Senèze: Life in Central France Around Two Million Years Ago

Vertebrate Paleobiology and Paleoanthropology Series

Edited by

Eric Delson
City University of New York and American Museum of Natural History
NY, USA
eric.delson@lehman.cuny.edu
http://www.nycep.org/ed/

Eric J. Sargis
Yale University, New Haven, USA
eric.sargis@yale.edu
http://www.yale.edu/anthro/people/esargis.html

Focal topics for volumes in the series will include systematic paleontology of all vertebrates (from agnathans to humans), phylogeny reconstruction, functional morphology, Paleolithic archaeology, taphonomy, geochronology, historical biogeography, and biostratigraphy. Other fields (e.g., paleoclimatology, paleoecology, ancient DNA, total organismal community structure) may be considered if the volume theme emphasizes paleobiology (or archaeology). Volumes in the series may either be monographic treatments (including unpublished but fully revised dissertations) or edited collections, especially those focusing on problem-oriented issues, with multidisciplinary coverage where possible. The two Series Editors are assisted by an Editorial Advisory Board. All contributions in the series (whether monographs or chapters in edited volumes) will be peer-reviewed by at least three readers, at the level of a journal submission.

Editorial Advisory Board

Senèze: Life in Central France Around Two Million Years Ago

Paleontology, Geochronology, Stratigraphy and Taphonomy

Edited by

Eric Delson

Department of Vertebrate Paleontology, American Museum of Natural History, New York, NY, USA;
Department of Anthropology, City University of New York, New York, NY, USA

Martine Faure

Université Lumière-Lyon 2, Lyon, France
Université de Lyon, UCBL, ENSL, CNRS, LGL-TPE, Villeurbanne, France

Claude Guérin

Université Claude Bernard-Lyon 1, Villeurbanne, France

Springer

Editors
Eric Delson
Department of Vertebrate Paleontology
American Museum of Natural History
New York, NY, USA

Department of Anthropology
City University of New York
New York, NY, USA

Martine Faure
Université Lumière-Lyon 2
Lyon, France

Université de Lyon, UCBL, ENSL, CNRS, LGL-TPE
Villeurbanne, France

Claude Guérin
Université Claude Bernard-Lyon 1
Villeurbanne, France

ISSN 1877-9077 ISSN 1877-9085 (electronic)
Vertebrate Paleobiology and Paleoanthropology
ISBN 978-3-031-64414-6 ISBN 978-3-031-64415-3 (eBook)
https://doi.org/10.1007/978-3-031-64415-3

Cover image: Left, view of Senèze area looking west, with Trench 1 in center of image and Trench 2 (with sunshade) to the right; to left along the road is the office bungalow, to right the hamlet of Senèze, background rising to rim of crater (photo E. Delson, larger image in Fig. 1.7). Right, partial skeleton of *Allohippus senezensis* excavated in 2005, specimen FSL SEN 05-0081+ (photo A. Monguillon, see also Fig. 11.24 and Fig. 2.7).

This Springer imprint is published by the registered company Springer Nature Switzerland AG
The registered company address is: Gewerbestrasse 11, 6330 Cham, Switzerland

In memory of Claude Guérin (1939–2016), colleague and friend, co-director of the Senèze project and of many other paleontological projects in France and abroad

Preface

The project to reopen the renowned paleontological site of Senèze began at a meeting of old friends. The investigators of the Vatera paleontological sites on the island of Lesvos, Greece, had invited a small group of colleagues to visit the area and participate in a short workshop from September 29 to October 2, 1999. The workshop "On Late Plio/Pleistocene Extinction and Evolution in the Palearctic: The Vatera Site" was held in Polichnitos, on Lesvos, and mainly included discussions of the mammalian fauna of the Villafranchian interval of Europe. Guérin and Faure presented a report on their recently completed study of the Saint-Vallier site and mammals, noting that it was older than the famous but less well-known locality of Senèze. Delson discussed his ongoing work on the large European cercopithecine primate *Paradolichopithecus*, which was known from Senèze and had also been recovered from Vatera-F.

During the meeting, Delson renewed his long acquaintanceship with Guérin and met Faure. He mused that it would be interesting to look again at Senèze, as Guérin and Faure had done at Saint-Vallier, bringing together a team of colleagues to investigate the geology, dating, paleontology, taphonomy and paleoenvironment. Guérin and Faure were thinking about a new project and were interested, so Delson suggested a joint venture, offering to seek funding in the US while Guérin and Faure directed the work on the ground. They agreed to start investigating the possibilities of a Franco-American team project.

Following numerous email exchanges later in 1999, Delson applied to NSF for a small grant, while Guérin and Faure requested a permit to work at Senèze. With the support of the Mayor of Domeyrat (the hamlet of Senèze is administratively part of the commune of Domeyrat), Faure submitted a request to the French Ministère de la Culture, Service Régional de l'Archéologie, Direction Régionale des Affaires Culturelles (Regional Archaeological Service of the Regional Office of the Ministry of Cultural Affairs); since 1945, Pleistocene paleontology in France falls under archaeological legislation. The Ministère de la Culture agreed to provide an authorization to survey and prospect for fossils. The NSF Physical Anthropology Program provided almost $9,000 in a "high risk exploratory" grant to Delson with Co-Investigators Guérin, Faure, geologist Evelyne Debard (University of Lyon) and dating specialist Bonnie Blackwell (Williams College). Blackwell agreed to wait until the following year when a full season would occur if the first work proved successful, and the other four were joined by Andrea Valli, then a Ph.D. student with Guérin at Lyon. Details of that first season (and the following years) are presented in Chap. 1 below. As a result of that "chance meeting" on Lesvos, 7 years of fieldwork were undertaken at Senèze, and the results of that project are presented in this volume. It was long delayed by a combination of other responsibilities and the need for detailed editing but the final product is an extensive analysis of this important Early Pleistocene locality, with implications for mammalian systematics and regional biochronology.

Acknowledgments We thank all our co-authors for their assiduous analyses of the data and fossils collected by our fieldwork as well as fossils recovered by earlier investigators. We thank all of the volunteer excavators who worked with us, the colleagues who assisted our studies and those who reviewed the manuscripts, the agencies that provided permits and funds, and the local landowners and other friends who made our work at Senèze possible. Each of them is named in the individual chapter acknowledgments, especially to Chap. 1. We thank Eric Sargis, co-editor of the VERT (Vertebrate Paleobiology and Paleoanthropology) series; Aaron Schiller, our Springer representative for VERT; and Henry Rodgers, Shalini Selvam and Jayanthi N. Narayanaswamy and their colleagues in the Springer production teams.

Great Neck and New York City, USA — Eric Delson
Lyon, France — Martine Faure
Villeurbanne, France — Claude Guérin

Contents

Contributors

Note: * indicates preferred address for correspondence

Luca Aprile Associazione Aditus in Rupe, Floridia (SR), Italy

Alain Argant Université Aix-Marseille, LAMPEA, UMR 7269 CNRS, MMSH, Aix-en-Provence, France

Jacqueline Argant Université Aix-Marseille, LAMPEA, UMR 7269 CNRS, MMSH, Aix-en-Provence, France

Jean-Jacques Bahain UMR 7194 HNHP (MNHN-CNRS-UPVD), Département Homme et Environnement, Muséum National d'Histoire Naturelle, Paris, France

Federica Candelato 3A soc.coop., via Campofiore 33, 37129, Verona, Italy

Evelyne Crégut-Bonnoure *Muséum Requien, Avignon, France;
TRACES, UMR 5608 (CNRS—Université Toulouse Jean-Jaurès), Toulouse, France

Evelyne Debard *25 rue Paul Chevrel, 69370 Saint-Didier-au-Mont-d'Or, France; Université de Lyon, UCBL, ENSL, CNRS, LGL-TPE, Villeurbanne, France

Eric Delson *Department of Vertebrate Paleontology, American Museum of Natural History, New York, NY, USA; Department of Anthropology, Lehman College of the City University of New York, Bronx, NY, USA; Ph.D. Program in Anthropology, The Graduate Center of the City University of New York, New York, NY, USA; New York Consortium in Evolutionary Primatology, New York, NY, USA; Institut Català de Paleontologia Miquel Crusafont, Universitat Autònoma de Barcelona, Cerdanyola del Vallès, Barcelona, Spain

Véra Eisenmann CR2P (Paléontologie)—Muséum national d'Histoire Naturelle, Paris, France

Christophe Falguères UMR 7194 HNHP (MNHN-CNRS-UPVD), Département Homme et Environnement, Muséum National d'Histoire Naturelle, Paris, France

Martine Faure Université Lumière-Lyon 2, Lyon, France; Université de Lyon, UCBL, ENSL, CNRS, LGL-TPE, Villeurbanne, France

Yolanda Fernández-Jalvo Museo Nacional de Ciencias Naturales (CSIC), Madrid, Spain

Jean Gaudant Université Paris 7, France, deceased 2015

Claude Guérin Université Claude Bernard-Lyon 1, France, deceased 2016

María Dolores Marín-Monfort Museo Nacional de Ciencias Naturales (CSIC), Madrid, Spain

Bernard Ménouret 3 rue de l'Amouroy, 13680 Lançon-de-Provence, France

Angélique Monguillon Réserve Naturelle Régionale du site géologique de Limay, Limay, France

Cécile Mourer-Chauviré Université de Lyon, UCBL, ENSL, CNRS, LGL-TPE, Villeurbanne, France

Luca Natali Istituto Italiano di Paleontologia Umana, Anagni (FR), Italy

Sébastien Nomade Laboratoire des Sciences du Climat et de l'Environnement, CEA Saclay, LSCE/IPSL, UMR CEA-CNRS-UVSQ 8212, Gif sur Yvette, France

Fabio Parenti *Universidade Federal do Paraná, Curitiba (PN), Brazil; Istituto Italiano di Paleontologia Umana, Anagni (FR), Italy

Jean-François Pastre *23, rue de l'Eglise, 93370 Montfermeil, France; Laboratoire de Géographie Physique, Environnements quaternaires et actuels, UMR 8591 CNRS, Thiais, France

María Dolores Pesquero-Fernández Museo Nacional de Ciencias Naturales (CSIC), Madrid, Spain

Vittorio Rioda Istituto Italiano di Paleontologia Umana, Anagni (FR), Italy

Sevket Sen CR2P (Paléontologie)—Muséum national d'Histoire Naturelle, Paris, France

Qingfeng Shao Nanjing Normal University, College of Geography Science, Nanjing, China

Andrea M. F. Valli *Société scientifique du Bourbonnais, Moulins (Allier), France; Istituto Italiano di Paleontologia Umana, Anagni (FR), Italy

Chapter 1
Introduction and History of Research in the Senèze Maar

Martine Faure, Eric Delson, and Claude Guérin

Abstract The Senèze paleontological site (Haute-Loire, France) is situated on the banks of a maar lake, in the Auvergne volcanic region. It was first reported by M. Boule in 1892, following the chance discovery of a proboscidean skeleton reported by local scholars. Pierre Philis, a local peasant, became interested in fossil mammals and collected them in the fields around the hamlet over more than 40 years, from the end of the nineteenth century until the start of World War II. He sold them to museum curators, especially to H. G. Stehlin (Naturhistorisches Musem Basel), C. Depéret (Faculté des Sciences de Lyon), C. Gaillard (Muséum d'Histoire naturelle de Lyon) and M. Boule (Muséum national d'Histoire naturelle, Paris). The resulting collections, including mounted skeletons exhibited in museums, made the site famous. New species (and genera) of ruminant artiodactyls, primate and carnivores were defined from Senèze, which became a biochronological reference site for the late Villafranchian. After the war, only a few short field campaigns (prospecting, survey, mapping and minimal excavation) were undertaken, and it became clear that renewed research was required to better understand this important site. The Franco-American fieldwork led by the authors began with a survey and request for excavation permits in 2000, followed by intensive mapping, prospection and excavation from 2001–2006. After a brief review of previous work at and about Senèze, each year of our research is summarized and illustrated. The succeeding chapters of this volume are also "previewed".

Résumé Le gisement paléontologique de Senèze (Haute-Loire, France) se situe sur le pourtour d'un ancien lac formé dans un maar, dans la région volcanique de l'Auvergne. Il a été signalé pour la première fois par M. Boule en 1892, à la suite de la découverte fortuite d'un squelette de proboscidien associé à d'autres restes de mammifères fossiles, signalés par des érudits locaux. Un paysan du hameau, Pierre Philis, s'est intéressé à ces ossements fossiles. Le gisement étant très riche, il en a récolté pendant une quarantaine d'années, de la fin du XIXe siècle jusqu'au début de la Deuxième Guerre mondiale. Il les a proposé aux conservateurs de musées et en a ainsi vendu une grande quantité à H. G. Stehlin (Naturhistorisches Museum Basel), C. Depéret (Faculté des Sciences de Lyon), C. Gaillard (Muséum d'Histoire naturelle de Lyon) et M. Boule (Muséum national d'Histoire naturelle, Paris). Ces collections, et entre autres les squelettes montés exposés dans les musées, ont fait la notoriété du site. Ces restes fossiles ont permis de définir de nouvelles espèces (et genres) ruminants, de primate et de carnivores. Senèze est devenu une référence biostratigraphique pour le Villafranchien supérieur. Après guerre, le gisement n'a fait l'objet que de courtes explorations de terrain (prospections, sondage géologique, prélèvements et fouille ponctuelle). Ce site majeur pour l'étude des faunes du Plio-Pléistocène

Claude Guérin—Deceased 2016.

M. Faure (✉)
Université Lumière-Lyon 2, Lyon, France
e-mail: fauremartine@free.fr

Université de Lyon, UCBL, ENSL, CNRS, LGL-TPE, Villeurbanne, France

E. Delson
Department of Vertebrate Paleontology, American Museum of Natural History, New York, NY, USA

Lehman College and the Graduate Center of the City University of New York, New York, NY, USA

New York Consortium in Evolutionary Primatology, New York, NY, USA

Institut Català de Paleontologia Miquel Crusafont, Barcelona, Spain

C. Guérin
Université Claude Bernard-Lyon 1, Lyon, France

E. Delson et al. (eds.), *Senèze: Life in Central France Around Two Million Years Ago*,
Vertebrate Paleobiology and Paleoanthropology, https://doi.org/10.1007/978-3-031-64415-3_1

nécessitait une reprise des recherches. En 2000 nous avons fait un sondage et obtenu une autorisation de fouille programmée qui a été renouvelée jusqu'en 2006. Dans cette introduction, après un historique et un survol des recherches antérieures, nous exposons nos propres travaux de prospection et de fouille. Les chapitres de ce volume sont ensuite présentés.

Keywords Fieldwork • Pierre Philis • Villafranchian

Mots-clés Fouilles programmées • Pierre Philis • Villafranchien

Work at Senèze in the Nineteenth and Twentieth Centuries

The Discovery of Senèze Mammals in the Late 19th Century

In the early 1890s, the remains of fossil mammals were discovered by chance in a farm field in the hamlet of Senèze, in the commune of Domeyrat (Haute-Loire Department, central France). Three antiquarian scholars of the region, H. Mosnier, A. Vernière and P. le Blanc, advised Marcellin Boule of the Muséum national d'Histoire naturelle in Paris (MNHN) of this find. Boule had worked on the geology and fossil mammals of the region (Boule 1892–1893), and he went to Senèze, accompanied by Albert Gaudry (the Professor of Paleontology at the MNHN). They excavated an almost complete skeleton identified as *Elephas* (now *Mammuthus*) *meridionalis*, associated with remains of (sic) *Equus stenonis*, *Bos elatus*, *Rhinoceros*, *Hyaena* and antlers of several cervids, buried in volcanic ash layers exposed by "storm rains" (Boule 1892).

A few years earlier, the MNHN had been deeply involved in the excavation and reconstruction of a mammoth (*M. meridionalis*) skeleton from Durfort (Gard Department, southeastern France). They also had to deal with the problems and costs of transporting, mounting and installing the skeleton for exhibit at MNHN (Gaudry 1893), at a cost estimated around 30,000 francs (Lascombe 1894). Gaudry was not prepared to initiate a similarly complicated and expensive project for the Senèze skeleton, which was even larger. Thus, the Senèze proboscidean skeleton was left in place, and only its molars and some rhino and cervid bones were transported to Paris and catalogued as MNHN-P 849, 1892-15. Gaudry (1896, p. 189, Fig. 190) published the drawing of one lower molar, and two teeth were exhibited in the new Galerie de Paléontologie, opened in 1898 (Gaudry 1898), where they remain (Fig. 1.1). The specimen was compared to other finds of this species and said to be similar to that from the British Crags, older than those from Durfort or the Cromer Forest Bed of Britain (Boule 1892; Gaudry 1896). In turn, Senèze was thought to be younger than the nearby volcanic sites of Chilhac or Le Coupet, which had yielded mastodons. These three sites were considered paleontological Pompeiis, preserving the fauna contemporaneous with their eruptions (Boule 1892, p. 626; Gaudry 1898, p. 822). The remainder of the Senèze proboscidean skeleton was left in place, from where Boule hoped it would be included in local collections, but in fact it deteriorated badly and was eventually salvaged in part by a team from Lyon: Depéret and Mayet (1912) roundly criticized this abandonment of the fossil, which led to its deterioration.

The discovery of the mammoth must have been an extraordinary event for the peasant inhabitants of the tiny Senèze hamlet. A local farmboy, then 12 years old, witnessed the extraction of the gigantic bones and became interested in vertebrate paleontology. That boy, Pierre Philis, was a central figure in the recovery (and sale) of fossils from Senèze over the next half century (1892–1942; Faure et al. 2022). Several years later, Philis discovered a partial skull of "*Rhinoceros etruscus*" and sold it to the local collector Vernière (Schaub 1943b). Vernière had kept in contact with Boule and offered him the specimen, which they donated to the MNHN (catalogued as MNHN-P 942, 1896-22; Faure et al. 2022: 342; Fig. 1.2). For the second time, a Senèze fossil entered the MNHN collection.

Fig. 1.1 Two upper molars of *Mammuthus meridionalis* MNHN-F 849, 1892-15 on exhibit in the Paleontology Hall of the Muséum national d'Histoire naturelle, Paris, part of the first fossil mammal collected at Senèze, in 1892 (© Sevket Sen, MNHN)

The Philis Era

Philis Starts to Contact Buyers. By 1906, Philis apparently began to search actively for fossils on his farmland, in whatever time he could spare from agriculture. Beginning in 1907, Philis corresponded with Hans Georg Stehlin (Curator of Osteology [i.e., Paleontology] at the Naturhistorisches Museum Basel [NMB]). Based on letters they exchanged in June 1907, Stehlin visited Senèze in late 1906 or early 1907 regarding an order for Senèze fossils. The first major discovery made by Philis was a complete skeleton of "*Machairodus*" (today *Megantereon cultridens*), found in 1909 and reported widely in the national press (Faure et al. 2022). This specimen was acquired by the NMB, where it is on exhibit today, following a new restoration in 2013.

Philis soon widened his circle of clients within France. In 1908, S. Rougier of the neighboring commune of La Chomette helped Philis contact Boule, who had succeeded Gaudry at MNHN in 1902 and was known for his work on the geology of the Velay area. Boule continued to be interested in Senèze fossils into the 1920s, and in August,

Fig. 1.2 Left, portrait of Pierre Philis, date unknown (after 1927; © Philis archives). Right, cranium of "*Rhinoceros etruscus*" from Senèze MNHN-F 1896-22, on exhibit in the Paleontology Hall of the Muséum national d'Histoire naturelle, Paris (© Sevket Sen, MNHN). This was the first fossil collected by Pierre Philis, then just 16 years old. It was sold to Antoine Vernière; he and M. Boule gave it to the MNHN collection in 1896

1921, he sent Pierre Teilhard de Chardin, who was staying at a Jesuit residence near Le Puy, to see P. Philis. Teilhard termed Philis a true "fossils hunter" in a letter to Boule (Faure et al. 2022: 333–334).

Depéret, Gaillard and the Senèze Collections in Lyon. The third of Philis' main customers was Charles Depéret, Professor of Geology in the Faculty of Sciences at Lyon (FSL; Fig. 1.3). He became interested in Senèze by 1908, when he sent his paleontological preparator, Laurent Maurette, to work there with Philis and train the young farmer how to prepare fossils and introduce him to the techniques of excavating fossil vertebrates (Faure et al. 2022: 335). With paleontologist Lucien Mayet (who taught the course on anthropology and human paleontology at FSL), Depéret described the results of the first intensive prospection and excavation at Senèze, mainly on the farmland of MM. Philis and Thomas (Depéret and Mayet 1911, 1912). In a short preliminary report, Depéret and Mayet (1911, p. 262) wrote that "le gisement de Senèze contient des squelettes complets, avec tous les os restés en connexion, condition des plus rares dans les gisements européens et qui ne se rencontrent guère que dans les magnifiques gisements de l'ouest des États-Unis." ("Senèze yielded complete skeletons, with all the bones in connection, which was rare in European localities and hardly occurred except in the magnificent localities of the American West"; our translation, also below). They also noted that a "savant étranger" (foreign scholar), Dr. Stehlin of Basel, who zealously followed discoveries of fossils in French localities, had obtained a quite important collection of Senèze fossils for the NMB. In light of an awareness to protect fossils from French deposits from being removed, they suggested that it was necessary for a French scientific institution to take over excavations at Senèze; they reported that it had become possible for the Faculté des Sciences Lyon (in part thanks to the support of the Association française pour l'avancement des Sciences) to begin careful exploration of the site. Depéret and Mayet (1912) reported

Fig. 1.3 Portrait of Charles Depéret in university regalia. Published in Multiple Authors (1929)

the results of their first year of intensive research at Senèze, including discussion of the regional geology and recognition of the volcanic origin of the depression around the hamlet. They did not use the term "maar", but it seems that they understood the concept of a volcanic lake. They discussed the local stratigraphy in several agricultural fields (giving the names of the landowners) and presented a preliminary sketch of a section based on natural exposures in a ravine where Philis had collected extensively, on the slopes above the volcanic valley. They also discussed each of the fossil species recovered, including one new cervid ("*Cervus senezensis*"), offering their thoughts about the possibility of eventually recovering evidence of early human occupation. A few years later, Depéret et al. (1923) wrote a major review of Pliocene elephantids, including the material from Senèze (see Ménouret and Guérin 2024). Meanwhile, C. Gaillard began to build a Senèze collection at the Muséum d'Histoire naturelle de Lyon.

Depéret's third and last paper about Senèze fossils was the description of *Dolichopithecus arvernensis* in 1929. As Faure et al. (2022: 358) pointed out, he had long hoped for remains of Pliocene humans, with primate fossils a close second option. In February 1924, Philis informed Depéret that he had found a small skull which he thought might be of a monkey. At Depéret's request, Philis sent the specimen to Lyon, where Depéret determined that it was indeed a cercopithecid monkey that he considered was a representative of *Dolichopithecus*, a genus which he had described from the early Pliocene of Perpignan in 1889. Depéret reported this find, under the new name of *Dolichopithecus arvernensis*, at the 16th International Geological Congress (Madrid, May 1926) and then named and described it briefly in the publication of his talk, a review of French Pliocene primates (Depéret 1928). A more detailed description and comparison was published a year later (Depéret 1929), just months before his death. The species was later transferred to the genus *Paradolichopithecus* Necrasov et al., 1961 (see Delson 2024). There was great interest in the find in the 1920s and 1930s, both among scientists and collectors: Claudius Côte, an antiquarian and major benefactor of the Lyon Muséum d'Histoire naturelle (MHNL), asked Philis numerous times to find him another such "beau crâne de singe" (Faure et al. 2022: Fig. 8).

Frédéric Roman, assistant professor in the FSL, succeeded Depéret as Professor after the latter's death in 1929. Jacques Dareste de la Chavanne had succeeded Maurette after his death in 1917. Together (Roman and Dareste de la Chavanne 1931) they announced the discovery of a Senèze elk skeleton identified as *Alces latifrons* Johnson, 1874. Azzaroli (1952) later determined that it typified the new genus and species *Libralces* (today *Cervalces*) *gallicus* (see Valli 2024).

In the early decades of the twentieth century, museums and university paleontology labs desired to acquire complete skeletons of fossils in order to mount them for exhibit purposes. Seven skeletons were obtained by the FSL (Gómez Llueca 1921; Roman 1931, Fig. 2, 1935), catalogued at the time as: *Cervus senezensis* (Roman 1935: Fig. 17), *Cervus philisi*, *Alces latifrons*, *Leptobos etruscus*, *Rhinoceros etruscus*, *Machairodus crenatidens* (Gómez Llueca 1929: Fig. 3; Ballesio 1963: pl. 4; Van Hooijdonk 2006: Fig. 9) and *Equus stenonis*. Up to the time of the move from the historic university building in central Lyon to the new campus of La Doua in Villeurbanne (at the time of the creation of the Université Claude Bernard-Lyon 1 or UCB-Lyon 1), these skeletons were among the flagship attractions of the paleontological collection (David 1962, p. 74). At the same time, the NMB had eight mounted skeletons of Senèze fossils excavated by Philis. Six of these are still on public exhibit in the permanent collection: *Megantereon cultridens* (Schaub 1925; Van Hooijdonk 2006: Fig. 12), *Stephanorhinus etruscus* (Schaub 1943b, Fig. 1; Wittmann 1980, Fig. 79; Van Hooijdonk 2006, Fig. 5), *Allohippus senezensis* (Schaub 1943b, Fig. 2), *Sus strozzii* (Schaub 1943b, Fig. 3), *Gallogoral meneghinii* (Schaub 1922, Fig. 2; Van Hooijdonk 2006: Fig. 4) and *Metacervoceros rhenanus philisi* (Schaub 1941, Fig. 1). Two others (*Gallogoral meneghinii* and *Eucladoceros ctenoides senezensis*) are kept in storage off limits to the public.

In 1920, W. D. Matthew of the American Museum of Natural History (AMNH) traveled through Europe to see the state of museums after World War I. In Basel, he saw "a Pliocene collection from Senèze that equals the Val d'Arno fauna or the Asti fauna in richness. Proboscideans are scarce, but all the smaller forms far better represented. Dr. Stehlin has already several fine skeletons mounted from this horizon–*Cervus*, *Machaerodus*, etc. Among the antelopes is one quite near to *Oreamnus* (our mountain goat). This is certainly one of the great faunas, and it is fine to see it in such competent hands". In Lyon he was impressed by various FSL collections, including their material from Senèze, writing that it is "a very splendid collection and I wish I could spend a couple of months studying it" (Matthew 1921: 189–190). Matthew was the first AMNH paleontologist to develop an interest in the fossil mammals of Senèze; it would take eighty years for a second (Delson) to realize the wish of his predecessor.

Stehlin, Helbing and Schaub—Building the Senèze Collection in Basel. Hans Georg Stehlin of the NMB continued to purchase fossils from Philis, but he only wrote one major paper about them, summarizing the Senèze fauna in 1923. He was joined after 1910 by Hermann Helbing and Samuel Schaub (Fig. 1.4; Faure et al. 2022: 339). Helbing worked at Senèze with Philis on several occasions. Schaub started as a volunteer at NMB in 1913, joined the staff in 1922 and succeeded Stehlin (who died in 1941) in 1943. Schaub worked on Senèze fossils sold by Philis starting in 1913 and first visited the site in 1920, after which he collaborated with Philis until the latter's death in 1942. He wrote numerous articles about this material, often in comparison with specimens from other Pliocene sites: among bovids (see Crégut-Bonnoure 2024), Schaub (1922) described and named *Nemorhoedus philisi* (a misspelling of *Nemorhaedus*, now *Gallogoral meneghinii*); he (Schaub 1923) also described and named *Procamptoceras brivatense* and *Megalovis latifrons*, as well as discussed *Tragelaphus* (now *Gazellospira*) *torticornis*, *Deperetia ardea* (now *Pliotragus ardeus*) and "*Antilope* sp." (now included in *Megalovis latifrons*). Among carnivorans (see Argant 2024), Schaub (1925) described a partial skeleton of *Machaerodus* (now *Megantereon*) *cultridens*; later, he (Schaub 1942) named *Brachyprosopus vireti* for some fragments which are now included in *Acinonyx pardinensis*. Finally, he (Schaub 1941) named *Cervus philisi*, now considered a subspecies of *Metacervoceros rhenanus* (see Valli 2024).

Schaub was among the first paleontologists to utilize screen-washing of sediment in the laboratory to enable searching for micromammal fossils. Screen-washing of material from Senèze led to the recovery of several small teeth. Stehlin (1923) included in his faunal list *Lepus*, *Sciurus*, and two species each of *Mimomys* and *Arvicola*. A few years later, he allowed the Hungarian paleontologist T. Kormos to review the arvicolids from Senèze, and Kormos (1931) identified three species of *Mimomys*. Maul (2004) re-evaluated the few teeth known and broadly agreed with Kormos, recognizing the *Mimomys pliocaenicus*/*M. ostramosensis* group and the *M. pitymyoides* group.

Schaub (1943) presented a major survey of the Senèze assemblage based on the Basel material. In addition to taxa recognized previously, he noted the presence of a sciurid close to the Asian *Eutamias* and the leporid *Oryctolagus* cf. *lacosti*; these taxa have not been re-examined recently.

Also in 1943 (but apparently unaware of Schaub's review), Masson (1943) submitted a Diplôme d'Etudes Supérieures on Senèze at the FSL. He reviewed the geology of the site and its fauna, with a special section on ruminant artiodactyls. In addition to birds and mammals, his faunal list included a turtle, a frog and molluscs of the genera *Limnea* and *Ancylus*.

Fig. 1.4 Portraits of (left to right): Hans Georg Stehlin, Samuel Schaub (© NMB archives)

Senèze After Philis and World War II

Fieldwork. After the war, the conduct of archaeology in France changed significantly. In 1941 a law required all people wishing to carry out excavations in sites which may have a relationship to prehistory to first have an excavation authorization. This law was validated in 1945. All Plio-Pleistocene deposits were affected by this law, which put an end to excavations without state control. This limited the possibility for amateurs like Philis to undertake the type of work that he did before the war (Faure et al. 2022: 364–368). Only a limited number of short field campaigns were undertaken at Senèze, while individual studies focused on specific taxa or included Senèze fossils in broader analyses.

In 1953, J. Roger and colleagues from the MNHN undertook several days' fieldwork at Senèze (Roger 1954; also Brébion et al. 1953). They examined the local geology and observed the presence of associated mammalian skeletons in one or two locations, while in others the elements were dissociated. It is unclear if they collected any fossils, but Roger (1954) noted that some property to the north of the "Philis ravine" had been purchased by the MNHN for that purpose. A map showed contour lines and the location of a number of fossiliferous points and observed sections, but without much detail. Roger (1954) also suggested that the faunal list included species which were adapted to widely different environments (forest, woodland, steppe) and proposed that a taphonomic study would be useful.

Bout (from 1960 onward) discussed the geology and stratigraphy of numerous French Villafranchian localities. In 1970, he described the geological setting of the Senèze maar and the sequence of events which formed it (Bout 1970b). As seen in his Fig. 1.1, he mapped the site as a large scoria cone to the (north)west of a deep infilled maar, about 500 m across. The scoria cone was formed first, at the source of a lava flow that covered the area of the site. Next, the maar formed in the crater of an explosive vent that penetrated the flow and left a deposit of explosion breccia against the western side of the scoria cone. As explosive activity died down, the maar lake slowly filled in with lacustrine deposits. Finally, offwash slope deposits built up on the inner (especially western) slopes of the maar. Bout (1970a, 1970b; also Couthures 1989, following Depéret & Mayet 1912 and Roger 1954) indicated that the mammalian fauna was derived from the upper levels of the maar infilling as well as from the slope deposits.

Elhaï and Grangeon (1963) briefly studied the pollen from the upper 60 m of a 100 m core drilled into the maar in 1962 (Grangeon 1962). Elhaï (1969) presented a more detailed analysis of the upper 120 m of a 175 m core taken in 1965. Elhaï reported that the lowest 52 m of the core was composed of solidified tuff and crater debris; this was overlain by 43 m of fluvio-lacustrine sediments, in turn underlying 80 m of diatomaceous clay, with a sandy layer between 28–22 m below the top. Elhaï described a general alternation between two main plant assemblages: a warm-climate forest of thermophile, generally leafy trees, both local and exotic; and a cooler-climate forest of conifers mixed with herbs, the latter sometimes expanding to cover most of the ground surface. Elhaï divided the core into a series of phases dominated by one of these two assemblages: from 120–94 m, a temperate phase with poorly diversified vegetation might signify the first "recovery" after a significant cold interval (for Elhaï, perhaps the start of the Pleistocene); this was followed by a cool and humid phase from 94–76 m. Between 76–43 m came the interval richest in thermophile plants and poorest in herbs, within which the earlier part was humid and temperate, while the later part was humid and cool; between 43–29 m, a cold and relatively dry phase occurred; and from 24 m to the top of the core came a complex alternation of three cycles, each with a temperate-humid subphase followed by a cool or cold subphase.

In studying the diatoms from Elhaï's core, Ehrlich (1968) distinguished three main intervals: from 107–97 m, littoral species of diatoms were more common than planktonic ones; between 96–20 m, littoral, benthic and epiphytic taxa disappeared to be replaced by a euplanktonic species; and from 19 m up, the littoral forms again dominated. Bout (1970a, 1970b) interpreted these three phases to imply an early period of slow lake filling, with shore diatoms washed in; followed by a deep lake where plankton flourished; and finally an interval when the lake was filling with sediment and only shallow-water diatom species flourished. Ehrlich (1968) also estimated that the number of laminations in the diatomite (which she interpreted as annual) was probably between $2–3 \times 10^5$. Bout (1970a, 1970b) combined that with an early estimate of 1.9 Ma for the underlying lava to suggest that the upper levels of the maar might have dated to 1.6 Ma and the mammalian fauna perhaps to 1.5 Ma.

Pelletier (1968) studied the sedimentology and mineralogy of three samples of sand in different levels of the core (see Debard 2024). In a discussion of petrology of the volcanic deposits at Senèze, Devis (1970) again noted a decline in lake depth over time and pointed out two levels with feldspar crystals. The partial skeleton of a fish found in the core was reported by Gaudant (1975) as representing *Tinca* (the tench). Another coring program was undertaken in 1989, with several short cores taken near the earlier one (see below, Roger et al. 2000).

Prévot and Dalrymple (1970) reported paleomagnetic data for the upper 132.5 m of the 1965 core. The analyzed section was entirely reversed, except for a 5–10 m normal zone from 17.5 to somewhere between 23.5 and 28 m below the top of the core. Prévot and Dalrymple suggested that the normal might correlate to the "Lower Olduvai" which they thought dated ca. 2.1 Ma. They also reported a whole-rock K–Ar age of 2.3 ± 0.15 Ma for samples of the Pié de Charenty basanite lava flow, assumed to predate the Senèze maar deposits. Couthures and Pastre (1983) obtained a date of 2.52 ± 0.06 Ma for the same flow, equivalent within the error ranges [Couthures (1989) reported the date as 2.48 ± 0.06 Ma]. Both Prévot and Dalrymple and later Couthures and Pastre reported a reversed polarity for the flow, correlated to the lower Matuyama chron. Prévot and Dalrymple (1970) further suggested that the normal magnetozone was the Réunion "Event" (then called the "Lower Olduvai"), implying that the Senèze fauna dated ca. 2.1–2.0 Ma.

C. Guth (Université de Poitiers) worked briefly at Senèze in the early 1960s, seeking fossils in two locations as indicated on the cadastral plan published by Parenti et al. (2024, Fig. 2.2). Guth (1975) reported finding a partial juvenile rhinocerotid (*Dicerorhinus etruscus*) along with specimens of *Canis etruscus*, *Eucladoceros* sp., "*Cervus philisi*" (now *Metacervoceros rhenanus philisi*), and indeterminate equid and bovid; he also reported work at Chilhac and Blassac-la-Girondie, southwest of Senèze.

The last fieldwork before our campaigns was undertaken by J. Couthures, who proposed the hypothesis that the death assemblage from Senèze was the result of intermittent CO_2 emission over a long time, leading to asphyxiation of mammals from diverse habitats (Couthures 1989). In part to test that hypothesis, Couthures (with D. Ablin and D. Hadjouis) undertook a single excavation campaign in August, 1991 (Couthures et al. 1991). They cleaned a section in or near the "Philis ravine" and recognized nine geological layers. They recovered four fossils: a horn core of *Procamptoceros* (figured in Lacombat et al. 2010, p. 77), hindlimb bones of a cervid, a cervid foot bone and a tooth of *Equus stenonis*. No further publication resulted from this work.

Analyses. In terms of systematic paleontology of ungulates, Richard (1944) wrote a short note about milk teeth of *Equus stenonis* from Senèze. Prat (1964) reviewed Villafranchian equids and described *Equus stenonis senezensis* for the same material (see Eisenmann and Delson 2024). Azzaroli (1952) named *Libralces gallicus* for two specimens of elk described earlier (see above); later he (Azzaroli 1982) included the species in the genus *Cervalces*. Breda (2001) suggested that the postcranial elements associated by Azzaroli with the holotype cranium were from a different individual and instead associated the postcranium of Azzaroli's paratype with that cranium. Heintz (1970) revised the Villafranchian deer from Spain and France, and he recognized four taxa from Senèze, including the new subspecies currently termed *Croizetoceros ramosus minor*. Pfeiffer (1999, 2005) reviewed the place of *Dama* among fossil cervids and discussed the Senèze taxa among others (see Valli 2024). Faure and Guérin (1979) examined supposed hippopotamids from Senèze and determined that none were correctly identified, but this idea has continued to appear in publications (see Crégut et al. 2024). Guérin (1980) studied the Senèze rhinocerotid in his doctoral thesis and attributed it to *Dicerorhinus* (*Brandtorhinus*) *etruscus etruscus*. Fortelius et al. (1993) moved this species into *Stephanorhinus*, but Guérin (e.g., 2024) always disagreed with that view and retained the former taxonomy. Duvernois and Guérin (1989) discussed bovids from Senèze as part of a broader work, and Duvernois (in that article) named *Leptobos furtivus* for bovine specimens from the site (see Crégut-Bonnoure 2024).

Numerous scholars have worked on Senèze carnivores (see A. Argant 2024). Ballesio (1963) studied the sabertooth cat *Homotherium crenatidens* in the FSL collection, while Perrot et al. (1972) discussed the paleopathology of a humerus of that species. Turner (1987) reviewed Old World *Megantereon*, including that from Senèze, and Antón and Werdelin (1998) discussed the skull of that cat. Both felids from Senèze were studied by Van Hooijdonk (2006), while Christiansen and Adolfssen (2007) focused on the osteology and ecology of *Megantereon cultridens*. Martin (1973) named three new Villafranchian canids, including *Canis senezensis*, and Mazza and Rustioni (1994) discussed the *Ursus* from Senèze.

The Senèze fauna was considered to belong to a later part of the Villafranchian "stage" (now land mammal age), by numerous researchers in the 1960's and 1970's. Heintz (1970) defined four successive phases based on cervids from France and Spain, of which the third (Villafranchien supérieur A) was the "zone de Senèze". Azzaroli (1970) broadly followed this subdivision, but he was uncertain whether the Senèze assemblage belonged in the earlier or later part of the upper Villafranchian. Mein (1975) proposed a system of numbered MN "zones" which divided Neogene time based on associations of mammalian taxa. The early Villafranchian was equated to MN 16, the middle Villafranchian to MN 17 and the late Villafranchian formed part of a subsequent, unnumbered unit. Guérin (1980, 1982) added MNQ 18 and 19 for the late Villafranchian (as well as several later units). Mein (1989) included Senèze in MN 17, whereas Guérin (1982) retained that site as one of the reference faunas in MNQ 18. We follow the latter concept, as will be discussed in greater detail in Chap. 17 (Crégut-Bonnoure et al. 2024), which makes the geochronological and biochronological position of Senèze important in a pan-European context.

In the meantime, Heintz et al. (1974) provided a complete faunal list of larger mammals for many French Villafranchian sites including Senèze. Azzaroli et al. (1988) proposed that the Senèze local fauna was composed of two associations of different age, on the basis of an analysis of Schaub's (1943b) faunal list plus original studies; these authors, however, did not cite and may not have been aware of the revision by Heintz et al. (1974). Azzaroli et al. (1988, p. 82) suggested that the larger and more speciose assemblage, including (in current taxonomy) *Eucladoceros ctenoides senezensis*, *Metacervoceros rhenanus philisi*, *Croizetoceros ramosus minor*, *Nyctereutes* and *Paradolichopithecus*, would fit in the middle Villafranchian. To the contrary, they considered that *Cervalces gallicus*, *Equus bressanus*, "a small equid which may possibly be *Equus stehlini*" and perhaps *Megalovis* "clearly point to a late Villafranchian age." They suggested that the former group, presumably from the upper levels of the maar, might date to ca. 2 Ma, in their middle Villafranchian, while the less extensive faunule might date to ca. 1 Ma, near the end of the Villafranchian as they conceived of it. These and later assessments will also be discussed in Chap. 17 (Crégut-Bonnoure et al. 2024).

Senèze in the Twenty-First Century

The Franco-American Research Project Year by Year

2000. The origin of this project is discussed above in the Preface. As noted there, Faure received a permit allowing survey and paleontological prospecting from the *Ministère de la Culture et de la Communication*. A small exploratory team (Faure, Guérin, Delson, Evelyne Debard and Andrea Valli [then a UCB-Lyon 1 PhD student]) spent a week in June 2000 in the Senèze area (Fig. 1.5; located at 45.241 N, 3.483 E; see Fig. 2.1). We began by examining the cadastral plan of the hamlet of Senèze (part of the commune of Domeyrat). This plan shows the ownership of land plots (parcels) as inherited and often subdivided over time. Thanks to the indispensable help of Mme R. Martin of

Domeyrat, we were able to locate the plots of land and meet their current owners. Checking the plan against local geomorphology, altitude and ground cover, the team selected a number of parcels which might be likely sites for excavation and then spoke with their proprietors, mainly landowners who were desended from those mentioned by Depéret and Mayet (1912) as owning fossiliferous properties. Many of the holdings had been divided through inheritance, but several of the proprietors were willing to allow excavation on their land the following year (see below).

Fig. 1.5 C. Guérin, M. Faure and E. Delson at the entrance to the hamlet of Senèze

2001. With the authorization and support of the Mayor of Domeyrat and permission from the selected landowners, Faure received a survey and prospecting permit and funding from the *Ministère de la Culture et de la Communication*. We (the three team leaders) and Debard were joined by a team of specialist colleagues and a group of mainly student volunteers from several institutions: Alain Argant (Research Associate of UMR 7269 CNRS (LAMPEA), Université Aix-Marseille, Aix en Provence, France; fossil carnivores); Jacqueline Argant (Research Associate of UMR 7269 CNRS (LAMPEA), Université Aix-Marseille, Aix en Provence, France; palynology); Fabio Parenti (Istituto italiano di Paleontologia Umana, Rome, Italy; topography, mapping, drafting of fossil placement); Jean-François Pastre (Laboratoire de Géographie Physique, CNRS Meudon, France; volcanology, tephrochronology). Sevket Sen (Laboratoire de Paléontologie, Muséum national d'Histoire naturelle, Paris; paleomagnetism); David Reddy (American Museum of Natural History; informatics); and John A. Van Couvering (Micropaleontology Press, American Museum of Natural History; geology and stratigraphy). Andrea Valli (UCB-Lyon 1) and Angélique Monguillon (PhD candidate, UCB-Lyon 1) were charged with technical supervision of the student excavators. Our team of volunteers included students from the PhD Program in Anthropology, City University of New York (CUNY) Graduate Center: Karen Baab, Terence Capellini, Steve Frost, Suzanne Hagell, Kieran McNulty and Tara Peburn (Fig. 1.6); and students and volunteers from Lyon universities (UCB-Lyon 1 and Lumière-Lyon 2 unless otherwise indicated): Amandine Alphonse, Célia Beaudoin, Loïc Costeur, Marie-Anne Héran, Hélène Jousse, Samuel Maillot, Bernard Ménouret and Laurent Servant. During the course of the season (July 2–21), the site was visited by M.-P. Aubry (Rutgers University, USA), W. A. Berggren (Rutgers and Woods Hole Oceanographic Institution, USA), M. Grandjean (Musée Crozatier au Puy-en-Velay, France), C. Lécuyer (UCB-Lyon 1) and P. Noyaret (Association Saint-Vallier, Histoire et Archéologie, France).

Fig. 1.6 American team in 2001; left to right: J. van Couvering, K. Baab, S. Frost, S. Hagell, E. Delson, K. McNulty, T. Peburn, T. Capellini

We concentrated on work in parcels 233 and 234, near the hamlet of Senèze (see Fig. 2.3). The landowner for parcel 233 (termed the "champ Thomas" by Depéret and Mayet 1912) was Ms. Marguerite Limagne; and for parcel

234 M. Jean-Jacques Chabrier. They graciously allowed the team to survey and excavate on their land, and this permission was renewed in following years (as was true for landowners mentioned in later seasons). The team did not seek permission to work on land owned by the Philis family for two reasons: On the one hand, the terrain had been excavated in the first half of the twentieth century by P. Philis, and we wished to avoid problems with his backfill; moreover, the current owner, his grandson M. Omer Philis, was not ready to grant us authorization to excavate on his land (see below).

Two rented trailers (in French, "bungalows") were placed along the road between Senèze and La Chomette to serve as an office and a storage area each season (Fig. 1.7), next to a small rock face termed Section A, which turned out to include the youngest deposits in the area (see Debard 2024, for details about this and other trenches and sections studied). Three trenches were excavated by a backhoe (Fig. 1.8) at the start of the season: T1, in parcel 234, was 3 m deep, 2.5 m wide and 40 m long; T2, in parcel 233, was 3 m deep, 3 m wide and 15 m long; T3, also in parcel 233, lying perpendicular to T1 and T2, was 3 m wide and 20 m long. T3 was essentially sterile, while a cervid mandible and teeth were found near the base of T1; T2, on the other hand, proved fossiliferous. The backhoe revealed part of the skeleton of a rhinoceros lacking the cranium in T2 (*Dicerorhinus etruscus etruscus*, see Guérin 2024), which was excavated throughout the season (Figs. 1.9 and 1.10). Debard trained the students to draw the stratigraphy of these trenches to scale by using a grid placed against the exposed trench surface and drawing the visible contents of each grid square (Figs. 1.11 and 1.12). Samples were taken for sedimentological and tephric analysis. Any fossils recovered were photographed in place, located in three dimensions with a plane table and alidade (Fig. 1.13) or total station (see Fig. 2.4) and drawn onto a base map under the supervision of Parenti (see Figs. 2.5, 2.6 and 2.7). Details were entered into an Excel spreadsheet and a GIS program and assigned a sequential field number of the form SEN 01-#. This methodology, developed by prehistoric archaeologists, is commonly used for Plio-Pleistocene deposits.

Blackwell supervised the collection of samples to estimate the radiation dose rate for analysis of ESR age, and four dosimeters were emplaced in auger holes in the wall of T2 to be retrieved in 2002. Sen took paleomagnetic samples in T1 and T2 and in underlying basalts (Fig. 1.14; see Delson et al. 2024b), and J. and A. Argant sampled for pollen in T1 and T2 (Fig. 1.15; see J. Argant 2024). At the end of each season, a geotextile was laid over excavated areas to mark the level for opening the following year (Fig. 1.16), and the backhoe refilled the excavations to prevent farm animals from harming themselves.

Fig. 1.7 View of Trench 1 and Trench 2 (with sunshade), in center of image; to left team cars and office bungalow, to right hamlet of Senèze, background rising to rim of crater

Fig. 1.8 Backhoe excavating Trench 1

Fig. 1.9 Excavation and cleaning of rhino skeletal elements, 2001

Fig. 1.10 Partial postcranial skeleton of *Dicerorhinus etruscus etruscus* found in connection in Trench 2, 2001. **A** Part of right forelimb and ribs; **B** Part of right hind foot; **C** left forefoot; **D** part of left hind foot

Fig. 1.11 Three teams of two students each making scaled drawings of Trench 2 stratigraphy

Fig. 1.12 Closeup of H. Jousse and L. Servant making a scaled drawing in Trench 2

Fig. 1.13 F. Parenti mapping fossil finds with plane table and alidade, K. McNulty taking notes

Fig. 1.14 S. Sen sampling for paleomagnetism in Trench 2

Fig. 1.15 Sampling for pollen in Trench 2. Left to right: A. Argant, J. Argant, C. Beaudoin

Fig. 1.16 Placement of geotextile to mark the excavation level for next year before the trench is refilled by the backhoe, 2001

2002. As in all later years, Faure and the team leadership submitted a report of the previous season's work to the DRAC early in the year, along with a request for continued permission and funding. The 2002 field season was short (June 29–July 9), given limited funding. We were joined by Debard, Pastre, Monguillon, Valli and Héran (Fig. 1.17). The dosimeters were retrieved from the former site of T2, and some localized coring was undertaken by Pastre (Fig. 1.18). Several new areas were prospected for potential detailed study in following years.

Fig. 1.17 French team in 2002. Left to right: E. Debard, M.-A. Héran, M. Faure, J.-F. Pastre, C. Guérin, A. Valli, A. Monguillon

Fig. 1.18 J.F. Pastre preparing to drill a core in parcel 233, 2000, observed by A. Valli, A. Monguillon and E. Debard

2003. The third field season again lasted four weeks, from June 30 to July 26. Faure was granted a permit for planned excavation (fouille programmée), which was renewed annually through 2006. We were joined by Debard, Pastre and Reddy, along with Luca Aprile (Associazione Aditus in Rupe, Floridia, Italy; topography and mapping); Elvira Martín Suárez (Universidad de Granada, Spain; micromammal paleontology); and Abel Prieur (UCB-Lyon 1; fossil collection and preparation). Monguillon and Valli continued technical supervision of the student excavators. Students and volunteers (from UCB-Lyon 1 and Lumière-Lyon 2 unless otherwise indicated) included: Catherine Albouy, Olivier Ambrosini, Anthony Bailly, Adama Cisse, Guillaume Colombeau, Camille Daujeard, Mikaël Dumas, Clément Fay, Marie-Anne Héran, Russell Hogg (CUNY), Pierre Joris, Bénédicte Kay, Bernard Ménouret, Fabien Méras, Emilie Pijolat, Caroline Ramirez, Caitlin Schrein (Arizona State University), Michelle Singleton (Midwestern University, Illinois) and Isabelle Sobis. Visiting colleagues included: Pierre Élouard, Pierre Hantzpergue and Pierre Mein (UCB-Lyon 1), Véra Eisenmann (CNRS and MNHN Paléontologie), Evelyne Crégut-Bonnoure (Musée Requien d'Histoire naturelle, Avignon), George Lyras (University of Athens) and Sandra van der Geer (University of Leiden).

A surface for excavation was exposed in parcel 233 near the place where T2 had been dug two years previously; this was termed zone H8-N9-L14-G11. Additional fragments of the 2001 rhino were recovered, along with teeth of a cheetah (*Acinonyx pardinensis*) and teeth and foot bones of bovids and cervids (especially *Metacervoceros rhenanus philisi*). A partial tusk of a mammoth (*Mammuthus meridionalis*) was found nearby and excavated, then covered in a plaster jacket for transport (Fig. 1.19). Trench T4 was dug in parcel 174 (landowner Ms. Marie Lonjon) but proved sterile of fossils. J. Argant sampled for additional pollen in the area of the rhino skeleton of T2 and also located a number of coprolites (presumably of a hyaenid) nearby. Martín Suárez supervised sampling for micromammals by passing several tons of sun-dried sediment through a specially-designed screen-washing system (Fig. 1.20), but no identifiable specimens were recovered. In light of this effort (also in 2004), we must note that Eugen Huber, who undertook screen-washing for Stehlin and Schaub from 1916–1940, had immense good luck to find even a few rodent teeth as discussed above.

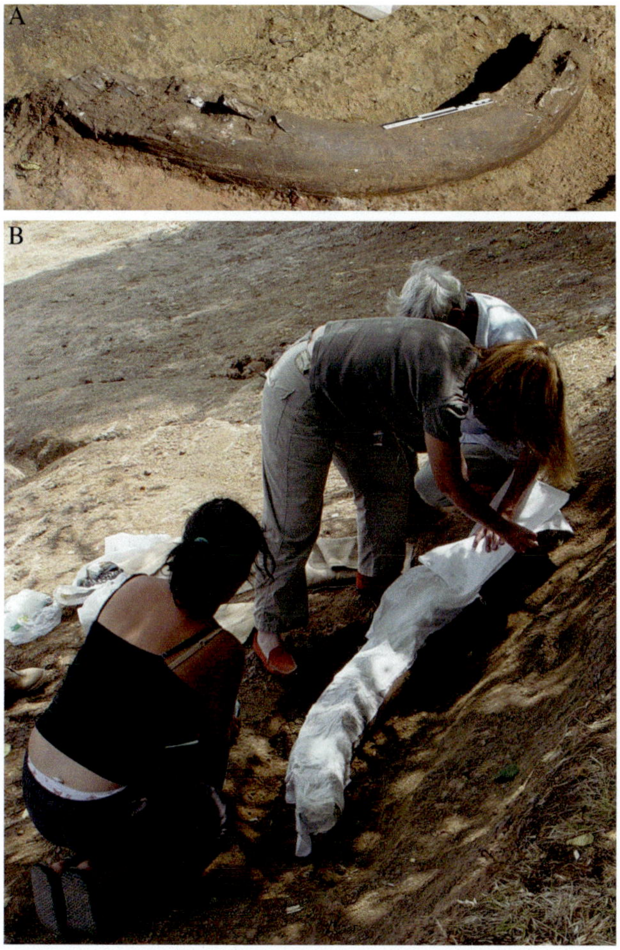

Fig. 1.19 Partial tusk of *Mammuthus meridionalis*. **A** After cleaning in place; **B** Jacketing in plaster, in order to protect it for transport (see Ménouret & Guérin 2024)

Fig. 1.20 E. Martín Suárez, assisted by student volunteers, washing dried sediment through screens in order to search for microfaunal remains

2004. The field season ran from July 1–31. Debard, Martín Suárez (and Matthijs Freudenthal, also from Granada and Naturalis/Leiden), Parenti (assisted by Federica Candelato, Vittorio Rioda and Maria Primicerio) and Pastre continued in their specialties. Students and other volunteers (from UCB-Lyon 1 and Lumière-Lyon 2) included: Audrey Crousilles, Julien Droz-Vincent, Véronique Fiastre, Stéphanie Guillerme, Marie-Pierre Guirado, Marie-Anne Héran, Mathieu Mazières, Bernard Ménouret, Guillaume Michel, Adrien Pozzi, Violette Ravel, Isabelle Sobis, Baptiste Sucheras, Jean-Alix Suy, Bérangère Tarka and Thi My Linh Vo. William Harcourt-Smith (AMNH postdoc) supervised training of three Lehman College (CUNY) undergraduate students (Randy Rampersaud, Roxanne Rivera and Jennifer Ross; see Fig. 1.21). Visiting colleagues included: Katherine St. John (Lehman College, CUNY), Kieran McNulty (Baylor University) and Tamara Munzner (University of British Columbia).

Excavation surface H8-N8-L14-G11 was reopened in parcel 233 and yielded most of the postcranial skeleton of a deer, *Eucladoceros ctenoides senezensis* (Fig. 1.22, also Fig. 2.6; see Valli 2024). The screen-washing operation yielded a porcupine incisor, representing a new family for the site (Hystricidae, see Mörs and Hugueney 2017) and also one molar of a small rodent (*Mimomys pitymyoides*), as well as some artiodactyl teeth. A major excavation was undertaken in parcel 172, where trench T5 was dug perpendicular to the slope toward the center of the maar. It was some 30 m long and 3 m wide, about 3 m deep at any one point but over 11 m in exposed elevation (Fig. 1.23). T5 yielded a partial skeleton and another partial cranium of *Eucladoceros*; articulated foot bones of horse (*Allohippus senezensis*, Fig. 1.24), rhino and artiodactyl; and teeth of bovids and cervids. Sevket Sen took samples for paleomagnetic analysis in T5, in zone H8-N8-L14-G11 and in a small pit about 1 m^2 in area and 2.5 m deep which had been dug along the northern edge of the same parcel (233) for that purpose. J. Argant sampled for pollen in T5.

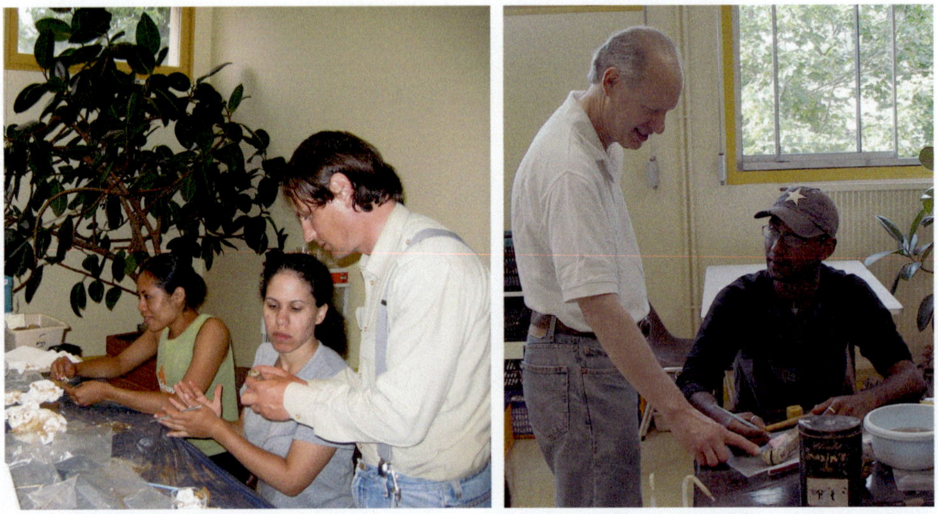

Fig. 1.21 Lehman College undergraduates supported by NSF UBM award in 2004. Left, A. Valli helping R. Rivera (left) and J. Ross wash fossils; right, E. Delson showing R. Rampersaud how to clean fossils

Fig. 1.22 Left, excavation surface H8-N8-L14-G11 in parcel 233, 2004; right, closeup of students excavating postcranial skeleton of *Eucladoceros ctenoides senezensis*

Fig. 1.23 Trench 5, 2004; lower (inf) segment, with E. Debard and J.-F. Pastre, "step" and excavation surface and upper (sup) segment in shadow; total drop from land surface at back of upper segment to lowest point about 11 m, length ca. 30 m

Fig. 1.24 Partial foot of *Allohippus senezensis* from T5 "step", 2004

2005. The field season ran from July 4–30, with the team leaders accompanied by Debard, Pastre, Valli and Monguillon (who also took responsibility for topography and mapping-in of fossils). Students and other volunteers (from UCB-Lyon 1 and Lumière-Lyon 2 unless otherwise indicated) included: Vivien Aubry, Siobhán Cooke (CUNY), Céline Domergue, Sarah Freidline (CUNY), Pauline Girard, Stéphanie Guillerme, Alexandra Houssaye, Bernard Ménouret, Isabelle Sobis and Michael Steiper (CUNY).

Excavation surface H8-N8-L14-G11 was reopened in parcel 233 in an unsuccessful search for the skull of the *Eucladoceros* skeleton recovered in 2004. Work concentrated in parcel 172 where a smaller trench (T6) was opened near where T5 had been excavated (Figs. 1.25 and 1.26). In addition to smaller finds, a nearly complete skeleton of *Allohippus senezensis* was recovered near the base of the section (Fig. 1.27; see Eisenmann & Delson 2024, Fig. 11.24; also Fig. 2.7). In parcel 228, the naturally

Fig. 1.25 Trench 6, 2005 (left, during excavation; right, at the end of the season)

Fig. 1.26 Excavating equid skeleton near T6, 2005

exposed "Philis ravine", where P. Philis had recovered many fossils in the first half of the twentieth century, was cleaned and studied stratigraphically thanks to the permission of landowner M. Omer Philis, but no additional fossils were found.

2006. The field season ran from July 3–28, under the direction of Guérin and Delson (Faure was only available for part of the season), assisted by Valli and Monguillon (who also took responsibility for topography and mapping-in of fossils). Debard and Pastre supervised geological excavation and tephra sampling; J. Argant sampled for pollen. Yolanda Fernández-Jalvo (Museo Nacional de Ciencias Naturales,

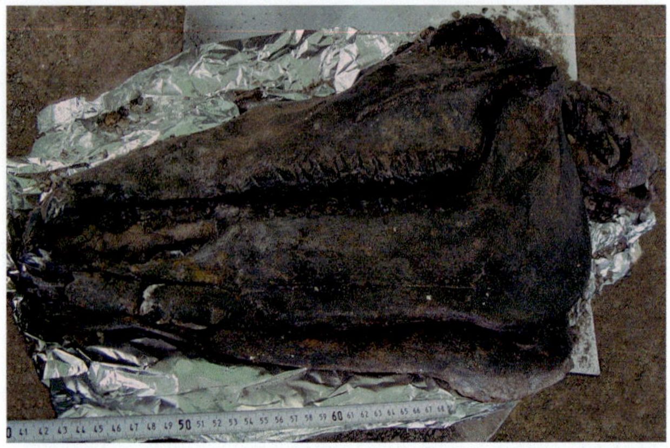

Fig. 1.27 Skull of *Allohippus senezensis* partial skeleton from area alongside T6

Madrid) studied the taphonomy, assisted by Valli. Students and other volunteers (from UCB-Lyon 1 and Lumière-Lyon 2 unless otherwise indicated) included: Florent Bartolucci, Hélène Bourget, Virginie Caup, Camille Clément, Audrey Crousilles, Erwan Ermel, Eva Garrett (CUNY), Christine Guingenaud, Hélène Lecas, Bernard Ménouret, Florence Olive, France Rousselières and Ilya Shmulenson (CUNY). In addition, James Delson (Eric's brother), his wife Jennifer and children Caroline and Robert spent several days volunteering on the site.

Another trench, T7, was emplaced in parcel 172 near where T5 and T6 had been (Figs. 1.28 and 1.29). A partial skeleton of *Allohippus senezensis* was found in the middle levels, where it had been held in place by large blocks; it was much less well preserved than the 2005 skeleton (see Eisenmann & Delson 2024, Fig. 11.25). To obtain additional stratigraphic data, trenches T8 and T9 were excavated nearby, but no fossils were recovered.

Fig. 1.28 Trench 7, 2006

Fig. 1.29 Team members cleaning and excavating in T7, 2006; **A** student volunteers; **B** E. Delson (photo by I. Shmulenson)

Other Senèze Research After 2000

A small number of publications focused on Senèze and its fossils have appeared since 2000, some of which have been mentioned above. Roger et al. (2000) sampled the 1989 cores for paleomagnetism and also discerned a thin (5 cm) volcanic tephra within the normal interval. This was dated by $^{40}Ar/^{39}Ar$ at 2.1 ± 0.1 Ma, which was suggested as the end of the Réunion subchron. The Senèze fauna was thought to lie above this level, following the work of Bout (1970a, 1970b) .

At the 2004 Weimar conference on Late Neogene and Quaternary faunas, Delson et al. (2004) reported the ongoing fieldwork at Senèze and Maul (2004) discussed Schaub's Senèze rodents (see above). Also in 2004, the site was visited by members of the Association Française pour l'Etude du Quaternaire during a field excursion on Quaternary and vulcanism of Auvergne and Velay (Argant 2004; Debard & Pastre 2004; Faure et al. 2004). Delson et al. (2006) published a long summary of the Franco-American work at Senèze (in the proceedings of the 2004 conference). Nishimura et al. (2009) analyzed a CT scan of the unique holotype cranium of Senèze *Paradolichopithecus arvernensis*, reporting the lack of a maxillary sinus as found otherwise only in *Macaca* among extant Cercopithecinae (see Delson 2024).

Nomade et al. (2014) reported the results of a major campaign of argon dating French Villafranchian faunal horizons. Five dates were obtained from tephras collected by our team, resulting in an age range between 2.21–2.09 Ma. These ages were recalculated to current standards and compared to paleomagnetic sequences in Delson et al. (2024b). Pastre et al. (2015) presented a range of geological data from the Franco-American fieldwork, including material developed at greater length in Parenti et al. (2024), Debard (2024), Pastre (2024) and several of the mammalian systematic chapters, as well as a brief discussion of Villafranchian biochronology. A partial upper incisor recovered during the 2004 field season was identified as *Hystrix refossa* by Mörs and Hugueney (2017). Paquette et al. (2021) reported several U–Pb dates on zircon from a number of French Villafranchian sites (see Delson et al. 2024b), including one of 2.100 ± 0.029 Ma from Senèze.

Plan of This Volume

The focus of this book is the work undertaken from 2001–2006 by the Franco-American field project. The first four chapters after this introduction cover aspects of the techniques employed and the geological background to the paleontological results. In Chap. 2, Parenti et al. (2024) discuss their surveying of the site and the basic excavation techniques employed; they also present several maps and measured drawings of partial skeletons recovered during the fieldwork. In Chap. 3, Debard (2024) presents the stratigraphical framework of the site and discusses aspects of correlation among the numerous trenches and other sampled sections. Pastre (2024, Chap. 4) describes the volcanic context of Senèze and details of the tephras identified. In Chap. 5, Delson et al. (2024b) discuss the age of the site, based on a series of argon-argon dates (originally reported by Nomade et al. 2014, but recalculated here) and paleomagnetic "columns", suggesting that the Senèze fauna was deposited between 2.20 and 2.08 Ma, with the fossils from our fieldwork coming from two intervals ca. 2.20–2.18 and 2.10–2.08 Ma.

Three chapters then present non-mammalian paleontology. In Chap. 6, J. Argant (2024) discusses palynological finds and their implications for paleoenvironment. Gaudant (2024) describes some fossil fishes recovered from our excavations in Chap. 7. In Chap. 8, Mourer-Chauviré (2024) describes the birds from Senèze and considers their importance for reconstructing the paleoenvironment; seventeen species are recognized, including four extinct ones: *Pavo bravardi*, *Tetrao* cf. *partium*, *Surnia robusta*, and *Corvus corax antecorax*.

The fossil mammals from Senèze are discussed in detail in the following eight chapters. A. Argant (2024, Chap. 9) revises the carnivores in detail. Twelve species are recognized: four canids (*Nyctereutes megamastoides*, *Vulpes alopecoides*, *Canis arnensis* and *Canis* sp.), one ursid (*Ursus etruscus*), three hyaenids (*Pachycrocuta perrieri*, *Chasmaporthetes lunensis*, and an indeterminate hyaenid), and four felids (*Acinonyx pardinensis*, *Megantereon cultridens*, *Homotherium crenatidens* and a new record for Senèze, *Dinofelis* sp.). Guérin (2024) reviews the rhino, *Dicerorhinus etruscus etruscus*, in Chap. 10. He describes the large sample of this form from Senèze, including a newly recovered partial skeleton, 17 crania and numerous other elements. Eisenmann and Delson (2024) distinguish three genera of monodactyl equids (*Equus*, *Plesippus* and *Allohippus*) and review all equid specimens known from Senèze in Chap. 11. The great majority of fossils belong to *Allohippus senezensis senezensis*, with less than 20 identified as *Allohippus major*. There are also a few specimens either larger or smaller than the former. Faure and Guérin (2024, Chap. 12) discuss the suid *Sus strozzii*. No new specimens were recovered, but the previously known sample includes a nearly complete skeleton; the species is a good indicator of a forested environment with a wet and probably warm climate. In Chap. 13, Valli (2024) discusses the newly recovered cervid material of *Eucladoceros ctenoides senezensis* and *Metacervoceros rhenanus philisi*; *Croizetoceros ramosus minor* and *Cervalces gallicus* are also known from Senèze, but no specimens of these taxa were recovered during our fieldwork. Chapter 14 is a detailed review by Crégut-Bonnoure (2024) of the bovid taxa known from Senèze. She recognizes *Gazellospira torticornis*, *Procamptoceras brivatense*, *Gallogoral meneghinii*, *Pliotragus ardeus*, *Megalovis latifrons*, cf. *Hemitragus* sp., *Ovis claudiusguerini* nov. sp., *Leptobos etruscus*, *Leptobos furtivus* and Bovidae indet. *Bison* (*Eobison*) sp. is known from a donated specimen which appears to derive from a significantly younger horizon; one specimen of *Megalovis* was also donated from a different unknown horizon. In Chap. 15, Ménouret and Guérin (2024) review the remains of *Mammuthus meridionalis meridionalis* from the site, including the one partial tusk found in 2004. Delson (2024, Chap. 16) discusses the two primate specimens recovered by Philis from Senèze: a partial ulna assigned to *Macaca* cf.

sylvanus and the holotype skull of *Paradolichopithecus arvernensis*. The latter species is differentially diagnosed, and the Senèze specimen (one of the youngest known for this taxon) is described and compared to those of other *Paradolichopithecus* samples.

The last three chapters are analytical summaries. Crégut-Bonnoure et al. (2024) review the biochronological position of Senèze within the Villafranchian in Chap. 17, comparing the mammalian assemblage to those of the Massif Central (and Saint-Vallier) in France as well as to selected localities in Spain, Italy, the Netherlands and Germany. Senèze is distinguished from older sites (in MNQ 17) and clarified as the earliest in MNQ 18, which is estimated to begin about 2.2 Ma. The end date of this unit is discussed, whether close to 2.0 or 1.7 Ma, depending on which other localities are included in it. In Chap. 18, Fernández-Jalvo et al. (2024) discuss the taphonomy of large mammals from Senèze. They conclude that most animals found as fossils died after drinking at the lake shore and falling into the water, from which they were unable to escape and drowned. In Chap. 19, Delson et al. (2024a) provide longer summaries of each chapter's results and broader conclusions about paleoenvironment, faunal composition and site formation.

Acknowledgments We honor the memory of our friend Paul Sondaar, who invited us in autumn 1999 to the International Workshop "On Late Plio/Pleistocene extinction and evolution in the Palearctic: The Vatera site" (on the island of Lesbos, Greece), where we began the discussion of resuming fieldwork at Senèze, in turn leading to this volume. Our new fieldwork at Senèze was only possible thanks to the authorizations for survey and excavation from the French *Ministère de la Culture et de la Communication*, for which we are extremely grateful. We thank the *Service Régional de l'Archéologie* of the *Direction Régionale des Affaires Culturelles* of Auvergne, in particular Messieurs P. Vergain and Frédérik Letterlé, then Conservateurs régionaux, Monsieur René Liabeuf, Conservateur, and Madame Isabelle Magy. Furthermore we thank Monsieur Gustave Courtet, Maire of the commune of Domeyrat (Haute-Loire) who authorized our fieldwork at the hamlet of Senèze, and his successor at the Mairie, Christian Poulet, who always supported us. We thank Mesdames Marguerite Limagne and Marie Lonjon, Messieurs J.J. Chabrier (of Rafayet), Maurice Oléon, Omer Pierre Philis, Gérard Vierne and the *Direction* of the Muséum National d'Histoire naturelle in Paris, the landowners who graciously gave us permission to undertake excavations and surveys and to seek geological data on their land.

We thank Madame Catherine Henri-Martin, then Vice-Présidente du Conseil Régional d'Auvergne; Monsieur Nicolas Di Giambattista, then Conseiller général and Président of the *Communauté de Communes du Pays de Paulhaguet;* the *Conseil Général de Haute-Loire*, its successive Présidents Messieurs Jacques Barrot and Gérard Roche; Monsieur Jean-Luc Barré, then Conservateur of the Conseil's *Direction de la Jeunesse, de la Culture et du Tourisme* (*Service Conservation et Patrimoine*); and Madame Karine Roche of its *Service Communication*. At the Université Claude Bernard-Lyon 1, we are grateful for support from Messieurs Christophe Lécuyer, then Directeur of the UMR 5125 *Paléoenvironnements et paléobiosphère*, and Serge Legendre, then Sous-directeur; Monsieur Abel Prieur, then head of the *Service des Collections* of the Geology Laboratory and his successor Monsieur Emmanuel Robert; and Madame Dominique Barbe, then in charge of the duplicating office, who with all her expertise dedicated much of her time to duplicating and editing our annual reports. We also thank M. Michel Philippe and M. Didier Berthet, successive Curators of the collections of fossil and modern vertebrates in the Muséum d'Histoire naturelle de Lyon (today the Musée des Confluences), for permitting us and our colleagues access to their collections.

We thank the many funding agencies who provided the financial support necessary for all of our field and laboratory efforts. The French *Ministère de la Culture et de la Communication* provided both field and laboratory funding from 2000–2006. The UMR "Paléoenvironnements et Paléobiosphère" of CNRS/UCB-Lyon 1 provided funding and loaned a field vehicle; the *Conseil Général de la Haute Loire* awarded field funds. The US National Science Foundation awarded a High Risk Exploratory Research grant (BCS 00-83166) in 2000 and a Small Grant for Exploratory Research (BCS 01-04781) in 2001 for work at Senèze; in addition, an Undergraduate Biology & Mathematics Initiative award (UBM 03-37561) supported training of Lehman College students and travel to Senèze for several in 2004. The National Geographic Society awarded grant numbers 7050-01 in 2001 and 7434-03 in 2003; and the Leakey Foundation provided funds in 2001. The PSC-CUNY Faculty Research Award program awarded funds each year: grants 63528, 2001; 64542-33, 2002; 65509-34, 2003; 66578-35, 2004; 67356-36, 2005; and 68016-37 in 2006.

We thank Monsieur A. Poitrasson-Rivière, Principal du Collège (high school) du Val-de-Sénouire at Paulhaguet (Haute-Loire), and his staff, notably Monsieur L. Redon (gestionnaire), who opened the doors of his school to provide us lodging and food-service. We acknowledge Monsieur Robert Amaduble, the cook at the Collège, who agreed to provide our morning and evening meals during his summer vacation; Madame Jacqueline Oléon, who served us our midday meals in her modest restaurant in Domeyrat, despite our sometimes large and rowdy group; Monsieur O. P. Philis and Madame Marinette Da Costa, the grandchildren of Pierre Philis, and Madame Sylvie Philis, his great-granddaughter, for access to their abundant family archives, and Angélique Monguillon (then a Ph.D. student), for scanning many of these documents; Madame Renée Martin of Domeyrat, who provided us with constant aid and encouragement, from our first steps with the cadastral plan; the *municipalité de Domeyrat* which gave us the possibility to store our equipment and our finds locally during the six years of our work; and Monsieur Vidal, from Collat, for the care with which he opened and closed our excavations with his backhoe.

We are grateful to all members of the Senèze project, especially Angélique Monguillon and Andrea Valli for their day-to-day supervision of student activities; David P. Reddy, for informatics maintenance; as well as our other colleagues and the many students and other volunteers listed above who participated in fieldwork. We especially acknowledge our friends Elvira Martín-Suarez and Matthijs Freudenthal, who did not hesitate to transport their ingenious screen-washing station more than 1400 km in order that we might profit from its use. Even if their efforts resulted in only two rodent teeth, they washed and sorted many tons of sediment with student assistance along the banks of the Doulon creek, in hopes of finding more micromammal remains, which are exceedingly rare at Senèze.

Finally, we thank Roberta Marx Delson and Eric Sargis for helpful comments on an earlier version of this manuscript.

References

Antón, M., & Werdelin, L. (1998). Too well restored? The case of the *Megantereon* skull from Senèze. *Lethaia, 31*, 165–243.

Argant, A. (2024). The carnivores of Senèze (Haute-Loire). In E. Delson, M. Faure, & C. Guérin (Eds.), *Senèze: Life in Central France two million years ago. Paleontology, geochronology, stratigraphy and taphonomy* (pp. 165–243). Cham (Switzerland): Springer.

Argant, J. (2004). Palynologie [du maar de Senèze]. In J.-F. Pastre (Ed.), *Quaternaire et volcanisme en Auvergne et Velay. Livret-guide de l'excursion de l'Association Française pour l'Etude du Quaternaire (mai, 2004)* (pp. 108–113).

Argant, J. (2024). Palynological analysis of the fossiliferous layers of the late Villafranchian deposit of Senèze (Domeyrat, Haute-Loire, France). Taphonomy and paleoenvironmental data. In E. Delson, M. Faure, & C. Guérin (Eds.), *Senèze: Life in Central France two million years ago. Paleontology, geochronology, stratigraphy and taphonomy* (pp. 123–140). Cham (Switzerland): Springer.

Azzaroli, A. (1952). L'Alce di Senèze. *Palaeontographia italica, 47*, n. ser., 17 (1950–1951), 133–141.

Azzaroli, A. (1970). Villafranchian correlations based on large mammals. *Giornale di Geologia, 35*, 1–21.

Azzaroli, A. (1982). On the quaternary and recent cervid genera *Alces, Cervalces, Libralces. Bollettino della Società Paleontologica Italiana, 20*, 147–154.

Azzaroli, A., De Giuli, C., Ficcarelli, G., & Torre, D. (1988). Late Pliocene to early Mid-Pleistocene mammals in Eurasia: Faunal succession and dispersal events. *Palaeogeography, Palaeoclimatology, Palaeoecology, 68*, 77–100.

Ballesio, R. (1963). Monographie d'un Machairodus du gisement villafranchien de Senèze: *Homotherium crenatidens* Fabrini. *Travaux du laboratoire de Géologie de la Faculté des Sciences de Lyon,* N.S. *9*, 1–129.

Boule, M. (1892). Découverte d'un squelette d'*Elephas meridionalis* dans les cendres basaltiques du volcan de Senèze (Haute-Loire). *Comptes rendus hebdomadaires des séances de l'Académie des sciences, 115*, 624–626.

Boule, M. (1892–1893). Description géologique du Velay. *Bulletin des Services de la Carte géologique de la France*, 4e année, *28*, 1–259.

Bout, P. (1960). Le Villafranchien du Velay et du Bassin hydrographique moyen et supérieur de l'Allier. Thèse de Doctorat ès Sciences naturelles, Université de Paris, Imprimerie Jeanne d'Arc édit., Le Puy.

Bout, P. (1970a). Absolute ages of some volcanic formations in the Auvergne and Velay areas and chronology of the European Pleistocene. *Palaeogeography, Palaeoclimatology, Palaeoecology, 8*, 95–106.

Bout, P. (1970b). Problèmes du volcanisme XV-le maar de Senèze (Haute-Loire). *Revue d'Auvergne, 84*, 54–68.

Brébion, P., Calas, P., Drot, J., Pimienta-Freneix, S., & Nazemi, L. (1953). Le gisement villafranchien de Senèze [Haute-Loire]; unpublished manuscript in the archives of the Laboratoire de Paléontologie, MNHN, Paris, 5 p.

Breda, M. (2001). The holotype of *Cervalces gallicus* (Azzaroli, 1952) from Senèze (Haute-Loire, France) with nomenclatural implications and taxonomical-phylogenetic accounts. *Rivista Italiana di Paleontologia e Stratigrafia, 107*, 439–449.

Christiansen, P., & Adolfssen, J. S. (2007). Osteology and ecology of *Megantereon cultridens* SE311 (Mammalia; Felidae; Machairodontidae), a sabre cat from the Late Pliocene-Early Pleistocene of Senèze, France. *Zoological Journal of the Linnean Society, 151*, 833–884.

Couthures, J. (1989). The Senèze maar (French Massif Central): Hypothesis regarding a catastrophe occuring about 1.5 million years ago. *Journal of Volcanology and Geothermal Research, 39*, 207–210.

Couthures, J., & Pastre, J. F. (1983). Chronostratigraphie du Plio-Pléistocène d'Auvergne et du Velay: nouveaux apports des datations radiométriques et du paléomagnétisme. *Bulletin de l'Association française pour l'étude du Quaternaire,* 1983-1, 9–18.

Couthures, J., Ablin, D., & Hadjouis, D. (1991). Senèze [Haute-Loire]. Rapport des fouilles 1991. Service Régional de l'Archéologie, Direction Régionale des Affaires Culturelles d' Auvergne, 55 p.

Crégut-Bonnoure, E. (2024). The Bovidae (Mammalia, Cetartiodactyla) from Senèze. In E. Delson, M. Faure, & C. Guérin (Eds.), *Senèze: Life in Central France two million years ago. Paleontology, geochronology, stratigraphy and taphonomy* (pp. 433–585). Cham (Switzerland): Springer.

Crégut-Bonnoure, E., Guérin, C., Argant, A., Argant, J., Debard, E., Delson, E., et al. (2024). Biochronology of the Senèze faunal assemblage. In E. Delson, M. Faure, & C. Guérin (Eds.), *Senèze: Life in Central France two million years ago. Paleontology, geochronology, stratigraphy and taphonomy* (pp. 633–652). Cham (Switzerland): Springer.

David, L. (1962). Rapport annuel 1962. *Documents des laboratoires de Géologie de la Faculté des Sciences de Lyon, 1*, 3–75.

Debard, E. (2024). Geological study of the Early Pleistocene site of Senèze (Domeyrat, Haute-Loire, France). In E. Delson, M. Faure, & C. Guérin (Eds.), *Senèze: Life in Central France two million years ago. Paleontology, geochronology, stratigraphy and taphonomy* (pp. 37–82). Cham, Switzerland: Springer.

Debard, E. & Pastre, J.-F. (2004). Géologie [du maar de Senèze]. In J.-F. Pastre (Ed.) *Quaternaire et volcanisme en Auvergne et Velay. Livret-guide de l'excursion de l'Association française pour l'étude du Quaternaire (mai, 2004)* (pp. 102–108).

Delson, E. (2024). Primates from Senèze. In E. Delson, M. Faure, & C. Guérin (Eds.), *Senèze: Life in Central France two million years ago. Paleontology, geochronology, stratigraphy and taphonomy* (pp. 607–632). Cham (Switzerland): Springer.

Delson, E., Faure, M., Guérin, C., Aprile, L., Argant, J., Blackwell, B., et al. (2004). Franco-American research at the Villafranchian locality of Senèze (Haute-Loire, France): Why are new enquiries needed? In L. C. Maul & R.-D. Kahlke (Eds.), *Conference volume for 18th International Senckenberg Conference—VI International Palaeontological Colloquium in Weimar. Late Neogene and Quaternary biodiversity and evolution: Regional developments and interregional correlations*. Weimar (Germany), 25–30 April, 2004. *Terra Nostra, 2004* (2), 96–97.

Delson, E., Faure, M., Guérin, C., Aprile, L., Argant, J., Blackwell, B., et al. (2006). Franco-American renewed research at the Late Villafranchian locality of Senèze (Haute-Loire, France). *Courier Forschunginstitut Senckenberg, 256*, 275–290.

Delson, E., Guérin, C., Faure,. M, Aprile, L., Argant, A., Argant, J., et al. (2024a). Summary and conclusions of the Senèze research project. In E. Delson, M. Faure & C. Guérin (Eds.), *Senèze: Life in Central France two million years ago. Paleontology, geochronology, stratigraphy and taphonomy* (pp. 683–699). Cham (Switzerland): Springer.

Delson, E., Nomade, S., Sen, S., Debard, E., Pastre, J-F., Bahain, J.-J., et al. (2024b). Geochronology of Senèze: $^{40}Ar/^{39}Ar$ dating and magnetostratigraphy, with notes on an ESR/U-series dating attempt. In E. Delson, M. Faure & C. Guérin (Eds.), *Senèze: Life in Central France two million years ago. Paleontology, geochronology, stratigraphy and taphonomy* (pp. 99–122). Cham (Switzerland): Springer.

Depéret, C. (1928). Les singes fossiles du Pliocène de France. *Comptes-Rendus du XIVe Congrès Géologique International, 1926* (pp. 3–4). Madrid: Gráficas Reunidas.

Depéret, C. (1929). *Dolichopithecus arvernensis* Depéret: Nouveau singe du Pliocène supérieur de Senèze (Haute-Loire). *Travaux du*

Laboratoire de Géologie de la Faculté des Sciences de Lyon, fascicule *15* (mémoire 12), 5–12.

Depéret, C., & Mayet, L. (1911). Le gisement de Senèze et sa faune paléomammalogique. *Association Française pour l'Avancement des Sciences, Compte Rendu de la 39e session Toulouse 1910. Notes et Mémoires, 2*, 261–263.

Depéret, C., & Mayet, L. (1912). Le gisement de mammifères pliocènes de Senèze (Haute-Loire). *Association Française pour l'Avancement des Sciences, Compte Rendu de la 40e session Dijon 1911. Mémoire hors série*, 138–151.

Depéret, C., Mayet, L., & Roman, F. (1923). Les éléphants pliocènes. *Annales de l'Université de Lyon, NS 1, 42*, 1–224.

Devis, G. (1970). Le volcanisme et le gisement fossilifère de Senèze (Haute-Loire). *Revue des Sciences naturelles d'Auvergne, 36*, 25–87.

Duvernois, M.-P., & Guérin, C. (1989). Les Bovidae (Mammalia, Artiodactyla) du Villafranchien supérieur d'Europe occidentale. *Geobios, 22*, 339–379.

Ehrlich, A. (1968). Les diatomées fossiles des sédiments villafranchiens de Senèze (Haute-Loire, Massif central français). *Bulletin de l'Association française pour l'étude du Quaternaire, 4*, 267–280.

Eisenmann, V., & Delson, E. (2024). The Senèze equids. In E. Delson, M. Faure, & C. Guérin (Eds.), *Senèze: Life in Central France two million years ago. Paleontology, geochronology, stratigraphy and taphonomy* (pp. 307–385). Cham (Switzerland): Springer.

Elhaï, H. (1969). La flore sporo-pollinique du gisement villafranchien de Senèze (Massif Central, France). *Pollen et Spores, 9*, 127–139.

Elhaï, H., & Grangeon, P. (1963). Nouvelles recherches sur le gisement villafranchien de Senèze (Haute-Loire, France). *Bulletin de la Société Géologique de France,* Série, *7*, 483–488.

Faure, M., & Guérin, C. (1979). Le problème de l'hippopotame villafranchien de Senèze (Haute-Loire, France) et ses implications biostratigraphiques. 7ème R.A.S.T. (Société géologique de France), Lyon, Avril 1979, p. 184.

Faure, M., & Guérin, C. (2024). *Sus strozzii* (Mammalia, Suidae) of the Late Villafranchian site of Senèze (Domeyrat, Haute-Loire, France). In E. Delson, M. Faure & C. Guérin (Eds.), *Senèze: Life in Central France two million years ago. Paleontology, geochronology, stratigraphy and taphonomy* (pp. 387–407). Cham (Switzerland): Springer.

Faure, M., Guérin, C., & Delson, E. (2004). Le gisement villafranchien de Senèze à Domeyrat (Haute-Loire). In J.-F. Pastre (Ed.) *Quaternaire et volcanisme en Auvergne et Velay. Livret-guide de l'excursion de l'Association française pour l'étude du Quaternaire (mai, 2004)* (pp. 114–122).

Faure, M., Sen, S., & Costeur, L. (2022). Pierre Philis *Fossil Hunter*: un demi-siècle de recherches de vertébrés fossiles (1892–1942). *Geodiversitas, 44*, 321–383.

Fernández-Jalvo, Y., Valli A. M. F., Marín-Monfort, M. D. & Pesquero-Fernández, M. D. (2024). The taphonomy of Senèze. In E. Delson, M. Faure, & C. Guérin (Eds.), *Senèze: Life in Central France two million years ago. Paleontology, geochronology, stratigraphy and taphonomy* (pp. 653–681). Cham (Switzerland): Springer.

Fortelius, M., Mazza, P., & Sala, B. (1993). *Stephanorhinus* (Mammalia: Rhinocerotidae) of the Western European Pleistocene, with a revision of *S. etruscus* (Falconer, 1868). *Palaeontographica Italica, 80*, 63–155.

Gaudant, J. (1975). Présence du genre *Tinca* Cuv. (Poisson Téléostéen, Cypriniforme) dans le Villafranchien de Senèze (Haute-Loire). *Geobios, 8*, 353–354.

Gaudant, J. (2024). The fishes from the Maar of Senèze. In E. Delson, M. Faure, & C. Guérin (Eds.), *Senèze: Life in Central France two million years ago. Paleontology, geochronology, stratigraphy and taphonomy* (pp. 141–144). Cham (Switzerland): Springer.

Gaudry, A. (1893). L'éléphant de Durfort, *in Centenaire de la fondation du Muséum d'Histoire naturelle, 10 juin 1793–19 juin 1893*. Volume commémoratif publié par les Professeurs du Muséum (pp. 327–347). Paris: Imprimerie nationale.

Gaudry, A. (1896). *Essai de paléontologie philosophique*. Paris: Masson.

Gaudry, A. (1898). Le nouveau musée de Paléontologie. *Revue des deux Mondes, 149*, 799–824.

Gómez Llueca, F. (1921). El Laboratorio de Geología de la Facultad de Ciencias de la Universidad de Lyon. Tomo extraordinario publicado con motivo del 50° aniversario de su fundación. *Real Sociedad española de Historia natural*, 379–387.

Gómez Llueca, F. (1929). El Profesor Charles Depéret (1854–1929). *Conferencias y reseñas científicas de la Real Sociedad española de Historia natural, 4*, 137–150.

Grangeon, P. (1962). Note préliminaire à l'étude géologique du volcan et du gisement fossilifère de Senèze (Haute-Loire). *Compte Rendu sommaire des séances de la Société géologique de France, 5*, 149–150.

Guérin, C. (1980). Les rhinocéros (Mammalia, Perissodactyla) du Miocène terminal au Pléistocène supérieur en Europe occidentale. Comparaison avec les espèces actuelles. *Documents des Laboratoires de Géologie de Lyon, 79*, 1–1185.

Guérin, C. (1982). Première biozonation du Pléistocène européen, principal résultat biostratigraphique de l'étude des Rhinocerotidae (Mammalia, Perissodactyla) du Miocène terminal au Pléistocène supérieur d'Europe occidentale. *Geobios, 15*, 593–598.

Guérin, C. (2024). *Dicerorhinus etruscus etruscus* from the Late Villafranchian site of Senèze (Domeyrat, Haute-Loire, Central France). In E. Delson, M. Faure, & C. Guérin (Eds.), *Senèze: Life in Central France two million years ago. Paleontology, geochronology, stratigraphy and taphonomy* (pp. 245–306). Cham (Switzerland): Springer.

Guth, C. (1975). Lagerstätten des Villafranchiums in der Umgebung von Brioude in Mittelfrankreich. *Quartärpaläontologie, 1*, 177–182.

Heintz, E. (1970). Les Cervidés villafranchiens de France et d'Espagne. *Mémoires du Muséum national d'Histoire naturelle, 22*, Volumes I and II, 1–317 and 1–206.

Heintz, E., Guérin, C., Martin, R., & Prat, F. (1974). Principaux gisements villafranchiens de France: listes fauniques et biostratigraphie. *Mémoires du Bureau des Recherches Géologiques et Minières, 78*, 169–182.

Hooijdonk, K. van (2006). De Sabeltandkatten *Homotherium* en *Megantereon* (Felidae, Carnivora) van de Plio-Pleistocene site van Senèze (Haute Loire, Fr.). *Cranium, 23*, 2, 25–38.

Kormos, Th. (1931). Oberpliocaene Wühlmäuse von Senèze und Val d'Arno. *Abhandlungen der Schweizerischen Palaeontologischen Gesellschaft, 51*, 1–14.

Lacombat, F., Mol, D., Borget, J. N., & Magne, E. (2010). Haute-Loire: des volcans, des mammouths & des hommes. Le Puy-en-Velay: Éditions Jeanne-d'Arc.

Lascombe, A. (1894). Lettre de M. Henry Mosnier au sujet du squelette fossile d'*Elephas meridionalis* trouvé à Senèze, canton de Paulhaguet. Séance du 3 novembre 1892. *Société agricole & scientifique de la Haute-Loire Mémoires et Procès-Verbaux de 1891, 1892, 1893, 7*, 86–88.

Martin, R. (1973). Trois nouvelles espèces de Caninae (Canidae, Carnivora) des gisements plio-villafranchiens d'Europe. *Documents des Laboratoires de Géologie de la Faculté des Sciences de Lyon, 57*, 87–96.

Masson, R. [1943 unpublished]. Le gisement de Mammifères de Senèze. Diplôme d'Etudes Supérieures, Faculté des Sciences de Lyon.

Matthew, W. D. (1921). Notes on the scientific museums of Europe. *Natural History, 21*, 185–190.

Maul, L. C. (2004). The arvicolids of the old collection of Senèze reconsidered. In L. C. Maul & R-D. Kahlke (Eds.), *Conference volume for 18th International Senckenberg Conference—VI International Palaeontological Colloquium in Weimar. Late Neogene and Quaternary biodiversity and evolution: Regional developments and interregional correlations*. Weimar (Germany), 25–30 April, 2004. *Terra Nostra, 2004*, 168–169.

Mazza, P., & Rustioni, M. (1994). The fossil bear from Senèze (Southern France). *Rendiconti Lincei. Scienze Fisiche e Naturali, 9* (5), 17–26.

Mein, P. (1975). Résultats du groupe de travail des vertébrés. In J. Senes (Ed.), *Report on Activity of the R.C.M.N.S. Working Groups* (pp. 78–81). Bratislava: Regional Committee on Mediterranean Neogene Stratigraphy.

Mein, P. (1989). Updating of MN Zones. In E. H. Lindsay, V. Fahlbusch, & P. Mein (Eds.), *European Neogene Mammal Chronology* (pp. 73-90). New York: Plenum Press.

Ménouret, B., & Guérin, C. (2024). The *Mammuthus meridionalis* of the Late Villafranchian site of Senèze (Domeyrat, Haute-Loire, Central France). In E. Delson, M. Faure, & C. Guérin (Eds.), *Senèze: Life in Central France two million years ago. Paleontology, geochronology, stratigraphy and taphonomy* (pp. 587–606). Cham (Switzerland): Springer.

Mörs, T., & Hugueney, M. (2017). A *Hystrix* Linnaeus, 1758 incisor (Mammalia, Rodentia) from the Early Pleistocene of Senèze, France. *Geodiversitas, 39*, 797–802.

Mourer-Chauviré, C. (2024). The birds (Aves) of the Early Pleistocene locality of Senèze (Haute-Loire, France). In E. Delson, M. Faure, & C. Guérin (Eds), *Senèze: Life in Central France two million years ago. Paleontology, geochronology, stratigraphy and taphonomy* (pp. 145–163). Cham (Switzerland): Springer.

Multiple Authors. (1929). Ch. Depéret 1854–1929. Discours prononcés aux obsèques de M. le Professeur Depéret. Lyon.

Necrasov, O., Samson, P., & Rădulesco, C. (1961). Sur un nouveau singe catarrhinien fossile, découvert dans un nid fossilifère d'Olté- (R. P. R.). *Analele Stiintifice Universitatii "Al. I. Cuza" din Iasi, Sec. II, Stiinte Naturale, 7*, 401–416.

Nishimura, T. D., Senut, B., Prieur, A., Treil, J., & Takai, M. (2009). Nasal architecture of *Paradolichopithecus arvernensis* (late Pliocene, Senèze, France) and its phyletic implications. *Journal of Human Evolution, 56*, 213–217.

Nomade, S., Pastre, J. F., Guillou, H., Faure, M., Guérin, C., Delson, E., et al. (2014). ^{40}Ar/^{39}Ar constraints on some French landmark Late Pliocene to Early Pleistocene large mammalian paleofaunas: Paleoenvironmental and paleoecological implications. *Quaternary Geochronology, 21*, 2–15.

Paquette, J.-L., Médard, E., Poidevin, J.-L., Barbet, P. (2021). Precise dating of middle to late Villafranchian mammalian paleofaunae from the Upper Allier River valley (French Massif Central) using U–Pb geochronology on volcanic zircons. *Quaternary Geochronology, 65*, 101198 (12p).

Parenti, F., Valli, A., Aprile, L., Candelato, F., Monguillon, A., Natali, L., et al. (2024). Surveying and excavations at Senèze: 2001–2006. In E. Delson, M. Faure, & C. Guérin (Eds.), *Senèze: Life in Central France two million years ago. Paleontology, geochronology, stratigraphy and taphonomy* (pp. 27–35). Cham (Switzerland): Springer.

Pastre, J.-F. (2024). The Senèze maar and paleontological site (Massif Central, France): Their volcanic context and studies of related tephras. In E. Delson, M. Faure, & C. Guérin (Eds.), *Senèze: Life in Central France two million years ago. Paleontology, geochronology, stratigraphy and taphonomy* (pp. 83–98). Cham (Switzerland): Springer.

Pastre, J.-F., Debard, E., Nomade, S., Guillou, H., Faure, M., Guérin, C., & Delson, E. (2015). Nouvelles données géologiques et téphrochronologiques sur le gisement paléontologique du maar de Senèze (Pléistocène Inférieur, Massif Central, France). *Quaternaire, 26*, 225–244.

Pelletier, H. (1968). Contribution à l'étude sédimentologique de l'ancien lac de Senèze (Haute-Loire, Massif Central français). *Revue des Sciences naturelles d'Auvergne, 34*, 39–50.

Perrot, R., Morel, P., & Ballesio, R. (1972). Les proliférations osseuses en paléopathologie, les exostoses isolées à propos de l'humérus d'un *Machairodus* découvert à Senèze (Haute - Loire). *Cahiers médicaux lyonnais, 48*(32), 3647–3658.

Pfeiffer, T. (1999). Die Stellung von *Dama* (Cervidae, Mammalia) im System plesiometacarpaler Hirsche des Pleistozäns. *Courier Forschungsinstitut Senckenberg, 211*, 1–218.

Pfeiffer, T. (2005). The position of *Dama* (Cervidae, Mammalia) in the system of fossil and living deer from Europe—Phylogenetical analysis based on the postcranial skeleton. In É. Crégut-Bonnoure (Ed.), *Les Ongulés holarctiques du Pliocène et du Pléistocène. Actes Colloque International (Avignon, 19–22 septembre 2000). Quaternaire*, Hors-série, *2*, 39–57.

Prat, F. (1964). Contribution à la classification des Equidés villafranchiens. *Procès-Verbaux de la Société Linnéenne de Bordeaux, 101*, 14–32.

Prévot, M., & Dalrymple, G. B. (1970). Un bref épisode de polarité géomagnétique normale au cours de l'époque inverse Matuyama. *Comptes Rendus hebdomadaires des Séances de l'Académie des Sciences Série D, Sciences Naturelles, 271*, 2221–2224.

Richard, L. (1944). Observations sur les dentitions de lait d'*Equus stenonis* Cocchi de Senèze (Haute-Loire). *Bulletin du Muséum national d'Histoire naturelle*, 2e série, *16*, 544–546.

Roger, J. (1954). Le gisement villafranchien de Senèze (Haute-Loire). *Bulletin du Museum national d'Histoire naturelle*, Paris, 2ème série., t. *XXVI*, 292–295.

Roger, S., Coulon, C., Thouveny, N., Féraud, G., Van Velzen, A., Fauquette, S., et al. (2000). ^{40}Ar/^{39}Ar dating of a tephra layer in the Pliocene Senèze maar lacustrine sequence (French Massif Central): Constraint on the age of the Réunion-Matuyama transition and implications on paleoenvironmental archives. *Earth and Planetary Science Letters, 183*, 431–440.

Roman, F. (1931). Un savant lyonnais, le Professeur Depéret. *La vie lyonnaise, 13*(543), 5–6.

Roman F. (1935). Les collections de Géologie et de Paléontologie de la Faculté des Sciences de Lyon. *Annales de la Société linnéenne de Lyon et de la Société botanique de Lyon, Société d'Anthropologie et de Biologie de Lyon réunies, NS 78*, 197–264. https://doi.org/10.3406/linly.1935.14764

Roman, F., & Dareste de la Chavanne, J. (1931). Sur la présence d'un élan (*Alces latifrons* Johnson) dans le Pliocène supérieur de Senèze (Haute-Loire). *Comptes rendus hebdomadaires des Séances de l'Académie des Sciences, 192*, 1256–1257.

Schaub, S. (1922). *Nemorhoedus philisi* nov. spec., ein fossiler Goral aus dem Oberpliocän der Auvergne. *Eclogae Geologicae Helvetiae, 16*, 558–563.

Schaub, S. (1923). Neue und wenig bekannte Cavicornier von Senèze. *Eclogae Geologicae Helvetiae, 18*, 281–295.

Schaub, S. (1925). Über die Osteologie von *Machaerodus cultridens* Cuvier. *Eclogae Geologicae Helvetiae, 19*, 255–266.

Schaub, S. (1941). Die kleine Hirschart aus dem Oberpliocaen von Senèze (Haute-Loire). *Eclogae Geologicae Helvetiae, 34*, 264–271.

Schaub, S. (1942). Un nouveau Félidé de Senèze (Haute-Loire). *Eclogae Geologicae Helvetiae, 35*, 183–189.

Schaub, S. (1943a). Osteologische Sammlung. *Verhandlungen der Naturforschenden Gesellschaft in Basel, 54*, 368–371.

Schaub, S. (1943b). Die oberpliocaene Säugetierfauna von Senèze (Haute-Loire) und ihre verbreitungsgeschichtliche Stellung. *Eclogae Geologicae Helvetiae, 36*, 270–289.

Stehlin, H. G. (1923). Die oberpliocaene Fauna von Senèze. *Eclogae Geologicae Helvetiae, 18*, 268–281.

Turner, A. (1987). *Megantereon cultridens* (Cuvier) (Mammalia, Felidae, Machairodontinae) from Plio-Pleistocene deposits in Africa and Eurasia, with comments on dispersal and the possibility of a New World origin. *Journal of Paleontology, 65*, 1256–1268.

Valli, A. M. F. (2024). New data on Cervidae (Mammalia, Artiodactyla) from the Villafranchian locality of Senèze (Haute-Loire, France). In E. Delson, M. Faure & C. Guérin (Eds.), *Senèze: Life in Central France two million years ago. Paleontology, geochronology, stratigraphy and taphonomy* (pp. 409–431). Cham (Switzerland): Springer.

Wittmann, O. (1980). *Raritäten und Curiositäten der Natur: Die Sammlungen des Naturhistorischen Museums Basel.* Basel: Springer.

Chapter 2
Surveying and Excavations at Senèze: 2001–2006: Evaluating and Comparing Techniques for the Recovery of Fossil Remains

Fabio Parenti, Andrea M. F. Valli, Luca Aprile, Federica Candelato, Angélique Monguillon, Luca Natali, and Vittorio Rioda

Abstract We provide the description of goals and techniques of surveying and drawing of fossils in the 2001–2006 excavation campaigns in the Lower Pleistocene deposits of Senèze (Haute-Loire, France). After a detailed survey at 1:1000 scale of each targeted parcel, a new topographic map of the whole site was established. During the excavations a total station was employed as the main topographic instrument. Detailed drawing and positioning of each fossil concentration was accomplished mainly by digital photographs followed by GIS treatment. From these images we have drawn each fossil on the general site map, linking it to the database. We illustrate here the three main fossil concentrations, consisting of partial skeletons of *Dicerorhinus etruscus*, *Eucladoceros ctenoides* and *Allohippus senezensis*. Finally, we provide an evaluation of different spatial data recording techniques in unconsolidated fossiliferous deposits.

F. Parenti (✉)
Universidade Federal do Paraná, Curitiba, PN, Brazil
e-mail: scriptoriumparentii@gmail.com

Istituto Italiano di Paleontologia Umana, Anagni, FR, Italy

A.M.F. Valli
Société Scientifique du Bourbonnais, Moulins (Allier), France

Istituto Italiano di Paleontologia Umana, Anagni, FR, Italy

L. Aprile
Associazione Aditus in Rupe, Floridia, SR, Italy

F. Candelato
Freelance Archaeologist, Verona, Italy

A. Monguillon
Réserve Naturelle Régionale du site géologique de Limay, Limay, France

L. Natali · V. Rioda
Istituto Italiano di Paleontologia Umana, Anagni, FR, Italy

Résumé Nous présentons les objectifs que nous nous sommes fixés et les techniques employées au cours des campagnes de fouille 2001–2006 dans le gisement du Pléistocène inférieur de Senèze (Haute-Loire). Après un relevé topographique détaillé au 1/1000ème de chaque parcelle fouillée, nous avons établi une nouvelle carte topographique de l'ensemble du gisement. Au cours de la fouille le relevé des données spatiales a été fait essentiellement à l'aide d'une station totale topographique. La description détaillée et la localisation de chaque amas fossilifère ont été réalisées à l'aide de photographies digitales, successivement redressées avec le module de géoréférentiation du logiciel ArcGIS. À partir de ces images nous avons pu dessiner chaque fossile sur le plan général du site et le relier à la base de données. Nous présentons les trois principales concentrations osseuses, constituées de squelettes partiels de *Dicerorhinus etruscus*, *Eucladoceros ctenoides*, *Allohippus senezensis*. Pour finir, nous évaluons l'efficacité et la valeur des différentes techniques de relevé dans divers types de gisements paléontologiques.

Keywords Large scale topography • Excavation of macrofossils • Evaluation of surveying techniques

Mots-clés topographie à grande échelle • fouille des grands mammifères • évaluation des techniques de relevé

Introduction

During the first years of the twenty-first century, paleontologists and prehistoric archaeologists have witnessed the widespread adoption of surveying and data recording techniques previously reserved to a few specialists: total stations and GPS in the field and, most of all, Geographic

E. Delson et al. (eds.), *Senèze: Life in Central France Around Two Million Years Ago*,
Vertebrate Paleobiology and Paleoanthropology, https://doi.org/10.1007/978-3-031-64415-3_2

Information Systems (GIS) software on laptops for the treatment of topographic data and georeferencing needs. The Senèze field team included some of the same people who worked on the 1995–1999 campaigns at the middle Villafranchian site of Saint Vallier (Faure & Guérin 2004); for this reason several procedures were maintained along a ten year span: careful direct drawing of each fossiliferous *locus* and attribution of a label number before documenting and removing each fossil in the field (Parenti & Valli 2004). Some technical improvements, however, have been adopted in the Senèze excavation cycle; these include automatic recording of polar coordinates by total station and GIS elaboration in the field by digital camera, for supporting the drawing of fossil remains or documenting stratigraphic sections. In this chapter we will briefly describe the procedures we adopted and the main methodological caveats in this specific situation.

Surveying Goals and Techniques

The Senèze maar, as for any site in France, is covered by a rich panoply of good topographic and cadastral maps (Figs. 2.1 and 2.2). The landscape is divided into parcels which belong to local landowners from whom permission must be obtained.

The available maps, however, are not adequately detailed with regard to some important items, such as: microrelief, vegetation cover, rural paths and local power lines. Moreover, the main stratigraphic and taphonomic research goals require a very careful description of each sedimentary body, both surficial and buried, with a centimetric tolerance. In order to prepare a good base for succeeding campaigns, we surveyed each targeted parcel at a 1:500 scale, drawing it at a 1:1,000 scale (Fig. 2.3).

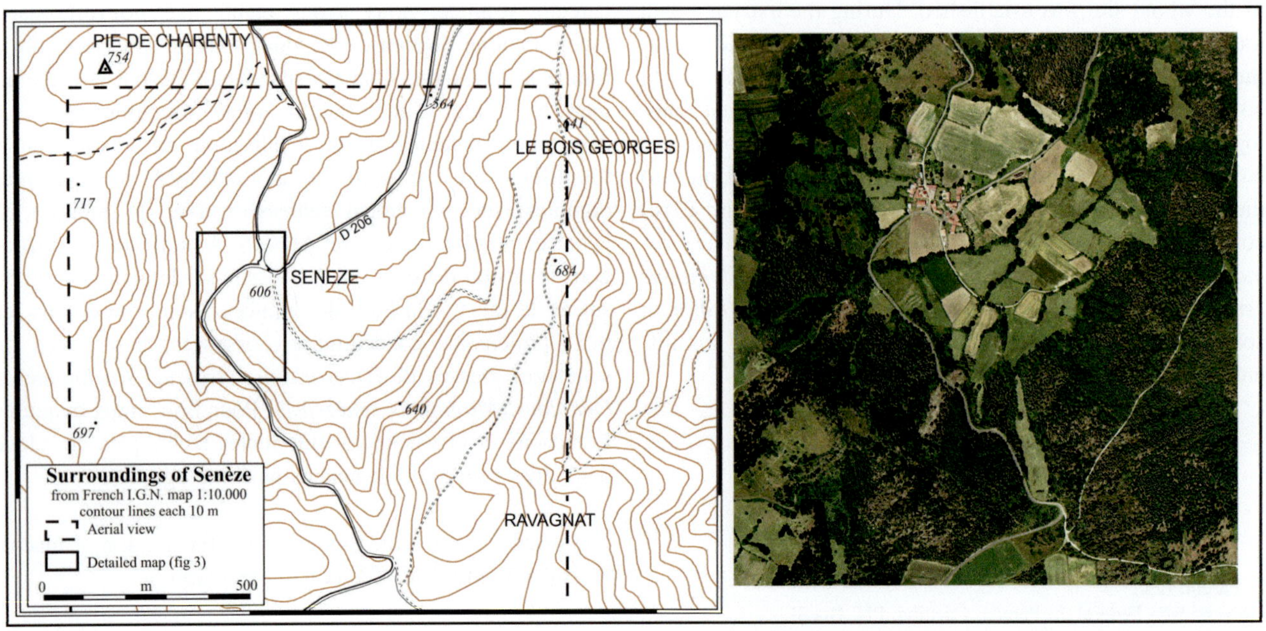

Fig. 2.1 Regional topographic map (left) and aerial view of Senèze maar (IGN 1995-FD-43–63/300, n° 633)

As an altimetric reference, owing to the impossibility to have a direct sightline to the I.G.N benchmark on Pié de Charenty (Fig. 2.1), we started from the 606 m a.s.l. point close to the village of Senèze (Fig. 2.1); from this point we transferred the elevation to parcel 233, where we placed a ferroconcrete benchmark as a *datum*. In the first campaign (2001), after having prepared a polygonal net with a Topcon GTS 229 total station, we surveyed with a Breithaupt transit alidade and plane table and placed more reference points for future years. A plane table, although now considered outdated, allows direct drawing in the field, and for this reason is the best device for recording topographic, morphologic and geological details (Wilford 2000). In 2001, parcels 233, 234, 235 (Eastern sector), Senèze village and the road to La Chomette were surveyed. In 2003 and 2004, the survey was completed by total station over the area illustrated in Fig. 2.3, including parcels 157, 172, 173, 174 and 238 (Fig. 2.2). For the digitization of the topographic data of the 2001 campaign and the drawing of contour lines, in the 2003 campaign we used the *Topografia & Strade* software from STS (Software Tecnico Scientifico). Concerning routine excavation procedures, both for fossils and samples, from 2003 onward we used total stations, providing fast and easy transfer of polar coordinates in *ArcView* (later *ArcGIS*, both from ESRI Corp.) software for the control of the whole excavation (Fig. 2.4B).

Fig. 2.2 Survey plot of land parcels (section AN), with locations of previous work (excavations, coring and trenches) and responsible researchers in red

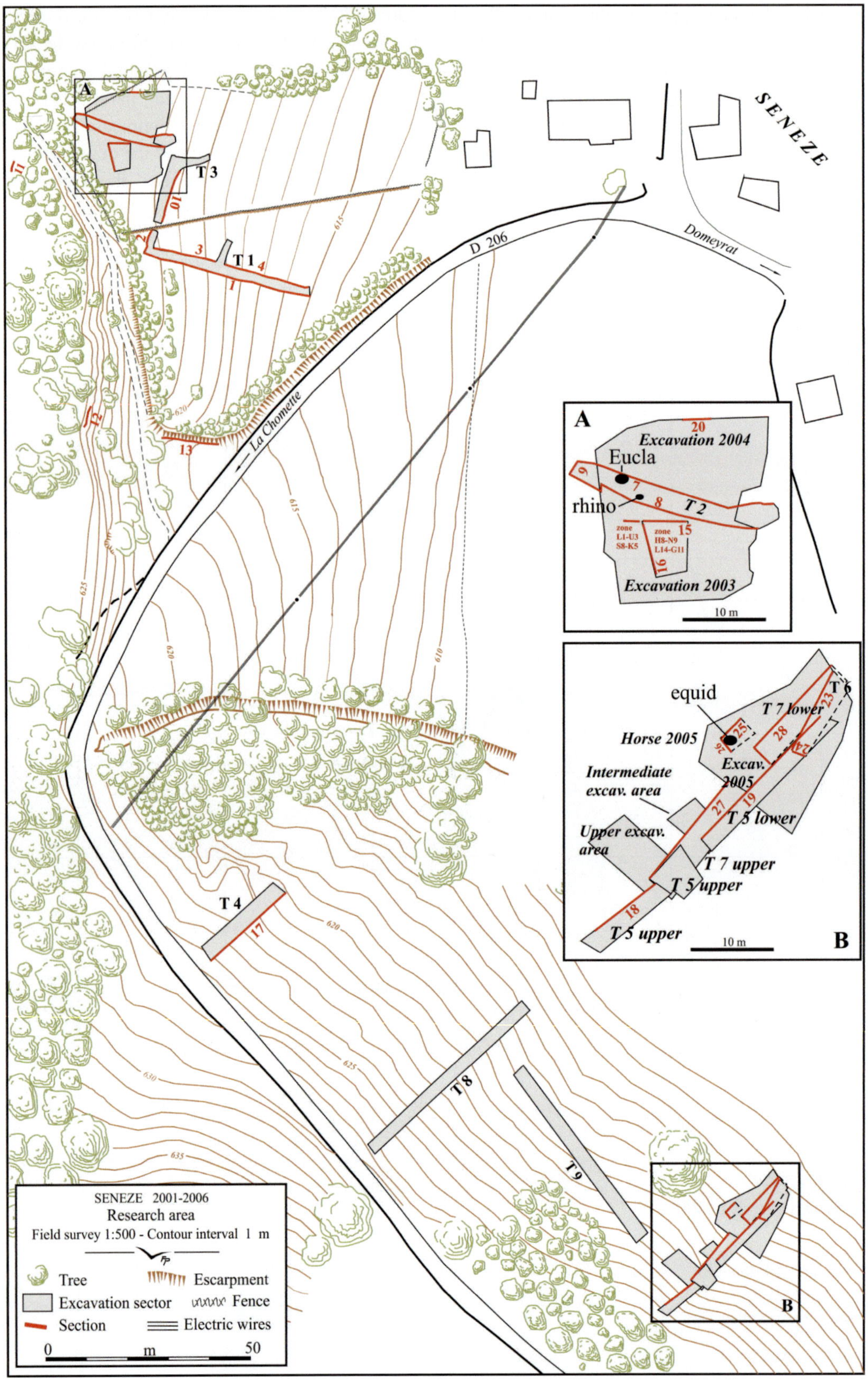
SENEZE
Domeyrat
D 206
La Chomette
T 1
T 3
T 4
T 8
T 9
A
B
Excavation 2004
Eucla
rhino
T 2
Excavation 2003
10 m
equid
Horse 2005
T 7 lower
T 6
Excav. 2005
Intermediate excav. area
Upper excav. area
T 5 lower
T 7 upper
T 5 upper
T 5 upper
10 m
SENEZE 2001-2006
Research area
Field survey 1:500 - Contour interval 1 m
Tree
Escarpment
Excavation sector
Fence
Section
Electric wires
0 m 50

◄ **Fig. 2.3** Senèze topography and 2001–2006 excavation sectors and trenches, 1:1000, north at top. Numbers in parentheses indicate location (chapter.figure) of detailed section views if any. (1) Trench 1, parcel 234, south face; (2) Trench 1, parcel 234, west face (3.2); (3) Trench 1, parcel 234, north face, upslope half (3.1); (4) Trench 1, parcel 234, north face, downslope half (3.1); (7) Trench 2, parcel 233, north face (3.3); (8) Trench 2, parcel 233, south face (3.4); (9) Trench 2, parcel 233, west face; (10) Trench 3, parcel 233, east face (3.5); (11) Section (profile) C, parcel 235, west face; (12) Section (profile) B, parcel 234, west face; (13) Section (profile) A, parcel 234, north face (3.8); (15) Zone [H8-N9-L14-G11] north face (3.6); (16) Zone [H8-N9-L14-G11] west face; (17) Trench 4, parcel 174, southeast face (3.10); (18) Trench 5 upper, parcel 172, northwest face (3.11); (19) Trench 5 lower, parcel 172, northwest face (3.12); (20) Pit at the northern boundary of parcel 233, north face (3.7); (23) Trench 6, parcel 172, northwest face (3.13); (24) Trench 6, parcel 172, southwest face; (25) 2005 equid excavation, parcel 172, northwest face; (26) 2005 equid excavation, parcel 172, southwest face; (27) Trench 7 upper, parcel 172, northwest face (3.14); (28) Trench 7 lower, parcel 172, northwest face (3.15). Inset A shows enlargement of Trench 2 area with location of profiles 7, 8, 9, 15, 16 and 20, as well as *Dicerorhinus* skeleton of 2001 ("rhino"; see Fig. 2.5) and *Eucladoceros* skeleton of 2004 ("Eucla"; see Fig. 2.6). Inset B shows enlargement of Trenches 5–7 area with locations of subregions and excavation areas

Detailed topographic recording of each fossil accumulation or remarkable isolated pieces was performed by direct drawing at a 1:10 scale and vertical digital photographs in which four identified points were measured by total station and therefore easily plotted in the GIS with its georeferencing module. In larger bone concentrations, as illustrated in Figs. 2.5, 2.6 and 2.7, a photomosaic was composed by the above technique. Single small fossils were plotted with coordinates by total station. Figures 2.5, 2.6 and 2.7 have been built on the basis of georeferenced mosaics. Finally each outlined (or just plotted) fossil was linked to the general database by an intermediate spreadsheet.

Excavation and Macrovertebrate Collecting

At the beginning of the four week annual average period of excavation at Senèze, topsoil was slowly cleared away by an earth-moving machine; the excavator also provided the first exploratory trenches of each sector, such as T1–3 of the 2001 campaign. During this delicate, unavoidable operation, all team members were looking for stratigraphic and textural changes as well as for the presence of macrofossils. This was the case, for instance, of rhino limbs, marginally bulldozed in T2 (2001) but essentially preserved (Figs. 2.4A and 2.5).

Fig. 2.4 Drawing and surveying procedures. **A** Excavation of rhino remains (2001); **B** survey by total station: the pole at right is placed on a reference point

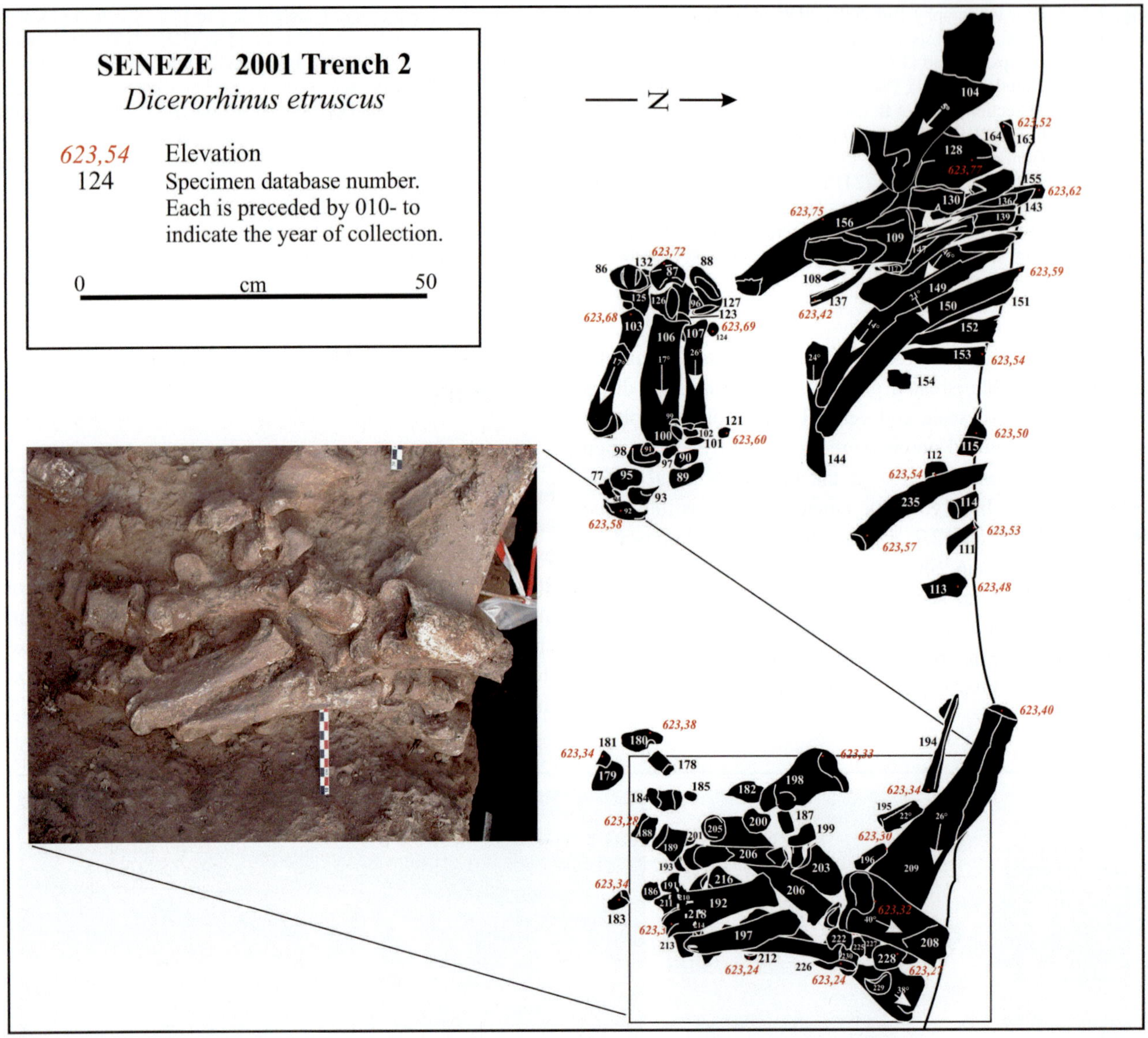

Fig. 2.5 *Dicerorhinus etruscus* partial skeleton, 2001: Photo of specimen during excavation, drawing and list of numbered elements in fossil database

Hand excavation in each sector followed survey and stratigraphic observations. Obviously, the so-called excavation by "natural layers" is simply a respectful managing of excavation according to visible sedimentary units and macrostructures, which is what we did at Senèze, where fossils are generally lying along a steep slope in proximity to basin borders (Debard 2024; Fernández-Jalvo et al. 2024). Because of the good preservation and mineralization of fossils, we usually removed short or massive bones without any additional treatment. In the (rare) case of long or fragmented bones, or especially fragile specimens such as the portion of elephant tusk, we provided a plaster jacket for transport to the laboratory.

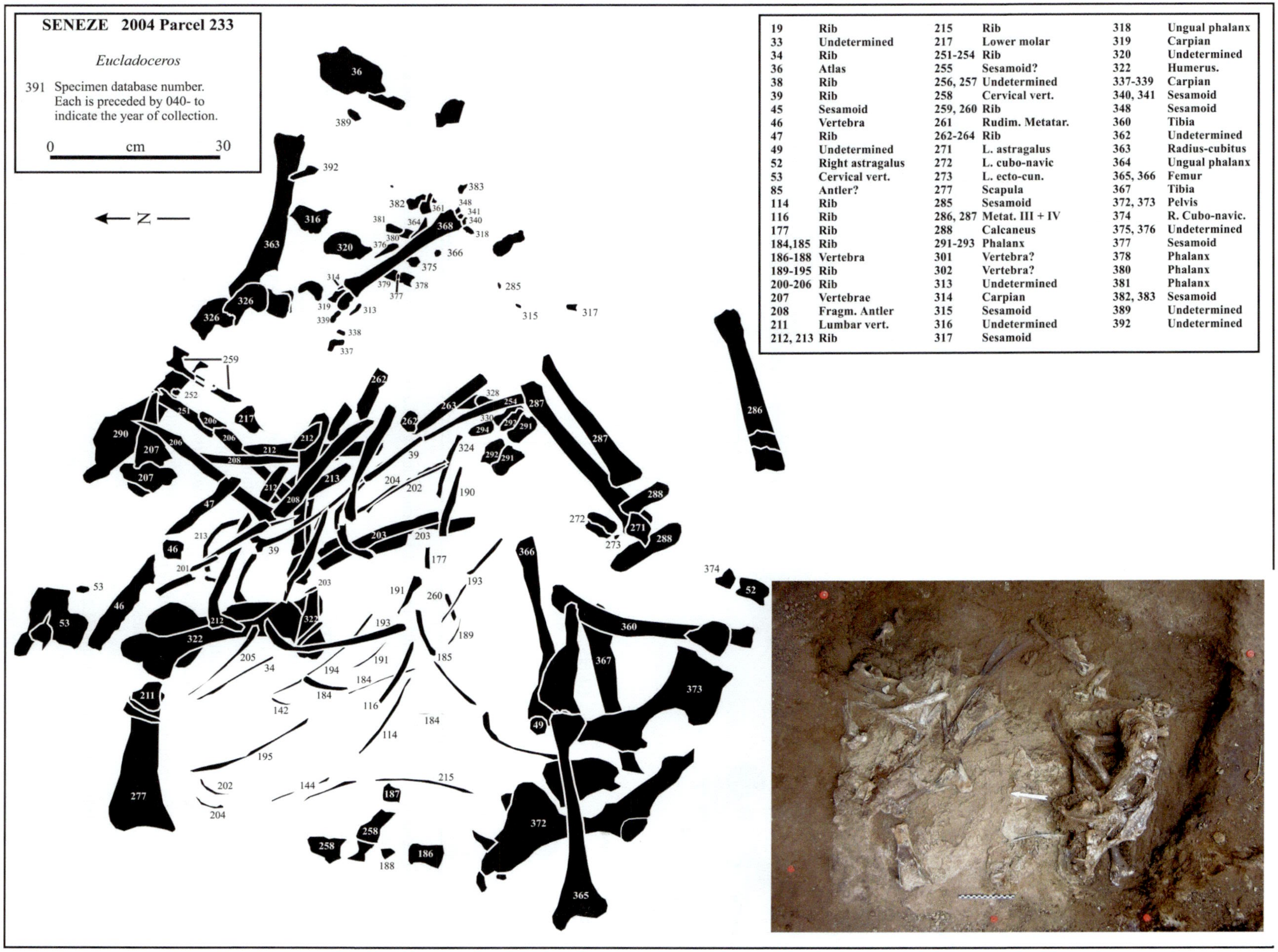

Fig. 2.6 *Eucladoceros ctenoides* partial skeleton, 2004: Photo of specimen during excavation, drawing and list of numbered elements in fossil database

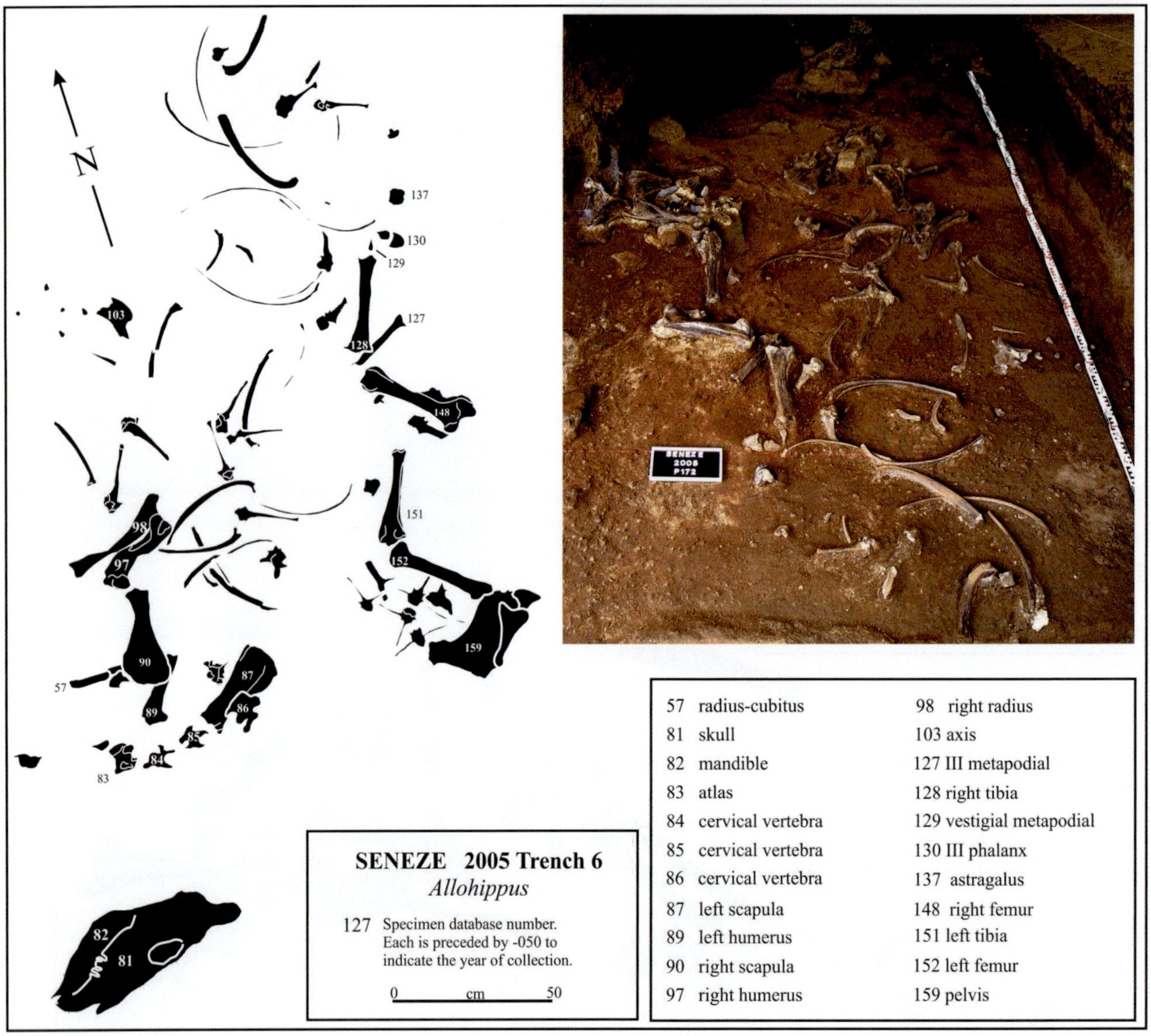

Fig. 2.7 *Allohippus senezensis* partial skeleton. 2005: Photo of specimen during excavation (view toward southeast); drawing (view toward northwest) and list of numbered elements in fossil database

The Case of Senèze in the Context of Different Recording Techniques

Since the excavations in southern England by Pitt-Rivers in the last decades of the nineteenth century, prehistoric archaeologists have been aware of the importance of a careful positioning of the remains they recovered (Trigger 1989). In vertebrate paleontology, this awareness has often been obscured by the intrinsic difficulties of extracting fragile (and very old) bones from their matrix (Leiggi et al. 1994; Whybrow 1985). After World War II, grid and 3-coordinate systems have been generally adopted, mainly for constricted deposits such as caves or rock-shelters (Barker 1977; Laplace-Jauretche & Méroc 1954; Lévêque 2002), but afterward for any kind of excavation. At Senèze, through five years of field recording practices, we experienced and evaluated pros and cons of multiple ways for attaining the same goal: i.e., the precise position of any piece of evidence, no matter its size, for a future understanding of its natural history. In reflecting on this history and in order to contextualize the choices we made, some *caveats* have been highlighted, and we review them for the reader's attention.

First of all, optical devices such as theodolites and levels are usually employed as recording tools in paleontological and archaeological excavations; this is because of their availability in academic departments and because they are (apparently) easy to manage, at least in the case of a level. However, they are often improperly used, because they were originally conceived for purely topographic tasks such as survey, geodesy or altimetric leveling for engineering. In fact these devices, and their modern descendants such as total stations,

Table 2.1 Evaluation of recording techniques (other than topographic surveying) in different unconsolidated fossil deposits

	Open: alluvial, lacustrine, fluvial/marine terraces	*Long and narrow*: channel, gully, canyon	*Restricted and lit:* rock-shelters	*Restricted and dark:* caves
Direct drawing	Slow	–	Good	Good
Compass/clinometer	–	Slow	Slow	Good
Transit alidade	Good, but slow	Good	Good	Difficult
Level, theodolite	Slow	Slow	Slow	Difficult
Total station	Good and fast	Good	Good	Difficult
Vertical pictures	For concentrated remains	–	Good	Good
GPS	Good and fast	Difficult	Impossible	Impossible
Laser scanning	–	–	Good	Good

are very precise and sophisticated but insufficient for the description of the landscape, which is the very basis of paleoenvironmental reconstruction. Paleontologists, archaeologists and taphonomists have their minds focused on looking backward, trying to describe what really occurred, rather than planning a new building or transforming the environment.

Secondly, the growing availability of computer-assisted drawing (also designed for engineering) is rapidly blurring any residual capacity of observation and understanding via the manual reproduction of forms and dimensions. As Sennett (2008) has admirably pointed out in a recent book on the history of craftsmanship, the human brain can really understand, and so describe, only what the hand is reproducing; because of this, the only possible reproduction of spatial relationship *during* excavation—excluding experimental taphonomy or archaeology—is drawing.

In Table 2.1 we present an evaluation of data recording techniques in different kinds of unlithified fossil deposits. We do not consider here the equipment for the preliminary topographic survey of the site, but only the devices—and consequently the techniques—for recording the position of macrofossils, clumped or sparse as they are, and describing their form before removing them. As for recovering microfossils—the difference between the two depending basically on visibility—the procedure should be oriented by the sampling strategy of each research project. In general terms, a grid system (materialized or virtual) is the best solution for regular sampling.

The drawing of sections, to record sampling for paleoenvironmental or geochronological studies, is a distinctly different matter. Such sections are, usually, regular vertical surfaces requiring careful detailed drawing, complemented by color pictures, whose correct position can be obtained by total station in the field and GIS analysis in the laboratory. Photogrammetry or laser scanning are today very fast and precise but, at the time of the Senèze fieldwork, they still were experimental and not so widespread. We expect that future methods and techniques will further expand the opportunities available for data collection and analysis in field paleontology projects.

Acknowledgments We thank Martine Faure, Eric Delson and the late Claude Guérin for inviting us to participate in the Senèze fieldwork and contribute to this volume. We thank William Harcourt-Smith, Eric Delson and an anonymous reviewer for comments on previous versions of this chapter. This article was originally written in 2012 and partially revised in 2021.

References

Barker, P. (1977). *The techniques of archaeological excavation*. London: Batsford.

Debard, E. (2024). Geological study of the Early Pleistocene site of Senèze (Domeyrat, Haute-Loire, France). In E. Delson, M. Faure & C. Guérin (Eds.), *Senèze: Life in Central France two million years ago. Paleontology, geochronology, stratigraphy and taphonomy* (pp. 37–82). Cham (Switzerland): Springer.

Faure, M., & Guérin, C. (Eds.) (2004). *Le Gisement pliocène final de Saint-Vallier (Drôme, France)*. Mémoire spécial 26, *Geobios, 37*.

Fernández-Jalvo, Y., Valli, A. M. F., Marín-Monfort, M. D., & Pesquero-Fernández, M. D. (2024). The taphonomy of Senèze. In E. Delson, M. Faure & C. Guérin (Eds.), *Senèze: Life in Central France two million years ago. Paleontology, geochronology, stratigraphy and taphonomy* (pp. 653–681). Cham (Switzerland): Springer.

Laplace-Jauretche, G., & Méroc, L. (1954). Application des coordonnées cartésiennes à la fouille d'un gisement. *Bulletin de la Société Préhistorique Française, 51*, 58–66.

Leiggi, P., Schaff, C.R., & May, P. (1994). Macrovertebrate collecting. Field organization and specimen collecting. In P. Leiggi & P. May (Eds.), *Vertebrate paleontological techniques* (Vol. 1, pp. 59–77). Cambridge: Cambridge University Press.

Lévêque, F. (2002). Méthodes de fouilles. In J. C. Miskovsky (Ed.), *Géologie de la Préhistoire: Méthodes, techniques, applications* (pp. 415–423). Paris: Géopré.

Parenti, F., & Valli, A. M. F. (2004). Méthodes de fouille, techniques de terrain et de laboratoire. In M. Faure & C. Guérin (Eds.), *Le Gisement pliocène final de Saint-Vallier (Drôme, France)*. Mémoire spécial 26, *Geobios, 37*, 66–76.

Sennett, R. (2008). *The Craftsman*. New Haven & London: Yale Univ. Press.

Trigger, B. G. (1989). *A history of archaeological thought*. Cambridge: Cambridge University Press.

Whybrow, P. J. (1985). A history of fossil collecting and preparation techniques. *Curator, 28*, 5–26.

Wilford, J. N. (2000). *The mapmakers*. New York: Alfred A. Knopf.

Chapter 3
Geological Study of the Early Pleistocene Site of Senèze (Domeyrat, Haute-Loire, France)

Evelyne Debard

Abstract The fieldwork carried out on the Senèze deposits between 2001 and 2006 allowed a detailed stratigraphic analysis of their nature and geometry. The emplacement of phreatomagmatic beds related to maar formation was followed by deposition of prograding lenses of more or less coarse sediments, originating from the reworking of outcrops located around the crater. These slope layers are covered by lacustrine layers, occurring between 625 and 620 m altitude. Marked differences in sedimentation found across sectors of the site are due to the morphological context (stiffness of the slopes of the maar), the nature of the underlying outcrops (gneissic basement, volcanic products), syn-sedimentary deformations (load casts, collapse faults, mass slip) and climatic changes leading to variations in the water level of the paleo-lake. This sedimentary context allows a better understanding of the genesis of the paleontological assemblages.

Résumé Les fouilles réalisées entre 2001 et 2006 ont permis une étude stratigraphique détaillée des formations mises au jour à Senèze et de préciser la nature et la géométrie des dépôts. Après la mise en place de produits phréatomagmatiques liés à la formation du maar, se déposent des lentilles progradantes de sédiments plus ou moins grossiers, issus du remaniement des affleurements situés sur le pourtour du cratère. Ces formations de versant sont recouvertes par des niveaux lacustres, apparaissant entre 625 et 620 m d'altitude. Les différences de sédimentation d'un secteur à l'autre sont marquées. Elles s'expliquent par le contexte morphologique (raideur des pentes du maar), la nature des formations affleurantes (socle gneissique, formations volcaniques), des déformations syn-sédimentaires (figures de charge, failles de tassement, glissement en masse) et des changements climatiques entraînant des variations du niveau d'eau du paléo-lac. Ce contexte sédimentaire permet de mieux appréhender la formation des assemblages paléontologiques.

Keywords Stratigraphy • Geomorphology • Sedimentology • Slope deposits • Lacustrine levels • Paleoshoreline • Climatic changes

Mots-clés Stratigraphie • géomorphologie • sédimentologie • formations de versant • niveaux lacustres • paléorivage • changements climatiques

Introduction

The Senèze site corresponds to a maar-crater in the gneissic basement that developed during a volcanic, phreatomagmatic event. The center of this depression was occupied by a lake that gradually filled with sediments over an estimated period of 200,000–300,000 years (Ehrlich 1968), in a combination of lacustrine layers and slope deposits on the exposed flanks. These slope formations as well as the upper levels of the lacustrine infill contained the main fossiliferous sites that were uncovered by the many previous excavations.

Although earlier fieldwork yielded a certain amount of data on the geology of the site, none provided a summary of the different formations visible around the maar. One goal of the new program of research at Senèze was to establish a detailed stratigraphy (and sedimentology) of this complex locality, where fossils have been recovered from both lacustrine and slope deposits whose relationships are unclear (Delson et al. 2006). The study presented here, based on a detailed stratigraphic analysis of the deposits exposed between 2001 and 2006, makes it possible to propose some correlations from one sector to another, to know the sedimentological context in which the fossils occur and to situate them in the dynamics of

E. Debard (✉)
Université de Lyon, UCBL, ENSL, CNRS, LGL-TPE, 69622 Villeurbanne, France
e-mail: evelyne.debard@free.fr

25 rue Paul Chevrel, 69370 Saint-Didier-au-Mont-d'Or, France

E. Delson et al. (eds.), *Senèze: Life in Central France Around Two Million Years Ago*,
Vertebrate Paleobiology and Paleoanthropology, https://doi.org/10.1007/978-3-031-64415-3_3

sediment placement along the edge of the paleolake. This study also makes it possible to place all the samples (palynology, tephrology, magnetostratigraphy, dates, most fossil concentrations, etc.) in their stratigraphic context since they have been rigorously indexed on the sections.

History of the Geological Data

The Slope Deposits

In 1892, Boule mentioned for the first time the presence of fossil mammals at Senèze "buried and scattered amongst volcanic ashes incised by storm rains". The approximate location of his and later excavations mentioned below are indicated on Fig. 2.2.

Depéret and Mayet (1912) documented the first geological profile of slope formations using a 60 m trench on the southern flank of what is here termed the Philis ravine, located northwest of Senèze village. These authors described, from upslope to downslope, three groups of formations, steeply dipping towards the center of the depression. The first and lowest unit, in contact with the gneissic bedrock, consists of hard and coarse-grained conglomerates, including more or less rounded gneiss, basalt and volcanic scoriae debris, mixed with sandy gravel containing volcanic, siliceous and micaceous elements. Downslope, less coarse gravels of ferruginous color include gneiss and basalt debris, alternating with fine red sands interbedded with thin clay layers that are hard and compact and look like ashy or vitric tuff. Further down slope, layers of more or less coarse red sands appear as well as conglomerates similar to the basal conglomerates higher on the slope. According to the authors, these slope formations extend to the southeastern part of the depression, where the volcanic debris becomes rare or disappears and only red silty clay layers mixed with gneiss debris, exceptionally including rounded basalt fragments, are present. Depéret and Mayet carried out five excavations in the southern, western and eastern part of the depression, finding fossils within the slope formations (coarse gravels, fine ferruginous sands, lower-slope conglomerates).

In 1953, Roger and colleagues of the Muséum national d'Histoire naturelle in Paris carried out field work (Brébion et al. 1953), and Roger (1954) compiled the first geological map of the Senèze formations. In the center of the maar they reported that the alluvial deposits of the creek that presently drains the infilled depression covered layers of ashy sediment. These latter are also visible farther north of Senèze as more or less fine-grained pyroclastic beds. These volcanic formations form a regular ring around the depression up to an altitude of 620 m in the east and the south and up to an altitude of 630 m in the northwest. From the south to the east they are characterized by increasing proportions of crystalline detritus. Collapsed basalt blocks can be observed within these beds in the northwestern part. In this area coarse ejectamenta (bombs, pozzolans, etc.) are visible up to the top of the flank. These slope formations, which are reworked from the summit of a volcanic complex, cover the gneissic bedrock in the form of longitudinal bands that crop out everywhere in the remainder of the depression. Roger mapped the ancient paleontological excavations and his profile sections, which were placed at altitudes between 610 m and 620 m, except for the section on the edge of the former road to La Chomette located at an altitude of nearly 600 m.

A decade later, Grangeon drilled boreholes in the center of the Senèze depression to further study its volcanic context (Elhaï & Grangeon 1963; Grangeon 1962). He stressed the variation of the slope deposits according to their position, with volcanic scoriae northwest and west of the village, breccia blocks at the base of the western and southwestern slopes, crystalline elements elsewhere, and more finely stratified and clearly lacustrine layers as one moves away from the crystalline banks. The upper limit of the slope deposits is at an average altitude of 625 m. The early excavations investigated the lacustrine sediments down to an altitude close to 595–600 m, from which complete skeletons were extracted, whereas rare bones were recovered from the slope deposits. Grangeon indicated that the wells formerly dug below the village constantly remained within "sands" to a depth of 15–20 m. This remark matches the observation made by Bout (1970c), who noted that during the building of the "new" road from La Chomette to Domeyrat in 1952–53, steeply dipping yellow loamy and sandy layers were found immediately downslope from Senèze at an estimated altitude of nearly 598 m.

Bout discussed the geology and stratigraphy of a large number of Villafranchian sites in the Velay and in the Allier basin (Bout 1969, 1970a, 1970b, 1972, 1975, 1978), and he was the first to refer to the depression of Senèze as a "maar", describing its geological composition and the succession of events in its origin (Bout 1970c). Specifically, using the data published by Elhaï and Grangeon (1963), Elhaï (1969), and Ehrlich (1968), he discussed the climatic processes that affected the lacustrine and slope deposits. These latter formed sublacustrine deltas which developed during a cooling event that followed the cold period reported by Elhaï (1969) in the upper third of the lacustrine filling.

Further excavations were carried out by Couthures (Couthures et al. 1991) with the aims of understanding the development dynamics of the slope deposits and discovering faunal elements in situ in order to verify his hypothesis of volcanic outgassing having been responsible for death by asphyxiation of the animals (Couthures 1989). He cleaned the profile of the "Philis ravine" along several meters and also the section located along the former path to La Chomette, while also documenting a profile at 700 m altitude at the foot of the Pié de Charenty.

In the profile of the ravine, he identified an initial sedimentary package or "set" formed by pyroclastic tuffs at the top, with more or less coarse sands and "block fields" below, overlying shales that exhibited volcanic airfall products at the bottom. Most of the sandy levels and the clays are fossiliferous. The second set, identified as being more than one meter thick, included red sands full of scoriae that weathered to clay. Couthures interpreted these deposits as being successive screes in a torrential environment that formed after the filling of the lake, as identified by coring (Elhaï & Grangeon 1963). He explained the absence of these deposits on the opposite flank of the maar as the result of slumps that carried them towards the center of the depression.

The cross-section of the former path to La Chomette, located at an altitude of 607 m, is exposed up to a height of 5 m and mainly shows azoic pyroclastic tuff containing basalt granules in a weathered gneiss matrix, interrupted by two beds of rounded basalt pebbles and a layer with a large number of unweathered bedrock and basalt blocks one meter above the base.

The section at Pié de Charenty revealed pyroclastic tuff, with underlying scoriae, lapilli and rare bombs, either flattened or appearing as closely set stripes, that were interpreted as being Strombolian ejecta reworked from the volcano that preceded the formation of the intracrateral lake.

In conclusion, all these investigations highlighted the great variability of the slope deposits around the depression. Most of the authors agree in associating most of the fossils in the slope formations to a level between 625 m and 610 m. Below this level lacustrine sediments occur that may intergrade laterally and locally with early slope deposits. These lacustrine levels are commonly fossiliferous, frequently containing complete skeletons.

The Lacustrine Filling

The center of the maar is occupied by a thick lacustrine filling that was identified in several drillings that reached or came close to gneissic substrate.

An initial, 60 m deep test hole was drilled in 1962, nearly 100 m east of the small village of Senèze (Elhaï & Grangeon 1963). Unfortunately, the sampling only represented some 20% of the section. Below 2 m of topsoil and scree, a sequence of mainly diatomaceous clays can be observed, more or less well bedded depending on the abundance of diatoms (whitish layers) compared to detrital material (biotite occurring in greyish layers, quartz in sandy, 0.5 mm to 1 mm thick layers). From 29.30 m to 31 m the investigators observed basalt and bedrock screes. All the samples contained lignitic fragments. Several layers between 2 m and 6 m thick yielded fossil bones, plant remains, freshwater crustaceans and small bivalve molluscs.

In 1965 a second core was drilled to a depth of c. 175 m., also from a base level of 590 m, with the following succession (Ehrlich 1968):

- from 0 m to 22 m, alternating clayey, sandy-clayey and sandy beige, brown or grey levels, sometimes compact, sometimes bedded;
- from 22 m to 28 m, waterlogged layers of medium-grained and coarse-grained sands;
- from 28 m to 100 m, fine clayey-sandy sediments, less frequently greyish sandy sediments, generally occurring as very thin layers revealed by alternating whitish folia, 0.1 mm to 1 mm thick and dark-colored micaceous, 0.2 mm to 2 mm thick folia, forming pseudo-varves;
- from 100 m to 123 m, sediments, bedded at the top and becoming increasingly sandy and micaceous towards the bottom; they frequently contain gravels and coarse detritic elements;
- from 123 m to 175 m, bedrock elements and lava pieces become more numerous.

According to Grangeon, who followed up the drilling in the field, the crater floor was probably reached, or is located not much farther down.

This drilling allowed analyses of the sediment (Pelletier 1968), diatoms (Ehrlich 1968), pollen (Elhaï 1969), palaeomagnetism (Prévot & Dalrymple 1970) and the description of a fish fossil (Gaudant 1975). Although the lacustrine filling reached at least 123 m in thickness, the water level in the hole never exceeded 40 m at most.

Pelletier (1968) carried out mineralogical and grain-size analyses of three sand samples from 28–25 m, 85–82 m and 173–172 m respectively. The lower sand is coarse (mean grain size 2.5 mm) and appears to be very well sorted. The two other samples are finer-grained (mean grain sizes 0.50 and 0.55 mm) and much less well sorted. Minerals of volcanic origin strongly predominate compared to minerals reworked from bedrock. However, a significant increase in garnet in the 28–25 m sample is probably related to heavier erosion of the bedrock.

In 1970 Devis published a petrographical study of the volcanic deposits of Senèze. He summarized the main data on the lacustrine sediments known from the works of Grangeon (Elhaï & Grangeon 1963 and unpublished notes), Pelletier (1968) and Ehrlich (1968). He explained the distinction made between the two sets (from 0 m to 114.80 m with relatively fine sediment; from 114.80 m to 174 m, with coarser deposits) as being due to a decline in the depth of the lake as sedimentation proceeded. The repeated alternating of sandy and clayey layers in the upper set was attributed to variation in rainfall. Devis highlighted the presence of two beds of feldspar crystals at 21 m, which correspond to the level dated by Roger et al. (2000).

In 1989, three drillings of 24 m were carried out 50 m south of the 1965 borehole with the aim of $^{40}Ar/^{39}Ar$ dating pyroclastic tephra trapped in the lacustrine layers (Roger et al. 2000), which resulted in a reported date of 2.10 ± 0.01 Ma (recalculated with current calibrations as 2.12 ± 0.01 Ma, see Delson et al. 2024) on a tephra at a depth of 20.75 m. Unfortunately, the cores were not described adequately for correlation to the drilling carried out in 1965.

The New Excavations

The fossiliferous deposits are currently not exposed except at the level of the older profile sections, as is the case for section A. Most of the observations below were made thanks to drilling, trenching and surface stripping.

The Investigated Sectors

Two sectors of the depression were investigated in detail through trial trenches and surface stripping (see Fig. 2.3), as follows:

- the sector located west of Senèze village including the trenches T1, T2, T3 and the surface excavation areas [H8-N9-L14-G11] and [L1-U3-S8-K5];
- the sector southeast of the village including the trenches T4 to T9 and several surface excavations.
- Observations were also made in the sections of the historic excavators, primarily through careful cleaning of the profile surfaces (see Fig. 2.2):
- section A (labeled "Bungalow section"), mentioned by Depéret and Mayet (1912) and excavated at the beginning of the 1960s by Guth (according to Bout 1970c);
- section B, described by Roger in 1954;
- section C (labeled "pine section"; Roger 1954);
- section D (labeled "Philis ravine section"; Depéret & Mayet 1912; Roger 1954; Couthures et al. 1991);
- section E, first excavated by the Muséum national d'Histoire naturelle (Brébion et al. 1953; Roger 1954) and then by Guth, and section F (labeled "section of the former road to La Chomette") studied by Brébion et al. (1953) and later by Couthures et al. (1991), showed only slope deposits with not a single fossil visible; these were neither documented nor sampled.

The Western Sector

In 2001 three trial trenches (T1, T2 and T3) were dug on parcels 233 and 234 located west of the village (see Figs. 2.2 and 2.3). Between 626.50 m and 614 m altitude, it was possible to uncover fossiliferous slope facies exhibiting interbedded lacustrine levels, the whole dipping towards the center of the maar, with an average slope of between 15° and 20°. Subsequent opening of excavation areas made it possible to complete the observations made in the trenches. Lastly, some of the sections of the ancient excavations were cleaned in order to provide an overview of the deposits in the western sector of the maar.

Trench 1

The trench extends over more than 40 m following the line of maximum gradient of parcel 234 (See Fig. 1.8). For technical reasons it was dug in two parts (T1 upslope and T1 downslope) separated by a perpendicular access trench (Figs. 3.1 and 3.2).

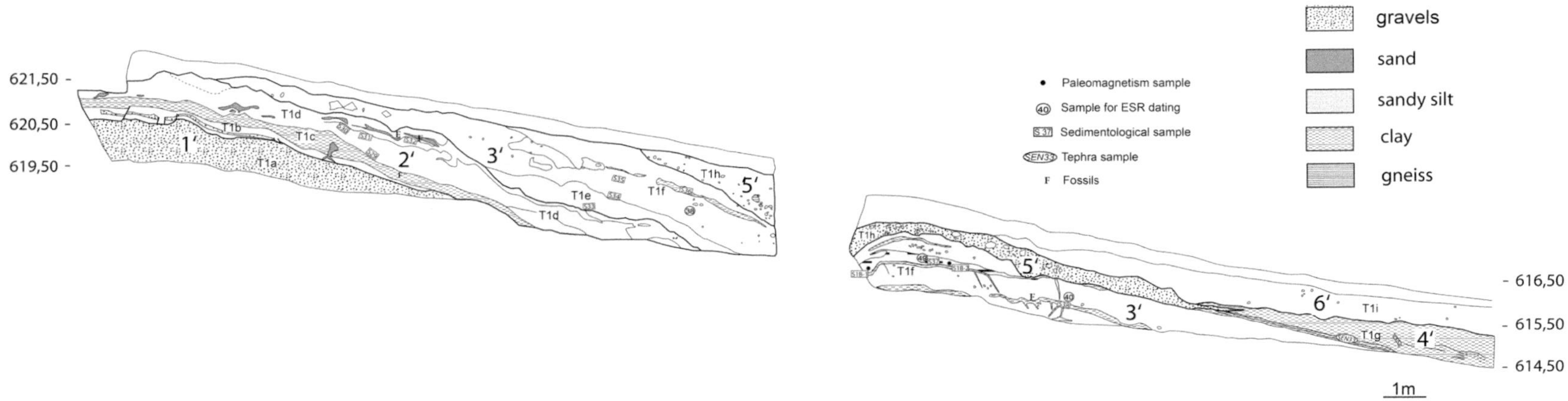

Fig. 3.1 Trench 1, north face, drafted at 1/10 scale. 40 m long, divided into two parts, it mainly shows slope deposits (sets 1', 3', 5', 6'), except for two lacustrine episodes (2', 4' partly reworked). It has been sampled for sedimentology, paleomagnetism and the study of tephras. This figure corresponds to segments 3 and 4 of Fig. 2.3; hereafter this will be indicated as [2.3:3,4]. (Figure 3.1: Tranchée 1, Coupe nord, dessinée à 1/10e. Longue de 40 m, divisée en deux parties, elle montre essentiellement des dépôts de versant (ensembles 1', 3', 5', 6'), excepté deux épisodes lacustres (2' et 4', ce dernier en partie remanié). Elle a fait l'objet de prélèvements pour la sédimentologie, le paléomagnétisme et l'étude de téphras. Cette figure correspond aux segments 3 et 4 de la Fig. 2.3; ci-après, on va l'indiquer comme [2.3:3,4])

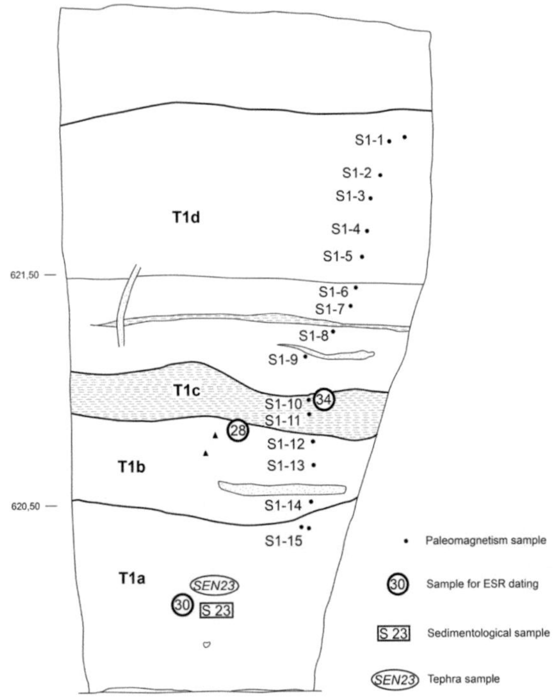

Fig. 3.2 Trench 1, west face, drafted at 1/10 scale. [2.3:2] (Figure 3.2: Tranchée 1, Coupe ouest, dessinée à 1/10e. [2.3:2])

Trench T1 makes it possible to observe the deposits located at altitudes between 622 m and 614 m.

The upslope part of the trench exposes four sedimentary sets, described here from bottom to top:

– Set 1' (T1a) has a thickness of 1 m in the uppermost part of the trench and consists of a polygenic microconglomerate with sandy cement, without grain sorting or clear stratification. Its upper limit is irregular with a corrugated boundary. The matrix of small, rounded granules is composed of bedrock, basalt and olivine crystal, associated with larger scoriae.

This first unit corresponds to a slope formation, incorporating elements ejected from the maar vent and scoriae reworked from the small enclosing cone (Pastre 2024), that accumulated in a massive deposit on the lakeshore. Its sedimentation was obviously favored by the absence of vegetation on the slope. The abrasion of the elements probably happened either during eruption or much later on the slopes of the maar. This formation was then eroded by wave action or lake currents as evidenced by its corrugated upper surface.

– Set 2' groups together several loamy units with enclosed sand or gravel lenses.

At its bottom, filling the corrugations of the upper surface of set 1', there is a grey-colored loamy level (T1b), which is laminated and mica-rich, cut by numerous faults and sharply truncated downslope. It has a maximum thickness of 40 cm and is truncated by a thin gravelly bed with irregular thickness (2 to 10 cm), containing fine elements identical to those of set 1'. T1b yielded cervid teeth in 2001.

The following unit (T1c), with a thickness of 15 to 20 cm on average and up to a maximum of 40 cm, is a laminated brown clayey loam that is increasingly dark and plastic towards the top. Fracture faces are coated by films of iron and manganese oxide. Surfaces with striated micro-cracks (N0 to N29) are highlighted by extremely thin deposits of yellow sand. In the upper part of the layer this medium-grained sand forms small lenses filling the depressions. The deposit is sub-horizontal in the topset part of the trench and plunges sharply downslope.

The next unit (T1d) has a basal layer nearly 25 cm in thickness of alternating laminae of light or dark-colored loam a few centimeters thick. Above this layer there is a set of interbedded lenses of grey brown and dark brown micaceous loamy sands with a strong downslope gradient. Some small pockets of yellow orange coarse gravelly sand can be observed in several places. This unit is between 60 cm and 100 cm thick, and it is strongly marked by ferric pedogenesis in the form of rust stains and films along the flow surfaces and cracks.

Set 2' corresponds to a lacustrine high level, more or less enriched with organic matter reworked from the slope, indicating the beginning of forest growth. Stages with very high energy remobilizing coarser matter on the lakeshore contributed to the formation of sand lenses.

- Set 3' begins with loamy sands (T1e), with a maximum thickness of 60 cm, enclosing scattered small scoriae and granule lenses (bedrock, scoriae, basalt). These are covered by more than 80 cm of micro-conglomerate (T1f) composed of small granules, gravels of bedrock and scoriae, and small pebbles scattered within a sandy loamy matrix, with flow surfaces covered by red manganese clayey pellicles. A discontinuous loamy chestnut brown layer 10 cm thick with prismatic jointing and without lamination and pedologic differentiation is intercalated.

This set corresponds to a succession of mudflows with coarser matter reworked from the slope (coarse-grained material with veins of "fine colluvium").

- Set 4' (T1g) is not visible in this part of trench 1.
- Set 5' (T1h) consists of coarse conglomerate with massive stone blocks, cutting into set 3'. This is a deposit of the "heads" type, which developed most probably during a glacial stage.

The downslope part of trench 1 intersects sets 3' to 5' as well as new formations that were uncovered there. The deposits clearly reveal upslope deformation, the origin of which remains uncertain because of an observational gap caused by the access trench: was it a fracture?, a slump?, or a compression due to the massive deposit of set 5'?

- Set 3' measures between 80 cm and 110 cm at the bottom of the trench. Only the micro-conglomerate (T1f), with discontinuous veins of unstructured brown loam can be observed. Loam or yellow sand lenses occur randomly. This unit is affected by open fractures filled with red clay, which results in a block structure.
- Set 4' (T1g) is present only in the lower part of the trench. It starts with alternating thin levels of yellow sandy loams and brown loams between a few millimeters and one centimeter thick. This banded unit is discontinuous and thickens towards the east from a few centimeters upslope to about 15 cm at the bottom end of trench 1. It is covered by an unstructured brown loamy level, which is also discontinuous. This level is not very thick on the west side and is between 80 cm and 1 m thick at the eastern end of the trench. It encompasses groups of banded disturbed loamy levels associated with yellow-greenish sandy loams. We are probably dealing with a colluvium covering and reworked lacustrine levels.
- Set 5' (T1h) is truncated and overlain by fine matter colluvium (set 6'= T1i), which fills the depressions of the underlying formations.

Trench 2

Trench 2 extends over 25 m following the line of the maximum gradient, cutting through the deposits between altitudes of 626.50 m and 620 m (Fig. 3.3; see also Figs. 1.7 and 1.12).

- Set 1' can be identified in the upslope part of the trench along a length of nearly 6 m. It is characterized by mainly

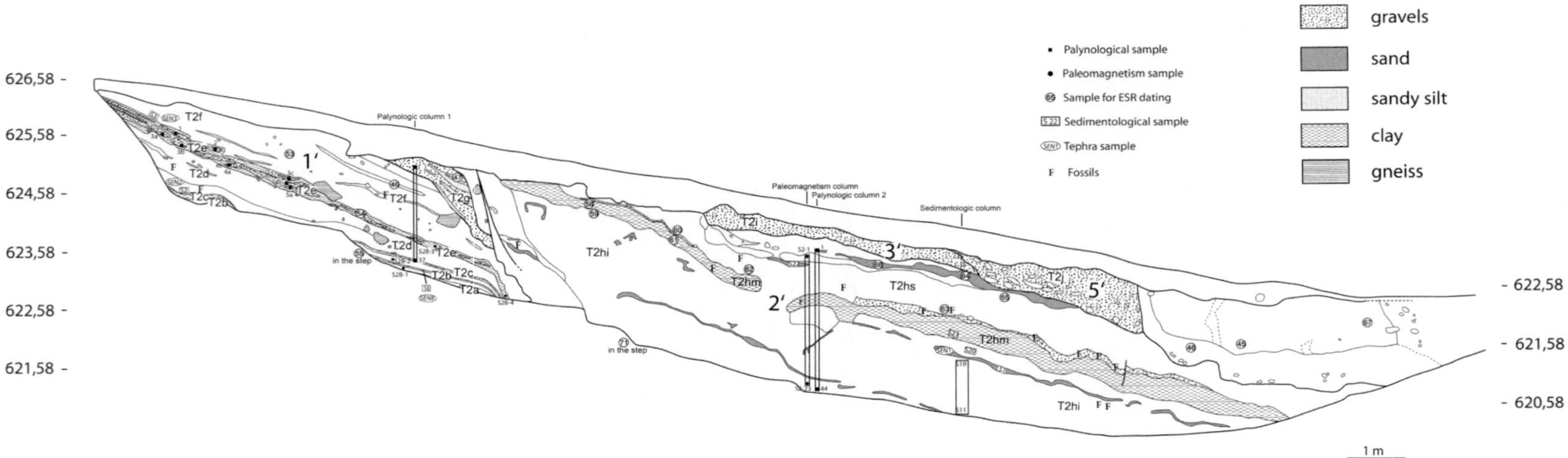

Fig. 3.3 Trench 2, north face, drafted at 1/10 scale. 25 m long, it is a reference for the stratigraphy of the western sector. It is characterized by the highest lake level (T2b). Upstream, slope deposits are predominant (set 1'). Downstream, in abnormal contact, a thick fossiliferous lake deposit (set 2') develops, overlain by slope deposits (sets 3' and 5'). Two tephras were dated 2.10 ± 0.03 Ma (SEN 1) and 2.16 ± 0.05 Ma (SEN 8; see Delson et al. 2024). Numerous samples were taken (sediments, tephras, pollen [see Fig. 1.15] and paleomagnetism [see Fig. 1.14]). The end of the trench shows traces of earlier excavations. [2.3:7] (Figure 3.3: Tranchée 2, Coupe nord, dessinée à 1/10e. Longue de 25 m, elle est une référence pour la stratigraphie du secteur ouest. Elle se caractérise par le niveau lacustre le plus haut (T2b). A l'amont, les dépôts de pente sont prépondérants (ensemble 1'). A l'aval, en contact anormal, se développe un épais dépôt lacustre (ensemble 2'), fossilifère, surmonté par des dépôts de pente (ensembles 3' et 5'). Deux niveaux de téphras ont été datés de 2,10 ± 0,03 Ma (SEN 1) et 2,16 ± 0,05 Ma (SEN 8; voir Delson et al. 2024). De nombreux prélèvements ont été effectués (sédiments, tephras, pollens [voir Fig. 1.15] et paleomagnetisme [voir Fig. 1.14]). L'extrémité de la tranchée montre des traces de fouilles anciennes. [2.3:7])

coarse sediments intersected by layers of fine sediment and can be subdivided into three units. The slope is about 25° towards the east.

The first unit, with a height of a dozen centimeters, corresponds to a micro-conglomerate (T2a) rich in small scoriae except for some isolated blocks. This is a typical slope formation.

The second unit is a heterogeneous association of several loamy layers more or less rich in organic matter and coarser levels. From bottom to top, the following layers are identified:

- a grey loam deposit (T2b), banded with parallelepipedic jointing, the cleavage faces covered by black ferro-manganese coating; well sorted micaceous yellow sand lens (SEN 8 = TS2 [Pastre 2024], and three thin layers composed of reddish scoriae granules and rare basalts. This deposit, with a maximum thickness of 20 cm, corresponds to a lacustrine period interbedded with reduced slope deposits and volcanic fallout;
- a dark grey loamy level (T2c) with red zones due to the presence of sands and scoriae granules, progressively passing to a black plastic level. This is a fine-textured colluvium less than 10 cm thick, enriched in organic matter;
- a coarse detritic deposit (T2d) 10-13 cm thick, composed of granules, gravels and small rounded pebbles, very closely packed within a loamy-sandy matrix, which is red colored in the lower part (altered scoriae) and yellow colored in the upper part; this is followed by about 20 cm of more scattered elements in a yellow sandy matrix including several gravel lenses; lastly, 65 cm of reddish sandy loam (altered scoriae) with included rounded granules and blunt pebbles (basalt, gneiss). T2d delivered bone fragments;
- a succession (T2e) of lenses from 5 to 25 cm thick, including granules and rounded gravels, grey loam lenses and organic layers.

The third and last unit of set 1' corresponds to a micro-conglomerate (T2f) containing granules, gravels, small pebbles and sparse stone blocks (basalt, scoriae) with a maximum visible thickness of 90 cm. Some lenses include coarser elements and occasional interbeds of thin loamy layers. At its top (T2g), there are alternating gravel and sandy loam layers.

The transition between set l' and set 2' is marked by the steep dip of the deposits of the former against a nearly abnormal vertical contact (apparent dip of 75-80°). This latter may correspond either to a fault with multiple slips, as seems to be indicated by a fossil which is deformed and extended along the contact surface, or to a sedimentary talus related to downslope slipping of the deposits of set 1'.

- Set 2' corresponds to a thick loamy deposit (T2h), in which are interbedded coarser, sandy or gravelly layers, also divided into three subunits.

At the bottom, the first unit (T2hi), visible at about 1.80 m, is grey sandy loam, locally affected by bioturbation,

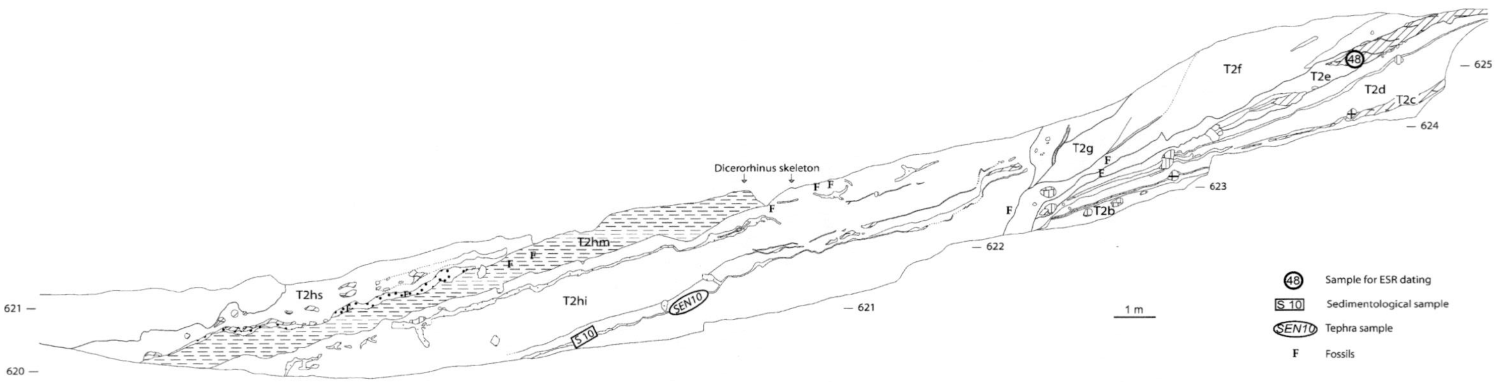

Fig. 3.4 Trench 2, south face, drafted at 1/10 scale. The T2hm layer yielded a skeleton of *Dicerorhinus etruscus etruscus*. [2.3:8](Figure 3.4: Tranchée 2, Coupe sud, dessinée à 1/10e. Le niveau T2hm a livré un squelette de *Dicerorhinus etruscus etruscus*. [2.3:8])

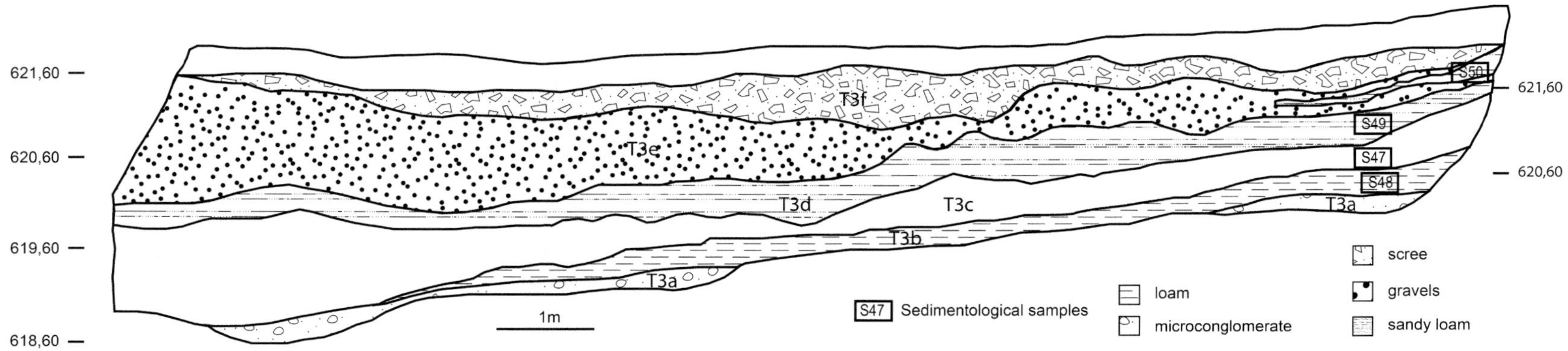

Fig. 3.5 Trench 3, east face, drafted at 1/50 scale. Perpendicular to the slope, it shows units defined in the upstream part of trench 1 (sets 1', 2', 3' and 5'). [2.3:10] (Figure 3.5: Tranchée 3, Coupe est, dessinée à 1/50e. Perpendiculaire à la pente, elle montre des unites définies dans la partie amont de la tranchée 1 (ensembles 1', 2', 3', 5'). [2.3:10])

marked by areas rich in organic matter and in iron oxides. It encloses two tephra levels: the lower level, quite continuous and up to 10 cm thick, corresponds to homometric green sand (SEN 10 = TS3 [Pastre 2024]), sampled in the southern section of trench 2 (Fig. 3.4) and matching level hi8 of zone [H8-N9-L14-G11]); the upper is yellow-colored fine sand distributed in discontinuous lenses and irregular sheets (SEN 1 = TS1 [Pastre 2024]). SEN 1 was dated by Ar-Ar to 2.10 ± 0.03 Ma (see Delson et al. 2024)

The second unit (T2hm) is laminated brown loam, locally discontinuous with a thickness varying from 10 cm to 30 cm. It is sealed by a gravelly level that thickens downslope (from 3 cm upslope up to 15 cm), which is affected by micro-cracks. A partial skeleton of *Dicerorhinus etruscus etruscus* was recovered in this unit in 2001 (Fig. 3.4; see also Fig. 2.5 and Guérin 2024).

The third unit (T2hs) is formed by brown grey to chestnut loam depending on the presence of zones rich in diffuse organic matter and the marbling of iron oxides. It becomes much redder in the upper part. This unit contains discontinuous lenses of yellow sand that can be as much as 10 cm to 15 cm thick.

- Set 3' (T2i) is a more or less indurated microconglomerate containing granules, gravels, pebbles and dispersed blocks.
- Set 5' (T2j) is a very coarse scree that cuts through the two previous sets. The sub-vertical contact in the downslope part of trench 2 may correspond to the limit of earlier excavations.

Trench 3

Trench 3 is 15 m long and runs perpendicularly to trenches 1 and 2. Its upper part is located at an altitude of nearly 622 m and its bottom reaches 618.80 m (Fig. 3.5). The formations outcropping in the upslope part of trench 1 can be identified here, but generally in lesser thicknesses.

- Set 1' (T3a): the upper part of this unit is seen as several centimeters of a microconglomerate at the bottom of the trench, including purplish scoriae, sometimes of large size (up to 18 cm). Very abundant granules are mostly highly altered powdery small scoriae associated with basalts. The matrix is dark reddish brown.
- Set 2' starts with a laminated dark brown yellowish loamy layer (T3b) between 20 cm and 25 cm thick; it breaks into small lumps, 2 cm to 3 cm high and 5 cm wide, with the sides covered by ferro-manganese oxides. At the base is a plastic loamy level (T3c), strongly marked by ferric pedogenesis: the oxides cover the discontinuity surfaces in a grid pattern with marbled appearance (moderate brown to moderate brown reddish color). The bottom of the level is marked by a thin layer of dark grey loam (1 to 2 cm thick). The upper boundary merges irregularly with coarse material reworked from the overlying unit; this boundary can easily be followed thanks to the discontinuous presence of small spots of organic matter.
- Set 3' is formed by two gravelly units, starting with a brown sandy loam layer (T3d) between 18 cm and 25 cm thick and mottled with numerous granules and gravels of altered brown red scoriae associated with rarer basalts. Occasionally larger pieces of altered gneiss and scoriae and sometimes concentrations of granules can be seen. Its upper boundary is irregular and wavy. This first unit is overlain by a microconglomerate (T3e), 25 cm to 50 cm thick, formed by gravels and granules (bedrock, basalt and first and foremost altered scoriae, providing a reddish color to the clayey matrix). Near trench 1 a non-laminated, brown loamy lens with prismatic jointing is interbedded. The upper part of set 3 is irregular and heavily eroded.
- Set 5' (T3f) corresponds to heterometric scree visible over a length of 60 cm. The generally rounded components are mainly of volcanic origin (vesicular or non-vesicular basalt with zeolites). Less frequently they stem from the bedrock (gneiss, vein quartz). The sizes range from large blocks (such as a basalt block extracted during the digging of the trench that measured 60 x 40 x 35 cm and a quartz block 40 x 26 x 16 cm) to granules. The loamy brown matrix is not very abundant and therefore makes the scree unstable.

The excavated areas

Two surface excavations were carried out in 2003 and 2004 in parcel 233 on either side of the location of trench 2.

A. Zone [H8-N9-L14-G11] = excavation 2003 upslope

The excavation of this area started at an altitude of 623.70 m and extended down to an altitude of 620.50 m (Fig. 3.6). This made it possible to recognize the three units of set 2', visible in trench 2, and to detail their composition.

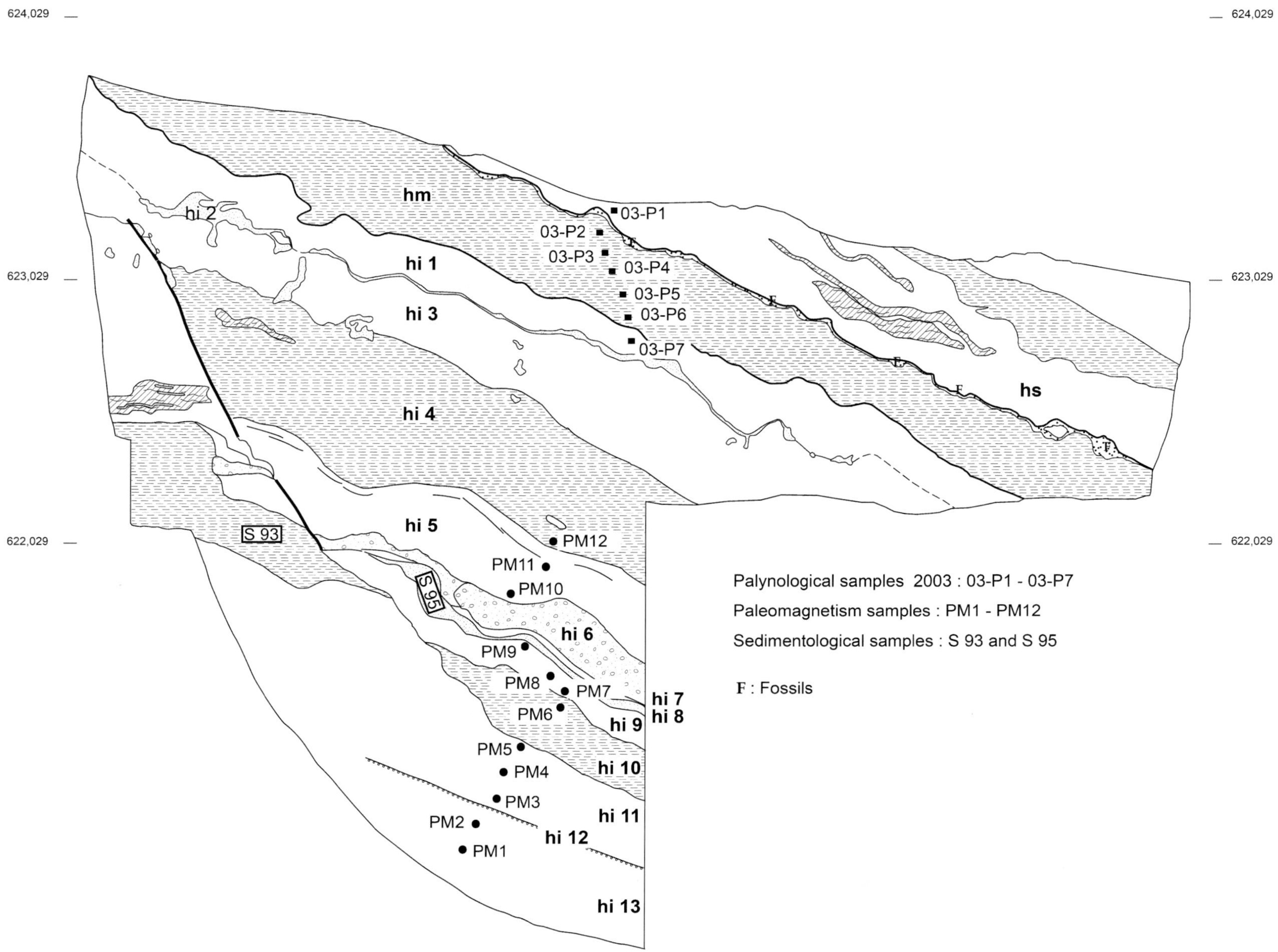

Fig. 3.6 Zone [H8-N9-L14-G11], north face, drafted at 1/10 scale. This excavation, near trench 2, allowed us to detail the units in set 2'. Three levels of tephras have been recognized. Sedimentological, paleomagnetic and palynological samples were taken. [2.3:15] (Figure 3.6: Zone [H8-N9-L14-G11], Coupe nord, dessinée à 1/10e. Cette fouille, proche de la tranchée 2, a permis de détailler les unites de l'ensemble 2'. Trois niveaux de téphras ont été reconnus. Des prélèvements sédimentologiques, paléomagnétiques et palynologiques ont été effectués. [2.3:15])

Lower unit (hi): exposed on a face more than 2 m high, it displays several thin layers of loam associated with coarser input (sand, granules and pebbles). The loamy levels have discontinuous and wavy laminar structures about one decimillimeter thick, except at the bottom of the unit where they can reach 1 mm with alternating black and grey laminae. The detachment surfaces are coated with manganese and iron oxides. From bottom to top it is possible to identify:

- hi13 to hi11: light brown to greyish loam, visible over 60 cm, with crude platy jointing, interbedded with a row of altered scoriae granules (= hi12) ; scoriae granules scattered in the upper part of hi13;
- hi10: brown sandy loam, with thin platy jointing, very irregular thickness (from 0 to 40 cm);
- hi9: 7 cm to 10 cm of compact light brown loam with crude platy jointing;
- hi8: irregular layer (2 cm to 8 cm) of coarse homometric dark green sand (see Fig. 3.4 - SEN 10 = TS3 [Pastre 2024]);
- hi7: thin level of compact grey loam (2 to 3 cm);
- hi6: lens with small stones (elements reworked from the bedrock and scoriae smaller than 2 cm), unevenly scattered within a sandy gravelly to sandy matrix;
- hi5: 20 to 30 cm of light brown to greyish white compact and laminated loam, thick platy jointing (2 to 3 cm thick); very rich in iron and manganese oxides, in the upper 5 cm, discontinuous lenses of fine sand (SEN 60 ≅ TS7 [Pastre 2024]), interbedded with layers of clay; in 2004, a partial skeleton of *Eucladoceros* was excavated in hi5 (see Fig. 2.6 and Valli 2024);
- hi4: 40 cm of dark sandy loam, with thin jointing several millimeters thick; finely detrital (abundant muscovite); iron and manganese oxide mottling, sometimes red clayey clasts (= altered scoriae);
- hi3 to hi1: 30 to 45 cm of compact, finely laminated loams, interbedded with a yellow sand layer (hi2), drawn-out into discontinuous lenses or penetrating into cracks (SEN 51 = TS5 [Pastre 2024]).

These levels are cut by a minor fault that displaces them by about 25 cm (well visible for hi6) or that draws them out (see hi5 and hi10). The fault ends in the loams at the bottom of the section. This first unit exhibits an upper boundary that is clearly wavy, of sedimentary origin (slump), sloping about 25° towards the center of the maar. It corresponds to sedimentation processes on the lake shores as is indicated by the presence of fish remains in the basal levels.

Central unit (hm): grey loam 20 to 35 cm in thickness. The structure is polyhedral, cubic with brownish red faces. This level contains very large amounts of gravels and rounded scoria granules captured in the very numerous cracks and fissures. Its upper boundary is highlighted by an almost continuous level of heterometric scoriae granules up to 6 cm long that contains numerous, often broken fossiliferous remains. This central unit evidences the continuation of lacustrine sedimentation sealed by a slope deposit.

Upper unit (hs): sandy loam including black lenses and grey loamy clasts with angular polyhedral structure, probably reworked from the central unit. At the top there is a gravelly level, which is visible over several centimeters. The sedimentation is still lacustrine but slope deposits seem to be more strongly marked.

B. Zone [L1-U3-S8-K5] = excavation 2003 upslope

This zone corresponds to the excavation of a fossiliferous slope deposit. In the upslope part it is characterized by a homometric structure including angular stones with rounded edges associated with abundant rounded granules measuring 0.5 cm to 1 cm in diameter. Bedrock elements are dominant, accompanied by reddish scoriae, in a sandy brown yellow matrix. In the downslope part gravels prevail because the larger stones were blocked upslope by the fossils. In the upper part of this excavation, a mammoth tusk was found in 2003 (see Ménouret and Guérin 2024).

A trench laid out between this zone and the [H8-N9-L14-G11] zone uncovered a wide basalt block, elevated along a fault line, against which the slope deposits accumulated. First and foremost, this made it possible to identify, at an altitude of about 625 m, the downslope passage to the lacustrine sediments. This situation confirms the observations made in the upper part of trench 2, located several meters to the north.

C. Pit at the northern boundary of parcel 233

The section (designed to allow additional paleomagnetic sampling) is exposed in a pit over 2.50 m deep dug in 2004 (Fig. 3.7):

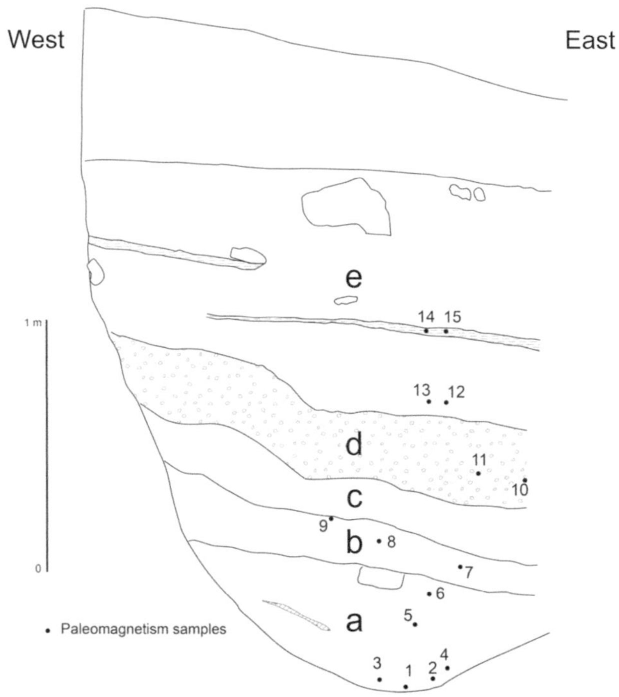

Fig. 3.7 North face of pit at the northern boundary of parcel 233. Located to the northwest of trench 2 and mainly designed to be sampled for paleomagnetism, this pit intersected the units of the upper part of trench 2 (sets 2' and 3'). [2.3:20] (Figure 3.7: Coupe nord du sondage à la limite nord de la parcel 233. Situé au nord ouest de la tranchée 2 et essentiellement réalisé pour des prélèvements pour le paleomagnétisme, ce sondage a recoupé les unites de la partie supérieure de la tranchée 2 (ensembles 2' et 3'). [2.3:20])

a upper part of a brown grey loam with reddish mottling, visible over about 60 centimeters; with a more sandy lens at its upper boundary;
b laminated grey loam; varying thickness (12 cm to 25 cm);
c red sandy loam including scattered gravels; wavy upper boundary; 15 cm to 20 cm thick;
d compact conglomerate composed of gravels and small pebbles (1 to 5 cm), in a brown grey matrix; the gravels are mainly scoriae accompanied by some bedrock fragments; irregular thickness (25-35 cm) and boundaries;
e conglomerate composed of small elements and a few basalt blocks dispersed in a yellow orange sandy matrix; small lenses of interbedded loams measure about 1 cm in thickness.

The units defined in this section can be linked with those of trench 2: the thin levels (a, b, c) correspond to the upper unit of set 2' (T2hs) and the conglomerate levels (d, e) to set 3'.

Stratigraphic Correlations of the Western Sector (Parcels 233 and 234)

The stratigraphic units identified in this sector belong to slope formations and lacustrine deposits. Detailed correlations are difficult to establish because of the numerous lateral and vertical variations and because of the displacement along fractures and slips. It is nonetheless possible to group the different units together into six sedimentary sets (Table 3.1), described from top down:

Table 3.1 Stratigraphic correlations of the western sector (parcels 233 and 234; trenches 1-3 and nearby) (Tableau 3.1 : Corrélations stratigraphiques dans le secteur ouest (parcelles 233 et 234; tranchées 1-3 et coupes voisines)

Set	Trench 1	Trench 2	Trench 3	L1-U3-S8-K5	H8-N9-L14-G11	Pit N P233	Section A
6'	T1i						
5'	T1h	T2j	T3f				Upper set
4'	T1g						
3'	T1f T1e	T2i	T3e T3d	Slope deposits		[d e]	Upper set
2'	T1d T1c T1b	T2hs T2hm T2hi	T3c T3b		hs hm hi	[a b c]	Central set
1'	T1a	T2g -T2f T2e - T2b T2a	T3a	Deposits in trench zone			Lower set

- set 6' (T1i): fine colluvium visible only in the downslope section of trench 1;
- set 5' (T1h - T2j - T3f): very coarse slope deposits (head or block field type), in erosional contact with the underlying sets;
- set 4' (T1g): colluvium from reworked lacustrine deposits, visible only in the downslope part of trench 1;
- set 3' (T1f and T1e - T2i - T3d and T3e – deposit of the zone [L1-U3-S8-K5] – layers d and e of the pit at the northern boundary of P233): slope deposits with varying grain sizes (conglomerate, microconglomerate, gravelly loams) interbedded with loamy levels, which rework lacustrine deposits at the end of trench 1.
- set 2' (T1d to T1b - T2hs to T2hi - T3c - T3b, hs to hi of zone [H8-N9-L14-G11], thin levels a, b, c of the pit at the northern boundary of P233): loamy lacustrine levels associated with sand lenses (volcanic fallout), interbedded with rather fine-grained slope deposits (gneissic gravels, small scoriae, colluvium).
- set 1' (T1a - T2g to T2a - T3a - trench zone [L1-U3-S8-K5]): slope deposits consisting of reworked ejecta from the maar vent and scoriae reworked from the small enclosing cone, with interbedded loamy levels (colluvium and lacustrine high level).

Sections Discussed by Previous Field Workers

A Profile section A

The section is located in parcel 234 beside the road leading from Senèze to La Chomette. It was termed the "bungalow" section because the team's daytime base camp, including a trailer/office ("bungalow"), was located in front of the section face. Along a length of 11.50 m it displays deposits at altitudes between 619.30 m and 617 m (Fig. 3.8). These can be grouped together into three sets:

Lower set: sequence of micro-conglomerates with coarse sandy matrix. The differences in size and the quantity of coarse elements and grain-size variations of the matrix causes color differences. The elements are mixtures of basalt, scoriae and bedrock. These micro-conglomerates are

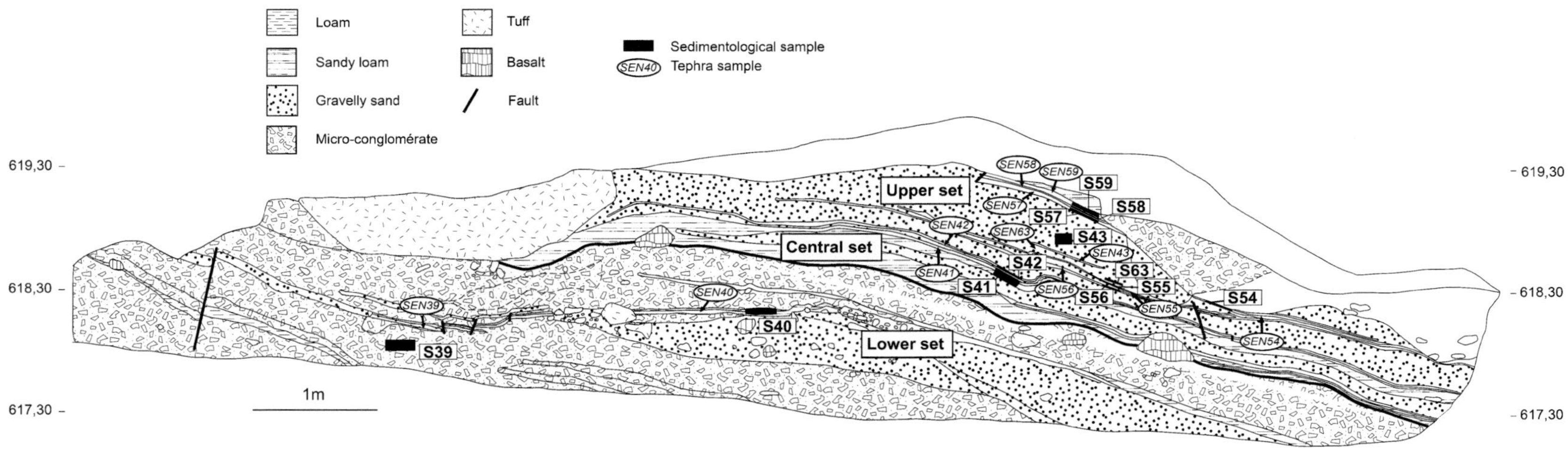

Fig. 3.8 North face of profile A, drafted at 1/10 scale.This section located about 30 meters south of trench 1 shows three sets. The lower one is characterized by slope deposits (with a megablock of tuff at the top), equivalent to those of set 1', as identified in trenches 1 and 2. The central one, formed mainly by lake levels alternating with slope deposits, corresponds to set 2'. The upper one, showing mainly slope deposits, is equivalent to set 3'; several tephras have been recognized, one of which (SEN 56) was dated 2.07 ± 0.02 Ma. [2.3:13] (Figure 3.8: Coupe A, ouest-est, dessinée à 1/10e. Cette coupe située à une trentaine de mètres au sud de la tranchée 1 montre trois ensembles. L'inférieur est caractérisé par des dépôts de pente (avec un mégabloc de tuff au sommet), équivalents à ceux de l'ensemble 1', identifié dans les tranchées 1 et 2. L'intermédiaire, formé essentiellement par des niveaux lacustres alternant avec des dépôts de pente, correspond à l'ensemble 2'. Le supérieur, montrant essentiellement des dépôts de pente, est à relier à l'ensemble 3'. Plusieurs téphras ont été reconnus dont l'un (SEN 56) a été daté de 2,07 ± 0,02 Ma. [2.3:13])

interbedded with thin bands of loams or sands as well as conglomerate lenses containing basalt and gneiss blocks, associated with scoriae in the upper levels. Distinct layers are shifted by small faults due to compaction caused by the impact of a displaced megablock of indurated tuff visible in the upper part of the section.

This lower set, which outcrops at a height of nearly 1.50 m, consists of slope deposits of reworked bedrock and the volcanic formations located on the western edge of the maar vent (basalt, scoriae, phreatomagmatic material). It is the equivalent of set 1' identified in the trenches.

Central set: about 50 cm of alternating more or less sandy loams and gravelly sands. The deposits are subhorizontal in the central part of the section and then abruptly dip towards the center of the maar. Some deposits are between 10 and 15 cm thick, others only between 1 and 3 cm. The light beige loamy beds display an angular polyhedral structure. The beds of sandy loams are compact; some of these enclose granules of scoriae. The yellow sands are coarse and generally homometric, containing reddish lava gravels with rounded and dispersed elements, sometimes associated with small pebbles of basalt and gneiss. The central set documents the interlayering of lacustrine loams and slope deposits of sands and gravel. It corresponds to set 2' of the trenches.

Upper set: banded, beige yellow gravelly sands interbedded with thin bands of loam and small lenses of rounded lava gravels. Distinct loamy or sandy beds correspond to volcanic fallout (Pastre 2024): TS 4 (= SEN 42) and TH 1-5 = (samples SEN 54, SEN 55, SEN 56, SEN 57 and SEN 58). SEN 56 was dated by Ar-Ar to 2.07 ± 0.02 Ma (see Delson et al. 2024). These layers mainly correspond to slope deposits interbedded with tephra beds (= set 3'). This upper set is incised by heterometric scree including gneiss and basalt blocks but no scoriae in a sandy matrix (= set 5').

B **Profile section B**

Section B is located at the top of the talus bordering the pathway between the parcels 234 and 235. This section is nearly 10 m long and exhibits slope deposits dipping eastwards, visible at altitudes between 626.90 m and 623.50 m. They are mainly formed by gravelly, more or less indurated sands including stones and dispersed blocks of scoriae, bedrock and rare basalts (probably equivalent of set 3'). The set is covered by heterometric more or less loose scree including big blocks of the same composition as those of the upper set of profile section A. This set probably matches set 5' identified in the trenches.

C. **Profile section C (labeled "Pine section")**

Located on parcel 235, at a higher level than the preceding section, it shows non-consolidated heterometric scree over a length of 1.50 m containing bedrock and basalt blocks, corresponding to set 5'.

D. **Profile section D (labeled "Philis ravine section")**

This section is visible on the left side of the ravine between parcel 231 on the one hand and parcels 230 and 228 on the other. This ravine, which yielded numerous fossils in the excavations carried out by P. Philis, was studied by Depéret and Mayet (1912) and later by Couthures, who cleaned a length of nearly 4 m and a height of 6 m in which he identified nine layers (Couthures et al. 1991) with fossils including a gazelle frontlet.

In 2005 the east-west trending section was cleaned and enlarged by an additional 4 m, which made it possible to observe deposits between 628.50 m and 624 m in altitude, grouped together into nine layers following the labelling of Couthures, identified from top to bottom as follows (Fig. 3.9):

Layer 1: heavily indurated, pale brown sandy formation with numerous coarse elements. It exhibits an overall massive structure except for the central part in which subtle bedding is developed. The upper part is progressively invaded by topsoil. The bed rests on an erosional surface in the eastern part of the section. The very abundant matrix is formed by coarse sand reworked from weathered bedrock.

The constituent elements are mostly small (granules to small pebbles). A few larger pebbles (up to 5 cm) and huge blocks, predominantly basalt, are dispersed at the bottom and the top of the layer. The larger stones are mainly angular to rounded pieces of basalt, either quite compact or

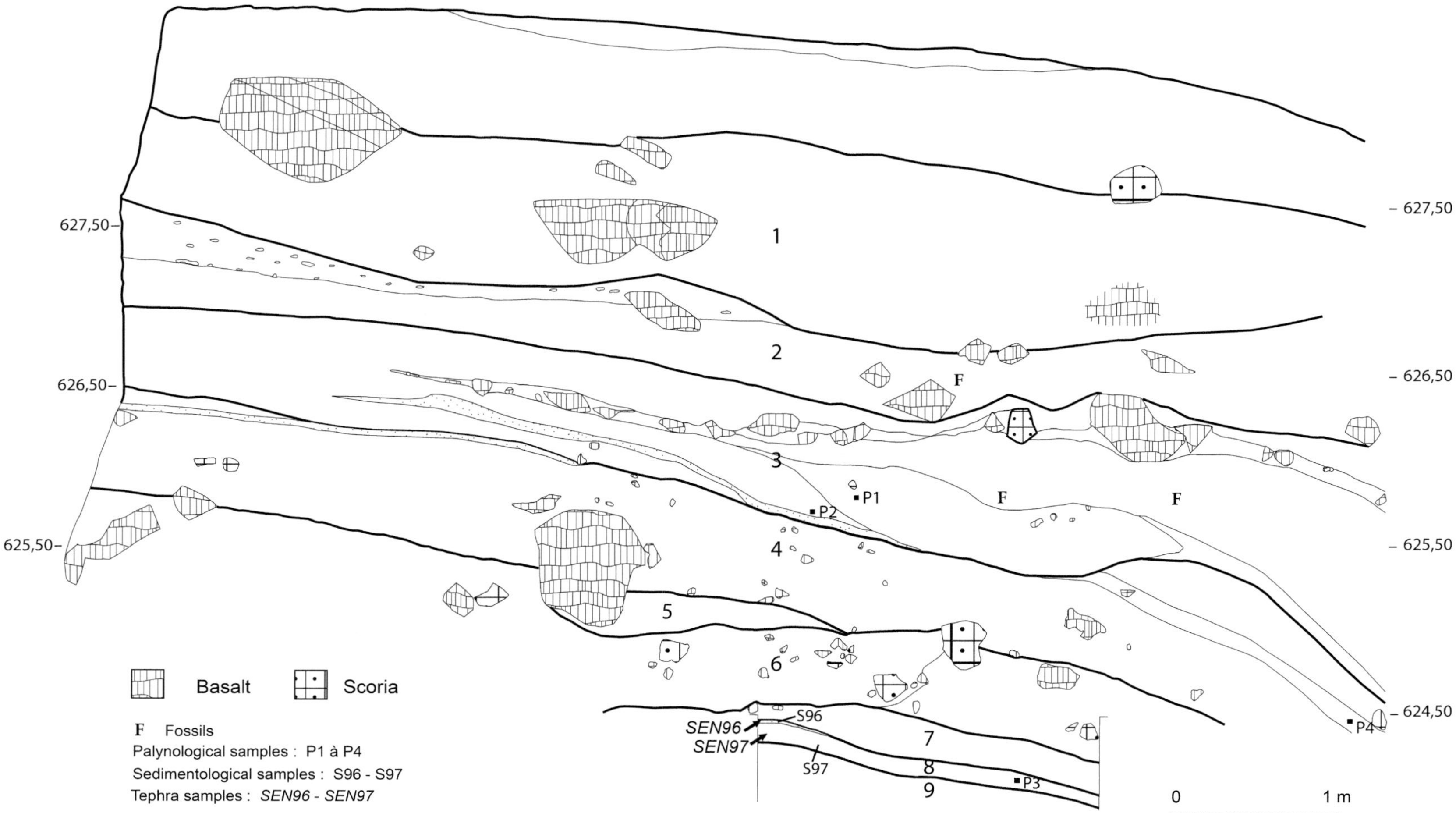

Fig. 3.9 North face of profile D ("Philis ravine"), drafted at 1/10 scale. On this section about 4 m long by 6 m high, eight slope deposits have been identified. At the base, a thin level of silt (bed 8) corresponds to a lacustrine episode, containing a tephra (SEN 96). Many fossils were unearthed here during the early excavations of Philis and then by Couthures. [see Fig. 2. 2, not on Fig. 2.3] (Figure 3.9: Coupe D ouest-est (dite "du ravin de Philis"), dessinée à 1/10e. Sur cette coupe de 4m de longueur sur 6m de hauteur, huit niveaux de dépôts de versant ont été identifiés. A la base, un mince niveau de limons (lit 8) correspond à un épisode lacustre, renfermant un tephra (SEN 96). De nombreux fossiles ont été dégagés lors des anciennes fouilles de Philis et puis par Couthures.) [voir Fig. 2.2; pas sur Fig. 2.3]

delaminated into small angular fragments. These clasts are associated with rare, brown-purplish weathered scoriae. A few big stones of pale gneiss, generally not weathered and rather angular, stem from the bedrock. The blocks of rather rounded basalts are frequently cracked or show delamination peeling like the layers of an onion. In the upper part of the layer a single block of reddish scoria can be observed. The gravels and granules are mostly composed of weathered or unweathered rounded scoriae, associated with quartzose granules.

Interpretation: head and pebble gravel in a sandy matrix, massively deposited in a southward-oriented channel. Based on its characteristics, this layer can be attributed to set 5'.

Layer 2: this layer is formed by crushed stones and gravel embedded in a dark yellow-brown sandy matrix, with a roughly laminated structure linked to grain-size variation. The bottom of the layer, enriched by sand and basalt granules, is more compact. In the eastern part of the section, the upper layers are incised by layer 1 and the lower layers are rich in large stones and blocks that are concentrated in a channel. The lower boundary of the bed is irregular.

The supporting matrix is composed of coarse sand, mainly basaltic with a few orangey scoriae grains. The coarse elements are rounded, slightly weathered basalt granules and cobbles grouped in beds with distinct characteristics: angular particles range from 1 to 2 cm and rounded particles range from 5 to 12 cm. The rare scoriae that could be identified are clearly weathered.

Interpretation: slope formation linked to wash, reworked elements of volcanic origin.

Layer 3: this layer is formed by several beds in a prograding pattern in the western part of the section. The beds, enclosed in a supporting matrix of brown yellowish sand reworked from weathered bedrock, can be distinguished by the grain sizes of the represented elements, i.e.:

- beds formed by small homometric pebbles (1 cm); primarily basalt, associated with reddish scoriae;
- beds of coarse sands with very rounded basalt grains and red scoriae associated with a few bedrock elements;
- a very large sand lens exposed from the center to the eastern part of the section formed mainly by basalt blocks and smaller stones; together with a single block of red scoria and a cobble of gneiss;
- light brown yellow sandy-loamy lens formed by bedrock elements and interbedded with two thinner laminae;
- massive, indurated lens, exposed in the center of the section, thickening just above the channels of layers beneath; the matrix is very abundant, containing dispersed granules, rounded gravels and small pebbles (1-2 cm up to 5 cm). These are mainly basaltic, but several red scoriae and a peridotite fragment were identified.

Interpretation: more or less channelled slope formation exhibiting various reworked materials: the coarser ones are of volcanic origin, the finer ones stem from the weathering of the bedrock.

Layer 4: this layer corresponds to a group of beds that are coarser than layer 3, with a compact to sometimes slightly loose structure (more particularly in the eastern part). The matrix is composed of light brown arenaceous fine sand overlain by a thin layer of coarse elements in the upper 30 cm of the layer. The beds are formed by:

- small homometric basalt pebbles and rounded scoriae associated with a few bigger elements (cobbles and small blocks);
- or rounded basalt granules and scoriae;
- or more indurated, brown yellowish arenaceous sands.

The central and eastern part of the section is clearly richer in larger stones.

Interpretation: slope formation that after its deposit underwent washing that locally dragged the matrix and even small granules.

Layer 5: restricted to the central part of the section, this layer corresponds to a lens composed of rounded, mainly basalt granules, deposited in beds with varying grain sizes. There are a few weathered basalt pebbles of medium size, around 4-5 cm. The matrix is unevenly distributed fine brown arenaceous sand. Distinct beds display weathered basalt granules and red scoriae embedded in an abundant matrix, others are loose with unweathered granules, overlain by a thin white yellowish deposit.

Interpretation: slope formation of small-sized reworked elements.

Layer 6: heterometric crushed stones embedded in highly compacted brown yellowish coarse sand. The stones, measuring 5 cm to 7 cm on average, do not exhibit any particular direction. They are mainly composed of basalt and of sometimes quite numerous scoriae, associated with blocks, some quite large. The smaller components are rounded granules and gravel, mainly weathered reddish scoriae, locally overlain by a very thin whitish layer similar to the one identified in distinct beds of layer 5. The abundant matrix is composed of fine, more or less indurated sand derived from weathering of bedrock and basalt. Several very compact areas correspond to loams.

Interpretation: slope formation.

Layer 7: this layer is formed by sands and granules that display graded bedding, with granules at the bottom followed by coarse sands and overlain by finer sands. The granules are rounded basalt and gneiss with red, severely altered scoriae embedded in the sandy levels above. The sands are also overlain by a very thin whitish layer as is the case in distinct levels of layers 5 and 6. Some rare pebbles are mainly weathered basalts. The matrix is formed by arenaceous fine sand enriched in loams.

Interpretation: slope formation, deposited by water flow.

Layer 8: 8 to 9 cm thick laminated loam with parallelepipedic jointing. It is overlain by a lens, visible only in the western part of the section, composed of three beds:

- at the base, a light grey brown 8 mm thick banded loam;
- overlain by 1 cm coarse orange greyish sand including some basalt granules;
- lastly, 8 mm brown yellowish sand derived from bedrock and basalt.

The bottom of layer 8 includes dispersed powdery whitish grains 2 to 5 mm in thickness. These are volcanic minerals (Couthures et al. 1991; Pastre 2024).

Interpretation: This layer corresponds to a lacustrine level containing pyroclastic ash from Mont-Dore (TH6 = SEN 96 [Pastre 2024]), overlain by fine detritic slope deposits.

Layer 9: this layer can be defined as an indurated unstructured microconglomerate. The represented elements are homometric: basalt granules and gravel as well as small pebbles of weathered scoriae 1-2 cm in diameter, associated with some bigger elements composed of more or less weathered vesicular basalt, weathered scoriae and slightly weathered bedrock elements. The whole is embedded in a sandy, arenaceous type matrix.

Interpretation: slope formation with mass deposit.

In conclusion, section D made it possible to identify formations reworked from the slope deposits, most often established in channels and dipping towards the center of the maar. A lacustrine level (layer 8) appears at an altitude of 624.50 m. It contains tephra derived from Mont-Dore at its base.

The Southeastern Sector

Six trenches were laid out between 2004 and 2006: trench 4 is in an isolated position on parcel 174, whereas the others are located on parcel 172 except for the upper end of trench 8, which extends slightly onto parcel 174 (see Figs. 2.2 and 2.3).

Trench 4

Trench 4 was dug for approximately 20 meters along the line of maximum gradient, oriented southwest/northeast (N 50°) and exposes three depositional sets at altitudes between 623.80 m and 619.80 m (Fig. 3.10):

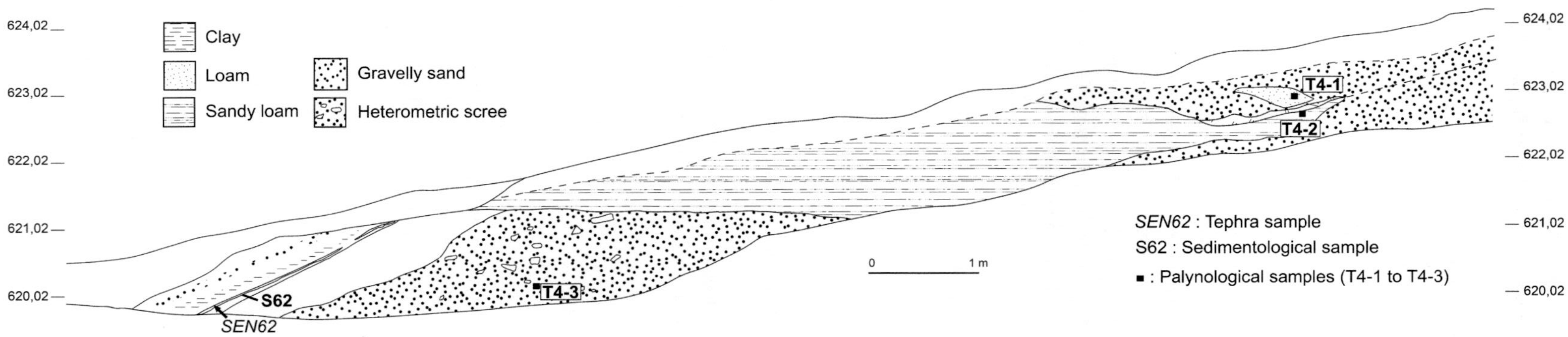

Fig. 3.10 Trench 4, southeast face, drafted at 1/20 scale. It only shows colluvium deposited in a channel. At its end is a level of reworked ashes (SEN 62). [2.3:17] (Figure 3.10: Tranchée 4, Coupe sud-est, dessinée à 1/20e. Elle montre seulement des colluvions déposées dans un chenal. A son extrémité affleure un niveau de cendres remaniées (SEN 62). [2.3:17])

At the bottom of the trench, unstructured heterometric scree can be recognized (from 30 cm blocks to gravel), in a matrix of arenaceous type coarse sand. The coarsest elements are heaped together mainly on the upslope side, primarily gneiss with large feldspars in weathered but well preserved forms, associated with a few quartzose clasts. Basalt scoriae are seen in the upper part of the scree. This first bed matches typical slope formations derived from weathered bedrock. Downslope this bed is incised by finer-grained stratified slope deposits of small stones and gravels, which contains a discontinuous bed of pale ash that corresponds to a reworked tephra from Mont-Dore (TS8 = SEN 62 [Pastre 2024]).

Above this bed is a level of dark orange sandy loam, rich in weathered gneiss granules, that fades upwards in the trench. It exhibits greyish reduction traces in vertical tongues at its base and in larger zones at the top. This is a slope deposit impacted by probably modern pedogenesis.

Towards the upper part of the trench the sediment becomes sandy and gravelly, with some isolated blocks, as well as loamy and sandy lenses and a level with scattered charcoal fragments. This bed also corresponds to a slope formation with reworked arenaceous and charcoal zones (hearths? charcoal kilns?). The presence of charcoal and several fragments of pottery at the top of the western section of the trench indicates modern colluvial dynamics. In conclusion, trench 4 only exposes colluviums filling a fossil dell. These colluviums are mainly reworked from weathered bedrock.

Trench 5

Trench 5 was dug in 2004, (see Fig. 1.23) along the line of maximum gradient and oriented southwest/northeast (N50°; see Fig. 1.23). It consists of two segments separated by a subhorizontal scouring surface about 7 m long. The upslope part (T5 sup) is slightly less than 10 m long, at altitudes between 624.70 m and 620 m (Fig. 3.11); the downslope part (T5 inf) is 15 m long between 618.80 m and 613.40 m (Fig. 3.12).

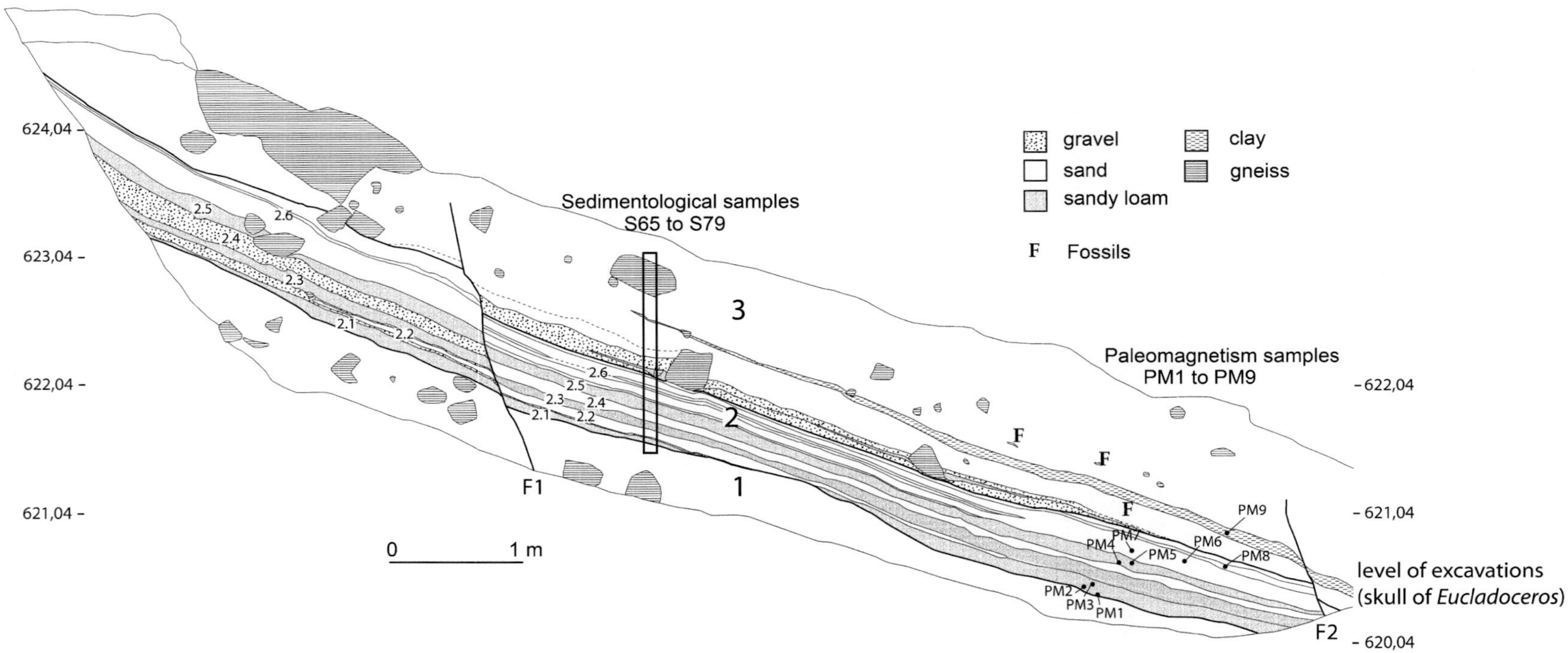

Fig. 3.11 Trench 5, upper part, northwest face, drafted at 1/10 scale. Trench 5 is the reference for the stratigraphy of the south-eastern sector. Its upper part, 10 m long, shows three sets of more or less coarse slope deposits with two microfaults (F1 and F2). Fossils have been found in set 1 (including a nearly complete skeleton of *Allohippus senezensis* recovered in this area in 2005) and at the base of set 3. Samples were taken for sedimentology and paleomagnetism. [2.3:18] (Figure 3.11: Tranchée 5, Partie supérieure: coupe nord-ouest, dessinée à 1/10e. La tranchée 5 est la reference pour le secteur sud-est. Sa partie supérieure, longue de 10 m, montre trois ensembles de dépôts de versant plus ou moins grossiers. Deux microfailles (F1 et F2) les affectent. Des fossiles ont été trouvés dans l'ensemble 1 (dont un squelette presque complet d'*Allohippus senezensis* trouvé dans cette aire en 2005) et à la base de l'ensemble 3. Des prélèvements ont été faits pour la sédimentologie et le paléomagnétisme. [2.3:18])

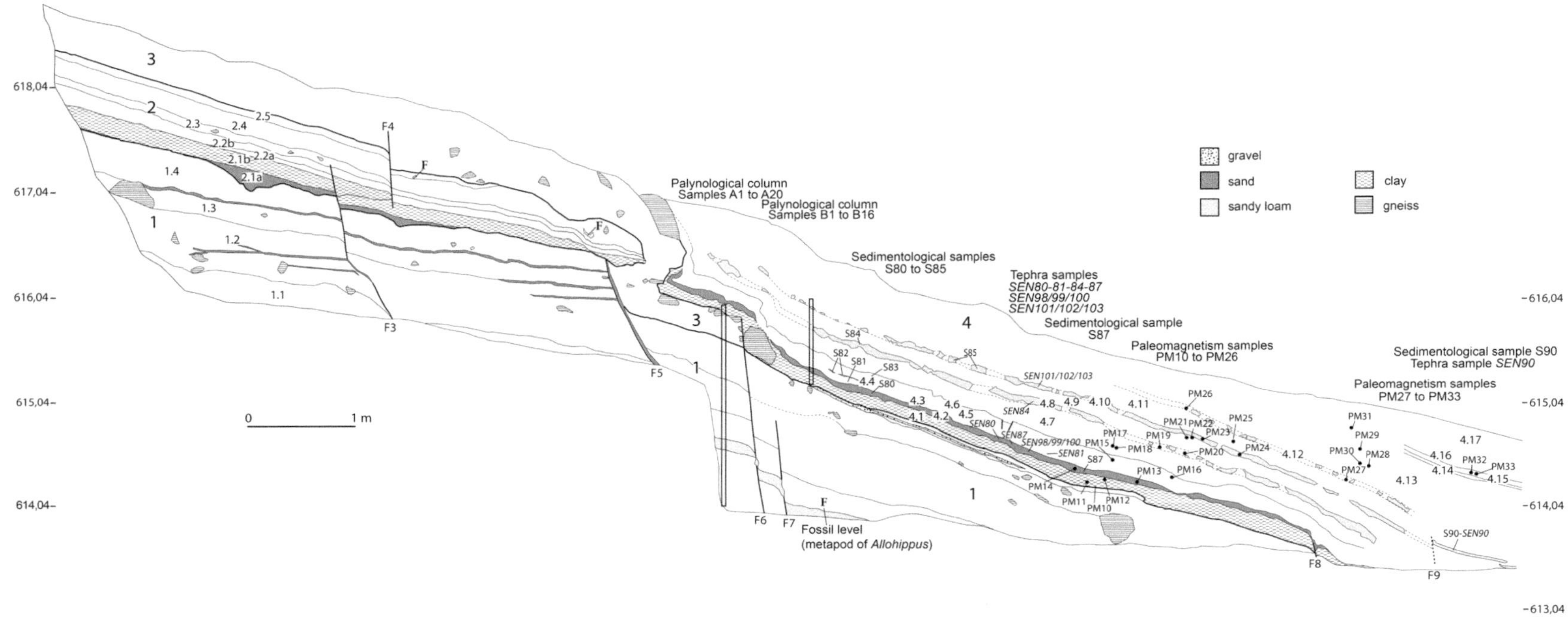

Fig. 3.12 Trench 5, lower part: northwest face, drafted at 1/10 scale. 15 m long, it intersects the sets visible on the T5 sup section (sets 1 to 3) and lake layers (set 4). The latter are formed by more or less coarse silts and sand levels containing tephras. Several slip cleavages are visible (F3 to F9) as well as ductile deformations (folds affecting sets 3 and 4, micro-injections and stretching in set 4). Samples were taken for palynology, sedimentology, paleomagnetism and tephras. SEN 98 of level 4.2 is dated 2.18 ± 0.03 Ma; SEN 101 from level 4.10 is dated 2.13 ± 0.04 Ma (Delson et al. 2024). Layer 1 yielded remains of *Allohippus*. [2.3:19] (Figure 3.12: Tranchée 5, partie inférieure: coupe nord-ouest, dessinée à 1/10e. Longue de 15 m, elle recoupe les ensembles visibles sur la coupe T5 sup (ensembles 1 à 3) et des niveaux lacustres (ensemble 4). Ces derniers sont formés par des limons plus ou moins grossiers et des niveaux de sables contenant des téphras. Plusieurs microfailles sont visibles (F3 à F9) ainsi que des déformations ductiles (pli affectant les ensembles 3 et 4, micro-injections et boudinage dans l'ensemble 4). Des prélèvements ont été faits pour la palynologie, la sédimentologie, le paléomagnétisme et les téphras. SEN 98 du niveau 4.2 est daté de 2,18 ± 0,03 Ma; SEN 101 du niveau 4.10 est daté de 2,13 ± 0,04 Ma (Delson et al. 2024). Le niveau 1 a livré des restes d'*Allohippus*. [2.3:19])

Trench 5 makes it possible to identify, from bottom to top, four important sedimentological units on an average slope of 20°, somewhat affected by fractures and sliding.

a Set 1

This set is 60 cm thick at the bottom of T5 sup, compared to more than 1.20 m in T5 inf. It is a coarse, heterometric conglomerate with supporting matrix. Its upper boundary is irregular due to erosion. The structure of the elements in beds with varying grain sizes make it possible to distinguish four units, identified from bottom to top:

– Unit 1.1

This unit is only identifiable in T5 inf. It is mainly yellowish gravel with alternating levels comprised either of inframillimetric fine granules and supporting matrix, or of coarser millimetric granules and supporting clastic support, with sandy lenses where the matrix becomes predominant. Globally, the grain size of the elements decreases downslope in the trench, where loam-sand lenses are present. Within the gravel levels the graded bedding is ordinary. The granules are mainly of basalt; with a few from the bedrock. Some rare gneiss pebbles can be identified in the upper 20–25 cm of this unit. An equid metatarsal and associated phalanges were found in this layer.

– Unit 1.2

This unit consists of pebble gravels associated with gravels. In the upslope part of the trench, the pebble gravels have a clastic support of elements less than 3 cm across and are accompanied by well-worn gravels and coarse sands reworked from the bedrock. Gravels of stream-polished quartz are notable.

In the downslope part the gravels predominate in superposed layers with differing grain sizes. Lateral sorting in these lenses can be observed with a downslope concentration of the coarser gravels. Blocks and pebbles of 1 to 2 cm width (mainly from the bedrock and basalt) are occasionally seen. The matrix is arenaceous gravelly sand. This stony unit is incised by two or three thinner beds with inverse graded bedding: sand at the bottom, overlain by inframillimetric granules, then by a bed of small gravelly pebbles. The top of the unit is enriched with coarse gravel of basaltic origin that provides a greyer color.

In 2005, at the base of this unit, many fossils were found, including a nearly complete skeleton of *Allohippus senezensis* (see Fig. 2.7 and Eisenmann and Delson 2024). These finds were not made in T5 inf itself, but in an area of excavation begun alongside T5 and continued in 2005 near T6.

– Unit 1.3.

This unit is very coarse in the upper part of the trench with various types of more or less weathered, rounded to angular blocks of medium-grained porphyroblastic gneiss and dark gneiss very rich in ferro-manganese minerals, reworked from the bedrock. These blocks are scattered amongst mainly basaltic gravels.

In T5 inf, the basement blocks are still present, some very large. They are associated with 3 to 4 cm pebbles, often lying flat according to the slope. The gravels are arranged in beds which do not extend very far. The upper boundary of unit 1.3 is highlighted by a bed of fine arenaceous sand.

– Unit 1.4

This is a set of gravel lenses with very loose clastic and sandy support. The gravel consists of rounded clasts of basalt and gneiss. Some bedrock cobbles are also present.

In the lower part of T5 inf the characteristics of the units 1.2, 1.3 and 1.4 tend to fade and it becomes difficult to distinguish them. Set 1 is a typical slope formation of reworked elements from the weathered bedrock (arenaceous, rounded blocks) and components of the tuff ring. The presence of rounded gravels, such as quartz in unit 1.2 may also indicate the reworking of eolian elements. Deposition of the successive units was prograded downslope. The presence of huge angular blocks in the downslope part of the trench suggests debris flow.

b Set 2

This set is characterized by alternating loamy and gravelly levels in T5 sup, changing to thinner levels in T5 inf. It is possible to identify six units from bottom to top:

– Unit 2.1

In T5 sup this unit corresponds to a loam lens that disappears at mid-slope. Its thickness varies between 2 to 3 cm at the upper part of the trench and up to 11 cm in the lower part. Regular bedding is apparent in the presence of thin bands (2 to 5 mm thick) of rounded basalt gravels dispersed in the loamy layers.

In T5 inf this unit is more sandy. It begins with micaceous sand lenses filling the depressions of set 1 (= 2.1a), with a bedding of varying grain sizes (coarse to fine, even loamy sands). The upper surface of the lenses sometimes shows small rippling. Above this is a level (= 2.1b) which is initially sandy and then more loamy with normal internal graded bedding. At its upper part micro-cracks can be seen as well as injections of unit 2.2 to a depth of 1 cm.

– Unit 2.2

In T5 sup this is a level with supportive matrix 5 to 6 cm thick, composed of millimetric to multimillimetric granules displaying inverse graded bedding; the matrix is sandy and becomes increasingly abundant downslope.

In T5 inf the transition from unit 2.1 to unit 2.2 is characterized by the progressive encroachment of clayey sand (2.1b) with multimillimetric gravels.

– Unit 2.3

This unit consists of sandy loam, 7 to 8 cm thick, thinning at the center of T5 sup and adding fine particles in the downslope direction. Microbedding clearly alternates sand, loam and gravel layers. In T5 inf, counterpart levels are arenaceous, 8 cm thick sand including bedrock stones at its bottom (6 cm to 7 cm).

– Unit 2.4

This unit exhibits upslope-downslope evolution characterized by the presence of basalt and gneiss gravels. In T5 sup the gravels are infra-millimetric to 5 mm, well arranged in beds with varying grain sizes, several centimeters thick with matrix or clastic support. The thickness of the unit grades from 15 cm upslope to 5-7 cm downslope. In T5 inf the gravels are coarser (up to 1 cm in diameter), concentrated in arenaceous sand 17 to 18 cm thick.

– Unit 2.5

In T5 sup this is a sandy loam level incised by beds of millimetric gravels in the upper part. It thickens (up to 10 cm) and becomes richer in fine particles towards the downslope part. In T5 inf it is loamy sand enclosing dispersed gravels.

– Unit 2.6

In the upper part of T5 sup, this unit displays alternating gravelly and sandy levels with matrix or clastic support, 31 to 33 cm thick. At the bottom of the unit millimetric gravels reworked from the bedrock, mixed with basalt gravels with pebbles 3 cm to 5 cm in diameter, are the only sediments represented. In the lower part of T5 sup the sandy levels increase whereas the gravel beds become thinner or even disappear. The unit is not very visible in T5 inf because it is located at the top of the section. The gravels are coarser than in the upper trench.

Set 2 corresponds to a slope formation. The grain sizes, which are finer compared to set 1, probably mirror a change of environmental conditions (less cold climate, plant development).

c Set 3

This is a heterometric conglomerate with matrix support visible measuring 80 cm thickness in T5 sup and nearly 40 cm in T5 inf. In T5 sup the blocks are large angular pieces of bedrock, up to 1.80 m in length, but smaller in T5 inf. They are accompanied by small, rounded pebbles, dominantly basalt and some gneiss. Basalt gravels are abundant, particularly in the lower 20-25 cm. The matrix is gravelly sand with grey color at the bottom due to abundant material reworked from the maar, with yellower material from the granitic sand more abundant above. Strongly rounded quartz grains of the eolian type are also found.

Set 3 is a typical formation linked to slope destabilization leading to a large debris flow from the bedrock and the tuff ring. In T5 sup, at the base of this set, many bones are present (including two partial cervid crania extracted in 2004, see Valli 2024).

d Set 4

This set is visible only in the downslope part of T5 inf where it is in direct contact with set 1. It is formed by more or less coarse loams and sands containing tephra particles. Sixteen units, from bottom to top, were identified according to the nature and grain size of their components:

– Unit 4.1

This unit consists of clayey loam with finely laminated polyhedral jointing, varying between 10 cm upslope and 14 cm downslope. Its bottom contains coarser particles, with fine elements progressively increasing towards the top. This unit conforms to the irregular surface of set 1 and displays a wavy upper boundary.

– Unit 4.2

This is a 3-5 cm thick orange homometric, more or less clayified sand, in which whitish millimetric to multimillimetric clay beds can be observed. This unit shows stretching and boudinage patterns. This is a tephra layer (TS9 = samples SEN 80, SEN 87, SEN 98, SEN 99 and SEN 100 [Pastre 2024]); SEN 98 was dated by Ar-Ar to 2.18 ± 0.03 Ma (see Delson et al. 2024).

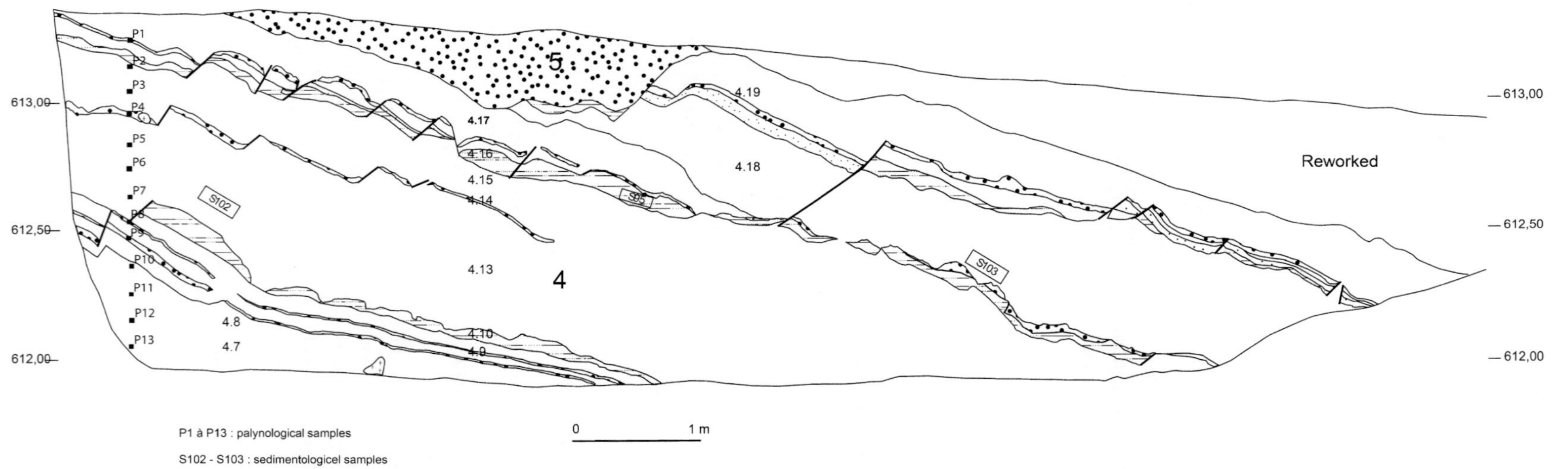

Fig. 3.13 Trench 6, northwest face, drafted at 1/20 scale. Excavated downstream from trench 5, it shows the lake deposits of set 4. They are intersected by a slope formation (set 5). Palynological samples were taken in its upstream part. [2.3:23] (Figure 3.13: Tranchée 6, Coupe nord-ouest, dessinée à 1/20e. Creusée à l'aval de la tranchée 5, elle montre les dépôts lacustres de l'ensemble 4. Ils sont recoupés par une formation de versant (ensemble 5). Des prélèvements palynologiques ont été réalisés dans sa partie amont. [2.3:23])

– Unit 4.3

This unit is 3-4 cm thick, comprised of brown fine laminated loam with polyhedral jointing. Its upper boundary is irregular.

– Unit 4.4

This unit corresponds to a thin deposit (2 cm at most) of heterometric arenaceous discontinuous sand, filling the irregularities and cracks of unit 4.3.

– Unit 4.5

This is a 4-5 cm thick, fine brown laminated loam with polyhedral jointing that includes sands and iron oxides towards the top.

– Unit 4.6

This unit displays a transition from fine light brown loam (3 cm thick) to light beige, increasingly sandy loam (3-4 cm thick), with a more subtle lamination compared to the preceding unit. The upper boundary is marked by a border of iron oxides (1 to 2 mm).

– Unit 4.7

This unit consists of fine laminated loam with polyhedral jointing, 16 to 25 cm thick. Its upper surface is irregular.

– Unit 4.8

This is a 7 to 8 cm thick bed of laminated light beige loam. Distinctly more compact zones evidence block jointing.

– Unit 4.9

This unit is a brown loamy bed with polyhedral jointing, 8 to 12 cm thick. It encloses four strata of coarser sand and basalt gravels.

– Units 4.10 to 4.13

These units are alternating thin beds (4-5 cm thick) of light beige fine loams (units 4.10 and 4.12) and thicker brown laminated beds with polyhedral jointing (12 to 17 cm thick as regards unit 4.11, 40 cm as regards unit 4.13). Unit 4.10 consists of tephra (TS10 = samples SEN 90, SEN 101, SEN 102 and SEN 103 [Pastre 2024]). SEN 101 was dated by Ar-Ar to 2.13 ± 0.04 Ma (see Delson et al. 2024).

The highest units (4.14 to 4.17) are visible at the end of T5 inf over a thickness of nearly 1 m. They are affected by modern weathering given their proximity to the surface, which makes their characterization difficult. They are described in trench 6.

In sum, Set 4 is composed of lacustrine, loamy and laminated beds that thicken downslope. They are traversed by thin detritic seams (mostly sandy loams, sands and gravels, occasionally small pebbles) marking the arrival of slope deposits. Tephra levels are present in the form of homometric sands (unit 4.2) or loams (unit 4.10).

e Deformations

Numerous deformations, linked with gravitational movements or mechanically induced compaction, affect the deposits of trench 5:

- Brittle deformations: eight ordinary slip cleavages, 10 to 20 cm slip, are visible in the documented sections. They generally end in coarse formations and are frequently highlighted by sand or clay infiltrations. One slip cleavage (number 9), visible over several centimetres at the end of T5 inf, is inverted.
- Ductile deformations appear in various ways, such as fold affecting sets 3 and 4 in the central part of T5 inf; micro-injections in the upper part of the units 2.1, 4.1; and stretching and boudinage of distinct levels of set 4 (4.2, 4.5).

Trench 6

This trench was laid out in 2005, downslope from trench 5. It is more than 6 m long, at an altitude between 613.30 m and 611.90 m (Fig. 3.13; also Fig. 1.25). It exposes mainly lacustrine deposits belonging to set 4, cut by a slope formation deposited on an erosion surface (= set 5).

The following units can be identified in Trench 6:

a Set 4

– Units 4.7 and 4.8:

At the bottom of the section, brown yellowish plastic loams that are micaceous and laminated are exposed (unit 4.7). The laminae are highlighted by ferro-manganese pellicles. The upper part (unit 4.8) is split into parallelepipedic orange greyish blocks exhibiting vertically striated surfaces covered by ferro-manganese deposits. The upper boundary is diffuse. Units measure 35 cm thickness at most.

– Unit 4.9

This unit is 12 cm to 14 cm of moderate-brown to yellowish laminated loam traversed by two ferruginous sandy seams. The laminae are highlighted by black mottling and/or iron oxide coating. The sandy levels display inverse graded bedding with orange greyish relatively thin (at most 1 cm) sandy loam at the bottom with a diffuse and more or less wavy lower boundary, overlain by 0.5 cm coarse ferruginous sand and lastly a row of small rounded weathered basalt granules.

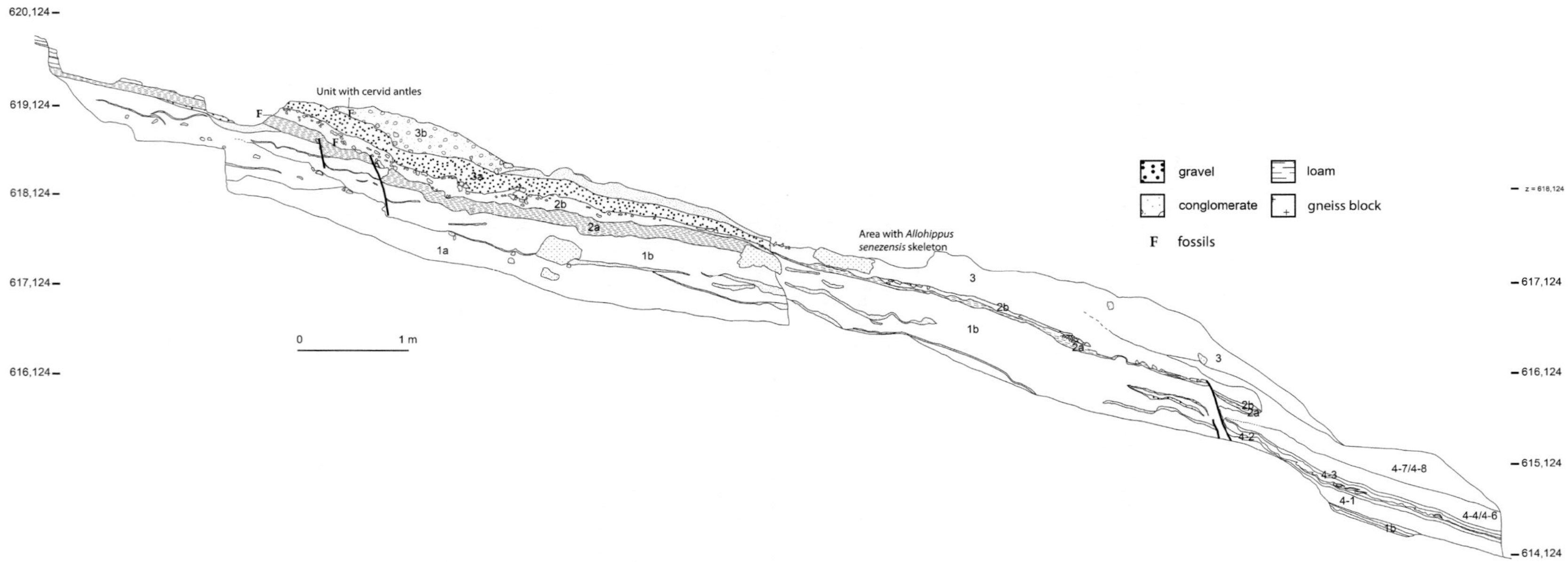

Fig. 3.14 Trench 7, upper part, northwest face, drafted at 1/10 scale. It shows deposits similar to those found in the T5 sup trench and the upper part of the T5 inf trench. Set 3 contains many fossils, including a second partial skeleton of *Allohippus senezensis*. [2.3:27] (Figure 3.14: Tranchée 7, Partie supérieure: coupe nord-ouest, dessinée à 1/10e. Elle montre des dépôts semblables à ceux décits dans la tranchée T5 sup et la partie supérieure de la tranchée T5 inf. L'ensemble 3 renferme de nombreux fossiles, y compris un deuxième squelette incomplet d'*Allohippus senezensis*. [2.3:27])

– Unit 4.10
Whitish sandy-loamy bed split into blocks with irregular thickness (from 0.8 cm to less than 0.5 cm) and a very irregular upper boundary.

– Unit 4.11
12 to 14 cm of loam with polyhedral jointing. Uniformly massive aspect but subtle lamination is sometimes revealed on detachment surfaces

– Unit 4.12
This unit is present over a thickness of 10 cm, with angular sandy-loamy blocks, stretched along the slope, that are separated by loam several millimetres thick.

– Units 4.13 to 4.15
This is a thick level of brown loam with polyhedral jointing, traversed by a thin coarse lens. Unit 4.13 is between 30 and 50 cm thick. Unit 4.14 is 3 to 5 cm thick, comprised of basalt granules and gravels associated with small basalt and gneiss pebbles (1-2 cm long) wrapped in a loamy-sandy matrix; some gneissic pieces can reach 5 cm. This lens is cut by slip cleavages with shifts of several centimeters, and disappears downslope in the trench. Unit 4.15, 18 to 25 cm thick, is progressively invaded by loamy sands towards the upper part.

– Unit 4.16
Detritic level with rather pale color cut into blocks by many slip cleavages. It is 23 cm thick on average, decreasing downslope, and composed of several beds with varying composition and grain sizes: at the bottom there is a bed of arenaceous sand of irregular thickness (0 to 4 cm), marked by oxide deposits and containing millimetric basalt granules; this is followed by a rather consistent level (1-2 cm) of very pale laminated ashes (sandy or loamy laminae). The upper part begins with a bed (1.5-2 cm) of slightly sandy orange greyish loam containing a very large number of carbonized root casts, overlain by 1 to 1.5 cm thick pale brown loam above the sandy lenses of the unit's bottom; this is followed by 2-3 cm of pale brown yellow fine loam and lastly a 1.5 to 2 cm thick bed of weathered and brittle basalt granules and unweathered gneiss granules associated with some bigger elements; this latter level is thicker but discontinuous in the downslope direction.

– Unit 4.17
Orange greyish laminated loam lens, with progressively uploaded fine elements in the downslope direction; lower boundary irregular because of the slip cleavages that affect unit 4.16; upper boundary eroded or wavy.

– Unit 4.18
Dark brown yellowish slightly sandy unstructured loamy layer, which expands downslope (40 cm thick at the end of the section).

– Unit 4.19
Alternating thin loamy and gravelly levels, 22 cm thick, cut by slip cleavages. Generally, these become richer in loams downslope. In more detail, the following succession can be identified: at the bottom, 4-5 cm of sandy loams with rounded and dispersed basalt granules, then rounded, ferruginous and more or less weathered homometric (2 mm on average) basalt gravels without matrix 1.5 to 2 cm thick, overlain by 2 to 4 cm thick brown yellowish plastic and slightly detritic loam; then a 2 to 3 cm thick layer of rounded, weathered basalt gravels (average diameter: 1 mm, maximum diameter: 5 mm) enclosed in an abundant loamy-sandy matrix; followed by 2.5 cm brown greyish plastic loam; and lastly 4cm of non-weathered basalt gravels, with sorted bedding (grain sizes ranging from 2-3 mm at the bottom to 1 mm at the top) and very pale orange abundant loamy-sandy matrix.

– Unit 4.20
Pale loam 15-20 cm, with dispersed gravels intersected by thin veins of water-worn basalt gravels.

b Set 5

This set is deposited on a distinct erosional surface, and consists of a massive concentration of water-worn basalt gravels and some weathered bedrock elements enclosed in an arenaceous matrix.

In conclusion, trench 6 exposes the upper units identified in trench 5, but its more downslope location makes it possible to identify other, more recent units. The sediments are mainly lacustrine with poorly defined slope deposits occurring as thin detritic levels with rather fine grain sizes.

Trench 7

Trench 7 was dug in 2006, close to trenches 5 and 6 (see Fig. 1.28). For safety the trench was dug in two parts

- the upper part (T7 sup), almost 16 m long makes it possible to identify detritic slope deposits and several lacustrine levels at an altitude of between 619.80 m and 614.10 m. These deposits are similar to those described in T5 sup and the upper half of T5 inf;

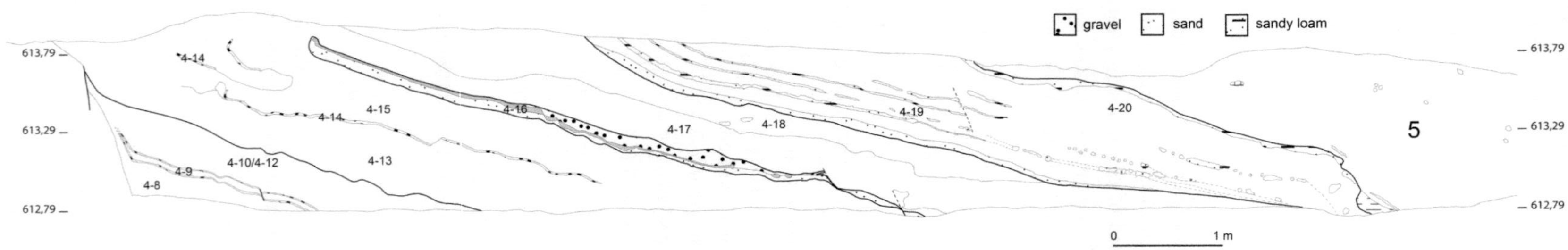

Fig. 3.15 Trench 7, lower part: northwest face, drafted at 1/10 scale. It mainly shows the lacustrine deposits described in T6. They are intersected in the downstream part by set 5. [2.3:28] (Figure 3.15: Tranchée 7, Partie inférieure: coupe nord-ouest, dessinée à 1/10e. Elle montre principalement les dépôts lacustres décrits en T6. Ils sont recoupés dans la partie avale par l'ensemble 5. [2.3:28])

- the lower part (T7 inf), about 10 m long, exposes mainly lacustrine deposits that are also identified in T5 inf and T6.

Deposits of T7 Sup

These deposits, with an average slope of 15° to 20° (N20°) are grouped in four large sets (Fig. 3.14). The first three, described from bottom to top, are exposed in most of T7 sup. Set 4 is only visible at its downslope end.

a Set 1

This set includes two units:

- Unit 1a

This unit is characterized by its heterometry (stones to sand) and the abundance of basalt elements. The bedrock elements (gneiss rich in black mica) are rarer, of varying sizes and weathered. Basalt grains make up some 80% of the matrix, the remaining part being bedrock minerals. The predominance of basaltic material provides a grey color. Unit 1a typically is a phreatomagmatic formation.

- Unit 1b

This unit displays significant variations as regards the grain sizes and its components. For the greater part, the components stem from the bedrock and include angular elements with rather fine grain sizes (small stones and sands). Some basalt pebbles and occasional very large blocks of gneiss are also included. Globally, the grain sizes become smaller from upslope to downslope. The debris is organized in prograding beds with a slope of 15° in the upper part of the trench, and of 30° to 40° downslope. These beds exhibit either normal (small stones-gravels-sands) or inverse graded bedding with a clastic or matrix support, in which the matrix is fine granitic sand. Sand lenses highlight the presence of micro-channels. Unit 1b typically is a detritic fan reworking deposits present on the sides of the maar.

b Set 2

This set has fine grain sizes and is comprised of two units that develop laterally from upslope to downslope:

- Unit 2a

This unit consists of compact, bedded and laminated fine loam that shows significant thickness variations between the top and the bottom of the trench; these modifications are due to both brittle (slip cleavings) and ductile deformations (ripples, slumps). The slip displacements measure between 8 and 12 cm, which affects the upper part of set 1 but unit 2b less so. This loam level is 10 cm thick in the upper part of the trench and reaches 16 cm behind a big gneiss block visible in the canter of T7 sup. Beyond the block, it thins out significantly (6 cm). These thickness variations and the ripples of this unit reveal its role as a "soap" layer in regard to the deposits of set 3, which clearly slipped down along the slope.

- Unit 2b

This unit, about 15 cm thick, was formed by the alternation of beds of fine, ferruginous (tephra?) sands and light, compact loamy-sandy layers. These latter contain the internal molds of small gastropods and root marks and show destructuration in the form of clasts and cracks. These clasts are arranged in thin beds or infiltrate along the fissure surfaces. Manganese oxides cover these surfaces and form dot-like patches. Downslope, within less than 2 m, these levels become less sandy and loamier. They exhibit deformation patterns (sparks, ripples) and complete destructuration in the form of small blocks lying along the slope before fading. Several hypotheses can be advanced to explain this destructuration:

animal trampling of loamy soil affected by desiccation cracks triggering the formation of clasts;

set of faults affecting the underlying deposits, at the origin of a slow downslope movement of consolidated loam;

deposition of the coarse-grained mudflow that forms set 3, which overloaded the loamy sediments;

combination of these factors.

At the limit with unit 2a, fossils were found in a thin bed of clasts.

In the downslope part of T7 sup, set 2 is affected by a slip cleaving that causes a 10 cm shift. It then disappears in the heart of the frontal lobe of the sliding that affected set 3.

The sedimentological characteristics of this set 2, including a merging of slope deposits with fine grain sizes and of loam deposits of lacustrine origin, and the presence of mollusc and root imprints, reveal a depositional environment most probably corresponding to the lake shore.

c Set 3

This set is characterized by coarse grain sizes; according to the mean size of the elements, it can be subdivided into two units, which are clearly visible, primarily in the upper half of the trench.

- Unit 3a (20 cm)

Gravels of granules and small bedrock and basalt pebbles, including some small pebbles, in abundant sandy matrix (quartz, feldspars, amphiboles); at the top of the trench, on the southern section, over 12 cm beds of rusty arenaceous sand alternate with gravels; a second partial skeleton of *Allohippus senezensis* was recovered in 2006 in a level corresponding to 3a (see Eisenmann and Delson 2024);

– Unit 3b (80 cm)
conglomerate of gravels and small (< 2 cm) rounded basalt pebbles, including dispersed angular blocks from the bedrock (10–15 cm); clastic to matrix support (arenaceous matrix); antlers have been recovered in this conglomerate.

Set 3 is typically a slope formation. It penetrated into the lacustrine deposits causing a drag fold. As is the case for the section, the slipping was favored by the "soap" layer represented by the loams of set 2.

d Set 4

This set is visible only in the downslope part of T7 sup. Directly overlying set 1, the identified units can be correlated with most of those described in trenches 5 and 6. From bottom to top:

- Unit 4.1 (15 to 17 cm thick):
  a bed l of fine, grey-colored laminated loam. The surfaces of the laminae are highlighted by mica and ferro-manganese oxides. The upper 2-4 cm has a darker color, clear polyhedral jointing, traces of modern roots between the laminae, and irregular boundaries;
- Unit 4.2 (3 to 9 cm thick), in three levels:
- basal discontinuous level of light laminated loam, frequently penetrating the overlying sand;
- rather fine-grained (1 to 0.5 mm) homometric orangish sand deformed into irregular thickness (from 2 to 7 cm) ;
- 1 cm thick level of whitish ash that slides over itself downslope.

Upslope, these levels are turned up in a drag fold. Their irregular internal structure shows that they slipped down within the lacustrine deposits, clearly at the moment when set 3 slumped. This unit 4.2 corresponds to a polyphase tephra (= TS9 [Pastre 2024]).

- Unit 4.3 (3 to 4 cm thick): a bed of dark laminated loam with polyhedral jointing and irregular upper boundary;
- Units 4.4 to 4.6 (16 to 18 cm thick): light loams, rich in fine detritus (mainly mica). The bottom encloses a discontinuous level of orange homometric sand (unit 4.4), with a maximum thickness of 1.5 cm. Sandier in the top 1 cm (unit 4.6) and marked by iron oxide precipitation;
- Units 4.7 to 4.8 (35 to 40 cm thick): brown, clearly laminated micaceous loams with polyhedral jointing. By contrast with Trench 6, a boundary between the two units is not discernible here.
- The following deposits are visible on the section at the height of the level separating T7 sup and T7 inf:
- Unit 4.9: two beds of rounded basalt granules, probably lapilli, transected by a light loamy bed 2.5 cm thick. The boundary with unit 4.8 is highlighted by a thin level (0.5 mm) of fine sand impregnated by iron oxides;
- Unit 4.12 (10 cm thick): light loam, fractured into polyhedral blocks;
- Unit 4.13 (60 cm thick): brown loam with polyhedral jointing.

These last units are found again at the bottom of T7 inf, thickened by several dozen centimeters due to slip cleavage.

Set 4 consists of lacustrine deposits with redeposited fine slope deposits and enclosing a tephra layer (Pastre 2024).

Deposits of T7 inf

This part of the trench was dug below the level corresponding to the lower locus of excavation. It cuts through most of the units identified in trench 6. The southern section exhibits the beds of unit 1b in abnormal contact with those of set 4. In the northwestern section the levels, sloping towards the center of the lake, such as those of T7 sup, are organized in two sets described from bottom to top (Fig. 3.15):

a Set 4:

The different units can be grouped together into five members:

- Units 4.8 to 4.12 (visible over more than 35 cm) are formed by fine, micaceous, homogeneous loam with subtle lamination. Unit 4.9 consists of rounded basalt granules (average diameter = 2 mm), closely to loosely packed. They are arranged in two layers separated by 1 mm or 0.5 mm of loam. The bottom of unit 4.8 is highlighted by deposits of bright rusty iron oxides. The first units correspond to lacustrine deposits, interbedded with two lapilli levels (similar to TS 10 of T5; see Pastre 2024).
- Units 4.13 to 4.15 (65 cm): dark laminated loam, rich in fine detritic sediment, split into polyhedral blocks coated with iron oxides. Locally light areas more heavily loaded in fine sediments and small lenticular amounts of granules and sands are isolated from one another. Unit 4.14 is a detritic bed 1 cm thick formed by granules, gravels and small pebbles in a slightly loamy sand matrix. The basalts predominate among the coarse elements, bedrock among the fine elements. The level is made up of concentrated iron oxides (rust color). Discontinuous and stretched downslope in the trench, it is broken up into several lenses upslope, which highlight plastic deformations (folds). These units are typical lacustrine deposits intersected by discontinuous detritic slope deposits. The presence of deformations (folds, stretch) probably results

from the silting of set 3 in these lacustrine deposits, triggering their slump.

- Unit 4.16: this unit is comprised of three levels:
  - at the bottom: orangish, medium to coarse sand (grain sizes from 1 mm to 0.5 mm), rich in feldspar; maximum thickness: 3–4 cm;
  - followed by 1 cm of laminated ashes, unstructured downslope;
  - lastly, a lens of predominantly basalt granules, gravels and small pebbles, associated with angular fragments of bedrock gneiss, micas, feldspars. Inverse graded bedding and downslope thickening can be noted, marking the accumulation of the material at the bottom of the slope.

This unit probably corresponds to tephra with successive differentiated deposits (Pastre 2024).

- Unit 4.17 and unit 4.18: laminated loam, rich in fine detritic material (micas). Unit 4.17 (about 30 cm) is greyish. Unit 4.18 (35 cm on average), brown, breaks up into lumps at the surfaces with ferro-manganese oxide coating. These are lacustrine formations.
- Unit 4.19 and unit 4.20: sequence of ten levels with medium (gravels and granules) to fine grain sizes (more or less sandy loams), with more or less sharp boundaries and varying thickness.

The coarsest levels are between 4 and 15 cm thick. They are composed of basalt bedrock elements, in equal proportions at the bottom of the unit with basalt predominant at the top. The bedrock gravels are frequently angular and the basalts more rounded. Normal graded bedding can be noted. The matrix is abundant, loamy and sandy, except for the upper level (15 cm of basalt and bedrock pebbles, in which the matrix is of arenaceous type. The fine thinner levels (1.5 to 3 cm thick) show very clear loam enrichment. They invariably include granules and sands of basalts and bedrock material, however, in smaller amounts. The bottom level is fine, homometric, 4 cm thick yellow orangish micaceous sand that may correspond to tephra. These units 4.19 and 4.20 are mainly formed by slope deposits, regularly intermixed with thin lacustrine deposits.

b Set 5:

This set corresponds to a large accumulation of small pebbles (1 to 2 cm maximum), gravels and granules in an abundant arenaceous matrix. Angular to rounded fragments with numerous large feldspar crystals reworked from the bedrock are predominant (80-90%). Basalt elements are mostly found among the gravels and granules; these are mostly rounded but some, however, are larger with an angular shape. The parallel arrangement of the deposited fragments reveals a diffuse though subtle structure in beds or lenses with varying grain sizes. This is a typical slope deposit that cuts across the preceding layers. The presence of a clayey level at its base may indicate a massive slipping.

Trenches 8 and 9

a Trench 8, spanning parcels 174 and 172, was laid out along the maximum gradient between 629 and 616 m in altitude. Below several dozen centimeters of topsoil, more or less lithified phreatomagmatic debris quickly halted the backhoe excavation of the trench.

b Trench 9 (parcel 172) was laid out perpendicularly to trench 8 at an altitude of between 618 and 621 m. Over nearly 50 m, it revealed traces of earlier excavation (perhaps that carried out by Boule). In the preserved areas and at the bottom of the trench, slope deposits with rather fine grain sizes (gravels and predominantly small pebbles), arranged in oblique beds, can be identified. Not a single fossil was identified therein. On the western side, phreatomagmatic debris identical to that of trench 8 was exposed at the bottom of the excavation.

Stratigraphic Correlations in the Southeastern Sector

While the correlations between trenches 5, 6 and 7 are quite easy, this is not the case as regards the other trenches: indeed, within a few meters there is a change from fossiliferous slope deposits covered by lacustrine clays to phreatomagmatic products ejected at the moment of the formation of the maar (trenches 8 and 9) or colluviums (trench 4). However, it is possible to link set 1 visible in the downslope part of the latter trench (heterometric scree covered by gravelly stratified deposits) with set 1 of trenches 5 and 7.

Sedimentological Study

Grain Sizes

Two types of grain size analyses were carried out on the samples taken in the less coarse levels of different sections or excavation loci.

The global grain size analysis was carried out on 26 samples that were sorted into pebbles (> 1 cm), gravels (1 cm-2 mm), sands (2 mm-50 μ), silts (<50 μ) (Table 3.2). Most of the samples evidence the predominance of fine fractions below 50 μ associated with sands to which a small amount of gravel can be sometimes added. Four samples (S1, S22, S60, S95) are mainly composed of sands. Six other samples (S2, S3, S23, S39, S91, S100) are clearly coarse with abundant gravels, sometimes associated with pebbles.

The study by laser diffraction analysis of the fractions below 2 mm of 30 samples was carried out by the platform "Environnement" of the Université Lyon 2. This analysis reveals the general predominance of sands (2000 μm to 63 μm) associated with silts (63 μm to 2 μm) with highly varying proportions (from 10 to 45%). One single sample (S101) evidences a slight predominance of silts compared to the sands. The clays (< 2 μm) are very scarcely represented (less than 2.5%).

The calculation of the sorting coefficient of Trask So indicates that these fractions below 2 mm are invariably very well sorted (coefficient slightly above 2.5). The skewness coefficient shows that the sorting is better for the coarse particles, except for four samples: S23 with better sorted fine particles; S35, S91 and S106, which display perfect symmetry. These results suggest that these elements stem from scarcely developed formations originating from the erosion of the rocks forming the perimeter of the maar. The low representation of clays indicates that the effects of soil weathering are hardly developed.

Table 3.2 Global grain size analysis (stones: >1 cm; gravels: 1 cm-2 mm; sands: 2 mm-50 μ; particles < 50 μ) and Laser grain size distribution (particles < 2 mm). (Tableau 3.2: Granulométrie globale (cailloux: >1 cm; graviers: 1 cm-2 mm; sables: 2 mm-50 μ; fins < 50 μ) et granulométrie Laser des fractions inférieures à 2 mm)

	Global grain size					Laser grain size distribution			
	Particles <50 μm	Sands >50 μm	Gravels	Stones	Definition	Clay	Silt <63 μm	Sand >63 μm	Definition
S1	39.0	61.0	0.0	0.0	Loamy sand	0.99	26.58	72.42	Loamy sand
S2	11.3	23.9	54.2	10.7	Sandy gravels	0.14	11.68	88.18	Loamy sand
S3	10.4	25.5	56.6	7.5	Sandy gravels	0.21	10.46	89.32	Sand
S8	84.7	11.0	4.3	0.0	Loam	0.36	13.39	86.26	Sand
S11						1.28	42.31	56.41	Loamy sand
S18						0.77	24.56	74.67	Loamy sand
S20						0.91	26.88	72.21	Loamy sand
S21						1.08	29.24	69.68	Loamy sand
S22	41.1	58.8	0.0	0.0	Loamy sand	1.09	29.80	69.11	Loamy sand
S23	4.3	26.7	42.5	26.5	Coarse sandy gravels	0.00	11.26	88.73	Sand
S29	97.0	3.0	0.0	0.0	Loam	0.64	22.38	76.98	Loamy sand
S30	97.0	3.0	0.0	0.0	Loam	1..35	44.31	54.34	Loamy sand
S31	92.3	7.7	0.0	0.0	Loam	0.79	23.56	75.65	Loamy sand
S33	63.3	36.7	0.0	0.0	Sandy loam	0.75	22.37	76.88	Loamy sand
S35	45.3	41.8	12.8	0.0	Sandy loam	0.79	24.52	74.69	Loamy sand
S39	3.8	34.5	35.8	25.9	Coarse sandy gravels	0.00	7.09	92.91	Sand
S60	29.7	70.1	0.2	0.0	Loamy sand	0.56	20.78	78.66	Loamy sand
S61	64.2	35.8	0.0	0.0	Loamy sand	2.53	43.69	53.78	Loamy sand
S91	3.5	29.9	48.3	18.3	Sandy gravels	0.27	11.19	88.54	Sand
S92	92.7	7.0	0.3	0.0	Loam	0.37	13.85	85.79	Sand
S93	98.5	1.5	0.0	0.0	Loam	0.59	22.64	76.77	Loamy sand
S94	88.7	11.3	0.0	0.0	Loam	0.67	23.80	75.53	Loamy sand
S95	27.5	71.3	1.2	0.0	Loamy sand	0.52	13.33	86.15	Sand
S98	52.7	47.3	0.0	0.0	Sandy loam	0.88	29.38	69.75	Loamy sand
S99						0.18	8.41	91.41	Sand
S101	94.7	5.3	0.0	0.0	Loam	1.93	51.61	46.46	Sandy loam
S102	95.0	5.0	0.0	0.0	Loam	1.73	41.99	56.29	Loamy sand
S103	85.5	14.5	0.0	0.0	Loam	1.61	44.83	53.56	Loamy sand
S105						0.82	28.59	70.59	Loamy sand
S106						0.53	15.10	84.36	Sand

Table 3.3 Composition of the clays (Tableau 3.3 Nature des argiles)

Location	Samples	On global samples	On clay	Illite	Chlorite	Nontronite	Volkonskoite	Interstratified mica-smectite	Kaolinite
T2 North section T2e	S5	x		x		xx	xx		
T2 North section T2b	S6	x		xxx	x				
T2 Step sup	S9	x		xxx					
T2 North section T2hi	S11	x		xx	x				
T2 North section	S14	x		xx	x	x?			
T2 North section T2hi	S18	x		xx	x				
T2 North section T2hm	S21	x		x	xx	x?			
T1 North section T1d	S29		x	x		xx		x	x
T1 North section T1e	S33	x		x		xx	xx		
[H8-N9-L14-G11] West section hi 4	S61	x		x		xx	xx		
[H8-N9-L14-G11] West section hi11	S92		x	x		xx			x
[H8-N9-L14-G11] North section	S93		x	x		xx			x
[H8-N9-L14-G11] West section hi 5	S94	x		x		xx	xx		
T6, level 4.1 (equid excavation)	S101		x	x		x		x	x
T6 4.13	S102		x	x		xx		x	x
T6 4.17	S103	x		x		xx	xx		
T7 sup	S105		x	x		xx			x
T7 sup	S105 bis		x	x		xx		x	x
Lower excavation	S106	x		x			x		

Nature of the Clays as Demonstrated by X-Ray Crystallography

The analyses were carried out at the Centre de Diffractométrie Longchambon de l'Université Lyon 1. They were conducted using two protocols depending on the amount of the clay fraction present in the samples: a global analysis on the fine fraction for those with the lowest clay content and a fraction analysis of less than 2 microns for the richest, following the usual protocol (natural passages, glycol, heating) (Table 3.3).

Illite is present in all the samples. It is associated with smectites, the most plentiful of which are nontronite, which results from the weathering of the basement biotite or from some basalts. Volkonskoite appears more rarely (samples S5, S33, S61, S94, S103, S106). Kaolinite also appears in some samples. Finally, we find interstratified mica-smectite (S29, S101, S102, S105bis), and chlorite was determined on some unoriented raw samples.

These results are difficult to explain because the obtained diagrams are dominated by the excessive amounts of quartz in the global analysis of fine fraction in the first case and the low representation of clays in the second case. They show however the heritage of the deposits around the crater and the absence of long term differentiated pedological profiles. The high frequency of smectites is consistent with the observations of Grangeon on the lacustrine filling (Elhaï & Grangeon 1963).

Micromorphology

A micromorphological analysis was carried out on two samples (S44 and S45) taken on the western section of the upper ledge located in the upstream part of Trench 2. They relate to layer T2b.

Description of S44 (upper part of T2b)

Sample of about 12 cm thickness comprised of five levels, from bottom to top:

- grey micaceous clay (4 cm thick);
- bed of reddish granules, including some black granules (3 to 5 mm thick) ;
- laminated clay (3.5 cm thick);
- sandier bed (3 to 5 mm thick);
- dark laminated clay (3.5 cm thick).

Microscopic analysis clearly shows lamination highlighted by the presence of many fissures/cracks separating the levels and by the orientation of elongated minerals (micas).

The basal layer is clearly detritic: at the clayey bottom, dispersed mineral grains can be identified everywhere, sometimes forming beds or lenses a few millimetres thick in which the grains are bigger. Quartz in angular crystals largely predominates; it is associated with micas (muscovite and weathered biotite), rare feldspars (alkali feldspars and plagioclases) and opaque oxides. Sillimanite occurs in fine more or less flexuous needles. One or two garnets were identified. This mineralogical range is typical in metamorphic bedrock. The overlying bed of granules consists of rounded lava elements, including both unwearthered black and weathered red granules. In the succeeding clayey beds, quartz grains decrease in number and in size. Sillimanite remains abundant, associated with numerous oxides and very small lava fragments. The sandier bed displays a composition that is similar to the basal level, although with a larger proportion of coarse detritus.

Description of S45 (lower part of T2b)

Sample of 8 cm thickness containing two levels:

- at the bottom, a 3 cm, heterogenous, poorly sorted sandy clay with high porosity. In the greater part of the clayey matrix, oxides and small dispersed lava grains can be identified. Areas characterized by stacking porosity are more detritic with abundant angular quartz grains and oxides. Some of these zones are crack fillings;
- above is a 5 cm very homogenous compact clay with a distinct decrease of the detritic grains in number and in size.

The micromorphological analysis of these two samples evidences the overall presence of detritic elements (lava granules, quartz grains, micas, feldspars, garnets, sillimanite, opaque oxides), even in the thinnest levels. The relatively homogenous results suggest that the components of the sediments have a common origin, reworked from freshly developed formations resulting from immediately preceding erosion of the rocks forming the perimeter of the maar. No soil weathering was detected

The grain size variations registered in the sediments reflect the evolution of environmental conditions such as the rainfall patterns, the level of flows, and the slumping resulting from the weight of more or less coarse debris on the slopes of the crater.

Formation of the Site

The shores of the lakes of the maar are characterized by the presence of a littoral prism (Bonifay & Truze 1987; Bonifay 1987; Bout 1970c; Raynal 1987; Truze 1990). This generally forms a continuous ring all around the lake and, in most cases, is dominated by a littoral slip cleaving of 0.5 to 1.5 m, the bottom of which corresponds to the mean water level. The littoral prism is comprised of:

- a littoral platform, the slightly sloping surface of which dips in the direction of the center of the lake up to a depth of 3 to 5 m;
- a talus slope at an angle of 35° to 40° that connects with the "deep" horizontal lacustrine sediments.

The littoral prism increases by prograding towards the center of the lake and shows the intrication of lacustrine levels with slope formations.

Examination of the modern Lake Issarlès maar makes it possible to complete these data and to better understand the organisation and structure of the deposits of the Senèze maar.

The Data From Lake Issarlès

Located at the boundaries of Haute-Loire, Lozère and Ardèche, Lake Issarlès formed in a maar about 1.5 km in diameter. It is nearly 140 m deep, and its waters are pumped during winter to supply a hydropower plant, significantly reducing the water level by about 40 to 50 m at maximum (Fig. 3.16A). The formations surrounding the maar are then partially uncovered, making it possible to study their nature and their structure.

On the western side, phreatomagmatic products are clearly visible in the form of more or less consolidated outcrops, into which were dug troglodyte dwellings. The beds consist of angular heterometric bedrock blocks eroded during the initial stage of the maar, that are below outcrops of finer-grained well bedded phreatomagmatic deposits, slabs of which which are locally collapsed towards the center of the maar. On this side the bedrock is not very visible.

By contrast, on the eastern side, at the level of the pumping station, the bedrock is well exposed and delimits the subvertical margin of the maar (Fig. 3.16B). Between the rock outcrops, lenses of stratified phreatomagmatic deposits were deposited (Fig. 3.16C). These beds display wide variations with regard to their thickness and the nature of the components, with sandy levels mainly formed by angular material reworked from the bedrock (gneiss and migmatite debris, crystals (Fig. 3.16G), and layers of gravels and small rounded pebbles of compact or scoriaceous basalt. Some large bedrock blocks can be identified (Fig. 3.16D). Locally, a scree of basalt blocks covers these stratified formations. All the deposits bank steeply towards the lake, but they display a large number of benches in the form of 40 cm to 2 m wide terraces, which are more or less continuous over some distance (Fig. 3.16E, F). The internal structure of the lenses, which is visible thanks to the cutting of a creek, reveals that the benches correspond to prograding shallowly-dipping beds that dip towards the lake with a slope of 30° to 45° (Fig. 3.16E).

Coûteaux (1984) highlighted the presence of numerous reworkings in the lacustrine filling resulting from repetition or inversion of the vegetation phases, gaps, and a mixture of pollen of different ages. These particularities are obviously due to slumps and collapses of littoral sediments. A section in the most elevated berm located below the talus cone of a creek shows a sequence of transgressive deposits and regressive erosion, thus of stabilization leading to the consolidation of the bench by peat. Coûteaux linked the variations of the lake level with the climate fluctuations between the end of the Würm glacial and the Holocene.

The observations made on the Issarlès maar reveal lateral variation of the formations visible on the slopes of this type of volcanic system, making it possible to better understand the structure of the deposits of the paleontological site.

The Deposits of the Senèze Paleontological Site

The variability of the sedimentary units identified all around the Senèze maar makes their correlations difficult. Nonetheless, distinct regularity can be described with regard to the large sets uncovered during the 2001–2006 excavations.

None of these excavations reached the bedrock. However, Depéret and Mayet (1912) identified bedrock in the trench of the Philis ravine, and Grangeon noted that the contact with the slope formations is located at an altitude of about 625 m (Elhaï & Grangeon 1963; Grangeon 1962). Roger (1954) indicated that the bedrock is overlain by slope formations in the entire northwestern part of the maar, while it crops out everywhere else. As was observed at Issarlès, the bedrock probably presents locally as subvertical rock faces. This is suggested by the drillings carried out in March 2005 by the BRGM at the request of Jean-François Pastre, at the limit of parcels 234 and 235 (see Fig. 2.2), at the height of the terrace next to the road along which is located section A. These drillings, which respectively reached depths of 22.50 m and 28.50 m, went through only fine detritic deposits without hitting the bedrock.

The formations uncovered by the excavations were deposited in four episodes of sedimentation.

The phreatomagmatic products

The first deposits covering the uplifted bedrock are volcanically erupted phreatomagmatic layers linked to the formation of the maar, such as were uncovered in the trenches T1, T5, T7, T8 and T9. They were also identified in the section of the Champ de Bos ravine (boundary of parcel 163 and the old pathway from Senèze to La Chomette) and in the

A
B
C
D
E
F
G

◀ **Fig. 3.16** Lake Issarlès. A: general view to the west of the partially emptied lake; B: basement outcrop and basalt block scree, near the pumping station; note the steep banks of the lake and the line of reeds indicating the lake level high water mark; C: general view of the bank at the pumping station: the light-colored deposits correspond to an accumulation of phreatomagmatic products, the dark outcrops to basement more or less covered with basaltic scree; D: view of the accumulation of phreatomagmatic products, arranged in tiers, and strewn with large angular blocks of basement; E: internal organization of the phreatomagmatic products, arranged in prograding layers on a slope of 30 to 40°; F: stepped levels separated by a few tens of cm ; in the background, the pumping station on a sub-vertical basement outcrop; G: detail of the constituents of the phreatomagmatic products: dominantly angular basement debris associated here with rounded elements of basalt. (Figure 3.16: Lac d'Issarlès. A : vue générale vers l'ouest du lac en partie vidé; B : affleurement du socle et éboulis de blocs de basalte, à proximité de la station de pompage ; on remarquera l'abrupt des berges du lac et la ligne de roseaux matérialisant le niveau du lac en hautes eaux; C : vue générale de la berge au niveau de la station de pompage : les dépôts clairs correspondent à une accumulation de produits phréatomagmatiques, les affleurements sombres, au socle plus ou moins recouverts d'éboulis basaltiques; D : vue de l'accumulation des produits phréatomagmatiques, disposés en gradins, et parsemée de gros blocs de socle anguleux; E : organisation interne des produits phréatomagmatiques, disposés en couches progradantes selon une pente de 30 à 40°; F : disposition en gradins de quelques dizaines de cm de large ; au fond, la station de pompage sur un affleurement de socle sub-vertical; G : détail des constituants des produits phréatomagmatiques : débris anguleux de socle dominants associés ici à des éléments arrondis de basalte.)

form of big collapsed blocks in section A and next to sections D and E. These deposits display an irregular structure: massive lenses without bedding or graded bedding cut by channels more or less floored with sands. The nature of the volcanic elements and the weathering of the bedrock fragments document deposition clearly linked with the initial phases of the formation of the maar.

The slope formations

Those initial formations are overlain by slope deposits from the reworking of the preceding layers of bedrock and volcanic ejecta around the growing maar, accumulated in lenses prograding towards the center of the crater along a slope of 15°-20°. Depending on the supply zones, the nature and size of the components differ: for example, the red scoriae, reworked from a small volcanic cone (Pastre 2024), are abundant in the western sector (trenches 1 and 2, sections A and D) and absent elsewhere; the gneiss elements, generally exhibiting coarser grain sizes, are clearly predominant in the southern sector (trenches 5 and 7). The structure and the characteristics of these formations show that they were deposited by debris flows, or in some cases by solifluction, below a shallow water level with differentiated fluvial deposits: generally coarse at the bottom of the sedimentary sets and finer at the top. The largest elements were apparently included in the bulk of the deposit because they do not cause deformation of the underlying levels (see the block in section A). The thickness of these formations varies from one sector to the other according to the local topography and the existence of a channelled water flow. The present-day morphology still mirrors their presence on the slopes of the crater: while they display a slope with a 35% grade, a rupture can be observed at an altitude of between 630 m and 620 m, marking the upper limit of the sedimentary prism formed by these detritic formations. This prism formed mainly in the western and southern part of the depression, the other sides being steeper.

The lacustrine levels

The topographically highest lacustrine deposits are located in the western sector at an altitude of 625 m, at the level of T2, becoming rapidly thicker (set 2 = T2h), and they can be identified down to the bottom of this trench, i.e., at an altitude of 620 m. Section D contains a lacustrine level (layer 8) at 624.40 m, an altitude that matches the top of T2h. Trench 1 mainly exhibits slope deposits, and the lacustrine levels first appear at an altitude of about 621 m and are less thick than in T2 (less than one meter); they extend widely only at an altitude of 616 m (Fig. 3.17).

In the southeastern sector, at the height of the lacustrine levels of trench 2, only slope deposits are visible (in T4, T5 sup, T8 and T9). The most elevated lacustrine level was identified in T7 sup at an altitude of 619.50 m, i.e., almost at the height of the bottom of T2. The lacustrine sedimentation becomes continuous at an altitude of about 617.50 m, which is nearly at the same height as in trench 1. Trenches T6 and T7 inf only display lacustrine deposits, as they were dug below this altitude, thus marking the spreading out of these latter.

The differences in elevation of the lacustrine deposits can be explained by the reworking of the initial lacustrine sediments into debris flows that are fully developed in the southeastern sector (trenches 5 and 7) and appear only locally in the western part (trench 1). These differences can also reveal a sedimentation shifted in time according to variations of lacustrine level (see below 5.2.6). Close to the ancient lake shore, the lacustrine deposits invariably include a significant amount of reworked slope detritus. These sediments at times certainly reached as far as the center of the maar, as is suggested by the presence of sand beds in the core of 1965 (Ehrlich 1968; Elhaï 1969). The first levels crossed by this drilling and the other boreholes are clearly below the highest lacustrine levels visible in the upper trenches. The difference in altitude is about 35 m in trench 2 and 29.50 m in trench 7. These differences perhaps reveal that a significant thickness of sediments was eroded between the top of the cores and the level of the fossiliferous sites.

Nonetheless, erosion was certainly less extensive than indicated by these values given the geometry of the deposits of the lake shore: in fact, the shoreline prism is connected to horizontal deep lacustrine sediments by a talus with an average slope of 30° and a thickness of up to several meters. This talus is the site of sub-lacustrine slides whose importance depends on the bathymetry of the central part of the lake (Bonifay & Truze 1987).

The block fields and the colluviums

The last type of deposits overlies the preceding formations. These are coarse conglomerates including large fragments together with more or less abundant sandy matrix, deposited in channels that cut through the underlying units. These are ubiquitous in the western sector where they form set 5' identified at the tops of trenches T1, T2 and T3 and of the profile sections A, B, C and D (layer 1). In the southeastern sector, set 5 of trenches 6 and 7 may be equivalent, however with clearly smaller grain sizes. The deposit of these block fields probably occurred during a glacial climate.while later colluviums were deposited locally, filling the depressions of the underlying formations, as is the case in trench 1 (set 6'), or a thalweg (trench 4). The age of their deposition cannot be determined more precisely, but some of these seem to be very recent (trench 4).

Variability of the sedimentation from one sector to another

From one sector to another and even from one trench to another, great variation can be found in the structure and the nature of the deposits. Several hypotheses can be proposed to explain these facies changes (Pastre et al. 2015).

- The morphological context obviously conditioned the sedimentation on the crater rim. Thus, the formation of the littoral prism took place in the less steep sectors while the steeper flanks made such a structure impossible. Subvertical faces of the bedrock, similar to those identified on the eastern side of Lake Issarlès, possibly exist as revealed by the drill coring carried out by the BRGM close to section A. The steep slope possibly prevented detritic sedimentation on the lake shore or triggered the instability of the deposits, inducing their collapse towards the center of the crater. Another example of the role played by topography is illustrated by the formations of section D in channels or the formations of trench 4, located at the end of a little-marked thalweg, which regularly channelled the colluviums.
- The formations outcropping on the crater rim also played a role as regards the nature of the deposits located below. Those of the western sector were derived from the topset scoriae cone, which provided the basaltic pebbles, gravels and red grains present in all the layers. The phreatomagmatic tuffs produced much of the fine material but also blocks such as the one visible in the western part of section A. By contrast, the greater part of the sediments originated from the erosion of the gneissic bedrock.
- Topography and the nature of the crater slopes possibly combined during the meteoric evolution of the crater. Thus it seems that the instability of its southeastern flank may have started prematurely triggering the displacement and the deposit of large blocks during the formation of set 3, whereas on the western flank these reworkings occurred later in the form of block fields.
- Synsedimentary and postsedimentary reworking resulted in various modifications to the deposits. Thus, load, stretch and boudinage patterns can be identified in many sections revealing obviously slow-moving subaquatic slipping of distinct units (for example the sand levels of T2hi or units 4.2 and 4.5 of trench T5 inf). In the areas of trenches 2 and 5, such movements, perhaps linked with postdepositional desiccation of the sediments, introduced growth faults responsible for the slab collapse of the deposits towards the center of the maar. Other collapses appear to have been larger. Thus the deposits of the region of section A seem to have been affected by massive detachment, revealed by the absence of deformation of the sediments under the tuff block and by the present-day morphology delimiting a well-marked depression between the foot of the slope and the beginning of the outcrop. The hypothesis of a mass sliding is confirmed by the date obtained on tephra SEN 56 (2.07 ± 0.02) which is the youngest age of all dated tephras, whereas altitudinally, it is lower than the older tephras in trench 2 (see below 5.2.6.). The slipping of the upper sets of trench T5 caused their folding. These reworkings may have reached the lacustrine sediments deposited in the center of the maar. Thus, Bout (1970c) hypothesized disruptions of the upper part of the lacustrine filling to explain the triple recurrence of the same warm phase evidenced by the pollen spectra in the core (Elhaï 1969). Those gravitational slidings were also identified in the deposits of Lake Issarlès and can be linked with several processes, such as break-offs of sediments deposited thickly on the terraces, earth tremors and fluctuations of the water level (Coûteaux 1984; Jouve et al. 2023). These hypotheses can be considered in explaining the

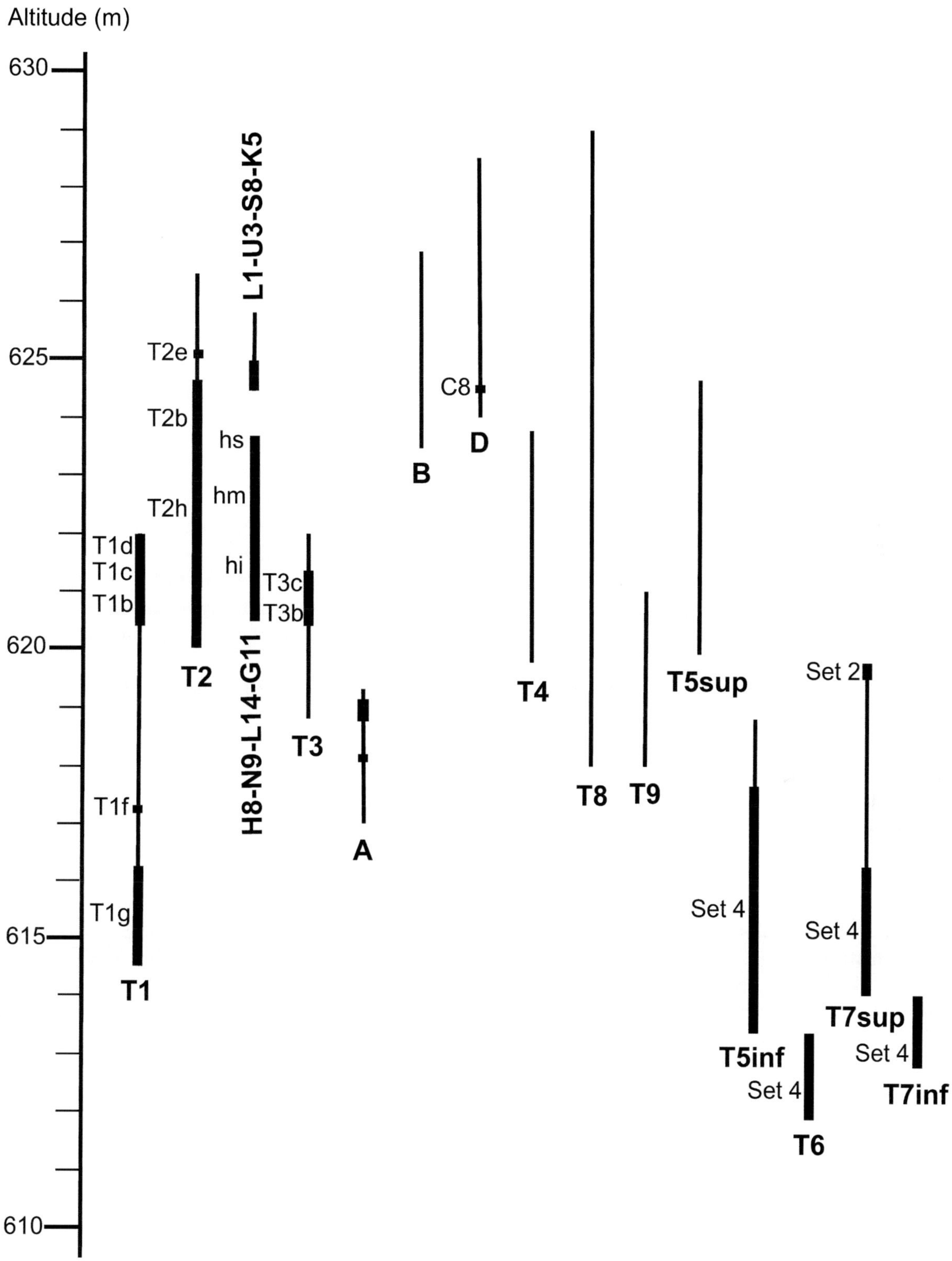

Fig. 3.17 Altitudinal distribution of trenches, sections and excavations. Location of lacustrine levels indicated by thicker lines. (Figure 3.17 – Répartition altitudinale des tranchées, coupes et excavations. Localisation des niveaux lacustres indiquée par des traits plus épais)

reworking of the deposits of Senèze. However, an alternative interpretation is also suggested by reference to the sublacustrine collapses of Lake Pavin (Chapron et al. 2010, 2010–2011) that was caused by the rapid draw-down of the lake level that followed the incision of the outlet. Nowadays, the start of the creek draining the depression of Senèze is located at an altitude of 570 m, i.e. more than 55 m below the fossiliferous slope deposits. It may be possible that the incision of the outlet and emptying of the lake may have started as early as the beginning of the sedimentation related to the evolution, 2 Ma, of the Sénouire water course into which the Senèze creek flows (Grangeon 1967; Pastre et al. 1997).

– The presence of lacustrine deposits interstratified with slope formations indicates a deposit at the height of the zone of the water level variations of the paleolake. This imbrication may reflect climate changes. Thus the introduction of slope deposits might rather indicate reduced plant cover on the slopes of the crater, linked to the periglacial conditions favoring solifluction flows and runoff during ground thaw. Taking into account the dates obtained on tephra that place most of the deposits within the time range 2.19 to 2.07 Ma (Nomade et al. 2014; Pastre et al. 2015; Pastre 2024; Delson et al. 2024), this climate deterioration may correspond to the well-marked cooling evidenced within the Tiglian (Popescu et al. 2010; Westerhoff et al. 2020; Zagwijn 1975). The lacustrine levels would mark lake rises because of significant rainfall triggering a rising water level. This change, related to wetter climate fluctuations, would have initiated the transition to the warmer conditions of the later part of the Tiglian.

The formation of the paleontological site: A complex history

The phreatomagmatic products visible at the base of several trenches in the site are linked to the initial phase of maar formation. As indicated in the oldest deposits crossed by the borehole of 1965 (Ehrlich 1968), they mark the beginning of the filling of the maar. This continues with a thick sequence of lacustrine sediments, often fine and diatomaceous, interspersed with coarser deposits, which occur especially in the upper 28 m of the borehole. Based on the thickness of the diatomaceous sediments with a fine seasonal bedding (around 110m), Ehrlich estimated the duration of lake filling at 200,000–300,000 years.

The fossiliferous deposits located on the periphery of the maar developed after this long period of sedimentation. They are typical of a littoral prism, marking the shore and progressively connecting with the lacustrine deposits of the center of the maar. The prism is formed by slope deposits gradually covered by lake sediments. In the two sectors explored by the Franco-American excavations, these develop at different altitudes (Fig. 3.17). The highest lacustrine levels are in the western sector (see T2, Fig. 3.3) while the levels in the southeastern sector are generally 6 - 7 m lower (see T5 inf, Fig. 3.12 and T7 sup, Fig. 3.14). This différence between the sectors reveals a diachronic construction of the prisms, depending on the dynamics of the establishment of slope deposits and variations in the water level. The mineralogical study of tephras with different compositions (Pastre 2024) and the ^{40}Ar / ^{39}Ar dates obtained on five of these tephras (Nomade et al. 2014, Delson et al. 2024) support this hypothesis of diachronic deposits, linked to several episodes of sedimentation between ca. 2.2 and 2.1 Ma. The oldest dates were obtained from the tephras sampled in the southeastern sector (trench T5 inf, Fig. 3.12): SEN 98 dated at 2.18 ± 0.03 Ma and SEN 101 at 2.13 ± 0.04 Ma. In the western sector, the dates obtained are more recent with SEN 8 at 2.16 ± 0.05 Ma and SEN 1 at 2.10 ± 0.03 Ma (Fig. 3.3). The SEN 56 tephric horizon of section A (Fig. 3.8) yielded the most recent age (2.07 ± 0.02 Ma). This young age was confirmed by the U/Pb dating of zircons collected from a level of section A slightly lower than the SEN 56 level and giving an age of 2.10 ± 0.03 Ma (Paquette et al. 2021). Located between 619.10 m and 617.60 m, the SEN 56 level is clearly altitudinally below the two tephras of trench 2. This apparent chronological inversion is explained by the fact that all the deposits in section A are part of a mass detachment which must have been at a higher altitude originally (see above Sect. 5.2.5). The littoral prisms were affected by numerous landslides which induced stresses in the sediments in place and caused deformations which sometimes were synsedimentary. This is particularly visible in the southeastern sector. The slipping of set 3 (trench T5 inf, Fig. 3.12 and trench T7 sup, Fig. 3.14) caused breaks and the formation of a fold that was injected into the lacustrine deposits, inducing the overturning of certain layers or their fragmentation. In the western sector, such fragmentation of lake levels occurred at the base of trench T1 (Fig. 3.1). After this long lacustrine period, only slope deposits (block fields, colluvium) of undetermined age are found.

Conclusion

The facies changes observed from one point to another of the excavated areas are characteristic of sedimentation at the paleo-shore level. Arranged in units prograding towards the center of the maar, the deposits form two sedimentary prisms that show important variability in their width as well as in their composition: these characteristics depend on the nature and morphology of the watershed, the importance of detrital inputs from the slope and the development of sub-lacustrine landslide slumps on the internal slope of the prisms that remobilized and redistributed the sediments. The imbrication of slope formations and lake levels also reflects climate changes: coarse deposits attest to significant erosion on bare surfaces that may agree with more or less marked cooling phases, whereas lacustrine sediments correspond to rises in lake level related to climate warming. The $^{40}Ar/^{39}Ar$ ages obtained on the tephras, mainly between 2.2 and 2.1 Ma (Nomade et al. 2014; Pastre et al. 2015; Pastre 2024; Delson et al. 2024) show that the formation of the paleontological site occurred in a minimum duration of about 100 kyr. This period covers several climatic phases corresponding to marine isotope stages (MIS) 83 to 79. During this time interval, variations in environmental conditions are seen in the taphonomic characteristics of fossil assemblages (Fernández-Jalvo et al. 2024). The remains can be included in fine-grained levels as well as in coarser deposits. Thus, in the western sector, they were mainly fossilized in clay or loamy levels corresponding to lake sediments, while in the south-eastern sector they are instead associated with slope deposits. The geomorphological and sedimentological context provides a better understanding of the formation conditions of the fossiliferous assemblages. During the cold phases, dead or still living animals could be dragged by solifluction to be trapped in the icy waters of the lake. The association of remains with pebbles and blocks (frequent in the south-eastern sector) may reflect very cold conditions where the surface of the lake, partly frozen and covered with stony pavement, could yield under the weight of heavier animals, causing their death by drowning. During the warmer phases, as seen in the western lake beds, they could simply slip on the slope while coming to drink; because the stiffness of the sedimentary layers resulted in bluffs 0.5 to 1.5 m high at the edge of the lake and due to the instability of the waterlogged sediments, the animals could not get back up onto the bank and drowned. Subsequently, the collapse of sediment panels, delimited by faults, could further break up the resulting skeletons. This interpretation helps to explain why the fossils are found in sediments whose depositional conditions differ from one area to another and cover several climatic periods.

Acknowledgments I am grateful to Claude Guérin, Martine Faure and Eric Delson for inviting me to undertake this study and funding my research. My thanks also go to Andrea Valli, Angélique Monguillon and Fabio Parenti and his colleagues for technical assistance in the field and to all the students for help in preparation of the sections and their plotting. I also thank John Van Couvering and Eric Delson for assistance with language editing, and Jean-François Pastre, John Van Couvering and an anonymous reviewer for their constructive comments on the manuscript.

References

Bonifay, E. (1987). La paléolimnologie en Velay : état d'avancement des travaux en 1985 et programmes internationaux. *Documents du Centre d'Etudes et de Recherche sur les lacs, anciens lacs et tourbières, 1*, 23–28.

Bonifay, E., & Truze, E. (1987). Dynamique sédimentaire et évolution des lacs de maars : l'exemple du Velay. *Documents du Centre d'Etudes et de Recherche sur les lacs, anciens lacs et tourbières, 1*, 29–64.

Boule, M. (1892). Découverte d'un squelette d'*Elephas meridionalis* dans les cendres basaltiques du volcan de Senèze (Haute-Loire). *Comptes Rendus hebdomadaires des Séances de l'Académie des Sciences, 115*, 624–626.

Bout, P. (1969). Datations absolues de quelques formations volcaniques d'Auvergne et du Velay, et chronologie du Quaternaire européen. *Revue d'Auvergne, 83*, 267–280.

Bout, P. (1970a). Absolute ages of some volcanic formations in the Auvergne and Velay areas and chronology of the European Pleistocene. *Palaeogeography, Palaeoclimatology, Palaeoecology, 8*, 95–106.

Bout, P. (1970b). Maars et cônes de scories d'Auvergne et du Velay. *Revue d'Auvergne, 84*, 29–42.

Bout, P. (1970c). Problèmes du volcanisme XV: le maar de Senèze (Haute-Loire). *Revue d'Auvergne, 84*, 54–68.

Bout, P. (1972). Problèmes du volcanisme VII. *Revue d'Auvergne, 86*, 53–76.

Bout, P. (1975).The contribution of the volcanic Massif Central of France to European Quaternary chronology. In K.W. Butzer & G. Ll. Isaac (Eds.), *After the Australopithecines* (pp. 73–98). The Hague: Mouton.

Bout, P. (1978). *Problèmes du volcanisme en Auvergne*. Brioude: Watel.

Brébion, P., Calas, P., Drot, J., Pimienta-Freneix S. & Nazemi L. (1953). Le gisement villafranchien de Senèze (Haute-Loire). 5 p., unpublished manuscript in the archives of the Institut de Paléontologie, MNHN].

Chapron, E., Albéric, P., Jézéquel, D., Versteeg, W., Bourdier, J.-L., & Sitbon, J. (2010). Multidisciplinary characterisation of sedimentary processes in a recent maar lake (Lake Pavin, French Massif Central) and implication for natural hazards. *Natural Hazards and Earth System Sciences, 10*, 1815–1827.

Chapron, E., Albéric, P., Jézéquel, D., Ledoux, G., & Massault, M. (2010–2011). Les archives sédimentaires de l'histoire du lac Pavin. *Revue des Sciences naturelles d'Auvergne,, 74–75*, 57–65.

Coûteaux, M. (1984). Recherches pollenanalytiques au lac d'Issarlès (Ardèche, France): évolution de la végétation et fluctuations lacustres. *Bulletin de la Société royale de botanique belge, 117*, 197–217.

Couthures, J. (1989). The Senèze maar (French Massif Central): hypothesis regarding a catastrophe occurring about 1.5 million years ago. *Journal of Volcanology & Geothermal Research, 39*, 207–210.

Couthures, J., Ablin, D., & Hadjouis, D. (1991). Senèze (Haute-Loire) Rapport de fouilles 1991. Direction Régionale des Affaires Culturelles – Auvergne, Clermont-Ferrand, 55 p., unpublished manuscript.

Delson, E., Faure, M., Guérin, C., Aprile, L., Argant, J., Blackwell, B., et al. (2006). Franco-American renewed research at the Late Villafranchian locality of Senèze (Haute-Loire, France). *Courier Forschunginstitut Senckenberg, 256*, 275–290.

Delson, E., Nomade, S., Sen, S., Debard, E., Pastre, J-F., Bahain, J-J. et al. (2024). Geochronology of Senèze: Argon Dating and Magnetostratigraphy, with Notes on an ESR/U-Series Dating

Attempt. In E. Delson, M. Faure & C. Guérin (Eds), *Senèze: Life in Central France two million years ago. Paleontology, geochronology, stratigraphy and taphonomy* (pp. 99–122). Cham (Switzerland): Springer.

Depéret, C., & Mayet, L. (1912). Le gisement de mammifères pliocènes de Senèze (Haute-Loire). Association française pour l'avancement des Sciences, *Compte Rendu de la 40e session Dijon 1911. Mémoire hors volume*, 138–151.

Devis, G. (1970). Le volcanisme et le gisement fossilifère de Senèze. *Revue des Sciences Naturelles d'Auvergne, 36*, 25–87.

Eisenmann, V., & Delson, E. (2024). The Senèze equids. In E. Delson, M. Faure & C. Guérin (Eds), *Senèze: Life in Central France two million years ago. Paleontology, geochronology, stratigraphy and taphonomy* (pp. 307–385). Cham (Switzerland): Springer.

Ehrlich, A. (1968). Les diatomées fossiles des sédiments villafranchiens de Senèze. *Bulletin de l'Association française pour l'Etude du Quaternaire, 4*, 267–280.

Elhaï, H. (1969). La flore sporo-pollinique du gisement villafranchien de Senèze (Massif central, France). *Pollen et Spores, 11*, 127–139.

Elhaï, H., & Grangeon, P. (1963). Nouvelles recherches sur le gisement villafranchien de Senèze (Haute-Loire). *Bulletin de la Société géologique de France*, 483-488

Fernández-Jalvo, Y., Valli A. M. F., Marín-Monfort, M. D., & Pesquero-Fernández, M. D. (2024). The taphonomy of Senèze. In E. Delson, M. Faure & C. Guérin (Eds), *Senèze: Life in Central France two million years ago. Paleontology, geochronology, stratigraphy and taphonomy* (pp. 653–681). Cham (Switzerland): Springer.

Gaudant, J. (1975). Présence du genre *Tinca* Cuv. (Poisson Téléostéen, Cypriniforme) dans le Villafranchien de Senèze (Haute-Loire). *Geobios, 8*, 353–354.

Grangeon, P. (1962). Note préliminaire à l'étude géologique du volcan et du gisement fossilifère de Senèze. *Compte Rendu sommaire des Séances de la Société géologique de France, 5*, 149–150.

Grangeon, P. (1967). Le cours villafranchien de la Sénouire dans le Bassin de Paulhaguet et quelques remarques sur certains phénomènes volcaniques de la région au S-E de Cerzat (Haute-Loire – Massif Central français). *Bulletin de l'Association Française pour l'Etude du Quaternaire, 4*, 145–163.

Guérin, C. (2024). *Dicerorhinus etruscus etruscus* from the Late Villafranchian site of Senèze (Domeyrat,Haute-Loire, Central France). In E. Delson, M. Faure, & C. Guérin (Eds), *Senèze: Life in Central France two million years ago. Paleontology, geochronology, stratigraphy and taphonomy* (pp. 245–306). Cham (Switzerland): Springer.

Jouve, G., Chapron, E., Chassiot, L., Foucher, A., Raynal, J.-P., & Defive, E. (2023) What triggered mass movement deposits in the maar Lake Issarlès (Ardèche, France)? *28th Réunion des Sciences de la Terre.* sciencesconf.org:rst2023-rennes: 484653.

Nomade, S., Pastre, J.-F., Guillou, H., Faure, M., Guérin, C., Delson, E., et al. (2014). $^{40}Ar/^{39}Ar$ constraints on some French landmark Late Pliocene to Early Pleistocene large mammalian paleofaunas: Paleoenvironmental and paleoecological implications. *Quaternary Geochronology, 21*, 2–15.

Paquette, J.-L., Médard, E., Poidevin, J.-L., & Barbet, P. (2021). Precise dating of middle to late Villafranchian mammalian paleofaunae from the Upper Allier River valley (French Massif Central) using U-Pb geochronology on volcanic zircons. *Quaternary Geochronology, 65*, 1–12.

Parenti, F., Valli, A., Aprile, L., Candelato, F., Monguillon, A., Natali, L., et al. (2024). Surveying and excavations at Senèze: 2001-2006. In E. Delson, M. Faure & C. Guérin (Eds), *Senèze: Life in Central France two million years ago. Paleontology, geochronology, stratigraphy and taphonomy* (pp. 27–35). Cham (Switzerland): Springer

Pastre, J.-F. (2024). The Senèze maar and paleontological site (Massif Central, France): their volcanic context and studies of related tephras. In E. Delson, M. Faure & C. Guérin (Eds), *Senèze: Life in Central France two million years ago. Paleontology, geochronology, stratigraphy and taphonomy* (pp. 83–98). Cham (Switzerland): Springer.

Pastre, J.-F., Defive, E., Gablier, F., & Lageat, Y. (1997). Changements hydrographiques et volcanisme plio-quaternaire dans les bassins de la Loire et de l'Allier (Massif central, France). *Géographie Physique et Quaternaire, 51*, 295–314.

Pastre, J.-F., Debard, E., Nomade, S., Guillou, H., Faure, M., Guérin, C., & Delson, E. (2015). Nouvelles données géologiques et téphrochronologiques sur le gisement paléontologique du maar de Senèze (Pléistocène inférieur, Massif Central, France). *Quaternaire, 26*, 225–244.

Pelletier, H. (1968). Contribution à l'étude sédimentologique de l'ancien lac de Senèze (Haute-Loire, Massif Central français). *Revue des Sciences naturelles d'Auvergne, 34*, 39–50.

Popescu, S. M., Biltekin, D., Winter, H., Suc, J.-P., Melinte-Dobrinescu, M. C., Klotz, S., et al. (2010). Pliocene and Lower Pleistocene vegetation and climate changes at the European scale: Long pollen records and climatostratigraphy. *Quaternary International, 219*, 152–167.

Prévot, M., & Dalrymple, G. B. (1970). Un bref épisode de polarité géomagnétique normale au cours de l'époque inverse Matuyama. *Comptes Rendus hebdomadaires des Séances de l'Académie des Sciences, 271*, 2221–2224.

Raynal, J.-P. (1987). Evolution comparée des lacs de maars en Auvergne et Velay (France). Datation et contribution à la connaissance des climats pléistocènes. *Documents du Centre d'Etudes et de Recherche sur les lacs, anciens lacs et tourbières, 1*, 65–96.

Roger, J. (1954). Le gisement villafranchien de Senèze (Haute-Loire). *Bulletin du Museum national d'Histoire naturelle*, Paris, 2ème série., t. *XXVI*, 292–295.

Roger, S., Coulon, C., Thouveny, N., Feraud, G., Van Velzen, A., Fauquette, S., et al. (2000). $^{40}Ar/^{39}Ar$ dating of a tephra layer in the Pliocene Senèze maar lacustrine sequence (French Massif Central): constraint on the age of the Reunion-Matuyama transition and implications on paleoenvironmental archives. *Earth and Planetary Science Letters, 183*, 431–440.

Truze, E. (1990). Etude sédimentologique et géochimique des dépôts du maar du Bouchet (Massif Central, France). Evolution d'un système lacustre au cours du dernier cycle climatique (0 -120 000 ans). Thèse de Doctorat, Université Aix-Marseille II.

Valli, A. M. F. (2024) New Data on Cervidae (Mammalia, Artiodactyla) from the Villafranchian Locality of Senèze (Haute-Loire, France). In E. Delson, M. Faure & C. Guérin (Eds), *Senèze: Life in Central France two million years ago. Paleontology, geochronology, stratigraphy and taphonomy* (pp. 409–431). Cham (Switzerland): Springer.

Westerhoff, W. E., Donders, T. H. , Trabucho Alexandre, N., & Busschers, F. S. (2020). Early Pleistocene Tiglian sites in the Netherlands: A revised view on the significance for Quaternary stratigraphy. *Quaternary Science Reviews, 242.* https://doi.org/10.1016/j.quascirev.2020.106417

Zagwijn, W. H. (1975). Variations in climate as shown by pollen analyses, especially in the Lower Pleistocene of Europe. In Wright, A. E., Mosely, F. (Eds.). *Ice ages: Ancient and modern* (Vol. 6). *Geological Journal, Special Issue,* 137–152

Chapter 4
The Senèze Maar and Paleontological Site (Massif Central, France): Their Volcanic Context and Studies of Related Tephras

Jean-François Pastre

Abstract The Senèze volcano is composed of a basanite flow, scoriae deposits, phreatomagmatic tuffs and a maar crater. Recent fieldwork and mapping clarify the extension and the succession of these volcanic products. A first phreatomagmatic episode precedes the lava flow which seems accompanied by the deposition of scoriae. In a later phase, the maar crater is formed and the main phreatomagmatic tuffs are emitted. The comparative study of the fossiliferous deposits exposed in newly excavated sections and of two other sections reveals the interstratified occurrence of 10 tephras and six reworked tephric layers issued from the Mont-Dore strato-volcano, 60 km north-westwards. These mainly trachytic products are attributed to the Early Pleistocene pyroclastic activity of the strato-volcano, *ca* 2.1 Ma. Their distinct composition in the different outcrops evidences the polyphased deposition of the fossiliferous site.

Résumé Le volcan de Senèze est composé d'une coulée de basanite, de dépôts de scories, de tufs phréatomagmatiques et d'un cratère de maar. Un travail de terrain récent et une cartographie précisent l'extension et la succession de ces produits volcaniques. Un premier épisode phréatomagmatique précède la coulée de lave qui semble accompagnée par le dépôt de scories. Dans une phase plus tardive le cratère de maar est formé et les principaux tufs phréatomagmatiques sont émis. L'étude comparative des dépôts fossilifères exposés dans les nouvelles coupes et dans deux coupes antérieures révèle l'occurrence de 10 téphras et de six niveaux téphriques remaniés issus du strato-volcan du Mont-Dore situé 60 km au nord-ouest. Ces produits à dominance trachytique sont attribués à l'activité pyroclastique du strato-volcan il y a environ 2,1 Ma. Leur composition distincte dans les différents affleurements met en évidence le caractère polyphasé des dépôts du site fossilifère.

Keywords Devès • Mont-Dore • Mineralogy • Tephrostratigraphy

Mots-clés Devès • Mont-Dore • Contexte volcanique • Mineralogie • Téphrostratigraphie

Introduction

The alkaline volcanism in the French Massif Central is characterized by a major activity period during the Early Pleistocene (Brousse & Lefèvre 1990; Maury & Varet 1980). This activity concerns two volcanic provinces: the basaltic Devès and the Mont-Dore, a differentiated strato-volcano which is the source of numerous tephras (Nomade et al. 2014a; Pastre & Cantagrel 2001). The Senèze volcano, which is situated at the northwestern edge of the Devès, reflects the activity of these two provinces. Several volcanic stages characterize its eruptive activity, while several tephras of the Mont-Dore are interbedded in the fossiliferous sediments following the construction of the volcano. This paper will focus on these two aspects. It complements one already published (Pastre et al. 2015).

The volcanic context of the Senèze maar

General Data

The geological setting and volcanic context of the Senèze volcanic system have been thoroughly studied by Bout (1960), Grangeon (1962), Devis (1969, 1970), and Lasnier et al. (1982). These authors proposed various interpretations

J.-F. Pastre (✉)
23, rue de l'Eglise, 93370 Montfermeil, France
e-mail: adline.pastre@wanadoo.fr

Laboratoire de Géographie Physique, Environnements quaternaires et actuels, UMR 8591 CNRS, Thiais, France

E. Delson et al. (eds.), *Senèze: Life in Central France Around Two Million Years Ago*,
Vertebrate Paleobiology and Paleoanthropology, https://doi.org/10.1007/978-3-031-64415-3_4

which will be reviewed here, in the frame of a new geological map of the area.

Like other phreatomagmatic craters, the Senèze maar does not constitute a strictly monogenetic volcano. It is associated with scoriae emissions and with an important basaltic flow, the Les Laniers (or Senèze) lava flow. It is also fringed by a pyroclastic crescent contemporaneous with its formation (Figs. 4.1 and 4.2).

In all previous studies, the Les Laniers lava flow is regarded as having occured during a first phase of activity. However, the views differ about the relative position of the scoriae and maar products, the accompanying maps and fieldwork data giving different interpretations of the volcanic formations and of their extension.

The Les Laniers Lava Flow and Related Deposits

The Les Laniers basanite flow (or Senèze flow) lies to the west of the maar crater (Figs. 4.1 and 4.2). It flowed over an irregular topography, masking most of its unevenness. Its origin is clearly situated at the "Pié de Charenty", but this hill does not correspond to the scoria cone described by most of the authors and mapped by the geological map "Brioude 1:50,000" (Lasnier et al. 1982). It is a massive basanite intrusion directly continuing the lava flow.

Most limits between the flow and the basement are masked, and only a few outcrops expose the contact. A complete section can be observed along national road 102, near Les Laniers hamlet, exposing:

1. The metamorphic basement (gneiss), of which the upper 6 m are visible.
2. A more or less quartz-rich colluvium including a large amount of basement elements: about 100 cm.

3a. Clayey tuff with strongly weathered basalt grains and rare basement xenoliths: 55 cm.

3b. Clayey tuff, ocher-brown colored with numerous small basalt grains < 1 cm: 30 cm.

4. Basanite flow: large weathered prisms; the texture of the base of the prisms is vitreous: 280 cm

The hypothesis that the Les Laniers lava flow preceded the maar is not unquestionable. It lies alongside the maar which destroyed part of it. If subsequent, it would have flowed into the maar. On the other hand, the tuff which precedes the lava flow proves the occurrence of volcanic activity prior to the flow. The small number and the distant location of other volcanoes in the area do not fit with the type and thickness of this tuff. It probably resulted from the initial phase of the system activity. This hypothesis of a short phreatomagmatic activity prior to the lava flow emission seems the most likely. This lava flow is dated by K–Ar at 2.28 ± 0.15 and 2.30 ± 0.15 Ma (Dalrymple in Prévot 1975) and 2.52 ± 0.06 Ma (Couthures & Pastre 1983).

The Scoriae

The cartography of the 1/50,000 geological map "Brioude" (Lasnier et al. 1982) is largely erroneous. It attributes an excessive size to the scoriae situated west of the maar and figures a non-existent gneiss outcrop north of La Chomette.

The Pié de Charenty is interpreted and mapped by most authors as a scoria-cone corresponding exactly to the hill of the same name. This interpretation, due to the misleading convex morphology of the hill, must be rejected. Indeed, the SE and SW flanks of the hill show only basanite, while its northern slope is shaped in the metamorphic basement (Fig. 4.1). Only some welded scoriae exposed north of the 754 m summit indicate a small lava-fountain activity.

The only true scoriae outcrop at Senèze is of a small size. It occurs south of the Pié de Charenty and is clearly distinct from the basanite forming the hill. Composed of red scoriae, it fringes the explosion crater 450 m W of Senèze hamlet and is directly bordered by the maar tuff to the south and the north (Figs. 4.1 and 4.2). The internal scree exhibits scoriae which are reworked in the fossiliferous slope deposits (e.g., in trenches T1 and T2). This observation documents their chronological place in the volcanic system.

On the one hand, they may precede the phreatomagmatic tuffs of the maar. These tuffs, which do not cover the scoriae, are eroded. The solution of a scoria-cone cut by the maar fits this hypothesis. On the other hand, the scoriae may follow the tuffs, thus explaining their disappearance and the rapid contact between them.

Actually, this question cannot be definitely answered, but the hypothesis of a scoriae cone synchronous with the lava-flow and partially destroyed by the younger maar is the most probable. We never observed the stratigraphic relationship between both formations in any of the various outcrops. According to Couthures et al. (1991), one section shows the scoriae covering the phreatomagmatic tuff. However, we never observed such a section; on the contrary, we think that the scoriae are possibly colluvial. The succession of maar, basaltic flow, scoria-cone occurs frequently in the Devès (Boivin & Gourgaud 1978; Mergoil & Boivin 1993; Pastre 1987, 2001); in most cases however, cones totally cover the maar after the emission of one, two, or even three lava-flows. The case of a scoria-cone clearly preceding a maar occurence is also known (e.g., maar of the "Marais de Limagne").

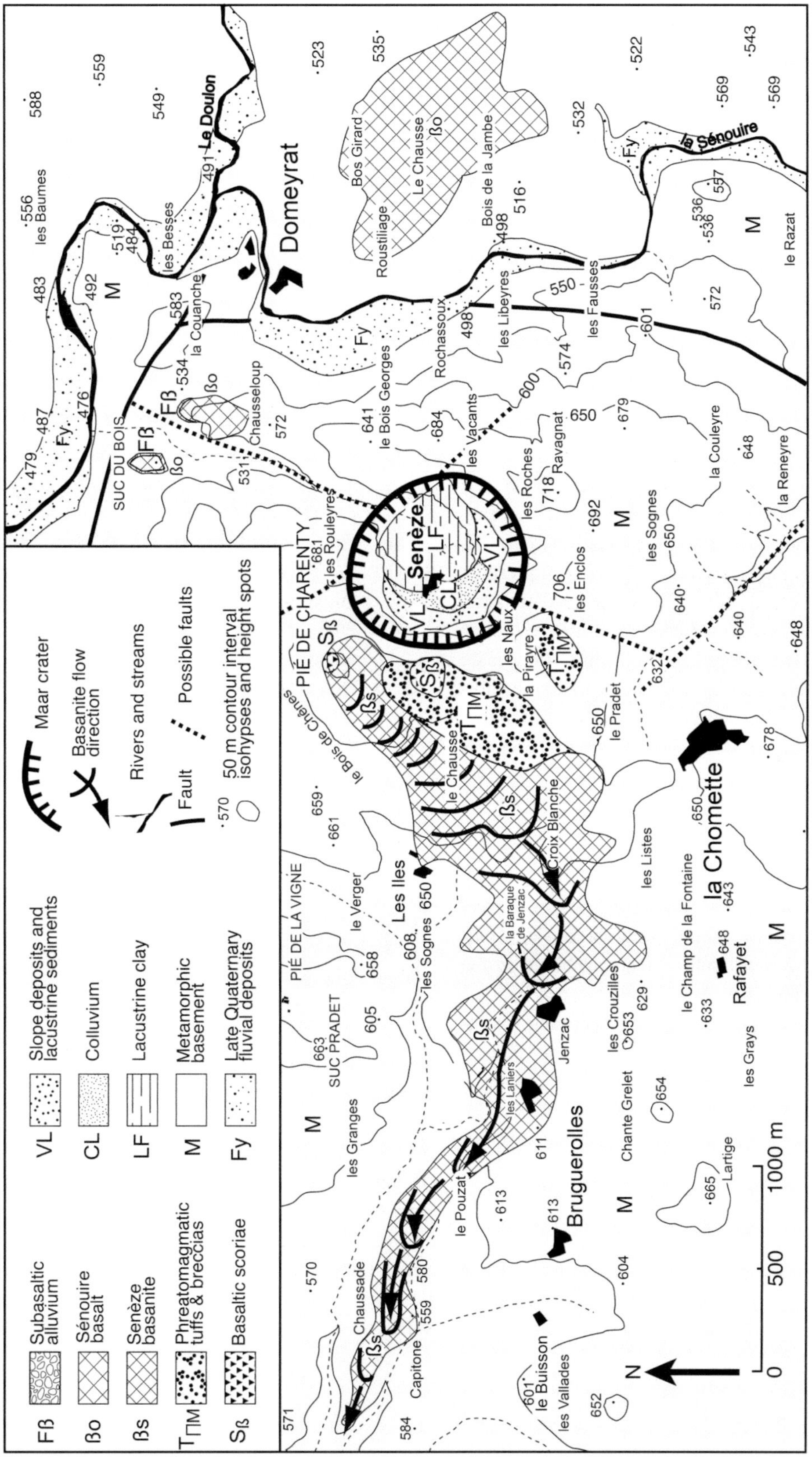

Fig. 4.1 Geological map of the Senèze maar area

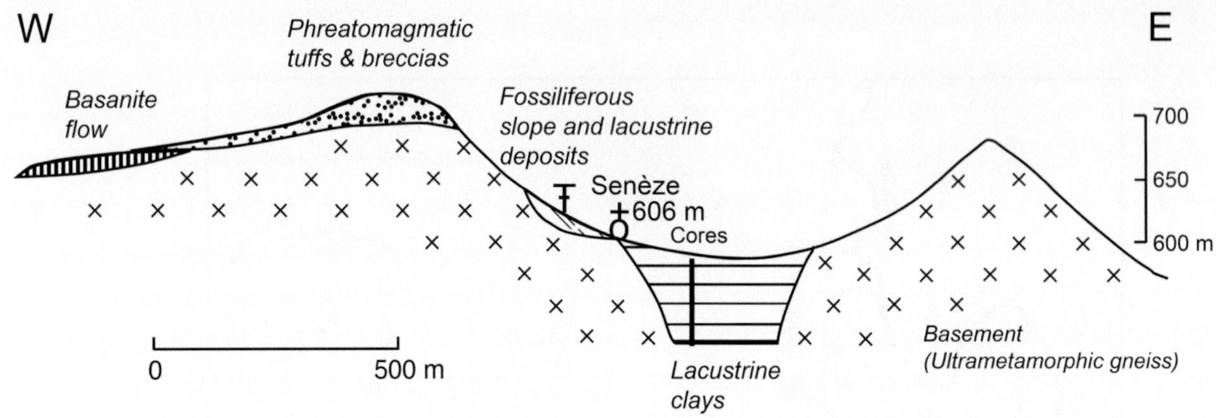

Fig. 4.2 W-E cross section of the Senèze maar and surroundings

The Phreatomagmatic Tuff

The phreatomagmatic tuff forms a crescent around the W and SW limit of the maar crater. In the map presented in Fig. 4.1, its extension is largely different from the one figured on the 1:50,000 geological map (Lasnier et al. 1982). It presents two distinct outcrops: (1) at 750 m NE from La Chomette, south of the 722 m hill; and (2) between the northern foot of the 697 m hill (S) and the talweg located south of the Pié de Charenty 717 m hill (N). The central part of this latter outcrop is interrupted by the scoriae previously discussed. These outcrops are quarried at "la Vieille Pireyre" (N) and "la Jeune Pireyre" (S).

In the "la Vieille Pireyre" quarry, about 5 m of the phreatomagmatic tuffs are visible. Its type-section is the following:

1. Indurated cindery tuff with a cindery brown matrix coating feldspars, quartz, small amphiboles and biotite < 1 mm, small basalt lapilli < 1 cm scattered in the matrix; 1 m of visible thickness.
2. Tuff with blocks and lapilli with graded bedding (thickness of the beds < 20 cm), less abundant cindery matrix, numerous lithic fragments: vitreous juvenile basalt (= more than 70%), gneiss (< 30%), basement blocks (< 20 cm), one big block of about 1 m^3, a few amphiboles; thickness ca. 2 m.
3. Indurated cindery tuff, homogenous, not bedded, fine-grained, poor in lithics; thickness about 2 m.

The succession exposed at the "la Jeune Pireyre" quarry is different. The typical (average) section, as measured mainly from the eastern part of the quarry, is the following:

1. Phreatomagmatic tuff, composed of five sub-units.
   (a) Indurated tuff with numerous lapilli of basalt < 2 cm = about 90% + basement = about 10%; not bedded, visible for a 2.50 m thickness.
   (b) Fine, welded, brown ash: 0.10 m thickness.
   (c) Tuff with lapilli similar to (a): about 1.50 m thickness.
   (d) Fine brown tuff, very poor in lithics: about 1 m thickness.
   (e) Graded bedded tuff (in the central part of the quarry: about 1.5 m thickness.
2. Graded bedded tuff (beds from < 1 cm to 5 cm), sandy to gravely/sandy texture: 0.90 m.
3. Massive tuff, both welded and not bedded, with sandy texture, rare lithic fragments: about 2 m thickness.

These formations are characterized by their richness in blocks and their indurated facies which corresponds probably to a primary induration. The absence of bedding in several horizons evokes a flow facies instead of the classical bedded base-surges which are usually found in most maars. These observations raise the question of the genesis of such phreatomagmatic products. The induration points to either a thermic or a hydromagmatic origin. The flow facies can be either phreatomagmatic or epiclastic. The high number of blocks indicates a strong explosive activity. The frequency of amphibole megacrystals (deep origin xenoliths) must be stressed.

These formations were channelized over an irregular paleomorphology. The subvertical position of the « la Jeune Pireyre» tuff over the basement, suggests a paleoslope of a small paleovalley. The « la Jeune Pireyre» tuff thickness spans over 50 m, from 670 to 722 m, possibly resulting from this topographic control.

Other phreatomagmatic tuffs can be observed in an internal position in the southern part of the crater. Represented by welded tuffs with no bedding, they are very rich in basement xenoliths < 2 cm, with rare hardened basalt bombs. Either this tuff was emplaced in the crater during the eruptive phase or it corresponds to a collapse of the tuff ring during a mass-movement. In any case, a more or less important collapse of the phreatomagmatic tuff occurred during the deposition of the fossiliferous beds, as seen in section A (see below) where a big block of tuff outcrops within the deposits.

The Tephras at the Senèze Fossiliferous Site

Introduction

The pyroclastic activity of the Mont-Dore strato-volcano during the Plio-Pleistocene constitutes the origin of numerous tephras scattered over the Massif Central and its surroundings (Pastre 1987; Teulade 1989). In particular, these tephras are known in the lacustrine fill of several maar craters (Degeai et al. 2013; Nomade et al. 2010; Pastre et al. 2007; Roger et al. 1999, 2000; Teulade 1989). Their in situ or slightly reworked occurrence is known in several fossiliferous sites (Debard & Pastre 1988, 2008; Nomade et al. 2014b; Pastre 2004; Pastre et al. 1996). At Senèze, a tephra bed was found at a 20.75 m depth in the lacustrine fill and dated at 2.10 ± 0.01 Ma by $^{40}Ar/^{39}Ar$ (Roger et al. 2000). Moreover, the location of the site 60 km SE of the strato-volcano constitutes a favorable condition for finding tephras in the fossiliferous sequence bordering the shore of the paleo-lake. The discovery of ten different tephras and six tephric horizons (slightly reworked tephras) in the various excavations fits with this perspective and provides an exceptional opportunity to characterize the tephrostratigraphical context and to date the site.

Location and Facies of the Tephras and Tephric Layers

The tephras (TS) and tephric layers (THS) of the fossiliferous site of Senèze were discovered in the new paleontological excavations (trenches T2 to T5, Zone [H8-N9-L14-G11] (west face; see Fig. 3.8 for north face)) and in two previously known sections (section A, also named “the bungalow section”; and section D, “the Philis ravine section”). The precise locations of the sections are given by Parenti et al. (2024) while the positions of tephras in each section are delineated by Debard (2024) and are summarized in Table 4.1. The sandy or silty-clayey texture of the samples results from weathering, which caused a complete disappearence of the vitreous material (cindery fraction/micro-pumices). TS 1 tephra (sample SEN 1) is a fine yellow sand interbedded in the lower clay of the middle part of trench T2 (bed hi2). TS 2 tephra (SEN 8) is a fine homometric sand layer at the base of the northern section of the T2 trench (bed b). TS 3 tephra (SEN 10) is also a fine grey sand, 1 cm thick; it was found in the southern part of the T2 trench (unit f, bed hi8). TS 4 tephra (SEN 42) is a white powdery clayey silt resembling weathered ash; it was identified in the upper part of section A. TS 5 (SEN 51), TS 6 (SEN 53) and TS 7 (SEN 60) are fine grained sands found in the excavated 233 West parcel. TS 8 tephra (sample SEN 62) is a white silt occurring in set 1 of the unfossiliferous T4 section. TS 9 (samples SEN 80, 87 and 98) and TS 10 (samples SEN 90, 95, 101, 102 and 103) tephras were found in trench T5, where they occur in units 4.2 (TS 9) and 4.10 (TS 10). TS 9 tephra presents a sandy texture; in some parts of the outcrop it can be divided into three layers: a (0–2 cm thickness, sandy); b (2–4 cm, sandy); and c (0–0.5 cm, white silt). TS 10 tephra resembles a sandy layer (a, 2 to 5 cm thick) which can be overlain by a white sandy silt (b, 0 to 4 cm) and a silty layer with small basalt lapilli (c, 0 to 10 cm).

The tephric horizons THS 1–5 (SEN 54, 55, 56, 57, 58) are white fine slightly reworked clayey-silty layers (weathered ash) occuring in the upper part of section A. The tephric horizon THS 6 (SEN 96, 97) is a clayey sand located in section D.

Table 4.1 Location and types of the studied samples

Sample	Label	Section	Bed	Type
SEN 1	TS 1	Trench T2 N	hi2	Fine sand (tephra)
SEN 2		Trench T2 N	d	Gravel sand
SEN 3		Trench T2 N	f	Gravel sand
SEN 8	TS 2	Trench T2 N	b	Fine sand (tephra)
SEN 10	TS 3	Trench T2 S	Set f hi8	Fine sand (tephra)
SEN 23		Trench T1 W		Microconglomerate
SEN 25		Trench T1 S		Gravel sand
SEN 33		Trench T1 N	Base T1g	Sandy silt
SEN 39		A (bungalow)	Lower part	Gravel sand
SEN 40		A (bungalow)	Middle part	Silt
SEN 41		A (bungalow)	Middle part	Silt
SEN 42	TS 4	A (bungalow)	Upper part	Clayey silt (tephra)
SEN 43		A (bungalow)	Upper part	Gravel sand
SEN 51	TS 5	Zone [H8-N9-L14-G11] W face	Lower locus	Fine sand (tephra)
SEN 53	TS 6	Zone [H8-N9-L14-G11] W face	hi	Sandy silt (tephra)
SEN 54	THS 5	A (bungalow)	Upper part	Clayey silt
SEN 55	THS 4	A (bungalow)	Upper part	Clayey silt
SEN 56	THS 3	A (bungalow)	Upper part	Clayey silt
SEN 57	THS 2	A (bungalow)	Upper part	Clayey silt
SEN 58	THS 1	A (bungalow)	Upper part	Silt with lava grains
SEN 59		A (bungalow)	Upper part	Heterometric sand
SEN 60	TS 7	Zone [H8-N9-L14-G11] W face	Lower locus	Fine sand (tephra)
SEN 62	TS 8	Trench T4	Set 1	Silt (tephra)
SEN 63		A (bungalow)	Upper part	Sand
SEN 80	TS 9	Lower trench T5	Unit 4.2	Fine sand (tephra)
SEN 81		Lower trench T5	Unit 4.3	Fine sand
SEN 84		Lower trench T5	Unit 4.8	Clayey silt
SEN 86		Lower trench T5		Gravel sand
SEN 87	TS 9	Trench T5	Unit 4.2	Sandy silt (tephra)
SEN 88		Scoria cone (W)		Ash & scoriae
SEN 90	TS 10	Lower trench T5	Unit 4.10	Sandy silt (tephra)
SEN 96	THS 6	D (Philis)	Unit 8	Sand
SEN 97		D (Philis)	Unit 8	Sand
SEN 98	TS 9	Lower trench T5	Unit 4.2	Sand a (tephra)
SEN 99	TS 9	Lower trench T5	Unit 4.2	Sand b (tephra)
SEN 100	TS 9	Lower trench T5	Unit 4.2	Silt c (tephra)
SEN 101	TS 10	Lower trench T5	Unit 4.10	Sand a (tephra)
SEN 102	TS 10	Lower trench T5	Unit 4.10	Silt b (tephra)
SEN 103	TS 10	Lower trench T5	Unit 4.10	Silt c lapilli (tephra)

Mineralogy

Because of the absence of glass, the tephras can only be characterized by their feldspars and heavy mineral content (ferromagnesians, oxides and accessory minerals). The feldspars and clinopyroxenes have been studied using an electron WDS microprobe (Cameca SX 100, 15 kV, 10 to 12 nA, counting time 10 s). The heavy minerals separated with bromoform (d = 2.8) in the 50 µm–2 mm granulometric fraction have been counted under the microscope (silicates and apatite; Fe-Ti oxides = mainly magnetite, some ilmenite not counted; biotite not counted).

Table 4.2 Heavy mineral percentages (silicates and apatite) of the tephras and tephric horizons in the Senèze fossiliferous site

Mineral/sample	SEN 01	SEN 08	SEN 10	SEN 42	SEN 51	SEN 53	SEN 54	SEN 55	SEN 56	SEN 57	SEN 58	SEN 60
Tephra/tephric horizon	TS 1	TS 2	TS 3	TS 4	TS 5	TS 6	THS 5	THS 4	THS 3	THS 2	THS 1	TS 7
Brown amphibole	7.9	20.1	41	13.2	2.7	78.2	19.7	15.3	10.5	17.8	13.1	60.5
Enstatite						0.5	1.1		0.4	2.5	5.4	
Chromian diopside			0.7				0.7	0.9	0.2	1.3	2.5	
Brown clinopyroxene		13.7	28.4	3.7	1.3	7.1	4.9	8.1	2.6	9.4	14.1	21.8
Green clinopyroxene	29.3	46.6	20.2	60.1	60.3	10.4	34.9	45.5	60.1	36.8	35.4	8.1
Olivine			6.7				3.6	8.8	1.6	3.8	2.9	4.8
Zircon	1.4	0.3		0.6	1.3		0.7	0.5	0.4	0.5	0.2	
Titanite	41.6	16.4	2.2	18.7	25.2	3.3	11.6	13.5	17.8	10.4	10.1	4.8
Apatite	19.8	2.7	0.7	3.7	9.1	0.5	1.6	1.2	2	3.8	1.6	
Sillimanite							1.6	0.7	1.6	7.4	6.3	
Staurolite							0.9	0.5				
Garnet							18.8	4.9	3	6.3	8.4	

Mineral/sample	SEN 62	SEN 80	SEN 87	SEN 90	SEN 96	SEN 98	SEN 99	SEN 100	SEN 101	SEN 102	SEN 103
Tephra/tephric horizon	TS 8	TS 9	TS 9	TS 10	THS 6	TS 9	TS 9	TS 9	TS 10	TS 10	TS 10
Brown amphibole	59.2	24.5	45.8	17.6	54.3	50.4	41.8	40.4	22.7	21	5.3
Enstatite										1	1.9
Chromian diopside										0.5	1.1
Brown clinopyroxene	9.2	47.3	8.6	8.6		23.8	33.6	23.1	30.9	27	31.7
Green clinopyroxene		16.8	0.5	27.3		5.2	8.2	9.6	32	25.5	9.5
Olivine										5.5	46.8
Zircon			0.8	0.6		0.8					
Titanite	31.6	11.5	43.6	41.5	45.7	19.8	16.4	26.9	13.7	19.5	3.1
Apatite			0.8	4.5					0.8		0.6
Sillimanite											
Staurolite											
Garnet											

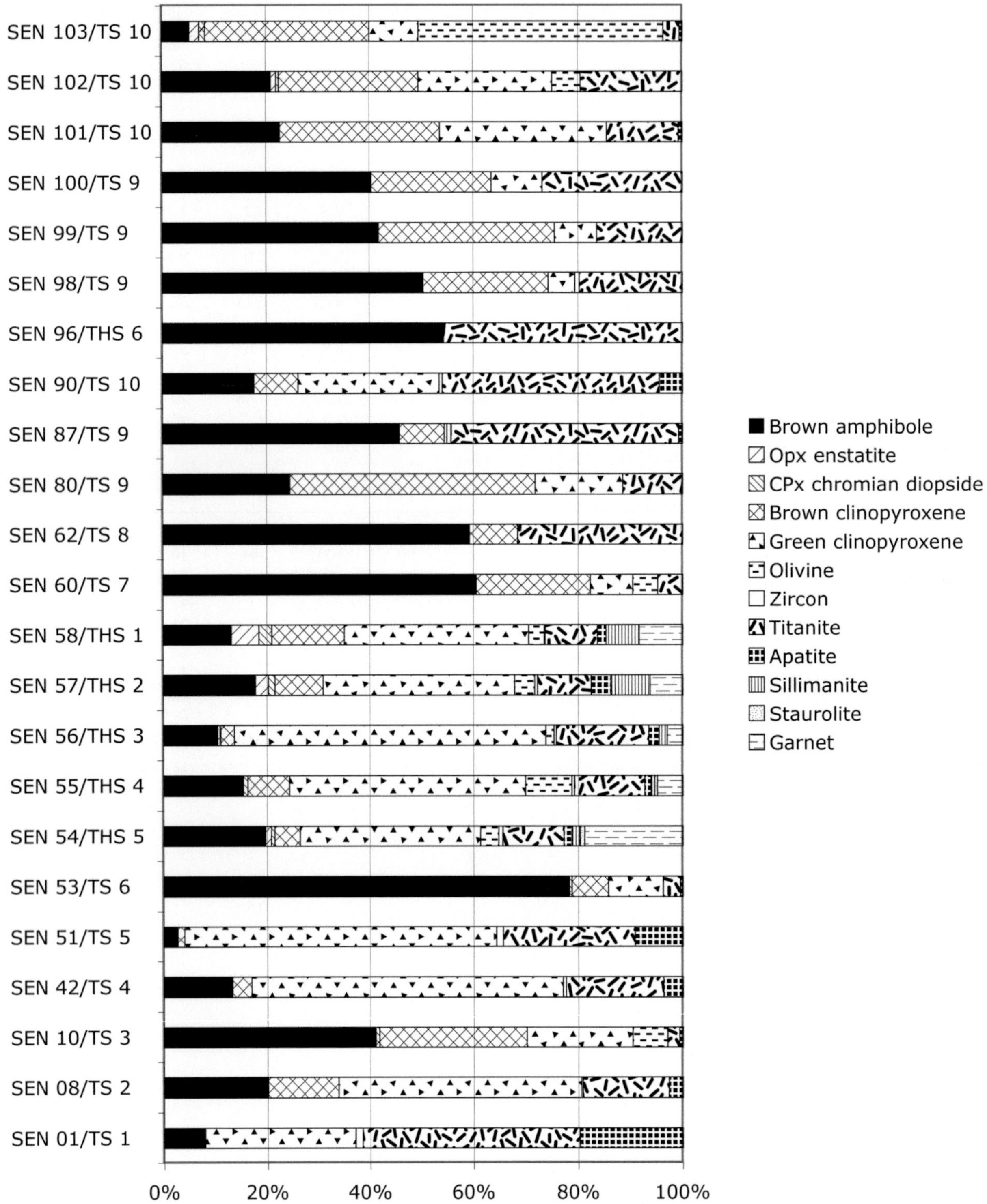

Fig. 4.3 Heavy mineral assemblages (silicates and apatite) of Senèze tephras and tephric horizons

The Heavy Minerals

The heavy mineral percentages are reported in Table 4.2 and illustrated in the diagram of Fig. 4.3. Spectra are dominated either by brown amphiboles (mainly kaersutite and pargasite: Ménard 1979; Pastre & Cantagrel 2001), brown clinopyroxenes (diopside $Wo_{45.4}$ $En_{41.9}$ $Fs_{12.7}$ to $Wo_{49.8}$ En_{38} $Fs_{12.2}$, Al_2O_3 = 4.15–10.08%, TiO_2 = 1.08–3.64%), green clinopyroxenes (diopside $Wo_{46.7}$ En_{31} $Fs_{22.3}$ to $Wo_{48.7}$ $En_{26.8}$ $Fs_{24.5}$, Al_2O_3 = 1.02–2.08%, TiO_2 = 0.24–1.18%), or titanite. Among the accessory minerals, zircons and apatite are present in low percentages in most of the samples. All these minerals are typical for the Mont-Dore pyroclastites (Pastre & Cantagrel 2001).

TS 1 tephra (sample SEN 1) is dominated by titanite (41.5%) followed by green clinopyroxene (29.3%) and shows a high percentage of apatite (19.8%). This assemblage characterizes a trachytic composition. Tephra TS 2 (sample SEN 8) presents a spectrum dominated by green clinopyroxenes (46.6%) again signaling a trachytic composition (Pastre & Cantagrel 2001). TS 3 (sample SEN 10) is dominated by brown amphiboles (41%) followed by brown and green clinopyroxenes. Such a composition could point to a tephritic composition (hyaloclastite/phreatomagmatic tuff ?; Pastre & Cantagrel 2001). TS 4 (SEN 42) contains green clinopyroxenes (60.1%) and titanite (18.7%), indicating a trachytic composition (Pastre & Cantagrel 2001). TS 5 (SEN 51) is dominated by green clinopyroxenes (60.3%), followed by titanite (25.2%) and apatite (9.1%), a composition typical of trachytic pyroclastites. TS 6 (SEN 53) is dominated by brown amphiboles (78.2%). TS 7 (SEN 60) mainly contains brown amphiboles (60.5%) and brown clinopyroxenes (21.8%). TS 8 (SEN 62) comprises brown amphiboles (59.2%) and titanite (31.6%). Samples SEN 80, 87, 98, 99 and 100, all of them representing the TS 9 tephra, show various compositions dominated by brown amphiboles and brown clinopyroxenes followed by titanite. Samples SEN 90, 95, 101, 102 and 103 representing the TS 10 tephra yield various spectra with brown amphiboles, brown clinopyroxenes, green clinopyroxenes and titanite. Sample SEN 103 delivered olivine crystals indicating a basaltic contamination during the later part of the eruption as also indicated by small basalt clasts (micro-lapilli).

The tephric horizons THS 1–5 (SEN 54, 55, 56, 57, 58) are mainly dominated by green clinopyroxene (34.9–60.1%), followed by brown amphiboles and titanite with some zircons and apatite (Table 4.2). They are slightly reworked trachytic tephras, with a small amount of basement (sillimanite, garnet and staurolite) and basaltic/mantellic minerals (brown clinopyroxene, olivine, enstatite and Cr-diopside) indicating their contamination by local material. The tephric horizon THS 6 (SEN 96) contains brown amphiboles (54.3%) and titanite (45.7%). The frequency of red cinders probably indicates its contamination by the scoriae dominating the maar.

Among the other minerals, biotite/phlogopite is common in most of the samples. Fe-Ti oxides (mainly magnetite) are frequent in most of the samples (except in SEN 10/TS 3, SEN 53/TS 6, SEN 60/TS 7 and SEN 96/THS 6 where they are rare).

Table 4.3 Selected microprobe analyses of feldspars from the tephras sampled in the Senèze fossiliferous site

Sample	SEN 1/1	SEN 1/2	SEN 8/1	SEN 8/2	SEN 10/1	SEN 10/2	SEN 42/1	SEN 42/2	SEN 53/1	SEN 53/2
Na_2O	6.037	7.275	5.14	5.875	7.787	7.332	6.04	7.984	6.9828	3.9951
MgO	0.003	0.008	0	0	0	0.001	0	0	0.064	0.0262
Al_2O_3	20.02	20.211	19.581	19.495	23.931	25.255	27.771	23.078	25.7633	30.1362
SiO_2	64.301	65.882	64.666	64.881	62.985	60.172	56.238	61.826	58.287	51.6043
K_2O	7.244	5.682	8.821	7.556	1.592	1.163	0.519	1.645	0.7661	0.2472
CaO	0.747	0.775	0.386	0.54	4.718	6.136	8.735	4.017	7.0882	12.8851
TiO_2	0.016	0.033	0.044	0.035	0.052	0.006	0.02	0.054	0.0363	0.0566
MnO	0	0.012	0	0	0.005	0.062	0	0.051	0.047	0
FeO	0.0196	0.212	0.264	0.236	0.315	0.317	0.379	0.384	0.4494	0.5153
Total oxides	98.3876	100.09	98.902	98.618	101.385	100.444	99.702	99.039	99.4841	99.466
Na	0.53368	0.62952	0.45491	0.51904	0.66277	0.63272	0.52792	0.69577	0.61049	0.35435
Mg	0.0002	0.00053	0	0	0	0.00007	0	0	0.00043	0.0179
Al	1.07581	1.06311	1.05343	1.04694	1.23812	1.32478	1.47547	1.22251	1.36917	1.62482
Si	2.93176	2.94033	2.9518	2.95636	2.76489	2.67812	2.53519	2.77886	2.62825	2.36072
K	0.42132	0.32349	0.51364	0.4392	0.08915	0.06603	0.02985	0.09432	0.04407	0.01443
Ca	0.03649	0.03706	0.01888	0.02636	0.2219	0.29261	0.42189	0.19345	0.34245	0.63155
Ti	0.00055	0.00111	0.00151	0.0012	0.00172	0.0002	0.00068	0.00183	0.00123	0.00195
Mn	0	0.00045	0	0	0.00019	0.00234	0	0.00194	0.00018	0
Fe	0.00747	0.00791	0.01008	0.00899	0.01156	0.0118	0.01429	0.01443	0.01695	0.01971
Total cations	5.00729	5.00351	5.00425	4.99809	4.99029	5.00866	5.00528	5.0031	5.01321	5.00931
An	3.68	3.74	1.91	2.68	22.79	29.52	43.07	19.67	34.35	63.13
Or	42.49	32.67	52.02	44.61	9.15	6.66	3.05	9.59	4.42	1.44
Ab	53.83	63.58	46.07	52.72	68.06	63.82	53.89	70.74	61.23	35.42

Sample	SEN 80/1	SEN 80/2	SEN 87/1	SEN 87/2	SEN 87/3	SEN 90/1	SEN 90/2	SEN 95/1	SEN 95/2	SEN 95/3
Na_2O	6.1983	6.0948	6.0363	6.9069	5.7642	4.8408	5.0235	4.981	5.0053	4.783
MgO	0	0.003	0.0067	0.0053	0	0.0177	0	0.0042	0	0.0125
Al_2O_3	19.5959	19.5245	19.669	19.5402	19.427	19.3818	19.1315	18.9958	19.5045	19.1325
SiO_2	65.1738	65.6326	65.1666	65.5854	64.9495	64.4624	64.5304	65.2116	64.4813	64.3684
K_2O	7.5007	7.832	7.5526	6.4772	8.1288	9.1882	9.0049	9.1555	8.5957	9.3753
CaO	0.8521	0.5752	0.9216	0.8785	0.7282	0.5015	0.5885	0.5031	0.7158	0.5058
TiO_2	0.0285	0.0044	0.0298	0.0363	0.026	0.0259	0	0.0408	0.017	0.0606
MnO	0	0.0213	0.033	0	0	0	0	0.006	0	0.0435
FeO	0.2162	0.2202	0.1552	0.2071	0.2466	0.2886	0.2994	0.1671	0.2307	0.2386
Total oxides	99.5655	99.908	99.5708	99.6369	99.2703	98.7069	98.5782	99.0651	98.5503	98.5202
Na	0.54342	0.53255	0.52905	0.60282	0.50788	0.43008	0.44684	0.44042	0.44424	0.42621
Mg	0	0.0002	0.00045	0.00036	0	0.00121	0	0.00029	0	0.00086
Al	1.04433	1.03703	1.0479	1.03668	1.04049	1.04673	1.03445	1.02098	1.05228	1.03636
Si	2.94704	2.95783	2.94581	2.95231	2.95155	2.95383	2.9605	2.97388	2.95169	2.95836
K	0.43266	0.45025	0.43552	0.37194	0.47123	0.53708	0.527	0.53261	0.50193	0.54966
Ca	0.04128	0.02777	0.04464	0.04237	0.03546	0.02462	0.02.893	0.02458	0.03511	0.02491
Ti	0.00097	0.00015	0.00101	0.00123	0.00089	0.00089	0	0.0014	0.00059	0.00209
Mn	0	0.00081	0.00126	0	0	0	0	0.00023	0	0.00169
Fe	0.00818	0.0083	0.00587	0.0078	0.00937	0.01106	0.01149	0.00637	0.00883	0.00917
Total cations	5.01787	5.01491	5.01151	5.0155	5.01687	5.00549	5.0092	5.00075	4.99467	5.00.931
An	4.06	2.75	4.42	4.17	3.49	2.48	2.88	2.46	3.58	2.49
Or	42.53	44.55	43.15	36.57	46.45	54.15	52.55	53.39	51.15	54.92
Ab	53.41	52.7	52.42	59.27	50.06	43.36	44.56	44.15	45.27	42.59

The Feldspars

Feldspars are abundant in all samples. Microprobe analysis reveals a classical composition with anorthite (An), albite (Ab) and orthose (Or), the percentages of which is a useful means for distinguishing tephras. Among the 236 crystal analyses performed, only some typical results are reported in Table 4.3. Three triangular diagrams, presented in Fig. 4.4, allow the identification of differences in the composition spectra.

The feldspars of TS 1 tephra (sample SEN 1) are alkali feldspars (anorthoclase–sodic sanidine). Their composition varies between $Or_{19.2}$ $Ab_{69.1}$ $An_{11.7}$ and $Or_{45.1}$ $Ab_{51.9}$ An_3. The feldspars of TS 2 tephra (SEN 8) are mainly sodic sanidine $Or_{32.5}$ Ab_{61} $An_{6.5}$ to Or_{52} $Ab_{46.1}$ $An_{1.9}$. TS 3 (SEN 10) is mainly composed of plagioclases (oligoclase-andesine $Ab_{49.9}$ $An_{46\ 0.7}$ $Or_{3.4}$ to $Ab_{68.1}$ $An_{22.8}$ $Or_{9.1}$) with rare alkali feldspars (sodic sanidine $Or_{40.6}$ $Ab_{57.1}$ $An_{2.3}$). TS 4 (SEN 42) is composed of a mixture of plagioclases (oligoclase-andesine $Ab_{51.4}$ $An_{45.9}$ $Or_{2.7}$ to $Ab_{69.6}$ $An_{19.5}$ $Or_{10.9}$) and alkali feldspars (sodic sanidine $Or_{49.2}$ $Ab_{46.8}$ An_4 to $Or_{52.1}$ $Ab_{44.9}$ An_3). TS 5 (SEN 51) mainly contains anorthoclase $Or_{13.2}$ $Ab_{73.3}$ $An_{13.5}$ to $Or_{33.2}$ Ab_{61} $An_{5.8}$ and few sodic sanidines (1 grain analysed $Or_{46.6}$, $Ab_{50.5}$ $An_{2.9}$). TS 6 (SEN 53) contains alkali feldspars (anorthoclase–sodic sanidine $Or_{24.8}$ $Ab_{69.8}$ $An_{5.4}$ to $Or_{57.8}$ $Ab_{40.3}$ An_{19}) and plagioclases (labrador–oligoclase $Ab_{35.5}$ An_{63} $Or_{1.5}$ to $Ab_{64.7}$ $An_{29.8}$ $Or_{5.5}$. TS 7 (SEN 60) is mainly composed by plagioclases (oligoclase-andesine $Ab_{63.8}$ $An_{30.4}$ $Or_{5.8}$ to $Ab_{70.3}$ An_{19} $Or_{10.7}$) and contains also some alkali feldspars $Or_{25.4}$ $Ab_{64.2}$ $An_{10.4}$ to $Or_{55.2}$ $Ab_{41.9}$ $An_{2.9}$. TS 8(SEN 62) is composed of plagioclases (oligoclase $Ab_{83.2}$ $An_{15.3}$ $Or_{1.5}$ to $Ab_{77.6}$ $An_{19.5}$ $Or_{2.9}$) and alkali feldspars (anorthoclase–sodic sanidine $Or_{28.5}$ $Ab_{64.1}$ $An_{7.4}$ to $Or_{42.9}$ $Ab_{55.3}$ $An_{1.8}$). The feldspars of the TS 9 tephra (SEN 80, 87 and 98) are alkali feldspars (anorthoclase–sodic sanidine $Or_{23.3}$ $Ab_{65.6}$ $An_{11.1}$ to $Or_{51.5}$ $Ab_{45.3}$ $An_{3.2}$. The TS 10 tephra (SEN 90, 95, 101 and 103) is composed of sodic sanidine with a very homogeneous composition ($Or_{49.3}$ $Ab_{46.5}$ $An_{4.2}$ to $Or_{55.6}$ $Ab_{41.8}$ $An_{2.6}$, 52 analysis).

The tephric horizon THS 2 (SEN 56) is composed by an homogeneous sodic sanidine (Or_{50} $Ab_{46.6}$ $An_{3.4}$ to $Or_{56.8}$ $Ab_{39.6}$ $An_{3.5}$) and by oligoclase $Ab_{67.1}$ $An_{25.9}$ Or_7 to $Ab_{70.1}$ An_{21} $Or_{8.9}$. The tephric horizons THS 1, 3, 4 and 5 have not been analysed as they are probably issued from the same tephra. THS 6 (SEN 96), heterogeneous in composition, contains sodic sanidine $Or_{38.8}$ $Ab_{58.1}$ $An_{3.1}$ to $Or_{56.8}$ $Ab_{39.6}$ $An_{3.5}$ and anorthoclase $Or_{11.4}$ $Ab_{71.8}$ $An_{16.8}$ to Or_{28} $Ab_{66.4}$ $An_{5.6}$. All these feldspar compositions and mineralogical assemblages are typical of the Mont-Dore strato-volcano pyroclastic products (Pastre & Cantagrel 2001, Pastre unpublished analysis).

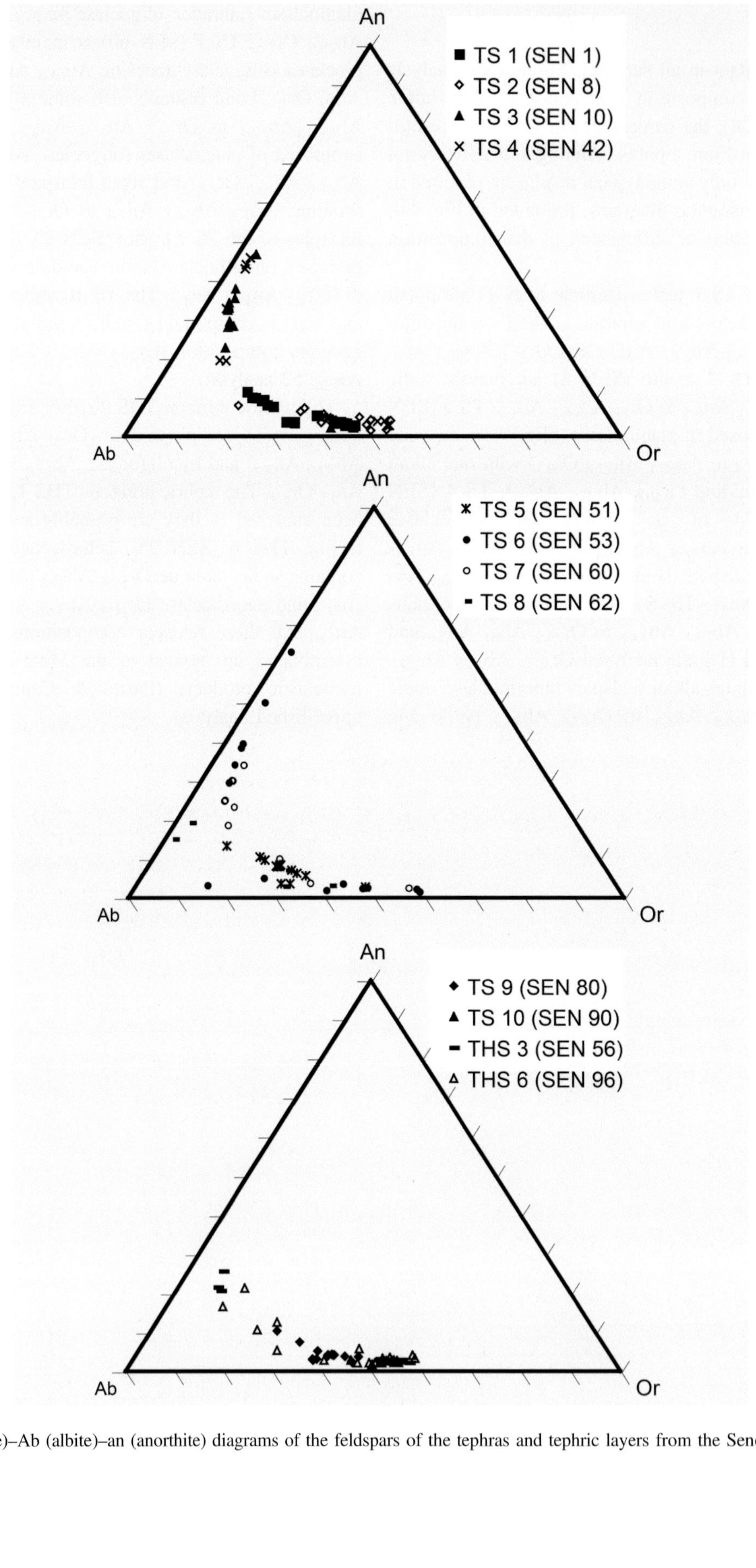

Fig. 4.4 Or (orthose)–Ab (albite)–an (anorthite) diagrams of the feldspars of the tephras and tephric layers from the Senèze fossiliferous site

Discussion, Tephrostratigraphic Implications

The different mineralogical compositions of the tephras studied indicate distinct ages for the deposit sequences in which they occur. These sequences characterize several evolutionary stages of the lacustrine shore and of the crater slopes featured by the various stratigraphic sequences observed in the paleontological excavations (Debard 2024). This fact implies the diachrony (rather than contemporaneity) of the fauna found in the various paleontological excavations (T1 + T2, 233W parcel and T5), and also shows that sediments in the A and T4 sections do not have the same age.

The nature of these tephras can be compared with the composition of the tephras emitted during the Early Pleistocene activity phase of the Mont-Dore strato-volcano around 2 Ma (Pastre & Cantagrel 2001; Pastre 1987). If no known pyroclastic unit of the Mont-Dore can be directly correlated with a Senèze tephra bed, several may be contemporaneous.

In the Guéry area, in the northeast part of the Mont-Dore, the Guéry pumice flow displays a mineralogical composition with anorthoclase-sodic sanidine, Fe-Ti oxides, biotite, brown amphibole, brown clinopyroxene, green clinopyroxene, titanite and zircon (Pastre & Cantagrel 2001; Pastre 1987) which is close to several tephras of Senèze. It is associated with other minor pyroclastites (pumice flows and falls) which are probably close in age. It was first dated by $^{40}Ar/^{39}Ar$ to 2.04 ± 0.03 Ma (Féraud et al. 1990) and recently redated to 2.09 ± 0.01 Ma (Nomade et al. 2014a). In the northern part of this area, the pyroclastites of the western border of the Puy de l'Ouire dome show several units mainly dominated by feldspars, Fe-Ti oxides and amphiboles (Pastre 1987). Several tephras are also associated with the unsaturated activity (tephrite-trachyphonolite) recorded in this area around ca. 2 Ma. As for most of the Senèze tephras, the joint occurence of brown amphibole, brown clinopyroxene and green clinopyroxene is typical of many pyroclastites of this period. Results of this comparison fit with the $^{40}Ar/^{39}Ar$ ages of the Senèze tephras comprised between 2.18 ± 0.03 and 2.07 ± 0.02 Ma (Nomade et al. 2014a, 2014b; Pastre et al. 2015; Delson et al. 2024).

At Blassac, in the upper Allier valley, the heavy mineral content of the fluvial sequence associated with local basaltic products shows the influence of four types of contemporaneous Mont-Dore tephra falls: (a) with green clinopyroxene; (b) with brown amphiboles and titanite; (c) with green clinopyroxene and titanite; and (d) with brown amphiboles, green clinopyroxene and titanite (Pastre 1987). These various tephras with heavy mineral assemblages dominated by either brown amphiboles or green clinopyroxene and ages of 2.04 ± 0.05 (Couthures 1982; Couthures & Pastre 1983) and 2.14 ± 0.06 Ma (Fouris 1989) for the succeeding lava-flows could point to a close comparison with Senèze tephras.

The Heavy Mineral Assemblages of the Sediments of the Paleontological Sequence: Composition and Implications for the Origin of Sedimentation

The different sequences excavated for paleontological research show several beds of coarse silty sands with gravels (e.g., SEN 2, SEN 3, and trench T2; see Table 4.1 and Debard (2024) for precise locations). The light sandy fraction of these beds is mainly constituted by quartz, feldspars and rock grains (weathered basalt, polymineral grains of quartz, feldspars and biotite = gneiss). Some sandy silts dominated by quartz, like SEN 33 (trench T1, see Table 4.1) or silts like SEN 40 (section A, see Table 4.1) can also be observed. The heavy mineral assemblages of these beds (Table 4.4) clarify their origin. The brown clinopyroxenes and olivine which dominate several samples (e.g., SEN 2 and SEN 3 in trench T2; SEN 25 in trench T1; SEN 39 in section A) are mainly basaltic minerals. The olivine, enstatite and chromian diopside which are common in many samples are mantellic minerals issued from lherzolite (peridotite) nodules. The basement minerals are dominated by garnets and sillimanite, associated with some staurolite and andalusite. If a small amount of these minerals could originate from surrounding hard rocks (gneiss of the crater, basalt of the lava-flow), most of them derive from the phreatomagmatic tuffs dominating the crater where they have been pulverized by phreatomagmatic explosions. The brown amphiboles could also originate from the maar tuffs, where they are present in megacrysts, or could have been reworked from Mont-Dore tephras. Some important concentrations of brown clinopyroxene and olivine as in SEN 59 (sand, upper part of section A) can probably be partly related to basaltic scoriae source dominating this area (SEN 88). The green clinopyroxene (automorphic diopside) and/or titanite which are present in several samples (e.g., SEN 3, 23, 25 and 41) show the reworking of Mont-Dore tephras fallen on the crater-rim. They are especially frequent in the upper part of the A section (gravelly sand SEN 43 and sand SEN 63) where they follow the trachytic TS 4 tephra and THS 1–5 tephric horizons. The fine sand SEN 81 (lower part of T5) which contains brown amphiboles, brown CPx and titanite are probably reworked from a Mont-Dore tephra. The gravel sand SEN 86 (lower part of T5), which is dominated by olivine (71.3% of volcanic minerals), probably represents a reworked phreatomagmatic tuff or perhaps an in situ tuff.

These mineralogical compositions agree with slope wash or gravity processes mainly reworking the phreatomagmatic tuffs of the upper part of the crater.

and five tephric horizons which are typical of the activity at Mont-Dore, 60 km to the NW. These weathered tephras, mainly trachytic in composition, correspond to products

Table 4.4 Heavy mineral percentages (volcanic and basement minerals) of the sediments in the Senèze fossiliferous site

Mineral/sample	SEN 2	SEN 3	SEN 23	SEN 25	SEN 33	SEN 39	SEN 40	SEN 41
Brown amphibole	5.3	9.2	8.1	16.8	20.1	7.8	17.6	16.4
Enstatite	11.7	10.4	11.9	14.5	5.5	7.8	12.6	4.9
Chromian diopside	4.6	7.2	6	6.2	4.8	5.4	10.1	1.6
Brown clinopyroxene	43.6	23.7	18.2	15.9	14.6	17.6	16.8	11.5
Green clinopyroxene		2.8	5	0.3				
Basaltic green clinopyroxene	0.2		1.6	0.9				
Olivine	32.1	28.1	6	36.9	7.8	51.2	18.5	
Zircon								
Titanite		0.4	2.2	0.3	4.9			6.6
Apatite			1.9					
Andalusite		0.8						
Sillimanite	2.4	9.6	6.6	6.5	37.2	2.2	18.5	21.3
Staurolite						2.4		3.3
Garnet		7.6	32.6	1.8	5.2	5.6	5.9	34.4
Other								
Mineral/sample	SEN 43	SEN 59	SEN 63	SEN 81	SEN 84	SEN 86	SEN 88	SEN 97
Brown amphibole	13.4	5.9	13	40.9	51.7	14.2		6.2
Enstatite	1.1	2.8	1		5	3.9		3.7
Chromian diopside	0.8	2.7	0.7		2.5	3		1.8
Brown clinopyroxene	8.6	31.2	4.1	10	29.2	4.5	66.7	32.1
Green clinopyroxene	49.7	10.4	56.5	1	2.5			
Basaltic green clinopyroxene							5.9	
Olivine	6.3	33.1	10.4			64.2	22.5	56.2
Zircon	0.3		0.1					
Titanite	12.6	3	10.4	7.2				
Apatite	3.2	0.3	2.2	0.9	1.7			
Andalusite						0.6		
Sillimanite	1.1	5.2	0.4	24.8	6.7	6.4	1	
Staurolite								
Garnet	3	5.4	1.1	15.2	0.8	3.1	1	
Other							2.9	

Conclusions

The Senèze volcano is characterized by the association of one basanite lava-flow, scoriae deposits and the phreatomagmatic tuff fringing the maar crater. The map in Fig. 4.1 presents a new view of their distribution. A first phreatomagmatic episode, characterized by sub-basanitic tuffs, precedes the lava-flow emission, which is probably contemporaneous with the scoriae. Afterwards, the main phreatomagmatic episode is recorded by the later western slope tuffs occurrence and the formation of the maar crater.

Slope deposits, lacustrine deposits and slope wash deposits containing the mammal fauna comprise ten tephras emitted during the Early Pleistocene activity of the strato-volcano *ca* 2.1 Ma. The difference in mineral content of these tephras, each of which occurs in different paleontological excavations and outcrops, shows the polyphased deposition of the fossiliferous deposits.

The mineralogical composition of the sands and silts interbedded in the different sections of the lacustrine shore indicate a major sedimentological contribution of phreatomagmatic tuffs dominating the crater and several inputs from local basaltic scoriae and Mont-Dore tephras. The widespread occurrence of the tephras provides several possibilities for dating the fauna and offers a great opportunity to clarify the tephrochronology of the Mont-Dore strato-volcano during this period.

References

Boivin, P., & Gourgaud, A. (1978). Synchronisme de dynamismes de type « maar » et « strombolien » près de la Sauvetat (Sud Devès, Massif central français). *Comptes-rendus sommaires de la Société géologique de France, 1*, 24–27.

Bout, P. (1960). *Le Villafranchien du Velay et du bassin hydrographique moyen et supérieur de la Loire*. Le Puy: Imprimerie Jeanne d'Arc.

Brousse, R., & Lefèvre, C. (1990). *Le volcanisme en France. Guides géologiques régionaux*. Masson.

Couthures, J. (1982). Contribution à la chronostratigraphie de formations plio-pléistocènes du Massif Central (France). Thèse de 3ème cycle, Université Paris VI.

Couthures, J., & Pastre, J. F. (1983). Chronostratigraphie du Plio-Pléistocène d'Auvergne et du Velay: Nouveaux apports des datations radiométriques et du paléomagnétisme. *Bulletin de l'Association française pour l'étude du Quaternaire, 1983–1*, 9–18.

Couthures, J., Ablin, D., & Hadjouis, D. (1991). *Senèze (Haute-Loire) Rapport de fouilles 1991*. Service Régional de l'Archéologie, DRAC Auvergne (unpublished).

Debard, E. (2024). Geological study of the Early Pleistocene site of Senèze (Domeyrat, Haute-Loire, France). In E. Delson, M. Faure & C. Guérin (Eds.), *Senèze: Life in Central France two million years ago. Paleontology, geochronology, stratigraphy and taphonomy* (pp. 37−82). Cham (Switzerland): Springer.

Debard, E., & Pastre, J. F. (1988). Un marqueur chronostratigraphique du Pléistocène moyen à la périphérie du Massif Central : La retombée à clinopyroxène vert du Sancy dans le site acheuléen d'Orgnac III (Bas Vivarais, S-E France). *Comptes-rendus de l'Académie des Sciences, Série II, 306*, 1515–1520.

Debard, E., & Pastre, J. F. (2008). Nouvelles données sur les téphras pléistocènes piégés dans les remplissages karstiques ardéchois (SE France). *Quaternaire, 19*, 107–116.

Degeai, J. P., Pastre, J. F., Gauthier, A., Robert, V., Nomade, S., Bout-Roumazeilles, V., & Guillou, H. (2013). La séquence lacustre du maar d'Alleret (Massif Central, France): Téphrochronologie et évolution paléoenvironnementale en Europe occidentale au début du Pléistocène moyen. *Quaternaire, 24*, 443–459.

Delson, E., Nomade, S., Sen, S., Debard, E., Pastre, J.F., Bahain, J.-J., et al. (2024). Geochronology of Senèze: $^{40}Ar/^{39}Ar$ dating and magnetostratigraphy, with notes on an ESR/U-series dating attempt. In E. Delson, M. Faure & C. Guérin (Eds), *Senèze: Life in Central France two million years ago. Paleontology, geochronology, stratigraphy and taphonomy* (pp. 99−122). Cham (Switzerland): Springer.

Devis, G. (1969). Le volcanisme du Brivadois. Diplôme d'Etudes Supérieures, Clermont-Ferrand.

Devis, G. (1970). Le volcanisme et le gisement fossilifère de Senèze. *Revue des Sciences naturelles d'Auvergne, 36*, 25–87.

Féraud, G., Lo Bello, P., Hall, C. M., Cantagrel, J. M., York, D., & Bernat, M. (1990). Direct dating of plio-quaternary pumices by $^{40}Ar/^{39}Ar$ step-heating and single-grain laser fusion methods: The example of the Monts-Dore massif (Massif Central, France). *Journal of Volcanology and Geothermal Research, 40*, 39−53.

Fouris, M. (1989). Les sites villafranchiens du Devès et les basaltes de la vallée de l'Allier: Application de la méthode K-Ar. Thèse de doctorat Université Blaise Pascal, Clermont-Ferrand.

Grangeon, P. (1962). Note préliminaire à l'étude géologique du volcan et du gisement fossilifère de Senèze. *Comptes-rendus sommaires de la Société géologique de France, 5*, 149–150.

Lasnier, B., Marchand, J., Bouiller, R., Cornen, G., Burg, J. P. Leyreloup, A. & Forestier, F. H. (1982). *Carte géologique de la France à 1/50 000ème, feuille 166 Brioude*.

Ménard, J. J. (1979). Contribution à l'étude pétrogénétique des nappes de ponces du massif volcanique du Mont-Dore (Massif Central français), Thèse de 3ème cycle. Université de Paris-sud, Orsay.

Maury, R. C., & Varet, J. (1980). *Le volcanisme tertiaire et quaternaire en France*. In A. Autran & J. Dercourt (Eds.), *Evolutions géologiques de la France, 26e Congrès géologique international, Colloque C7, Mémoires du Bureau de Recherches Géologiques et Minières* (vol. 107, pp. 138–159).

Mergoil, J., & Boivin, P. (1993). Le Velay. Son volcanisme et les formations associées. *Géologie de la France, 3*, 3–96.

Nomade, S., Gauthier, A., Guillou, H., & Pastre, J. F. (2010). $^{40}Ar/^{39}Ar$ temporal framework for the Alleret maar lacustrine sequence (French Massif Central): Volcanological and paleoclimatic implications. *Quaternary Geochronology, 5*, 20–27.

Nomade, S., Pastre, J. F., Nehlig, P., Guillou, H., Scao, V., & Scaillet, S. (2014a). Tephrochronology of the Mont-Dore volcanic Massif (Massif Central, France): New $^{40}Ar/^{39}Ar$ constraints on the late Pliocene and early Pleistocene activity. *Bulletin of Volcanology*, 76, 798. https://doi.org/10.1007/s00445-014-0798-6

Nomade, S., Pastre, J. F., Guillou, H., Faure, M., Guérin, C., Delson, E., et al. (2014b). $^{40}Ar/^{39}Ar$ constraints on some French landmark Late Pliocene to Early Pleistocene large mammalian paleofaunas: Paleoenvironmental and paleoecological implications. *Quaternary Geochronology, 21*, 2–15.

Parenti, F., Valli, A., Aprile, L., Candelato, F., Monguillon, A., Natali, L., et al. (2024). Surveying and excavations at Senèze: 2001–2006. In E. Delson, M. Faure & C. Guérin (Eds), *Senèze: Life in Central France two million years ago. Paleontology, geochronology, stratigraphy and taphonomy* (pp. 27–35). Cham (Switzerland): Springer.

Pastre, J. F. (1987). Les formations plio-quaternaires du bassin de l'Allier et le volcanisme régional (Massif Central, France). Thèse de doctorat, Université Pierre et Marie Curie (Paris VI), Mémoire 87–32, p. 753.

Pastre, J. F. (2001). L'impact du volcanisme du Devès sur l'évolution morpho-sédimentaire plio-quaternaire de la haute vallée de l'Allier (Massif Central, France). *Géomorphologie: Relief, Processus, Environnement, 2001–2*, 127–136.

Pastre, J.F. (2004). Le cryptotéphra de Saint-Vallier: comparaison avec d'autres retombées pyroclastiques du Mont-Dore issues de gisements mammaliens plio-pléistocènes du Massif Central. Implications chronostratigraphiques. In M. Faure & C. Guérin (Eds), *Le gisement pliocène final de Saint-Vallier (Drôme, France). Geobios, 37*, Mémoire spécial n° 26, 44–57.

Pastre, J. F., Billard, A., Debard, E., Faure, M., & Guérin, C. (1996). Un nouvel horizon téphrique du Mont-Dore repéré dans les loess plio-pléistocènes du gisement fossilifère de Saint-Vallier (Drôme, France). *Comptes-rendus de l'Académie des Sciences, Série IIa, 323*, 607–614.

Pastre, J. F., & Cantagrel, J. M. (2001). Téphrostratigraphie du Mont-Dore (Massif Central, France). *Quaternaire, 12*, 249–267.

Pastre, J. F., Gauthier, A., Nomade, S., Orth, P., Andrieu, A., Goupille, F., et al. (2007). The Alleret maar (Massif Central, France): A new lacustrine sequence of the early middle Pleistocene in western Europe. *Comptes Rendus Geoscience, 339*, 987–997.

Pastre, J. F., Debard, E., Nomade, S., Guillou, H., Faure, M., Guérin, C., & Delson, E. (2015). Nouvelles données géologiques et téphrochronologiques sur le gisement paléontologique du maar de Senèze (Pléistocène inférieur, Massif Central, France). *Quaternaire, 26*, 225–244.

Prévot, M. (1975). Magnétisme et minéralogie magnétique de roches néogènes et quaternaires, contribution au paléomagnétisme et à la géologie du Velay. Thèse d'Etat, Université Pierre et Marie Curie, Paris VI.

Roger, S., Coulon, C., Thouveny, N., Féraud, G., Van Velzen, A., Fauquette, S., et al. (2000). $^{40}Ar/^{39}Ar$ dating of a tephra layer in the Pliocene Senèze maar lacustrine sequence (French Massif Central): Constraint on the age of the Reunion-Matuyama transition and implications on paleoenvironmental archives. *Earth and Planetary Science Letters, 183*, 431–440.

Roger, S., Féraud, G., de Beaulieu, J. L., Thouveny, N., Coulon, C., Cochemé, J. J., et al. (1999). $^{40}Ar/^{39}Ar$ dating on tephra of the Velay maars (France): Implications for the late Pleistocene proxy-climatic record. *Earth and Planetary Science Letters, 170*, 287–299.

Teulade, A. (1989). Téphrologie des formations cendro-ponceuses en milieux lacustres quaternaires. Méthode d'étude et applications au Massif Central français (Velay) et aux Carpathes orientales roumaines (dépression de Brasov). Thèse de doctorat, Université Aix-Marseille.

Chapter 5

Geochronology of Senèze: $^{40}Ar/^{39}Ar$ Dating and Magnetostratigraphy, with Notes on an ESR/U-Series Dating Attempt

Eric Delson, Sébastien Nomade, Sevket Sen, Evelyne Debard, Jean-François Pastre, Jean-Jacques Bahain, Qingfeng Shao, and Christophe Falguères

Abstract The age of the Senèze mammalian fauna has been discussed since it was first reported in 1892. $^{40}Ar/^{39}Ar$ ages reported by Nomade et al. (2014) and recalculated here to agree with current standards placed the deposits between 2.20 and 2.07 Ma. Paleomagnetic data collected in 2001 and 2004 help to narrow the age range, especially of the levels yielding fossils in 2001–2006. In the western sector, fossils can be securely dated between 2.10 and 2.08 Ma, while in the southeastern sector, they are slightly older, between 2.20 and 2.18 Ma. Senèze is one of the few later Cenozoic European sites dated by both argon geochronometry and paleomagnetism, which makes these ages so precise. Experiments with ESR/U-series dating on teeth proved unsuccessful as a result of the early U-uptake and high natural dose rate in the sediments of Senèze.

Résumé L'âge de la faune mammalienne de Senèze fait débat depuis sa découverte en 1892. Les âges $^{40}Ar/^{39}Ar$ obtenus par (Nomade et al. 2014) sont ici recalculés selon les calibrations les plus récentes et placent définitivement les dépôts de ce site paléontologique majeur entre 2.20 et 2.07 Ma. Les données paléomagnétiques recueillies en 2001 et 2004 permettent d'affiner encore la fourchette d'âge, en particulier pour les niveaux ayant livré des fossiles entre 2001–2006. Dans le secteur occidental, les fossiles peuvent être datés avec une grande précision entre 2.10 et 2.08 Ma, tandis que dans le secteur sud-est, ils sont légèrement plus anciens, entre 2.20 et 2.18 Ma. Senèze est l'un des rares gisements européens du Cénozoique supérieur à être daté à la fois par radio-isotopie et paléomagnétisme, ce qui rend sa datation si précise. Nous notons enfin que les tentatives de datation de dents par ESR/U-Th se sont révélées infructueuses en raison de l'absorption précoce de l'U et du débit de dose naturel élevé des sédiments de Senèze.

Keywords Argon-argon ages • Biochronology • Feni subchron • Huckleberry Ridge excursion • Late Villafranchian • Mammal fauna • MNQ 18 • Paleomagnetism

Mots-clés Ages argon-argon • Biochronologie • Faune mammalienne • Paléomagnétisme • Villafranchien supérieur

E. Delson (✉)
Division of Paleontology, the American Museum of Natural History, New York, NY, USA
e-mail: eric.delson@lehman.cuny.edu

Department of Anthropology, Lehman College of the City University of New York, Bronx, NY, USA

PhD Program in Anthropology, The Graduate Center of the City University of New York, New York, NY, USA

New York Consortium in Evolutionary Primatology, New York, NY, USA

Institut Català de Paleontologia Miquel Crusafont, Barcelona, Spain

S. Nomade
Laboratoire des Sciences du Climat et de l'Environnement, CEA Saclay, 7LSCE/IPSL, UMR CEA-CNRS-UVSQ 8212, Gif sur Yvette, France

S. Sen
CR2P (Paléontologie)—Muséum National d'Histoire Naturelle, 8 Rue Buffon, 75005 Paris, France

E. Debard
Université de Lyon, UCBL, ENSL, CNRS, LGL-TPE, 69622 Villeurbanne, France

J.-F. Pastre
23, rue de l'Eglise, 93370 Montfermeil, France

Laboratoire de Géographie Physique, Environnements quaternaires et actuels, UMR 8591 CNRS, Thiais, France

J.-J. Bahain · C. Falguères
UMR 7194 HNHP, Département Homme et Environnement, Muséum National d'Histoire Naturelle, Paris, France

Q. Shao
College of Geography, Nanjing Normal University, Nanjing, China

E. Delson et al. (eds.), *Senèze: Life in Central France Around Two Million Years Ago*, Vertebrate Paleobiology and Paleoanthropology, https://doi.org/10.1007/978-3-031-64415-3_5

Introduction

The age of the Senèze mammalian fauna has been subject to discussion since it was first reported by Boule (1892). He noted that the proboscidean he recovered was *Elephas meridionalis*, similar to that found in the British Crags, but older than the same taxon at Durfort and Cromer, although younger than the mastodons of Chilhac and Le Coupet. Over the following decades, work by Depéret (1929), Depéret and Mayet (1912), Stehlin (1923), Schaub (1943) and others discussed the fauna as Late Pliocene in age, without suggesting actual dates. Heintz (1970) and Azzaroli (1970) both included Senèze in their Late Villafranchian mammal unit, younger than the Saint-Vallier assemblage, which Guérin thought might date to ca. 2.0–1.9 Ma (Guérin et al. 2004). As summarized in Delson et al. (2006) and Faure et al. (2024), age estimates for the Senèze fauna (not considering the suggestion of Azzaroli et al. (1988) that there was a younger assemblage somehow mixed in) ranged between ca. 1.5 and ca. 2.1 Ma, near the then-accepted Plio-Pleistocene boundary.

Prévot and Dalrymple (1970) provided the first chronometric date for the Senèze complex, a whole-rock potassium-argon date of 2.30 ± 0.15 Ma for samples of the Pié de Charenty lava flow assumed to predate the maar deposits. Couthures and Pastre (1983) and Couthures (1989) reported a roughly comparable date of 2.48 ± 0.06 Ma. All of these works reported a reversed polarity for the flow, correlated to the lower Matuyama. Prévot and Dalrymple (1970) also reported paleomagnetic data for the upper 132.5 m of a core drilled into the lake beds in 1965. The section was entirely reversed, except for a normal zone near the core top, between 17.5 and 23.5 m down from the surface (between 22–23 m and 23.5–28 m, sandy layers did not allow paleomagnetic sampling). This normal interval was correlated to the "Lower Olduvai" which they thought dated ca. 2.1 Ma. Prévot and Dalrymple (1970) estimated that the maar lake filled in about 0.25 Myr (from the 2.35 Malava flow to the 2.1 Ma core top), which agreed with an estimate by Ehrlich (1968) of 200–300 kyr for the age span of the core based on the number of annual diatomite layers in a portion she counted. Delson and Plopsor (1975) discussed these geochronological elements and agreed with Prévot and Dalrymple (1970) that the short normal interval correlated with the Réunion normal magnetosubchron and the Senèze mammalian fauna likely dated between 2.1 and 2.0 Ma, as it was then assumed to derive from levels altitudinally above the lake beds. Roger et al. (2000) reported a $^{40}Ar/^{39}Ar$ age of 2.10 ± 0.01 Ma on a thin tephra within the normal magnetozone near the top of a shorter core, drilled in 1989. They considered that this normal was the Réunion subchron and estimated that the subchron's top would have been close to 2.09 Ma.

Bonnie Blackwell was part of the original Senèze field team and collected a range of data in order to undertake ESR analysis of the age of the site. A preliminary abstract (Mei et al. 2001) reported results from a blind test of the ESR method on teeth which did not in fact come from Senèze. Little further work was done with the Senèze data, in part because it became necessary to include U-series analyses, which was growing more common as time passed. Finally, Blackwell et al. (2020) published another abstract before her untimely death, providing some analytical details: TL dosimetry, γ (gamma) spectrometry and neutron activation analyses measured the sedimentary dosimetry. Sedimentary water concentrations ranged from 11 to 36% by weight. Volumetrically and time-averaged sedimentary and time-averaged cosmic dose rates ranged from 1.5 to 2.0 mGy/yr above and below the Trench 2 layer yielding *Dicerorhinus etruscus* (see below), but up to 3.0–4.0 inside it. For each tooth studied, 5–8 subsamples were independently dated by standard ESR, while U uptake rates were analyzed by isochron analyses. Enamel U concentrations averaged 3–6 ppm, dentine 70–100 ppm, and bone 60–80 ppm depending on the tooth and subsample. Accumulated doses ranged from 2799 to 4721 Grays. Dates for two teeth suggested that bioturbation had mixed younger fauna into the Villafranchian deposits but preliminary dates for *D. etruscus* teeth were 1.5–2.1 Ma, assuming p = 4–8. It is likely, however, that the two teeth which indicated a possibly younger age may have been blind test samples from other localities. Moreover, the subsequent ESR/U-series analyses reported here in the Appendix demonstrated that the assumed value of p (linked to a recent uptake of uranium) was unlikely thus vitiating the date obtained.

Nomade et al. (2014) reported a number of $^{40}Ar/^{39}Ar$ ages from several French Villafranchian faunal sites, including five from Senèze. These tephra derived from the Mont-Dore stratovolcano (see Pastre 2024). Tephra samples SEN 1 (dated 2.13 ± 0.02 Ma) and SEN 8 (2.17 ± 0.02 Ma) derived from levels below the main faunal horizons in parcel 233 (see below and also Parenti et al. (2024), Debard (2024) and Pastre (2024). SEN 56 (2.09 ± 0.01 Ma) was collected in parcel 234 but is not connected directly to any faunal horizon. SEN 98 (2.21 ± 0.02 Ma) and SEN 101 (2.16 ± 0.02 Ma) were collected in the large trench excavated in parcel 172 and were said to bracket a major fossiliferous level. Details on these samples are presented below, but in brief, one faunal level in parcel 172 appeared to be roughly 2.18 Ma while one in parcel 233 was younger than 2.13 Ma. The overall implication was that Senèze mammals dated between 2.20 and 2.10 Ma. The older Chilhac fauna was dated ca. 2.37 Ma, suggesting that the still older Saint-Vallier assemblage might be at least 2.40 Ma.

Paquette et al. (2021) reported several U–Pb dates on zircon from a number of French Villafranchian sites (see Crégut-Bonnoure et al. 2024), including one of 2.100 ± 0.029 Ma from Senèze. The location given for that sample does not agree with the local geography, but communication with one of the authors (J.-L. Poidevin) revealed that the sample in fact came from our "section A" (see below), which also produced the argon date on sample SEN 56. This level appears to be stratigraphically higher than any recovered fossils.

In addition to collecting tephra samples for argon dating, the team invited Sevket Sen to sample the deposits for evidence of magnetozones at Senèze. The results of this work have not yet been published in detail, although Delson et al. (2006) and Nomade et al. (2014) noted that Sen suggested a short interval of normal polarity occurred high in the section, above the SEN 1 tephra.

In this chapter, Nomade and Delson provide recalculated argon ages based on the most recent standards (Schaen et al. 2021), Delson and Sen interpret paleomagnetic data from five separate sequences, and we place these results into the stratigraphic context provided by Debard (2024). Combining these sources of data, we offer an overall interpretation of the age of the Senèze deposits and faunal horizons. In addition, Bahain, Shao and Falguères discuss (in the Appendix) their unsuccessful attempt to obtain a comparable date using an ESR/U-series approach.

Stratigraphic Context

In order to place the dated samples and paleomagnetic "columns" in the stratigraphic context of Senèze, it is important to briefly consider the positions of the sections yielding the samples and then examine each in detail. Parenti et al. (2024, Fig. 2.3) provide a plan of all the major excavated areas and some additional sections. Our team's work at Senèze can be divided into two main geographic sectors, the western sector with excavations and sections mainly located in cadastral parcels 233 and 234 (these are official land divisions which were passed down through family ownership, see Faure et al. 2024) and the southeastern sector mainly in parcel 172. Because these parcels were used by their owners for pasturage, all excavations and trenches were refilled at the end of each season's work.

In the western sector, trenches 1 and 2 were excavated in 2001 and sampled for magnetostratigraphy and tephra (see Debard 2024, Figs. 3.2 and 3.3 for more detail). Samples SEN 1 and SEN 8 were recovered from trench 2 near the team's first major paleontological find, the partial skeleton of *Dicerorhinus etruscus* (see Guérin 2024). Debard (2024) has described the sediments in these sections (and others nearby) and defined small- and medium-scale units in each, which have been correlated as shown in Table 5.1. The rhino skeleton derived from layer T2hm, and sample SEN 1 was located slightly below it. The upper part of trench 2 (layers T2h–T2j) had slipped downslope so that it now lies at the same altitude as the lower part (layers T2a–T2g), with a small fault between the two parts. Sample SEN 8 derived from one of the stratigraphically lowest levels of this trench (T2b). Note that the tephra samples were extracted from sedimentological samples named S1, S8, etc. on the diagrams. A small cliff face termed section A at the southern end of parcel 234 (near the trailer used as an office) was studied in 2003 and sampled for sedimentology and tephra; sample SEN 56 was recovered near the middle of this section (Debard 2024, Fig. 3.8). The whole of Section A also slipped downslope from its original higher position in the maar, so that this tephra was high in the overall sequence of the site. Also in 2003, an area slightly south of the original trench 2 (termed Zone [H8-N9-L14-G11], hereafter "Zone H8") was excavated. In 2004, a section in Zone H8 was reopened and sampled for magnetostratigraphy (Debard 2024, Fig. 3.6), as was a pit dug for that purpose at the northern edge of parcel 233 (Debard 2024, Fig. 3.7). Thus, three dated tephra and four sets of magnetostratigraphic samples ("columns") derived from the western sector, as summarized in Fig. 5.1. SEN 8 is the stratigraphically lowest (and oldest) sample, followed up-section by several cervid fossils, SEN 1, a partial cervid skeleton, the rhino and other fossils, and then SEN 56 above.

In the southeastern sector, trench 5 was opened in 2004 over a depth of more than 11 m in two segments with a step

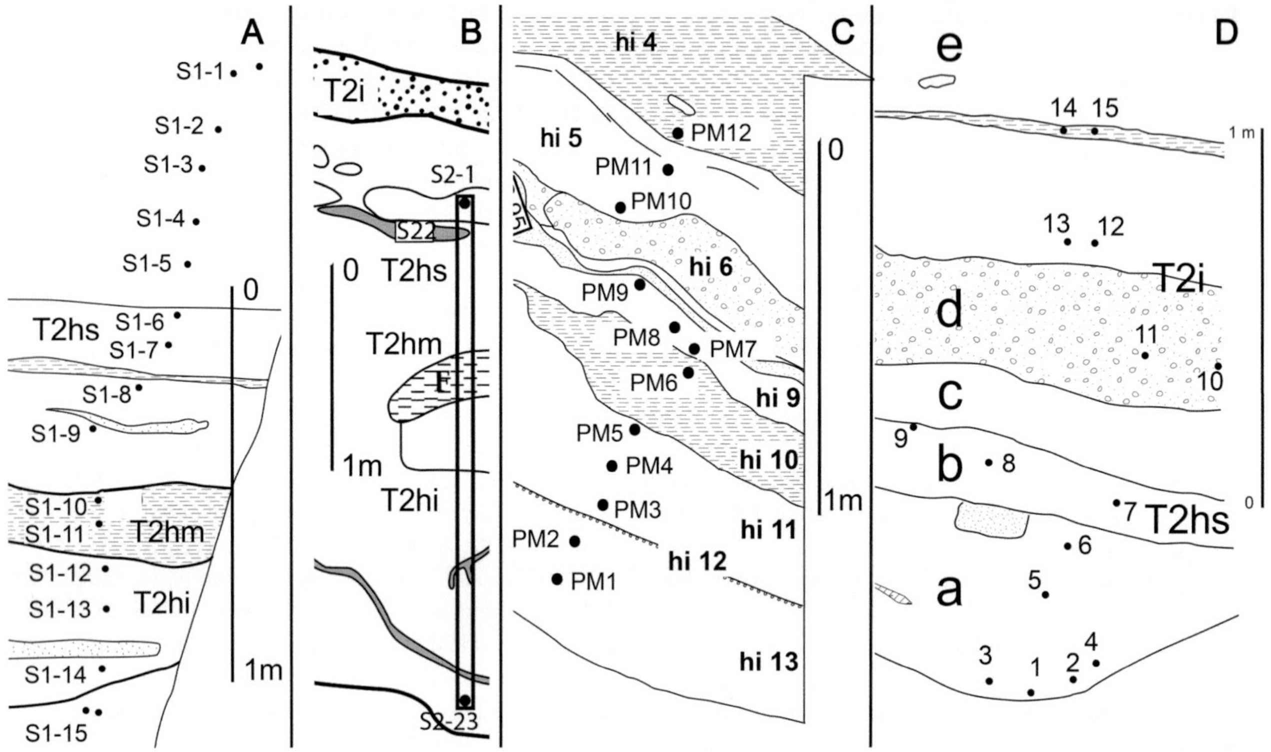

Fig. 5.1 Close-up views of paleomagnetic sampling positions in western sector sections. Scale bars = 1 m. Trench 2 horizon equivalents (T2h, T2i) in A and D approximate, after correlations in Debard (2024). A, Trench 1 (2001), from Debard (2024), Fig. 3.2; paleomagnetic samples S1-1–S1-15. B, Trench 2 (2001), from Debard (2024), Fig. 3.3; T2hm is the sandy level with the *Dicerorhinus* fossil; paleomagnetic samples S2-1–S2-23 shown as a column with intermediate samples not individually labeled; approximate elevation of argon dating sample SEN 1 just above letters "T2hi", although the tephra is actually located about 3 m to the right; sample SEN 8 located below the base of the figure (see Fig. 3.3). C, Zone H8 (2004), from Debard (2024), Fig. 3.6; hi horizons equivalent to those of Trench 2 (T2hi); paleomagnetic samples PM1–PM12. D, Pit at north end of P 233 (2004), from Debard (2024), Fig. 3.7; layers a-c correspond to T2h, while layers d-e correlate with T2i; paleomagnetic samples 1–15 (tabulated as T2-1 through T2-15 in Table 5.4). Note that the SEN 56 tephra is located in section A, not among those shown here; see Fig. 3.8

Table 5.1 Stratigraphic correlations of the western sector (parcels 233 and 234; trenches 1–2 and nearby)

SSet	Trench 1	Trench 2	Zone H8-N9-L14-G11	Pit N P233	Section A
6′	T1i				
5′	T1h	T2j			Upper set
4′	T1g				[U/Pb 2.1 + 0.029][1]
3′	T1f T1e	T2i		d-e [T2i; PM 10–15]	Upper set [SEN 56] (2.07 ± 0.02 Ma)
2′	T1d (PM S1-1–9) T1c (PM S1-10–11) T1b (cervid, PM S1-12–14)	T2hs T2hs (PM S2-1–7) T2hm (fossils, incl rhino; PM S2-8–10) T2hi (SEN 1 T2hi 2, 2.10 ± 0.03 Ma; PM S2-11–23)	[T2]hs [T2]hm [T2]hi (4: PM 12; 5: 10–11 cervid; 9: 7–9; 10: 6; 11: 3–5; 13: 1–2)	b-c [T2hs; PM 7–9] a [T2hs; PM 4–6] a [T2hs; PM 1–3]	Central set
1′	T1a (PM S1-15)	T2g –f [older level in place:] T2e-b (PM S2B-2B-3A) T2b (SEN 8, 2.16 ± 0.05 Ma) T2a (PM S2-B1B)			Lower set

Notes Modified after Debard (2024), Table 3.1; SEN # = argon dated tephra; PM # = paleomagnetic sample numbers; in the column for Zone H8-N8-L14-G11, data for [T2]hi are presented only briefly in the table—for example "4: PM 12" means "sample PM 12 is found in layer hi4", "5: 10-11 cervid" means "samples PM 10 and 11 and cervid fossils are found in layer hi5"

[1]Zircon date on unknown level of Section A from Paquette et al. (2021).

in between, total length ca 25 m. Sets of paleomagnetic samples were taken in each segment, and tephra samples SEN 98 and SEN 101 were collected, although due to a numbering change, these tephra samples do not correspond to sedimentological samples S98 and S101 (Fig. 5.2; see Debard 2024, Figs. 3.11 and 3.12 for more detail). In 2005, an area equivalent to lower trench 5 was reopened as trench 6, and the greater part of an equid skeleton was recovered (see Eisenmann and Delson 2024). Another partial equid skeleton was found in 2006 in the nearby trench 7. Correlations among the samples and equids is complex, but there is clear overlap between the two argon-dated samples and the magnetostratigraphic "columns". Table 5.2 is an attempt to show such correlations. Broadly, the 2005 *Allohippus* skeleton was found in layer 1.2, near the base of the sequence, followed upsection by additional fossils including two partial cervid crania recovered in 2004 and the 2006 *Allohippus* skeleton, with the SEN 98 and SEN 101 tephra above these.

Following Debard (2024), no direct correlation is possible between the two sectors, but the argon dating results reveal broad similarity in age.

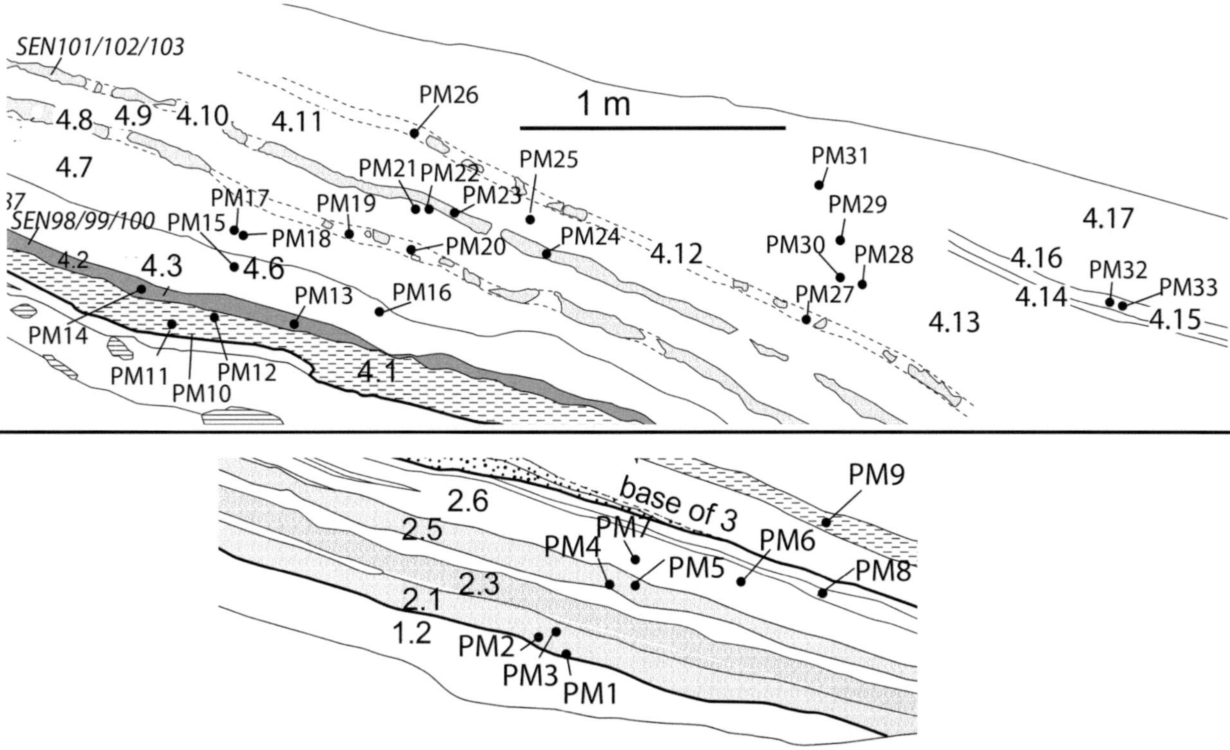

Fig. 5.2 Close-up views of paleomagnetic sampling positions in southeastern sector Trench 5. Above, Trench 5 lower part (inf) with layers 4.1–4.17 and paleomagnetic samples PM10-PM33; the positions of tephra samples SEN 98 and SEN 101 (each as part of a group of nearby samples) are shown at left. Below, Trench 5 upper (sup, stratigraphically below T5 inf), with layers 1.2–3 and paleomagnetic samples PM1-PM9. The 1 m scale bar is accurate for both parts

Table 5.2 Stratigraphic correlations of the southeastern sector (parcel 172; trenches 5–7)

Set	Trench 5 sup	Trench 5 inf	Trench 6	Trench 7 sup	Trench 7 inf
6					
5 (slope)			5 (gravels on erosional surface)		5 (slope, slippage at base?)
4 (lacustrine with some slope)		4.16–4.1	4.20–4.7	("step" between T7 sup & inf)	4.20–4.8
		(PM T5 10–31 in 4.13–4.1)		4.13, 4.12, 4.9	
		(4.10 SEN 101, 2.13 ± 0.04 Ma)		(downslope part only)	
		(4.2 SEN 98, 2.18 ± 0.03 Ma)		4.8–4.1	
		(set 4 visible and overlies set 1 in downslope part only)		(set 4 overlies set 1)	
3 (slope)	3	3		3	
	(PM T5 1–9)			(antlers in conglomerate)	
	Fossils at base			(*Allohippus* skeleton 2006)	
2 (slope)	2.6–2.1	2.6–2.5		2b (fossils) Lake shore	
		2.4 (fossils)		2a	
		2.3–2.1			
1 (slope)		1.3–1.4		1b	
		1.2			
		(*Allohippus* skeleton 2005)		1a	
		1			

Notes SEN # = argon dated tephra; PM # = paleomagnetic sample numbers

$^{40}Ar/^{39}Ar$ Dating

Samples SEN 1, SEN 8, SEN 56, SEN 98 and SEN 101 were analyzed at the $^{40}Ar/^{39}Ar$ facility hosted at the LSCE (Gif-sur-Yvette, France). In order to check the accuracy of these measurements and test possible inter-laboratory bias, SEN 1 was also analyzed at the Berkeley Geochronology Center (USA). In both laboratories, pristine sanidine crystals ranging from 160 to 400°m in size were handpicked from a K-feldspars concentration and slightly leached for 5 min in a 5% HF acid solution. A total of 25–40 grains for each sample were loaded in aluminum disks along with standard reference minerals and irradiated in a nuclear reactor. All basic data from both labs were reported in Supplementary online Tables T1 and T2 of Nomade et al. (2014) and are not repeated here.

Gif sur Yvette Laboratory Analytical Method

Samples were irradiated for 90 (Irr. 1b, SEN 1, SEN 8 and SEN 56) and 60 min (Irr. 20, SEN 98 and SEN 101) in the B1 tube of the OSIRIS reactor (CEA Saclay, France). After irradiation, samples were transferred to a stainless steel sample holder and then into a differential vacuum Cleartran© hublot. Single grains were fused (with 9% of the laser power) using a focused 25 W CO_2 laser (Synrad©). Ar isotopes were analyzed using a micromass 5400 mass spectrometer equipped with a single ion counter (Balzers© SEV 217 SEN). Depending on the xenocrystic contamination, 7 to 15 sanidine single crystals were individually dated for each sample. Neutron fluence (J) was monitored by co-irradiation of Alder Creek Sanidine (ACs-2, Nomade et al. 2005) or Fish Canyon Sanidine (FCs, Renne et al. 1998) placed in the same pit as the sample or in three positions around the sample. The J value for each sample was determined from fusion analyses of 3–6 ACs-2 or FCs single grains. Corresponding J values were calculated using an age of 1.193 Ma (Nomade et al. 2005) and 28.02 Ma (Renne et al. 1998) for ACs-2 and FCs respectively, as well as the total decay constant of Steiger and Jäger (1977). J uncertainty is calculated using the standard deviation of all measurements and excludes age uncertainty on the standard age as well as the decay constant. Procedural blanks were measured every two or three grains depending on the unknown bean size previously measured. For example with a 9 min static blank, typical backgrounds are about 2.0×10^{-17} and 5.0×10^{-19} mol for ^{40}Ar and ^{36}Ar, respectively. Each Ar isotope measurement consists of 20 cycles by magnetic switching of the argon spectrum and baselines. The precision and accuracy of the mass discrimination correction was monitored by periodic measurements of air argon. This monitoring is performed using a dedicated air calibration system featuring a 6 l tank filled with purified atmospheric argon. This tank is connected to the mass spectrometer through the laser extraction system via two pneumatically-actuated air pipettes of approximately 0.1 and 1.0 cc. This system allows for 1 cc (e.g., ~600,000 counts/second [cps] on ^{40}Ar) and 0.1 cc (e.g., ~70,000 cps on ^{40}Ar) atmospheric aliquots to be delivered into the mass spectrometer and permits a careful monitoring of the mass discrimination over a wide dynamic range with a precision better than 0.2% (2σ)for any given bean size measured (see Nomade et al. 2010 for more details). This level of precision allows the user (in theory) to measure with confidence samples with <1% $^{40}Ar^*$ (* indicates radiogenic argon). As the measurements were done during two periods in 2008 and 2010, two discrimination calibration curves (see Nomade et al. 2010 for a calibration curve) were used to correct measurements.

Berkeley Geochronology Center Analytical Method

The SEN 1 aliquot was irradiated for 30 min in the CLICIT facility of the TRIGA reactor at Oregon State University. After irradiation, 34 single grains of sanidine were fused with a defocused argon laser (Synrad) then analyzed with a MAP 215C mass spectrometer at the Berkeley Geochronology Center using methods described by Renne et al. (1998). Procedural blanks were measured between every three grains, and typical values were $3.3 \pm 0.5 \times 10^{-16}$ for ^{40}Ar and $2.26 \pm 0.3 \times 10^{-18}$ for ^{36}Ar moles. Neutron fluence (J) was monitored by Alder Creek sanidine (ACs-2, 1.193 Ma, Nomade et al. 2005) grains in four positions in the Al disk. The J value for each disk was determined from individual analyses of 6 separate ACs-2 crystals in each position fused with the CO_2 laser (total of 24 grains) per disk. The corresponding J values and associated uncertainties were calculated with the total decay constant of Steiger and Jäger (1977) and exclude age uncertainties on the ACs-2 age and the decay constant. Mass discrimination was monitored by automated analysis air pipettes, interspersed with the unknowns.

Recalculation of Ages

Since the dates were published (Nomade et al. 2014), revised calibrations were recommended by the Ar/Ar community for the ACs-2 flux standard (Jicha et al. 2016; Niespolo et al. 2017; Schaen et al. 2021) as well as the decay constant (Min et al. 2000; Renne et al. 2011). Overall these new calibrations insure a better accuracy for the $^{40}Ar/^{39}Ar$ ages. The use of these new calibrations affects the weighted mean age calculations made previously by <1%, therefore well within the uncertainties reported at the time. However, and to comply with these new recommendations, hereafter all weighted mean ages are recalculated according to ACs-2 at 1.1851 Ma (Schaen et al. 2021) and using the K total decay constant proposed by Min et al. (2000). The recommended revised age for FCs is 28.21 Ma, indistinguishable from the value used in several runs at Gif, and thus these dates are not modified.

$^{40}Ar/^{39}Ar$ Results

Full analytical details for individual crystal experiments corrected for backgrounds and radioactive decay were reported in Supplementary online Tables T1 and T2 of Nomade et al. (2014). Age-probability density spectra with individual single crystal age (2 σ) for each ash layer are presented in Fig. 5.3. Detailed results for each tephra layer are discussed below and summarized in Table 5.3. Uncertainties are reported below at the 2 σ level (corresponding to a 95% probability of accuracy) including J value uncertainties and the K decay constant.

Table 5.3 Summary of $^{40}Ar/^{39}Ar$ dated samples from Senèze

Sample	Lab	n/N	Age[a] (weighted mean) (ka)	Error (2 s) internal (ka)	Error (2 s) external (ka)	MSWD	Isochron age[a] (ka)	Error (2 s) (ka)	$^{40}Ar/^{36}Ar$ in	Error (2 s)	MSWD	"Best age" (Ma)	Error (2 s) (Ma)
SEN 1	GIF (France)	5//7	2119	24	49	1.0	2080	76	310	23	0.8	N/A	
SEN 1	BGC (USA)	18//34	2100	13	26	0.5	2100	31	316	42	0.5	N/A	
SEN 1 combined		27//41	2110	13	26	1.0						2.10	0.03
SEN 8	GIF (France)	15//15	2159	22	45	1.7	2163	45	291	4	1.2	2.16	0.05
SEN 56	GIF (France)	10//10	2066	9	21	1.4	2074	21	292	8	0.8	2.07	0.02
SEN 98	GIF (France)	9//9	2177	16	32	1.0	2177	33	296	9	1.2	2.18	0.03
SEN 101	GIF (France)	9//12	2133	14	42	0.5	2130	44	311	24	0.3	2.13	0.04

Notes [a]Ages recalculated according to ACs-2 at 1.1851 Ma (Schaen et al. 2021) and K decay constant from Min et al. (2000)
Basic data were reported in Supplementary online Tables T1 and T2 of Nomade et al. (2014)
n/N = number of grains used to calculate age/total number of grains analyzed. MSWD = Mean Square Weighted Deviation

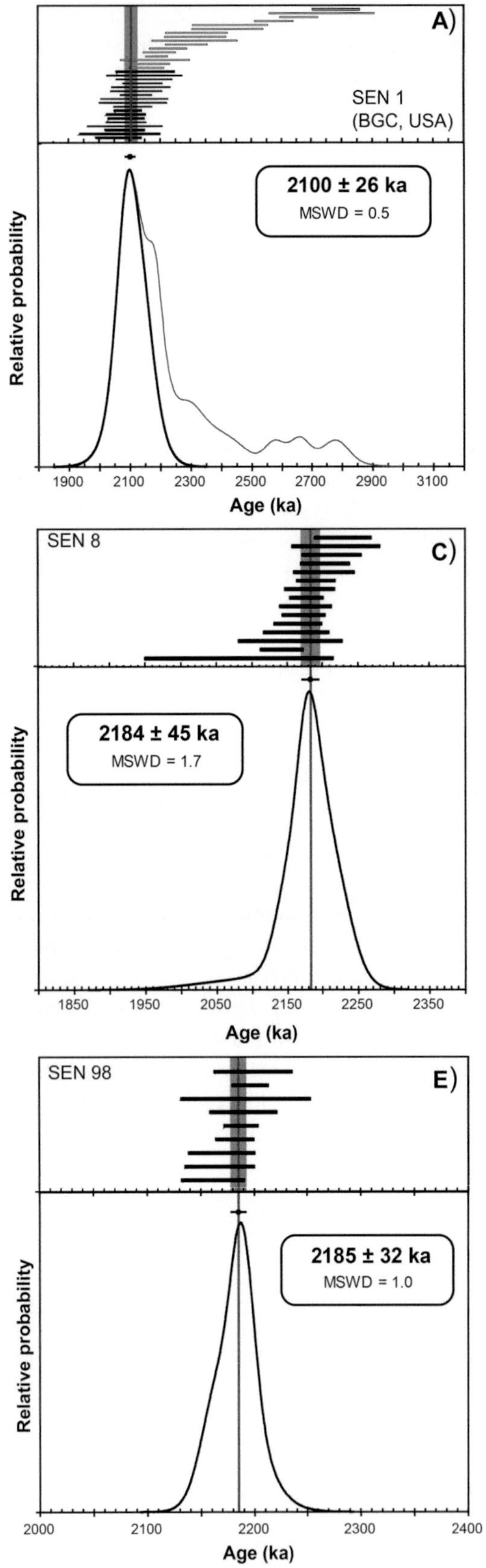

Fig. 5.3 Age probability diagrams of the $^{40}Ar/^{39}Ar$ experiments on Senèze tephra. Black boxes are juvenile crystals whereas white ones correspond to xenocrystals and were not used for the age calculation. Full external uncertainties (1 sigma) are indicated for each weighted mean and isochron ages. A and B indicate results for the US and French analyses on SEN 1, respectively. The dates shown were not recalculated, but those in the text and Table 5.3 were

SEN 1. A total of 7 and 34 sanidine crystals were individually analyzed at Gif and the BGC, respectively, for this ash layer. The corresponding probability spectra are complex with a strong asymmetry toward older ages (Fig. 5.3A, B). This complex distribution can be due to partial incorporation of extraneous argon and/or the presence of xenocrysts from older eruptive events incorporated during the explosive phase. Despite this complex behavior, probability spectra display a dominant mode including the majority of the analyzed sanidine allowing the calculation of robust eruptive ages of 2.10 ± 0.03 Ma (n = 18/34) and 2.12 ± 0.05 Ma (n = 5/7) for the Berkeley and Gif experiments, respectively. Because all crystals analyzed belong to the same tephra layer, a mean age of 2.11 ± 0.03 Ma (n = 27/41) is calculated by combining all single crystal ages. Because of the relatively low spread for each laboratory, the calculated $^{40}Ar/^{36}Ar$ initial ratios have large uncertainties and values above 310 (ca. 310 ± 23 and 316 ± 42) but suggest within error an atmospheric ratio (Table 5.3).

SEN 8. Fifteen sanidine crystals were analyzed for this ash layer. The age probability spectrum exhibits a unimodal pattern leading to a robust eruptive age of 2.16 ± 0.05 Ma (n = 15/15) with a MSWD of 1.7 (Fig. 5.3C) for this tephra layer. The inverse isochron yields a precise initial $^{40}Ar/^{36}Ar$ ratio. The age calculated from this isochron is 2.16 ± 0.05 Ma (MSWD = 1.2), identical to the weighted mean age calculated using all the single crystal ages.

SEN 56. Ten sanidine crystals extracted from this tephra layer were dated, and all of them share the same age within uncertainties, resulting in an almost Gaussian probability diagram allowing us to calculate a robust weighted mean age of 2.07 ± 0.02 Ma (MSWD = 1.4, n = 10/10) (Fig. 5.3D). The corresponding inverse isochron displays a relatively low spread but allows calculation of an age of 2.07 ± 0.02 Ma (MSWD = 0.8) and an atmospheric initial $^{40}Ar/^{36}Ar$ intercept of 292 ± 8.

SEN 98. Nine crystals were analyzed for this tephra horizon. The sanidine population probability spectrum exhibits a simple almost Gaussian distribution allowing calculation of a straightforward mean age of 2.18 ± 0.03 Ma including all analyzed single crystals (n = 9/9) (MSWD = 1) (Fig. 5.3E). The corresponding inverse isochron calculated with this sanidine population displays a nice spread with an initial $^{40}Ar/^{36}Ar$ ratio of 296 ± 9, identical with an atmospheric ratio. The age of 2.18 ± 0.03 Ma (MSWD = 1.2) is identical to the weighted mean age calculated using all the single crystal ages.

SEN 101. Twelve sanidine crystals were analyzed. The probability spectra obtained is here bimodal with a main mode corresponding to 9 crystals. The secondary older peak is probably the result of contamination by older xenocrysts. The main mode allows us to calculate a weighted mean age of 2.13 ± 0.04 Ma (n = 9/12) with a MSWD of 0.5 (Fig. 5.3F). The corresponding inverse isochron displays an initial $^{40}Ar/^{36}Ar$ ratio of 311 ± 24, identical within error to the atmospheric ratio. The age calculated from this isochron is 2.13 ± 0.04 Ma (MSWD = 0.3), in agreement with the weighted mean age calculated using the primary crystal population.

Magnetostratigraphic Analysis

Magnetostratigraphic Sampling

In 2001, S. Sen sampled trenches 1 and 2 for paleomagnetic analysis. The apparently older Jeune Pireyre volcanoclastic deposit was also sampled, but these deposits are unrelated to the mammalian fossils and entirely reversed, so they are not further discussed here. All samples were drilled and the cores cut in the laboratory to obtain cylindrical blocks of standard volume: diameter 25 mm, height 22 mm. Analyses were performed by S.S. in the Laboratoire de Paléomagnetisme of IPG, Paris, using a three-axis 2G cryogenic magnetometer. Bulk susceptibility was measured at room temperature, and samples were thermally demagnetized from room temperature (20 °C) to 350 or 400 °C in 50° increments; a selection of orthogonal projection (Zijderveld) diagrams for 9 samples from the upper part of the Trench 2 section are presented in Fig. 5.4. Fourteen samples were analyzed from trench 1, and a soft secondary component of magnetization was removed before 250 °C, after which there is a stable component. Twenty-two samples were analyzed from trench 2, with a soft component removed by 200° C. Three additional samples were taken in the block which enclosed dating sample SEN 8. Basic data are provided in Table 5.4, and positions of samples are shown above in Fig. 5.1.

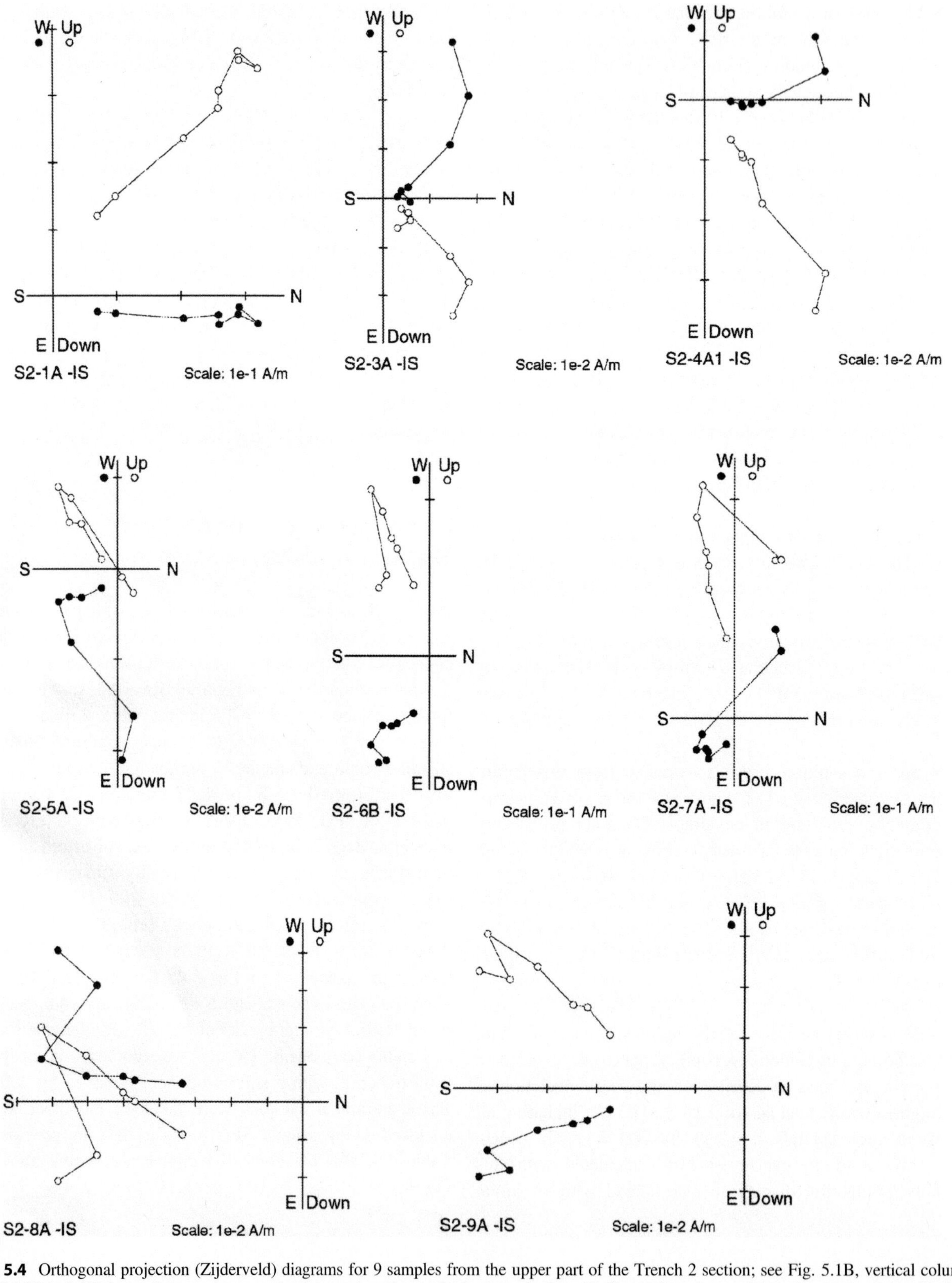

Fig. 5.4 Orthogonal projection (Zijderveld) diagrams for 9 samples from the upper part of the Trench 2 section; see Fig. 5.1B, vertical column headed S2-1. Solid data points are vector end points projected onto the horizontal plane (declination), and open data points are vector end points projected onto a north–south oriented vertical plane (inclination). Thermal demagnetization proceeded from 20° C (room temperature) to 50° and thence to 350° or 400° in 50° steps

Table 5.4 Data for paleomagnetic samples from Senèze "sections"

Location	Sample no.	Height in cm	Declination	Inclination	VGP lat	Polarity
Trench 1	S1-15A	154	172	−61	−83.5	R
(2001)	S1-14A	143	180	−22	−56.4	R
Parcel 234	S1-13A2	128	200	−20	−51.3	R
	S1-12B	115	178	−30	−61.1	R
	S1-11A	107	200	−47	−66.9	R
	S1-10A	100	158	−36	−59.1	R
	S1-9A	83	152	−53	−65.7	R
	S1-8A	72	190	−62	−82.6	R
	S1-7A	65	175	−60	−84.5	R
	S1-6A	57	202	−40	−61.4	R
	S1-5A	47	185	−34	−63.3	R
	S1-4B	37	245	−20	−24.9	R
	S1-2A	15	330	−32	−21.9	R
	S1-1B	0	152	−35	−55.4	R
Trench 2	S2-1A	240	7	−49	14.8	N
(2001)	S2-3A	210	338	30	55.7	N
Parcel 233	S2-4A	199	7	54	78.2	N
	S2-5A	194	150	−45	−59.8	R
	S2-6B	182	120	−57	−45.4	R
	S2-7A	176	140	−76	−61	R
	S2-8A	165	187	−14	−51.6	R
	S2-9A	160	169	−29	−59.1	R
	S2-10A	153	172	−45	−70.5	R
	S2-11A	138	148	−8	−40.4	R
	S2-12A	126	155	−35	−57	R
	S2-13A	117	75	−50	−11.8	R
	S2-14A	108	185	-43	−69.6	R
	S2-15A	96	160	−39	−61.8	R
	S2-16A	84	150	−50	−62.6	R
	S2-17A	73	180	−50	−75.8	R
	S2-18A	61	182	−53	−78.5	R
	S2-19A	49	185	−61	−85.3	R
	S2-20A	41	195	−49	−70.8	R
	S2-21A	29	175	−52	−77	R
	S2-22A	10	181	−57	−82.6	R
	S2-23A	0	151	−47	−61.5	R
Trench 2						
Below fault	S2-B3A	33	175.7	−73.1	−76.2	R
	S2-B2B	12	181	−57.5	−82.9	R
	S2-B1B	0	153.7	−41.4	−59.8	R
Zone H8	P12	140	145	−30	−49.4	R
(2004)	P11 B	125	180	−50	−75.8	R
Parcel 233	P10 A	115	175	−50	−75.3	R
	P9	95	175	−42	−68.8	R
	P8	89	175	−60	−84.5	R
	P7	77	175	−40	−67.4	R
	P6	71	155	−15	−46.8	R
	P5	51	160	−40	−62.4	R
	P4	36	155	−38	−58.7	R
	P3	24	170	−52	−75.4	R
	P2	12	170	−50	−73.8	R
	P1	0	170	−45	−69.9	R
Pit N P233	T2-15	149	100	−45	−25.2	R
(2004)	T2-14	149	215	18	−27.3	R

(continued)

Table 5.4 (continued)

Location	Sample no.	Height in cm	Declination	Inclination	VGP lat	Polarity
Parcel 233	T2-13	124	165	−60	−78.3	R
	T2-12	124	140	−45	−53.2	R
	T2-11	94	260	−10	−10.6	R
	T2-10	91	263	8	−2.1	R
	T2-9	71	340	69	74.9	N
	T2-8	59	190	−50	−73.8	Trans?
	T2-7	49	0	70	81.1	N
	T2-6	34	340	70	74.3	N
	T2-5	24	350	65	82.8	N
	T2-4	14	320	36	48.6	N
	T2-3	7	188	−39	−66.1	R
	T2-2	4	189	−57	−80	R
	T2-1	0	195	−62	−79.1	R
Trench 5	T5-31	250	184.1	−70.2	−80.4	R
Lower/inf	T5-30	250	60.2	−51.0	−4.2	R
(2004)	T5-29	237	29.3	−30.8	22.9	N
Parcel 172	T5-28	222	105.2	−60	37.1	N
	T5-27	202	87.1	−35.8	−11.9	Trans?
	T5-26A	202	97.2	−52	−27.1	Trans?
	T5-25	192	75.2	−30.2	−1.5	Trans?
	T5-24B	180	64.2	−15.8	11.7	Trans?
	T5-23	178	75.3	−15.4	5.8	Trans?
	T5-22	174	205.3	−25.4	−51.5	R
	T5-21	174	167.5	−17.5	−52.1	R
	T5-20	160	50.00	−64.3	−11.5	Trans?
	T5-19	155	89.60	46.8	19.8	Trans?
	T5-18	145	112.3	−8.60	−18.6	Trans?
	T5-17	145	105.0	3.50	−9.2	Trans?
	T5-16	130	89.40	−60.0	−27.4	R
	T5-15	130	126.0	−58.1	−50.1	R
	T5-13	113	212.2	−71.2	−67.3	R
	T5-12	105	146.2	−55.6	−63	R
	T5-11	105	130.4	−53.2	−50.7	R
	T5-10A	97	169.3	−55.7	−77.9	R
Trench 5	T5-9	77	190.9	−23.2	−55.5	R
Upper/sup	T5-8	47	207.6	−48.9	−63.4	R
(2004)	T5-7	41	196.7	−36.1	−61.2	R
Parcel 172	T5-5	28	187.0	−65.2	−84.8	R
	T5-4B	24	167.0	−41.2	−65.9	R
	T5-3	17	117.2	−55.3	−42.4	R
	T5-2	5	182.3	−60.6	−85.9	R
	T5-1	0	190.4	−70.3	−78.7	R

Notes Height in cm above base of sampled sequence; VGP lat., virtual geomagnetic pole latitude; Polarity, estimated from VGP lat.: R, reversed; N, normal; trans?, possibly transitional

In 2004, S.S. sampled three more areas. Twelve samples were collected in Zone H8 and 15 in the pit dug for this purpose at the north edge of parcel 233 (see Fig. 5.1). In trench 5, nine samples were collected in the upper portion (T5 sup), but one was discarded; 24 were collected in the lower portion (T5 inf), but three were lost. As a result of slippage of the upper layers, T5 inf is in fact stratigraphically higher than T5 sup. Dated tephra SEN 98 and SEN 101 were collected in horizons that had been sampled for paleomag (see Fig. 5.2). Samples were thermally demagnetized from 20 °C up to 650 °C, with increments of 50° after the 150° step. These samples were found to have no reliable secondary magnetization or just a very soft one that is removed before the 200 °C step. Thereafter, a stable component was defined in all samples with a very low angle error. Some samples showed an unstable viscous magnetization behavior after the 500 °C step, when the direction became random. In such samples, the results at steps over 500° were discarded. Basic data are provided in Table 5.4.

Magnetostratigraphic Results

In parcels 233 and 234, four separate but correlated "sections" yielded paleomagnetic signal (Fig. 5.5). Three magnetozones can be distinguished. A reversed zone is recorded in the two upper layers (d and e) of the pit at the northern edge of parcel 233 (Fig. 5.5D), correlated to T2i at the top of Trench 2; there is a gap between layers b and d, with the lowest sample in layer d (T2-10) weakly reversed, as might be expected for a transition to reversed from an underlying normal. Most of the samples from the lower layers a and b (all equivalent to T2hs) are normal, but the lowest three at the base of layer a are reversed. Three higher samples in a and one at the base of b are normal. Above these is a single sample (T2-8) which seems definitely reversed, but perhaps it is transitional or represents a minor "flip" within the magnetozone, and above it is sample T2-9, which is clearly normal. Trench 1 is entirely reversed, spanning most of Layer T2h (T2hs-T2hm-T2hi), but with an isolated sample at its base correlated to T2g (Fig. 5.5A).

In Trench 2 itself (Fig. 5.5B), the highest level sampled for magnetic signal is T2hs, and its upper part is normal. The lower part of T2hs, T2hm and T2hi down to sublayer 9 is all reversed. The SEN 1 tephra dated to 2.12 ± 0.03 Ma derives from the upper part of T2hi. The section in Zone H8 (Fig. 5.5C), correlated to T2hi 4–5 and 9–13, overlaps the lower part of Trench 2 and is similarly reversed. The small segment of Trench 2 in place (T2a-e) enclosing the SEN 8 tephra dated to 2.18 ± 0.05 Ma is also reversed. This results in a simple pattern of reversed above short normal (in layer T2hs) above reversed.

There are two questions about this inferred pattern: In the pit at the north of P233, sample T2-8 seems to be strongly reversed within the normal magnetozone; could this reflect transitional behavior of the field or a minor reversal? Trench 1 spans most or all of the T2hs equivalent where the normal magnetozone occurs in Trench 2 and the pit, but it is all clearly reversed. If the normal magnetozone is a short-lived interval (as we suggest below), perhaps it was not recorded in the sediments of Trench 1 as a result of stochastic processes.

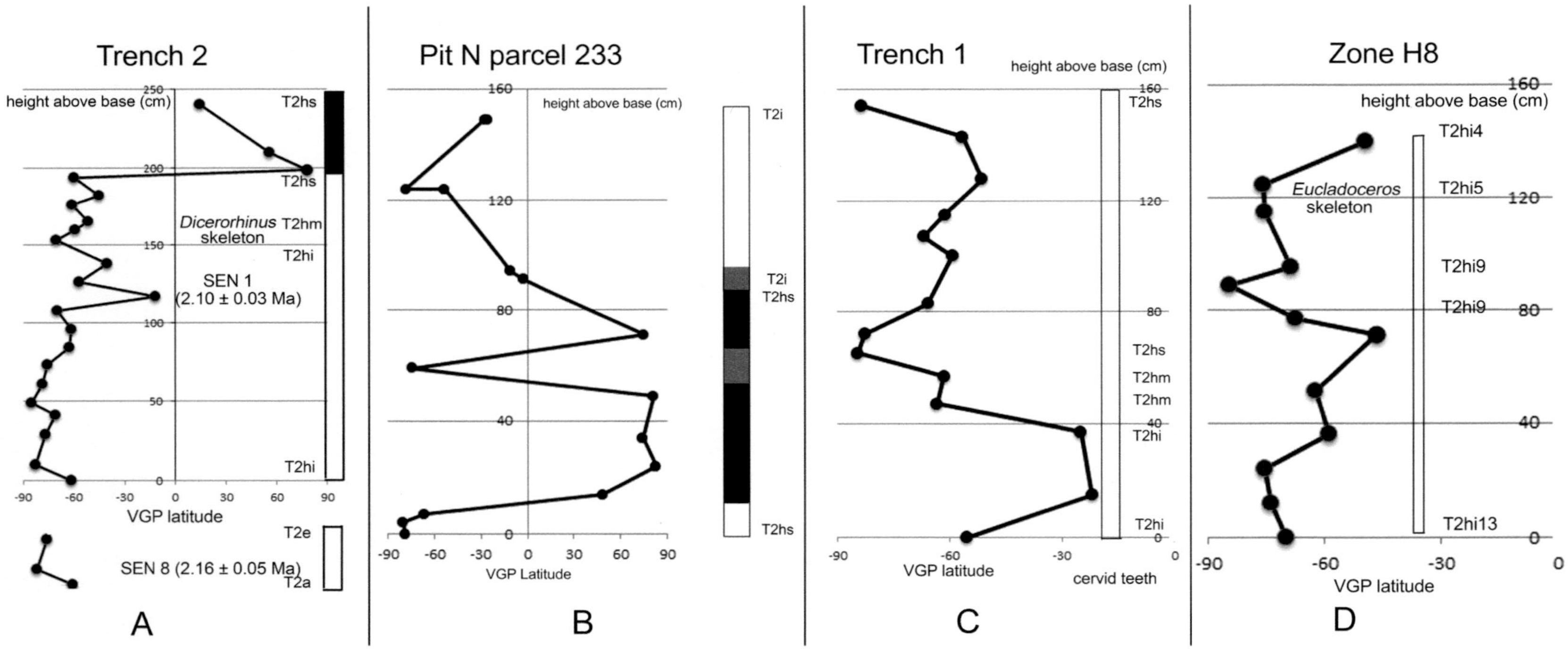

Fig. 5.5 Paleomagnetic "columns" of the western sector with vertical scale, calculated VGP latitudes and correlations of layers to those of Trench 2. A, Trench 2 (2001), normal at top and reversed below, with *Dicerorhinus* skeleton in T2hm and dated tephra SEN 1 high in T2hi; a short reversed paleomagnetic section was sampled in a small block from the base of the trench, enclosing dated tephra SEN 8. B, pit at the north end of parcel 233 (2004), with upper and lower reversed magnetozones bounding normal zone, with strong and weak reversed samples suggesting possible transitional behavior gray). C, Trench 1 (2001), entirely reversed, with cervid teeth recovered below lowest paleomagnetism sample. D, Zone H8 (2004), entirely reversed through most of T2hi equivalent, with cervid skeleton in the upper part

In parcel 172, two segments of Trench 5 yielded signal (Fig. 5.6). The highest two samples in T5 inf were reversed; eleven of the next 13 samples were probably normal, with two apparently reversed samples (T5-21 and especially T5-22) in the middle. The remaining samples in T5 inf and all those in T5 sup (stratigraphically lower) were reversed. A possible interpretation is that the field was reversing during the time represented here, with an early transitional zone, followed by a short fully(?) reversed interval and then a later transitional zone before the field stabilized as normal. The pattern appears broadly similar to that from the western sector, with a short normal and transitional zones between two reversed segments. The dated tephra SEN 101 and SEN 98, however, indicate that the majority of this section is older than what is known from the western sector.

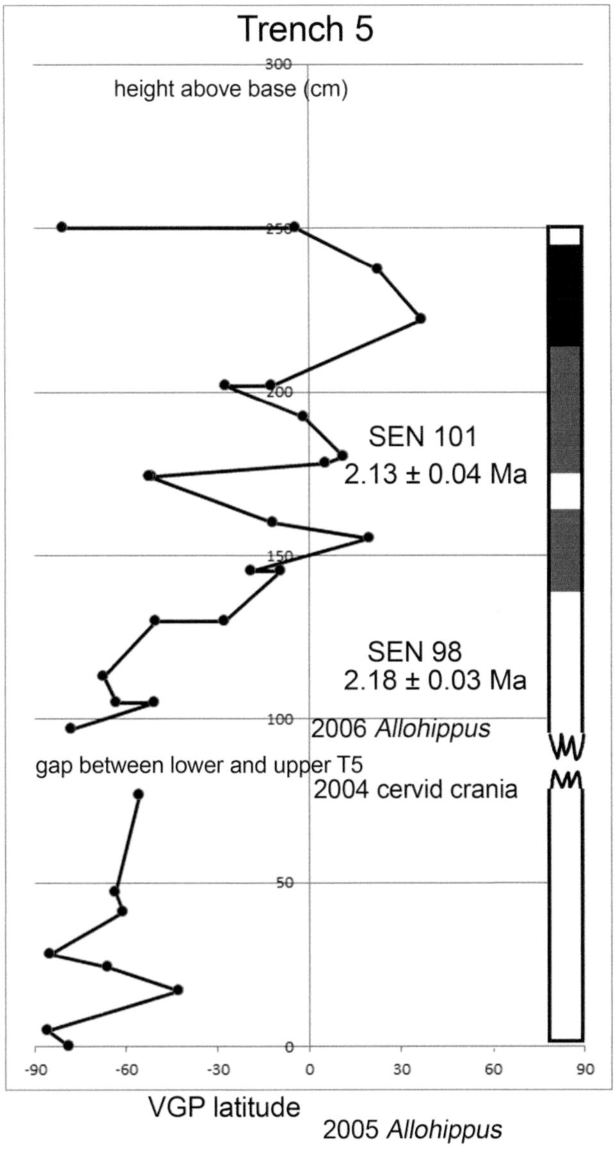

Fig. 5.6 Paleomagnetic "column" of Trench 5 in the southeastern sector (2004) with vertical scale, calculated VGP latitudes and approximate positions of dated tephra and major fossils. The normal magnetozone near the top is preceded by an apparently transitional zone (in gray); the long reversed zone below may be "interrupted" by another transitional zone. Note that the lower segment (inf) of T5 is stratigraphically above the upper segment (sup) due to local dip. The two segments have been plotted together with a gap of 20 cm placed between them; this gap is indicated on the paleomagnetic column by ragged edges

Combined ESR and U-series Analyses on Teeth

A combined Electron Spin Resonance (ESR) and Uranium-series (U-series) approach was also attempted on several teeth from parcels 172 and 233. Unfortunately, the obtained ESR/U-series ages are severely underestimated by comparison with both paleontological and ^{40}Ar/^{39}Ar data, supported by paleomagnetic correlation (see details of the ESR/U-series protocol and results in the Appendix). At Senèze, both relatively early U-uptake history reconstructed from the analytical data and high dose rates linked to the volcano-sedimentary context seem to prohibit the use of the ESR/U-series method.

Integrated Geochronology

Channell et al. (2020) have provided the most up to date review of geomagnetic reversals and excursions in the Early Pleistocene. In brief, for the time frame between 2.20 and 2.00 Ma, where the Senèze ages fall, they interpret data from sedimentary records and argon ages on volcanics to suggest three intervals of normal polarity, with some uncertainty due to complex correlation across great distances. The oldest of these is documented in lava flows on Réunion Island dated by ^{40}Ar/^{39}Ar to 2.19 ± 0.01 Ma (recalculated). This was previously known as the Réunion II (or "older Réunion") subchron, but Channell et al. (2020: 20) term it the Réunion excursion. Slightly younger in the record is a longer normal interval which was previously termed the Réunion I ("later" or "younger" Réunion) subchron, but is now renamed the Feni subchron (Singer 2014: 29; Channell et al. 2020: 20; Chron C2A.n1). Its age is slightly unclear, depending on the deep-sea drill site from which the dates are derived: either 2.14–2.09 or 2.137–2.116 ± 0.005 Ma. Ogg (2020) suggested 2.140–2.116 ± 0.005 Ma, which is accepted here. Finally, an even younger normal excursion, the Huckleberry Ridge, dates to 2.071 ± 0.003 Ma (recalculated); or 2.07 ± 0.01 Ma if typical argon uncertainties are included. With this terminology, Roger et al.'s (2000) date, recalculated to 2.12 ± 0.01 Ma, would have been within the Feni (and near its end, as they thought).

Using the argon dates from Senèze as a framework, the above results indicate that the youngest horizon, well above any fossiliferous level, dates to about 2.07 Ma (sample SEN 56). Unfortunately no paleomagnetic samples were taken nearby, but the level involved is correlated to layer T2i in trench 2, in turn correlated to layer e high in the pit at the northern edge of parcel 233, which has yielded a reversed signal. In that pit and at the top of trench 2 some 30–50 cm above the *Dicerorhinus* skeleton, there is a normal signal which probably equates to the Huckleberry Ridge excursion. It is above SEN 1, which is located about 30–50 cm below the rhino and dated at 2.10 ± 0.03 Ma in a reversed interval. Given this error range, SEN 1 must actually lie between 2.07 and 2.13 Ma, which is too young to fall below the Feni normal Subchron, but it must fit above the Feni and below the Huckleberry Ridge excursion; its real age would lie between 2.116 and 2.070 Ma. Therefore, the normal magnetozone must correlate to the Huckleberry Ridge (Fig. 5.7), despite its approximately 40 cm thickness in Trench 2 (Table 5.4). The complex alternation of normal, reversed and possible transitional geomagnetic intervals in the pit north of 233 makes it difficult to correlate zones precisely but suggests that the main normal segments equate to the Huckleberry Ridge as well.

Lower in the pit and in trenches 1 and 2, the remaining sediments are reversed. Fossils including a *Eucladoceros* skeleton, *Metacervoceros* teeth and other cervid remains were recovered below that dated tephra. The small segment of in situ sediment at the base of trench 2 is also reversed and dated 2.16 ± 0.05 Ma (SEN 8), thus presumably older than the Feni (Fig. 5.7). On the other hand, it is just possible (within error) that this block is younger than the Feni. The parcel 233–234 sequence as a whole (above T2a-T2e) therefore probably dates between about 2.15 and 2.07 Ma, with known fossils postdating the Feni subchron and predating the Huckleberry Ridge excursion; they would thus date between 2.11 and 2.07 Ma, or probably ca. 2.10–2.08 Ma if we allow some time on either side of this interval for the "empty" part of the column.

In parcel 172, as discussed above, the section is reversed at the top (in the upper part of T5 inf, see Fig. 5.8). Lower in T5 inf, there is a normal magnetozone above tephra sample SEN 101, dated 2.13 ± 0.04 Ma. This normal interval would equate to part of the Feni subchron. Below the fully

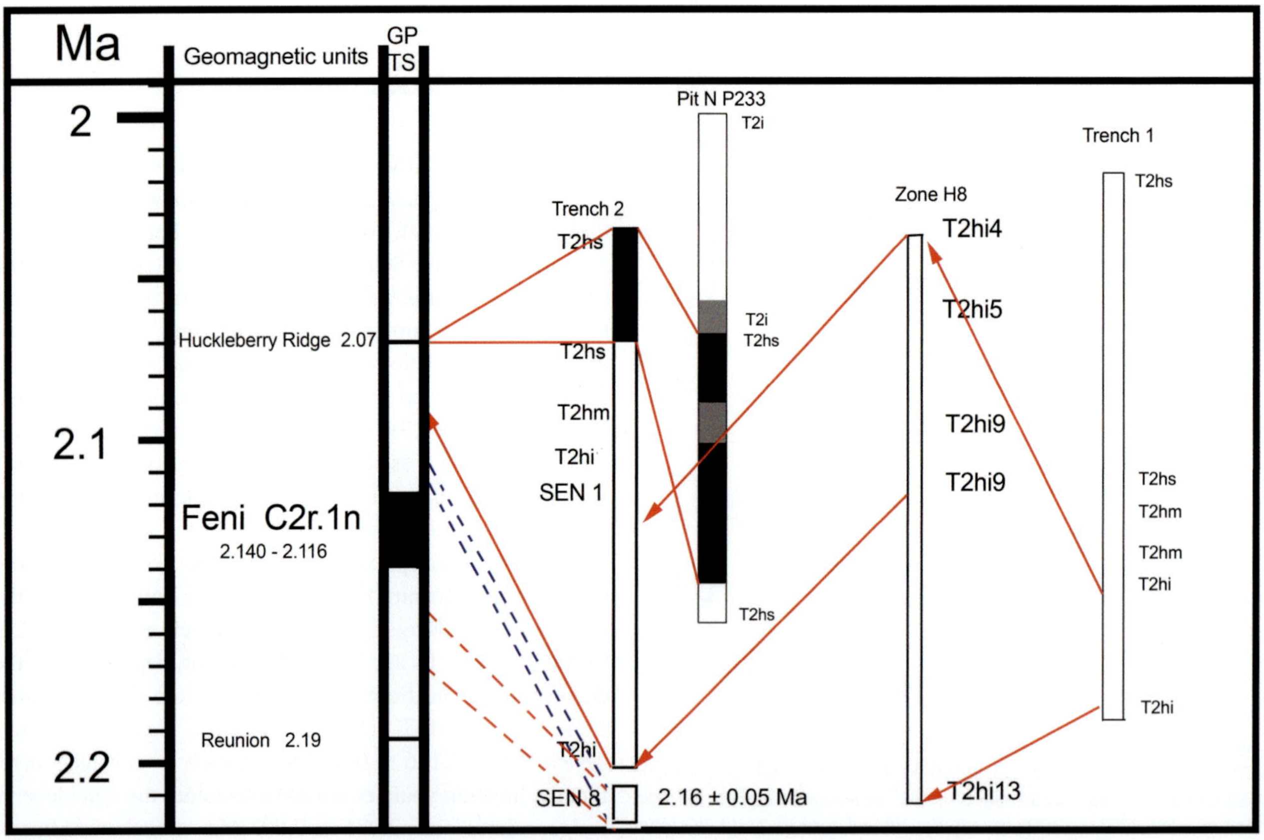

Fig. 5.7 Summary of western sector geochronology, on the Geomagnetic Polarity Time Scale (GPTS) from 2.2 to 2.0 Ma. Paleomagnetic polarity columns from Fig. 5.5, with suggested correlations to the GPTS (and each other). Solid red lines indicate approximate correlations; arrowheads indicate very broad correlation in that area. For lower block of Trench 2, dashed red lines indicate preferred possible correlations, while dashed blue lines represent a less likely option

normal interval, we have suggested that there is a zone of transition from reversed to normal with several questionable (and one clear) reversed samples in the middle (see Fig. 5.6 and Table 5.4). The upper transitional interval, which includes SEN 101, could represent the earlier part of the Feni, which begins ca. 2.14 Ma, continuing into the normal interval. It is possible that this transitional plus normal zone might instead equate to the Huckleberry Ridge excursion if the actual age of SEN 101 were between 2.11 and 2.09 Ma (at the limit of the error range). This seems unlikely, especially given that the upper transitional plus normal magnetozone(s) spans about 0.67 m of sediment, nearly twice the thickness of the Trench 2 normal (see Table 5.4 and Figs. 5.5 and 5.6). Farther down section is the lower transitional interval and then a long reversed magnetozone interpreted to be (part of) the reversed interval (C2r.2r) between the Feni and the (unobserved) Réunion excursion (2.19 Ma).

Near the top of this reversed magnetozone is the SEN 98 tephra, dated 2.18 ± 0.03 Ma. The true age of SEN 98 must lie between 2.21 and 2.15 Ma, thus older than the Feni no matter to which global polarity zone the overlying normal is equated. Even lower in the reversed zone are the two partial *Allohippus* skeketons (and also a partial cervid cranium and other specimens), whose exact ages are unknown but

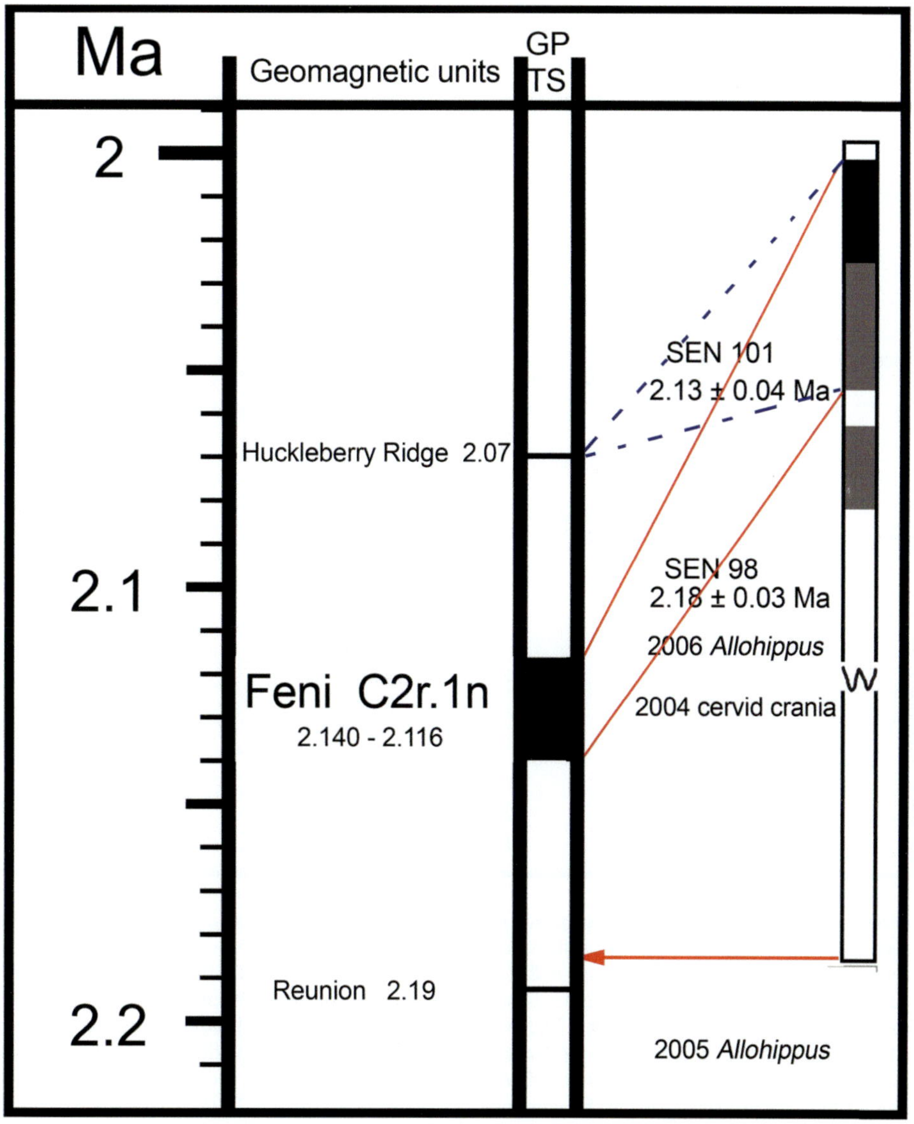

Fig. 5.8 Summary of southeastern sector (parcel 172, Trench 5) geochronology, on the Geomagnetic Polarity Time Scale (GPTS) from 2.2 to 2.0 Ma. Paleomagnetic polarity column, argon-argon ages and major fossil finds from Fig. 5.6, with suggested correlation to GPTS in red; arrowhead indicates a very broad correlation in that area; dashed blue lines indicate possible but unlikely alternative correlation. The upper middle part of the sequence is interpreted as a transition from reversed to normal, with the lower transitional zone part of the reversed interval, the upper transitional zone representing the earlier part of the Feni subchron and the fully normal zone later in the Feni. The gap between the lower (inf) portion of Trench 5 above and the upper (sup) portion below is indicated by the wavy line

probably lie between 2.20 and 2.18 Ma. The earlier suggestion (Delson et al. 2006; Nomade et al. 2014) that the younger of the skeletons was bracketed by the SEN 101 and SEN 98 ages was a misinterpretation of the preliminary analysis. This fossiliferous interval must be somewhat older than the oldest stratigraphic level in parcel 233. The parcel 172 section thus spans from about 2.20–2.11 Ma (possibly 2.08 Ma), with fossils at least between ca. 2.20 and 2.18 Ma, but probably not above this.

Combining data from the two sectors, sediments from the Senèze maar span ca. 2.20–2.07 Ma. Broadly speaking, the Senèze local fauna derives from the time interval between 2.20 and 2.10 Ma.

In our fieldwork, there seem to be two main fossiliferous horizons ca. 2.20–2.18 and 2.10–2.08 Ma (Fig. 5.9). These may represent actual intervals of faunal presence or they may reflect local lenses or clusters of fossils which we recovered, with no fossils located between them. But we don't know if the material recovered by Philis corresponds to these intervals, to the entire time span from 2.20 to 2.08 Ma or an even longer span not sampled by our work. No faunal change can be observed between these two levels, nor is any expected, mainly because there are few fossils involved and few taxa found in both "horizons", but also because the time separating them is only 100 ka. Crégut-Bonnoure (2024) does note some variation in several bovid species within the

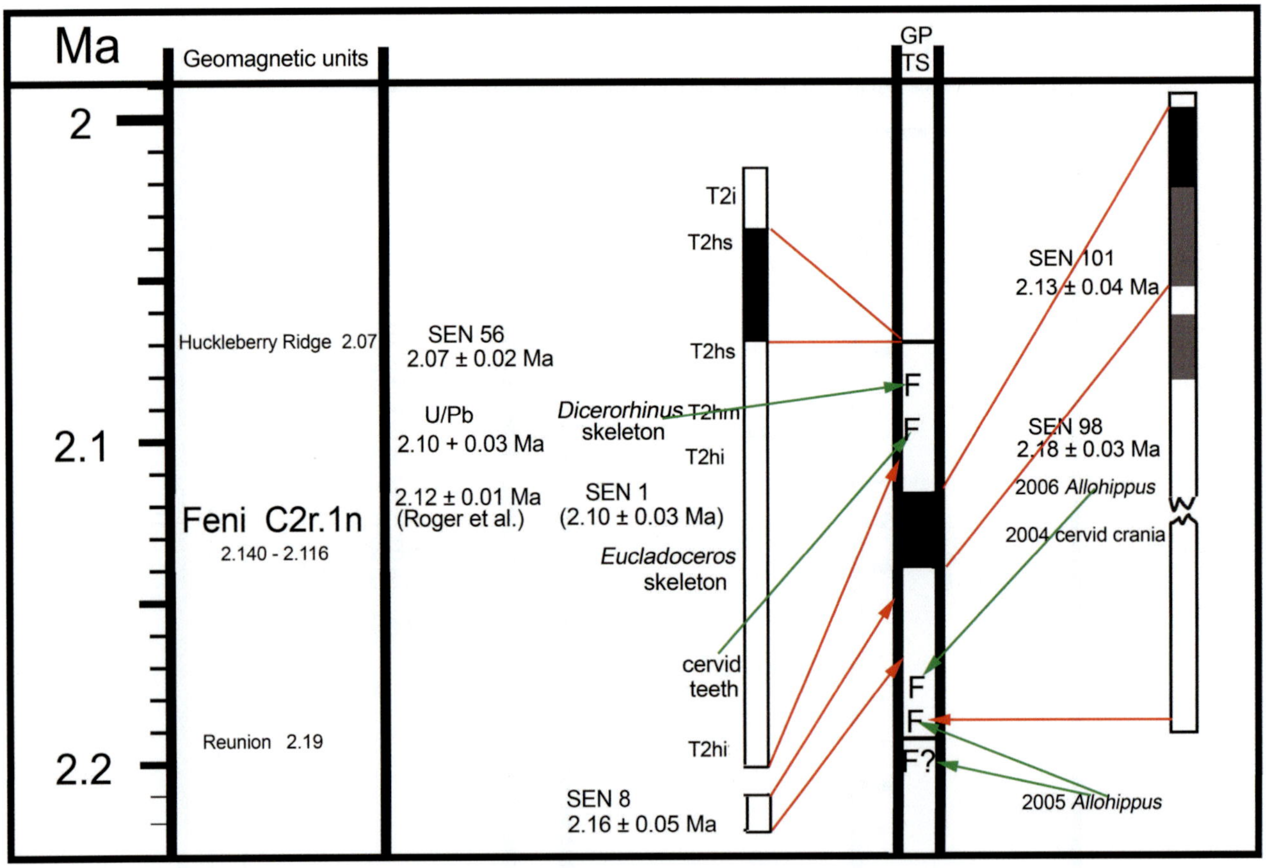

Fig. 5.9 Overview of Senèze geochronology, on the Geomagnetic Polarity Time Scale (GPTS) from 2.2 to 2.0 Ma; names of normal intervals in the GPTS column indicated at far left. The right column repeats the southeastern sequence of Trench 5 from Fig. 5.8, with suggested correlations to the GPTS in red; the arrowhead indicates a very broad correlation in that area; alternative correlation of the normal/transitional interval to the Huckleberry Ridge excursion indicated in Fig. 5.8 is not repeated here, as it is thought far less likely. The gap between the lower (inf) portion of Trench 5 above and the upper (sup) portion below is indicated by the wavy line. The green lines (with arrowheads indicating uncertainty) show possible correlation of individual fossil finds to the GPTS. The "F"s in the GTPS column indicate the fossiliferous interval, the lowest showing the possible position of the 2005 *Allohippus* skeleton below the Réunion excursion. Names of dated tephra indicate approximate position in Trench 5 section. The middle column is a composite of the four western sector sequences in Fig. 5.7 with suggested correlations to the GPTS in red (arrowheads indicate broad correlations). The green lines and F's in the GPTS column as for the southeastern sector above. Names of dated tephra indicate approximate position in Trench 2 section. At far left are three additional dates (broadly within the western sector) in their approximate chronological positions. SEN 56 is a tephra from section A, correlated to layer T2i at the top of Trench 2 (and the pit at the northern edge of parcel 233); most of section A has slid down from the top of the local sequence to a lower altitude, but the date confirms its young age. The U/Pb date was published by Paquette et al. (2021) for a horizon said to be near the top of the local sequence, but it fact it derives from section A, although from an unknown level. The argon-argon age of 2.12 Ma (recalculated here) was obtained by Roger et al. (2000) on a tephra layer within a normal magnetozone in the 1989 core. It would appear to correlate with part of the Feni subchron. All of these dates agree well with their stratigraphic positions.

Senèze sample collected by Philis, but the cause is not clear. Moreover, one specimen of *Bison* (*Eobison*) sp. (Crégut-Bonnoure 2024) and one of *Allohippus major* (Eisenmann & Delson 2024) were recovered from parcel 164 (as surface finds) and donated to the team by local collector A. Consigny; the *Bison* probably indicates the presence of a significantly younger fossil horizon (Epivillafranchian?) not otherwise sampled.

Three other ages may be added to the picture (Fig. 5.9). Tephra SEN 56, from section A in parcel 234, derives from high in the Senèze sequence (Debard 2024), as confirmed by the date of 2.07 ± 0.02 Ma. From an uncertain level in section A, Paquette et al. (2021) obtained a U/Pb date (rounded to two decimal places) of 2.10 ± 0.03 Ma. These are equivalent within error. Finally, Roger et al. (2000) dated a tephra within a normal magnetozone in the 1989 core, and we have recalculated that age to 2.12 ± 0.01 Ma, within the Feni and presumably equivalent to the normal zone in the Trench 5 sequence. Most of the pollen samples from the 1965 core must predate most or all of the Senèze fauna collected by our team (see also Argant 2024).

Summary and Conclusions

The Senèze maar (Haute-Loire, central France) has yielded fossil mammals since 1892. It has long been considered to date to the Late Pliocene, modified to Early Pleistocene following the redefinition of the Plio-Pleistocene boundary to 2.58 Ma. The first chronometric date for Senèze was provided by Prévot and Dalrymple (1970), who published a K–Ar age of 2.30 Ma for the underlying basalt marking the opening of the maar. The normal magnetozone they discerned in the 1965 core was equated to the Réunion “1” (now Feni) subchron. The $^{40}Ar/^{39}Ar$ ages on tephra from the Senèze sequence published by Nomade et al. (2014) were recalculated to conform to current standards. Samples for paleomagnetic analysis were collected by S.S. in 2001 and 2004 and are analyzed here.

The investigation by Bahain et al. (Appendix) has revealed that dating by ESR and U-series methods is especially problematic in the earlier Early Pleistocene. Only in cases where favorable conditions of late uptake and low environmental radioactive dose rate prevail would such dating succeed. Both these conditions are unfavorable at Senèze, resulting in ESR dates younger than or close to 1 Ma, far younger than those obtained via $^{40}Ar/^{39}Ar$ dating and confirmed with paleomagnetic correlation.

The Senèze deposits probably span most of the time between the Huckleberry Ridge and Réunion excursions, ca. 2.20–2.07 Ma (see Fig. 5.9). The western sector in parcels 233 and 234 (especially Trench 2) spans roughly 2.15–2.07 Ma. A normal magnetozone high in the sequence is correlated to the Huckleberry Ridge excursion at 2.07 Ma. The Feni subchron was apparently not sampled, but a date of 2.16 ± 0.05 Ma on a lower segment of Trench 2 suggests that it occurred within the time interval represented in the sector. Major fossil finds include a partial skeleton of *Dicerorhinus etruscus etruscus*, a partial skeleton of *Eucladoceros ctenoides senezensis* and teeth of *Metacervoceros rhenanus philisi*, which derive from a fossiliferous level dated ca. 2.10–2.08 Ma.

The southeastern sector in parcel 172 (especially Trench 5) appears to span from ca. 2.20–2.11 (or 2.08) Ma. A short reversed interval at the top of the paleomagnetic “column” postdates the Feni subchron (2.140–2.116 Ma). A normal magnetozone is preceded by a transitional complex which may include part of the Feni at the top and then the pre-Feni reversed interval (C2r.2r). There is no indication of the Réunion excursion, which may predate the entire sequence or might possibly be located below the lowest paleomagnetic sample. The stratigraphically constrained faunal remains in this sector (including two partial skeletons of *Allohippus senezensis* from Trenches 6 and 7) range from just below the SEN 98 tephra dated 2.18 ± 0.03 Ma through the lower reversed magnetozone and below it, thus ca. 2.20–2.18 Ma, but no younger than 2.15 Ma (the youngest age of SEN 98 within error). Those two equid skeletons were not located between dated samples SEN 98 and SEN 101, as incorrectly suggested by Delson et al. (2006) and Nomade et al. (2014). Instead both are older than these dated tephra, and the older of them lies below the oldest paleomagnetic sample. The position of that skeleton with regard to the Réunion excursion is unknown: it could lie above, within or below that (unsampled) short normal interval.

Senèze is a pivotal site in the sequence of western European mammalian faunal evolution. It has been recognized as the reference locality for both the Late Villafranchian land mammal age and the MNQ 18 regional mammal unit (see Crégut-Bonnoure et al. 2024). The age of this site is now known accurately, because Senèze is one of the few European later Cenozoic localities with a tightly developed stratigraphic context which has yielded both $^{40}Ar/^{39}Ar$ ages and paleomagnetic data which help to tie down the dating. Study of additional such well-documented fossil sites will further establish precise dating of the European Plio-Pleistocene mammalian biochronology.

Acknowledgments We thank Martine Faure and Claude Guérin for field support during sampling and for organization of the entire Senèze project. We thank Messieurs Didier Miallier et Jean Fain, of the Laboratoire de Physique Corpusculaire de l'Université Blaise Pascal—Clermont Ferrand II, for the loan of a gamma spectrometer; and Madame Hélène Valladas, then of the Centre des Faibles Radioactivités, Laboratoire mixte CNRS-CEA, Gif sur Yvette, for the loan of several TL dosimeters; these devices were used by Bonnie Blackwell in her attempt to obtain an ESR age for Senèze. Advice on the interpretation of paleomagnetic data was given to Delson by the late Frank Brown, John Kappelman and Chris Lepre. We used the TimeScale Creator program (https://timescalecreator.org/index/index.php) to make several of the figures. We thank Eric Sargis, an anonymous reviewer and especially Isaac Casanovas-Vilar for comments on earlier drafts of this manuscript.

Appendix: Attempted Combined ESR and U-series Dating of Tooth Enamel from Senèze

Jean-Jacques Bahain, Qingfeng Shao and Christophe Falguères.

A combined Electron Spin Resonance (ESR) and Uranium-series (U-series) approach was applied at the Muséum national d'Histoire naturelle (MNHN), Paris, on several teeth excavated from the different areas of Senèze to investigate the success of this approach on fossils of a known Early Pleistocene age. The combination of these two methods allows a better evaluation of post-mortem uranium-uptake into the dental tissues and associated variations of the dose rate over time. This phenomenon, which depends on the nature of the site and the age of the sample, makes the determination of the ESR ages (strongly linked to uranium content) particularly delicate, and it is necessary to use mathematical models to describe the evolution of this parameter over time. In order to calculate the age of a given sample, the various data acquired through ESR and U-series analyses are used together to model the post-mortem U-uptake in the different tissues (US model with "p" U-uptake parameter, Grün et al. 1988) and in some cases a late loss of uranium (leaching, AU model with "n" U-uptake parameter, Shao et al. 2012; see Fig. 5.A1). One of the main interests of this approach is to allow the determination for each dental tissue of a U-uptake parameter calculated from the whole set of analytical data and hence the description of the uranium-uptake kinetics into the considered tissue.

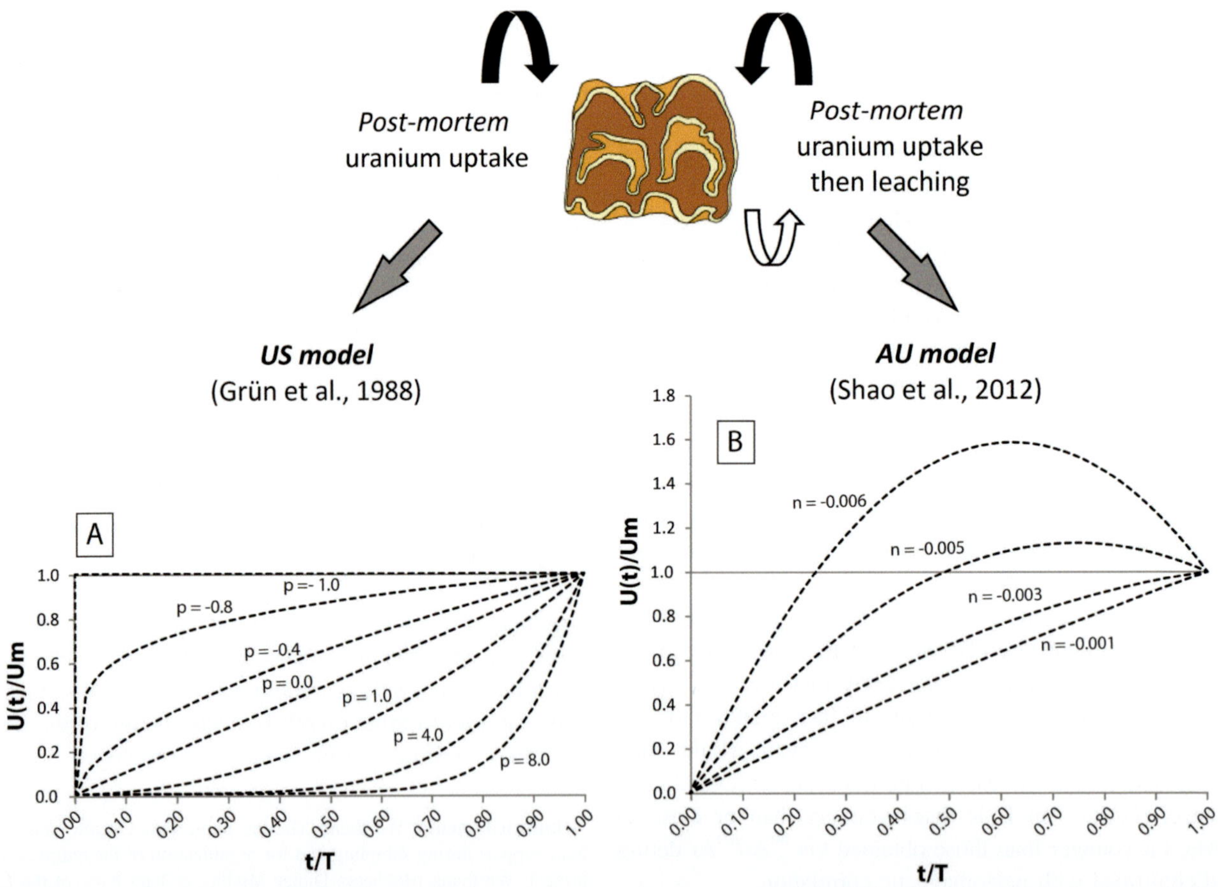

Fig. 5.A1 Kinetics of the uranium uptake in a given paleontological tissue as a function of the determined U-uptake parameter (A: incorporation only, p-parameter, US model, Grün et al. 1988; B: incorporation then leaching, n-parameter, AU model, Shao et al. 2012). U(t) is the uranium content in a tissue at time t, Um is the current uranium content in this tissue, T is the age of the sample

MNHN Laboratory Analytical Method

The Senèze teeth were prepared following the protocol described by Bahain et al. (2020). After mechanical separation of the dental tissues, the enamel was cleaned using a dental drill to remove any contamination of sediment or dentine. For the calculation of the beta contribution, the enamel thickness was measured before and after enamel preparation on both sides. The enamel was then ground and sieved. ESR study was then conducted by the additive method. The 100–200 μm enamel fraction was split in ten aliquots, nine of them were irradiated by a γ ^{60}Co source at doses ranging from 260 to 11,700 Gy. The ESR intensity of each aliquot was then measured using a Bruker© EMX ESR spectrometer, and a dose–response curve was built from the data set for the tooth. Palaeodoses, D_e, were determined using an exponential + linear function (E + L) (Shao et al. 2015). Radioelement contents of the dental tissues and associated sediments were measured by γ-ray spectrometry, and U-series analyses were performed on each dental tissue. The γ dose was recorded on the field by TL dosimeters *in situ* at different points in Trench 2 around the level of the rhino skeleton. Additional laboratory gamma measurements were performed on several sediment samples from the same level.

U-series analyses were then realized at Nanjing Normal University, China, on a Neptune Multi-Collector Inductively Coupled Plasma Mass Spectrometer (MC-ICPMS) (see details of the chemical and analytical protocols in Shao et al. 2019). Lastly, dose rates (D_a) contributions, U-uptake parameters and ESR/U-series ages were calculated from both ESR and U-series data applying either the US model (Grün et al. 1988) or the AU model (Shao et al. 2014), according to their isotopic characteristics.

Results of ESR/U-Series Analyses

The results are presented in Appendix Table 5.A1 (U-series data) and Appendix Table 5.A2 (ESR/U-series data). The U content in enamel and dentine ranges from 0.8 to 5.0 ppm and 52.5 to 82.1 ppm, respectively. These relatively high U-content values are associated with systematically high ^{230}Th/^{234}U ratios, mainly higher than 1.00, often too high to permit the use of the US model. On the other hand, the equivalent doses (D_e) are surprisingly relatively low, between 2000 and 3100 Gy, and the combination of these parameters leads to dental tissue dose being the most important factor in the age calculation.

The ESR/U-series ages, ranging from 686 ± 59 ka to 965 ± 75 ka, were obtained using the AU model for three teeth. Unfortunately, the whole set of ages is severely underestimated by comparison with both paleontological and ^{40}Ar/^{39}Ar data. Even if the values are relatively low, the higher the enamel U-content, the lower the age estimate.

As the same environmental dose was used for all the teeth and in order to evaluate if the age underestimation could be related to a poor dosimetric reconstruction, an isochron plot was fitted to the obtained data (Fig. 5.11). Despite a relatively good fit indicating that the dose reconstruction seems correct, the isochron age estimate is older (around 1047 ka), but it is still significantly lower than the ^{40}Ar/^{39}Ar age estimate. The dosimetric reconstruction therefore seems correct but cannot explain the age underestimation. It appears instead that the observed age underestimation of the Senèze teeth could be linked to the determination of the equivalent dose rather than to the dosimetric history of the samples.

This confirms the observations of Joannès-Boyau and Grün (2011) who argued that the enamel ESR signal would be composite with at least one unstable component leading to such age underestimation. The dating of earlier Early Pleistocene deposits by this method would thus be compromised except in favorable conditions (late uptake history in the dental tissues and low environmental radioactivity context), such as at Longgupo, China (Han et al. 2017). At Senèze, both relatively early U-uptake history and high dose rates seem to prohibit the use of the ESR/U-series method.

Table 5.A1 U-series data (MC-ICP-MS) obtained on fossil teeth from Senèze

Sample	Tissue	U (ppm)	Isotopic ratios				U-series apparent ages (ka)
			$^{234}U/^{238}U$	$^{230}Th/^{232}Th$	$^{230}Th/^{234}U$	$^{222}Rn/^{230}Th$	
SEN12 (P233-2004–04-0012)	Enamel	3.583 ± 0.015	1.153 ± 0.004	>100	0.960 ± 0.005	1.00	280.2 ± 7.5
	Dentine	82.140 ± 0.291	1.268 ± 0.004	>1000	1.043 ± 0.006	1.00	382.8 ± 17.4
SEN102a (P172-2006–06-0102)	Enamel	2.321 ± 0.009	1.128 ± 0.005	>1000	0.951 ± 0.005	1.00	275.8 ± 7.5
	Dentine	52.513 ± 0.195	1.121 ± 0.003	>1000	1.001 ± 0.006	0.92	377.3 ± 18.4
SEN102b (P172-2006–06-0102)	Enamel	0.807 ± 0.001	1.233 ± 0.002	>100	1.022 ± 0.003	1.00	350.9 ± 7.0
	Dentine	57.648 ± 0.150	1.120 ± 0.002	>1000	1.036 ± 0.005	1.00	535.1 ± 57.9
SEN253 (P233-2004–04-0253)	Enamel	2.190 ± 0.005	1.177 ± 0.003	>100	0.939 ± 0.004	1.00	252.4 ± 3.8
	Dentine	68.211 ± 0.272	1.277 ± 0.005	>1000	1.091 ± 0.007	0.68	681.3 ± 12.7
SEN346 (P172-2004–04-0346)	Enamel	4.983 ± 0.011	1.205 ± 0.003	>1000	0.987 ± 0.004	1.00	301.3 ± 5.4
	Dentine	78.815 ± 0.268	1.206 ± 0.004	>1000	1.034 ± 0.006	0.87	395.9 ± 18.9

Notes SEN12 and SEN253 are cheek teeth of *Dicerorhinus etruscus*, SEN102a and SEN102b are incisors of *Allohippus senezensis* and SEN346 is a cheek tooth of *Eucladoceros ctenoides*; P172 and P233 refer to the parcels from which the teeth were recovered; the 6-digit number is the field catalog number, formally FSL SEN 04-0012, etc. Uncertainties for isotopic ratios and ages are given with ±2σ. The eventual radon (Rn) losses were determined for each tissue by combining α and γ spectrometry data (Bahain et al. 1992)

Table 5.A2 ESR/U-series data obtained on fossil teeth from Senèze

Sample	Tissue	Equivalent doses D_e (Gy)	U uptake parameters p (US) or *n* (*AU*)	Annual dose rate contributions (μGy/a)				ESR/U-series ages–US or *AU* (ka)
				α	β	γ + cosmics	Total	
SEN12 (P233-2004-04-0012)	Enamel	2843 ± 184	−0.563 ± 0.038	1319 ± 292	763 ± 147	1120 ± 55	3202 ± 332	***888 ± 72***
	Dentine		*−0.0012 ± 0.0001*					
SEN102a (P172-2006-06-0102)	Enamel	2097 ± 124	−0.529 ± 0.042	800 ± 208	477 ± 96	1120 ± 55	2397 ± 236	**875 ± 69**
	Dentine		−0.851 ± 0.003					
SEN102b (P172-2006-06-0102)	Enamel	2078 ± 79	−0.891 ± 0.003	385 ± 110	649 ± 139	1120 ± 55	2154 ± 186	**965 ± 75**
	Dentine		−0.001 ± 0.001					
SEN253 (P233-2004-04-0253)	Enamel	2334 ± 68	−0.408 ± 0.042	726 ± 125	815 ± 129	1120 ± 55	2659 ± 186	***878 ± 56***
	Dentine		*−0.0012 ± 0.0001*					
SEN346 (P172-2004-04-0346)	Enamel	3050 ± 235	−0.757 ± 0.026	2161 ± 469	1169 ± 206	1120 ± 55	4446 ± 514	***686 ± 59***
	Dentine		*−0.0016 ± 0.0001*					

Notes Sample identifications, see Table 5.A1. Ages were calculated taking into account the effects of sample preparation on the beta dose rate contributions (Brennan et al. 1997). Uranium uptake parameters and ESR/U-series ages are presented in italics if n (for the AU model) or not bold if p (for the US model). Dose rate contributions (alpha, beta, gamma + cosmic and total) were calculated using conversion factors from Guérin et al. (2011). The γ-dose used for the age calculation corresponds to the mean value of the doses recorded by TL dosimeters. The β-sediment dose rate was calculated from the uranium, thorium and potassium contents of the paleontological layer, and the cosmic dose rate was determined from the depth of this level using the equations given by Prescott and Hutton (1994). A k-value of 0.13 ± 0.02 was used in the α-dose rate calculation (Grün and Katzenberger-Appel 1994). The water content was estimated to be 0 wt% in the enamel, 7 wt% in the dentine and 18 wt% in the sediment. The ESR/U-series ages were calculated with the "ESRUS" and "ESRAU" Matlab routines written by Qingfeng Shao, and the ages are given with ± 1 sigma

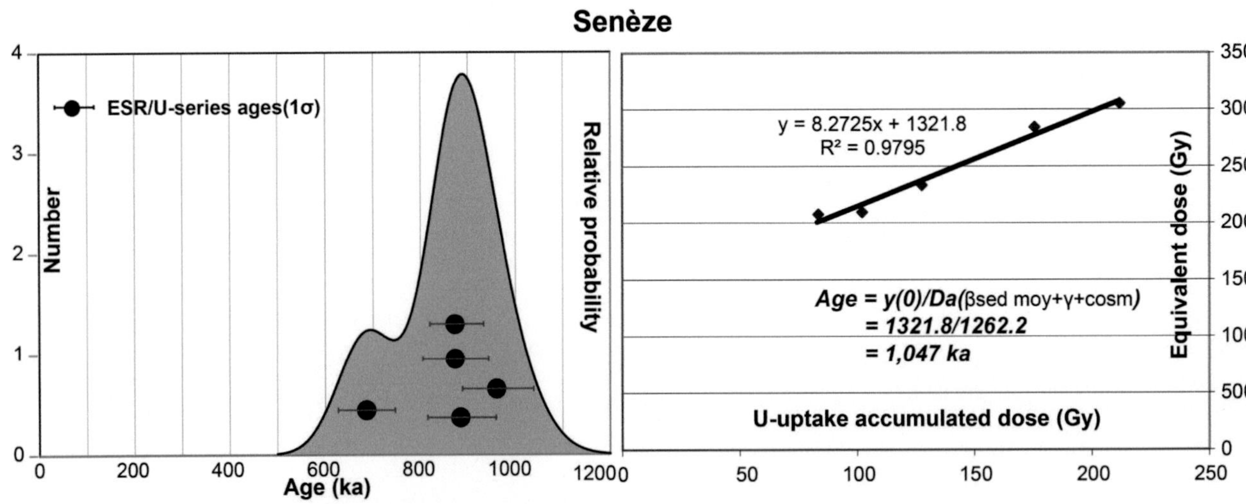

Fig. 5.A2 Probability plot (left) and isochron plot (right) of the ESR/U-series data obtained on fossil teeth from Senèze

References

Azzaroli, A. (1970) Villafranchian correlations based on large mammals. *Giornale di Geologia, 35*, 1–21.

Azzaroli, A., de Giuli, C., Ficcarelli, G., & Torre, D. (1988). Late Pliocene to early Mid-Pleistocene mammals in Eurasia: Faunal succession and dispersal events. *Palaeogeography, Palaeoclimatology, Palaeoecology, 68*, 77–100.

Bahain, J.-J., Yokoyama, Y., Falguères, C., & Sarcia, M. N. (1992). ESR dating of tooth enamel: A comparison with K-Ar dating. *Quaternary Science Reviews, 11*, 245–250.

Bahain, J.-J., Duval, M., Voinchet, P., Tissoux, H., Falguères, C., Grün, R., et al. (2020). ESR and ESR/U-series chronology of the Middle Pleistocene site of Tourville-la-Rivière (Normandy, France)—A multi-laboratory approach. *Quaternary International, 556*, 66–78

Blackwell, B. A. B., Delson, E., McNulty, K., Capellini, T. D., Blickstein, J. I. B., Skinner, A. B., et al. (2020). Dosimetry and ESR analyses at Senèze, France: dating Villafranchian monkeys and associated fauna (abstract). *American Journal of Physical Anthropology, 171*(S69), 29.

Bouchez, R., Lopez Carranza, E., Ma, J. L., Amosse, J., Piboule, M., Cornu, A., et al. (1984). Datation par spectrometrie RPE d'émail dentaire fossile dans le domaine de 50,000 ans à plusieurs millions d'années. *Revue d'Archéométrie, 8*, 70–79.

Boule, M. (1892). Découverte d'un squelette d' *Elephas meridionalis* dans les cendres basaltiques du volcan de Senèze (Haute-Loire). *Comptes Rendus hebdomadaires des Séances de l'Academie des Sciences, 115*, 624–626.

Brennan, B. J., Rink, W. J., Mcguirl, E. L., Schwarcz, H. P., & Prestwich, W. V. (1997). Beta doses in tooth enamel by "One Group" theory and the Rosy ESR dating software. *Radiation Measurements, 27*, 307–314.

Channell, H. E. T., Singer, B. S., & Jicha, B. R. (2020). Timing of Quaternary geomagnetic reversals and excursions in volcanic and sedimentary archives. *Quatermary Science Reviews, 228*, 106–114.

Couthures, J. (1989) The Senèze maar (French Massif Central): Hypothesis regarding a catastrophe occurring about 1.5 million years ago. *Journal of Volcanology and Geothermal Research, 39*, 207–210.

Couthures, J., & Pastre, J.F. (1983). Chronostratigraphie du Plio-Pléistocène d'Auvergne et du Velay: Nouveaux apports des datations radiométriques et du paléomagnétisme. *Bulletin de l'Association française pour l'étude du Quaternaire, 1983-1*, 9–18.

Crégut-Bonnoure, E. (2024). The Bovidae (Mammalia, Cetartiodactyla) from Senèze. In E. Delson, M. Faure, & C. Guérin (Eds.), *Senèze: Life in Central France two million years ago. Paleontology, geochronology, stratigraphy and taphonomy* (pp. 433–585). Cham (Switzerland): Springer.

Crégut-Bonnoure, E., Guérin, C., Argant, A., Argant, J., Debard, E., Delson, E. et al. (2024). Biochronology of the Senèze faunal assemblage. In E. Delson, M. Faure & C. Guérin (Eds.), *Senèze: Life in Central France two million years ago. Paleontology, geochronology, stratigraphy and taphonomy* (pp. 633–652). Cham (Switzerland): Springer.

Debard, E. (2024). Geological study of the Early Pleistocene site of Senèze (Domeyrat, Haute-Loire, France). In E. Delson, M. Faure & C. Guérin (Eds.), *Senèze: Life in Central France two million years ago. Paleontology, geochronology, stratigraphy and taphonomy* (pp. 37–82). Cham (Switzerland): Springer.

Delson, E., & Plopsor, D. (1975). *Paradolichopithecus*, a large terrestrial monkey (Cercopithecidae, Primates) from the Plio-Pleistocene of southern Europe and its importance for mammalian biochronology. In *Proceedings of the VIth session, Regional Committee on Mediterranean Neogene Stratigraphy*, Bratislava, 1975 (pp. 91–96). Bratislava: Slovak Academy of Sciences.

Delson, E., Faure, M., Guérin, C., Aprile, A., Argant, J., Blackwell, B., et al. (2006). Franco-American renewed research at the Late Villafranchian locality of Senèze (Haute-Loire, France). *Courier Forschunginstitut Senckenberg, 256*, 275–290.

Depéret, C. (1929). *Dolichopithecus arvernensis* Dépéret: Nouveau singe du pliocène supérieur de Senèze (Haute-Loire). *Travaux du Laboratoire de Géologie de la Faculté des Sciences de Lyon 15*, 5–12.

Depéret, C., & Mayet, L. (1912). Le gisement de mammifères pliocènes de Senèze (Haute-Loire). *Association française pour l'avancement des Sciences–Compte rendu de la 40me session–Dijon–1911–Notes et Mémoires. Mémoire hors volume*, 138–151.

Ehrlich, A. (1968). Les diatomées fossiles des sédiments villafranchiens de Senèze (Haute-Loire, Massif Central français). *Bulletin de l'Association française pour l'Étude du Quaternaire 4*, 267–280.

Eisenmann, V., & Delson, E. (2024) The Senèze equids. In E. Delson, M. Faure, & C. Guérin (Eds.), *Senèze: Life in Central France two million years ago. Paleontology, geochronology, stratigraphy and taphonomy* (pp. 307–385). Cham (Switzerland): Springer.

Faure, M., Delson, E., & Guérin, C. (2024). Introduction and history of research in the Senèze maar. In E. Delson, M. Faure & C. Guérin (Eds.), *Senèze: Life in Central France two million years ago. Paleontology, geochronology, stratigraphy and taphonomy* (pp. 1–25). Cham (Switzerland): Springer.

Grün, R., & Katzenberger-Apel, O. (1994). An alpha irradiator for ESR dating. *Ancient TL, 12*, 35–38.

Grün, R., Schwarcz, H. P., & Chadam, J. M. (1988). ESR dating of tooth enamel: Coupled correction for U-uptake and U-series disequilibrium. *Nuclear Tracks and Radiation Measurements 14*, 237–241.

Guérin, C. (2024). *Dicerorhinus etruscus etruscus* from the Late Villafranchian site of Senèze (Domeyrat, Haute-Loire, Central France). In E. Delson, M. Faure & C. Guérin (Eds.), *Senèze: Life in Central France two million years ago. Paleontology, geochronology, stratigraphy and taphonomy* (pp. 245–306). Cham (Switzerland): Springer.

Guérin, C., Faure, M., Argant, A., Argant, J., Crégut-Bonnoure, E., Debard, E., et al. (2004). Le gisement pliocène supérieur de Saint-Vallier (Drôme, France): synthèse biostratigraphique et paléoécologique. In M. Faure & C. Guérin (Eds.), *Le gisement-pliocène terminal de Saint-Vallier (Drôme, France), Geobios*, Mémoire spécial 26, 37, 349–360.

Guérin, G., Mercier, N., Adamiec, G. (2011). Dose-rate conversion factors: Update. *Ancient TL, 29*, 5–8.

Han, F., Bahain, J.-J., Deng, C., Boëda, É., Hou Y., et al. (2017). The earliest evidence of hominid settlement in China—Combined electron spin resonance and uranium series (ESR/U-series) dating of mammalian fossil teeth from Longgupo cave. *Quaternary International, 434*, part A, 75–83.

Heintz, E. (1970). Les cervidés Villafranchiens de France et d'Espagne. *Mémoires du Muséum National d'Histoire Naturelle*, Paris, *série C*, 22, volumes 1 and 2, 1–317, 1–206.

Jicha, B., Singer, B., & Sobol, P. (2016). Re-evaluation of the ages of $^{40}Ar/^{39}Ar$ sanidine standards and supereruptions in the western U.S. using a Noblesse multi-collector mass spectrometer. *Chemical Geology, 431*, 54–66

Joannes-Boyau, R., & Grün, R. (2011). A comprehensive model for CO_2-radicals in fossil tooth enamel: Implications for ESR dating. *Quaternary Geochronology, 6*, 82–97.

Mei, T. D. G, Blackwell, B. A. B., Delson, E., Frost, S., Blickstein, J. I. B., & Divjak, M. N. (2001). ESR dating the Villafranchian in France and elsewhere: A "handicapped" study? (abstract) GSA *Program with Abstracts*, 2001, poster BTH 58, A-199.

Min, K., Mundil, R., Renne, P. R., & Ludwig, K. R. (2000). A test for systematic errors in $^{40}Ar/^{39}Ar$ geochronology through comparison with U/Pb analysis of a 1.1-Ga rhyolite. *Geochimica et Cosmochimica Acta, 64*, 73–98.

Niespolo, E. M., Rutte, D., Deino, A. L., & Renne, P. R. (2017). Intercalibration and age of the Alder Creek sanidine $^{40}Ar/^{39}Ar$ standard. *Quaternary Geochronology, 39*, 205–213.

Nomade, S., Renne, P. R., Vogel, N., Deino, A. L., Sharp, W. D., Becker, T. A., et al. (2005). Alder Creek Sanidine (ACs-2): A Quaternary $^{40}Ar/^{39}Ar$ standard. *Chemical Geology, 218*, 319–342.

Nomade, S., Gauthier, A., Guillou, H., & Pastre, J. F. (2010). $^{40}Ar/^{39}Ar$ temporal framework for the Alleret maar lacustrine sequence (French Massif Central): Volcanological and paleoclimatic implications. *Quaternary Geochronology, 5*, 20–27.

Nomade, S., Pastre, J.-F., Guillou, H., Faure, M., Guérin, C., Delson, E., et al. (2014). $^{40}Ar/^{39}Ar$ constraints on some French landmark Late Pliocene to Early Pleistocene large mammalian paleofaunas: Paleoenvironmental and paleoecological implications. *Quaternary Geochronology, 21*, 2–15.

Ogg, J. (2020). Geomagnetic polarity time scale. In F. M. Gradstein, J. G. Ogg, M. D. Schmitz, & G. M. Ogg (Eds.), *Geologic Time Scale 2020* (pp. 159–192). Amsterdam: Elsevier.

Paquette, J.-L., Médard, E., Poidevin, J.-L., & Barbet, P. (2021). Precise dating of middle to late Villafranchian mammalian paleofaunae from the Upper Allier River valley (French Massif Central) using U–Pb geochronology on volcanic zircons. *Quaternary Geochronology, 65*, 101198.

Parenti, F., Valli, A., Aprile, L., Candelato, F., Monguillon, A., Natali, L. et al. (2024). Surveying and excavations at Senèze: 2001–2006. In E. Delson, M. Faure & C. Guérin (Eds.), *Senèze: Life in Central France two million years ago. Paleontology, geochronology, stratigraphy and taphonomy* (pp. 27–35). Cham (Switzerland): Springer.

Pastre, J.-F. (2024). The Senèze maar and paleontological site (Massif Central, France): Their volcanic context and studies of related tephras. In E. Delson, M. Faure & C. Guérin (Eds.), *Senèze: Life in Central France two million years ago. Paleontology, geochronology, stratigraphy and taphonomy* (pp. 83–98). Cham (Switzerland): Springer.

Prescott, J. R., & Hutton, J. T. (1994). Cosmic ray contributions to dose rates for luminescence and ESR dating: Large depths and long-term time variations. *Radiation Measurements, 23*, 497–500.

Prévot, M., & Dalrymple, G. B. (1970). Un bref épisode de polarité géomagnétique normale au cours de l'époque inverse Matuyama. *Comptes Rendus de l'Académie des Sciences, 271 D*, 2221–2224.

Renne, P. R., Swisher, C. C., Deino, A. L., Karner, D. B., Owens, T. L., & DePaolo, D. J. (1998). Intercalibration of standards, absolute ages and uncertainties in $^{40}Ar/^{39}Ar$ dating. *Chemical Geology, 145*, 117–152.

Renne, P. R., Balco, G., Ludwig, K. R., Mundil, R., & Min, K. (2011). Response to the comment by W.H. Schwarz et al. on "Joint determination of K-40 decay constants and Ar-40*/K-40 for the Fish Canyon sanidine standard, and improved accuracy for Ar-40/Ar-39 geochronology" by P.R. Renne et al. (2010). *Geochimica et Cosmochimica Acta, 75*, 5097–5100.

Roger, S., Coulon, C., Thouveny, N., Féraud, G., Van Velzen, A., Fauquette, S., et al. (2000). $^{40}Ar/^{39}Ar$ dating of a tephra layer in the Pliocene Senèze maar lacustrine sequence (French Massif Central): constraint on the age of the Réunion-Matuyama transition and implications on paleoenvironmental archives. *Earth and Planetary Science Letters, 183*, 431–440.

Schaen, A. J., Jicha, B. R., Hodges, K. V., Vermeesch, P., Stelten, M. E., Mercer, C. M., et al. (2021). Interpreting and reporting $^{40}Ar/^{39}Ar$ geochronologic data. *GSA Bulletin, 133*, 461–487.

Schaub, S. (1943). Die oberpliocaene Säugetierfauna von Senèze (Haute-Loire) und ihre verbreitungsgeschichtliche Stellung. *Eclogae Geologicae Helvetiae, 36*, 270–289.

Shao, Q., Bahain, J.-J., Dolo, J.-M., & Falguères, C. (2014). Monte Carlo approach to calculate US-ESR ages and their uncertainties. *Quaternary Geochronology, 22*, 99–106.

Shao, Q., Bahain, J. J., Falguères, C., Dolo, J.-M., & Garcia, T. (2012). A new U-uptake model for combined ESR/U-series dating of tooth enamel. *Quaternary Geochronology,10*, 406–441.

Shao, Q., Bahain, J.-J., Wang, W., Jin, C., Wang, Y., Voinchet, P., et al. (2015). Combined ESR and U-series dating of early Pleistocene *Gigantopithecus* faunas at Mohui and Sanhe Caves, Guangxi, southern China. *Quaternary Geochronology, 30*, 524–528.

Shao, Q., Li, C., Huang, M., Liao, Z., Arps, J., Huang, C., et al. (2019). Interactive programs of MC-ICPMS data processing for 230Th/U Geochronology. *Quaternary Geochronology, 51*, 43–52.

Singer, B. S. (2014). A Quaternary geomagnetic instability time scale. *Quaternary Geochronology, 21*, 29–52.

Stehlin, H. G. (1923). Die oberpliocaene Fauna von Senèze (Haute-Loire). *Eclogae Geologicae Helvetiae, 18*, 268–281.

Steiger, R. H., & Jäger, E. (1977). Subcommission on geochronology: Convention on the use of decay constants in geo- and cosmochronology. *Earth and Planetary Science Letters, 36*, 359–362

Chapter 6
Palynological Analysis of the Fossiliferous Layers of the Late Villafranchian Deposit of Senèze (Domeyrat, Haute-Loire, France). Taphonomy and Paleoenvironmental Data

Jacqueline Argant

Abstract Following the first paleobotanical analysis done in the 1960s on a core drilled in the center of the maar of Senèze (Domeyrat, Haute-Loire), a palynological study of the fossil layers situated 30 m higher altitudinally remained to be undertaken. The aim of our new pollen analysis was to know more about the landscape and climate contemporaneous with the late Villafranchian fauna. Unfortunately the sediments (mainly either colluvial slope deposits or clays) were not very favorable to the preservation of pollen grains. However, qualitative information was obtained from hyena coprolites and their embedding sediments. Those samples indicate the coexistence of leafy trees, conifers and herbs in an open landscape, under cool-temperate climatic conditions.

Résumé La première analyse pollinique réalisée à Senèze est due à H. Elhaï (1969) qui a étudié entre 5 m et 120 m un carottage de 175 m effectué dans les dépôts lacustres du centre du maar (altitude 590 m). Ce carottage était situé à une trentaine de mètres en dessous des couches contenant la faune villafranchienne. Les cinq premiers mètres étaient stériles. A la faveur de la reprise des fouilles, il était donc nécessaire de procéder à un échantillonnage systématique en liaison avec les nouvelles découvertes paléontologiques. Un total de deux cent huit échantillons ont été prélevés de 2001 à 2006, afin de tester tous les types de sédiments: sables et graviers grossiers des dépôts de pente, sables fins et argiles des dépôts lacustres recueillis en colonnes sur les coupes des tranchées T1 à T6 ou à proximité des os au fur et à mesure de leur dégagement par les fouilleurs. Onze coprolithes de Hyénidés, dont quatre découverts à proximité du squelette de *Dicerorhinus etruscus* ont également été analysés ainsi que, pour ces derniers, le sédiment sur lequel ils reposaient.

Les échantillons ont tous été traités suivant la méthode de concentration en liqueur dense et seuls 24 ont conservé des grains de pollen, généralement en excellent état, mais malheureusement souvent peu abondants. Les autres échantillons sont stériles. C'est le cas en particulier des colluvions riches en matériaux volcaniques altérés issues du versant. Les argiles lacustres sont un peu plus favorables mais restent pauvres en matériel pollinique. Une végétation clairsemée, la pente importante des abords du lac favorisant le glissement des dépôts et l'existence de courants emportant rapidement les particules les plus légères au large en sont peut-être responsables.

Les données restent donc pour l'essentiel qualitatives. Quarante-deux taxons ont été identifiés sur l'ensemble: 18 arbres, 22 herbacées, un Ptéridophyte (spore monolète) et une mousse (*Sphagnum*). Les feuillus sont constamment présents: *Quercus*, *Corylus*, *Alnus* et, plus sporadiquement *Carpinus*, *Castanea*, *Fagus*, *Tilia*, *Betula*, *Fraxinus*, *Juglans*, *Ligustrum* et *Platanus*. *Hedera* est noté une fois. Les conifères sont *Pinus*, *Abies*, *Cedrus* et *Juniperus*. La présence de *Alnus*, *Fraxinus* et *Ligustrum* évoque les substrats les plus humides en bordure du lac. Bien que le nombre de grains de pollen d'arbres soit souvent supérieur à celui des herbacées, le rôle des herbacées ne doit pas être sous-estimé, comme en témoignent les Poaceae toujours présentes, *Plantago*, *Rumex* et les arbres héliophiles tels *Betula*, *Juniperus*, *Alnus*, *Platanus* qui profitent de l'ouverture du paysage. Les plantes hygrophiles sont rarement perceptibles. Seules *Filipendula*, Cyperaceae et Typhaceae sont notées régulièrement mais les bords du lac semblent cependant avoir été défavorables à l'installation d'une mégaphorbiaie étendue.

Si des variations climatiques ont probablement existé au cours du dépôt des couches contenant les vestiges paléontologiques, il est impossible de les mettre clairement en évidence et de les quantifier par un diagramme. La

J. Argant (✉)
LAMPEA UMR 7269, Université Aix-Marseille, CNRS, MMSH, Aix-en-Provence, MCC, France
e-mail: j.argant@wanadoo.fr

E. Delson et al. (eds.), *Senèze: Life in Central France Around Two Million Years Ago*, Vertebrate Paleobiology and Paleoanthropology, https://doi.org/10.1007/978-3-031-64415-3_6

comparaison avec les résultats d'Elhaï reste délicate en raison de la stérilité des sédiments de surface des 5 premiers mètres supérieurs du sondage profond. Cependant, il apparaît que les spectres polliniques des sédiments fossilifères ne peuvent qu'être plus récents que ceux de la partie supérieure du diagramme d'Elhaï. Les conditions climatiques rencontrées par les animaux qui vivaient à proximité du maar de Senèze montrent une tendance tempérée fraîche.

Keywords Palynology • Pollen taphonomy • Coprolites • Landscape

Mots-clés Coprolithes • Palynologie • Paysage • Taphonomie des pollens

Introduction

The hamlet of Senèze (Domeyrat, Haute-Loire, France) is located in a depression where a tributary of the Sénouire River, itself a tributary of the Allier, has its source. This depression is currently devoted to crops and livestock, while the slopes and summits are forested (mainly *Pinus silvestris, Quercus pedunculata, Fraxinus excelsior, Salix caprea, Prunus avium*). It corresponds to a Plio-Pleistocene phreatomagmatic explosion crater filled secondarily by a lake that gradually filled in during the Villafranchian period. A paleontological deposit was formed on the edges of this maar and the filling of the paleolake has yielded numerous fossils (at least 39 species of mammals and 17 birds), collected sporadically and without precise location, from 1892 until the end of the 1930s. After a survey campaign in 2000, a programmed excavation operation was carried out each year from 2001 to 2006. At the end of this work, the exact position of at least one major mammalian fossiliferous level, affected by a steep slope, is known; a certain number of stratigraphic sections have been widely sampled from sedimentological, palynological and paleomagnetic points of view (Debard 2024). They also made it possible to identify several tephras used for dating (Guérin 2006; Delson et al. 2024a).

The lacustrine filling that occupies the central part of the maar (590 m a.s.l.) has been sampled by several boreholes. The deepest, carried out in 1965, reached a depth of 175 m. The first studies of pollen (Elhaï 1969) and diatoms (Ehrlich 1968) were undertaken on this long core situated at a distance of some hundred meters from the Villafranchian fossil site and about 30 m below it (see Faure et al. 2024).

The upper 120 m of the core were analyzed with a spacing of 40 to 60 cm between the samples. The pollen sum used to construct the diagram referred only to tree pollen (100 to 650 pollen grains per sample), and the five upper meters of this core were sterile in pollen. Elhaï's interpretation was based on the ecology of four main plant groups:

(a) relict thermophilous trees: *Taxodium* type or *Sequoia, Sciadopitys, Pinus haploxylon, Nyssa, Liquidambar, Eleagnus, Symplocos, Carya, Pterocarya, Juglans,* and *Eucommia*;
(b) leafy trees native in the area today: *Quercus, Ulmus, Corylus, Tilia, Fraxinus, Acer,* and *Alnus*;
(c) conifers: *Pinus, Abies, Picea, Cedrus,* and *Tsuga*;
(d) herbs, with distinction among terrestrial or aquatic taxa and ferns.

According to the variation of these four groups, Elhaï (1969) described a general alternation between two main plant assemblages: a warm-climate forest of thermophilous, generally leafy trees, both local and relict; and a cooler-climate forest of conifers mixed with herbs, the latter sometimes expanding to cover a large surface. He divided the core into a series of phases dominated by one of these assemblages (Table 6.1). These findings are consistent with those from the study of the diatoms (Ehrlich 1968).

Ablin (1991) recalculated the pollen data from Elhaï on the basis of all the pollen grains determined. A synthetic ecological diagram was reconstructed, excluding the poorest 17 samples at the bottom, and compared with the pollen sequences from Ceyssac (Ablin 1991). It is assumed that Senèze documented an alternation of leafy forest (with poor diversity among conifers, and with rare Mediterranean taxa) and open steppe with *Artemisia* and Poaceae. This interpretation is close to that given by Elhaï (1969) but insists more on a steppic character for the coolest periods. Furthermore, Ablin (1991) correlated the upper part of the pollen diagram with the Eburonian, while Elhaï (1969) assigned it to the Waalian.

These early dating assumptions need to be reconsidered in light of new $^{40}Ar/^{39}Ar$ data obtained since then (Nomade et al. 2014; Roger et al. 2000). Chronological hypotheses have to be integrated with the pollen analyses of this long sequence and the fauna discovered thirty meters above during the 2001–2006 excavations directed by Faure, Guérin and Delson.

Table 6.1 Biostratigraphic units in the 1965 Senèze drilled core according to pollen analysis. Suggested dating after Elhaï (1969), Ablin (in Couthures et al. 1991) and Roger et al. (2000). (Tableau 6.1. Unités biostratigraphiques dans la carotte de Senèze fondées sur les résultats polliniques. Hypothèses d'interprétation et de datation d'après Elhaï (1969), Ablin (in Couthures et al. 1991) et Roger et al. (2000))

Biostratigraphic units (depth below top of core)		Climate	Hypothetical dating	
			Elhaï	Ablin
	S I a	Cool or cold		
	S I b	Temperate and humid (warmer than today)		
	S I c	Cool or cold		
S I (24–5 m)	S I d	Temperate and humid	Waalian	Eburonian
	S I e	Cool or cold		
	S I f	Temperate and humid		
S II (43–29 m)		Climate cold and relatively dry, first major Deterioration	Eburonian	Eburonian
S III (75–43 m)	S III a	Cool and humid	Tiglian ?	Tiglian
	S III b	humid and temperate		
S IV (94–76 m)		Cool and humid		
S V (120–94 m)		Temperate with poorly diversified vegetation perhaps the first recovery after a significant cold phase (the start of the Pleistocene ?)		

The links between the fossiliferous deposits and those studied by Elhaï (1969) were not clearly established at the outset. Thus, palynological investigation of the fossil layers remained to be undertaken to provide information about the vegetation directly contemporaneous with the excavated fauna (Delson et al. 2006).

Materials and Methods

Sampling of the Fossil Layers

Sediments

Pollen analysis was carried out on samples taken from the different sections as the excavation progressed. Since 2001, 208 pollen samples were collected in trenches T1, T2, T4, T5 and T6. Samples were taken in columns or isolated places (Table 6.2, Fig. 6.1).

The whole diversity of sediments was sampled, from coarse colluvium to fine-grained sediment but the analysis was not emphasized in sections where the samples tested were found to be sterile or substerile (e.g. T2). Sediments in contact with the bones were prioritized, especially in the case of *Dicerorhinus etruscus*, *Allohippus senezensis* (Fig. 6.2), cervids and fishes. These are almost entire skeletons found in connection. Therefore, even if they were moved after the death of the animal, they did not undergo any significant disturbance that would have dispersed the bones. The sediments brought by runoff and blocked at their level are therefore a priori likely to contain pollen material contemporary with their death. Moreover, the skeletons may themselves be a source of pollen originally retained in the fur or in the digestive tract of the animal, for example.

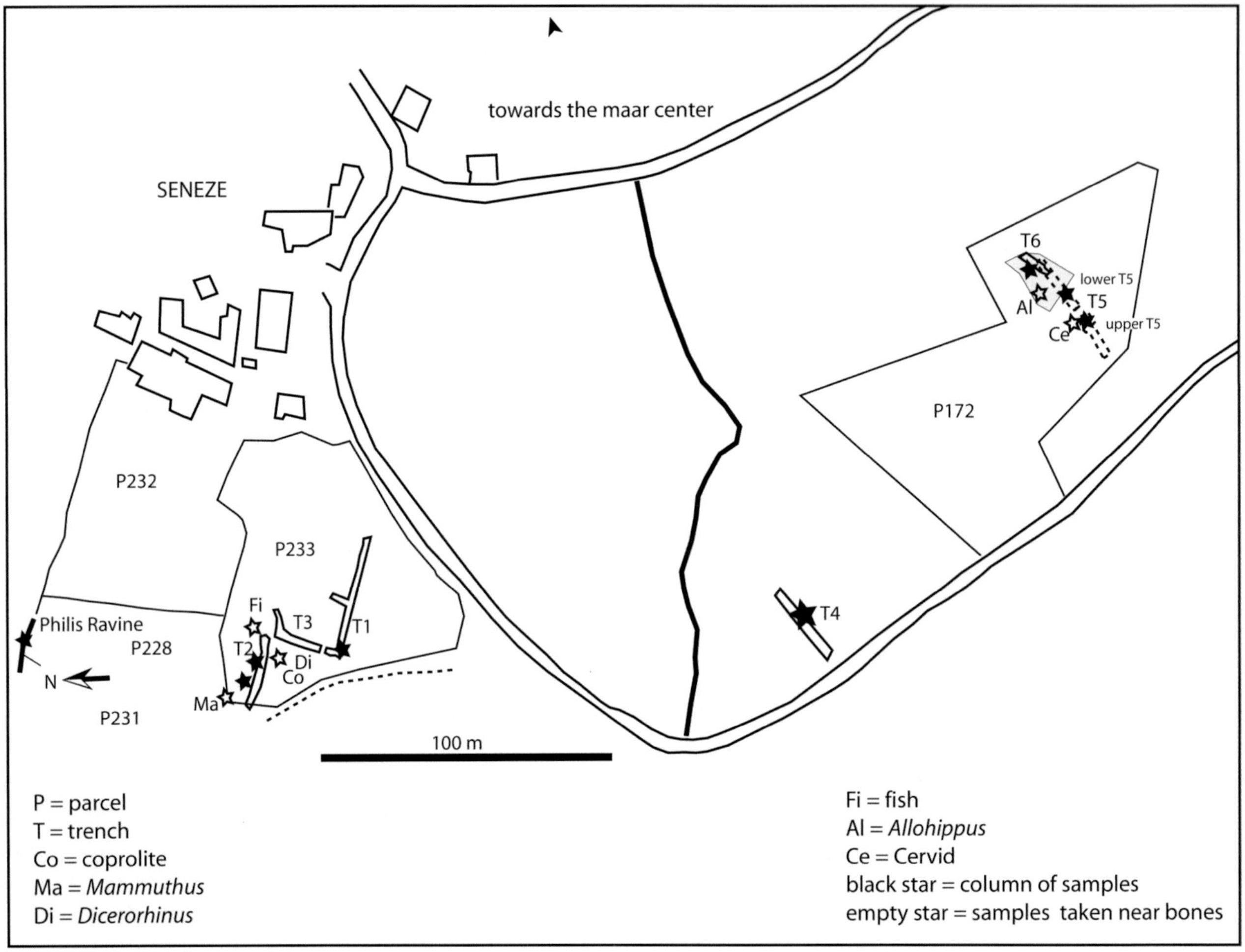

Fig. 6.1 Senèze, Domeyrat (Haute-Loire, France). Location of the main samples taken for palynological analysis. (Fig. 6.1. Senèze, Domeyrat (Haute-Loire, France). Localisation des principaux échantillons prélevés pour l'analyse pollinique.)

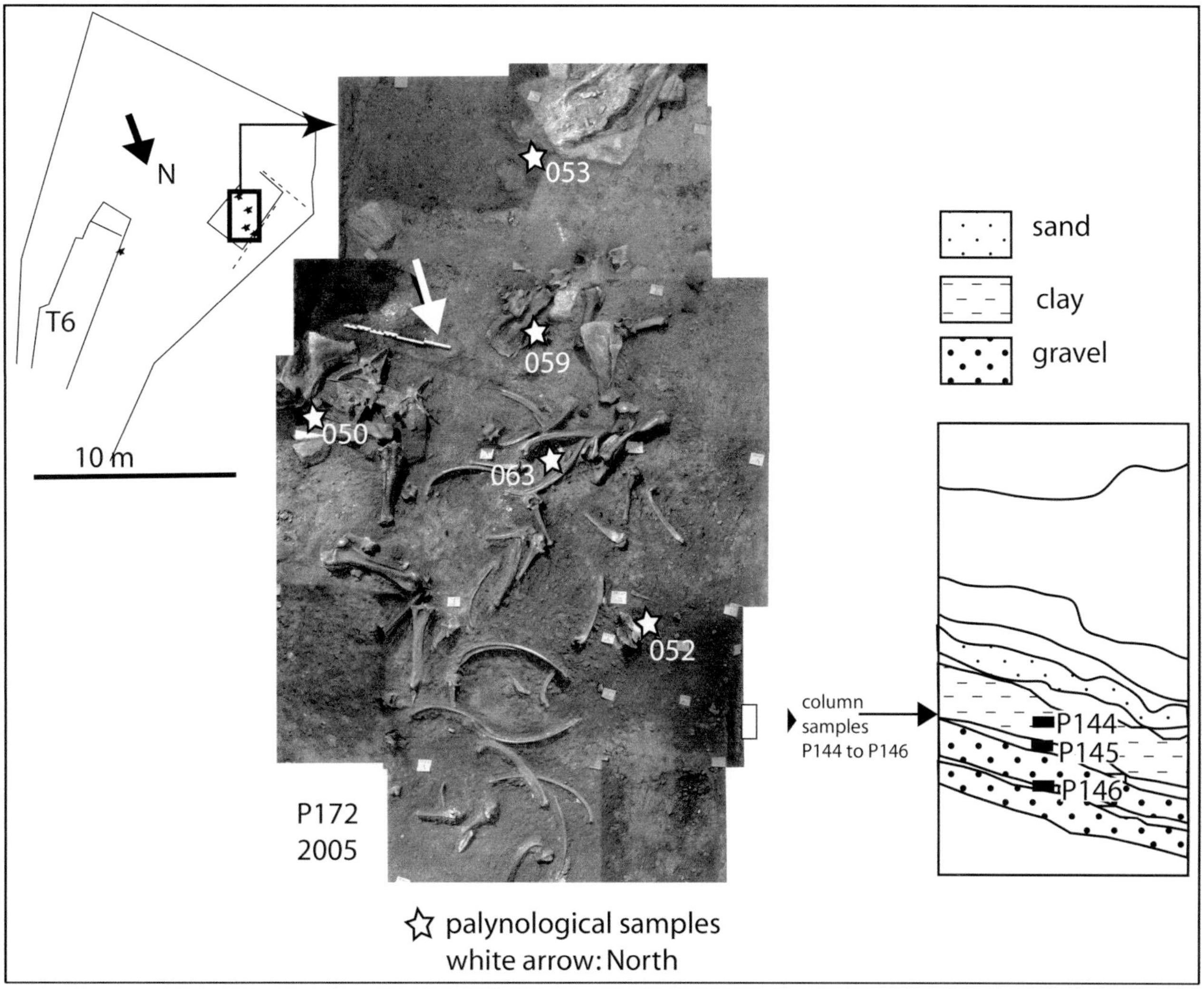

Fig. 6.2 Senèze, Domeyrat (Haute-Loire, France). Location of the samples taken around the 2005 *Allohippus* skeleton. Top left, wider view of position of skeleton northwest of Trench 6; right, local stratigraphic column with samples P144-P146. (Fig. 6.2. Senèze, Domeyrat (Haute-Loire, France). Localisation des échantillons prélevés près du squelette de *Allohipus* en 2005. En haut à gauche, vue d'ensemble du squelette au nord-ouest de la tranchée 6; à droite, colonne avec échantillons P144-P146.)

Table 6.2 Samples taken at Senèze for palynological analysis. (Tableau 6.2. Echantillonage à Senèze pour analyses palynologiques.)

Prefix (Year)	Parcel/Trench	Column or skeleton	Number of samples	Elevation (meters)	Remarks
SEN 01 (2001)	P233 / T2	C1	45	624.47 /622.78	Colluvium and clay
	P233/T2 lower part	C2	44	622.90 / 620.75	Colluvium and clay
	P234 / T1	C3	24	621.12 / 619.45	Colluvium and clay
	P233 / T2	Isolated samples	8	> 624.47	Dark matter
	P233 /T2	*Dicerorhinus*	1	623.50	Clay
SEN 03 (2003)	P174 / T4	Isolated samples	3	623.25 / 620.31	Colluvium
	P233 /	Tusk *Mammuthus*	1	–	Colluvium
	P233 /	Fish	1	–	Clay
	P233 /	Rib	1	–	Sandy clay
	P233 /	*Dicerorhinus*	1	–	Sandy clay
	P233 /	Northern Section [N9-L14]	8	623.35 / 622.86	Colluvium to sandy silt
	P233 /	Coprolites (see Table 6.3)			
SEN 04 (2004)	P172 / T5	Cervid	3	616.04 / 614.24	Gravels, sand
	P172 / T5 lower part	C A	19	616.06 / 615.20	Gravels, sand
	P172 / T5 lower part	C B	16	621.68 / 621.38	Gravels, sand
	P172 / T5 upper part	C4	9		Gravels, sand
SEN 05 (2005)	P172	*Allohippus*	8	613.12 / 612.11	Gravels, beds of sandy silt or sand
	P172	NW section, locus *Allohippus*	3		Clay, sandy silt, gravels
	P172	T6	13		

Coprolites and Embedding Sediments:

In 2003, ten coprolites were discovered near the skeleton of *Dicerorhinus etruscus* excavated in 2001 in trench T2 and form a relatively homogeneous series by their shape and size (Table 6.3 A). They differed from the embedding sediment by their hard consistency and their yellowish color. Breaks reveal many small vesicles of variable form which sometimes evoke the mold of an element which has disappeared. For example, the form in Fig. 6.3 E looks like a small fragment of stem of a plant. Although it is often deeply transformed, bone can also be observed in very small fragments recognizable by their vesicular structure (Fig. 6.3 F). They show that these are coprolites of bone-crushing carnivores. Their shape and large size make it easy to recognize coprolites of Hyaenidae, clearly different from those of canids for instance. They confirm the presence of hyenas near the rhinoceros carcass. Their species cannot be specified because three hyaenids are represented on the site by their skeletal parts: *Pachycrocuta perrieri*, *Chasmaporthetes lunensis* and a third undetermined species (A. Argant 2024). Five of these coprolites, the best preserved, were used for pollen analysis, as well as nine isolated others collected at different points across the excavation area (Fig. 6.3, Table 6.3 B). They are the only samples from Senèze to contain a little calcium carbonate and to effervesce with hydrochloric acid during preparation. The reacting carbonate can only come from animal bone crushed by the carnivore while eating. As in trench 2, the embedding sediment was also sampled, so it is possible to compare the two types of contemporaneous materials.

The mass of the treated samples is obviously highly variable according to the composition of the sediments (stony or not), ranging here between 8.5 g and 40.6 g, and most often between 20 and 30 g. It reaches 94.3 g in the case of coprolites.

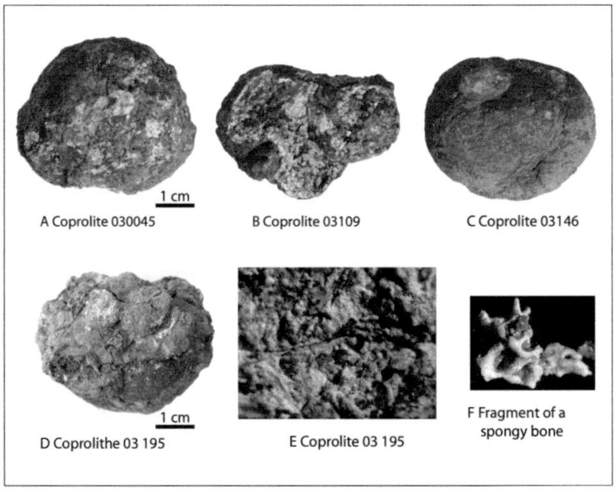

Fig. 6.3 Senèze, Domeyrat (Haute-Loire, France). Hyaenid coprolites (**A–D**; **E**, stem trace; **F**, bone found in one of them). (Fig. 6.3. Senèze, Domeyrat (Haute-Loire, France). Coprolithes d'hyène (**A–D**; **E**, trace de tige; **F**, os trouvé dans l'un d'eux).)

Sample Processing

The process is based on the principle of dense liquor concentration (d = 2), which must be carried out when the treated material no longer contains aggregates that can trap any pollen grains. This method (now ancient, e.g. Pokrovskaia 1950; Frenzel 1964; Girard et Renault-Miskovsky 1969; Goeury et Beaulieu 1979) is particularly efficient if a very powerful diffluent (sodium hexametaphosphate, 122 g/l) is used at the beginning, before a mixing that does not damage the pollen in any way—probably due to cavitation—if there are pollen and spores of course (Argant 1990). In the case of coprolites their surface needs to be precisely cleaned, by intensive brushing under a jet of water in order to remove potentially polluted material, especially from modern rainfall, and to ensure that only the content of the coprolite is treated. After being crushed and decarbonated, they are prepared by the same method as other mineral sediments.

Table 6.3 Group A, coprolites (C) and their embedding sediment samples (S) collected in 2003 near the 2001 skeleton of *Dicerorhinus etruscus* (trench T2, P233); Group B, coprolites collected in different places (P233 and P172). The analyzed samples are in bold. (+) = contain some pollen grains (cf. Table 6.7). (Tableau 6.3. Groupe A, coprolithes (C) et les échantillons des sédiments encaissants (S) récoltés en 2003 près du squelette de *Dicerorhinus etruscus* (2001, tranchée T2, P233); Groupe B, coprolithes collectés en différents endroits (P233 et P172). Les échantillons analysés sont en gras. (+) = contient des grains de pollen (cf. Table 6.7).)

Group A: near *Dicerorhinus etruscus* skeleton (P233)		
Number	Dimensions in mm	S = corresponding sediment
C030149	34 × 27,4 × 26,3	S030153
C030045	41 × 35,6 × 31	S030042
C030114	32 × 31 × 32	S030115
C030141	Not measured	S030034
C030175	Not measured	**S030176** (+)
C030055	38 × 28 × 30	**S030055** (+)
C030091 (+)	Not measured	**S030092** (+)
C030195 (+)	Not measured	**S030196** (+)
C030173 (+)	Not measured	**S030174** (+)
C030120	Not measured	**S030121**
Group B, other locations		
Number	Dimensions in mm	Parcel
C040172 (2 pieces)	47.3 × 47.3 × 40 + 59.4 × 50.2 × 50.3	P172 – T5 S040173
C030073	33 × 27.5 × 24.3	P233
C030141	Not measured	P233
C060064	37.7 × 37.7 × 52.5	P172
C030031	44 × 44 × 55.6	?
C050060	27.3 × 25.4 × 21.4	P233
C040017	Not measured	P172
C040120	52.7 × 43.5 × 48.4	P172
No Number	36.7 × 29.7 × 36.7	?
C030146	63 × 51 × 33	–

- The sieving (mesh size 1 mm) is done before mixing in the case of mineral sediments, afterward in the case of peat. Then, the following operations are performed:
- a cold decarbonation (hydrochloric acid 28%) if the carbonate test is positive. It is done from the start for coprolites,
- then, after rinsing and centrifugation (3000 rpm for 5 min.), cold hydrofluoric acid (40%) acting for about 24 h,
- centrifugation and then removal of the fluorosilicates with hot hydrochloric acid (10 min. in a water bath),
- rinsing, then removal of the organic matter with heated potassium hydroxide (10%) for 10 min.
- concentration by density (Thoulet liquor d = 2) becomes possible under good conditions after mechanical stirring for 10 min.
- the supernatant is collected. It is then diluted in distilled water (twice) to obtain the final pellet, rinsed, centrifuged and stored in glycerine. The number of drops of glycerin added is proportional to the volume of the pellet. This is the basis for the calculation of the pollen concentration in relation to the starting mass (volumetric method), which remains a simple indication because of the many uncertainties that mark the different stages of the treatment (mass of the sieve rejects, rejected surpluses, etc.). The main thing is to always proceed in the same way in order to be able to compare.

A Leica M3200 light microscope with a magnification of 500 × and phase contrast was used for routine pollen analysis. Pollen types were identified using our own reference collection, identification key (Moore et al. 1991) and atlas (Reille 1992).

Results of Pollen Analysis

The Colluvium of Trench T2 (Columns C1 and C2)

The volume of the residues obtained at the end of preparation is generally very low. Each of them contain almost exclusively mineral elements ranging from 5 to 150 µm translucent crystals of rectangular form, showing rectilinear lines of cleavage parallel to the length, or translucent crystals of irregular, angular form, and black elements of irregular form.

Some fungal ascospores are the only elements that can be observed in the samples P6 and P8, closest to the surface. They are indeterminable, and no information can be deduced from their presence.

The Very Fine-Grained Clays of Trench T1 (Column C3)

Among the 24 samples, only two contained some pollen grains (Table 6.4):

- T1-C3-15 contained six pollen grains representing five taxa (*Pinus, Castanea, Plantago, Rumex*, Cichorioideae)
- T1-C3-21 is the richest sample. The totality of the residue was observed. It contained 63 pollen grains, largely dominated by *Pinus* (49 grains). The six other taxa were *Betula* and herbs (Poaceae, Cichorioideae, *Plantago, Rumex,* Rosaceae). One large sized pollen grain of Poaceae might suggest modern pollution by cereal agriculture, but this sample is 1.07 m below the surface, out of the plough zone. Furthermore, among the wild forms of Poaceae, several species such as *Glyceria* type from wetland ecosystems have coarse pollen grains (> 40 µm) similar to those of cereals. Such pollen grains sometimes occur in Tardiglacial sediments, and some were also found in the hardened loess of Saint-Vallier (Leroi-Gourhan 1973). Thus, we can exclude modern pollution.

The number of grains for each of these samples is too low to allow an interpretation of the vegetation cover, though only taxa indicative of cold periods are represented (*Pinus, Betula*, herbs). The contents of samples T1-C3-15 and T1-C3-21 could thus indicate a cool episode.

Table 6.4 Results of the pollen analysis of samples from trench T1 at Senèze. (Tableau 6.4. Résultats de l'analyse pollinique des échantillons de la tranchée T1 de Senèze.)

SENEZE 2001	T1-C3-P15	T1-C3-P21
Betula		2
Castanea	1	
Pinus	2	49
AP	**3**	**51**
NAP	**3**	**10**
Cichorioideae	1	
Poaceae		4
Large Poaceae		1
Rosaceae		1
Plantago lanceolata type	1*	2
Rumex	1*	2
Monolete spore		1
Indeterminate		1
TOTAL	**6**	**63**
* with fresh aspect		

Sediments in Contact with the Skeletons

Sediments in Contact with the Skeleton of *Dicerorhinus etruscus*

The sediment taken near trench T2 in contact with the skeleton of *Dicerorhinus etruscus* (Fig. 6.1) delivered only pollen of *Pinus silvestris* type (28 grains). No interpretation can be given from these results.

Sediments in Contact with the Large Cervid (*Eucladoceros*)

Sediments were sampled in parcel 172, trench T5, near a tibia (lower T5) and a metapodial (upper T5) of this large cervid (Fig. 6.4 A). Five samples from a 35 cm sequence (column C4, in the upper trench T5 section above the skeleton) were also treated. The pollen grains and spores were very well preserved (Fig. 6.4 B-D), but again far from abundant (maximum of 20 in a single sample), and the taxa were also limited in number (maximum of 12 in the most diverse sample—sample 7—and only 20 overall), which once more prohibited statistical analysis (Table 6.5).

By considering all samples we find *Quercus*, *Alnus*, *Fagus*, *Betula* and *Corylus*, *Pinus*, *Cedrus*, these last three appearing in five samples. The herb taxa are mainly represented with Poaceae, Cyperaceae, Ranunculaceae and *Rumex*. They appeared a little more diversified in sample 5 (column C4), where, in particular, *Calluna* is recorded. A spore of *Sphagnum*, a bog moss, was found in three samples of column C and those taken in contact with the bones; perhaps it implies the existence of small marshy zones favorable to its development near the lake or of ancient peat reworked after destruction by erosion. However, no modern sediment of this kind was observed on the excavations carried out in the site.

Analysis of the Sediments Taken in 2005 in Contact with the Skeleton of *Allohippus senezensis* (FSL SEN 05–0081) (See Section "Sediments" in Materials and Methods and Fig. 6.2)

In every sample, the volume of the residue obtained at the end of the chemical process was very small and almost totally observed under the microscope. When they are preserved, the pollen grains appear very fresh. Table 6.6 shows the heterogeneity of the results according to the sampling points around the skeleton of the equid (050 to 063). Silica crystals remained in all cases, except in the sample taken close to the ribs where fine microscopic organic remains dominated. The angular silica grains were sometimes relatively coarse (30 to 40 μm). However, there is no clear relation between their size and the presence or absence of pollen.

The bones of the horse "are in mostly basaltic gravel lenses, with clastic support, associated to limono-sandy beds. The bones were blocked by very large angular blocks…" (Debard *in* Faure et al. 2005). Two samples (053, 059) located in the higher part of the locus (Fig. 6.2) can be regarded as substerile. Thus it should be assumed that the sedimentation of pollen could be disturbed for example by strong rain washing down steep slopes, carrying away the lightest particles or blocking them on obstacles. The carcass of the horse could also trap them locally (on the level of the axis vertebra 052 in particular) or be itself a source of pollen, from the contents of the stomach for example. This may explain why sample 052 is the richest of all the samples analyzed from Senèze, with 108 pollen grains and 28 taxa.

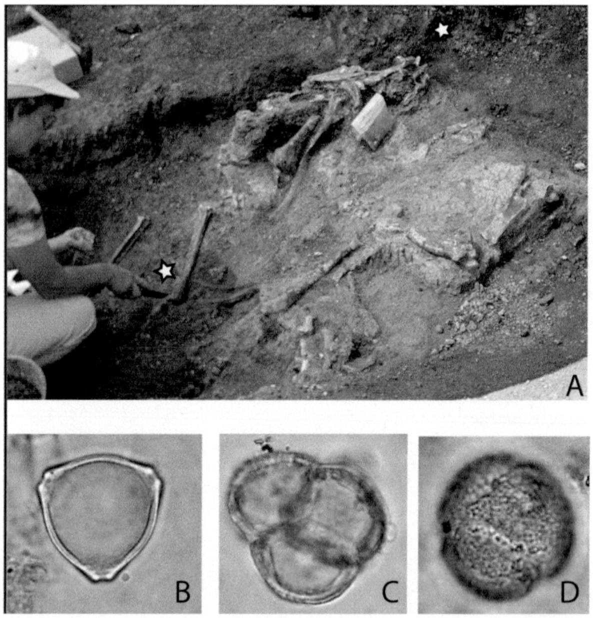

Fig. 6.4 Senèze, Domeyrat (Haute-Loire, France). A, Sampling of sediment near the cervid skeleton in upper T5. Pollen grains: B, *Corylus;* C, *Calluna;* D, Ranunculaceae. The stars show the positions of samples. Scale bar: 20 μm. (Fig. 6.4. Senèze, Domeyrat (Haute-Loire, France). A, Echantillonnage près du squelette de cervidé dans la tranchée T5 supérieure. Grains de pollen: B, *Corylus;* C, *Calluna;* D, Ranunculaceae. Les étoiles montrent la localisation des échantillons. Echelle: 20 μm.)

Table 6.5 Results of the pollen analysis of Senèze samples from trench T5, column C4 and near cervid bones. (Tableau 6.5. Résultats de l'analyse pollinique des échantillons de la tranchée T5 de Senèze, colonne C4 et près des os de cervidés).)

SENEZE 2004	Upper T5. col C4					Cervid	
						metapodial	tibia
Sample number	5	6	7	8	9		
Anacardiaceae					1		
Alnus	1			1		1	1
Betula							2
Cedrus	1	1			1	1	1
Corylus		1	2		3	2	6
Fagus			1		1		1
Pinus	2	2	3	2		1	
Quercus	1	1				2	4
AP	**5**	**5**	**7**	**3**	**6**	**7**	**15**
NAP	**12**	**3**	**5**	**1**	**2**	**4**	**4**
Caryophyllaceae			1	1			
Lamiaceae	2						
Poaceae	3	3	2		2	1	2
Ranunculaceae	1					1	
Calluna	3						
Polygonum bistorta	1						
Rumex	1		1			1	1
Cyperaceae	1		1			1	1
TOTAL (AP + NAP)	**17**	**8**	**12**	**4**	**8**	**12**	**19**
Sphagnum	1	1	1			1	1
Monolete spores			1				
Indét		1	1				

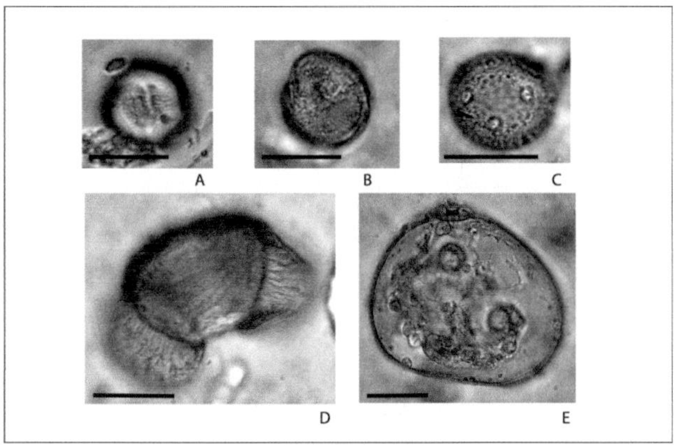

Fig. 6.5 Senèze, Domeyrat (Haute-Loire, France). Examples of pollen grains found near the skeleton of *Allohippus*: A-B, *Platanus*; C, *Plantago*; D, *Cedrus*; E, Poaceae. Scale bars: 20 µm. (Fig. 6.5. Senèze, Domeyrat (Haute-Loire, France). Exemples de grains de pollen trouvés près du squelette d'*Allohippus*: A-B, *Platanus*; C, *Plantago*; D, *Cedrus*; E, Poaceae. Echelle: 20 µm.)

In this sample:

- The pollen of trees accounts for 53.7% of the total pollen sum. Two taxa clearly dominate: *Quercus* (23.15%) and *Pinus* (10.2%). Many other leafy trees and some conifers (*Abies, Cedrus, Juniperus*) are also represented, but they only appear in a small quantity (from 1 to 3 grains only). Some are trees or shrubs requiring edaphic moisture such as *Alnus*, *Ligustrum* or *Platanus* (Fig. 6.5 A-B). The presence of *Platanus*, also observed in the coprolite C030091, is thus confirmed, as are those of *Cedrus* (Fig. 6.5 D) (observed in the T5 trench) and *Castanea*. Thus the locality was partially wooded, but heliophilous trees (*Betula, Juniperus, Corylus*) can develop in open spaces, in which herbs are also well represented (46.3%), especially with *Plantago lanceolata* type (Fig. 6.5 C), *Rumex*, Asteraceae and Cichorioideae, Apiaceae and Poaceae. Among these - as in sample T1-C3-21 – there is one grain of large size (approx. 65 µm, Fig. 6.5 E). Grains of this type are also abundant (15 grains) in sample 063 taken near the ribs of the horse. They might relate to the proximity of the guts of the animal which could contain remains of food at the moment of its death.

If we assume that the 5 samples taken close to the bones likely represent contemporaneous deposits, it is then possible to add together all the pollen grains that they contain. The total pollen sum reaches 155 grains, with 14 taxa of trees and 14 of herbs. The percentages obtained are logically

Table 6.6 Results of the pollen analysis of Senèze samples from around the 2005 *Allohippus* skeleton. (Tableau 6.6. Résultats de l'analyse pollinique des échantillons pris autour du squelette d'*Allohippus* de Senèze 2005.)

SENEZE 2005	*Allohippus*								E. section/*Allohippus*	
	Skull	**Axis**		Scapula	Ribs	Pelvis	**Sum**		Upper layer	Lower layer
Sample number	53	52		59	63	50	Samples 50–52-53–59-63		64	67
	n	n	%	n	n	n	n	%	n	n
Abies		2	1.85			1	3	1.94		
Alnus		3	2.78				3	1.94		
Betula		2	1.85				2	1.29		
Castanea		3	2.78				3	1.94		
Cedrus		3	2.78				3	0 02		
Corylus		2	1.85			2	4	2.58		
Hedera		1	0.93				1	0.65		
Juniperus		1	0.93				1	0.65		
Ligustrum		2	1.85				2	1.29		
Pinus	2	11	10.19	1	3	7	24	15.48	2	
Platanus		1	0.93				1	0.65		
Quercus	1	25	23.15				26	16.77	1	1
Undeterminate conifer		1	0.93			1	2	1.29	1	
Tilia		1	0.93				1	0.65		
AP	**3**	**58**	**53.70**	**1**	**3**	**11**	**76**	**49.03**	**4**	**1**
NAP	**0**	**50**	**46.30**	**0**	**22**	**7**	**79**	**50.97**	**3**	**0**
Apiaceae		4	3.70				4	2.58		
Asteraceae undet		4	3.70				4	2.58		
Caryophyllaceae		1	0.93				1	0.65		
Chenopodiaceae		1	0.93				1	0.65		
Cichorioideae		4	3.70			3	7	4.52		
Fabaceae		3	2.78				3	1.94		
Large Poaceae		1	0.93		15	1	17	10.97		
Poaceae		8	7.40		7	3	18	11.61	1	
Plantago lanceolata-type		11	10.19				11	7.10		
Ranunculaceae		1	0.93				1	0.65		
Rubiaceae		2	1.85				2	1.29		
Rosaceae		6	5.56				6	3.87		
Rumex		3	2.78				3	1.94		
Scrophulariaceae		1	0.93				1	0.65	1	
Total	**3**	**108**		**1**	**25**	**18**	**155**		**7**	**1**

Note The sample from the axis element of this skeleton is the richest at the site, thus it is emphasized in bold and its percentages are provided

close to those of the richest sample 052, except for the large size Poaceae because that pollen is concentrated in a single place (063). No pollen grain of any hygrophilous plant was registered. Was the shore of the lake unfavorable to the development of wetland shrubs and herbs? A steep slope and coarse sediments which often slumped could explain this, but it is only an assumption that cannot be easily verified for now.

This analysis makes it possible to think that the horse moved in an open milieu. There were grassy lands covered with Poaceae, various Asteraceae, Apiaceae, *Plantago lanceolata* type, *Rumex*… scattered with groups of trees distributed between the banks of the lake and the nearest summits.

Sediments in Contact with Fish Bones

Near trench 2, fish bones were embedded in very thin dark grey clastic clays. Unfortunately these clays were totally sterile.

Coprolites

The residues collected at the end of preparation of the coprolites were very limited. Only three samples among those taken near the skeleton of *Dicerorhinus etruscus* (trench T2) contained some well preserved pollen grains (Table 6.7, Fig. 6.6).

Table 6.7 Results of pollen analysis of coprolites and their embedding sediment samples collected near the skeleton of *Dicerorhinus etruscus* (trench T2, 2001); the sediment sample numbers are one higher than those of the associated coprolites (e.g., C030091 and S030092, dark grey; C030195 and S030196, white; C030173 and S030174; light grey). (Tableau 6.7. Résultats de l'analyse pollinique des coprolithes et des échantillons des sédiments encaissants recoltés près du squelette de *Dicerorhinus etruscus* (tranchée T2, 2001); les numéros des échantillons de sédiment sont plus grands d'une unité que ceux des coprolithes associées (ex: C030091 et S030092, gris foncé; C030195 et S030196, blanc; C030173 et S030174; gris clair).)

SENEZE 2003	(C) Coprolites			(S) Sediment associated with the coprolites				
N° of sample	C030091	C030195	C030173	S030092	S030196	S030174	S030055	S030176
	n	n	n	n	n	n	n	n
Abies	1	2	1	4		1	1	
Alnus	1							1
Betula	1				1			
Carpinus	1			3				
Castanea	1	1						
Corylus	1	1	2	2	2			
Fagus	2	2	1	3				
Fraxinus				1				2
Juglans				1				
Juniperus								1
Platanus	1			1				3
Pinus	1	2	1	3			1	15
Quercus	1		4	2	4			11
Salix								1
Tilia	1				1			1
AP	**12**	**8**	**9**	**20**	**8**	**1**	**2**	**35**
NAP	**4**	**6**	**2**	**19**	**14**	**0**	**0**	**4**
Apiaceae		1		1				
Poaceae	4	4		13	12			2
Plantago lanceolata - type		1						
Polygonum bistorta - type				3				
Rumex				2	1			
Cyperaceae					1			
Filipendula			1					2
Typha			1					
Monolete spore		1						
Indeterminate	2							1
All TOTAL (AP + NAP + SP + Undet)	**18**	**15**	**11**	**39**	**22**	**1**	**2**	**40**

Seventeen taxa were identified: eleven trees, five herbs and one fern. Figure 6.6 shows the good state of conservation of pollen grains. Qualitatively, the trees are dominant, the leafy trees being the most abundant and the most diversified with nine taxa: *Quercus*, *Fagus*, *Alnus*, *Tilia*, *Betula*, *Corylus*, *Carpinus*, *Castanea* and *Platanus*. The conifers are *Pinus* and *Abies*. Poaceae, Apiaceae, *Rumex*, and *Plantago lanceolata* type suggest open spaces; while *Filipendula*, Cyperaceae, and Typhaceae indicate a wet environment, probably connected with the bank of the lake.

The sediments associated with the coprolites yield six more taxa: *Fraxinus*, *Salix*, and *Juglans* for the trees, and *Rumex*, Cyperaceae and *Polygonum* for the herbs. From a qualitative point of view, their pollen content is reminiscent

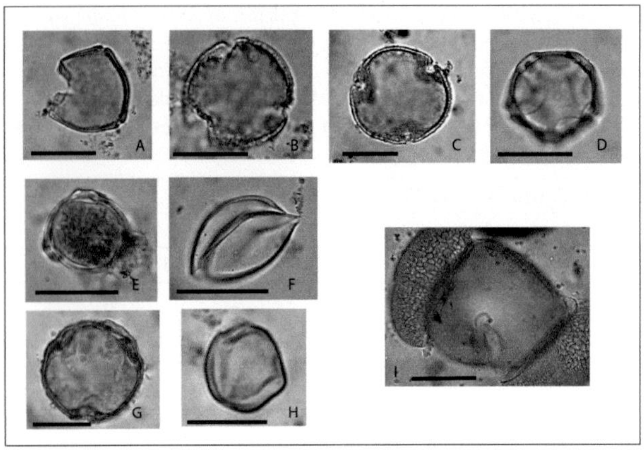

Fig. 6.6 Senèze, Domeyrat (Haute-Loire, France). Pollen grains extracted from coprolites: A, *Corylus*; B, *Quercus*; C, *Tilia*; D, *Alnus*; E, *Betula*; F, *Juniperus*; G, *Fagus*; H, *Filipendula*; I, *Abies*. Scale bars: 20 µm. (Fig. 6.6. Senèze, Domeyrat (Haute-Loire, France). Grains de pollen extraits des coprolithes: A, *Corylus*; B, *Quercus*; C, *Tilia*; D, *Alnus*; E, *Betula*; F, *Juniperus*; G, *Fagus*; H, *Filipendula*; I, *Abies*. Echelle: 20 µm.)

of that of the coprolites, in particular by the presence of the same leafy trees.

The isolated coprolites analyzed from other sections were unfortunately sterile.

Discussion

Taphonomic Observations: Relations Between Presence and Aspect of Pollen and Sediments

It should be remembered that most of the samples were taken from the slope deposits of the crater. Many of them were found to be sterile or very poor in pollen and spores whatever the sediment and in spite of the concentration by heavy liquid used for their preparation. Similarly, Carrión et al. (2009) point out the frequent lack of pollen material in Plio-Pleistocene paleontological sites, such as Fonelas and Mencal (Spain), where, as at Senèze, very diverse sediments including clays, silts and sands formed on the margins of a paleo lake environment are found.

However, at Senèze some differences can be observed according to the type of sediment analyzed.

The Colluvium

Composed of volcanic products, they are always almost sterile in pollen material. Their emplacement on steep slopes, their composition and the importance of oxidation, can easily explain the paucity of pollen in the sediment. Some samples here are exceptions to this rule when they are in contact with skeletons (for example near the *Allohippus* skeleton).

The Polyhedric Clays (T1)

They are poor in pollen grains but those present are very well preserved. Some even have a fresh aspect and one might suggest it represents pollution by modern pollen. This hypothesis can be refuted because of the compactness of the clay and the deep position of the samples. The same observation is done on very compact clays interbedded with gypsum on the banks of Lake Jabbul in Syria (J. Argant, personal observation). The absence of corrosion, degradation or thinning as described by Havinga (1984) can be due to a very fast sinking of grains in the mud. On the other hand, the scarcity of the grains would be due to a limited input of pollen within the clays, because of sparse local vegetation with only pollen of long distance origin (essentially pine) arriving and/or the existence of lake currents quickly carrying pollen offshore. One can also invoke introduction of pollen by various animals (e.g., insects or insectivores such as fishes or birds) as demonstrated in some Quaternary lake deposits (Bińka 2003).

Coprolites

It is known that pollen is incorporated into coprolites in different ways, absorbed with food (meat and stomach content of the prey), with water, by the licking of fur or paws and, in some cases, even by ingesting vegetable matter (Argant 1990, 2004, 2014; Carrión et al. 2001; Leroi-Gourhan 1966; Moe 1983; Scott 1987; Yll et al. 2006). Taphonomic observations on fresh hyena dung show that the pollen spectra obtained from coprolites give relatively unbiased pictures of the landscape (Scott et al. 2003). However, while pollen conservation is generally good in hyena coprolites of the genera *Crocuta* or *Hyaena*, it seems that the coprolites of *Chasmaporthetes* and *Pachycrocuta* are generally very poor or even sterile (Carrión et al. 2009; Argant 2004; Argant et Dimitrijevic 2007). This is confirmed by the analyzed Senèze coprolites. Fossilization processes acting in this case over a very long time may be the cause (Scott 1987).

Nevertheless, pollen encompassed in the mass of the three coprolites near the skeleton of *Dicerorhinus etruscus* (trench T2) provides available information about the plants growing around the lake. Found in the same vicinity and on the same level, they are of comparable age, and thus the results given by each of them can be regarded as representative of the same environment. In addition, they make it possible to validate the results obtained on sediments where the same list of taxa is found—in particular that of leafy trees.

Interpretation of the Different Results

If we consider all the results obtained in the analyzed samples (sediments and coprolites), 42 taxa were identified: 18 are trees, 22 herbs, one type of pteridophyte spore (monolete) and one of moss (*Sphagnum*). The taphonomic processes acting on the pollen assemblages are obviously intense and may affect the reliability of the pollen spectra. We will consider these results globally and only from a qualitative point of view.

The taxa recorded in the samples from Senèze mostly indicate the presence of leafy trees: *Quercus*, *Corylus*, *Alnus* and, more sporadically *Carpinus*, *Castanea*, *Fagus*, *Tilia*, *Betula*, *Fraxinus*, *Juglans*, *Ligustrum* and *Platanus*. *Hedera* occurs one time. The conifers are *Pinus*, *Abies*, *Cedrus* and *Juniperus*. Most of these taxa are European leafy and coniferous trees native to the region. Alongside these, *Juglans* (1), *Platanus* (2), *Castanea* (3), as well as *Cedrus* (4) are among the taxa considered to have disappeared from most of the European continent either at the end of the Early Pleistocene or during more recent periods of the Quaternary. In order to date the disappearance of relict taxa from the end of the Early Pleistocene, Magri et al. (2017) compiled the most significant pollen data from about 100 sites in Southern Europe. For *Castanea*, *Juglans* and *Platanus*, the data collected at these sites are not sufficient to conclude. One can only note that they are recorded in the Early Pleistocene, and that, depending on the region, some of them last until the Holocene and even the present time: this is the case, for example, for *Juglans* in Spain at Atapuerca (Gran Dolina). In this site we also find mention of *Platanus* for the last time between 0.8 and 0.6 Ma (TD5-TD8). This tree is also present at Saint-Vallier around 2.4 Ma (Argant 2004; pollen in a coprolite). For all of these taxa, the percentages recorded are generally very low (< 2%). The most abundant data concern *Cedrus* (Magri 2012). In the Early Pleistocene it is notably found at Bernasso (Leroy & Roiron 1996) and Nogaret (Leroy & Seret 1992), to mention only the sites closest to Senèze, both geographically and chronologically (around 2 Ma; Table 6.8). It persists sporadically until today in Spain and southern Europe.

Successive changes in vegetation are now clearly recognized in southern Europe for the Gelasian and Calabrian Stages, from 2.6 Ma onwards (Leroy 2007; Leroy et al. 2011). The stages of these changes during a glacial-interglacial cycle forced by obliquity (41 ka) are: open vegetation (steppe or tundra); dry open woodland; deciduous forest; mixed forest; and conifer forest (often *Picea* or *Abies*). These reflect climate changes from dry and cold to warmer but still dry, to warm and humid, and finally cool and still humid. Based on the taxa found in the fossiliferous layers of Senèze, where leafy trees are commonly present, one can logically consider that the paleontological deposit is located in an interglacial period. However, it is more difficult to assess the importance of afforestation and to say how dry or humid this period was. One can only note the rarity of edaphic moisture indices and think of a transitory

Table 6.8 Paleoenvironmental comparison with other sites in the Massif Central and its borders dated between 2.2 and 1.2 Ma. (Tableau 6.8. Comparaison de paléoenvironnement avec d'autres sites du Massif Central et de ses bordures datés entre 2. 2 et 1.2 Ma)

Site (Author)	Department	Layer	Approximate age of the fauna after the authors	Vegetation type	Occurrences of
Ceyssaguet (Argant & Bonifay 2011)	Haute-Loire	Coprolites from the fossiliferous layer B2	1.4- 1.2 Ma	Temperate vegetation, leafy trees	*Cedrus*
Peyrolles (Valli et al. 2006)	Puy-de-Dôme	Fossiliferous layer (square BW)	1.5–1.4 Ma	Wooded zones with leafy trees	*Cedrus*
Saint-Vallier (Argant 2004; Leroi-Gourhan, 1973)	Drôme	Hardened loess LD 3 *Hyaena* coprolite	2.2 Ma	Cool temperate vegetation with leafy trees and conifers	*Platanus*, *Cedrus*
La Rochelambert (Defive et al. 2005)	Haute-Loire	IC (pollen zones Rlb-2 to Rlb-4)	2.2–1.8 Ma (Tiglian)	Two phases (Rlb3-Rlb-5b) with open landscape alternating with a more wooded phase (Rlb-5a)	
Ceyssac (Ablin 1991)	Haute-Loire	Cey-1b et 3b	Tiglian / Bavelian 2.0–1.5 Ma	Alternation of wooded phases and steppic phases	*Cedrus*

situation between dry open woodland and deciduous forest, reflecting a climate change to warmer but still dry. In the Le Puy-en-Velay graben (Table 6.8), pollen and sedimentological records show several cold intervals due to cooling phases during the Tiglian (Gauthier in Defive et al. 2005). It seems then that the faunas lived in alternatively open and more or less wooded landscapes with both leafy trees and conifers—with *Cedrus* among them (except at La Rochelambert).

What is the Relationship Between the Core Pollen Units (Elhaï 1969) and the Paleontological Samples?

The diagram obtained by Elhaï shows that *Juglans*, *Castanea* and *Cedrus* persisted during the lake filling period, but that they appear only in small proportions as in the fossil layers, for some of them very sporadically, and it is noted that the plane tree was not encountered in the study. *Eucommia* is one of the relict taxa best represented in the Elhaï diagram where its highest rates occured in the S3 pollen zone, at the time when the oak forest dominated, and it reappeared in notable quantities in the most temperate episodes of the S1 zone. On the other hand, it is not found in the fossiliferous layers of the recent excavations, which do not contain *Ulmus* pollen either, although it is quite abundant throughout the diagram.

This link is difficult to establish because of the distance and obviously the difference in altitude between them. The highest lake level was observed at 625 m elevation in trench T2, and pollen analyses of the fossiliferous levels are distributed between 613 and 624 m (see Debard 2024).

In the context of punctual analyses, as is the case here for paleontological sediments, the absence of a taxon in a single sample does not mean its absence in the environment. However, based on all 24 samples and the 605 counted pollen grains and spores related to the faunal assemblages of the recent excavations, the diversity of the analysed environments allow us to consider that the absence of relicts on the one hand and that of *Ulmus* on the other hand distinguish the spectra of the paleontological sediments from those of the terminal part of the deep core.

The first five meters of the drilling starting at elevation 590 m are pollen sterile. Therefore, placing the analyses of the paleontological levels in relation to the palynological results of Elhaï becomes difficult. Although their spectra are reminiscent of those of the S.1 b pollen zone or even of the S1.d, due to their European leafy tree content and a significant frequency of herbaceous plants, the pollen reported here differ from the latter by the absence of *Ulmus* and *Eucommia* and the presence of *Platanus*. They are even more clearly distinguished from the latest core pollen zone S.1a, which ends the sequence by evoking colder conditions.

How can this be explained? The absence of pollen in the terminal part of the core contrasts with the continuous presence of numerous pollen grains in the remainder which documents a lacustrine environment (Elhaï 1969). If the last five meters correspond to the final lake deposits, there are no objective reasons for this change. Furthermore, the diatom study does not record the ultimate phase of lacustrine filling (Ehrlich (1968: 276). Numerous changes have been revealed in the maar by the sedimentological study of fossiliferous deposits with layer shifts toward the center of the maar, sometimes in connection with rapid drops in the lake level due to the incision of the outlet (Pastre et al. 2015). After the disappearance of the lake following these ineluctable drops in level, the basin found itself in a terrestrial environment. It seems obvious that erosion and collluviation played an important role inside the crater framed by steep slopes. Their erosion could only be amplified by humans in the Holocene through deforestation and an early and rather intensive agropastoral use of these particularly fertile volcanic soils. For example, in archaeology we often note an accumulation of colluvium of about 5 m since the Roman period, always sterile in pollen. No precise description of the upper 5 m of the core is given in the literature, other than the mention of clays of various colors, more or less compact and sandy, and the presence of diatoms limited to a single sample one meter below the surface. There is no argument for linking them to the end of a lake infill rather than to deposits formed from the reworking of the upper part of the infill truncated by post-lake erosion. Colluviation has created this layer in the terrestrial environment, a deposit where pollen is generally absent. This hypothesis may explain the inexact correspondance between the pollen spectra of the fossiliferous layers and those from the core, as well as the absence of pollen in the upper five meters of the core. Thus it appears that the vegetation of the fossiliferous layers from the new excavations can only be more recent that those of the Elhaï diagram, which has clearly been truncated by erosion. In turn, the correlations proposed by Elhaï and Ablin (Table 6.1) cannot be correct, as the levels they suggest to be Tiglian, Eburonian and even perhaps Waalian are in fact older than the 2.2 Ma date of the base of the paleontological deposits. Westerhoff et al. (2020, Fig. 1) date the base of the Tiglian around 2.25 Ma, the base Eburonian around 1.77 Ma (top of the Olduvai subchron) and the base of the Waalian around 1.5 Ma. In the Praetiglian (ca. 2.6–2.25 Ma), there are several climatic fluctuations which might correlate to those seen near the top of the Senèze core.

Conclusion

The excavations programmed in Senèze from 2001 to 2006 under the direction of M. Faure, C. Guérin and E. Delson aimed to collect as much data as possible in all fields on this paleontological site, and in particular all the possible remains by carefully locating them in space (Delson et al. 2006). Pollen, a microscopic fossil witness of the vegetation, was naturally integrated into this research. In the context of the complex slope deposits of the crater, one could of course not expect to obtain a record comparable to what can be obtained from purely lacustrine deposits or peat bogs. It is therefore not surprising to find numerous sterile or insufficiently rich samples, but the elements collected, especially in contact with skeletons, confirmed by the content of hyaenid coprolites, make it possible to imagine an environment with both open spaces and wooded areas, under relatively temperate or cool temperate conditions. Paleoenvironmental analysis (in Delson et al. 2024b) reaches the same conclusion. This leads one to think that part of the fauna expanded in an interglacial warming phase of the Gelasian Stage around 2.2–2.0 Ma, considering the $^{40}Ar/^{39}Ar$ dating (Nomade et al. 2014; Delson et al. 2024a). If climatic variations existed during the formation of the deposits in the fossil site, it is impossible to highlight and to quantify them because not enough pollen was preserved.

Acknowledgments First of all, I would like to thank Martine Faure, Claude Guérin and Eric Delson for entrusting me with the pollen analysis of Senèze. This analysis benefited from funding allocated by the DRAC of Auvergne for the excavation of the paleontological site. Samples were prepared by Alain Argant in the laboratory of A.R.P.A. My thanks also go to Agnès Gauthier, Suzanne Leroy and an anonymous reviewer for their valuable comments and suggestions for improving the manuscript. Finally, I am grateful to Eric Delson for the improvement of the English text.

References

Ablin, D. (1991). Analyse pollinique des dépôts lacustres de Ceyssac, Plio-Pléistocène du Velay (Massif Central, France). *Cahiers de Micropaléontologie, 6*, 21–38.

Argant, A. (2024). The Carnivores of Senèze (Haute-Loire). In E. Delson, M. Faure and C. Guérin (eds.) *Senèze: Life in central France two million years ago. Paleontology, geochronology, stratigraphy and taphonomy* (pp. 165–243). Cham, Switzerland: Springer.

Argant, J. (1990). Climat et environnement au Quaternaire dans le Bassin du Rhône d'après les données palynologiques. *Documents du Laboratoire de Géologie de Lyon, 111*, 1–199.

Argant, J. (2004). Le gisement pliocène final de Saint-Vallier (Drôme, France): palynologie. In M. Faure & C. Guérin (Eds,), *Le gisement pliocène final de Saint-Vallier (Drôme, France)*. Mémoire spécial 26, *Geobios* (Vol. 37, pp. 81–90).

Argant, J. (2014). La copropalynologie: Coprolithes et paléoenvironnement. *Bulletin du Musée d'Anthropologie préhistorique de Monaco, 54*, 41–46.

Argant, J., & Bonifay, M.-F. (2011). Les coprolithes de hyène (*Pachycrocuta brevirostris*) de la couche 2 du site villafranchien de Ceyssaguet (Lavoûte-sur-Loire, Haute-Loire, France): Analyse pollinique et indications paléoenvironnementales. *Quaternaire, 22*, 3–11.

Argant, J., & Dimitrijevic, V. (2007). Pollen analyses of Pleistocene hyaena coprolites from Montenegro and Serbia. *Annales Géologiques de la Péninsule Balkanique, 68*, 73–80.

Bińka, K. (2003). Palynological evidence for plant-animal interaction in the late Holocene. *Vegetation History and Archaeobotany, 12*, 37–47.

Carrión, J. S., Fernández, S., González-Sampériz, P., Leroy, S. A. G., Balley, G. N., López-Sáez, J. A., et al. (2009). Quaternary pollen analysis in the Iberian Peninsula : the value of negative results. *Internet Archaeology, 25*.

Carrión, J. S., Riquelme, J. A., Navarro, C., & Munuera, M. (2001). Pollen in hyaena coprolites reflects late glacial landscape in Southern Spain. *Palaeogeography, Palaeoclimatology, Palaeoecology, 176*, 193–205.

Couthures, J., Ablin, D., & Hadjouis, D. (1991). *Senèze (Haute-Loire)*. Rapport de fouilles 1991, Direction Régionale des Affaires Culturelles Auvergne, 55 p. (Unpublished).

Debard, E. (2024). Geological study of the paleontological site of Senèze (Domeyrat, Haute-Loire, France). In E. Delson, M. Faure and C. Guérin (eds.) *Senèze: Life in Central France two million years ago. Paleontology, geochronology, stratigraphy and taphonomy* (pp. 37–82). Cham, Switzerland: Springer.

Defive, E., Gauthier, A., & Pastre, J.-F. (2005). L'évolution plio-quaternaire du Bassin du Puy (Massif central, France): Rythmes morphosédimentaires et volcanisme. *Quaternaire, 16*, 177–190.

Delson, E., Faure, M., Guérin, C., Aprile, L., Argant, J., Blackwell, B. A. B., et al. (2006). Franco-American renewed research at the Late Villafranchian locality of Senèze (Haute-Loire, France). *Courier Forschungs-Institut Senckenberg, 256*, 275–290.

Delson, E., Nomade, S., Sen, S., Debard, E., Pastre, J-F., Bahain, J.-J., et al. (2024a). Geochronology of Senèze: argon dating and magnetostratigraphy, with notes on an ESR/U-series dating attempt. In E. Delson, M. Faure & C. Guérin (Eds.), *Senèze: Life in Central France two million years ago. Paleontology, geochronology, stratigraphy and taphonomy* (pp. 99–122). Cham (Switzerland): Springer.

Delson, E., Guérin, C., Faure, M., Aprile, L, Argant, A., Argant, J., et al. (2024b). Summary and Conclusions of the Senèze Research Project. In E. Delson, M. Faure & C. Guérin (Eds.), *Senèze: Life in Central France two million years ago. Paleontology, geochronology, stratigraphy and taphonomy* (pp. 683–699). Cham (Switzerland): Springer.

Elhaï, H. (1969). La flore sporo-pollinique du gisement villafranchien de Senèze (Massif Central, France). *Pollen et Spores, 11*, 127–139.

Ehrlich, A. (1968). Les diatomées fossiles des sédiments villafranchiens de Senèze (Haute Loire, Massif Central français). *Bulletin de l'Association française pour l'étude du Quaternaire, 4*, 267–280.

Faure, M., Guérin, C., & Delson, E. (2005). Rapport 2005 sur l'opération de fouille programmée annuelle dans le gisement paléontologique villafranchien supérieur de Senèze à Domeyrat, Haute-Loire, Ministère de la Culture.

Faure, M., Delson, E., & Guérin, C. (2024). Introduction and history of research in the Senèze maar. In E. Delson, M. Faure & C. Guérin (Eds.), *Senèze: Life in Central France two million years ago.*

Paleontology, geochronology, stratigraphy and taphonomy (pp. 1–23). Cham (Switzerland): Springer.

Frenzel, B. (1964). Zur Pollenanalyse von Lössen. Untersuchung der Lössprofile von Oberfellabrunn und Stillfried (Niederösterreich). *Eiszeitalter und Gegenwart, 15*, 5–39.

Girard, M., & Renault-Miskovsky, J. (1969). Nouvelles techniques de préparation en palynologie appliquées à trois sédiments du Quaternaire final de l'abri Cornille (Istres, Bouches-du-Rhône). *Bulletin de l'Association française pour l'étude du Quaternaire, 1969–4*, 275–283.

Goeury, C., & de Beaulieu, J. (1979). A propos de la concentration du pollen à l'aide de la liqueur de Thoulet dans les sédiments minéraux. *Pollen et Spores, 21*(1–2), 239–251.

Guérin, C. (2006). « Domeyrat – Senèze » [notice archéologique], *ADLFI. Archéologie de la France - Informations* [En ligne], Auvergne-Rhône-Alpes, mis en ligne le 01 mars 2006, URL : http://journals.openedition.org/adlfi/4661

Havinga, A. J. (1984). A 20-year experimental investigation into the differential corrosion susceptibility of pollen and spores in various soil types. *Pollen et Spores, 26*, 541–558.

Leroi-Gourhan, A. (1966). L'analyse pollinique des coprolithes. *Bulletin de la Société préhistorique française, 63*, 163–164.

Leroi-Gourhan, A. (1973). Les cèdres villafranchiens de Saint-Vallier (Drôme). *Bulletin de l'Association française pour l'Etude du Quaternaire, 34*, 25–30.

Leroy, S. A. G. (2007). Progress in palynology of the Gelasian-Calabrian Stages in Europe: Ten messages. *Revue de Micropaléontologie, 50*, 293–308.

Leroy, S. A. G., Arpe, K., & Mikolajewicz, K. (2011). Vegetation context and climatic limits of the Early Pleistocene hominin dispersal in Europe. *Quaternary Science Reviews, 30*, 1448–1463.

Leroy, S. A. G., & Roiron, P. (1996). Latest Pliocene pollen and leaf floras from Bernasso palaeolake (Escandorgue Massif, Hérault, France). *Review of Palaeobotany and Palynology, 94*, 295–328.

Leroy, S.A.G., & Seret, G. (1992). Duration and vegetation dynamic of the Nogaret Interglacial (1.9 Ma, S. France). Tentative correlation with stage 75. In G. Kukla and E. Went (Eds.), *Start of a Glacial, Proceedings of the Mallorca NATO ARW, NATO ASI.* (pp. 113–125). Heidelberg: Springer Verlag.

Magri, D. (2012). Quaternary history of *Cedrus* in Southern Europe. *Annali di Botanica, 2*, 57–66.

Magri, D., Di Rita, F., Aranbarri, J., Fletcher, W., & González-Sampériz, P. (2017). Quaternary disappearance of tree taxa from Southern Europe: Timing and trends. *Quaternary Science Reviews, 163*, 23–55.

Moe, D. (1983). Palynology of sheep's faeces: Relationship between pollen content, diet and local pollen rain. *Grana, 22*, 105–113.

Moore, P. D., Webb, J. A., & Collinson, M. E. (1991). *Pollen analysis*. Blackwell.

Nomade, S., Pastre, J.-F., Guillou, H., Faure, M., Guérin, C., Delson, E., et al. (2014). ^{40}Ar/^{39}Ar constraints on some French landmark Late Pliocene to Early Pleistocene large mammalian paleofaunas: Paleoenvironmental and paleoecological implications. *Quaternary Geochronology, 21*, 2–15.

Pastre, J.-F., Debard, E., Nomade, S., Guillou, H., Faure, M., Guérin, C., et al. (2015). Nouvelles données géologiques et téphrochronologiques sur le gisement paléontologique du maar de Senèze (Pléistocène inférieur, Massif central, France). *Quaternaire, 26*(3), 225–244.

Pokrovskaia, I. M. (1950). *Analyse pollinique*. Annales du service d'information du B.R.G.M., éditions du C.N.R.S.

Reille, M., (1992). *Pollen et Spores d'Europe et d'Afrique du nord.*. Marseille: Laboratoire de Botanique Historique et de Palynologie.

Roger, S., Coulon, C., Thouveny, N., Féraud, G., Van Velzen, A., Fauquette, S., et al. (2000). ^{40}Ar/^{39}Ar dating of a tephra layer in the Pliocene Senèze maar lacustrine sequence (French Massif central): Constraint on the age of the Réunion-Matuyama transition and implications on paleoenvironmental archives. *Earth and Planetary Science Letters, 183*, 431–440.

Scott, L. (1987). Pollen analysis of hyena coprolites and sediments from *Equus* Cave, Taung, Southern Kahalari (South Africa). *Quaternary Research, 28*, 144–156.

Scott, L., Fernández-Jalvo, Y., Carrión, J. S., & Brink, J. (2003). Preservation and interpretation of pollen in hyaena coprolites: Taphonomic observations from Spain and southern Africa. *Palaeontologica Africa, 39*, 83–91.

Valli, A. M. F., Caron, J.-B., Debard, E., Guérin, C., & Argant, J. (2006). Le gisement paléontologique villafranchien terminal de Peyrolles (Issoire, Puy-De-Dôme, France): Résultats de nouvelles prospections. *Geodiversitas, 28*, 297–317.

Westerhoff, W. E., Donders, T. H., Trabucho Alexandre, N., & Busschers, F. S. (2020). Early Pleistocene Tiglian sites in the Netherlands: A revised view on the significance for quaternary stratigraphy. *Quaternary Science Reviews, 242*, 106417.

Yll, R., Carrión, J. S., Marra, A. C., & Bonfiglio, L. (2006). Vegetation reconstruction on the basis of pollen in Late Pleistocene hyena coprolites from San Teodoro Cave (Sicily). *Palaeogeography, Palaeoclimatology, Palaeoecology, 237*, 32–39.

Chapter 7
The Fishes from the Maar of Senèze

Jean Gaudant

Abstract The late Villafranchian fish fauna from the maar of Senèze (Haute-Loire, France) is restricted to only two species. The first is a tench, *Tinca* sp., the incomplete skeleton of which was found in the core of the research drilling carried out in 1965. The second is the recent perch, *Perca fluviatilis*. It is represented by one incomplete skeleton, a disarticulated head, and isolated remains (incomplete dentary, abdominal vertebra and scales) found during the recent excavations.

Résumé L'ichthyofaune du Villafranchien supérieur du maar de Senèze (Haute-Loire) ne comporte que deux espèces. La première est une tanche, *Tinca* sp., dont un squelette incomplet a été trouvé dans la carotte du sondage de recherche foré en 1965. La seconde est la perche actuelle, *Perca fluviatilis*. Elle est représentée par un squelette incomplet, une tête désarticulée et des restes isolés (dentaire incomplet, vertèbre abdominale et écailles) découverts pendant les fouilles récentes.

Keywords Teleosts • Cyprinids (tench) • Percids (perch) • Villafranchian • Anatomy

Mots-clés Téléostéens • Cyprinidae (tanche) • Percidae (perche) • Lac de cratère • Villafranchien • Anatomie

There is little available information concerning the fish fauna from the maar of Senèze: the material at hand is restricted to three incomplete fish skeletons and some isolated bony remains and scale fragments.

Family Cyprinidae
Genus *Tinca* Cuvier
Tinca sp.

An incomplete skeleton of a tench was found as a result of drilling a research borehole in the sediments filling the volcanic crater of Senèze. This drilling was carried out by the CNRS in 1965 under the supervision of Pierre Grangeon and reached a depth of 174 m (see Faure et al. 2024). The fish was found at −12.10 m in a greyish clayey sediment, preserved in part and counterpart, and first reported by Devis (1970). This incomplete skeleton was examined by Gaudant (1975), who figured the right side (here Fig. 7.1A) and recognized that it belongs to the recent genus *Tinca*. It was kept in the Pierre Grangeon collection, in the "Magmas et Volcans" laboratory of Université Blaise Pascal, Clermont-Ferrand, now Université Clermont-Auvergne.

The specimen lacks both the head and the caudal region, as the drill core went through the middle part of the body (Fig. 7.1B). The cleithrum is visible anteriorly. The pectoral fin, which consists of at least 14 or 15 rays, is preserved under the ventral arm of the cleithrum. It seems to have been rather short as its estimated length was about 2/3 of the pectoro-pelvic distance.

In the anterior part of the vertebral column, the transversal process of the fourth vertebral component of the Weberian apparatus is preserved; this confirms that the fish is an otophysan. The number of free abdominal vertebral centra is about 20. Long and robust pleural ribs are present in the abdominal region.

The dorsal fin begins above the pelvic fins. Only two anterior short rays and the base of the articulated unbranched long ray are present on the left side of the fish, whereas seven branched rays are visible behind the long unbranched ray on its right side. The endoskeleton of the dorsal fin

J. Gaudant—Deceased 2015

J. Gaudant (✉)
Université Paris 7, Paris, France

E. Delson et al. (eds.), *Senèze: Life in Central France Around Two Million Years Ago*,
Vertebrate Paleobiology and Paleoanthropology, https://doi.org/10.1007/978-3-031-64415-3_7

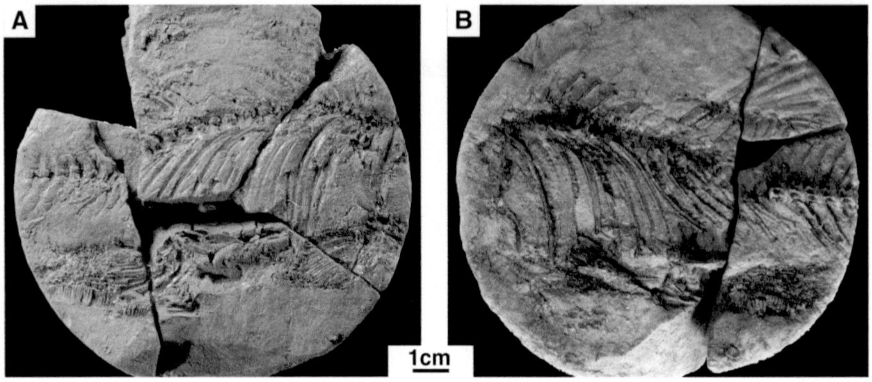

Fig. 7.1 A, B. *Tinca* sp. Part and counterpart of the male specimen found in the research borehole drilled at Senèze by the CNRS (1965)

includes nine pterygiophores. The first of them is situated behind the eighth long vertebral neurapophysis borne by the abdominal vertebrae.

The pelvic fin is clearly larger than the pectoral. It is supported by a long and robust pelvic bone, the length of which equals about 6/10 of the pectoro-pelvic distance. It consists of nine rays: a short lateral one is situated against the enlarged one which is thickened and exhibits short and wide elements, characterizing a male tench. The basal elements of seven branched rays are present against it. The scales are not preserved.

Male specimens of tenches are already known in the Pliocene, as they have been described from the Late Ruscinian of Willershausen am Harz (Lower Saxony, Germany) under the name *Tinca pliocaenica* Gaudant, 1997. This species is characterized by its caudal fin being divided, instead of paddle shaped as in the recent species *Tinca tinca*. Unfortunately, it is impossible to know the shape of the caudal fin of the Senèze fossil tench. In addition, other European Early Pleistocene localities have yielded tench remains: Tegelen, in the Netherlands (Gaudant, 1979); and in Italy, San Giovanni Valdarno (Lawley, 1876), Montetermini Valdarno (Pieragnoli, 1921), and Montecarlo Valdarno (Ghinassi et al., 2005).

Were the head preserved, the morphology of the pharyngeal teeth might have helped to determine the species. Without it, we can only identify this fish as *Tinca* sp.

Family Percidae
Genus *Perca* Linnaeus
Perca fluviatilis Linnaeus

Fig. 7.2 *Perca fluviatilis*. **A**: Head and part of body showing the anterior part of the dorsal fin. **B**: Disarticulated remains of a head. Key to abbreviations: Hmd: hyomandibular; Neur: neurocranium; Op: operculum; Q: quadratum; vert: vertebrae; fin sp.: fin spines

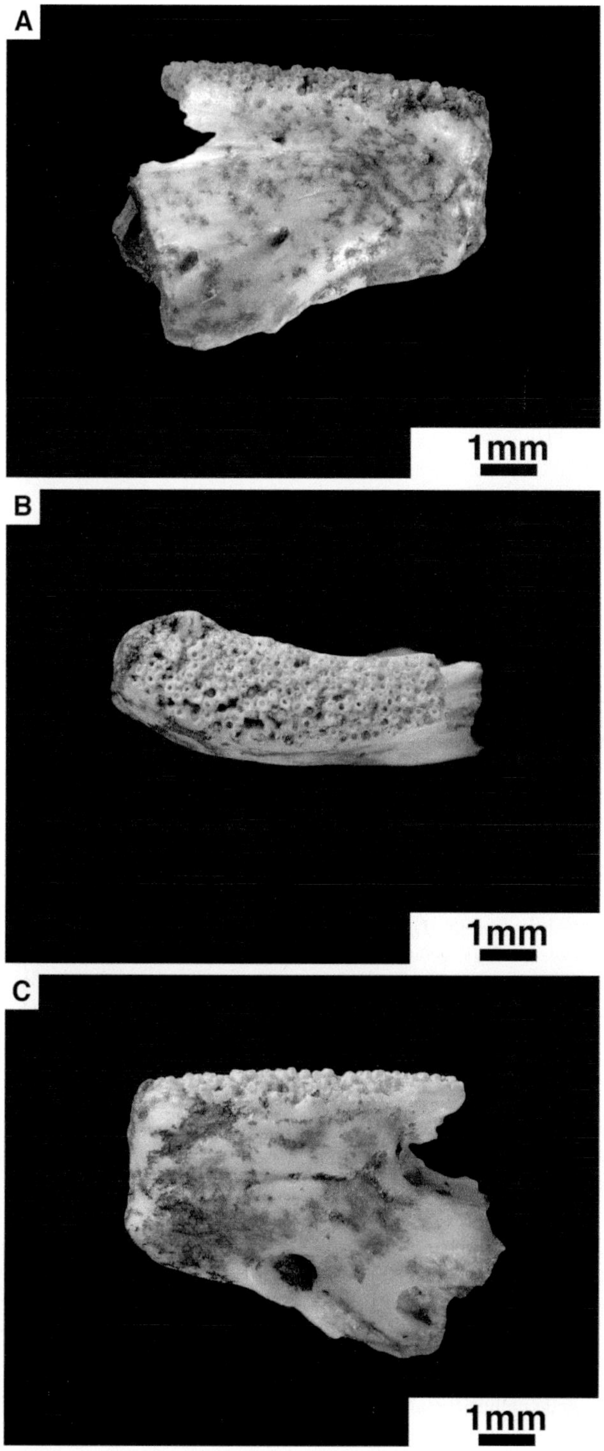

Fig. 7.3 *Perca fluviatilis*. Symphyseal region of a right dentary. **A**: lingual view; **B**: occlusal surface; **C**: labial view

Two more or less disarticulated skeletons, isolated vertebrae and scales recovered by the Franco-American fieldwork are indicative of the occurrence of percoid fishes in the maar of Senèze.

The best preserved specimen shows the head and the anterior part of the dorsal fin and several disarticulated vertebrae (Fig. 7.2A). The head has a length of 63 mm, corresponding to a standard length of about 210 mm. The remains of a lacrymal are present in front of the orbit. The maxillary and the premaxillary are also present, as well as remains of the lower jaw. The most characteristic and most diagnostic component is the preoperculum, which has its lower arm ventrally ornamented with four rather large spines that are oriented both downwards and forwards, as in the recent *Perca fluviatilis*. Three additional spines are borne by the postero-ventral part of the bone. They are oriented downwards and slightly backwards. Although it is incompletely preserved, the shape of the operculum, which is rather wide, fully agrees with that of the recent *Perca fluviatilis*. Although the dorsal fin is more or less disarticulated, it includes at least 11 spines, compared with the 13 to 17 spines of the recent *Perca fluviatilis*. The abdominal vertebrae are characterized by the presence of longitudinal ridges delimiting rather deep elongated fossae.

The second specimen is an argillaceous block bearing the disarticulated remains of a head, several vertebrae and four fin spines (Fig. 7.2B). Unfortunately, the state of preservation of these bones is rather poor, but they include a neurocranium, an operculum, a quadratum, a hyomandibular and other bony fragments. A bone situated near the operculum has a denticulated edge.

Although it is impossible to identify these percoid fish remains precisely, the find, in parcel 233, of an isolated anterior part of a dentary (Fig. 7.3) showing an oral surface covered with small round sockets which supported only small villiform teeth suggests that this percoid probably also belongs to the genus *Perca*.

An isolated abdominal centrum bearing the basal part of the neurapophysis (Fig. 7.4A) was also found in parcel 233. As in vertebrae of *Perca*, it is characterized by its lateral side bearing a longitudinal ridge above which a series of thin bony ridges delimit small alveoli. Finally, two scale fragments were also found in parcel 233, level Hi (Fig. 7.4B). Both exhibit four radii diverging forward from the nucleus. Although their posterior part is unknown, these scales look

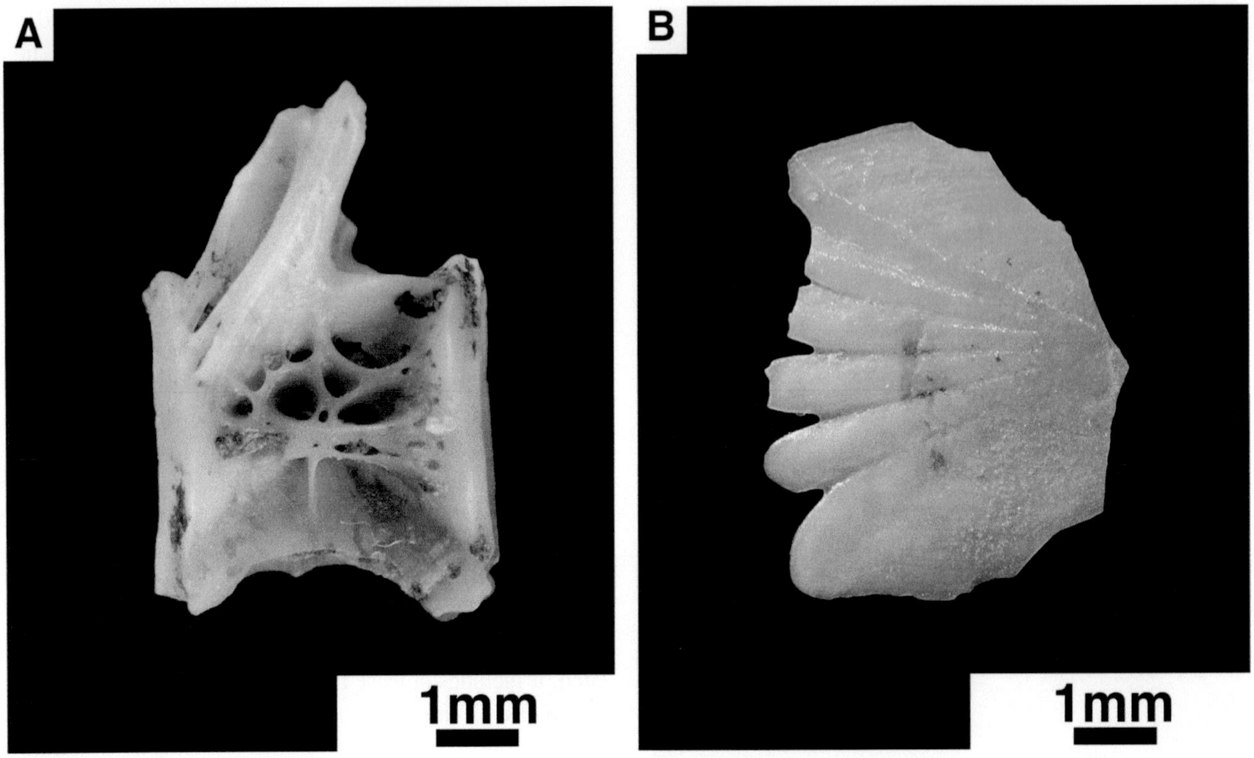

Fig. 7.4 *Perca fluviatilis*. **A**: lateral view of an abdominal vertebra; **B**: view of an incomplete scale

similar to those of *Perca*, because the distal ends of the radii are separate, giving a scalloped shape to the scale margin.

The occurrence of *Perca fluviatilis* in the late Villafranchian of Senèze is not surprising, because this genus is already known in the European Pliocene and Early Pleistocene. It was identified by Weiler (1965) in the interglacial clay of Bilshausen, near Göttingen (Germany), and is present in the late Ruscinian of Willershausen am Harz (Gaudant, 1997).

From a paleoenvironmental point of view, *Tinca tinca* preferentially lives on muddy and grassy bottoms in quiet rivers and ponds, whereas *Perca fluviatilis* dwells in water plant communities of rivers and lakes.

Acknowledgments The editors thank John Maisey for a number of helpful editorial suggestions which improved the clarity of the manuscript. Madelaine Boehme and Eric Delson also provided editorial comments.

References

Devis, G. (1970). Le volcanisme et le gisement fossilifère de Senèze (Haute-Loire). *Revue des Sciences naturelles d'Auvergne, 36*, 25–87.

Faure, M., Delson, E., & Guérin, C. (2024). Introduction and history of research in the Senèze maar. In Delson, E., Faure, M., & Guérin, C. (Eds.), *Senèze: Life in Central France two million years ago. Paleontology, geochronology, stratigraphy and taphonomy* (pp. 1–23). Cham (Switzerland): Springer.

Gaudant, J. (1975). Présence du genre *Tinca* CUV. (Poisson téléostéen, Cypriniforme) dans le Villafranchien de Senèze. *Geobios, 8*, 353–354.

Gaudant, J. (1979). L'ichthyofaune tiglienne de Tegelen (Pays-Bas): Signification paléoécologique et paléoclimatique. *Scripta Geologica, 50*, 1–16.

Gaudant, J. (1997). L'ichthyofaune pliocène de Willershausen am Harz (Basse Saxe, Allemagne)—un réexamen. *Stuttgarter Beiträge zur Naturkunde, (B), 257*, 1–51.

Ghinassi, M., Abbazzi, L., Esu, D., Gaudant, J., & Girotti, O. (2005). Facies analysis, stratigraphy and palaeontology (Molluscs and Vertebrates) in the Upper Pliocene sandy flood-basin deposits of the Upper Valdarno basin (Northern Apennines). *Rivista Italiana di Paleontologia e Stratigrafia, 111*, 467–487.

Lawley, R. (1876). Nuovi studi sopra ai pesci ed altri vertebrati fossili delle colline toscane (cf. pp. 84–85). Firenze: Tipografia dell'Arte della Stampa.

Pieragnoli, L. (1921). Due pesci pliocenici. *Rivista Italiana di Paleontologia, 27*, 11–14.

Weiler, W. (1965). Die Fischfauna des interglazialen Beckentons von Bilshausen bei Göttingen. *Neues Jahrbuch für Geologie und Paläontologie, Abhandlungen, 123*, 202–219.

Chapter 8
The Birds (Aves) from the Early Pleistocene Site of Senèze (Domeyrat, Haute-Loire, France)

Cécile Mourer-Chauviré

Abstract The Senèze avifauna includes four extinct taxa, *Pavo bravardi*, *Tetrao* cf. *partium*, *Surnia robusta*, and *Corvus corax antecorax*. *Pavo bravardi*, closely related to *Pavo muticus*, the Green Peafowl of South-East Asia, has been reported in seven European localities, of which Senèze is the most recent one. The other elements can be referred to Recent species, many of which either live on water, or on water banks, or in forested regions with lakes. The Senèze avifauna shows some similarities with the slightly older one of Saint-Vallier, in the same area, but at Senèze the environmental conditions indicated by birds correspond to a more humid climate and to a more wooded paleohabitat.

Résumé Le gisement de Senèze correspond à la biozone mammalienne MNQ 18. L'avifaune comporte les taxons suivants: *Ciconia nigra*, Cigogne noire; *Tadorna* cf. *ferruginea*, cf. Tadorne casarca; *Anas clypeata*, Canard souchet; *Aythya* sp., Fuligule; *Mergus* cf. *merganser*, cf. Harle bièvre; *Mergellus* sp., taille de *Mergellus albellus*, Harle piette; Anatidae indet.; *Haliaeetus albicilla*, Pygargue à queue blanche; *Buteo* sp., Buse; *Tetrao* cf. *partium*, forme primitive de Tétras-lyre; *Pavo bravardi*, Paon de Bravard; *Alectoris* cf. *graeca*, cf. Perdrix bartavelle; *Otis* sp. taille *Otis tarda*, Outarde de la taille de l'Outarde barbue; *Bubo* sp., forme proche du Grand-duc; *Surnia robusta*, grande forme de Chouette épervière; Turdidae indéterminé; et *Corvus corax antecorax*, forme ancienne de Grand Corbeau. Les formes éteintes sont *Pavo bravardi*, *Tetrao* cf. *partium*, *Surnia robusta* et *C. corax antecorax*. *Pavo bravardi* est une forme très proche de *Pavo muticus*, le Paon spicifère actuel d'Asie du Sud Est. Il a été identifié dans sept gisements du Plio-Pléistocène de France, Grèce, Moldavie et Bulgarie, et Senèze est jusqu'à présent le plus récent d'entre eux. *Tetrao partium* est probablement l'ancêtre direct de *T. tetrix*. Cette espèce a été signalée dans de nombreux gisements du Pliocène supérieur, et du Pléistocène inférieur et moyen d'Europe centrale et orientale, mais Senèze est le premier gisement où elle est signalée en Europe occidentale. *Surnia robusta* est représentée uniquement par une phalange du pied. Cette espèce est caractérisée par une très grande taille, beaucoup plus grande que celle de l'unique espèce actuelle de ce genre, *Surnia ulula*. *C. corax antecorax* est caractérisé par une taille un peu plus faible que celle de la majorité des sous-espèces actuelles de Grand Corbeau et par des proportions différentes. *Surnia robusta* est une espèce très rare. Durant tout le Pléistocène de la Province paléarctique, elle n'a été signalée que dans quatre gisements du Sud de la Hongrie, dont les âges s'étendent de MN 16/17 à MNQ 20, ainsi que dans le gisement de Ahl al Oughlam, au Maroc, sur la côte atlantique, daté d'environ 2,5 MA. Dans trois de ces gisements hongrois elle est associée à *Tetrao partium*, comme à Senèze. Parmi les différentes avifaunes de la fin du Pliocène et du début du Pléistocène, c'est de l'avifaune de Saint-Vallier que celle de Senèze se rapproche le plus. Elles ont en commun *Ciconia nigra*, *Haliaeetus albicilla*, *Pavo bravardi*, une grande outarde, et *C. corax antecorax*. Les oiseaux de Senèze comportent un grand nombre d'espèces vivant sur ou au bord de l'eau et d'autres indiquent la présence de forêts avec des lacs, ou de cours d'eau bordés de forêts. *Tetrao* cf. *partium* et *Surnia robusta* vivaient sans doute en limite de forêts ou dans des forêts de basse ou moyenne altitude. *Alectoris* cf. *graeca* et *Otis* sp. correspondent à des milieux d'espaces découverts, grandes plaines herbeuses ou couvertes de buissons pour l'outarde, milieux plus rocheux pour *Alectoris*, tandis que le Grand-duc ou le Grand Corbeau peuvent avoir des habitats très variés. L'environnement indiqué par les oiseaux de Senèze est plus

C. Mourer-Chauviré (✉)
Université de Lyon, UCBL, ENSL, CNRS, LGL-TPE, Villeurbanne, France
e-mail: cecile.mourer@gmail.com

E. Delson et al. (eds.), *Senèze: Life in Central France Around Two Million Years Ago*, Vertebrate Paleobiology and Paleoanthropology, https://doi.org/10.1007/978-3-031-64415-3_8

humide et plus boisé que celui indiqué par les oiseaux de Saint-Vallier. *Pavo bravardi* est une relicte tropicale. L'avifaune ne comporte pas d'éléments correspondant à un climat plus froid que le climat actuel.

Keywords Pavo • Tetrao • Surnia • Corvus • Systematics • Paleoecology

Mots-clés Pléistocène inférieur • Oiseaux • Pavo • Tetrao • Surnia • Corvus • Systématique • Paléoécologie

Introduction

The fossil birds examined here come from the old excavations carried out in the locality of Senèze. They were briefly reported by Stehlin (1923), and since that time, only the remains of peafowl and raven have been revised (Mourer-Chauviré 1990, 2004). Most of the material comes from the collection of the Naturhistorisches Museum Basel, and the numbers bear the prefix NMB Se. Only two bones come from the collection of the Université Claude Bernard, Lyon, and they bear the numbers FSL 91636 and FSL 91637.

Senèze is a reference locality for Mammal Neogene unit 18 (Delson et al. 2006; Crégut-Bonnoure et al. 2024). Several tephra layers, mostly bracketing the fossiliferous beds, have been dated between 2.2 Ma and 2.1 Ma (Nomade et al. 2014; Delson et al. 2024). The biochronological framework is based on Guérin (1982), Mein (1990), and Tyrberg (1998, 2022). The anatomical terminology follows Baumel and Witmer (1993) and when necessary Howard (1929).

Systematic Study

The Senèze material has been compared with skeletons of recent species from the collections of the Muséum national d'Histoire naturelle (Laboratoire d'Anatomie comparée and Institut de Paléontologie Humaine), Paris; the Université Claude-Bernard, Lyon; the Natural History Museum, Tring (UK); the Pierce Brodkorb collection, Gainesville, Florida (USA); and the Institute of Paleontology of the Russian Academy of Science, Moscow.

Order Ciconiiformes

Family Ciconiidae

Ciconia nigra (Linnaeus, 1758), Black Stork.

Material: Right tarsometatarsus, distal part, very juvenile, NMB Se 1562.

Description and comparisons

Morphologically and metrically this tarsometatarsus corresponds perfectly to the recent species *Ciconia nigra*. The distal part is less widened than in *Ciconia ciconia*. The fossa metatarsi I is situated closer to the trochlea metatarsi II and closer to the medial side of the bone in *C. nigra* than in *C. ciconia*. The morphological characteristics distinguishing the two species indicated by Gruber (1990) are not visible due to the very juvenile state of the bone.

An extinct species, *Ciconia lucida* Kurochkin (1982), has been described in the Early and Middle Pliocene of Mongolia (Kurochkin 1982, 1985), but the dimensions of the tarsometatarsus are unknown. However, the dimensions of the carpometacarpus indicate that this species was larger than the two recent species, *C. ciconia* and *C. nigra* (Mourer-Chauviré 2004).

Another extinct species, *Ciconia stehlini*, has been described by Janossy (1992) from the (Hungarian) localities of Beremend 15 (MN 16) and Beremend 16 (MNQ 21). The distal width of the tibiotarsus of that species is included in the variation range of the two recent species (Gruber 1990). The proximal width of the tarsometatarsus is slightly larger than in *C. nigra* but included in the variation range of *C. ciconia* (Gruber 1990). The distal part of the tarsometatarsus illustrated by Janossy (1992, Fig. 8.4, n° 3) does not resemble the genus *Ciconia*. The trochlea metatarsi II is very narrow, strongly displaced plantarly, and much shorter than the trochlea metatarsi III, while in the genus *Ciconia* it is wide, not strongly displaced plantarly, and almost as elongated as the trochlea metatarsi III. Janossy (1992, p. 16) considers the condition where "the hypotarsus in the proximal epiphysis of the tarsometatarsus is narrower as in *C. nigra*" as a distinguishing characteristic; this, however, is not visible on the Senèze material, nor are the other characteristics of the distal part of the tarsometatarsus, the identification of which now seems doubtful.

The recent species *Ciconia nigra* has been reported from the Early Pleistocene of Saint-Vallier, France (MNQ 17) (Mourer-Chauviré 2004) and Torre di Picchio, Italy (Bedetti 2003).

Measurements: Table 8.1

Table 8.1 *Ciconia ciconia* and *C. nigra*. Measurements (in mm) of the distal part of the tarsometatarsus. Recent specimens after Gruber (1990)

Tarsometatarsus	*Ciconia nigra* Senèze NMB Se 1562	*Ciconia ciconia* Recent, juveniles	*Ciconia nigra* Recent, juveniles
Distal width	17.7		
Males, range		18.6–20.6	
Mean (n)		19.80 (8)	17.5 (1)
Females, range		18.0–21.0	
Mean (n)		18.92 (6)	18.0 (1)
Sex unknown, range		18.4–20.0	–
Mean (n)		19.22 (22)	–

Order Anseriformes
Family Anatidae
Tadorna cf. *ferruginea* (Pallas, 1764), cf. Ruddy Shelduck.
Figures 8.1 and 8.2A.
Material: Left humerus, largely restored shaft and distal part, NMB Se 492.
Description and comparisons

The distal part of this specimen is incomplete, and the processus flexorius is missing. Morphologically this humerus corresponds to the genus *Tadorna*, and by its size it is closer to the species *T. ferruginea*. Janossy (1972) said that it is practically impossible to find morphological differences between *T. tadorna* and *T. ferruginea*. On average the dimensions of *T. ferruginea* are slightly larger than those of *T. tadorna*, but there is a considerable overlap between the two species. However Soergel (1966) has noted some points on the humerus that make it possible to distinguish *T. ferruginea* from *T. tadorna*. In particular in *T. tadorna* the impressio brachialis (fossa m. brachialis in Baumel & Witmer 1993) is elliptical and well delimited on all sides while in *T. ferruginea* "nur auf den Seiten deutlich markiert ist, nach oben aber allmählich spitz ausläuft" ("[it] is only clearly marked on the sides, but gradually tapers upward": Soergel 1966, p. 95) In addition, on the single examined skeleton of *T. ferruginea* one can see a very thin ridge, which starts from the ectepicondylar prominence (Howard 1929) and runs obliquely towards the fossa m. brachialis. This ridge is strongly indicated on the Senèze fossil and does not exist on my comparative material of *T. tadorna*. However, due to the small amount of comparative material, it is not possible to tell whether this character is constant or not in these two species.

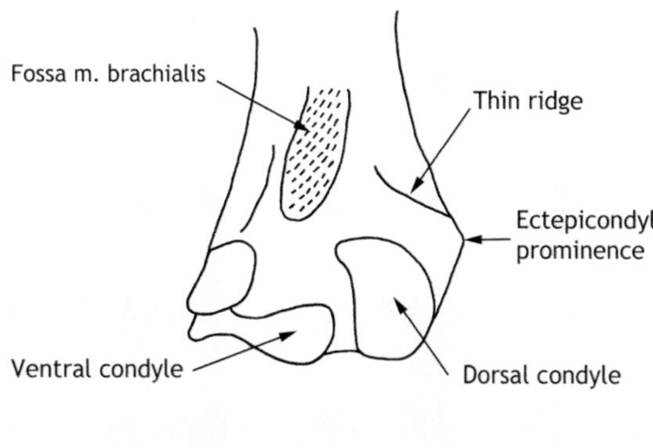

Fig. 8.1 Drawing of the distal part of the humerus, in cranial view, of a recent *Tadorna ferruginea*, showing the presence of a thin ridge, which starts from the ectepicondylar prominence and runs obliquely towards the fossa m. brachialis

The recent species *Tadorna ferruginea* has been reported from the Early Pleistocene of Somssichhégy 1, Hungary (Tyrberg 2022), Pirro Nord, Italy (MNQ 19, age between 1.7 and 1.3 Ma, Bedetti 2003; Bedetti & Pavia 2013) and Dursunlu, Turkey (MNQ 20, age between 1.0 and 0.9 Ma, Louchart et al. 1998). It has also been reported in Middle Pleistocene localities in Azerbaidjan and Spain (Sánchez Marco 1989, 2004; Tyrberg 1998, 2022).

Table 8.2 *Tadorna tadorna* and *T. ferruginea*. Measurements (in mm) of the distal partof the humerus. Recent specimens after Woelfle (1967)

Humerus	*Tadorna* cf. *ferruginea* Senèze NMB Se 492	*Tadorna tadorna* Recent	*Tadorna ferruginea* Recent
Distal width	Superior to 18.0		
Males, range		13.7–17.0	16.4–17.2
Mean (n)		16.11 (18)	(2)
Females, range		14.4–15.6	
Mean (n)		15.05 (19)	16.6 (1)

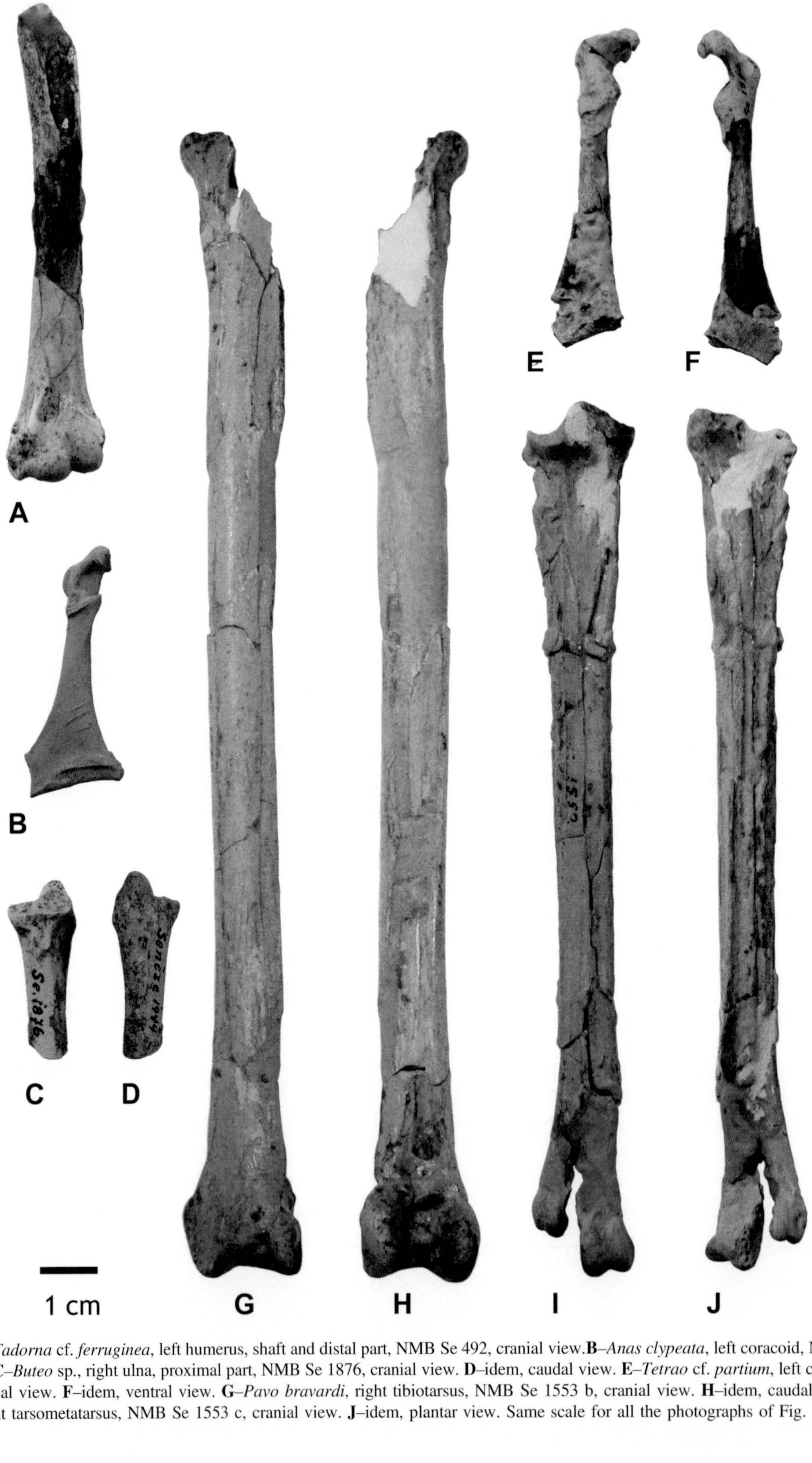

Fig. 8.2 A–*Tadorna* cf. *ferruginea*, left humerus, shaft and distal part, NMB Se 492, cranial view.**B**–*Anas clypeata*, left coracoid, NMB Se 1556, dorsal view. **C**–*Buteo* sp., right ulna, proximal part, NMB Se 1876, cranial view. **D**–idem, caudal view. **E**–*Tetrao* cf. *partium*, left coracoid, NMB Se 1555, dorsal view. **F**–idem, ventral view. **G**–*Pavo bravardi*, right tibiotarsus, NMB Se 1553 b, cranial view. **H**–idem, caudal view. **I**–*Pavo bravardi*, right tarsometatarsus, NMB Se 1553 c, cranial view. **J**–idem, plantar view. Same scale for all the photographs of Fig. 8.2

Measurements: Table 8.2 (after Woelfle 1967).
Anas clypeata Linnaeus, 1758, Northern Shoveler
Fig. 8.2B
Material: Left coracoid, NMB Se 1556; right humerus, distal part, NMB Se 1557; right coracoid, omal part, NMB Se 1565a; right coracoid, sternal part, NMB Se 1565b.
Description and comparisons.

The distal part of humerus NMB Se 1557 had been linked (with reservations) to other specimens referred to "*Caccabis* sp." (= *Alectoris* sp.) by Stehlin (1923), but it actually belongs to Anseriformes. According to its morphological characteristics and dimensions, it corresponds to the recent species *Anas clypeata*. The complete left coracoid corresponds to the smallest of the two or three forms of Anatidae spec. div. reported by Stehlin (1923). The two fragments of the right coracoid, NMB Se 1565 a and b, probably come from the same bone and from the same individual as the left coracoid NMB Se 1556.

A form close to the recent species *Anas clypeata* is known as early as the Late Miocene (MN 11–13) of Ukraine (Mlíkovsky 2002a) and a form identified sometimes as *A.* aff. *clypeata*, *A.* cf. *clypeata*, or sometimes as *A. clypeata*, is known from numerous Early Pleistocene localities in Bulgaria, Hungary, Italy, Romania, Spain, Turkey, and Ukraine (Tyrberg 1998, 2022).

Measurements: Table 8.3

Aythya sp., indeterminate Pochard
Material: Right humerus, proximal part, NMB Se 495
Description and comparisons

This humerus is not well preserved and it is not possible to take measurements. It corresponds to an *Aythya* sp., of the size of *A. ferina*, the Pochard, or *A. fuligula*, the Tufted Duck.

Two extinct species of *Aythya* have been described from the Early Pleistocene of Valdarno, Italy, namely *Aythya aretina* (Portis 1891), and *A. sepulta* (Portis 1891). These two species have been transferred by Cheneval (1987) to the genus *Anas*, but Bedetti (2003) and Bedetti and Pavia (2004) have shown that the type actually belongs to the genus *Aythya* and that these two species are valid. The Senèze form is comparable in size to *Aythya sepulta* but due to its condition it is not possible to ascribe it to this species which so far is known only from the Early Pleistocene of Central Italy (Bedetti 2003; Tyrberg 2022).

Recent species of the genus *Aythya* have been reported in the Late Pliocene: *A. marila* at Las Higueruelas, Spain (MN 16) (Sánchez Marco 2005), *A. ferina*, *A. nyroca* and *A. fuligula* in several Early Pleistocene localities of Romania, Spain, and Turkey (Tyrberg 2022).

Mergus cf. *merganser* Linnaeus, 1758, cf. Goosander.
Material: Right radius, NMB Se 494 + NMB Se 496
Description and Comparisons

Stehlin (1923, p. 279) wrote: "Ein Radius… nähert sich in den Proportionen dem von *Mergus merganser*" (a radius is similar in its proportions to those of…). This radius morphologically corresponds to the genus *Mergus*, and its size is included in the variation range of the recent *M. merganser* males (Woelfle 1967).

Janossy (1972) has described the extinct species *Mergus connectens*, which is present in the Early Pleistocene of Betfia 5, Romania, and the Middle Pleistocene of Stránská Skála, Czech Republic (Janossy 1972, 1979a; Mlíkovsky 1995, 2002b). Janossy (1972) does not give measurements for the radius but *Mergus connectens* is smaller than the two recent species *M. merganser* and *M. serrator;* therefore, the Senèze form cannot be referred to this extinct species.

Table 8.3 *Anas clypeata*. Measurements (in mm) of the coracoid, after Bacher (1967). Recent specimens after Woelfle (1967)

Coracoid	*Anas clypeata* Senèze NMB Se 1565a and NMB Se 1565b	*Anas clypeata* Recent	
		Males	Females
Maximum length			
Range		42.5–47.2	42.1–47.1
Mean (n)	45.7	45.03 (24)	43.38 (23)
Medial length			
Range		38.9–42.9	37.6–43.5
Mean (n)	41.2	41.10 (24)	39.78 (23)
Distal width			
Range		17.7–19.3	17.0–19.3
Mean (n)	18.1	18.70 (25)	17.59 (23)
Basal width			
Range		16.2–18.1	15.1–17.9
Mean (n)	17.4	16.92 (25)	16.47 (23)

The recent species *Mergus merganser* is reported from the Early Pleistocene of Pietrafitta, Italy (Bedetti 2003), whose age corresponds to MNQ 18 (Zucchetta et al. 2003).

Mergellus sp. size *M. albellus* (Linnaeus, 1758), Smew
Material: Left carpometacarpus, proximal part, incomplete, FSL 91637.

It is not possible to measure this fragment because both the os metacarpale alulare and the tip of the processus pisiformis are broken, but its size corresponds to that of *M. albellus*.

The recent species *Mergellus albellus* has been reported in the Late Pliocene (MN 16) of Las Higueruelas, Spain (Sánchez-Marco 2005) and in the Early Pleistocene of Dursunlu, Turkey (Louchart et al. 1998).

Indeterminate Anatidae
Material: phalanx of posterior digit, NMB Se 1565c.
Order Accipitriformes
Family Accipitridae
Haliaeetus albicilla (Linnaeus, 1758), White-tailed Sea-eagle Fig. 8.4A and B
Material: left coracoid, incomplete, NMB Se 1788.
Description and comparisons

This bone was not mentioned by Stehlin (1923). The tip of the processus acrocoracoideus, the tip of the processus procoracoideus, a part of the lateral side of the shaft, a part of the facies articularis sternalis, and the processus lateralis are all missing. This bone shows the morphological characteristics of the genus *Haliaeetus* but differs from the genus *Aquila*. It has been compared with different species of the genus *Haliaeetus*, such as *H. leucocephalus*, *H. albicilla*, *H. leucoryphus*, *H. vocifer*, and *H. leucogaster*. The average dimensions of *H. leucocephalus* are smaller than those of *H. albicilla*. In these two species, females are larger than males, and there is an overlap in their dimensions, while *H. leucogaster*, *H. leucoryphus* and *H. vocifer* are smaller. The width in the middle of the coracoid, 14.0 mm, corresponds equally well to an *H. albicilla* male as to *H. leucocephalus*, but the latter only lives in North America, while *H. albicilla* is widespread all over the Eurasian continent (del Hoyo et al. 1994).

Comparisons with fossil forms

From the Middle Pliocene of Mongolia, Kurochkin (1985) has described the extinct species *Haliaeetus fortis*, which is known only by an omal part of coracoid and two posterior phalanges. This species is approximately the same size as *H. leucoryphus*, or *H. vocifer*, and thus is smaller than *H. albicilla*. From the Late Pliocene of Tologoi, in the Buriat Republic (eastern Siberia, Russia), a proximal part of carpometacarpus has been referred to *Haliaeetus* sp. (Kurochkin 1985). This fragment is almost indistinguishable from the recent species *H. albicilla*. Janossy (1983a) described the extinct species *Haliaeetus angustipes* from the Middle Pleistocene of Prezletice, in the Czech Republic, but Mlíkovsky (1997), with a very large amount of comparative material, has shown that this form did not differ from the recent species *H. albicilla*. In addition, *H. albicilla* has been reported in the Early Pleistocene (MNQ 17) of Saint-Vallier, France (Mourer-Chauviré 2004); in the Early Pleistocene of Tegelen, Netherlands (Junge 1953), which corresponds to the mammalian biozone MNQ 18 (Crégut-Bonnoure et al. 2024); in the Early Pleistocene of the "Sima del Elefante", Spain (Rosas et al. 2001); and in the Early Middle Pleistocene of Casal Selce, Italy (Bedetti 2003; Pavia et al. 2018).

Measurements (in mm): Internal length from the proc. acrocor. to the angulus medialis, as preserved, 78.6; depth of pars omalis, as preserved, 20.7; width of facies articularis humeralis, 12.2; width of shaft in the middle, 14.0; depth of shaft in the middle, 10.8; depth of the sternal facet on the medial side, 11.0.

In spite of its name, the Sea-eagle, *Haliaeetus albicilla* is not confined to sea coasts. Its habitat is made up of large rivers, lakes, ponds, estuaries, and sea coasts. During the Middle and Late Pleistocene it is often reported in inland localities (Tyrberg 2022).

Buteo sp., size of *Buteo buteo* (Linnaeus, 1758), Eurasian Buzzard,
or *Buteo lagopus* (Pontoppidan, 1763), Rough-Legged Buzzard. Fig. 8.2C and D
Material: Right ulna, proximal part, NMB Se 1876.
Description and comparisons.

This bone was not reported by Stehlin (1923). It is complete except for a small break on the cranial part of the olecranon. Its measurements are: proximal width, 12.0; proximal depth, 9.5. This proximal depth corresponds to the measurement called "Tiefe proximal" in Otto (1981). These measurements are included in the variation range of *Buteo buteo*, males and females, and of *Buteo lagopus* males (Otto 1981).

The recent species *B. buteo* is known as early as the Early Pleistocene in Cueva Negra del Estrecho del Rio Quipar, Spain (Rosas et al. 2001) and *Buteo* cf. *buteo* at Kujal'nitzkogo limana, Ukraine (Tyrberg 2022), and the recent species *B. lagopus* is also known from the Early Pleistocene at Somssichhégy 1, Hungary (Tyrberg 2022). In addition Harrison (1979) reported *Buteo*? *buteo*, or *B. buteo/lagopus/rufinus* superspecies in the Cromer Forest Bed Series of England.

Order Galliformes
Family Tetraonidae
Tetrao cf. *partium* (Kretzoi, 1962), ancestor or extinct relative of the Black Grouse. Figs. 8.2E and F, and 8.3
Material: left coracoid, almost complete, NMB Se 1555
Description and comparisons

This bone has been shattered and restored, but in the course of the restoration the shaft has undergone a rotation and the omal part is in a plane which forms an angle of about 45° compared to the plane of the sternal part. However it is still possible to measure the total length and the internal length. Stehlin (1923) identified this bone as ?*Lyrurus tetrix* and has written that it corresponds so well to that of the Black Grouse that if it had been found in a Pleistocene locality, it would doubtless have been placed in this species. The total length (58.4 mm) is included in the variation range of the recent *Tetrao tetrix* males (Erbersdobler 1968). However, this bone shows some morphological differences compared to the corresponding bone of the recent *Tetrao tetrix*. In a recent *T. tetrix* the acrocoracoid forms an obtuse angle compared to the axis of the shaft, while in the Senèze coracoid the acrocoracoid forms an angle which is less open and closer to a right angle (Fig. 8.3). On the medial face, the facies articularis clavicularis shows two tubercles directed sternally, one on the ventral side and the other on the dorsal side. The tubercle situated on the dorsal side is much more projecting medially than in all the coracoids of recent and fossil *T. tetrix* that I have been able to examine. The sternal part has a pneumatic foramen, but it is almost completely filled by the product used for the restoration. The angulus medialis, as well as the processus lateralis, are broken. However, even taking into account its incomplete preservation, this sternal part of the coracoid is strikingly narrower than in the recent *T. tetrix*. The total length is 58.4 mm, and the estimated width of the sternal facet is 13.5 mm. The ratio Length of the sternal facet × 100/Total length is 23.1. According to the measurements given by Erbersdobler (1968), the same ratio is on average 25.9 for the recent females (range 24.9–27.4, n = 12), and 25.6 for the recent males (range 24.6–27.0, n = 50). This difference in the proportions of the bone is an argument to attribute it to a different species.

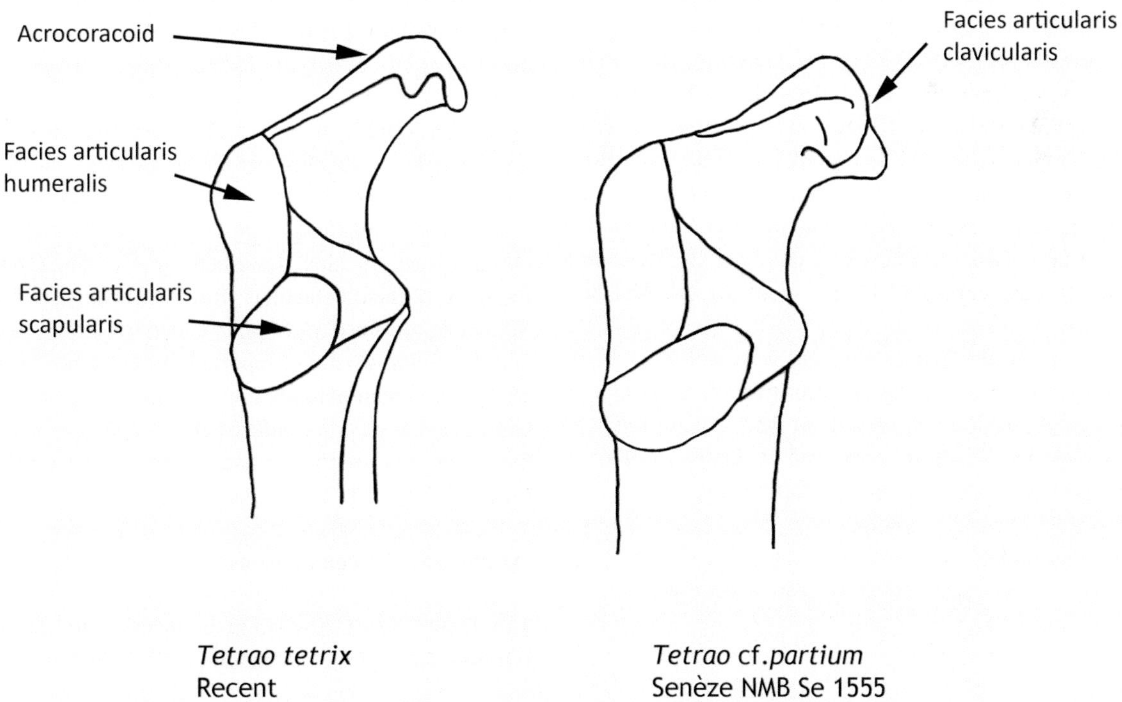

Fig. 8.3 Drawing of the omal part of the coracoid, dorsal view, showing the differences between *Tetrao tetrix*, Recent, and *Tetrao* cf. *partium*, NMB Se 1555, from Senèze

The extinct species *Tetrao partium* has been described from the Early Pleistocene of Betfia, Romania (Kretzoi 1962), age MNQ 19, and it has been reported in numerous localities of Late Pliocene and Early and Middle Pleistocene age, in Bulgaria, Czech Republic, France, Germany, Hungary, Romania, and Slovakia, (Boev 1999; Mlíkovsky 2002a; Tyrberg 2022). *Tetrao partium* is characterized by proportions which differ from those of *T. tetrix*, with some bones such as the tarsometatarsus and the distal part of the tibiotarsus being more elongate, while others, such as the humerus, the carpometacarpus, and the first phalanx of the major digit of the wing, are more robust (Janossy 1976). Sánchez Marco (2009) has placed *Tetrao partium* into synonymy with the Recent species *Tetrao tetrix*, arguing that the dimensions given by Janossy (1976) are included in the variation range of the *T. tetrix* males and females given by Erbersdobler (1968). This is inaccurate. The distal width of the Villany 3 tibiotarsus (7.8 mm) is smaller than that of the recent females (8.0–8.9 mm, n = 12) and the width in the middle of the Osztramos 2 humerus (8.2 mm) is larger than that of the recent males (7.2–8.0 mm, n = 14). The total length of the Osztramos 8 carpometacarpus (41.7 mm) is larger than that of the recent females (38.9–40.8 mm, n = 14) and smaller than that of the recent males (43.0–46.2, n = 5).

Boev (1999) has reported *Tetrao partium* from the Early Pleistocene locality of Varshets, Bulgaria, the age of which corresponds to MNQ 17 and is, therefore, close to the age of Senèze. Boev has also indicated, for his material, some morphological characteristics differing from *T. tetrix*. The measurements of the coracoid, according to Boev (1999), are indicated in Table 8.4. In its omal part, the Senèze coracoid is slightly larger than those of Varshets, but the facies articularis clavicularis is narrower. In its sternal part, the Senèze coracoid is very close to the Varshets examples.

In conclusion, the Senèze coracoid is identified as *T.* cf. *partium* because it is slightly different from *T. tetrix* and its chronological position corresponds to the temporal range of the fossil species, which is considered by Janossy (1976) to be the direct ancestor of the recent form *T. tetrix*.

Table 8.4 Tetrao cf. *partium* and *Tetrao partium*. Measurements (in mm); Varshets after Boev (1999)

Coracoid Omal part	a	b	c	d
Tetrao cf. *partium* (Senèze)	20.1	12.6	7.5	5.3
Tetrao partium (Varshets, n = 2)	18.2	11.1	8.3	5
	18.1	11.1	8	5.2
Coracoid Sternal part	e	f		
Tetrao cf. *partium* (Senèze)	est. 13.5	est. 4.5		
Tetrao partium (Varshets, n = 2)	–	4.9		
	14	4.4		

Omal part
a: length of the omal part, from the base of facies articularis scapularis to the top of the acrocoracoid. b: width of the omal part, from the lateral border of facies articularis humeralis to the medial border of facies articularis clavicularis. c: depth of facies articularis clavicularis. d: length of facies articularis clavicularis in omalo-sternal direction.
Sternal part
e: length of the sternal facet. f: depth of the sternal facet.

Family Phasianidae
Pavo bravardi (Gervais, 1849), Bravard's Peafowl
Fig. 8.2G and J.

Material: Right femur, distal part, NMB Se 1553a; right tibiotarsus, almost complete, NMB Se 1553b; right tarsometatarsus, almost complete, NMB Se 1553c; metatarsal I, NMB Se 1565d; phalanx 1 pedal dig. I, NMB Se 1565e; phalanx 2 pedal dig. I, NMB Se 1565f; phalanx 1 pedal dig. III, NMB Se 1565 g; phalanx 2 pedal dig. III, NMB Se 1565 h; phalanx 1 pedal dig. IV, NMB Se 1565i; phalanx 3 pedal dig. IV, NMB Se 1565j; phalanx 4 pedal dig. IV, NMB Se 1565 k.

Stehlin (1923) indicated that the tibiotarsus and tarsometatarsus were close to *Pavo cristatus* and *Meleagris gallopavo* but did not completely correspond to any of them. This material was designated as? *Gallus bravardi* to bring it closer to the species described by Gervais (1849) for a fragment of tarsometatarsus from the locality of Ardé in the Montagne de Perrier. Mourer-Chauviré (1989) has shown that the *Gallus bravardi* reported in the Early Pliocene (MN 15) of Perpignan actually corresponds to a Peafowl, closely related to the recent southeast Asian species *Pavo muticus*, the Green Peafowl; accordingly, she transferred the species *Gallus bravardi* to the genus *Pavo*.

Description and comparisons

The femur NMB Se 1553a is incomplete and has been restored with plaster. The only measurement that can be made is the depth of the medial condyle, here 21 mm. On the tibiotarsus NMB Se 1553b, the proximal part is incompletely preserved. On the distal part the lateral condyle extends further proximally than the medial condyle, and the supratendinal bridge is wide and orientated perpendicularly to the long axis of the bone, as in the genus *Pavo*; in the genus *Gallus* the lateral condyle does not extend proximally farther than the medial condyle, and the supratendinal bridge is obliquely oriented. It seems that the different remains found at Senèze come from the same individual, which was probably a female since there is no spur on the tarsometatarsus. The measurements of the tibiotarsus are slightly larger than those of recent *Pavo muticus* females (Mourer-Chauviré 1990). In the locality of Megalo Emvolon, the tibiotarsus is also larger than those of the recent *Pavo cristatus* and *P. muticus* males (Boev & Koufos 2000) (Table 8.6). The tarsometatarsus NMB Se 1553c is also heavily restored. The distal part is incomplete, the medial part of the shaft and the trochlea metatarsi II are missing, but there is a contact between the lateral part of the shaft and the trochlea metatarsi IV which makes it possible to have an estimate of the total length. On the plantar face there is an ossified tendon, which follows the medial side of this face to the level of the break of the distal part, where it is abruptly interrupted, although it probably extended a little farther distally. On the recent *Pavo* females, it is on this distal part of the ossified tendon that a rudiment of spur is found. This rudiment of spur is totally absent on the Senèze tarsometatarsus.

Peafowl have also been reported in the Pliocene of Lucheshti, Moldova; Megalo Emvolon, Macedonia, Greece (Boev & Koufos 2000); Muselievo, Bulgaria (Boev 2001); and in the Early Pleistocene of Saint-Vallier, France (Mourer-Chauviré 1990, 2004). In summary, *Pavo bravardi* is known in the localities of Perpignan, Megalo Emvolon, and Muselievo (MN 15), Ardé and Lucheshti (MN 16), Saint-Vallier (MNQ 17) and Senèze (MNQ 18). Large-sized peafowl are presently absent from Africa, but they occurred there in the early Pliocene. A *Pavo* sp., related to the recent *Pavo muticus* and to the extinct *Pavo bravardi*, has been reported from the Ethiopian localities of Aramis 1, 6, and 17 (Louchart 2003; Louchart et al. 2009) and from the Kenyan locality of Mabaget (Pickford et al. 2004).

Measurements: Tables 8.5, 8.6, and 8.7.

Table 8.5 *Pavo bravardi* and recent peafowl. Measurements of the tibiotarsus and tarsometatarsus (in mm)

	Pavo bravardi Senèze	*Pavo muticus* Recent		*Pavo cristatus* Recent	
		Males	Females	Males	Females
Tibiotarsus	NMB Se 1553 b				
Total length					
Range	ca. 194	197–232	179–192	187–198	161–173
Mean (n)		214.60 (5)	185.33 (3)	190.50 (4)	165.00 (3)
Distal Width					
Range	20	17.5–20.5	16.6–18.5	17.4–20.2	14.8–16.6
Mean (n)		19.14 (5)	17.60 (3)	18.43 (4)	15.90 (3)
Distal depth					
Range	19	17.2–20.0	16.0–16.8	17.0–19.1	15.4–17.1
Mean (n)		18.82 (5)	16.40 (3)	18.20 (4)	16.43 (3)
Width at midpoint					
Range	11.9	10.3–12.0	9.1–10.1	10.5–12.5	9.8–10.3
Mean (n)		11.16 (5)	9.73 (3)	11.25 (4)	10.03 (3)
Tarsometatarsus	NMB Se 1553 c				
Total length					
Range	148	139.4–171	126.8–133.4	126–138	106.4–112.4
Mean (n)		157.63 (7)	130.73 (3)	130.01 (8)	109.70 (4)
Proximal width					
Range	ca. 20.3	19	18.4–20.4	20.6–21.3	17.6–19.0
Mean (n)		(1)	(2)	21.05 (4)	18.43 (3)
Width at midpoint					
Range	10.3	8.2–9.8	8.2–9.0	8.8–9.7	7.9–8.6
Mean (n)		9.06 (5)	8.60 (3)	9.13 (4)	8.30 (3)

Table 8.6 *Pavo bravardi*. Measurements of the tibiotarsus (in mm)

	Pavo bravardi Senèze	cf. *Pavo bravardi* Saint-Vallier	*Pavo bravardi* Megalo Emvolon
Tibiotarsus	NMB Se 1553 b	FSL 497,095	
Width at midpoint	11.9	11.1	ca. 15[a]
Depth at midpoint	9.9	9.2	ca. 12[a]
Distal width	20	–	21.3
Distal depth	19	–	20.6

[a] After the Fig. 8.2 in Boev and Koufos (2000)

Table 8.7 *Pavo bravardi* and recent peafowl. Measurements (in mm) of metatarsal I and pedal phalanges

	Pavo bravardi Senèze	*Pavo bravardi* Muselievo NMNHS	*Pavo muticus* Recent Females (n = 2)	*Pavo cristatus* Recent	
				Males (n = 3)	Females (n = 3)
Metatarsal I	NMB Se 1565 d	–			
Length	14.1	–	15.6–16.2	15.5–16.7	13.0–15.0
Distal width	5.7	–	5.0–6.0	6.5–7.2	5.4–7.0
Ph. 1 D I	NMB Se 1565 e	–			
Length	20.4	–	19.3–21.1	19.6–21.7	18.0–19.0
Prox. width	6.9	–	5.2–6.9	7.7–8.5	5.8–6.8
Dist. width	4.6	–	3.6–5.1	5.9–6.2	4.0–4.9
W. at midpoint	4.1	–	3.2–4.0	4.1–4.4	3.0–3.6
Ph. 2 D I	NMB Se 1565 f	n° 421			
Length	ca. 15	11	11.0–15.0	14.0–16.1	11.0–13.0
Prox. width	4.6	4.7	3.4–4.8	6.4–7.5	3.7–5.0
Ph. 1 D III	NMB Se 1565 g	–			
Length	30.2	–	26.6–30.1	28.1–31.6	26.0–27.0
Prox. width	9.8	–	8.2–10.0	ca.10–ca.115	7.5–10.9
Dist. width	7.4	–	7.0–7.5	7.4–8.0	6.1–8.0
W. at midpoint	5.6	–	4.7–5.2	5.5–5.7	4.3–5.4
Ph. 2 D III	NMB Se 1565 h	n° 419			
Length	23.6	ca. 19.5	22.0–22.0	ca.22–23.9	19.8–21.5
Prox. width	–	–	6.4–9.0	ca.8–ca.10	6.1–8.6
Dist. width	5.7	5.5	5.2–6.5	6.0–6.7	4.7–6.3
W. at midpoint	–	–	4	4.6–4.9	3.2–4.4
Ph. 1 D IV	NMB Se 1565 i	–			
Length	20.2	–	18.3–19.8	18.4–20.4	17.6–18.0
Prox. width	7.8	–	6.0–7.6	7.1–8.1	6.0–7.3
Dist. width	5.5	–	5.1–5.3	5.3–5.8	4.6–5.4
W. at midpoint	4.1	–	3.4–3.9	3.7–3.9	3.2–3.6
Ph. 3 D IV	NMB Se 1565 j	n° 420			
Length	10.8	9.7	10.8–11.1	10.1–10.5	9.3–9.5
Prox. width	5	4.5	4.4–5.4	5.2–5.4	4.3–5.0
Dist. width	4.3	4.4	4.2–4.5	4.3–4.6	3.9–4.3
W. at midpoint	4	–	3.3–4.0	3.9–4.1	3.4–3.8
Ph. 4 D IV	NMB Se 1565 k				
Length	13.1		12.8–13.1	12.4–13.3	11.9–12.3
Prox. width	4.2		3.8–4.9	4.4–4.7	3.5–4.1
Dist. width	3.9		3.7–4.4	4.0–4.5	3.3–3.9
W. at midpoint	3.3		3.0–3.6	3.0–3.2	2.7–2.9

Alectoris cf. *graeca* (Meisner, 1804), cf. Rock Partridge
Fig. 8.4C and D.
Material: Cranial part of sternum, NMB Se 1538; left femur, almost complete, NMB Se 291; left tibiotarsus, slightly incomplete, NMB Se 1538.
Description and comparisons.

According to Stehlin (1923) this material corresponds very precisely to *Caccabis petrosa* (now *Alectoris graeca*) but does not completely attain its dimensions. The fragment of humerus tentatively attributed to this species actually belongs to an anseriform (*Anas* cf. *clypeata*).

The femur corresponds, in all its details, to the femur of the recent *Alectoris graeca*, and its dimensions are very close to the mean dimensions of a recent population of 44 individuals (Erbersdobler 1968). The tibiotarsus length is relatively small but it is included in the variation range of the recent forms, and its other measurements are very close to the mean dimensions of recent Rock Partridges. The sternum also perfectly agrees with that of *Alectoris graeca*.

Harrison (1980) has described a new species of *Alectoris*, *A. sutcliffei*, from the late Middle Pleistocene of Glutton Stratum, Tornewton Cave, England; it is known only from a distal part of the tibiotarsus. However, Stewart (1996), who has revised the material, considers it as a junior synonym of *Perdix perdix*.

Tchernov (1980) has described a new species of *Alectoris*, *A. baryosefi*, from the Early Pleistocene (MNQ 19) of 'Ubeidiya, Israel. It is characterized by morphological differences when compared to the recent species *A. graeca*, *A. rufa*, and *A. chukar*, and by a relatively small size; on the other hand, the total length of the holotype carpometacarpus is included in the variation range of the recent *A. chukar*.

Mlíkovsky (2002a) has placed *Alectoris baryosefi* in synonymy with the species *Palaeocryptonyx donnezani*, from the Early Pliocene of Perpignan, France, where he also places the genus *Alectoris*. The type material of *Palaeocryptonyx donnezani* (Depéret 1892; Pavia et al. 2012) includes an almost complete carpometacarpus (UCBL-FSL 92891–3) which differs strongly from the holotype of *Alectoris baryosefi*. The total length of this carpometacarpus is 21.5 mm (in *A. baryosefi* it is 29.3 mm), while the proximal depth, from the tip of the os metacarpale alulare to the caudal border of the trochlea carpalis, is 5.8 mm (in *A. baryosefi*, 8.5 mm). In addition, in *A. baryosefi* the processus intermetacarpalis is situated very close to the proximal junction of both major and minor metacarpals, while in *Palaeocryptonyx donnezani* this process is situated further distally compared to the junction. *Alectoris baryosefi* therefore cannot be put into the same genus as *Palaeocryptonyx donnezani* and, consequently, this species must be retained in the genus *Alectoris* until the material can be revised. It is not possible to compare the Senèze form with the species *A. baryosefi* because the femur and tibiotarsus are unknown in the extinct Israeli species.

Alectoris cf. *graeca* is found in the Early Pleistocene of Italy, at Monte Argentario, Tuscany (Bedetti 2003; Siori et al. 2014), but in other Early Pleistocene localities of Central Italy, *Alectoris* is represented either by *Alectoris rufa* or by *Alectoris* sp. (Bedetti 2003). This species is also known in Greece, occurring in the Middle Pleistocene, at Petralona (Tyrberg 2022). In Spain *Alectoris* sp. is reported in the Early Pleistocene localities of Quibas and Cueva Victoria, and in the Middle Pleistocene locality of Huéscar 1 (Sánchez Marco 2009).

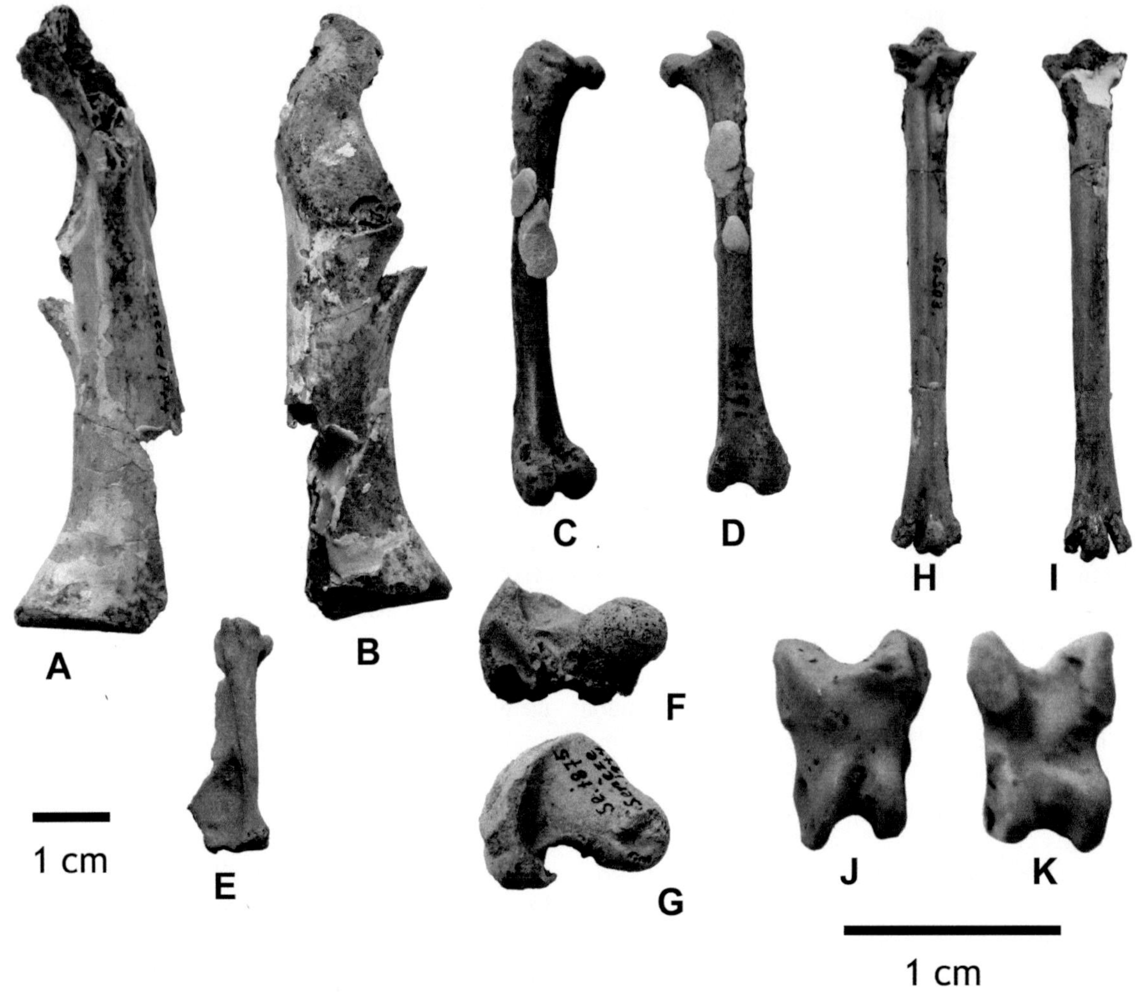

Fig. 8.4 **A**–*Haliaeetus albicilla*, left coracoid, NMB Se 1788, dorsal view. **B**–idem, ventral view. **C**–*Alectoris* cf. *graeca*, left femur, NMB Se 291, caudal view. **D**–idem, cranial view. **E**–*Bubo* sp., phalanx 1 of the major digit of the right wing, NMB Se 1563, dorsal view. **F**–*Otis* sp., right femur, proximal part, NMB Se 1875, cranial view. **G**–idem, proximal view. **H**–*Corvus corax antecorax*, right tarsometatarsus, NMB Se 503, cranial view. **I**–idem, plantar view. **J**–*Surnia robusta*, Phalanx 1 of left pedal digit II, NMB Se 1565 l, cranial view. **K**–idem, plantar view.Same scale for photographs a-i; scale × 3 for photographs j-k

Order Gruiformes

Family Otididae

Otis sp. size *Otis tarda* Linnaeus, 1758, female, Great Bustard Fig. 8.4F and G.

Material: Right femur, proximal part, NMB Se 1875.

Description and comparisons

This femur exhibits the very characteristic shape of the Otididae. The crista trochanteris describes an arc which rises proximally above the proximal articular surface. This crista trochanteris also extends along the cranial face of the proximal part and delimits a fossa inside of which a large pneumatic foramen opens. The caput femoris is situated in the same plane as the proximal articular surface, and the collum femoris is very slightly indicated. The fovea lig. capitis is situated on the proximal surface of the caput femoris. The dimensions of the proximal part (width 25.5 mm; depth 19.0 mm) are included in the variation range of the recent *Otis tarda* females and are smaller than the recent *Otis tarda* males.

Comparisons with fossil forms

The extinct species *Otis lambrechti* was described by Kretzoi (1941) from the locality of Betfia 5, Romania, age MNQ 19. The type is a tarsometatarsus which has never been described, nor illustrated, and the dimensions of which are not given. It has only been said that its size is similar to that of *Otis tarda* but that its proportions are slightly different. A few other remains have been referred to this species from different localities of Hungary, Romania, and Austria, dating from MNQ 18 to MNQ 22 (Janossy 1974, 1979b; Tyrberg

2022). However, these remains do not include any femora, and therefore the Senèze material cannot be compared with them. Their dimensions are not different from those of the recent *Otis tarda* males and females. Due to the very small quantity of material attributed to *Otis lambrechti*, it is difficult to ascertain whether it is a species that is distinct from *Otis tarda*. It does not seem justified to assign the Senèze material to this species only because it has been found in the same period of time.

In the locality of Chishmikioy, Moldova, age MN 16 (Mlíkovsky 1996), Bochenski and Kurochkin (1987) have reported an *Otis* sp. (group of *Otis tarda*), represented by the distal part of a radius. This indicates that large forms of bustard were already present at the beginning of the Late Pliocene. Another very large form of bustard, *Iorotis gabunii* (Burchak-Abramovich & Vekua 1981), has been described from the locality of Kvabebi, eastern Georgia, the age of which is MN 16 (Mein 1990). It is larger than the recent *Otis tarda* males and thus larger than the Senèze form.

Another extinct species, *Chlamydotis mesetaria* Sánchez Marco (1990), has been described from the Early Pliocene of Layna, Spain (age MN 15). It is placed in the recent genus *Chlamydotis*, which includes only one species, *Chlamydotis undulata*, the Houbara Bustard. It is only known by a distal part of tibiotarsus which is a little larger than in the largest *C. undulata*, but is smaller than in the recent *Otis tarda* females and, thus, smaller than the Senèze form.

Otis sp., size of *O. tarda*, is known in the Early Pleistocene of Saint-Vallier, France (MNQ 17) (Mourer-Chauviré 2004), while the extant species *O. tarda* is present in the Early Pleistocene of Pirro Nord, Italy (MNQ 19) (Bedetti 2003; Bedetti & Pavia 2013).

Order Strigiformes

Family Strigidae

Bubo sp., relative of the Eurasian Eagle-owl Fig. 8.4E

Material: Left radius, shaft without extremities, NMB Se 1594; phalanx 1 of the right wing major digit, NMB Se 1563.

The radius has been referred by Stehlin (1923) to a Gruidae close to *Anthropoides virgo*, and the phalanx to *Bubo* sp., showing only slight differences compared to *Bubo bubo*. The radius is broken at both extremities but the shaft shows the point of departure of a small osseous arch situated distally to the proximal extremity, a feature which is present in most modern Strigidae (Peters 1992) but which does not exist in the Gruidae. This osseous arch is extended by a ridge which continues over the proximal two-thirds of the bone and gives a particular form to its section. In addition the shaft is strongly incurved and corresponds, in all morphological details, to the recent genus *Bubo*. Its dimensions (largest width, 5.8 mm; depth measured perpendicularly to the width, 4.3 mm) are included in the variation range of the recent *Bubo bubo*.The measurements of the wing phalanx are (in mm): Total length, 30.9; proximal width, 7.8; distal width, 6.3. Compared to recent *B. bubo*, in the Senèze phalanx the proximal width is relatively large in relation to the total length but, otherwise there are no morphological differences.

The recent species *Bubo bubo* had been reported in several Early Pleistocene localities, but in these localities it was only known by few and fragmentary remains. A revision of these remains (Meijer et al. 2017) led to the description of a new extinct species, *Bubo ibericus* Meijer et al. (2017) from Cal Guardiola, Spain, age latest Early Pleistocene, and of *Bubo* sp. nov. indet. from Soave, Cava Sud, Italy, age Early Pleistocene (Pavia & Bedetti 2013), a very large form which has not been formally named. All the remains from the other localities are considered too fragmentary and are designated as *Bubo* sp. The Senèze material is also very incomplete and, as it is not possible to compare it either with *Bubo ibericus*, or with *Bubo* sp. nov. indet., it is preferable to designate it as *Bubo* sp.

Surnia robusta Janossy (1977), Large Hawk-owl Fig. 8.4J and K.

Material: Phalanx 1 of left pedal digit II, NMB Se 1565 l.

Description and comparisons

This species was described by Janossy (1977, 1983b) on the basis of material from four Early Pleistocene localities of Hungary. It has never been reported again from any other European localities, but it was recently identified in the Early Pleistocene Moroccan locality of Ahl al Oughlam, near Casablanca, dated at about 2.5 Ma (Mourer-Chauviré & Geraads 2008, 2010). In the Moroccan locality it is represented by several bones including an almost complete tarsometatarsus that is very similar to the holotype of the species (Mourer-Chauviré & Geraads 2010). *Surnia robusta* is characterized by its size, which is about 30% larger than that of the single recent species of that genus, *Surnia ulula*, the Northern Hawk-owl.

The Senèze phalanx is very short and robust. It does not correspond to the recent genera *Bubo*, *Strix*, and *Asio*. Its proportions are similar to that of the recent species *Bubo scandiacus*, but it is much smaller. Morphologically it corresponds perfectly to the genus *Surnia*, but it is conspicuously larger than *Surnia ulula*. Its dimensions (Table 8.8) are slightly larger than those of two corresponding phalanges of *Surnia robusta* from Villány 3, but the ratio between the width in the middle × 100 and the total length is very close.

In the four Hungarian localities where *Surnia robusta* has been recorded, it is the only species identified in Beremend 4, age MN 16/17, but in the other localities, Villany 3 (MN 17), Nagyharsány-hegy 1–4 (MNQ 19), and Somssichhégy 2

Table 8.8 Measurements (in mm) of phalanx 1 of pedal digit II in *Surnia robusta* and recent *Surnia ulula*

Phalanx 1 of pedal digit II	*Surnia robusta* Senèze NMB Se 1565 l	*Surnia robusta* Villány 3	*Surnia ulula* Recent		
			Range	Mean (n)	s^{n-1}
1. Total length	9.6	8.5–9.0	6.6–7.6	7.06 (8)	0.346
2. Prox. width	7.2	–	5.0–5.7	5.39 (7)	0.219
3. Dist. width	5.1	–	3.2–3.9	3.57 (7)	0.229
4. W. at midpoint	5.2	4.6–4.8	3.5–3.9	3.63 (8)	0.149
Ratio 4 × 100/1	54.2	54.1–53.3	49.3–53.0	51.35 (8)	1.522

(MNQ 20), it is associated with *Tetrao partium*, as in Senèze (Tyrberg 2022).

The occurrence of *Surnia robusta* at Senèze is also interesting from a paleobiogeographical perspective, because Senèze is midway between the two geographically very distant areas where this species has been recorded, i.e., the southern part of Hungary, on the one hand, and the Atlantic coast of Morocco, near Casablanca, on the other.

Order Passeriformes
Cf. Turdidae
Cf. Turdidae, size *Turdus merula* Linnaeus, 1758, Blackbird.
Material: Right carpometacarpus, NMB Se 1565 m.
Measurements: Total length, 20 mm.

Family Corvidae

Corvus corax antecorax Mourer-Chauviré (1975), primitive form of the Common Raven Fig. 8.4H and I.

Material: Right tarsometatarsus, almost complete, NMB Se 503. Only the hypotarsus and a part of the trochlea metatarsi II are missing.

Measurements (in mm): Total length, 68.3; proximal width, 11.6; distal width, ca. 8.7; distal depth, 5.6; width in the middle, 5.0; depth in the middle, 5.5.

Description and comparisons.

Stehlin (1923) attributed this tarsometatarsus to the species *Corvus hungaricus*, described by Lambrecht (1916) from the Early Pleistocene of Nagyharsány-hegy in Hungary, MNQ 19, age about 1.1 Ma. It is also present in the Early Pleistocene of Beremend 16–17, Hungary, age about 1 Ma (Tyrberg 2022). This species is characterized by the slenderness of the shaft of its tarsometatarsus. As has been previously indicated for the Saint-Vallier specimen NMB St. V. 287 (Mourer-Chauviré 2004), the dimensions of these tarsometatarsi are included in the size range of *Corvus corax antecorax* from La Fage, and the width in the middle of both of them, 4.7 and 5.0 mm respectively, is larger than that of *Corvus hungaricus* (4.0 mm).

The extinct form was first described as the species *Corvus antecorax*, smaller than the recent species *Corvus corax* (Mourer-Chauviré 1975); it was later considered only as an extinct subspecies (Brodkorb 1978) *C. corax antecorax*. More recently, Stewart (2007) has resumed the study of this taxon and has compared it, in particular, with the small extant subspecies *C. corax hispanus*, which lives in the Iberian Peninsula and the Balearic islands. He has shown that *C. corax antecorax* is significantly smaller than this modern subspecies. In addition, using measurements from Hernandez et al. (1993), Mourer-Chauviré (2004) has demonstrated that *C. corax tingitanus* (from the Canary Islands and Morocco) is significantly smaller in several dimensions than *C. corax antecorax*. Moreover, analysis of mtDNA indicates that the subspecies *C. corax tingitanus* is different from the Holarctic clade of the Common Raven (Baker & Omland 2006), suggesting that it could belong to a distinct species.

C. corax antecorax also differs from the recent subspecies of *Corvus corax* by its proportions. Its humeri and tarsometatarsi are more robust than those of the recent forms, both the large Scandinavian ones and the small Spanish ones (Stewart 2007). The same author has also shown that the coefficients (R^2 values) of the regression lines of the elements of the post-cranial skeleton are different in *C. corax antecorax* and in the recent subspecies of *Corvus corax*. In spite of these differences, Stewart (2007, p. 195) does not consider *C. corax antecorax* as a valid species, or even as a valid subspecies. From my point of view the differences from the extant subspecies are strong enough to justify a subspecific status and, unlike Stewart, I consider *C. corax antecorax* as a valid fossil subspecies. Comparisons with the extinct species *Corvus shufeldti* and *C. neomexicanus* have been performed by Mourer-Chauviré (2004).

C. corax antecorax is known from the beginning of the Early Pleistocene (Saint-Vallier), to the end of the Middle Pleistocene, in France, Spain, and Portugal (Sánchez Marco 2004; Tyrberg 1998, 2022). This subspecies is found during the relatively mild periods of the Early and Middle Pleistocene. During the very cold periods, for example in the locality of L'Escale (at Saint-Estève-Janson, MNQ 22), from the end of MNQ 24, and during MNQ 26, larger forms are found, comparable in size to the recent subspecies *C. corax corax* (Mourer-Chauviré 2004).

Indeterminate birds.

Material: Pelvis, NMB Se 1559; sacrum, NMB Se 1560; alar phalanx, incomplete, NMB Se 1595; right femur, partly embedded in matrix, FSL 91636; pedal phalanx, NMB Se 1565n.

Comparisons of the Senèze Avifauna with Other Approximately Coeval Avifaunas

The avifauna which is most similar to that of Senèze is the slightly older one of Saint-Vallier (MNQ 17) (Mourer-Chauviré 2004). They have in common *Ciconia nigra*, *Haliaeetus albicilla*, *Pavo bravardi*, a great bustard the size of *Otis tarda*, and *Corvus corax antecorax*. However the Senèze avifauna differs from that of Saint-Vallier by the presence of a larger number of aquatic forms: *Tadorna* cf. *ferruginea*, *Anas clypeata*, *Aythya* sp., *Mergus* cf. *merganser*, and *Mergellus* sp., while in Saint-Vallier the only aquatic form is *Cygnus* cf. *cygnus*.

The Senèze avifauna also differs by the presence of a tetraonid, *Tetrao* cf. *partium*, while this family is absent from Saint-Vallier. The tetraonids, with the different species of the genus *Lagopus*, were to become extremely abundant during the Middle and Late Pleistocene, in almost all the European avifaunas. Species of Tetraonidae are present in Central Europe as early as the Early Pliocene (MN 15) (Mourer-Chauviré 1993) but, in France, Senèze (MNQ 18) is their first appearance.

The occurrence of a peafowl is an important chronological marker, but *Pavo bravardi* has not been found in any other localities of MNQ 18. These localities are mainly karstic fillings, and most bird remains come from the pellets of raptors; peafowl were probably too large to be captured by these predators.

The Senèze avifauna is even more different from that of the Italian localities Pietrafitta and Pirro Nord (Bedetti 2003; Bedetti & Pavia 2013). Pietrafitta includes mainly aquatic forms (Podicipedidae, Phalacrocoracidae, Ardeidae, Anatidae, and Rallidae), and *Gallus* sp. The Pirro Nord avifauna is more diverse, including in particular two extinct forms: the genus *Palaeocryptonyx*, described from Perpignan (MN 15) and known in France at Montoussé 5 (MNQ 18), and *Corvus pliocaenus*, described from the Early Pleistocene of Valdarno (MNQ 18) and present in France until MNQ 22 (Mourer-Chauviré 1993). The Pirro Nord avifauna has in common with that of Senèze *Tadorna ferruginea*, *Anas clypeata*, *Aythya* sp., *Alectoris* sp., *Otis tarda* and *Bubo* sp. But, it differs from Senèze because it includes an ibis, *Geronticus eremita*, more numerous Anseriformes, many Charadriiformes, and several mid-sized and small Passeriformes (Bedetti 2003; Bedetti & Pavia 2013).

In Hungary the Villány 3 avifauna (MNQ 17, age 2.0 Ma, Tyrberg 2022) shows affinities with that of Senèze by the presence of the genera *Tadorna*, *Otis*, *Bubo*, and the species *Tetrao* cf. *partium* and *Surnia robusta* (Janossy 1986; Tyrberg 1998). In Bulgaria the Slivnitsa avifauna is referred to MNQ 18, age 1.95–2.0 Ma (Boev 2002; Tyrberg 2022). It contains extinct forms such as an ibis, *Geronticus balcanicus*, a small galliform, *Chauvireria balcanica*, and a primitive form of raven, *Corvus* cf. *praecorax*. This avifauna is quite different from that of Senèze and moreover includes numerous middle- or small-sized Passeriformes, which are practically absent from Senèze.

The rich avifauna of Las Higueruelas (Ciudad Real Province, Spain) is older than that of Senèze, as it is referred to MN 16 (Sánchez Marco 2005, 2009). This avifauna includes a very large number of aquatic forms (Podicipedidae, Ardeidae, Threskiornithidae, Anatidae, Rallidae, and Scolopacidae), but does not contain large-sized species such as peafowl and great bustards. It includes only two extinct forms, *Palaeocryptonyx novaki* and *Corvus* cf. *antecorax*. There are at least three species which nest in northern regions, *Cygnus cygnus*, *Aythya marila*, and *Mergellus albellus*, none of which is commonly reported today in the Iberian Peninsula. Sánchez Marco (2005) considers that the Las Higueruelas avifauna, although dating from ca. 3.5 Ma, is a Quaternary-like avifauna. The Senèze avifauna shares with Las Higueruelas the taxa *Tadorna*, *Aythya*, *Mergus*, and *Corvus* cf. *antecorax*, but differs from it by the presence of the extinct species *Pavo bravardi*, *Tetrao* cf. *partium*, and *Surnia robusta*.

Paleoecological and Paleoclimatic Signification

In addition to the extinct taxa, the avifauna from Senèze includes a number of species which still occur in this part of Western Europe whereas others have disparate distribution. For these recent species it is possible to draw information about the paleoenvironment and the paleoclimate. For extinct species information can be inferred from the recent species which are related to them.

Ciconia nigra, the Black Stork, has practically disappeared from Western Europe, as a consequence of the destruction of its habitat in its breeding area, except in the center of the Iberian Peninsula (Voous 1960). In Eastern

Europe and Asia, its breeding area extends to about latitude 60° north. It is migratory and winters mainly in Africa, India, and southern China. It lives principally in old undisturbed forests, interspersed with rivers, pools, swampy patches and other humid areas. It feeds mainly on fish taken from shallow waters (Cramp & Simmons 1977).

Tadorna ferruginea, the Ruddy Shelduck, breeds in Central Asia and winters in India, the Indochinese Peninsula, southern China and Eastern Africa. Its distribution area is discontinuous since it also breeds and winters in Western Africa. It was formerly present in Europe. It lives on often brackish water bodies, in open regions, feeding on plants but also on small aquatic animals (del Hoyo et al. 1992).

Anas clypeata, the Northern Shoveler, breeds in all the Palearctic and Nearctic provinces, except in their coldest parts. It is migratory and winters in the south of Europe and in the tropical regions of the Northern Hemisphere. Its habitat is made up of a wide variety of shallow, freshwater wetlands, and in winter it can be found on brackish lagoons and tidal mudflats. Its diet is composed of small aquatic invertebrates, seeds, and plant remains (del Hoyo et al. 1992).

Mergus merganser, the Goosander, breeds in the north and center of the Palearctic and Nearctic provinces and winters in the south of its breeding area. It breeds on freshwater lakes, pools, and upstream of rivers, mainly in wooded regions, and winters on larger lakes and rivers, lagoons and brackish marshes (del Hoyo et al. 1992; Voous 1960). It feeds mainly on fish and also on small aquatic animals (invertebrates and vertebrates).

Haliaeetus albicilla, the White-tailed Sea-eagle, was formerly present in the whole of the Palearctic province, as well as in Greenland. It had been practically exterminated by humans in Central and Western Europe, but it is now slowly recovering thanks to protection. Its habitat consists of wide rivers, lakes, pools, estuaries and seacoasts, in wooded or open regions, but not in mountains. It feeds mainly on fish and aquatic birds, but also on terrestrial and aquatic mammals and carrion. Its distribution area is very extensive, from the tundra to the Mediterranean and desert climatic zones (Voous 1960).

Tetrao partium is considered as the direct ancestor of *Tetrao tetrix*, the Black Grouse. *Tetrao tetrix* lives in the northern part of the Palearctic province and in the Alps, but not in the tundra zone. Its habitat is typically the transition area between forests and open environments (del Hoyo et al. 1994). It is sedentary and feeds exclusively on vegetable matter, shoots and buds of leaves and flowers, needles, catkins, seeds and berries (Voous 1960).

Pavo bravardi is closely related to the recent species *Pavo muticus*, the Green Peafowl, which lives in Northeast India, Bangladesh, Burma, Thailand, South China, Laos, Vietnam, and Java. In Java it occurs in open woodland and forest edge, from sea level up to 3,000 m. Its food is made up of vegetable matter and insects. It is believed to be sedentary (del Hoyo et al. 1994).

Alectoris graeca, the Rock Partridge, lives in the south east of Europe, in the Alps, Italy, Sicily, former Yugoslavia, Bulgaria, and Greece. Its habitat consists of rocky hills and mountain slopes with short grass and sparse shrub cover, usually above 900 m a.s.l. It is generally sedentary and predominantly vegetarian (del Hoyo et al. 1994).

The family Otididae is represented by a fragment of bone the size of *Otis tarda*, the Great Bustard. The bustards live in grassy plains, temperate or tropical, always in areas that are at least seasonally dry. Their habitat varies from near-desert regions and salted steppes to dense bushlands and thorny country. They are most diverse in Africa, and only two species are found in Europe today. Their food includes mainly plant material and invertebrates (del Hoyo et al. 1996).

The different species of the recent genus *Bubo* live in varied habitats (del Hoyo et al. 1999).

Surnia robusta is probably not the direct ancestor of the recent species *Surnia ulula*, the Northern Hawk-owl, which lives in the Holarctic province, in the forest tundra and boreal taiga (del Hoyo et al. 1999). Outside the four Early Pleistocene localities from which it was described in Hungary, it has been identified in the Early Pleistocene of Ahl al Oughlam, Morocco. This locality was on the seashore of the Atlantic Ocean at the time of the deposit of the fossil material; there are no high mountains nearby which might have housed high-altitude forests (Mourer-Chauviré & Geraads 2010). Janossy (1977, p. 14) had asked whether *Surnia robusta* had the same ecological significance as *S. ulula* "in view of the fact that most of the remains originate from the submediterranean region of the Villány mountains". But it seems instead that *Surnia robusta* was not linked to an environment of boreal or mountain forests, rather perhaps to an environment of middle or low altitude forests.

Corvus corax, the Common Raven, has a very wide distribution in almost all the Holarctic province, and in the north of the Ethiopian province, in climatic zones ranging from tundra to desert and savanna. In Western Europe it had been practically exterminated by humans, but it is now recovering in most of its former range. Its habitat and food are extremely varied: its diet varies according to the available resources but can include a high proportion of carrion (Voous 1960).

The birds from Senèze include a certain number of species living on or along water, such as *Tadorna* cf. *ferruginea*, *Anas clypeata*, *Aythya* sp., and *Haliaeetus albicilla*, which is not surprising since the fossils were deposited in a lake. Other species indicate the presence of forests with water bodies, such as *Ciconia nigra*, *Mergus* cf. *merganser*, and *Pavo bravardi*. *Tetrao* cf. *partium* and perhaps also *Surnia robusta* lived probably at the edges of forests. The occurrence of *Alectoris* cf. *graeca* and of a large bustard, of the size of *Otis tarda*, indicates the proximity of more or less rocky open regions, grassy plains or bushlands. Finally, *Bubo* sp. and *Corvus corax antecorax* can have very diverse habitats.

As indicated by the avian fossils, the paleoenvironment of Senèze was more humid and more wooded than that of Saint-Vallier. Previous palynological studies (Elhaï 1969) were based on material recovered from a lake core sampling lower layers than the fossiliferous level. New palynological studies, from the layers which have yielded the fossil material, indicate that the vegetation was characterized by leafy trees. The area was partly wooded, and the trees were mainly mesothermophilous. The forest was not dense, as pollen of herbs and light-demanding trees were also found (J. Argant 2024). *Pavo bravardi* is a relict tropical form, Senèze is the last occurrence of the genus *Pavo* in Europe. The avifauna at Senèze does not include any element corresponding to a climate colder than the present one.

Acknowledgments For the loan of fossil material, I thank B. Engesser and L. Costeur, from the Naturhistorisches Museum, Basel, and A. Prieur, from the University Claude Bernard, Lyon. For providing me the opportunity of studying comparative material, I thank the late Professor P. Brodkorb, Gainesville, USA, and the curators of the collections of the Muséum national d'Histoire naturelle (Laboratoire d'Anatomie comparée and Institut de Paléontologie Humaine), Paris; the Université Claude Bernard, Lyon; the Natural History Museum, Tring, United Kingdom; and the Institute of Paleontology of the Russian Academy of Sciences, Moscow, Russia. I thank Samuel Mailliot and Jean-Paul Bourseau for their welcome help in the preparation of the figures. I thank Eric Delson for language suggestions and the reviewers, Gerald Mayr, Marco Pavia, and Antonio Sánchez Marco, for their helpful comments.

References

Argant, J. (2024). Palynological analysis of the fossiliferous layers of the late Villafranchian deposit of Senèze (Domeyrat, Haute-Loire, France). Taphonomy and paleoenvironmental data. In E. Delson, M. Faure & C. Guérin (Eds), *Senèze: Life in Central France two million years ago. Paleontology, geochronology, stratigraphy and taphonomy* (pp. 123–140). Cham (Switzerland): Springer.

Bacher, A. (1967). Vergleichend morphologische untersuchungen an Einzelknochen des postcranialen Skeletts in Mitteleuropa vorkommender Schwäne und Gänse. Inaugural Dissertation, Institut für Paläoanatomie, Domestikationsforschung und Geschichte der Tiermedizin, München.

Baker, J. M., & Omland, K. E. (2006). Canary Island Ravens *Corvus corax tingitanus* have distinct mtDNA. *Ibis, 148*, 174–178.

Baumel, J. J., & Witmer, L. M. (1993). Osteologia. In: Baumel, J. J. (Ed.), *Handbook of Avian Anatomy. Nomina Anatomica Avium* (2nd Ed.). *Nuttall Ornithological Club, 23*, 45–132.

Bedetti, C. (2003). Le avifaune fossili del Plio-Pleistocene italiano: sistematica, paleoecologia, ed elementi di biocronologia. Ph. D. Dissertation, Università "La Sapienza" di Roma, Roma.

Bedetti, C., & Pavia, M. (2004). The validity of the fossil *Aythya aretina* (Portis) and *Aythya sepulta* (Portis) (Aves, Anatidae) on the basis of new finding in the Italian early Pleistocene. In *Sixth International Meeting of the Society of Avian Paleontology and Evolution, Quillan, France, 28th September-3rd October 2004*. Abstracts, 4.

Bedetti, C., & Pavia, M. (2013). Early pleistocene birds from pirro nord (Puglia, southern Italy). *Palaeontographica, 298*, 31–53.

Bochenski, Z., & Kurochkin, E. N. (1987). Pliocene bustards (Aves: Otididae and Gryzajidae) of Moldavia and S. Ukraine. In: C. Mourer-Chauviré (Coord.), L'évolution des oiseaux d'après le témoignage des fossiles. *Documents des Laboratoires de Géologie de Lyon, 99*, 173–187.

Boev, Z. N. (1999). On the presence of *Tetrao partium* (Kretzoi, 1962) (Aves: Tetraonidae) in the Late Pliocene of Bulgaria. *Historia Naturalis Bulgarica, 10*(1999), 85–96.

Boev, Z. N. (2001). Early pliocene avifauna of Muselievo (C. Northern Bulgaria). *Acta Zoologica Cracoviensia, 44*, 37–52.

Boev, Z. N. (2002). Neogene Avifauna of Bulgaria. In Z. Zhou & F. Zhang (Eds.), *Proceedings of the 5th Symposium of the Society of Avian Paleontology and Evolution, Beijing, 1–4 June 2000* (pp. 29–40). Beijing, China: Science Press.

Boev, Z. N., & Koufos, G. (2000). Presence of *Pavo bravardi* (Gervais, 1849) (Aves, Phasianidae) in the Ruscinian locality of Megalo Emvolon, Macedonia, Greece. *Geologica Balcanica, 30*, 69–74.

Brodkorb, P. (1978). Catalogue of fossil birds. Part 5 (Passeriformes). *Bulletin of the Florida State Museum, Biological Sciences, 23*, 139–228.

Burchak-Abramovich, N. I., & Vekua, A. K. (1981). Fossil-bustard-runner of East Georgian Akchagil. *Izvestia Akademii Nauk Gruzinskoy SSR, Seria Biologitcheskaia, 7*, 53–60. (in Russian).

Cheneval, J. (1987). Les Anatidae (Aves, Anseriformes) du Miocène de France. Révision systématique et Evolution. In C. Mourer-Chauviré (Coord.), *L'évolution des oiseaux d'après le témoignage des fossiles. Documents des Laboratoires de Géologie de Lyon* (Vol. 99, pp. 137–157).

Cramp, S., & Simmons, K. E. L. (1977). *The birds of the Western Palearctic* (Vol. I). Oxford University Press.

Crégut-Bonnoure, E., Guérin, C., Argant, A., Argant, J., Debard, E., Delson, E., et al. (2024). Biochronology of the Senèze faunal assemblage. In E. Delson, M. Faure & C. Guérin (Eds), *Senèze: Life in Central France two million years ago. Paleontology, geochronology, stratigraphy and taphonomy* (pp. 633–652). Cham (Switzerland): Springer.

Del Hoyo, J., Elliott, A., & Sargatal, J. (1992). *Handbook of the birds of the world. Ostrich to Ducks* (Vol. 1). Barcelona: Lynx edicions.

Del Hoyo, J., Elliott, A., & Sargatal, J. (1994). *Handbook of the Birds of the World. New World Vultures to Guineafowl* (Vol. 2). Barcelona: Lynx edicions.

Del Hoyo, J., Elliott, A., & Sargatal, J. (1996). *Handbook of the Birds of the World.* Hoatzin to Auks (Vol. 3). Barcelona: Lynx edicions.

Del Hoyo, J., Elliott, A., & Sargatal, J. (1999). *Handbook of the Birds of the World.* Barn-owls to Hummingbirds (Vol. 5). Barcelona: Lynx edicions.

Delson, E., Faure, M., Guérin, C., Aprile, L., Argant, J., Blackwell, B. A. B., Debard, E., Harcourt-Smith, W., Martin-Suarez, E., Monguillon, A., Parenti, F., Pastre, J.-F., Sen, S., Skinner, A. R., Swisher, C. C., III., & Valli, A. M. F. (2006). Franco-American renewed research at the late Villafranchian locality of Senèze (Haute-Loire, France). *Courier Forschungsinstitut Senckenberg, 256*, 275–290.

Delson, E., Nomade, S., Sen, S., Debard, E., Pastre, J.-F., & Bahain, J.-J. (2024). Geochronology of Senèze: $^{40}Ar/^{39}Ar$ dating and magnetostratigraphy, with notes on an ESR/U-series dating attempt. In E. Delson, M. Faure & C. Guérin (Eds.), *Senèze: Life in Central France two million years ago. Paleontology, geochronology, stratigraphy and taphonomy* (pp. 99–122). Cham (Switzerland): Springer.

Depéret, C. (1892). Les animaux pliocènes du Roussillon. *Mémoires de la Société géologique de France. Paléontologie, 3*, 117–136.

Elhaï, H. (1969). La flore sporo-pollinique du gisement villafranchien de Senèze (Massif-Central-France). *Pollen et Spores, 11*, 127–139.

Erbersdobler, K. (1968). Vergleichend morphologische Untersuchungen an Einzelknochen des postcranialen Skeletts in Mitteleuropa vorkommender mittelgrosser Hühnervögel. Inaugural Dissertation, Institut für Paläoanatomie, Domestikationsforschung und Geschichte der Tiermedizin, München.

Gervais, P. (1849). Oiseaux et Reptiles fossiles de France. *Mémoires de l'Académie des Sciences de Montpellier, 1*, 220–222.

Gruber, A. (1990). Vergleichend morphologische Untersuchungen an Einzelknochen in Ägypten vorkommender Ciconiidae. Inaugural-Dissertation zur Erlangung der tiermedizinischen Doktorwürde der Tierärztlichen Fakultät der Ludwig-Maximilians-Universität München.

Guérin, C. (1982). Première biozonation du Pléistocène européen, principal résultat biostratigraphique de l'étude des Rhinocerotidae (Mammalia, Perissodactyla) du Miocène terminal au Pléistocène supérieur d'Europe occidentale. *Geobios, 15*, 593–598.

Harrison, C. J. O. (1979). Birds of the cromer forest bed series of the East Anglian pleistocene. *Transactions of the Norfolk and Norwich Natural Society, 24*, 277–286.

Harrison, C. J. O. (1980). Pleistocene bird remains from tornewton cave and the Brixham windmill cave in South Devon. *Bulletin of the British Museum (Natural History), Geology, 33*,, 91–100.

Hernández, E. C., Martín, M., & Rando, J. C. (1993). Estudio osteológico comparado de dos subspecies de *Corvus corax* (Aves: Passeriformes). *Archaeofauna, 2*, 181–190.

Howard, H. (1929). The avifauna of the Emeryville shellmound. *University of California Publications in Zoology, 32*, 301–394.

Janossy, D. (1972). Die mittelpleistozäne Vögel fauna der Stránská Skála. *Anthropos, Studia Musei Moraviae, 20*, 35–64.

Janossy, D. (1974). Die mittelpleistozäne Vogelfauna von Hundsheim (Niederösterreich). In *Sitzungsberichte der Österreichischen Akademie der Wissenschaften, Mathematisch-Naturwissenschaftliche Klasse* (Vol. 182, pp. 211−257). Wien, Abteilung I.

Janossy, D. (1976). Plio-pleistocene bird remains from the Carpathian basin. II. Galliformes. 2. Phasianidae. *Aquila, 83*, 29−42.

Janossy, D. (1977). Plio-Pleistocene bird remains from the Carpathian basin. III. Strigiformes, Falconiformes, Caprimulgiformes, Apodiformes. *Aquila, 84*, 9−36.

Janossy, D. (1979a). Plio-pleistocene bird remains from the Carpathian Basin. IV. *Anseriformes, Gruiformes, Charadriiformes, Passeriformes. Aquila, 85*, 11−39.

Janossy, D. (1979b). Plio-Pleistocene bird remains from the Carpathian basin. V. Podicipediformes, Ciconiiformes, Otidiformes, Columbiformes. *Piciformes. Aquila, 86*, 19–33.

Janossy, D. (1983a). Die mittelpleistozäne Vogelfauna von Prezletice bei Prag. *Schriftenreihe Für Geologische Wissenschaften, 19*(20), 247–269.

Janossy, D. (1983b). Lemming-remain from the Older Pleistocene of Southern Hungary (Villány, Somssich-hegy 2). *Fragmenta Mineralogica et Palaeontologica, 11*, 55–60.

Janossy, D. (1986). *Pleistocene Vertebrate Faunas of Hungary. Developments in Palaeontology and Stratigraphy* (Vol. 8). Amsterdam: Elsevier.

Janossy, D. (1992). Lower pleistocene bird remains from Beremend, (S-Hungary, LOC. 15. and 16.). *Aquila, 99*, 9–25.

Junge, G. C. A. (1953). Fossil bones of a white-tailed eagle, *Haliaëtus albicilla* (L.) from the Tegelen clay. *Koninklijke Nederlandse Akademie van Wetenschappen, 56, ser. B*, 285–290.

Kretzoi, M. (1941). Die mittelpleitozäne Säugetierfauna von Betfia bei Nagyvarad. *Földtani Közlöny, 71*, 308–335.

Kretzoi, M. (1962). Vogelreste aus der altpleistozänen Fauna von Betfia. *Aquila, 67–68*, 167–174.

Kurochkin, E. N. (1982). Novie ptitzi iz crednievo Pliotzena Mongolii. *Ornithologia, 17*, 150–154. (in Russian).

Kurochkin, E. N. (1985). *Ptitzi tzentralnoi Asii v Pliotzenie*. Birds of the central Asia in Pliocene. The joint Soviet-Mongolian paleontological expedition. *Transactions, 26* (in Russian). Moskva: Naouka.

Lambrecht, K. (1916). Die Erste Ungarische Präglaciale Vogelfauna. *Aquila, 22*, 168−175.

Louchart, A. (2003). A true peafowl in Africa. *South African Journal of Science, 99*, 368–371.

Louchart, A., Mourer-Chauviré, C., Güleç, E., Howell, F. C., & White, T. D. (1998). L'avifaune de Dursunlu, Turquie, Pléistocène inférieur: Climat, environnement et biogéographie. *Comptes rendus de l'Académie des Sciences, 327*, 341–346.

Louchart, A., Wesselman, H., Blumenschine, R. J., Hlusko, L. J., Njau, J. K., Black, M. T., Asnake, M., & White, T. D. (2009). Taphonomic, Avian and small-vertebrate Indicators of *Ardipithecus ramidus* Habitat. *Science, 326*, 66 and 66e1–4.

Meijer, H. J. M., Pavia, M., Madurell-Malapeira, J., & Alba, D. M. (2017). A revision of the fossil eagle owls (Aves: Strigiformes: *Bubo*) from Europe and the description of a new species, *Bubo ibericus*, from Cal Guardiola (NE Iberian Peninsula). *Historical Biology, 29*, 822–832.

Mein, P. (1990). Updating of MN zones. In E. H. Lindsay, V. Fahlbusch, & P. Mein (Eds.), *European Neogene Mammal Chronology* (pp. 73–90). Plenum Press.

Mlíkovsky, J. (1995). Early pleistocene birds of Stránská Skála hill, Czech Republic: 1. Musil's Talus Cone. In R. Musil (Ed.). *Stránská Skála Hill. Excavations of open air sediments 1964–1972. Moravian Museum Brno, Anthropos Series, 26 (N. S. 18)*, 111–126.

Mlíkovsky, J. (1996). Tertiary Avian localities of Holland. In J. Mlíkovsky (Ed.). *Tertiary Avian Localities of Europe. Acta Universitatis Carolinae, 39(1995)*, 3–4, 685–691.

Mlíkovsky, J. (1997). Taxonomic identity of *Haliaeetus angustipes* JANOSSY, 1983 (Aves: Accipitridae) from the early Pleistocene of the Czech Republic. *Buteo, 9*, 51–56.

Mlíkovsky, J. (2002a). *Cenozoic birds of the world. Part 1: Europe*. Praha: Ninox Press.

Mlíkovsky, J. (2002b). Early pleistocene birds of Stránská Skála, Czech Republic: 2. Absolon's Cave. *Sylvia, 38*(2002), 57–65.

Mourer-Chauviré, C. (1975). Les Oiseaux du Pléistocène moyen et supérieur de France. *Documents des Laboratoires de Géologie de la Faculté des Sciences de Lyon, 64*, 1–624.

Mourer-Chauviré, C. (1989). A peafowl from the Pliocene of Perpignan, France. *Palaeontology, 32*, 439–446.

Mourer-Chauviré, C. (1990). Les paons du Pliocène et du Pléistocène inférieur d'Europe. Biostratigraphie et Paléoécologie. *Journée Louis David. Documents des Laboratoires de Géologie de Lyon H, s., 9*, 77–91.

Mourer-Chauviré, C. (1993). The Pleistocene Avifaunas of Europe. *Archaeofauna, 2*, 53−66.

Mourer-Chauviré, C. (2004). Les oiseaux du Pliocène supérieur de Saint-Vallier (Drôme, France). In M. Faure & C. Guérin (Eds), *Le gisement pliocène final de Saint-Vallier (Drôme, France)*. Geobios, Mémoire spécial n° 26, 37, 97−114.

Mourer-Chauviré, C., & Geraads, D. (2008). The Struthionidae and Pelagornithidae (Aves: Struthioniformes, Odontopterygiformes) from the late Pliocene of Ahl al Oughlam, Morocco. In *Proceedings of the Sixth International Meeting of the Society for Avian Paleontology and Evolution, Quillan, France. Oryctos, 7,* 169–194.
Mourer-Chauviré, C., & Geraads, D. (2010). The Upper Pliocene avifauna of Ahl al Oughlam, Morocco. Systematics and biogeography. In *Proceedings of the 7th International Meeting of the Society of Avian Paleontology and Evolution, Sydney. Records of the Australian Museum, 62,* 157–184.
Nomade, S., Pastre, J. F., Guillou, H., Faure, M., Guérin, C., Delson, E., Debard, E., Voinchet, P., & Messager, E. (2014). $^{40}Ar/^{39}Ar$ constraints on some French landmark Late Pliocene to Early Pleistocene large mammalian paleofaunas: Paleoenvironmental and paleoecological implications. *Quaternary Geochronology, 21*, 2–15.
Otto, C. (1981). Vergleichende Morphologische Untersuchungen an Einzelknochen in Zentraleuropa vorkommender mittelgrosser Accipitridae. 1. Schädel, Brustbein, Schultergürtel and Vorderextremität. Inaugural-Dissertation zur Erlangung der tiermedizinischen Doktorwürde der Tierärztlichen Fakultät der Ludwig-Maximilians-Universität München.
Pavia, M., & Bedetti, C. (2013). Early pleistocene fossil birds from Cava Sud, Soave (Verona, North-Eastern Italy). In U. B. Göhlich, & A. Kroh (Eds.) *Palaeornithological Research. Proceedings of the 8th International Meeting of the Society of Avian Paleontology and Evolution* (pp. 171–183).
Pavia, M., Bedetti, C., & Carrera, L. (2018). A new middle pleistocene bird assemblage from Cava di Breccia di Canal Selce (Ponte Galeria, Rome, Italy). *Avocetta, 42*, 31–38.
Pavia, M., Göhlich, U. B., & Mourer-Chauviré, C. (2012). Description of the type-series of *Palaeocryptonyx donnezani* Depéret, 1892 (Aves: Phasianidae) with the selection of a lectotype. *Comptes Rendus Palevol, 11*, 257–263.
Peters, D. S. (1992). A new species of Owl (Aves, Strigiformes) from the middle eocene messel oil shale. *Natural History Museum of Los Angeles County, Science Series, 36*, 161–169.
Pickford, M., Senut, B., & Mourer-Chauviré, C. (2004). Early pliocene Tragulidae and peafowls in the rift Valley, Kenya: Evidence for rainforest in East Africa. *Comptes Rendus Palevol, 3*, 179–189.
Portis, A. (1891). Gli ornitoliti del Valdarno superiore e di alcune altre località plioceniche di Toscana. *Memorie Regio Istituto di Studi Superiore e Practici (Firenze), 1889*, 1–20.
Rosas, A., Pérez-González, A., Carbonell, E., van der Made, J., Sánchez, A., Laplana, C., Cuenca-Bescós, G., Parés, J. M., & Huguet, R. (2001). Le gisement pléistocène de la "Sima del Elefante" (Sierra de Atapuerca, Espagne). *L'Anthropologie, 105*, 301–312.
Sánchez Marco, A. (1989). Huéscar-1 (Granada, España): Avance de la lista de aves y consiguientes conjeturas sobre paleoambiente y paleogeografia. In M. T. Alberdi, F. P. Bonadonna, (Eds.), *Geologia y Paleontologia de la Cuenca de Guadix-Baza. Trabajos sobre el Neogeno-Cuaternario, 11,* 175–183.
Sánchez Marco, A. (1990). A new bustard (Otididae, Aves) from the early pliocene of Layna (Soria, Spain). *Paleontologia i Evolució, 23*, 223–229.
Sánchez Marco, A. (2004). Avian zoogeographical patterns during the Quaternary in the Mediterranean region and paleoclimatic interpretation. *Ardeola, 5*, 91–132.
Sánchez Marco, A. (2005). Aves del Plioceno superior de la meseta sur ibérica: Una asociación ornítica aparentemente cuaternaria. *Revista Española de Paleontología, 20*, 143–157.
Sánchez Marco, A. (2009). New Iberian Galliforms and reappraisal of some pliocene and pleistocene Eurasian taxa. *Journal of Vertebrate Paleontology, 29*, 1148–1161.
Siori, M. S., Boero, A., Carnevale, G., Colombero, S., Delfino, M., Sardella, R., & Pavia, M. (2014). New data on early pleistocene vertebrates from Monte Argentario (Central Italy). Paleoecological and Biochronological Implications. *Geobios, 47*, 403–418.
Soergel, E. (1966). Die Teufels-oder Fuchsenlucken bei Eggenburg (NÖ). *Denkschriften der Österreichischen Akademie der Wissenschaften, Mathematisch-Naturwissenschaftliche Klasse, 112,* 93–107.
Stehlin, G. (1923). Die oberpliocaene Fauna von Senèze (Haute-Loire). *Eclogae Geologicae Helvetiae, 23*, 268–281.
Stewart, J. R. (1996). Quaternary birds from the Torbryan Valley. In D. J. Charman, R. M. Newnham & D. G. Croot (Eds.), *The Quaternary of Devon and East Cornwall: Field Guide* (pp. 190–191). London, England: Quaternary Research Association.
Stewart, J. R. (2007). An evolutionary study of some archaeologically significant Avian Taxa in the Quaternary of the Western Palaearctic. *British Archaeological Reports International Series, 1653*, 1–272.
Tchernov, E. (1980). *The Pleistocene birds of 'Ubeidiya, Jordan Valley*. Jerusalem: The Israel Academy of Sciences and Humanities.
Tyrberg, T. (1998). Pleistocene birds of the Palearctic: A Catalogue. *Publications of the Nuttall Ornithological Club, 27*, 1–720.
Tyrberg, T. (2022). *Pleistocene Birds of the Palearctic* (2nd Ed.). Privately pubished by the author.
Voous, K. H. (1960). *Atlas of European Birds*. Nelson.
Woelfle, E. (1967). Vergleichend morphologische untersuchungen an Einzelknochen des postcranialen Skeletts in Mitteleuropa vorkommender Enten, Halbgänse und Säger. Inaugural Dissertation, Institut für Paläoanatomie, Domestikationsforschung und Geschichte der Tiermedizin, München.
Zucchetta, G., Gentili, S., & Pavia, M. (2003). A new early Pleistocene Bird association from Pietrafitta (Perugia, Central Italy). *Rivista Italiana di Paleontologia e Stratigrafia, 109*, 527–538.

Chapter 9
The Carnivores of Senèze (Haute-Loire)

Alain Argant

Abstract The recently excavated carnivore fossils at Senèze are rather limited in number, in contrast with the material of earlier decades. A global review of the earlier material is presented. Twelve species are identified: four canids (*Nyctereutes megamastoides, Vulpes alopecoides, Canis arnensis, Canis* sp.), one ursid (*Ursus etruscus*), three hyaenids (*Pachycrocuta perrieri, Chasmaporthetes lunensis*, Hyaenidae indet.), and four felids (*Acinonyx pardinensis, Megantereon cultridens, Homotherium crenatidens* and, for the first time, *Dinofelis* sp.). The large-scale excavation carried out by Faure, Guérin and Delson provided the first precise knowledge of the stratigraphy corresponding to the paleontological remains and paleoenvironmental data linked to the fauna. The carnivore taxa of Senèze range over a long time span, but they don't indicate the existence of two chronologically distinct periods. Their study places Senèze in the second part of the middle Villafranchian, around 2.1 Ma.

Résumé À Senèze, les restes de Carnivores fossiles exhumés lors des fouilles récentes (2001–2006) sont des plus réduits, contrairement au matériel paléontologique du siècle dernier. L'étude de ce matériel récent a été l'occasion de faire le point sur l'ensemble des vestiges de Carnivores de Senèze et d'essayer de dater le gisement par la biochronologie. Neuf espèces ont été identifiées, et trois autres restent imprécises: quatre Canidés (*Nyctereutes megamastoides, Vulpes alopecoides, Canis arnensis, Canis* sp.), un Ursidé (*Ursus etruscus*), trois Hyénidés (*Pachycrocuta perrieri, Chasmaporthetes lunensis*, Hyaenidae indét.) et quatre Félidés (*Acinonyx pardinensis, Megantereon cultridens, Homotherium crenatidens* et *Dinofelis* sp.). Trois éléments nouveaux par rapport à la liste de faune antérieure sont à relever: la présence possible de *Canis etruscus*, pour la première fois celle d'un *Dinofelis* et, peut-être, d'une autre espèce de Hyénidé. La fouille à grande échelle réalisée par M. Faure, C. Guérin et E. Delson livre de précieuses données sur des niveaux fossilifères en place et permet de préciser leur stratigraphie et de connaître leurs paléoenvironnements. Les Carnivores de Senèze s'échelonnent probablement sur une longue durée mais rien ne permet d'évoquer deux périodes chronologiques distinctes. Cette étude place la faune des Carnivores de Senèze au Villafranchien. Elle est chronologiquement postérieure à celle du gisement de Saint-Vallier et antérieure à celles du Val d'Arno et de Dmanisi, aux environs de 2,1 Ma.

Keywords Villafranchian • Biochronology • *Dinofelis* • Canidae • Ursidae • Felidae • Hyaenidae

Mots-clés Villafranchien • Biochronologie • *Dinofelis* • Canidae • Ursidae • Felidae • Hyaenidae

Introduction

The carnivore material recently excavated at Senèze by Faure, Guérin and Delson (2001–2006) is rather limited in extent, in contrast to the fossils recovered over the course of earlier decades. However, carnivores have never been very abundant at Senèze, and their remains have been found scattered over a wide area bordering the body of water in the maar, which attracted animals to drink. It is not surprising that carnivore fossils remain rare.

A. Argant (✉)
Aix-Marseille Université, CNRS, Minist. Culture, LAMPEA, UMR 7269, Aix-en-Provence, France
e-mail: a.argant@wanadoo.fr

E. Delson et al. (eds.), *Senèze: Life in Central France Around Two Million Years Ago*, Vertebrate Paleobiology and Paleoanthropology, https://doi.org/10.1007/978-3-031-64415-3_9

Notes and Abbreviations

All the measurements in this study are in mm, and this unit does not appear in the tables.

The height of the crown (HC) of a cheek tooth is measured on the external side in a direct line, and it is only given in the case of teeth that are not very worn.

* about (after a measurement).
A: measure of the alveolus of a tooth, (A) measure between the alveoli for a series of teeth
APD = antero-posterior diameter = length or *DAP = diamètre antéro-postérieur*
CAN = Canidae in the Romain-la-Roche collection
Dentition: I, i = incisor; C, c = canine; P, p = premolar; M, m = molar. Upper teeth are indicated by uppercase (capital) letters, lower teeth by lowercase
DPD = dorso-palmar or dorso-plantar diameter = height or *DDP = diamètre dorso-palmaire ou dorso-plantaire (= hauteur)*
fragt = fragment
HC = height of the crown
inf.; sup.; ant.; post.; med.; art. = inferior; superior; anterior; posterior; median; articular
juv. = juvenile
L or R = left or right
mand. = mandible
MD = mesio-distal diameter = length (for a tooth)
min., max. = minimum; maximum
MNI = minimum number of individuals
Mc, Mt = metacarpal, metatarsal bone; Mt 3 = 3rd metatarsal bone
NR = number of remains (complete or fragments)
PEMAC (Plan d'Enregistrement Morphométrique Argant pour Canidés) = Argant system for morphometry of canids.
PFEMA (Plan Félidés d'Enregistrement Morphométrique Argant) = Argant system for morphometry of felids
PUEMA (Plan Ursidés d'Enregistrement Morphométrique Argant) = Argant system for morphometry of ursids
prox. ext. = proximal extremity; dist. ext. = distal extremity
TD = transverse diameter or *DT = diamètre transverse*
TN = transfer number (in a table, a figure, a caption)
U = degree of wear (Argant 1991: 61–62): 0 = no wear; 1 = very slightly worn; 2 = moderately worn, enamel has sometimes disappeared in places; 3 = heavily worn but the relief of the crown is still visible; 4 = crown almost flattened; 5 = crown completely flattened
VL = vestibulo-lingual diameter = width (for a tooth)
Δ l = measurement range
σ n − 1 = standard deviation

Institutional Abbreviations

CCEC = Center of Conservation and Study of the Collections, MHNL
FSL = Faculté des Sciences, Lyon (used as collection acronym with specimen numbers, located in UCB-Lyon 1)
GNM = Georgian National Museum, Tbilisi
IGF = Istituto Geologico Firenze (now Dipartimento di Scienze della Terra, Università di Firenze, Florence. Italy)
IIPU = Istituto Italiano di Paleontologia Umana (Rome, now Anagni, Italy)
IQW = Institut für Qartärpaläontologie, Weimar (Germany)
LGPUT = Laboratory of Geology and Paleontology, University of Thessaloniki (Greece)
MHNL = Museum of Natural History of Lyon (used as collection acronym with specimen numbers)
MNPE = Musée national de préhistoire des Eyzies (Les Eyzies, France)
NMB = Naturhistoriches Museum Basel
RSU = Rostov State University, Rostov-on-Don (Georgia)
SBAU: Soprintendenza per i Beni Archeologici dell'Umbria (Italy)
SEN = Senèze specimen from recent excavations directed by Martine Faure, Claude Guérin and Eric Delson (2001–2006); field number in the FSL collection
UCB-Lyon 1 = Paleontological Collection of Claude Bernard University - Lyon 1 (used to indicate the collection location)
ZIN = Zoological Institute, Russian Academy of Science, St. Petersburg, Russia.

Carnivore Finds During Recent Fieldwork Directed by Faure, Guérin and Delson (2001–2006)

The recent excavations at the Senèze site carried out by Faure, Guérin and Delson from 2001 to 2006 have provided a large scale localization of the paleontological remains, a precise knowledge of the corresponding stratigraphy and the first paleoenvironmental data linked to the fauna. Although very rare, the remains of carnivores found then must be treated separately from those of earlier discoveries impossible to relate to the stratigraphy.

Family Hyaenidae

Hyaenidae Indet.

Material:

- FSL SEN 01–0028 (Trench 2—excavated soil), fragment of a right demi-mandible with the two p2 roots in place and the complete p3, old breaks.

The fragment corresponds to part of the right mandibular body of a hyaenid (Fig. 9.1). The bone and the tooth appear to be very strongly mineralized. The mandibular body shows the specific characteristics of hyena mandibles. The upper and lower edges are sub-parallel. The corpus is thick and is 43.5 mm high under the p2. The circular mental foramen, unique, important, is visible on the external surface directly beneath the middle of the p2, the classic position for hyenas. The crown of the p2 has disappeared and only the two roots remain, barely showing above the level of the mandibular bone. On the other hand the p3 is complete, brownish-black in color. Its posterior root, visible in the break, is very elongated (Fig. 9.1). It is a narrow premolar with almost parallel sides. A small cuspid is evident at the front, at the base of the anterior ridge of the protoconid, leaning slightly away from its lingual side. This protoconid is high, powerful, elongated. It is tilted towards the rear and hardly worn at its tip (indicating a young adult). It occupies almost the whole of the tooth except for a short talonid. A small groove at the base of the posterior ridge of the protoconid marks its separation from the talonid and a second ridge with a small cuspid finishes the back edge of the tooth. A narrow slightly raised convex band, not really a cingulum, encircles the crown on its lingual side.

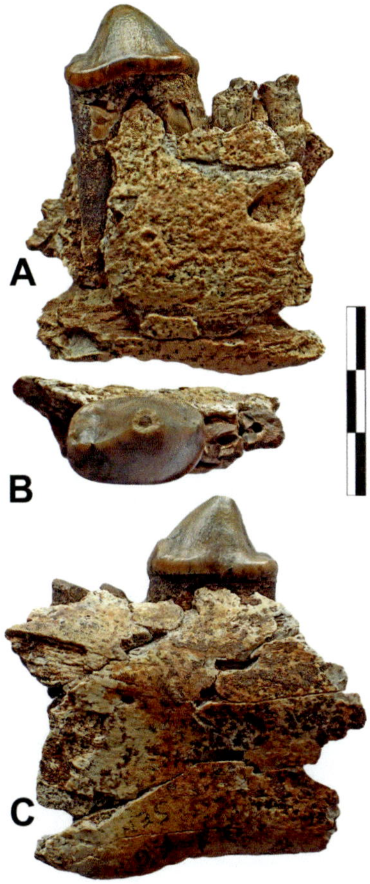

Fig. 9.1 Senèze, 2003. Hyaenidae indet., FSL SEN 01-0028, fragment of a right half mandible with p3, lateral occlusal and lingual views. Scale bar equals 3 cm

Its dimensions indicate a hyaenid of small size:

- Length (MD) = 19.8 mm
- Width (VL) = 12.2 mm
- Height of the crown (HC) = estimated about 16.1 mm (very little wear U = 0.5 on a scale of 5)
- Length (MD) of the protoconid = 13.3 mm

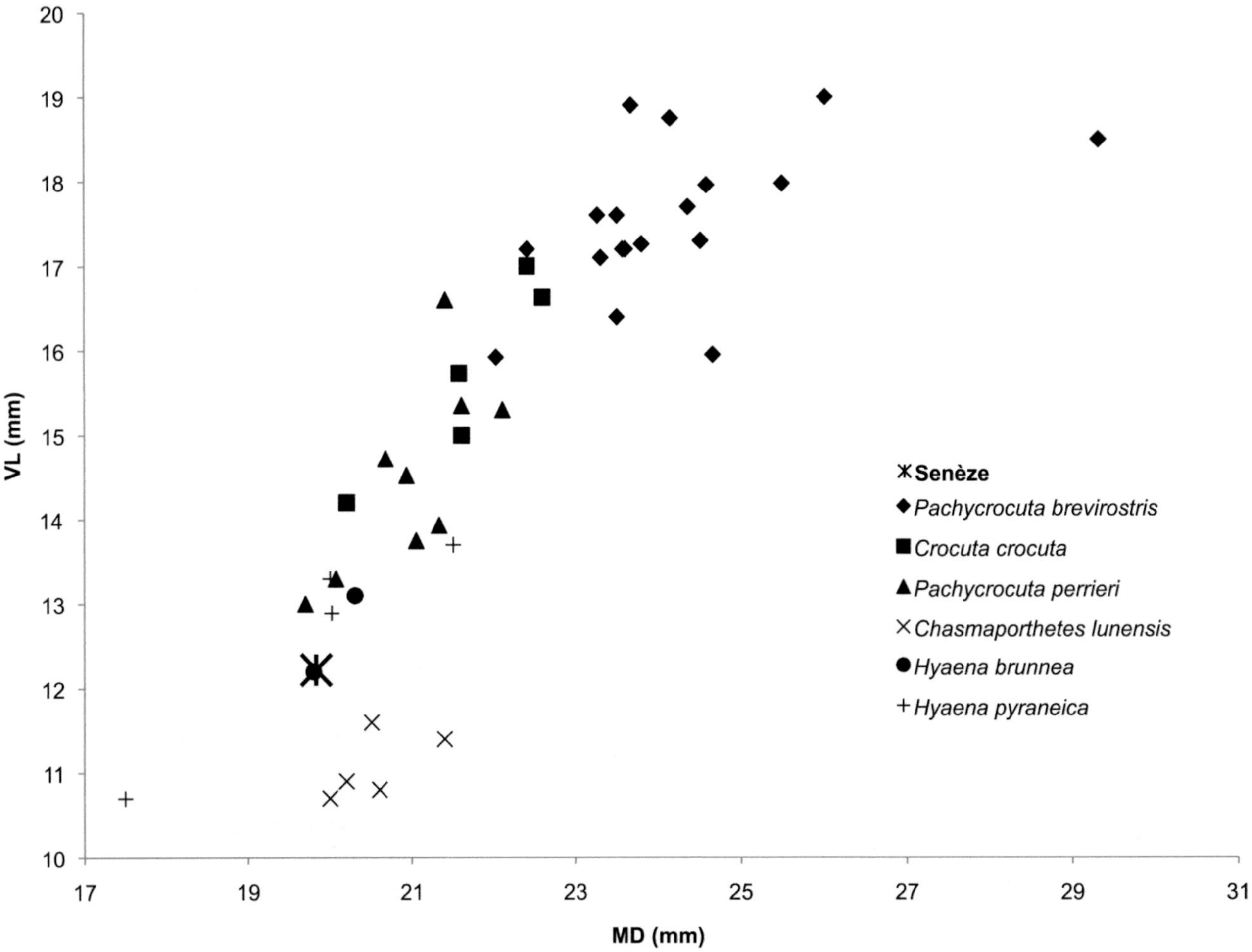

Fig. 9.2 Scatterplot of length (MD) versus width (VL) of the p3 of western European hyaenids. Data from Howell and Petter (1980), Turner and Antón (1996), Antón et al. (2006), Arribas and Garrido (2008). Senèze and Saint-Vallier, author's measurements on the original material

Figure 9.2, which places this p3 from Senèze (MD × VL) among the potential species of western European Hyaenidae, makes clear its small size. It allows us to easily rule out the "giant hyena" *Pachycrocuta brevirostris*. *Pachycrocuta perrieri* and *Crocuta spelaea* are also not possible candidates because of their clearly greater dimensions and obvious morphological differences. This Senèze p3 with a tall and elongated protoconid (trenchant function) belongs probably to the "hunting hyaena" group, which is composed of the genera *Lycyaena*, *Hyaenictis* and *Chasmaporthetes*. This premolar has the same type of structure and general proportions of the extant *Hyaena hyaena* but without the development of the anterior denticle, nor of the talonid.

I have compared the material from Senèze to Schaub's (1941) holotype of *Euryboas bielawskyi* (= *Chasmaporthetes lunensis* for Werdelin & Solounias 1991), a complete mandible from Perrier-Rocca-Neyra (FSL 211221 "Collection Clermont-Ferrand"). The p3 from Senèze is slightly shorter, proportionally a little wider, and has a larger and more elevated paraconid. The protoconid of Schaub's specimen of *Chasmaporthetes lunensis* and of that from Saint-Vallier (Viret 1954) does not cover the greater part of the premolar as in Senèze and is clearly isolated in the center of the tooth. The talonid at Perrier and Saint-Vallier appears to be much larger and higher (Viret 1954, pl. 8, 3a and 3b). A good resemblance can be observed between the Saint-Vallier and Perrier specimens, which confirms the differences with the p3 from Senèze. The latter does not belong to the species *Chasmaporthetes lunensis*.

Thus, the question of the existence of a third species of Hyenidae in Senèze arises. The presence of *Hyaena brunnea* must be considered. It is usually absent from the European Plio-Pleistocene, but is reported at Fonelas P-1 (Spain; Arribas & Garrido 2008). Their Figs. 6 and 7 (pp. 216–217) of two hemi-mandibles found in this site will be discussed later. The two p3 of Fonelas P-1 have the same dimensions as that of Senèze (Fig. 9.2). Morphologically, according to Fig. 6, which in my opinion is quite different from Fig. 7, we do not exactly find the morphology of the anterior part, nor of the talonid. The resemblance is closer to *Hyaena*

hyaena than to Senèze. The ancestor of *Hyaena brunnea*, *Parahyaena howelli* Werdelin, 2003, right mandible KNM-KP 10033 from Kanapoi (Kenya), shows a p3 which is rather similar morphologically but without a slight paraconid at its anterior part. It is moreover strictly African, and its relationship with *Hyaena brunnea* is not certain. It seems to me that *Hyaena brunnea* can be ruled out for the identification of the p3 from Senèze.

I compared the Senèze p3 to the photo of the right half-mandible of *Hyaena borissiaki* (= *Chasmaporthetes borissiaki fide* Werdelin & Solounias 1991; see Khomenko 1931, plate 5–3) in external lateral view. The profile of the p3 corresponds fairly well but an occlusal view is missing to confirm the correspondence, and it is not possible to be sure. This hyena is attributed to the Ruscinian, therefore much older and incompatible with the fauna of Senèze. It presents intermediate characters between *Hyaenictis* sensu lato and *Chasmaporthetes* and presages the characters of *Chasmaporthetes* (Werdelin & Soliounas 1991).

The paleontology of hyaenids is particularly complex. Practically every paleontologist has his own synonymies and ideas about phylogeny. Apart from the frequent *Crocuta spelaea* and the slightly more abundant *Pachycrocuta perrieri*, the remains of the other fossil hyenas are very rare and it becomes almost impossible to distinguish between individual variations and geographical differences, especially since we are chronologically outside the field of DNA and ^{14}C. The great morphological variability of the teeth of hyaenids has been repeatedly stressed by many authors (e.g., Baryshnikov 2012; Fourvel 2012). In these conditions, the determination of the p3 FSL SEN 01-0028, becomes difficult. It is the only vestige of Hyaenidae from the recent excavations in Senèze, thus the only one with a stratigraphic position. In my opinion, it is neither *Pachycrocuta perrieri* nor *Chasmaporthetes lunensis*, which are already known from the site. There is therefore another species of Hyaenidae in Senèze. A number of questions arise. Does *Hyaena brunnea* or *Parahyaena brunnea* exist in Senèze? No, it seems not. Is there another Eurasian form, not yet determined, similar to *Hyaena*? In my opinion, a single specimen cannot be used to create a new species until the morphometric range of individual variations can be approached. Is there an ancient form that would suggest the presence of two faunas of different ages at Senèze? The hypothesis has sometimes been put forward in the past, but it is impossible to verify because, except for recent excavations, no location or stratigraphic survey has been able to determine contemporary assemblages for the faunas discovered The persistence and extension of a Eurasian form such as *C. borrisiaki* is not impossible. However, the comparison in lateral view with *Chasmaporthetes borissiak*i is not supported by an occlusal view and does not bring any certainty. Finally I would retain the determination "Hyaenidae indet." for this third right lower premolar FSL SEN 01-0028.

The Hyaenid Coprolites

Material:

Because of their shape and abundant bone fragments, hyaenid coprolites are easily recognizable. On the other hand it is difficult to attribute them to species. The criterion of size is not reliable because this can be affected by the age of particular individuals. At Senèze, the coprolites collected from around the remains of the *Dicerorhinus etruscus* discovered in Trench 2 are sufficient to confirm the presence of hyenas, which had come to feed on the carcass. Argant (2024) has been able to analyse several of them and provide paleoenvironmental information. Additional hyaenid coprolites have come from other locations at this site. A number of them are conserved in the NMB collection, including some of impressive size (Fig. 9.3). Hyenas were therefore often present on the banks around the Senèze lake, whether *Pachycrocuta perrieri* or *Chasmaporthetes lunensis*.

Fig. 9.3 Senèze, hyaenid coprolites preserved in the Natural History Museum of Basel

Family Felidae

Acinonyx pardinensis **(Croizet & Jobert, 1828)**

Material:

- FSL SEN 03-0058, fragment of a right demi-mandible bearing its p3 and p4 (Fig. 9.4),
- FSL SEN 03-0210, lower right canine probably from the same individual.

This very small quantity of material was referred to in the Rapport de Fouilles 2003 (Faure et al. 2003: 71), which included a photograph of the mandible fragment. It has not been possible to find this material again in the FSL collection, however, despite several exhaustive searches. It was therefore impossible to estimate measurements of the two premolars shown in the only photograph of them that exists, because there is no scale. Their morphology indicates without doubt that they belong to the cheetah *Acinonyx pardinensis* already known at Senèze from previously excavated material.

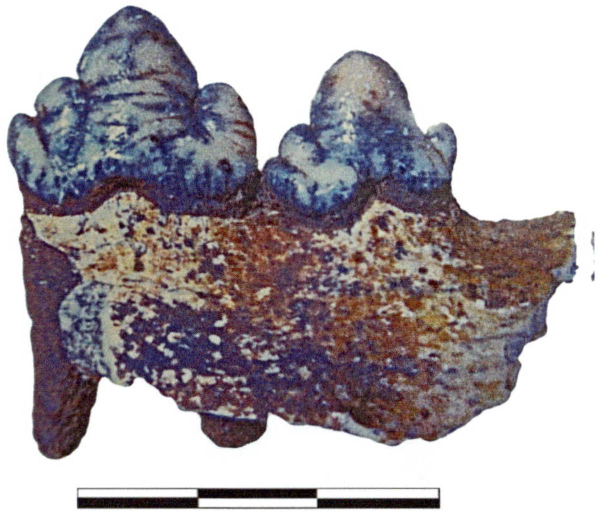

Fig. 9.4 Senèze, 2003, FSL SEN 03-0058. *Acinonyx pardinensis*, fragment of a right hemi-mandible with p3 and p4, lateral view (from Faure et al. 2003: 71), specimen now lost. Scale bar equals 3 cm

State of Knowledge of the Senèze Carnivores—Contribution of Earlier Excavations

The three new carnivore remains discovered during recent excavations at Senèze have not really brought to light any new element, other than the possible third hyaenid. On the other hand they do provide the opportunity to reassess earlier discoveries which are no longer important, in the light of current knowledge, in determining the age of the Senèze deposit.

Family Canidae

I have examined the earlier material from Senèze in the NMB and the UCB-Lyon 1 collections). I was readily able to recognize the three species previously identified as *Nyctereutes megamastoides*, *Vulpes alopecoides* and an intermediate canid the size of a small wolf. This raised again the taxonomic problem of *Canis arnensis* or *Canis senezensis*, as well as the possible attribution of one of these remains to another *Canis*.

Nyctereutes megamastoides (Pomel, 1842)

Material:

- NMB Se 313: damaged cranium largely restored and with practically complete dentition (Fig. 9.5, Table 9.1);
- NMB Se 1779: the most complete cranium, without incisors, but with L and R P4-M2, and the two demi-mandibles of the same individual bearing R p1-p4 and m1 and L m1-m2 (Figs. 9.5 and 9.6, Table 9.1).

These specimens undoubtedly belong to *Nyctereutes megamastoides*, whose cranial features include a thin, pointed, but not very elongated muzzle, a convex profile and slightly rounded frontal bones. The ramus of NMB Se 1779 appears quite massive (Fig. 9.6, Table 9.2). The sub-angular lobe exhibits the angled indentation characteristic of *Nyctereutes*, forming a very pronounced preangular notch. The squat, solid, globular form of the condyle of this mandible confirms the crushing role of the dentition.

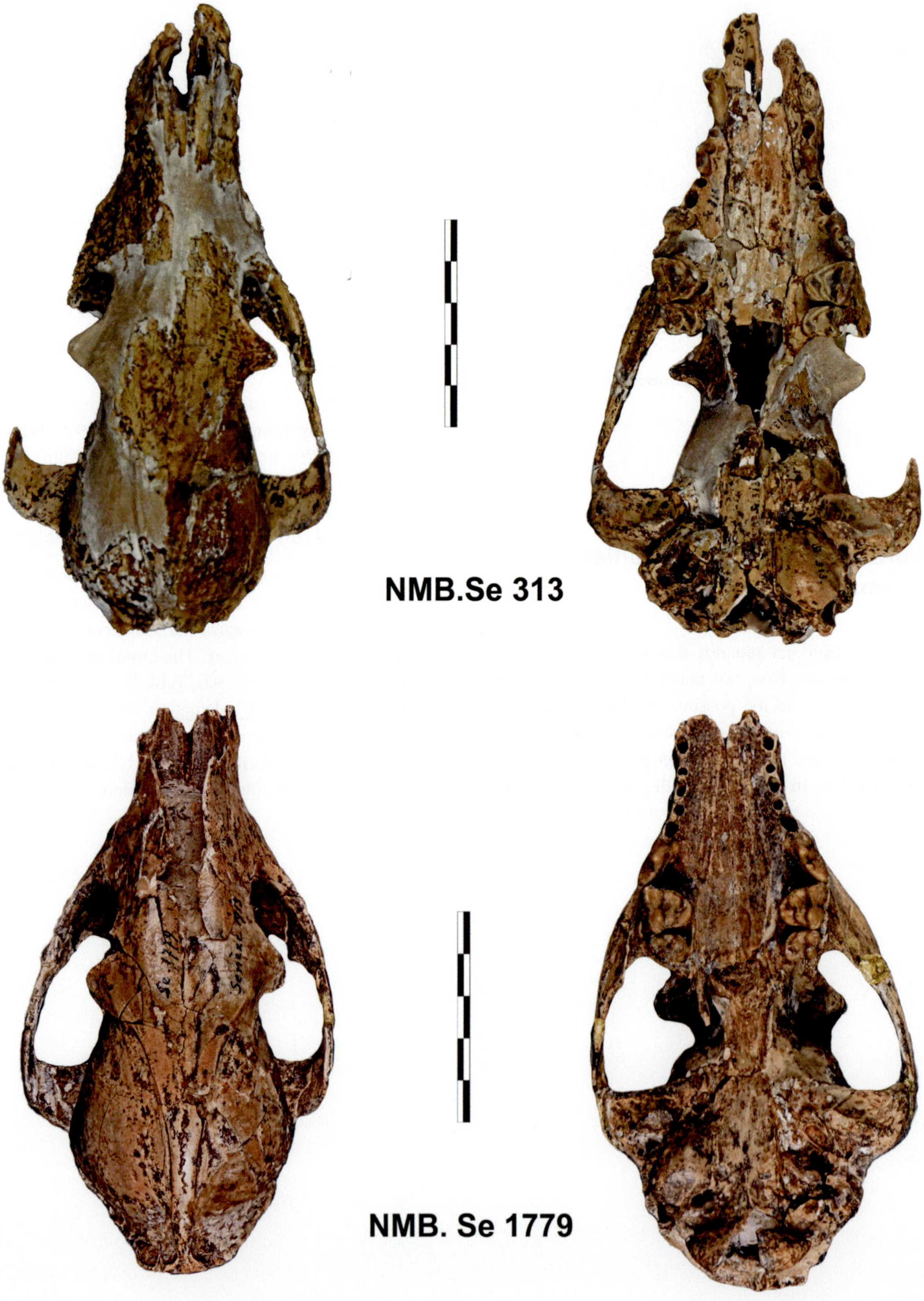

Fig. 9.5 Senèze, *Nyctereutes megamastoides*, crania NMB Se 313 and NMB Se 1779, superior view on the left, inferior view on the right. Scale bar equals 5 cm

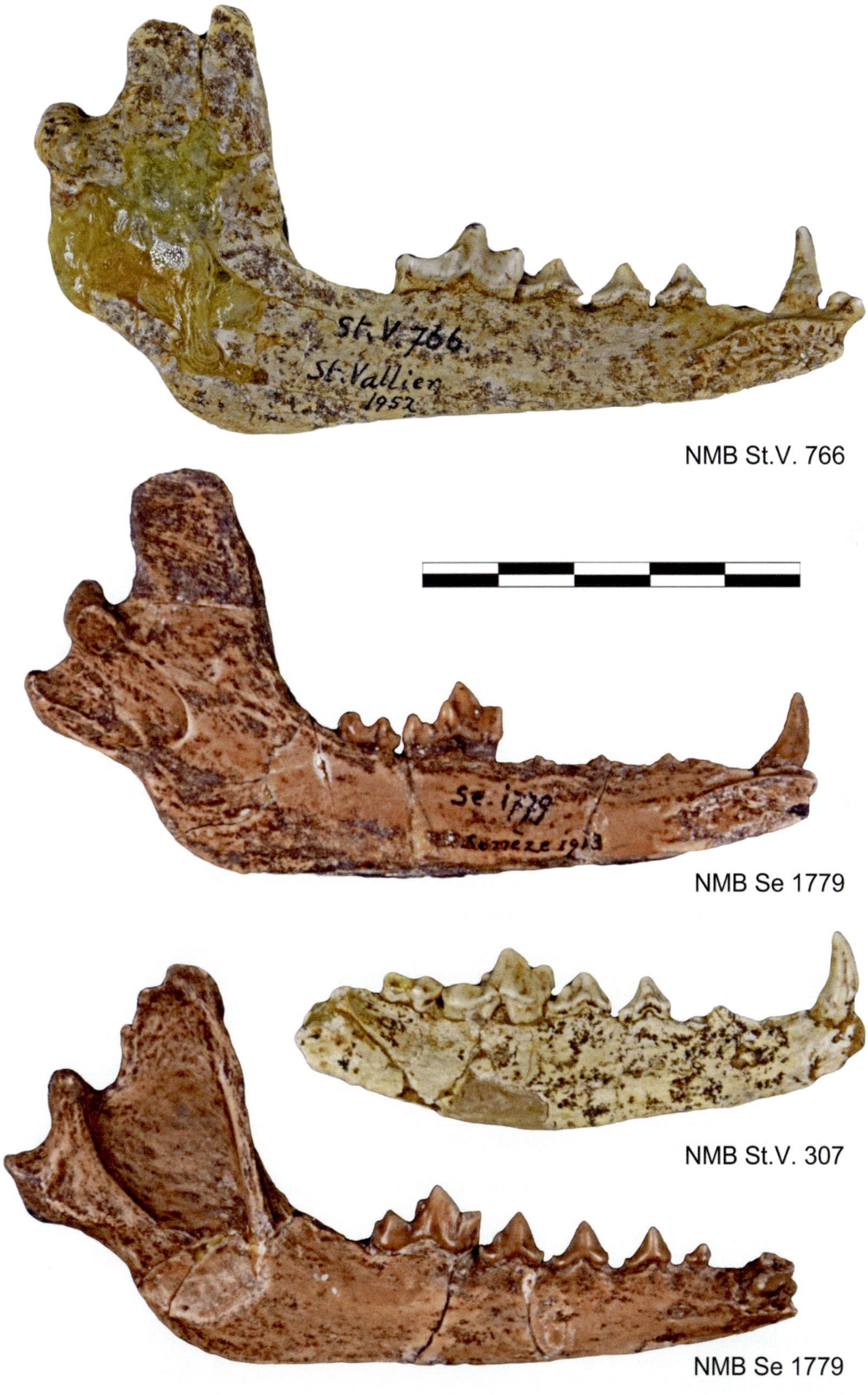

Fig. 9.6 Senèze, *Nyctereutes megamastoides*, mandible NMB Se 1779. Comparison with *Nyctereutes vulpinus* from Saint-Vallier: left NMB SV 766 and right NMB SV 307; above, lingual views; below, lateral views. Scale bar equals 5 cm

Direct comparison of the material from Senèze with the *Nyctereutes* from Saint-Vallier conserved in the NMB reveals some notable differences in their dentition.

Material from Saint-Vallier in the NMB:

- NMB StV 308, fragment of left maxilla with L M1-2 (U = 0);
- NMB StV 252, fragment of cranium with R M1 and M2 and initial part of the zygomatic arch;
- NMB StV 307, left demi-mandible with c-m2;
- NMB StV 766, left demi-mandible with i2-c, p2-m1, without apparent alveoli for the missing m2 and m3;
- NMB StV 256, isolated L p4;
- NMB StV 255, isolated m1;
- NMB StV 254, isolated m1.

The dental morphology is clearly that of *Nyctereutes* for the material from Senèze just as it is for the material from Saint-Vallier, but a number of differences are apparent:

- the Senèze M1 (NMB Se 313, Se 1779) for example is longer (MD) and less elongated transversely (Fig. 9.7, Table 9.3), with parallel sides, without an internal sharp-edged talon, and not forming an arc of a circle as at Saint-Vallier. The paracone and the metacone lack the development that we find at Saint-Vallier.
- the m1 from Senèze is not as long as the one from Saint-Vallier (Table 9.4). The two very developed tubercles of the talonid (entoconid and hypoconid) add to its width. The dentition of the Senèze *Nyctereutes* appears to be more fully developed for crushing than that of the more carnivorous *Nyctereutes vulpinus* from Saint-Vallier.
- the anterior part of the m2 from Senèze is relatively more developed than in the m2 of the Saint-Vallier *Nyctereutes*.

Soria and Aguirre (1976) created the sub-species *Nyctereutes megamastoides vulpinus* for the *Nyctereutes* from Saint-Vallier, a deposit where it is particularly well represented (MHNL, NMB and UCB-Lyon 1 collections) (A. Argant 2004; Soria & Aguirre 1976; Viret 1954). This subspecies has subsequently been considered a separate species, *Nyctereutes vulpinus* (Monguillon et al. 2004).

The *Nyctereutes* from Senèze differs from the *Nyctereutes vulpinus* from Saint-Vallier. We attribute it to *Nyctereutes megamastoides* in agreement with the study by Soria and Aguirre (1976). The morphology of the ramus of the mandibles of NMB Se 1779 and NMB StV 766 is not identical. In spite of the poor general condition of the latter hemi-mandible StV 766 (Fig. 9.6) and a probable poor restoration of the lower part of the ramus, one can observe in particular its greater development and its position nearly perpendicular to the corpus, as on NMB QSV 99 (Fig. 2a, Plate 1, Viret, 1954; Fig. 3, p. 143, A. Argant 2004); on the contrary, there is a clear rounding at Senèze. The angular process has disappeared and cannot be observed. The position of the condyle appears to be consistent and the top of the coronoid process has disappeared (Fig. 9.6). An examination of Tables 9.1, 9.2, 9.3 and 9.4 reveals the slightly smaller size of the *N. megamastoides* from Senèze in comparison with the Saint-Vallier *Nyctereutes* in almost every respect, but it nevertheless remains within the range of dimensions from Saint-Vallier. However there is one notable exception, the size of the m2 relative to the size of the m1. The m1 from Senèze (NMB Se 1779, L) is clearly smaller (MD lower than the lowest value from Saint-Vallier) but the m2 from Senèze (NMB Se 1779) corresponds to the highest values from Saint-Vallier.

Table 9.1 Comparative dimensions of the Senèze *Nyctereutes megamastoides* crania

Cranial variables (code 03)		*Nyctereutes megamastoides*					*Nyctereutes vulpinus*				
		NMB		Kurtèn and Crusafont-Pairó (1977)			Saint-Vallier MHNL (Coll. Viret)				FSL
		Senèze Se 1779	Senèze Se 313	Puebla Valverde	Layna	Villarroya	n	Mean	ΔI	σ n − 1	SV 97483
1	Total length: acrocranion- prosthion	–	144.1*	–	–	–	2	155.0	154.6–154.4	0.5	–
2	Condylobasal length	137.0*	141.0*	142.0*	–	140.0*	2	147.5	147.4–147.6	0.1	–
3	Length, prosthion-basion	–	135.2*	–	–	–	2	138.2	138.1–138.3	0.1	–
4	Basicranial length, synsphenion-basion	41.5	–	–	–	–	–	–	–	–	–
5	Basifacial length, prosthion-synsphenion	–	–	–	–	–	–	–	–	–	–
6	Neurocranial length, basion-nasion	75.1	–	–	–	–	–	–	–	–	–
7	Neurocranial length, point F-acrocranion	73.6	67.6*	–	–	–	3	75.9	72.8–79.1	3.1	71.7
8	Viscerocranial length, prosthion nasion	–	–	–	–	–	2	66.2	65.7–66.6	0.6	–
9	Facial length, prosthion-point F	–	82.9	–	–	–	4	79.5	71.4–84.4	6.0	–
10	Max length of nasal: nasion-rhinion	–	–	–	–	–	4	52.9	50.0–56.4	2.7	–

(continued)

Table 9.1 (continued)

	Cranial variables (code 03)	*Nyctereutes megamastoides*					*Nyctereutes vulpinus*				
		NMB		Kurtèn and Crusafont-Pairó (1977)			Saint-Vallier MHNL (Coll. Viret)				FSL
		Senèze Se 1779	Senèze Se 313	Puebla Valverde	Layna	Villarroya	n	Mean	Δl	σ n − 1	SV 97483
11	Length of neurocranium	–	–	–	–	–	–	–	–	–	–
12	Length of muzzle	–	–	–	–	–	4	60.8	59.8–61.2	0.7	–
13	Median palatinal length, prosthion-staphylion	–	–	–	–	–	4	78.6	74.4–84.2	4.1	–
14	Palatal length	–	–	–	67.0*	–	3	75.0	71.4–78.1	3.4	–
15	Length staphylion-palatinooral	–	–	–	–	–	4	25.3	23.6–27.3	1.8	–
16	Horizontal length of the palate	–	–	–	–	–	4	24.8	23.0–25.4	1.2	–
17	Breadth of tympanic bulla	21.8	19.2	–	–	–	6	15.7	13.8–20.4	2.5	–
18	Breadth, otion–otion	39.3*	47.5	–	–	–	3	52.2	49.0–56.3	3.7	–
19	Breadth at external auditory meatus	–	37.5	–	–	–	5	48.7	44.2–56.0	4.4	–
20	Max breadth at condylus occipitalis	27.8	–	27.2	–	26.7	5	28.8	27.3–31.1	1.5	31.0
21	Max breadth at paraoccipital process	49.7	–	–	–	–	3	43.0	41.0–47.1	3.5	–
22	Max breadth of occipital foramen	12.1	–	–	–	–	4	15.2	14.8–15.3	0.2	–
23	Height of occipital foramen, basion-opisthion	14.4	–	–	–	–	4	12.4	10.7–14.6	1.6	–
24	Max breadth of neurocranium: euryon–euryon	52.4	49.1	–	–	–	7	52.5	50.5–54.3	1.3	54.0
25	Bizygomatical breadth, zygion - zygion	73.5*	77.0	–	–	69.0*	4	81.0	80.0–82.2	0.9	80.0*
	Cranial variables (code 04)										
1	Min breadth of neurocranium	35.9	32.0	–	–	–	–	–	–	–	–
2	Frontal breadth, ectorbital-ectorbital	48.8*	47.2	–	–	–	–	–	–	–	–
3	Min breadth at orbits, entorbital-entorbital	30.3	29.7	–	–	–	–	–	–	–	–
4	Max breadth of palate	41.1	45.4	–	–	–	–	–	–	–	–
5	Min breadth of palate	27.4	27.5	–	–	–	–	–	–	–	–
6	External breadth between the roots of canines	–	26.8*	–	–	–	–	–	–	–	–
7	Max height of orbit	27.4	24.0	–	–	–	–	–	–	–	–
8	Height of cranium	43.8	44.0*	–	–	–	–	–	–	–	–
9	Height of cranium without sagittal crest	38.5	–	–	–	–	–	–	–	–	–
10	Height of occipital triangle, acrocranion -basion	38.2	40.4	–	–	–	–	–	–	–	–
11	Capacity of neurocranium	–	–	–	–	–	–	–	–	–	–
	Maxillary variables (Code 06)										
1	Length of cheek teeth (or alveoli)	49.1	49.6	–	–	–	–	–	–	–	–
2	Length from C (ant. edge) to M2/ (post. edge)	–	61.4	–	–	–	–	–	–	–	–
3	Length of molars	16.2	17.1	–	–	–	–	–	–	–	–
4	Length of premolars	35.8	35.7	–	–	–	–	–	–	–	–
5	Length of P4	12.3	12.4	–	–	–	–	–	–	–	–
6	Max width of P4	5.9	6.1	–	–	–	–	–	–	–	–
7	Length of P4 alveolus	7.6	–	–	–	–	–	–	–	–	–
8	Length of M1	10.2	9.6	–	–	–	–	–	–	–	–
9	Width of M1	11.1	11.3	–	–	–	–	–	–	–	–
10	Length of M2	6.7	7.0	–	–	–	–	–	–	–	–
11	Width of M2	7.6	8.7	–	–	–	–	–	–	–	–
12	Height of the crown of the canine (external face)	–	–	–	–	–	–	–	–	–	–

Notes Point F: frontal mid-point between ectorbitals. Variables after OSTEO (Desse et al. 1986). * indicates approximate value (in all tables)

Table 9.2 Comparative dimensions of the Senèze *Nyctereutes megamastoides* hemi mandible. Measures after OSTEO (Desse et al. 1986)

Mandibular variables (code 10)		*Nyctereutes megamastoides*					*Nyctereutes vulpinus*						
		NMB		Kurtèn & Crusafont-Pairó (1977)			Saint-Vallier MHNL (Coll. Viret)				Saint-Vallier UCB - Lyon 1 (Coll. Faure-Guérin)		
		Senèze Se 1779 L	Senèze Se 1779 R	Puebla Valverde	Layna	Villarroya	n	Mean	Δl	σ n − 1	SV 98799	SV 99686	SV 991168
1	Length infradental-mid condylus post	102.8	–	–	–	–	3	109.2	103.1–116.8	7.0	107.8	–	–
2	Length infradental-angular apophysis	106.3	–	–	–	–	2	112.1	106.2–117.9	8.3	–	–	–
3	Length infradental-angular incisure	101.3	–	–	–	–	3	107.5	101.4–115.1	7.0	106.0	–	–
4	Length canine post-mid condylus post	92.5	94.0	–	–	–	2	95.3	93.8–96.8	2.1	96.9	–	–
5	Length canine post-angular incisure	93.7	94.0	–	–	–	2	93.3	91.5–95.1	2.5	95.2	–	–
6	Length canine post-angular apophysis	98.8	–	–	–	–	1	96.5	–	–	–	–	–
7	Length canine post-m3 post	60.8*	62.5	–	–	–	6	61.5	55.7–65.3	3.3	63.5	–	–
8	Length p1 ant-m3 post	55.6	58.0	–	–	–	7	58.2	50.9–61.9	4.0	60.0	62.0	–
9	Length p2 ant-m3 post	52.0	52.6	–	–	–	3	52.5	45.9–56.0	5.7	55.8	56.0	–
10	Length of molars	26.9	26.1*	22.0	–	–	3	28.0	27.0–28.6	0.9	28.6	28.4	–
11	Length of premolars p1 ant-p4 post	29.4*	31.5	–	–	–	4	32.2	29.3–34.2	2.1	32.6	32.8	–
12	Length p2 ant-p4 post	25.7*	26.1	23.3	–	–	5	26.8	24.8–28.2	1.6	27.8	27.8	–
13	Length of m1	14.1	14.2	16.30	15.40	15.98	11	15.5	14.5–16.0	0.4	16.0	15.9	–
14	Breadth of m1	5.9	5.7	–	–	–	11	6.5	5.9–6.8	0.2	6.4	6.5	–
15	Length of m1 alveolus	–	–	–	–	–	–	–	–	–	–	–	–
16	Length of m2	8.9	–	8.4	8.16	–	7	8.0	7.1–8.9	0.6	8.6	8.4	8.6
17	Breadth of m2	5.8	–	–	–	–	6	5.8	5.7–6.3	0.3	5.6	6.0	5.7
18	Length of m3	–	–	–	–	–	5	3.9	3.4–4.4	0.4	3.5*	–	3.0*
19	Breadth of m3	–	–	–	–	–	6	3.0	2.2–3.7	0.6	2.4*	–	2.3*
20	Transversal diameter of mandible behind m1	7.8	7.7	–	–	–	11	7.5	6.3–8.2	0.6	7.2	–	7.5
21	Height angular apophysis-coronion	39.5	40.1	–	–	–	2	54.4	52.9–55.8	2.1	–	–	37.0
22	Horizontal ramus height behind m1	16.1	16.0	–	–	–	11	17.1	16.3–18.1	0.8	16.9	18.1	17.4
23	Horizontal ramus height between p2 – p3	13.4	14.2	–	–	–	5	13.0	11.9–14.3	1.1	12.5	11.9	–
24	Height of the crown of the canine (external face)	11.4	–	–	–	–	5	10.9	10.0–11.8	0.9	–	–	–

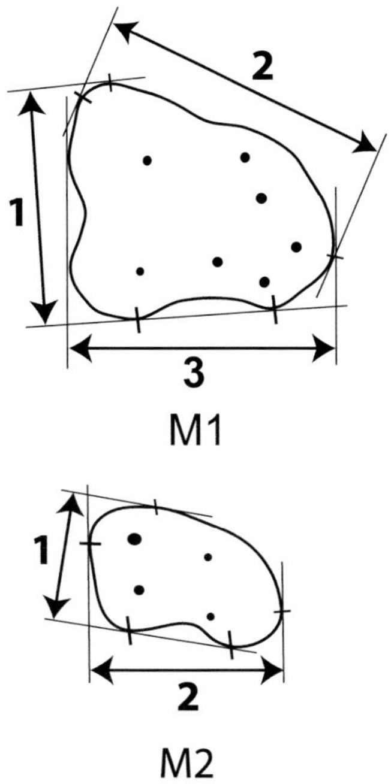

Fig. 9.7 Measurements of the upper molars of canids: M1 (PEMAC 17); M2 (PEMAC 18)

Table 9.3 Comparative dimensions of the Senèze *Nyctereutes megamastoides* upper dentition. See Fig. 9.7

Upper teeth		*Nyctereutes megamastoides*				*Nyctereutes vulpinus*									
		Senèze NMB				Saint-Vallier UCB-Lyon 1 (Coll. Faure-Guérin)						Saint-Vallier MHNL (Coll. Viret)			
		Se 1779		Se 313		SV 308	SV 252	SV 99141	SV 991144	SV 96185	SV 97075	n	Mean	Δl	σ n − 1
		L.	R.	L.	R.										
P4	MD	12.3	12.7	12.4	12.4	–	–	13.2	–	–	13.9	15	13.4	12.4–15.3	0.9
	VL max	5.9	5.9	6.1	6.3	–	–	6.5	–	–	7.0	14	6.7	6.2–7.7	0.5
	VL talon	–	–	–	–	–	–	5.4	–	–	–	11	5.2	4.8–5.5	0.2
	HC	–	–	–	–	–	–	8.2	–	–	9.5	–	–	–	–
M1 (Fig. 9.7)	MD (1)	10.2	10.2	9.6	9.9	11.1	10.9	10.3	10.5	11.0	–	16	10.5	10.0–11.4	0.2
	VL ant max (2)	–	–	–	–	–	–	–	13.2	13.2	13.8	7	13.6	12.8–14.2	0.6
	VL post (3)	11.1	11.6	11.3	11.1	11.7	11.6	11.5	11.4	11.6	–	16	11.8	11.0–12.6	0.2
	HC	5.2	4.6	–	–	5.4	4.7	4.8	4.9	5.1	–	–	–	–	–
M2 (Fig. 9.7)	MD (1)	6.7	6.8	7.0	7.0	–	–	–	6.8	6.8	–	11	6.6	5.7–7.3	0.5
	VL (2)	7.6	8.5	8.7	9.0	–	–	–	8.0	8.0	–	12	8.0	7.1–8.9	0.5

Table 9.4 Comparative dimensions of the Senèze *Nyctereutes megamastoides* lower dentition

Lower teeth		*Nyctereutes megamastoides*		*Nyctereutes vulpinus*						
		NMB		Saint-Vallier UCB - Lyon 1 (Coll. Faure-Guérin)			Saint-Vallier MHNL Coll. Viret			
		Se 1779		SV 98799	SV 99686	SV 991168	n	Mean	Δl	σ n − 1
		L	R							
p1	MD	–	3.6	4.2	3.9	–	7	3.8	3.3–4.0	0.2
	VL	–	2.5	2.5	2.4	–	7	2.5	2.1–3.0	0.3
	HC	–	2.9	3.7*	2.9	–	–	–	–	–
p2	MD	–	6.5	7.4	7.5	–	7	6.7	6.2–7.3	0/4
	VL	–	3.0	3.0	3.3*	–	7	3.0	2.7–3.3	0.2
	HC	–	4.7	4.4	4.3	–	–	–	–	–
p3	MD	–	7.6	9.2	9.1	–	6	8.5	8.1–9.0	0.4
	VL	–	3.1	3.4	3.4	–	7	3.3	3.0–3.7	0.2
	HC	–	5.2	5.0	4.8	–	–	–	–	–
p4	MD	–	8.8	10.1	10.0	–	9	9.7	8.7–10.3	0.5
	VL	–	3.7	4.1	4.3	–	9	4.2	3.7–4.5	0.3
	HC	–	5.8	5.9	5.0*	–	–	–	–	–
m1	MD	14.1	14.2	16.0	15.9	–	12	15.6	14.4–16.8	0.6
	VL	5.9	5.7	6.4	6.5	–	12	6.5	5.8–6.7	0.2
	MD trigonid	9.4	9.8	10.9	10.8	–	–	–	–	–
	MD talonid	–	–	6.0	5.9	–	–	–	–	–
	HC	7.8	7.4	7.9	7.5*	–	2	7.7	7.4–8.0	0.4
m2	MD	8.9	–	8.6	8.4	8.6	7	8.0	7.1–8.9	0.6
	VL	5.8	–	5.6	6.0	5.7	6	5.8	5.6–6.3	0.2
	HC	4.7	–	–	–	–	–	–	–	–
m3	MD	–	–	3.5*	–	3.0*	5	3.9	3.4–4.4	0.4
	VL	–	–	2.4*	–	2.3*	5	3.2	2.6–3.7	0.4
	HC	–	–	–	–	–	–	–	–	–

The genus *Nyctereutes* is known by several species across Eurasia since the Early Pliocene. *Nyctereutes donnezani* from the later Ruscinian was found at Layna (Spain, MN 15, ca. 3.9 Ma, Bartolini Lucenti et al. 2018) and at Perpignan (France) (Depéret 1890). *N. sinensis* (Schlosser, 1903) is known from Nihowan (China) and *N. tingi* (Tedford & Qiu, 1991) from the Yushe Basin, to which must be added the extant species *N. procyonoides*. Of the fossil forms, only *N. sinensis* survived in Asia, until the Middle Pleistocene. In Europe, *Nyctereutes donnezani* was the ancestor of *Nyctereutes megamastoides* at the end of the Pliocene and at the beginning of the Early Pleistocene. Monguillon et al. (2004) raised the question of the origin of the *N. vulpinus* from Saint-Vallier; was it through the evolution of *N. megamastoides* from *N. donnezani,* or the evolution of populations of *N. tingi* coming from Asia? The presence of *N.* cf. *tingi* at Varshets in Bulgaria (Spassov 2000, 2003) makes the latter hypothesis more likely (Monguillon et al. 2004).

N. megamastoides is identified at Senèze, just as it is at Perrier-Etouaires and Chilhac (France), and at Puebla de Valverde, Layna and Villarroya in Spain. This provides us with an argument to distinguish the Senèze deposit from that of Saint-Vallier. Has this difference come about because of evolution at each site over a sufficiently long period of time, or because over a shorter period of time there has been an influx of populations from different origins (southern Europe in one case, Asia in the other), or because of the adaptation (whether synchronous or not) of two biologically similar species to different biotopes? We note for example the significant variations in dental morphology apparent in specimens of the extant *Nyctereutes procyonoides* held in the NMB, evidence of real individual variability and very probably of adaptations to different biotopes requiring different modes of feeding. It is difficult to draw conclusions with any certainty, but in my opinion the last hypothesis cannot be disregarded.

Are we able to use our knowledge of the evolution of *Nyctereutes megamastoides* to help date the Senèze deposit? In my opinion the only factor we could make use of is the difference between the two species at Senèze and Saint-Vallier. *N. vulpinus* of Saint-Vallier has a greater carnivorous tendency than the *N. m. megmastoides* from Senèze. These differences make it unlikely that the two deposits are strictly contemporaneous. Assuming the living species allows a sense of evolutionary trend, the Senèze *Nyctereutes* is more recent than Saint-Vallier.

***Vulpes alopecoides* (Forsyth Major, 1877)**

Material:

- NMB Se 315: 1 femur L, 1 Mt 2 L, 1 Mt 5 L;
- NMB Se 316: 1 femur R, 1 tibia L and 1 tibia R;
- FSL 210,980: 1 tibia L, without distal part.

This group of long bones clearly belongs to a very small canid (Fig. 9.8). The length and gracility of these bones are incompatible with *Nyctereutes megamastoides* and even more so with *Canis arnensis.* The extant *Vulpes vulpes* is clearly much larger (Table 9.5). The tibias (NMB Se 316, FSL 210,980) have the morphology of *Vulpes* but are much smaller and are proportionally more elongated. The morphology, slenderness and proportions closely resemble those of the arctic fox *Vulpes lagopus (= Alopex lagopus)* which is a little smaller, however (Table 9.5). Without hesitation I attribute this material from Senèze to *Vulpes alopecoides*.

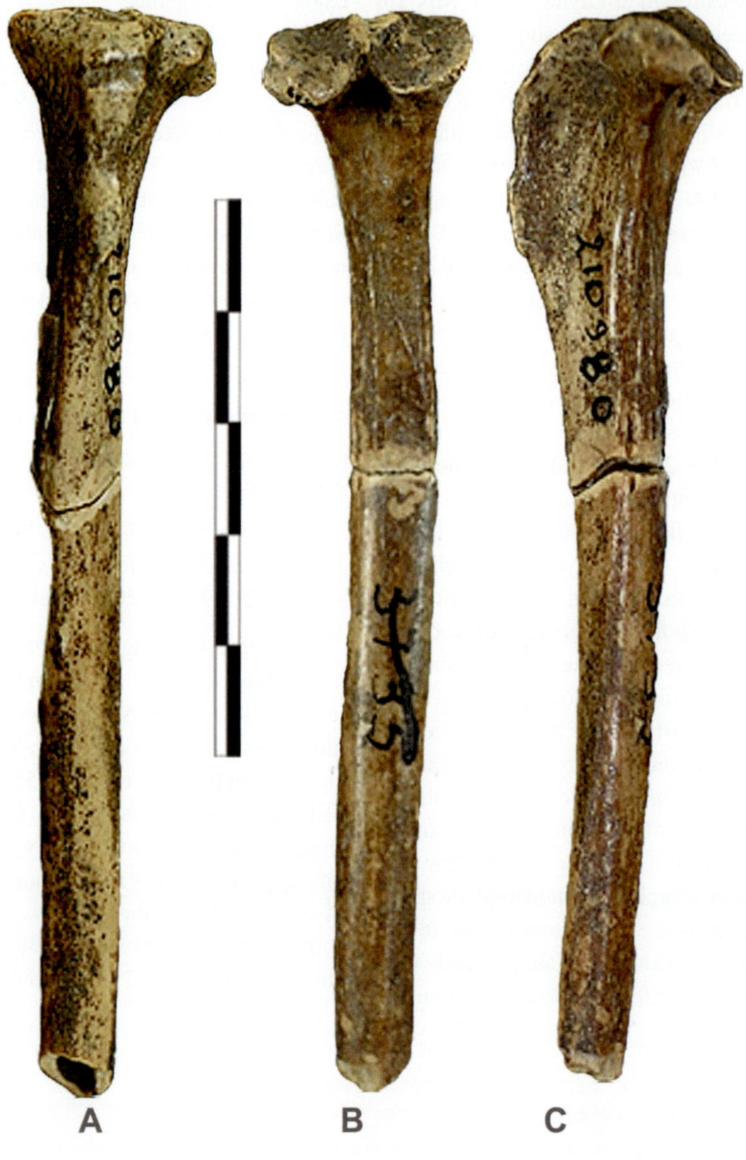

Fig. 9.8 Senèze, *Vulpes alopecoides*, left tibia (FSL 210980). Anterior (**A**), posterior (**B**) and lateral (**C**) views. Scale bar equals 5 cm

Table 9.5 Comparative dimensions of the Senèze *Vulpes alopecoides* post cranial skeleton

		Fossil					Extant	
		Vulpes alopecoides			*Nyctereutes megamastoides*	*Vulpes lagopus*	*Vulpes vulpes*	*Alopex lagopus*
		Senèze		Senèze	Saint-Vallier	Romain-la-Roche		
		NMB Se 315	NMB Se 316	FSL 210980	FSL SV 99107	CAN. 787		
		L	R	L	L			
Femur	L	110.8	119.6		–	–	140.6	–
	Prox TD	–	22.7		–	–	27.71	–
	Prox APD	10.9*	10.6*		–	–	13.0	–
	Med TD	7.8	8.3		–	6.9	9.8	–
	Med APD	7.0	7.2		–	–	8.7	–
	Dist TD	18.5	18.3		–	16.8	23.2	–
	Dist APD	21.8	21.5		–	–	24.8	–
	Art TD	17.8	16.8		–	–	21.7	–
	Caput art TD	10.5	10.6		–	–	13.0	–
	Trochlea L	–	–		–	–	20.7	–
	Neck L	–	11.9		–	–	12.3	–
Tibia	L	–	130.6	–	149.1	–	153.5	118.4
	Prox TD	–	19.2	19.5	–	–	25.7	19.6
	Prox APD	–	–	20.8	27.0*	–	27.7	20.8
	Med TD	–	7.5	7.7	9.3	–	9.1	6.9
	Med APD	–	8.0	7.0	10.2	–	10.9	7.0
	Dist TD	13.9	–	–	17.8*	–	17.6	13.6
	Dist APD	9.4	8.8*	–	13.2		12.0	9.5
	Art TD	10.5	10.3	–	–	–	13.2	10.1
Mt 2	L	–	–		–	–	65.3	46.1
	Prox TD	–	–		–	–	7.9	5.0
	Prox APD	–	–		–	–	10.3	7.8
	Med TD	4.0	–		–	–	5.1	3.5
	Med APD	3.9	–		–	–	4.3	3.2
	Dist TD	5.8	–		–	–	7.5	5.6
	Dist APD	5.4					6.5	5.0
	Art TD	5.5	–		–	–	6.9	5.6
Mt 5	L	54.1	–		–	–	66.1	47.2
	Prox TD	6.2	–		–	–	8.6	5.5
	Prox APD	7.0	–		–	–	7.1	6.9
	Med TD	3.5	–		–	–	4.8	2.9
	Med APD	3.9	–		–	–	4.2	3.5
	Dist TD	5.2	–		–	–	6.7	5.3
	Dist APD	4.9	–		–	–	6.4	5.1
	Art TD	5.1					6.8	5.3

Lacking dental remains from Senèze, dental comparison with other deposits remains impossible. On the other hand post-cranial remains have been found at other deposits though these have been rare. In the case of *Vulpes alopecoides* it has therefore not been possible to make use of biochronology. However, based on what we know of its "first known appearance and its last known presence", its chronological range was MNQ 16 to 19. *Vulpes alopecoides* is the first representative of the genus *Vulpes*. It first appears at Villarroya (Spain) (Villalta 1952, p. 23, pl. 3, Figs. 1, 2), zone MNQ 16, about 2.7 Ma, in a fluvio-paludal context (Garrido 2008). Its last known presence was at Ceyssaguet (France; Tsoukala 2004) about 1.2 Ma (MNQ 19), at the end of the Villafranchian. Its phyletic position also remains imprecise because of the scarcity of material but the evolutionary line *Vulpes alopecoides* …*Vulpes praeglacialis* … *Vulpes lagopus* seems likely (Argant 2010).

Canis arnensis Del Campana, 1913

Material:

- FSL 210987: right maxilla, type of *Canis senezensis* Martin, 1973, illustrated in Martin (1973, Fig. 5 p. 93), with P1-M2, the teeth slightly worn (the M2 visible in the figure is currently missing on the specimen, only its roots are left);
- NMB Se 318: 1 incisor of a canid clearly much larger than *Vulpes vulpes*;
- NMB Se 1780: 1 fragment of a left maxilla with P1-M2, teeth slightly worn.

These fossils all belong to canids. The dimensions and the dental morphology of the two maxillae are very similar. There is no doubt that they come from the same species and one could even think that they come from the left and right sides of the same individual but this is not the case.

They cannot belong to *Vulpes alopecoides* which is clearly smaller, and they are also much larger than those of the extant *Vulpes vulpes*. On the other hand, direct comparison (Table 9.6) shows that they are smaller than those of *Canis etruscus*. It is clear that these fossils come from a canid of intermediate size similar to a small wolf. Martin (1973) made the maxilla FSL 210987 (formerly FSL 3736), the holotype of a new species, *Canis senezensis* (Fig. 9.9), in which he saw the ancestral form of *Canis arnensis*. Numerous authors now accept the synonymy of the two species (Garrido 2006; Garrido & Arribas 2008; Rook & Torre 1996). Boudadi-Maligne (2010) and Brugal and Boudadi-Maligne (2011) have recently carried out a statistical analysis of the metrics of *Canis* fossils from the Early Pleistocene using data from both dental and post-cranial remains. None of the variables examined there showed any significant difference among *Canis arnensis*, *Canis senezensis* and *Canis occitanus*. Differentiation remains more difficult for *Canis appoloniensis* whose measurements fall between this first group and those of *Canis etruscus*. In my view, *Canis arnensis* Del Campana is synonymous with *C. senezensis* and has priority. I therefore attribute this medium-sized canid from Senèze to *Canis arnensis*. On the other hand I believe it would be quite problematic to try to link *Canis arnensis* to an extant species of canid, whether a jackal or even a coyote. It is true that *Canis arnensis*, at least based on its dentition, resembles *Canis priscolatrans* as Kurtén (1974, Fig. 1, p.6) thought. However we must not forget that these are ancestral forms which have undergone almost 2 Myr of evolution and which may even have produced forms as a result of hybridisation. For *Canis arnensis* we are still not able to specify the range of individual or sexual variation, nor the effects of adaptations to different environmental conditions, and of course we are even less able to use genetic techniques for such ancient periods.

Comparison of the dental material (NMB Se 1780 and FSL 210987) of *Canis arnensis* from Senèze (Table 9.6) with that from other sites and with *Canis etruscus* shows that it is appropriate for morphology to play a determining role because comparisons based solely on metrics can be misleading. If it is true that on the whole the dental dimensions for all the *Canis arnensis* remains are similar, they are sometimes very close to those of *Canis etruscus* so that in certain cases it is difficult to know which of these two species is the larger.

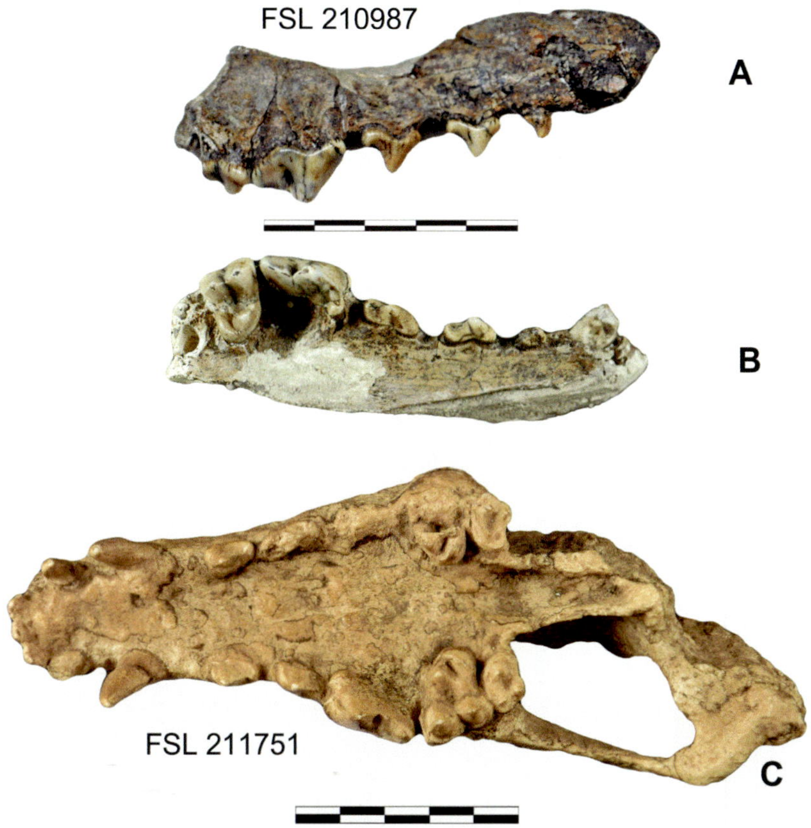

Fig. 9.9 Senèze, *Canis arnensis*, right maxilla (FSL 210987, formerly FSL 3736), type of *Canis senezensis* Martin, 1973. Lateral view (**A**) and occlusal view (**B**). Comparison with *Canis etruscus* (**C**) (cast FSL 211751; Montevecchio, Val d'Arno, Italy). Scale bar equals 5 cm

Table 9.6 Comparative dimensions of the Senèze *Canis arnensis* upper teeth. See Fig. 9.7

Upper teeth		*Canis arnensis*								*Canis etruscus*							
		Senèze	Senèze	Val d'Arno sup	Khapry	Fonelas P.1	Kudaro 1	Apollonia (LGPUT)		Ceyssaguet	Pantalla				Monte-vecchio	Gerakarou (LGPUT)	Fonelas P.1
		FSL	NMB	Rook (1993)	Sotnikova et al. (2002)	Garrido (2008)	Baryshnikov (2012)	Koufos (1992)		Tsoukala (2004)	Cherin et al. (2013)					Koufos (1992)	Garrido (2008)
		210987	Se 1780		RSU-778	FP-1	ZIN 31889	APL.16	APL.1	MNPE CEY (range)	SBAU 337628		SBAU 337646		NMB	GER-168–169	FP-1
											dext	sin	dext	sin			
C	MD	10.9	–	–	–	–	8.5–9.6	9.8	–	9.3–11.8	–	10.8	11.8	10.0	8.5–9.5	10.7–10.8	9.8
	VL	6.4	–	–	–	–	5.4–5.9	6.2	–	6.0–7.6	–	7.6	7.2	7.0	6.3–6.4	7.1–7.2	6.3
	HC	–	–	–	–	–	–	–	–	17.4*–25.7	–	–	–	–	18.4–19.6	–	–
P1	MD	5.6	5.8	–	–	–	–	7.4	–	5.9–7.0	6.4	6.5	6.6	7.0	–	–	6.2–6.8
	VL	3.7	3.5	–	–	–	–	4.7	–	3.7–4.3	4.1	4.3	3.8	3.9	–	–	4.1–4.4
	HC	6.9	5.5	–	–	–	–	–	–	–	–	–	–	–	–	–	–
P2	MD	10.5	10.2	–	–	10.9	10.8	11.4–11.8	–	10.9–13.0	11.6	12.0	12.1	11.8	11.8–12.6	–	10.9–11.0
	VL	4.0	4.0	–	–	4.2	4.5	4.8–5.0	–	4.1–5.1	4.8	4.8	4.1	4.2	6.1–7.2	–	4.5–4.6
	HC	6.6	6.2	–	–	–	–	–	–	–	–	–	–	–	8.8–9.5	–	–
P3	MD	12.0	12.0	–	–	11.6	12.2	13.1	–	11.7–14.4	13.7	13.4	13.5	14.0	12.0*–13.7	–	12.5
	VL	4.5	4.7	–	–	4.5	4.6	5.3	–	4.4–5.4	5.3	5.3	4.8	4.8	6.3	–	5.0
	HC	7.1	8.0*	–	–	–	–	–	–	–	–	–	–	–	9.8	–	–
P4	MD	18.4	19.0	17.8–22.9	19.0	18.5	18.3	20.7	–	19.3–22.8	23.3	22.7	–	21.0	22.5	22.0	20.6–21.7
	VL	9.8	9.6	–	–	8.6	9.7	10.4	–	8.6–11.5	11.2	11.2	–	10.7	11.2	10.0	10.1–10.5
	HC	11.5*	12.0*	–	–	–	–	–	–	–	–	–	–	–	14.7	–	–
	paracone MD	12.2	14.0	–	–	–	11.6	–	–	–	–	–	–	–	15.3	–	–
	metacone MD	7.8	8.0	–	–	–	–	–	–	–	–	–	–	–	9.4	–	–
	metacone TD	7.5	6.6	–	–	–	–	–	–	6.2–8.2	8.4	8.8	–	8.0	7.9	–	–
M1	MD	12.2	12.1	11.7–15.3	12.6	12.5	11.6	13.8	13.0	13.4–15.7	14.8	14.8	13.7	13.7	15.9	14.5*	14.0–14.4
	VL	14.7	14.6	–	–	14.4	15.1	17.0	16.3	14.5–18.0	16.2	16.6	15.6	16.2	18.7	17.7	16.7–17.1
	HC	8.3	8.6*	–							–				9.7		–
	paracone MD	6.7	–	–	–	–	6.0	–	–	–	–	–	–	–	7.7		–
M2	MD	7.4	7.6	–	6.8	7.2	7.5	8.0	–	7.6–8.7	7.9	8.1	8.5	8.3	8.4–8.7	8.0*	8.3
	VL	11.0	9.9	–	10.3	9.5	10.2	11.5	–	9.2–11.1	12.0	12.4	11.3	11.4	11.1–11.8	13.0	10.3
	HC	–	4.4	–	–	–	–	–	–	–	–	–	–	–	5.2–5.6		–

The canid from Ceyssaguet (Haute-Loire, France) is an example of this ambiguity. It is well represented among the fauna of this important site from the end of the Villafranchian. Tsoukala (2004) attributed the remains to *Canis etruscus*. Boudadi-Maligne (2010) re-examined the material: "it has all the characters used to identify *Canis etruscus* (well-developed sagittal crest, curvature of the lower row of teeth, and paraconid of the lower carnassial at the same height as the p4). However it has all the morpho-functional characteristics needed [to] link it to *Canis arnensis*" (p.151). In the end, Boudadi-Maligne decided on *Canis arnensis* because of the greater significance of these latter characteristics. I agree with her assessment that the two species need to be redefined on the basis of other criteria, not least because such adaptations need not be genetic, but may be due to phenotypic plasticity.

When we examine the upper teeth of the *Canis* from Senèze, the only known specimens, we can see significant differences from *Canis etruscus*. I directly compared *Canis arnensis* NMB Se 1780 and FSL 210,987 with *Canis etruscus* from the Italian sites: Inferno (original); Figline (cast); and Montevecchio (cast). The teeth of this latter species are larger and their morphology, especially of the P4, M1 and M2 can be easily distinguished from those of *Canis arnensis*:

- P4: in *Canis arnensis* this tooth is narrower, and the paracone is more slender, more tilted towards the rear and more sharply pointed. In *Canis etruscus* the paracone is wide, massive and relatively lower, the antero-internal talonid less separated from it and the metacone clearly wider.
- M1: in *Canis arnensis* this tooth is elongated, narrower, strongly curved. The paracone and the metacone are moderately developed and their external side is almost vertical in relation to the two points. In *Canis etruscus* the M1 is a wide tooth, more triangular, with the paracone and metacone strongly developed, round, and separated from the external edge by an oblique slope. The paracone is clearly more developed than the metacone, unlike in *Canis arnensis.* The structure of the internal talonid is different, because of the position and the development of the protocone and its articulation with the hypocone, the latter much more distinct and developed in *Canis etruscus.*
- M2: in *Canis arnensis* it is narrow, curved and shorter but in *Canis etruscus* it is wider.

The species *Canis etruscus* has recently been redefined on the basis of new cranio-dental criteria made possible by the rich material found at Pantalla (Italy) (Cherin et al. 2013). It is clearly distinct from *Canis arnensis* because of a number of characters that completely justify the species distinction.

NMB Se 1780 consists of the maxilla connected to the left palatine (Fig. 9.10). The anterior part is missing but all the cheek teeth are there, hardly worn. In occlusal view, the caudal nasal spine, which marks the middle of the opening of the choanae, clearly indicates the medial line of the palate. Apart from one minor and readily observable exception, this is the axis along which the specimen has been broken, following for the most part the still visible suture line of a young animal. Specimen FSL 210987 of an equivalent size (Table 9.6) provides the dimension of the length of the canine. Any ideas about the degree of elongation of the muzzle remain subjective. On the other hand, through the use of computer reconstruction by symmetry, the shape and proportions of the palate of the *Canis arnensis* from Senèze can be determined with some confidence.

The reconstruction allows us to quantify the external TD between the two canines ([PEMAC] variable 1), the maximum TD between the two contact points of the P4-M1 (variable 2) corresponding to the maximum width of the palate, and the length (variable 3) between the front of the canine and the rear of the M2 (Fig. 9.10). Three indices can be constructed from combinations of these variables. The first index (1/3 × 100) indicates the relative minimum width of the anterior part in relation to the length, the second (2/3 × 100) the relative maximum width of the palate in relation to the length. The sum of the two widths provides a third index (or ratio) (1 + 2)/ 3 characterizing the degree of elongation. Table 9.7 places the *Canis arnensis* from Senèze among the extant and fossil canids in terms of these indices.

Three groups are apparent according to their ratio of relative elongation:

A = ratio from 0.97 to 1.07; long and narrow muzzle, extant *Vulpes* type;

B = ratio from 1.23 to 1.40; muzzle of intermediate length and width, extant *Canis lupus* type;

and C = ratio from 1.52 to 1.62; short and wide muzzle, extant *Lycaon* type.

The *Canis arnensis* from Senèze clearly belongs to Group A, as does the *Canis etruscus* from Val d'Arno (Montevecchio) and the canid from Ceyssaguet. It was a canid with an elongated muzzle and about the size of a modern wolf (*Canis lupus*, females: 18–55 kg, males: 20 to 79 kg; Hunter and Barret 2011). I estimate the weight of the *Canis arnensis* from Senèze (NMB Se 1780) to be approximately 16 to 24 kg based on length of its cranium (prosthion-acrocranion) of at least 200 mm (extrapolation from the reconstruction of NMB Se 1780). For comparison, the length of the cranium of the modern dhole (*Cuon alpinus*) is 180 mm. The weight of this canid ranges from 16 to 30 kg for females and from 24 to 41 kg for males (Hunter & Barret 2011). The cranium of *Canis arnensis* of Senèze indicates a canid at least as big as a dhole.

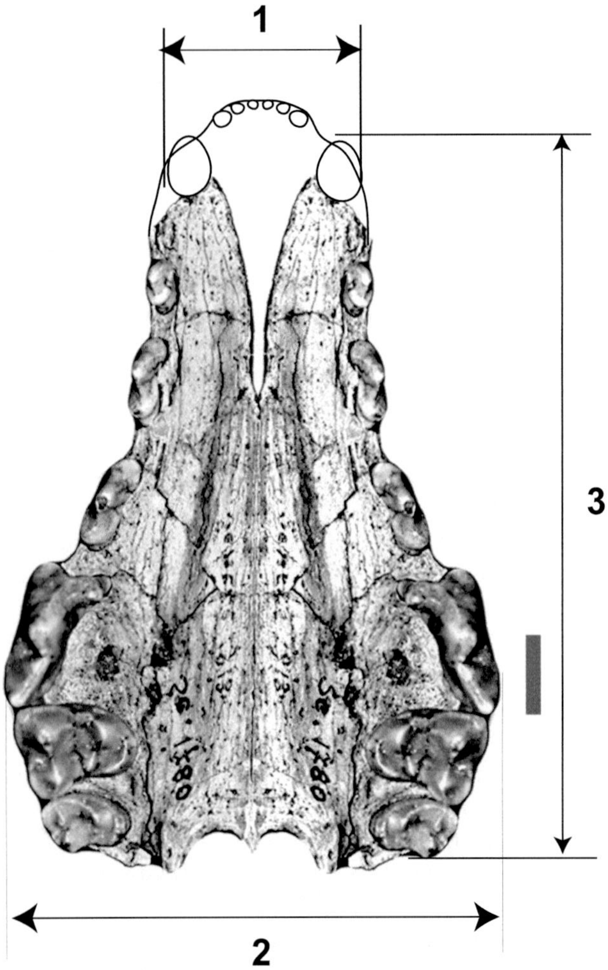

Fig. 9.10 Senèze, *Canis arnensis*, reconstruction of the palate of the individual NMB Se 1780. See Table 9.7 for variable names and proportions. Scale bar equals 1 cm

Canis sp.

- NMB Se 543: fragment of tooth corresponding to the talonid of a left m1 (Fig. 9.11);
- FSL 210979: 1 right complete radius (Fig. 9.12).

The fragment of tooth NMB Se 543 is labelled "Ursus?" in H.G. Stehlin's handwriting. The caution of that author must be emphasized in indicating that it could be the talonid of a left m1 of *Ursus*. The fragment is small, the talonid especially so. Signs of significant wear of the hypoconid and of the entoconid appear in black. The break in the tooth at the front of the talonid, also stained black (tooth greatly fossilized), forms a curved transverse barrier that occupies practically the full width of the fragment and cuts off from it a small part of the central basin at the anterior end of the fragment. In occlusal view, the rear of the tooth is in the form of an asymmetric curved surface with a steeply inclined slope. It does not have the rounded symmetrical form of the posterior edge of the talonid of an ursid m1, nor does it have the verticality of the edges of the entoconid and hypoconid. The transverse break which has removed the most forward part of the talonid basin, shows that the base of the almost vertical surface corresponds to the breaking of the lower part of a protoconid. The complete posterior root does not have the shape of a bear's m1, and it has a longitudinal groove on its anterior edge, which does not occur in ursids. In reality, this is the talonid of a canid m1. The transverse diameter of this talonid (the only measurement possible) is 9.95 mm.

Table 9.8 compares NMB Se 543 to the m1 of fossil and extant *Canis*. The only way any conclusions can be drawn from this comparison is by evaluation of the dimensions of this m1 (MD x VL) knowing that the transverse diameter of

Table 9.7 Comparative dimensions and proportions of the Senèze *Canis arnensis* palate

	Fossil Canidae			Extant Canidae									
Palate proportions (see Fig. 9.10)	Se 1780	*C. etruscus* FSL 211,751 cast Montevecchio	*C. etruscus* MNPE CEY. 5971	*V. vulpes*	*C. latrans*	*N. procyonoides*	*Cerdocyon thous*	*Canis lupus*			*Cuon alpinus*	*Lycaon pictus*	
1 TD at canines	28.5	29.96	35.5	23.8	10.55	11.69	12.23	55.4	42.03	47.11	40.8	45.1	47.3
2 TD P4-P4	65.5	55.14	62.0	40.6	17.1	22.87	22.01	87.5	75.3	85.56	66.8	74.5	82.9
3 Length C ant. /M2	94.5	90.3	100.0	61.7	27.09	37.75	31.85	116	82.3	101.34	76.5	78.6	80.25
4 (1) × 100 /(3)	30.16	33.18	35.5	38.57	38.94	30.97	38.40	47.76	51.07	46.49	53.33	57.38	58.94
5 (2) × 100 /(3)	69.31	61.06	62.0	65.80	63.12	60.58	69.11	75.43	91.50	84.43	87.32	94.78	103.30
6 Muzzle length index: (1 + 2)/(3)	**0.99**	0.94	0.98	1.04	1.02	0.92	1.08	1.23	1.43	1.31	1.41	1.52	1.62

Measurements of extant canids by author from various collections.

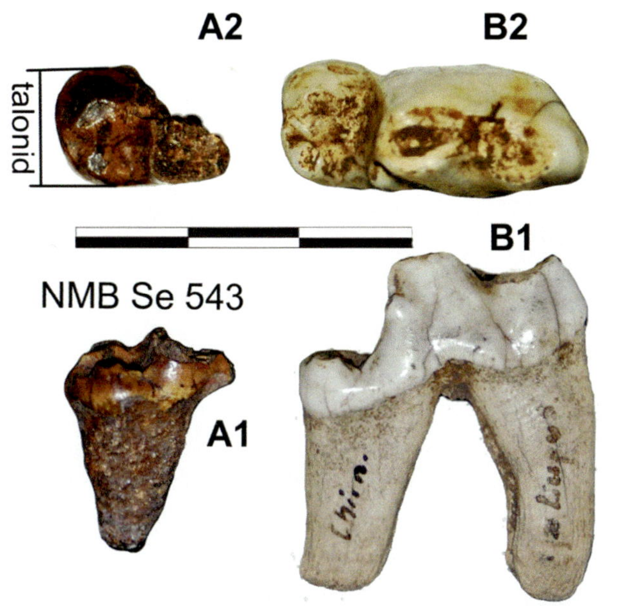

Fig. 9.11 Senèze, *Canis* sp. (NMB Se 543), m1, **A1**, lateral view; **A2**, occlusal view; *Canis lupus* (NMB), m1; **B1**, lateral view; **B2**, occlusal view. Comparison. Scale bar equals 3 cm

the talonid is close to but not quite the width of the complete tooth. The size of NMB Se 543 seems relatively large, on the order of an m1 of fossil or modern wolves. It is clearly larger than the m1 of the canid from Ceyssaguet attributed to *Canis etruscus* by Tsoukala (2004) or to *Canis arnensis* by Boudadi-Maligne (2010). If we take, as our basis, the dimensions of the m1 of the *Canis etruscus* from Val d'Arno (Table 9.8), the dimensions of the unbroken m1 of NMB Se 543, extrapolated from the width of the talonid, would be about 26 mm (MD) and 10.3 mm (VL). It could therefore quite possibly belong to the species *Canis etruscus*, but this is by no means certain given the fragmentary nature of the specimen. In addition, the wolf of Dmanisi, originally determined as *C. etruscus*, could belong to a new species *Canis borjgali* presaging the morphology of *C. mosbachensis* and the extant wolf *Canis lupus* (Bartolini Lucenti et al. 2020).

The complete right radius (Fig. 9.12) was in a box labelled FSL 210979, "Carnassier", together with the proximal part of a tibia (Fig. 9.20, B) and one femur without its proximal extremity (Fig. 9.20, A), the last two belonging to *Chasmaporthetes lunensis*. These three remains are all catalogued under the same number FSL 210979, but in fact the radius belongs to a canid. It is black, light in weight, with epiphysal edges worn by rolling and almost as big as the radius of the *Canis lupus* from Châtillon-Saint-Jean. It differs from the latter by the slenderness of its diaphysis (whose edges are perfectly parallel) and by some morphological details of the slightly raised ridge running along the palmar face of the diaphysis. This ridge is central on the canid radius from Senèze and parallel to the two edges, while it is shifted nearer to the median side in *Canis lupus*. Its dimensions are (in mm): L = 193.0; TD prox. = 21.4; prox APD = 14.3; med TD = 16.4; med APD = 11.0; dist TD = 28.9; dist. APD = 26.4.

Is this radius *Canis arnensis* or *Canis etruscus*? The post-cranial skeleton of *Canis arnensis* is unknown, and that of *Canis etruscus* is still poorly known and little studied, so comparisons are difficult. Size seems to be insufficient to distinguish them. We are far from knowing the extent of individual variation for these two species, and sometimes the teeth of *Canis arnensis* seem to be as large as those of *Canis etruscus* or even larger. The former species is present at Senèze, but the second has not yet been encountered there except in a simple list of fauna (unaccompanied by any real study of the material) extracted during the unpublished excavation by Guth (1975). In the case of this radius, *Canis etruscus* cannot be totally ruled out without any reliable morphological element of comparison.

Table 9.8 Comparative dimensions of the Senèze *Canis* sp. m1

	Talonid TD (Fig. 9.11)	MD	VL	HC	Trigonid MD	Talonid MD	Source
***Canis* sp. m1 NMB Se 543**	**10.0**	(25.0)	(10.3)	–	–	–	**NMB**
Canis adoxus, Saint–Estève (Pyrénées orientales, France)	–	18.1	7.6	–	–	–	Martin (1973)
Canis arnensis, LGPUT APL.17	–	21.7	8.6	–	–	–	Koufos (1992)
Canis arnensis Ceyssaguet (MNPE)	n = 17	n = 17	n = 17	–	n = 15	–	Boudadi-Maligne (2010)
Mean	8.42	22.61	9.26	–	15.39	–	
σ n − 1	0.48	1.16	0.47	–	0.97	–	
Canis arnensis Ceyssaguet (MNPE)	–	n = 9	n = 9	–	n = 8	n = 9	Tsoukala (2004)
Mean	–	22.88	9.36	–	16.73	6.44	
σ n − 1	–	1.71	0.62	–	1.24	0.66	
Canis etruscus NMB Va 165-Tasso 1900	9.3	23.8	9.6	–	–	–	NMB
Canis etruscus NMB Va 1669-1912	9.4	25.5	10.3	–	–	–	NMB
Canis etruscus LGPUT GER-166	–	25.2	10.5	–	–	–	Koufos (1992)
GER-167	–	25.0	10.4	–	–	–	
Extant *Canis lupus*	–	n = 17	n = 17	–	–	–	Baryschnikov (2012)
Mean	–	27.82	11.20	–	–	–	
σ n − 1	–	1.85	0.89	–	–	–	
Extant *Canis lupus americanus* c. 1362	10.1	26.7	10.7	–	–	–	NMB
Extant *Canis lupus*—Canada—male	9.4	30.4	12.2	17.8	–	–	Cast-Skull Unlimited

() estimation by comparison with other canids

Fig. 9.12 Senèze, *Canis* sp. (FSL 210979), right radius, **1**, plantar view; **2**, dorsal view; **3**, external lateral view; **4**, internal lateral view. Scale bar equals 5 cm

Family Ursidae

The ursids have not left many remains at Senèze.

Ursus etruscus Cuvier, 1823

Material:

- NMB Se 1505: 1 complete left C, not worn, and 1 complete thoracic vertebra;
- NMB Se 1506: 1 complete thoracic vertebra;
- NMB Se 1581: fragment of left C;
- NMB Se 1660: fragment of a damaged C;
- NMB Se 1714: left demi mandible with c, empty alveolus of p1, crowns of p2-m2 and empty alveolus of m3;
- FSL 210966: cranial fragment preserving part of the right maxilla with I3 (mistakenly positioned several centimeters behind the correct position when it was restored), M1 and M2, as well as the right demi-mandible with p3-m3. These two pieces certainly belong to the same individual, sharing the same color, degree of fossilization and stage of wear, as well as a perfect occlusal match between the teeth. Unfortunately it is an old individual with very worn teeth, and occlusal details are not clearly observable (Figs. 9.13 and 9.14).

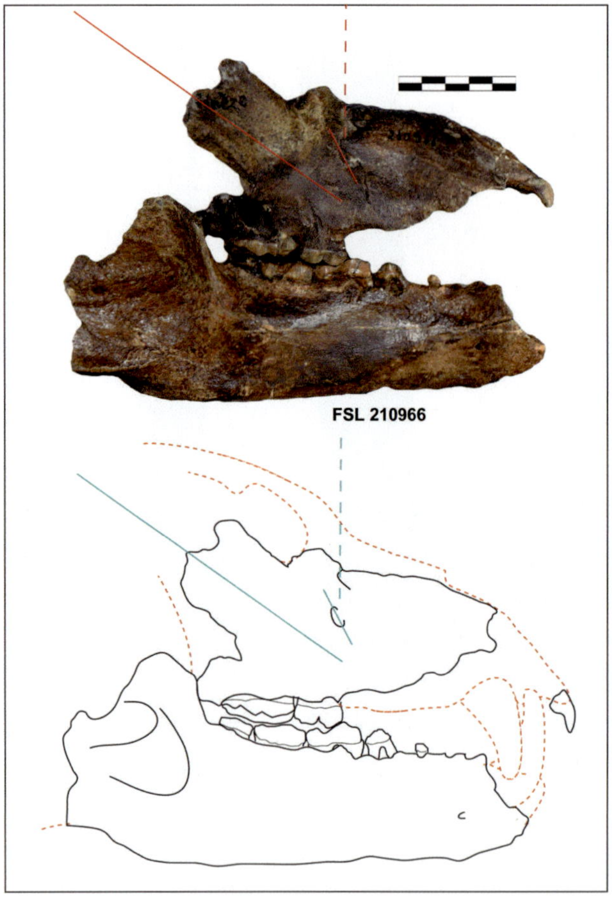

Fig. 9.13 Senèze, *Ursus etruscus*; **A**, fragments of cranium and mandible (FSL 210966); **B**, reconstruction of the skull and correction of the position of the I3. Scale bar equals 5 cm

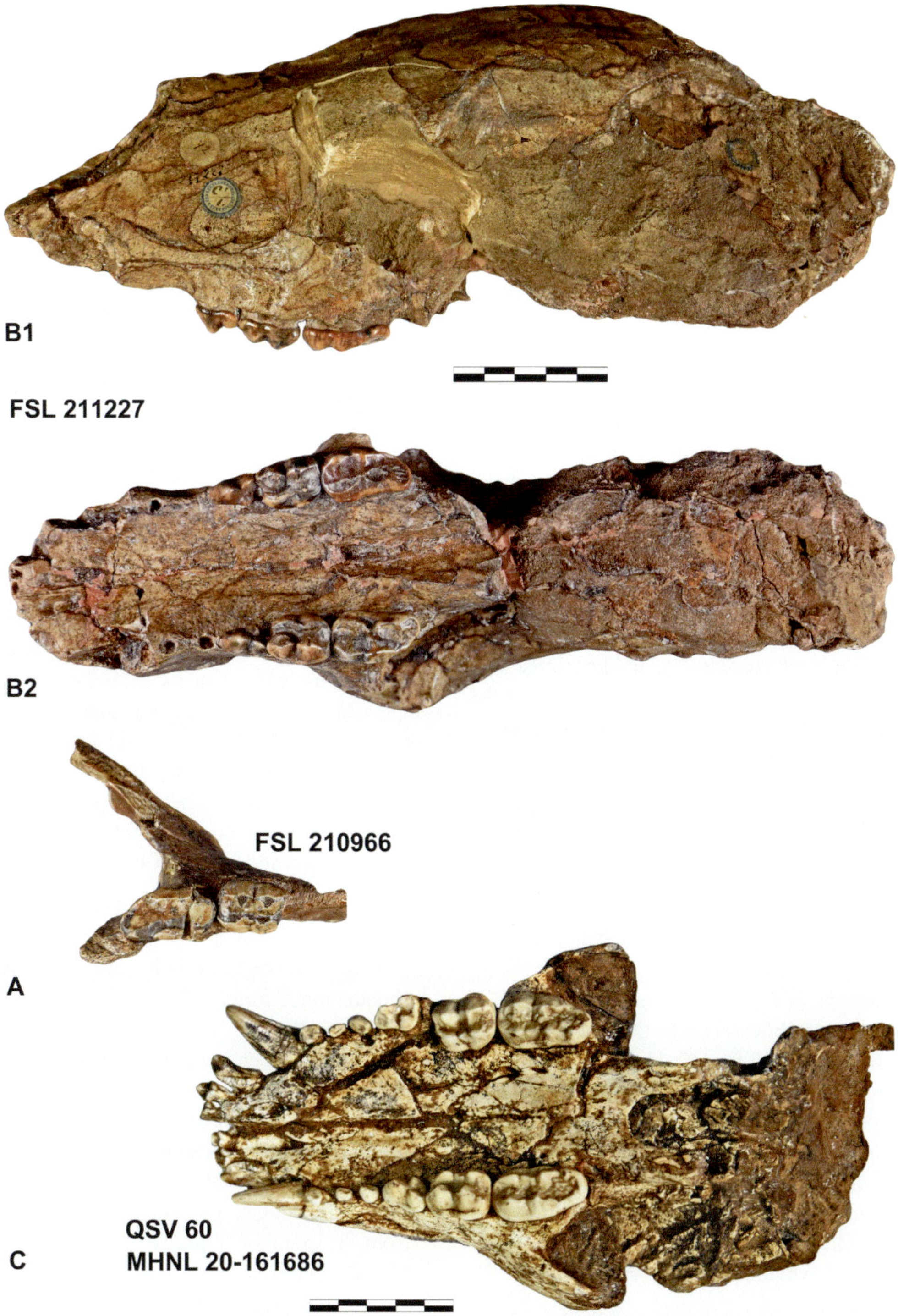

Fig. 9.14 **A**, *Ursus etruscus*, Senèze cranial fragment (FSL 210966) occlusal view. Comparison with **B**, holotype of *Ursus minimus* (FSL 211227) from Perrier-Etouaires France), **1**, left lateral view; **2**, occlusal view; and **C**, *Ursus etruscus* (MHNL 20-161686) from Saint-Vallier (France), occlusal view. Scale bar equals 5 cm

Stehlin (1923) and Schaub (1943) identified *Ursus etruscus* at Senèze based on the material in the NMB. Jung (1947) published a list of fauna at Senèze including (after information from J. Piveteau) *Ursus arvernensis* (= *Ursus minimus*, see below) instead of *Ursus etruscus*. Subsequently *Ursus etruscus* has been the only taxon recognized (e.g. Bout 1960, 1970; Delson et al. 2006; Devis 1970; Heintz et al. 1974).

However, after re-examining the FSL specimen, Mazza and Rustioni (1994, p. 22) attributed the Senèze bear to the *Ursus minimus-thibetanus* group. In this they were followed by Wagner (2006). Some of their arguments merit further attention. They observed that FSL 210966 has:

- "a particularly short muzzle";
- "the persistence of the premolars in front of the P4/4, which tend to progressively disappear from *Ursus etruscus* during the course of its evolution": however, all the Etruscan bears that I have studied had all four premolars or their corresponding empty alveoli, including the bear from Ceyssaguet (Haute-Loire) which I consider, as does Tsoukala (2004), to be the last known *Ursus etruscus* (at about 1.2 Ma, MNQ 19 biozone);
- "moderately prominent cuspids …, unlike *Ursus etruscus*". However, considering the degree of wear on the teeth of FSL 210966 and even of NMB Se 1714, it seems very difficult to be definitive;
- "the M2 with a reduced talonid": this is true, although it is difficult to define its morphology because of its heavy wear (U = 5). The talonid is relatively small at Saint-Vallier too, although a little less so;
- "the lower teeth rather small and laterally compressed": the lower teeth are relatively narrow; but they closely match the ones from Saint-Vallier;
- metric comparisons, in the form of diagrams of the distribution of molar length/width ratios (Mazza & Rustioni 1994, Fig. 2 for M1; Fig. 3 for M2; Fig. 4 for m1; Fig. 5 for m2; Fig. 6 for m3), show that the bear from Senèze is almost always situated in the middle of the range of *Ursus etruscus* and *Ursus* aff. *etruscus*. The one exception is the m3 which is in the middle of the range of the *Ursus minimus-thibetanus* group, indicating a small third lower molar, which could reflect either an individual variation or a different stage of evolution (a single specimen).

Mazza and Rustioni (1994) suggested that the very small parastyle and metastyle of the first upper molar of the bear from Senèze is unlike *Ursus etruscus*. From Senèze there is only one upper first molar, very worn at both ends (FSL 210966). It is true that there are no lateral styles on either side of the paracone and metacone, but if they had been there before the wear occurred, they would not have been well developed. The parastyle often present at the front of the M1 of *Ursus etruscus* is always quite small, whereas the metastyle is often better developed. However, the situation varies with different individuals. At Saint-Vallier, NMB StV 293 for example, has only a very small parastyle whereas on NMB StV 292 it occurs quite idiosyncratically on a cingulum. Individual variation is also evident on MHNL QSV 60, QSV 63, QSV 84 and QSV 85 in the Viret Collection of Saint-Vallier, and later, on the abundant *Ursus deningeri-spelaeus* group material.

In 1933 Stehlin was already discussing the validity of the distinction between *Ursus minimus* (= *U. arvernensis*) and *Ursus etruscus* (Dubois & Stehlin 1933, p. 63). I have compared Senèze FSL 210966 to the skull of the bear from Perrier-Etouaires (FSL 211227), the holotype of *Ursus arvernensis* Croizet & Jobert, 1828 (Fig. 9.14). *Ursus minimus* Devèze and Bouillet, 1827 was named one year earlier and everyone has agreed for a long time that the two taxa are synonymous, thus today only the name *U. minimus* is used. The skull from Perrier is complete and very well fossilized. The zygomatic arches are missing, but we can still see the marks left by their removal. It has been crushed laterally on its right side, while the left side seems less deformed. The muzzle appears short in relation to the overall length of the skull. The palate is complete and has a medial depression which is more marked in its posterior part because of the compression. The teeth are almost all present and well preserved (Fig. 9.14). The skull and teeth of the bear from Senèze FSL 210966 are slightly (about 13%) larger than those of the *Ursus minimus* from Perrier (Fig. 9.15, Table 9.9). From a morphological point of view several important observations can be made:

- the plaster restoration of the third right upper incisor of FSL 210966, which attaches it to the fragment of the maxilla, has not been well done. This tooth should be situated much further forward and a little lower than it is now (Fig. 9.13);
- the imprint left on the Perrier cranium by the removal of the zygomatic arch, clearly visible on the right side, runs from the facial crest of the orbit to the base of its lower edge. This imprint is short (TD = 34.5 mm) in comparison with its equivalent on the Senèze cranium (TD = 45.3 mm) and in particular, its direction of departure from the cranium is similar to that of *Ursus thibetanus*, whose zygomatic arch has a relatively short anterior attachment and a more abrupt angle of departure. On the Senèze cranium the length of the attachment is much greater and the departure has an upward direction like that of *Ursus spelaeus* (e.g. FSL 95,660–95,661, mounted skeleton from Vallon Pont d'Arc). This arrangement is also found on the *Ursus etruscus* cranium from Saint-Vallier (Viret 1954, pl.3, Fig. 1c);
- the plane of the infraorbital foramen at the front of the orbit is vertical on the bear from Perrier (*Ursus minimus*),

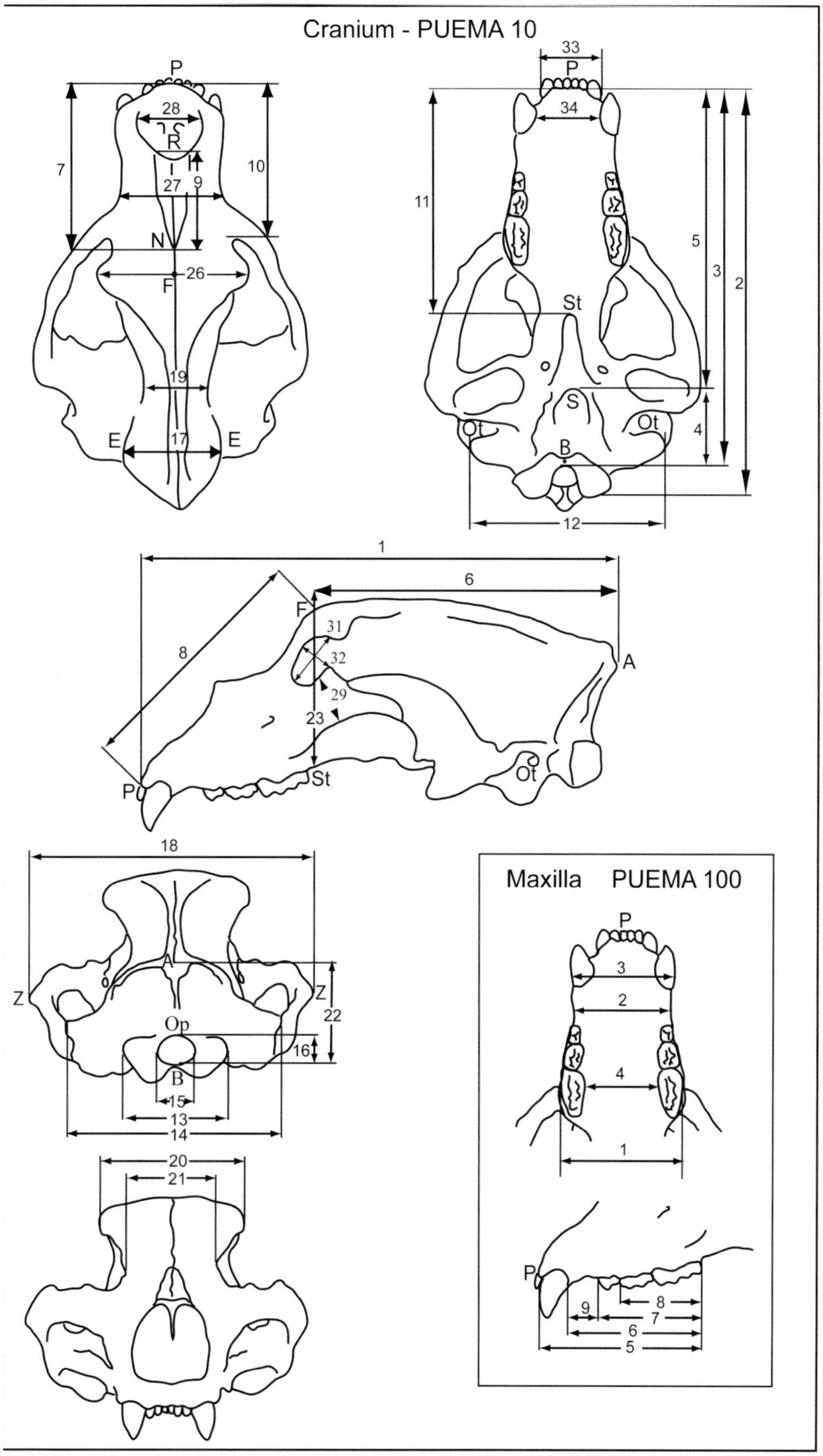

Fig. 9.15 Ursids, measurements of the skull (PUEMA 10) and of the maxilla (PUEMA 100). See Table 9.11

and on the extant *Ursus thibetanus*, whereas it tilts outwards at about 45° on the *Ursus etruscus* from Saint-Vallier, as well as on the Senèze bear, and later on *Ursus spelaeus*;

- the proportions of the M1 of the *Ursus minimus* from Perrier can be determined in spite of the great wear of this tooth. The ratio of width/length is 76.3% for Perrier and 75.3% for the fossil *Ursus thibetanus* from Blanot 2 (OIS 5 or early OIS 4, Argant 1991). The ratio on the bear from Senèze (FSL 210966) is lower at 72.7%, which is distinct from that of the *Ursus minimus-thibetanus* group.

metrical data for this comparison, and Figs. 9.14 and 9.16 show the morphology.

The mandible from Senèze is rather elongated and narrow (Figs. 9.16 and 9.17, Table 9.10). The start of the ramus with its characteristic anterior edge rises steeply (almost vertically) and borders a relatively deep masseteric fossa as in the *Ursus etruscus* from Saint-Vallier (Viret 1954, pl. 3, Fig. 2a) and in *Ursus spelaeus*. In *Ursus thibetanus* the ramus rises progressively in a regular curve and much less abruptly, and the fossa is shallow and almost imperceptible. No mandible has been found at Perrier, but the similarities between *Ursus minimus* and *Ursus thibetanus* are accepted and are the basis of the case put forward by Mazza and Rustioni.

Table 9.9 Comparative dimensions of the Senèze *Ursus etruscus* cranium

Template	Var. No.	Cranial variables (See Fig. 9.15)	*Ursus etruscus* Senèze FSL 210966 R	*Ursus minimus* Saint-Vallier (MNHL) 20 161 689	Perrier-Etouaires 20 161 739	Perrier-Etouaires 20 161 686	Perrier-Etouaires FSL 211227 (holotype) L	FSL 211227 (holotype)	FSL 211227 (holotype) R
PUEMA 10	1	Length Prosthion-Acrocranion		–	–	–		285.0	
	6	Neurocranial length: F-Acrocranion		–	–	–		161.5	
	7	Viscerocranial length: Prosthion-Nasion		120.1	125.0*	131.1		[94.0]	
	8	Facial length: Prosthion-F		161.8	–	149.8*		138.4	
	10	Lateral length of muzzle		119.7	116.7	106.0*	102.6		
	28	Internal breadth of nasal opening		46.9	49.6	41.6*		37.3	
	31	Max. length of orbit	–	43.8	–	–	42.9	–	–
	33	Internal breadth of incisors or alveoli	–	–	37.8	38.8	–	37.1	–
	34	Internal breadth of canines or alveoli	–	42.3 (A)	39.4 (A)	40.1	–	42.3	–
	36	TD of initial part of zygomatic arch	43.5	–	–	–	34.3	–	34.5
PUEMA 100	1	Max. breadth of the palate						[64.5]	
	7	Length of cheek teeth P2-M2	–	–	–	–	61.1		61.1
	9	Length of diastema: C-P2	–	–	–	–	–		19.8

Values in square brackets ([94.0] and [64.5]) are rough estimates due to the lateral flattening of the cranium.

I then compared the bear from Senèze to the specimens from Saint-Vallier in NMB, MHNL, and FSL (A. Argant 2004; Viret 1954). Tables 9.10, 9.11 and 9.12 provide the

Fig. 9.16 Senèze, *Ursus etruscus*, left mandible (NMB Se 1714). **A** lateral view; **B** lingual view. Scale bar equals 5 cm

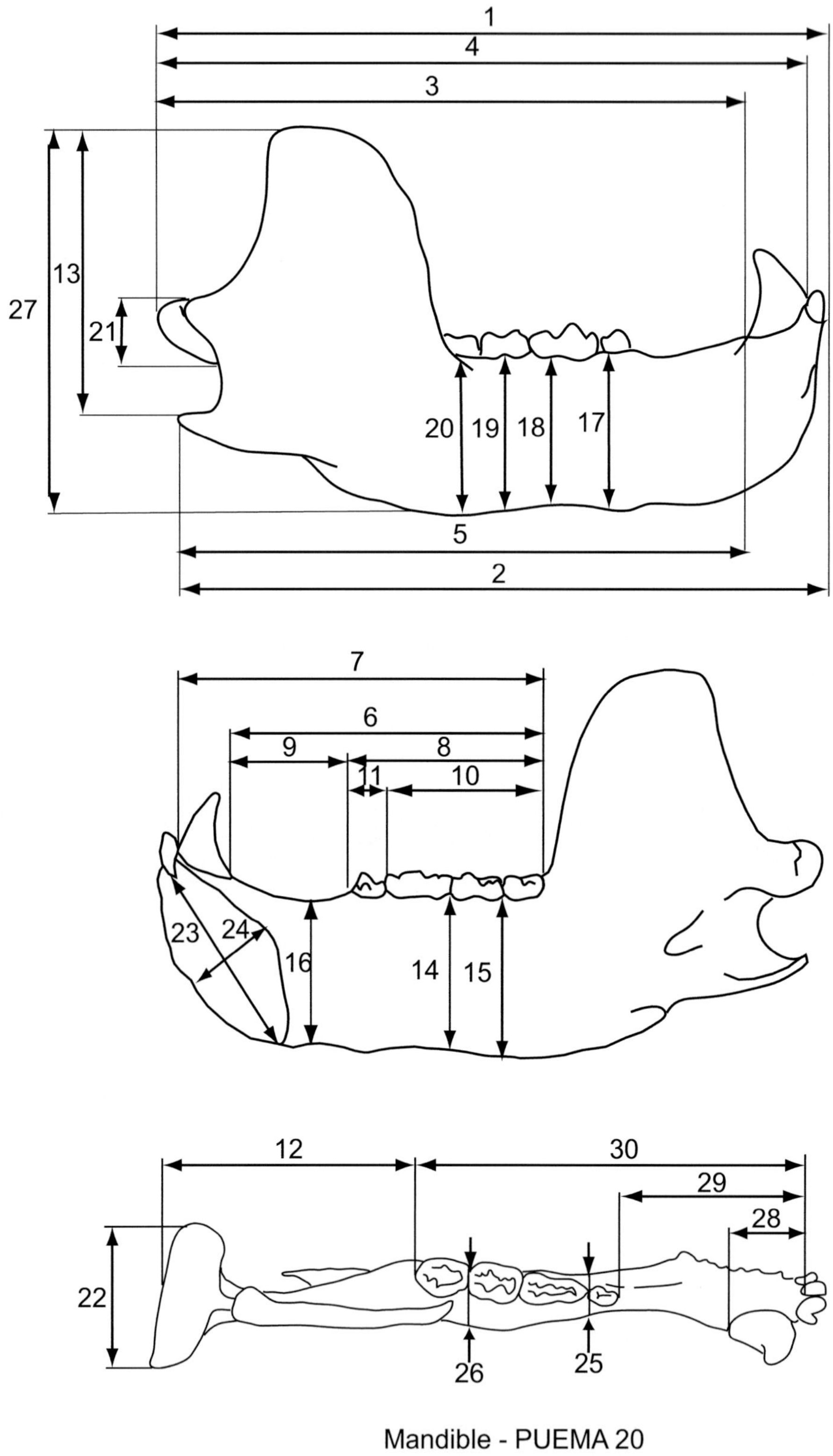

Fig. 9.17 Ursids, measurements of the mandible (PUEMA 20). See Table 9.10

Table 9.10 Comparative dimensions of the Senèze *Ursus etruscus* mandible

Mandibular variables		Senèze		Saint-Vallier	Dmanisi (Vekua 1995)	
		FSL 210966	NMB Se 1714	NMB StV 291	GNM D.355, D.1277	
PUEMA 20 (See Fig. 9.17)						
1	Total length infradental—condylus	–	>208.2	–	248.0	285.0
2	Length infradental—angular apophysis	–	–	–	–	–
3	Length condyles—posterior of canine	–	–	–	–	–
4	Length condyles—anterior of canine	–	–	–	–	–
5	Length angular apophysis—posterior of canine	–	–	–	–	–
6	Length posterior of canine—posterior of last molar	123.9	117.5	115.4	–	–
7	Length anterior of canine—posterior of last molar	–	145.0*	141.6	–	–
8	Length of the cheek teeth	78.3	77.9	74.2	–	–
9	Length of diastema	40.3	40.3	43.9	–	–
10	Length of molars	64.8	65.1	64.2	75.0	75.5
11	Length of premolars	11.2*	12.4	11.0	–	–
12	Length condyles—posterior of last molar	–	–	–	–	–
13	Height angular apophysis—coronion	–	–	–	–	–
14	Horizontal ramus height behind m1, lingual side	51.7	52.0	45.6	–	–
15	Horizontal ramus height behind m2, lingual side	54.4	53.7	48.1	–	–
16	Ramus height at distema	45.9	47.8	40.3	–	–
17	Ramus height under p4, labial side	44.1	46.3	42.8	–	–
18	Ramus height under m1, labial side	47.4	48.2	43.6	–	–
19	Ramus height under m2, lingual side	50.5	50.7	45.6	–	–
20	Ramus height under m3, lingual side	59.2	57.0	52.0	50.0	55.0
21	Height of condylus	–	–	–	–	–
22	Breadth of condylus	–	–	–	–	–
23	Maximum diameter of symphysis	–	68.5*	66	74.0	82.0
24	Minimum diameter of symphysis	–	34.5	30.4	–	–
25	Breadth of ramus, between p4-m1	17.2	16.1	18.6	–	–
26	Breadth of ramus, between m2-m3	21.8	22.2	20.3	–	–
27	Maximum height of mandible	–	–	–	–	–
28	Length infradental—posterior of canine	–	–	–	–	–
29	Length infradental—anterior of p4	–	–	–	–	–
30	Length infradental—posterior of last molar	–	–	–	–	–

Table 9.11 Comparative dimensions of the Senèze *Ursus etruscus* upper teeth

Upper tooth variables		*Ursus etruscus* Senèze		*Ursus etruscus* Saint-Vallier				*Ursus minimus* Perrier		*Ursus etruscus* Valdarno				*Ursus etruscus* Dmanisi			
		NMB Se 1505	FSL 210966	MHNL				FSL 211227		Garcia, 2003				Vekua, 1995			
				n	Δl	Mean	σ n − 1	L	R	n	Δl	Mean	σ n − 1	n	Δl	Mean	σ n − 1
I3	MD	–	13.7	–	–	–	–	–	–	–	–	–	–	–	–	–	–
	VL	–	10.0	–	–	–	–	–	–	–	–	–	–	–	–	–	–
	HC	–	15.3	–	–	–	–	–	–	–	–	–	–		–	–	–
C	MD	22.3	–	7	18.9–24.3	20.58	1.91	–	–	–	–	–	–	2	21.5–24.0	22.8	1.76
	VL	15.9	–	7	11.1–16.2	13.45	1.57	–	–	–	–	–	–	2	14.0–16.0	15.0	1.41
	HC	37.4	–	5	30.8–36.2	33.06	2.69	–	–	–	–	–	–	–	–	–	–
	PC	64.2	–	5	–	55.32	2.68	–	–	–	–	–	–	–	–	–	–
P4	MD	–	–	8	14.7–18.4	16.42	1.32	15.0	15.0	10	15.2–17.7	16.47	0.88	3	17.0–18.0	17.6	0.52
	VL	–	–	8	10.0–13.0	11.32	0.97	9.7	9.5	9	10.2–12.0	11.28	0.66	3	10.9–13.0	12.0	1.05
	HC	–	–	–	–	–	–	9.2	9.1	–	–	–	–	–	–	–	–
M1	MD	–	23.1	9	17.7–22.0	20.74	1.39	19.8	19.2	15	19.1–23.7	21.79	1.24	3	23.0–24.8	23.8	0.91
	VL max	–	16.8	7	14.5–17.9	16.75	1.5	15.1	15.3	15	14.2–17.4	16.11	1.26	3	18.0–19.3	18.43	0.75
	VL ant	–	16.6	8	14.1–18.3	16.22	1.21	14.1	14.0	15	14.4–17.2	15.92	0.8	–	–	–	–
	HC	–	–	–	–	–	–	9.0	9.4	–	–	–	–	–	–	–	–
M2	MD	–	31.3	10	28.1–33.7	30.40	1.91	27.5	27.7	29	31.0	32.68	0.93	3	35.0–40.0	37.66	2.21
	VL	–	16.7	10	15.4–18.9	17.43	1.00	15.9	15.9	10	16.5–19.3	18.09	0.8	3	18.6–21.6	20.3	1.53
	HC	–	–	–	–	–	–	7.4	7.4	–	–	–	–	–	–	–	–

Table 9.12 Comparative dimensions of the Senèze *Ursus etruscus* lower teeth

Lower tooth variables		*Ursus etruscus*															
		Senèze	Senèze	Saint-Vallier MNHL				Dmanisi Vekua, 1995 GNM		Val d'Arno Garcia, 2003				Ceyssaguet (MNPE) Tsoukala, 2004			
		FSL 216966	NMB Se 1714	n	Δl	Mean	σ n − 1	D.1278	D.355	n	Δl	Mean	σ n − 1	n	Δl	Mean	σ n − 1
c	MD	–	21.3	–	–	–	–	28.6	–	–	–	–	–	3	19.1–24.6	22.75	3.17
	VL	–	14.6	–	–	–	–	18.6	–	–	–	–	–	4	12.1–16.8	15.07	2.06
	HC	–	–	–	–	–	–	–	–	–	–	–	–	–	–	–	–
	U	–	–	–	–	–	–	–	–	–	–	–	–	–	–	–	–
p1	MD	–	4.2 (A)	–	–	–	–	8.6	–	–	–	–	–	2	7.8–8.4	8.1	0.44
	VL	–	4.3 (A)	–	–	–	–	–	–	–	–	–	–	2	5.6–5.7	5.6	0.07
p2	MD	5.4 (A)	5.6	–	–	–	–	–	–	–	–	–	–	–	–	–	–
	VL	3.2 (A)	4.3	–	–	–	–	–	–	–	–	–	–	–	–	–	–
	HC	–	3.6	–	–	–	–	–	–	–	–	–	–	–	–	–	–
	U	–	0.5	–	–	–	–	–	–	–	–	–	–	–	–	–	–
p3	MD	6.1	6.1	–	–	–	–	6.9	–	–	–	–	–	–	–	–	–
	VL	4.0	4.5	–	–	–	–	–	–	–	–	–	–	–	–	–	–
	HC	2.6	4.1	–	–	–	–	–	–	–	–	–	–	–	–	–	–
	U	0	0.5	–	–	–	–	–	–	–	–	–	–	–	–	–	–
p4	MD	12.7*	12.4	8	10.9–14.1	12.78	1.02	14.8	14.8	11	11.6–16.0	13.17	1.42	3	13.6–14.2	13.95	0.32
	VL	7.0	7.3	8	7.2–8.2	7.53	0.32	8.9	8.1	11	6.7–8.5	7.36	0.62	3	7.4–7.8	7.58	0.17
	HC	–	7.3	–	–	–	–	–	–	–	–	–	–	–	–	–	–
	U	3	0.4	–	–	–	–	–	–	–	–	–	–	–	–	–	–
m1	MD	26.1	24.6*	8	21.0–24.6	23.31	1.14	27.7	26.4	13	21.2–25.4	24.01	1.47	4	23.9–27.1	25.52	1.64
	VL max	10.9	11.2	7	10.2–11.8	11.07	0.6	14.0	14.1	12	9.7–12.1	10.93	0.76	4	11.1–12.4	11.74	0.67
	VL ant	–	9.9	7	8.3–12.3	9.76	1.22	–	–	13	8.3–10.3	9.32	0.6	–	–	–	–
	U	2.5	2	–	–	–	–	–	–	–	–	–	–	–	–	–	–
m2	MD	23.8*	23.8	8	20.9–24.3	22.75	1.04	27.1	28.0	13	21.6–25.0	23.55	1.8	3	23.1–26.3	25.18	1.81
	VL max	14.5	14.7	5	13.5–14.7	14.10	0.58	17.5	17.6	13	13.4–15.9	14.33	1.14	3	14.3–15.8	15.23	0.82
	VL ant	–	13.8	7	13.4–14.8	13.84	0.53	–	–	13	13.2–15.2	14.02	0.76	–	–	–	–
	HC	–	–	–	–	–	–	–	–	–	–	–	–	–	–	–	–
	U	2	3	–	–	–	–	–	–	–	–	–	–	–	–	–	–
m3	MD	15.7	15.6 (A)	7	16.7–17.7	17.23	0.41	20.9	18.3	11	16.0–20.4	18.42	2.17	3	17.6–19.0	18.55	0.8
	VL	13.4	9.4 (A)	7	13.4–14.8	13.80	0.53	16.0	17.0	10	12.5–15.5	13.65	1.53	3	14.4–16.9	16.03	1.38
	HC	–	–	–	–	–	–	–	–	–	–	–	–	–	–	–	–
	U	3	–	–	–	–	–	–	–	–	–	–	–	–	–	–	–

For me *Ursus etruscus* is a species in the broad sense, which can always be subdivided by the creation of subspecies if the material allows that. Thus we have *U. etruscus saintvallierensis* Baryshnikov, 2007 and *U. etruscus verescagini* Sarapov, 1986 with a high degree of polymorphism, in accordance with what Erdbrink has said (1953, Fig. 28, p. 261) and with what we know about fossil bears when based on statistically sufficient series (*U. deningeri*, *U. spelaeus*, *U. arctos*), as well as with what we know about extant ones (*U. arctos*, *U. thibetanus*). I consider, along with many other authors, that *Ursus etruscus* is the ancestral form from which the cave bear lineage *Ursus deningeri-spelaeus* and the brown bear lineage *Ursus arctos* originated, the divergence occurring probably around 1.5 Ma (Argant 2001). The bear from Senèze closely resembles the bear from Saint-Vallier. It already displays the characteristics that we find again in its descendants in the cave bear lineage. It does not in any case display the short muzzle of the *Ursus thibetanus* type as indicated by Mazza and Rustioni (1994). In conclusion, the bear from Senèze is undoubtedly *Ursus etruscus*, and the *Ursus minimus-thibetanus* group must be completely ruled out.

The analyses of tooth microwear and dental morphometrics performed on the specimens of *U. etruscus* of Dmanisi (Georgia) (Medin et al. 2019) suggest that the bear species had an omnivorous behavior, feeding on both plant resources and vertebrate, depending upon availability. Isotopic data from Venta Micena indicates that *U. etruscus* recorded high levels of δ15N, which suggests that a regular consumption of fish from the moderately saline lacustrine systems present in the Orce basin of the Baza and Guadix depression could have played a significant role in the diet of *U. etruscus* (Medin et al. 2017).

Family Hyaenidae

Earlier excavations demonstrated the existence of two species of hyenids, *Pachycrocuta perrieri* and *Chasmaporthetes lunensis*, occurring together at many sites. The large number of coprolites found during earlier excavations (housed in the NMB) as well as during the recent excavations, demonstrates that the presence of hyenas at Senèze was more significant than the remains listed below would suggest. I did not always take in account the rare post-cranial remains because, given the coexistence of the two species, attribution to one species or the other is often very difficult, and moreover the remains allow practically no comparison.

Pachycrocuta perrieri (Croizet & Jobert 1828): Figs. 9.18, 9.19 and 9.20, Tables 9.13, 9.14 and 9.15

Material

- NMB Se 279: two fragments of upper jaw, with P2 L, P3 L and R, P4 L and R, and M1 L;
- NMB Se 312: left and right mandible of the same individual, incisors absent, with p2-m1, complete left C and damaged right C;
- NMB Se 312: isolated I2 R;
- NMB Se 1627: left complete calcaneus;
- FSL 210969: left demi-mandible, with p2-p4, broken m1 (cast, De Brun collection).

Earlier excavations at Senèze demonstrated the presence of *Pachycrocuta perrieri*, recognizable because of its wide and short "crocutoid" type p3 (NMB Se 312). The teeth can be easily distinguished from those of *Chasmaporthetes lunensis* (Fig. 9.18).

The study of hyaenids is far from achieving the degree of clarity we would like. Each author tends to have his/her own view of the phylogeny, undoubtedly because of the number of fossil species, but also because of the rarity of Early Pleistocene fossils, which prevents us from knowing the extent of individual, sexual or adaptive variations. As for me, following Howell and Petter (1980), I consider that *Pachycrocuta pyrenaica* corresponds to the ancestral form of the Villafranchian Eurasian species *Pachycrocuta perrieri*, from which originated *Pachycrocuta brevirostris*, the last species of the lineage.

Table 9.13 Comparative dimensions of *Chasmaporthetes lunensis*, *Pachycrocuta brevirostris* and *Pachycrocuta perrieri* upper teeth from Senèze and other localities

Upper tooth variables		*Chasmaporthetes lunensis*					*Pachycrocuta perrieri*				*Pachycrocuta brevirostris*			
		Senèze	St-Vallier	St-Vallier	La Puebla de Valverde		Senèze	Europe			Val d'Arno	Europe		
		FSL 210970	MHNL	Argant	Kurtén and Crusafint Pairó (1977)		NMB	Howell and Petter (1980)			NMB	Howell and Petter (1980)		
			Coll. Viret	2004	L	R	Se 279	n	Mean	Range	VA. 364	n	Mean	Range
C	MD	–	17.5	22.4	–	–	–	5	17.8	15.9–18.8	–	2	22.05	20.1–24.0
	VL	–	13.2	14.8	–	–	–	5	13.0	11.9–14.2	–	–	–	–
	HC	–	31.1	–	–	–	–	–	–		–	–	–	-
P2	MD	–	18.3	19.1	17.0	16.5	18.1	9	17.8	17.0–18.9	20.5	2	21.15	20.0–22.0
	VL	–	11.6	12.0	10.3	10.2	12.4	8	11.7	11.7–13.1	13.9	2	15.00	14.0–16.0
	HC	–	–	15.8*	–	–	–	–	–		14.3*	–	–	–
P3	MD	24.5	20.9	21.7	–	19.4	22.4	11	23.7	22.3–25.8	26.0	2	26.4	25.8–27.0
	VL	11.5	13.2	12.9	–	12.1*	17.5	10	16.6	13.6–18.1	18.2	2	19.6	18.2–21.0
	HC	–	–	17.3*	–	–	–	–	–		22.1*	–	–	–
P4	MD	36.0	32.1	32.9	30.3*	29.9	35.2	11	34.8	31.0–37.8	39.0	2	42.25	40.0–44.5
	VL max	19.1	18.0	17.8	–	16.0	20.5	10	20.5	18.5–22.9	24.0	2	24.85	25.0–28.7
	HC	18.9	–	16.5	–	–	22.2*	–	–	–	22.4*	–	–	
	MD protocone	10.5		–	–	–	12.2	–	–	–	–	–	–	–
	MD paracone	12.6	11.2	12.0	9.9	10.2	15.2	–	–	–	13.8	–	–	–
	MD metacone	15.3	14.3	16.3	–	13.6	14.4	9	13.2	11.0–15.4	15.4	2	15.5	14.0–17.0
M1	MD	–	6.7	9.2 (A)	–	6.0	6.0	5	5.7	5.2–6.1	6.1	–	–	–
	VL	–	16.3	13.3 (A)	–	14.9	13.9	6	14.3	14.3–14.7	15.3	–	–	–

Table 9.14 Comparative dimensions of the Senèze hyaenid lower teeth

Chasmaporthetes		Hyaenidae indet	*Chasmaporthetes lunensis*			*Pachycrocuta perrieri*					*Pachycrocuta brevirostris*			*Hyaena brunnea*	
Pachycrocuta		Senèze	Perrier-Rocca-Neyra	St-Vallier	St-Vallier	Senèze	Senèze	Europe			Europe			Fonelas P-1	
Lower teeth		Faure-Guérin-Delson	type Schaub 1941	A. Argant (2004)	A. Argant (2004)			Howell and Peter (1980)			Howell and Peter (1980)			Arribas and Garrido (2008)	
		FSL SEN	FSL	MHNL	UCB-Ly1	FSL	NMB	n	Mean	Range	n	Mean	Range	549	452
		01-0028	212804	QSV.52 161817	SV. 99.814	210969	Se 312								
c	MD	–	15.6	18.6	–	–	21.1	10	18.9	17.0–21.0	4	22.0	20.9–24.0	17.3	16.1
	VL	–	11.5	12.4	–	–	16.5	8	14.5	13.0–16.0	1	17.9	–	11.9	11.8
	HC	–	28.6	27.4	–	–	–	–	–	–	–	–	–	–	–
p2	MD	–	15.8	16.1	16.3	16.8	17.6	14	15.63	14.0–17.7	4	18.57	18.0–19.0	15.1	14.3
	VL	–	9.4	8.9	9.2	12.1	11.6	13	10.73	9.3–12.2	4	13.2	12.3–14.5	10.1	9.5
	HC	–	11.2	12.7	12.0*	–	–	–	–	–	–	–	–	–	–
p3	MD	19.8	20.6	21.4	20.5	23.2	22,0	16	20.93	19.5–22.9	4	24.35	22.5–26.0	20.3*	19.8
	VL	12.2	10.8	11.4	11.6	16.3	16.4	13	14.52	13.3–16.6	4	17.7	15.6–19.0	13.1	12.2
	HC	16.1	13.8	14.5	–	–	–	–	–	–	–	–	–	–	–
p4	MD	–	22.9	25.8	–	23.5	24.4	19	23.46	21.4–25.5	7	27.27	25.5–28.1	23.3	23.0
	VL max	–	11.3	12.2	–	16.4	15.9	16	14.43	11.8–16.1	6	16.77	16.2–17.8	13.8	12.5
	HC	–	15.4	17.2	–	–	–	–	–	–	–	–	–	–	–
	MD paraconid	–	11.8	–	–	–	–	–	–	–	–	–	–	–	–
m1	MD	–	24.7	–	–	–	27.0	22	25.44	23.4–26.9	4	29.35	27.5–31.2	–	23.8
	VL	–	11.4	–	–	–	14.4	18	12.80	10.8–14.6	4	14.42	13.2–15.0	–	12.0
	HC	–	11.0	–	–	–	–	–	–	–	–	–	–	–	–
	MD paraconid	–	12.7	–	–	–	–	–	–	–	–	–	–	–	–
	MD protoconid	–	11.4	–	–	–	–	–	–	–	–	–	–	–	–
	MD talonid	–	5.0	–	–	–	–	19	4.0	4.0–4.1	4	4.8	4.5–5.4	–	5.2

I compared the mandible from Senèze (NMB Se 312) to that of the type of "*Hyaena perrieri*" from Perrier-Etouaires, Croizet & Jobert, 1828 (FSL 211223, left demi-mandible). As with the material from Saint-Vallier, it shows the specific characteristics of *Pachycrocuta perrieri* described by Viret (1954), then in detail by Howell and Petter (1980, pp. 598–599) which I will not repeat.

There is no reason to suspect this specimen is *Pachycrocuta brevirostris* (Fig. 9.20) or even *Hyaena brunnea*, which Arribas and Garrido (2008) thought they had found for the first time out of Africa at Fonelas P.1 (Spain). However the plot points of Senèze and those of Fonelas P.1 are all in the same position on Fig. 9.2. The comparison of the mandible from Senèze (NMB Se 312) with the illustration of the two demi-mandibles of *Hyaena brunnea* from this site (Arribas & Garrido 2008, p. 215–216, Figs. 6 and 7) reveals some morphological differences on the p3—wide, short, and "crocutoid" type at Senèze—but longer and narrower at Fonelas of the *Hyaena* or *Parahyaena* type of their Fig. 6. The second Fonelas figure (Arribas & Garrido 2008, p. 216, Fig. 7), however, shows a p3 with a rather "crocutoid" form, quite different from that in their Fig. 6. This indicates at least a significant individual variability, if not two different species. Compared to Saint-Vallier, the demi-mandible from Fonelas FP.1-2001-0452 (Arribas & Garrido 2008, Fig. 6) is quite similar to that of the *Chasmaporthetes lunensis* (= *Euryboas lunensis*) of Viret (1954, Plate 8-3b, QSV.52). Its p3 (Specimen 0452) measures 19.8 × 12.2 mm. These are exactly the same measurements as for the p3 from Senèze (recent excavations), but slightly different from those (20.5 × 11.6 mm) of the p3 from Saint-Vallier (SV 99814, excavation Faure/Guérin). The specimen FP.1 2001-0549 from Fonelas shown in Fig. 7 (Arribas & Garrido 2008) could correspond to the other hyaenid from Saint-Vallier (Viret 1954, Plate 8-2, *Pachycrocuta perrieri*), on the basis of the morphology of its full row of teeth. The association of both species at the Fonelas site would agree with our current knowledge. The m1 of *Pachycrocuta perrieri* from Senèze (NMB Se 312) has no metaconid on the internal face of its protoconid, nor does the m1 from Saint-Vallier. There is one on the first m1 from Fonelas (FP.1 2001-0452), but it is not visible on the other m1 (FP.1 2001-0549), which is damaged (Arribas & Garrido 2008). However, I have not seen the material on which this study is based, nor the excavation data, which are probably essential. I simply regret that no use has been made of data from the post-cranial skeleton because these would have been a welcome source of additional information.

Pachycrocuta perrieri, which first appears in the early Villafranchian at Perrier-Etouaires, is found at many sites (see Werdelin & Solounias 1991, p. 20): Layna, Villarroya, Puebla de Valverde, Saint-Vallier, Senèze, Valdarno, and Blassac, for example. It is still present in the median Middle Pleistocene at Mauer and Mosbach in Germany. Werdelin and Solounias (1991, p. 21, Fig. 11) have established "a ratio diagram of various samples of hyaenids referred to as *P. pyrenaica* and *P. perrieri*", the reference standard having been based on the material from Villarroya. Figure 9.19 makes use of this analysis for the mandible. By comparing the tooth profiles of hyenas from the three sites, Villarroya, Saint-Vallier, and Senèze, as well as the tooth profile of *Pachycrocuta brevirostris*, it allows us to place Senèze relative to the others. Werdelin and Solounias (1991, p. 20) highlight the considerable variation among the samples in their Fig. 11: "Some, such as at Saint-Vallier, are robust and have broad premolars, whereas others, such as at La Puebla de Valverde, are small, with relatively slender premolars". There appears to be some indication of geographic patterning between the samples, as the two Spanish samples (Villarroya and La Puebla de Valverde) are very similar and differ from the French samples (Etouaires, Saint-Vallier and Senèze). Nevertheless the three profiles compared to Senèze (NMB Se 312) provide in my opinion a biochronological indication. For the mandible, the profiles from Villarroya (reference standard) and from Saint-Vallier are the closest, the profile of *P. brevirostris* is the furthest. The profile of the mandible from Senèze indicates an intermediate degree of evolution, sightly more evolved than the hyena from Saint-Vallier. If we complete this "ratio diagram" by using the values for the upper teeth of NMB Se 279, the profile of the hyena from Senèze (an individual different from the one providing the mandible) corresponds exactly to the *Pachycrocuta perrieri* of Saint-Vallier except that the characteristics of the upper teeth are a little more exaggerated. In any case it seems clear that the hyenas at Senèze and Saint-Vallier are both *Pachycrocuta perrieri*, and that, once again, the two sites are not strictly contemporaneous, though the evolutionary gap between them does not appear great. It is nevertheless true that our reasoning is based on very few individuals.

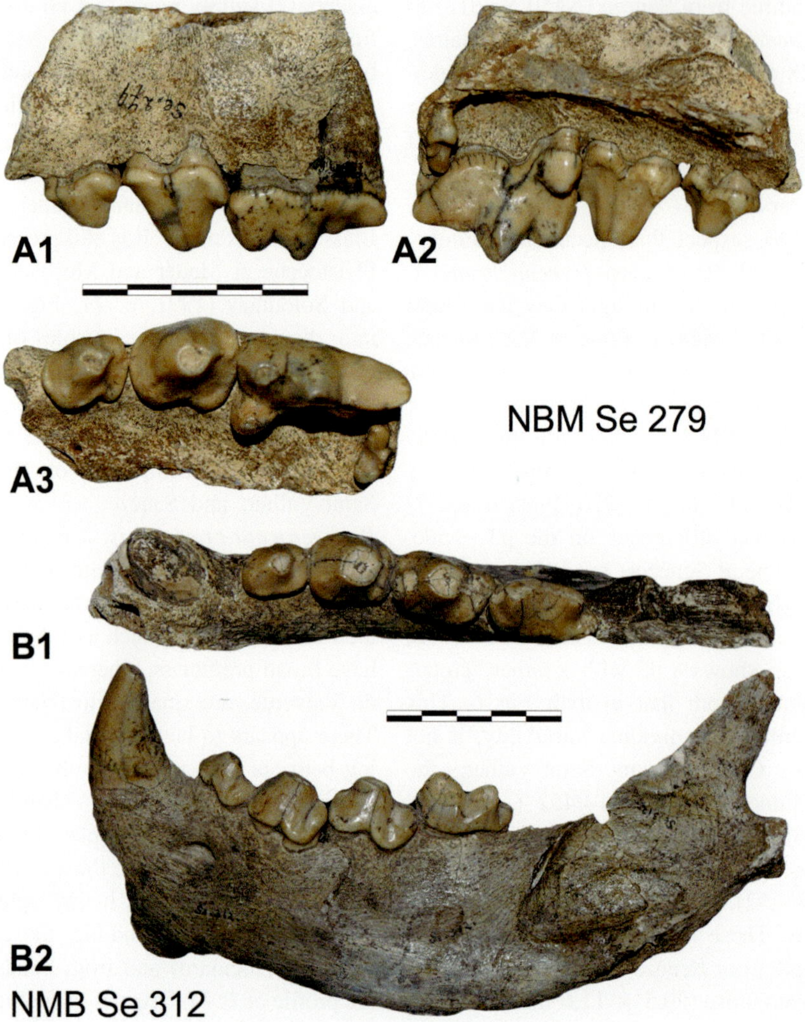

Fig. 9.18 Senèze, *Pachycrocuta perrieri*. A, left fragment of cranium (NMB Se 279) with maxilla, 1, lateral view; 2, lingual view; 3, occlusal view; and B, left mandible (NMB Se 312), 1, occlusal view; 2, external lateral view

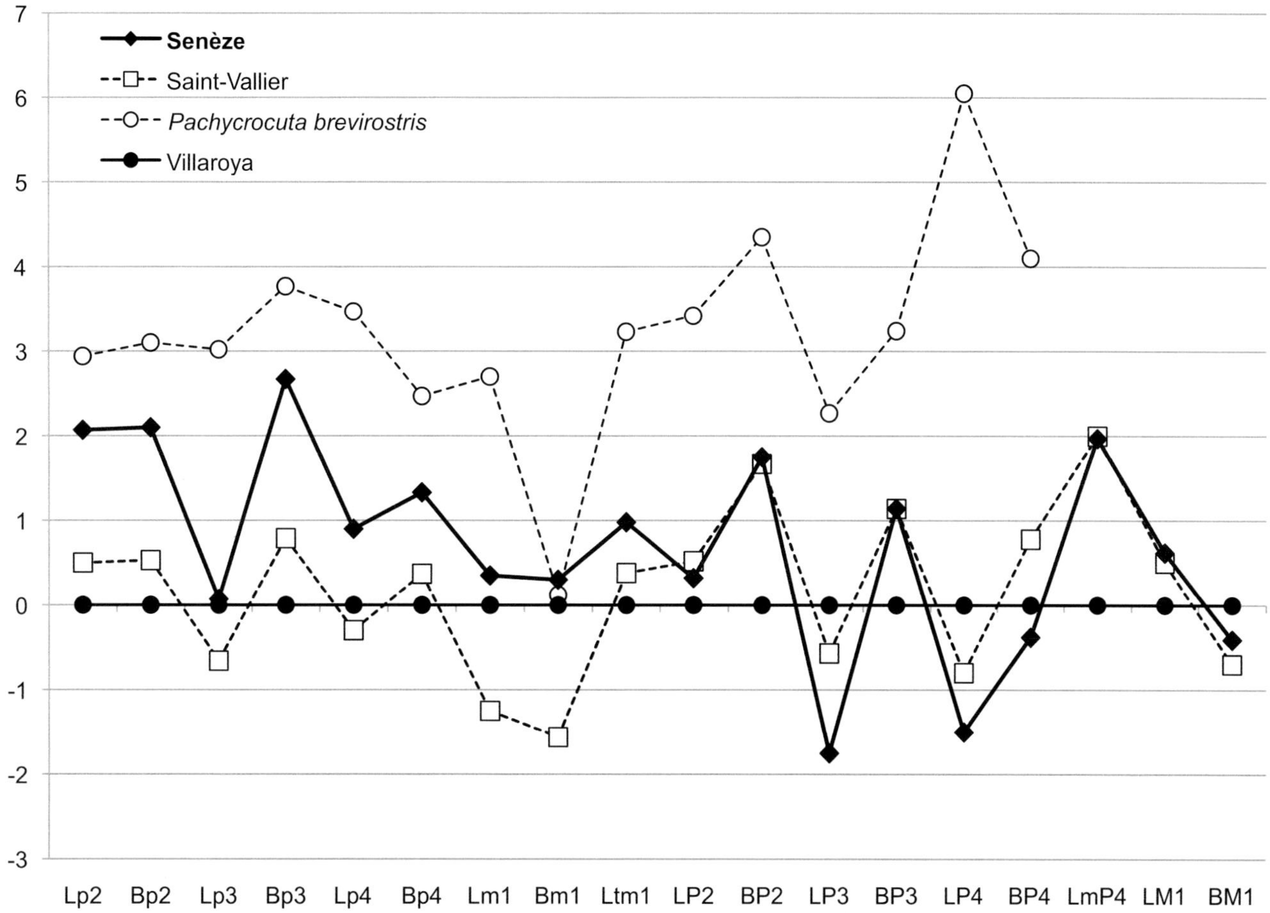

Fig. 9.19 Senèze, *Pachycrocuta perrieri*, ratio diagram of various samples of hyaenids: *Pachycrocuta pyraneica* (Villarroya), *Pachycrocuta perrieri* (Senèze, Saint-Vallier) and *Pachycrocuta brevirostris* (European samples, Howell & Petter 1980), after Werdelin and Solounias (1991: Fig. 11). L = length = MD; B = breadth = VL

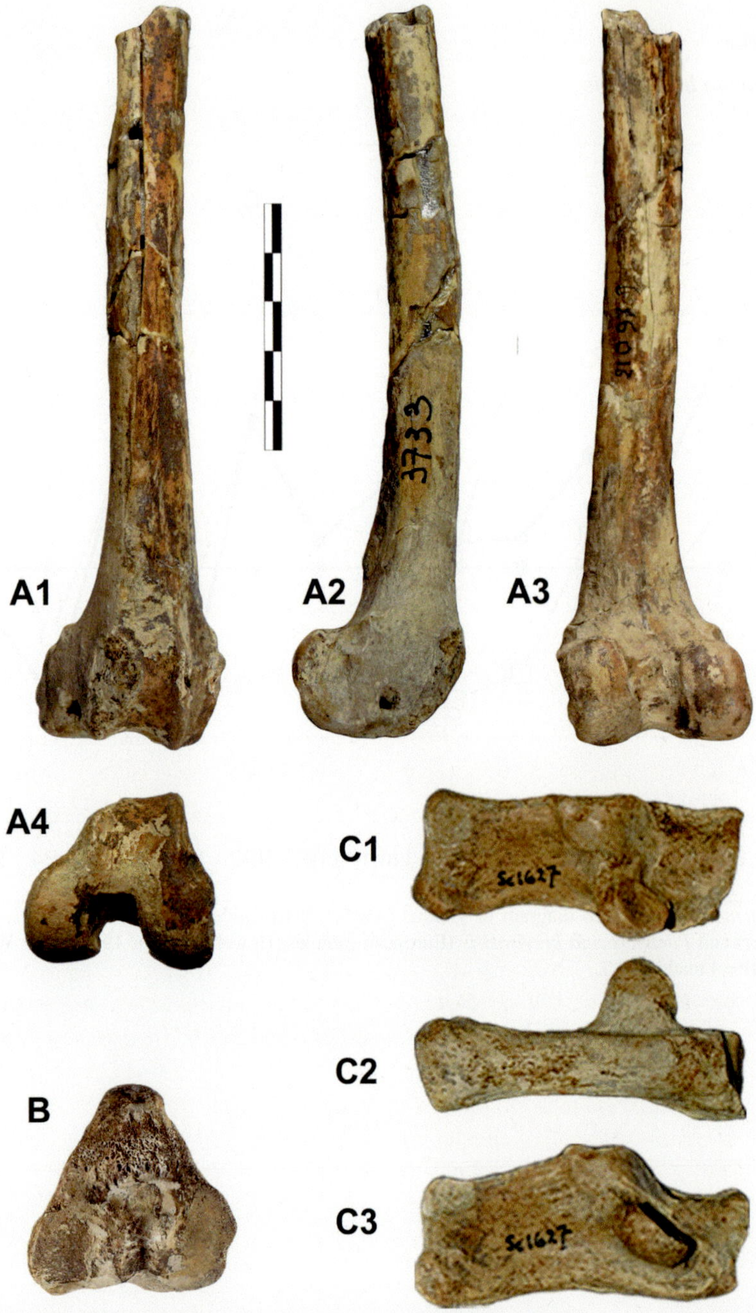

Fig. 9.20 Senèze, *Chasmaporthetes lunensis* and *Pachycrocuta perrieri*, post-cranial elements. *Chasmaporthetes lunensis*: **A**, Left femur, distal part (FSL 210,979): 1, dorsal view; 2, lateral internal view; 3, plantar view; 4, articular view; **B**, left tibia proximal part (FSL 210979). *Pachycrocuta perrieri:* **C**, Left calcaneus (NMB Se 1627), 1, dorsal view; 2, plantar view; 3, lateral internal view

Chasmaporthetes lunensis (Del Campana, 1914): Figs. 9.20 and 9.21, Tables 9.13, 9.14, 9.15 and 9.16

Material:

- FSL 210970: fragment of left upper jaw, cast (original "Collection Clermont-Ferrand"), Senèze 1953, with P3 and P4, illustrated in Viret 1954, Fig. 5, p. 55, corresponding to the cast NMB Se 1882;
- FSL 210979: a half distal part of a left femur and the proximal extremity of a left tibia, same side, of compatible size, could belong to the same individual (Fig. 9.20);
- FSL 210975: a set of bones from the left hind limb of a single individual consisting of the distal half of a tibia, the calcaneus, the astragalus, the cuboid, the navicular, the lateral (third) cuneiform, 3 metatarsals (Mt 2, Mt 3 and Mt 5) and one phalanx 1 (Fig. 9.21).

I have compared the material from Senèze to Schaub's (1941) holotype, a complete mandible from Perrier-Rocca-Neyra (FSL 211221 "Collection Clermont-Ferrand") and *Chasmaporthetes lunensis* of Saint-Vallier The morphology of the p3 (Hyaenid indet.) from the recent Senèze excavations studied above does not correspond.

The tibial plateau and the femur FSL 210979 (Fig. 9.20, Table 9.15) were given a label with the single inscription "Carnivores". They clearly come from the hind limb of a carnivore about the size of a small wolf or a small extant *Hyaena hyaena* of the Middle East. The femur is a distal half with its complete epiphysis. The two condyles, medial and lateral, have the slightly oblique position of hyaenids. In between, the trochlea appears wide and less hollowed than in canids. The diaphysis profile is curved, which eliminates the felids. There is a supracondylar ledge in antero-external position, elongated oval in shape, which covers the bottom quarter of the diaphysis. Contrary to the extant canids, the popliteal surface has almost no supracondylar tuberosity, and a longitudinal line bordered by a slight groove in the middle of the caudal part of the diaphysis rises a few centimeters above the epiphysis. Any one of these features can be regarded as sufficient, because we are not dealing here with extant or recent fossil species but with long extinct species, so that the comparative data are practically non-existent. The articular surfaces of the femur match well with the tibial plateau. The dimensions are perfectly compatible and even if we cannot demonstrate that they are from the same individual (slightly different colors), we can confirm their belonging to the same species.

The tibial plateau is not from a canid because it does not have the indentation of the extensor tendon groove on its antero-external edge and because its general shape is distinct from that of canids. The medium sized felids, such as *Felis issiodorensis*, can also be ruled out. The living carnivore which most closely resembles this material from Senèze is *Hyaena hyaena* (Collections MHNL) which has a nearly similar morphology (Fig. 9.20). There is no doubt that this material from Senèze belongs to a hyaenid. This Senèze hyena is small and has slender limbs similar to those of a small extant *Hyaena hyena*. It can be distinguished from the extant *Crocuta crocuta* because the morphology of its tibial plateau is different. *Pachycrocuta perrieri* has a tibia with a crocutoid morphology and must be ruled out. The only possibility is a hyaenid of the genus *Chasmaporthetes.* Berta's (1981, p. 348) description of the post-cranial material of *Chasmaporthetes ossifragus* provides further support for this conclusion: "The proximal articular surface of the head of the tibia near its median sector is pierced by numerous foramina. In anterior view, the proximal end displays a transversely broad tibial tuberosity for attachment of the patella. The tibial crest extends distally and has a pronounced medial inclination." These characteristics are also evident on the tibia FSL 210979 from Senèze. There are a number of morphological differences between this hyaenid from Senèze and the extant striped hyena. The triangular shape of the tibial plateau is proportionally higher and narrower at its base on *Hyaena hyaena* and the initial section of the tibial tuberosity above the nutrient foramina of the medial fossa is much wider and more substantial on the *Chasmaporthetes* from Senèze.

A set of bones (listed above) from the left hind limb of a single individual (FSL 210975; Fig. 9.21, Table 9.16) bears the manuscript label "*Canidé de la taille d'un petit loup*" (canid of the size of a small wolf). It is clearly the posterior limb of a carnivore but it presents some differences of morphology and proportions with those of canids. The living carnivore which most closely resembles this material from Senèze is *Hyaena hyaena*, which has a similar overall morphology in spite of some noticeable discrepancies:

- calcaneus FSL 210975: this calcaneus shows epiphyseal fusion and is in good general condition except for the sustentaculum tali, which has lost the extremity of its protruding part. Compared to the calcaneus of the extant *Canis lupus*, the calcaneus from Senèze is more massive, higher at the manubrium, more quadrangular and seems more laterally compressed and less elongated. In medial view its shape is relatively high in relation to its length. The articular facet for the astragalus seems more elongated and less hollowed out. The contact between the anterior edge of the sustentaculum tali and this articular facet occurs nearer to the extremity and forms a right angle, while it begins more to the rear. In sum, this calcaneus is not a canid but presents the characteristics of a hyaenid;

- astragalus FSL 210975: complete astragalus in good condition. There is a relatively longer neck and a more prominent head on this specimen than on the astragalus of *Canis lupus*. The articular pulley is less hollow and wider. In its proportions it is more similar to those of hyaenids;
- navicular FSL 210975: complete navicular in good condition. Its distal articular part connecting with the cuneiform is not so clearly widened as in *Canis lupus* or *Vulpes vulpes*, but looks more similar to the navicular of felids or hyaenids;
- cuboid FSL 210975: complete cuboid in good condition. It seems relatively a little more elongated than in *Canis lupus* and the morphology of the lateral external view is quite different with an oblique relief originating in the middle part and keeping a constant width toward the inferior part (Fig. 9.21 I);
- lateral (3rd) cuneiform FSL 210975: the short plantar peduncle is in median position on the Senèze individual as opposed to distal in wolf and fox;
- 3 metatarsal bones (Mt 2, Mt 3, Mt 5) FSL 210975: complete metatarsals, except Mt 3 whose proximal end is slightly damaged;
- 1 phalanx 1 FSL 210975: complete phalanx, broad, rather massive. Compared to the phalanx 1 of the wolf, the diaphysis is relatively more hourglass shaped and the articular pullies are much bigger.

All these elements and their comparison with those of the fossil *Canis lupus* from Châtillon-Saint-Jean (France), a small wolf dated to the end of the Middle Pleistocene, confirm without ambiguity that these remains do not correspond to a canid of similar size but to a hyaenid.

Table 9.15 Comparative dimensions of the Senèze *Chasmaporthetes lunensis* and *Pachycrocuta perrieri* post-cranial skeleton

Hind limb variables			1	2	3	4	5	6	7
			L	Prox TD	Prox APD	Med TD	Med APD	Dist TD	Dist APD
Femur	*Chasmaporthetes lunensis*	Senèze FSL 210979	–	–	–	15.7	14.5	38.9	35.3
	Crocuta crocuta	Extant (n = 7) Ballesio (1979)	244.3	–	–	21.0	–	50.0	–
Tibia	*Chasmaporthetes lunensis*	Senèze FSL 210979	–	39.9	41.6	–	–	–	–
		Senèze FSL 210975	–	–	–	15.0	15.2	–	20.6*
		Perrier Berta (1981)	247.0	–	–	17.0	–	30.5	–
	Chasmaporthetes ossifragus	Inglis 1A (Florida) Berta (1981)	273.0	53.4	65.7	23.5	–	40.7	38.3
	Chasmaporthetes borissiaki	Khomenko (1931)	220.5	40.5	39.5	15.5	–	27.5	28.5
	Pachycrocuta perrieri	Senèze NMB Se 1496	–	51.2	61.1*	19.6*	23.3*	–	–
		Senèze NMB Se 1697	–	–	–	–	–	39.6	26.6
	Crocuta spelaea	Le Blanc (France)	196.0	58.9	62.0	21.9	26.5	42.9	30.4
	Crocuta crocuta	Extant (n = 10) Berta (1981)	184.0–206.0 198.1	39.9–50.2 40.5	44.1–57.2 51.4	13.8–19.8 17.5	17.6–23.4 20.9	28.6–37.1 33.7	19.3–25.5 23.7
		Extant (n = 2/6) Cardoso (1993)	201.2	–	50.5	17.1	–	37.0	–
	Hyaena hyaena	Extant MNHL 50 000 557	198.2	40.9	45.9	14.5	18.2	29.6	20.3
		Extant MNHL 50 000 561	178.3	39.9	42.6	14.0	16.5	28.7	18.5

(continued)

Table 9.15 (continued)

Hind limb variables			1	2	3	4	5	6	7
			L	Prox TD	Prox APD	Med TD	Med APD	Dist TD	Dist APD
Calcaneum		Calcaneal variables	L	max TD	max DPD	min med TD	med DPD	Tuber calcanei TD	Tuber calcanei DPD
	Chasmaporthetes lunensis	FSL 210975	45.8	22.4	21.9	10.1	16.6	13.7	18.8
	Pachycrocuta perrieri	NMB Se 1627	63.8	30.7	28.9	12.2	–	18.6	25.0
	Hyaena hyaena	Extant MNHL 50 000 557	48.0	20.7	–	14.9	18.7	–	–
		Extant MNHL 50 000 561	48.6	22.4	22.4	14.6	18.0	–	–

Table 9.16 Dimensions of the Senèze *Chasmaporthetes lunensis* postcranial skeleton (hind limb of a single individual FSL 210975)

Chasmaporthetes lunensis	1	2	3	4	5	6	7
Tibia	L	Prox TD	Prox APD	Med TD	Med APD	Dist TD	Dist APD
	–	–	–	15.0	15.2	–	20.6*
Astragalus	L	Max TD	Max DPD	Trochlea TD	Neck TD	Caput TD	Caput APD
	29.2	24.0	15.7	18.7	13.0	16.6	–
Calcaneum	L	Max TD	Max DPD	Med TD	Med DPD	Tuber calcanei TD	Tuber calcanei DPD
	45.8	22.4	21.9	10.1	16.6	13.7	18.8
Navicular	L	Max TD	Max DPD				
	12.0	14.8	18.5				
Cuboid	L	Max TD	Max DPD				
	19.8	15.5	14.1				
Cuneiform 3	L	Max TD	Max DPD				
	13.6	10.2	18.5*				
	L	Prox TD	Prox APD	Med TD	Med APD	Dist TD	Dist APD
Mt 2	73.5	9.1	–	8.4	7.5	10.4	10.1
Mt 3	82.3	11.0	–	8.6	7.3	11.6	11.0
Mt 5	71.0	7.6	12.4	5.8	6.9	8.4	9.3
Phalanx 1 post	24.5	9.0	8.6	5.5	6.0	8.6	6.3

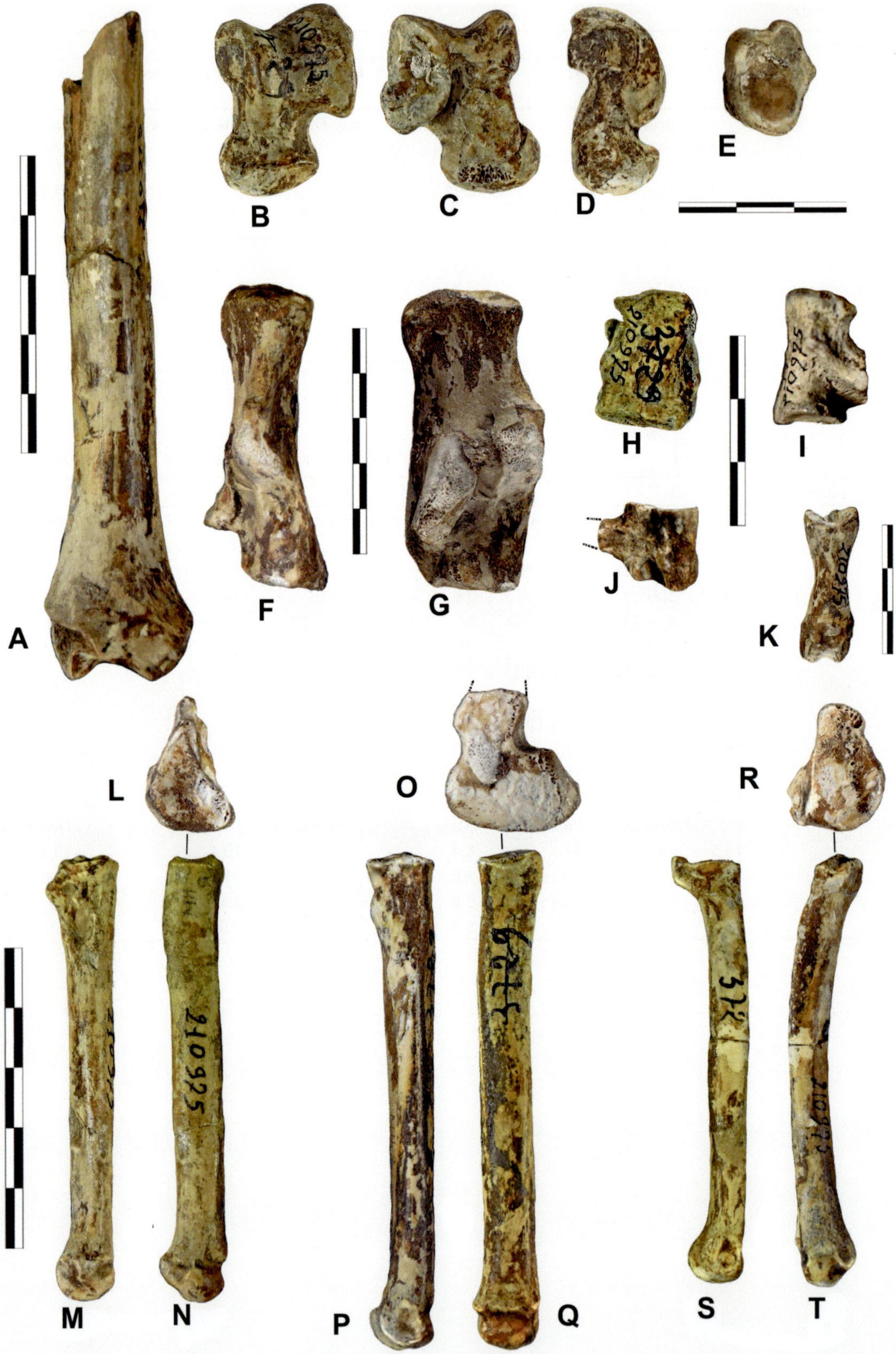

Fig. 9.21 Senèze, *Chasmaporthetes lunensis*, left hind limb of a single individual (FSL 210975): **A**, distal half of a tibia; **B–D**, astragalus, **B**, dorsal view; **C**, plantar view; **D**, internal lateral view; **E**, navicular, proximal view; **F–G**, calcaneus, **F**, dorsal view; **G**, internal lateral view; **H–I**, cuboid, **H**, dorsal view; **I**, internal lateral view; **J**, damaged lateral (3rd) cuneiform, external lateral view; **K**, 1st posterior phalanx, dorsal view; **L–N**, Mt 2: **L**, proximal view; **M**, internal lateral view; **N**, dorsal view; **O–Q**, Mt 3: **O**, proximal view; **P**, internal lateral view; **Q**, dorsal view; **R–T**, Mt 5: **R**, proximal view; **S**, internal lateral view; **T**, dorsal view. Scale bar equals 3 or 5 cm

The next problem is to determine the species of this material. The post-cranial skeleton of *Chasmaporthetes* remains practically unknown, except for that of *Chasmaporthetes ossifragus* from North America (Berta 1981) and *Chasmaporthetes borissiaki* (Khomenko 1931). The latter are clearly larger than the small *Chasmaporthetes* from Senèze and can be ruled out.

Recently Tseng et al. (2013) have described a new cursorial hyena *Chasmaporthetes gangsriensis* from the Zanda Basin (Tibet), with long, slender limbs and smaller size than indicated by other Plio-Pleistocene Eurasian records of the genus. This hyaenid was probably the ancestor of *Chasmaporthetes lunensis*, which became widespread in Europe and was undoubtedly at the origin of the American forms. But *Chasmaporthetes gangsriensis* corresponds to a clearly more ancient form than the population from Senèze.

The posterior limb FSL 210975 from Senèze also indicates a small animal with slender limbs, close to the extant *Hyaena hyaena* in spite of some morphological discrepancies, whereas *Crocuta crocuta* appears to be very different. It seems possible to attribute this post-cranial material to a *Chasmaporthetes* while knowing that in the case of the presence of a third species of Hyenidae—a possible case for the p3 of the recent Faure-Guérin-Delson excavations—we also have the presence of an individual of small size.

The right tibia of *Chasmaporthetes lunensis* from the Spanish Pliocene site of Las Higueruelas (Arribas & Antón 1997, Fig. 2, p. 115) has lost its proximal epihysis but the diaphysis and the distal part also show that *Chasmaporthetes lunensis* had elongated slender limbs. This species is well adapted to running, and we know that it is often associated with the cheetah *Acinonyx pardinensis* as at Perrier-Etouaires, Pardines, Saint-Vallier and La Puebla de Valverde. Pollen analyses at Saint-Vallier and Senèze indicate open landscapes with some wooded areas (A. Argant, 2004, J. Argant 2004, 2024).

The genus *Chasmaporthetes* clearly differs from other hyaenids. Probably we must look for its origins in Asia where the most ancient fossils have been recorded. The earliest records of *Chasmaporthetes* from Africa and Europe are slightly younger (5.2 Ma for *Chasmaporthetes australis* from Langbaanweg; MN13 for *Chasmaporthetes 'bonisi'* from Greece). The earliest geologic occurrence is that of *C. exitelus* from the late Miocene Qingyang locality in Gansu Province, China. It is also in Asia that we find the greatest number of species of this genus. In Europe, the hunting hyena (or cursorial hyena) occurs at many Villafranchian sites (see Werdelin & Solounias 1991, p. 30). It appears at Layna at the end of the Ruscinian (MN 15) and is still present in the late Villafranchian (MNQ 18). The first complete skull known came from Saint-Vallier (A. Argant 2004). Since then, a second skull has been discovered at La Puebla de Valverde (Antón et al. 2006). The Senèze fossils allow us to describe for the first time the morphology and the size of the posterior limb of a single individual. It is even possible that the proximal end of the left tibia (FSL 210979), studied separately, corresponds to the distal end of the tibia of the FSL 210975 set. They are both left, their proportions and size are similar, and we are missing only a small piece of the diaphysis to be sure.

Given the small number of elements compared, it would be unrealistic at this stage to draw biochronological conclusions about the species *Chasmaporthetes lunensis.*

Family Felidae

The felids are well represented at Senèze by the same species as those at Saint-Vallier except for *Lynx issiodorensis* and *Puma pardoides.* This is a classic grouping of large Villafranchian felids found in open spaces and close to streams or lakes where herds of herbivores came to drink.

Acinonyx pardinensis (Croizet & Jobert, 1828)

Some remains of *Acinonyx pardinensis* found during earlier excavations exist, but they are rare. The data from the recent excavations confirm its presence along with *Dicerorhinus etruscus* and a hyaenid:

- fragment of right demi-mandible with p3 and p4 (FSL SEN 03-0058) and right c (FSL SEN 03-0210), probably of the same individual (discussed above);
- fragment of a right maxilla with P3-4 and one left C (FSL 210986); this is the type of *Brachyprosopus vireti* Schaub, 1942 [a synonym of *Acinonyx pardinensis* (Croizet & Jobert, 1828)], which corresponds to the casts NMB Se 1782 for the maxilla and NMB Se 1763 for the left C (Fig. 9.22);

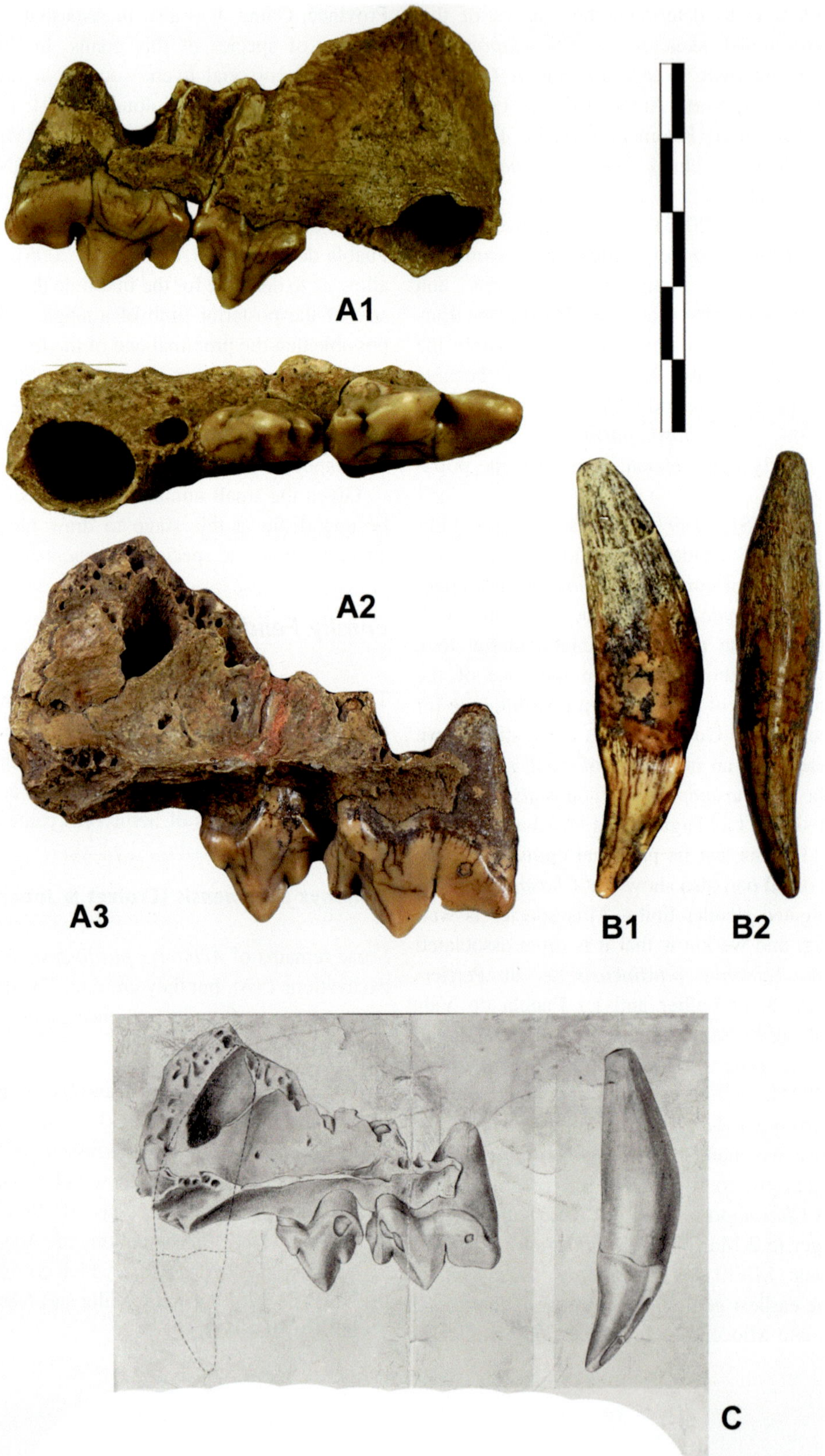

Fig. 9.22 Senèze, *Acinonyx pardinensis*. **A**, Right maxilla (FSL 210986), **1**, external view; **2**, occlusal view; **3**, lingual view. **B**, Left upper canine (FSL 210986), **1,** lateral lingual view; **2**, posterior view; **C**, Drawing of the type specimen of *Brachyprosopus vireti* (synonym of *Acinonyx pardinensis*), extracted from Schaub (1942, pl. V)

Table 9.17 Comparative dimensions of the Senèze and Saint-Vallier *Acinonyx pardinensis* cranium and mandible

Acinonyx pardinensis			Senèze	Saint-Vallier
			FSL	MHNL, Collection Viret
			210986	20 161 826
Cranium	Length before alveolus C - behind P4		68.2	72.97
	Length before alveolus P2-behind P4		47.6	52.38
	Empty alveolus of P2	MD	4.7	4.28
		VL	3.2	4.09
	Empty alveolus of C	MD	13.1	15.42
		VL	12.7	11.08
		HC	-	28.15*
Mandible			Senèze	Saint-Vallier
			NMB	NMB
			Se 1497	SV 777 (cast)
	Empty alveolus of p3	MD	12.5	16.7
		VL	8.5	9.4
	Empty alveolus of p4	MD	17.4	21.1
		VL	8.8	9.7
	DT mandible between p3 and p4		13.8	14.7
	Alveolus of c	MD	-	14.6
		VL	-	10.4
	Length of diastema c - p3		-	7.3

- NMB Se 1499: one damaged right astragalus, labeled "Senèze, juin 1909", "*Cynailurus* astragalus" another synonym of *Acinonyx pardinensis* (Fig. 9.23);
- NMB Se 1497: fragment of right demi-mandible, Senèze 1909, empty alveoli.
- Morphometric data: Tables 9.17, 9.18 and 9.19.
- demi mandible: the fragment of mandible from Senèze (NMB Se 1497) has no teeth but has their alveoli, so the length of the diastema between the posterior part of the canine and the p3, and the TD of the mandible between the p3 and p4 can be measured. Comparison with the cast NMB StV 777 (original MHNL QSV 113) shows the same concave curve of the mandible in both cases and the same type of breakage of the teeth at the level of the mandible, the roots remaining in place. We can see that these roots are fully developed, confirming that they belong in both cases to adults. Except for the difference in the length of the diastema, shorter at Senèze than at Saint-Vallier (Table 9.17), these two specimens are identical.

Table 9.18 Comparative dimensions of the Senèze *Acinonyx pardinensis* upper and lower teeth

Acinonyx pardinensis			Senèze			Saint-Vallier					Montopoli	Perrier-Etouaires	Liventsovka Russia	Ceyssaguet	Untermassfeld	
			FSL	MNB	FSL	Faure-Guérin	Faure-Guérin	MHNL Collection Viret			Lower Valdarno				IQW 1980/16.350	IQW 1980/15.503
			210986	Se 1763	SEN 03–0058	SV 96 449; 98 624; 98 624	isolated tooth SV99 477	161822	161825	161826			RSU-669	MNPE CEY 1918		
						A. Argant (2004)		Viret (1954)			Ficcarelli (1984)	Schaub (1949)	Sotnikova et al. (2002)	Tsoukala (2004)	Hemmer (2001)	
Upper teeth	C	MD	13.6	13.3	–	–	–	–	–	–	14.2	–	–	–	15.4	–
		VL	10.9	10.4	–	–	–	–	–	–	10.1	–	–	–	12.1	–
		HC	25.4	27.6	–	–	–	–	–	–	–	–	–	–	> 27.0	–
		HT	59.3	59.7	–	–	–	–	–	–	–	–	–	–	–	–
		PC	39.5	–	–	–	–	–	–	–	–	–	–	–	–	–
	P3	MD	16.6	–	–	15.4	–	16.7	16.0	17.9	17.0	–	–	–	18.1	–
		VL	8.3	–	–	8.7	–	9.5	8.0	9.5	8.2	–	–	–	9.8	–
		HC	11.4	–	–	–	–	–	–	–	–	–	–	–	> 13.0	–
		HT	30.3	–	–	–	–	–	–	–	–	–	–	–	–	–
		MD paracone	8.2	–	–	7.1	–	–	–	–	–	–	–	–	8.0	
	P4	MD	25.3	–	–	27.1	–	28.8	26.1	28.5	26.7	–	30.1	–	29.7	–
		VL max	11.5	–	–	12.4	–	12.4	11.6	13.4	12.1	–	12.6	–	12.5	–
		HC	14.5	–	–	14.4	–	15.4	14.1	14.0	–	–	–	–	> 15.0	–
		HT	29.9	–	–	–	–	–	–	–	–	–	–	–	–	–
		MD protocone	6.7	–	–	–	–	–	–	–	–	–	–	–	–	–
		MD paracone	10.4	–	–	10.5	–	11.2	10.3	10.9	–	–	12.6	–	11.7	–
		MD metacone	10.3	–	–	11.8	–	10.4	–	11.9	–	–	12.3	–	12.6	–
		DT metacone	9.3	–	–	–	–	–	–	–	–	–	10.2	–	10.0	–
Lower teeth	p3	MD	–	–	16.3*	13.7	15.5	14.1	–	–	14.2	13.0	–	15.1	–	16.0
		VL	–	–	–	7.6	8.1	8.4	–	–	7.6	–	–	9.0	–	9.7
		MD protoconid	–	–	–	7.7	8.5	8.2	–	–	–	–	–	–	–	8.0
		HC	–	–	–	10.6	–	10.7	–	–	–	–	–	12.8	–	> 12.0
	p4	MD	–	–	20.6*	17.4	–	19.2	19.2	–	17.5	17.0	–	19.6	–	19.8
		VL	–	–	–	8.7	–	9.5	–	–	8.5	–	–	9.9	–	9.7
		HC	–	–	–	12.6	–	13.7	–	–	–	13.0	–	15.7	–	> 13.0
		MD protoconid	–	–	–	8.5	–	9.5	–	–	–	–	–	–	–	9.0

– astragalus (NMB Se 1499): unfortunately, it is a specimen which is too damaged to allow all the necessary observations. The trochlear groove is deep and the articular pulley (damaged) seems narrow (Fig. 9.23). The morphology of this astragalus closely matches that shown in different views of the astragalus of *Acinonyx pardinensis* from the middle Villafranchian at Varshets (Bulgaria) (Spassov 2011, Fig. 1, p. 248), especially the view D.3 showing the plantar articulation facets. The flat internal articular facet, laterally bordering the facet connecting with the sustentaculum tali is characteristic of *Acinonyx pardinensis*. The specimen from Senèze shows a wide and shallow fossa covering the full width of the lower part of the neck which borders the articular head. This specimen is undoubtedly from *Acinonyx pardinensis*, a species which disappeared during the Middle Pleistocene.

Table 9.19 Comparative dimensions of the Senèze *Acinonyx pardinensis* astragalus

Acinonyx pardinensis Astragalar variables	*Acinonyx pardinensis*			*A. p. pleistocaenicus*	extant *Acinonyx jubatus*		
	Senèze	Olivola	Puebla de Valverde	Untermassfeld			
PFEMA 642	NMB Se 1499	IGF 2613	Kurtén and Crusafont Pairó (1977) Tab. 22 p.32	IQW 1980/16354 (Mei.15865) Hemmer, 2001	Kurtén and Crusafont Pairó (1977)	NMB 5115	NMB 3770
APD	41.5	41.2	41.0	53.0	37.0	32.2	28.5
TD	–	33.5	39.1	–	–	29.9	26.9
DPD	–	–	–	–	–	19.1*	16.7
DPD min. central	14.0	–	–	–	–	17.1	14.6
TD max. trochlea	–	–	24.5	28.0	19.7	18.4	16.6
TD neck	16.8	–	20.1	–	15.7	14.3	12.9
TD max. caput	–	–	25.3	–	19.6	18.0	15.7
Max. med. length	–	–	–	47.0	–	37.4	34.6

Fig. 9.23 Senèze, *Acinonyx pardinensis*, damaged right astragalus (NMB Se 1499), plantar view

Acinonyx pardinensis is not the same species as the extant *Acinonyx jubatus*, and there is no direct phyletic link between them, as demonstrated by numerous differences. *A. pardinensis* appears without doubt at the end of the Ruscinian at Layna (Spain) in MN 15 (cf. *Acinonyx pardinensis*, Moralès 1985), and is present during the whole of the Villafranchian at Villarroya, Perrier-Etouaires, La Puebla de Valverde, Saint-Vallier, Senèze, Olivola, and Dmanisi (Georgia) for example. It is found in the late Early Pleistocene at Untermassfeld and Vallonnet and until the Middle Pleistocene at Mosbach and Hundsheim. It is now known to have occurred in more than sixteen localities in France, Spain, Italy, Austria, Germany, Bulgaria, and Georgia. The history of cheetahs is beginning to become clear (Hemmer et al. 2008, 2011; Spassov 2011) but some gray areas still exist. The rarity of remains at each site and the lack of studies of the post-cranial skeleton explain the reservations. *Acinonyx pardinensis*, a species in the broad sense, had a very wide distribution. It is still unclear whether it is of American (*Miracinonyx* Adams, 1979), African or Eurasian origin (Spassov 2011).

The Saint-Vallier deposit (A. Argant 2004; Viret 1954) has provided sufficient material to act as a reference for characterizing *A. pardinensis pardinensis* of the middle Villafranchian (MNQ 17). On the other hand the material from Senèze is very limited, and comparisons are only possible for the two upper premolars of FSL 210986. The dimensions of the Senèze teeth are almost the same as those from Saint-Vallier, although they fall at the lower end of the range of values (Table 9.18). Their morphology is very similar, especially for the P4: paracone elevated, pointed and not very elongated; metacone a horizontal blade, relatively elongated; parastyle with a small secondary prominence in the form of a simple bulge at the base of the crown, characteristic of *Acinonyx pardinensis* (Fig. 9.22). The Senèze specimen corresponds to the nominal subspecies *A. pardinensis pardinensis* that inhabited Eurasia during the beginning of the Pleistocene in the late early and middle Villafranchian. This indicates a degree of chronological proximity for the Saint-Vallier and Senèze deposits. On the other hand the morphology of the P4 of the *A. pardinensis pardinensis* from Varshets appears more ancient. This premolar is wider in the posterior part and less triangular. It should be emphasized however that these observations have been made on only a small number of specimens. The large size of the intermediate form, the subspecies *Acinonyx pardinensis pleistocaenicus* (Zdansky, 1925) from the Epivillafranchian (Untermassfeld and Le Vallonnet), which probably originated in Asia (Hemmer et al. 2008), is incompatible with the specimen from Senèze (Table 9.20).

Dinofelis Zdansky, 1924: *Dinofelis* sp.

Material:

– NMB Se 320: left complete calcaneus.

The specimen NMB Se 320 (Fig. 9.24, Table 9.20) has presented me with a problem. It is a complete left calcaneus of a carnivore with fused epiphyses. The original label, handwritten by Stehlin, bears the inscription "*Felis* sp. am nächsten bei *Felis uncia*, [*Felis* sp. near *Felis uncia*] Senèze, Feb. 1911". A catalog check revealed an error in numbering on the piece (now NMB Se 320, it was formerly Se 319). As for all the old Senèze material, the exact place of origin of this specimen remains unknown.

In plantar view, the calcaneus NMB Se 320 appears elongated, relatively narrow and perfectly rectilinear, with parallel sides. At the proximal extremity the tuber calcanei is not very wide, only slightly concave, and in dorsal view, it terminates at an oblique angle to the central axis. At the distal extremity the very prominent sustentaculum tali has a shallow groove on its upper surface parallel to the central axis. The height of the manubrium decreases progressively from front to rear. In lateral view, the external face is in the form of a concavity, bounded by an oblique carinate rim, a little below the distal articular part, which connects to the astragalus. The distal extremity has a lateral external tuber, clear but moderately developed, without any marked groove (Fig. 9.24B). This distal part lacks any navicular facet.

There is no apparent resemblance between this unique well-preserved adult specimen and its equivalents from *Acinonyx pardinensis*, *Megantereon cultridens* and *Homotherium crenatidens* found at the same site:

- compared to the calcaneus of *Acinonyx pardinensis* from Saint-Vallier and that of the extant *Acinonyx jubatus*, the calcaneus from Senèze has neither their elongated proportions, nor their very characteristic morphology;
- comparison with the calcaneus from the complete skeleton of *Megantereon cultridens* from Senèze (NMB Se 311), an animal of much larger size, shows that its morphology is incompatible with NMB Se 320. *Megantereon cultridens* from Pardines (Puy-de-Dôme) NMB LP 19, a small individual, has a calcaneus with almost the same dimensions as NMB Se 320, but its morphology is clearly different (Fig. 9.25, B1, B2, B3);
- comparison with the calcaneus of *Homotherium crenatidens* shows that it is incompatible with NMB Se 320 both because of its much larger size and its morphology.

There could have been at Senèze a fourth species of felid, one obviously too large to be *Lynx issiodorensis*. In the Senèze context, one possible candidate is *Puma pardoides*, which occurs at Saint-Vallier for example. Remains of *Puma pardoides* are rare and post-cranial remains little known, apart from those at Puebla de Valverde (Teruel, Espagne), Varshets (Bulgaria) (Spassov 1997) and Saint-Vallier (A. Argant 2004), which don't include a calcaneus. The only calcaneus I know is that of the *Puma pardoides* from Untermassfeld (Hemmer 2001, Fig. 13, p. 732). While the size of NMB Se 320 may be compatible, there is no satisfactory morphological resemblance. The Senèze animal is not a puma. Another possibility is *Panthera gombaszogensis*[1] *toscana*, which appeared at the end of the middle Villafranchian and was smaller than *Panthera gombaszogensis gombaszogensis* of the Middle Pleistocene. *Panthera gombaszogensis* can be ruled out because of the morphology of the small articular facet connecting with the sustentaculum tali, and the morphology of the plantar part of the neck, in spite of some similarities, particularly in relation to the astragalar groove. The extant *Panthera uncia* (NMB 3917) is the closest in terms of its dimensions, but if the calcaneus of the extant snow leopard has the massiveness and the rectilinear aspect of NMB Se 320, it does not have the shape of its sustentaculum tali nor of its tuber calcanei. All these felids must be ruled out.

[1]This species name is sometimes rendered "*gombaszoegensis*", but Wagner (2011, p. 245) wrote: "The species-group name [originally] '*gombaszögensis*' is based on the Hungarian name of the locality Gombaszög, that is itself composed from two Hungarian words: 'gomba' ('mushroom' in English) and 'szög' ('corner' in English). Therefore the species-group name "*gombaszögensis*" is not to be treated as based upon a German word (confirmed by L. Kordos in verb, 2005). Based on the above mentioned facts, the spelling '*gombaszogensis*' (with deleted umlaut sign only) is the correct subsequent spelling according to the ICZN."

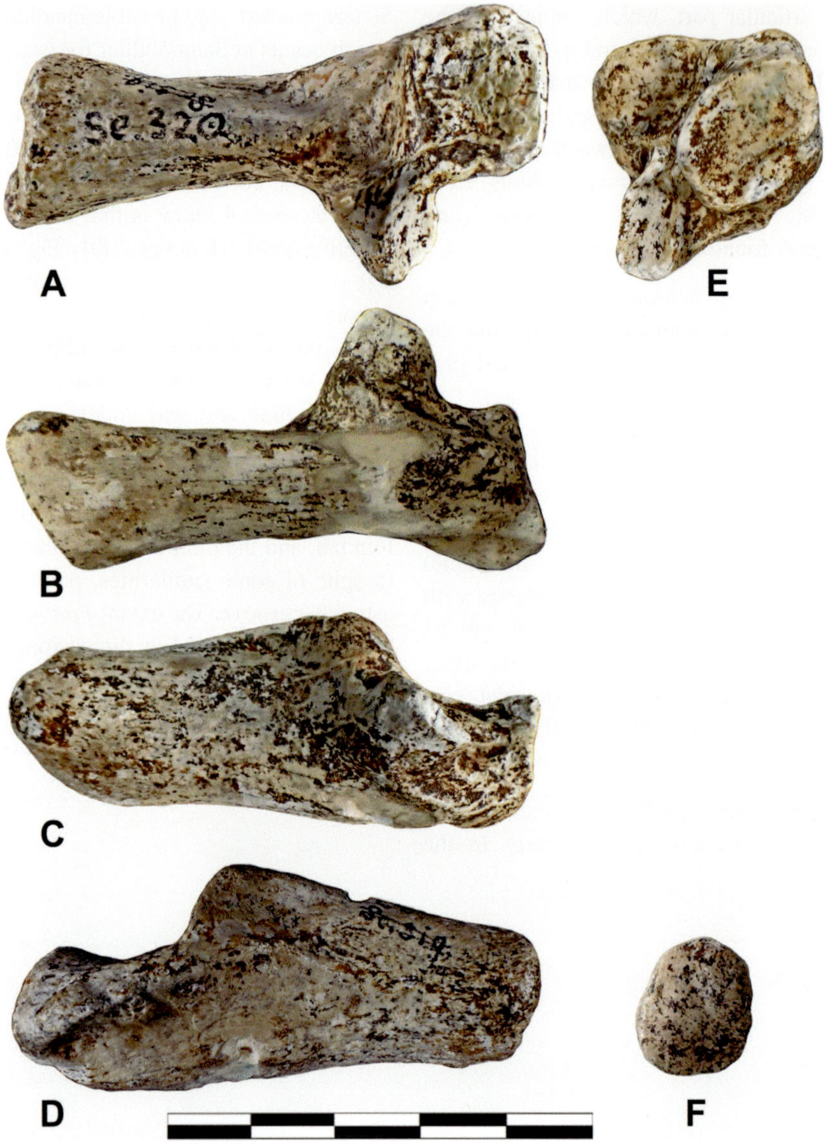

Fig. 9.24 Senèze, *Dinofelis* sp. Left calcaneum (NMB Se 320), **A**, dorsal view; **B**, plantar view; **C**, internal lateral view; **D**, external lateral view; **E**, distal articular view; **F**, tuber calcanei. Scale bar equals 5 cm

Table 9.20 Comparative dimensions of the Senèze *Dinofelis sp.* (NMB Se 320) calcaneus

Dinofelis sp. Calcaneal variables				1 Length	2 Dist TD max	3 Dist DPD max	4 Tuber calcanei TD	5 Tuber calcanei DPD	6 L. manubrium	7 Med. TD	8 Med DPD
Dinofelis sp.		NMB Se 320		64.2	29.5	26.8	19.0	18.5	48.7	13.0	27.3
Hyaenids	Fossil	*Chasmaporthetes borissiaki*	Khomenko (1931)	64.5	22.5	–	–	–	–	–	–
		Chasmaporthetes ossifragus	Hodnett et al. (2011)	76.4*	–	29.9*	21.6*	–	–	13.3*	–
		Pachycrocuta perrieri	NMB Se 1627	63.8	30.7	28.9	18.6	25.0	48.2	12.2	–
	Extant	*Hyaena hyaena*	MHNL 50 000 557	48.0	20.7	–	14.9	18.7	–	–	–
			MHNL 50 000 561	48.6	22.4	22.4	14.6	18.0	–	–	–
Felids	Fossil	*Puma pardoides*	Untermassfeld Hemmer (2001)	69.0	–	–	–	–	–	–	–
		Acinonyx pardinensis	Saint-Vallier NMB SV 779	92.1	34.2	36.8	22.0	21.1	67.4	13.1	32.5
		Megantereon cultridens	Senèze NMB Se 311	85.8	40.0	34.5	25.4	26.6	67.0	16.0	34.5
		Megantereon cultridens	Pardines NMB LP 19	69.3	32.9	29.1	19.6	20.4	51.1	11.5	27.1
		Homotherium crenatidens	Senèze Ballesio (1963)	85.0	38.0	45.0	–	–	–	–	–
		Panthera gombaszogensis	Argant and Argant (2011) CHA.1–00-F.8–144	96.3	44.6	41.5	–	28.8	73.6	20.6	39.4
	Extant	*Acinonyx jubatus*	NMB	75.1	25.3	25.0	16.5	20.2	59.9	8.9	22.9
		Panthera pardus	NMB	73.7	36.5	31.7	19.6	22.2	57.7	12.4	30.3
		Panthera uncia	NMB	59.8	28.0	25.6	16.6	18.6	42.5	10.9	25.3
		Panthera onca	NMB 2044	68.1	32.9	27.9	18.9	21.1	50.0	11.2	–
		Puma concolor	MNHL 50 000 431	66.6	29.9	24.7	16.7	17.4	–	–	–

There is certainly another potential felid, *Dinofelis,* which had been rejected at first because it is not present among the known material from Senèze and mostly because it is generally much bigger. The revision of the genus *Dinofelis* (Werdelin & Lewis, 2001) provides most of the available data. The calcanei of *Dinofelis* are still rare, and they are all larger than NMB Se 320 (L = 64.2 mm), sometimes very much so (L: n = 7; mean = 79.2; 71.0–84.9; sd 5.04, from Werdelin & Lewis 2001). However the morphology of the calcaneus of *Dinofelis* does indeed correspond to that of NMB Se 320: articular facet connected to the astragalus, longitudinal groove crossing over this articular part, sustentaculum tali, tuber calcanei, lateral external distal tuber not strongly developed and without any clear groove. The width, the massiveness of the manubrium and the large size of *Dinofelis piveteaui* from the Okote Member of the Koobi Fora Formation (Werdelin & Lewis 2001: Fig. 20) cannot be observed at Senèze. But NMB Se 320 and KNM-LT 23696 from the Apak Member of the Nachukui Formation at Lothagam (Kenya) have the same proportions and same elongated aspect, the same thin manubrium and after the pictures (Werdelin & Lewis 2001: Fig. 22) they have the same morphology. The machairodont from Lothagam is determined as *Dinofelis* sp. and the Apak Member is dated to the interval 5.0–4.2 Ma (Werdelin & Lewis 2001).

The morphology of NMB Se 320 is close to that of *Megantereon cultridens* from Hajnáčka (Slovakia) SNM-MNH Z 26377 (Fejfar & Sabol 2004), which is similar in size to Se 320 but small in relation to the complete specimen NMB Se 311. However, three differences can be noticed: the Hajnáčka specimen has a well developed external lateral tuber, characteristic of *Megantereon cultridens* with a marked sliding passage of a tendon; on the other hand the Senèze calcaneum has a wide oblique furrow all along the external part, which doesn't appear either on the Hajnáčka specimen or on the other calcanei of *Megantereon cultridens*; finally, an articular facet connecting with the navicular on the distal part of the calcaneus is seen on the bone from Hajnáčka (Sabol, 2017, personal communication). The latter is quite similar to the calcaneus of *Megantereon cultridens* from Pardines NMB LP 19, also very small compared to the conspecific complete specimen Se 311. The specimen from Pardines is definitely a small *Megantereon cultridens,* as confirmed by the skull and an abundant post-cranial material; could this imply evolution or a different species?

Hajnáčka and Pardines are much older than Senèze, which could explain the differences in the size of *Megantereon cultridens*, which is much smaller at these two sites than at Senèze. The calcaneus NMB Se 320 must also belong to a machairodontine, but instead to a *Dinofelis.*

The calcaneus NMB Va 552 from Val d'Arno, labeled "*Megantereon cultridens*", has the same morphology as NMB Se 320, with an external tuber present but not well developed, much less so than on the second calcaneus from Val d'Arno (NMB Va 1200) also labeled "*Megantereon cultridens*" (Fig. 9.25). NMB Se 320 is smaller, while the two calcanei from Val d'Arno (NMB Va 552 and Va 1200) have the larger size of the calcanei of the complete skeleton from Senèze (NMB Se 311).

In external lateral view the calcaneum of Senèze (Se 320, Fig. 9.24D) shows a ridge (surrounded by grooves) where the accessory flexor muscle originates, as on the calcaneum of *Pseudaelurus,* a 20 Ma cat (Turner & Antón 1997, Fig. 4. 32, p. 139). This feature infers a greater grasping ability in the foot of this animal better adapted to climbing than to running. It can be assumed that the *Dinofelis* of Senèze was not particularly adapted to running and could climb trees.

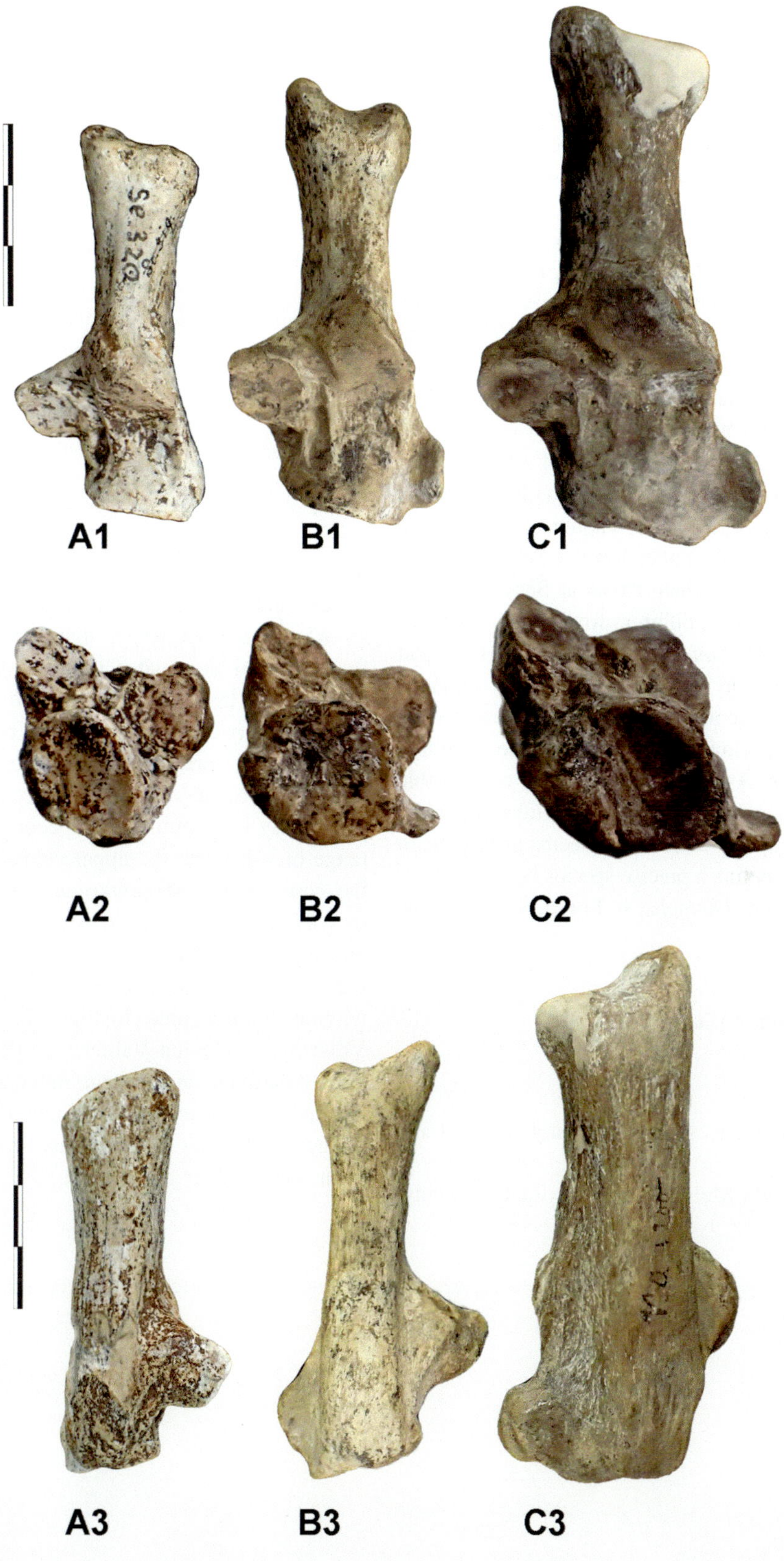

Fig. 9.25 Left calcanei of Machairodontinae: **A**, *Dinofelis* sp. from Senèze (NMB Se 320); **B**, *Megantereon cultridens* from Pardines (NMB LP 19); **C**, *M. cultridens* from Val d'Arno (NMB Va 1200); **1**, dorsal view; **2**, distal articular view; **3**, plantar view. Scale bar equals 3 cm

The rare *Dinofelis* samples encountered in Europe [*Dinofelis diastemata* (Astre, 1929) from the Ruscinian of Serrat d'en Vaquer (Astre 1929; Piveteau 1948), and *Dinofelis* sp. from Balaruc II (Beaumont 1983), from Novala Etulia 2 in Moldavia (Averianov & Baryshnikov 1999), and from Zemo Melaani (Werdelin & Lewis 2001: Fig. 32) and Kvabebi (Vekua 1972) in Georgia] have provided no calcaneus. The left calcaneus of the *Dinofelis* sp. from Venta del Moro in Spain (Morales & Aguirre 1976: Fig. 5b) seems similar to the calcaneus of *Dinofelis* from Lothagam, but about 10–15% smaller. Venta del Moro dates to the Late Miocene (6.23 Ma; Gibert et al. 2013). The *Dinofelis* from Balaruc II and from Etulia 2 are more recent than the previous ones and smaller (Balaruc II early MN 16, ca. 3 Ma; Etulia 2, MN 17, middle Villafranchian).

Venta del Moro indicates the beginning of a long-lasting presence of Machairodontinae of the Metalurini tribe in Europe, continuing at Serrat d'en Vaquer, Etulia 2 and Balaruc II and eventually a smaller form at Senèze. In any case, we have to admit that there exists at Senèze another species of Machairodontinae, different from the two famous "complete" skeletons, *Megantereon cultridens* in Basel and *Homotherium crenatidens* at the University Claude Bernard-Lyon 1. The genus *Dinofelis* is represented at Senèze by only one specimen, smaller than those generally encountered in Europe, Asia or Africa. It is not yet possible to tell if this small size reflects sexual, individual, specific, evolutionary or geographic variability. It seems unrealistic to distinguish much less define a precise species because of the scarcity of the calcanei of *Dinofelis*, so I prudently retain the name *Dinofelis* sp.

Megantereon cultridens (Cuvier, 1824)

Material:

- NMB Se 311: complete skeleton, greatly restored and mounted;
- NMB Se 321: 1 right Mc 3 without proximal extremity, Senèze 1911;
- NMB Se 1693: damaged extremity of a left Mt 3 (prox. TD = 14.18 mm);
- FSL 210981: fragment of a right demi-mandible with m1, original, corresponding to the cast NMB Se 1784;
- FSL 210976: distal half of a right tibia.

Morphometric data: Tables 9.21 and 9.22.

The complete skeleton from Senèze NMB Se 311 has been recently studied (Christiansen & Adolfssen 2007; Hooijdonk 2006). I do not believe it is useful for me to go back over the same ground and instead will concentrate on those aspects necessary to make comparisons.

Table 9.21 provides the dimensions of the teeth (Fig. 9.26) of the complete individual NMB Se 311 as measured directly on the original, measurements that will not always be exactly the same as those taken from casts. I used the dimensions MD $\times$ VL of the upper canines to check if there was evidence of an evolutionary trend of the species that might allow us to clarify the chronological relationship between the *Megantereon cultridens* from Senèze and the one from Saint-Vallier (Fig. 9.27). In doing so we should not forget that the number of individuals from each of these sites is very small (maximum n = 4). The general trend of a progressive increase in the size of individuals over time seems to be confirmed, with the lower values corresponding to the older forms from Villarroya and Saint-Vallier. However when the data is from a sufficiently large number of specimens, we can see that the range of values for the upper canines is quite wide. This is the case for the *Megantereon cultridens* from Villarroya (n = 4) and from Val d'Arno (n = 4), which probably indicates a significant degree of sexual dimorphism. The individual from Senèze NMB Se 311, of large size, is without doubt a male. In Fig. 9.27, its values appear after Villarroya, Saint-Vallier, Perrier-Etouaires and Perrier-Pardines, in the same range as Val d'Arno. While it is important not to place too much weight on the position of these dots, they do allow us to separate the two forms from Saint-Vallier and Senèze.

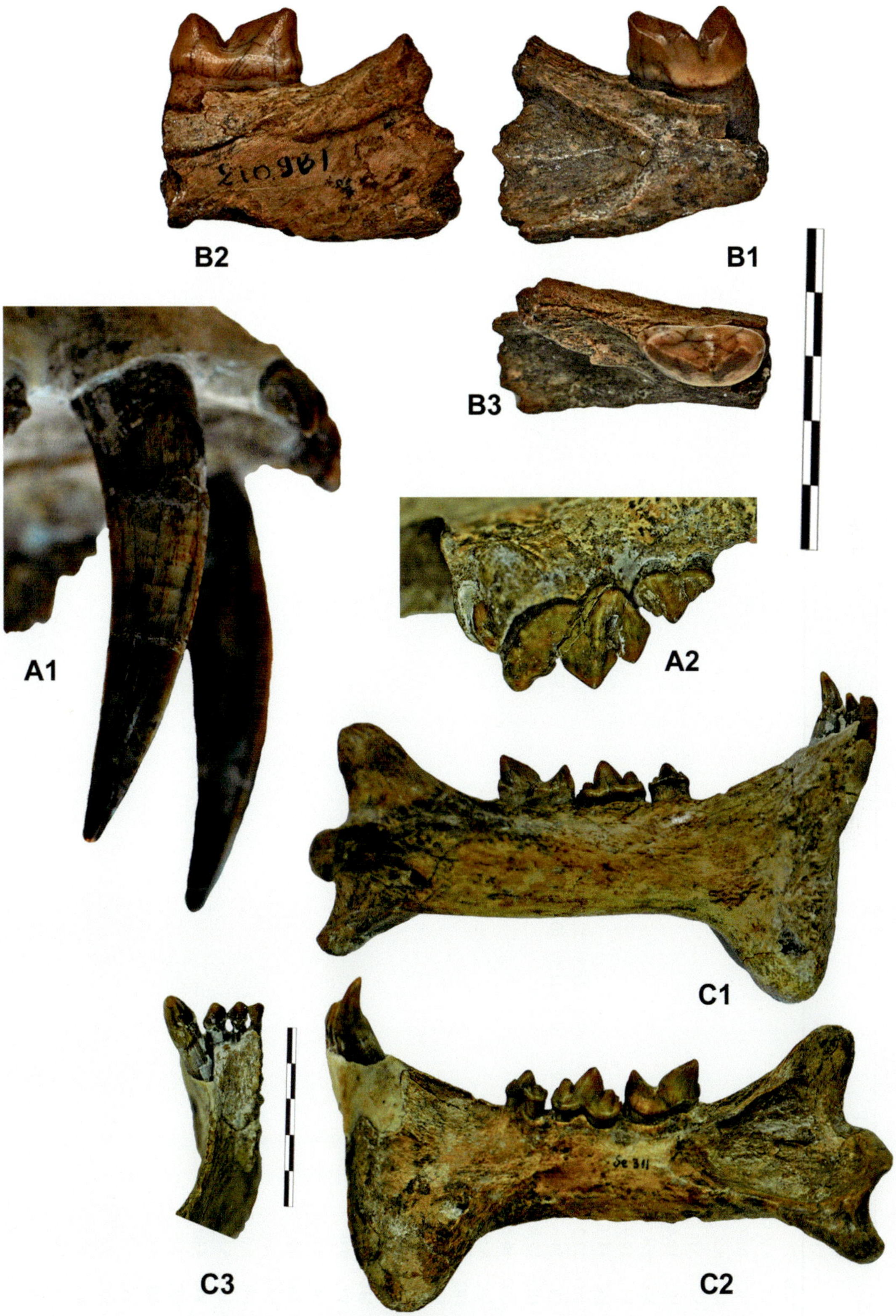

Fig. 9.26 Senèze, *Megantereon cultridens*. **A**, Cranium (NMB Se 311), **1**, anterior part; **2**, lingual view of P3-M1 in left maxilla; **B**, Fragment of right mandible (FSL 210981) with m1: **1**, buccal view; **2**, lingual view; **3**, occlusal view; **C**, Left hemi mandible (NMB Se 311), **1**, lingual view; **2**, buccal view; **3**, anterior teeth: i1-c. Scale bar equals 5 cm

Table 9.21 Comparative dimensions of the Senèze *Megantereon cultridens* upper and lower teeth

Dental variables			*Megantereon cultridens*																			
			Senèze		Villarroya*				Pardines*	Perrier-Etouaires*	Saint-Vallier*			Kuruksay*		Val d'Arno*				Fonelas P1*	Puebla de Valverde*
			NMB	FSL	1	2	3	4			1	2	3	3120	73459	1	2	3	4		
			Se 311	210981																	
Upper teeth	I1	MD	7																		
		VL	4.8																		
		HC	10.7																		
	I2	MD	8.2																		
		VL	5.6																		
		HC	10.4																		
	I3	MD	9.4																		
		VL	10.9																		
		HC	15.3																		
	C	MD	22.2		18	17	19.1	21.5	22	22.4	17	20.7	19.6	19.9		23.7	21.2	26	23.8	25.2	
		VL	11.6		9.3	9.9	10	11.5	10.7	10.7	8.5	10.5	11.3	11.1		12.7	10	12.3	12.1	9.3	
		HC	77.8													–	–	–	–	–	
	P3	MD	16.8																		
		VL	8.5																		
		HC	8.3																		
	P4	MD	30.6																		
		VL	15.3																		
		HC	16.1																		
	M1	MD	4.3																		
		VL	9.9																		
		HC	4.9																		
Lower teeth	i1	MD	4.8																		
		VL	3.2																		
		HC	6.6																		
	i2	MD	5.8																		
		VL	4.8																		
		HC	7.8																		
	i3	MD	6.4																		
		VL	7.2																		
		HC	9.4																		
	c.	MD	10.7												10.8						
		VL	7.8												–						
		HC	17.9												–						
	p3	MD	11.8								11			11	10.4						
		VL	6.2								–			–	4.9						
		HC	6.7								–			–	–						

(continued)

Table 9.21 (continued)

Dental variables		*Megantereon cultridens*																		
		Senèze		Villarroya*				Pardines*	Perrier-Etouaires*	Saint-Vallier*			Kuruksay*		Val d'Arno*				Fonelas P1*	Puebla de Valverde*
		NMB	FSL	1	2	3	4			1	2	3	3120	73459	1	2	3	4		
		Se 311	210981																	
p4	MD	19.5								17.8				18.6	19.6	18.8	19.2	19		18.8
	VL	9.4								8.6				7.2	9.1	8	8.5	8		8
	HC	11.8								–				–	–	–	–	–		–
m1	MD	21.4	19.1											19.1	22.9	23	18.4	20		
	VL	10.1	9.2											8.7	11.2	10.8	9.3	9.5		
	HC	12.9	12.1											–	–	–	–	–		

Dental variables			*Megantereon whitei*						
			Dmanisi* GNM		Venta Micena*	Apollonia*		Untermassfeld*	
			1341	1340		APL 12	APL 13	IQW 1992	IQW 1980
								24037	16002
Upper teeth	I1	MD	8						
		VL	4						
		HC	–						
	I2	MD	8.5						
		VL	4.5						
		HC	–						
	I3	MD	9.5						
		VL	8						
		HC	–						
	C	MD	22.2	23.4			24.6		
		VL	10.8	11.3			11.3		
		HC	85.8	84			–		
	P3	MD	13.5	11.2					
		VL	6.2	5.3					
		HC	–	–					
	P4	MD	28	29				31.1	
		VL	11	11				8.7	
		HC	–	–				15.1	
	M1	MD	4	8					
		VL	8	9					
		HC	–	–					
Lower teeth	i1	MD							
		VL							
		HC							

(continued)

Table 9.21 (continued)

Dental variables		*Megantereon whitei*						
		Dmanisi* GNM		Venta Micena*	Apollonia*		Untermassfeld*	
		1341	1340		APL 12	APL 13	IQW 1992	IQW 1980
							24037	16002
i2	MD							
	VL							
	HC							
i3	MD							
	VL							
	HC							
c.	MD				9.2			
	VL				6.2			
	HC				–			
p3	MD	6	6		5.6			
	VL	–	–		3.4			
	HC	–	–		–			
p4	MD	16	16	14.5	14.4			19.0*
	VL	–	–	6.5	6			8.1
	HC	–	–	–	–			
m1	MD	19.2	19		18.1			
	VL	–	–		8.5			
	HC	–	–		–			

* Measurement sources: Villarroya, Pardines, Perrier-Etouaires, Saint-Vallier, Val d'Arno, Venta Micena (Palmqvist et al. 2007); Kuruksay (Sotnikova 1989); Dmanisi (Vekua 1995); Fonelas P1 (Garrido 2006); Puebla de Valverde (Kurtén and Crusafont Pairó 1977); Untermassfeld (Hemmer 2001); Apollonia 1 (Koufos 1992); Senèze, author's measurements of the original complete specimen NMB Se 311

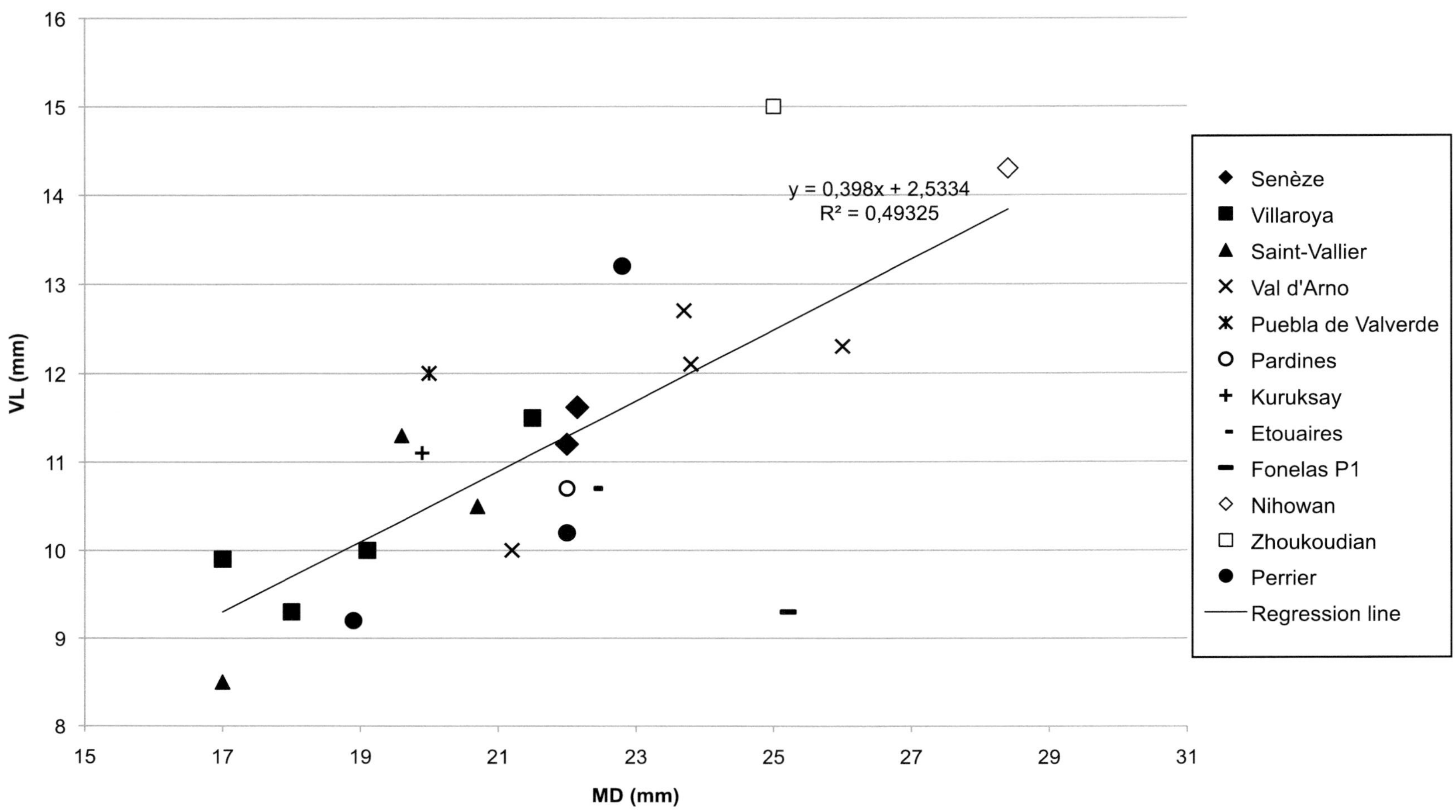

Fig. 9.27 Scatter diagram of the dimensions (MD x VL) of the upper canines of *Megantereon cultridens* from European sites, position of *M. cultridens* from Senèze

Comparison of the metacarpal bones of *Megantereon cultridens* (Table 9.22) confirms a true variability, probably sexual. The complete individual NMB Se 311 from Senèze is large, with a much bigger Mc 4 than that of Saint-Vallier (SV 96617). On the other hand, the Mc 5 SV 97544 from Saint-Vallier has equivalent dimensions and proportions to those of NMB Se 311. Likewise, the measurable parts of the Mc 3 from another Senèze individual (NMB Se 321) have much lower values than NMB Se 311. *Megantereon whitei* from Pirro Nord G.1 (Italy) and Monte Argentario (Italy) (Sardella et al. 2008) is smaller than the *Megantereon cultridens* from Senèze but one Mc 3 from Monte Argentario is particularly long in comparison to Pirro Nord and is even longer than the one from Senèze. Until we have statistically sufficient series, the size of the metapodials will continue to be unusable for the purposes of biochronology.

The genus *Megantereon* appeared in America about 4.5 Ma in the Hemphillian (Bone Valley Formation, Florida; Berta & Galiano 1983) and spread throughout Eurasia during the Villafranchian. *Megantereon cultridens*, initially the only representative of the genus, persisted in China until the Middle Pleistocene (= *Megantereon inexpectatus*).

Table 9.22 Comparative dimensions of the Senèze *Megantereon cultridens* metacarpals

Megantereon cultridens		Senèze		Saint-Vallier		Pirro-Nord (Apulia)*	Argentario*	Villa San Faustino*	Ceyssaguet* (MNPE)				
Metacarpal Variables		NMB	NMB	FSL	FSL	IIPU	IIPU	IIPU	CEY	CEY	CEY	CEY	CEY
		Se 311	Se 321	SV 96617	SV 97544	G12–G15	ARG-30–31	BT 16 (Mc3)-17 (Mc 5)	13911	13597	13596	13595	13594
Mc 2	L	77.8				75.0				74.2			
	TD prox	20.2				18.0				18.9			
	APD prox	24.0				21.0				22.8			
	TD med	11.4				11.5				12.9			
	APD med	12.0				11.0				13.0			
	TD dist	18.0				18.0				19.5			
	APD dist	17.4				14.0				15.6*			
Mc 3	L	84.6	–			81.5	86.0	91.0	89.2		80.5		
	TD prox	20.7	–			20.0	–	22.0	18.7		19.9		
	APD prox	21.4	–			20.0	–	22.7	19.4*		21.5		
	TD med	12.7	11.5			12.8	–	12.8	15.4		14.2		
	APD med	10.8	9.7			10.0	–	12.4	12.3		12.5		
	TD dist	19.4	15.3			18.5	–	21.5	19.0		20.8		
	APD dist	17.9	13.2			16.0	–	17.5	17.0		16.2		
Mc 4	L	83.4		78.3		77.0	78.7					75.0	
	TD prox	14.8		14.5		15.0	–					15.9	
	APD prox	18.6		17.0		17.5	–					19.5	
	TD med	10.6		9.5		10.5	–					12.9	
	APD med	10.3		8.6		10.2	–					13.9	
	TD dist	17.3		15.6		16.5	–					19.3	
	APD dist	13.9		15.0		15.5	–					17.6	

(continued)

Table 9.22 (continued)

Megantereon cultridens Metacarpal Variables		Senèze		Saint-Vallier		Pirro-Nord (Apulia)*	Argentario*	Villa San Faustino*	Ceyssaguet* (MNPE)				
		NMB	NMB	FSL	FSL	IIPU	IIPU	IIPU	CEY	CEY	CEY	CEY	CEY
		Se 311	Se 321	SV 96617	SV 97544	G12–G15	ARG-30–31	BT 16 (Mc3)-17 (Mc 5)	13911	13597	13596	13595	13594
Mc 5	L	69.0			68.6*	65.0	68.0	72.0					64.7
	TD prox	16.8			14.1	16.8	–	15.2					17.9
	APD prox	17.4			16.9	16.5	–	18.0					19.7
	TD med	10.2			8.9	11.0	–	10.0					12.2
	APD med	10.5			11.3	10.0	–	10.6					12.2
	TD dist	16.0			15.0*	16.0	–	17.0					17.7
	APD dist	16.5			–	13.8	–	16.0					15.4

* Measurement sources: Saint-Vallier (A. Argant 2004); Pirro-Nord, Argentario and Villa San Faustino (Sardella et al. 2008); Ceyssaguet (Tsoukala 2004); Senèze: author's measurements of the original complete specimen NMB Se 311

Martínez-Navarro and Palmqvist (1995, 1996) thought that the Early Pleistocene form in Europe, *Megantereon cultridens,* was replaced by *Megantereon whitei* (Broom 1937) originating in Africa. The spread of African populations, coming through the Middle East (Isthmus of Suez or Strait of Mandeb), or through Gibraltar, reached Europe after about 1.8–1.6 Ma, occurring in: Spain (Venta Micena); Georgia (Dmanisi); Greece (Appolonia 1); Italy (Monte Argentario); Israel ('Ubeidiya); Romania (Bugiulesti); Hungary (Urkùt); and Germany (Untermassfeld) (Martínez-Navarro & Palmqvist 1995, 1996; Palmqvist et al. 2007; Sardella et al. 2008). *Megantereon cultridens* and *Megantereon whitei* are both valid species because the differences between the sizes of their teeth are greater than the variation due to sexual dimorphism normally found in the carnivores (comparison with large populations of *Panthera pardus* and *Panthera leo*) (Palmqvist et al. 2007). *Megantereon cultridens* has on average bigger teeth (with different proportions) than *Megantereon whitei*, but the latter's upper canines are generally larger.

Several studies (Christiansen & Adolfssen 2007; Sardella 1998; Schaub 1925) link the material from Senèze to the species *Megantereon cultridens*. This attribution confirms the assignment of the Senèze site to the middle Villafranchian as shown by the diagram in Fig. 9.27. However, the *Megantereon cultridens* from Senèze seems to be more recent than the one from Saint-Vallier and closer to the one from Val d'Arno.

Homotherium crenatidens (Fabrini, 1890):

Material:

- NMB Se 1698: complete left C (in two parts);
- NMB Se 1789: proximal extremity of a damaged right radius and proximal extremity of a left radius;
- FSL 210991: complete skeleton, mounted and then later dismantled (Ballesio 1963);
- FSL 210972: 1 Mc 3, R;
- FSL 210973: 2 metapodials and 2 sesamoids;
- FSL 210974: 2 phalanges 1 and 1 phalanx 2;
- FSL 210971: 1 complete radius from a second individual, R.

The genus *Homotherium* is known to have existed since the Ruscinian in Europe and Africa. It was widespread in Eurasia and North America during the Villafranchian. As far as we know a single species, *Homotherium crenatidens*, lived in Western Europe after the early Villafranchian. The species occurred throughout the late Villafranchian and even at the beginning of the early Middle Pleistocene, for instance at Ceyssaguet around 1.2 Ma, at Untermassfeld (Germany) about 1.1 Ma, and even much later at Mauer and Mosbach (Germany). The species has been found at many sites (n $\approx$ 100 in Europe) but usually with few remains.

Senèze was the source of the complete skeleton held by the UCB-Lyon 1, which today is the best reference material for the species (Figs. 9.28, 9.29 and 9.30). It has been carefully studied by Ballesio (1963). I do not believe it is useful for me to go back over the same ground and instead will concentrate on those aspects necessary to make comparisons.

Table 9.23 Comparative dimensions of the Senèze *Homotherium crenatidens* upper and lower teeth

Dental variables			*Homotherium crenatidens*									*Homotherium latidens*		
			Senèze			Kvabebi*	Kuruksay*	Rocca-Neyra*	Val d'Arno Sup*	Ceyssaguet*	Untermassfeld*	Incarcal*	Westbury*	Hundsheim*
			FSL		NMB									
			210991		Se 1698	Georgia	Tadjikistan							
			L	R										
Upper teeth	I2	MD	12.3	12.6								11.1–12.1		
		VL	11.2	10.5								8.7–11.1		
	C	MD					n = 4		n = 5	n = 3				
			36.8	35.1	32.8	31.0	28.0–33.2	38.0–38.9	31.6–33.5	22.9–30.0	37.7	29.5–36.0	33.1	33.0
							31.2		32.2	25.4				
		VL					n = 4		n = 5	n = 3				
			13.6	13.7	13.3	15.0	11.0–12.6	17.0–17.5	11.7–14.0	11.2–15.0	14.7	12.2–13.5	14.3	12.5
							12.0		13.0	12.6				
		HC					n = 4		n = 3	n = 2				
			>53.0	–	71.1	110.0	73.5–96.0	108.0	81.0–95.0	56.7–68.5	100.8	71.4	–	73.5
				–			82.7		88.7					
		HT								n = 2				
			–	–	133.6	–	125.5		142.0*	117.6–128.4	158.2	–		122.0
	P4	MD					n = 3					n = 8		
			40.4	42.6		37.0	39.0–41.0		41.5–45.0			22.0–42.7		
							40.2					37.3		
		VL										n = 9		
			>10.6	>13.8		–	13.1–12.0		11.8–13.3			6.9–13.0		
												10.9		
Lower teeth	i1									n = 3				
		MD	9.3	9.0						8.5–9.1		7.5–8.1		
										8.8				
										n = 3				
		VL	6.6	6.5						6.2–6.6		4.3–6.0		
										6.4				
	i2									n = 3				
		MD	10.6	–						10.6–11.5		9.3–9.8		
										11.1				
										n = 3				
		VL	8.7	–						9.0–9.4		6.4–9.0		
										9.3				
	i3	MD								n = 3		n = 7		
			10.9	–						11.4–12.0		10.2–11.2		
										11.7		10.8		
		VL								n = 3		n = 7		
			11.7	–						11.1–12.3		8.6–12.2		
										11.7		9.7		

(continued)

Table 9.23 (continued)

Dental variables		*Homotherium crenatidens*									*Homotherium latidens*		
		Senèze			Kvabebi*	Kuruksay*	Rocca-Neyra*	Val d'Arno Sup*	Ceyssaguet*	Untermassfeld*	Incarcal*	Westbury*	Hundsheim*
		FSL		NMB									
		210991		Se 1698	Georgia	Tadjikistan							
		L	R										
c	MD								n = 3		n = 6		
		17.4	17.8		14.0		15.0		15.5–18.0	13.8–15.1	14.6–16.3		
									16.4		15.1		
	VL								n = 3		n = 6		
		11.1	11.3		10.0		11.0		10.3–12.1	10.0–10.1	9.4–10.5		
									11.0		9.9		
									n = 3				
	HC	–	–		–		–		24.3–25.7	> 20.0			
									24.9				
p3	MD					n = 3		n = 3					
		7.9			9.0	7.9–12.0		7.9–12.0	9.1–9.2				
						9.8		9.8					
	VL					n = 3		n = 3					
		5.6			–	4.5–6.4		4.5–6.4	5.6–6.2				
						5.5		5.5					
	HC	6.7			–	–		–	8.4–8.8				
p4	MD					n = 4		n = 4	n = 4				
		>18.3	22.5		20.0–21.0	20.4–24.7	18.3	20.4–24.7	21.1–23.6	19.4–21.0	19.0–23.5		19.7–21.1
						22.8		22.8	22.2				
	VL					n = 4		n = 4	n = 4		n = 3		
		10.1	>10.6		10.0	9.2–10.4	9.8	9.2–10.4	9.5–10.7	8.5–8.6	9.2–10.8		8.5–9.0
						10.0		10.0	10.2		10.0		
									n = 4				
	HC	14.7	–		–	–			16.3–17.8	–			
									16.5				
m1	MD					n = 5		n = 5	n = 4	n = 4	n = 5		
		32.7	32.4		25.0–26.0	30.0–33.9	32.4	30.0–33.9	27.5–32.5	30.4–34.0*	27.2–33.8		
						32.0		32.0	29.8	32.2	31.0		
	VL					n = 5		n = 5	n = 4	n = 4	n = 5		
		12.9	13.2		10.0	12.7–13.5	12.6	12.7–13.5	12.6–13.3	11.9–13.1	11.7–13.1		
						13.1		13.1	12.9	12.6	12.6		
									n = 4				
	HC	–	–		–	–	–	–	19.4–21.2	–			
									19.7				

* Measurement sources: Kvabebi (Vekua 1972); Kuruksay (Sotnikova 1989); Perrier-Rocca-Neyra (Bonis 1976); Val d'Arno sup. (Ficcarelli 1979); Ceyssaguet (Tsoukala 2004); Untermassfeld (Hemmer 2001); Incarcal (Galobart et al. 2003); Westbury (Bishop 1982); Hundsheim (Freudenberg 1914); Senèze: author's measurements of the original complete specimen FSL 210991

Table 9.24 Comparative dimensions of the Senèze *Homotherium crenatidens* long bones and metapodials

Homotherium crenatidens Postcranial variables			Lat	1 Length	2 Prox TD	3 Prox APD	4 Med TD	5 Med APD	6 Dist TD	7 Dist APD	8 Dist articular TD	9 See below	10 See below
Humerus	Senèze	FSL 210991	L	354.0	64.0	97.0	28.0	43.0	85.0	57.0	–	65.0	–
	Senèze	FSL 210991	R	354.0	65.0	97.0	29.0	44.0	86.0	56.0	–	65.0	–
	Ceyssaguet	MNPE CEY 7834	R	–	–	–	30.0*	42.2	76.2	50.8	56.9	–	–
	Ceyssaguet	MNPE CEY 2309	L	–	–	–	28.5	42.1	–	–	–	–	–
Radius	Senèze	FSL 210991	L	317.0	34.0	29.0	28.0	24.0	55.0	39.0	44.4	27.0	21.0
	Senèze	FSL 210991	R	317.0	34.0	29.0	28.0	23.0	55.0	38.0	–	27.0	21.0
	Senèze	FSL 210971	R	329.0	34.1	27.4	25.8	19.8	56.7	36.2	42.5	24.2	20.0
	Ceyssaguet	MNPE CEY 3664	L	–	34.9*	28.9*	26.8*	18.2	–	–	–	29.5	21.6
Ulna	Senèze	FSL 210991	L	372.0	–	–	19.0	33.0	24.0	35.0	–	–	59.0
	Senèze	FSL 210991	R	370.0	–	–	21.0	34.0	23.0	35.0	–	35.0	59.0
Femur	Senèze	FSL 210991	L	356.0	89.0	–	33.0	29.0	70.0	70.0	–	41.0	69.0
	Senèze	FSL 210991	R	357.0	90.0	–	32.0	28.0	70.0	70.0	–	41.0	69.0
Tibia	Senèze	FSL 210991	L	315.0	–	–	30.0	34.0	51.0	40.0	–	–	–
	Senèze	FSL 210991	R	308.0	69.0	80.0	29.0	34.0	50.0	41.0	–	–	–
	Ceyssaguet	MNPE CEY 6030	L	318.5	73.4	84.0	28.0	34.5	54.8	43.0	53.6	–	–
Fibula	Senèze	FSL 210991	R	280.0*	28.0	–	–	–	26.0	28.0	–	–	–
Mc 1	Senèze	FSL 210991	R	42.3	20.6	14.5	15.8	12.0	16.2	12.6	–		
Mc 2	Senèze	FSL 210991	L	109.7	27.1	29.3	14.3	16.4	22.7	–	–		
Mc 3	Senèze	FSL 210991	L	122.5	22.5	24.2	16.0	13.6	23.4	22.1	19.9		
	Senèze	FSL 210972	R	126.4	21.8	24.1	15.0	13.2	22.5	20.9	18.9		
Mc 4	Senèze	FSL 210991	L	119.3	21.0	26.8	14.2	14.3	21.9	22.3	20.0		
Mc 5	Senèze	FSL 210991	L	103.0	23.3	27.4	13.3	16.1	20.9	20.4	20.4		
Mt 2	Senèze	FSL 210991	L	108.5	19.0	–	12.9	14.8	20.9	19.6	18.2		
Mt 3	Senèze	FSL 210991	L	121.5	22.6	–	16.1	13.0	22.2	20.6	18.7		
	Senèze	FSL 210991	R	121.5	22.3	–	17.1	14.8	21.9	–	–		
Mt 4	Senèze	FSL 210991	L	119.9	19.8	27.5	14.2	12.5	20.6	20.4	17.6		
	Senèze	FSL 210991	R	118.1	15.6*	27.0	13.0	–	20.5	20.7	18.0		
Mt 5	Senèze	FSL 210991	L	108.3	16.8	19.5	9.5	13.8	19.1	17.6	17.3		
	Senèze	FSL 210991	R	110.8	–	21.4	9.4	14.2	21.4	18.9	18.7		

Ceyssaguet measurements from Tsoukala 2004. Senèze: author's measurements of the original complete skeleton FSL 210991. Variables for numbers 9 and 10 for specific elements: **Humerus**, 9 = proximal articular APD; **Radius**, 9 = neck TD; 10 = neck APD; **Ulna**: 9 = olecranon TD; 10 = olecranon APD; **Femur**: 9 = APD head; 10 = distance between greater and lesser trochanters.

Morphometric data: Tables 9.23 and 9.24.

The complete left upper canine (NMB Se 1698) belongs to a young individual (U = 0.2) (Fig. 9.29). Visible along the anterior and posterior ridges are the serrations characteristic of the species, although these are much less evident on the anterior ridge because of greater wear. Its dimensions correspond to those of the complete individual from Senèze and to the canines of Early Pleistocene forms. The more recent specimens seem larger (Table 9.23).

The mandible of the complete skeleton of *Homotherium crenatidens* from Senèze (FSL 210991) is particularly robust and massive. It has all its incisors, which are relatively tall, massive and quite worn. The two lower canines, the L p3, the L and R p4 and the L and R m1 are also still in place (Table 9.23). Completing the mandible at the rear are the two articular condyles, massive and not very elongated transversally. The ramus is not very robust and is typical of a machairodont. I have compared this mandible from Senèze to the one from Liventsovka (Russia; Khapry Faunal Assemblage; Sotnikova et al. 2002, pl. II-9, RSU 94). In the case of the p3, the specimen from Senèze has a similar dental morphology to that of Liventsovka RSU 94, short with a prominent cuspid surrounded by a band that entirely encircles the crown. A long diastema separates the p3 from the canine, as is usual for the species, and there is a gap between it and the p4. However, because of their very distinctive morphology, the p4 and m1 from Senèze are difficult to use for comparisons. Ballesio (1963: 50) drew attention to the very extensive wear on the posterior lobe of the p4 that had been worn away because of the rubbing of its external face against the paracone and the parastyle of the upper carnassial. For the same reason most of the enamel on the external face of the m1 has been worn away. The wear is particularly deep in the middle of this face where a deep vertical groove has been cut, such that the V-shaped indentation in the cutting edge of the two blades is more accentuated than it would be on an unworn tooth. Because of this wear, it is difficult to make any comparison with an unworn tooth. The m1 from Khapry, also very worn on its external edge, seems longer in lateral view and has a pronounced talonid: a shape more reminiscent of the *Homotherium crenatidens* from Ceyssaguet (CEY 2083 cast) than the one from Senèze.

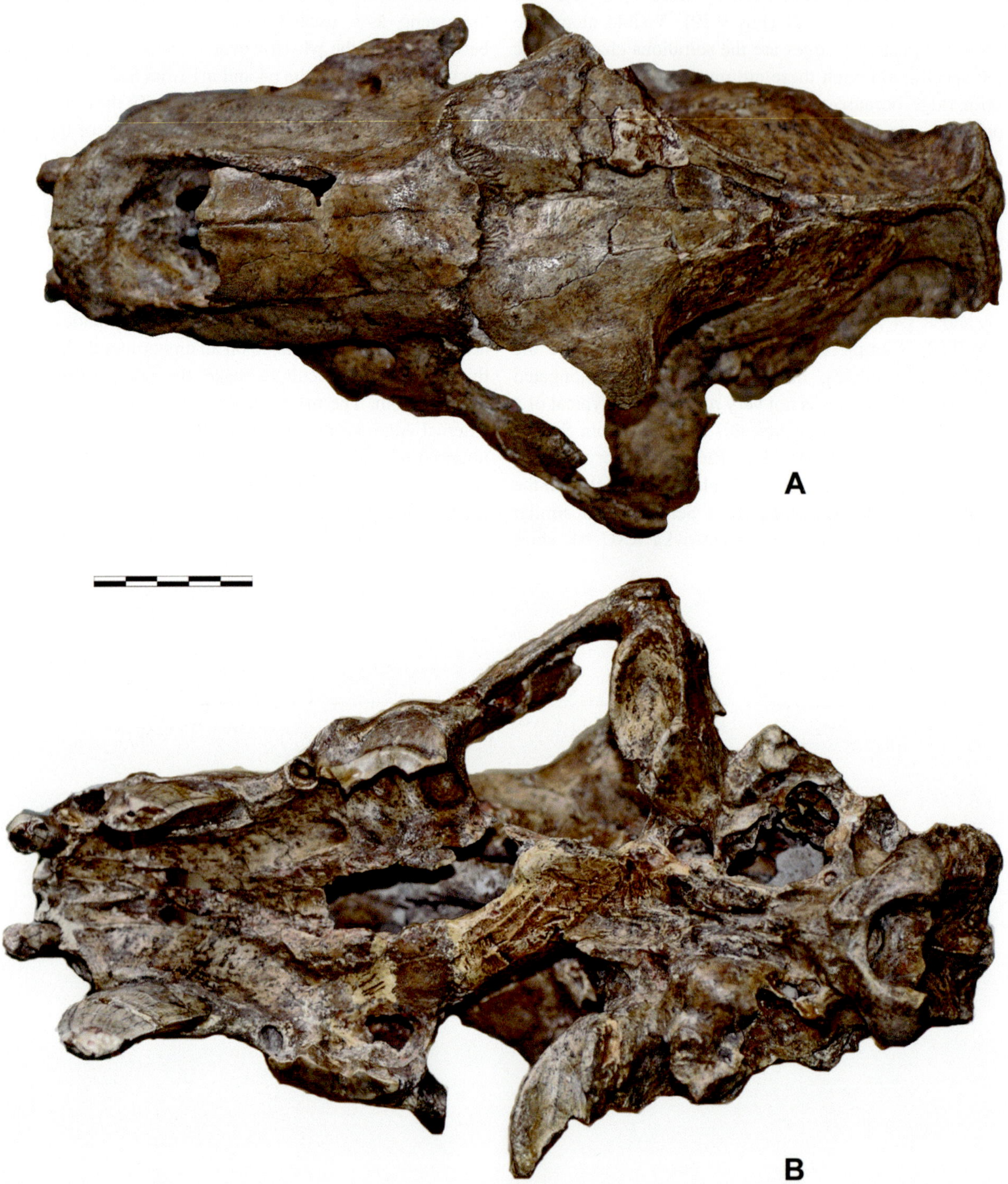

Fig. 9.28 Senèze, *Homotherium crenatidens*, cranium of the complete individual FSL 210991; **A,** dorsal view; **B**, basal view. Scale bar equals 5 cm

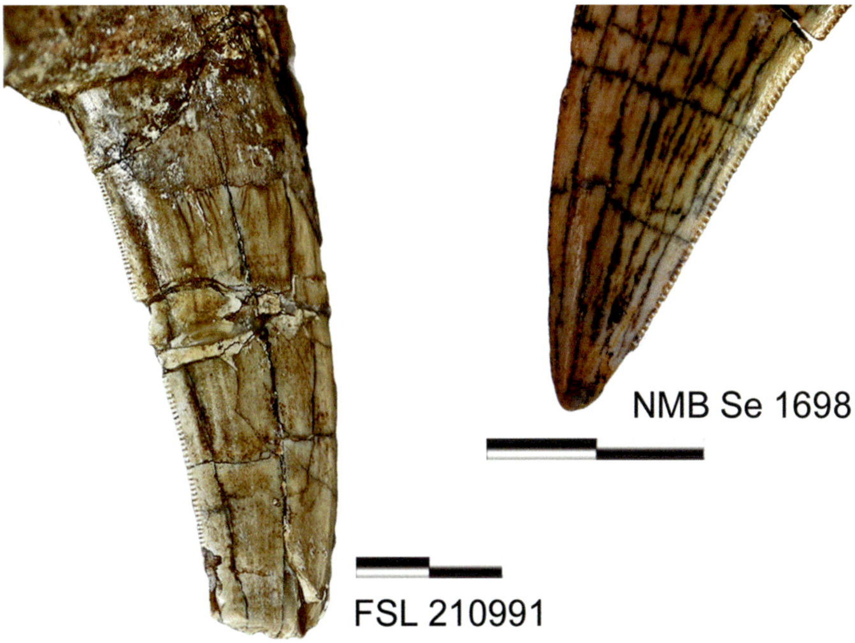

Fig. 9.29 Senèze, *Homotherium crenatidens*, right C of the complete individual FSL 210991, lingual view, and left C NMB Se 1698, labial view with clearly visible serrations on the posterior edge of the crown. Scale bar equals 2 cm

Fig. 9.30 Senèze, *Homotherium crenatidens*, mandible of the complete individual FSL 210991, superior view. Scale bar equals 5 cm

Figure 9.31, an adaptation of Fig. 4 of Sotnikova et al. (2002: 387), compares the length of the p4 to the width of m1. This ratio separates the older populations of the late Ruscinian-early Villafranchian from those of the middle—upper Villafranchian. As in the case of *Megantereon cultridens*, Senèze lies close to the specimens from the Upper Val d'Arno, while Liventsovka seems more recent.

I also tried to compare the metapodials, in this case the metacarpals, of the *Homotherium crenatidens* from Senèze (Table 9.24). However, it should be recognized that it is not currently possible to take these comparisons very far and they remain quite limited. Nevertheless, they can provide more reliable information on the size of individuals than dental data.

Authors have often emphasized the morphological variability of the teeth of *Homotherium crenatidens*. It is clear that the degree of dental wear often encountered with felids, particularly machairodonts, can easily distort essential morphological observations. In my opinion this is indicative of the difficulties arising from the rarity of well-dated sites with sufficient material (including relatively unworn teeth), difficulties compounded by a lack of consideration of post-cranial material. In my opinion, the taxonomy needs to be clarified, and in the future better account needs to be taken of the evolution of the species.

Signs of injury are visible on the diaphysis of the humerus of the complete *Homotherium crenatidens* from Senèze (Ballesio 1963; Perrot et al. 1972) as well as on the one from Saint-Vallier (A. Argant 2004). These indications of trauma confirm the physical engagement of these animals in the capture of their prey. I saw a good example of this type of physical engagement in a television documentary filmed in infra-red at night in a South African game preserve. A female elephant was dying at the edge of a wood. She was on the ground, forelegs upright, but no longer able to move. A group of four lionesses attacked her. Three of them clung to her hindquarters in order to keep her on the ground and prevent her from moving while the fourth, more daring, climbed onto her back. This lioness, with its mouth wide open, bit the elephant a number of times at the base of the neck, using its upper canines to try to stab her. However it did not succeed and the lionesses gave up when, with one last effort, the elephant was able to stand up and walk slowly away. This behavior helped me better understand how *Homotherium crenatidens* would have used its very long and narrow canines when attacking its prey. It would probably have been able to finish off the elephant quite easily.

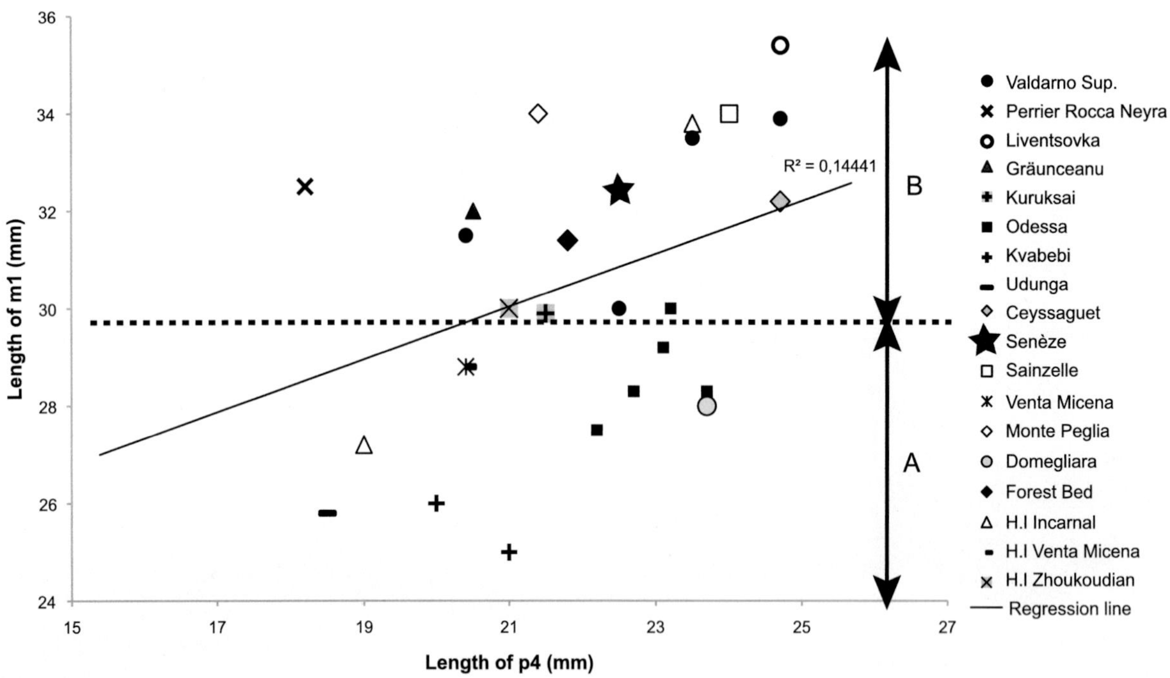

Fig. 9.31 Scatter diagram of length of the p4 versus length of m1 of some *Homotherium crenatidens*, and position of Senèze FSL 210991 (after Sotnikova et al., 2002: **Fig. 4**). The dotted horizontal line divides the early *Homotherium* into A = Terminal Ruscinian-Early Villafranchian and B = Middle-Late Villafranchian. Senèze clearly belongs to group B. The low correlation coefficient of the regression line reveals the heterogeneity of Eurasian *Homotherium crenatidens*

Paleoecological Assessment

A 175 m deep core obtained in 1965 at the center of the former maar at Senèze is sufficient to demonstrate how long it took for the lake to fill up with sediment (Elhaï 1969). The diatoms analysis by Ehrlich (1968) demonstrate that the 110 m of laminated sediment correspond to a long period: 200 to 300 ka. The pollen analysis of the paleontological sediments and coprolites by Argant (2024) indicates a temperate period. The bank of the lake attracted herds of herbivores to drink and often of course their predators. This occurred over a very long period of time since the environment of the lake provided biotopes favorable to the progressive establishment of life on the original volcanic substratum. Of course, we only know about the more recent paleontological remains. These were found because they were near the surface, or because they were rapidly covered. On such a large open-air site it is not surprising to find remains which seem to be the result of a variety of isolated events spread out over a period long enough to sometimes record several climatic alternations.

The association of different fauna and even the evolutionary record of some species can be understood without much difficulty. In my opinion it is unrealistic to attribute too precise an age to all the fauna discovered during more than a century or to conclude that there are two distinct groups with different ages. The only reasonable interpretation is that the Senèze remains are the result of a series of sporadic events reflecting life at the edge of a steep-sided lake over a long period of time. We would be able to conclude that there were two chronologically distinct periods only if there were groups of remains that have remained enclosed and undisturbed since their origins, something that seems quite impossible to claim in the case of the Senèze deposit.

The sporadic events spread out over time which led to the formation of fossilised remains at Senèze were likely to have occurred because of one of the following scenarios (see also Debard 2024; Fernández-Jalvo et al. 2024):

(1) animals coming to drink were sometimes the victims of accidents as they hurriedly descended the steep slopes bordering the lake. It is also likely that predators attacked them during their descent, by lying in wait beside the usual access routes. Scavengers feeding on the remains would have contributed to their dispersal, as would subsequent frequent trampling and slippage on unstable or sloping terrain.

(2) animals which had reached the water's edge and were already drinking were once again vulnerable to attack. Particularly for herbivores this was a critical time. With heads close to the water and forelegs wide apart, they were less able to detect predators using their vision, sense of smell or hearing, and they were dependent on the alert signals of other members of their herd to avoid attack. Predators know very well how to take advantage of these sorts of opportunities.

(3) some herbivores taking fright fled instinctively down the slopes and into the water. Unlike in Africa, there was no fear of crocodiles here to deter them. Some would have swum to safety, others would have drowned. After a while their corpses would have risen again to the surface and drifted according to the direction of wind and current. Some of these would end up on the shore and accessible to scavengers. Others would have sunk once again to the bottom of the lake and become perfectly protected. This I think would have been the origin of most of the complete skeletons found at Senèze, both those found in the past and in recent excavations.

When I examined the lists of the carnivores found at comparable sites (Fig. 9.32), I noticed that the sites yielding the greatest number of species, particularly of Felidae, were those which were close to large streams, lakes or fluvio-paludal environments, for example Villarroya, Saint-Vallier, Senèze, Ceyssaguet, and Untermassfeld. This convergence obviously demonstrates the importance of these types of biotopes for the development of life but also the role of these environments in facilitating contact between predators and their prey, in increasing the risks and concentrating the traps for prey, as well as providing favorable geological conditions for the rapid covering and hence fossilisation of remains.

Conclusions

Current List of Carnivores Found at Senèze:

Canids

- *Nyctereutes megamastoides megamastoides*
- *Vulpes alopecoides*
- *Canis arnensis* (= *Canis senezensis*)
- *Canis* sp. (*Canis arnensis* or *Canis etruscus*?)

Ursids:

- *Ursus etruscus*

Hyaenids:

- *Pachycrocuta perrieri* (= *Pliocrocuta perrieri*)
- *Chasmaporthetes lunensis*
- Hyaenidae *indet.*

Site n° 1 2 3 4 5 6 7 8 9 10 11 12 13 — 3.0 2.5 2.0 1.5 1.0 My

Carnivores Families Species (● species present; ○ possibly present)	Carnivores of Senèze	1 Villaroya	2 Perrier-Etouaies	3 Varshets	4 Saint-Vallier	5 Puebla Valverde	6 Dmanissi	7 Gerakarou	8 Slivnista	9 Olivola	10 Ubedieyh	11 Ceyssaguet	12 Venta Micena	13 Untermassfeld
CANIDS														
Nyctereutes megamastoides	●	●	●			●								
Nyctereutes vulpinus				○	●									
Vulpes alopecoides	●	●		●	●	●			○			●		
Canis arnensis	●							●		●	●			
Canis sp. ou C. etruscus	○	○			○	○	●	●	●	●		●	●	
URSIDS														
U. minimus-thibetanus		○	●	●										
Ursus etruscus	●	?			●	●	●			●	●	●	●	
Ursus sp.							●							
Ursus rodei														●
HYAENIDS														
Pachycrocuta perrieri	●	●	●	●	●	●	●	●	○					
Pachycrocuta brevirostris							○	●		●		●	●	
Chasmaporthetes lunensis	●	●			●	●			○	●				
FELIDS														
Acinonyx pardinensis	●	●	●	●	●	●				●		●		●
Puma pardoides		●	●	●	●	●								●
Panthera gombaszogensis							●	●	●	●	●	●	●	●
Megantereon cultridens	●	●	●		●	●					●	●		
Megantereon whitei							●				●		○	○
Homotherium crenatidens	●	●	●		●	●	●		●	●		●	●	●
Dinofelis sp.	●													

Fig. 9.32 Checklist of European Villafranchian carnivores, indicating the biochronological position of Senèze. Solid circles: species present; empty circles: possibly present

Felids:

- *Acinonyx pardinensis pardinensis*
- *Dinofelis* sp.
- *Megantereon cultridens*
- *Homotherium crenatidens*

Readily apparent is the absence, not significant in paleontology, of members of the mustelid family and of *Lynx issiodorensis* and *Puma pardoides* among the Felidae. For mustelids, their small size has evidently acted against their presence. Perhaps this has come about because of excavations in the past, which have favored discoveries regarded as more important or spectacular, or because of the paleontological conditions of a very open lakeside site and the large area over which potential remains could have been spread. This list of fauna (see Fig. 9.32) enables us to place the carnivores from Senèze in the Villafranchian context:

- the genus *Canis* is present at Senèze, but is rare, represented only by *Canis arnensis* and perhaps by *Canis etruscus*.
- *Chasmaporthetes lunensis*, the hunting hyena, is present at Senèze. The last known occurrence of this species corresponds to the middle of the "late Villafranchian" at about 1.8 Ma at Olivola.
- the hyena *Pachycrocuta perrieri* has been found at Senèze but *Pachycrocuta brevirostris*, a large hyena which appeared only in the middle of the "late Villafranchian", has never been found at the site.
- the genus *Dinofelis* is present at Senèze. It is a small individual like those of Balaruc II and Etulia 2. The probable conspecific individual of Val d'Arno-Fornacella (NMB Va 552) is clearly bigger. The calcaneus NMB Se 320 has neither the dimensions nor the proportions of the African *Dinofelis* or of *D. diastemata*. It could belong to an evolved form that continued in Europe until the end of the Villafranchian.
- *Puma pardoides* is not recorded at Senèze but was present at Saint-Vallier, La Puebla de Valverde and Varshets. It is true that the species was still present much later at Untermassfeld in Germany at about 1.1 Ma, but it was no longer encountered during the whole of the "late Villafranchian", when *Panthera gombaszogensis* became the large feline present, accompanied by *Megantereon cultridens*, *Homotherium crenatidens* and *Acinonyx pardinensis*.
- *Panthera gombaszogensis*, the European jaguar, appeared in Eurasia only at Dmanisi (Georgia) dated at 1.80–1.76 Ma. This species has not been found before this date, and it is not present at Senèze.
- the *Megantereon cultridens* from Senèze has been replaced later, in Europe, by the African form *Megantereon whitei* at about 1.8 Ma.
- the *Homotherium crenatidens* from Senèze corresponds to the forms from the middle or late Villafranchian (Sotnikova et al. 2002). It is different from the form of Ceyssaguet (1.2 Ma) but not very different from the form found at Olivola (Olivola unit: 2.0–1.8 Ma) (Fig. 9.31).

The conclusions to be drawn from an examination of the carnivores seem clear. The carnivores have been found to be good biostratigraphic markers, particularly for the middle-late Villafranchian. Chronologically the list of fauna places the Senèze deposit around 2.0 Ma. "The earliest Late Villafranchian fauna of Dmanisi is of predominantly Eurasian character. Several of the faunal elements are common in the Middle Villafranchian context of Western Asia and Europe…" (Hemmer et al. 2011, p. 2705; see also Bartolini Lucenti et al. 2020). The Dmanisi site, well dated by the convergence of several methods and yielding an abundant and varied fauna, marks the transition very well. At this site we find the following carnivores: *Canis etruscus*, *Vulpes alopecoides*, *Ursus etruscus*, *Pachycrocuta perrieri*, *Homotherium crenatidens* and *Acinonyx pardinensis*. *Puma pardoides* is missing but *Panthera gombaszogensis* is present. The *Megantereon* is *M. whitei*. Therefore the site of Senèze is prior to Dmanisi. It is close in age to the Saint-Vallier site, but the few data which allow a biochronological approach show that Senèze and Saint-Vallier are not contemporaneous: *Nyctereutes megamastoides megamastoides* at Senèze, *Nyctereutes vulpinus* at Saint-Vallier; *Ursus etruscus* more evolved at Senèze; *Megantereon cultridens* and *Homotherium crenatidens* in both sites but the forms from Senèze seem to be nearer to the Olivola unit (2.0–1.8 Ma). To conclude, the study of the carnivores chronologically places Senèze around 2.1–2.0 Ma.

Finally, what are the arguments that can be used to date the carnivores of Senèze? It is necessary to take into account data from the biochronology of carnivores but also from pollen analyses and the different physical dating methods available:

(1) the lava flow of the Pié de Charenty, located on the northern edge of the maar and occuring before its formation, is dated at 2.3 ± 0.15 Ma (Prévot & Dalrymple 1970) to 2.53 ± 0.06 Ma (Couthures 1989);
(2) in 1989, close to Elhaï's core, a shorter core included a tephra layer interbedded in the sediments at 20.75 m below the surface. This tephra was dated at 2.10 ± 0.01 Ma by the $^{40}Ar/^{39}Ar$ method (Roger et al. 2000) and occurred within a short normal polarity zone (correlated with the Feni [formerly Réunion 1] subchron?);
(3) the pollen data of the fossiliferous layers show that these latter are more recent than the upper part of

Elhaï's core and therefore posterior to the normal polarity (Feni) subchron;

(4) the $^{40}Ar/^{39}Ar$ ages obtained on tephras by Nomade et al. (2014) are of interest because they relate to fossils (see also Delson et al. 2024 for recalculated ages and notes):

- in trench T2, parcel 233: 2.10 ± 0.03 Ma, some decimeters under *Mammuthus meridionalis*, *Dicerorhinus etruscus*, *Acinonyx pardinensis*, Hyaenidae indet., coprolites of hyaenids;
- in trench T5, parcel 172: 2.18 ± 0.03 Ma and 2.13 ± 0.04 Ma; these two dates were thought to bracket the fossiliferous layer with *D. etruscus*, *Allohippus senezensis* and other taxa, but it has now been shown that both dates overlie these fossils (Debard 2024);

(5) the biochronology, for me that of the carnivores.

All these data indicate an age around 2.1 to 2.0 Ma, and considering the uncertainty of the different methods, including the physical dating, they are ultimately very consistent. It is obvious that the paleontological record of the maar of Senèze lasted a long time, in relation to changes in the water level of the lake due to climatic conditions and progressive erosion at the level of the outlet. I retain for the fauna of carnivores in Senèze an age of 2.0 to 2.1 Ma, a little more recent than that of Saint-Vallier. The date of 2.4 Ma advanced for the latter site by Nomade et al. (2014, p. 10, Fig. 7) seems to me to be too old because it is not based on new dates and on the other hand because of the palynological data (absence of exotic taxa in LD3). On my side, I place the Saint-Vallier deposits in the middle Villafranchian around 2.4–2.2 Ma.

The recent excavations at the Senèze site carried out by Faure, Guérin and Delson from 2001 to 2006 have provided a large scale localisation of the paleontological remains, a precise knowledge of the corresponding stratigraphy and the first paleoenvironmental data linked to the fauna. We can only regret once more the lack of stratigraphic and taphonomic data from the old excavations, which would have allowed taking a global view of the total material of Senèze in this study. Nevertheless this is still a major site, invaluable for the carnivores. It has yielded two complete reference skeletons, *Megantereon cultridens* (NMB Se 311) and *Homotherium crenatidens* (FSL 210991).

Acknowledgments It is a pleasure for me to thank Claude Guérin and Martine Faure for having entrusted me with the study of the carnivores of Senèze. I have always been given very warm welcome and excellent working conditions at the Museum of Natural History in Basel (Switzerland) where Loïc Costeur and Martin Schneider were particularly helpful. The same is true for the assistance provided at the Collections of the Université Claude Bernard-Lyon 1 by Emmanuel Robert and Abel Prieur. I am grateful to Lars Werdelin whose advice has been essential in the difficult determination of *Dinofelis*. John Paul Hodnett helpfully allowed me to compare the calcaneum from the David Formation to *Chasmaporthetes lunensis*. Robert Norington from Canberra (Australia) has contributed extensively to the English translation of the text, for which I thank him sincerely. Eric Delson has also assisted with language editing. Finally, I thank the three reviewers, Lars Werdelin, Bienvenido Martínez-Navarro and Xiaoming Wang, for their valuable advice that helped improve this article.

References

Adams, D. B. (1979). The Cheetah: Native American. *Science, 205*, 1155–1158.

Antón, M., Turner, A., Salesa, M. J., & Morales, J. (2006). A complete skull of *Chasmaporthetes lunensis* (Carnivora, Hyaenidae) from the Spanish Pliocene site of La Puebla de Valverde (Teruel). *Estudios Geológicos, 62*, 375–388.

Argant, A. (1991). Carnivores quaternaires de Bourgogne. *Documents des Laboratoires de Géologie de Lyon, 115*, 1–301.

Argant, A. (2001). Cave bear ancestors. *Cadernos Lab. Xeolóxico De Laxe Coruña, 26*, 341–348.

Argant, A. (2004). Les Carnivores du gisement Pliocène final de Saint-Vallier (Drôme, France). In M. Faure & C. Guérin (Eds.), *Le gisement pliocène final de Saint-Vallier (Drôme, France). Geobios*, Mémoire spécial n° 26, *37*, 133–182.

Argant, A. (2010). Carnivores (Canidae, Felidae et Ursidae) de Romain-la-Roche (Doubs, France). *Revue de Paléobiologie, 29*, 495–601.

Argant, A., & Argant, J. (2011). The *Panthera gombaszogensis* story: the contribution of the Château Breccia (Saône-et-Loire, Burgundy, France). In A. Argant & J. Argant (coord.), *Proceedings of the 16th International Cave Bear and Lion Symposium, Azé (Saône-et-Loire, France). Quaternaire*, Hors Série n°4, 201–208.

Argant, J. (2004). Le gisement pliocène final de Saint-Vallier (Drôme, France): palynologie. In M. Faure & C. Guérin (Eds.), *Le gisement pliocène final de Saint-Vallier (Drôme, France). Geobios*, Mémoire spécial n° 26, 37, 81–90.

Argant, J. (2024). Palynological analysis of the fossiliferous layers of the late Villafranchian deposit of Senèze (Domeyrat, Haute-Loire, France). Taphonomy and paleoenvironmental data. In E. Delson, M. Faure & C. Guérin (Eds), *Senèze: Life in Central France two million years ago. Paleontology, geochronology, stratigraphy and taphonomy* (pp. 123–140). Cham (Switzerland): Springer.

Arribas, A., & Antón, M. (1997). Los carnivoros de los yacimientos pliocenos de Las Higueruelas y Piedrabuena (Ciudad real, España). *Boletín del Instituto geológico y minero de España, 108*, 111–120.

Arribas, A., & Garrido, G. (2008). Hiénidos [*Pachycrocuta brevirostris* (Aymard, 1846) y *Hyaena brunnea* Thunberg, 1820] del yacimiento de Fonelas P-1 (Cuenca de Guadix, Granada). In A. Arribas (Ed.), *Vertebrados del Plioceno superior terminal en el suroeste de Europa: Fonelas P-1 y el Proyecto Fonelas*. Madrid: Instituto Geológico y Minero de España, serie Cuadernos del Museo Geominero, *10*, 201–230.

Astre, G. (1929). Sur un félin à particularités ursoïdes des limons Pliocènes du Roussillon. *Bulletin de la Société géologique de France, 29*, 199–204.

Averianov, A. O., & Baryshnikov, G. F. (1999). *Dinofelis* sp. (Carnivora, Felidae) from the Late Pliocene locality Etulia in Moldavia. *Neues Jahrbuch für Geologie und Paläontologie-Monatshefte, 1999*, 531–540.

Ballesio, R. (1963). Monographie d'un *Machairodus* du gisement villafranchien de Senèze: *Homotherium crenatidens* Fabrini. *Travaux du laboratoire de Géologie de la Faculté des Sciences de Lyon,* N.S. *9*, 1–129.

Ballesio, R. (1979). Le gisement pléistocène supérieur de Jaurens à Nespouls, Corrèze, France : Les carnivores (Mammalia, Carnivora). I: Canidae et Hyaenidae. *Nouvelles archives du Muséum d'histoire naturelle de Lyon, 17*, 25–56. https://doi.org/10.3406/mhnly.1979.1041

Bartolini Lucenti, S., Bukhsianidze, M., Martínez-Navarro, B., & Lordkipanidze, D. (2020). The wolf from Dmanisi and augmented reality: Review, implications, and opportunities. *Frontiers in Earth Science, 8*, 131. https://doi.org/10.3389/feart.2020.00131

Bartolini Lucenti, S., Rook, L., & Morales, J. (2018). *Nyctereutes* (Mammalia, Carnivora, Canidae) from Layna and the Eurasian raccoon-dogs: An updated revision. *Rivista Italiana di Paleontologia e Stratigrafia, 124*, 597–616.

Baryshnikov, G. F. (2012). Pleistocene Canidae (Mammalia, Carnivora) from the Paleolithic Kudaro caves in the Caucasus. *Russian Journal of Theriology, 11*, 77–120.

Beaumont, G. de. (1983). Une mandibule de *Dinofelis* (Mammifère, Carnivore, Félidé) du Villafranchien inférieur de Balaruc II, Hérault, France. *Archives des Sciences*, Genève, *36*, 469–478.

Berta, A. (1981). The Plio-Pleistocene hyaena *Chasmaporthetes ossifragus* from Florida. *Journal of Vertebrate Paleontology, 1*, 341–356.

Berta, A., & Galiano, H. (1983). *Megantereon hesperus* from the late Hemphillian of Florida with remarks on the phylogenetic relationships of machairodonts (Mammalia, Felidae, Machairodontinae). *Journal of Paleontology, 57*, 892–899.

Bishop, M. J. (1982). The mammal fauna of the early Middle Pleistocene cavern infill site of Westbury-Sub-Mendip, Somerset. *Special Papers in Paleontology, 28*, 1–108.

Bonis, L. (de) (1976). Un félidé à longues canines de la colline de Perrier (Puy-de-Dôme): ses rapports avec les félinés machairodontes. *Annales de Paléontologie* (Vertébrés) *62*, 159–198.

Boudadi-Maligne, M. (2010). Les *Canis* pléistocènes du sud de la France : approche biosystématique, évolutive et biochronologique. Thèse de Doctorat, Université de Bordeaux.

Bout, P. (1960). Le Villafranchien du Velay et du Bassin hydrographique moyen et supérieur de l'Allier. Thèse de Doctorat d'Etat, Université de Paris.

Bout, P. (1970). Absolute ages of some volcanic formations in the Auvergne and Velay areas and chronology of the European Pleistocene. *Palaeogeography, Palaeoclimatology, Palaeoecology, 8*, 95–106.

Broom, R. (1937). On some new fossil mammals from limestone caves of the Transvaal. *South African Journal of Science, 33*, 750–769.

Brugal, J.-P., & Boudadi-Maligne, M. (2011). Quaternary small to large canids in Europe: Taxonomic status and biochronological contribution. *Quaternary International, 243*, 171–182.

Cardoso, J.L. (1993). La Hyène des « Oubliettes» de Gargas, *Crocuta crocuta spelaea* (Mammalia, Carnivora). *Bulletin du Muséum National d'Histoire naturelle*, Paris, 4ème sér., *15*, C, 79–104.

Cherin, M., Bertè, D. F., Rook, L., & Sardella, R. (2013). Re-defining *Canis etruscus* (Canidae, Mammalia): A new look into the evolutionary history of Early Pleistocene dogs resulting from the outstanding fossil record from Pantalla (Italy). *Journal of Mammalian Evolution, 21*, 95–110.

Christiansen, P., & Adolfssen, J. S. (2007). Osteology and ecology of *Megantereon cultridens* SE311 (Mammalia; Felidae; Machairodontidae), a sabre cat from the Late Pliocene-Early Pleistocene of Senèze, France. *Zoological Journal of the Linnean Society, 151*, 833–884.

Couthures, J. (1989). The Senèze maar (French Massif Central): Hypothesis regarding a catastrophe occuring about 1.5 million years ago. *Journal of Volcanology and Geothermal Research, 39*, 207–210.

Croizet, J.-B., & Jobert, A. C. G. (1828). *Recherches sur les ossemens fossiles du département du Puy-de-Dôme*. Paris: Clermont-Ferrand.

Debard, E. (2024). Geological study of the paleontological site of Senèze (Domeyrat, Haute-Loire, France). In E. Delson, M. Faure, & C. Guérin (Eds.), *Senèze: Life in Central France two million years ago. Paleontology, geochronology, stratigraphy and taphonomy*. (pp. 37–82). Cham (Switzerland): Springer

Delson, E., Faure, M., Guérin, C., Aprile, L., Argant, J., Blackwell, B. A. B., et al. (2006). Franco-American renewed research at the Late Villafranchian locality of Senèze (Haute-Loire, France). *Courier Forschungs-Institut Senckenberg, 256*, 275–290.

Delson, E., Nomade, S., Sen, S., Debard, E., Pastre, J-F., Bahain, J.-J., et al. (2024). Geochronology of Senèze: $^{40}Ar/^{39}Ar$ dating and magnetostratigraphy, with notes on an ESR/U-series dating attempt. In E. Delson, M. Faure & C. Guérin (Eds.), *Senèze: Life in Central France two million years ago. Paleontology, geochronology, stratigraphy and taphonomy* (pp. 99–122). Cham (Switzerland): Springer.

Depéret, C. (1890). Les animaux pliocènes du Roussillon. *Mémoires de la Société géologique de France, 3*, 5–195.

Desse, J., Chaix, L., & Desse-Berset, N. (1986). *"Ostéo", Base-réseau de données ostéométriques pour l'archéozoologie. Procédures, codages, exploitation des données, organisation du réseau.* Valbonne: Editions du C.N.R.S.

Devis, G. (1970). Le volcanisme et le gisement fossilifère de Senèze (Haute-Loire). *Revue des Sciences naturelles d'Auvergne, 36*, 25–87.

Dubois, A., & Stehlin, H. G. (1933). La grotte de Cotencher, station moustérienne. *Mémoires de la Société paléontologique Suisse, 52*, 1–178; *53*, 179–292.

Ehrlich, A. (1968). Les diatomées fossiles des sédiments villafranchiens de Senèze (Haute-Loire, Massif central français). *Bulletin de l'Association française pour l'étude du Quaternaire, 4*, 267–280.

Elhaï, H. (1969). La flore sporo-pollinique du gisement villafranchien de Senèze (Massif Central, France). *Pollen et Spores, 11*, 127–139.

Erdbrink, D. P. (1953). *A review of fossil and recent bears of the Old World. With remarks on their phylogeny, based upon their dentition.* Utrecht: Geologisch Instituut, Rijks Universiteit Utrecht.

Faure, M., Guérin, C., & Monguillon, A. (2003). Rapport sur la campagne de fouilles 2003 dans le gisement villafranchien supérieur de Senèze (commune de Domeyrat, Haute-Loire). Submitted to DRAC Auvergne, unpublished.

Fejfar, O., & Sabol, M. (2004). Pliocene carnivores (Carnivora, Mammalia) from Ivanovce and Hajnáčka (Slovakia). *Courier Forschungs-Institut Senckenberg, 246*, 15–53.

Ficcarelli, G. (1979). The Villafranchian Machairodonts of Tuscany. *Palaeontographia Italica, 71*, 17–26.

Ficcarelli, G. (1984). The Villafranchian cheetahs from Tuscany and remarks on the dispersal and evolution of the genus *Acinonyx*. *Palaeontographia Italica, 73*, 94–103.

Fourvel, J.-B. (2012). Hyénidés modernes et fossiles d'Europe et d'Afrique: Taphonomie comparée de leurs assemblages osseux. Thèse de Doctorat, Université Toulouse le Mirail.

Freudenberg, W. (1914). Die Säugtiere des älteren Quartärs von Mitteleuropa mit besonderer Berücksichtigung der Fauna von Hundsheim und Deutschaltenburg in Niederösterreich nebst Bemerkungen über verwandte Formen anderer Fundorte. *Geologische und Paläontologische Abhandlungen (Neue Folge), 12*, 455–670.

Galobart, À., Pons-Moyà, J., Antón, M., & Maroto, J. (2003). Descripción del material de *Homotherium latidens* (Owen) de los yacimientos del Pleistoceno inferior de Incarnal (Girona, NE de la Península Ibérica). *Paleontologia i Evolució, 3*, 90–141.

Garcia, N. (2003). *Osos y otros carnivoros de la Sierra de Atapuerca.* Fondation oso de Asturias.

Garrido, G. (2006). Paleontología sistemática de grandes mamiferos del yacimiento del Villafranquiense superior de Fonelas P-1 (Cuenca de Guadix, Granada). Tesis Doctoral Universidad Complutense de Madrid.

Garrido, G. (2008). El registro de *Vulpes alopecoides* (Forsyth Major, 1877), *Canis etruscus* Forsyth Major, 1877 y *Canis cf. falconeri* Forsyth Major, 1877 (Canidae, Carnivora, Mammalia) en Fonelas P-1 (Cuenca de Guadix, Granada). In A. Arribas (Ed.), *Vertebrados del Plioceno superior terminal en el suroeste de Europa: Fonelas P-1 y el Proyecto Fonelas* (pp. 159–186). Madrid: Instituto Geológico y Minero de España, serie Cuadernos del Museo Geominero, *10*.

Garrido, G., & Arribas, A. (2008). Hiénidos [*Pachycrocuta brevirostris* (Aymard, 1846) y *Hyaena brunnea* Thunberg, 1820] del yacimiento de Fonelas P-1 (Cuenca de Guadix, Granada). In A. Arribas (Ed.), *Vertebrados del Plioceno superior terminal en el suroeste de Europa: Fonelas P-1 y el Proyecto Fonelas* (pp. 201–230). Madrid: Instituto Geológico y Minero de España, serie Cuadernos del Museo Geominero, *10*.

Gibert, L., Scott, G. R., Montoya, P., Ruiz-Sánchez, F. J., Morales, J., Luque, L., et al. (2013). Evidence for an African-Iberian mammal dispersal during the pre-evaporitic Messinian. *Geology, 41*, 691–694.

Guth, C. (1975). Lagerstätten des Villafranchiums in der Umgebung von Brioude in Mittelfrankreich. *Quartärpaläontologie, 1*, 177–182.

Heintz, É., Guérin, C., Martin, R., & Prat, F. (1974). Principaux gisements villafranchiens de France: listes fauniques et biostratigraphie. *Mémoires du Bureau de Recherches Géologiques et Minières, 78*, 169–182.

Hemmer, H. (2001). Die Feliden aus dem Epivillafranchium von Untermassfeld. In R.-D. Kahlke, *Das Pleistozän von Untermassfeld bei Meiningen (Thüringen)*, Teil 3. *Monographien des Römisch--Germanischen Zentralmuseums Mainz, 40*, 699–782.

Hemmer, H., Kahlke, R.-D., & Keller, T. (2008). Geparde im Mittelpleistozän Europas: *Acinonyx pardinensis* (sensu lato) *intermedius* (Thenius, 1954) aus den Mosbach-sanden (Wiesbaden, Hessen, Deutschland). *Neues Jahrbuch für Geologie und Paläontologie, Abhandlungen, 249*, 345–356.

Hemmer, H., Kahlke, R.-D., & Vekua, A. K. (2011). The cheetah *Acinonyx pardinensis* (Croizet et Jobert, 1828) s. l. at the hominin site of Dmanisi (Georgia)—A potential prime meat supplier in Early Pleistocene ecosystems. *Quaternary Science Reviews, 30*, 2703–2714.

Hodnett, J-P., White, R., & Werdelin, L. (2011). The second record of a calcaneum for *Chasmaporthetes* (Mammalia; Carnivora; Hyaenidae); evidence from the St. David formation, Southeastern Arizona. In *Proceedings of the Arizona-Nevada Academy of Science*, fifty-fith annual meeting, 09/04/2011.

van Hooijdonk, K. (2006). De Sabeltandkatten *Homotherium* en *Megantereon* (Felidae, Carnivora) van de Plio-Pleistocene site van Senèze (Haute Loire, Fr.). *Cranium, 23*, 25–38.

Howell, F. C., & Petter, G. (1980). The *Pachycrocuta* and *Hyaena* lineages (Plio-Pleistocene and extant species of the Hyaenidae). Their relationships with Miocene Ictitheres: *Palyhyaena* and *Hyaenictitherium*. *Geobios, 13*, 579–623.

Hunter, L., & Barrett, P. (2011). *A field guide to the carnivores of the World*. Totnes, Devon: New Holland Publishers (UK).

Jung, J. (1947). *Géologie de l'Auvergne et de ses confins bourbonnais et limousins*. Mémoires du Service de la Carte géologique de France.

Khomenko, I. P. (1931). *Hyaena borissiaki* n. sp. aus der Russillion Fauna Bessarabiens. *Travaux de l'Institut de Paléozoologie de l'Académie des Sciences d'URSS, 1*, 81–136 [in Russian, abstract in German].

Koufos, G. D. (1992). The Pleistocene Carnivores of the Mygdonia Basin (Macedonia, Greece). *Annales de Paléontologie, 78*, 205–257.

Kurtén, B. (1974). A history of coyote-like dogs (Canidae, Mammalia). *Acta Zoologica Fennica, 140*, 1–38.

Kurtén, B., & Crusafont Pairó, M. (1977). Villafranchian Carnivores (Mammalia) from La Puebla de Valverde (Teruel, Spain). *Commentationes Biologicae, Societas Scientiarum Fennica, 85*, 3–39.

Martin, R. (1973). Trois nouvelles espèces de Caninae (Canidae, Carnivora) des gisements pliovillafranchiens d'Europe. *Documents des Laboratoires de Géologie de la Faculté des Sciences de Lyon, 57*, 87–96.

Martínez-Navarro, B., & Palmqvist, P. (1995). Presence of the African Machairodont *Megantereon whitei* (Broom, 1937) (Felidae, Carnivora, Mammalia) in the Lower Pleistocene site of Venta Micena (Orce, Granada, Spain), with some considerations on the origin, evolution and dispersal of the genus. *Journal of Archaeological Science, 22*, 569–582.

Martínez-Navarro, B., & Palmqvist, P. (1996). Presence of the African saber-toothed felid *Megantereon whitei* (Broom, 1937) (Mammalia, Carnivora, Machairodontinae) in Appolonia 1 (Mygdonia Basin, Macedonia, Greece). *Journal of Archaeological Science, 23*, 869–872.

Mazza, P., & Rustioni, M. (1994). The fossil bear from Senèze (Southern France). *Rendiconti Lincei. Scienze Fisiche e Naturali, 9* (5), 17–26.

Medin, T., Martínez-Navarro, B., Rivals, F., Madurell-Malapeira, J., Ros-Montoya, S., Espigares, M.-P., Figueirido, B., Rook, L., & Palmqvist, P. (2017). Late Villafranchian *Ursus etruscus* and other large carnivorans from the Orce sites (Guadix-Baza basin, southern Spain): Taxonomy, biochronology, paleobiology, and ecogeographical context. *Quaternary International, 431*, 20–41. https://doi.org/10.1016/j.quaint.2015.10.053

Medin, T., Martínez-Navarro, B., Madurell-Malapeira, J., Figueirido, B., Kopaliani, G., Rivals, F., Kiladze G., Palmqvist, P., & Lordkipanidze, D. (2019). The bears from Dmanisi and the first dispersal of early *Homo* out of Africa. *Scientific Reports, 9*, 17752. https://doi.org/10.1038/s41598-019-54138-6

Monguillon, A., Spassov, N., Argant, A., Kauhala, K., & Viranta, S. (2004). *Nyctereutes vulpinus* comb. et stat. nov. (Mammalia, Carnivora, Canidae) du Pliocène terminal de Saint-Vallier (Drôme, France). In M. Faure & C. Guérin (Eds.), *Le gisement pliocène final de Saint-Vallier (Drôme, France). Geobios*, Mémoire spécial n° 26, *37*, 183–188.

Morales, J. (1985). El registro fósil de los Carnívoros ibéricos. In R. García-Perea, R. A. Baquero, R. Fernández-Salvador & J. Gisbert (Eds.), *Carnívoros, evolución, ecología y conservación* (pp. 117–135). Madrid: Museo Nacional de Ciencias Naturales.

Nomade, S., Pastre, J.-F., Guillou, H., Faure, M., Guérin, C., Delson, E., et al. (2014). $^{40}Ar/^{39}Ar$ constraints on some French landmark Late Pliocene to Early Pleistocene large mammalian paleofaunas: Paleoenvironmental and paleoecological implications. *Quaternary Geochronology, 21*, 2–15.

Morales, J., & Aguirre, E. (1976). Carnivoros de Venta del Moro. *Trabajos Sobre Neogeno-Cuaternario, 5*, 31–81.

Palmqvist, P., Torregrosa, V., Pérez-Claros, J. A., Martínez-Navarro, B., & Turner, A. (2007). A re-evaluation of the diversity of *Megantereon* (Mammalia, Carnivora, Machairodontinae) and the problem of species identification in extinct Carnivores. *Journal of Vertebrate Paleontology, 27*, 160–175.

Perrot, R., Morel, P., & Ballesio, R. (1972). Les proliférations osseuses en paléopathologie, les exostoses isolées à propos de l'humérus d'un *Machairodus* découvert à Senèze (Haute - Loire). *Cahiers Médicaux Lyonnais, 48*(32), 3647–3658.

Piveteau, J. (1948). Un Félidé du Pliocène du Roussillon. *Annales de Paléontologie, 34*, 3–28.

Prévot, M., & Dalrymple, G. B. (1970). Un bref épisode de polarité géomagnétique normale au cours de l'époque inverse Matuyama. *Comptes Rendus de l'Académie des Sciences Série D, Sciences Naturelles, 271*, 2221–2224.

Roger, S., Coulon, C., Thouveny, N., Féraud, G., Van Velzen, A., Fauquette, S., et al. (2000). ^{40}A/^{39}/A dating of a tephra layer in the Pliocene Senèze maar lacustrine sequence (French Massif Central): Constraint on the age of the Reunion-Matuyama transition and implications on paleoenvironmental archives. *Earth and Planetary Science Letters, 183*, 431–440.

Rook, L. (1993). I cani dell'Eurasia dal Miocene superiore al Pleistocene medio. Ph.D. dissertation, Universities of Modena, Bologna, Firenze, Roma "La Sapienza".

Rook, L., & Torre, D. (1996). The wolf-event in Western Europe and the beginning of Late Villafranchian. *Neues Jahrbuch für Geologie und Paläontologie, Monatshefte, 8*, 495–501.

Sardella, R. (1998). The Plio-Pleistocene dirk-toothed cat *Megantereon* ex gr. *cultridens* (Mammalia, Felidae, Machairodontinae) with comments on taxonomy, origin and evolution. *Neues Jahrbuch für Geologie und Paläontologie, Abhandlungen, 207*, 1–36.

Sardella, R., Petrucci, M., & Rook, R. (2008). The African species *Megantereon whitei* from the Early Pleistocene of Monte Argentario (South Tuscany, Central Italy). *Comptes Rendus Palevol, 7*, 601–606.

Schaub, S. (1925). Über die Osteologie von *Machaerodus cultridens* Cuvier. *Eclogae Geologicae Helvetiae, 19*, 255–266.

Schaub, S. (1942). Un nouveau Félidé de Senèze (Haute-Loire). *Eclogae Geologicae Helvetiae, 35*, 183–189.

Schaub, S. (1943). Die oberpliocaene Säugetierfauna von Senèze (Haute-Loire). *Eclogae Geologicae Helvetiae, 36*, 270–289.

Schaub, S. (1949). Révision de quelques carnassiers villafranchiens du niveau des Etouaires (Montagne de Perrier, Puy-de-Dôme). *Eclogae Geologicae Helvetiae, 42*, 492–506.

Schlosser, M. (1903). Die fossilen Säugthiere Chinas nebst einer Odontographie de recenten Antilopen. *Abhandlungen der Bayerischen Akademie der Wissenschaften, 22*, 1–220.

Soria, D., & Aguirre, E. (1976). El Canido de Layna: revision de los *Nyctereutes* fosiles. *Trabajos Sobre Neogeno-Cuaternario, 5*, 83–114.

Sotnikova, M. S. (1989). Late Pliocene-early Pleistocene Carnivora stratigraphic significance. *Academy of Sciences of the USSR, Transactions, 140*, Moskva (in Russian).

Sotnikova, M. V., Baigusheva, V. S., & Titov, V. V. (2002). Carnivores of the Khapry faunal assemblage and their stratigraphic implications. *Stratigraphy and Geological Correlation, 10*, 375–390.

Spassov, N. (1997). Villafranchian succession of mammalian megafaunas from Bulgaria and the biozonation of South-East Europe. *Mémoires et Travaux de l'Institut de Montpellier, Ecole Pratique des Hautes Etudes, 21*, 669–676.

Spassov, N. (2000). Biochronology and zoogeographic affinities of the Villafranchian Faunas of Bulgaria and South Europe. *Historia naturalis Bulgarica, 12*, 89–128.

Spassov, N. (2003). The Plio-Pleistocene vertebrate fauna in South-Eastern Europe and the megafaunal migratory waves from the east to Europe. *Revue de Paléobiologie, 22*, 197–229.

Spassov, N. (2011). *Acinonyx pardinensis* (Croizet et Jobert) remains from the Middle Villafranchian locality of Varshets (Bulgaria) and the Plio-Pleistocene history of the cheetah in Eurasia. *Estudios Geológicos, 67*, 245–253.

Stehlin, H. G. (1923). Die oberpliocaene Fauna von Senèze. *Eclogae Geologicae Helvetiae, 18*, 268–281.

Tedford, R. H., & Qiu, Z. (1991). Pliocene *Nyctereutes* (Carnivora, Canidae) from Yushe, Shanxi, with comments on Chinese fossil racoondogs. *Vertebrata PalAsiatica, 29*, 176–189.

Tseng, Z. J., Li, Q., & Wang, X. (2013). A new cursorial Hyena from Tibet, and analysis of biostratigraphy, paleozoogeography, and dental morphology of *Chasmaporthetes* (Mammalia, Carnivora). *Journal of Vertebrate Paleontology, 33*, 1457–1471.

Tsoukala, E. (2004). The early pleistocene carnivores (Mammalia) from Ceyssaguet (Haute-Loire). *Paleo, 16*, 193–242.

Turner, A., & Antón, M. (1996). The giant hyaena *Pachycrocuta brevirostris* (Mammalia, Carnivora, Hyaenidae). *Geobios, 29*, 455–468.

Turner, A., & Antón, M. (1997). *The big cats and their fossil relatives. An illustrated guide to their evolution and natural history*. New York: Columbia University Press.

Vekua, A. K. (1972). *The Kvabebi Fauna of Akchagylian Vertebrates*. Moscow: Nauka.

Vekua, A. K. (1995). Die Wirbeltierfauna des Villafranchium von Dmanisi und irhe biostratigraphische Bedeutung. *Jahrbuch des Römisch-Germanischen Zentralmuseums Mainz, 42*, 77–180.

Villalta, J.-F. (de). (1952). Contribución al conocimiento de la fauna de mamiferos fósiles del Plioceno de Villarroya (Logroño). *Boletín del Instituto geológico y minero de España, 64*, 1–201.

Viret, J. (1954). Le loess à bancs durcis de Saint-Vallier (Drôme) et sa faune de mammifères villafranchiens. *Nouvelles Archives du Muséum d'Histoire naturelle de Lyon, 4*, 1–200.

Wagner, J. (2006). A list of craniodental material of Pliocene Ursids (genus *Ursus*) in the collection of Naturhistorisches Museum Basel. *Scientific Annals, School of Geology Aristotle University of Thessaloniki, 98*, 127–139.

Wagner, J. (2011). A short nomenclature note to correct spelling of the species-group name "*Gombaszögensis*". In A. Argant & J. Argant (coord.), *Proceedings of the 16th international cave bear and lion symposium, Azé (Saône-et-Loire, France). Quaternaire*, Hors Série n°4, 245.

Werdelin, L., & Solounias, N. (1991). The Hyaenidae: Taxonomy, systematics and evolution. *Fossils and Strata, 30*, 1–104.

Werdelin, L., & Lewis, M. E. (2001). A revision of the genus *Dinofelis* (Mammalia, Felidae). *Zoological Journal of the Linnean Society, 132*, 147–258.

Zdansky, O. (1925). Quartäre Carnivoren aus Nord-China. *Palaeontologica Sinica, C, 2*(2), 1–29.

Chapter 10
Dicerorhinus etruscus etruscus from the Late Villafranchian Site of Senèze (Domeyrat, Haute-Loire, Central France)

Claude Guérin

Abstract Many rhino remains were found in the fossiliferous site of Senèze, all belonging to *Dicerorhinus etruscus* (Falconer, 1859). The new excavations yielded a mandible fragment, two proximal radial epiphyses, 11 complete or subcomplete carpal bones, six metacarpals, 13 tarsal bones, seven metatarsals and numerous phalanges and sesamoids, pertaining for the most part to one individual. Including the numerous specimens gathered during the old excavations, but excluding the heavily fragmented or damaged pieces as well as the vertebrae and ribs, the grand total for the Senèze rhino remains is 17 crania, 19 mandibles, four isolated upper tooth rows, about 20 isolated cheek teeth, 11 scapulae, 14 humeri, 18 radii, 12 ulnae, 60 carpals, 45 metacarpals, 15 femora, 27 patellas, 12 tibiae, 60 tarsals, 43 metatarsals, 105 phalanges and 70 sesamoids. Such a quantity of fossils makes Senèze by far the richest locality for *Dicerorhinus etruscus*. These remains are compared anatomically and biometrically with the homologous material of *Dicerorhinus etruscus etruscus* of other European Villafranchian sites, the Middle Pleistocene *D. etruscus brachycephalus* and the Early Villafranchian *Dicerorhinus jeanvireti*. Owing to its size and proportions, all the Senèze material is identified as belonging to evolutionary stage II of *Dicerorhinus etruscus etruscus*, typical of the Late Villafranchian. Its presence therefore helps to date the site and to identify its paleoenvironment as a wooded savanna with a rather wet climate. The osteological and biometrical differences between *D. etruscus etruscus* and the younger and somewhat larger *D. etruscus brachycephalus* do not justify a full species status for the latter.

Résumé De nombreux restes de Rhinocerotidés ont été découverts dans le site fossilifère de Senèze, tous attribués à *Dicerorhinus etruscus* (Falconer, 1859). Les nouvelles fouilles ont livré du nouveau matériel. La quantité de fossiles en fait de loin le plus riche gisement à *D. etruscus*. Ces restes sont comparés du point de vue anatomique et biométrique avec du matériel homologue de *D. etruscus etruscus* d'autres gisements villafranchiens d'Europe, avec *D. etruscus brachycephalus* du Pléistocène moyen, et *D. jeanvireti* du Villafranchien ancien. Compte tenu de sa taille et de ses proportions, les restes de rhinocéros de Senèze correspondent au stade évolutif II de *D. etruscus etruscus*, caractéristique du Pléistocène ancien. Sa présence contribue à la datation du site et indique un environnement de savane arborée et un climat plutôt humide. Les différences ostéologiques et biométriques entre *D. etruscus etruscus* et le plus récent et un peu plus grand *D. etruscus brachycephalus,* ne justifient pas une différenciation spécifique de ce dernier.

Keywords Rhinocerotidae • *Stephanorhinus* • Anatomy • Biometry • Biochronology • Paleoenvironment

Mots-clés Anatomie • Biométrie • Biochronologie • Paléoenvironnement

Introduction

The Late Villafranchian (Early Pleistocene) site of Senèze (Haute-Loire) has been known since the extraction, during roadwork in 1892, of a complete *Mammuthus meridionalis* skeleton. Boule (1892) published this discovery and pointed out the presence of rhinoceros remains in the site, but without indication of a species.

Four years later the Muséum national d'Histoire naturelle in Paris acquired a well preserved rhino skull from Senèze

Claude Guérin—Deceased 2016.

C. Guérin (✉)
Université Claude Bernard-Lyon 1, Lyon, France

E. Delson et al. (eds.), *Senèze: Life in Central France Around Two Million Years Ago*,
Vertebrate Paleobiology and Paleoanthropology, https://doi.org/10.1007/978-3-031-64415-3_10

(Catalogue MNHN.F.1896–22), donated by Boule and Vernière (Schaub 1943a). This fossil, the first rhino specimen recovered at Senèze, had been sold to amateur collector A. Vernière by Pierre Philis, a young local peasant. It is exhibited in the Galerie de Paléontologie of the MNHN (see Fig. 1.2 in Faure et al. 2024). In the following decades, most of the material found at Senèze was excavated between 1907 and 1942 by Philis, who gathered a huge quantity of fossils which was sold to the NMB, the FSL and the MNHN (see Faure et al. 2022). Depéret and Mayet (1912) reported that they found numerous cheek teeth, small limb bones, fragments of long bones and a posteriorly damaged cranium and mandible. For the first time the rhino remains were identified as *Rhinoceros etruscus*. In an activity report, the *Bulletin du Muséum national* mentions the making of a cast of "un crâne de jeune *Rhinoceros etruscus*" exhibited to the public (Anonymous 1927). A complete mounted skeleton of "*Rh.*" *etruscus* has been preserved in the FSL collection since the 1920s (Roman 1931).

The fine state of preservation of many fossils allowed the first general paleontological publications of the Senèze fauna utilizing the Basel collection: Stehlin (1923) pointed out the presence in Basel of three crania of *Rhinoceros etruscus*: one juvenile and two adults, including one without the ossified nasal wall. Schaub (1943b Fig. 1; and later Wittmann, 1980, Fig. 79) illustrated the mounted skeleton NMB Se 1711 and noted that most of the spinal column and the rib cage are from another individual.

Several more recent scientific papers referred to the Senèze rhinos: Viret (1954) noted that the *Rhinoceros etruscus* from Senèze is very similar to that of Saint-Vallier (Drôme, France, Middle Villafranchian). Heintz et al. (1974) provided a critical revision of the faunal lists of the main Villafranchian deposits of France and confirmed the presence of *Dicerorhinus etruscus* in all but one of the classical French sites. More recently Lacombat (2005, 2006) published a part of the MNHN material.

Some small excavations were made by diverse scholars at Senèze between 1953 and 1991 (see Faure et al. 2024), but no new rhino remains were discovered with the exception of an incomplete juvenile cranium found by Guth and his team at the beginning of the sixties (Bout 1970; Guth 1975 Fig. 3). This specimen is stored in the Poitiers University collection. In his revision of the West European Late Miocene to Pleistocene rhinos, Guérin (1980) studied it and made extensive use of all the Senèze fossils for biometrical purposes but did not describe them individually.

The 2000–2006 excavations of the Franco-American team co-directed by Faure, Guérin and Delson represented the first real paleontological research ever made on a large scale at Senèze. During this excavation work, new fossil rhino specimens were found, especially the major part of an adult skeleton whose autopodials are complete and well preserved (see Parenti et al. 2024, Fig. 2.5). The purpose of the present work is to list and describe all the rhinoceros material found in Senèze.

The following abbreviations are used:
abs. = absolute; anat. = anatomical; APD = antero-posterior diameter; art. = articular; CV = coefficient of variation; D,d = deciduous premolar; dia. = diaphysis; dI, di = milk incisor; dist. = distal; h = height; I = incisor; L = length; M,m = molar; Mc = metacarpal; Mt = metatarsal; P,p = premolar; proc. = process; prox. = proximal; SD = standard deviation; TD = transverse diameter; w = width. Upper teeth are indicated by capital letters, lowers by lowercase. Institution acronyms: FSL, Université Claude Bernard-Lyon 1 (formerly Faculté des Sciences, Université de Lyon); MNHN, Muséum national d'Histoire naturelle, Institut de Paléontologie, Paris; NMB, Naturhistorisches Museum Basel. The ages of the fossiliferous sites are given following the MNQ biozones system (Guérin 1980, 1982, 1990, 2007). The measurements provided in Tables 1−29 were defined in Guérin (1980).

Systematic Paleontology

Order: Perissodactyla Owen, 1848
Suborder: Ceratomorpha Wood, 1937
Family: Rhinocerotidae Owen, 1845
Subfamily: Dicerorhininae Simpson, 1945
Genus: *Dicerorhinus* Gloger, 1841

Simplified synonymy: *Didermocerus* Brookes, 1828 (invalidated in 1977, Opinion 1080, by the International Commission on Zoological Nomenclature because it was published in a catalogue).

Ceratorhinus Gray, 1868, junior synonym.
Stephanorhinus Kretzoï, 1942, junior synonym.
Procerorhinus Kretzoï, 1942, junior synonym.
Brandtorhinus Guérin (1980), as a subgenus.
Type species *Dicerorhinus sumatrensis* (Fischer, 1814).

The genus *Dicerorhinus* Gloger, 1841 is still represented in the modern fauna by *D. sumatrensis*: it includes many Neogene and Quaternary species in Eurasia and Africa. McKenna and Bell (1997) estimated that *Dicerorhinus* is perfectly valid in this sense. The European Villafranchian and Quaternary species *D. jeanvireti*, *D. etruscus*, *D. mercki* and *D. hemitoechus* share three derived characters, namely the possession of a partially ossified nasal septum, a gradual increase of hypsodonty of the cheek teeth and the complete disappearance of the anterior teeth. These characters find their maximum expression in the "woolly" rhinoceros *Coelodonta antiquitatis* of the European Pleistocene.

Dicerorhinus etruscus (Falconer, 1859).

Four species of *Dicerorhinus* are known during the Ruscinian and the Villafranchian of Europe (Guérin 1980): the Ruscinian *D. megarhinus* (de Christol, 1834); the Late Ruscinian *D. miguelcrusafonti* Guérin & Santafe, 1978; the Early Villafranchian *D. jeanvireti* Guérin, 1972; and the Villafranchian and Middle Pleistocene *D. etruscus* (Falconer, 1859). The Middle Pleistocene representative of *D. etruscus* is best recognized as an evolved subspecies, *D. etruscus brachycephalus* (Schröder, 1903; see Guérin 1980). All rhinoceros remains from Senèze correspond to the nominal subspecies *D. etruscus etruscus*.

The holotype is the skull IGF 756, preserved in the Museum of Firenze University. It was collected in the Upper Valdarno Formation of Tuscany.

The dimensions and proportions of the bones and the teeth of *D. etruscus etruscus* of Senèze will be compared with those of the same subspecies found in other sites, and with those of the immediately preceding and succeeding European rhinos, *D. jeanvireti* and *D. etruscus brachycephalus*.

Editorial note: Due to the untimely death of Claude Guérin in 2016, his completed manuscript was edited by his collaborators Eric Delson and Martine Faure. C. Guérin disagreed with Fortelius et al. (1993) about including *D. etruscus* in *Stephanorhinus* Kretzoi, 1942, a genus which he considered invalid. He discussed this in Guérin (1980, p. 444) and repeated this view in several subsequent publications; we follow his opinion here.

Material

The recently excavated specimens have numbers beginning with SEN followed by the 2-digit year and a 4-digit sequence number in that year, e.g., FSL SEN 04–0276, and will be housed in the FSL. Most of the material discovered during previous excavations is stored in the NMB (numbers beginning with Se, often followed by the year of cataloguing), in the FSL and in the MNHN, plus the one fossil excavated during the 1960s in the University of Poitiers. The small quantity stored in Clermont-Ferrand University, originating mainly in the old Roux-Vernière private collection, was transferred during the 1970s to FSL. Those fossils have no formal FSL numbers (or no number at all), so we refer to them below and in the Tables as "ex-Clermont". The entire Seneze sample includes:

- 18 complete or partial skulls or crania, including 15 adult or subadult ones and two juveniles.

NMB: skull Se 561 (1913); cranium Se 1703; skull Se 1711 (1927); cranium Se 548 (adult but without nasal bony wall); cranium Se 1785 (1944), pathologic, and with a cervid humerus in the right zygomatic arch and an equid rib inside the nasal cavity; cranium with broken nasal bones Se 791 (1911); skull Se 560 (1913); the back part of three crania Se 743, Se 753 (both 1913) and Se 1406 (1912); palate with both tooth-rows Se 187.

Univ. Poitiers: the forepart of a juvenile cranium.

MNHN: 3 crania–1923–4 (figured in Lacombat 2005, p. 22 Fig. 9–2); 1896–22; 1922–15 (juvenile).

FSL: 3 crania–210959 (ex 96129); 210958 (ex 602 V); one specimen without number.

- Upper cheek-teeth:

FSL SEN: right P3 04-0276.

NMB: left D4 Se 1509 (1914); left M2 Se 334 (1911); P2-P4 Se 1416 (1909); right D2 and D3 Se 325 or 335 (1911).

FSL: right D4 (or 3) 210930 (ex 3707); left P1 210925 (ex 3707); left M3 3707; left D1 3707; right D4 3707; left isolated P1 (ex-Clermont); two reconstructed left and right upper tooth-rows (ex-Clermont, teeth set in plaster) 211118.

- 14 mandibles (plus 5 associated with crania listed above):

FSL SEN: left fragment 01–0004 with p2, p3 and m1.

NMB: left mandible with p3-m3 Se 1420 (1912); left juvenile mandible with d1-d3 Se 1862 (1944); fragment of a right mandible with p4-m3 (1920); left mandible with p2-4 and m3 Se 1413; fragment of a left mandible with m1-m3 Se 1863 (1944); left mandible with p3-m3 Se 188; right juvenile mandible with d2-d4 Se 1715 (1928); right juvenile mandible with d1-d4 Se 1682 (1920); mandible with d4-m3, both sides, without number (1920).

FSL: mandible 210957 (ex 604 V) with all the cheek-teeth; left corpus with p4-m3, ex-Clermont n° 3275; right corpus with p3-m3, ex-Clermont n° 3274; fragment of left corpus with m1-m3 ex-Clermont.

- Lower cheek-teeth: NMB: right m3 Se 330–333 (1909–1911); lower molar Se 88 (1908); right d4 Se 1509 (1914); Se 1422; left p2 Se 1416 (1909). FSL: right m2 3707.
- Atlas: NMB: Se 1409 (1911), Se 1410 (1908), Se 1412 and Se 1412 (both 1911), Se 1708 (1927). FSL: 210959 (ex 96129); 3712.
- Axis: NMB: Se 741 (1913); Se 1384 (1912); incomplete without number (1930); Se 1708 (1927); Se 565 (1913); incomplete Se 1756 (1933). FSL: 210959 (ex 96129).
- scapulae: NMB: right Se 1703 (1926); left proximal fragment Se 1703 (1926); right Se 747 (1913); right incomplete Se 1403 (1912); left Se 1401 (1909); left fragment Se 192 (1908); fragment Se 1756 (1933); right fragment Se 1764 (1933). Univ. Poitiers: 1. FSL: right and left 210959 (ex 96129).

- humeri: NMB: proximal fragment Se 1703 (1926); left distal half Se 1703 (1927); right distal fragment Se 1397 (1910); left diaphysis Se 1740 (1913); right Se 801; incomplete right Se 191 (1908); two distal fragments Se 1756 (1933); left and right Se 1711. MNHN: 1 right 1923–8; juvenile left (incomplete) and right 1922–5. FSL: left and right 210954 (ex 96129).
- radii: FSL SEN: two proximal fragments 01–0007, left, and 01–0109, right. NMB: right distal fragment Se 1703 (1926); left distal half Se 1703; left Se 756 (1933); juvenile left and right Se 1756 (1933); proximal two thirds of a right Se 1703; complete left Se 1703; left and right Se 1711. FSL: left and right 210959 (ex 96129); 1 complete 210936 (ex 594 V); 1 distal third 210931 (ex 3709); 1 ex-Clermont. MNHN: juvenile left and right 1922–5.
- ulnae: NMB: proximal fragment Se 1703 (1926); fragment Se 1399 (1909); Se 1765 (1933); complete Se 1703; distal fragment Se 1703; left and right Se 1711. MNHN: proximal fragment 1921–9; juvenile left and right 1922–5. FSL: left and right 210959 (ex 96129).
- scaphoids: FSL SEN: right 01–0086; left 01–0156. NMB: right incomplete without number (1930); left Se 1703 (1927); right Se 1703 (1927); right Se 1756 (1933); left Se 1756 (1933). MNHN: left and right 1922–5. FSL: left and right 210959 (ex 96129).
- lunars: FSL SEN: left 01–0125; right 01–0156; NMB: left Se 1703 (1927); right Se 1703 (1927); Se 1756 (1933); right Se 1388 (1911). MNHN: left and right 1922–5. FSL: left and right 210959 (ex 96129).
- pyramidals: FSL SEN: right 01–0087; left fused with a pisiform 01–0069. NMB: right Se 1703 (1927); left Se 1729 (1928). FSL: left and right 210959 (ex 96129); 1 ex-Clermont; 1 without number. MNHN: left 1923–8: left and right 1922–5.
- pisiforms: FSL SEN: incomplete right 01–0088; left fused with a pyramidal 01–0069. NMB; left and right Se 1703 (1927. FSL: left and right 210959 (ex 96129).
- trapezoideum: FSL SEN: right 01–0132. NMB: left and right Se 1703 (1927); right Se 1763 (1933); Se 1756 (1933). MNHN: 1922–5. FSL: left and right 210959 (ex 96129).
- magnums: FSL SEN: right 01–0126 + 01–0123. NMB: right Se 1703 (1927). MNHN: left 1923–8; 1 n° 1922–5. FSL: left and right 210259 (ex 96129).
- unciforms: FSL SEN: incomplete right 01–0127. NMB: incomplete left and right Se 1756 (1933); right Se 1387 (1909). MNHN: left 1923–8; right 1923–4; 1922–5. FSL: left and right 210959 (ex 96129).
- Mc II: FSL SEN: right 01–0103. NMB: left Se 1703 (1926); left Se 1756 (1933); left and right Se 1711. FSL: left and right 210959 (ex 96129); right 210946 (ex 3711); right ex-Clermont-Ferrand. MNHN: left 1923–8; juvenile left and right 1922–5.
- Mc III: FSL SEN: right 01–0106. NMB: right Se 1634; right Se 1703 (1927); right Se 1392; left Se 1736 (1905); left Se 1756; right Se 1756 (1933); left Se 1755 (1933); left and right Se 1711. MNHN: left 1923–8; juvenile left and right 1922–5. FSL: left and right 210959 (ex 96129).
- Mc IV: FSL SEN: right 01–0197. NMB: right Se 1703 (1927); right Se 1756 (1933); left Se 1706 (1927). MNHN: left 1923–8; juvenile left and right 1922–5; left and right Se 1711. FSL: left and right 210959 (ex 96129); right 210946 (ex 3711).
- Mc V: FSL SEN: 01–0124; 01–0156; 01–0131. NMB: right Se 1703 (1927). FSL: two without numbers.
- pelvis: NMB: Se 190 (1908); Se 1757 (1933); Se 1758 (1933); Se 565 (1914); Se 746 (1910); Se 1705 (1927); Se 1383; Se 1703; Se 1383 (1912).
- femora: NMB: right juvenile Se 1756 (1933); left fragment juvenile Se 1756 (1933); very juvenile diaphysis Se 1684 (1920); left diaphysis Se 1400 (1909); left distal fragment Se 1759 (1933); right fragment Se 739 (1913); right distal fragment Se 1707 (1927); left and right (incomplete) Se 1711. MNHN: left and right 1923–8; juvenile left and right 1922–5. FSL: left and right 210953 (ex 96129); right 591 V; left without number.
- patellae: FSL SEN: right 06-0080. NMB: Se 178; fragment Se 1510 (1911); Se 1386 (1910); fragment Se 1398 (1911); Se 1756 (1933); Se 787; Se 726; Se 1385 (1912); Se 788; Se 780 (1911); Se 782; Se 781; Se 786 (1908); Se 783 (1912). MNHN: left and right 1923–8; left and right 1922–5. FSL: 210959 (ex 96129); 5 all n° 3708; 2 without number.
- tibiae: NMB: juvenile right distal half Se 1756 (1933); left and right Se 1711. Univ. Poitiers: 1. MNHN: left and right 1923–8; juvenile left and right 1922–5. FSL: left

and right 210956 (ex 96129); left 210956 (589 V); right 210952.
- fibulae: MNHN: 1922–5. FSL: 1 left without number.
- astragali: FSL SEN: left 01–0203; right 01–0228. NMB: right without number (1930); right Se 1756 (1933); fragment of a right Se 1756 (1933); left Se 784 (1912). MNHN: left and right 1923–8; left 1922–5. FSL: left and right 210959 (ex 96129).
- calcanei: FSL SEN: right 01–0224. NMB: incomplete left juvenile Se 1756 (1933); juvenile right Se 1756 (1933). MNHN: left and right 1923–8; left 1922–5. FSL: left and right 210959 (ex 96129).
- cuboids: FSL SEN: left 01–0200; right 01–0230. NMB: incomplete Se 1756 (1933); Se 1756 (1933); Se 1686 (1920). FSL: left and right 210959 (ex 96129); left ex-Clermont. MNHN: left and right 1923–8; right 1923–4; left 1922–5.
- naviculars: FSL SEN: left 01–0199; right 01–0227. NMB: Se 1756 (1933). MNHN: left and right 1923–8; left 1922–5. FSL: left and right 210959 (ex 96129).
- lateral cuneiforms FSL SEN: right 01–0204; left 01–0222. NMB: Se 1756 (1933). MNHN: left and right 1923–8; 1923–4; left 1922–5. FSL: left 210959 (ex 96129).
- intermediate cuneiforms: FSL SEN: 01–0187; 01–0225. NMB: Se 1762 (1933); right Se 1763; Se 1687 (1920). MNHN: left 1922–5. FSL: right 210959 (ex 96129).
- medial cuneiforms FSL SEN: 01–0182; 01–0226. NMB: Se 1687 (1920). MNHN: left and right 1923–8. FSL: 1 without number.
- Mt II: FSL SEN: right 01–0192; left 01–0221; left 04–0015. NMB: juvenile left proximal third and complete right Se 1756 (1933); left and right Se 1711. MNHN: left and right 1923–8; juvenile left and right 1922–5. FSL: left (complete) and right (incomplete) 210959 (ex 96129); left 210942.
- Mt III: FSL SEN: left 01–0192; right 01–0206. NMB: right proximal half Se 191 (1908); distal half Se 1756 (1933); right Se 1634 (1920); left and right Se 1711. MNHN: left and right 1923–8; juvenile left and right 1922–5. FSL: left and right 210959 (ex 96129).
- Mt IV: FSL SEN: left 01–0197; right 01–0223 + 030199 (two parts of one bone). NMB: proximal half Se 1864 (1944); proximal fragment Se 1865; 2 right fragments Se 1761 (1933); proximal epiphysis Se 1865 (1944); left juvenile Se 1756 (1933); incomplete right Se 1756 (1933); left Se 1711. MNHN: 1923–8. FSL: left and right 210959 (ex 96129); left and right 210940 (ex 587 V).
- Axial proximal phalanges: FSL SEN: 01–0098; 01–0189; 01–0216. NMB: Se 1703 (1927); Se 737 (1913); 4 × Se 1756 (1933). FSL: 4 × 210959 (ex 96129); 1 587 V; 1 without number.
- Abaxial proximal phalanges: FSL SEN: 2 × 01–0101; 01–0076; 01–0220; 01–0205; 01–0184. NMB: Se 1703 (1927); 3 × Se 1756 (1933); Se 1390 (1909); Se 1394 (1909). FSL: 8 × 210959 (ex 96129); 2 × 587 V; 2 without number.
- Axial middle phalanges: FSL SEN: 01–0095; 01–0188; 01–0218. NMB: Se 737 (1913); Se 1393; juvenile Se 1772 (1936); Se 1703. FSL: 4 × 210959 (ex 96129); 2 without number; 587 V.
- Abaxial middle phalanges: FSL SEN: 01–0090; 01–0215; 01–0186; 01–0233; 01–0077; 01–0078. NMB: 2 × Se 1703 (1927); Se 1393; 2 × Se 1756 (1933); Se 1390 (1909); Se 1394 (1909). FSL: 2 × 587 V; 3 × 210959 (ex 96129); 7 without number.
- Axial distal phalanges: FSL SEN: 01–0092; 01–0179. NMB: Se 1703; Se 1391 (1910). FSL: 587 V; 1 without number; 2 × 210959 (ex 96129).
- Abaxial distal phalanges: FSL SEN: 01–0089; 01–0183; 01–0180; 01–0093. 01–0181. NMB: Se 1685 (1920); 2 × Se 1703; fragment Se 1635. FSL: 2 × 587 V; 7 × 210959 (ex 96129); 3 without number.
- Sesamoids: FSL SEN: 01–0091; 01–0097; 01–0193; 01–0212; 03–0187; 01–0211; 01–0185; 01–0102; 01–0201; 01–0100; 01–0099; 01–0121; 02–0186; 01–0190; without number. NMB: 6 × Se 1703 (1927); 12 × Se 1711 (1927); 1 Se 1387 (1909); 1 Se 1392 (1909). FSL: 11 × 3714; 12 without number; 2 × 587 V.

Some more or less complete skeletons were mounted, like NMB Se 1711 (1927) and FSL 210959 (ex 96129), the latter being dismounted. Such a quantity of fossils makes Senèze by far the richest locality for *Dicerorhinus etruscus*.

Description

The following description is built mainly upon the work of Guérin (1980).

Cranium

Fig. 10.1 *Dicerorhinus etruscus etruscus* cranium from Senèze, FSL 210959, right lateral view

In lateral view (Fig. 10.1) the main features of the cranium are those typical of all the European Plio-Pleistocene rhinos: it is lengthened with a very long facial part, the nasal bones are bulky with a large insertion for the anterior (nasal) horn while that of the posterior (frontal) one is much smaller but always obvious. The major occipital crest is high and shows a reduced overhang behind the occipital condyles. That overhang is much larger in the Middle and Late Pleistocene rhinos *Dicerorhinus hemitoechus* and especially *Coelodonta antiquitatis* because it is linked with the low-held head and the hypsodont cheek teeth of these two species, two characters involving the consumption of hard grasses (Gramineae). The zygomatic arch is thick, the orbit is low. The nasal notch is long. The post-glenoid and post-tympanic apophysis are fused, closing downwards the auditory pseudo-meatus. The ratio of the distances outer occipital bump (outer angle of the supraoccipital crest)-antorbital process (variable 8 in Table 10.1)/occipito-nasal length (variable 2) presents an average value of 0.55 for *D. etruscus etruscus*, 0.56 and 0.57 for two specimens of *D. jeanvireti* and 0.51 and about 0.5 for two specimens of *D. etruscus brachycephalus*.

The tip of the nasals protrudes forward to the premaxillae, and the ossified nasal septum protrudes forward to the tip of the nasals. The adult crania Se 548 and Se 1785 show no trace of ossified nasal septum; Thenius (1955) suggested that the few adult specimens of *D. etruscus* in which this wall is not ossified are females.

The level of the posterior end of the nasal notch is between the middle of P3 and the posterior part of P4, the most frequent position being between the posterior part of P3 and the anterior part of P4. This level is somewhat more anterior for three specimens of *D. jeanvireti*: from the middle of P3 to the interval P3-P4. For *D. etruscus brachycephalus* it is between the posterior part of P2 and the anterior part of M1, the most frequent position being between the middle of P4 and the interval P4-M1.

The level of the anterior border of the orbit is between M1-M2 and M2-M3, the most frequent position being above M2. For *D. jeanvireti* it is between the middle of M2 and M2-M3. For *D. etruscus brachycephalus* it is between the middle of M1 and the posterior part of M3, the most frequent position being above the posterior part of M2.

The level of the posterior end of the infraorbital foramen is between P3-P4 and P4-M1, the most frequent position being above P4. For *D. jeanvireti* it is between P3-P4 and the middle of P4. For *D. etruscus brachycephalus* it is between the anterior part of P3 and the posterior part of M1, the most frequent position being between the posterior part of P3 and the anterior part of M1. Antoine (2002) showed that generally in the Rhinocerotidae, evolutionary change reflects a receding of these relative positions.

The major occipital crest is lower than that of *D. jeanvireti*, but its dihedral is more acute. The occipital is tilted slightly backwards and downwards; for *D. jeanvireti* it is nearly vertical in its lower and middle parts, and tilted slightly backwards and upwards in its superior part.

The post-glenoid apophysis is very strong and slightly tilted forwards. The post-tympanic apophysis is reduced, it extends downwards then forwards and comes into contact with the post-glenoid one. The paroccipital apophysis is reduced and less extended ventrally than the post-typanic one. In *D. jeanvireti* the post-glenoid apophysis is strong, more vertically developed than the post-tympanic and paroccipital apophyses and tilted forward and inferiorly. The post-tympanic apophysis is very short, while the paroccipital apophysis is quite long, sub-vertical, much more extended ventrally than the post-tympanic one, and bent slightly forward at the end.

The "wedge-shaped" general form of the rhinocerotid cranium is evident from superior and inferior view, as is the strong post-orbital constriction. In superior view the front of the nasal bone is rounded with a small recess within the sagittal plane. Following Thenius (1955) the nasals of the males are broader and less pointed than those of the females. At the posterior edge of the cranium the major occipital crest is greatly depressed in the middle. The ante-orbital processes are strong, the supra-orbital ones are slightly prominent and the post-orbital ones are very low. Except for its much larger size *D. jeanvireti* is morphologically very similar to *D. etruscus* in dorsal view.

In ventral view, the basioccipital muscular tubercule is strong. The bottom of the post-palatal notch is between M1-M2 and M2-M3, the most frequent position being above M2. For two specimens of *D. jeanvireti* it is between M1-M2 and posterior to M2.

The occipital face is trapezoidal, the long base being at the level of the mastoid apophysys (roughly the axis of the auditory pseudo-meatus), and the short one constituted by the main occipital crest. The depression in the centre of the major occipital crest is clearly visible from this angle. Guérin (1980) noted that the occipital condyles are more oblique to the sagittal plane for *D. etruscus* than for *D. jeanvireti*.

The dimensions of the crania are given in Table 10.1. All are clearly inferior to those of *D. jeanvireti* and *D. etruscus brachycephalus*.

Table 10.1 Comparative dimensions of the Senèze *Dicerorhinus etruscus etruscus* crania

D. etruscus etruscus Senèze		Poitiers	FSL	FSL	FSL	NMB	NMB	NMB	NMB	NMB	NMB	NMB	NMB	MNHN.F	MNHN.F
	Variable number	No number	210959 (96129)	210958 (602 v)	No number	Se 791	Se 753	Se 743	Se 1703	Se 1711	Se 561	Se 1785	Se 548	1923–4	1896–22
Cranial measurement variables															
Occipito-nasal length	2		626		620				620	592	628	530	648	642	630[a]
Angle supraoccip. crest-nasal	3		655		630				621	582	636	530	632	650	
Length nasal notch	4		209.5	190	209				196	179	207	138	177	200	200[a]
Minimal braincase width	5		121		104.5	87	112	105	103.5	105[a]	94	86		97[a]	
Angle supraoccip. crest-postorbital proc	6		275		288.5	284			283	265	282	247[a]	247	290	260[a]
Angle supraoccip. crest-supraorbital proc	7		314		318.5						340				
Angle supraoccip. crest-antorbital proc	8		334.5		346.5										
Nasal notch-orbit	9	113.5	112	106	110	118			91	117	90	96		101.5	
Post M3-condyle	10		290.5		252.5	266			320	288	315	256	305	295	320[a]
Nasal-orbit	11		323	309.5	313				285	296	300	236			
TD major occipital crest	12		174[a]		146[a]	155[a]	142	141	125[a]	136	142	126		146	
Width at mastoid apophyses	13		228[a]		188[a]	192	190	207	182[a]	196	195	190[a]	163	212	
Minimal frontal width	14		60		41		60.5	48[a]	56	34	36.5	49		35	
Width at postorbital proc	15		199[a]		200[a]	162a									
Width at supraorbital proc	16		214[a]		192[a]				200					196[a]	
Width at antorbital proc	17		224		216[a]				213	212[a]	207	178[a]			
Bizygomatic width	18		267		253.5[a]	300[a]			291	296	287	256[a]	284[a]	324	
Width bottom nasal notch	19		145[a]	117[a]	118	103			126	117	114	110[a]	132	134	
Height occip. Face	20		133[a]		125	153[a]	129	134[a]	134	127	118	117		133	
Height cranium at P2 level	21		175	176.5	164				162	166	150	146		170	
Height cranium at P4-M1 level	22		193[a]	182.5	127	169			170	175	162	138		195	
Height cranium at M3 level	23		218[a]	185	166	171			206	165	175	138		183.5	
Palatine width between P2	24	61	49		54										
Palatine width between P4-M1	25	90	61.5		70										
Palatine width between M3	26		79		68										
TD occipital foramen	27				54.5	53	57.5	55	51	47	49	44	50	47	
Outer distance between occipital condyles	28				127.5	125	123.5	126	118	134	118	109	123	125	

(continued)

Table 10.1 (continued)

	Dicerorhinus jeanvireti						Other *Dicerorhinus etruscus etruscus*					
Cranial measurement variables	N	Mean	Min	Max	SD	CV	N	Mean	Min	Max	SD	CV
Occipito-nasal length	3	689.67	637	727			3	611.00	595	628		
Angle supraoccip. crest-nasal	3	726.00	692	750			2	625.50	606	645		
Length nasal notch	4	207.75	168	249			6	194.50	176	202	9.61	4.94
Minimal braincase width	2	123.50	109	138			6	100.33	79	126	15.80	15.75
Angle supraoccip. crest-postorbital proc	3	330.33	308	343			2	283.00	280	286		
Angle supraoccip. crest-supraorbital proc	3	347.00	308	370			1	322.00				
Angle supraoccip. crest-antorbital proc	2	407.50	400	415			0					
Nasal notch-orbit	2	122.75	117	128.5			9	105.44	92	123	12.60	11.95
Post M3-condyle	2	332.00	330	334			2	289.50	275	304		
Nasal-orbit	4	363.25	348	374			2	311.50	296	327		
TD major occipital crest	1	172.00					2	127.50	101	154		
Width at mastoid apophyses	2	241.00	232	250			4	184.63	158	203.5		
Minimal frontal width	2	72.25	71	73.5			2	51.50	43	60		
width at postorbital proc	2	226.75	224.5	229			1	188.00				
Width at supraorbital proc	2	249.25	244.5	254			2	207.50	188	227		
Width at antorbital proc	2	243.50	221	266			5	192.40	162	217	19.71	10.24
Bizygomatic width	2	356.00	320	392			4	293.25	255	313		
Width bottom nasal notch	3	142.00	108	161			5	115.20	100	133.5	12.01	10.43
Height occip. face	2	145.00	133	157			2	132.00	122	142		
Height cranium at P2 level	1	155.50					6	154.00	124	178	21.53	13.98
Height cranium at P4-M1 level	1	178.50					6	160.67	138	202	22.38	13.93
Height cranium at M3 level	1	162.00					6	165.00	138	198	20.77	12.59
Palatine width between P2	2	57.25	43.5	71			3	53.67	46	67		
Palatine width between P4-M1	2	87.50	82.5	92.5			3	80.33	68	89		
Palatine width between M3	2	101.00	100	102			3	80.33	75	88		
TD occipital foramen	1	55.00					2	44.25	40	48.5		
Outer distance between occipital condyles	2	141.50	141	142			4	118.75	103	132		

(continued)

Table 10.1 (continued)

		Dicerorhinus etruscus brachycephalus				
Cranial measurement variables	N	Mean	Min	Max	SD	CV
Occipito-nasal length	12	706.17	640	750	32.65	4.62
Angle supraoccip. crest-nasal	9	691.56	605	760	49.09	7.10
Length nasal notch	12	252.58	217.5	289	19.75	7.82
Minimal braincase width	8	123.81	107	145	13.99	11.30
Angle supraoccip. crest-postorbital proc	9	326.44	289	360	25.33	7.76
Angle supraoccip. crest-supraorbital proc	3	335.33	315	358		
Angle supraoccip. crest-antorbital proc	1	376.00				
Nasal notch-orbit	12	99.50	81	122	12.81	12.88
Post M3-condyle	11	334.45	293	370	26.67	7.97
Nasal-orbit	10	351.30	313	400	25.67	7.31
TD major occipital crest	9	154.61	132	175	15.10	9.77
Width at mastoid apophyses	4	237.63	223	254		
Minimal frontal width	8	58.94	35	88	15.70	26.64
Width at postorbital proc	4	216.13	172	263		
Width at supraorbital proc	4	241.38	228.5	256		
Width at antorbital proc	3	248.33	224	265		
Bizygomatic width	4	321.25	302	340		
Width bottom nasal notch	10	151.40	133	174	13.38	8.84
Height occip. face	8	163.44	139	191.5	17.55	10.74
Height cranium at P2 level	10	201.35	156	234	21.64	10.75
Height cranium at P4-M1 level	11	207.86	150	243	23.79	11.44
Height cranium at M3 level	10	207.10	152	232	22.41	10.82
Palatine width between P2	1	76.00				
Palatine width between P4-M1	2	93.00	86	100		
Palatine width between M3	2	103.00	93	113		
TD occipital foramen	4	55.88	49.5	63		
Outer distance between occipital condyles	4	140.00	133	146		

[a] Estimated value

Figure 10.2 is a Simpson diagram showing a general similarity of the cranial proportions of the three fossil European rhinos and the recent Asian *Dicerorhinus sumatrensis*. All but one of the dimensions of *D. etruscus brachycephalus* are larger than those of *D. etruscus etruscus*, the exception being the distance from nasal notch to orbit, shorter for *D. e. brachycephalus;* the proportions of the two subspecies are the same. For *D. jeanvireti* the general size is larger but the length of the nasal notch, the width at the level of the anteorbital process and the height of the facial part are relatively lower. The main proportional differences of *Dicerorhinus sumatrensis* are the relatively smaller general lengths, the narrower occipital part, the larger palatal widths, and the reduced lengths but larger widths of the post-orbital part of the cranium. The proportions of the reference taxon *Diceros bicornis* are quite different.

Evolutionary trends in the cranium of *Dicerorhinus etruscus etruscus* include the increasing distance from the nasal notch to the orbit and the enlargement of the cranial dome (Guérin 1980).

Figure 10.3 presents the left lateral and dorsal views of the very juvenile specimen Se 560. The premolars and the last two molars are unerupted, there is no trace of an ossified nasal wall, the occipital face is nearly vertical. Most of the cranial sutures are unfused. The level of the posterior end of the nasal notch is above the middle of D3 and the level of the anterior border of the orbit is above the middle of M1.

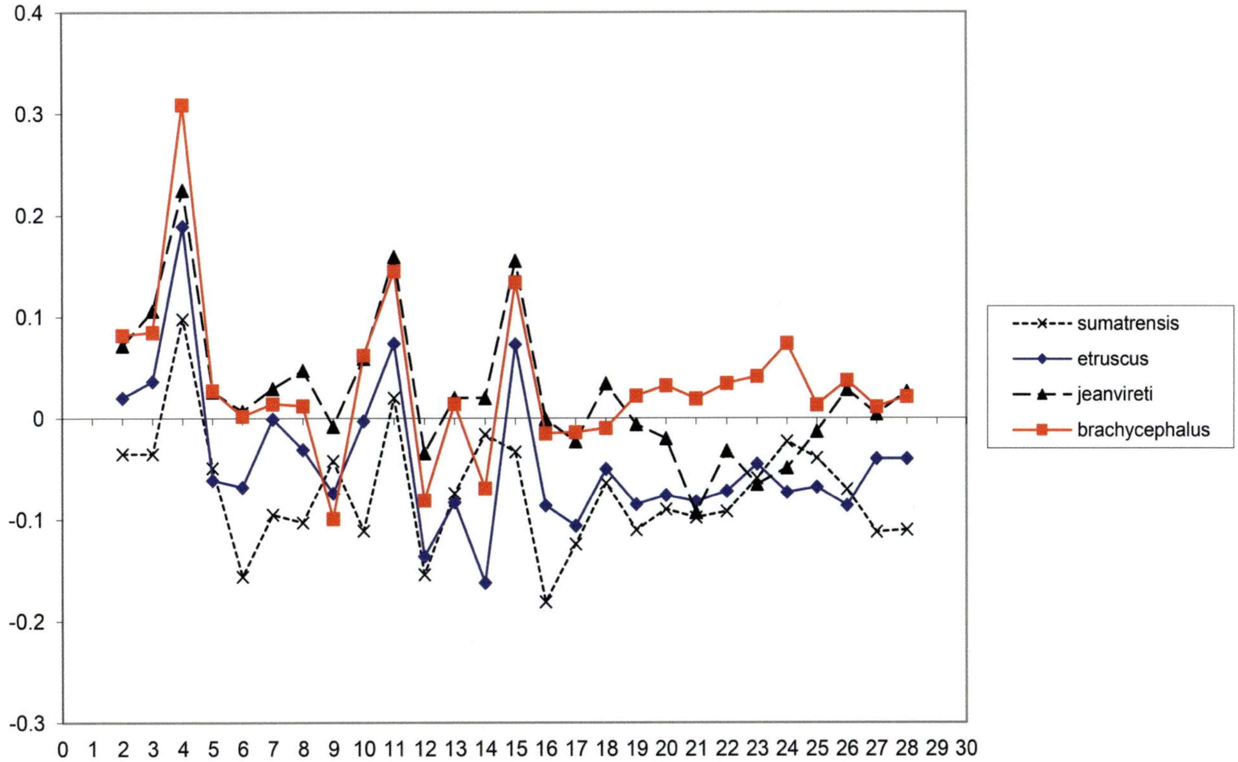

Fig. 10.2 Simpson diagram of crania: mean values of *Dicerorhinus etruscus etruscus*, *D. etruscus brachycephalus*, *D. jeanvireti*, and *D. sumatrensis*. Reference is the recent *Diceros bicornis*. The variables on the x-axis are the same and in the same order as in Table 10.1

Fig. 10.3 *Dicerorhinus etruscus etruscus* juvenile cranium from Senèze, NMB Se 560, left lateral view below dorsal view

Mandible

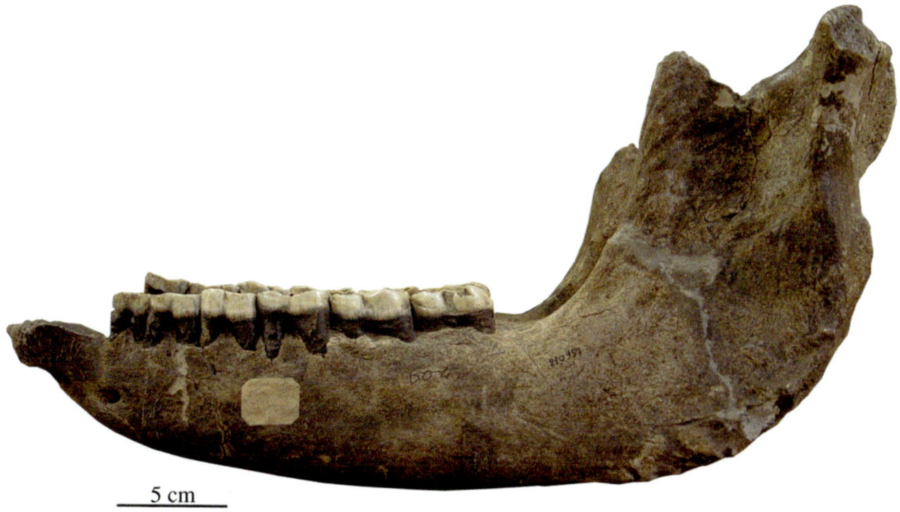

Fig. 10.4 *Dicerorhinus etruscus etruscus* mandible from Senèze, FSL 210957, left lateral view

Fig. 10.5 *Dicerorhinus etruscus etruscus* from Senèze, occlusal view of the symphyseal part of the mandible FSL 210957

FSL 210957 provides a good example of this element (Fig. 10.4). The maximum width of the symphyseal region (Fig. 10.5) is at the level of its anterior edge, while for *D. jeanvireti* it is in a more posterior position. In *D. etruscus etruscus* the anterior border bears two to four small alveoli more or less filled with bone, without visible incisors, while two very small and not functional but relatively well developed tusks corresponding to the i2s are present in *D. jeanvireti*. The ventral side of the symphysis generally bears no median crest. The posterior edge of the symphysis is between the anterior part of p2 and the posterior part of p3, the most frequent position being between the middle of p2 and the interval p2-p3. For *D. etruscus brachycephalus* the range of that position is the same, but the most frequent one is the level of p2. For *D. jeanvireti* it is between the middle of p2 and the anterior part of p3. Antoine (2002) showed that among Rhinocerotidae, the relative position of the posterior border of the symphysis usually tends to recede during the evolution of the family.

The rather deep horizontal branch (corpus) has a relatively straight ventral edge in lateral view (Fig. 10.4). The external face is convex and the internal one more or less depressed in its middle part. The mandibular foramina are extremely variable in size, number and position. The gonion is not prominent backwards, contrary to *D. jeanvireti* where it is extended up to a little beyond the posterior edge of the condyles. The anterior border of the ramus, like the coronoid apophysis, is tilted slightly backwards. For *D. jeanvireti* it is subvertical, barely oblique backwards and upwards in its superior part.

The dimensions of the mandibles are given in Table 10.2. The 11 specimens found in Senèze are all in the range of variation of 1 to 33 specimens of *D. etruscus etruscus*, while the mean values of those from 2 to 20 specimens of *D. jeanvireti* are stronger, like those from 4 to 65 specimens of *D. etruscus brachycephalus*.

Table 10.2 Comparative dimensions of the Senèze *Dicerorhinus etruscus etruscus* mandibles

Senèze		MNHN.F 1923–4	FSL 210958		FSL 210959		FSL 210957		FSL 604 V	NMB Se561	NMB Se1711	NMB Se1920
			Formerly 602 V		Formerly 96129		Formerly 604 V					
Mandibular measurement variables	Var. #		Right	Left	Right	Left	Left	Right				
Length	1	470[a]	494		474		473			509		
Posterior symphysis to angle	2	375	402		364[a]					421	385	
Height corpus between P2-P3	3		58				64	66		69		
Height corpus between P3-P4	4		67.5		68		71	68.5		73		
Height corpus between P4-M1	5		74		79		74	74		82	77	
Height corpus between M1-M2	6	82	82.5	82	87	91	79	79		85	82	
Height corpus between M2-M3	7	90	85.5	87		96.5	81	81.5		85	89	77
Height corpus behind M3	8	97	88	87		97.5	86	85		91	91	81
TD corpus between P4-M1	9	47.5	44.5	46.5		51	41	42		46	48	
TD corpus below ant. M3	10	47.5	48	49.5		52	44	45		48	44	47.5
Symphysis length	11		108	108	123	123	103.5	103.5		101		
AP D corpus at occlusal level	13	139		131	144	138	143.5	144	148	152	148	
TD condyle	14	89	85			91		90		107	90	
Condyle height	15	243	215.5			237	216	221		203	197	
Height coronoid apophysis	16										254	
		Dicerorhinus jeanvireti					Other *Dicerorhinus etruscus etruscus*					
	N	Mean	Min	Max	SD	CV	N	Mean	Min	Max	SD	CV
Length	2	516.00	497	535			7	471.00	430	499	25.59	5.43
Posterior symphysis to angle	4	399.25	380	416	17.08		7	389.57	367	402	11.22	2.88
Height corpus between P2-P3	6	69.25	61.5	77	6.46	9.33	15	66.87	55.5	80	7.46	11.15
Height corpus between P3-P4	11	75.41	70	82	3.77	5.00	20	72.23	62.5	83.5	5.58	7.73
Height corpus between P4-M1	16	79.72	74	85	3.28	4.12	24	76.71	66	85.5	5.49	7.16
Height corpus between M1-M2	11	84.23	71	92	5.31	6.30	26	82.79	65	91	7.00	8.46
Height corpus between M2-M3	20	89.55	77	104.5	6.96	7.77	30	86.15	70	95	6.71	7.79
Height corpus behind M3	10	97.50	90	105	5.34	5.48	29	91.74	79	105	7.56	8.24
TD corpus between P4-M1	17	51.82	45	61	4.91	9.47	26	49.04	41.5	60	4.57	9.32
TD corpus below ant. M3	20	52.15	43.5	69	6.38	12.24	33	49.50	41	60	4.59	9.27
Symphysis length	7	120.57	114	133	6.44	5.35	8	110.94	94	121	7.69	6.93
AP D corpus at occlusal level	6	143.58	135.5	163	9.93	6.92	8	143.00	115	152	12.50	8.74
TD condyle	8	102.56	88	115	9.05	8.83	6	88.92	77.5	96	7.53	8.46
Condyle height	7	236.71	220	270	17.83	7.53	3	230.67	218	238		
Height coronoid apophysis	2	302.00	286	318			1	268.00				
		Dicerorhinus etruscus brachycephalus										
	N	Mean	Min	Max	SD	CV						
Length	4	526.75	485	542	27.85							
Posterior symphysis to angle	8	424.38	385	452	23.05	5.43						
Height corpus between P2-P3	39	72.86	62	86	6.52	8.95						
Height corpus between P3-P4	59	77.96	63	90	6.70	8.60						
Height corpus between P4-M1	60	84.13	74	96.5	5.40	6.42						
Height corpus between M1-M2	62	89.75	78	105	6.80	7.58						
Height corpus between M2-M3	62	90.49	77.5	104	6.38	7.05						
Height corpus behind M3	53	94.32	79	108	7.28	7.71						
TD corpus between P4-M1	30	51.35	42	62.5	4.88	9.50						
TD corpus below ant. M3	65	53.50	43	68.5	4.95	9.26						
Symphysis length	7	104.21	91	126	12.20	11.71						
AP D corpus at occlusal level	11	153.82	128	169	12.02	7.82						
TD condyle	6	95.00	83	100	6.30	6.63						
Condyle height	10	249.80	230	292	23.61	9.45						
Height coronoid apophysis	–											

[a]Estimated value

Upper Teeth

No trace of anterior upper teeth was ever observed in *D. etruscus*. One example, identified as a dI2 or an I2, is known for *D. jeanvireti*, on a cranium from the Early Villafranchian site of Milia, Macedonia, Greece (Guérin & Tsoukala 2013).

Figure 10.6 is an occlusal view of a right upper tooth row, rather well worn, of the cranium FSL 210959. The dimensions of the upper cheek teeth are given in Table 10.3. The Senèze examples fall within the range of variation of *D. etruscus etruscus*, but there are no significant differences in upper cheek-tooth size among the three rhino taxa compared here. Neither are there important differences in the morphology of their upper cheek-teeth as they all belong to the same genus; the ectoloph profiles (by far the most distinguishing feature of these teeth for all kinds of rhinocerotids, see Guérin 1980) are only slightly different. In more detail:

Fig. 10.6 *Dicerorhinus etruscus etruscus*, right upper tooth row of Senèze cranium FSL 210959, occlusal view

Table 10.3 Comparative dimensions of the Senèze *Dicerorhinus etruscus etruscus* upper cheek teeth

		Dicerorhinus etruscus etruscus (Senèze)																		
	Collection	Univ. Poitiers		FSL		FSL		FSL		FSL		FSL	FSL	FSL	FSL SEN	NMB	NMB	NMB	NMB	NMB
	Number	Unnumbered		211118		210958		210959		Unnumbered		210925	210930	3707	04–0276	Unnumb	Se 1416	Se 334	Se 335	Se 1509
	Notes	Juvenile face		Formerly Clermont-Ferr		Formerly 602 V		Mounted skeleton, ex 96129		Damaged cranium		ex3707	ex3707						(325?)	
Dental measurement variables	Side	Left	Right	Left	Right	Right	Left	Right	Left	Left	Right	Left	Right					Left	Right	Left
P 1 L		23.5	23.5									27.5								
P 1 B		25.5	26									23.5								
P 2 L		35	35	29	31					30	29					30	35			
P 2 B		37	38.5	33.5	32	35	35.5	41.5		35	36					38	38			
P 3 L		40	41	37.5	37					35	37				33[a]	37	43.5			
P 3 B		46.5	47.5	43	42	44.5	44	54		47.5	53				50	46	49			
P 4 L				39.5	40					37						39	47			
P 4 B				45	49	49	49	57	56	52						52	56			
M 1 L		49	50	48	48											41[a]				
M 1 B		55	53.5	51		48	50	58	59	52						53.5				
M 2 L		56	57	47	45.5						47					48		53		
M 2 B		56.5	57.5	48	49	54	52	61	61		55					56		56.5		
M 3 L abs			45[a]		47					54	52	54								
M 3 L anat					40	52	53	54	54	52	46	47				51				
M 3 B			46		46	48.5	49.5	55	51	50	52	52.5				50				
L P2-M3 (maximum)							215	227.5		223	230					223				
L premolar row		137	132					100		101	103.5					102				
L P3 (parastyle) to P4 (metastyle)		79.5	75.5				68.5	70.5	67	71						73				
L M1 (parastyle) to M3 (distal end)			129[a]				126.5	132	133	133	131					126[a]				
D1L														27.5						
D1B														23.5						
D2L																				
D2B																				
D3L																			37.5	
D3B																			36	
D4L													45							45
D4B			43.5										43							45

(continued)

Table 10.3 (continued)

	Dicerorhinus jeanvireti						*Dicerorhinus etruscus etruscus* (not Senèze)					
	N	Mean	Min	Max	SD	CV	N	Mean	Min	Max	SD	CV
P 1 L	1	24.00					2	23.50	23	24		
P 1 B	1	20.00					2	23.25	21	25.5		
P 2 L	5	34.50	31	37	2.29	6.64	4	33.38	32.5	34	0.75	
P 2 B	5	38.50	35	41	2.18	5.66	12	39.38	33	42.5	2.81	7.14
P 3 L	7	41.79	40	43	1.25	3.00	7	40.36	39.5	41.5	0.80	1.99
P 3 B	8	49.88	45	53.5	2.83	5.66	17	49.35	43	53.5	3.13	6.34
P 4 L	12	42.88	40.5	46.5	1.75	4.07	7	43.14	38	49	3.82	8.84
P 4 B	14	54.14	46	58.5	3.94	7.28	16	56.59	52.5	63	3.15	5.57
M 1 L	10	50.60	48.5	56	2.28	4.51	11	49.23	45	54	2.52	5.13
M 1 B	10	54.60	50	58	3.27	5.99	16	56.72	50.5	60.5	3.10	5.47
M 2 L	9	53.67	51	56	1.71	3.19	6	50.17	49	52	1.29	2.57
M 2 B	11	59.36	51.5	63.5	3.98	6.70	10	59.10	53	65.5	3.70	6.25
M 3 L abs	8	57.75	53	61.5	2.36	4.09	13	54.35	49	59	2.79	5.13
M 3 L anat	6	52.58	45	58	5.96	11.33	13	44.62	38.5	51.5	3.52	7.90
M 3 B	9	51.16	46	59	4.73	9.25	13	53.62	51	56.5	2.07	3.87
L P2-M3 (maximum)	2	250.75	248	253.5			8	233.13	220	245	8.45	3.62
L premolar row	3	115.83	114	117	1.61		11	108.27	102	117	4.84	4.47
L P3 (parastyle) to P4 (metastyle)	4	80.13	71	84	6.14		11	77.09	72	83.5	3.83	4.97
L M1 (parastyle) to M3 (distal end)	5	138.60	120	149	11.82	8.53	9	137.39	128	145	6.25	4.55
D1L	2	26.00	25.5	26.5			9	28.17	27	30	0.90	3.20
D1B	2	20.75	20.5	21			9	24.06	22	25	1.07	4.46
D2L	1	34.50					12	41.75	36	46	2.83	6.78
D2B	1	33.00					12	41.92	36	46	3.14	7.49
D3L	3	41.83	39.5	46			8	37.25	35	40	1.81	4.87
D3B	3	41.50	38.5	46			8	36.31	35.5	37	0.59	1.64
D4L	2	45.00	45	45			14	46.39	43	48.5	1.46	3.14
D4B	2	41.00	40	42			11	44.55	42	47	1.42	3.19

(continued)

Table 10.3 (continued)

		Dicerorhinus etruscus brachycephalus				
	N	Mean	Min	Max	SD	CV
P 1 L	27	35.72	28.5	45	3.06	8.57
P 1 B	40	40.04	31	52.5	4.28	10.69
P 2 L	35	38.97	28.5	47.5	4.78	12.26
P 2 B	38	46.11	33	58	7.35	15.95
P 3 L	32	45.14	39	54	3.57	7.91
P 3 B	50	55.42	47.5	61.5	3.17	5.71
P 4 L	35	51.56	44	60	3.85	7.46
P 4 B	41	56.30	44.5	63.5	4.06	7.21
M 1 L	48	54.59	47	63	3.36	6.15
M 1 B	63	58.84	52	67	3.28	5.57
M 2 L	44	54.57	45	66	3.92	7.18
M 2 B	44	48.38	40.5	56	4.18	8.63
M 3 L abs	59	50.31	41.5	60	4.07	8.10
M 3 L anat	16	243.91	226	263.5	12.88	5.28
M 3 B	19	108.55	96	119.5	5.95	5.48
L P2-M3 (maximum)	22	77.59	69	87.5	4.98	6.42
L premolar row	18	142.67	128	159.5	9.03	6.33
L P3 (parastyle) to P4 (metastyle)	8	28.06	25.5	30	1.76	6.28
L M1 (parastyle) to M3 (distal end)	8	24.31	21	28	2.12	8.72
D1L	7	37.86	34	41	2.34	6.18
D1B	9	36.17	32.5	39	2.17	5.99
D2L	8	42.38	36	46.5	3.22	7.59
D2B	10	42.10	33.5	46.5	3.43	8.15
D3L	12	48.50	44	58	4.53	9.34
D3B	15	46.83	43.5	56.5	3.61	7.72
D4L						
D4B						

[a]Estimated value

D1: the ectoloph is convex and without any fold. The crochet is present or absent, the crista generally present, sometimes multiple; the anticrochet is rarely present. The medifossette is open or closed. The lingual cingulum is very variable.

D2: the ectoloph profile is rather variable, sometimes regularly convex without any fold, sometimes wavy with a clear mesostyle and a very small metacone fold. The crochet is always present, like the crista which is sometimes double. The medifossette is mainly closed, the anticrochet is rather rare. The protocone constriction is present or not. The lingual cingulum is always present, continuous or discontinuous.

D3: the ectoloph profile presents a paracone fold which is always clear but more or less prominent; this profile tends to become wavy with the beginning of a mesostyle and a very small metacone fold. The crochet is always present, sometimes double, the crista is generally present. The medifossette is more often open than closed, the anticrochet is generally absent. The protocone constriction is always present, sometimes very strong. The lingual cingulum is present or not, often discontinuous.

D4: the ectoloph profile is the same as for D3. The crochet is always present, sometimes double, the crista is present or not, the anticrochet rarely present. The medifossette is generally open. The protocone constriction is always present, often very strong. The lingual cingulum is present or not.

M1: the ectoloph profile tends to be wavy with a small but clear paracone fold, a beginning of a mesostyle and/or of a metacone fold. The crochet is fast always present, the crista generally absent, the anticrochet is rare; the medifossette is rarely closed. The protocone is generally constricted. The cingulum, if present, is very variable.

The M2 is constructed along the same lines as the M1, its ectoloph profile is similar but with a more bulky paracone fold.

M3: the ectometaloph is moderately convex with a distinct paracone fold. The crochet is always present, the crista frequently present, the anticrochet rare. The medifossette is sometimes closed. The protocone constriction is rather rare, the lingual cingulum, if present, is very variable.

The P1 is rarely found, and its place is sometimes taken by a persistent D1. The size is small; the convex ecoloph shows no fold; on the unworn specimens the protoloph is notched in the middle, insulating the protocone. The crochet is present or not, the crista generally present, the medifossette is generally open. The lingual cingulum is very variable.

P2: the ectoloph is convex and bears a very small and barely perceptible paracone fold. The protocone is generally isolated. The always present crochet is of varying spread. The crista is generally present, the anticrochet generally absent. The medifossette is rarely closed. The lingual cingulum is always present.

P3: the ectoloph profile is only slightly convex, the paracone fold being weak or very weak. The crochet is always present, the crista is present or not, the anticrochet generally absent. The protocone constriction is rarely seen. The lingual cingulum is generally present.

P4: the ectoloph profile is relatively flat, the paracone fold being little prominent even bulky; sometimes the profile tends to be wavy because the beginning of a depression appears at the mesostyle and a very weak metacone fold appears as well. The crochet is always present, sometimes multiple, the crista is present or not, the anticrochet rarely present. The medifossette is rarely closed. The protocone constriction is present or not. The lingual cingulum is generally present.

Lower Teeth

As noted above, two to four small empty alveoli at the front of the mandibular symphysis show that there were anterior lower teeth in the Senèze rhino. The presence of such teeth is known solely by these alveoli and they were probably totally regressed and nonfunctional; this is always the case for *D. etruscus*. Several specimens of *D. jeanvireti* possess small nonfunctional tusks corresponding to the i2 (Guérin & Tsoukala 2013).

Except for *Coelodonta*, the construction of the lower cheek-teeth of Pliocene and Pleistocene European rhinoceroses is very monotonous (Fig. 10.7 is an occlusal view of a typical *D. etruscus etruscus* left tooth row, rather well worn), and there are no anatomical characteristics that clearly distinguish the diverse species of *Dicerorhinus*. The dimensions of the lower cheek teeth are given in Table 10.4. The Senèze rhinoceros can only be specifically determined from their dimensions, and all fall within the range of variation of *D. etruscus etruscus*. The lower deciduous premolars show no special features. Other details:

Table 10.4 Comparative dimensions of the Senèze *Dicerorhinus etruscus etruscus* lower cheek teeth

Senèze specimens	Collection	FSL		FSL	FSL	FSL	FSL		FSL	FSL	FSL	FSL	FSL SEN	NMB	NMB	NMB	NMB	NMB	NMB	NMB
	Number	210957		Unnumb	3274	3275	210958		210959	Unnumb	"210927"	"210929"	01–0004	Se 1413	Se 1416	Se 330	Se 1682	Se 1862	Se 1509	Unnum
	Notes	ex 604 V		Ex clermont	Ex clermont	Ex clermont	ex 602 V		Mounted skel	Mounted skel	604 V	3707								Mand
Dental measurement variables	Side						R	L			R	R	L	L	L	R	R	L	R	
p1 L																				
p1 B																				
p2 L													27	29	30.5					
p2 B			16										15.5	17	21					
p3 L					35						32.5		33	37.5						
p3 B		22.5	21.5		22			21.5	22.5	22.5	22		22	23						
p4 L					37	38					35			38.5						
p4 B		24.5	24		25	26.5		26	26.5	26	24			22						
m1 L				43	43	41					39		38.5							41
m1 B		30	29.5	28	28	28.5		27.5	30	30	27		26.5							29
m2 L				45	44	44					40.5	46.5								46
m2 B		28	28	28	27	29	27.5	27	31	30.5	28	30.5								29.5
m3 L		41	41.5	44	43	46	46	46	45	43	43.5			49		46.5				45
m3 B		27	27.5	27		28.5	27.5	28	29	29	28			30		27				28.5
L p2-m3		211	210																	
L p2-4		88	87																	
L p3-4		64	63		72			68	68	67	67									
L m1-3		123	121		130	133		126	123.5	123.5	124									135
d1L																	19	19		
d1B																	11	11		
d2L																	29	30		
d2B																	16.5	16		
d3L																	38	40		
d3B																	21	22		
d4L																	40[a]		42	
d4B																	23.5		24.5	

(continued)

Table 10.4 (continued)

	Dicerorhinus jeanvireti						*Dicerorhinus etruscus etruscus*					
Dental measurement variables	N	Mean	Min	Max	SD	Coef. var	N	Mean	Min	Max	SD	Coef. var
p1 L												
p1 B												
p2 L	11	29.77	27	32	1.95	6.56	15	29.97	25	33	1.87	6.23
p2 B	11	18.77	16.5	21	1.60	8.54	15	19.17	16.5	22	1.69	8.80
p3 L	9	35.28	33	38	1.82	5.16	13	34.81	31.5	37.5	1.97	5.67
p3 B	10	25.85	23	34	3.08	11.93	12	24.42	23	29	1.61	6.58
p4 L	14	39.00	37	40.5	1.11	2.84	18	38.00	35	42.5	1.85	4.86
p4 B	14	28.39	24	31	2.18	7.67	20	27.23	22	31	2.25	8.27
m1 L	16	43.16	40	47	1.92	4.45	24	41.88	36	50	3.71	8.87
m1 B	19	30.50	25	34	2.32	7.59	31	29.23	26.5	33	1.72	5.89
m2 L	20	47.50	43	53.5	2.59	5.44	28	46.11	40	54	3.52	7.63
m2 B	22	30.95	27	34.5	1.81	5.85	33	30.03	27	34	2.00	6.65
m3 L	16	49.22	43.5	54	2.96	6.02	28	46.73	42	52	2.35	5.03
m3 B	15	29.93	26	33	1.56	5.20	28	28.89	26	33	1.76	6.07
L p2-m3	5	238.60	234	242	2.97	1.24	11	235.50	213	251.5	12.19	5.18
L p2-4	7	102.36	97	106.5	3.77	3.68	10	102.25	88	108	6.25	6.12
L p3-4	12	73.29	68.5	77	2.86	3.91	17	71.53	63.5	80.5	5.47	7.65
L m1-3	14	134.29	127	140	3.38	2.52	23	131.22	122	143	6.18	4.71
d1L	1	22.00					3	18.50	17.5	20		
d1B	1	11.00					3	11.17	10.5	12		
d2L							8	29.81	27	31.5	1.53	5.14
d2B							8	17.56	16.5	19	1.08	6.17
d3L	2	43.50	40	47			7	40.07	38	42	1.54	3.84
d3B	2	24.75	23	26.5			8	20.94	19	22.5	1.21	5.77
d4L	1	41.00					8	40.38	39	43	1.48	3.67
d4B	1	26.00					7	23.93	22.5	25	0.98	4.08

(continued)

Table 10.4 (continued)

		Dicerorhinus etruscus brachycephalus				
Dental measurement variables	N	Mean	Min	Max	SD	Coef. var
p1 L						
p1 B						
p2 L	36	30.25	27	39	2.48	8.20
p2 B	37	20.70	16	33	2.82	13.61
p3 L	51	36.25	32	42.5	2.29	6.30
p3 B	49	26.33	21	31	1.88	7.12
p4 L	64	39.88	33	49.5	3.05	7.66
p4 B	74	28.61	24	35	2.17	7.59
m1 L	77	44.13	37	57	4.01	9.09
m1 B	84	30.45	25	37	2.32	7.62
m2 L	102	48.06	42	57.5	3.13	6.52
m2 B	114	30.91	23.5	35	1.87	6.05
m3 L	72	49.87	44	60.5	3.24	6.51
m3 B	72	29.76	25	33	1.68	5.63
L p2-m3	15	244.63	228	263.5	11.06	4.52
L p2-4	26	105.62	99	116	5.13	4.86
L p3-4	51	74.63	63	84	4.51	6.04
L m1-3	54	138.62	119	163.5	7.67	5.53
d1L	4	19.00	17	23.5	3.03	
d1B	4	11.88	11	13	0.85	
d2L	9	31.78	27.5	34	2.06	6.49
d2B	9	18.50	17.5	20	0.87	4.68
d3L	16	41.88	30	47	3.93	9.39
d3B	16	22.66	18	28.5	2.53	11.19
d4L	16	42.38	36	47	2.97	7.01
d4B	16	25.22	22.5	31	2.28	9.04

[a]Estimated value

Fig. 10.7 *Dicerorhinus etruscus etruscus*, left lower tooth row of Senèze mandible FSL 210957, occlusal view

m1: the two lingual valleys are transversally V or U-shaped, and their level difference is medium. One example of a closed posterior valley is known. The lingual cingulum is generally absent, in some cases there are small traces of it. The labial cingulum is very variable, it can be continuous and rather strong, or present as discontinuous small traces, or totally absent.

The m2 is constructed along the same lines as the m1. The lingual cingulum is normally absent, the labial one is generally present but its development is very variable.

m3: the lingual valleys are transversally shaped in a broad V or a U, their level difference is variable. The lingual cingulum is generally absent, the labial cingulum generally present.

p2: the sole lingual valley is the posterior one. The lingual and labial cingula are frequently present.

p3: the two lingual valleys are V-shaped with a rather strong level difference. The labial cingulum is frequently present, the lingual one is normally absent.

p4: the two lingual valleys are V-shaped with a reduced or medium level difference. There is no lingual cingulum. The labial cingulum is frequently present but of variable strength.

Humerus

There are eight complete or sub-complete humeri from Senèze, all from the old excavations. Their main anatomical characters are those of *D. etruscus* (Fig. 10.8). The articular head is nearly rectangular. The trochanteric convexity is narrow and sinusoidal, the tip of the trochanter is very high relatively to the deltoid tuberosity. The bicipital groove is trough shaped with a flat bottom.

The dimensions of the humeri are given in Table 10.5. That of *D. etruscus* is rather small relative to *D. jeanvireti*. There are no significant differences in size or proportions of

Fig. 10.8 *Dicerorhinus etruscus etruscus* right humerus, Senèze, FSL 210954, anterior view

the two subspecies of *D. etruscus*, but in the older form the two diameters of the distal epiphysis show a tendency to reduce from the Early-Middle to the Late Villafranchian (Guérin 1980, 2004).

whole articulation is slightly undulating with a small medial re-entrant at the level of the coronoid process. The anterior border of the deep lateral facet is ahead of the anterior margin of the medial facet. The posterior border of the lateral facet is

Table 10.5 Comparative dimensions of the Senèze *Dicerorhinus etruscus etruscus* humeri

Senèze		FSL "586 V"	FSL 210954		MNHN.F 1923–8	MNHN.F 1922–5 (juv)		NMB Se 1711		NMB Se 1703	NMB Se 801
Measurement	Side		L	R		R	L	R	L	L	R
L		417	427	400	385	336		374	378[a]		390
TD proximal		151	146	147	145	125			126		140
APD proximal		140	149.5	150	153	130					
TD diaphysis		56	62	67.5	64	45.5	46	54.5	56	62	55.5
APD diaphysis		63	68	67.5	64.5	56	54	64	61		62
TD distal		126	125	130	128	112	114	120	121	127	130
APD distal		104	97	108	106.5	97		97	95	104	103
			Dicerorhinus jeanvireti								
	N	Mean	Min	Max	SD	CV					
L	5	436.20	428	457	11.84	2.71					
TD proximal	5	159.00	151	175	9.77	6.15					
APD proximal	5	178.80	164	184	8.41	4.70					
TD diaphysis	8	65.00	62	70	2.62	4.03					
APD diaphysis	8	72.50	69	77	3.25	4.48					
TD distal	6	147.50	142	152	3.45	2.34					
APD distal	9	120.22	117	126	2.99	2.49					
		Dicerorhinus etruscus etruscus (not Senèze)									
	N	Mean	Min	Max	SD	CV					
L	14	387.93	376	412	11.75	3.03					
TD proximal	12	144.38	128	202	20.07	13.90					
APD proximal	6	146.83	135	155	7.39	5.03					
TD diaphysis	31	56.39	52.5	65	3.06	5.43					
APD diaphysis	26	63.98	58	72	3.82	5.97					
TD distal	29	123.03	110	137	6.15	5.00					
APD distal	29	101.90	83.5	114	5.56	5.46					
		Dicerorhinus etruscus brachycephalus									
	N	Mean	Min	Max	SD	CV					
L	7	384.00	363	395	11.83	3.08					
TD proximal	4	143.63	140	150	4.42						
APD proximal	2	162.50	159	166							
TD diaphysis	22	59.77	50	70	5.14	8.60					
APD diaphysis	22	64.16	55	77	5.69	8.86					
TD distal	18	128.36	118	145	7.84	6.11					
APD distal	17	106.24	94	116.5	7.00	6.59					

[a] Estimated value

Radius and Ulna

The new excavations yielded only two proximal radial epiphyses, but 11 complete or subcomplete adult radii were found previously. The proximal articulation comprises a lateral facet and a medial one. The anterior border of the very oblique and slightly concave. In anterior view, notable features include the strong lateral tuberosity of the proximal epiphysis, which is well developed below the articulation, and the convex medial border of the same epiphysis, as well as the strong insertion of the biceps brachii, shifted clearly medially. The dimensions of the radii are given in Table 10.6.

Table 10.6 Comparative dimensions of the Senèze *Dicerorhinus etruscus etruscus* radii and ulnae

Senèze		FSL SEN	FSL SEN	FSL	FSL	FSL	FSL	FSL		MNHN.F		NMB	NMB		NMB	
		01–0109	01–0007	ex Clermt	210931	210935	210936	210959		1922–5		Se 1703	Se 1711		Se 1756	
	SIDE	R	L			R		L	R	R	L		R	L	L	R
Radius L				366.5			354	382	376	349	342[a]	389	362[a]	367[a]	364	366
Radius prox. TD		92	90				83	87	87	82	82	93	77.5		87	
Radius prox. APD		59		54.5			55	59	61.5	58		60		53	52[a]	56
Radius TD diaph				43.5			46	49.5	50	40	41	55.5	43	49	48.5	47
Radius APD diaph				36.5			36	38	39	31	30	37	35	37	37.5	32
Radius dist. TD				86	92		87	92	92.5	86	87	93			90.5	93.5
Radius dist. APD				58.5	58		60	62	63	63	62.5	62			62.5	62
Ulna L						501		477	479	432		476	437	433[a]		
Ulna artic. prox. TD						78		79	77	66		72	69	70		
			Dicerorhinus jeanvireti													
	N	mean	min	max	SD	CV										
Radius L	8	425.63	383	455	21.88	5.14										
Radius prox. TD	16	101.09	97	107	3.05	3.01										
Radius prox. APD	13	70.73	63	77	5.13	7.26										
Radius TD diaph	13	56.38	48	61	3.76	6.67										
Radius APD diaph	14	39	35	44	2.52	6.46										
Radius dist. TD	10	104.15	91	113	7.5	7.2										
Radius dist. APD	9	66.67	55	76	5.68	8.52										
Ulna L	6	526.17	508	560	18.21	3.46										
Ulna artic. prox. TD	8	89.69	79	97.5	5.62	6.27										
		Dicerorhinus etruscus etruscus														
	N	mean	min	max	SD	CV										
Radius L	16	379.59	351	415	17.58	4.63										
Radius prox. TD	34	86.46	78	94	3.51	4.06										
Radius prox. APD	29	57.78	50.5	65	3.18	5.51										
Radius TD diaph	25	47.98	42	56	3.58	7.45										
Radius APD diaph	25	36.34	31.5	42	3.08	8.47										
Radius dist. TD	18	85.5	81	89	2.4	2.81										
Radius dist. APD	18	55.17	49	61	2.86	5.18										
Ulna L	4	470.25	460	478	8.02											
Ulna artic. prox. TD	16	75.53	67	82	3.78	5.01										
		Dicerorhinus etruscus brachycephalus														
	N	mean	min	max	SD	CV										
Radius L	13	378.27	358	404	15.35	4.06										
Radius prox. TD	39	98.06	80	112	7.32	7.46										
Radius prox. APD	39	66.14	57	80	5.94	8.97										
Radius TD diaph	35	52.6	43.5	64	4.89	9.3										
Radius APD diaph	36	39.97	33	53	4.4	11.01										
Radius dist. TD	16	94.59	88	102	3.98	4.2										
Radius dist. APD	20	60.03	54	67	3.16	5.26										
Ulna L	11	470.18	432	505	21.12	4.49										
Ulna artic. prox. TD	29	76	66	86	5.04	6.64										

[a] Estimated value

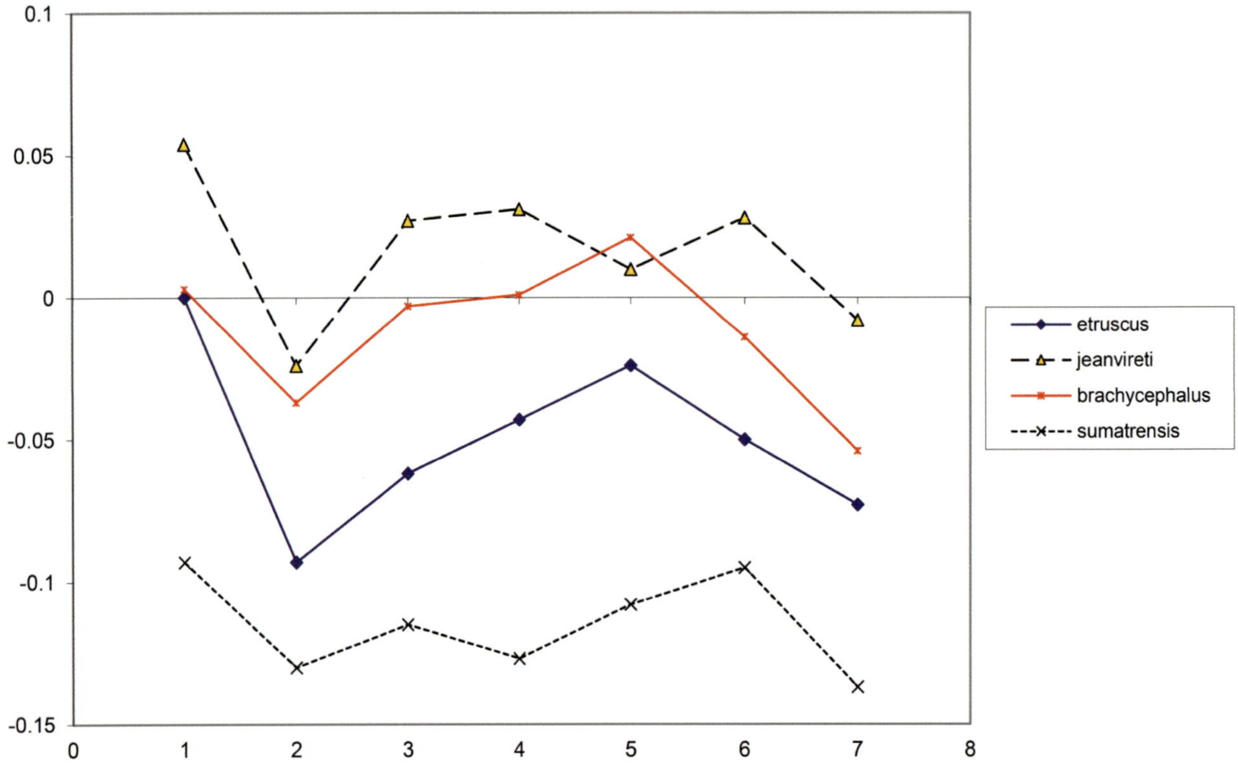

Fig. 10.9 Simpson diagram of the radius: mean values of *Dicerorhinus etruscus etruscus, D. etruscus brachycephalus*, *D. jeanvireti* and *D. sumatrensis*. Reference is the recent *Diceros bicornis*. The variables on the x-axis are the same and in the same order as in Table 10.6

Figure 10.9 is a Simpson diagram of the mean dimensions; the radius of *D. e. brachycephalus* is larger and stockier than that of *D. etruscus etruscus* but the proportions of the two subspecies are the same. That of *D. jeanvireti* is larger and slimmer, and the antero-posterior diameter of its epiphysis is relatively reduced. That of *D. sumatrensis* is smaller and relatively stockier. From the Early-Middle to the Late Villafranchian the length of the bone and the transverse diameter of the proximal epiphysis show a tendency to reduce (Guérin 1980, 2004).

The ulna of *D. etruscus* is normally not firmly fused with the radius. Seven almost complete ulnae were found during the old excavations. The olecranon is relatively long with a thick and oblique anterior border. The proximal articulation is narrow.

Scaphoid

Ten scaphoids, including two from the new excavations, were found at Senèze. In anterior view the medial border is regularly globular, and the lateral edge is rectilinear or slightly concave; the undulation of the proximal border is symmetrical. The medial height is superior to the lateral one (Fig. 10.10). The proximal articulation is trapezoidal. The dimensions of the scaphoid are given in Table 10.7. During the Villafranchian, the scaphoid of *D. etruscus etruscus* tends to shorten and to become lower.

Fig. 10.10 *Dicerorhinus etruscus etruscus* Senèze right scaphoid FSL SEN 01–0086, lateral view

Table 10.7 Comparative dimensions of the Senèze *Dicerorhinus etruscus etruscus* scaphoids

Senèze		FSL SEN	FSL SEN	FSL		MNHN.F		NMB		NMB	
Scaphoid		01–0086	01–0156	210959		1922–5		Se 1756		Se 1703	
Measurement	Side	R	L	L	R					R	R
L		77.5	77.5	78.5	75[a]	67	68	73	67	77	77
W		49	> 43	42	45	44	43	47.5	48	51	50[a]
H		62	63	64	65	56.5	56	57	55	58	55[a]
Prox. artic. APD		47.5	46	42	45	48		47.5	48	48	47
Prox. artic. TD		50		40[a]	46			47	47.5	48	47
Dist. artic. APD		65.5	66	64		55	56	61	61	62	63
Dist. artic. TD				30				26	27.5	31	31
			Dicerorhinus jeanvireti								
	N	Mean	Min	Max	SD	CV					
L	10	98	91.5	103	3.171	3.24					
W	8	58.94	53.5	62	3.19	5.41					
H	8	75.19	72	78	1.889	2.51					
Prox. artic. APD	7	55.29	51	60	2.87	5.19					
Prox. artic. TD	6	55.25	44	64	6.586	11.92					
Dist. artic. APD	4	73.5	68	77							
Dist. artic. TD	3	33.5	33	34							
		Dicerorhinus etruscus etruscus (not Senèze)									
	N	Mean	Min	Max	SD	CV					
L	11	73.86	67.5	82.5	4.675	6.33					
W	11	47.14	43	50	2.589	5.49					
H	10	58	52.5	64	4.72	8.14					
Prox. artic. APD	12	45.17	43	47.5	1.285	2.85					
Prox. artic. TD	9	45.72	40.5	49	2.938	6.43					
Dist. artic. APD	9	62.11	59	65	2.133	3.43					
Dist. artic. TD	7	26.5	25	28	1.155	4.36					
		Dicerorhinus etruscus brachycephalus									
	N	Mean	Min	Max	SD	CV					
L	17	78.15	71	82.5	3.196	4.09					
W	18	48.53	42	55.5	3.747	7.72					
H	21	60.86	55	68.5	3.705	6.09					
Prox. artic. APD	20	49.45	41	59	3.967	8.02					
Prox. artic. TD	18	45.36	39	51	3.796	8.37					
Dist. artic. APD	18	62.56	57.5	68	3.034	4.85					
Dist. artic. TD	17	28.76	25	34	2.195	7.63					

[a] Estimated value

Lunar

Nine lunars were found at Senèze, including two during the new excavations. The front side is proximally enlarged, and its distal end is rounded. The ulnar part of the articular hinge makes a marked angle with the proximal edge. In lateral view (Fig. 10.11) the proximal facet for the pyramidal is triangular, high and short; the distal facet is made of two successive lobes, the posterior being higher and shorter than the anterior one. In medial view the proximal articulation with the scaphoid is constituted by a very elongated anterior facet followed by a posterior one, more or less distinct; there is a distal articular facet, triangular, high and elongated, near the anterior border of the face. The dimensions of the lunar are given in Table 10.8. During the Villafranchian the lunar of *D. etruscus etruscus* tends to shorten, to enlarge and to become lower.

Fig. 10.11 *Dicerorhinus etruscus etruscus* Senèze left lunar FSL SEN 01–0125, lateral view

Table 10.8 Comparative dimensions of the Senèze *Dicerorhinus etruscus etruscus* lunars

Senèze		FSL SEN	FSL SEN	FSL		MNHN.F		NMB		NMB
		01–0156	01–0125	210959		1922–5		Se 1703		Se 1388
Measurement	Side	R	L	L	R			R	L	
L		69	70	62	62	56	57	64	64	60
W		44[a]	47	44	45.5	40	39	45	45	50
H		51	52	48	51	45	43.5	45		47
ant. H		49.5	51	52						
			Dicerorhinus jeanvireti							
	N	Mean	Min	max	SD	CV				
L	6	75.75	72	80	3.34	4.41				
W	7	53.86	50	60	3.89	7.23				
H	7	56.71	53	60	2.93	5.16				
ant. H	2	57.75	57	58.5						
		Dicerorhinus etruscus etruscus (not Senèze)								
	N	Mean	Min	Max	SD	CV				
L	11	63.18	59	68	2.87	4.54				
W	13	44.46	35	52.5	5.52	12.41				
H	12	48.75	45.5	51.5	2.27	4.66				
ant. H	1	52.00								
		Dicerorhinus etruscus brachycephalus								
	N	Mean	Min	Max	SD	CV				
L	23	67.13	60	74	3.54	5.28				
W	21	47.50	42	52.5	3.70	7.80				
H	23	49.02	43.5	56.5	3.43	6.99				
ant. H	21	50.10	43.5	58	3.25	6.48				

[a] Estimated value

Pyramidal

There are eleven pyramidals, including two found during the new excavations (Fig. 10.12); one of these, a left, is firmly fused with a pisiform (Fig. 10.13). Such a fusion was never previously described in the family Rhinocerotidae.

The antero-external side is slightly wider than high (Fig. 10.12). The postero-internal side bears two articular facets, a very large and rectangular proximal one, and a distal facet which is higher laterally than medially and shaped like a lying-down L. The proximal face is trapezoid shaped with a lateral edge much shorter than the medial one. The distal face, totally articular, is half or two thirds of an ellipse. The dimensions of the pyramidal are given in Table 10.9.

Pisiform

Six were found at Senèze, including two during the new excavations; as previously noted, one is fused with a pyramidal (Fig. 10.13). The general shape is that of a tennis racket, the proportions are very variable. The dimensions of the pisiform are given in Table 10.10.

Fig. 10.12 *Dicerorhinus etruscus etruscus* Senèze right pyramidal FSL SEN 01–0087, anterior view

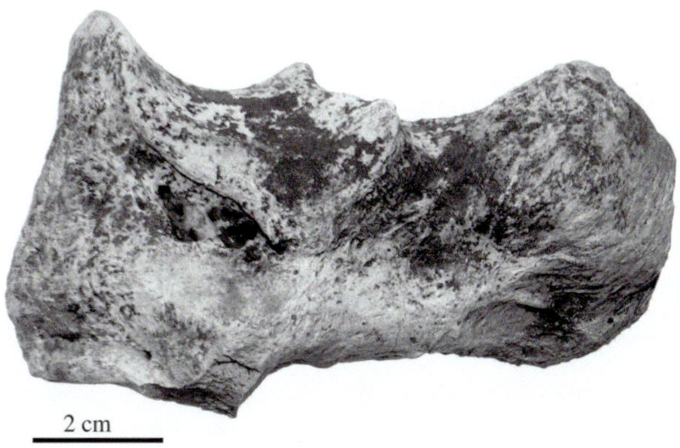

Fig. 10.13 *Dicerorhinus etruscus etruscus*, fused left pyramidal and pisiform, Senèze, FSL SEN 01–0069, anterior view

Table 10.9 Comparative dimensions of the Senèze *Dicerorhinus etruscus etruscus* pyramidals

Dicerorhinus etruscus etruscus (Senèze)												
		FSL SEN	FSL SEN	FSL	FSL		FSL	MNHN.F		MNHN.F	NMB	NMB
		01–0087	01–0069	210951	210959		(Ex-Clermont)	1922–5		1923–8	Se 1703	Se 1729
Measurement	Side	R	L		L	R					R	L
L		45			36	38	38.5	33.5	33.5	33	37	35
W		49[a]	58.5	45	52	50	55.5	51	51	50	53	48.5
H		49.5	52	42	50	50	47.5	49	50	50	51	50
		Dicerorhinus jeanvireti										
	N	Mean	Min	Max	SD	CV						
L	5	45.3	41	50	4.35	9.61						
W	6	57.75	51	65	5.02	8.69						
H	6	62.58	55.5	67	4.15	6.63						
		Dicerorhinus etruscus etruscus (not Senèze)										
	N	Mean	Min	Max	SD	CV						
L	11	36.32	32	40	2.9	7.99						
W	10	49.25	41	56.5	4.16	8.44						
H	12	49.46	47	55	2.29	4.63						
		Dicerorhinus etruscus brachycephalus										
	N	Mean	Min	Max	SD	CV						
L	18	36.58	31.5	40	2.38	6.5						
W	18	53.22	49	56.5	2.41	4.53						
H	20	49.75	45	54	3.1	6.23						

[a] Estimated value

Table 10.10 Comparative dimensions of the Senèze *Dicerorhinus etruscus etruscus* pisiforms

Senèze		FSL SEN	FSL SEN	FSL 210959		NMB Se 1703	
		01–0088	01–0069	Ex 96129			
		R	L	L	R	R	L
L		58.5		60	60	59	58[a]
W				27	27	26	25.5
H			42	37	35.5	40	40.5
		Dicerorhinus jeanvireti					
	N	Mean	Min	Max	SD		
L	3	68.00	65	70	2.65		
W	3	27.67	25	29	2.31		
H	3	47.67	42	53	5.51		
		Dicerorhinus etruscus etruscus (not Senèze)					
	N	Value					
L	1	58.00					
W	1	25.00					
H	1	38.00					
		Dicerorhinus etruscus brachycephalus					
	N	Mean	Min	Max	SD	CV	
L	7	60.71	46	68.5	7.58	12.49	
W	7	26.43	22	29	2.42	9.17	
H	7	39.79	35	46	3.74	9.40	

[a] Estimated value

Trapezoid

There are seven trapezoids, including one, a right, found during the new excavations (Fig. 10.14). On the medial side the articular facet for the trapezium is expanded on about the third of the length of the bone, up to its posterior border; the facet does not reach the distal edge. The proximal side is rectangular, as wide in front as it is behind. The distal side is elliptical. The dimensions of the trapezoid are given in Table 10.11.

Fig. 10.14 *Dicerorhinus etruscus etruscus* Senèze right trapezoid FSL SEN 01–0132, medial view

Table 10.11 Comparative dimensions of the Senèze *Dicerorhinus etruscus etruscus* trapezoids

Senèze		FSL SEN 01–0132	FSL 210959 (ex 96129)		MNHN.F 1922–5	NMB Se 1703		NMB Se 1763
		R	R	L	R	R	L	R
L		43	39	39	35	40	40	37.5
W		28	27.5	27	22	32	32	25
H		32	35.5	29.5	29.5	25.5	25.5	31.5
		Dicerorhinus jeanvireti						
	N	Mean	Min	Max	SD	CV		
L	8	44.25	40.5	47	2.24	5.05		
W	7	29.64	27.5	32	1.60	5.40		
H	8	36.00	24.5	40	5.06	14.07		
		Dicerorhinus etruscus etruscus (not Senèze)						
	N	Mean	Min	Max				
L	2	36.50	36	37				
W	2	27.25	26.5	28				
H	2	33.00	30	36				
		Dicerorhinus etruscus brachycephalus						
	N	Mean	Min	Max	SD	CV		
L	11	40.55	37.5	45	2.37	5.85		
W	9	26.00	24	28	1.39	5.35		
H	11	34.27	26	37.5	3.09	9.03		

Magnum

Seven were found in Senèze, including one right, during the new excavations (Fig. 10.15). The anterior side has a pentagonal outline and is rounded distally. The lateral side shows a rectangular facet for the unciform, higher than long. The distal articulation is triangular, lengthened, widened anteriorly, with a shallow notch in the middle of its lateral margin.

The dimensions of the magnum are given in Table 10.12. During the Villafranchian, the magnum of *D. etruscus etruscus* tends to shorten and to become lower.

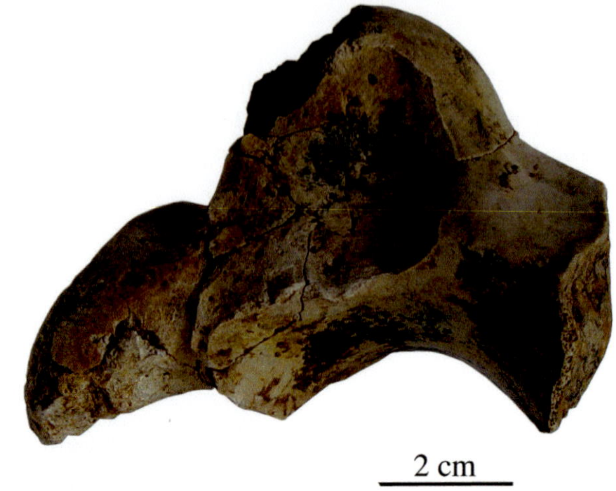

Fig. 10.15 *Dicerorhinus etruscus etruscus* Senèze right magnum FSL SEN 01–0123, lateral view

Table 10.12 Comparative dimensions of the Senèze *Dicerorhinus etruscus etruscus* magnums

Senèze		FSL SEN	FSL	FSL		MNHN.F	MNHN.F	NMB
		01–0123 + 01–0126	No number	210959 (ex 96129)		1922–5	1923–8	Se 1703
		R	R	L	R			R
L		87.5	77.5	84		77	83.5	84.5
W		48	45	43	43	40	39.5	48.5
H		62	54	55[a]	57	55		60.5
Artic. H		60		53	56.5			
		Dicerorhinus jeanvireti						
	N	Mean	Min	Max	SD	CV		
L	2	97.75	94	101.5				
W	4	53.75	47	58	4.99	9.29		
H	2	70.00	69	71				
Artic. H	2	65.50	61	70				
		Dicerorhinus etruscus etruscus (not Senèze)						
	N	Mean	Min	Max	SD	CV		
L	8	84.50	79	89	3.60	4.26		
W	11	45.05	38	48.5	2.93	6.50		
H	9	60.17	55	66	3.61	5.99		
		Dicerorhinus etruscus brachycephalus						
	N	Mean	Min	max	SD	CV		
L	13	88.42	75	95	5.50	6.22		
W	18	45.42	40	50	3.05	6.73		
H	16	60.44	55	65.5	3.26	5.39		
Artic. H	10	61.10	55	65.5	3.27	5.36		

[a] Estimated value

Unciform

There are nine unciforms including one found during the new excavations. The anterior face is somewhat higher laterally than it is medially. Its distal border is almost rectilinear, does not present any median depression and is convex solely in its lateral angle. The medial articulation is rather low, its shape is a rounded quadrilateral. In upper view the facets for the pyramidal and the metacarpal V are in contact. The dimensions of the unciform are given in Table 10.13. During the Villafranchian, the unciform of *D. etruscus etruscus* tends to shorten.

Table 10.13 Comparative dimensions of the Senèze *Dicerorhinus etruscus etruscus* unciforms

Senèze		FSL SEN	FSL		MNHN.F	MNHN.F	MNHN.F	NMB	NMB	Bâle
		01–0127	210959 (ex 96129)		1923–1	1923–8	1922–5	Se 1756	Se 1387	SE 1756
		R	L	R					R	
Abs. L			82	83.5	80.5	80.5	79		75	
Anat. L			59	60	61	60	55		53	
W		63.5	73	63	62	61	55.5	58.5	57	58.5
H		50.5	48	48	47	46	46	44	45	44.5
		Dicerorhinus jeanvireti								
	N	Mean	Min	Max	SD	CV				
Abs. L	2	97.00	97	97						
Anat. L	3	75.33	72	82						
W	5	71.60	70	76	2.61	3.64				
H	5	56.80	53	61	3.56	6.27				
		Dicerorhinus etruscus etruscus (not Senèze)								
	N	Mean	Min	Max	SD	CV				
Abs. L	13	82.27	75	87.5	3.58	4.35				
Anat. L	14	59.75	53	67	4.08	6.82				
W	14	60.18	54.5	64	2.81	4.66				
H	15	46.97	42.5	52	2.55	5.42				
		Dicerorhinus etruscus brachycephalus								
	N	Mean	Min	Max	SD	CV				
Abs. L	10	84.95	78	94	5.07	5.97				
Anat. L	12	63.67	55	71.5	4.50	7.07				
W	12	65.54	60	70	3.33	5.09				
H	15	51.83	46	55	2.87	5.54				

Metacarpal II

There are ten specimens from Senèze, with one, a right, found during the new excavations (Fig. 10.16). The Mc II is long and slim. The proximal epiphysis shows a well developed posterior tuberosity. The proximal articulation, with a more or less crescent shaped outline, is relatively wide and rather short; its most anterior point is located clearly behind the most anterior point of the epiphysis. The postero-lateral articulation is slightly notched in its distal part; it is constituted by two superposed facets, the distal being lower and much shorter than the proximal. On the medial face of the epiphysis the articular facet corresponding to the trapezium is nearly absent. The transverse section of the diaphysis is a rounded trapezoid whose posterior edge is sometimes slightly concave. The dimensions of the Mc II are given in Table 10.14. From the Early-Middle to the Late Villafranchian, the length of the bone and the two diameters of the diaphysis show a tendency to reduce (Guérin, 1980, 2004).

Fig. 10.16 *Dicerorhinus etruscus etruscus* Senèze right Mc II FSL SEN 01–0103, posterior view

Table 10.14 Comparative dimensions of the Senèze *Dicerorhinus etruscus etruscus* Mc IIs

Senèze		FSL SEN	FSL	FSL		FSL	MNHN.F	MNHN.F	NMB		NMB	NMB
		01–0103	210946 (ex 3711)	210459 (ex 96129)		ex Clermont	1923–8	1922–5	Se 1711		Se 1756	Se 1703
		R	L	L	R	R	L	L	R	L	L	R
L		183.5	181.5	180	180.5	168.5	175	169		162	175	175
Prox. TD		46.5	34	39	42	34.5	42.5	31	29	30	37	41
Prox. APD		44	38	46	46.5	43	43.5	35	40	39	41.5	44
TD dia		37.5	31	31	33.5	27		27		24	31	38
APD dia		27	24	21	22	22		17		28	22	20
Max. TD		45	43	43	44	41.5	47.5	38		39	41.5	46.5
Dist. TD		42	38	39.5	38	39.5	45.5	33		36	37	45.5
Dist. APD		40.5	36	38	38	36	40	37		36	39	40
		Dicerorhinus jeanvireti										
	N	Mean	Min	Max	SD	CV						
L	6	200.92	194	207	5.258	2.62						
Prox. TD	7	46.5	40.5	53	4.682	10.07						
Prox. APD	7	43.43	38	49.5	3.856	8.88						
TD dia	8	41.25	38	43	1.753	4.25						
APD dia	7	24.93	23	27.5	1.858	7.45						
Max. TD	5	51.4	49.5	53.5	1.475	2.87						
Dist. TD	8	43.5	42	45.5	1.102	2.53						
Dist. APD	8	44.44	40	48	2.945	6.63						
		Dicerorhinus etruscus etruscus (not Senèze)										
	N	Mean	Min	Max	SD	CV						
L	11	174.27	139.5	186	13.024	7.47						
Prox. TD	17	37.91	27	47	4.621	12.19						
Prox. APD	17	39.76	34	46	2.796	7.03						
TD dia	12	33.75	31	38	1.865	5.53						
APD dia	12	22.71	19	25	1.777	7.82						
Max. TD	9	42.06	38	48	3.302	7.85						
Dist. TD	10	37.35	30	45	4.13	11.06						
Dist. APD	10	37.9	36	41	1.524	4.02						
		Dicerorhinus etruscus brachycephalus										
	N	Mean	Min	Max	SD	CV						
L	11	181.59	174	194	7.283	4.01						
Prox. TD	11	43.95	39.5	49	3.061	6.97						
Prox. APD	12	42.92	37	45.5	2.12	4.94						
TD dia	11	35.59	30	41	2.948	8.28						
APD dia	11	23.27	20	28	2.621	11.26						
Max. TD	11	45.23	42	49	2.172	4.8						
Dist. TD	11	38.59	34	43	2.386	6.18						
Dist. APD	11	39.45	36	41	1.65	4.18						

Metacarpal III

Eleven complete or sub-complete Mc III were found at Senèze, including one during the new excavations. The Mc III is long and slender (Fig. 10.17). The proximal articulation is wide, triangular with a convex anterior margin. On the lateral surface of the proximal epiphysis there are two articular facets; the anterior one is almost circular and subdivided into two parts of which the distal one is very reduced; the posterior facet is about as high as the anterior one but more elongated, with an elliptical or rounded quadrate outline. On the medial surface of the proximal epiphysis, there is a small articular facet in the shape of an S which is variable in height. There is a strong individual variability in the shape of the outlines of all the lateral and medial facets of the proximal epiphysis, and in the surface and the spacing of the medial ones. The transverse section of the diaphysis is elliptical, with the posterior border straight or slightly depressed. The dimensions of the Mc III are given in Table 10.15. From the Early-Middle to the Late Villafranchian the length of the bone and the width of its distal epiphysis show a clear tendency to reduce, while the antero-posterior diameters of the diaphysis and especially of the distal epiphysis tend to increase (Guérin 1980, 2004).

Fig. 10.17 *Dicerorhinus etruscus etruscus* Senèze right Mc III FSL SEN 01–0106, anterior view

Table 10.15 Comparative dimensions of the Senèze *Dicerorhinus etruscus etruscus* Mc IIIs

Senèze	FSL SEN	FSL		MNHN.F	NMB		NMB		NMB	NMB	NMB	NMB
	01–0106	210459 (ex 96129)		1923–8	Se 1711		Se 1756		Se 1634	Se 1703	Se 1392	Se 1736
	R	R	L		R	L	L	R	R	R	R	L
L	207.5	205.5	205	196		197	197.5	192[a]	194	197.5	208	210
Prox. TD	58	56	56	56		54.5	51		52	54.5	55	55.5
Prox. APD	49[a]	49	47	48.5			44		44	48	47.5	46.5
TD dia	57.5	49	50	53	46.5	45	45	46	47.5	51.5	48	53.5
APD dia	20	21.5	22	21.5			20	20	20.5	20.5	20	24.5
Max. TD	58.5	57	57	58	49	50[a]	59	60	55	57	57.5	61
Dist. TD	51.5	49	48	48		43	46.5	46.5	45	47.5	47	46.5
Dist. APD	44	40	42	41			42	42.5	40	41	38	41

(continued)

Table 10.15 (continued)

	N	Mean	Min	Max	SD	CV
	Dicerorhinus jeanvireti					
L	10	228.95	215	239.5	8.7	3.8
Prox. TD	14	63.79	57	69	3.662	5.74
Prox. APD	13	54.15	48	58	3.478	6.42
TD dia	14	57.18	51	63	3.635	6.36
APD dia	11	22.4	20	27.5	1.99	8.88
Max. TD	11	66.41	60	74.5	4.592	6.92
Dist. TD	13	53.52	48	58.5	3.001	5.61
Dist. APD	13	47.86	42	53.5	3.582	7.48
	Dicerorhinus etruscus etruscus (not Senèze)					
	N	Mean	Min	Max	SD	CV
L	11	202.73	194	220.5	7.019	3.46
Prox. TD	17	52.79	48.5	58	2.574	4.88
Prox. APD	15	45.53	42.5	51	2.083	4.57
TD dia	12	49	45.5	52.5	1.883	3.84
APD dia	11	20.32	18	22	1.27	6.25
Max. TD	11	56.55	54	59	1.781	3.15
Dist. TD	11	45.73	43	50	2.09	4.57
Dist. APD	8	40.5	35	43	2.577	6.36
	Dicerorhinus etruscus brachycephalus					
	N	Mean	Min	Max	SD	CV
L	17	204.59	188.5	221	7.942	3.88
Prox. TD	23	55.11	51	63	2.78	5.04
Prox. APD	17	47.32	44	51.5	2.106	4.45
TD dia	18	51	46	58	3.382	6.63
APD dia	18	21.31	18.5	24	1.646	7.73
Max. TD	15	58.8	52.5	65.5	3.688	6.27
Dist. TD	17	48.21	43	54.5	3.103	6.44
Dist. APD	15	43.23	39	46.5	2.625	6.07

Figure 10.18 is a scatter diagram of the supra-articular proximal width of the Mc III (Y axis) as a function of length (abscissa). All the Senèze Mc III plot near the Mc III of *D. etruscus etruscus* from other Villafranchian sites. The Mc III of *D. etruscus brachycephalus* are somewhat larger, and those of *D. jeanvireti* clearly longer.

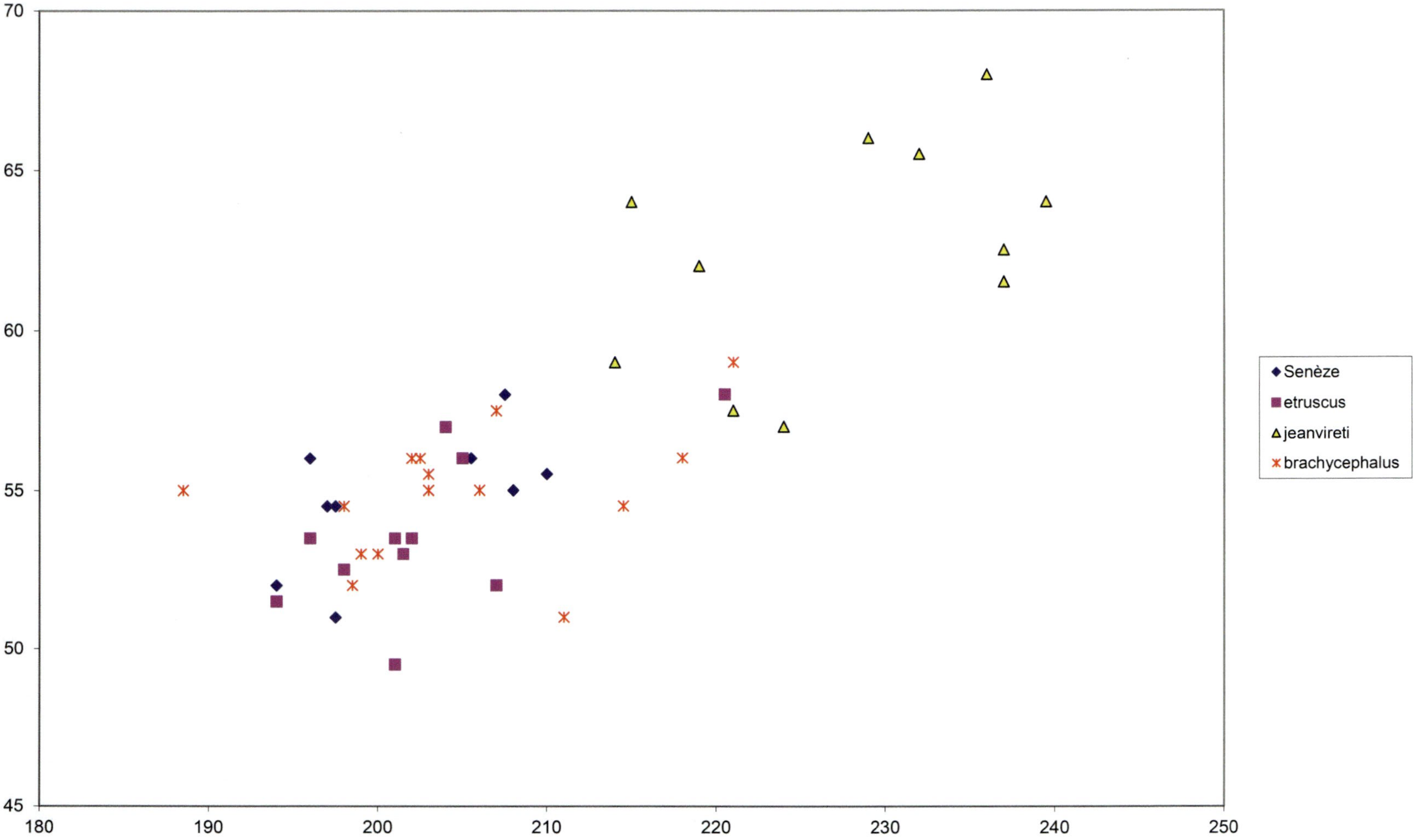

Fig. 10.18 Scatter diagram of proximal supra-articular transverse diameter (y-axis) compared to length (x-axis) of the Mc III of the Senèze rhino, other *Dicerorhinus etruscus etruscus*, *D. etruscus brachycephalus* and *D. jeanvireti*

Figure 10.19 is a Simpson diagram of the Mc III of *D. etruscus etruscus* (mean values of all available specimens, including those from Senèze) compared with those of *D. etruscus brachycephalus* and *D. jeanvireti*, with recent *Diceros bicornis* as the reference. It shows very strong similarities in general proportions of the three fossil *Dicerorhinus*, while *D. bicornis* is quite different. *D. sumatrensis* is more reminiscent of *D. bicornis* but with a relatively reduced antero-posterior diameter of its diaphysis.

small and quite indistinct, merging with the posterior boundary of the proximal articulation. The medial surface of the proximal epiphysis has two articular facets, whose spacing is rather variable; the anterior one is clearly longer than the posterior, and its proximal margin is moderately oblique; its vertical extension corresponds to the third of the height of the posterior facet. The latter is sub-circular or elliptical and can be as high as wide or slightly higher than wide. The transverse section of the diaphysis is slender, of triangular or

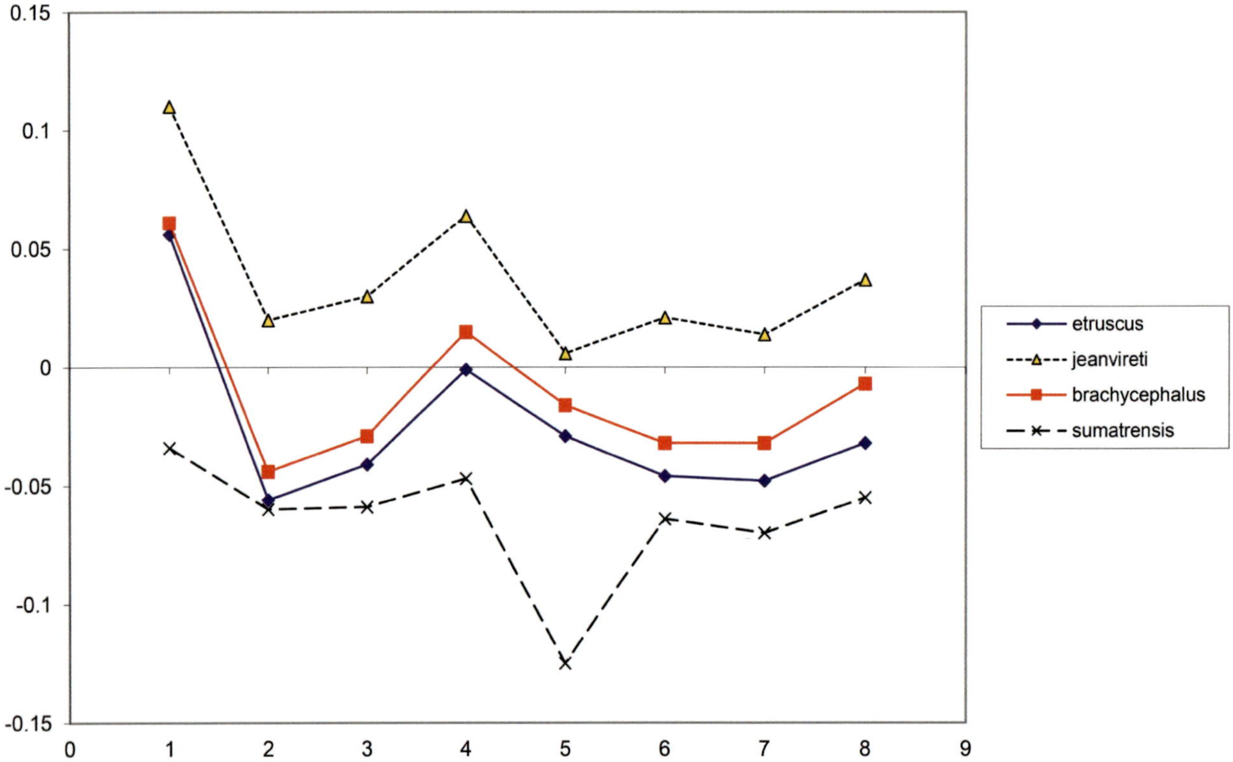

Fig. 10.19 Simpson diagram of the Mc III: mean values of *Dicerorhinus etruscus etruscus, D. etruscus brachycephalus*, *D. jeanvireti*, and *D. sumatrensis*. Reference is the recent *Diceros bicornis*. The variables on the x-axis are the same and in the same order as in Table 10.15

Metacarpal IV

There are ten complete or subcomplete Mc IV, including a right specimen found during the new excavations (Fig. 10.20). The proximal articulation is triangular, less enlarged posteriorly than for *D. jeanvireti*. Its posterior margin is sinuous. The articular facet for the Mc V is very pentagonal midsection; in the latter case the medial border is not rectilinear but presents an obtuse dihedral shape. The dimensions of the Mc IV are given in Table 10.16. During the Villafranchian, the length of the bone, the two diameters of the proximal epiphysis and the antero-posterior diameter of the distal epiphysis tend to reduce and the antero-posterior diameter of the diaphysis tends to increase.

Fig. 10.20 *Dicerorhinus etruscus etruscus* Senèze right Mc IV FSL SEN 01–0107, medial view

Table 10.16 Comparative dimensions of the Senèze *Dicerorhinus etruscus etruscus* Mc IVs

Senèze		FSL SEN	FSL		FSL		MNHN.F	NMB		NMB	NMB
		01–0197	210946 (ex 371)		210959 (ex 96129)		1890–5	Se 1711		Se 1703	Se 1706
		R	L	R	L	R		R	L	R	L
L		167.5	167	169	171	170	158	157	157.5	158.5	160
Prox. TD		42.5	40.5	40	45		40			41	42
Prox. APD		38.5	33	35	36	37	38	32	35		37
TD dia		32	29.5	28.5	32.5	31	34	27[a]	28	33.5	32
APD dia		25	21.5	21	23.5	24	21	23[a]	22	23	23.5
Max. TD		44	41	40	45	45.5	42		38	40	41.5
Dist. TD		44	35	34	39	38	39		35	37.5	36.5
Dist. APD		41	37	37	38.5	38	39	31.5	36.5	38.5	36
		Dicerorhinus jeanvireti									
	N	Mean	Min	Max	SD	CV					
L	6	194.75	185	201	6.243	3.21					
Prox. TD	9	51.17	45.5	57	3.857	7.54					
Prox. APD	9	43.39	41	46.5	2.205	5.08					
TD dia	9	39.94	37	45	2.338	5.85					
APD dia	8	24.25	20	28	3.012	12.42					
Max. TD	4	47.5	43	50	3.317						
Dist. TD	9	45.17	38	53	3.873	8.57					
Dist. APD	9	46.72	44.5	48	1.176	2.52					
		Dicerorhinus etruscus etruscus (not Senèze)									
	N	Mean	Min	Max	SD	CV					
L	10	169.1	158	177	6.244	3.69					
Prox. TD	15	42.2	40	45.5	1.771	4.2					
Prox. APD	14	36.75	32	40	2.471	6.72					
TD dia	11	33	30	38	2.54	7.7					
APD dia	10	21.65	19	26.5	2.427	11.21					
Max. TD	10	42.6	38	46	2.307	5.42					
Dist. TD	10	38.5	35	42	2.333	6.06					
Dist. APD	9	38.22	36	43	2.238	5.85					
		Dicerorhinus etruscus brachycephalus									
	N	Mean	Min	Max	SD	CV					
L	16	172.16	151	182	8.08	4.69					
Prox. TD	13	45.35	41	48	2.249	4.96					
Prox. APD	14	38.75	32	48	4.108	10.6					
TD dia	17	34.74	29.5	39	2.646	7.62					
APD dia	17	23.47	17	27	2.503	10.66					
Max. TD	17	45.29	40	49	2.398	5.29					
Dist. TD	17	39.68	35.5	46	2.867	7.22					
Dist. APD	16	39.84	32	43.5	2.779	6.98					

[a] Estimated value

Femur

Seven complete or subcomplete femora were found at Senèze, all during the old excavations. The articular head presents a well marked ligamentary fossa. The moderately elongated neck is rather thick. The convex lesser trochanter is thick and clearly distinct from the head. The tip of the greater trochanter is lower than the top of the head (Fig. 10.21). The third trochanter is situated in the middle of the diaphysis, it is more expanded upwards than downwards. Its axis is oblique upwards. The distal articulation is strongly dissymetric, with a very oblique sagittal plane. The lateral lip is much less developed than the medial one, the articular pulley is relatively broad, its cross section is at a slightly obtuse angle. The dimensions of the femur are given in Table 10.17. During the Villafranchian, there are no important changes in the relative proportions of the femur of *D. etruscus etruscus*, except for the reduction of the transverse diameter of the proximal epiphysis.

Fig. 10.21 *Dicerorhinus etruscus etruscus* Senèze right femur FSL 210953, anterior view

Table 10.17 Comparative dimensions of the Senèze *Dicerorhinus etruscus etruscus* femora

Senèze		FSL	FSL	FSL		MNHN.F		NMB	
		591 V	no number	210953 (ex 96129)		1923–8		Se 1711	
		R	L	L	R	R	L	L	R
L		424	424	474	453	455	460	449	
Head TD		84	80	76	80	80		70[a]	
Head APD		80	78	79	83	75.5			
Prox. TD		170	169		178	163		145[a]	
TD dia		63	62	69	61	65	67.5	50	52
APD dia		50	50	63	60	55	52	57	59
Dist. TD		128	124	135	129	128	130	118	
Dist. APD		139	140	141	150		136		
		Dicerorhinus jeanvireti							
	N	Mean	Min	Max	SD	CV			
L	10	510.75	494	519	9.11	1.78			
Head TD	12	94.42	89	99	3.70	3.92			
Head APD	13	87.58	81	96	4.74	5.42			
Prox. TD	11	195.64	190	208	5.41	2.76			
TD dia	13	69.27	65	78	4.24	6.12			
APD dia	12	61.17	53	72	6.06	9.90			
Dist. TD	13	150.54	143	160.5	5.32	3.53			
Dist. APD	10	171.35	167	176	2.73	1.59			

(continued)

Table 10.17 (continued)

	Dicerorhinus etruscus etruscus (not Senèze)					
	N	Mean	Min	Max	SD	CV
L	6	437.17	406	468	27.77	6.35
Head TD	8	79.25	68	89	5.81	7.33
Head APD	9	75.06	66	81	4.65	6.19
Prox. TD	8	162.31	148	187	12.53	7.72
TD dia	16	59.50	48	65	5.37	9.03
APD dia	16	52.94	47	65.5	4.83	9.12
Dist. TD	14	122.43	115	135	6.52	5.32
Dist. APD	9	147.22	139.5	161	8.12	5.51
	Dicerorhinus etruscus brachycephalus					
	N	Mean	Min	Max	SD	CV
L	7	448.21	412.5	472	21.87	4.88
Head TD	13	85.81	74	95	6.51	7.59
Head APD	13	81.04	72	88	5.37	6.62
Prox. TD	8	177.50	162	186	9.59	5.40
TD dia	19	62.87	53	72	5.09	8.10
APD dia	19	52.11	45	61.5	4.82	9.26
Dist. TD	16	127.59	120	140.5	6.16	4.83
Dist. APD	15	150.23	140	163	8.31	5.53

[a] Estimated value

Tibia

Nine complete or subcomplete tibiae were found in Senèze, all during the old excavations. The proximal epiphysis bears a relatively weak tibial tuberosity clearly bounded backwards by a clear but variable depression. The spacing of the edges of the tibial spine is rather reduced, the medial edge is shifted forwards and slightly higher than the lateral one. On the posterior face of the diaphysis the poplital notch is strongly excavated. The distal epiphysis is moderately developed transversally. The fibular notch is rather high, slightly and irregularly excavated. The medial malleolus is only slightly prominent and its medial angle is obtuse (Fig. 10.22). The dimensions of the tibia are given in Table 10.18. Figure 10.23 is a Simpson diagram of the mean dimensions; the tibia of *D. e. brachycephalus* is larger and stockier than that of *D. etruscus etruscus*, but the proportions of the two subspecies are roughly the same. The *D. jeanvireti* tibia is larger but its proportions are once again the same. That of *D. sumatrensis* is smaller and relatively stockier, with a reduced antero-posterior diameter of its diaphysis. During the Villafranchian, the sole modification observed for the tibia is a reduction of its distal antero-posterior diameter.

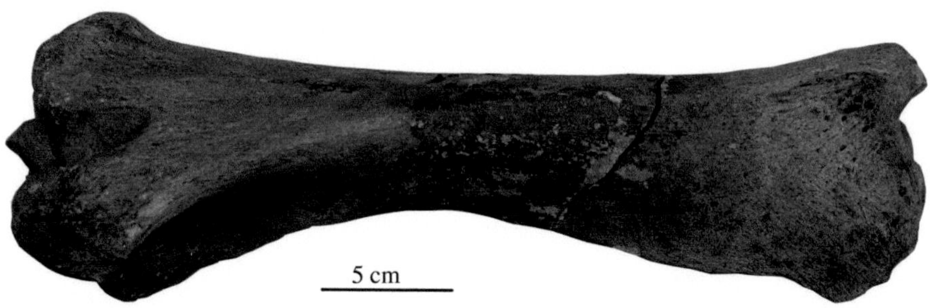

Fig. 10.22 *Dicerorhinus etruscus etruscus* Senèze right tibia FSL 210956, anterior view

Table 10.18 Comparative dimensions of the Senèze *Dicerorhinus etruscus etruscus* tibiae

Senèze		FSL	FSL	FSL		MNHN.F		MNHN.F	NMB		NMB
		210952	210956 (589 V)	210959 (ex 96129)		1923–8		1922–5 juv	Se 1711		Se 1756
		R	L	L	R	L	R	L	L	R	L
L		352	354	364	378	378	380	348	350[a]	353	
Prox. TD		110	110	111	115	115	115	102	108	105	
Prox. APD		109	110	104	116	108	107	95[a]	109.5		
TD dia		53.5	52	60	61.5	56	57	43.5	51.5		
APD dia		49	49	63	57	52	50	43	52	54	
Dist. TD		94	94	104	105	94.5		83		87.5	90
Dist. APD		65.5	65.5		69	67	67	63.5		58	64
		Dicerorhinus jeanvireti									
	N	Mean	min	max	SD	CV					
L	9	410.22	375	431	18.85	4.60					
Prox. TD	10	134.90	123	140	5.34	3.96					
Prox. APD	9	131.22	123	137.5	5.78	4.40					
TD dia	14	65.21	61	70	2.72	4.16					
APD dia	13	62.69	56	69	3.45	5.50					
Dist. TD	13	108.38	102	118	5.34	4.93					
Dist. APD	13	78.65	71	83.5	3.42	4.35					
		Dicerorhinus etruscus etruscus (not Senèze)									
	N	Mean	Min	Max	SD	CV					
L	13	353.62	334	389	14.12	3.99					
Prox. TD	11	109.73	105	114	3.14	2.86					
Prox. APD	9	109.28	100	117.5	5.88	5.39					
TD dia	20	53.10	46	57.5	3.26	6.14					
APD dia	19	51.92	44	58	4.39	8.45					
Dist. TD	23	89.76	80	99	5.93	6.61					
Dist. APD	24	64.21	58	72	4.15	6.47					
		Dicerorhinus etruscus brachycephalus									
	N	Mean	Min	Max	SD	CV					
L	16	370.91	343	394.5	14.95	4.03					
Prox. TD	21	113.60	103	123.5	6.00	5.28					
Prox. APD	14	111.75	104	126	5.68	5.08					
TD dia	32	57.22	50	68	4.19	7.32					
APD dia	29	52.53	43	65.5	5.35	10.19					
Dist. TD	34	92.99	81	103.5	5.32	5.72					
Dist. APD	36	67.44	60.5	73	3.53	5.23					

[a] Estimated value

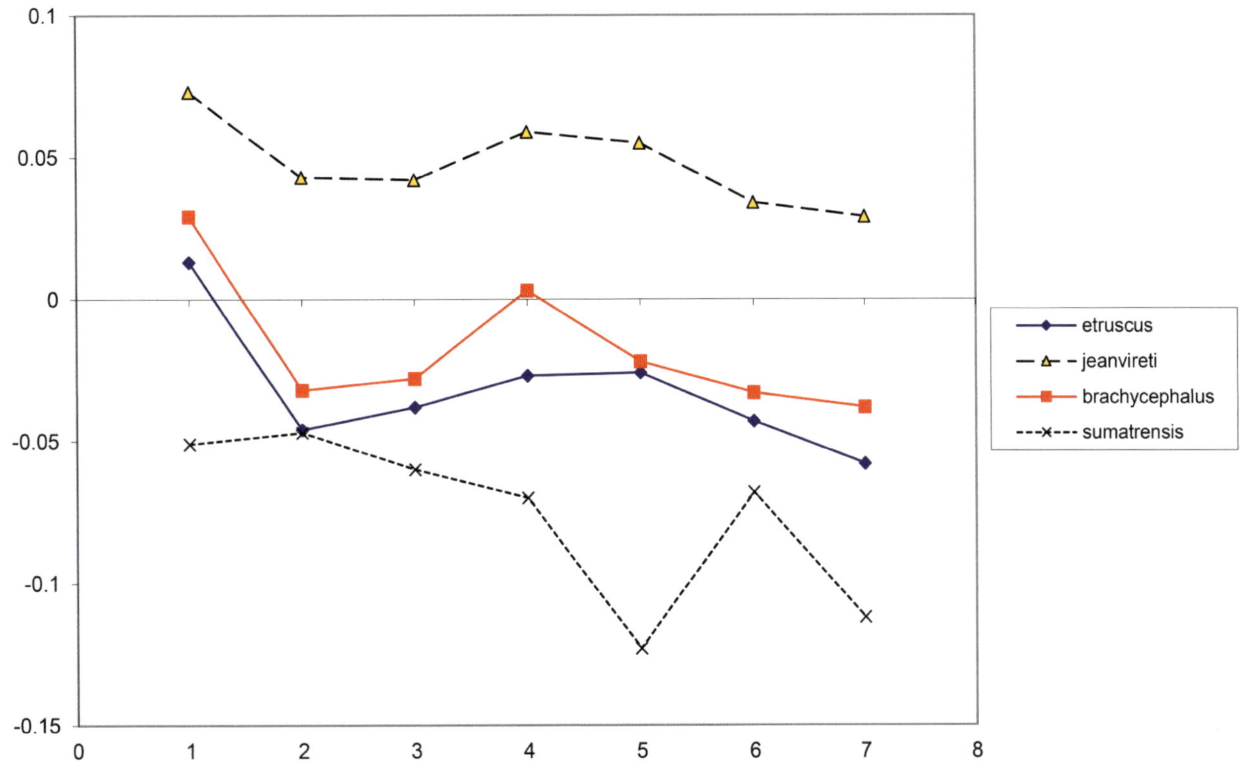

Fig. 10.23 Simpson diagram of tibia: mean values of *Dicerorhinus etruscus etruscus, D. etruscus brachycephalus* and *D. jeanvireti*. Reference is the recent *Diceros bicornis*. The variables on the x-axis are the same and in the same order as in Table 10.19

Astragalus

There are eleven rhino astragali from Senèze, two of them found during the new excavations. In anterior view (Fig. 10.24) the pulley appears broad and shallow, and also asymmetrical owing to its very little sheer leteral lip. The distal boundary of the articulation is a slightly carved low depression. In medial view, the tubercle of the inferior part of the medial surface is positioned far from the distal margin and nearer to the posterior margin. In distal view, the anterior margin of the whole articulation is straight or very slightly concave. A small notch is sometimes present between the anterior border of the facets for the cuboid and the navicular. The anterior margin of the cuboidal facet may be shifted ahead of the anterior margin of the navicular facet. The dimensions of the astragalus are given in Table 10.19. During the Villafranchian, the transverse diameter tends to reduce but only at its end; the bone height is greater during the Early Villafranchian; the anteroposterior diameter decreases at the end of MNQ 16 (end of the Early Villafranchian), remains relatively stable up to the beginning of the Late Villafranchian and decreases again late in that subage.

Fig. 10.24 *Dicerorhinus etruscus etruscus* Senèze left astragalus FSL SEN 01–0203, anterior view

Table 10.19 Comparative dimensions of the Senèze *Dicerorhinus etruscus etruscus* astragali

Senèze		FSL SEN	FSL SEN	FSL	FSL	FSL		MNHN.F		MNHN.F	NMB	NMB
		01–0228	01–0203	"210959"	"210959" (587 V)	210959 (ex 96129)		1923–8		1922–5	Se 1756 b	Se 784
		R	L			L	R	L	R	L	R	L
Max TD		88	89	83	80	87	86	83	83	73	82.5	80
H		85	80[a]	73	76	82.5	81	77	78	75	76	77
Lat APD		56	57	55	54.5	56	58	56	57	50	52.5	54
Art. dist.TD		70	70	71	66	73	73.5	67	64.5	61.5		
Art. dist. APD		43.5		41	41	44	42	41.5	45	39		43
Trochlear width		59	60.5	58	57	53	50	59	57	52	54.5	55
Dist.TD		76.5	75	71	66	73	73.5	68.5	67	65	72	66.5
		Dicerorhinus jeanvireti										
	N	Mean	Min	Max	SD	CV						
Max TD	18	99.17	92	107.5	5.128	5.17						
H	14	90.75	87	104	4.322	4.76						
Lat APD	13	64.42	56.5	72	3.662	5.68						
Art. dist.TD	11	77.86	70	85	5.348	6.87						
Art. dist. APD	10	48.3	46	52	2.098	4.34						
Trochlear width	14	65.36	62	69	2.742	4.19						
Dist.TD	13	85.31	77	94	5.739	6.73						
		Dicerorhinus etruscus etruscus (not Senèze)										
	N	Mean	Min	Max	SD	CV						
Max TD	40	80.61	74	88	3.333	4.13						
H	40	77.16	71	84	3.333	4.32						
Lat APD	38	53.07	47	58	2.475	4.66						
Art. dist.TD	38	65.21	57	75	3.755	5.76						
Art. dist. APD	37	40.38	36	45.5	2.373	5.88						
Trochlear width	41	55.7	49	62	3.199	5.74						
Dist.TD	36	70.24	60	78	3.777	5.38						
		Dicerorhinus etruscus brachycephalus										
	N	Mean	Min	Max	SD	CV						
Max TD	39	86.38	76	107	6.522	7.55						
H	39	80.47	72	89	5.113	6.35						
Lat APD	35	55.69	46	68	4.683	8.41						
Art. dist.TD	33	71.41	59.5	85	6.036	8.45						
Art. dist. APD	31	41.4	35	52	3.953	9.55						
Trochlear width	38	60.41	52	71	4.853	8.03						
Dist.TD	36	73.85	61	82	5.253	7.11						

[a] Estimated value

Calcaneum

Eight complete or subcomplete calcanea were found at Senèze, including a right one during the new excavations (Fig. 10.25). In lateral view the distal section of the tuber is located somewhat proximal to the beak (which is in many rhino species the most anterior part of the bone). There is a strong difference in the anteroposterior height between the most proximal point of the bone and the most distal part of the anterior tuber, which is much more posterior. The posterior edge of the facet is very slightly concave, the distal one is poorly elongated. In posterior view the axis of the sustentaculum tali is slightly oblique downwards, making a weakly obtuse angle with the long axis of the body of the bone. The medial extension of the sustentaculum is reduced, its medial end is thick and rounded. In distal view the cuboid facet is triangular shaped and rather short. The dimensions of the calcaneum are given in Table 10.20. During the Villafranchian, there are few biometrical modifications, mainly a significant decrease of the antero-posterior diameter of the beak and a slight decrease of the transverse diameter at the sustentaculum tali.

Table 10.20 Comparative dimensions of the Senèze *Dicerorhinus etruscus etruscus* calcanei

Senèze		FSL SEN	FSL	FSL	FSL		MNHN.F		MNHN.F
		01–0229	587 V	no number	210959 (ex 96129)		1923–8		1922–5
		R	R	L	L	R	R	L	L
H		122.5	111	111	129	129	119	121	118
APD tuber		70.5	59	59.5	68	68	61.5	63.5	58
APD beak		63.5	56	54	62	61	55	56.5	54
Sust. TD		76	66	67	62	66	71.5	70	65
TD tuber		47	44	43	49	48.5	46	45	40
Middle TD		39.5	36.5	37	39	37.5	38	37	33
		Dicerorhinus jeanvireti							
	N	Mean	Min	Max	SD	CV			
H	13	139.08	128	149.5	7.664	5.51			
APD tuber	14	74.08	66	83.5	4.037	5.45			
APD beak	13	73.35	67	83	4.17	5.69			
Sust. TD	12	81.25	75	89	4.24	5.22			
TD tuber	15	53.77	48	61	4.423	8.23			
Middle TD	11	38.59	35	43.5	2.558	6.63			
		Dicerorhinus etruscus etruscus (not Senèze)							
	N	Mean	Min	Max	SD	coef. var			
H	16	118.28	109	130	5.819	4.92			
APD tuber	18	64.08	57.5	71	3.528	5.51			
APD beak	18	61.11	55	70	3.563	5.83			
Sust. TD	16	67.97	60	75	4.33	6.37			
TD tuber	16	43.38	39.5	46	2.217	5.11			
Middle TD	14	34.04	31	37	1.926	5.66			
		Dicerorhinus etruscus brachycephalus							
	N	Mean	Min	Max	SD	CV			
H	28	119.16	110	131	6.149	5.16			
APD tuber	23	68.41	55.5	77	4.673	6.83			
APD beak	24	60.75	54	68	4.019	6.62			
Sust. TD	27	72.94	64	80	4.382	6.01			
TD tuber	29	48.6	43	58	3.764	7.74			
Middle TD	28	38.29	32.5	44	3.489	9.11			

Fig. 10.25 *Dicerorhinus etruscus etruscus* Senèze right calcaneus FSL SEN 01–0224, anterior view

Cuboid

There are thirteen rhino cuboids from Senèze, two of them found during the new excavations. The anterior face is trapezoidal and somewhat broader than tall; its lateral border is clearly higher than the medial one. In medial view (Fig. 10.26) the cuboid is remarkably high at the level of the posterior edge of the proximal and the distal articulations. The anterior border of the medial face is slightly oblique. The articular facets are fused in two clusters: one, continuous, for the four upper and median ones, and another one, antero-distal, whose length is variable. The proximal articulation is strongly enlarged forwards. It is constituted by two facets, the lateral being no longer than the medial one but much more broader. The anterior and posterior margins of the lateral facet are shifted forwards. The shape of the distal articulation is very variable. The dimensions of the cuboid are given in Table 10.21. The length of the cuboid tends to decrease progressively during the Villafranchian. The height of the anterior face tends to increase from the Early

Fig. 10.26 *Dicerorhinus etruscus etruscus* Senèze left cuboid FSL SEN 01–0200, medial view

Table 10.21 Comparative dimensions of the Senèze *Dicerorhinus etruscus etruscus* cuboids

Senèze	FSL SEN	FSL SEN	FSL	FSL	FSL	FSL	FSL	MNHN. F	MNHN. F	MNHN. F	NMB	NMB
	01–0230	01–0200	Ex-Clermont	210959	210959	210959	210959	1923–8	1923–4	1922–5	Se 1756	Se 1686
	R	L	L									
L	67	70.5	58	56	57	57	60	64 61	54	55	59	58.5
W	43.5	44	41	48	44	43.5	44	45 42	38.5	36	42	41
H	56.5	55.5	55	55	57.5	54.5	54	57 57	51	52	57	54
Prox art. L	42	40	34	44	41.5	40	41.5	42 42.5	39	41	39	37
Prox art. W	46.5	46.5	41	41.5	43	40	40.5				40.5	40.5
H ant. face	41	43	42.5	42.5	42	36.5	36	42 40	37.5	39	39.5	39.5

(continued)

Table 10.21 (continued)

		Dicerorhinus jeanvireti				
	N	Mean	Min	Max	SD	CV
L	8	66.91	63	71	2.742	4.1
W	9	50.31	47.8	54	2.164	4.3
H	10	63.75	56	73	5.969	9.36
Prox art. L	8	45.42	36	51	4.48	9.86
Prox art. W	3	46.33	40	55	7.767	16.76
H ant. face	5	47.96	44.8	52	3.132	6.53
		Dicerorhinus etruscus etruscus (not Senèze)				
	N	Mean	Min	Max	SD	CV
L	13	59.31	53.5	67	3.903	6.58
W	12	40.25	36	45	3.187	7.92
H	10	57	51	62	3.009	5.28
Prox art. L	11	39.59	36	47	2.905	7.34
Prox art. W	6	39	31.5	42.5	3.912	10.03
H ant. face	11	39.86	36	42.5	2.44	6.12
		Dicerorhinus etruscus brachycephalus				
	N	Mean	Min	Max	SD	CV
L	12	63.79	59	70	3.271	5.13
W	9	40.94	37	44	2.364	5.77
H	11	57.45	52	70.5	5.846	10.17
Prox art. L	7	43.71	39	48	3.638	8.32
Prox art. W	5	41.5	38.5	43	1.871	4.51
H ant. face	8	40.88	36	43	2.475	6.05

Villafranchian to the Middle Villafranchian, then to decrease from the Middle Villafranchian to the Late Villafranchian.

Navicular

Ten complete naviculars were found at Senèze, including two during the new excavations. The proximal face (Fig. 10.27) is slightly wider than long; it presents a regular outline, with a medio-anterior edge well rounded, a weakly protruding antero-external tip and a slightly expanded postero-medial tuberosity. In lateral view all the articular facets are fused. The dimensions of the navicular are given in Table 10.22. The length and the breadth of the bone tend to decrease from the Early Villafranchian to the Middle Villafranchian, then to increase from the Middle Villafranchian to the Late Villafranchian.

Fig. 10.27 *Dicerorhinus etruscus etruscus* Senèze left navicular FSL SEN 01–0199, upper view

Table 10.22 Comparative dimensions of the Senèze *Dicerorhinus etruscus etruscus* naviculars

Senèze		FSL SEN	FSL SEN	FSL	FSL	FSL		MNHN.F	MNHN.F		NMB
		01–0227	01–0199	No number	587 V	210259 (ex 96129)		1922–5	1923–8		Se 1756
		R	L	L	R	L	R	L			
L		63	62	55	56	61	60	55	53	55	53.5
W		46	45.5	42	41	44	45	39	43	42	41
H		31.5	30	32	31	33	32	30	31	30.5	26.5
		Dicerorhinus jeanvireti									
	N	Mean	Min	Max	SD	CV					
L	8	66.38	63	69	2.57	3.88					
W	7	53.93	50.5	59	2.92	5.42					
H	8	33.81	27	39	3.7	10.95					
		Dicerorhinus etruscus etruscus (not Senèze)									
	N	Mean	Min	Max	SD	CV					
L	7	54.57	52	58	2.05	3.76					
W	7	45.5	37	49	4.17	9.17					
H	6	28.83	24.5	31.5	2.66	9.22					
		Dicerorhinus etruscus brachycephalus									
	N	Mean	Min	Max	SD	CV					
L	17	58.03	53.5	63	3.1	5.34					
W	17	46.53	41	55	4.16	8.95					
H	15	29.9	27	34	1.72	5.77					

Lateral Cuneiform

There are nine lateral cuneiforms from Senèze, two of them discovered during the new excavations. In anterior view the latero-inferior tip is acute and the upper edge depressed. The medial face shows three clearly distinct articular facets, and the anterior one is well developed. Figure 10.28 is a superior view of the right lateral cuneiform FSL SEN 01–0204. The dimensions of the lateral cuneiform are given in Table 10.23.

Fig. 10.28 *Dicerorhinus etruscus etruscus* Senèze right lateral cuneiform FSL SEN 01–0204, superior view

Table 10.23 Comparative dimensions of the Senèze *Dicerorhinus etruscus etruscus* lateral cuneiforms

Senèze		FSL SEN	FSL SEN	FSL	FSL	FSL	MNHN.F	MNHN.F	MNHN.F	
		01–0222	01–0204	587 V	no number	210959 (ex 96129)	n° 1922–5	n° 1923–4	1923–8	
		L	R	R	L	R	L			
L		48.5	48	47.5	49	45	39	37	45	47
W		45.5	44	46	45	40	38	37	44	44
H		25.5	26.5	27	27	31	25	25	27.5	25
		Dicerorhinus jeanvireti								
	N	Mean	Min	Max	SD	CV				
L	4	51.63	49	55	2.50	4.83				
W	9	49.61	45.5	54	3.21	6.47				
H	9	28.50	25	30.5	1.66	5.82				
		Dicerorhinus etruscus etruscus (not Senèze)								
	N	Mean	Min	Max	SD	CV				
L	8	44.94	42	50.5	2.92	6.50				
W	8	40.06	36	43	2.16	5.40				
H	8	25.63	22.5	28	1.75	6.82				
		Dicerorhinus etruscus brachycephalus								
	N	Mean	Min	Max	SD	CV				
L	10	47.20	44	51	2.36	5.00				
W	10	43.85	41.5	47	1.78	4.06				
H	11	26.59	23	30	2.02	7.61				

Intermediate Cuneiform

Seven intermediate cuneiforms were found at Senèze, including one recovered during the new excavations. In lateral view the proximal and the distal border have the same length. The proximal articulation has a semi-circular outline. The shape of the distal face is a rather narrow elongated ellipse. The dimensions of the intermediate cuneiform are given in Table 10.24.

Medial Cuneiform

There are six medial cuneiforms from Senèze, two of them found during the new excavations. This bone (Fig. 10.29) is very variable in shape and morphology. The dimensions of the medial cuneiform are given in Table 10.25.

Table 10.24 Comparative dimensions of the Senèze *Dicerorhinus etruscus etruscus* intermediate cuneiforms

Senèze		FSL SEN	FSL SEN	FSL	MNHN.F	NMB	NMB
		01–0187	01–0225	210959 (ex 96129)	1922–5	Se 1763	Se 1687
				R	L	R	
L		35	35	33	30	30.5	28.5
W		21.5	21.5	19	16	20	18
H		18.5	17.5	20	25	21	17
		Dicerorhinus jeanvireti					
	N	Mean	Min	Max			
L	2	37.00	34	40			
W	2	22.75	21.5	24			
H	2	19.75	19.5	20			
		Dicerorhinus etruscus etruscus (not Senèze)					
	N	Mean	Min	Max	SD	CV	
L	5	30.30	27	34.5	3.13	10.34	
W	5	20.40	19	22	1.19	5.85	
H	4	19.88	14.5	30	6.91		
		Dicerorhinus etruscus brachycephalus					
	N	Mean	Min	Max	SD	CV	
L	7	33.79	30	36.5	2.46	7.29	
W	7	20.21	18	22	1.35	6.68	
H	7	17.64	14	19	1.77	10.05	

Metatarsal II

Eleven complete or subcomplete Mt II were found in Senèze including three from the new excavations. The proximal articulation is elliptical, elongated and rather narrow. Owing to the relatively reduced development of its anterior tuberosity, the most anterior point of the articular surface is located behind the most anterior point of the epiphysis. The lateral surface of the proximal epiphysis presents two well separated articular facets, both with a more or less elliptical outline. The posterior facet, situated more distally than the anterior one, is not clearly subdivided; it is as high but a bit narrower than the anterior facet. On the posterior surface of the proximal epiphysis there is a small, subcircular and well delimited facet for the lateral cuneiform. The transverse mid-section of the diaphysis is a rounded polygon. Figure 10.30 is a front view of the left Mt II FSL SEN 01–0221. The dimensions of the Mt II are given in Table 10.26. There is no clear difference in the size and proportions between the Mt II of the Early and Middle Villafranchian and those of the Late Villafranchian, just a slight small tendency to the thinning down of the diaphysis and to a decrease of the transverse diameters of the distal epiphysis.

Fig. 10.29 *Dicerorhinus etruscus etruscus* Senèze medial cuneiform FSL SEN 01–0226

Table 10.25 Comparative dimensions of the Senèze *Dicerorhinus etruscus etruscus* medial cuneiforms

Senèze		FSL SEN 01–0226	FSL SEN 01–0182	FSL 587 V	FSL No number	MNHN.F 1923–8	
L		72.5	69.5	62	66	67	66
W		32		28	33.5	31	31
H		20		18	22	21	18.5
		Dicerorhinus jeanvireti					
	N	Mean	Min	Max			
L	2	65.50	65	66			
W	2	36.25	36	36.5			
H	2	22.25	22	22.5			
		Dicerorhinus etruscus etruscus (not Senèze)					
	N	Mean	Min	Max			
L	4	63.13	53	73			
W	4	32.50	30.5	36			
H	3	18.00	16	19			

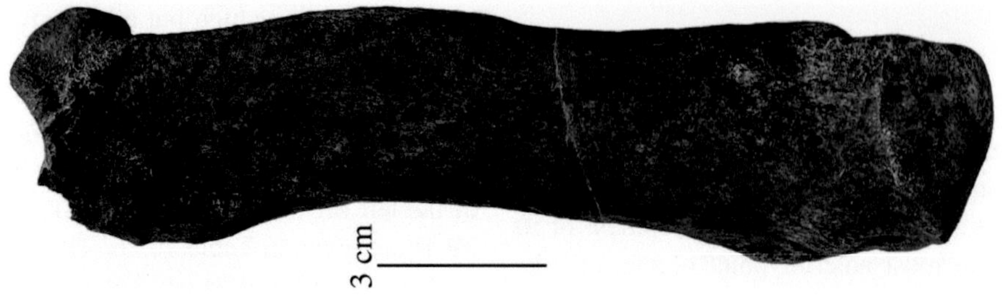

Fig. 10.30 *Dicerorhinus etruscus etruscus* Senèze left Mt II FSL SEN 01–0221, anterior view

Table 10.26 Comparative dimensions of the Senèze *Dicerorhinus etruscus etruscus* Mt IIs

Senèze	FSL SEN	FSL SEN	FSL SEN	FSL	FSL	FSL		MNHN.F		MNHN.F	NMB		NMB
	01–0221	01–0202	04–0015	"210942" (587 V)	210942	210959 (ex 96129)		1923–8		1922–5 (juv)	Se 1711		Se 1756 (juv)
	R	L	L	R	L	L	R	L	R	L	L	R	R
L	172	170	175	150	150	160.5		158	157.5	151.5	150[a]	159[a]	160
Prox. TD	36	32	41	26	24	28	27	27	27.5	25	21		27.5
Prox. APD	40	37.5	40	36.5	37	37	38	42	42		32.5	33.5	36[a]
TD dia	27.5	30.5	31.5	25.5	24	27	27	27	26.5	20[a]	23		23.5
APD dia	28.5	28	22.5	26.5	24	26.5	25	26	25	22[a]	24		24
Max. TD	38.5	39	41	37	37			35.5	34.5	30[a]	35		34
Dist. TD	38.5	37.5	36.5	35	33	31		33	32.5	29[a]	32		31
Dist. APD	37	38.5	38.5	34	34	37		36	36	34[a]	32		35

Dicerorhinus jeanvireti

	N	Mean	Min	Max	SD	CV
L	6	182.75	179	189	3.71	2.03
Prox. TD	8	35.13	30	40	3.31	9.43
Prox. APD	7	46.21	45	50	2.00	4.32
TD dia	8	30.00	22	33	3.46	11.55
APD dia	8	28.81	23	34	3.23	11.21
Max. TD	7	43.07	39	46	2.32	5.38
Dist. TD	9	38.72	36	40	1.20	3.10
Dist. APD	9	43.44	41	47	1.76	4.05

Dicerorhinus etruscus etruscus (not Senèze)

	N	Mean	Min	Max	SD	CV
L	8	155.13	145	162	5.94	3.83
Prox. TD	11	28.09	25	31	1.80	6.41
Prox. APD	11	38.50	35	42	2.55	6.62
TD dia	10	26.30	22.5	29	2.02	7.67
APD dia	10	23.80	21	27.5	1.96	8.24
Max. TD	9	38.17	34	41	2.56	6.71
Dist. TD	9	33.17	32	35	1.27	3.84
Dist. APD	9	35.39	32.5	39	2.20	6.23

Dicerorhinus etruscus brachycephalus

	N	Mean	Min	Max	SD	CV
L	13	160.96	151	171	5.87	3.65
Prox. TD	12	28.25	26	32	1.67	5.92
Prox. APD	14	41.86	38	47	2.68	6.40
TD dia	14	27.29	23	32.5	2.35	8.62
APD dia	14	24.07	20	28	2.63	10.93
Max. TD	11	37.45	33.5	41.5	2.46	6.58
Dist. TD	11	31.68	28.5	37	2.66	8.39
Dist. APD	11	38.14	35.5	42	2.36	6.18

[a] Estimated value

Metatarsal III

There are eleven rhino Mt III in Senèze, two of them found during the new excavations. The proximal articulation is triangular with a regularly convex anterior margin. The antero-lateral angle is rounded. The medial border is convex in its anterior part, then notched, and once more convex in its posteriormost part. The lateral face of the proximal epiphysis has two articular facets, the anterior one being a vertically elongated ellipse and the posterior one, more rounded, being positioned somewhat lower. Individual variation concerns mainly the outlines and the spacing of the two lateral facets. The medial face of the proximal epiphysis bears two triangular or semi-circular articular facets of about the same area. The medium transverse section of the diaphysis is rather variable; it can be regularly elliptical, ellipsoidal with a straight posterior border or a rounded triangle resulting from the shortening of the medial border of the ellipse. The dimensions of the Mt III are given in Table 10.27.

Figure 10.31 is a scatter diagram of the supra-articular proximal width of the Mt III (Y axis) as a function of length (X axis). All the Senèze Mt III plot near the Mt III of *D. etruscus etruscus* from other Villafranchian sites. The Mt III of *D. etruscus brachycephalus* are somewhat larger, and those of *D. jeanvireti* clearly longer. Figure 10.32 is a Simpson diagram of the Mt III of *D. etruscus etruscus* (mean values of all the available specimens, including those from Senèze) compared with those of *D. etruscus brachycephalus* and *D. jeanvireti*, recent *Diceros bicornis* being the reference. It shows very strong similarities in general proportions of the three fossil *Dicerorhinus*, the diaphysis and the distal epiphysis of *D. jeanvireti* being wider, while *D. bicornis* is quite different. *D. sumatrensis* is smaller and wider. During the Villafranchian, the Mt III shows a small tendency to shorten and to increase the width of its proximal epiphysis and its diaphysis.

Metatarsal IV

Nine complete or subcomplete Mt IV were found at Senèze including two from the new excavations. Figure 10.33 is a front view of the left specimen FSL SEN 04–0015. The proximal articulation is pentagonal or hexagonal in shape, with a posterior border sometimes notched and a variable anterior angle. The articulation is somewhat larger anteriorly, but its transverse extent does not exceed its antero-posterior one. The proximo-medial articulation bears two facets, the anterior one being located higher than the posterior. The anterior facet is trapezoidal in shape and its length is variable; the more rounded posterior facet is generally smaller than the anterior, and their espacement is rather small. The cross section of the diaphysis is thin, it can be a more or less flattened ellipse or a parallelogram. The dimensions of the Mt IV are given in Table 10.28. At the end of the Villafranchian, the length of the bone clearly tends to reduce, as does the antero-posterior development of its proximal epiphysis and of its diaphysis.

Table 10.27 Comparative dimensions of the Senèze *Dicerorhinus etruscus etruscus* Mt IIIs

Senèze		FSL SEN	FSL SEN	FSL	FSL	FSL		MNHN.F		MNHN.F	NMB		NMB
		01–0192	01–0206	210940 (587 V)	210941	210959 (ex 96129)		1923–8		1922–5	Se 1711		Se 191
		L	R	R	L	L	R	R	L	R	L	R	R
L		186.5	187	167.5	165	180	179	175	175	169	175	172	
Prox. TD		51	52.5	56	54	46	43	51	52	45	43	43	46
Prox. APD		44	44	44.5	46	40	40	47	44	37[a]			38
TD dia		45	46.5	42.5	42	41.5	42	42	41	35[a]	37	35	
APD dia		21.5	23.5	22	22	24.5	23.5	23	23	20,5[a]			
Max. TD		51	53.5	51	50.5	51	51	54	54.5	43[a]			
Dist. TD		46.5	47.5	43	43.5	42	43	45	45	38[a]	39.5	39	
Dist. APD		37	38	35	35	39	40	38	38	35			
		Dicerorhinus jeanvireti											
	N	Mean	Min	Max	SD	CV							
L	8	204.63	200	212	4.47	2.18							
Prox. TD	12	57.13	53	60.5	2.40	4.21							
Prox. APD	7	47.86	44	52	2.97	6.20							
TD dia	12	49.54	46	53	2.32	4.68							
APD dia	9	24.39	21	27	1.65	6.78							
Max. TD	7	61.50	58	65.5	2.61	4.25							
Dist. TD	10	50.15	45	54	2.39	4.77							
Dist. APD	10	43.70	40	46	1.89	4.32							
		Dicerorhinus etruscus etruscus (not Senèze)											
	N	Mean	Min	Max	SD	CV							
L	9	178.78	169	187	5.70	3.19							
Prox. TD	14	49.00	45	54	2.29	4.67							
Prox. APD	11	41.27	39	46	2.27	5.51							
TD dia	11	40.05	33.5	45	3.30	8.23							
APD dia	9	22.83	21.5	25	1.25	5.47							
Max. TD	10	52.20	48	59	3.25	6.23							
Dist. TD	10	42.10	40	46	2.13	5.06							
Dist. APD	7	36.57	32	38.5	2.26	6.19							
		Dicerorhinus etruscus brachycephalus											
	N	Mean	Min	Max	SD	CV							
L	14	183.79	164	197	10.99	5.98							
Prox. TD	18	50.39	37	54.5	4.04	8.01							
Prox. APD	18	43.67	39.5	49	2.89	6.61							
TD dia	18	40.97	33	47	4.00	9.76							
APD dia	18	23.94	20	29	2.19	9.14							
Max. TD	14	53.50	46	57.5	3.42	6.39							
Dist. TD	14	45.21	39	49.5	2.90	6.41							
Dist. APD	15	40.37	33.5	45	2.86	7.09							

[a] Estimated value

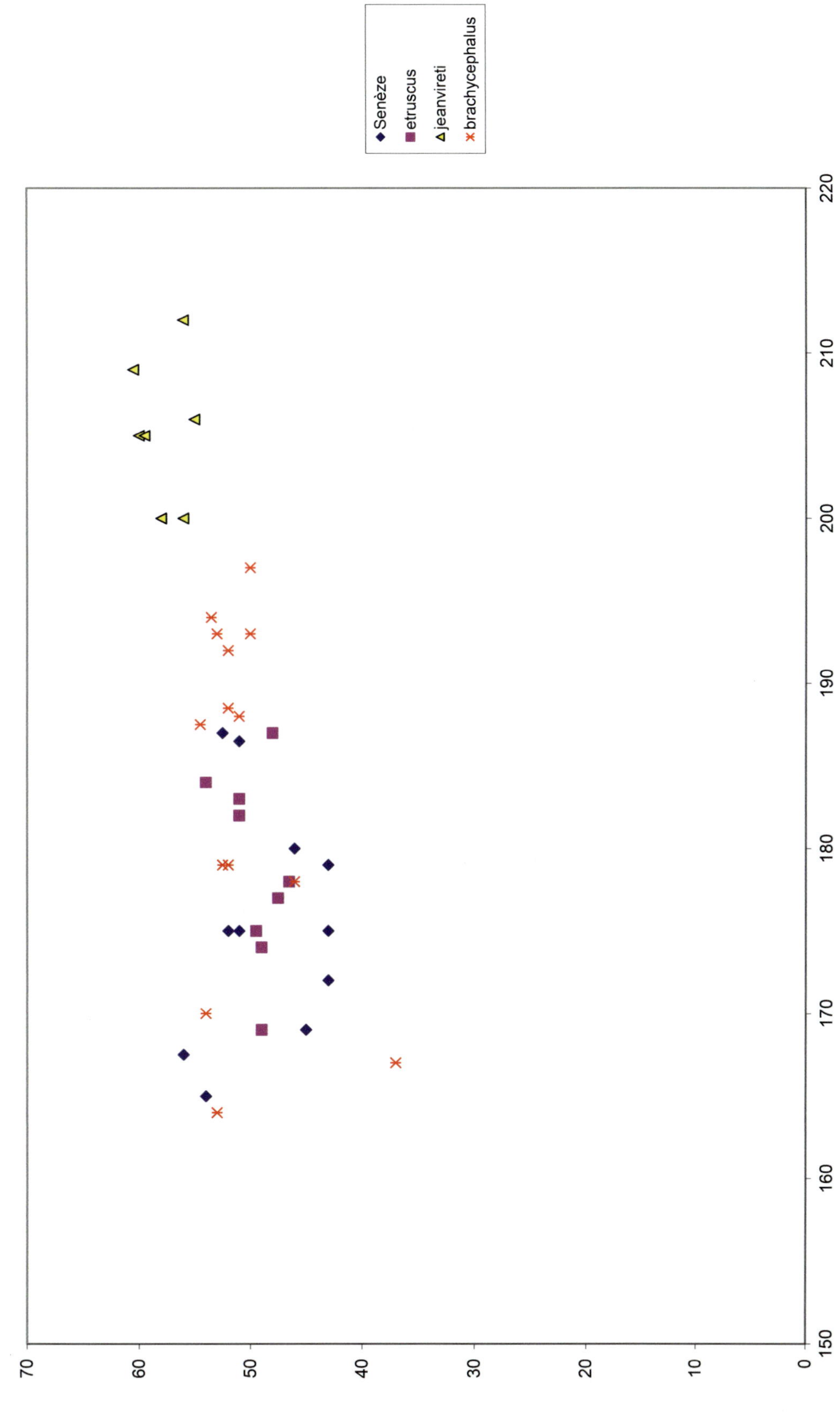

Fig. 10.31 Scatter diagram of proximal supra-articular transverse diameter (y-axis) compared to length (x-axis) of the Mt III of the Senèze rhino, other *Dicerorhinus etruscus etruscus*, *D. etruscus brachycephalus* and *D. jeanvireti*

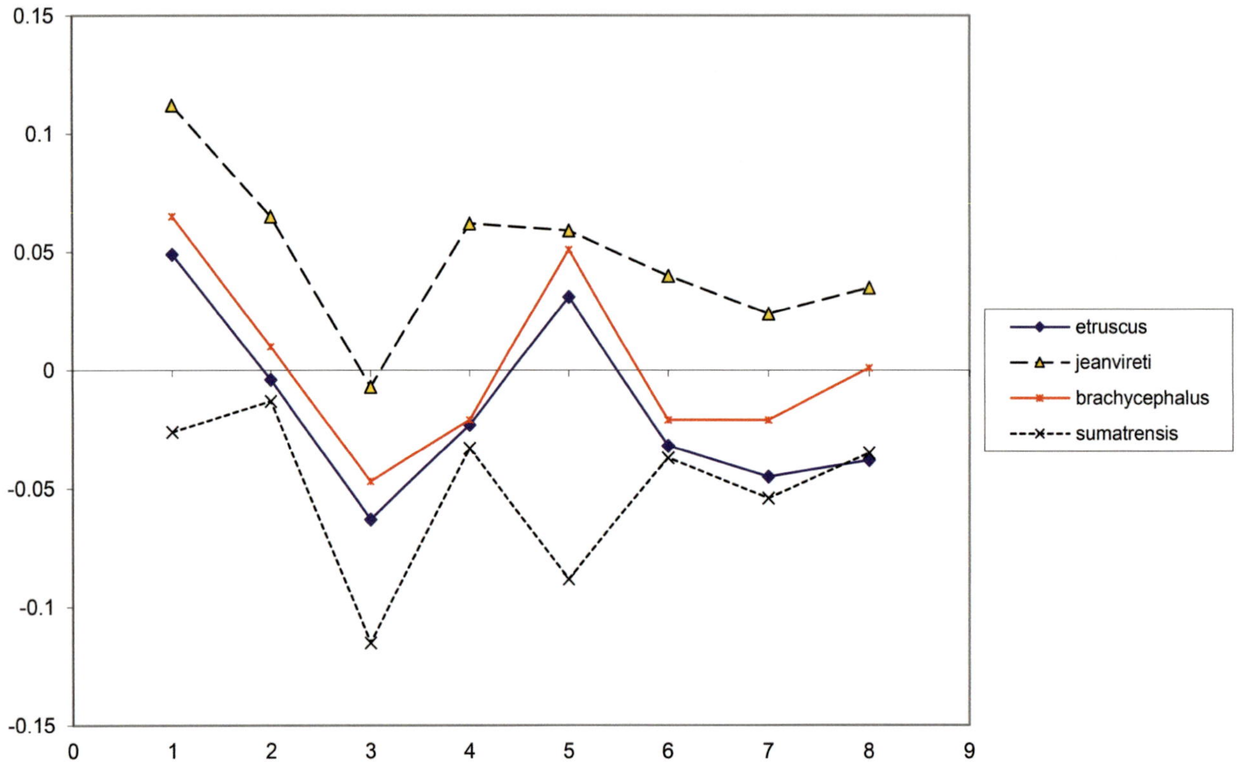

Fig. 10.32 Simpson diagram of the Mt III: mean values of *Dicerorhinus etruscus etruscus, D. etruscus brachycephalus, D. jeanvireti* and *D. sumatrensis*. Reference is the recent *Diceros bicornis*. The variables on the x-axis are the same and in the same order as in Table 10.27

Fig. 10.33 *Dicerorhinus etruscus etruscus* Senèze left Mt IV FSL SEN 04-0015, anterior view

Phalanges

In most cases rhinoceros phalanges are not specifically nor even generically determinable. Moreover it is very difficult to definitively distinguish the anterior from the posterior ones, and still more difficult to separate the lateral from the medial phalanges. The dimensions of the phalanges from Senèze are given in Table 10.29.

Table 10.28 Comparative dimensions of the Senèze *Dicerorhinus etruscus etruscus* Mt IV

Senèze		FSL SEN	FSL SEN	FSL		FSL		MNHN.F	NMB	NMB	NMB	NMB
		01–0197	01–0223 + 03–0199	210940 (ex 587 V)		210959 (ex 96129)		1923–8	Se 1711	Se 1756 (juv)	Se 1865	Se 1864
		L	R	R	L	L	R		L			
L		164	167	148.5	147	161.5	161.5	152.5	153	157		
Prox. TD		38.5	38	37	39	38.5	41	37	30	33.5	37	35
Prox. APD		41.5	40	38.5	38	35	38	41		40.5	37	36
TD dia		29	28.5	25.5	27	29	28	25	25	22.5		
APD dia		30	29.5	22.5	22	24	24.5	27	25.5	25.5		
Max. TD		37	38	32	31.5		35	32		31.5		
Dist. TD		37.5	38	30	31	33	33	33		31.5		
Dist. APD		37.5	37.5	34	33	39	39	35	33.5	35.5		
		Dicerorhinus jeanvireti										
	N	Mean	Min	Max	SD	CV						
L	8	180.19	168	188.5	6.63	3.68						
Prox. TD	11	49.64	44	56	3.23	6.51						
Prox. APD	11	44.82	40.5	51.5	3.59	8.00						
TD dia	11	35.14	31	38	2.01	5.73						
APD dia	10	29.25	23	36	4.87	16.66						
Max. TD	6	41.83	40	44	1.47	3.52						
Dist. TD	8	40.25	37	45	2.66	6.61						
Dist. APD	8	45.00	41	49.5	2.46	5.48						
		Dicerorhinus etruscus etruscus (not Senèze)										
	N	Mean	Min	Max	SD	CV						
L	9	153.11	146	163	5.61	3.66						
Prox. TD	16	38.22	32	42	2.73	7.15						
Prox. APD	14	35.36	33	37.5	1.45	4.09						
TD dia	11	26.41	22	31	2.80	10.60						
APD dia	11	23.77	20	28	2.51	10.57						
Max. TD	9	33.83	29	41.5	3.80	11.23						
Dist. TD	9	32.22	28.5	35	2.72	8.43						
Dist. APD	8	36.25	33	38.5	2.05	5.66						
		Dicerorhinus etruscus brachycephalus										
	N	Mean	Min	Max	SD	CV						
L	9	160.22	151.5	166.5	5.77	3.60						
Prox. TD	9	41.00	36.5	47.5	3.76	9.17						
Prox. APD	10	39.05	33.5	46	3.52	9.00						
TD dia	10	27.55	22	35.5	3.56	12.93						
APD dia	10	25.90	19	32.5	3.94	15.20						
Max. TD	6	32.67	29	36	2.50	7.66						
Dist. TD	7	31.21	28.5	35	2.23	7.16						
Dist. APD	8	38.88	35	46	3.48	8.96						

Table 10.29 Dimensions of the Senèze *Dicerorhinus etruscus etruscus* phalanges

	Axial proximal phalanges				Axial middle phalanges		
	FSL SEN 01–0189	FSL SEN 01–0216	FSL SEN 01–0098		FSL SEN 01–0188	FSL SEN 01–0218	FSL SEN 01–0095
L	53.5	52.5	48		37	37	36.5
Prox. TD	51	50.5	53		48	49	48.5
Prox. APD	37	36	37		27	27	28
TD dia	43	43	47.5		42.5	41	42
APD dia	20.5	21	22		23.5	22	23.5
Dist. TD	41.5	42	44		41	42	41
Dist. APD	24	24.5	27.5		25	25	26
	Abaxial proximal phalanges						
	FSL SEN 01–0101	FSL SEN 01–0101	FSL SEN 01–0076	FSL SEN 01–0220	FSL SEN 01–0205	FSL SEN 01–0184	
L	46.5	53.5	53.5	48.5	48	48	
Prox. TD	35	40	40.5	38	35	37	
Prox. APD	32	37	37	37	32.5	35.5	
TD dia	32	32.5	36	32	31.5	33	
APD dia	23	28	25	24.5	24.5	25.5	
Dist. TD	30.5	34.5	36	33	31.5	32.5	
Dist. APD	26	28.5	29	26.5	26.5	25	
	Abaxial middle phalanges						
	FSL SEN 01–0090	FSL SEN 01–0215	FSL SEN 01–0186	FSL SEN 01–0233	FSL SEN 01–0077	FSL SEN 01–0078	
L	36.5	32	34	34	40		
Prox. TD	36	32.5	32	34	39.5	35	
Prox. APD	27.5	24.5	26.5	25.5	28.5	25	
TD dia	30.5	27.5	29	31	30		
APD dia	26	22.5	25	23.5	26.5		
Dist. TD	30	27	30	28	30.5		
Dist. APD	27.5	25.5	31	27.5	30		
	Axial distal phalanges			Abaxial distal phalanges			
	FSL SEN 01–0092	FSL SEN 01–0179		FSL SEN 01–0089	FSL SEN 01–0183	FSL SEN 01–0180	FSL SEN 01–0093
L	33	35		32	33.5	34	34
Prox. TD	72	70.5		60	56.5	62	66
Prox. APD	23	24		25	25.5	23.5	25.5

Conclusion of the Anatomical Study of the Senèze Rhino

Rhino remains are very numerous in Senèze, and frequently in anatomical connection, as exemplified by a right manus found during the 2001 campaign (Fig. 10.34; see also Fig. 2.5 in Parenti et al. 2024).

The Senèze rhino presents all the anatomical and biometric characteristics of *D. etruscus etruscus*; within that subspecies the Senèze biometric data are nearer to its more derived evolutionary stage, typical of the Late Villafranchian. *D. etruscus* and *D. jeanvireti*, as well as the Middle and Late Pleistocene *D. mercki* and *D. hemitoechus*, possess several common derived cranial and dental characters, namely the possession of a partially ossified nasal septum, a gradual increase of hypsodonty of the cheek teeth and the complete disappearance of the anterior teeth. These characteristics find their maximum expression in the "woolly" rhinoceros *Coelodonta antiquitatis* of the Eurasian Pleistocene.

Figure 10.35 is a comparative Simpson diagram of the lengths of the limb bones for the three taxa of European rhinoceros considered in this chapter, and for *D. sumatrensis*. It shows an evident similarity among the European *Dicerorhinus*, which (like the Middle and Late Pleistocene European species, see Guérin 2010) present some distinctive common characteristics, such as the remarkable lengthening of the metapodials, especially of the abaxial ones, compared to *Diceros bicornis*, which serves as the reference taxon. In this feature, *D. sumatrensis* is nearer to *Diceros bicornis* than to *Dicerorhinus jeanvireti* and *D. etruscus*, but distinguished from *D. bicornis* by the reduction of the lengths of the radius and the tibia. Owing to the fact that *D. bicornis* is an "unspecialized" form, such a metapodial lengthening (like the shortening of the humerus, femur and tibia) indicates that *D. etruscus* was a cursorial species, as was *D. jeanvireti*. Such a type of cursorial rhino is not known at the present time. However, *D. etruscus* is clearly different from *D. jeanvireti*, which is always much taller and whose general proportions are different. The differences between *D. etruscus etruscus* and *D. etruscus brachycephalus* are less important, confined mainly to the larger average size of *D. e. brachycephalus* and some slight modifications of skeletal proportions, tending to be stockier. Such a difference does not justify more than a subspecific distinction.

Fig. 10.34 Right manus of *Dicerorhinus etruscus etruscus* found in anatomical connection during the 2001 Senèze fieldwork

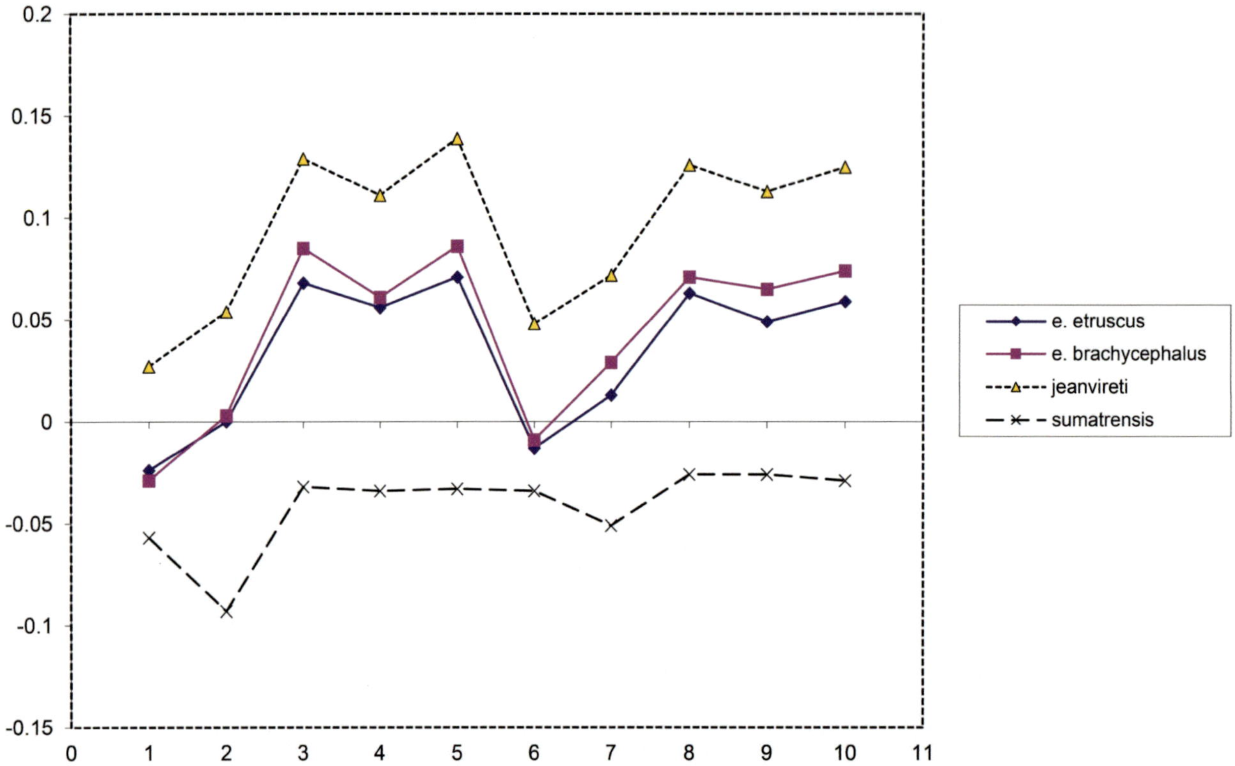

Fig. 10.35 Simpson diagram of mean lengths of the limb segments of *Dicerorhinus etruscus etruscus, D. etruscus brachycephalus, D. jeanvireti* and *D. sumatrensis*. Reference is the recent *Diceros bicornis*. The variables on the x-axis are, in order, the average lengths of: 1 = humerus, 2 = radius, 3 = Mc II, 4 = Mc III, 5 = Mc IV, 6 = femur, 7 = tibia, 8 = Mt II, 9 = Mt III, 10 = Mt IV

Biostratigraphic Significance of *D. etruscus etruscus*

D. etruscus first appeared in Europe in the Early Villafranchian (later part of MNQ 16). It is absent from the oldest sites of that unit, such as Vialette and Triversa (Fornace RDB), when the sole European rhino was *D. jeanvireti*. The two species are sympatric at Perrier-Les Étouaires and Montopoli in MNQ 16b, which is characterized in part by the appearance of *D. etruscus etruscus*. Guérin (2007) suggested that the age of Vialette and Triversa (MNQ 16a) is around 3 Ma and that of Les Étouaires and Montopoli (16 b) about 2.6–2.5 Ma, but Nomade et al. (2014) provided a date of 2.78 Ma for the earliest levels at Perrier-Les Étouaires, suggesting that the MNQ 16a/b transition may be closer to 2.8 Ma (see also Crégut-Bonnoure et al. 2024).

Following Guérin (1980) there are two evolutionary stages of the nominal subspecies *D. etruscus etruscus*, stage I corresponding to the Early and Middle Villafranchian (MNQ 16 and 17), and stage II corresponding to the Late and Final Villafranchian (MNQ 18 and 19). From stage I to stage II, evolutionary changes include the following:

- for the cranium, receding of the orbit and widening of the posterior part;
- for the mandible, thinning and shallowing of the corpus;
- for the teeth, reduction of size, namely the width of the upper cheek-teeth, and slight increase of hypsodonty;
- for the post-cranial skeleton, a general tendency to a slight reduction of the length of the long bones and the metapodials, slendering of the distal epiphysis of the humerus and the tibia, and slight changes in the proportions of the carpal and tarsal bones.

A new and more evolved subspecies, *D. etruscus brachycephalus* (Schröder, 1903: Guérin 1980) replaces the nominal one in Europe at the beginning of the Middle Pleistocene and lasts up to about 0.5 Ma, and even later in the Caucasus (Guérin & Baryschnikov 1987). *D. e. brachycephalus* is larger than *D. e. etruscus*, with a bulkier cranium whose facial part is lengthened, a deeper and more robust mandible, taller cheek-teeth, stockier long bones and metapodials. The abaxial metapodials are relatively slightly more elongated.

D. e. etruscus is abundant in the Villafranchian sites of western Europe (Guérin 1980), namely in France (Guérin 2004; Heintz et al. 1974), Italy (Pandolfi et al. 2017), Spain (Cerdeño 1993), and Great Britain (Stuart 1982). It is frequent in Central and South-Eastern Europe, in the Caucasus (Vekua 1995) and in the Middle East (Guérin 1986). The main sites with *D. etruscus etruscus* stage I are Perrier-Les Etouaires (Puy-de-Dôme; in this site it is associated with *D. jeanvireti*), Triversa/Villafranca d'Asti (Piedmont, Italy) and Villaroya (Logroño, Spain) in MNQ 16; and Saint-Vallier (Drôme, France), Olivola (Val di Magra, Italy), La Puebla de Valverde (Teruel, Spain) and the English Red Crag in MNQ 17. Among the main sites with *D. etruscus etruscus* stage II are Senèze (Haute-Loire), Cueva Victoria (Murcia, Spain: Santafé & Casanovas 1987), Lodesana (Parma, Italy) and at least part of the Upper Valdarno Formation (Tuscany, Italy). *D. e. brachycephalus* can be abundant also, as for example in Le Vallonnet and Lunel-Viel in France, or in Mauer and Mosbach in Germany (associated in the two last sites with *Dicerorhinus mercki*).

Paleoecological Significance of *D. etruscus etruscus*

The high-carried head and brachyodont cheek teeth indicate the consumption of soft foods like leaves, buds, young bark and twigs. The lengthening of the metapodials shows that *D. etruscus* was much more cursorial than the recent rhino species, all of which are more or less graviportal. The faunas and floras associated with *D. etruscus* are a mixture of steppic and forest species, and even aquatic taxa are not rare. Many of the associated plant species are thermophilous. *D. etruscus* thus appears to be a ubiquitous species living in variable habitats, the wooded savanna being probably its preferred one. Its paleoenvironment was rather moist, but the temperature was variable.

Conclusion

The rhinoceros of Senèze clearly represents *Dicerorhinus etruscus etruscus*. It is very abundant in the site and can be identified as belonging to the second evolutionary stage of the subspecies, typical of the Late Villafranchian. Its presence helps to identify the paleoenvironment of the site as a wooded savanna. The osteological and biometrical differences between *D. etruscus etruscus* and the younger and somewhat larger *D. etruscus brachycephalus* do not justify a full specific status for the latter.

Acknowledgments I thank Abel Prieur, Curator of the Paleontological collections of the University Claude Bernard-Lyon I, and Loïc Costeur, Curator of the Vertebrate collections of the Natural History Museum of Basel, for access to specimens and photographs of many of these. I thank my colleague Martine Faure for help during the measurements and in making preparations for the photographs of the fossils. Bernard Ménouret gave me great help for the preparation of the photographs. Eric Delson assisted with language editing.

References

Anonymous. (1927). Paléontologie. Travaux de Laboratoire et dans la Galerie. *Bulletin du Muséum national d'Histoire naturelle*, 235e réunion des naturalistes du Muséum, 27 janvier, n° 1, p. 33.

Antoine, P.-O. (2002). Phylogénie et évolution des Elasmotheriina (Mammalia, Rhinocerotidae). *Mémoires du Muséum national d'Histoire naturelle, 188*, 1–359.

Boule, M. (1892). Découverte d'un squelette d'*Elephas meridionalis* dans les cendres basaltiques du volcan de Senèze (Haute-Loire). *Comptes Rendus hebdomadaires des séances de l'Académie des sciences, 115*, 624–626.

Bout, P. (1970). Problèmes du volcanisme XV: Le maar de Senèze (Haute-Loire). *Revue d'Auvergne, 84*, 54–68.

Cerdeño, E. (1993). Remarks on the Spanish Plio-Pleistocene *Stephanorhinus etruscus* (Rhinocerotidae). *Comptes Rendus de l'Académie des Sciences, 317*, 1363–1367.

Crégut-Bonnoure, E., Argant, A., Argant, J., Debard, E., Delson, E., Eisenmann, et al. (2024). Biochronology of the Senèze faunal assemblage. In E. Delson, M. Faure & C. Guérin (Eds.), *Senèze: Life in Central France two million years ago. Paleontology, geochronology, stratigraphy and taphonomy* (pp. 633–652). Cham (Switzerland): Springer.

Depéret, C., & Mayet, L. (1912). Le gisement de mammifères pliocènes de Senèze (Haute-Loire). *Association française pour l'Avancement des Sciences, Compte Rendu de la 40e session, Dijon, 1911, Mémoire hors série*, 138–151.

Falconer, H. (1859). Mammalia. *Rhinoceros*. In: Ansted, T.D. (Ed.), On the geology of Malaga and the Southern part of Andalusia. *Quarterly Journal of the Geological Society of London, 15*, 602–603

Faure, M., Delson, E., & Guérin, C. (2024) Introduction and history of research in the Senèze maar. In E. Delson, M. Faure & C. Guérin (Eds.), *Senèze: Life in Central France two million years ago. Paleontology, geochronology, stratigraphy and taphonomy* (pp. 1–25). Cham (Switzerland): Springer.

Faure, M., Sen, S., & Costeur, L. (2022). Pierre Philis *Fossil Hunter*: Un demi-siècle de recherches de vertébrés fossiles (1892–1942). *Geodiversitas, 44*, 321–383.

Fortelius M., Mazza P.P. & Sala B. (1993). *Stephanorhinus* (Mammalia: Rhinocerotidae) of the Western European Pleistocene, with a revision of S. *etruscus* (Falconer, 1868). *Palaeontographia Italica, 80*, 63–155.

Guérin, C. (1972). Une nouvelle espèce de rhinocéros (Mammalia, Perissodactyla) à Vialette (Haute-Loire, France) et dans d'autres gisements du Villafranchien inférieur européen: *Dicerorhinus jeanvireti* n. sp. *Documents des Laboratoires de Géologie de la Faculté des Sciences de Lyon, 49*, 53–150.

Guérin, C. (1980). Les rhinocéros (Mammalia, Perissodactyla) du Miocène terminal au Pléistocène supérieur en Europe occidentale. Comparaison avec les espèces actuelles. *Documents des Laboratoires de Géologie de Lyon, 79*, 1–1185.

Guérin, C. (1982). Première biozonation du Pléistocène européen, principal résultat biostratigraphique de l'étude des Rhinocerotidae (Mammalia, Perissodactyla) du Miocène terminal au Pléistocène supérieur d'Europe occidentale. *Geobios, 15*, 593–598.

Guérin, C. (1986). Le rhinocéros du Pléistocène ancien d'Oubeidiyeh (Israël). In: E. Tchernov & C. Guérin (Eds.), *Les mammifères du*

Pléistocène inférieur de la vallée du Jourdain à Oubeidiyeh. Mémoires et Travaux du Centre de Recherche français de Jérusalem (Association Paléorient, Paris), *5*, 183–190.

Guérin, C. (1990). Biozones or mammal Units? Methods and Limits in Biochronology. In E. H. Lindsay, V. Fahlbusch & P. Mein (Eds.), *European Neogene Mammal Chronology* (pp. 119–130). New York: Plenum Press.

Guérin, C. (2004). Les rhinocéros (Mammalia, Perissodactyla) du gisement villafranchien moyen de Saint-Vallier (Drôme). In M. Faure & C. Guérin (Eds.), *Le gisement pliocène terminal de Saint-Vallier (Drôme, France). Geobios*, Mémoire spécial 26, 37, 259–278.

Guérin, C. (2007). Biozonation continentale du Plio-Pléistocène d'Europe et d'Asie occidentale par les mammifères: état de la question et incidence sur les limites Tertiaire/Quaternaire et Plio/Pléistocène. *Quaternaire*, *18*, 23–33.

Guérin, C. (2010). *Coelodonta antiquitatis praecursor* (Rhinocerotidae) du Pléistocène moyen final de l'Aven de Romain-la-Roche (Doubs, France). In C. Guérin & T. Malvesy (Eds), L'aven pléistocène moyen final de Romain-la-Roche (Doubs, France), *Revue de Paléobiologie*, Genève, *29*, 697–746.

Guérin, C., & Barychnikov, G. F. (1987). Le rhinocéros acheuléen de la grotte de Koudaro I (Géorgie, URSS) et le problème des espèces relictes du Pléistocène du Caucase. *Geobios, 20*, 389–396.

Guérin, C., & Tsoukala, E. (2013). The Tapiridae, Rhinocerotidae and Suidae (Mammalia) of the Early Villafranchian site of Milia (Grevena, Macedonia, Greece). *Geodiversitas, 35*, 447–489.

Guérin, C., Faure, M., Argant, A., Argant, J., Crégut-Bonnoure, E., Debard, E., et al. (2004). Le gisement pliocène supérieur de Saint-Vallier (Drôme, France): synthèse biostratigraphique et paléoécologique. In M. Faure & C. Guérin (Eds.), *Le gisement pliocène terminal de Saint-Vallier (Drôme, France), Geobios*, Mémoire spécial 26, *37*, 349–360.

Guth, C. (1975). Lagerstätten des Villafranchiums in der Umgebung von Brioude in Mittelfrankreich. *Quartärpaläontologie, 1*, 177–182.

Heintz, E., Guérin, C., Martin, R., & Prat, F. (1974). Principaux gisements villafranchiens de France: Listes fauniques et biostratigraphie. *Mémoires du Bureau de Recherches Géologiques et Minières, 78*, 169–182.

Lacombat, F. (2005). Les rhinocéros fossiles des sites préhistoriques de l'Europe méditerranéenne et du Massif Central. Paléontologie et implications biochronologiques. *British Archaeological Reports International Series, 1419*, 1–175.

Lacombat, F. (2006). Pleistocene Rhinoceroses in Mediterranean Europe and in Massil Central (France). *Courier Forschungsinstitut Senckenberg*, *256*, 57–69.

McKenna, M. C., & Bell, S. K. (1997). *Classification of Mammals above the species level*. New York: Columbia University Press.

Nomade, S., Pastre, J.-F., Guillou, H., Faure, M., Guérin, C., Delson, E., et al. (2014). $^{40}Ar/^{39}Ar$ constraints on some French landmark Late Pliocene to Early Pleistocene large mammalian paleofaunas: Paleoenvironmental and paleoecological implications. *Quaternary Geochronology*, *21*, 2–15.

Pandolfi, L., Cerdeño, E., Codrea, V., & Kotsakis, T. (2017). Biogeography and chronology of the Eurasian extinct rhinoceros *Stephanorhinus etruscus* (Mammalia Rhinocerotidae). *Comptes Rendus Palevol, 16*, 762–773.

Parenti, F., Valli, A., Aprile, L., Candelato, F., Monguillon, A., & Natali, L. et al. (2024). Surveying and excavations at Senèze: 2001–2006. In E. Delson, M. Faure & C. Guérin (Eds), *Senèze: Life in Central France two million years ago. Paleontology, geochronology, stratigraphy and taphonomy* (pp. 27–35). Cham (Switzerland): Springer.

Roman, F., & Dareste de la Chavanne, J. (1931). Note sur la présence d'un élan (*Alces latifrons* Johnson) dans le Pliocène de Senèze (Haute-Loire). *Comptes Rendus des séances de l'Académie des Sciences, 192*, 1256–1257.

Santafé, J.V. & Casanovas, M.L. (1987). *Dicerorhinus etruscus brachycephalus* (Mammalia, Perissodactyla) de los yacimientos pleistocénicos de la cuenca Guadix-Baza (Venta Micena y Huéscar) (Granada, España). *Paleontologia i Evolució*, mem. esp., *1*, 237–254.

Schaub, S. (1943a). Osteologische Sammlung. *Verhandlungen der Naturforschenden Gesellschaft in Basel, 54*, 368–371.

Schaub, S. (1943b). Die oberpliocäne Säugetierfauna von Senèze (Haute-Loire) und ihre verbreitungsgeschichtliche Stellung. *Eclogae Geologicae Helvetiae, 36*, 270–289.

Stehlin, H. G. (1923). Die oberpliocäne Fauna von Senèze (Haute-Loire). *Eclogae Geologicae Helvetiae, 18*, 268–281.

Stuart, A.J. (1982). *Pleistocene Vertebrates of the British Isles*. London: Longman.

Thenius, E. (1955). Die Verknoecherung der Nasenscheidewand bei Rhinocerotiden und ihr systematischer Wert: zum Geschlechtsdimorphismus fossiler Rhinocerotiden. *Schweizerisches Paläontologische Abhandlungen*, *71*, 1–18.

Vekua, A. (1995). Die Wirbeltierfauna des Villafranchium von Dmanisi und ihre biostratigraphische Bedeutung. *Jahrbuch des Römisch-Germanischen Zentralmuseums Mainz*, 42, 77–180.

Viret, J. (1954). Le loess à bancs durcis de Saint-Vallier (Drôme) et sa faune de mammifères villafranchiens. *Nouvelles Archives du Muséum d'Histoire naturelle de Lyon, 4*, 1–200.

Wittmann, O. (1980). *Raritäten und Curiositäten der Natur: Die Sammlungen des Naturhistorischen Museums Basel*. Basel: Springer/Birkhäuser.

Chapter 11
The Senèze Equids

Véra Eisenmann and Eric Delson

Abstract Three genera of Plio-Pleistocene monodactyl equines are recognized and distinguished by cranial proportions: *Equus*, *Plesippus* and *Allohippus*. The rich material of equid fossils from Senèze is not homogeneous. In addition to a few caballine teeth and limb bones (possibly cataloguing errors or intrusive specimens), there is evidence of at least two species. The bulk of the material may be referred to *Allohippus senezensis senezensis*, which by size and proportions is intermediate between *A. senezensis guthi* of La Puebla de Valverde (Spain) and *A. senezensis mygdoniensis* of Gerakarou (Greece). Its relatively short muzzle and deep, not very robust, metapodials are usually found in equids living in rather dry conditions. Two partial skeletons and a few other possibly associated hindlimb elements were recovered in the new excavations close to the bottom of the local sequence, thus ca 2.2 Ma. One upper cheek tooth and 16 limb bones belong to a very large *Allohippus*. There are moreover a few fossils larger than the average of *A. stenonis vireti* of Saint-Vallier (France) and a few others as small as ?*Allohippus* of Pyrgos (Greece). The affinities and ages of various Pliocene and Pleistocene equid species are discussed and illustrated.

Résumé Grâce aux proportions crâniennes il est possible de reconnaître trois genres d'Equinés monodactyles: *Equus*, *Allohippus* et *Plesippus*. Si l'on prend pour modèles de variation interspécifique les *Equus* actuels, *Allohippus stenonis* et *A. senezensis* sont deux espèces distinctes, chacune comprenant plusieurs sous-espèces, les premières à museau long, les secondes à museau court.

Les fossiles d'équidés de Senèze, pour la plupart issus d'anciennes fouilles et rassemblés dans les collections de Lyon, Bâle et Paris, comprennent de nombreux crânes, squelettes et os des membres isolés sans provenance stratigraphique précise. Le matériel est fortement hétérogène. La présence de quelques spécimens caballins (notamment mandibule et métacarpien du squelette monté FSL 210993, ex 96132) s'explique sans doute par leur caractère intrusif ou par des erreurs de catalogue. Les autres fossiles appartiennent à au moins deux espèces. La plupart peuvent être attribués à *Allohippus senezensis senezensis*, intermédiaire par sa taille et ses proportions entre les *A. senezensis guthi* de La Puebla de Valverde (Espagne) et *A. senezensis mygdoniensis* de Gerakarou (Grèce). La brièveté relative de son museau et les proportions de ses métapodes suggèrent un environnement plutôt sec. Deux squelettes incomplets et quelques os de membre postérieur probablement associés proviennent de fouilles récentes proches du bas de la séquence locale daté d'environ 2,2 Ma. Une P3 supérieure et 16 os des membres (humérus, radius, métacarpien, tibias, astragale, métatarsiens et premières, deuxièmes et troisièmes phalanges) appartiennent à un très grand équidé. Sur la P3 le protocône est court et ne porte pas de sillon, le pli caballin est petit. Sur les métapodes la largeur sus-articulaire est plus grande que la largeur articulaire. Ces caractères rendent plausible une attribution à *Allohippus major*. Deux premières phalanges dépassent la taille moyenne d'*A. stenonis vireti* de Saint-Vallier (ca 2,5 Ma). Un MC III ressemble à *A.*

V. Eisenmann (✉)
CR2P (Paléontologie), Muséum National d'Histoire Naturelle, Département Histoire de la Terre, Paris, France
e-mail: vera.eisenmann@orange.fr

E. Delson
Department of Vertebrate Paleontology, American Museum of Natural History, New York, NY, USA

Department of Anthropology, Lehman College of the City University of New York, Bronx, NY, USA

PhD Program in Anthropology, The Graduate Center of the City University of New York, New York, NY, USA

New York Consortium in Evolutionary Primatology, New York, NY, USA

Institut Català de Paleontologia Miquel Crusafont, Universitat Autònoma de Barcelona, Cerdanyola del Vallès, Barcelona, Spain

E. Delson et al. (eds.), *Senèze: Life in Central France Around Two Million Years Ago*,
Vertebrate Paleobiology and Paleoanthropology, https://doi.org/10.1007/978-3-031-64415-3_11

stehlini du Valdarno Supérieur (ca. 1,7 Ma). Un MT III et une première phalange ressemblent à *?Allohippus* sp. de Pyrgos (MNQ 18 ou MNQ 19, ca 1,75 Ma). Ces cinq spécimens pourraient provenir d'horizons différents de Senèze ou même d'autres sites. L'âge et les affinités de différentes espèces d'équidés pliocènes et pléistocènes sont discutés. La Fig. 11.51 présente une vue générale de la distribution temporelle des huit taxons ou morphes d'Equidés en question.

Keywords *Allohippus* • *A. senezensis* • *A. major* • *Equus (Suessemionus) suessenbornensis* • *Plesippus* • Late Pliocene • Early Pleistocene • Middle Pleistocene

Mots-clés *Allohippus* • *A. senezensis* • *A. major* • *Equus (Suessemionus) suessenbornensis* • *Plesippus* • Pliocène supérieur • Pléistocène inférieur • Pléistocène moyen

Preliminaries

Before describing and discussing the Senèze equids, it is useful to summarize and explain the general background used here. Figs. 11.1 and 11.2 describe the system of cranial measurements and data used in diagrams illustrating this article. All of the data and almost all of the opinions in this chapter (especially those indicated by "I" or "my") are those of the senior author.

In the Old World, there are three groups of equine crania belonging to the genera *Plesippus*, *Allohippus*, and *Equus*. They have been discussed at length previously (Eisenmann 2004; Eisenmann & Baylac 2000; Eisenmann & Deng 2005; Forsten & Eisenmann 1995; Samson 1975). Although Eisenmann (2022) has used *Allohippus* as a subgenus of *Equus*, we here recognize it as a distinct genus, given the distinguishing features detailed below. Barrón-Ortíz et al. (2019) also accepted all three equids as full genera, following a detailed cladistic analysis. Other authors (e.g., Cirilli et al. 2021, 2024) include species here considered as *Allohippus* or *Plesippus* in *Equus* without infrageneric division, but importantly, they recognize *senezensis* and *major* as full species present at Senèze.

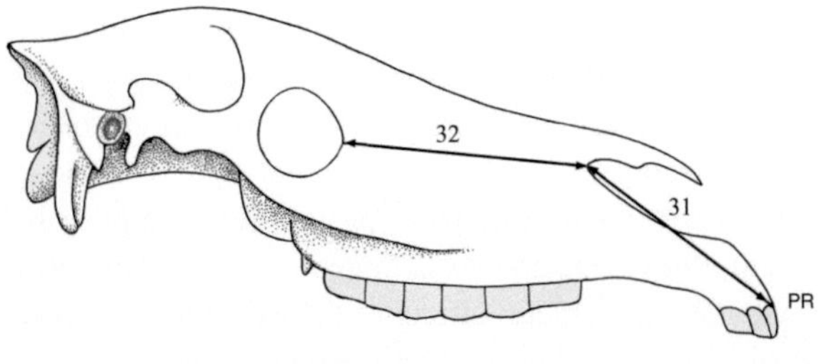

Fig. 11.1 Schematic representation of the profile of an *Equus* cranium. PR: prosthion. 31: length of the naso-incisival notch. 32: cheek length. Measurements defined at https://vera-eisenmann.com/skulls-system-of-measurements

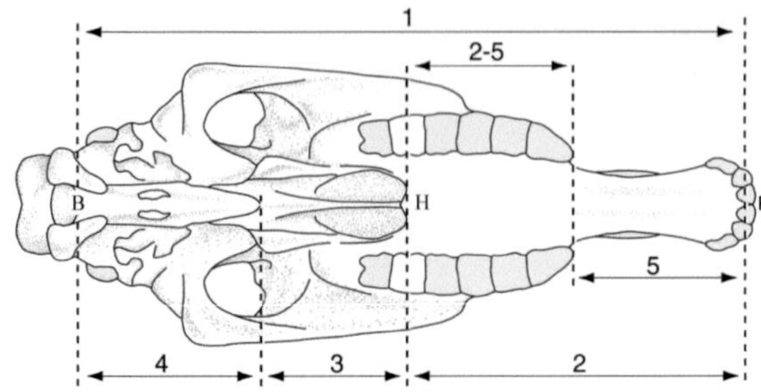

Fig. 11.2 Schematic occlusal view of an *Equus* cranium. B: basion. H: hormion. PR: prosthion. 1: basilar length. 2: overall palatal lengh. 2–5: palatal length sensu stricto. 3: vomerine length. 4: post-vomerine length. 5. muzzle length. Measurements defined at https://vera-eisenmann.com/skulls-system-of-measurements

Equus have a shorter naso-incisival notch relative to cheek length than *Allohippus* (Figs. 11.1, 11.2, 11.3, Table 11.A1). The same is true also in general for *Plesippus* (Fig. 11.4, Table 11.A1). According to this character, the crania of Valdarno (Italy), Saint-Vallier, Senèze, Ceyssaguet (France), La Puebla de Valverde (Spain), Gerakarou (Greece), Kuruksai (Tajikistan), Nihowan, SE Shansi, Locality D (China), East Turkana (Kenya), and Grandview (USA) all belong to *Allohippus*.

To distinguish *Allohippus* and *Plesippus* the Palatal Index (vomerine length relative to the palatal length sensu stricto, Fig. 11.2, higher in *Plesippus*) may be of some help: the ratio of vomerine length to palatal length is generally lower in *Allohippus* (and *Equus caballus*) than in *Plesippus* (and other *Equus*). There is overlap among all three genera, but they are mostly distinguishable (Fig. 11.5). There is some evidence of intermediate forms between *Plesippus* and *Allohippus*, at least at Longdan, China (Qiu et al. 2004 and Wang & Deng 2011).

Equus have longer post-vomerine length relative to overall palatal length than *Plesippus* and *Allohippus* (Fig. 11.6, Table 11.A1). As far as I know it is the only differential diagnostic character of the genus *Equus*. It may be hypothesized that this longer post-vomerine length is related to a larger brain.

There is an unfortunate tendency to refer to *Dolichohippus* any *Equus* with a large cranium. Thus Skinner and Hibbard (1972) referred "*Equus*" *simplicidens* to *Dolichohippus*; Forsten and Eisenmann (1995) showed that it was unacceptable. Churcher and Richardson (1978) and Churcher (1986, 1993) considered that *Equus capensis* was congeneric with *Dolichohippus* while cranial morphology (Eisenmann and Baylac 2000), confirmed later by molecular biology (Orlando et al. 2009), clearly shows its affinity with Quaggas. Following Churcher and Richardson (1978),

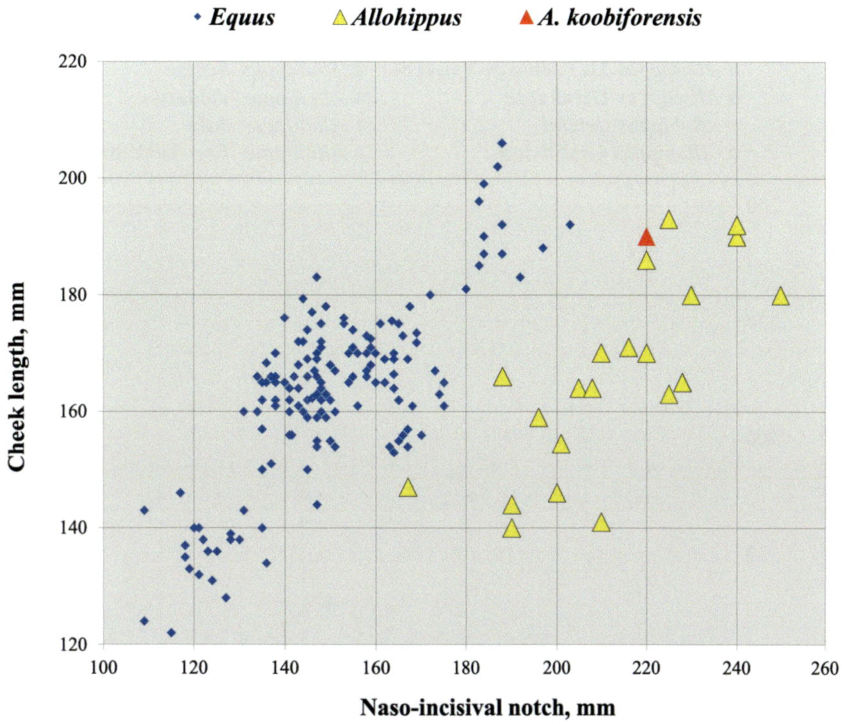

Fig. 11.3 Scatter diagram of the cheek length versus the length of the naso-incisival notch in crania of *Equus* and *Allohippus*. The naso-incisival notch is longer in *Allohippus* than in *Equus*. The East Turkana cranium KNM-ER 1484 (type of *A. koobiforensis*) plots with *Allohippus*, not with *Equus* and therefore should not be included in the subgenus *Dolichohippus*

Bernor et al. (2010) referred '*Equus' koobiforensis* to *Dolichohippus* although its naso-incisival notch is much deeper (red triangle on Fig. 11.3 and Fig. 11.4, Table 11.A1).

The concept of species in paleontology is problematic. Should it be based on osteology, chronology, or geography, or to what exact combination of these factors? My primary guide is the osteological variability in extant wild species of *Equus*. In the case of *Allohippus,* differences in osteological cranial characters, apart from size, are mostly found in the muzzle length (variable 5 in Fig. 11.2). The crania of *A. stenonis* from Valdarno have a long and narrow muzzle (Fig. 11.7, Table 11.A1). So do the crania of Saint-Vallier and Ceyssaguet (France), Liventsovka (Russia), the larger species of Kuruksai (Tajikistan), Nihowan and Fan Tsun, SE Shansi (China). Although similar morphology may indicate conspecificity, I do not think that being as far away in distance and in time as they are, they must all be referred to the single species *A. stenonis*. We refer them to the *A. stenonis* species group. A long muzzle does not give any indication about the ages of taxa, since Saint-Vallier is about 2.4 Ma (Nomade et al. 2014) and Ceyssaguet around 1.2 Ma (Aouadi 1999).

Other crania usually also referred to '*A. stenonis'* have very different proportions: muzzles are shorter and wider at Senèze (France), La Puebla de Valverde (Spain), Gerakarou (Greece), in the smaller species of Kuruksai (Tajikistan), and Localities A and D in China (Fig. 11.8, Table 11.A1). In extant *Equus*, these crania would never be considered conspecific with the preceding. We refer them to the *A. senezensis* species group. According to the available data, short-muzzled skulls seem more restricted in time than long-muzzled ones: between about 2.2–2.1 Ma (Senèze: Nomade et al. 2014; Delson et al. 2024) and the Olduvai subchron (Valdarno). Both morphs seem to coexist at Kuruksai, which may be a little older than Senèze. (Table 11.A2).

In the genus *Allohippus,* there are also differences in limb bone size and proportions, and in relative lengths of limb bone segments. It is not in the scope of this paper to illustrate all relevant data, so I shall only illustrate differences in two characters. The *A. stenonis* species group have more robust and flatter metacarpals (MC) than the *A. senezensis* species group. In extant *Equus* such characters usually correspond respectively to more humid or drier environments (Fig. 11.9, Table 11.A3). Limb segment proportions are also different (Fig. 11.10, Table 11.A4a). Although I do not know why, there again the long-muzzled forms seem to differ from the short-muzzled ones. In consequence I propose to use the following classification and distribution pattern of Table 11.A5 for the taxa discussed in this chapter.

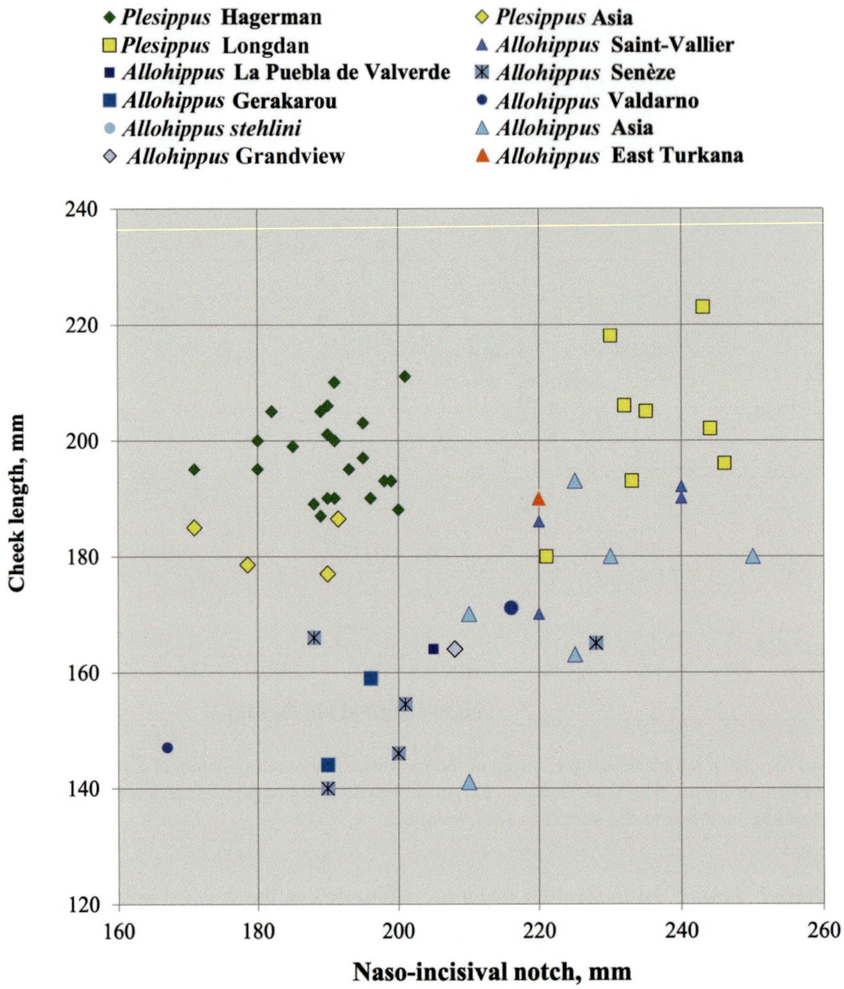

Fig. 11.4 Scatter diagram of the cheek length versus the length of the naso-incisival notch in crania of *Plesippus* and *Allohippus*. The naso-incisival notch is longer in *Allohippus* than in *Plesippus*. The Longdan crania and the East Turkana cranium KNM-ER 1484 plot with *Allohippus*

Material and Methods

Most of the fossil equids from Senèze were collected long ago and are preserved in several institutions: the Université Claude Bernard-Lyon I (UCB-Lyon 1; specimens indicated by the prefix FSL) and Muséum d'Histoire naturelle (now Musée des Confluences, Centre de Conservation et d'Etude des Collections) in Lyon (MHNL Sen), the Laboratoire de Paléontologie (MNHN.F or MNHN.F-AC) of the Muséum national d'Histoire Naturelle in Paris, and the Naturhistorisches Museum in Basel (NMB Se; see, e.g., Schaub 1943, fig. 2) are the richest. Specimens newly recovered by the Franco-American team have not yet been catalogued or deposited formally in any institution (they are stored now in FSL), but they are indicated here by SEN followed by the year of discovery and a field sequence number (e.g., FSL SEN 05–0081). The Senèze sample (combining old collections and new) is rich and comprises crania, mandibles, associated teeth and limb bones, and isolated fossils. The preservation of long bones and teeth is usually good; the crania, however, are often crushed, and some were heavily reconstructed. Abbreviations used for skeletal elements and other terms include: MC, metacarpal; MT, metatarsal; Ph, phalanx. Upper teeth are indicated by uppercase letters, lower teeth by lowercase. A list of other acronyms for museums preserving specimens used in comparative data tables and figures is provided below:

AM: Zoologisch Museum, Amsterdam, Netherlands.

AMNH-P FM (Paleontology, fossil mammals), **-M** (Mammalogy)–American Museum of Natural History, New York, USA;

BE: Naturhistorisches Museum, Bern, Switzerland.

BO: Museum Koenig, Bonn, Germany.

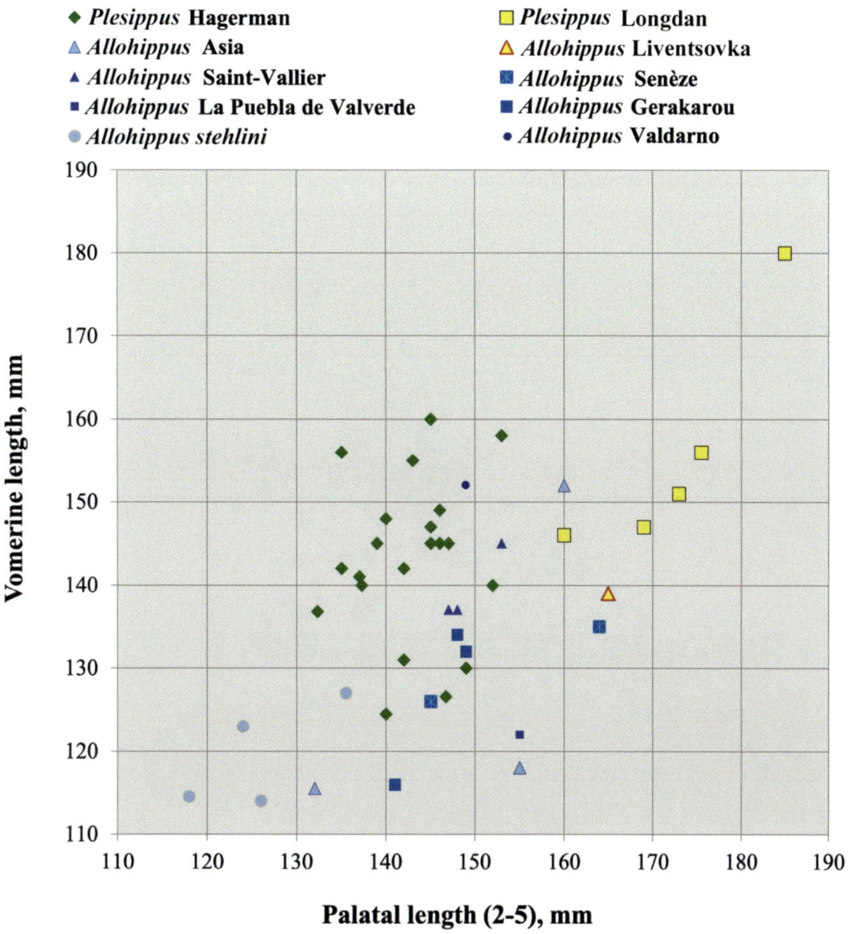

Fig. 11.5 Scatter diagram of the length of the vomer (hormion to posterior border of palate, variable 3) versus the length of the palate (posterior border of palate to anterior borders of P2, variable 2-variable 5) in crania of *Plesippus* and *Allohippus*. The length of the vomer is usually, but not always, larger in *Plesippus*

CRA: Centre de Recherches Archéologiques, Compiègne, France.

EV: Ecole Vétérinaire, Maisons-Alfort, France.

FMNH: Field Museum of Natural History, Chicago, USA;

GIN: Geological Institute, Moscow, Russia.

HL: Museum für Haustierkunde, Halle, Germany.

HPM: Hezheng Paleozoological Museum, Hezheng, China.

HUJ-ESE: Section of Ecology, Systematics & Evolution, Hebrew University, Jerusalem, Israel.

IA: Geological Institute, Iakutsk, Russia.

IGF: Istituto di Geologia, Firenze, Florence, Italy.

ISER: Institutul de Speologie Emil Racovitsa, Bucharest, Romania.

IVPP: Institute of Vertebrate Paleontology and Paleoanthropology, Beijing, China.

KI: Institut für Haustierkunde, Kiel, Germany.

KNM-P (Palaeontology), **-O** (Osteology): Kenya National Museums, Nairobi, Kenya.

LGPUT: Laboratory of Geology and Paleontology, Aristotle University Thessaloniki, Greece.

MB-Z: Zoologisches Museum der Humboldt Universität, Berlin, Germany.

MGRI: Moscow Geological Research Institute, Moscow, Russia.

MNCS-CSI: Museo Nacional de Ciencias Naturales, Madrid, Spain.

MNHN-ZM-MO: Laboratoire des Mammifères et Oiseaux du MNHN, Paris, France.

MNP-Bonifay: Musée National de Préhistoire, collection of M.F. Bonifay, Les Eyzies, France.

MS: Zoological Museum of Moscow University, Moscow, Russia.

MTA-M: Direction de Recherches Géologiques et Minières, Musée d'Histoire naturelle, Ankara, Turkey.

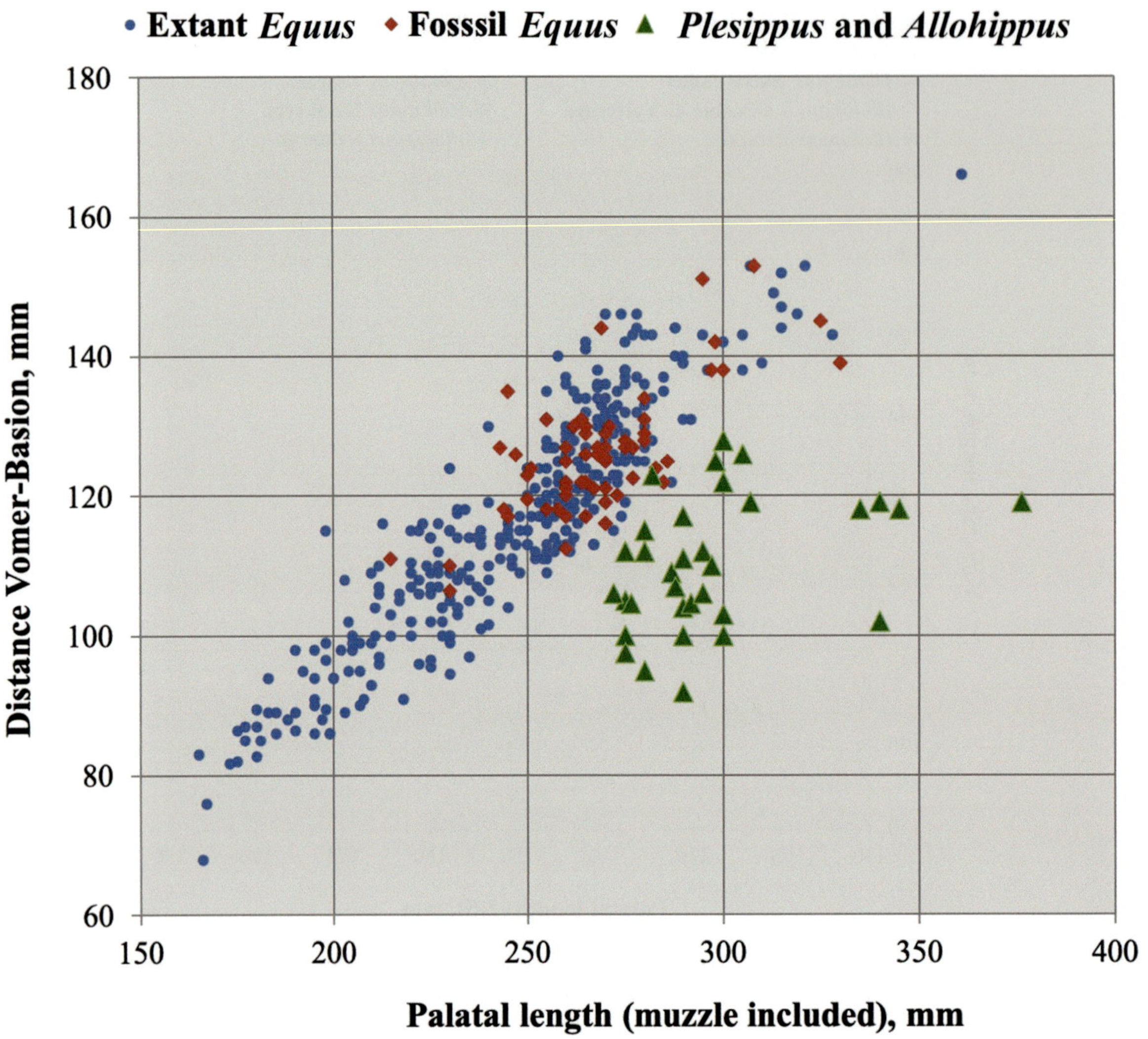

Fig. 11.6 Scatter diagram of the post-vomerine length (hormion to basion, variables 3 + 4) versus the overall palatal length (from posterior border of palate to the I1, variable 2) in crania of *Equus*, *Plesippus*, and *Allohippus*. In spite of some overlaps, *Equus* crania have larger post-vomerine lengths

NHCV: Natural History Collection, Museum of Vrissa, Lesvos, Greece (now mostly located in Athens Museum of Paleontology and Geology).

NHMUK -P (Palaeontology), **-ZD** (Zoology): Natural History Museum [formerly British Museum (Natural History)], London, Great Britain.

NM: Natural History Museum, Windhoek, Namibia.

NMB: Naturhistorisches Museum, Basel, Switzerland.

NMP: Narodni (National) Museum (Natural History), Prague, Czech Republic.

PIN: Paleontological Institite, Moscow, Russsia.

PMU: Paleontological Museum Uppsala University, Sweden.

RGU: State University of Rostov, Rostov on Don, Russia.

RNH-L: Naturalis (formerly Rijksmuseum van Natuurlijke Historie), Leiden, Nederlands.

ROMK: Azov Regional Museum, Rostov on Don, Russia.

SAM: Iziko South African Museum, Cape Town, South Africa.

SMF: Naturmuseum und Forschungsinstitut Senckenberg, Frankfurt, Germany.

SMNS: Staatliches Museum für Naturkunde, Stuttgart, Germany.

TMB: Magyar Természettudiomanyi Muzeum, Budapest, Hungary.

TMP: Ditsong (formerly Transvaal) Museum, Pretoria, South Africa.

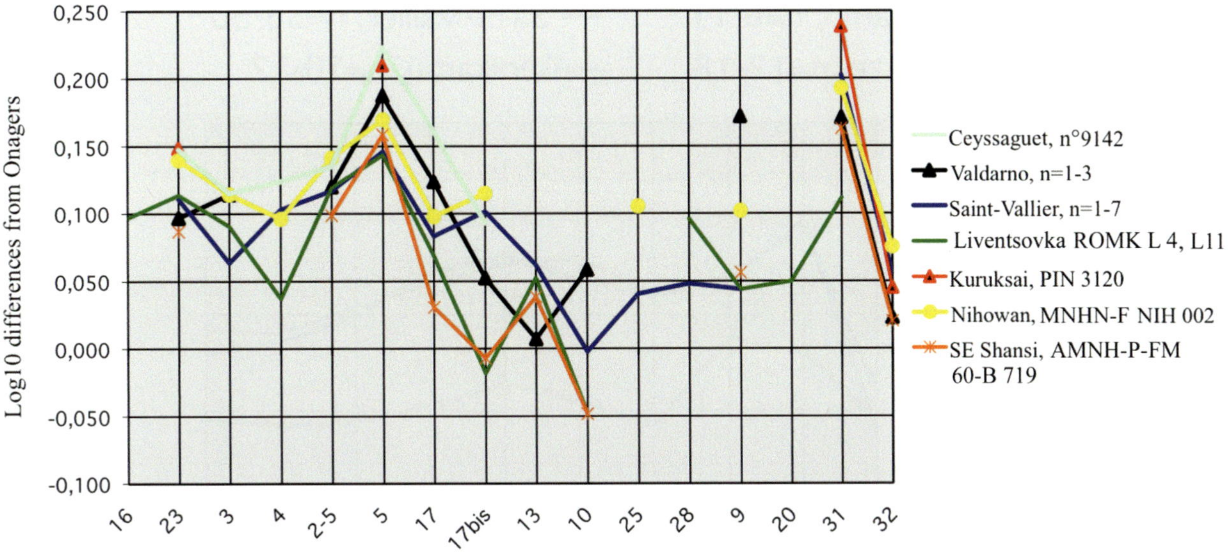

Fig. 11.7 Ratio diagrams of *Allohippus* crania with long muzzles (variable 5). 16: breadth of the supra-occipital crest; 23: anterior ocular line; 3: distance from palate to hormion; 4: distance from hormion to basion; 2–5: palatal length (sensu stricto, without the muzzle); 5: muzzle length; 17: muzzle breadth at the posterior border of the I3; 17bis: least muzzle breadth between the interalveolar borders; 13: frontal breadth: 10: greatest choanal breadth; 25: facial height in front of P2; 28: cranial height behind the orbits; 9: choanal length; 20: height of the external auditory meatus; 31: length of the naso-incisival notch (from prosthion to the back of the narial opening); 32: cheek length (from the back of the narial opening to the most anterior point of the orbit). n: number of specimens. Technique of measurement illustrated defined at https://vera-eisenmann.com/skulls-system-of-measurements

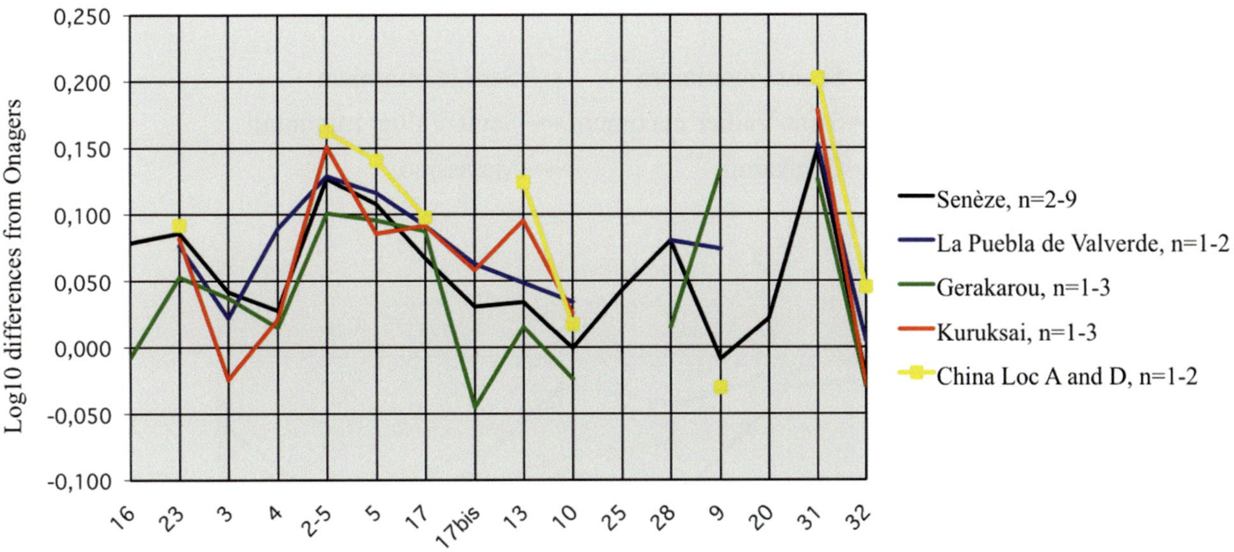

Fig. 11.8 Ratio diagrams of *Allohippus* crania with short muzzles (variable 5). Same variables as in Fig. 11.7

USNM: United States National Museum of Natural History [NMNH], Smithsonian, Washington, DC, USA.

Windhoek: Windhoek Sciences Museum, Namibia.

YPM: Yale Peabody Museum, New Haven, USA.

ZIN: Zoological Institute, Saint Petersburg, Russsia.

ZSM: Zoologische Sammlung des Bayerischen Staats, Munich, Germany.

ZU: Zoologisches Museum der Universität, Zurich, Switzerland.

The detailed system of measurements used in this chapter may be found on my web site at:

https://vera-eisenmann.com/-system-of-measurements-for-equus-bones-and-teeth-english

Click on the element for which measurement details are needed.

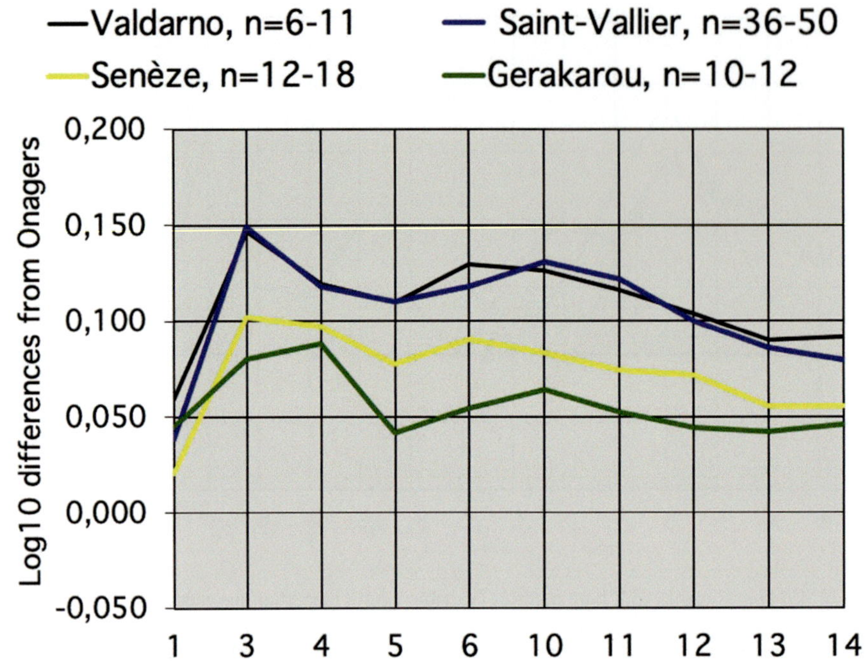

Fig. 11.9 Ratio diagrams of *Allohippus* third metacarpals. 1: maximal length; 3: minimal breadth of the diaphysis; 4: depth of the diaphysis at the same level; 5: proximal articular breadth; 6: proximal articular depth; 10: distal supra-articular breadth; 11: distal articular breadth; 12: depth of the keel; 13: minimal depth of the medial condyle; 14: maximal depth of the medial condyle. Technique of measurement illustrated at https://vera-eisenmann.com/metapodials-system-of-measurements

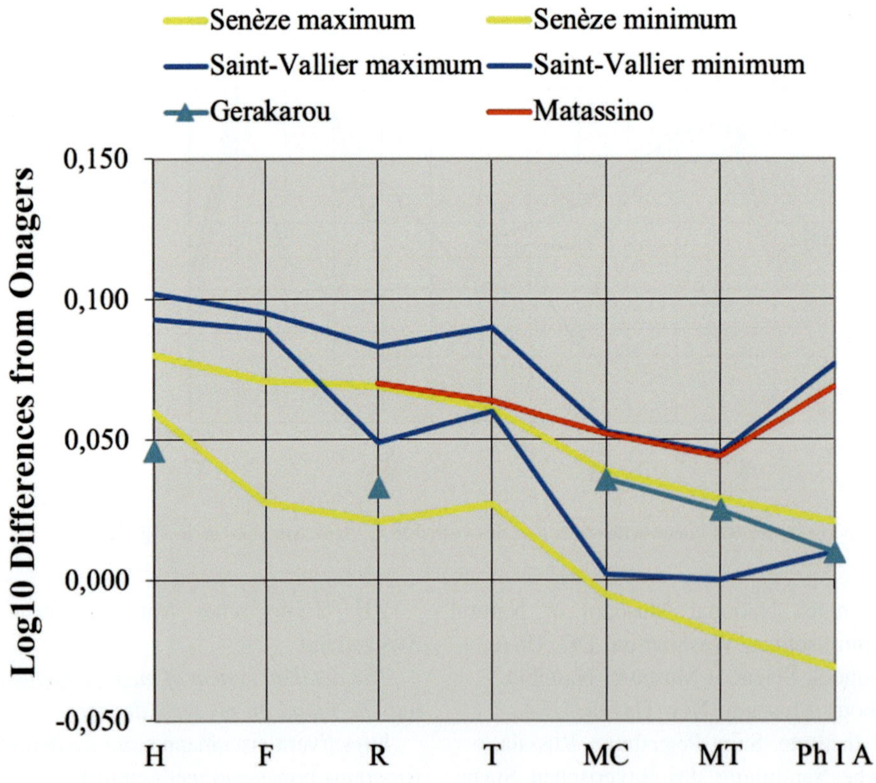

Fig. 11.10 Ratio diagrams of *Allohippus* limb bone segments. Maximal lengths of humerus (H), femur (F), radius (R), tibia (T), third metacarpal (MC), third metatarsal (MT) and first anterior phalanx (Ph1 A). First anterior phalanges are relatively shorter in *A. senezensis senezensis* from Senèze and *A. senezensis mygdoniensis* from Gerakarou than in *A. stenonis stenonis* from Matassino and *A. stenonis vireti* from Saint-Vallier

Comparisons in this article are based mostly on personal observations and occasionally on data communicated and/or published by: Samson (1975; Oasele, Romania); Mäuser (1992; Würzburg-Schalksberg, Germany); Koufos (1992; Gerakarou, Greece); Azzaroli and Voorhies (1993; Grandview, Idaho, USA); Aouadi (1999; Ceyssaguet, France); Athanassiou (2001; Sesklo, Greece); Koufos personal communication 1990 (Dafnero, Greece); and van Kolfschoten personal communication 1985 (Pyrgos, Greece; Tegelen, Netherlands). Simpson's ratio diagrams, scatter diagrams, and variability size indices (see below) are given to illustrate differences and similarities between fossil teeth and bones. All tables are placed in an Appendix at the end of the chapter.

The Variability Size Index (VSI) is one of the size index scaling techniques used by archeozoologists (Meadow 1999; Uerpmann 1982). A sample including all the bones of one taxon is chosen as the reference. Mean and standard deviation are calculated for each measurement of this sample. The comparisons are made using the following formula:

$$\mathrm{VSI} = 25^{*}(\mathrm{x} - \mathrm{m})/\mathrm{s}$$

where s is the standard deviation of the mean (m) of the reference measurements to which another measurement (x) is being compared. The obtained values are plotted on a histogram graduated in one, two, three, or more standard deviations from the reference. As phrased by Meadow (1986), "Using this formula, the standard dimension is set at zero; a measurement one standard deviation larger than the standard (reference) dimension will be plotted at 25, one standard deviation smaller at –25, etc." Some additional details are given at https://vera-eisenmann.com/variability-size-index-vsi.

Occurrence of Several Equids at Senèze

I have noted previously (Eisenmann 1981, 1985) that the mandible of the mounted skeleton FSL 210993 (ex 96132) cannot belong with the cranium. In consequence, the caballine pattern of the lower cheek teeth noted by Viret (1954, p. 145) and commented on by Azzaroli (1965, pp. 2–3) does not mean that *Allohippus* may have caballine teeth. Instead, it appears that some caballine specimens (e.g., the lower premolars MNHL Sen 3982) are present in the Senèze collections, either because they came from as yet unrecognized younger levels or because they were erroneously thought to come from Senèze. Here I will only illustrate the proportions of two probably caballine metapodials. On the MC of the same mounted skeleton FSL 210993, the relative flatness of the proximal end is unusual but a modern Shetland Pony MC (although smaller) has the same proportions (Fig. 11.11, Table 11.A3a). This skeleton thus contains at least two extraneous elements, and the relevance of other bones must be examined with caution. The MT NMB Se 821 although smaller has the same proportions as the "Mindel-Riss" caballine of Lunel-Viel (Fig. 11.12, Table 11.A6a).

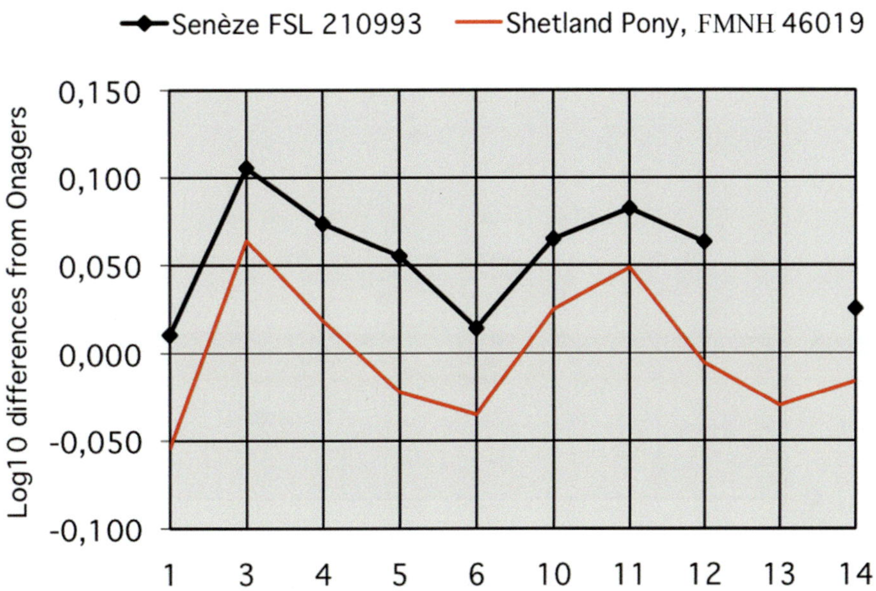

Fig. 11.11 Ratio diagrams of third metacarpals. Same variables as in Fig. 11.9

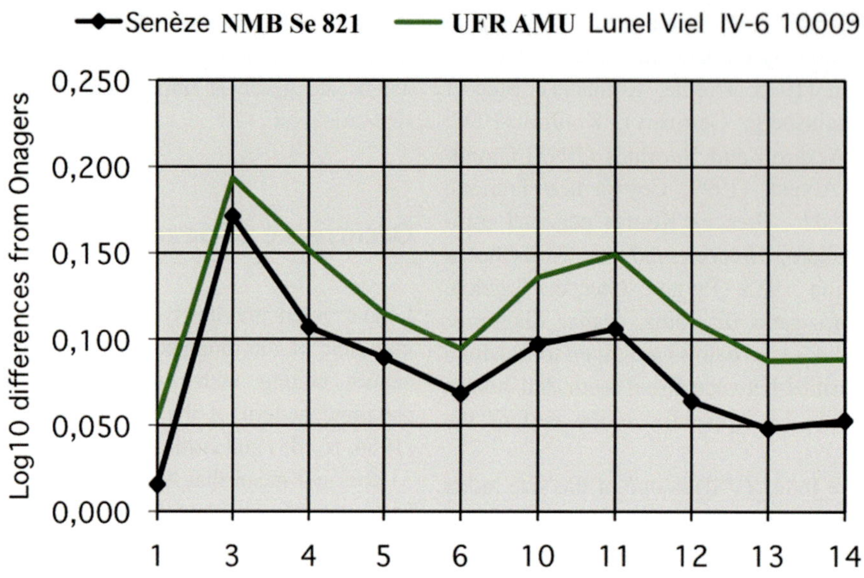

Fig. 11.12 Ratio diagrams of third metatarsals. Same variables as in Fig. 11.9

Cheek Teeth

With their short protocones, all the upper cheek teeth from Senèze are quite typical for *Allohippus*. The development of plis caballin, however, is very variable (see below). One P3, NMB Se 338, is larger than the rest (Fig. 11.13, Table 11. A7). and may be referred to the very large equid discussed below.

Limb Bone Size (Breadths)

Although there are differences in size within the *A. senezensis* sample, almost all fossils (excluding the "caballines" mentioned above) may belong to a single species. This is not the case of 16 considerably larger specimens (see Introduction to "Very Large Villafranchian Equid" below). The difference in size is well illustrated by the Variability

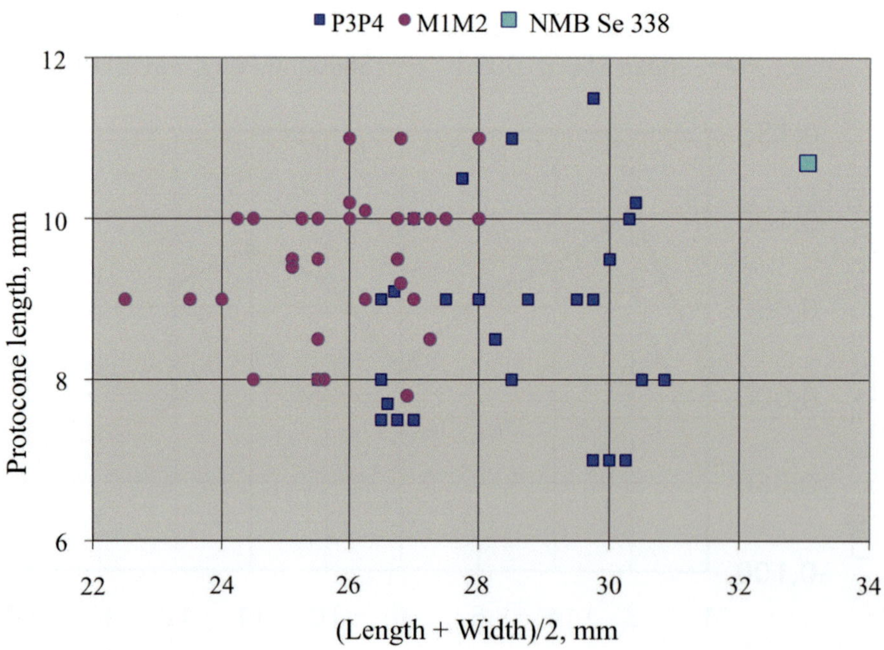

Fig. 11.13 Scatter diagram of protocone lengths versus average occlusal dimensions [(occlusal length + occlusal width)/2] in upper cheek teeth of Senèze

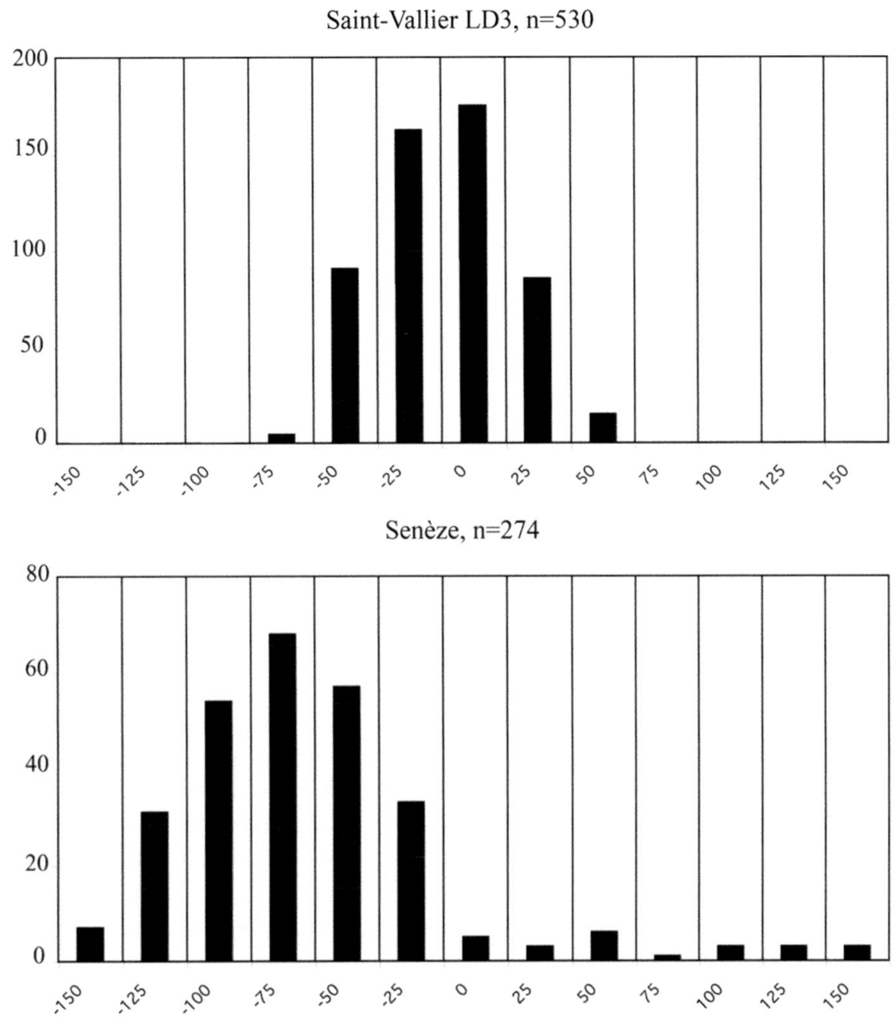

Fig. 11.14 Variability Size Index for 530 breadths of Saint-Vallier LD3 (younger level) and 274 breadths of Senèze limb bones

Size Index (VSI; see the end of the Materials and Methods section above). I have chosen the sample of *A. stenonis vireti* of Saint-Vallier as the reference. I considered only the bone breadths (excluding juvenile specimens). Means and standard deviations for the Saint Vallier reference standard used in Fig. 11.14 are given in Table 11.A8a. Naturally the histogram for Saint-Vallier is centered on 0; it is normally distributed, three standard deviations on each side of 0.

On the histogram for Senèze (Fig. 11.14, Table 11.A8a):

- Most of the breadths appear on the left side of the histogram, left of one standard deviation from 0, showing that most bones are thinner than those of Saint-Vallier. Their breadths seem normally distributed but less concentrated on the mean than at Saint-Vallier;
- The breadths of 16 bones (enumerated in Material section of "Very Large Villafranchian Equid") are on the right side of the diagram, all between one and six standard deviations from 0. They may be referred to a very large equid (see "Very Large Villafranchian Equid" below).

Limb Bone Size and Proportions

There is clearly a considerable size difference between *A. senezensis senezensis* and the very large equid Fig. 11.15, Table 11.A4b). I do not know what to make of the other large phalanges.

Very Large Villafranchian Equid

Introduction

The identity of the very large Villafranchian equid which Alberdi et al. (1998) have thoroughly discussed and decided to refer to as *E. major* Depéret, 1893 (rather than *E. robustus* Pomel, 1853 or *E. bressanus* Viret, 1954) is far from resolved, since its precise age is not known, and its main characteristic is its "large" size. I believe that part of the

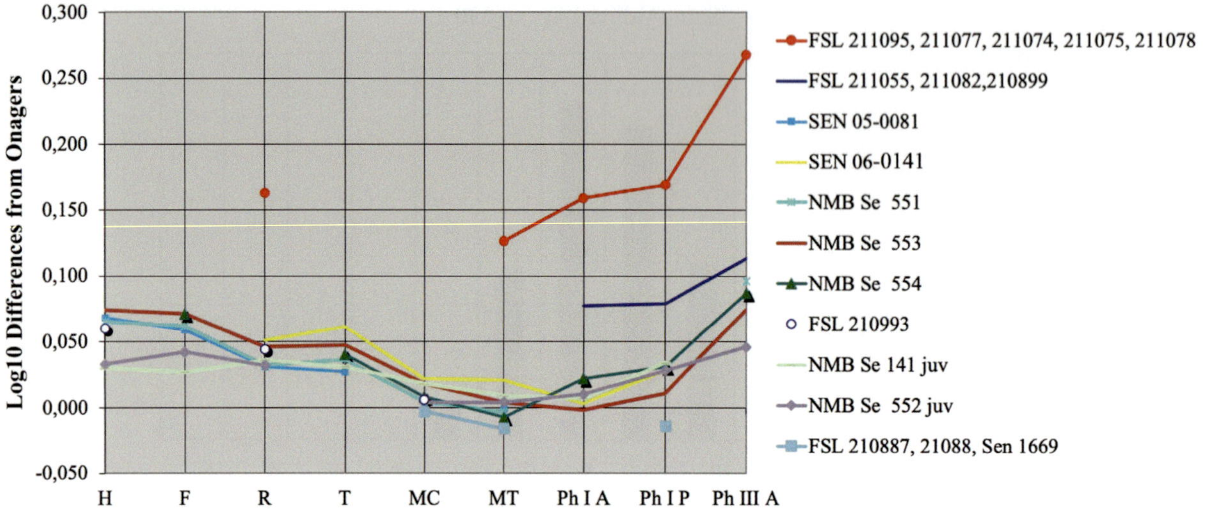

Fig. 11.15 Ratio diagrams of maximal lengths of humerus (H), femur (F), radius (R), tibia (T), first anterior and posterior phalanges (PhIa and PhIp) and plantar width of third anterior phalanx (Ph3a width) from Senèze

fossils referred to *E. major* (Alberdi et al. 1998; Boulbes & Asperen 2019) may belong either to *Allohippus* or to *Equus (Sussemionus)*. Others, like the equid of Tegelen illustrated and discussed below, are even more problematic. This "large" size overlaps with two other groups of large and better-documented equids: *Allohippus* (most species) and the large *Equus (Sussemionus)* (from Süssenborn, Akhalkalaki, and Northeastern Siberia, see Eisenmann 2006, 2010). As will be shown below, the large-bodied sample from Senèze can be identified as *Allohippus* on the basis of P3 enamel pattern and metapodial proportions, and due to its size it can be termed *Allohippus major*. (Fig. 11.16).

Material

At Senèze, the large equid is represented by a single tooth, the P3 NMB Se 338 illustrated in Fig. 11.17 below. The other 16 fossils are: humerus FSL 211091; radii FSL 210860 and 211095; MC III FSL 211079; tibiae NMB Se 812 and 813; tali FSL 211073; MT III FSL 211077 and FSL SEN 02–0006 (collected in parcel 164 and presented to the team by the amateur collector A. Consigny, thus essentially of unknown stratigraphic position); first phalanges FSL 211074 and 211075; second phalanges FSL 210919, 211075, and 211078; and third phalanges FSL 211078 and 211082. Their measurements are included in the general tables.

Combined Limb Bone Widths

Based on 41 widths (all available widths) of limb bones from Ceyssaguet (only the largest ones), Chagny, Gannat, Loubières de Pardines, Senèze, Solilhac (France), Schernfeld, Würtsburg-Schalksberg (Germany), Vatera (Greece), Kislang (Hungary), Oesterschelde (Netherlands), Oasele (Romania) and Morozovka (Russia), the VSI histogram of this combined sample overlaps the one of *Allohippus stenonis vireti* (Saint-Vallier) and even more the one of *Allohippus stenonis* subsp. of Ceyssaguet (Fig. 11.16, Table 11.A8, see details at https://vera-eisenmann.com/variability-size-index-vsi).

Upper Cheek Teeth

Figure 11.17 shows the enamel pattern of NMB Se 338 and two other P3s from Senèze compared to patterns of upper premolars of Sussemiones and one upper premolar from Chagny. *Equus (Sussemionus)* teeth from Süssenborn, Akhalkalaki, Liventsovka, and Chukochya have long and grooved protocones, plicated enamel, and very developed plis caballin (Fig. 11.17E, F, H, I). In contrast, NMB Se 338 (Fig. 11.17D) and other upper cheek teeth from Senèze like NMB Se 336 and MNHL Sen 1855 (Fig. 11.17A, G), as well as from Chagny (Fig. 11.17B) have very short, ungrooved protocones, as is usual in *Allohippus*. In *Allohippus* plis caballin are in general small but may be very developed (MNHL Sen 1855). The P3 from Kolyma (Fig. 11.17C), in spite of not having a groove on its rather small protocone, is closer to *E. (Sussemionus)* than to *Allohippus*. In that case, however, the attribution to an *Equus* is certain because of the cranium to which it belongs.

Third Metapodials

In MCs as well as in MTs, the distal articular width (variable 11 in Fig. 11.18 and Fig. 11.19, Table 11.A3a, Table 11.

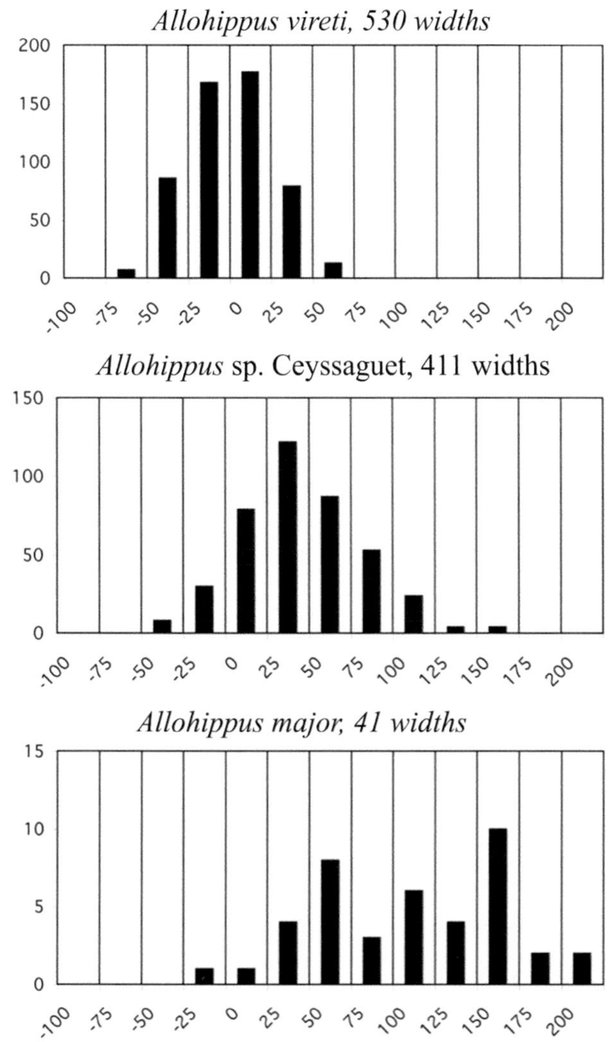

Fig. 11.16 Variability Size Index for Saint-Vallier LD3, Ceyssaguet, and for the *Allohippus major* of Senèze, Gannat, Würzburg-Schalksberg, Oasele, and Kislang

A6a, b, c) is usually larger than the supra-articular width (variable 10) in the large *E.* (*Sussemionus)* from Akhalakalaki, Georgia, and Northeastern Siberia, as opposed to what is seen in *Allohippus*. Using this criterion, I consider that the MCs of Senèze, Gannat (France), Würzburg-Schalksberg (Germany), and Oasele (Romania) belong to *Allohippus*. For the same reason I tentatively refer the gigantic MC from Overstrand (Great Britain) (Fig. 11.19) and the fragmentary MT from Feldioara (Romania) to the large *E. suessenbornensis* group. Although there are several MC morphs at Liventzovka (Fig. 11.20, Table 11.A3b), none qualifies to be *Allohippus major* or *Sussemionus*.

Since the same distinguishing characters seem as true for the third metatarsals as for the metacarpals, I refer the MTs from Kislang (Hungary), Würtzburg-Schalksberg, Gannat, and Senèze (FSL 211,077) to *Allohippus major*. Although smaller, I believe that the MTs from Chagny, Vatera (Lesvos, Greece), Liventsovka L-778 and Tataourova (Russia) may also belong to *Allohippus major* (Fig. 11.21, Table 11. A6c).

First Phalanges

Apart from a larger size, anterior first phalanges of the large equid do not seem to differ from those of Sussemiones (Fig. 11.22, Table 11.A12a). The morphology of the posterior first phalanges is more distinct (Fig. 11.23, Table 11. A12b). In specimens from both Chagny and Senèze, the distal supra-articular width is small relative to the proximal depth (variables 6 and 5 on the diagram, respectively). In Sussemiones from Süssenborn and from NE Siberia, the distal supra-articular width is large relative to the proximal depth. At Akhalkalaki, however, only a few phalanges have these proportions, so that the mean for Akhalkalaki is not typical.

Other Limb Bones

According to my data, there are no important differences in the proportions of the other limb bones of the large equids, whether they belong to *Allohippus major* or to the *E.* (*Suessemionus*) group.

Allohippus senezensis senezensis: Material, Description, and Comparisons

Skeletons

–FSL 210993 (ex 96,132) is a mounted skeleton at UCB-Lyon 1. As is usual for mounted skeletons, the data for FSL 210993 are poor. Moreover as mentioned before this skeleton is composite with a caballine mandible and MC.

–FSL SEN 05–0081 + is a nearly complete skeleton of an adult male with a crushed skull (Fig. 11.24, see also Fig. 2.7), excavated in 2005. There are many elements with different field numbers, but as 0081 is the cranium, the specimen can be termed 0081 + .

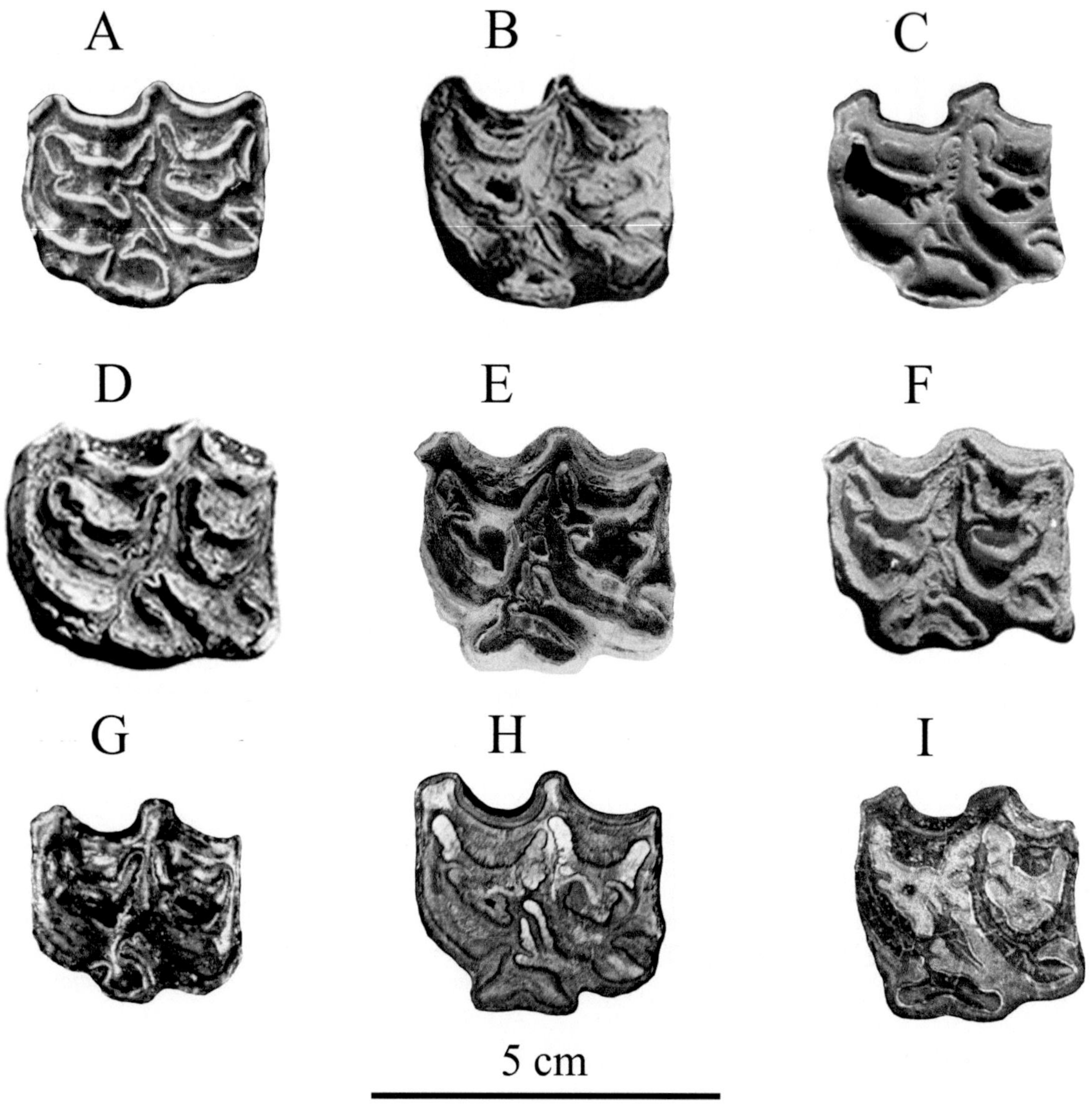

Fig. 11.17 Occlusal views of upper premolars. A. *Allohippus senezensis senezensis,* MNHL Sen 1855. B. *A. major*, Chagny no number. C. *E. (Sussemionus) coliemensis*, Kolyma IA 1741. D. *A. major*, Senèze NMB Se 338. E. *E. (Sussemionus) suessenbornensis,* Süsenborn P4 of the type series figured by Wüst (1910 Plate IV-9). F. *E. (Sussemionus)* cf. *suessenbornensis,* Akhalkalaki AKHA 1289. G. *A. senezensis senezensis,* NMB Se 336. H. *E. (Sussemionus)* sp., Liventsovka ROMK L 131. I. *E. (Sussemionus) verae,* Chukochya PIN 2998–243

–FSL SEN 06–0137 + , recovered in 2006, is less complete, including radius, tibia, metapodials, and first phalanges among other elements (Fig. 11.25).
–NMB Se 141 is a juvenile skeleton.
–NMB Se 551 is a mounted skeleton (the data are therefore poor) of an old female.
–NMB Se 552 is a juvenile skeleton.
–NMB Se 553 is the complete mounted skeleton of a male a little less than 4 years old.
–NMB Se 554 is a nearly complete skeleton of an old male.

Cranial Material

–NMB Se 336, very well preserved cranium of an adult male Fig. 11.26.
–NMB Se 551 is the skull of the above-mentioned mounted skeleton (Fig. 11.27). The cranium shows a malar depression, probably due to the shortness of the very worn crowns of the upper cheek teeth.
–NMB Se 553 is the fragmentary and badly preserved cranium (Fig. 11.28A, B) of another mounted skeleton.

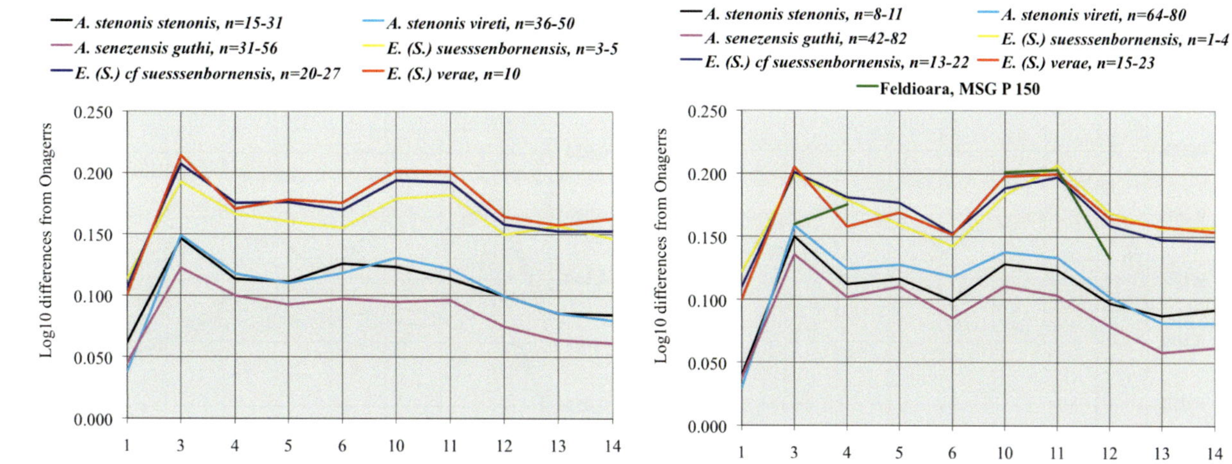

Fig. 11.18 Ratio diagrams of third metacarpals (left panel) and metatarsals (right panel) of *Allohippus* compared to *E. (Sussemionus)*. n: number of specimens. Same variables as in Fig. 11.9. In *Allohippus* the distal articular breadths are relatvely smaller

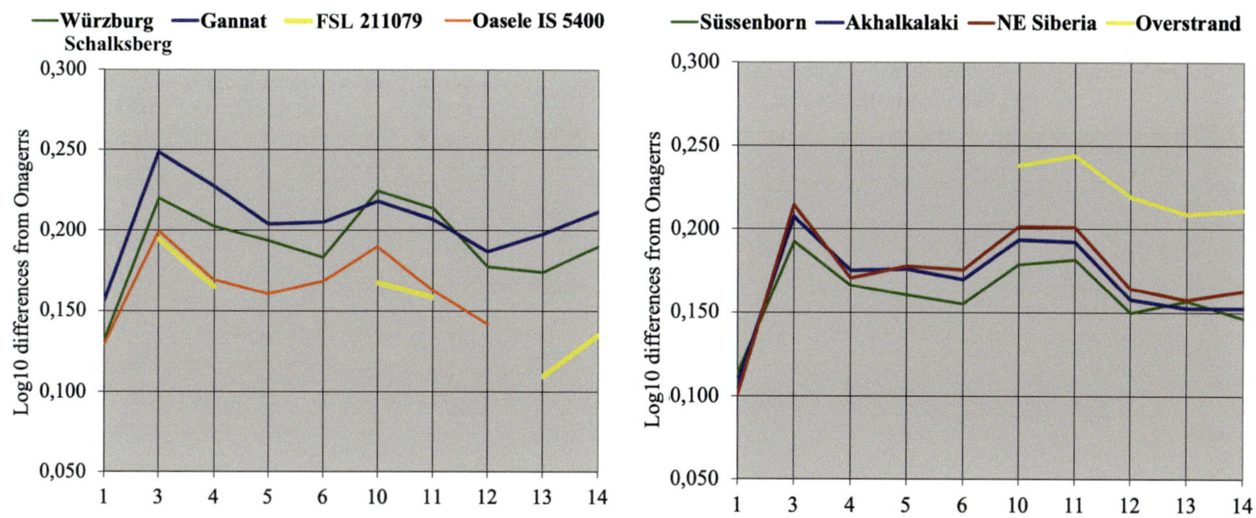

Fig. 11.19 Ratio diagrams of third metacarpals of *Allohippus major* (left panel) compared to Sussemiones (right panel). Same variables as in Fig. 11.9

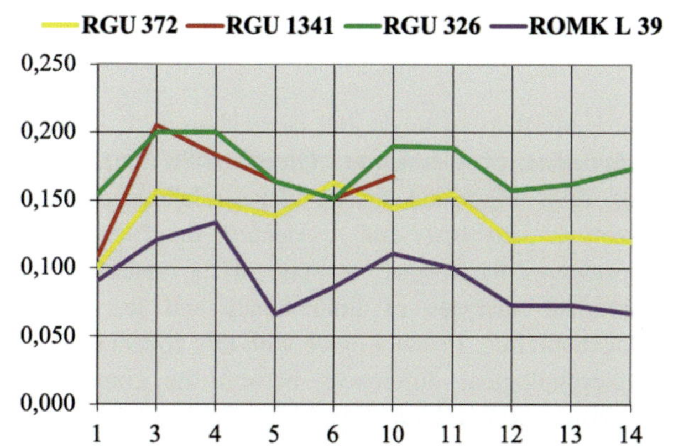

Fig. 11.20 Ratio diagram of some third metacarpals from Liventsovka. Same variables as in Fig. 11.9

–NMB Se 554 is the very badly preserved cranium (Fig. 11.28C) of a nearly complete skeleton.

–NMB Se 796 is the skull of a ca. 1 year old juvenile (Fig. 11.29).

–NMB Se 803, very old and very damaged fragmentary cranium.

–MNHL Sen 5233, adult male skull (Fig. 11.30).

–FSL 210887, crushed cranium of an adult female (Fig. 11.31)

–FSL 210993 (ex 96,132) is the cranium (Fig. 11.32) of the mounted skeleton discussed in Sect. 5.1.

–FSL SEN 05–0081/82 is the skull (Fig. 11.33) of the skeleton excavated in 2005 (see Fig. 1.27 for specimen before cleaning).

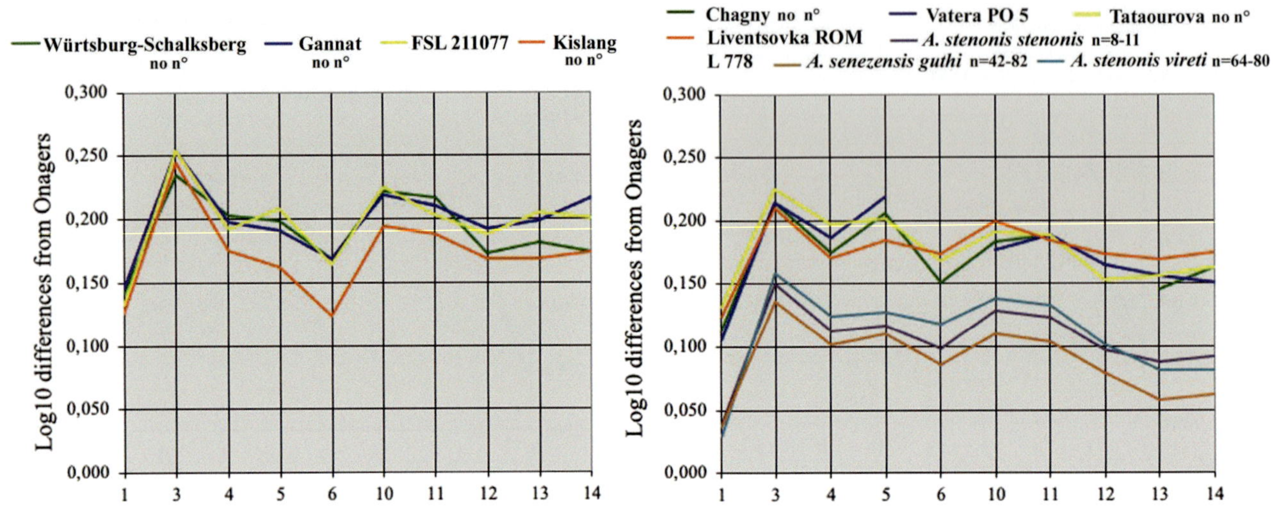

Fig. 11.21 Ratio diagrams of some third metatarsals of *Allohippus major* (left panel) compared to other *Allohippus* (right panel). n: number of specimens. Same variables as in Fig. 11.9

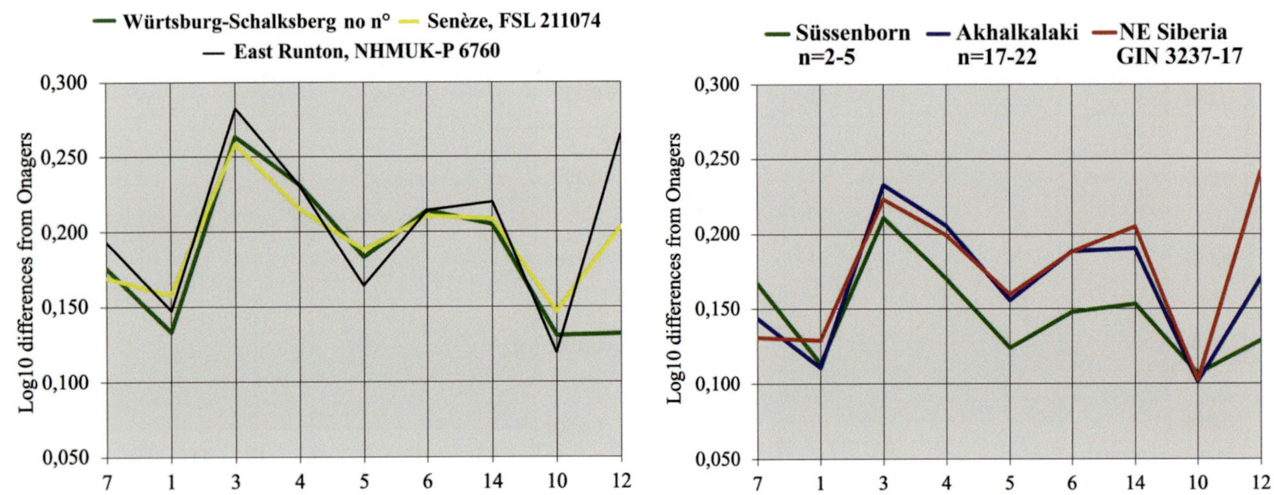

Fig. 11.22 Ratio diagrams of first anterior phalanges of *Allohippus major* from Senèze and other sites (left panel) compared to Sussemiones (right panel). n: number of specimens. 7. Maximal length of trigonum phalangis, 1. Maximal length, 3. Minimal breadth, 4. Proximal breadth, 5. Proximal depth, 6. Distal breadth at the tuberosities, 14. Distal articular breadth, 10. Medial supratuberosital length, 12. Medial infratuberosital length. Technique of measurement illustrated at https://vera-eisenmann.com/first-phalanges-system-of-measurements

Crania Description and Comparisons

In spite of some differences in size and shape, all crania (Table 11.A2) may be referred to a single form. As noted before they have relatively shorter muzzzles and are smaller in size than the 'long-muzzled' *Allohippus*: the average basilar length of Senèze skulls (523 mm) is smaller than in many other samples [545 mm in *Allohippus senezensis guthi* of La Puebla de Valverde (Spain), 554 mm in *A. stenonis vireti* of Saint-Vallier (France), 560 mm in *A. stenonis stenonis* of Olivola (Italy)], but larger than in *A. senezensis mygdoniensis* of Gerakarou, Greece (488 mm). Alberdi et al. 1998 suggested that a close relationship existed between *A. senezensis* and *A. stehlini*, the latter being a subspecies of the former. In spite of the results of their multivariate analyses of limb bones and the resulting phylogenetic tree, I cannot agree with this conclusion given the morphological differences betwen the crania of *A. senezensis* and *A stehlini* (Fig. 11.34, Table 11.A9a).

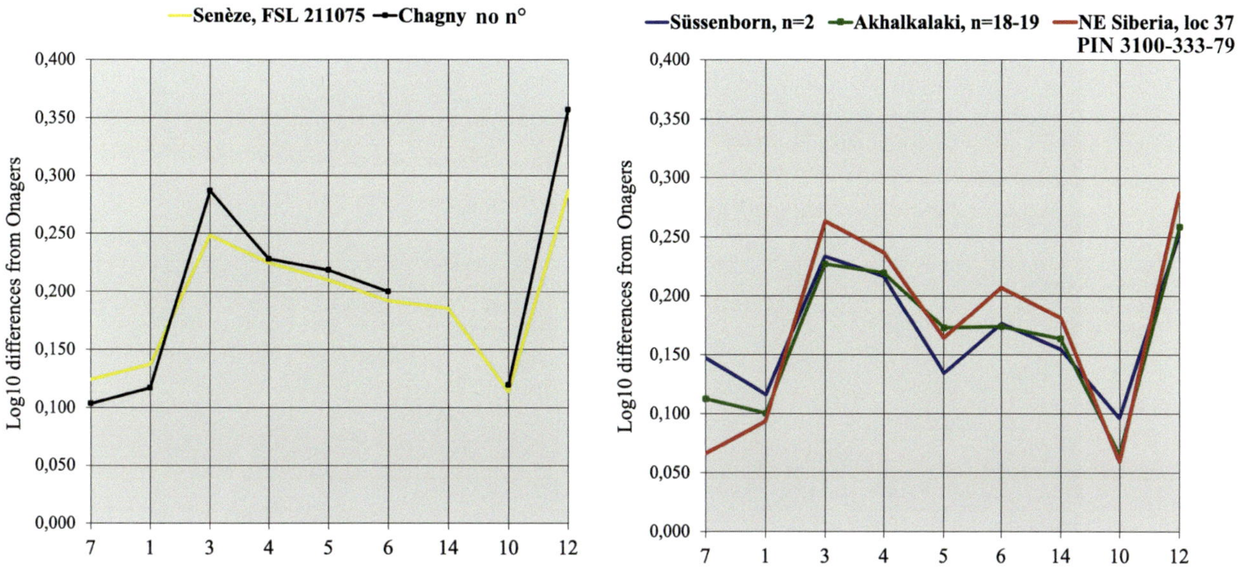

Fig. 11.23 Ratio diagrams of first posterior phalanges of *Allohippus major* (left panel) compared to Sussemiones (right panel). Same variables as in Fig. 11.22

Mandibles (Figs. 11.27, 11.29, 11.30, 11.33)

Mandibles are not very informative elements. Usually they just mirror cranial characters such as size and muzzle proportions (Table 11.A10). Figure 11.35 shows differences between extant *Equus* mandibles and among *A. stenonis vireti*, *A. senezensis senezensis* and *A. senezensis guthi*. The three *Allohippus* mandibles do not differ very much; they resemble those of extant *E. grevyi*.

Cheek Teeth–Specimens

–NMB Se 336, upper cheek teeth (Fig. 11.36A), associated with lowers.
–NMB Se 554, upper cheek teeth (Fig. 11.36F), associated with lowers.
–NMB Se 804, upper cheek teeth (Fig. 11.36B), associated with lowers.
–NMB Se 806, upper cheek teeth (Fig. 11.36C), associated with lowers.
–MNHL Sen 5233, upper cheek teeth (Fig. 11.36D), associated with lowers.
–FSL 210993 (ex 96132), upper cheek teeth (Fig. 11.36E).
–MNHL Sen 1855, upper cheek teeth associated with lowers (Fig. 11.37).
–FSL SEN 05–0081 upper cheek teeth associated with lowers -0082 (Figs. 11.38, 11.39).
–NMB Se 336, lower cheek teeth (Fig. 11.40A), associated with uppers.
–NMB Se 554, lower cheek teeth (Fig. 11.40F), associated with uppers.
–NMB Se 804, lower cheek teeth (Fig. 11.40B). associated with uppers.
–NMB Se 806, lower cheek teeth (Fig. 11.40C) associated with uppers.

Upper Cheek Teeth

As noted above, the short protocones of the upper cheek teeth are quite typical for *Allohippus stenonis*, but the development of the plis caballin is very variable. It is exceptionally long and wide at its base in MNHL Sen 1855 (Fig. 11.37); this is not a usual character. Figure 11.39 shows that most of the Senèze teeth are smaller than *Allohippus stenonis stenonis* from Olivola and Matassino (see measurements in Table 11.A7 and Table 11.A13).

Fig. 11.24 *A. senezensis senezensis* from Senèze. Skeleton SEN 05–0081 +

Fig. 11.25 *A. senezensis senezensis* from Senèze. Partial skeleton SEN 06–0137 +

Lower Cheek Teeth and Incisors

They are typical for *Allohippus*: rounded metaconid and metastylid, pointed liguaflexid, deep ectoflexid on the molars (Figs. 11.37, 11.38, 11.40); cups on incisors, at least on I1 and I2. Measurements are given in Table 11.A11.

Third Metacarpals

FSL 210887 is absolutely similar to *A. stehlini* from Valdarno (Fig. 11.41, Table 11.A3 a and c.). Even excluding the *E. caballus* mentioned above (Fig. 11.11) and the *A. major* FSL 211079, the MCs apppear very polymorphic (Figs. 11.42, Table 11.A3c). Many have deep proximal ends, not usual in *Allohippus stenonis stenonis*. The average MC is smaller and has a deeper diaphysis and proximal epiphysis than *A. senezensis guthi* (La Puebla de Valverde) (Fig. 11.43, Tables 11.A3, 11.A14, 11.A15).

Third Metatarsals

The smallest of them (FSL 210868b) resembles metatarsals of ?*Allohippus* sp. from Pyrgos, Greece (Fig. 11.44, Table 11.A16 a). Apart from the *Allohippus major* specimens and the small FSL 210868b, the morphology (Fig. 11.45 Table 11.A6d) of the MTs seems more homogeneous than for the MCs (even so, the coefficients of variation are larger than usual for one species). Their mean is intermediate in size between the larger *A. senezensis guthi* (La Puebla de Valverde) and the smaller *A. senezensis mygdoniensis* (Gerakarou, Greece) (Fig. 11.46, Table 11.A6b), differing from all of these by a deeper proximal end (measure 6).

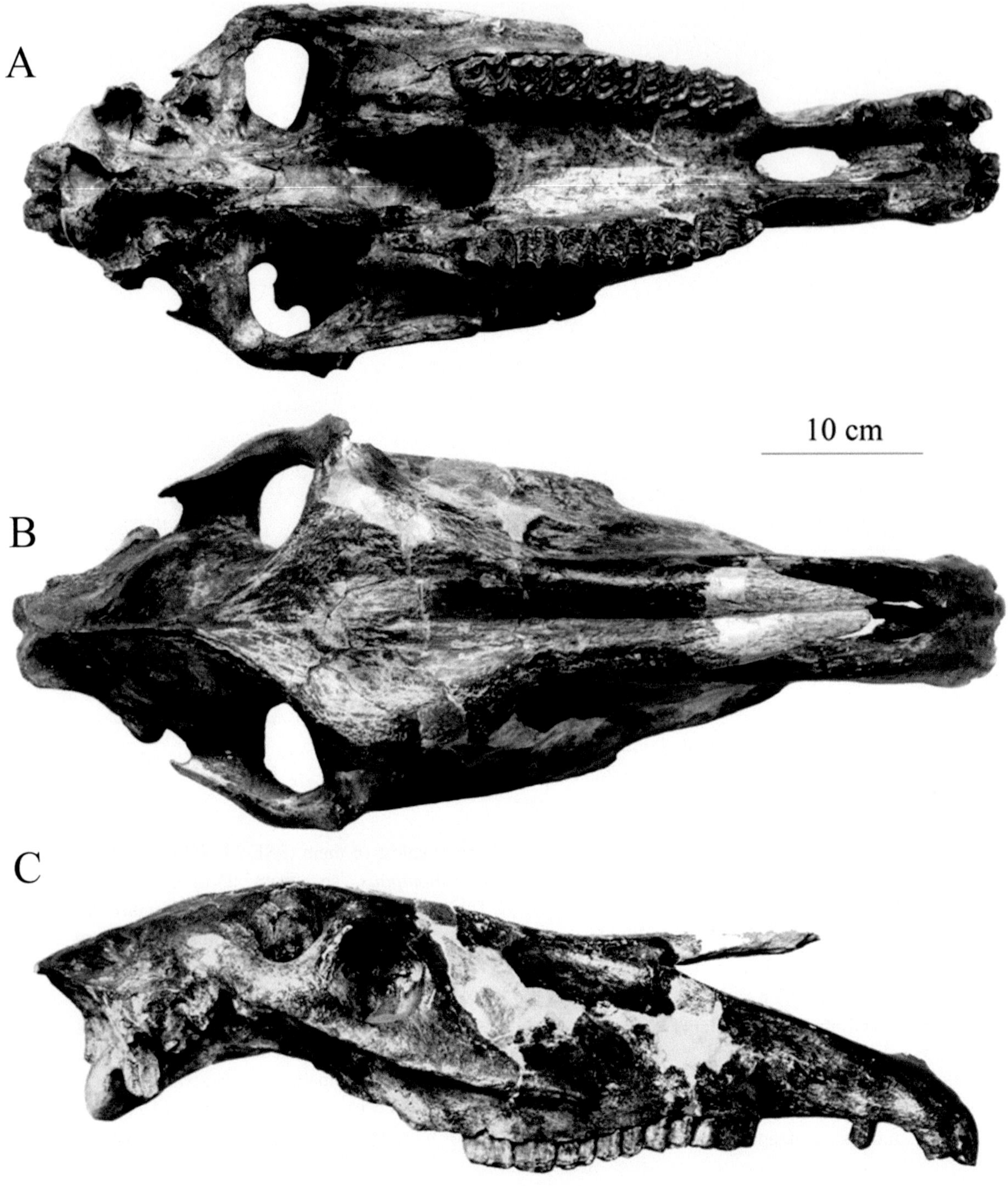

Fig. 11.26 *A. senezensis senezensis* from Senèze. Cranium NMB Se 336. A. Occlusal view. B. Dorsal view. C. Lateral view

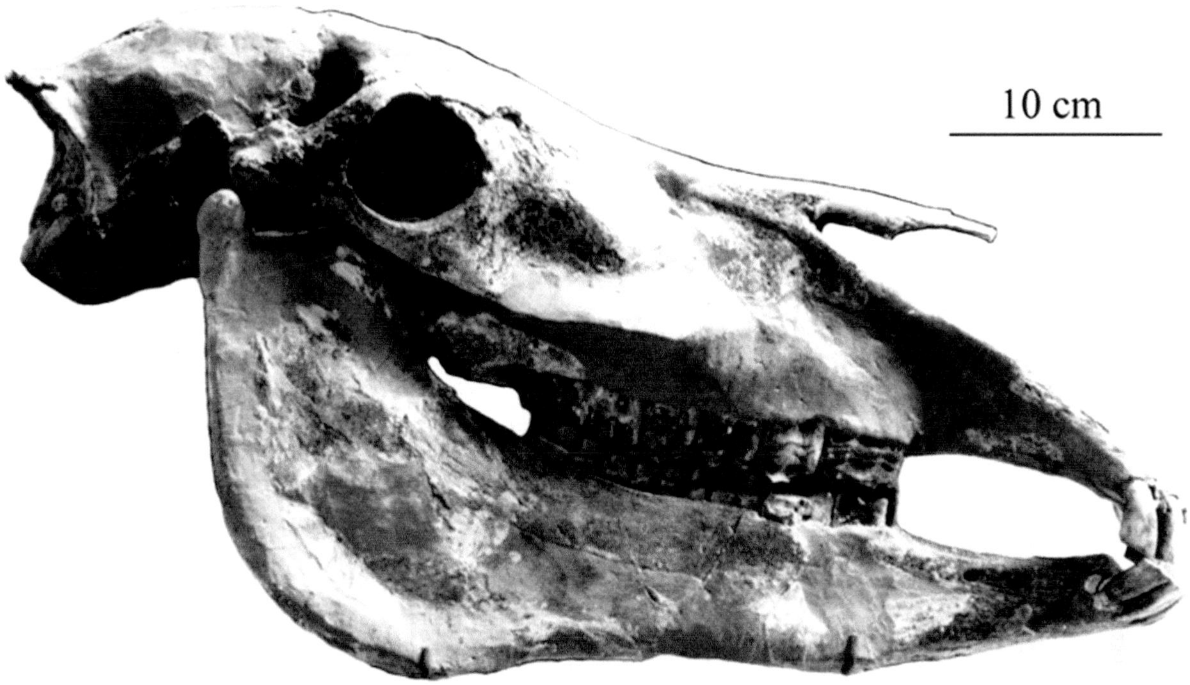

Fig. 11.27 *A. senezensis senezensis* from Senèze. Skull NMB Se 551, occlusal view

First Phalanges

First anterior phalanx FSL 211055, first posterior FSL 211082, and third anterior FSL 210899 are probably associated with each other. They are quite a bit larger than the rest but not as large as those of *Allohippus major* (Fig. 11.15). Ten anterior first phalanges which are alike in size and proportions are referred to *A. senezensis senezensis*. They are slightly smaller than those of *A. senezensis guthi* of La Puebla de Valverde. Three others are quite a bit larger: NMB Se 554, NMB Se 828, and FSL 211055, the last being even longer (variable 1) than the average of *A. stenonis vireti* (Fig. 11.47, Table 11.A12).

Among the first posterior phalanges, nine are relatvely similar. Two (NMB Se 554 and FSL 211090) are close to the average of *A. stenonis vireti* and two (MNHL Sen 1689 and FSL 211083) are as small as the specimens from Pyrgos (Fig. 11.48, Table 11.A12). The average dimensions of the nine posterior Ph1s which can be referred to *A. senezensis senezensis* are slighly smaller than in *A. senezensis guthi*; one of them (FSL 211085bis) overlaps the mean of the latter.

Other Limb Bones (Tables 11.A14, 11.A15, 11.A16, 11.A17, 11.A18, 11.A19, 11.A20, 11.A21, 11.A22).

Except for the large and small specimens already discussed, other Senèze elements are intermediate in size between those of *A. senezensis guthi* and smaller forms like *A. senezensis mygdoniensis*.

Conclusions

Tentative Identification of Selected Villafranchian Equids

A better understanding of Villafranchian equids cannot be achieved without a combination of detailed morphology and dating, at least for key localities, i.e., where the material is varied, abundant, homogeneous, and well preserved and/or where fossils are especially interesting by their "originality"

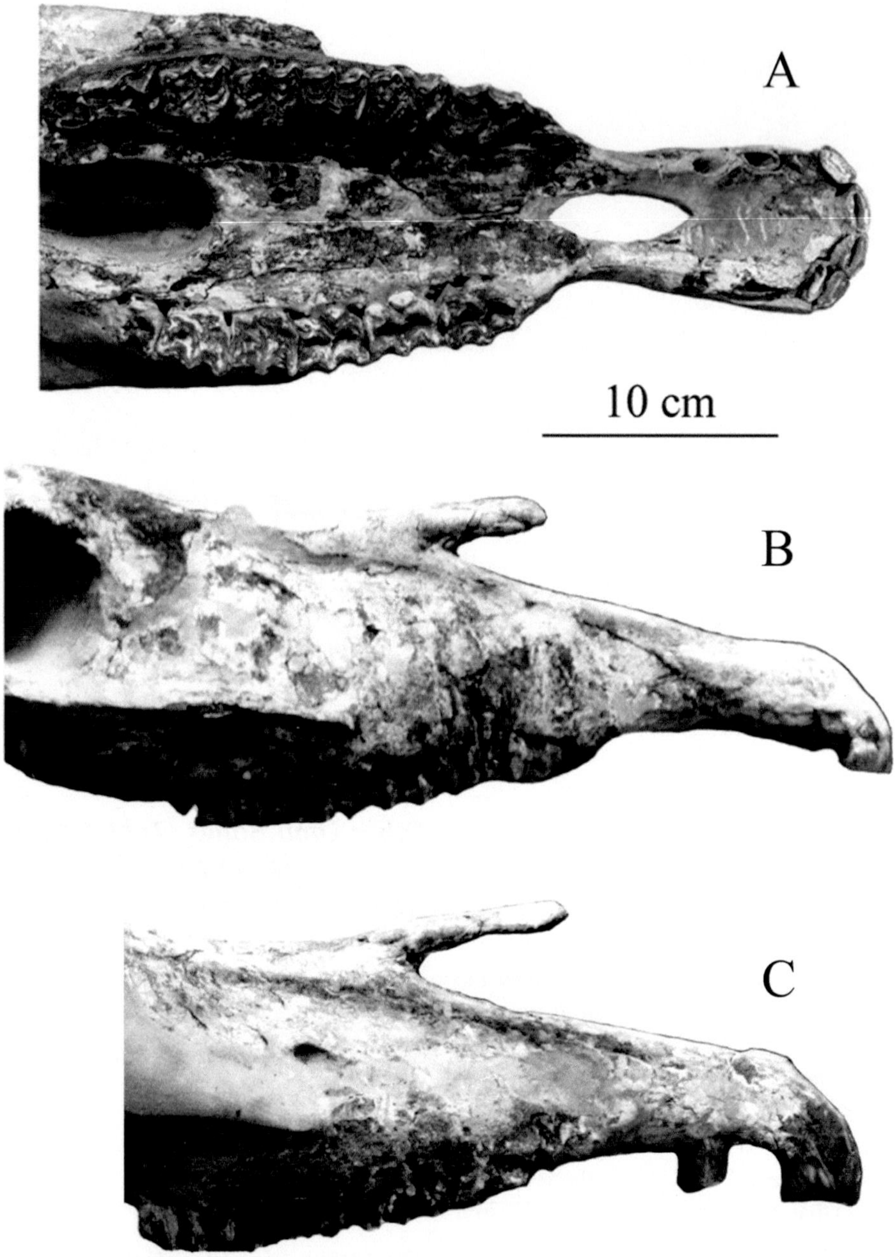

Fig. 11.28 *A. senezensis senezensis* from Senèze. Anterior fragment of cranium NMB Se 553. A. Occlusal view. B. Lateral view. Anterior fragment of cranium NMB Se 554. C. Lateral view

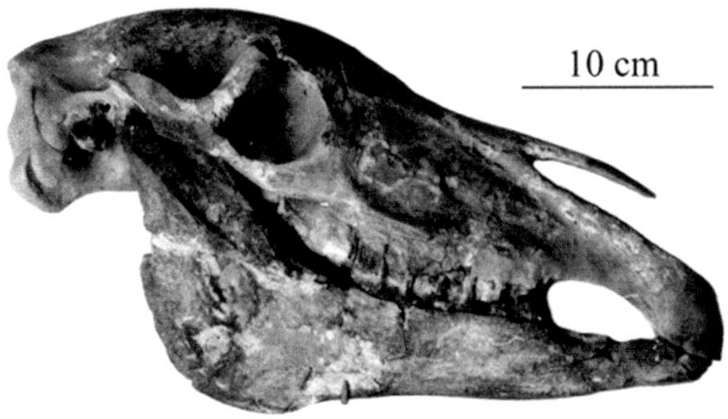

Fig. 11.29 *A. senezensis senezensis* from Senèze. Juvenile skull NMB Se 796, lateral view

(unusual size or morphology). The age may help to distinguish morphologically similar populations which differ widely in time. But progress also depends very much on the discovery of crania associated with limb bones.

The diversity of sizes and proportions of limb bones as well as their overlaps discourages attempts at any definitive classification. The size of an equid cannot be used as chronological evidence: large equids occur just above the Gauss-Matuyama boundary (Montopoli) as well as around the Brunhes-Matuyama boundary (Würzburg-Schalksberg). The first may belong to *Allohippus* and/or *Plesippus*; the second to *Allohippus* and/or *Equus*.

In fact, when crania are missing, geologically old fossils may possibly belong to *Plesippus* as well as to *Allohippus*: A few examples will show some of the problems in these cases (Fig. 11.49, Table 11.A3d).

- Figure 11.49A. ?*Allohippus* sp. 1 from Montopoli (Italy), Sarikol Tepe (Turkey), and Loubières de Pardines (France). Metacarpals are robust and have shallow distal keels (measures 12, 13, 14) as in *Allohippus*, but the distal articular breadth (11) is large relative to the supra-articular one (10), which is not typical for this genus.
- Figure 11.49B.?*Allohippus* sp. 2: two metacarpals from Vatera (Greece) and Huelago (Spain) are slender, with distal supra-articular breadth large relative to articular breadth (variables 10 and 11), as is usually the case in *Allohippus*. Their slenderness, however, is not typical for *Allohippus*, their proportions being more onager-like. Both localities are believed to be about 2 Ma or more (Alberdi & Ruiz-Bustos 1989; Alberdi et al. 1998; Eisenmann 2002).
- Figure 11.49C. ?*Allohippus* sp. 3: four metacarpals are larger than the preceding and very similar in size and morphology, being slender and with relatively large proximal depths and distal articular breadths (measures 6, 11). One is from Tegelen (Netherlands) believed to be 1.8 Ma (Westerhoff et al. 1998); the other three are from Morskaja, Liventzovka, and Khapry (from the Khapry complex, Russia) believed to be ca. 2.5 Ma.
- *Allohippus stenonis* ssp. from Ceyssaguet: According to the data kindly provided by N. Aouadi, the sample of MCs from Ceyssaguet is very rich but not homogeneous. Three specimens belong to some kind of slender and large equid. Four more (especially 9012) resemble a MC from El Rincon (12440). Another 15 are even larger, robust specimens with variable morphologies. The bulk of the material (n = 41–47) probably belongs to a large, robust *Allohippus*, with deep diaphyses and epiphyses. By their proportions, they resemble the average MC from Sainzelles (France) but are quite a bit larger (Fig. 11.50). Sainzelles is believed to be about 1.3 Ma (Lacombat 2005), which fits well with the supposed age of Ceyssaguet, about 1.2 Ma (Aouadi 1999).

Keeping all those caveats in mind, Fig. 11.51 presents the various taxa (and selected populations or individuals) and their distributions in time.

A. senezensis senezensis

- As illustrated above (Fig. 11.6, Table 11.A9 a), the average skull of *A. senezensis senezensis* belongs to the short-muzzled group and is intermediate in size between *A. senezensis guthi* and *A. senezensis mygdoniensis*.
- The average MC III of *A. senezensis senezensis* is smaller than *A. senezensis guthi* and larger than *A. senezensis mygdoniensis* (Fig. 11.43, Table 11.A3). As in *A. senezensis mygdoniensis*, the diaphysis is deep relative to its width (variables 4 and 3).
- Short muzzles and deep, not very robust, metapodials are usually found in equids living in rather dry conditions like extant *Equus hemionus* and *E. przewalskii* (Eisenmann 1984; Eisenmann & Guérin 1984).

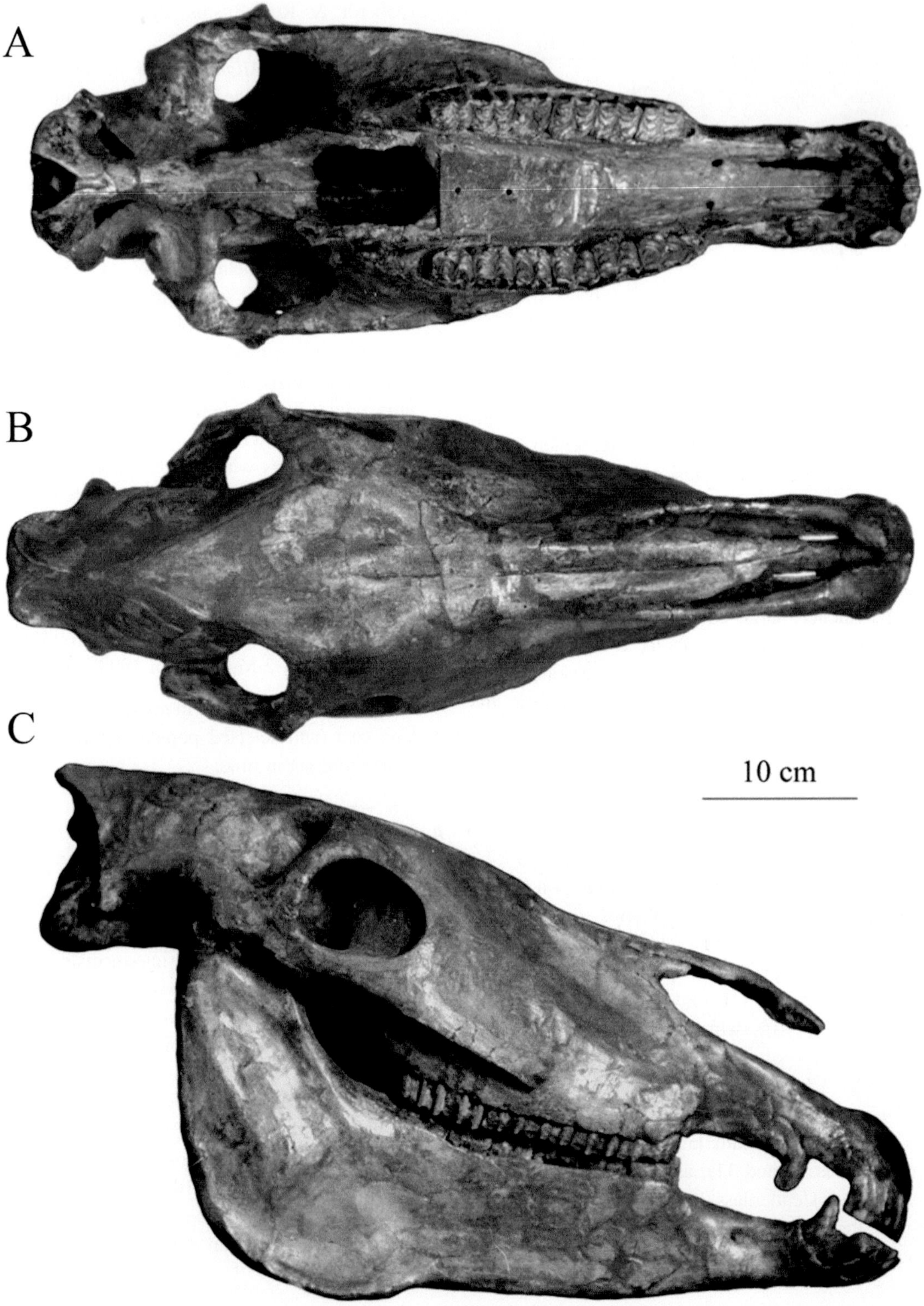

Fig. 11.30 *A. senezensis senezensis* from Senèze. Skull MHNL Sen 5233. A. Occlusal view. B, Dorsal view. C. Lateral view

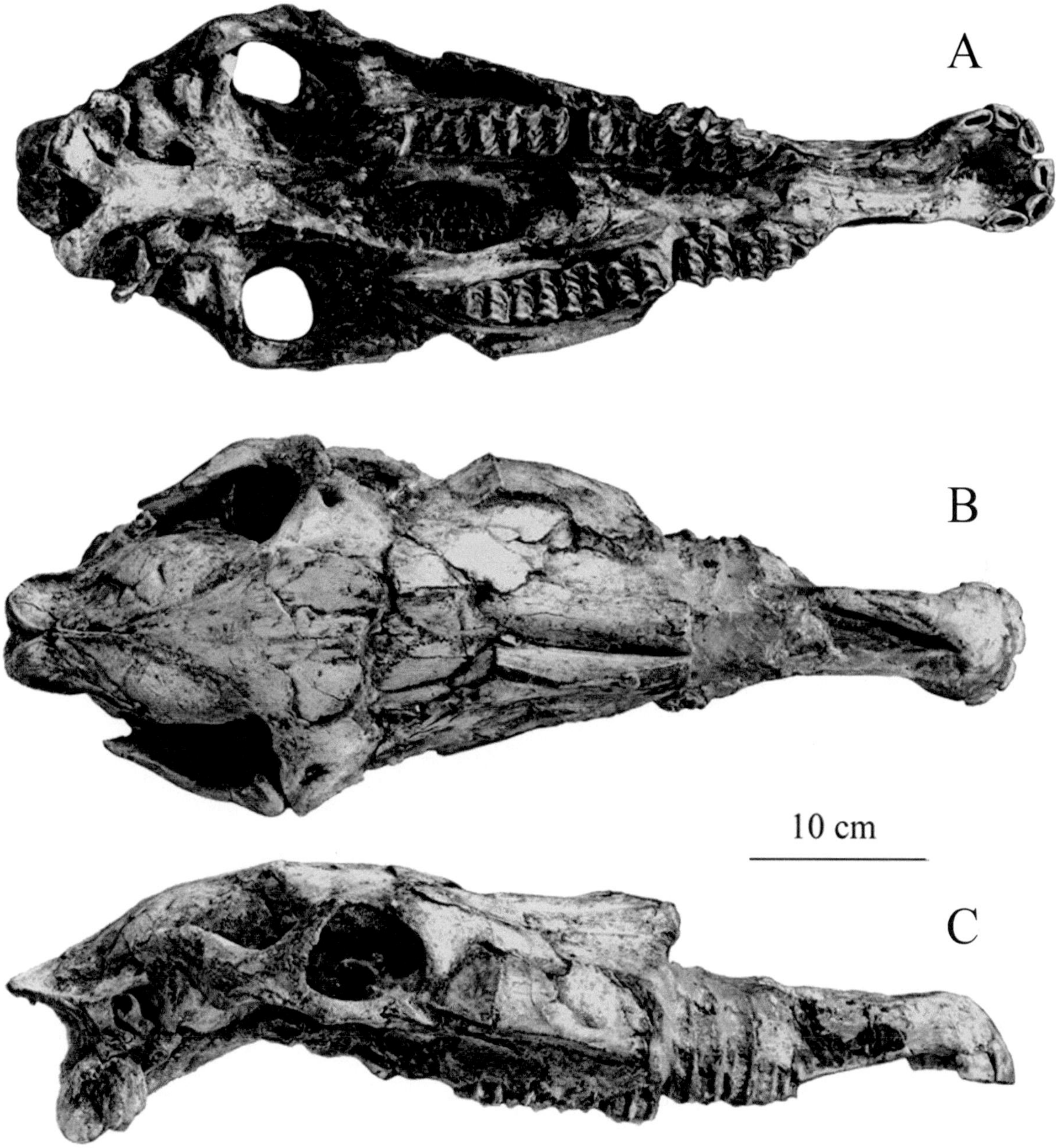

Fig. 11.31 *A. senezensis senezensis* from Senèze. Cranium FSL 210887. A. Occlusal view. B. Dorsal view. C. Lateral view

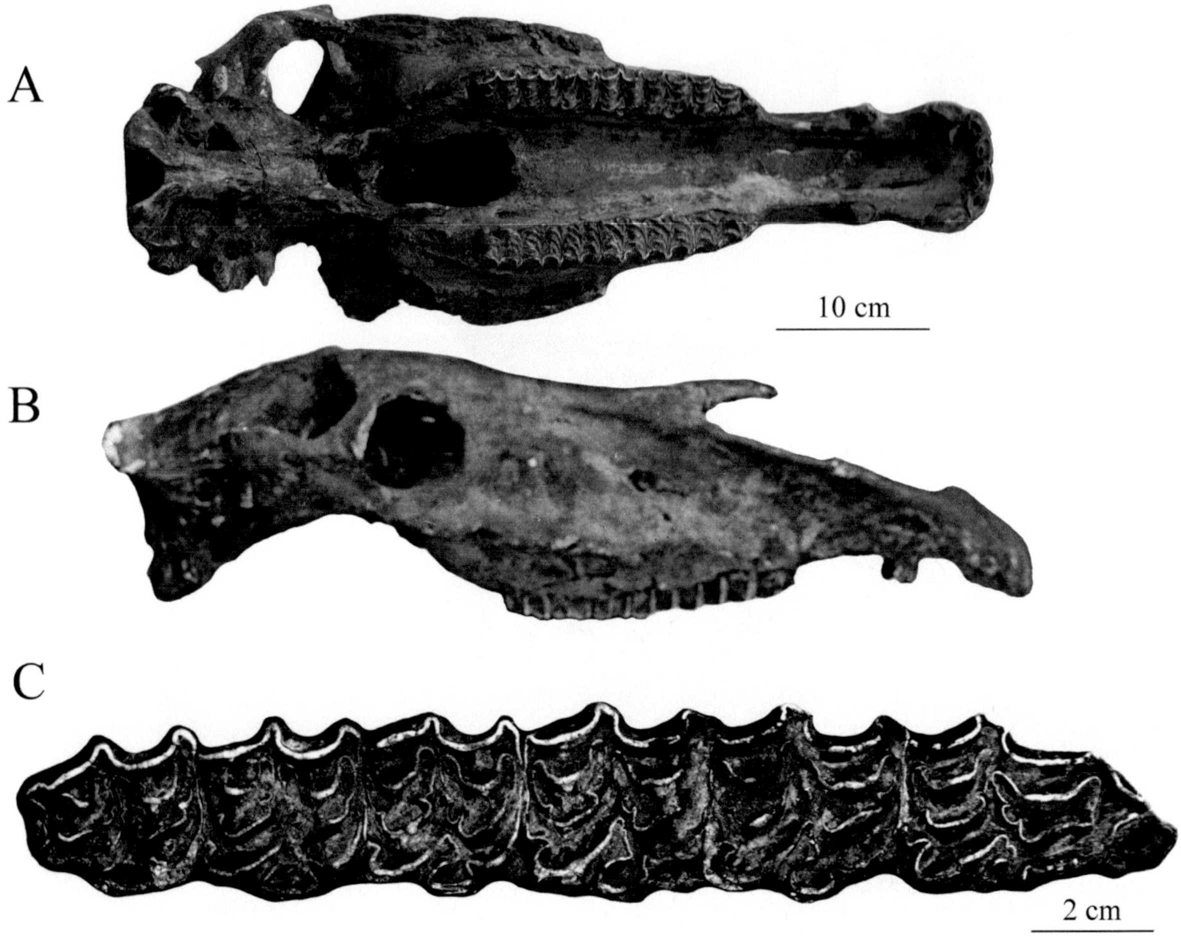

Fig. 11.32 *A. senezensis senezensis* from Senèze. Cranium FSL 210993. A. Occlusal view. B. Lateral view. C. Right P3-M3, occlusal view

Allohippus major *from Senèze and Other Localities*

From the enamel pattern of the P3 as well as the proportions of the metapodials, the large equid from Senèze is almost certainly an *Allohippus*, although we cannot be certain so long as its cranial proportions remain unknown. It can be referred to *A. major* although the upper cheek series from Chagny chosen as lectotype by Alberdi et al. (1998) is not completely typical for *Allohippus.* According to the prprtions of metapodials and first phalanges, the similarly-sized specimens from Gannat (France), Würtsburg-Schalksberg (Germany), East Runton (Great Britain), Vatera PO 5 (Greece), Kislang (Hungary), Oasele (Romania), and Tataourova (Russsia) also belong to *Allohippus major*.

Are there Age Indications for the Senèze Equids?

The occurrence of caballine fossils has already been discussed above. (See "Occurrence of several equids at Senèze"). Assuming that they do not actually belong to the primary Senèze collections, there still remains evidence for more than one species of equid at Senèze.

- *A. senezensis senezensis* is the most frequent equid. Both partial skeletons and other elements were found below the oldest dated level (SEN 98, 2.18 ± 0.03 Ma in Trench 5); the more complete FSL SEN 05–0081 + may be one of the oldest of all Senèze fossils, around 2.2 Ma (Delson et al. 2024). *A. senezensis senezensis* is similar to *A senezensis guthi* of La Puebla de Valverde (Spain),

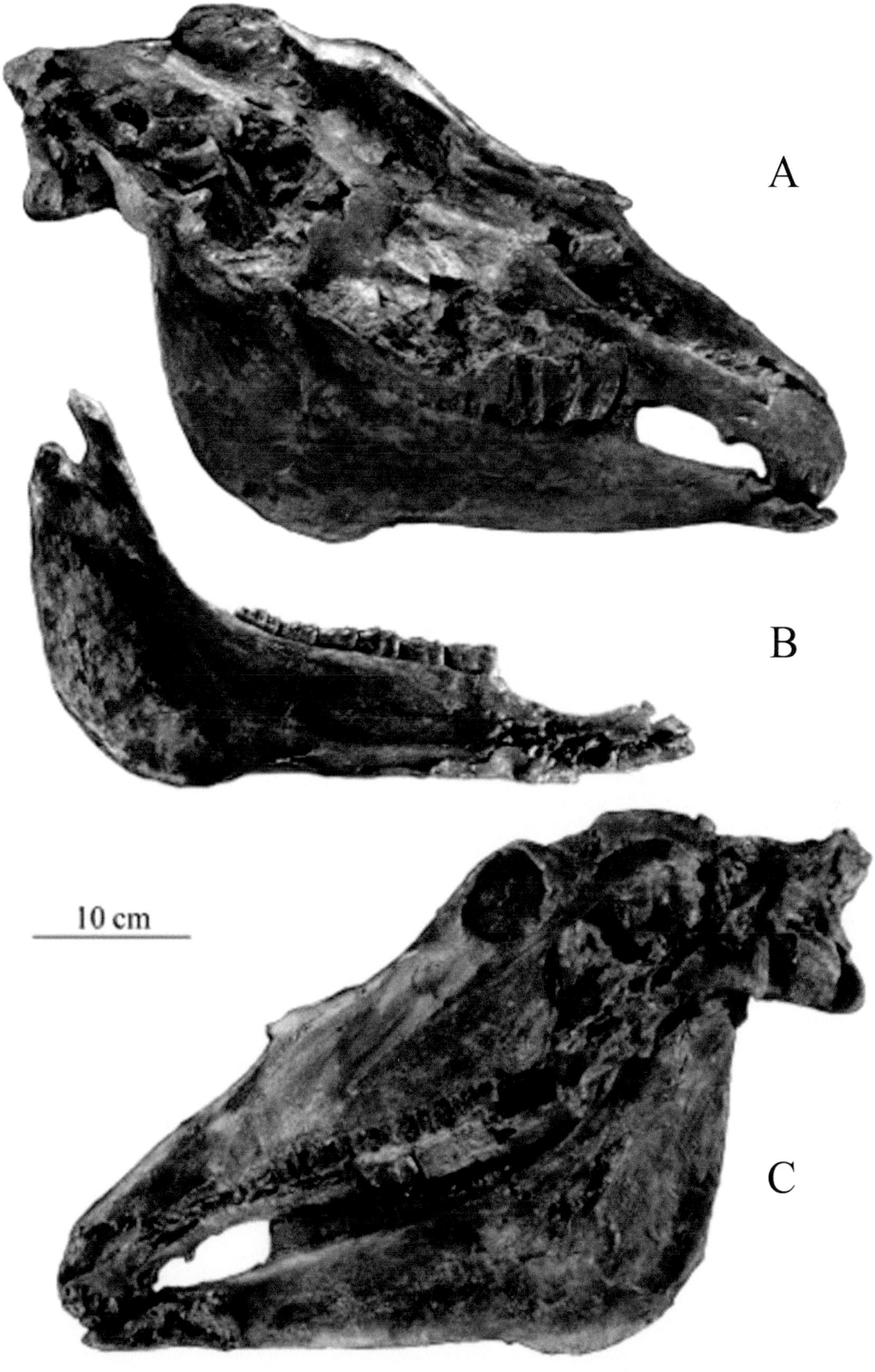

Fig. 11.33 *A. senezensis senezensis* from Senèze. Skull SEN 05–0081/82. A. Right lateral view. B. Internal view of the left mandibular corpus (05-0082). C. Left lateral view

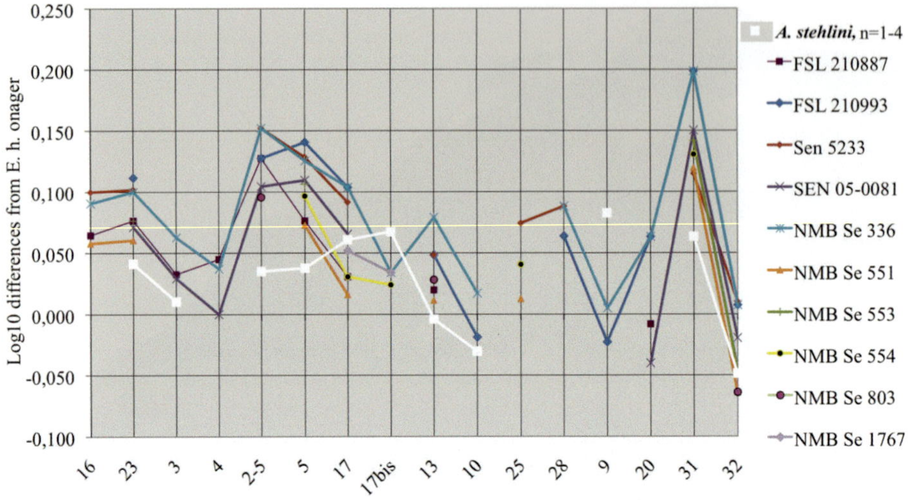

Fig. 11.34 Ratio diagrams of *Allohippus senezensis senezensis* (10 individuals) and *Allohippus stehlini* crania. Same variables as in Fig. 11.7. In addition to their smaller size, the crania of *A. stehlini* differ from those of *A. senezensis senezensis* by a wider muzzle breadth between the interalveolar borders (variable 17bis) and by longer choanae (variable 9)

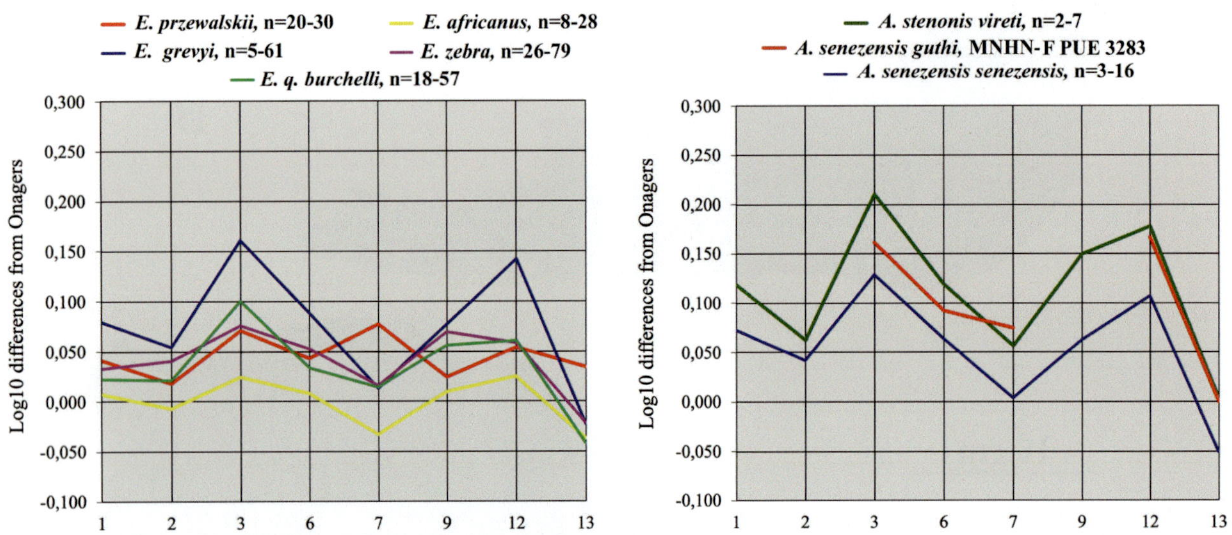

Fig. 11.35 Ratio diagrams of extant *Equus* (left panel) and *Allohippus* (right panel) mandibles. 1. Maximal length; 2. Radius of the mandibular angle; 3. Diastema; 6. Symphysis length; 7. Breadth at the posterior borders of i3; 9. Height in front of p2; 12. Muzzle length; 13. Least symphysis breadth. n: number of specimens. System of measurements illustrated at https://vera-eisenmann.com/mandibles-system-of-measurements

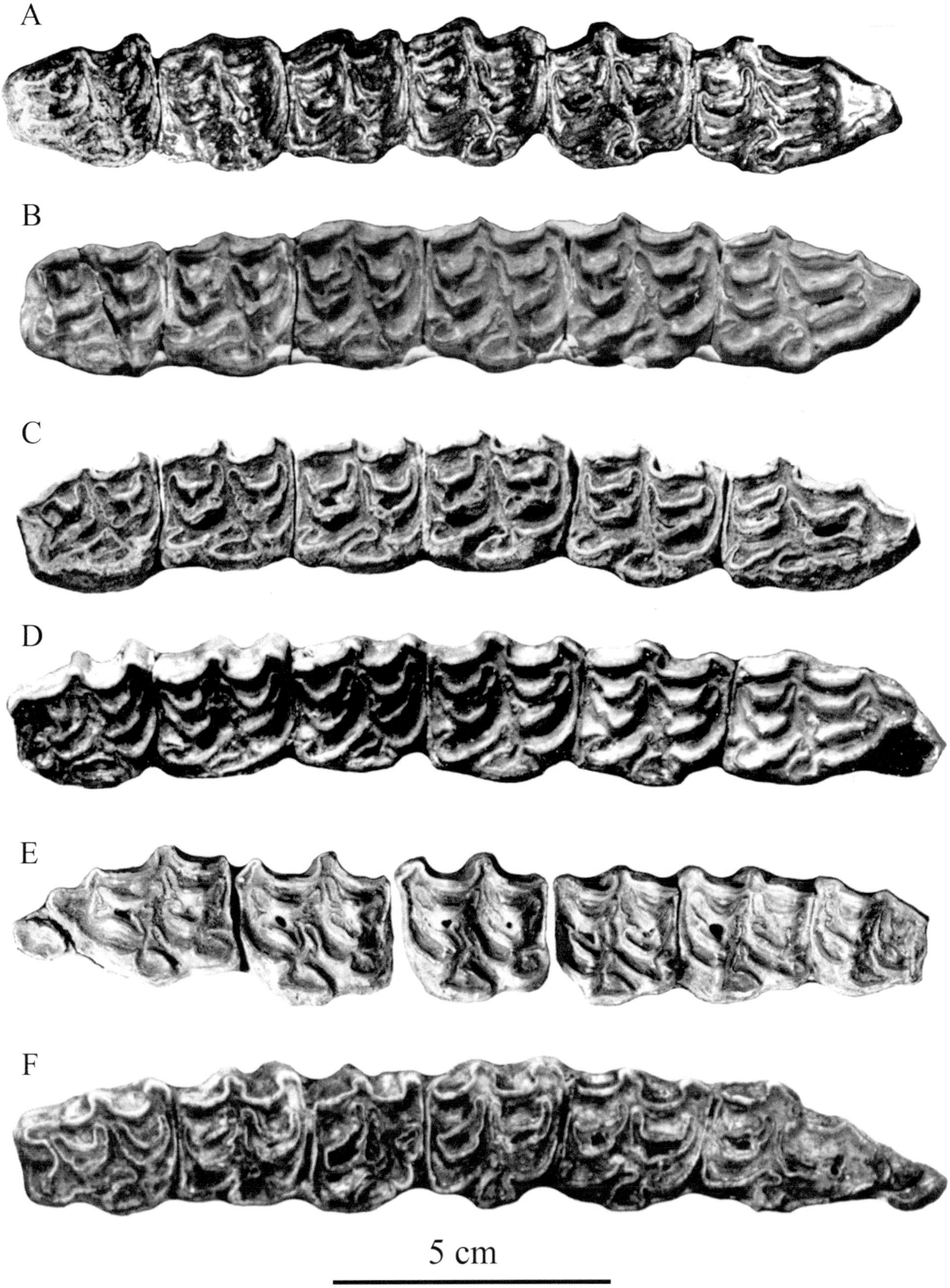

Fig. 11.36 *A. senezensis senezensis* from Senèze. Uppper cheek teeth: A. Right P2-M3 NMB Se 336. B. Right P2-M3 NMB Se 804. C. Right P2-M3 NMB Se 806. D. Right P2-M3 MHNL Sen 5233. E. Left P2-M3 FSL 210887. F. Right P2-M3 NMB Se 554

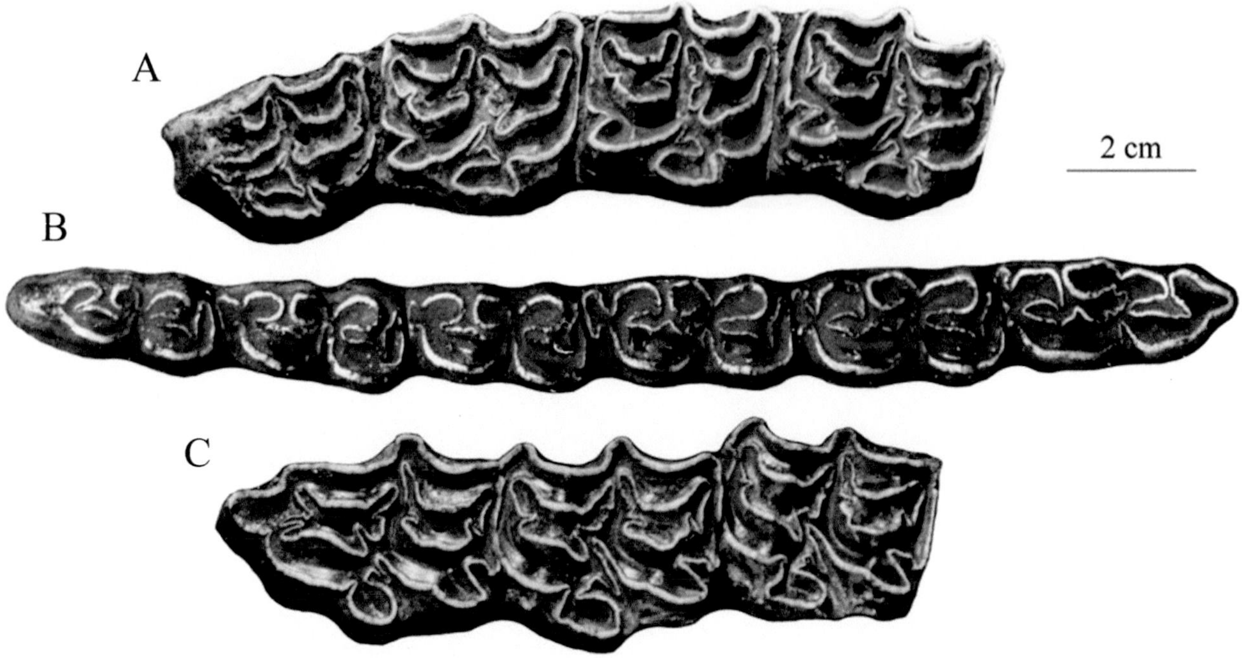

Fig. 11.37 *A. senezensis senezensis* from Senèze. Upper and lower cheek teeth of MNHL Sen 1855. A. Right P4-M3. B. Right p2-m3. C. Left P2-P4

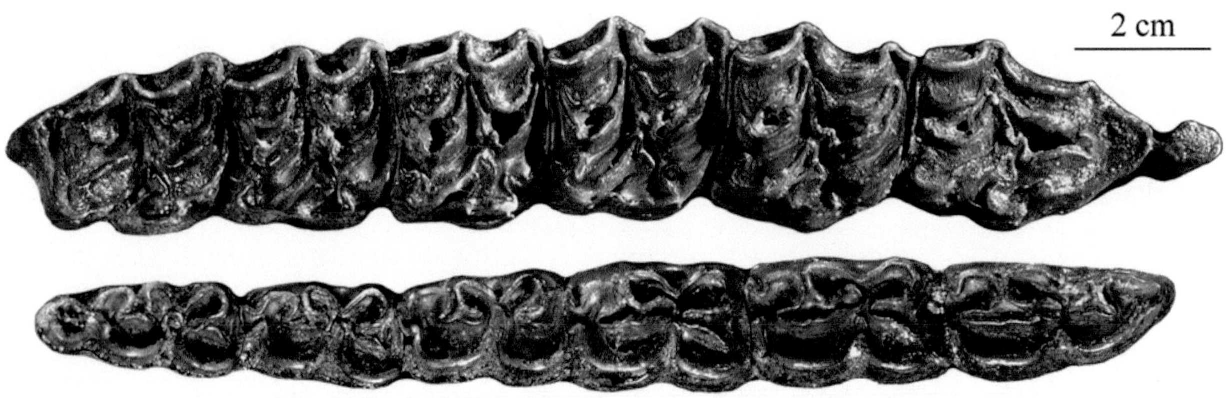

Fig. 11.38 *A. senezensis senezensis* from Senèze. A. SEN 05–0081 right P2-M3. B. SEN 05–0082 right p2-m3

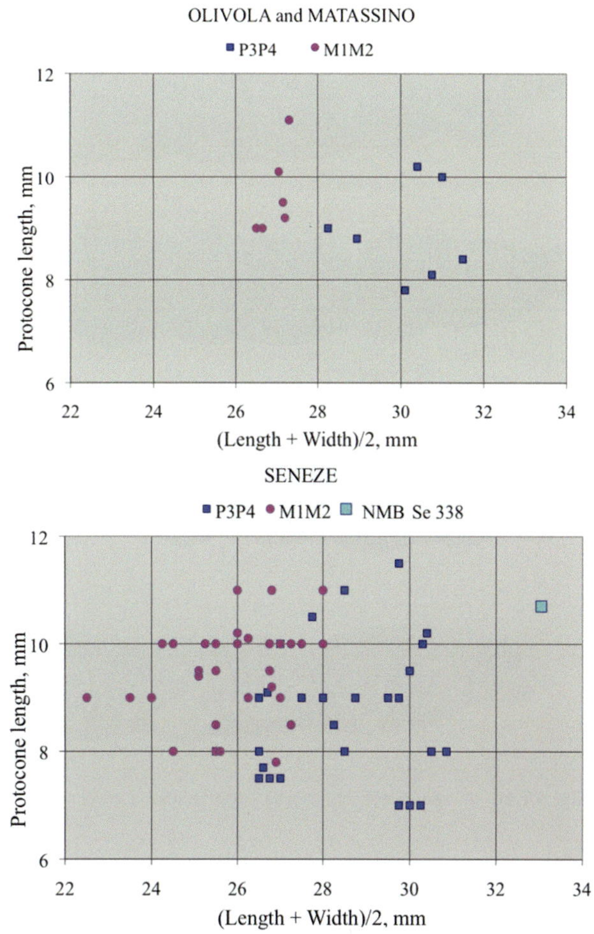

Fig. 11.39 Scatter diagrams of protocone lengths versus average occlusal dimensions [(occlusal length + occlusal width)/2] in *A. senezensis* of Senèze and *A. stenonis* of Olivola and Matassino

which is generally considered to be biochronologically closer to St. Vallier than to Senèze, suggesting that La Puebla de Valverde is older than 2.2 Ma. *A. senezensis mygdoniensis* of Gerakarou (Greece) is also similar and may thus be older than its estimated age of ca. 1.8 Ma shown on Fig. 11.51.

- One upper cheek tooth and 16 limb bones belong to a very large Villafranchian equid. I refer them to *Allohippus major* (see above the comparison of metapodials and first phalanges). They could belong to an earliest Pleistocene species like the one from Oasele (Montopoli faunal unit?) or to a mid-Early Pleistocene form as at Gannat (where it is associated with a Merck's Rhinoceros and a peculiar *Megaceros* [C. Guérin, 1995, pers. comm.]).
- Two first phalanges are larger than the average for *A. stenonis vireti* (ca. 2.5 Ma).
- One MC III resembles *A. stehlini* of Valdarno (younger than the Olduvai subchron).
- One MT III and one first phalanx resemble the small ?*Allohippus* sp. of Pyrgos (late MNQ 18 or early MNQ 19).

The last five specimens seem to be significantly different from typical *A. senezensis senezensis*. They might indicate additional rare equids at Senèze, in which case they would extend the known ranges of these taxa, or they might potentially be intrusive or erroneous (as with the caballine material).

Summary

Three genera of Plio-Pleistocene monodactyl equines are recognized and distinguished by cranial proportions: *Equus*, *Plesippus* and *Allohippus*. *Equus* (and generally *Plesippus*) have a shorter naso-incisival notch relative to cheek length than *Allohippus*, while the ratio of vomerine length to palatal length is generally lower in *Allohippus* than *Plesippus* (and *Equus caballus*). All *Equus* also have longer post-vomerine length compared to overall palatal length than either *Plesippus* or *Allohippus*. Using variation in extant wild species of *Equus* as a model, *Allohippus stenonis* is distinguished from *A. senezensis* at the species level based on relative

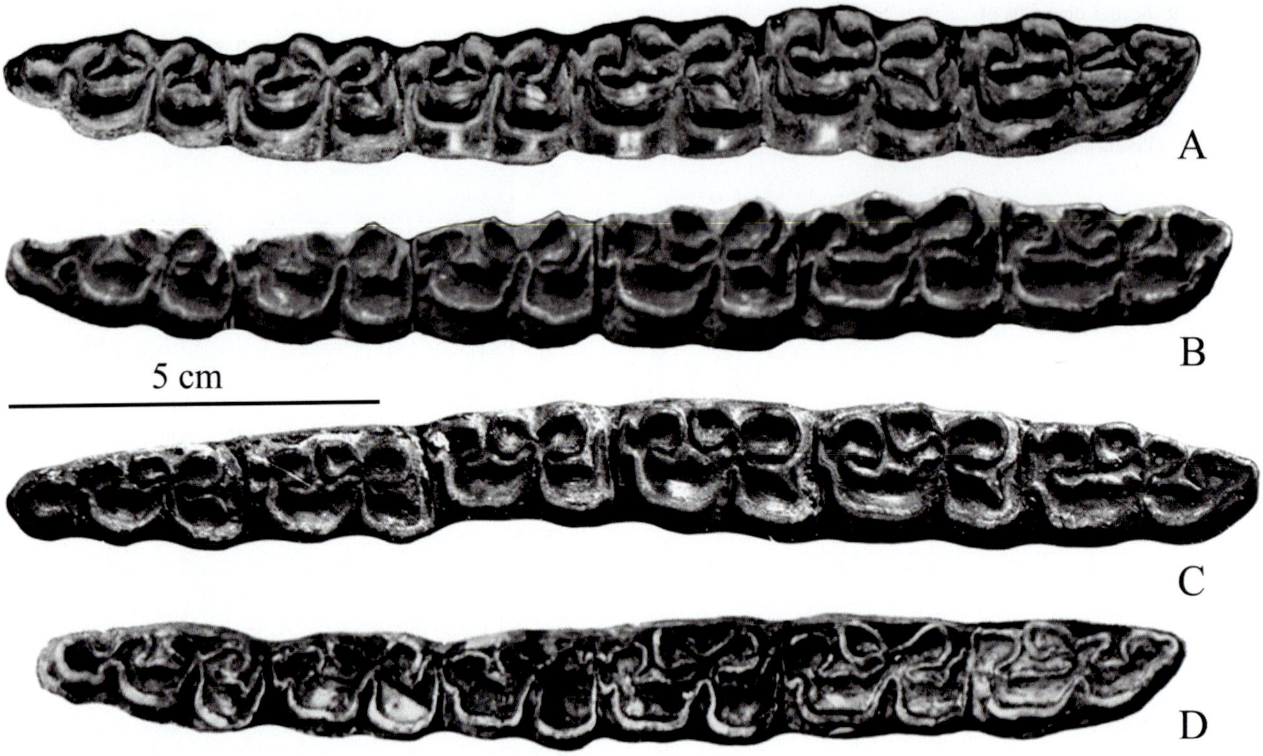

Fig. 11.40 *A. senezensis senezensis* from Senèze. Lower cheek teeth. A. NMB Se 804. B. NMB Se 806. C. Sen 5233. D. NMB Se 554

Fig. 11.41 Ratio diagrams of third metacarpals of *Allohippus stehlini* from Valdarno and of FSL 210,887 from Senèze. n: number of specimens

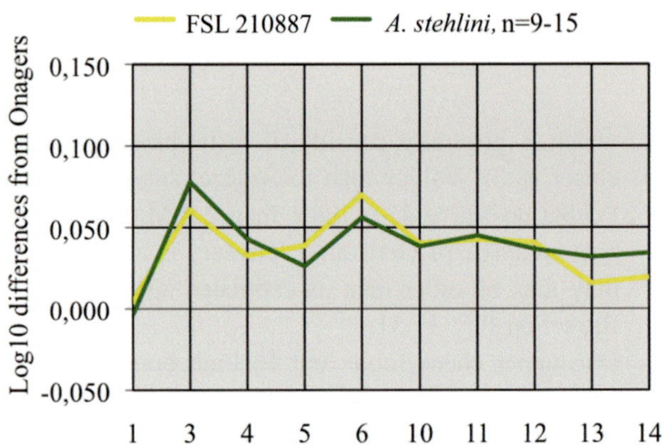

muzzle length (and some postcranial features), with several subspecies in each (see Table 11.A5).

The numerous equid fossils from Senèze were mostly collected in the first half of the twentieth century without detailed stratigraphic provenance. At least two and possibly up to six different taxa can be distinguished in the material catalogued as coming from Senèze in the Lyon, Basel and Paris collections. The mandible and metacarpal of a mounted skeleton, an isolated metatarsal and some isolated lower cheek teeth are clearly caballine and may be erroneously labeled or derived from an otherwise unknown younger horizon at Senèze.

The vast majority of the finds are identified as *Allohippus senezensis senezensis*, which is intermediate in size and proportions between *A. senezensis guthi* of La Puebla de Valverde (Spain) and *A. senezensis mygdoniensis* of

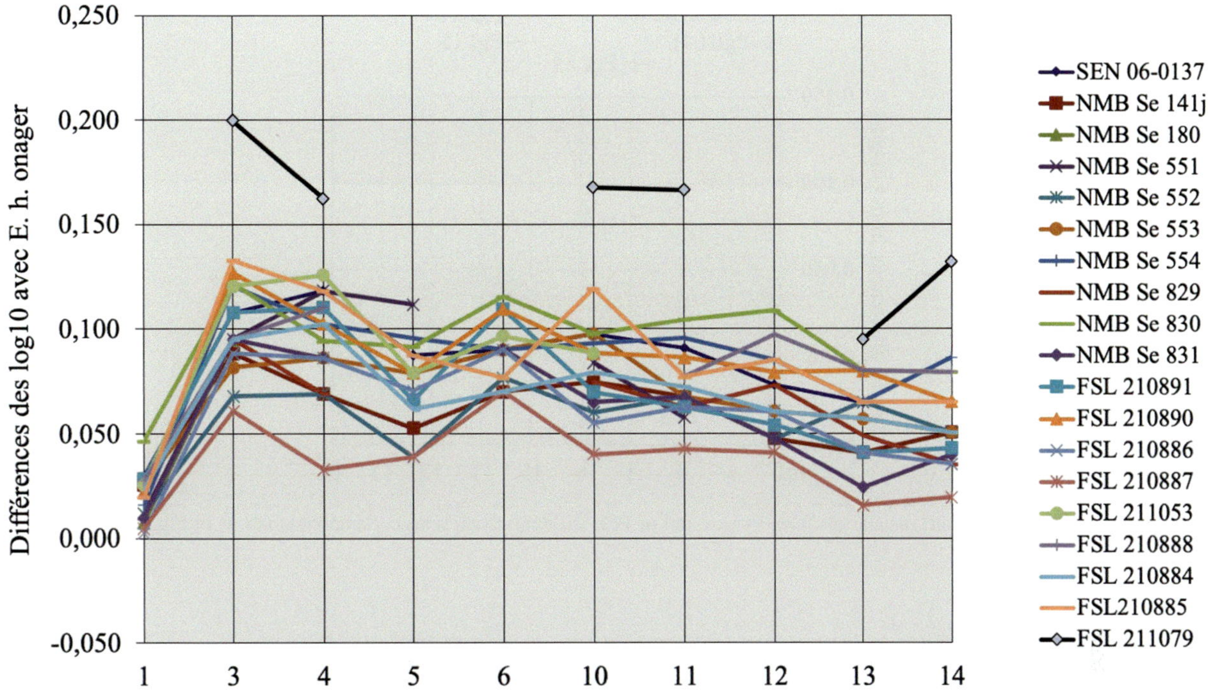

Fig. 11.42 Ratio diagrams of third metacarpals from Senèze. Same variables as in Fig. 11.9

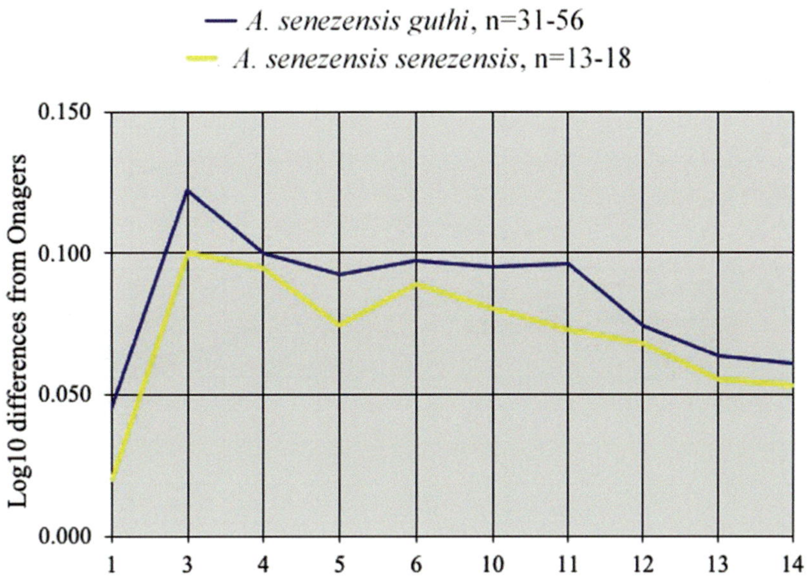

Fig. 11.43 Ratio diagrams of mean of third metacarpals of *A. senezensis senezensis* from compared to those of other equids. Same variables as in Fig. 11.9

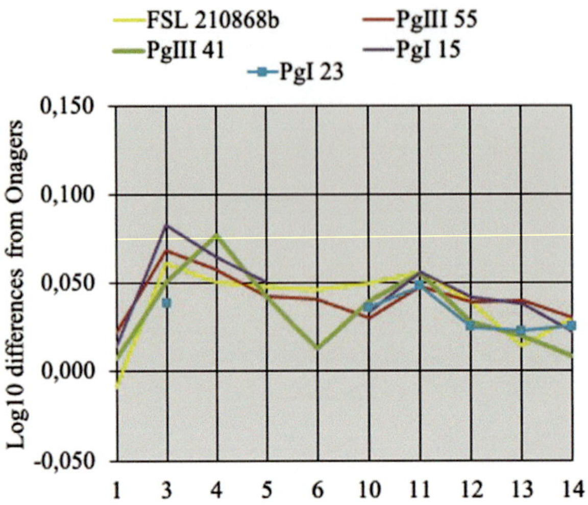

Fig. 11.44 Ratio diagrams of third metatarsals from Pyrgos and of FSL 210868b from Senèze. Same variables as in Fig. 11.9

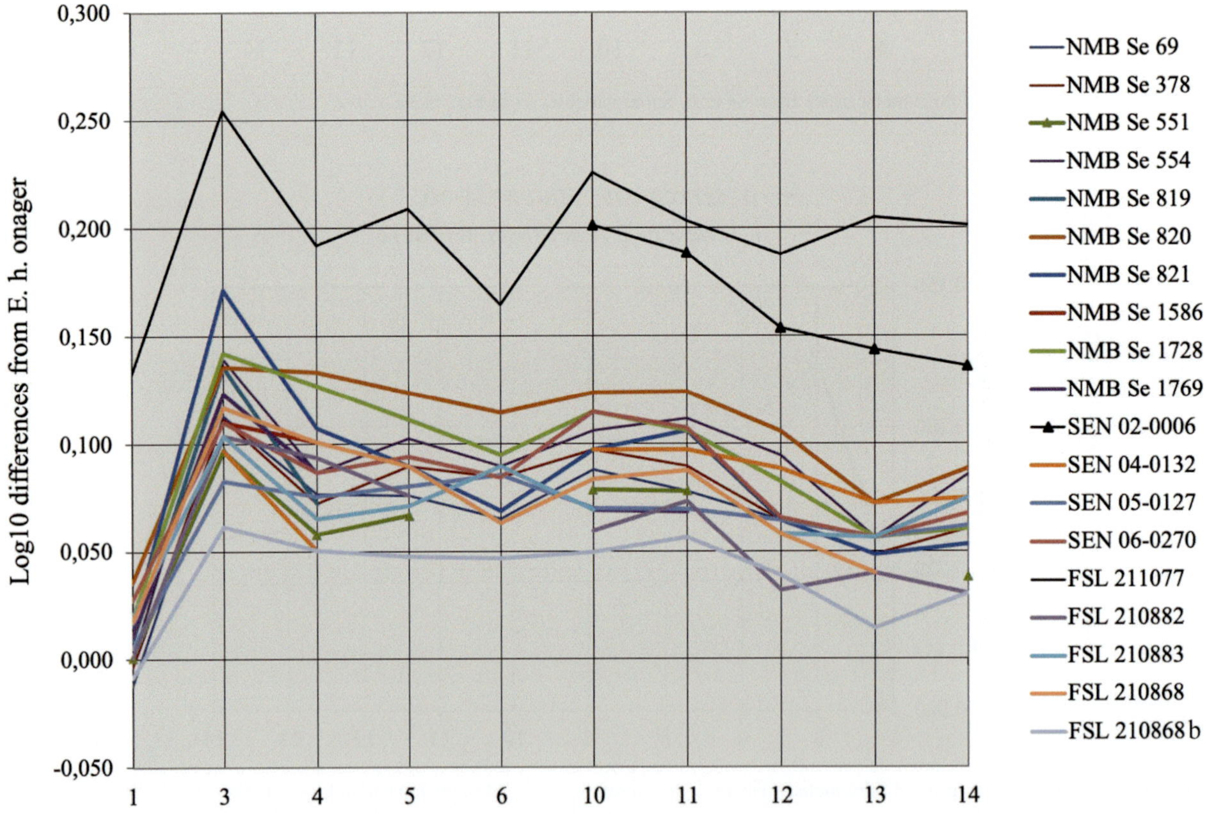

Fig. 11.45 Ratio diagrams of third metatarsals from Senèze. Same variables as in Fig. 11.9

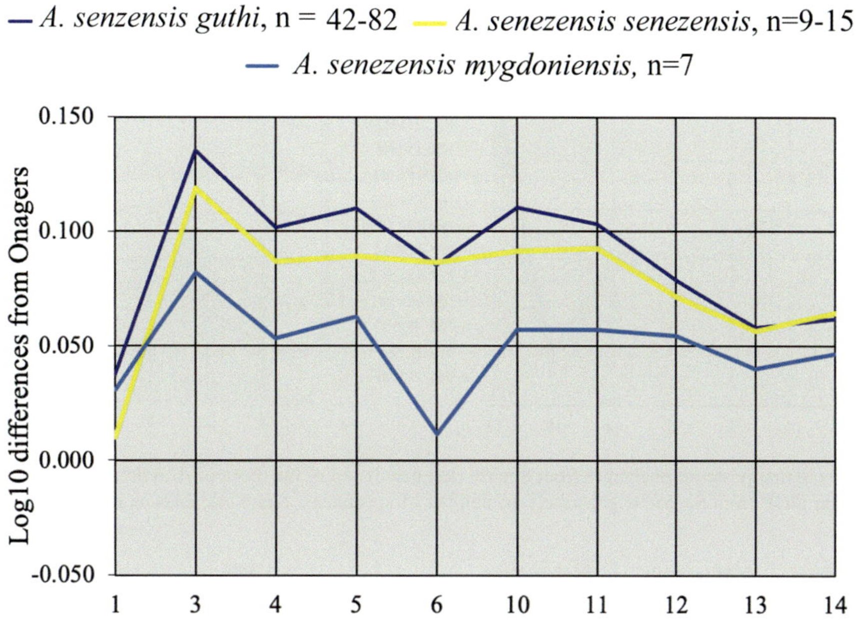

Fig. 11.46 Ratio diagrams of mean of third metatarsals of *A. senezensis senezensis* from Senèze compared to those of other equids. n: number of specimens. Same variables as in Fig. 11.9

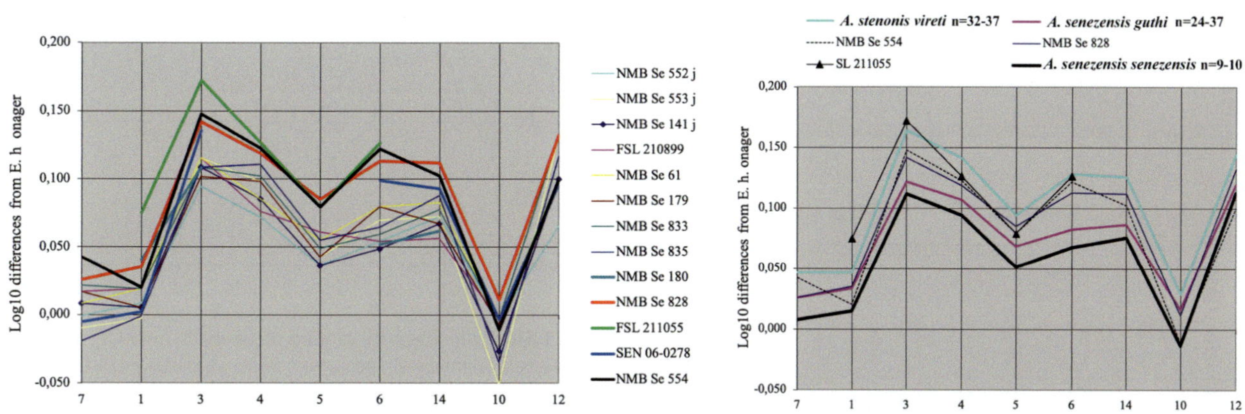

Fig. 11.47 Ratio diagrams of first anterior phalanges from Senèze (left panel) and of the mean of *A. senezensis senezensis* compared to those of other equids and three large Ph1A from Senèze (right panel). n: number of specimens. Same variables as in Fig. 11.22

Gerakarou (Greece). It has a relatively short muzzle and deep, not very robust, metapodials, comparable to those of extant equids living in rather dry conditions. Two partial skeletons and a few other possibly associated hindlimb elements were recovered in the new excavations close to the bottom of the local sequence, thus ca 2.2 Ma.

One isolated P3 and 16 limb bones (humerus, radii, metacarpal, tibiae, tali and first, second and third phalanges) belong to a very large ?*Allohippus*. The tooth has a short, ungrooved protocone and small pli caballin; on the metapodials, the distal articular width is larger than the supra-articular width; on the anterior first phalanx the distal supra-articular width is small relative to the proximal depth; these features are common for *Allohippus*. This sample can be identified as *Allohippus major*, and samples from other localities referred to this species are also discussed.

In addition to these clearly distinguishable samples, there are five specimens which appear to be beyond the variation range of either *Allohippus* population and might possibly represent additional taxa which are similar to larger samples

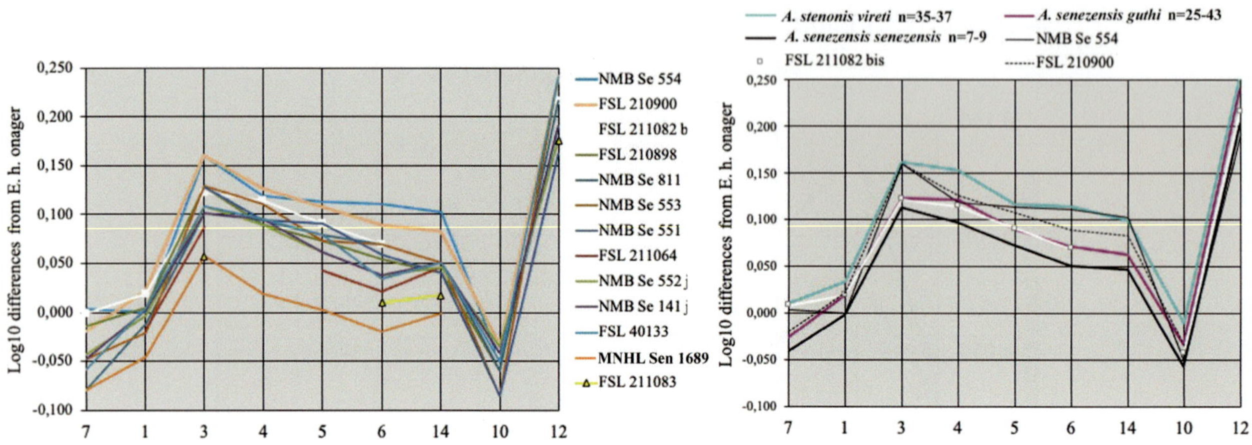

Fig. 11.48 Ratio diagrams of first posterior phalanges from Senèze (left panel) and of the mean of *A. senezensis senezensis* compared to those of other equids and three large Ph1P from Senèze (right panel). n: number of specimens. Same variables as in Fig. 11.22

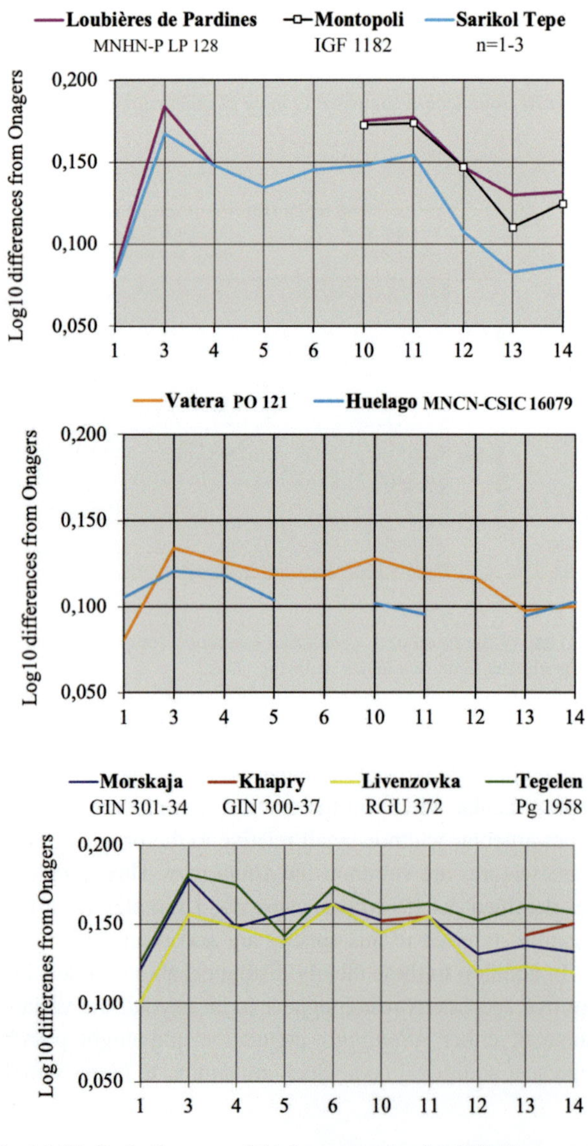

Fig. 11.49 Ratio diagrams of third metacarpals of A ?*Allohippus* sp.1, B ?*Allohippus* sp. 2, C ?*Allohippus* sp. 3. Same variables as in Fig. 11.9

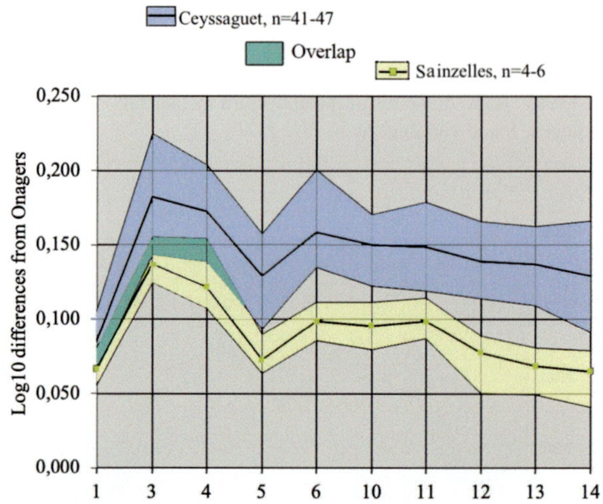

Fig. 11.50 Ratio diagrams of third metacarpals from Ceyssaguet and Sainzelles. n: number of specimens. Same variables as in Fig. 11.9

from sites older or younger than Senèze: two first phalanges are larger than the average for *A. stenonis vireti* (ca. 2.5 Ma); one MC III resembles *A. stehlini* from the Upper Valdarno (ca. 1.7 Ma); and one MT III and one first phalanx resemble the small ?*Allohippus* sp. of Pyrgos (late MNQ 18 or early MNQ 19, ca 1.75 Ma). These five specimens could document rare Senèze taxa or could be either cataloguing errors from another site or derived from different Senèze horizons, as noted for the caballine specimens.

Various other equid fossils from Plio-Pleistocene sites are identified where possible. Figure 11.51 presents an overview of the distribution of eight taxa or morphs of mainly Villafranchian equids from Old World localities.

chrons	Ma	MNQ units	*EQUUS* (*Suessemionus*)	*?ALLOHIPPUS* LARGE and ROBUST	*?ALLOHIPPUS* LARGE and SLENDER	*?ALLOHIPPUS* SMALL,SLENDER	*ALLOHIPPUS major*	*ALLOHIPPUS* LONG MUZZLES	*ALLOHIPPUS* SHORT MUZZLES	*PLESIPPUS*
								A. stenonis stenonis *A. stenonis vireti* *A. koobiforensis* *A. sanmeniensis*	*A. senezensis senezensis* *A. senezensis guthi* *A. senezensis mygdoniensis* *A. stehlini*	
BRUNHES	0.6		Overstrand Süssenborn Akhalkalaki							
Matuyama										
Jaramillo	1		NE Siberia Liventsovka (L-131?)				Würtzburg			
MATUYAMA	1.2							Ceyssaguet		
		MNQ 19				Libakos				
						Pyrgos?	Gannat?			
		?			Tegelen		Kislang? Tataourova?	Nihowan Upper Valdarno	Upper Valdarno Gerakarou	
Olduvai										
	2	MNQ 18					Chagny?			
	2.1									
Feni	2.2						Senèze		Senèze	
MATUYAMA		MNQ 17b		Sarikol Tepe	Vatera F (PO 121) Huelago (16079) St Vallier LD2		Vatera (PO 5 Electric site)	Fuentenueva Kuruksai Tetoiu St Vallier LD3 (classic)	Dafnero Kuruksai, Sesklo Chilhac, La Puebla de Valverde	Bajiazui
	2.5	MNQ 17a		Montopoli Loubières de Pardines	El Rincon Liventzovka (RGU 372) Morskaja, Khapry (300-37)		Liventzovka (L778) L. de Pardines Oasele	Liventzovka (L4-11)		Loc 32
	2.6									

Fig. 11.51 Selected *Plesippus*, *Allohippus*, and *Equus* taxa and samples from the Early and Middle Pleistocene and their approximate temporal distribution. "?" indicates uncertainty about age. Individual specimens include: Liventsovka, ROMK L-131 P3 or P4; Sarikol Tepe, mean of three MC; Montopoli, distal fragment of MC IGF 1182; Loubières de Pardines, distal fragment of MC MNHN.F LP 128; Vatera, MC NHCV PO 121; Huelago, MC MNCS-CSI 16079; Liventsovka MC RGU 372; Morskaja, MC GIN 302–34; Khapry, MC fragment GIN 300–37; Tegelen, MC Pg 1958; Oasele, MC ISER 5400; Liventsovka, MT ROMK L-178; Vatera, MT NHCV PO 5; Senèze, upper premolar and various limb bones of *Allohippus major*; Chagny, upper cheek teeeth, MT and Ph1 post no n°s; Kislang, MT no n°; Gannat, MC and MT no n°s; Würtsburg-Schalksberg, MC 76–500, MT no n°; Liventsovka, cranium ROMK L 4/RGU 11; Tetoiu, MC n = 6–9; Kuruksai, cranium PIN 3120-no n° (long-muzzle); Sesklo, cranial fragment LGPUT SES (Σ) 2003; Dafnero, MC and MT; Kuruksai, crania PIN 3120–319, 320, 360 (short-muzzled)

Acknowledgments First of all I (V.E.) wish to mention Claude Guérin, who was one of my oldest colleagues and friends. Without him, Martine Faure and Eric Delson, the whole project on Senèze would not have come to fruition. I thank them most sincerely for that. Of course I am indebted to all the curators of Lyon and Basel collections without whose help I could not have studied the fossils of Senèze. We thank Athanassios Athanassiou, George Koufos and an anonymous reviewer for comments which improved an earlier version.

Appendix

Table 11.A1 Cranial measurements of equids (in mm)

Origin or Breed	Locality	Species	Specimen number	Prosthion-Hormion	P2 to Hormion	Vomer length	Vomer-Basion	Muzzle length	N-i notch length	Cheek length
			Variable number	2	2–5	3	4	5	31	32
USA	Hagerman	*P. shoshonensis*	USNM 16991	280	139	145	112			
USA	Hagerman	*P. shoshonensis*	USNM 13841	275	149	130	105			
USA	Hagerman	*P. shoshonensis*	USNM 16982	288	146	149	107			
USA	Hagerman	*P. shoshonensis*	USNM 13835	300	153	158	100			
USA	Hagerman	*P. shoshonensis*	USNM 12538	280	135	142	115			
USA	Hagerman	*P. shoshonensis*	USNM 12528	295	147	145	112			
USA	Hagerman	*P. shoshonensis*	USNM 12576	275	137	141	112			
USA	Hagerman	*P. shoshonensis*	USNM 11989	275	135	156	100			
USA	Hagerman	*P. shoshonensis*	USNM 11988	290	145	160	100			
USA	Hagerman	*P. shoshonensis*	USNM 14560	290	145	145	92			
USA	Hagerman	*P. shoshonensis*	USNM 12155	300	146	145	103			
USA	Hagerman	*P. shoshonensis*	USNM 12501	290	142	142	117			
USA	Hagerman	*P. shoshonensis*	USNM 12535	280	152	140	95			
USA	Hagerman	*P. shoshonensis*	USNM 32555	287	140	148	109			
USA	Hagerman	*P. shoshonensis*	USNM 32553	290	145	147	111			
USA	Hagerman	*P. shoshonensis*	USNM 12543	288	143	155	107			
USA	Crawfish	*P. simplicidens*	AMNH-P 20077	290			104			
China	Bajiazui	*P. qingyangensis*	IVPP B 61249	275	132.3	136.8	97.6			
China	Longdan	*P.-A.. eisenmannae*	HPM 0978		185	180				
China	Longdan	*P.-A.. eisenmannae*	IVPP 13552	[376	199]	[135]	119			
China	Longdan	*P.-A.. eisenmannae*	HPM C 0055	345	175.5	156	118			
China	Longdan	*P.-A.. eisenmannae*	HPM C 0052	340	169	147	119			
China	Longdan	*P.-A.. eisenmannae*	HPM D 0014	340	173	151	102			
China	Longdan	*P.-A.. eisenmannae*	HPM EXPO 1	335	160	146	118			
China	Loc 32	*A. yunnanensis*	PMU M 1324–25 ?	292	146.7	126.6	104.5			
China	Madahai	*A.? yunnanensis*	IVPP 4250	272	140	124.5	106			
China	Nihowan	*A. sanmeniensis*	MNHN-F NIH 002	305	160	152	126	155		
China	Fan Tsun	*A.* sp.	FAM 60-B 719					151		
China	Loc. D	*A.* sp.	PMU M 1418					145		
Tajikistan	Kuruksai	*A. bactrianus*	n = 3					127.7		
Tajikistan	Kuruksai	*A. bactrianus*	GIN Kur 3120/320	295	155	118	106	134		
Tajikistan	Kuruksai	*A.* sp.	GIN no nb					170		
Russia	Liventsovka	*A. livenzovensis*	ZIN 31078		165	139		146		
Italy	Olivola	*A. stenonis stenonis*	IGF 11023		149	152		156		
Italy	Valdarno	*A. stehlini*	n = 4					114.4		
Italy	Valdarno	*A. stehlini*	IGF 563		118	114.5		113		
Italy	Valdarno	*A. stehlini*	IGF 581		135.5	127		113		
Italy	Valdarno	*A. stehlini*	IGF 582		126	114		114		
France	Saint-Vallier	*A. stenonis vireti*	MHNL QSV 222	300	147	137	128	153		
France	Saint-Vallier	*A. stenonis vireti*	n = 3					146.8		
Spain	La Puebla de Valverde	*A. senezensis guthi*	MNHN-P PUE 3280	298	155	122	125	137		
France	Ceyssaguet	*A.* sp.	9142					175.2		
France	Senèze	*A. senezensis senezensis*	n = 7					134.4		
France	Senèze	*A. senezensis senezensis*	FSL 210993 ex 96132	307			119	145		
France	Senèze	*A. senezensis senezensis*	Sen 5233	300			122	141		
France	Senèze	*A. senezensis senezensis*	FSL 210857	280			112	125		
France	Senèze	*A. senezensis senezensis*	NMB Se 336	297	164	135	110	140		

(continued)

Table 11.A1 (continued)

Origin or Breed	Locality	Species	Specimen number	Prosthion-Hormion	P2 to Hormion	Vomer length	Vomer-Basion	Muzzle length	N-i notch length	Cheek length
Greece	Gerakarou	*A. senezensis mygdoniensis*	n = 3	276.4			104.5	130.7		
Greece	Gerakarou	*A. senezensis mygdoniensis*	LGPUT-GER 8		141	116		130		
Greece	Gerakarou	*A. senezensis mygdoniensis*	LGPUT-GER 9					135		
Greece	Gerakarou	*A. senezensis mygdoniensis*	LGPUT-GER 31		148	134				
Greece	Gerakarou	*A. senezensis mygdoniensis*	LGPUT-GER 122		149	132		127		
Kenya	Koobi Fora	*A. koobiforensis*	KNM-ER 1484					150		
Russia	Mongolia	*E. nalaikhaensis*	GIN 3747/500	275			128			
Russia	NE Siberia	*E. cf. scotti*	SI 160–455 (BET 55)	325			145			
Algeria	Tighennif	*E. mauritanicus*	MNHN-F TER 1542	270			121			
Algeria	Tighennif	*E. mauritanicus*	MNHN-F TER 1543	260			122			
S. Africa	Elandsfontein	*E. capensis*	SAM E 21025	298			142			
Russia	Bug	*E. khomenkoi*	ZIN 1283–1	247			126			
Russia	N Yakutia	Fossil horse	ZIN 32300	258			118			
Russia	Liakhov	Fossil horse	ZIN 3965	264			131			
Russia	Lena	Fossil horse	IA 33	277			122.5			
Russia	N Yakutia	Fossil horse	IA 5059	265			126			
Russia	N Yakutia	*E. coliemensis*	IA 1741	282			123			
Russia	Kotelny	Fossil horse	PIN 301–539	250			119.5			
Russia	Kolyma	Fossil horse	MS 3752–25	268			126			
Russia	North Siberia	Fossil horse	Sher PIN no n°	268			127			
Russia	Toungouze	*E. chosricus* type	PIN 113–165	271			130			
Russia	Senguilei	Fossil horse	PIN 113–166	269			144			
Russia	Missy	*E. missi*	PIN 113–167	243			127			
Russia	Missy	*E. missi*	PIN 113–168	250			123			
Russia	Missy	*E. missi*	PIN 113–169	251			124			
Russia	Missy	*E. missi*	PIN 113–170	262			130			
Russia	Taburishche	Fossil horse	PIN 113–171	270			125			
Russia	Ukraine	Fossil horse	PIN 113–172	275			127			
Russia	North Siberia	*E. alaskae*	PIN 113–173	260			112.5			
Russia	North Siberia	*E. alaskae*	PIN 113–174	255			118			
Russia	Kotelny	Fossil horse	PIN 113–175	267			121			
France	Siréjol	*E. gallicus*	PIN 113–176	260			127			
France	Siréjol	*E. gallicus*	PIN 113–177	270			127			
Italy	ValdiChiana	Fossil horse	PIN 113–178	280			131			
Italy	ValdiChiana	Fossil horse	PIN 113–179	255			131			
Yukon		*E. lambei*	UNSM 8426	245			117			
Alaska		*E. alaskae*	UNSM 7700	265			122			
USA	Hay Springs	*E. niobrarensis*	UNSM 4999	285			122			
USA	Dawson	Fossil horse	NMC 9924–12	270			119			
USA	Gold Run	Fossil horse	NMC 13485	244			118			
USA	Gold Run	Fossil horse	NMC 17254	260			121			
USA	Gold Run	Fossil horse	NMC 34803 A	260			120			
USA	Dawson	Fossil horse	NMC 17905	265			129			
USA	Fairbanks	Fossil horse	AMNH-P 60026	265			117			
USA	Fairbanks	Fossil horse	AMNH-P 30702	270			116			
USA	Gilliland	*E.* cf. *scotti*	AA 46899	300			138			
USA	Rock Creek	*E. scotti*	NMC 2381	330			139			
USA	Rock Creek	*E. scotti*	AMNH-P 10612	297			138			
USA	Hay Springs	*E.* cf. *occidentaliss*	UNSM 5978	277			127			
USA	Hay Springs	Large caballine	UNSM 1346	308			153			
USA	Hay Springs	*E. calobatus*	AMNH-P "13"	260			117			
USA	Cedar Meadow	*E. fraternus*	AMNH-P FM 116143 + 3770	295			151			
USA	Channing	*E. semiplicatus*	AMNH-P FM 18–399	230			110			
USA	Lissie Fm	*E. (Amerhippus) francisci*	TAMU 2518	215			111			
USA	Rancho La Brea	*E. (Amerhippus) occidentalis*	AMNH-P 14396	264			122			
USA	Rancho La Brea	*E. (Amerhippus) occidentalis*	LACM 3500–14	270			129			
USA	Rancho La Brea	*E. (Amerhippus) occidentalis*	LACM 3500–26	283			124			
USA	Rancho La Brea	*E. (Amerhippus) occidentalis*	LACM 3500–5	286			125			

(continued)

Table 11.A1 (continued)

Origin or Breed	Locality	Species	Specimen number	Prosthion-Hormion	P2 to Hormion	Vomer length	Vomer-Basion	Muzzle length	N-i notch length	Cheek length
USA	Rancho La Brea	*E. (Amerhippus) occidentalis*	LACM 3500–21	273			120			
USA	Rancho La Brea	*E. (Amerhippus) occidentalis*	LACM 3500–1	260			125			
USA	Rancho La Brea	*E. (Amerhippus) occidentalis*	UCMP no n°	270			125.5			
USA	Rancho La Brea	*E. (Amerhippus) occidentalis*	UCMP 2051–12269	280			134			
USA	Rancho La Brea	*E. (Amerhippus) occidentalis*	UCMP 2051–21001	280			128			
Ecuador		*E. (Amerhippus) andium*	AMNH-P "14'	230			106.5			
Turkey		*E. asinus*	KI 3244	175			82			
Turkey		*E. asinus*	KI 3245	183			94			
?		*E. asinus*	KI 33	195			90			
?		*E. asinus*	KI 26192	180			82.7			
?		*E. asinus*	KI 4221	167			76			
Socotra		*E. asinus*	HL Eaw2	188			88			
Socotra		*E. asinus*	HL Eaw1	185			86			
Socotra		*E. asinus*	HL Eaw4	203			89			
?		*E. asinus*	HL Ea2	205			98			
?		*E. asinus*	HL Ea11	211			100			
?		*E. asinus*	HL E13	166			68			
?		*E. asinus*	HA no n°	207			95			
Abyssinia		*E. asinus*	HL Ea abs						135	140
		E. asinus	MCZ 14539						121	132
		E. asinus	MCZ 8272						118	135
		E. asinus	MS 3008						106	117
		E. asinus	MS 74781						130	138
		E. asinus	MS 74776						120	140
		E. asinus	MS 1747						115	122
		E. asinus	MS 102074						127	128
Kitengela		*E. burchelli granti*	KNM 2360	235			114			
Kitengela		*E. burchelli granti*	KNM 2383	213			116			
Kitengela		*E. burchelli granti*	KNM 2397	225			109			
Kitengela		*E. burchelli granti*	KNM 2399	235			97			
Kitengela		*E. burchelli granti*	KNM 2401	238			114			
Kitengela		*E. burchelli granti*	KNM 2402	228			104			
Kitengela		*E. burchelli granti*	KNM 2403	225			102			
Kitengela		*E. burchelli granti*	KNM 2407	237			107			
Kitengela		*E. burchelli granti*	KNM 2408	233			109			
Kitengela		*E. burchelli granti*	KNM 2409	230			106			
Kitengela		*E. burchelli granti*	KNM 2412	220			107			
Kitengela		*E. burchelli granti*	KNM 2413	235			105			
Kitengela		*E. burchelli granti*	KNM 2415	225			106			
Kitengela		*E. burchelli granti*	KNM 2418	225			114			
Kitengela		*E. burchelli granti*	KNM 2419	222			115			
Kitengela		*E. burchelli granti*	KNM 2421	240			101.5			
Kitengela		*E. burchelli granti*	KNM 2422	230			100			
Kitengela		*E. burchelli granti*	KNM 2423	220			115			
Kitengela		*E. burchelli granti*	KNM 2426	230			94.5			
Kitengela		*E. burchelli granti*	KNM 2428	225			95.5			
Kitengela		*E. burchelli granti*	KNM 2431	230			109			
Kitengela		*E. burchelli granti*	KNM 2432	227			107			
Kitengela		*E. burchelli granti*	KNM 2435	227			112			
Kitengela		*E. burchelli granti*	KNM 2436	228			100			
Kitengela		*E. burchelli granti*	KNM 2438	225			114			
Kitengela		*E. burchelli granti*	KNM 2439	245			116			
Kitengela		E. burchelli granti	KNM 2440	230			115			
Kitengela		*E. burchelli granti*	KNM 2441	222			108			
Kitengela		*E. burchelli granti*	KNM 2448	232			108			
Kitengela		*E. burchelli granti*	KNM 2449	220			109			
Kitengela		*E. burchelli granti*	KNM 2450	210			109			
Kitengela		*E. burchelli granti*	KNM 2451	227			110			
S. Africa		*E. burchelli burchelli*	NMBI 420						147	169
S. Africa		*E. burchelli burchelli*	NMBI 421						153	176
S. Africa		*E. burchelli burchelli*	NMBI 422						151	160
S. Africa		*E. burchelli burchelli*	NMBI 423						144	160

(continued)

Table 11.A1 (continued)

Origin or Breed	Locality	Species	Specimen number	Prosthion-Hormion	P2 to Hormion	Vomer length	Vomer-Basion	Muzzle length	N-i notch length	Cheek length
S. Africa		*E. burchelli burchelli*	NMBI 424						146	177
S. Africa		*E. burchelli burchelli*	NMBI 425						150	162
S. Africa		*E. burchelli burchelli*	NMBI 426						143	161
S. Africa		*E. burchelli burchelli*	NMBI 427						151	154
S. Africa		*E. burchelli burchelli*	NMBI 428						147	155
S. Africa		*E. burchelli burchelli*	NMBI 429						147	183
S. Africa		*E. burchelli burchelli*	NMBI 430						138	165
S. Africa		*E. burchelli burchelli*	NMBI 431						150	168
S. Africa		*E. burchelli burchelli*	NMBI 432						144	172
S. Africa		*E. burchelli burchelli*	NMBI 433						145	174
S. Africa		*E. burchelli burchelli*	NMBI 434						149	178
S. Africa		*E. burchelli burchelli*	NMBI 435						135	157
S. Africa		*E. burchelli burchelli*	NMBI 436						148	175
S. Africa		*E. burchelli burchelli*	NMBI 437						140	176
S. Africa		*E. burchelli burchelli*	NMBI 438						145	169
S. Africa		*E. burchelli burchelli*	NMBI 439						154	165
S. Africa		*E. burchelli burchelli*	NMBI 440						149	163
S. Africa		*E. burchelli burchelli*	NMBI 441						141	164
S. Africa		*E. burchelli burchelli*	NMBI 442						136	165
S. Africa		*E. burchelli burchelli*	NMBI 443						140	165
S. Africa		*E. burchelli burchelli*	NMBI 444						159	171
S. Africa		*E. burchelli burchelli*	NMBI 445						148	162
S. Africa		*E. burchelli burchelli*	NMBI 446						149	159
S. Africa		*E. burchelli burchelli*	NMBI 447						138	170
S. Africa		*E. burchelli burchelli*	NMBI 448						148	172
Turkmenistan		*E. hemionus kulan*	ZIN 19046	230			109			
Badkhyz		*E. hemionus kulan*	MS 49096	228			104			
Badkhyz		*E. hemionus kulan*	MS 49098	211			104			
Badkhyz		*E. hemionus kulan*	MS 74785	235			108			
Badkhyz		*E. hemionus kulan*	MS 74787	215			103			
Badkhyz		*E. hemionus kulan*	MS 74788	223			116			
Badkhyz		*E. hemionus kulan*	MS 74790	228			102			
Badkhyz		*E. hemionus kulan*	MS 74791	222			96			
Badkhyz		*E. hemionus kulan*	MS 74802	217			106			
Badkhyz		*E. hemionus kulan*	MS 74792	230			100			
Badkhyz		*E. hemionus kulan*	MS 74793	222			106			
Badkhyz		*E. hemionus kulan*	MS 74794	232			104			
Badkhyz		*E. hemionus kulan*	MS 74799	210			99			
Badkhyz		*E. hemionus kulan*	ZIN 32047	220			102			
Badkhyz		*E. hemionus kulan*	ZIN 32113	212			110			
Badkhyz		*E. hemionus kulan*	ZIN 2277	232			104			
Badkhyz		*E. hemionus kulan*	ZIN 32279	212			96			
Badkhyz		*E. hemionus kulan*	ZIN "49"	217			105			
Badkhyz		*E. hemionus kulan*	ZIN "47"	207			99			
Badkhyz		*E. hemionus kulan*	ZIN "50"	220			100			
?		*E. hemionus kulan*	HA 7682	225			96.5			
Zoo		*E. zebra*	ZIN 111	224			110		147	144
S Africa		*E. zebra*	NHMUK-ZD 1846.3.23.10	240			110			
S Africa		*E. zebra*	NHMUK-ZD 1847.1.27.2	240			105			
Zoo		*E. zebra*	MB-Z 8558	250			115			
Zoo		*E. zebra*	MB-Z 13415	245			104			
Kapland		*E. zebra*	MB-Z 47489	243			111			
Kapland		*E. zebra*	MU 542	220			110.5			
S Afrika		*E. zebra*	HA 1396	238			101			
Cape Prov		*E. zebra*	AMNH-M 83602	232			117.5		155	171
S Africa		*E. zebra*	AMNH-M 81775	227			109		148	160
?		*E. zebra*	AMNH-M 99700						156	170
?		*E. zebra*	AMNH-M 42753						155	171
Zoo		*E. zebra*	AMNH-M 90240	230			124		156	161
S Africa		*E. zebra*	NMBI 10918						155	166
S Africa		*E. zebra*	NMBI 7444	246			110		158	166
S Africa		*E. zebra*	NMBI 8702	232			114		158	173
S Africa		*E. zebra*	NMBI no n°	240			130		158	170
S Africa		*E. zebra*	NMBI 6026	230			110		148	164
S Africa		*E. zebra*	SAM 33855						169	171.8

(continued)

Table 11.A1 (continued)

Origin or Breed	Locality	Species	Specimen number	Prosthion-Hormion	P2 to Hormion	Vomer length	Vomer-Basion	Muzzle length	N-i notch length	Cheek length
S Africa		*E. zebra*	SAM 38645						163.5	175.5
S Africa		*E. zebra*	SAM 38646						164	166.4
S Africa		*E. zebra*	TMP no n°						167.5	178
Namibia		*E. zebra*	Windhoek CM 285						150	155
Namibia		*E. zebra*	Windhoek CM 361						170	156
Namibia		*E. zebra*	Windhoek C 459						166	173
Namibia		*E. zebra*	Windhoek XV 101						151	167
Namibia		*E. zebra*	Windhoek XV 102						160	165
Namibia		*E. zebra*	Windhoek XV 99–97						155	174
Namibia		*E. zebra*	Windhoek XIV 1082						168	161
Namibia		*E. zebra*	Windhoek XIV 88						167	169
Namibia		*E. zebra*	Windhoek XIV 80						175	161
Namibia		*E. zebra*	Windhoek XIV 86						164	169
Namibia		*E. zebra*	Windhoek XII 68						162	165
Namibia		*E. zebra*	Windhoek XII 60						165	162
Namibia		*E. zebra*	Windhoek XIV 81						161	175
Namibia		*E. zebra*	Windhoek XIV 92						172	180
Namibia		*E. zebra*	Windhoek XIV 83						159	172.5
Namibia		*E. zebra*	Windhoek XIV 78						158	167
Namibia		*E. zebra*	Windhoek XIV 87						175	165
Namibia		*E. zebra*	Windhoek XIV 82						162	169
Namibia		*E. zebra*	Windhoek XIII 74						153	175
Namibia		*E. zebra*	Windhoek XVI 108						160	170
Zoo		*E. przewalskii*	MNHN.Z-AC 1941.322	261			121			
Chaffanjon 1896–3		*E. przewalskii*	MNHN.Z-MO 1964–106	261			113			
Chaffanjon 1896–4		*E. przewalskii*	MNHN.Z-MO 1964–107	256			117			
"Kiang, Tibet"		*E. przewalskii*	MNHN.Z-AC 1986.269	257			120.3			
Zoo		*E. przewalskii*	MNHN.Z-AC 1973.109	250			115			
Gobi, Bedford		*E. przewalskii*	NHMUK-ZD 1907.5.15.1	255			118			
Zoo, Bedford		*E. przewalskii*	NHMUK-ZD 1945.6.11.1	275			119			
Mongolia	Bijsk	*E. przewalskii*	RNH-L 359	248			109			
Hagenbeck 1926		*E. przewalskii*	RNH-L 1534	267			113			
Zoo 1969		*E. przewalskii*	AM 11–913	253			119			
Gooilust 1938		*E. przewalskii*	AM 981	272			115			
?		*E. przewalskii*	MB-Z 60363	248			117			
Münich 1949		*E. przewalskii*	SMF 35389	260			122			
Bessie		*E. przewalskii*	ZSM 1951.173	255			109			
Neville		*E. przewalskii*	ZSM 1953.147	268			130			
Zoo melioïdose		*E. przewalskii*	MNHN.Z-MO 1977.55	270			122			
Bedford, Chubb 209		*E. przewalskii*	AMNH-M 32686	267			118			
Chubb 136		*E. przewalskii*	AMNH-M 16234	264			124			
?		*E. przewalskii*	AMNH-M 21523	255			113			
Zoo		*E. przewalskii*	AMNH-M 80062	251			117			
Sidor		*E. przewalskii*	NMB 10881	262			114			
?		*E. przewalskii*	MCZ 51058	261			112			
Roborovski and Kozlov		*E. przewalskii*	ZIN 5213	262			119			
Roborovski and Kozlov		*E. przewalskii*	ZIN 5214	263			116			
Roborovski and Kozlov		*E. przewalskii*	ZIN 5216	267			113			
Roborovski and Kozlov		*E. przewalskii*	ZIN 5218	265			119			
Zoo 1933		*E. przewalskii*	ZIN 17591	287			122			
Askania Nova		*E. przewalskii*	ZIN 27031	274			117			
Askania 1903–1909		*E. przewalskii*	ZIN 27089	255			119			
Zoo Tallin 1983		*E. przewalskii*	ZIN 31877	270			122			
Zoo 1987		*E. przewalskii*	ZIN 32578	255			116			
Zoo 1983		*E. przewalskii*	ZIN 32050	255			111			
Zoo London 1973		*E. przewalskii*	MS 95921	247			113			
Zoo London 1974		*E. przewalskii*	MS 133806	266			120			

(continued)

Table 11.A1 (continued)

Origin or Breed	Locality	Species	Specimen number	Prosthion-Hormion	P2 to Hormion	Vomer length	Vomer-Basion	Muzzle length	N-i notch length	Cheek length
Zoo London 1975		*E. przewalskii*	MS 1772	245			114			
Zoo London 1976		*E. przewalskii*	MS 110476	250			113			
?		*E. przewalskii*	HL wld 1	263			121			
Mongolia	Bijsk	*E. przewalskii*	HL wld 2	250			124			
?		*E. przewalskii*	HL wld 4	258			122			
Zoo Köln		*E. przewalskii*	BO 92324	254			117			
Vassienievitch 1892, Mongolia		*E. przewalskii*	ZIN 5230	255			112			
Prague 1976		*E. przewalskii*	NMP 47161	258			113			
Artemis		*E. przewalskii*	NMP 46585	262			118			
Uran		*E. przewalskii*	NMP 24688	255			122			
Hera		*E. przewalskii*	NMP 47160	257			112			
?		*E. przewalskii*	NMP 22772	250			120			
?		*E. przewalskii*	NMP 47165	260			118			
?		*E. przewalskii*	NMP 47167	260			115			
Nepal		*E. caballus*	NMUK-ZD 1858.6.24.119	243			113			
Nepal		*E. caballus*	NMUK-ZD 1858.6.24.150	232			118			
Norway		*E. caballus*	MNHN.Z-ACA 541	270			128			
France	Quercy	*E. caballus*	MNHN.Z-AC 1973.14	290			139			
Percheron		*E. caballus*	MNHN. Z-MO1977.84	315			152			
France		*E. caballus*	EV no n°	305			138			
France		*E. caballus*	MNHN.Z-AC 1880.744?	280			128			
France		*E. caballus*	MNHN.Z-AC 1926.301	280			133			
Tonkin		*E. caballus*	MNHN.Z-AC 1911.145	245			118			
Tonkin		*E. caballus*	MNHN.Z-AC 1900.108	243			114			
France		*E. caballus*	MNHN.Z-AC 1937–51	190			98			
Norway		*E. caballus*	MNHN.Z-AC 1925.78	265			124			
Flamand		*E. caballus*	MNHN.Z-AC 1930.28	307			153			
Boulonnais		*E. caballus*	MNHN.Z-AC 1930.32	310			139			
Russia	Irkoutsk	*E. caballus*	MNHN.Z-AC 1902.822	270			126			
France		*E. caballus*	MNHN.Z-AC 1930.31	275			142			
France		*E. caballus*	MNHN.Z-AC 1940.399	288			144			
France		*E. caballus*	MNHN.Z-AC 1930.30	255			131			
Scotland		*E. caballus*	MNHN.Z-AC 2319	230			124			
France		*E. caballus*	MNHN.Z-AC 1926.125	275			138			
France		*E. caballus*	MNHN.Z-AC 1950–8	180			87			
France		*E. caballus*	MNHN.Z-AC 1880.232.1	290			139			
France		*E. caballus*	MNHN.Z-AC 1930.27	275			136			
France		*E. caballus*	MNHN.Z-AC 1964–197	225			107			
Flamand		*E. caballus*	MNHN.Z-AC 1930.29	295			143			
France		*E. caballus*	MNHN.Z-AC 69	296			138			
Tarpan		*E. caballus*	ZIN 521	252			121			
France		*E. caballus*	MNHN.Z-MO 1977.72	270			132			
France		*E. caballus*	MNHN.Z-MO 1977.88	265			122			
Iceland		*E. caballus*	MNHN.Z-AC 1873–385	212			106			
Shetland		*E. caballus*	AMNH-M 204044	215			100			
'Jumper'		*E. caballus*	AMNH-M 204127	285			135			

(continued)

Table 11.A1 (continued)

Origin or Breed	Locality	Species	Specimen number	Prosthion-Hormion	P2 to Hormion	Vomer length	Vomer-Basion	Muzzle length	N-i notch length	Cheek length
Draft		*E. caballus*	AMNH-M 16274	361			166			
Arab?		*E. caballus*	YPM 1636	268			131			
Draft		*E. caballus*	AMNH-M 99	300			142			
Boulonnais		*E. caballus*	MNHN.Z-AC 1891.107	321			153			
Chubb 70		*E. caballus*	AMNH-M Chubb 70	258			125			
Shetland		*E. caballus*	FMNH 46019	185			89			
France		*E. caballus*	MNHN.Z-AC 1937–59	177			87			
Iceland		*E. caballus*	MNHN.Z-AC 1975.98	258			119			
Shetland		*E. caballus*	MNHN.Z-AC 1945–27	205			99			
Iceland		*E. caballus*	MNHN.Z-AC 1880.233	245			115			
Iceland		*E. caballus*	MNHN.Z-AC 1891–44	218			91			
France		*E. caballus*	EV no n°	232			103			
Arab		*E. caballus*	AMNH-M 204200	260			128			
Arab		*E. caballus*	AMNH-M 204191	260			137			
Arab		*E. caballus*	AMNH-M 204184	270			125			
Arab		*E. caballus*	AMNH-M 204176	270			146			
Arab		*E. caballus*	AMNH-M 204210	260			136			
Norway		*E. caballus*	AMNH-M 204174	270			130			
Norway		*E. caballus*	AMNH-M 204135	257			117			
Norway		*E. caballus*	AMNH-M204147	255			131			
Mongolia	Gobi	*E. caballus*	AMNH-M 204192	254			111			
Russia	Iakutsk	*E. caballus*	IA 1955	258			123			
Russia	Iakutsk	*E. caballus*	IA 1957	260			119			
Russia	Iakutsk	*E. caballus*	IA 1959	260			113			
Russia	Iana	*E. caballus*	IA 1944	261			117			
France		*E. caballus*	AA sans n°	319			146			
France		*E. caballus*	AMNH-M 1218	274			146			
Draft		*E. caballus*	AMNH-M 8178 A	315			147			
?		*E. caballus*	MCZ 16890	195			91			
Konik		*E. caballus*	PIN 657	230			105			
Israel		*E. caballus*	HUJ-ESE no n°	258			140			
France		*E. caballus*	MNHN.Z-AC 1971.325	248			115			
France		*E. caballus*	MNHN.Z-AC 1880.232.6	288			140			
France		*E. caballus*	MNHN.Z-AC 1880.232.7	277			143			
France		*E. caballus*	MNHN.Z-AC 1880.747	292			131			
France		*E. caballus*	MNHN.Z-AC 1880–232-5	315			144			
France	Gournay	*E. caballus*	CRA Gournay	290			140			
Welsh		*E. caballus*	MCZ 52978	212			97			
Tarpan		*E. caballus*	MS 94535	257			127			
Pushcha 1		*E. caballus*	MS 96873	255			124			
Pushcha 2		*E. caballus*	MS 96872	270			128			
Mongolia		*E. caballus*	MS 110478	240			119			
?		*E. caballus*	MS 106944	195			86			
?		*E. caballus*	MS 106942	230			105			
Shetland		*E. caballus*	KI 16449	197			88			
Shetland		*E. caballus*	KI 20253	208			91			
Shetland		*E. caballus*	KI 20214	199			86			
Shetland		*E. caballus*	KI 1661	204			102			
Iceland		*E. caballus*	Ki 16719	240			108			
Iceland		*E. caballus*	KI 18146	227			116			
Iceland		*E. caballus*	KI 2217	198			115			
Hannover		*E. caballus*	KI 1760	305			143			
Konik		*E. caballus*	HL CA 2	255			135			
?		*E. caballus*	HL Pol 1	199			86			
Iceland		*E. caballus*	HL 4	230			115			
Welsh		*E. caballus*	HL Epon2	203			108			
Welsh		*E. caballus*	HL mgl 2	234			118			
Iceland		*E. caballus*	HL mgl 1	258			118			

(continued)

Table 11.A1 (continued)

Origin or Breed	Locality	Species	Specimen number	Prosthion-Hormion	P2 to Hormion	Vomer length	Vomer-Basion	Muzzle length	N-i notch length	Cheek length
Iceland		*E. caballus*	HL mgl 3	238			115			
Arab		*E. caballus*	BO 1939	269			133			
Arab		*E. caballus*	BO 79629	272			131			
Arab		*E. caballus*	TMB no n°	253			124			
?		*E. caballus*	ZIN 518	238			113			
Mongolia		*E. caballus*	ZIN 18036	253			117			
Mongolia		*E. caballus*	ZIN 24504	258			125			
Mongolia		*E. caballus*	MS 102019	247			113			
'Przewalski'		*E. caballus*	NMP 14/60	267			123			
'Tarpan'		*E. caballus*	KI 1661	273			130			
Arab		*E. caballus*	Ki 16719	271			130			
Arab		*E. caballus*	KI 18146	260			126			
Kighize		*E. caballus*	KI 2217	265			141			
Russia	Carelia	*E. caballus*	KI 1760	313			149			
?		*E. caballus*	HL CA 2	273			135			
?		*E. caballus*	HL Pol 1	328			143			
?		*E. caballus*	NMP 2485.9884	282			143			
Turkmenistan		*E. caballus*	ZIN 18039	270			129			
Turkmenistan		*E. caballus*	ZIN 18046	273			123			
Turkmenistan		*E. caballus*	ZIN 18047	282			128			
Turkmenistan		*E. caballus*	ZIN 18052	280			129			
Berelekh		*E. caballus*	ZIN 31168	273			130			
Kighize		*E. caballus*	ZIN 18038	272			129			
Mongolie		*E. caballus*	ZIN 18055	255			119			
Turkmenistan		*E. caballus*	ZIN 18040	263			125			
Turkmenistan		*E. caballus*	ZIN 18041	260			129			
Turkmenistan		*E. caballus*	ZIN 18042	261			126.5			
Turkmenistan		*E. caballus*	ZIN 18043	262			118			
Turkmenistan		*E. caballus*	ZIN 18045	265			121			
Turkmenistan		*E. caballus*	ZIN 18048	255			120			
Turkmenistan		*E. caballus*	ZIN 18049	261			129			
Mongolia		*E. caballus*	ZIN 18056	250			113			
Russia	Selenga	*E. caballus*	ZIN 20135a	246			111			
Russia	Arctic	*E. caballus*	ZIN 29697	260			130			
Mongolia		*E. caballus*	ZIN 18057	245			114			
Mongolia		*E. caballus*	ZIN 18058	252			111			
Mongolia		*E. caballus*	ZIN 18059	238			106.5			
Kighiz		*E. caballus*	ZIN 18034	257			114			
Russia	Tobol	*E. caballus*	ZIN 4050	250			124			
Russia	Adycha	*E. caballus*	ZIN 4051	260			115			
Russia	Kolyma	*E. caballus*	ZIN 5233	237			114			
Russia	Kazakstan	*E. caballus*	ZIN 11485	257			127			
Russia	Iakoutsk	*E. caballus*	ZIN 12614	262			119			
Russia	Iakoutsk	*E. caballus*	ZIN 12615	252			112			
Russia	Iakoutsk	*E. caballus*	ZIN 12616	255			128			
Togo		*E. caballus*	NMP 47158	235			110			
Togo		*E. caballus*	NMP 47159	230			115			
?		*E. caballus*	NMP 4.1.1977	256			127			
?		*E. caballus*	NMP 1975	263			121			
Nesvera		*E. caballus*	NMP no n°	230			99			
Zoo		*E. grevyi*	MNHN.Z-AC 1913–58	276			126			
Expo coloniale		*E. grevyi*	MNHN.Z-AC 1931–392	265			134			
Zoo Vincennes		*E. grevyi*	MNHN.Z-AC 1932–123	273			122			
Zoo Vincennes		*E. grevyi*	MNHN.Z-AC 1939–75	278			125			
Zoo		*E. grevyi*	MNHN.Z-MO 1977–66	280			125		183	185
Somali		*E. grevyi*	ZIN 5237	278			125			
Ost Afr		*E. grevyi*	RNH-L 885	265			132			
?		*E. grevyi*	AM 980	288			144			
?		*E. grevyi*	AM 1923	275			137			
?		*E. grevyi*	AM 1963	265			142			
Kenya	Awash	*E. grevyi*	NMB 5463	270			136			
Kenya	Field	*E. grevyi*	Kenya field A	275			129			
Kenya	Field	*E. grevyi*	Kenya field B	278			144			

(continued)

Table 11.A1 (continued)

Origin or Breed	Locality	Species	Specimen number	Prosthion-Hormion	P2 to Hormion	Vomer length	Vomer-Basion	Muzzle length	N-i notch length	Cheek length
Kenya		*E. grevyi*	KNM 3965	255			127			
Kenya		*E. grevyi*	KNM 3967	270			119			
Kenya		*E. grevyi*	KNM 3968	272			123			
Kenya		*E. grevyi*	KNM 2488	268			135.5			
Kenya		*E. grevyi*	KNM 2489	290			131			
Somali		*E. grevyi*	NMUK-ZD 1893.12.1.2	280			127			
Juba		*E. grevyi*	NMUK-ZD 1923.10.20.16	268			134			
Kenya		*E. grevyi*	NMUK-ZD 1962.8.174	270			131			
Zoo		*E. grevyi*	SMLS 17490	272			132.5			
Ethiopia	Lake Stephanie	*E. grevyi*	NME ss n°	280			143			
Zoo		*E. grevyi*	NMB 10876	275			125			
Zoo		*E. grevyi*	NMB 10873	268			123			
Kenya		*E. grevyi*	BE 1923–177	270			129.5			
Ethiopia		*E. grevyi*	ZU 19018	270			127.5			
Zoo		*E. grevyi*	ZU 16656	270			125			
Kenya	Wamba	*E. grevyi*	HA 6879	278			132			
Kenya	Wamba	*E. grevyi*	HA 7196	275			132			
Kenya	Wamba	*E. grevyi*	HA 7197	270			134			
Kenya	Wamba	*E. grevyi*	HA 7198	268			136			
Kenya	Wamba	*E. grevyi*	HA 7201	262			135			
Kenya	Wamba	*E. grevyi*	HA 7202	282			134			
Kenya	Wamba	*E. grevyi*	HA 7203	273			125			
Kenya	Wamba	*E. grevyi*	HA 7204	272			126			
Kenya	Wamba	*E. grevyi*	HA 7205	265			127			
Kenya	Wamba	*E. grevyi*	HA 7206	268			138			
Kenya	Wamba	*E. grevyi*	HA 7207	263			134			
Kenya	Wamba	*E. grevyi*	HA 7209	275			138			
Kenya	Wamba	*E. grevyi*	HA 7210	272			131			
Kenya	Wamba	*E. grevyi*	HA 7211	262			128			
Kenya	Wamba	*E. grevyi*	HA 7212	280			130			
Kenya	Wamba	*E. grevyi*	HA 7213	280			136			
Kenya		*E. grevyi*	HA 6747	275			125			
Zoo		*E. grevyi*	HA 7111	268			123			
Kenya	Arussi	*E. grevyi*	SMF 657	278			146			
Zoo		*E. grevyi*	ZSM 1953–96	278			137			
Zoo		*E. grevyi*	ZMS 1965–113	275			126			
?		*E. grevyi*	AMNH-M 54247	285			137		188	206
?		*E. grevyi*	AMNH-M 82038	273			133		184	190
?		*E. grevyi*	NMBl 10899						192	183
?		*E. grevyi*	ZIN 32049						184	187
?		*E. grevyi*	HA 10.07.95						183	196
?		*E. grevyi*	NMP 6293						180	181
?		*E. grevyi*	NMP 46430						187	202
?		*E. grevyi*	NMP 46431						188	187
?		*E. grevyi*	SAM 39802						188	192
?		*E. grevyi*	FMNH 26987						184	199

[11.1] *A.* = *Allohippus*; *E.* = *Equus*; *P.* = *Plesippus*; *"P.-A."* = *Plesippus* or *Allohippus*

Table 11.A2 Cranial measurements (in mm) of Senèze equids

	FSL 210993	MHNL Sen 5233	NMB Se 336	FSL SEN 05–0081	FSL 210887	NMB Se 551	NMB Se 553	NMB Se 554	NMB Se 803
Sex	M	M	M	F	F	F	?	M	?
Age	adult	adult	old	adult	ca. 5 years	old	ca. 2 years	old	old
1	540	540	535	505	495				
2	307	300	297	282	280		270		
2–5	155	[159]	164	147	[155]		[150]		144
3		127	135	125	126				
4		122	110	101	112				
5	145	141	140	[135]	125	124	134	131	
6	112	110	97	99	95	94		104	
7	103	96	104	94	99	91	104	91.5	92
7bis	81	80	83	78	78	81		74	81
8	184	178	187	173	177	170	[188]	164	178
9	[60]	short	short		long		[70]		short
10	46		50				[45]		
10bis	38		42.5						
11	160	158	166		130	144	144		[164]
12	392	400	400		350				[393]
13	[220]	220	236		206	202	196		210
14		192	229		192	[200]			193
15	112	[100]	108		104	108			
16	[73]	70.5	69	[51]	65	64			
17	71	69	71	65–70	60	58		60	[62]
17bis	[51]	[53]	44				[40.5]	43	
18	[580]	605	600	580		560			
19	15	[11.5]	18	12	12.5	12	16	11	12
20	16.5			13	14				
21	65	[81]	67	58	62	58			63
22	62	[60]	60	54	55	56			57
23	450	440	438	410	415	400	415		
24	[205]	205	232	215	202	200			
25	[128]	121	[124]			105	[116]	112	
26	136	132	128		120	110	136	122	136
27	140	152	144		120	110			148
28	104	110	110		90				[103]
29	87	88	93	75	86.5	84	75	81	79
30	40	38	39	33	39		38	35.5	34
31	228	188	228	198–204		190	200	195	
32	165	166	165	154.5		140	146		140

M: male; F: female. Approximate measurements between brackets. Measurements as in Eisenmann, 1986 and http://www.vera-eisenmann.com: 1: basilar length; 2: overall palatal length; 2–5: palatal length (without the muzzle); 3: distance from palate to hormion; 4: distance from hormion to basion; 5: muzzle length; 6: diastema length; 7: occlusal length of P2-P4; 7bis: occlusal length of M1-M3; 8: length of P2-M3; 9: choanal length; 10: greatest choanal breadth; 10bis: choanal breadth between the pterygoid processes; 11: breadth between the foremost points of the facial crests; 12: distance between sesion and anterior ends of P2; 13: frontal breadth: 14: bizygomatic breadth; 15: greatest cranial breadth; 16: breadth of the supra-occipital crest; 17: muzzle breadth at the posterior border of the I3; 17bis: least muzzle breadth between the interalveolar borders; 18: vertex length; 19: infra-orbital height; 20: height of the external auditory meatus; 21: antero-posterior diameter of the orbit; 22: dorso-ventral diameter of the orbit; 23: anterior ocular line; 24: posterior ocular line; 25: facial height in front of P2; 26: facial height between P4 and M1; 27: facial height behind M3; 28: cranial height behind the orbits; 29: breadth of the occipital condyles; 30: breadth of the foramen magnum; 31: length of the naso-incisival notch (from prosthion to the back of the narial opening); 32: cheek length (from the back of the narial opening to the most anterior point of the orbit)

Table 11.A3 (a) Measurements (in mm) of third metacarpals of the reference *E. hemionus onager* and various *Allohippus* and a Shetland Pony. (b) Measurements (in mm) of third metacarpals of various equids (c) Measurements (in mm) of third metacarpals of *Allohippus senezensis senezensis* (d) Measurements (in mm) of some *?Allohippus* third metacarpals (e) Measurements (in mm) of *?Allohippus* third metacarpals from Sainzelles

	E. hemionus onager	Valdarno	*A. stenonis vireti*	*A. senezensis senezensis*	*A. senezensis mygdoniensis*	Shetland Pony
(a) Measurements (in mm) of third metacarpals of the reference *E. hemionus onager* and various *Allohippus* and a Shetland Pony.						
	n=29	n=6-11	n=36-50	n=12-18	n=10-12	FMNH 46019
Greatest length	210.2	241	229.6	220.4	233	187
Minimal breadth	26.5	37.2	37.4	33.5	31.9	30
Depth at same level	21.3	28.1	28.0	26.7	26.1	22
Proximal articular breadth	42.5	54.8	54.8	50.8	46.8	41
Proximal articular depth	26.8	36.1	35.2	33	30.4	25
Distal supra-articular breadth	38.8	50.3	51.0	45.7	43.5	41
Distal articular breadth	38.5	37.6	37.2	34.9	32.8	43
Depth of sagittal crest	29.6	29.7	29.4	27.4	26.6	29
Smallest depth of medial condyle	24.1	31.9	31	29.3	28.7	22.5
Greatest depth of medial condyle	25.8	50.3	51	45.7	43.5	25
(b) Measurements (in mm) of third metacarpals of various equids						
	Equus (S.) suesssenbornensis Süssenborn	*E. (S.) cf suesssenbornensis* Akhalkalaki	*E. (S.) verae* NE Siberia	*Allohippus stenonis stenonis*	*A. stenonis guthi*	*A. stehlini*
	n=3-5	n=20-27	n=10	n=15-31	n=31-56	n=9-15
Greatest length	273.2	268.8	265.2	242.4	233.2	208.5
Minimal breadth	41.3	42.7	43.4	37.2	35.2	31.7
Depth at same level	31.3	31.9	31.6	27.7	26.9	23.5
Proximal articular breadth	61.5	63.8	64.1	54.9	52.7	45.2
Proximal articular depth	38.3	39.6	40.2	35.8	33.6	30.5
Distal supra-articular breadth	58.5	60.5	61.6	51.5	48.2	42.3
Distal articular breadth	58.5	60	61.2	50.1	48.1	42.7
Depth of sagittal crest	41.8	42.6	43.2	37.2	35.1	32.2
Smallest depth of medial condyle	34.6	343	34.7	29.4	27.9	26.0
Greatest depth of medial condyle	36.2	367	37.5	31.4	29.7	27.9

(continued)

Table 11.A3 (continued)

	Würzburg Schalksberg	Gannat	Senèze	Oasele	Overstrand	Liventsovka			
	78-500	no nb	FSL 211079	IS 5400	NHMUK -P 19242	RGU 372	RGU 1341	RGU 326	ROMK L 39
Greatest length	285	301		283		265	270	300	259
Minimal breadth	44	47	42	42		38	42.5	42	35
Depth at same level	34	36	31	31.5		30	32.5	33.8	29
Proximal articular breadth	66.4	68		61.5		58.5	62	62	49.5
Proximal articular depth	40.9	43		39.5		39	38	38	32.7
Distal supra-articular breadth	65	64	57.0	60	67	54	57	60	50
Distal articular breadth	63	62	55.5	56	67.5	55		59.5	48.5
Depth of sagittal crest	44.5	45.5		41	49	39		42.5	35
Smallest depth of medial condyle	36	38.0	30		39	32		35	28.5
Greatest depth of medial condyle	40	42	35	35.5	42	34		38.5	30.1

(c) Measurements (in mm) of third metacarpals of *Allohippus senezensis senezensis*

	FSL 210884	FSL 210885	FSL 210886	FSL 210887	FSL 210888	FSL 210889j	FSL 210890	FSL 210891
Greatest length	222	220	211.5	212.5	224	216	221	224.5
Lateral length	215	213	205	204.5		209	213	217
Minimal breadth	33	36	32.5	30.5	33	28.5	35.5	34
Depth at same level	27	28	26	23	27.5		27	27.5
Proximal articular breadth	49	52	50	46.5			51	49.5
Proximal articular depth	31.5	32	33	31.5			34.5	34.5
Facet for Carpale 3	40.5	40	39	39			41	41
Anterior facet for Carpale 4	16.5	16.5	15	13.5			15.5	15
Posterior facet for Carpale 4	10	9	9	9			8	9.5
Facet for Carpale 2	0	0	[6]				0	0
Distal supra-articular breadth	46.5	51	44	42.5		41	47.5	45.5
Distal articular breadth	45.5	46	44.5	42.5	46	40	47	44.5
Depth of sagittal crest	34	36	34	32.5	37		35.5	33.5
Smallest depth of medial condyle	27.5	28	26.5	25	29		29	26.5
Greatest depth of medial condyle	29	30	28	27	31		30	28.5

(continued)

Table 11.A3 (continued)

	FSL 210993	FSL 211053	FSL 211079	FSL SEN 04-0132	FSL SEN 04-0062j	FSL SEN 06-0137	NMB Se 141j	NMB Se 180
Greatest length	[217]	223				225	223	[214]
Lateral length	210	216				220	217	
Minimal breadth	33	35	41.5	33.5	26	34	32.5	33
Depth at same level	25	28.5	31.2	27.4	17	28	25	
Proximal articular breadth	49	51			44.5	52	48	
Proximal articular depth	28	33.5			30	33	31.5	
Facet for Magnum (Carpale 3)	[40]	41			36	42	39	42
Anterior facet for Unciform (Carpale 4)	14.5	16			13.2	17.1	17	
Posterior facet Unciform (Carpale 4)	7	10.5			7	8.1		
Facet for Trapezoid (Carpale 2)		4				0	0	
Distal supra-articular breadth	45	47.5	57	49		48.5	46	
Distal articular breadth	46.5		55.5	46.9		47.5	45	
Depth of sagittal crest	34		40	36.8		35	33	
Smallest depth of medial condyle			31	27.9		28	26.5	
Greatest depth of medial condyle	27.5		35.2	30.9		30	29	
Greatest depth of lateral condyle						26.1		

	NMB Se 551	NMB Se 552j	NMB Se 553	NMB Se 554	NMB Se 829	NMB Se 830	NMB Se 831
Greatest length	216	216	223	218		234	215
Minimal breadth	33	31	32	35	33	35.1	33
Depth at same level	28	25	26	27	25	26.5	26
Proximal articular breadth	55	46.5	51	53		52.5	
Proximal articular depth		32	33	33		35	33
Facet for Magnum (Carpale 3)		38	41	40.5		43	39
Anterior facet for Unciform (Carpale 4)		15.5	16	17		17	
Posterior facet Unciform (Carpale 4)		7.5	7	11		7	9.5
Facet for Trapezoid (Carpale 2)		3	3.5	0		0	
Distal supra-articular breadth	47	44.5	48.5	48	46	48.5	45
Distal articular breadth	44	45	45	48	44.5	49	45
Depth of sagittal crest		33	34	36	35	38	33
Smallest depth of medial condyle		28	27.5	28	27	29	25.5
Greatest depth of medial condyle	28	29	29	31.5	28	31	28.3

(continued)

Table 11.A3 (continued)

(d) Measurements (in mm) of some ?*Allohippus* third metacarpals

	Montopoli IGF 11282	Sarikol Tepe MTA-M n=1-3	Loubières de Pardines MNHN-F LP 128	Vatera NHCV PO 121	Huelago MNCS-CSI 16079	Tegelen Pg 1958
Greatest length		253	254.5	253	268	281
Minimal breadth		39	40.5	36.1	35	40.3
Depth at same level		30	30	28.5	28	31.9
Proximal articular breadth		58		55.9	54	59
Proximal articular depth		37.5		35.2		40
Distal supra-articular breadth	57.7	54.5	58	52	49	56
Distal articular breadth	57.5	55	58	50.7	48	56
Depth of sagittal crest	41.5	37.9	41.5	38.7		42
Smallest depth of medial condyle	31.1	29.2	32.5	38.7		35
Greatest depth of medial condyle	34.4	31.6	35			37.1

	Morskaya MGRI 301-34	Khapry MGRI 300-27	Ceyssaguet *Allohippus* sp. n=41-47	Ceyssaguet *Allohippus* sp. minimum	Ceyssaguet *Allohippus* sp. maximum
Greatest length	278		257.9	246	270
Minimal breadth	40		39.4	36	43.4
Depth at same level	30		31.4	29	33.7
Proximal articular breadth	61		58.1	53.5	62
Proximal articular depth	39		39.1	37	43
Distal supra-articular breadth	55	55	54.7	51.3	57.3
Distal articular breadth	55	55	54.2	50.6	58
Depth of sagittal crest	40		40.5	38.2	43
Smallest depth of medial condyle	33	33.5	[33.1]	[31]	[35]
Greatest depth of medial condyle	35	36.5	34.9	32	38

(e) Measurements (in mm) of ?*Allohippus* third metacarpals from Sainzelles

	Sainzelles ?*Allohippus* sp. n=4-6	Sainzelles ?*Allohippus* sp. minimum	Sainzelles ?*Allohippus* sp. maximum
Greatest length	246.9	241	255
Minimal breadth	35.5	34.5	37
Depth at same level	27.9	27	30.1
Proximal articular breadth	51	50	53
Proximal articular depth	34	33	35
Distal supra-articular breadth	48.2	46.5	50
Distal articular breadth	48.3	47	50
Depth of sagittal crest	35.1	33	36
Smallest depth of medial condyle	28.2	27	29
Greatest depth of medial condyle	30.1	28.5	31.1

j = juvenile

Table 11.A4 Means of measurements (in mm) of limb bones lengths and of maximal plantar breadths of anterior Ph3 of (a) various *Allohippus*. (b) equids from Senèze

	E. hemionus onager	Senèze maximum	Senèze minimum	Saint-Vallier maximum	Saint-Vallier minimum	Gerakarou	Matassino		
(a) various *Allohippus*									
	n=8-10								
Humerus	241.3	290	277	305	299	268			
Femur	329.7	388	352	410	405				
Radius	293.5	344	308	355	328.5	317	344.5		
Tbia	313.0	360	333	385	359		362.5		
MC	214.1	234	211.5	242	215	232.7	241.6		
MT	250.8	268	240	278	251	265.9	277.5		
Ph1 A	76.3	80	71	91	78	78	89.5		
Ph1 P	71.2	80.5	68	86.5	80	74.5	89.5		
Ph3 A breadth	54	67.3	63	83.0	72		70		
	Humerus	Femur	Radius	Tbia	MC	MT	Ph1 A	Ph1 P	Ph3 A breadth
(b) equids from Senèze									
* FSL 211095, 211077, 211074, 211075, 211078			427			335	110	105	ca. 100
** FSL 211055, 211082, 210899							91	85.5	ca. 70
FSL SEN 05-0081+	282	378	315	333		250			
FSL SEN 06-0137+			330	360	225	263	76.9	76	
NMB Se 551	280	380	ca. 316	ca. 340	ca. 216	249	ca. 71	ca. 68	67.3
NMB Se 553	286	388	326	349.0	223	253	76	73	64
NMB Se 554		388		343.0	218	247	80.2	76.5	66
FSL 210993	ca. 277		ca. 325		ca. 217				
NMB Se 141 juv	ca. 258.6	351	319	336.0	223	256	77.5	77	ca. 50
NMB Se 552 juv	260.5	363	316		216	253	78	76	ca. 60
** FSL 210887, 210868, MNHL Sen 1669					212.5	242		69	

* The measurements in this row were taken from separate unassociated bones (Radius FSL 211095, MT FSL 211077, etc.). ** The measurements in these rows were taken from separate but probably associated bones. juv: juvenile

Table 11.A5 Taxonomy and partial distribution (for fossil taxa) of Plio-Pleistocene equine species

GENUS	Subgenus/Species group	SPECIES	SUBSPECIES	Localities	Countries
Plesippus Matthew, 1924		*simplicidens* Cope, 1892	*simplicidens*		
			shoshonensis Gidley, 1930		
Allohippus Kretzoi, 1938	*stenonis*	*stenonis* Cocchi, 1867	*stenonis*	Valdarno, Olivola, Matassino	Italy
			vireti Prat, 1964	Saint-Vallier, Ceyssaguet	France
			athanasiui Samson, 1975	Tetoiu	Romania
		livenzovensis Bajgusheva, 1978		Liventsovka	Russia
		sp. Kuruksai (GIN-KU no number)		Kuruksai	Tajikistan
		sanmeniensis Teilhard de Chardin and Piveteau, 1930		Nihowan	China
		sp. SE Shansi (AMNH 60-B 719)		Fan Tsun, Taigu	China
		koobiforensis Eisenmann, 1983		East Turkana	Kenya
	senezensis	*senezensis* Prat, 1964	*senezensis*	Senèze	France
			guthi Bœuf,1986	Chilhac	France
				La Puebla de Valverde	Spain
			mygdoniensis Koufos, 1992	Gerakarou	Greece
		stehlini Azzaroli,1965		Valdarno	Italy
		bactrianus Zhegallo, 1998*		Kuruksai	Tajikistan
		sp. China, locs A & D, M 1357, M 1418		Honan	China
Equus Linnaeus, 1758	(*Sussemionus*) Eisenmann, 2010	*suessenbornensis* Wüst, 1901	*suessenbornensis*	Süssenborn	Germany
			cf. *suessenbornensis*: Vekua, 1962	Akhalkalaki	Georgia
			verae Sher, 1971	NE Siberia	Russia
		hipparionoides Vekua, 1962		Akhalkalaki	Georgia
		coliemensis Lazarev, 1980		Kolyma	Russia
		granatensis Marin, 1987		Venta Micena	Spain
		wuesti Musil, 2001		Untermassfeld	Germany
		ovodovi Eisenmann and Vasiliev, 2011		Proskuriakova Cave	Russia
	(*Dolichohippus*) Heller, 1912	*grevyi* Oustalet, 1882			
	(*Quagga*) Shortridge, 1934	*quagga* Boddaert, 1785			
		burchelli Gray, 1824	numerous subspecies		
		mauritanicus Pomel, 1897			
		capensis Broom, 1909			
	(*Hippotigris*) Smith, 1841	*zebra* Linnaeus, 1758			
	(*Asinus*) Gray, 1824	*africanus* Heuglin and Fitzinger, 1866	and other species		
	(*Hemionus*) F. Cuvier, 1823	*hemionus* Pallas, 1775	numerous subspecies		
	(*Equus*) Linnaeus, 1758	*caballus* Linnaeus, 1758	and other species		

* in Vangengeim et al. 1998

Table 11.A6 (a) Measurements (in mm) of third metatarsals of various equids. (b) Measurements (in mm) of third metatarsals of various *Allohippus* and Equus (*Sussemionus*). (c) Measurements (in mm) of some *Allohippus major* third metatarsals (d) Measurements (in mm) of third metatarsals of Senèze equids

	Equus hemionus onager	Lunel Viel MNP-Bonifay	?*Allohippus* Pyrgos	?*Allohippus* Pyrgos	?*Allohippus* Pyrgos	?*Allohippus* Pyrgos	*A. senezensis senezensis*
(a) Measurements (in mm) of third metatarsals of various equids							
	n = 32	IV-6 10009	Pg II 55	Pg II 41	Pg I 15	Pg I 23	n = 9–15
Greatest length	246.9	280	260	251	255		252.7
Minimal breadth	25.6	40	30	28.8	31	28	33.7
Depth at same level	25.4	36	29	30.3	29.5	[26]	31
Proximal articular breadth	39.9	52	44	44	44.8		49
Proximal depth	34.6	43	38	35.6			42.2
Distal supra-articular breadth	38.4	52.5	41.1	42	41.5	41.7	47.4
Distal articular breadth	37.6	53	42	42.7	42.8	42	46.5
Depth of sagittal crest	30.2	39	33	32.2	33.2	32	35.6
Smallest depth of medial condyle	23.7	29	26	24.8	25.9	25	27
Greatest depth of medial condyle	26.1	32	28	26.6	27.5	27.7	29.6
	A. stenonis stenonis	*A. stenonis vireti*	*A. senezensis guthi*	*A. senezensis mygdoniensis*	*E. (S.) suesssen bornensis*	*E. (S.)* cf *suesssen bornensis*	*E. (S.) verae*
(b) Measurements (in mm) of third metatarsals of various *Allohippus* and *Equus (Sussemionus)*							
	n = 8–11	n = 64–80	n = 42–82	n = 7	n = 1–4	n = 13–22	n = 15–23
Greatest length	270.6	264.4	269.3	265.2	327	317.9	310.8
Minimal breadth	36.2	36.9	35	30.9	40.5	40.7	41.1
Depth at same level	329	33.8	32.1	28.7	38.3	38.5	36.5
Proximal articular breadth	52.2	53.5	51.4	46.1	57.5	59.9	58.8
Proximal depth	43.4	45.4	42.1	35.5	48	49.1	49
Distal supra-articular breadth	51 6	52.7	49.5	43.8	58.5	59.2	60.5
Distal articular breadth	49.9	51.1	47.7	42.9	60.5	59.2	59.5
Depth of sagittal crest	37.8	38.2	36.2	34.9	44.5	43.5	44.1
Smallest depth of medial condyle	29	28.6	27.1	26	34	33.3	34.1
Greatest depth of medial condyle	32.3	31.5	30.1	29.1	37.5	36.6	37.2

(continued)

Table 11.A6 (continued)

	A. major Würtsburg-Schalksberg no number	*A. major* Gannat no number	*A. major* Senèze FSL 211077	*A. major* Kislang no number	*A. major* Chagny no number	*A. major* Vatera NHCV PO 5	*A. major* Tataourova GIN 920	*A. major* Liventsovka ROMK L-778
(c) Measurements (in mm) of some *Allohippus major* third metatarsals								
Greatest length	342.7	345	335	330	321	315	334	329
Minimal breadth	44	46	46	45	42	42	43	41.5
Depth at same level	40.5	40	39.5	38	38	39	40	37.6
Proximal articular breadth	63	62	64.5	58	64	66	63.5	61
Proximal articular depth	52	51	50.5	46	49		51	51.6
Distal supra-articular breadth	64	63.5	64.5	60	58.5	57.7	59.5	60.7
Distal articular breadth	62	61	60	58	58	58	58	57.5
Depth of sagittal crest	45	47	46.5	44.5		44.1	43	45
Smallest depth of medial condyle	36	37.5	38	35	33.2	34	34	35
Greatest depth of medial condyle	39	43	41.5	39	38	37	38	39
	FSL SEN 02-0006 parcelle 164	FSL SEN 04–0132	FSL SEN 06–0270	FSL SEN 06–0249	FSL SEN 05–0127	FSL 210868	FSL 210868b	
(d) Measurements (in mm) of third metatarsals of Senèze equids								
Greatest length			263		244	257	242	
Lateral length			253		239	252	237.5	
Minimal breadth		32	33	33	31	33.5	29.5	
Depth at same level		[29]	31	32.5	30	32	28.5	
Proximal articular breadth			49.5	49	46	49	44.5	
Proximal depth			42	42.5	42	40	38.5	
Facet for Large Cuneiform (Tarsale 3)			45.5	45	44.7	42.5	41.5	
Facet for Cuboid (Tarsale 4)			[14]	10.5	10.7	12	11	
Facet for Small Cuneiform (Tarsale 2)			7.1	6	7.8	8	6.5	
Distal supra-articular breadth	61	48	48.1	50	45	46.5	43	
Distal articular breadth	[58]	47	48	48	44	46	42.8	
Depth of sagittal crest	[43]	37	35.1	36	35.1	34.5	33	
Smallest depth of medial condyle	33	28	27	26	26.2	26	24.5	
Greatest depth of medial condyle	35.7	31	30.5	30	30.1		28	
Greatest depth of lateral condyle	32	27	25.5		25.5			

(continued)

Table 11.A6 (continued)

	FSL 210882	FSL 210883	FSL 211077	MNHN.F (Sen) 69	NMB Se 141j	NMB Se 185j	NMB Se 378
Greatest length	247.5	258.5	335	240	256	241	245
Lateral length	244	252		237	251	237	240
Minimal breadth	32.5	32.5	46	33.2	30	30	33.1
Depth at same level	31.5	29.5	39.5	30.3	29	28	30
Proximal articular breadth	47.5	47	64.5	47.5	48.5	45	49
Proximal depth		42.5	> 50	40.2	40	40	42
Facet for Large Cuneiform (Tarsale 3)	42.5	43	58	43	44	42.5	44
Facet for Cuboid (Tarsale 4)	10	9	16	10	11	11	12
Facet for Small Cuneiform (Tarsale 2)	7	7.5		6	7	6.5	6.5
Distal supra-articular breadth	44	45	64.5	47	48	46	48
Distal articular breadth	44.5		60	45	46	44	46.2
Depth of sagittal crest	32.5	34.5	46.5	35	34	32	35
Smallest depth of medial condyle	26	27	38	27	25.5	24	26.5
Greatest depth of medial condyle	28	[31]	41.5	30.5	29	28	30

	NMB Se 551	NMB Se 552j	NMB Se 553j	NMB Se 554	NMB Se 811j?	NMB Se 819	NMB Se 820
Greatest length		253	253	247	242	252	268
Lateral length	245	245.5	248	242			261
Minimal breadth	32	29.5	31.5	35.3	32	35	35
Depth at same level	29	29.9	31	31	30	[30]	34.5
Proximal articular breadth	46.5	47	48.5	50.5			53
Proximal depth		40.8	40.5	42.5			45
Facet for Large Cuneiform (Tarsale 3)		43	44	44			49
Facet for Cuboid (Tarsale 4)		9.2	10	12			12
Facet for Small Cuneiform (Tarsale 2)		7.5	7.5	8			7
Distal supra-articular breadth	46	45.5	49	49	[45]		51
Distal articular breadth	45	45.5	45.5	48.6	44.2		50
Depth of sagittal crest		34.5	35	37.5	34		38.5
Smallest depth of medial condyle		26	27	27	26		28
Greatest depth of medial condyle	28.5	29.5	30.5	31.8	29		32

(continued)

Table 11.A6 (continued)

	NMB Se 821	NMB Se 822j?	NMB Se 1586	NMB Se 1728	NMB Se 1769
Greatest length	256			258	253
Lateral length	250			254	
Minimal breadth	38	32.5	33	35.5	34
Depth at same level	32.5	[32]	32	34	31
Proximal articular breadth	49	48.5		51.5	
Proximal depth	40.5	43.5		43	
Facet for Large Cuneiform (Tarsale 3)		43.5		44	
Facet for Cuboid (Tarsale 4)		11		9.5	
Facet for Small Cuneiform (Tarsale 2)	6	6		7.5	
Distal supra-articular breadth	48			50	[45]
Distal articular breadth	48			48	[44]
Depth of sagittal crest	35			36.5	
Smallest depth of medial condyle	26.5			27	27
Greatest depth of medial condyle	29.5			30	31

j = juvenile

Table 11.A7 (a) Upper cheek teeth measurements (in mm) of Senèze equids. (b) Isolated upper cheek teeth measurements (in mm) of Senèze equids

		MNHL Sen 5233	FSL 210887	FSL 210993	FSL SEN 05–0081	NMB Se 336	NMB Se 553	NMB Se 554	NMB Se 803
(a) Upper cheek teeth measurements (in mm) of Senèze equids									
Age		adult	ca. 5 years	adult	adult	old	ca. 2 years	old	old
P2	L	39	39	41.5	35.2	43			39
	Lp	7.5	8		6.5	7			
	W	25	26.5	28	25.2	28			28
P3	L	30	31	31	25.7	31		27	27
	Lp	9	11.5	8	7.7	7		7.5	
	W	27.5	28.5	30.7	27.4	29		27	28
P4	L	29.5	29	30	26.3	30		26	29
	Lp	11	10.5	9	9.1	7		7.5	
	W	27.5	26.5	29.5	27.1	29.5		27.5	30
M1	L	25.5	25.5	26	22.8	25	29	22	22
	Lp	9.5	9	7.8	8	10	10	9	
	W	25.5	27	27.8	26.2	26	26	25	27
M2	L	26	27	26	24.9	26	30	23.5	25
	Lp	11	10.1	10	9.4	10	13	10	10.2
	W	26	25.5	27.5	25.2	26	25.5	25	27
M3	L	27.5	24.5	28	26.7	30			34
	Lp	10	10	10	8	10.5			12
	W	22.5	20	24	22.4	23			25

(continued)

Table 11.A7 (continued)

		NMB Se 804	NMB Se 806	NMB Se 890	NMB Se 1855	NMB Se 1858
Age		adult	old	adult	adult	adult
P2	L	38	35		40	
	Lp	7	6.5		7.5	
	W	26.5	27		26	
	H		23			
P3	L	29	26.5		32	30
	Lp	8.5	7.5		9.5	8
	W	27.5	26.5		28	31
	H		28			49
P4	L	27	26		31	30.5
	Lp	10	8		9	7
	W	27	25		28	30
	H		29		ca. 60	50
M1	L	24	22.5	25.5	28	28
	Lp	10	9	8	9.5	10
	W	26.5	24.5	25	25.5	28
	H		27	57		44
M2	L	24	24	26	29	26.5
	Lp	10	9	8.5	10	8.5
	W	25	24	25	25.5	28
	H		36	55		40
M3	L	28	25	24	27	30.5
	Lp	10	8.5	9	12	12
	W	22	20	20	20	23
	H		27	44		36

(b) Isolated upper cheek teeth measurements (in mm) of Senèze equids

		L	Lp	W	H
P2	MNHN.F (Sen 41)	40.2	7	26	32
P2	MNHN.F-AC 1938–373	37	7	23.5	25
P3	NMB Se 338	34.1	10.7	32	27
P ?	MNHN.F-AC 1938–373	27	9	26	35
P ? young	MNHN.F-AC 1938–373	31.5	10	29	
P	MNHN.F (Sen 37)	30	9	29	32
P	MNHN.F (Sen 40)	29	9	[27]	34
P	MNHN.F (Sen 45)	28	8	25	29
P	MNHN.F (Sen 46)	29	[9]	[27]	25
P	MNHN.F (Sen 51)	31.3*	10.2*	29.5*	63
P	MNHN.F (Sen 70)	29*	8*	28*	77
M	MNHN.F (Sen 33)	27*	11*	26.5*	67
M	MNHN.F (Sen 34)	26*	9.2*	27.5*	62
M	MNHN.F (Sen 35)	28*	11*	28*	69
M	MNHN.F (Sen 36)	25	9.5	25.2	36
M	MNHN.F (Sen 39)	26	8	25.2	40
M	MNHN.F (Sen 49à	26*	10*	25*	
M ?	MNHN.F-AC 1938–373	24	10	27	18
M ? young	MNHN.F-AC 1938–373	29	10	25	
M ? young	MNHN.F-AC 1938–373	28	9	26	68
M3	MNHN.F (Sen 31)	27	10.1	21	63
M3	MNHN.F (Sen 38)	27	10	23	10

L: occlusal length; Lp: protocone length; W: occlusal width without cement; H: crown height

* measurements taken at mid-crown; approximate measurements between brackets

Table 11.A8 (a) *Allohippus vireti*: statistics of limb bone breadths (b) *Allohippus senezensis senezensis*: statistics of limb bone breadths (c) *Allohippus major*: statistics of limb bone breadths (d) *Allohippus* sp. of Ceyssaguet: statistics of limb bone breadths

	n	x	s	v
(a) *Allohippus vireti*: statistics of limb bone breadths				
Scapula, minimal breadth at the neck	7	64.2	5.99	9.33
Scapula, maximal breadth of articular process	5	100.7	4.02	3.99
Humerus, minimal breadth	27	39.2	1.44	3.66
Humerus, distal articular breadth	22	82.3	2.23	2.71
Femur, minimal breadth	19	43.9	2.56	5.84
Radius, proximal articular breadth	20	81.0	3.95	4.87
Radius, distal articular breadth	29	68.8	2.61	3.79
Third metacarpal, proximal articular breadth	44	54.8	1.82	3.33
Third metacarpal, distal articular breadth	42	51.0	1.56	3.06
Tibia, minimal breadth	20	48.2	1.44	2.99
Tibia, distal breadth	21	80.5	3.00	3.73
Talus, distal articular breadth	102	54.9	1.77	3.21
Calcaneum, proximal maximal breadth	20	36.7	1.91	5.20
Calcaneum, distal maximal breadth	25	56.2	1.88	3.34
Third metatarsal, proximal articular breadth	70	53.5	1.92	3.59
Third metatarsal, distal articular breadth	68	51.1	1.48	2.89
First phalanges, minimal breadth	74	35.8	1.24	3.46
Second phalanges, minimal breadth	54	47.4	2.21	4.67
Third phalanges, articular breadth	27	49.5	2.50	5.05
	n	*x*	*s*	*v*
(b) *Allohippus senezensis senezensis*: statistics of limb bone breadths				
Scapula, minimal breadth at the neck	16	55	7.18	13.06
Scapula, maximal breadth of articular process	15	89.2	4.11	4.61
Humerus, distal articular breadth	13	76.2	6.93	9.09
Femur, minimal breadth	7	38.6	2.07	5.38
Radius, proximal articular breadth	15	76.5	8.28	10.81
Radius, distal articular breadth	15	64.9	8.11	12.50
Third metacarpal, proximal articular breadth	15	50.3	2.66	5.29
Third metacarpal, distal articular breadth	18	46.3	2.79	6.03
Tibia, minimal breadth	15	44.5	3.02	6.79
Tibia, distal breadth	16	73	5.10	9.98
Talus, distal articular breadth	11	52.3	3.95	7.54
Calcaneum, proximal maximal breadth	8	34.1	3.13	9.20
Calcaneum, distal maximal breadth	8	51.9	1.90	3.65
Third metatarsal, proximal articular breadth	17	49.2	4.51	9.16
Third metatarsal, distal articular breadth	19	47.2	4.56	9.66
First phalanges, minimal breadth	27	33.2	3.70	11.16
Second phalanges, minimal breadth	29	42.6	5.81	13.64
Third phalanges, articular breadth	18	43.6	4.74	10.87
	n	*x*	*s*	*v*
(c) *Allohippus major*: statistics of limb bone breadths				
Humerus, distal articular breadth	1	49.00		
Humerus, distal maximum breadth	1	98.00		
Radius, proximal articular breadth	2	96.00	4.24	4.42
Radius, distal articular breadth	2	83.75	3.89	4.64
Third metacarpal, proximal articular breadth	2	62.20	5.94	9.55
Third metacarpal, distal supra-articular breadth	4	61.68	3.87	6.27
Third metacarpal, distal articular breadth	4	59.13	3.92	6.64
Tibia, minimal breadth	2	50.75	1.06	2.09
Tibia, distal breadth	2	81.55	2.19	2.69
Talus, distal articular breadth	2	62.50	1.41	2.26
Third metatarsal, proximal articular breadth	5	62.30	2.59	4.15
Third metatarsal, distal supra-articular breadth	5	62.10	2.68	4.31

(continued)

Table 11.A8 (continued)

	n	x	s	v
Third metatarsal, distal articular breadth	5	58.80	1.79	2.99
First phalanges, minimal breadth	3	45.17	2.08	4.61
Second phalanges, minimal breadth	3	57.17	3.69	6.45
Third phalanges, articular breadth	2	54.00	0.00	0.00
	n	*x*	*s*	*v*
(d) *Allohippus* sp. of Ceyssaguet: statistics of limb bone breadths				
Radius, proximal articular breadth	1	87		
Radius, distal articular breadth	2	77.5	2.12	2.74
Third metacarpal, proximal articular breadth	66	58.29	2.43	4.17
Third metacarpal, distal supra-articular breadth	69	54.89	2.16	3.93
Tibia, minimal breadth	2	55	1.41	2.57
Tibia, distal breadth	2	92.5	3.54	3.82
Talus, distal articular breadth	2	61.25	1.77	2.89
Third metatarsal, proximal articular breadth	5	59	2.87	4.87
Third metatarsal, distal articular breadth	77	55.8	2.75	4.93
First phalanges, minimal breadth	45	38.54	2.25	5.84

n: number of measurements, *x:* mean (in mm), *s:* standard deviation, *v:* coefficient of variation

Table 11.A9 (a) Cranial measurements (in mm) of *Allohippus senezensis senezensis* and *A. stehlini*. (b) Cranial measurements (in mm) of *Allohippus stenonis* species group (c) Cranial measurements (in mm) of *Allohippus senezensis* species group and the reference *Equus hemionus onager*

	FSL 210887	FSL 210993	MNHL Sen 5233	FSL SEN 05–0081	NMB Se 336	NMB Se 551	NMB Se 553	NMB Se 554
(a) Cranial measurements (in mm) of *Allohippus senezensis senezensis* and *A. stehlini*								
16	65		70.5	[51]	69	64		
23	415	450	440	410	438	400	415	
3	126			125	135			
4	112			101	110			
2–5	155	155	164	147	164		[150]	
5	125	145	141	135	140	124	134	131
17	60	71	69	65	71	58		60
17bis		[51]	[53]		44		[40.5]	43
13	206	220	220		236	202	196	
10		46			50			
25		[128]	121		[124]	105	[116]	112
28		104	110		110			
9		60			64			
20	14	16.5		13	16.5			
31		228	188	204	228	190	200	195
32		165	166	155	165	140	146	

	NMB Se 803	NMB Se 1767	*A. stehlini* n = 1–4
16			
23			382.8
3			119.7
4			
2–5	144		125.3
5			114.4
17	[62]	63	64.3
17bis		44	47.5
13	210		195
10			44.8
25			
28	[103]		
9			76.5
20			
31			167
32	140		145

(continued)

Table 11.A9 (continued)

	A. stenonis stenonis	*A. stenonis vireti*	*A. stenonis livenzovensis*	*A. stenonis* sp.	*A. sanmeniensis*	*A. stenonis* sp.	*A. stenonis* sp.	*A. koobiforensis*
(b) Cranial measurements (in mm) of *Allohippus stenonis* species group								
	Valdarno n = 1–3	Saint-Vallier n = 1–7	Liventsovka ROMK L 4, L 11	Kuruksai PIN 3120	Nihowan MNHN-F NIH 002	Fan Tsun AMNH-P 60-B 719	Ceyssaguet MNP-Bonifay 9142	East Turkana KNM- ER 1484
16			70					
23	435	449	452	490	480	425	489	472
3	152	135.3	144		152		152.3	
4		128	110		126		134.3	
2–5	152.3	151.2	152		160	145	157.7	
5	161.5	146.8	146	170	155	151	175.2	150
17	74.3	67.7	65.5		70	[60]	81	[64]
17bis	45.9	51.4	[39]		53	[40]	50.4	[46]
13	200	227	222			215		225
10		47.8	43			43		53
25		112			130			
28		103.3	112					
9	94	70	70		80	72		
20			16					
31	214	230	186.5	250	225	210		195
32	171	184.5		180	193	170		162
	E. hemionus onager	*A. senezensis senezensis*	*A. senezensis guthi*	*A. senezensis mygdoniensis*	*A. bactrianus*	*A.* sp.	*A. stehlini*	
(c) Cranial measurements (in mm) of *Allohippus senezensis* species group and the reference *Equus hemionus onager*								
	n = 30	Senèze n = 2–9	La Puebla de Valverde n = 1–2	Gerakarou n = 1–3	Kuruksai n = 1–3	China Loc A and D PMU 1357, 1418 n = 1–2	Valdarno n = 1–4	
16	56	67.1		55				
23	348.1	424	415	393	420	430	382.8	
3	116.9	128.7	123	127.3	110.5		119.7	
4	101	107.7	124	104.5	106			
2–5	115.6	154.8	155.5	145.8	163.7	168	125..3	
5	104.9	134.4	137	130.7	127.7	145	114.4	
17	55.9	65.2	69	68.3	69	70	64.3	
17bis	40.7	43.7	47	36.7	46.5		47.5	
13	196.8	212.9	220	203.8	245	262	195	
10	48.1	48	52	45.5	51	50	44.8	
25	102	112.7						
28	89.8	108	108	93				
9	63.3	62						
20	14.3	15						
31	143.3	203.9	205	193	217.5	230	167	
32	162.2	153.9	164	151.5	152	180	145	

Approximate measurements between brackets. Measurements as in Eisenmann (1986) and http://www.vera-eisenmann.com: 16: breadth of the supra-occipital crest; 23: anterior ocular line; 3: distance from palate to hormion; 4: distance from hormion to basion; 2–5: palatal length (without the muzzle); 5: muzzle length; 17: muzzle breadth at the posterior border of the I3; 17bis: least muzzle breadth between the interalveolar borders; 13: frontal breadth; 10: greatest choanal breadth; 25: facial height in front of P2; 28: cranial height behind the orbits; 9: choanal length; 20: height of the external auditory meatus; 31: length of the naso-incisival notch (from prosthion to the back of the narial opening); 32: cheek length (from the back of the narial opening to the most anterior point of the orbit)

Table 11.A10 (a) Means of measurements (in mm) of mandibles of *Allohippus* and extant *Equus*. (b) Measurements (in mm) of mandibles of Senèze equids

	variable:	1	2	3	6	7	9	12	13
(a) Means of measurements (in mm) of mandibles of *Allohippus* and extant *Equus*									
A. stenonis vireti n = 2–7	Saint-Vallier	505	133	116.5	97.9	57.6	73	141.5	38.3
A. senezensis guthi MNHN-F PUE 3283	La Puebla de Valverde			104	92	60		138	38
A. senezensis senezensis n = 4–15	Senèze	451.9	127	95.9	85.6	48.6	60.6	115.5	33.7
E. hemionus onager n = 4–22		384.5	115.1	71.7	74.4	50.5	51.8	94	38
E. przewalskii n = 20–30		422.5	119.9	84.4	82.1	60.3	54.7	106.4	41.1
E. africanus n = 8–28		*390.8*	113.2	75.8	75.7	46.8	52.9	99.5	34.9
E. grevyi n = 5–61		461.2	130.5	103.9	91.1	52	61.8	130.4	36
E. zebra n = 26–79		414.3	126.3	85.5	83.9	52.3	60.7	107.5	36.2
E. quagga n = 18–57		404.7	120.8	90.4	80.3	52.2	58.9	108.1	34.5
	MNHL Sen 5233	FSL 210845	FSL 210847	FSL 210849	FSL 210850	FSL 210851	FSL 210852	FSL 210859	
(b) Measurements (in mm) of mandibles of Senèze equids									
Sex	M				M	F	M	F	
Age	adult	adult		adult	adult	old	ca. 5 years	old	
1. Greatest length	460							480	
2. Mandibular angle radius	132							[135]	
3. Length of diastema	103		90	92	90			109	
4. Occlusal length of P2-P4	91	98	86	88				88	
4 bis. Occlusal length of M1-M3	87.5	84	80	82				85	
5. Occlusal length of P2-M3	179	180	164	168				174	
6. Length of symphysis	94		82	87	85 or 90	85	94	95	
7. Breadth at posterior borders of I3	56		54	51	52 or 55	48.5	66	[50]	
8. Height of ascending ramus								230	
9. Height in front of P2		62		62				55.5	
10. Height between P4 and M1								80	
11. Height behind M3									
12. Muzzle length				112	114			135	
13. Least symphysis breadth	[38]			34	39	29.1	34		

(continued)

Table 11.A10 (continued)

	FSL SEN 05–0081	NMB Se 551	NMB Se 553	NMB Se 554	NMB Se 804	NMB Se 805	NMB Se 1776	NMB Se 1855
(b) Measurements (in mm) of mandibles of Senèze equids								
Sex	F	F		M			F	F
Age	adult	old	ca. 3 years	old	adult	old		ca. 5 years
1. Greatest length	435	440	445	438			465	
2. Mandibular angule radius	126	120	130	119			129	
3. Length of diastema	91	90	90	96	84	102	116	97
4. Occlusal length of P2-P4	91	86	102	85.5	89	74	81	100
4 bis. Occlusal length of M1-M3	78	83		80	80	77	80	88
5. Occlusal length of P2-M3	170	170		163.5	170	151	162	187
6. Length of symphysis	[79]	72	87	92 or 121	80	79	90	84
7. Breadth at posterior borders of I3		51	52	51	[35]	[38]	53	53
8. Height of ascending ramus	210							
9. Height in front of P2	63							
10. Height between P4 and M1	79							
11. Height behind M3	107							
12. Muzzle length	101							
13. Least symphysis breadth				32				30

Variables listed by number at top of table and defined on left in section b. For NMB Se 374, Female, Length of symphysis = 83.5; breadth at posterior borders of I3 = 50

Table 11.A11 (a). Lower cheek teeth measurements (in mm) of Senèze equids. (b) Isolated lower cheek teeth measurements (in mm) of Senèze equids

		MNHL Sen 5233	FSL 210834	FSL 210835	FSL 210841	FSL 210842	FSL 210843	FSL 210844	FSL 210845
(a) Lower cheek teeth measurements (in mm) of Senèze equids									
Age		adult	< 4 years	adult	old	old	adult	adult	adult
p2	L	34	37	30.4	34	32.5	32	31	[35.5]
	Lpf	14.5	14.7	11.5	15.3	13	12.5	15.6	15
	W	15.5	14.3	14.5	15	15.5	14	15	15.7
	H		60	[43]		32			
p3	L	30	32.3	27.5	29	27	28.5	28	30
	Lpf	15	13.8	11.4	10	9	14	15	14.5
	W	16	16.5	15.7	15	15.3	15	16	16
	H		70						
p4	L	29	31	26	27	25.2	27.5	28	29.5
	Lpf	12.5	12	10.5	8.7	7.5	13.5	13	13
	W	16	15.5	16.5	16.3	15.5	15.5	16	15.7
m1	L	26.5	29	25.2	24	21	25.5	25.5	26
	Lpf	12	9.8	7.5	0	0	8	8	9.5
	W	14	14.8	15	15.4	14.5	13.5	15	15.5
	H			[31]					
m2	L	27	30.5			24	26.5	26	26.5
	Lpf	11.5	9.7			4.5	8	7.8	9
	W	14	13.6			13.5	13	14.5	14.8
	H		68			[18]			
m3	L	32			37		30	30	29.5
	W	13			12.7		12.5	13	13
		FSL 210847	FSL 210849	FSL 210857	FSL 210877	FSL 210887	NMB Se 142	NMB Se 553	NMB Se 554
Age		adult	adult	adult	old	ca. 5 years	ca. 5 years	ca. 2 years	old
p2	L	33	34.5	32		36.5	36.5	[38]	30.5
	Lpf	14	15.5	14.5		16.5	14.5		12
	W	14.5	14.8	15.5		14	14		13.5
p3	L	28.5	29.5	27	26.5	32	30	[33]	27.5
	Lpf	13.5	13.5	9.5		15	15		10.3
	W	16	16.5	16.5	15	16	16		15
	H		52						
p4	L	27.5		27	26.5	30.5	28.5		27
	Lpf	12		10.5	7.5	13.5	13		10
	W	16		17.5	14	13	16		15.5
m1	L	24.5		23.5	22.5	29.5	25.5	30.5	23
	Lpf	6.5		0	0	11	11	13	5
	W	15.5		16	15	14	15	13	14
m2	L	24		25	25	28	25.5	34	24.5
	Lpf	7.5		0	4.5	11	11	12	6
	W	13.5		15	14	12	15	12.5	14
m3	L	31		31		29	30		31
	W	12		12.5		9.5	14		12

(continued)

Table 11.A11 (continued)

		NMB Se 694	NMB Se 804	NMB Se 806	NMB Se 918	NMB Se 1496	NMB Se 1855	NMB Se 1857	MNHN.F (Sen 68)
Age		4 years	adult	old	4 years	adult	ca. 5 years	adult	adult
p2	L	37.5	34	31	33	34	35.5		31
	Lpf	17.5	12	13.5	16	14.5	17		14.5
	W	15.5	16	14	15.5	16	15		15
p3	L	32	28.5	26.5	29	30	32	28	29
	Lpf	14	13	12.5	13.5	15	15	12	14
	W	16.5	16.5	15	15	17	16	16	16.5
	H								[40]
p4	L	31	27	27	31	29	31	28	
	Lpf	12.5	13	12	13	13	13	11	
	W	15	16	15.5	15	16.5	14.5	16.5	
m1	L	29	24.5	24.5	27.5	27	27.5	25	
	Lpf	9	7.5	7	10	10.5	10.5	8	
	W	15	15	14		15	14.5	[14]	
m2	L	31	25	24.5	29	27.5	28.5	25	
	Lpf	10.5	9	7	12	9.5	10	7.5	
	W	13	15	13.5	14	15	14	14	
m3	L	31	30	30	31.5	33	30.5	32	
	W	10	13	12.5	11	13.5	11.5	13	
		MNHN.F-AC 1938–373	FSL SEN 05–0081						
Age		adult	adult						
p2	L	32	32.3						
	Ldn		13.5						
	Lpf	13	15						
	W	13	14.4						
p3	L	28	28.9						
	Ldn		17.3						
	Lpf	13.2	13.4						
	W	14.5	15.1						
p4	L	27	26.8						
	Ldn		15.8						
	Lpf	12.5	12.1						
	W	15.1	16.1						
m1	L	24.5	24.2						
	Ldn		14						
	Lpf	8	7						
	W	14	14						
m2	L	26	23.9						
	Ldn		12.8						
	Lpf	8	7.4						
	W	13	13.7						
m3	L		30.1						
	Ldn		12.6						
	W		12.4						

(continued)

Table 11.A11 (continued)

		L	Lpf	W	H
(b) Isolated lower cheek teeth measurements (in mm) of Senèze					
P2 unworn	MNHN.F-AC no n°	[39]			[56]
P2	MNHN.F-AC no n°		15	15.5	[46]
P2	MNHN.F-AC no n°	34	14	15	39
P2	MNHN.F-AC no n°	[35]	15	15	38
P2	MNHN.F (Sen 16)	[30]	10.5	14	16
P2	MNHN.F (Sen 17)	32	15	[14.5]	27
P2	MNHN.F (Sen 18)	32	14.5		45
P	FSL 211060	30.5	13.5	17.2	36
P	MNHN.F-AC no n°	31.5	13	16	51
P	MNHN.F-AC no n°	32	15	17	54
P	MNHN.F-AC no n°	27	13	17	42
P	MNHN.F-AC no n°	29.5	13.5	16	34
P	MNHN.F (Sen 19)	30.5	15	15	46
P	MNHN.F (Sen 20)	29.5	15	17	64
P	MNHN.F (Sen 21)	30.5	13.5	16.5	49
P caballine	MNHL Sen 3982	28.5	12	17	66
P caballine	MHNL Sen 3982	28	12	17	52
P ?	MNHN.F (Sen 9)	27.5	9.5	15.5	23
P ?	MNHN.F (Sen 10)	27	11	15	24
M young	MNHN.F-AC no n°	[34]	12		> 60
M young	MNHN.F-AC no n°	[33]	10		72
M young	MNHN.F-AC no n°	[31]	10		> 63
M	MNHN.F-AC no n°	27	10	14	50
M	MNHN.F-AC no n°	26.3	8.2	12.5	39
M	MNHN.F-AC no n°	26.5	10	13.5	32
M	MNHN.F-AC no n°	26	10	13	31
M unworn	MNHN.F (Sen 1)	[35]			> 64
M young	MNHN.F (Sen 2)	29	11.5	12	> 63
M young	MNHN.F (Sen 3)	29.5	10.5	12	> 55
M young	MNHN.F (Sen 4)	[33]	11.5		> 61
M	MNHN.F (Sen 5)	27.5	9.5	13.5	60
M	MNHN.F (Sen 6)	26	8.5	14.5	30
M unworn	MNHN.F (Sen 7)	[35]			
M	MNHN.F (Sen 11)	30	10.5	[14]	68
M	MNHN.F (Sen 12)		7	12.5	
M young	MNHN.F-AC 1938–373	29	11.5	13.2	60
M3	MNHN.F-AC no n°	32	13	13	33
M3	MNHN.F (Sen 22)	33	13	13	32
M3	MNHN.F (Sen 23)	31.5	13.5	13.5	34

L: occlusal length; Ldn: double knot length; Lpf: postfossette length; W: occlusal width without cement; H: height. Approximate measurements between brackets

Table 11.A12 (a) Measurements (in mm) of first anterior phalanges of *Allohippus major*, Sussemiones, and *E. hemionus onager*. (b) Measurements (in mm) of first posterior phalanges of *Allohippus major* and Sussemiones (c) Measurements (in mm) of first anterior phalanges of Senèze equids (d) Measurements (in mm) of first posterior phalanges of Senèze equids

	Würtsburg Schalksberg no number *A. major*	Senèze FSL 211074 *A. major*	East-Runton NHMUK -P 6760 *A. major*	Süssenborn n = 2–5	Akhalkalaki n = 17–22	NE Siberia GIN 3722–17	*E. hemionus onager* n = 14–15
(a) Measurements (in mm) of first anterior phalanges of *Allohippus major*, Sussemiones, and *E. hemionus onager*							
Greatest length	104	110	107.5	99.4	98.8	103	76.5
Smallest breadth	45	44.5	47	39.9	42	41	24.6
Proximal breadth	70	67.5	70	60.8	65.9	65	41.1
Proximal depth	47	47.5	45	41	44.1	44.5	30.8
Supra-articular distal breadth	60	59.5	60	51.5	56.5	56.5	36.7
Greatest length of trigonum phalangis	72	71	75	70.6	66.9	65	48.1
Medial supratuberosital length	74.8	73.9	74	74.8	73.9	74	58.5
Medial infratuberosital length	14	16.5	19	13.9	15.3	18	10.3
Distal articular breadth	57	57.5	59	50.6	55.1	57	48.1
	Senèze FSL 211075 *A. major*	Chagny FSL no number *A. major*	Süssenborn n = 2	Akhalkalaki n = 18–19	NE Siberia Loc 37 PIN 3100–333-79		
(b) Measurements (in mm) of first posterior phalanges of *Allohippus major* and Sussemiones							
Greatest length	105	100.2	100	96.5	95		
Smallest breadth	43.5	47.5	42	41.4	45		
Proximal breadth	69	69.5	67.7	68.2	70.9		
Proximal depth	50	51	42	45.9	45		
Supra-articular distal breadth	57	58.1	55	54.7	59		
Greatest length of trigonum phalangis	64	61	67.5	62.3	56		
Medial supratuberosital length	76	77	73	68	67		
Medial infratuberosital length	20	23.5	18.5	18.7	20		
Distal articular breadth	54.5		50.8	51.8	54		

(continued)

Table 11.A12 (continued)

	FSL 210899	FSL 211055	FSL 211074	FSL SEN 06–0279	MNHN.F (Sen 61)	NMB Se 141j	NMB Se 179	NMB Se 180
(c) Measurements (in mm) of first anterior phalanges of Senèze equids								
Greatest length	80	91	110	76.9	80	77.5	77.5	86
Anterior length	70		101	70.1		70	70	
Smallest breadth	32	36.5	44.5	33.5	32	31.5	31	31.5
Proximal breadth	49	55	67.5		50	50	51.5	
Proximal depth	35.5	37	47.5		35	33.5	34	
Supra-articular breadth	41.5	49	59.5	46	44	41	44	41.3
Greatest length of trigonum phalangis	50		71	47.5	49	49	50	
Smallest length of trigonum phalangis	45		67	42	43	43	44	
Posterior length	70		99	67.5		67	68.5	
Medial supratuberosital length	58		82	58	58	55	57.5	
Lateral supratuberosital length			83	59	58	55	55.5	
Medial infratuberosital length	14		16.5	12.9	13	13	13	
Lateral infratuberosital length			15.5	11	14	13	13.5	
Distal articular breadth	40.5		57.5	44	43	41.5	41.5	[41]
	NMB Se 552j	NMB Se 553j	NMB Se 554	NMB Se 828	NMB Se 833	NMB Se 835	NMB Se 1689jjj	
Greatest length	78	76	80.2	83	80	76.3	71	
Anterior length	71	67	71.5		72	67.5	65	
Smallest breadth	30.5	32	34.5	34	31.5	31.5	26	
Proximal breadth	48.5	51.5	54.5	54	52	53	42	
Proximal depth	33.5	34	37	37.5	34.5	35	29	
Supra-articular breadth	[41.5]	43	48.5	47.5	42	42.5	36	
Greatest length of trigonum phalangis	48	47	53	51	50.5	46		
Smallest length of trigonum phalangis	41	42	48		45	40		
Posterior length	68	66	71		70	66	63	
Medial supratuberosital length	58	52	57	60	58	54		
Lateral supratuberosital length	58	52	57.5		57	54		
Medial infratuberosital length	12	14	13	14	14	13.5		
Lateral infratuberosital length	11	14	12.5			13		
Distal articular breadth	[42]	42	45	46	42.5	43.5	37	

(continued)

Table 11.A12 (continued)

	FSL 210898	FSL 210900	FSL 211075	FSL 211082	FSL 211082bis	FSL 211083j	FSL SEN 04–0133	FSL SEN 06–0267
(d) Measurements (in mm) of first posterior phalanges of Senèze equids								
Greatest length	77.5	80.5	105	85.5	80		77.5	76
Anterior length	68	71	97	78	71.5		69	68
Smallest breadth	33	35.5	43.5	33.5	32.5	28	31.5	34
Proximal breadth	50.5	55	69	54	53.5		51.2	53.7
Proximal depth	36.5	39.5	50	39	38		37	37.2
Supra-articular breadth	41.5	45	57	45	[43]	37.5	39.6	44.1
Greatest length of trigonum phalangis	46.5	46	64	51	48		42	45
Smallest length of trigonum phalangis	40	42	57	45	42		35.5	38
Posterior length	66.5	71	94	74	70.5		67	65
Medial supratuberosital length	51	54	76	61	53		52	52
Lateral supratuberosital length			76	61			53	52
Medial infratuberosital length	17	18	20	16	17	15.5	18	17
Lateral infratuberosital length			19	15			16	17
Distal articular breadth	39	43	54.5	42		37	40	42
	NMB Se 141j	NMB Se 551	NMB Se 552j	NMB Se 553j	NMB Se 554	NMB Se 811	NMB Se 1689jjj	
Greatest length	77		76	73	76.5	74.5	69	
Anterior length	69		70	65.5	68.8	67	63.5	
Smallest breadth	31	33.5	31.5	33.1	35.5	31	28	
Proximal breadth	51.2	51	50.5	53	54	51	43	
Proximal depth	35.5	37	35.5	36.5	40	37	31	
Supra-articular breadth	40	42.5	[40]	43	47.3	43	35	
Greatest length of trigonum phalangis	43		43.5	43	48.5	40	40	
Smallest length of trigonum phalangis	37		36.6	37	42	34.5	34	
Posterior length	66		65.5	63	66	65	61	
Medial supratuberosital length	53	[48]	54	48	52	51		
Lateral supratuberosital length	54		52	48	52.5	50		
Medial infratuberosital length	16	[15]	15.5	18	16	17		
Lateral infratuberosital length	15		15	16	15.5	17		
Distal articular breadth	40	39.5	39.5	40	45	40	35.5	

j = juvenile

Table 11.A13 Upper cheek teeth measurements (in mm) of *Allohippus stenonis stenonis* from Olivola and Matassino

		IGF 11023	IGF 11025	IGF 1432	IGF 1431
Age		old	young adult	adult	adult
P3	L	30.2	33	30.5	32
	Lp	7.8	8.4	8.1	10
	W	30	30	31	30
P4	L	28.4	31.5	27.5	
	Lp	8.8	10.2	9	
	W	29.5	29.3	29	
M1	L	25.8	26.8	25	
	Lp	10.1	9.2	9	
	W	28.3	27.6	28	
M2	L	26.9	27.6	26	
	Lp	11.1	9.5	9	
	W	27.3	26.7	27.3	

L: occlusal length; Lp: protocone length; W: occlusal width without cement

Table 11.A14 Measurements (in mm) of scapula and pelvic acetabulum of Senèze equids

	Maximal length	Minimal breadth at the neck	Maximal breadth of articular process	Articular maximal breadth	Articular maximal depth	Pelvic acetabulum diameter
FSL 210993		64	92	54	43	[62]
MNHN.F-AC 1921–9		58	84	51	46.5	
MNHN.F (Sen 55)		60	90	56	49	
NMB Se 141j		50.5	87.5	55	46	
NMB Se 181		58	95	59	49	
NMB Se 551	345	58	91	56	47.5	58
NMB Se 552j	290	51.5	90	55	[47]	60
NMB Se 553j	[320]	59	91.5	56	51.5	62
NMB Se 570		51	85	53	45.5	
NMB Se 862		49.5	86	[53]	43	
NMB Se 863			91.5	58	42	
NMB Se 864				59.5	41.5	
NMB Se 864bis		< 58				
NMB Se 865		64			51	
NMB Se 866					50.5	
NMB Se 868		59.5	95.3	57	49	
NMB Se 869		40				
NMB Se 870		43		52	42.5	
NMB Se 923			90	55		
NMB Se 1471		60	89	54.5	43.5	
NMB Se 1471bis					48	
NMB Se 1511		47.5	80	51	43.5	
NMB Se 696						60
NMB Se 854						60
NMB Se 857						[64]
NMB Se 859						63

j = juvenile

Table 11.15 Measurements (in mm) of humeri of Senèze equids

	Greatest length	Posterior length	Smallest breadth	Proximal breadth	Proximal depth	Distal articular breadth	Distal medial depth	Trochlear smallest height
FSL 210864j	262	244	32.5		[80]	71	79	32
FSL 210869–70			35			73		34
FSL 210871			36			76	84	37
FSL 211091			[49]			98	107	43
FSL 210993	[277]	[263]	[33]	[99]	97	73	76	32
FSL SEN 05–0097	282	263	35	96	98	74	79	36
MNHN-F (sen 54)		260	34.5			71	80	36.5
NMB Se 141j		238	31.5		[90]	75	82	35
NMB Se 551	280	250	34	92	102	73	80	
NMB Se 552j	260.5	240	30.1			70	80	36.5
NMB Se 553j	286	261	34.5	92	103	76	82.5	36
NMB Se 554						78		37.2
NMB Se 824	290	268	38			78	87.5	39
NMB Se 825			34.5			74.5	82.5	34.5

Table 11.A16 Measurements (in mm) of radii of Senèze equids

	FSL 210860	FSL 210861	FSL 210862	FSL 210863	FSL 211093	FSL 211095	FSL 211101
Greatest length	398	315	344	308	[325]	427	
Lateral length	383.5	296		295	312	412	
Smallest breadth	58	39	44.5	40	38		
Proximal breadth	103.5	81		79	80.5	110	82.5
Proximal articular breadth	93	72.5		70	72	99	75
Proximal articular depth	49.5	38	39	33	[36]	46.5	38.5
Distal breadth	100	72	78	70	71.5	102	
Distal articular breadth	81	61	66	60	57	86.5	
Greatest distal articular depth	49	38	41	36.5	[35]	47	
Breadth of radial condyle	31	24.5	25.5	22.5	25	33	
Breadth of ulnar condyle	21	14.5	17	15	14.5	23	

(continued)

Table 11.A16 (continued)

	FSL SEN 05–0098	FSL SEN 06–0288	FSL SEN 06–0297	MNHN.F (Sen 66)	NMB Se 141j	NMB Se 551	NMB Se 552j
Greatest length	315	[330]		330	319		316
Lateral length	295			316.5	300	302	302
Smallest breadth	36.6	40	40	43.5	38	38	35.2
Proximal breadth	81	84		83.5	82	78	81
Proximal articular breadth	72.5	72.5		72.5	75	70	72
Proximal articular depth	37	40			38		35
Distal breadth	71		77	73	73	69.5	71
Distal articular breadth	59		62	61	64	60	59.5
Greatest distal articular depth	34		[42]	35	38		36.5
Breadth of radial condyle	23		27	26	26		26
Breadth of ulnar condyle	16.7		16	16.5	16		17
	NMB Se 553	NMB Se 554	NMB Se 827	NMB Se 828	NMB Se 1768		
Greatest length	326			327	340		
Lateral length	306			306	320		
Smallest breadth	40		44.5	43	43		
Proximal breadth	84	83		84	85.5		
Proximal articular breadth	76	75		75	78		
Proximal articular depth	38	39		37.5	40		
Distal breadth	76	76		74	79		
Distal articular breadth	63	63		64	66		
Greatest distal articular depth	37	38		38	41		
Breadth of radial condyle	26.5	26		27	28		
Breadth of ulnar condyle	17	18		16	17		

j = juvenile

Table 11.A17 Measurements (in mm) of femora of Senèze equids

	Greatest length	Medial length	Smallest breadth	Proximal breadth	Proximal depth	Depth of caput femoris	Distal breadth	Trochlear breadth	Distal medial depth
MNHN.F Sen 1223–4							[86]	[62]	118
NMB Se 551	380	340	38	117	81	51	89	60	114
NMB Se 554	388	347	39	123.5	88	58	93	64	115.5
NMB Se 552j	363	333	33.5	107	[76]	53	87	59	110.5
NMB Se 808		367	42			57	93	66	120
NMB Se 785	352	321	38	113	87	51	85.5	55.5	109
NMB Se 553j	388	349	39			56	92	63	116
NMB Se 141j	351	325	35	108	80	54	89		
FSL SEN 05–0148	378	340	39.1	112	86	53	89	65	112

j = juvenile

Table 11.A18 Measurements (in mm) of tibiae of Senèze equids

	FSL SEN 05–0151	FSL SEN 06–0274 right	FSL SEN 06–0278 left	MNHN.F (Sen56)	MNHN.F (Sen 57)	MNHN- AC 1938–373	NMB Se 107a	NMB Se 141j	
Greatest length	333	360					[335]	336	
Lateral length	300							321	
Smallest breadth	42	45	45	[46]		[43.5]	40	41	
Smallest depth	31	31	32	30		30	26	28	
Proximal breadth	[95]	[102]						[94]	
Proximal depth	[82]							80	
Distal breadth	69.5	76	80	67	76	70	65	71	
Distal depth	48		51	44	50	48		49	
Length of fossa digitalis	54							58	
Breadth of fossa digitalis	22							18	
Greatest length		349	343	338	355			348	
Lateral length	330	335	325	325	339			337	[290]
Smallest breadth	43	43.5	45	44	50	51.5	43	44.5	39
Smallest depth	31	30	32.5	31	35	33.5	30	31	27
Proximal breadth	94	98	100	[91]	104			96.5	
Proximal depth	80	83	89	80	87			84	
Distal breadth	70	75	74	69	80	83.1	[69]	74	[60]
Distal depth	47	47	47	46.5	50	56	47	47	
Length of fossa digitalis	58	60	54	54	64			54	
Breadth of fossa digitalis	19	19	18	[22]	20			19	

j = juvenile

Table 11.A19 Measurements (in mm) of tali of Senèze equids

	Gratest length	Length	Medial length of trochlea	Greatest breadth	Trochlear breadth	Distal articular breadth	Distal articular depth	Medial depth
FSL 210854–55	59.5		58.5	60	27	52	33.5	49.5
FSL 210856	63.5		64	61	27.5	53.5	36	51
FSL 211073	79		76.5	79	36	63.5	45.5	62.5
FSL 211082	61		60.5	61	29	51	36	52.5
FSL SEN 05–0137	57	55	56.2	59	27	49	33.1	50
FSL SEN 06–0278	63	61	62	62	28.5	52	35.1	
MNHN.F (Sen 58)	[63]		60	[63]	[30]	52.5	37.5	52.5
NMB Se 141j	64		62	61.5	26.5	50	36	52
NMB Se 181	64.5			63.5	27.5	[52]	34	
NMB Se 198			59.5		27			
NMB Se 551				59	27.5			
NMB Se 552j	61.5		59	55	27	49	35	49
NMB Se 553j	59		60	62	27	50	33	52
NMB Se 554	62		61.5	61	27	52	35	51
NMB Se 692	63		60	60		50	35.5	

j = juvenile

Table 11.A20 Measurements (in mm) of calcanei of Senèze equids

	Geatest length	Length of proximal part	Smallest breadth	Proximal breadth	Proximal depth	Distal breadth	Distal depth
FSL SEN 06–0255	114	77	22	36	50	52	55.1
FSL SEN 06–0261	114	75		35.5	50.5	55	54
MNHN.F (Sen 59)	113	76	20	31.5	47.5	53	52
NMB Se 141j	110	71.5	22.5	33	49	50	51
NMB Se 551	107	73	22	28	48		
NMB Se 552	104	68	20.5	32	47	49	50
NMB Se 553j	111	75	22	35.5	52	51	51
NMB Se 554	113	80	23	38	51	53	55
NMB Se 811	108		19			49	51
NMB Se 843	123	80	23	35	55		58
NMB Se 844j	118		20			52.5	45

j = juvenile

Table 11.A21 (a) Measurements (in mm) of second anterior phalanges of Senèze equids. (b) Measurements (in mm) of second posterior phalanges of Senèze equids

	Greatest length	Anterior length	Smallest breadth	Proximal breadth	Proximal depth	Distal articular breadth
(a)						
FSL 210899	45.5	33	40	47	32	43.5
FSL 210919	57.5	43.5	58.5	68.5	40.5	65.5
FSL 211067	48.5	36.5	42.5	51.5	32.5	48
FSL 211075	57.5	43	60	66	43	62.5
FSL SEN 06–0276 left	45	34	44	51	33	46.6
FSL SEN 06–0312 right	45.7	34	44	51.5	33	47
MNHN.F (Sen 62)	45	33	42.5	50.5	30.5	44.5
NMB Se 141j	44	33	41.5	47	31	44
NMB Se 180	47		42	49		47
NMB Se 551	39		43	47.1	31	45.5
NMB Se 552j	46	34	40	46.5	30.5	44.5
NMB Se 553j	46	33	42	50	32	45.2
NMB Se 554	47.5	34	46	53	33	50
	Greatest length	Anterior length	Smallest breadth	Proximal breadth	Proximal depth	Distal articular breadth
(b)						
FSL 210898	46.5	36	37	45	32.5	39.5
FSL 210900	46	35.5	37.5	45.5	32	40
FSL 211064		34	38			40
FSL 211075	59	43.5	48.5	58	38.5	52.5
FSL 211078	59	46	53	63	42.5	56
FSL 211083	45	34.5	37	43.5	30	38.5
FSL SEN 04–0134	47.2	35	38.7	46.2	32	41
FSL SEN 06–0264	47	35	41.5	50	33	43
MNHN.F (Sen 60)	47	37	41	50	32.5	42
NMB 141j	47	33.5	40	45	32.5	40.5
NMB Se 377	43	31	37	44	29	39
NMB Se 551	45		40	47	33	42
NMB Se 552j	47	37	39	46	32	41.5
NMB Se 553j	46.5	34.5	40	48.5	32	42
NMB Se 554	49	37	43	52	35	46.5
NMB Se 838	47	[33]	39	46	32	40.5
NMB Se 839	47	35	38	47	32	39.5
NMB Se 845	46.5	33	40	47.2	32	39.5

j = juvenile

Table 11.A22 (a) Measurements (in mm) of third anterior phalanges of Senèze equids. (b) Measurements (in mm) of third posterior phalanges of Senèze equids

	Anterior length	Greatest antero-posterior diameter	Height	Greatest breadth	Articular breadth	Articular antero-posterior diameter	Distal "circumference"
(a)							
FSL 210899	52.5	62	[42]	[70]		26	[170]
FSL 210901	[62]	80.5		[100]			[210]
FSL 211082	66	73	[44]	[98]	[54]	32	[200]
NMB Se 115, ant.?			39		[48]	29	
NMB Se 141j	44	46	40	[50]	44	25	120
NMB Se 551	52	67	41.5	67.3	44	25	160
NMB Se 552j	[45]	[53]	37	[60]	44	[24]	[130]
NMB Se 553j	47	60	42	64	[44]	25	145
NMB Se 554, ant.?	52	62	40.5	66	46	29	[150]
NMB Se 917jjj	42		28.5			24	
NMB Se 1861, ant.?	52	65	39	63	46	25	155
	Anterior length	Greatest antero-posterior diameter	Height	Greatest breadth	Articular breadth	Articular antero-posterior diameter	Distal "circumference"
(b)							
FSL 210898	51.5	59		[62]	[40]	25	[150]
FSL 210900	51	58.5	40	60	39	25.5	145
FSL 211078, post.?	66	78	[44.5]	[90]	54	31	
FSL 211083	40.5	43	33.5	43	38	24	[110]
FSL SEN 04–0135	51	[54]	40	58	44	28	> 136
FSL SEN 05–0130	50	59	40	57.2	41	27	147
NMB Se 141j	49	50	40	51.5	41	26	120
NMB Se 375	48	57.5	39	57.5	37	25	140
NMB Se 551	51	63	42	62.5	42	27	150
NMB Se 552j	46	50.5	34	54	40	24	125
NMB Se 553j	49.5	53	40	[59]	41	24	130
NMB Se 811	[50]	54	35.5	59	42	25	[143]
NMB Se 917jjj	43	39	35.5	42	40	26	103

j = juvenile

References

Alberdi, M.-T., & Ruiz-Bustos, A. (1989). Taxonomia y Biostratigrafia de Equidae (Mammalia, Perissodactyla) en la Cuenca de Guadix-Baza (Granada). *Geología y Paleontología de la Cuenca de Guadix-Baza, Trabajos sobre el Neógeno-Cuaternario, 11*, 239–270.

Alberdi, M. T., Caloi, L., & Palombo, M. R. (1998). Large mammal associations from the Early Pleistocene: Italy and Spain. In Van Kolfschoten, T., & Gibbard, P. L. (Eds.), *The Dawn of the Quaternary* (pp. 521–532). *Proceedings of the SEQS-EuroMam symposium 1996. Mededelingen Nederlands Instituut voor Toegepaste Geowetenschappen TNO.*

Alberdi, M. T., & Palombo, M. R. (2013). The Early to early Middle Pleistocene stenonoid horses from Italy. *Quaternary International, 288*, 25–44.

Aouadi, N. (1999). Etude préliminaire des restes crâniens de chevaux villafranchiens (Ceyssaguet, Haute-Loire). *Bulletin du Musée d'Anthropologie préhistorique de Monaco, 40*, 23–42.

Athanassiou, A. (2001). New data on the *Equus stenonis* from the Late Pliocene locality of Sesklo (Thessaly, Greece). *Geodiversitas, 23*, 439–469.

Azzaroli, A. (1965). The two Villafranchian Horses of the Upper Valdarno. *Palaeontographia Italica, 59* (n.s. *29*), 1–12.

Azzaroli, A. (1990). The genus *Equus* in Europe. In E. H. Lindsay, V. Fahlbusch, & P. Mein (Eds.), *European Neogene Mammal Chronology* (pp. 339–355). New York: Plenum Press.

Azzaroli, A., & Voorhies, M. (1993). The Genus *Equus* in North America. The Blancan species. *Palaeontographia Italica, 80*, 175–198.

Azzaroli, A., Colalongo, M. L., Nakagawa, H., Pasini, G., Rio, D., Ruggieri, G., et al. (1997). The Pliocene-Pleistocene boundary in Italy. In J. A. Van Couvering (Ed.), *The Pleistocene Boundary and the Beginning of the Quaternary* (pp. 141–155). Cambridge: Cambridge University Press.

Barrón-Ortiz, C. I., Avilla, L. S., Jass, C. N., Bravo-Cuevas, V. M., Machado, H., & Mothé, D. (2019). What is *Equus*? Reconciling taxonomy and phylogenetic analyses. *Frontiers in Ecology and Evolution, 7*, 343. https://doi.org/10.3389/fevo.2019.00343

Bennami, M., Aidona, E., Merceron, G., Koufos, G., & Kostopoulos, D. S. (2020). Magnetostratigraphy and chronology of the Lower Pleistocene primate bearing Dafnero fossil site. N. Greece. *Quaternary, 3*, 1–15.

Bernor, R. L., Armour-Chelu, M., Gilbert, H., Kaiser, T. M., & Schulz, E. (2010). Equidae. In L. Werdelin & W. L. Sanders (Eds.), *Cenozoic Mammals of Africa* (pp. 685–721). Berkeley: University of California Press.

Bernor, R. L., Cirilli, O., Wang, S. Q., & Rook, L. (2018). *Equus* cf. *livenzovenzis* from Montopoli, Italy (early Pleistocene; MN16b; ca. 2.6 Ma). *Bolettino della Società Paleontologica Italiana, 57*, 203–216.

Bernor, R. L., Cirilli, O., Jukar, A. M., Potts, R., Buskanidze, M., & Rook, L. (2019). Evolution of Early *Equus* in Italy, Georgia, the Indian Subcontinent, East Africa, and the Origins of African Zebras. *Frontiers in Ecology and Evolution, 7*, 166.

Boeuf, O. (1986). L'Equidé du Site Villafranchien de Chilhac (Haute-Loire, France): *Equus stenonis guthi* nov. subsp. *Annales de Paléontologie, 72*, 29–67.

Bonifay, M.-F., & Brugal, J.-P. (1996). Biogéographie et biostratigraphie des grandes faunes du Pléistocène inférieur et moyen d'Europe du Sud : Apport des gisements français. *Paléo, Revue d'Archéologie Préhistorique, 8*, 19–30.

Boulbes, N., & Asperen, E. (2019). Biostratigraphy and Paleoecology of European *Equus*. *Frontiers in Ecology and Evolution*, 7:301. https://doi.org/10.3389/fevo.2019.00301.

Churcher, C. S., & Richardson, M. L. (1978). Equidae. In V. J. Maglio & H. B. S. Cooke (Eds.), *Evolution of African Mammals* (pp. 372–422). Cambridge: Harvard University Press.

Churcher, C. S. (1986). The extinct Cape zebra. *Sagittarius, 1*(4), 4–5.

Churcher, C. S. (1993). *Equus grevyi. Mammalian Species, 453*, 1–9.

Cirilli, O., Saarinen, J., Pandolfi, L., Rook, L., & Bernor, R. L. (2021). An updated review on *Equus stenonis* (Mammalia, Perissodactyla): New implications for the European early Pleistocene *Equus* taxonomy and paleoecology, and remarks on the Old World *Equus* evolution. *Quaternary Science Reviews, 269*, 107155.

Cirilli, O., Saarinen. J., & Bernor, R. L. (2024). Lost in the collections. A critical re-appraisal on *Equus major* provides a new perspective on the paleobiogeography of the Plio-Pleistocene European equids and on the *Equus* Datum. *Quaternary Science Reviews, 323*, 108428.

Debard, E. (2024). Geological study of the Early Pleistocene site of Senèze (Domeyrat, Haute-Loire, France). In E. Delson, M. Faure, & C. Guérin (Eds.), *Senèze: Life in Central France two million years ago. Paleontology, geochronology, stratigraphy and taphonomy* (pp. 37–82). Cham (Switzerland): Springer.

De Giuli, C. (1972). On the type form of *Equus stenonis* Cocchi. *Palaeontographia Italica, 68*, 35–49.

Delson, E., Nomade, S., Sen, S., Debard, E., Pastre, J. -F., Bahain, J. -J, et al. (2024) Geochronology of Senèze: ^{40}Ar/^{39}Ar Dating and Magnetostratigraphy, with Notes on an ESR/U-Series Dating Attempt. In E. Delson, M. Faure, & C. Guérin (Eds.), *Senèze: Life in Central France two million years ago. Paleontology, geochronology, stratigraphy and taphonomy* (pp. 99–122). Cham (Switzerland): Springer.

Deng, T. (2009). Late Cenozoic environmental changes in the Linxia Basin (Gansu, China) as indicated by cenograms of fossil mammals. *Vertebrata Palasiatica, 10*, 282–298.

Eisenmann, V. (1981). Etude des dents jugales inférieures des *Equus* actuels et fossiles. *Palaeovertebrata, 10*, 127–226.

Eisenmann, V. (1984). Sur quelques caractères adaptatifs du squelette d'*Equus* et leurs implications paléoécologiques. *Bulletin du Muséum national d' Histoire naturelle, Paris, 4ème série, 6, section C, 2*, 85–195.

Eisenmann, V. (1985). Les Equidés des gisements de la vallée de l'Omo en Ethiopie (collections françaises). In: *Les Faunes plio-pléistocènes de la basse vallée de l'Omo (Ethiopie), T. 1: Périssodactyles, Artiodactyles (Bovidae). Cahiers de Paléontologie, Travaux de Paléontologie est-africaine* (Paris: CNRS), pp. 13–55.

Eisenmann, V. (1986). Comparative osteology of modern and fossil horses, halfasses and asses. In R.H. Meadow & H.P. Uerpmann (Eds.) *Equids in the ancient world. Beihefte zum Tübinger Atlas des Vorderen Orients, Reihe A*, 67–116.

Eisenmann, V. (2002). The primitive horses of the Vatera Formation (Lesbos, Greece). Proceedings of the 1st International Workshop "On Late Plio/Pleistocene extinction and evolution in the Palearctic. The Vatera site". *Annales Géologiques des Pays Helléniques, 1ère série, 39 (A)*, 131–153.

Eisenmann, V. (2004). Les Equidés (Mammalia, Perissodactyla) de Saint-Vallier (Drôme, France) et les Equidés Plio-Pleistocènes d'Europe. In M. Faure, & C. Guérin (Eds.), *Le gisement pliocène final de Saint-Vallier (Drôme, France). Geobios 37*, Mémoire spécial 26, 279–305.

Eisenmann, V. (2006). Pliocene and Pleistocene Equids: Paleontology versus Molecular Biology. In R. D. Kahlke, L. C. Maul, & P. Mazza (Eds.) *Late Neogene and Quaternary biodiversity and evolution:*

Regional developments and interregional correlations. Proceedings volume of the 18th International Senckenberg Conference (VI International Palaeontological Colloquium in Weimar), 25th-30th April 2004. Courier Forschungsinstitut Senckenberg, *256*,71–89.

Eisenmann, V. (2010). *Sussemionus*, a new subgenus of *Equus* (Perissodactyla, Mammalia). *Comptes Rendus de l'Academie des Sciences, Biologies, 333*, 235–240.

Eisenmann, V. (2022). The equids from Liventsovka and other localities of the Khaprovskii Faunal Complex, Russia: A revision. *Geobios, 70*, 17–33.

Eisenmann, V., & Baylac, M. (2000). Extant and fossil *Equus* (Mammalia, Perissodactyla) skulls: A morphometric definition of the subgenus *Equus*. *Zoologica Scripta, 29*, 89–100.

Eisenmann, V., & Deng, T. (2005). *Equus qingyangensis* (Equidae, Perissodactyla) of the Lower Pleistocene of Bajiazui, China: evidence for the North American origin of an Old World lineage distinct from *E. stenonis*. In Crégut, E. (Ed.) *Les Ongulés Holarctiques du Pliocène et du Pléistocène,* Actes du Colloque international d'Avignon, 19–22 Septembre 2000, *Quaternaire, hors-série 2, 2005,* pp. 113–122.

Eisenmann, V., & Guérin, C. (1984). Morphologie fonctionnelle et environnement chez les Périssodactyles. *Geobios, mémoire spécial n°8. Congrès International de Paléoécologie, Lyon, 1983*, 69–74.

Eisenmann, V., & Vasiliev, S. (2011). Unexpected finding of a new *Equus* species (Mammalia, Perissodactyla) belonging to a supposedly extinct subgenus in late Pleistocene deposits of Khakassia (southwestern Siberia). *Geodiversitas, 33*, 519–530.

Forsten, A. (1998). The fossil horses (Equidae, Mammalia) from the Plio-Pleistocene of Liventsovka near Rostov-Don, Russia. *Geobios, 31*, 645–657.

Forsten, A., & Eisenmann, V. (1995). *Equus (Plesippus) simplicidens* (Cope), not *Dolichohippus*. *Mammalia, 59*, 85–89.

Forsten, A., & Sharapov, S. (2000). Fossil equids (Mamalia, Equidae) from the Neogene and Pleistocene of Tadjikistan. *Geodiversitas, 22*, 193–314.

Franzen, J. L. (2010). *The rise of horses: 55 million years of evolution*. Baltimore: Johns Hopkins University Press.

Guérin, C., Faure, M., Argant, A., Argant, J., Crégut-Bonnoure, E., Debard, E., et al. (2004). Le gisement pliocène supérieur de Saint-Vallier (Drôme, France): synthèse biostratigraphique et paléoécologique. In M. Faure & C. Guérin (Eds.), *Le gisement pliocène final de Saint-Vallier (Drôme, France)*, *Geobios*, Mémoire spécial 26, 37, 349-360..

Heintz, E. (1970). Les Cervidés villafranchiens de France et d'Espagne. *Mémoires du Muséum national d'Histoire naturelle, Nouvelle Série, Série C, 32*, 2 volumes, 1-303, 1-206.

Hemmer, H., Kahlke, R.-D., & Vekua, A. K. (2001). The Jaguar - *Panthera onca gombaszoegensis* (Kretzoi, 1938) (Carnivora: Felidae) in the late Lower Pleistocene of Akhalkalaki (South Georgia; Transcaucasia) and its evolutionary and ecological significance. *Geobios, 34*, 475–486.

Kahlke, R. -D. (2006–2007). Untermassfeld - A late Early Pleistoce (Epivillafranchian) fossil site near Meiningen (Thuringia, Germany) and its position in the development of the European mammal fauna. *British Archaeological Reports, International Series 1578*, 1–141.

Kostopoulos, D. S., & Sen, S. (1999). Late Pliocene (Villafranchian) mammals from Sarikol Tepe, Ankara, Turkey. *Mitteilungen der Bayerischen Staatssammlung für Paläontologie und Historiche Geologie, 39*, 165–202.

Koufos, G. (1992). Early Pleistocene equids from Mygdonia basin (Macedonia, Greece). *Palaeontographia Italica, 79*, 167–199.

Koufos, G. D., & Kostopoulos, D. S. (1997). Biochronology and succession of the Plio-Pleistocene macromammalian localities of Greece. In J.P. Aguilar, S. Legendre, & J. Michaux (Eds.), *Actes du Congrès BiochroM'97. Mémoires et travaux de l'Institut de Montpellier de l'École Pratique des Hautes Etudes, 21*, 619-634.

Koufos, G. D., & Kostopoulos, D. S. (2016). The Plio-Pleistocene Large Mammal Record of Greece: Implications for Early Human Dispersals into Europe. In K. Harvati & M. Roksandic (Eds.), *Paleoanthropology of the Balkans and Anatolia* (pp. 269–280). Cham (Switzerland), Springer.

Kretzoi, M. (1954). Bericht über die Calabrische - (Villafranchische)—fauna von Kislang. Kom. Féjer. *Jahrbuch Ungarisches Geologischen Anstalt, 1*, 239–265.

Lacombat, F. (Ed) (2005). Les grands mammifères fossiles du Velay. Les collections paléontologiques du Plio-Pléistocène du Musée Crozatier, Le Puy-en-Velay. *Annales de la Société des Amis du Musée Crozatier,* 13/14, 1–208.

Maul, L. C., Heinrich, W. -D., Parfitt, S. A., & Paunescu, A. -C. (2007). Comment on the correlation between magnetostratigraphy and the evolution of *Microtus* (Arvicolidae, Rodentia, Mammalia) during the Early and early Middle Pleistocene. In: R.D. Kahlke, L.C. Maul, & P. Mazza (Eds.) *Late Neogene and Quaternary biodiversity and evolution: Regional developments and interregional correlations. Proceedings volume of the 18th International Senckenberg Conference (VI International Palaeontological Colloquium in Weimar). Volume II. Courier Forschungsinstitut Senckenberg 259,* 243–263.

Mäuser M. (1992). Zur Taphonomie der altpleistozänen Säugetier--Fundstelle Würzburg-Schalksberg. *Kaupia—Darmstädter Beiträge zur Naturgeschichte 1,* 89–98.

Meadow, R. (1999). The use of size index scaling techniques for research on archaeozoological collections from the Middle East. Historia Animalium ex Ossibus. Festschrift für Angela von den Driesch zum 65. Geburtstag. *Internationale Archäologie, Bd 8: Studia honoraria*, 285–300.

Nomade, S., Pastre, J. F., Guillou, H., Faure, M., Guérin, C., Delson, E., et al. (2014). ^{40}Ar/^{39}Ar constraints on some French landmark Late Pliocene to Early Pleistocene large mammalian paleofaunas: Paleoenvironmental and paleoecological implications. *Quaternary Geochronology, 21*, 2–15.

Orlando, L., Metcalf, J. L., Alberdi, M. T., Telles-Antunes, M., Bonjean, D., Otte, M., et al. (2009). Revising the recent evolutionary history of equids using ancient DNA. *Proceedings of the National Academy of Sciences USA, 106*, 21754–21759.

Rădulesco, C., & Samson, P. (1990). The Plio-Pleistocene Mammalian succession of the Olteţ Valley, Dacic Basin, Romania. *Quartärpaläontologie, 8*, 225–232.

Samson, P. (1975). Les Equidés fossiles de Roumanie. *Geologica Romana, 14*, 165–352.

Schaub, S. (1943). Die oberpliocaene Säugetierfauna von Senèze (Haute-Loire) und ihre verbreitungsgeschichtliche Stellung. *Eclogae Geologicae Helvetiae*, 36, 270-289.

Sher, A. V. (1987). Olyorian land mammal age of Northeastern Siberia. *Palaeontographia Italica, 74*, 97–112.

Sher, A. V. (1992). Beringian fauna and Early Quaternary mammalian dispersal in Eurasia: Ecological aspects. *Courier Forschungsinstitut Senckenberg, 153*, 125–133.

Skinner, M. P., & Hibbard, C. W. (1972). Early Pleistocene pre-glacial and glacial rocks and faunas of north-central Nebraska. *Bulletin of the American Museum of Natural History, 148*, 1–148.

Sotnikova, M. V. (1989). Khishchnye Mlekopitaiushchie Pliotsena - Rannego Pleistotsena. *Geologicheskij Institut, Trudy 440*, 1–121.

Qiu, Z., Deng, T., & Wang B. (2004). Early Pleistocene Mammalian Fauna from Longdan, Dongxiang, Gansu, China. *Palaeontologica Sinica*, New Series C, *121* (27), 1–198.

Torre, D. (1987). Pliocene and Pleistocene marine-continental correlations. *Proceedings of the VIIIth RCMNS Congress, Annales Instituti Publici Geologici Hungarici, 30*, 71–77.

Torre, D., Albianelli, A., Bertini, A., Ficcarelli, G., Masini, F., & Napoleone, G. (1996). Paleomagnetic calibration of Plio-Pleistocene mammal localities in central Italy. *Acta Zoologica Cracoviensa, 3*, 559–570.

Uerpmann, H. P. (1982). Faunal remains from Shams ed-din Tannira, a Halafian Site in Northern Syria. *Beyrutus, 30*, 3–52.

Van den Meulen, A. J., & Van Kolfschoten, T. (1986). Review of the late Turolian to early Biharian mammal faunas from Greece and Turkey. *Memorie della Società Geologica Italiana, 31*, 201–211.

Vekua, A. K. (1986). The Lower Pleistocene Mammalian Fauna of Akhalkalaki (Southern Georgia, USSR). *Palaeontographia Italica, 74*, 63–96.

Wang, S.-Q., & Deng, T. (2011). Some evolutionary trends of *Equus eisenmannae* (Mammalia, Perissodactyla) in the stratigraphic sequence of Longdan, China, in comparison to modern *Equus*. *Journal of Vertebrate Paleontology, 31*, 1356–1365.

Westerhoff, W. E., Cleveringa, P., Meijer, T., van Kolfschoten, T., & Zagwijn, W. H. (1998). The Lower Pleistocene fluvial (clay) deposits in the Maalbeek pit near Tegelen. In T. Van Kolfschoten & P. L. Gibbard (Eds.), *The Dawn of the Quaternary* (pp. 35–69). *Proceedings of the SEQS-EuroMam symposium 1996. Mededelingen Nederlands Instituut voor Toegepaste Geowetenschappen TNO*.

Wüst, E. (1900). Untersuchungen über das Pliozän und das älteste Pleistozän Thüringens nördlich vom Thüringer Walde und westlich der Saale. *Abhandlungen Naturforschende Gesellschaft Halle, 23*, 1–352.

Chapter 12
Sus strozzii (Mammalia, Suidae) of the Late Villafranchian Site of Senèze (Domeyrat, Haute-Loire, France)

Martine Faure and Claude Guérin

Abstract No remains of *Sus strozzii* were found during the new excavations in Senèze, but the nearly complete skeleton of a young male was found before World War II and is preserved in the Basel Natural History Museum. It was briefly studied and well illustrated by Azzaroli in 1954. The species is a large one, anatomically closer to the recent Southeast Asian long-snouted *Sus verrucosus* and *S. barbatus* than to the recent Eurasian *Sus scrofa*. Its cheek teeth are wider and lower, its molars less complicated and covered with thicker enamel than those of *S. scrofa*. Its limb bones are stockier than those of *S. scrofa*. *S. strozzii* is known in Western and Southeastern Europe and in the Middle East from the Early Villafranchian (biozone MNQ 16) onwards up to the early Middle Pleistocene (biozone MNQ 20). The presence of *Sus strozzii* is a clear indication of a rather forested area under a wet and probably warm climate.

Résumé Aucun reste de *Sus strozzii* n'a été découvert lors des nouvelles fouilles à Senèze, mais un squelette presque complet d'un jeune mâle avait été trouvé avant la Deuxième Guerre mondiale. Conservé au Muséum d'Histoire naturelle de Bâle, il a été brièvement étudié et bien figuré par Azzaroli en 1954. C'est un grand sanglier, anatomiquement plus proche des actuels *Sus verrucosus* et *S. barbatus* au groin long d'Asie du Sud-Est que de l'actuel *Sus scrofa* d'Eurasie. Ses dents jugales sont plus larges et basses, ses molaires sont plus simples, avec un émail plus épais que chez *S. scrofa*. Les os des membres sont aussi plus trapus que ceux de *S.scrofa*. *S. strozzii* est connu en Europe de l'Ouest et du Sud-Est, et au Moyen-Orient, depuis le Villafranchien inférieur (biozone MNQ 16) jusqu'au début du Pléistocène moyen (biozone MNQ 20). La présence de *Sus strozzii* indique clairement un environnement plutôt forestier, sous un climat humide et probablement chaud.

Keywords Anatomy • Biometry • Biochronology • Paleoenvironment

Mots-clés Anatomie • Biométrie • Biochronologie • Paléoenvironnement

Introduction

The late Villafranchian site of Senèze (Domeyrat, Haute-Loire) has been known since the publication of Boule (1892) dealing with the discovery of a proboscidean skeleton. The site was extensively excavated between 1907 and 1942 by the local peasant Pierre Philis who gathered a huge quantity of fossils which he sold to several scientific institutions, mainly the Naturhistorisches Museum Basel, the Muséum national d'Histoire naturelle in Paris and the Faculté des Sciences de Lyon (Faure et al. 2022). A nearly complete skeleton of a young male suid was found by Philis and is preserved in the Naturhistorisches Museum Basel. The 2000-2006 scientific excavations by the Franco-American team co-directed by Faure, Guérin and Delson recovered no suid remains.

The first indication of the occurence of *Sus strozzii* at Senèze was given by Stehlin (1923: 275) who pointed out an incomplete scapula described as too large for *Sus arvernensis* and somewhat small for *Sus strozzii* (*S. arvernensis* and *S. strozzii* were the sole Villafranchian suid species known in Europe at that time). The subcomplete young male skeleton was discovered afterwards and sent to the Naturhistorisches

Claude Guérin—Deceased 2016

M. Faure (✉)
Université Lumière-Lyon 2, Lyon, France
e-mail: fauremartine@free.fr

Université de Lyon, UCBL, ENSL, CNRS, LGL-TPE, Villeurbanne, France

C. Guérin
Université Claude Bernard-Lyon 1, Villeurbanne, France

E. Delson et al. (eds.), *Senèze: Life in Central France Around Two Million Years Ago*,
Vertebrate Paleobiology and Paleoanthropology, https://doi.org/10.1007/978-3-031-64415-3_12

Museum Basel, where Stehlin acknowledged receipt of the bones on February 11, 1937. The skeleton was prepared and mounted in 1937. It is still exhibited in the museum's permanent exhibition (Fig. 12.1). Schaub took a photograph of the mounted skeleton which he sent to Philis on January 1, 1938 (Letters of Stehlin and Schaub to Philis, Philis family archives; Faure et al. 2022). He published this picture (Schaub 1943-44, figure 3), with a very brief description including some measurements of the skull and, surprisingly enough, some measurements of the long bones, which are clearly juvenile with their unfused epiphyses and thus without real biometrical significance. Although, on the contrary, the cranium, the mandible, the teeth and the preserved metapodials are paleontologically much more interesting for comparison, Schaub did not really study them.

In a general paper dealing with the European Villafranchian suids, Azzaroli (1954) presented a rather short but well illustrated study comparing some cranial, dental and osteological characters of the *Sus strozzii* of Senèze with those of the Olivola and especially the Upper Valdarno specimens. He pointed out some differences, for example the persistence of the I3 in the Senèze skull (lost in the Valdarno specimens), but failed to recognize if such characters could be considered as evolutionary indicators or not.

Heintz et al. (1974) gave a critical revision of the faunal lists of the main Villafranchian deposits of France and confirmed the presence of *Sus strozzii* at Senèze and probably at Le Coupet (Haute-Loire). Lastly the presence of *S. strozzii* at Saint-Vallier, questioned by Viret (1954), was established by Faure (2004).

The following abbreviations are used:

abs. = absolute; anat. = anatomical; ant. = anterior; APD = antero-posterior diameter; art. = articular; C,c = canine; CV = coefficient of variation; dia. = diaphysis; dist. = distal; h = height; I,i = incisor; L = length; M,m = molar; min, minimum value in measured sample; max, maximum value in measured sample; Mc = metacarpal; Mt = metatarsal; N, number of specimens in measured sample; P, p = premolar; post = posterior; prox. = proximal; SD = standard deviation; TD = transverse diameter; v, variance; w = width. Upper teeth are indicated by uppercase letters, lowers by lowercase.

The ages of the fossiliferous sites follow the MNQ biozones system (Guérin 1980, 1990, 2007).

For comparative material we used the *Sus strozzii* remains stored in the paleontological collections of Firenze University (IGF), Montevarchi Museum, Naturhistorisches Museum Basel (NMB), Lyon Musée des Confluences (ML, formerly Muséum d'Histoire naturelle de Lyon), Université Claude Bernard-Lyon 1 (FSL), Muséum national d'Histoire naturelle Paris (MNHN), Natural History Museum London (NHMUK), Nationaal Natuurhistorisch Museum (Naturalis) in Leiden and Teyler Museum in Haarlem.

For the Simpson diagrams we used as reference the mean values of a large sample of the recent *Sus scrofa*. We also studied numerous remains of European and Middle Eastern Pleistocene *Sus scrofa* housed in all these collections, as well as some recent *Sus verrucosus* and *Sus barbatus* from NHMUK and NMB.

Systematic Paleontology

Order ARTIODACTYLA Owen, 1848
Suborder SUIFORMES Jaeckel, 1911
Superfamily SUOIDEA Gray, 1821
Family SUIDAE Gray, 1821
Subfamily SUINAE Gray, 1821
Genus *Sus* Linnaeus, 1758
Type species: *Sus scrofa* Linnaeus, 1758.

The genus is presently known in Europe, Asia and Northern Africa. It probably appeared during the Late Miocene in Southeast Asia. In Europe during the Pliocene, suids were represented by two species of the genus *Potamochoerus*: the Ruscinian *P. provincialis* (Gervais, 1848-1852) and the Villafranchian *P. magnus* Arribas & Garrido, 2008, both more or less sympatric with the genus *Sus*. The first European *Sus* species was *S. arvernensis* (Croizet & Jobert, 1828), with two successive subspecies, *S. arvernensis minor* Depéret, 1890 of the late Ruscinian (biozone MNQ 15) and *S. arvernensis arvernensis*, of the early Villafranchian (biozone MNQ 16). Guérin et al. (1998) estimated the mass of *Sus arvernensis* between 45 and 100 kg. Its cheek teeth are still relatively simple and the last molars usually end with a single tubercle.

Sus arvernensis is a small species anatomically closer to the present-day Southeast Asian boars with a long rostrum of the *Sus verrucosus* type than to the recent Eurasian wild boar *Sus scrofa*. Following Groves (1981), the main species of these long-snouted Asian boars are *Sus verrucosus* Müller, 1840 of Java, Madura and Bawean; *Sus barbatus* Müller, 1838 of Borneo, Sumatra, Malaysia, Philippines and neighboring islands; *Sus bucculentus* Heude, 1892 of Vietnam; and *Sus celebensis* Müller, 1840 of Sulawesi and neighboring islands.

The next European *Sus* was *Sus strozzii* Meneghini, 1858. That large species appeared during the early Villafranchian (biozone MNQ 16) and survived locally up to the latest Early Pleistocene (biozone MNQ 20). Azzaroli (1975) wrote that *Sus strozzii* is almost exactly a larger version of "*Sus minor*" (i.e., *Sus arvernensis minor*) and differs very slightly from *Sus verrucosus* in general proportions. He considered that "*Sus minor*" is a direct forerunner of *Sus strozzii*. The recent *Sus scrofa* has been known since the end of the Early Pleistocene (biozone MNQ 20) with the large primitive subspecies *Sus scrofa priscus* (Goldfuss 1823; Guérin & Faure 1997).

There are two main groups within the genus *Sus*: a long snouted one today inhabiting Southeast Asia with two main lineages, those of *S. verrucosus* and *S. barbatus*; and the shorter snouted *Sus scrofa* group presently found in Eurasia and Northern Africa. Both groups show evolutionary changes in body size and proportions, reduction of the enamel thickness of the cheek teeth, and progressive complication of the molars whose accessory tubercles become higher and more numerous.

Species: *Sus strozzii* Meneghini, 1858.

Sus strozzii was defined on the basis of fossils from the Upper Valdarno Formation in Italy, but the precise locality for the lectotype (skull IGF 424 with the mandible and some vertebrae) is unknown (Azzaroli 1954; Rook et al. 2013, Figure 16). The specific name was created by Meneghini in a catalogue that can not be found today. In a very short text, Forsyth Major (1881) was the first to observe analogies with *Sus verrucosus*. Stehlin (1899–1900) provided a more detailed description of the Valdarno teeth, published the main measurements of the lectotype, and confirmed these analogies.

Diagnosis following Koufos (1986): *Sus* of large size, with narrow and elongated palate; nasals separated by a prominence of the bones across them; elongated and parallel tooth rows. Supraorbital groove well developed.

Van der Made and Moyà-Solà (1989) characterized the species as follows: very large species. The male canine is verrucosic, the alveolar crests above it are well developed, and the base of the crown of the P4 is lower relative to the base of the crown of the M1 than in *Sus scrofa*. Usually the last molar has a simple talon. The shorter molars (i.e., the first and the second ones) can readily be differentiated from those of *S. scrofa* by their relatively great width. We do not follow the proposal of Berdondini (1992) to place *S. strozzii* in *Dasychoerus*, a genus created by J.E. Gray (1873) for the species *Sus verrucosus* and *S. celebensis*. This taxonomic change did not convince paleo-mammalogists then or after Pickford (2012) championed it 20 years later; see the recent detailed analysis by Iannucci (2024, pp. 46–47).

Subspecies: no subspecies were defined.

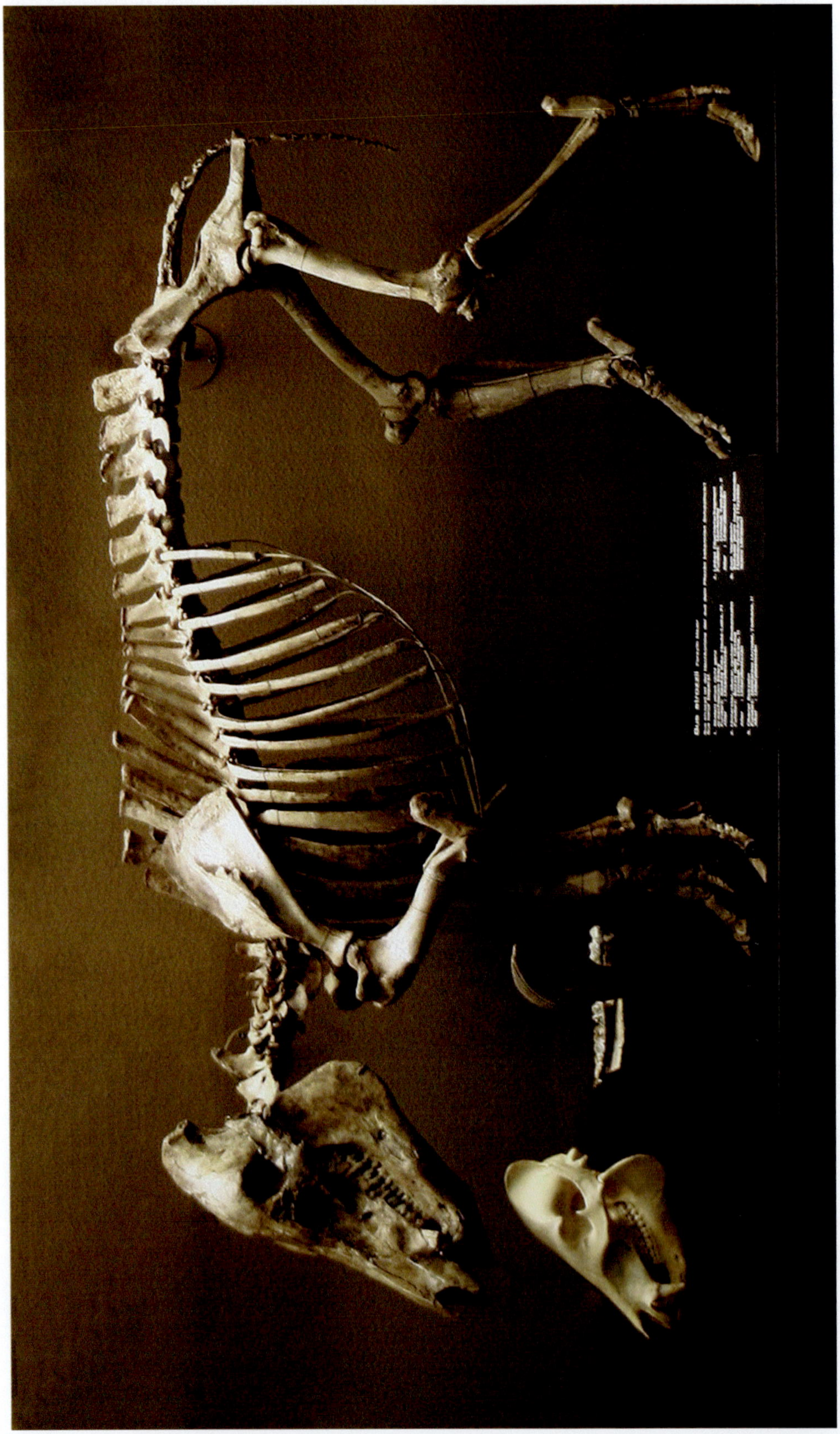

Fig. 12.1 Nearly complete young male skeleton of *Sus strozzii* from Senèze NMB Se 1775, mounted in 1937. Naturhistorisches Museum Basel (courtesy L. Costeur)

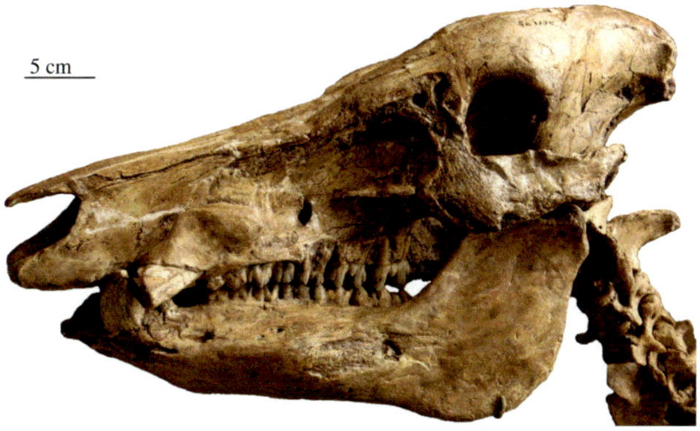

Fig. 12.2 Cranium and mandible NMB Se 1775 of *Sus strozzii* from Senèze, left lateral view; Naturhistorisches Museum Basel (courtesy L. Costeur)

Material

The Senèze *Sus* material is preserved in the Naturhistorisches Museum Basel. It corresponds to two individuals and consists of:

- the nearly complete skeleton of a young male, NMB Se 1775 (Figs. 12.1-12.2);
- an isolated incomplete scapula, probably female, NMB Se 1500 (registered April 1910).

Description

Cranium. The Senèze cranium was illustrated by Azzaroli (1954, pl. XV (VII) fig. 4 a, b and c) in dorsal, ventral and left lateral views. Comparative dimensions are provided in Table 12.1. The Senèze cranium is smaller than the lectotype of *S. strozzii*, larger than the mean values of a sample of 24 to 45 recent *Sus scrofa* but smaller than the mean values of a sample of 12 to 13 recent *Sus barbatus*.

Table 12.1 Cranial dimensions of *Sus strozzii* (Senèze and elsewhere) and recent *Sus scrofa*, *S. verrucosus* and *S. barbatus*

		Sus strozzii			*Sus scrofa* recent					
		Senèze	Olivola	Valdarno						
Variables	Variable Number	NMB Se 1775	NHMUK M 5634	IGF 424[1]	N	Mean	Min	Max	SD	v
Middle supra-orb. crest—premax. tip	1	431		479*	39	378.19	288	445.5	39.39	10.41
Lateral angle supra-orb. crest—premax. tip	2	430.5		483	41	384.18	295	446	36.35	9.46
Occipital condyle—premax. tip	3	370		401	37	340.47	260.5	391	32.26	9.48
ant. orbital border—premax. tip	4	273.5		317	43	245.77	189.5	295	24.89	10.13
Middle supra-orbital crest—nasal tip	5	420		450*	37	368.34	283.5	429	36.18	9.82
ant. orbital border—bottom nasal notch	6	222.5	221	259	43	202.72	128	245	25.73	12.69
Length of the temporal fossa	7	115		98	42	101.73	74	118	10.78	10.60
Facial height in front of P2	8	84		79	41	58.06	44	76.5	7.56	13.02
Facial height between P4 and M1	9	93.5	98	91.5	46	69.01	53.5	89.5	7.56	10.95
Facial height between M2 and M3	10	109	115	103	45	82.14	65	105	8.82	10.73
Breadth of the supraoccipital crest	11	98*		118*	40	73.51	53	99.5	10.89	14.82
Occipital breadth at the ext. auditory meatus level	12	131.5		146*	43	117.02	92	149	12.61	10.78
Middle supra-orb. crest—foramen magnum	13	118,5*		112	41	89.99	65	119	12.47	13.85
Middle supra-orb. crest—basion	14	157*		154	39	119.01	87	148	14.35	12.06
Breadth of the occipital condyles	15	57		60	41	59.38	45	71	5.79	9.76
Frontal breadth at level of post-orb. processes	16	125			43	105.93	86	126	9.90	9.34
Bizygomatic breadth	17	157		188	43	149.31	120	177.5	14.65	9.81
Palate width between P4 and M1	18		38	38	45	37.79	29	47	3.95	10.46
Post-palatal notch—premaxillary tip	19	250*	254	282	41	233.11	177	273	23.20	9.95
Muzzle breadth at the lateral border of the canines	20	96*	96	116.5	42	76.87	54	115.5	17.95	23.36
Muzzle breadth at the lateral border of the I3	21	61*	43*	48	43	45.63	33	58.5	6.61	14.50
lat. angle supra-orb. crest—ant. orb. Border	22	160		163.5	29	136.67	106	158.5	14.32	10.48

(continued)

Table 12.1 (continued)

		Sus strozzii			*Sus scrofa* recent					
		Senèze	Olivola	Valdarno						
Variables	Variable Number	NMB Se 1775	NHMUK M 5634	IGF 424[1]	N	Mean	Min	Max	SD	v
Premax. tip - anterior border of P2	23	118.5	122.5	139	28	110.34	77	139	14.69	13.31
Breadth of the post-orbital constriction	24	60		56*	31	54.48	42.5	82.5	7.81	14.33
Frontal breadth at the M3 level	25	82	70*	58.5	24	70.96	43.5	80.5	8.45	11.91

	Sus verrucosus recent						*Sus barbatus* recent					
Variables	N	Mean	Min	Max	SD	v	N	Mean	Min	Max	SD	v
Middle supra-orb. crest—premax. tip	8	408.25	373	430	18.31	4.49	12	477.00	434	500	20.25	4.24
Lateral angle supra-orb. crest—premax. tip	8	405.50	365	423	19.52	4.81	13	480.69	436	519	23.59	4.91
Occipital condyle—premax. tip	8	351.00	326	363	12.56	3.58	12	423.83	388	445	19.05	4.49
ant. orbital border—premax. tip	8	260.63	238	271	10.84	4.16	12	316.33	285	334	14.44	4.57
Middle supra-orbital crest—nasal tip	6	394.83	365	417	18.03	4.57	13	474.08	421	519	25.24	5.32
ant. orbital border—bottom nasal notch	8	221.00	198	233	10.85	4.91	13	271.46	245	293	14.84	5.47
Length of the temporal fossa	8	95.75	84	106	6.32	6.60	12	114.33	100	125	8.05	7.04
Facial height in front of P2	8	64.88	58	73.5	5.10	7.86	12	70.13	64	74.5	3.53	5.03
Facial height between P4 and M1	8	77.31	71.5	81	3.49	4.52	13	81.69	76	86.5	3.82	4.68
Facial height between M2 and M3	8	93.56	89	99	3.41	3.64	13	93.00	86.5	100.5	4.50	4.84
Breadth of the supraoccipital crest	8	93.13	89.5	96.5	2.62	2.81	12	77.75	70	88	5.19	6.68
Occipital breadth at the ext. auditory meatus level	8	128.88	122.5	135.5	5.33	4.14	13	135.38	123.5	148	6.57	4.85
Middle supra-orb. crest—foramen magnum	8	112.44	105	122.5	5.80	5.16	13	117.73	106.5	132	7.50	6.37
Middle supra-orb. crest—basion	8	138.06	132.5	145.5	4.16	3.01	13	146.92	134	157	6.60	4.49
Breadth of the occipital condyles	8	55.00	49.5	59.5	3.42	6.22	13	63.54	58	69	3.41	5.36
Frontal breadth at level of post-orb. processes	8	99.13	93.5	105.5	4.02	4.06	13	108.85	103.5	115	3.67	3.37
Bizygomatic breadth	8	166.50	161	172	3.85	2.32	13	169.92	156	182	7.84	4.61
Palate width between P4 and M1	8	36.13	28.5	43	4.39	12.15	13	36.65	33	43	3.27	8.93
Post-palatal notch—premaxillary tip	7	250.57	237	265	10.47	4.18	13	325.54	294	345	16.88	5.18
Muzzle breadth at the lateral border of the canines	8	93.69	84	100.5	5.78	6.16	13	92.31	84.5	99	4.96	5.37
Muzzle breadth at the lateral border of the I3	8	51.31	46.5	57	3.05	5.94	13	53.69	48	60	3.60	6.70
lat. angle supra-orb. crest—ant. orb. Border	8	138.31	105	153.5	15.19	10.98	12	156.33	117	170	14.66	9.37
Premax. tip—anterior border of P2	8	121.63	109	136	8.16	6.71	12	157.83	137.5	180	12.78	8.10
Breadth of the post-orbital constriction	8	63.81	60	67	2.39	3.74	12	57.50	50	78.5	7.94	13.81
Frontal breadth at the M3 level	8	70.31	66	76	3.27	4.65	12	76.21	70	81	3.43	4.50

[1]cast, observed as NHMUK M 35089

* Value estimated

Figure 12.3 is a Simpson diagram of the cranium of *Sus strozzii* (mean values of Senèze, the Vadarno lectotype and an incomplete cranium from Olivola) compared with that of *Sus verrucosus* and *Sus barbatus*, recent *Sus scrofa* being the reference. *S. strozzii* shows strong similarities in general proportions with the two Southeast Asian *Sus*, such as the reduced length of the temporal fossa, the low external width of the occipital condyles, the reduced width and the great length of the palate.

Sus strozzii also presents some distinguishing cranial proportions, such as the great height and the narrowness of the anterior part of the cranium, the height of the occipital face, a strong postorbital width, and a more advanced position of the orbit.

Lateral view (Fig. 12.2):

The "wedge-shaped" general form of the suid cranium is evident in lateral and superior views: the dorsal profile of the cranium is straight, the anterior angle between upper and lower profiles is acute. As observed by Azzaroli, the malars are inflated and stand out sharply behind the sides of the rostrum. The zygomatic arches are high and powerful. The occiput clearly overhangs the nuchal surface and the condyles. The anterior rostrum is very elongated. The posterior border of the infraorbital foramen is above the middle of M1. The frontal bones sweep backwards, inserting acutely into the parietals, the naso-frontal suture is saw-edged. The parietal region is nearly flat, with a small longitudinal convexity. The frontal region is flat transversely and longitudinally. Old males, for instance the lectotype of *Sus strozzii*, show very strong supra-alveolar crests (bony crests over the alveoli of the canines); these crests are present but less developed on the Senèze cranium owing to the youth of the specimen.

The rounded orbit is placed far posteriorly, its anterior border being above the posterior end of M3. Azzaroli (1954) pointed out that the lacrymal bone presents the same shape as

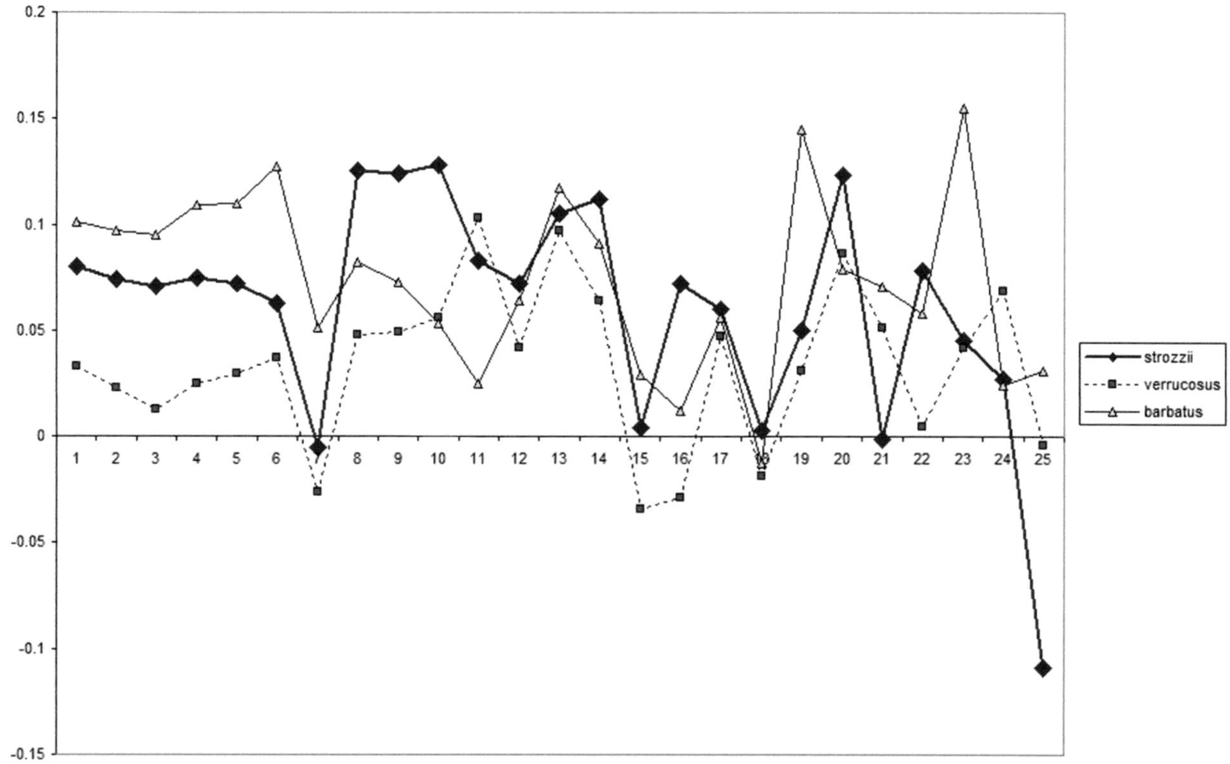

Fig. 12.3 Simpson diagram of the cranium: mean values of *Sus strozzii*, recent *Sus verrucosus* and recent *Sus barbatus*. The reference is mean values of recent *Sus scrofa*. The variables on the X axis are the same and in the same order as in Table 12.1

in *S. verrucosus* with an upper edge clearly longer than the lower one. The shape of the preorbital fossa of the Senèze cranium corresponds well to the description given by Groves (1981) for the recent long-snouted Southeast Asian boars: it is very deep, long and dorso-ventrally narrow, sharply overhung by a thickened shelf of bone postero-superiorly, similarly bordered behind and below by a sharp horizontal crest which runs forward from the zygomatic root and gradually fades anteriorly. Groves noted that the preorbital fossa is the origin of three snout muscles special to Suidae, levator rostri, dilator naris lateralis and depressor rostri. Azzaroli (1954) noted that for *S. strozzii* there is some individual variation in the importance of the insertion of these muscles and the crest which runs longitudinally on the lower part of the occipital scale is very strong.

Superior view

The narrow braincase and the strongly divergent zygomatic arches are characters of the *Sus verrucosus* group. The occiput is broad and convex. The zygomatic bones are bulky, very high and pneumatised. The roots of the zygomatic arches extend forward as far as the posterior part of the M2, the arch widens backwards with an outward slope. Azzaroli (1954) noted a strong individual variability for the zygomatic arch. The long and deep supraorbital grooves, rather divergent posteriorly, end anteriorly at the fronto-nasal suture. The nasal bones are wide, convex, rugose and inflated in the midline in old males but not especially rugose nor inflated on the young Senèze cranium.

Ventral view

The palate is narrow and very long, its posterior border is well behind the M3. The maximal width of the premaxilla is at the I3 level. The tooth-rows are parallel. The post-orbital constriction is narrowest at the position of the condyle.

Mandible. The mandible was illustrated by Azzaroli (1954, pl. XII (IV) figures 3 a and b) in dorsal and left lateral views. Table 12.2 provides comparative dimensions. The Senèze specimen fits well with four mandibles from Valdarno. Nearly all the mean dimensions of a sample of 21 to 42 *Sus scrofa* mandibles are inferior to those of *S. strozzii*, as is true for the mean of 6 *S. verrucosus* but that of 12 to 13 *S. barbatus* is sometimes larger than in the fossil species.

Table 12.2 Mandibular dimensions of *Sus strozzii* from Senèze and Valdarno, recent and fossil *S. scrofa*, recent *S. verrucosus* and *S. barbatus*

Sus strozzii		Senèze	Valdarno	Valdarno	Valdarno	Valdarno
		NMB	NMB	NMB	NMB	NHMUK
	Variable Number	Se 1775	VA 1962	VA 1960	VA 1961	M 35089b
Mandibular tip - angle.	1	320				347
Symphysis length.	2	96	89	99		107
Mandib. tip - coronoid apophysis.	3					328
APD of the ramus.	4					96
Corpus height between p4-m1.	5	58	54	54.5	66	56
Corpus height between m2-3.	6	58	59.5	60	61	60
TD between p4-m1.	7	41*	42.5	45.5	40	42
TD between m2-3.	8	23	39	39		40
Height of the coronoid apophysis.	9					163,5*
Height of the condyle.	10					149*
TD of the condyle.	11					34
Between lateral borders of the condyles.	12					138*
Mandibular breadth across canines	13					91.5
Breadth of symphyseal constriction.	14					80

	Sus scrofa recent						*Sus scrofa* fossil					
	N	Mean	Min	Max	SD	v	N	Mean	Min	Max	SD	v
Mandibular tip - angle.	37	296.05	225	355	30.36	10.25	2	350.00	335	365		
Symphysis length.	37	86.16	55	122.5	14.86	17.25	9	98.83	73.5	133.5	19.51	19.74
Mandib. tip - coronoid apophysis.	36	268.69	203	315	26.53	9.87	2	313.00	295	331		
APD of the ramus.	36	76.11	63	86	6.28	8.25	3	84.67	73	91		
Corpus height between p4-m1.	42	47.54	33.5	66	6.84	14.39	12	51.08	41	62	6.96	13.62
Corpus height between m2-3.	41	46.50	35	67	6.30	13.55	8	48.50	38	59.5	7.36	15.18
TD between p4-m1.	42	26.52	18.5	36	3.91	14.74	12	27.96	19	35.5	4.57	16.35
TD between m2-3.	38	30.63	23.5	38	3.65	11.91	9	32.06	23.5	39	5.19	16.18
Height of the coronoid apophysis.	35	121.66	96	140	10.61	8.72	2	145.00	138	152		
Height of the condyle.	37	112.07	89	130.5	10.46	9.33	2	134.00	125	143		
TD of the condyle.	36	28.31	23	36.5	2.75	9.71	2	33.25	32.5	34		
Between lateral borders of the condyles.	31	127.10	106	151	11.63	9.15	1	140.00				
Mandibular breadth across canines	21	60.81	42	82.5	10.83	17.80	2	63.50	63	64		
Breadth of symphyseal constriction.	25	47.36	33	60	7.03	14.85	2	51.75	47.5	56		

	Sus verrucosus recent						*Sus barbatus* fossil					
	N	Mean	Min	Max	SD	v	N	Mean	Min	Max	SD	v
Mandibular tip - angle.	6	291.50	280	305	8.41	2.88	13	341.50	313	381.5	18.13	5.31
Symphysis length.	6	97.08	90	105	5.04	5.20	13	113.85	93.5	136.5	12.00	10.54
Mandib. tip - coronoid apophysis.	6	285.00	270	294	8.56	3.00	12	312.08	229	346	30.92	9.91
APD of the ramus.	6	87.83	83	93.5	4.30	4.89	12	94.46	89	100	4.00	4.24
Corpus height between p4-m1.	6	52.33	50	57	2.58	4.93	13	56.08	49.5	63	3.80	6.78
Corpus height between m2-3.	6	52.33	49	55.5	2.60	4.97	12	54.88	49	61.5	3.59	6.55
TD between p4-m1.	6	32.50	30.5	36	2.21	6.81	13	27.81	25.5	30	1.71	6.17
TD between m2-3.	6	33.00	30	36.5	2.43	7.36	12	30.50	28	33.5	1.77	5.81
Height of the coronoid apophysis.	6	134.00	130	138.5	3.45	2.57	13	136.85	117	152	11.18	8.17
Height of the condyle.	6	120.00	111.5	128.5	6.55	5.46	13	125.50	108.5	140	9.59	7.64
TD of the condyle.	6	37.08	33	40.5	2.48	6.68	13	34.54	30	43	3.31	9.59
Between lateral borders of the condyles.	6	138.50	134	143	3.87	2.80	12	135.17	125	144	6.24	4.61
Mandibular breadth across canines	6	71.17	67.5	74.5	2.64	3.71	13	77.46	65	87.5	6.19	7.99
Breadth of symphyseal constriction.	6	52.08	51	54	1.20	2.31	12	50.04	46	54.5	3.02	6.03

* Estimated value

The mandible is bulky, with an obvious inflation on the external surface of males coresponding to the root of the canine; such an inflation does not exist on the female mandibles. The corpus mandibulae is long, with a slightly depressed ventral border; at the symphysis, this border leans obliquely anteriorly from bottom to top. The symphysis is not very long, its posterior border is at the level of the contact between p2 and p3. The gonial region is massive. The ramus is long, with an anterior border strongly oblique postero-superiorly. The coronoid process is not higher than the articular condyle. The mental foramen is situated below the diastema, in front of the anterior border of p2. On the internal border of the symphysis, male specimens present two rugose fossae for the insertion of the geniohyoid muscle; these fossae are less strong in the other *Sus* species. Kostopoulos & Athanassiou (2005) pointed out the following characters for the *S. strozzii* mandible: the large and massive mandibular corpora diverge strongly; the symphysis is elongated and strong, its posterior border is at the level of p2; the mandibular gonial angle is almost vertical.

Figure 12.4 is a Simpson diagram of the mandible of *Sus strozzii* (mean values of Senèze and 4 more or less complete mandibles of Upper Valdarno) compared with those of *Sus verrucosus* and *Sus barbatus*, recent *Sus scrofa* being the

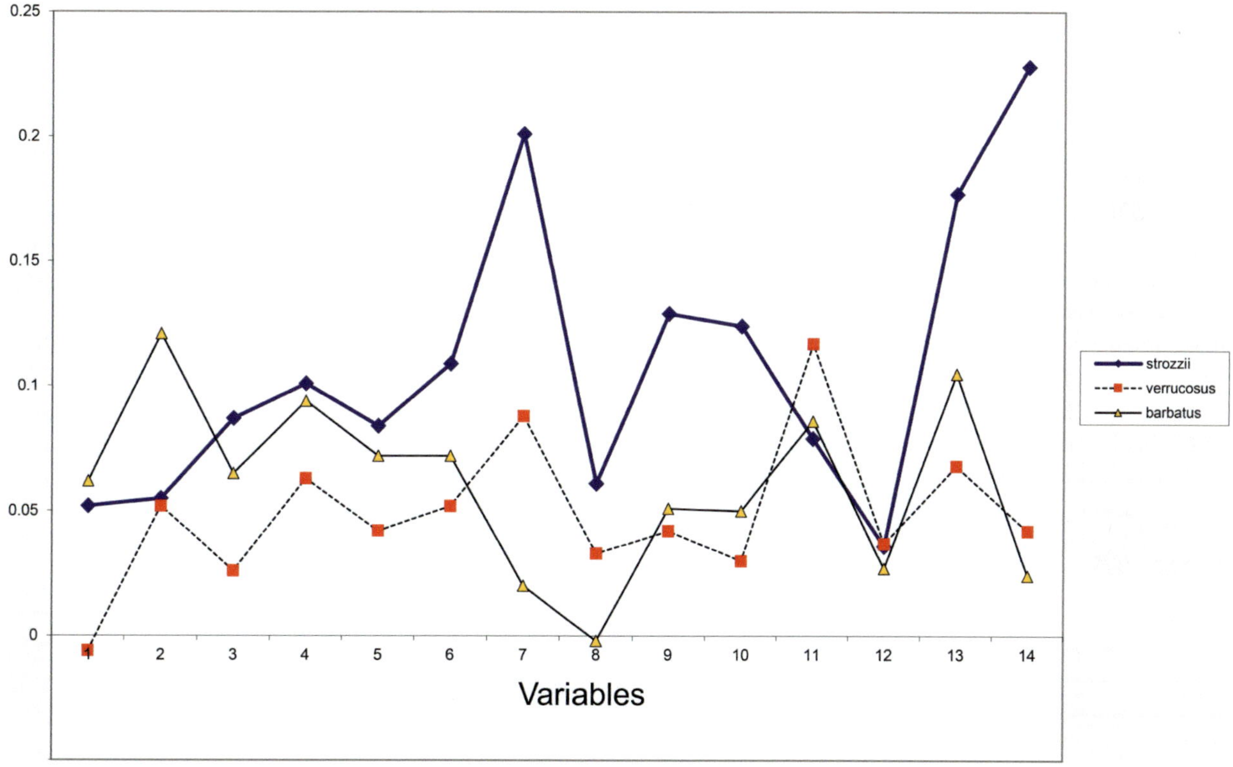

Fig. 12.4 Simpson diagram of the mandible: mean values of *Sus strozzii*, recent *Sus verrucosus* and recent *Sus barbatus*. Reference is mean values of recent *Sus scrofa*. The variables on the X axis are the same and in the same order as in Table 12.2

reference. The fossil shows strong analogies in general proportions with the two Southeast Asian *Sus*, especially with the first one: relative antero-posterior diameter of the ramus, height of the corpus, width of the corpus at the m2-3 level, height of the coronoid and the articular condyle, width at the condyle and at the symphyseal constriction.

Distinguishing features of the *Sus strozzii* mandible include its general length, the relative shortness of the symphysis, the width of the corpus at p4-m1, the weak transverse extension of the condyle, the strong breadth at the canine root level and the great width of the symphyseal constriction.

Teeth

Anterior teeth. The upper incisors are not preserved but their alveoli are clearly observable. The alveoli of the I3s show that these teeth were present and well developed on the Senèze cranium; they are missing on all the Upper Valdarno crania, whose geological age is younger, but a rudiment of alveolus can be seen on the Olivola face preserved in the NHMUK. As Olivola is often thought to be older than Senèze, it may be that the disappearance of the I3 is an evolutionary trend of the species, but the number of observations is not sufficient to support such a claim.

The upper canines of the males are very strong, with a broad and flattened crown; the enamel of their inferior face is longitudinally striated. The antero-posterior diameter of the Senèze specimen is 28 mm, the dorso-ventral diameter is 22 mm. For two upper canines from Tegelen preserved in the Teyler Museum (Haarlem), these dimensions are respectively 33 x 20 mm and 32 x about 22 mm. The large size of one of these upper canines from Tegelen led to its misidentification as *Hippopotamus,* a genus which figured sometimes afterwards in the Tegelen faunal list; this error was rectified by Bernsen (1931).

The six lower incisors are not preserved on the Senèze mandible but their alveoli are present. The Valdarno ones show strong longitudinal grooves. The robust lower canines of the males are of the *verrucosus* type, as described by Groves (1981): the inferior surface is as broad as the enamel-less postero-dorsal surface; the antero-lingual surface, the broadest, forms a rounded hypotenuse opposite the right angle which is subtended between the inferior and posterior surfaces. The inferior angle is acute. Azzaroli (1954, fig. 7) pointed out an important individual variation in the shape of their cross section, which can be triangular or heart-shaped. The transverse diameter of the Senèze lower canine is 15 mm, the dorso-ventral diameter is 24.5 mm.

Upper cheek teeth. They are illustrated in labial and occlusal views by Azzaroli (1954, pl. XIV (VI) figures 1a and b). The dimensions of the upper cheek teeth are given in Table 12.3. Those of Senèze fall within the range of variation of a sample of 5 to 23 specimens of *Sus strozzii*; the lengths and the widths of the Senèze teeth are generally close to the minimum or to the mean values in our

Table 12.3. Dimensions of the upper cheek teeth of *Sus strozzii* from Senèze and other sites, recent and fossil *S. scrofa*, recent *S. verrucosus* and *S. barbatus*

	Sus strozzii						
	NMB Se 1775						
	Senèze	N	Mean	Min	Max	SD	v
P1 L	10	6	11.00	10	12	0.89	8.13
P 1 w	5	6	6.00	5	10.5	2.21	36.89
P2 L	13	10	13.65	13	15	0.63	4.58
P2 w	9	8	8.94	5.5	10	1.47	16.50
P3 L	14	13	14.23	12.5	16.5	0.99	6.97
P3 w	13	13	13.23	10	15	1.28	9.71
P4 L	12	13	13.77	12	15.5	1.25	9.09
P4 w	17	13	17.38	15	19	1.23	7.06
M1 L	17.5	17	18.94	17	22	1.69	8.90
M1 ant. w	17	9	17.06	14.5	21	2.02	11.86
M1 post. w	17	10	18.15	16	21.5	1.93	10.63
M2 L	26.5	17	27.38	23.5	30.5	1.69	6.18
M2 ant. w	23	18	22.61	19.5	24.5	1.33	5.90
M2 post. w	22	19	22.42	19	25	1.40	6.23
M3 L	33.5	23	42.91	33.5	49	3.33	7.75
M3 ant. w	25.5	24	26.54	24	29	1.37	5.15
M3 median w	22	23	23.89	21	27	1.46	6.12
M3 post. w		19	17.26	13	21	1.90	11.02

(continued)

Table 12.3. (continued)

	Sus strozzii						
	NMB Se 1775						
	Senèze	N	Mean	Min	Max	SD	v
Tooth-row L	135	7	137.86	132	145	4.71	3.41
P2-M3 L	123	2	124.25	123	125.5		
P row L	53	6	51.25	48.5	54	1.99	3.89
Two last P L	29	8	27.06	25	29	1.32	4.88
M row L	80	10	86.60	80.5	92	3.33	3.85

	Sus scrofa recent						*Sus scrofa* fossil					
	N	Mean	Min	Max	SD	v	N	Mean	Min	Max	SD	v
P1 L	31	9.23	7	11	0.96	10.46	12	9.54	7.5	12.5	1.21	12.73
P 1 w	31	4.53	3.5	7.5	0.91	20.13	12	4.54	3.5	7.5	1.08	23.69
P2 L	41	13.28	11	15	0.92	6.89	29	13.00	10.5	17	1.46	11.26
P2 w	41	8.11	6	15	1.41	17.36	29	7.50	5	9.5	0.98	13.09
P3 L	45	14.00	10.5	15.5	0.93	6.64	39	13.96	11	17	1.16	8.31
P3 w	45	11.27	7.5	13.5	1.12	9.95	40	10.63	6.5	13	1.51	14.21
P4 L	46	13.35	10.5	18	1.30	9.73	67	13.76	11	16	1.11	8.03
P4 w	46	14.38	12	16.5	0.96	6.65	66	14.01	8	18	2.15	15.38
M1 L	46	17.30	13	20	1.36	7.84	80	17.82	13.5	21.5	1.72	9.65
M1 ant. w	46	14.69	11.5	17	0.95	6.49	77	15.41	10.5	19	1.65	10.71
M1 post. w	46	15.35	10	18.5	1.32	8.63	79	15.56	10	19	1.76	11.29
M2 L	46	23.13	18	26.5	1.61	6.95	83	23.97	17	29	2.45	10.20
M2 ant. w	46	19.07	14	21	1.39	7.31	80	19.44	14	23	2.24	11.51
M2 post. w	46	18.98	13.5	22	1.79	9.43	80	19.31	14	24	2.53	13.10
M3 L	45	35.92	28	46	3.38	9.42	96	38.60	27	49	5.46	14.16
M3 ant. w	45	21.27	15	25	1.79	8.43	95	21.67	16	27	2.63	12.12
M3 median w	43	18.26	11	21.5	2.00	10.95	92	19.10	13.5	27.5	2.88	15.08
M3 post. w	32	12.84	7.5	20	2.30	17.92	91	14.30	9	19	2.31	16.13
Tooth-row L	40	122.48	100	136	7.99	6.52	11	122.95	106	137	10.61	8.63
P2-M3 L	42	120.43	100	136	8.29	6.88	14	117.71	96.5	139.5	12.40	10.54
P row L	30	48.53	40	54.5	3.61	7.44	17	44.59	39	53	4.95	11.11
Two last P L	48	26.41	22	32	1.77	6.70	33	26.56	22	31.5	2.03	7.63

	Sus verrucosus recent						*Sus barbatus* fossil					
	N	Mean	Min	Max	SD	v	N	Mean	Min	Max	SD	v
P1 L	3	8.67	8	9.5			12	9.46	8.5	11	0.84	8.86
P 1 w	3	3.67	3.5	4			12	4.67	3.5	6	0.69	14.69
P2 L	4	12.13	11.5	12.5			13	13.62	12.5	15	0.92	6.73
P2 w	4	7.13	6.5	8			13	8.04	6.5	9	0.72	8.96
P3 L	4	13.75	13	14.5			13	14.69	13.5	16	0.78	5.30
P3 w	4	10.88	10	12			13	11.92	10.5	13.5	0.98	8.18
P4 L	4	12.63	12	14			13	13.42	11.5	15	0.95	7.11
P4 w	4	14.00	14	14			13	14.92	14	16.5	0.79	5.27
M1 L	4	15.63	13.5	17			13	18.00	15.5	19.5	1.24	6.90
M1 ant. w	4	14.13	13.5	14.5			13	15.73	15	16.5	0.63	4.02
M1 post. w	3	14.00	13.5	15			13	16.35	15	17.5	0.59	3.62
M2 L	4	22.13	19.5	24			13	23.46	21.5	25	1.05	4.48
M2 ant. w	4	17.63	15.5	19			13	19.77	19	20.5	0.63	3.20
M2 post. w	4	17.00	16.5	17.5			13	20.19	19	22	0.72	3.58
M3 L	4	37.00	34	40			13	36.54	33	39	1.90	5.19
M3 ant. w	4	20.38	20	21.5			13	22.15	21	23.5	0.69	3.11
M3 median w	4	18.38	17.5	19.5			13	18.96	16.5	21	1.23	6.50
M3 post. w	4	14.25	12	16			13	13.58	11	17	1.54	11.34
Tooth-row L	4	118.50	102.5	126			12	129.71	123.5	134	3.72	2.87
P2-M3 L	4	111.38	102.5	116.5			12	117.75	112	122	4.06	3.45
Two last P L	4	25.38	24.5	26.5			13	27.04	25.5	29	1.16	4.30
M row L	4	74.00	65.5	79			13	76.50	72	81	2.90	3.79

comparative sample. The incompletely erupted M3 does not present its full length.

There is no diastema in the upper tooth-row of the Seneze cranium. P1 and P2 show no traces of wear, and M3, although in place, is unworn. The wear on P4 and M2 is very weak. All the cheek teeth present thick and feebly folded enamel. P1 is double rooted, elongated and narrow, with a main central cusp and two much smaller cuspules, one anteriorly and the other one posteriorly. P2 is similat to P1 but much larger and broader distally. P3 shows the same design but is enlarged posteriorly. Azzaroli (1954) thought that a protocone and a hypocone had begun to develop on it, and on P2 as well, but much less evidently; the anterior part of the cingulum of P3 is rudimentary. P4 is larger and especially wider than the other premolars, its outline nearly circular and its occlusal surface rounded. Geraads et al. (1986) pointed out that the 'Ubeidiya P4 shows a more complicated structure than that of *Sus scrofa*: the tooth is constituted by three main cusps (paracone, metacone and protocone), with a main crest joining paracone and protocone; a secondary crest links this main crest to the higher of the tubercles of the anterior cingulum, near its labial end. Another main crest links the protocone to the metastyle and to the median tubercle of the posterior cingulum. The anterior cingulum is strong and bears 3 tubercles; the posterior cingulum is well developed. There is a vertical groove in the middle of the labial side.

M1 and M2 are typical for Suinae, with their four principal tubercles disposed in two successive pairs where two cusps are situated side by side along the axis of the tooth; these teeth have a peripheral cingulum which for the M2 is slightly folded distally. These molars are broad. On the M1, the posterior cingulum is less developed than in *S. scrofa*, and there is no tubercle in the middle of the lingual cingulum. On the M2 the metacone and the metaconule are much less subdivided than for *S. scrofa*, especially on their anterior face; posterior to these cusps, there are no tubercles other than those of the posterior cingulum; the anterior cingulum is more developed; the posterior cingulum is shallower; there is no tubercle in the middle of the lingual cingulum (Geraads et al., 1986). Figure 12.4 is a scatter diagram plotting the anterior breadth of the M2 against its length. Although of relatively moderate size, the Senèze M2 plots in the cloud of *S. strozzii*, and there is no clear overlap between this cloud and that of fossil *S. scrofa*. Of the latter, the plot points corresponding to *Sus scrofa priscus* fall near the border between the two species.

The M3 also has four principal cusps, but they are followed by a large posterior axial tubercle which makes the tooth trilobate. There are many accessory cuspules on the anterior cingulum, on the distal parts of the paracone and metacone and around the metaconule; this tooth resembles the homologous tooth from *S. scrofa* but is more narrowed posteriorly and less complicated.

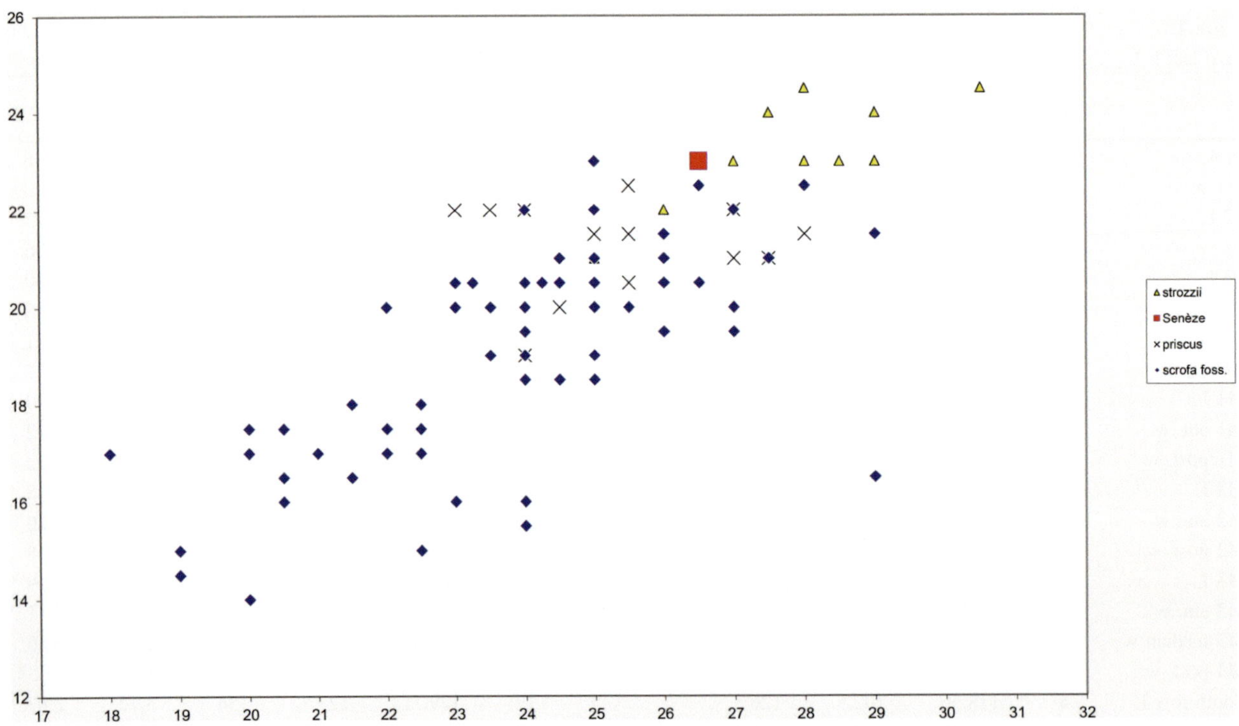

Fig. 12.5 Scatter diagram of anterior transverse diameter as a function of length (X axis) of the M2 of the Senèze *Sus*, other *Sus strozzii*, late Middle Pleistocene and Late Pleistocene *S. scrofa*, and early Middle Pleistocene *Sus scrofa priscus*

Lower cheek teeth. They are illustrated by Azzaroli (1954, pl. XIII (V) figures 7a and b) in labial and occlusal views. There is a long diastema between p1 and p2. The wear of the premolars and the m2 is light. The m3, although in place, is unworn. As for the upper cheek teeth the enamel is thick and little folded.

Figure 12.5 is a scatter diagram of the length of the lower molar row (Y axis) vs. the length of the p3-p4 segment for the Senèze *Sus*, *Sus strozzii*, fossil *S. scrofa*, *S. verrucosus* and *S. barbatus*. The Senèze *Sus* fits well with ten lower tooth-rows from Valdarno and Olivola. There are no significant differences between *S. barbatus* and the largest *S. scrofa*, nor between *S. verrucosus* and the medium sized *S. scrofa*. There is only a weak overlap between *S. strozzii* and *S. scrofa*, the premolar segment of the latter being relatively longer; *S. barbatus* appears more similar to *S. strozzii*.

The dimensions of the lower cheek teeth are given in Table 12.4, compared to a sample from 11 to 23 specimens of *S. strozzii*;, the lengths and widths of Senèze are somewhat lower than the observed averages of our sample. The Senèze teeth are larger than the mean values of those of recent and fossil *S. scrofa* and recent *S. barbatus*.

Table 12.4. Dimensions of the lower cheek teeth of *Sus strozzii* from Senèze and other sites, recent and fossil *S. scrofa*, recent S. *verrucosus* and *S. barbatus*

	Sus strozzii						
	NMB Se 1775						
	Senèze	N	Mean	Min	Max	SD	v
Tooth-row L	127	11	129.27	124	134	3.37	2.61
m row L	84	13	86.46	82	92	3.09	3.58
p row L	43.5	12	43.96	42	47	1.78	4.04
Two last p L	31	15	30.67	29	33	1.03	3.36
m3 L	40	23	43.78	40	48	2.52	5.76
m3 ant. w	20.5	25	21.30	19	23.5	1.00	4.69
m3 median w	18.5	25	19.96	15	22	1.75	8.77
m3 post. w	14	24	17.21	13	19	1.43	8.30
m2 L	25.5	22	26.18	24.5	28	1.08	4.11
m2 ant. w	16	22	17.66	16	19	0.76	4.31
m2 post. w	16.5	22	18.52	16.5	20	0.76	4.12
m1 L	17.5	22	19.36	17.5	22	1.36	7.05
m1ant. w	11	17	13.41	11	15	0.97	7.25
m1 post. w	12.5	18	13.72	11	15	1.00	7.31
p4 L	15	17	15.65	15	17	0.68	4.34
p4 w	13	18	13.00	12	14.5	0.66	5.11
p3 L	14.5	18	14.92	13	17.5	0.86	5.78
p3 w	9.5	17	9.65	8	11	0.86	8.93
p2 L	13	12	12.71	10.5	14	0.94	7.40
p2 w	6	12	6.92	6	8	0.56	8.06

	Sus scrofa recent						*Sus scrofa* fossil					
	N	Mean	Min	Max	SD	v	N	Mean	Min	Max	SD	v
Tooth-row L	64	113.55	86	148	10.77	9.48	25	117.42	97	127.5	9.21	7.85
m row L	63	74.14	54	85.5	6.83	9.21	54	79.61	61	91	6.34	7.96
p row L	54	40.28	31	55	4.15	10.30	19	41.03	33.5	45.5	3.48	8.48
Two last p L	65	28.10	22	31.5	2.34	8.34	46	29.39	22.5	34	2.55	8.68
m3 L	62	37.45	25	44.5	4.25	11.34	156	39.98	30	56	4.91	12.29
m3 ant. w	62	17.26	12	21	1.90	10.99	153	18.62	12.5	25.5	2.28	12.24
m3 median w	60	16.21	9.5	20	2.69	16.60	127	17.79	12	21	2.02	11.34
m3 post. w	35	13.54	9.5	21	2.04	15.07	143	14.52	10	19	1.81	12.44
m2 L	63	21.59	14	25	2.16	9.99	111	23.50	17	28.5	2.04	8.69
m2 ant. w	63	14.68	10.5	17.5	1.39	9.46	107	15.85	11.5	19	1.57	9.91
m2 post. w	62	15.59	11	22.5	1.80	11.52	109	16.52	11	20	1.95	11.79
m1 L	61	16.28	13	19	1.60	9.81	88	17.39	13.5	23.5	1.57	9.01
m1ant. w	59	11.08	8.5	15	1.07	9.70	81	11.51	9	15	1.10	9.56
m1 post. w	60	12.13	9.5	18	1.32	10.89	84	12.60	10	16.5	1.29	10.25
p4 L	63	14.63	8.5	17.5	1.63	11.14	75	15.65	12.5	18	1.26	8.05
p4 w	63	9.76	7	12.5	1.11	11.41	74	10.22	7.5	13	1.31	12.86
p3 L	63	13.68	7.5	16	1.50	10.95	51	14.25	11	17.5	1.41	9.91
p3 w	63	7.55	5.5	13.5	1.23	16.30	51	7.80	6	17.5	1.71	21.94
p2 L	60	11.99	5	14.5	1.64	13.70	30	12.72	10	17	1.52	11.98
p2 w	59	5.46	4	7.5	0.68	12.42	29	5.72	2	9	1.18	20.56

(continued)

Table 12.4. (continued)

	Sus verrucosus recent						*Sus barbatus* recent					
	N	Mean	Min	Max	SD	v	N	Mean	Min	Max	SD	v
Tooth-row L	4	119.13	113	123	4.29		13	124.92	116	134	5.81	4.65
m row L	4	79.88	74	83	4.01		13	80.04	75	87	4.29	5.36
p row L	4	29.50	27.5	30.5	1.41		13	30.81	29	33	1.30	4.22
Two last p L	4	43.50	39	45.5	3.03		12	41.33	36.5	51	4.31	10.42
m3 L	4	18.50	17.5	19.5	0.82		12	18.38	17.5	19.5	0.61	3.31
m3 ant. w	4	16.50	15.5	17	0.71		12	17.46	16.5	18.5	0.50	2.85
m3 median w	4	15.50	14	16.5	1.08		12	14.63	13.5	16	0.96	6.54
m3 post. w	4	22.88	21.5	23.5	0.95		12	23.21	21	25.5	1.18	5.07
m2 L	4	14.88	13	15.5	1.25		12	16.00	15	17.5	0.74	4.62
m2 ant. w	4	15.13	15	15.5	0.25		12	16.71	14.5	18.5	1.12	6.69
m2 post. w	4	15.63	14.5	16	0.75		12	17.67	17	19	0.62	3.48
m1 L	4	10.88	9.5	11.5	0.95		11	11.86	10.5	13	0.78	6.55
m1ant. w	4	11.63	11	12	0.48		12	13.29	12	16	1.03	7.77
m1 post. w	4	15.75	14	17	1.26		12	15.63	14	17	0.88	5.65
p4 L	4	10.25	10	11	0.50		12	10.54	9.5	11.5	0.54	5.14
p4 w	4	13.75	13.5	14	0.29		12	15.08	13.5	16	0.76	5.06
p3 L	4	7.13	6.5	8	0.63		12	8.13	7.5	9.5	0.53	6.49
p3 w	4	12.13	12	12.5	0.25		12	14.04	12.5	16	0.92	6.52
p2 L	4	12.12	12	12.5	0.25		12	14.04	12.5	16	0.92	6.52
p2 w	4	6.38	6	7	0.48		12	6.54	5.5	8.5	0.87	13.22

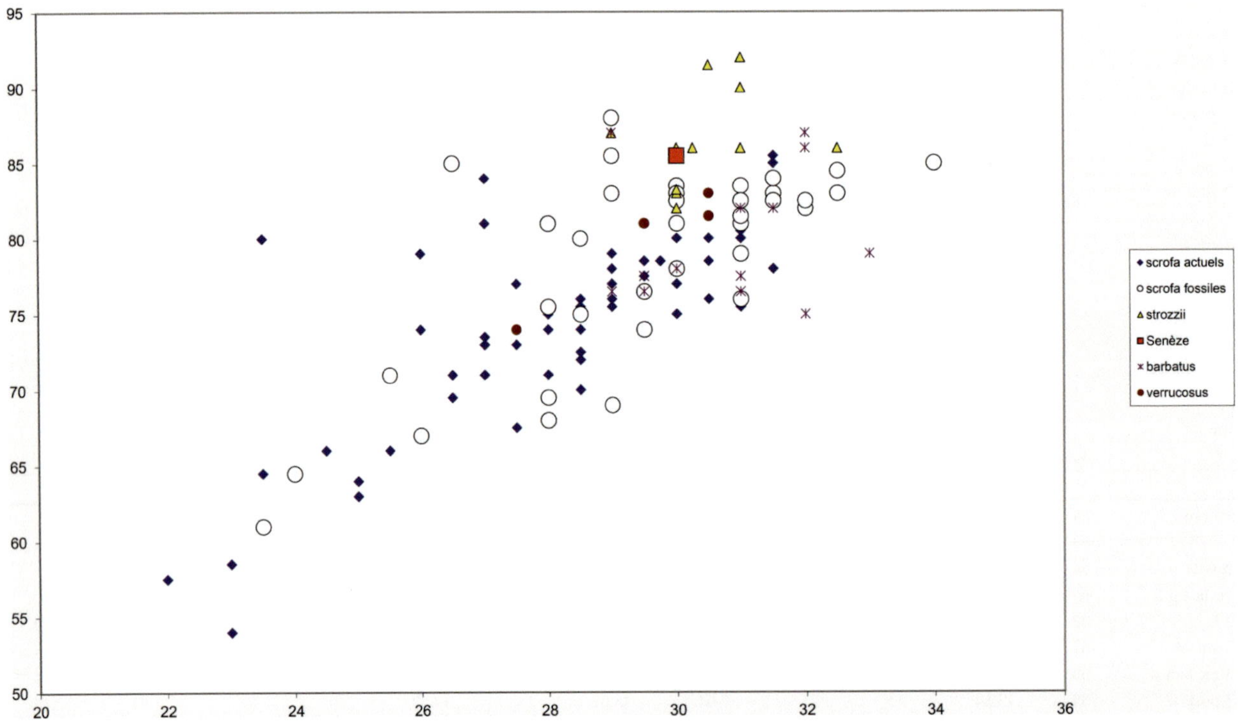

Fig. 12.6 Scatter diagram of the length of the three lower molars as a function of the length of the two last inferior premolars (X axis) for the Senèze *Sus*, other *Sus strozzii*, fossil *S. scrofa*, *S. verrucosus* and *S. barbatus*

The p1 is small and cone-shaped. p2 is blade-shaped; Kostopoulos & Athanassiou (2005) described this tooth as small, with a strong antero-central cuspid and shallow anterior and posterior cingulids. p3 is wider posteriorly; Kostopoulos & Athanassiou (2005) described it as similar to the p2, with a strong anterior cuspid. The postero-labial cingulid is thicker and higher than for *S. scrofa* (Geraads et al. 1986). The p4 possesses a rather low talonid, and the one from Senèze, close to the average length and width of 16 to 17 specimens of *S. strozzii,* has a bifid cuspid; an Olivola specimen shows a cuspid with two partially distinct tubercles, the second clearly offset medially. Kostopoulos & Athanassiou (2005) described the p4 as having two central cuspids, the buccal one situated more anteriorly. The anterior cingulid is shallow, the posterior one is more developed but not significantly raised.

The m1 possesses a posterior cingulid less developed than that of *S. scrofa*; its length corresponds to the lower limit of a sample of 19 *S. strozzii*, but its widths are both somewhat below the minimum. The m2 is very weakly worn and its morphology requires no special comment; its length is close to the average of 21 specimens of *S. strozzii* but it is narrower, especially for its distal part (Table 12.4). Figure 12.6 is a scatter diagram of anterior transverse diameter compared to length of m2s of the Senèze *Sus*, *Sus strozzii*, Late Middle Pleistocene and Late Pleistocene *S. scrofa*, and late Early to early Middle Pleistocene *Sus scrofa priscus*. The Senèze *Sus* plots near the inferior boundary of the *S. strozzii* cloud; there is a partial overlap of *S. strozzii* and *S. scrofa priscus* specimens. On m3 there are three transverse ridges, with two major cusps and a varying number of accessory cupules making up each ridge; the triangular posterior talonid has up to four cusps, more or less rounded, broad and pointed. Geraads et al. (1986) described the talonid as constituted by three terminal tubercles separated by an inverted Y shaped groove and pointed out that *S. strozzii* does not possess a postero-external tubercle as observed in *S. scrofa*. Figure 12.7 is a scatter diagram of anterior width versus length of the m3 of the Senèze *Sus*, *Sus strozzii*, Late Middle Pleistocene and Late Pleistocene *S. scrofa*, and late Early to early Middle Pleistocene *Sus scrofa priscus*. This diagram shows the same results as that of Fig. 12.6.

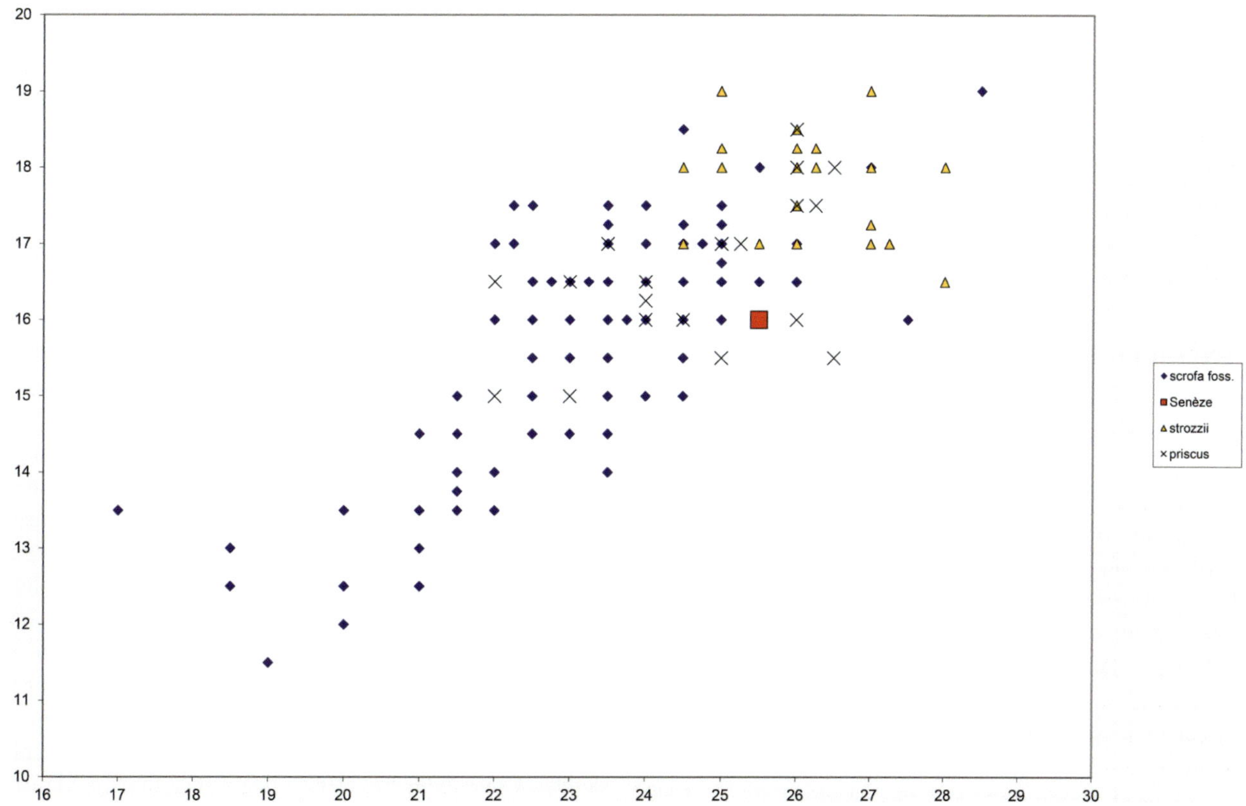

Fig. 12.7 Scatter diagram of anterior transverse diameter as a function of length (X axis) of the m2 of the Senèze *Sus*, other *Sus strozzii*, late Middle Pleistocene and Late Pleistocene *S. scrofa*, and early Middle Pleistocene *Sus scrofa priscus*

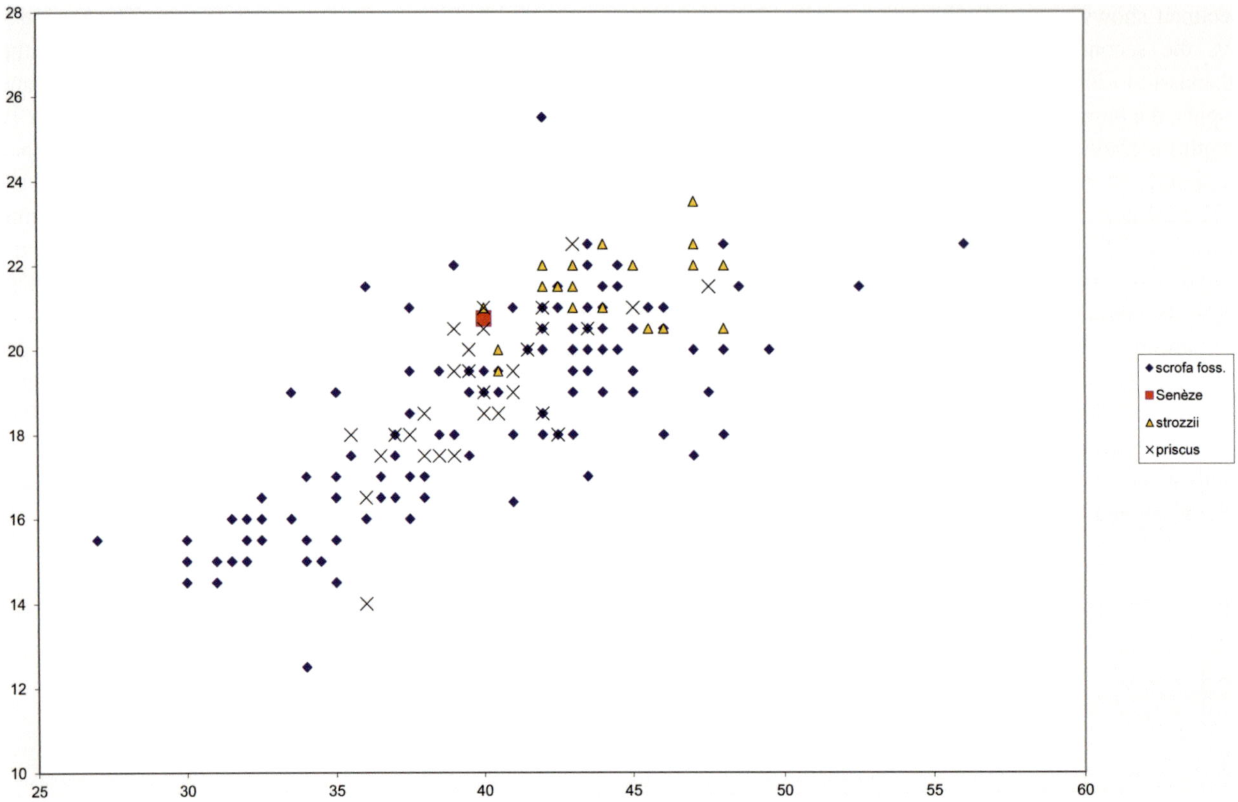

Fig. 12.8 Scatter diagram of anterior transverse diameter as a function of length (X axis) of the m3 of the Senèze *Sus*, other *Sus strozzii*, late Middle Pleistocene and Late Pleistocene *S. scrofa*, and early Middle Pleistocene *Sus scrofa priscus*

Postcranial Skeleton

Other than the Senèze specimen, the postcranial skeleton of *S. strozzii* is practically unknown, the sole exceptions (for complete or nearly complete adult bones) being a radius and an astragalus from Tegelen, a tibia from Olivola and two Mc III and one Mc IV from Valdarno. Therefore Azzaroli (1954) described the postcranial skeleton from Senèze briefly but illustrated it in detail (pl. XV (VII), XVI (VIII), XVII (IX), XVIII (X), XIX (XI), and XX (XII) fig. 2), pointing out several osteological characters. The first two cervical vertebrae present the characters of *S. verrucosus*. The articular surface of the scapula is perfectly circular. The humeral head is rounded, and the proximal third of the diaphysis very strong. The distal articulation of the radius recalls that of *S. verrucosus*, and the olecranon is strongly developed. The rather badly preserved carpus is broad, the astragalus massive. The metacarpals and metatarsals are short. Azzaroli gave some dimensions of the Senèze limb bones, especially the lengths, which are not significant owing to the incompletely fused epiphysis of the long bones, with the exception of the metapodials.

On each hand and foot, the carpals, tarsals and metapodials are mounted together and cannot be separated, so we were unable to take more than a handful of measurements. The length of the Mc III is 86 mm, the proximal transverse diameter is 23 mm, the transverse diameter of the diaphysis 18 mm and the supra-articular distal width 23 mm. Figure 12.8 is a plot of the supra-articular distal width of the Mc III compared to length. The Senèze Mc III falls near the two Mc III of Valdarno *S. strozzii*. Recent *S. scrofa* are clearly shorter, some fossil *S. scrofa* are longer but generally slender. Faure & Guérin (1984) established that the Mc III of *Sus strozzii* is on average slightly longer than that of recent *Sus scrofa*, but with a much more robust diaphysis (which is especially wider) and a bigger distal epiphysis. On the other hand, it is shorter than the Mc III of the Late Middle and Late Pleistocene *S. scrofa*, whose diaphysis is slender transversally but deeper anteroposteriorly (Figure 12.9).

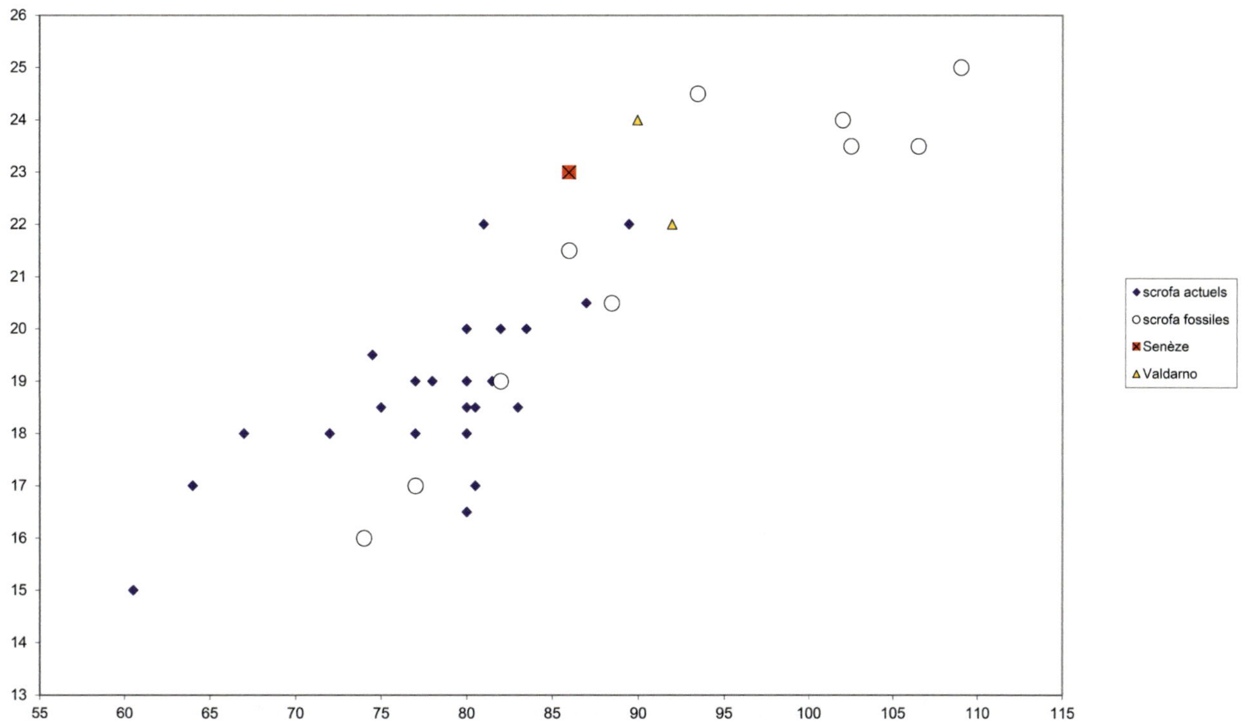

Fig. 12.9 Scatter diagram of distal supra-articular transverse diameter as a function of length (X axis) of the Mc III of the Senèze *Sus*, Valdarno *Sus strozzii*, recent *S. scrofa*, and fossil *Sus scrofa*

The sole other Senèze metapodial we were able to measure is the Mt III. Length is 90 mm, the proximal transverse diameter 19.5 mm, the transverse diameter of the diaphysis 17 mm and the supra-articular distal width 22 mm. The Senèze Mt III is no longer than the largest recent *S. scrofa*, and shorter than many fossil *S. scrofa*, but always stockier. Recent *S. verrucosus* and *S. barbatus* are indistinguishable from recent *S. scrofa*.

The Tegelen radius is 195 mm long, with a proximal transverse diameter of 40 mm and a diaphyseal transverse diameter of 26 mm; for the incompletely grown Senèze radius these measurements are respectively 179, 31.5 and 26 mm. For 11 recent *S. scrofa*, they are 126 to 191 mm (mean 163), 24 to 36.5 mm (mean 31,5), and 15.5 to 25 mm (mean 21.8); for four Pleistocene *S. scrofa*, 143.5 to 200 mm (mean 168.9), 27.5 to 40 mm (mean 32.6) and 18.5 to 29.5 mm (mean 23.1); a recent *S. barbatus* is 194 mm long, the two transverse diameters being 32.5 and 24 mm; radii of two recent *S. verrucosus* are smaller.

We did not see the Olivola tibia but Azzaroli (1954) reported its length as 236 mm and its distal transverse diameter 30.5 mm; owing probably to its relatively small size, he attributed it to a female. The subadult Senèze tibia is 245 mm long, and its distal transverse diameter is 36 mm. For 11 recent *S. scrofa* the observed range of the lengths is 166 to 259 mm (mean 216.9 mm), and that of the distal TD is 25 to 38 mm (mean 32.1 mm); two Pleistocene ones are 201.5 and 255 mm long. A tibia of *S. barbatus* is 240 mm long with a distal TD of 33.5 mm; two tibiae of *S. verrucosus* are much smaller.

Radius, tibia and axial metacarpals (but not the shorter third metatarsal) of *S. strozzii* appear longer than those of recent *S. scrofa* and shorter than those of the late Middle and Late Pleistocene ones, but they are always stockier. Such an observation gives the impression of relative limb shortness. Azzaroli (1954) suggested that this limb shortness could be an adaptation to soft and marshy soils.

Conclusion of the Descriptive Study

The Senèze *Sus* presents all the characters of *S. strozzii*, especially the large size, the *S. verrucosus*-like lower canines, many of the cranial and mandibular proportions of the *S. verrucosus*-*S. barbatus* group, and the thick-enameled molars of more simple construction and greater relative width than those of *Sus scrofa*. Many parts of the postcranial skeleton of the species are known only from the Senèze specimen. The large size and stocky construction of *S. strozzii* allow us to estimate its weight. Recent *S. scrofa* includes very large subspecies, such as *S. scrofa attila* Thomas, 1912 of Central and Eastern Europe, Central Asia, Caucasus, and Iraq, whose head and body length is up to 1.6

m, shoulder height 1 m and average weight of 200-250 kg, reaching up to 350 kg for large males (Groves 1981; Hainard 1988). The maximal values we observed for the radial diaphysis are, for recent *S. scrofa*, TD and APD of 25 and 22 mm; for fossil *S. scrofa*, 29.5 and 22 mm; for the Tegelen radius, 26 and 24 mm. Owing to these values, it seems reasonable to assign to *S. strozzii* a weight comparable to that of large *S. scrofa*, with a mean between 180 and 250 kg.

Despite the lack of comparative material, the Senèze specimen presents some differences with the homologous Valdarno remains, whose age is younger. These differences will be discussed in the next part.

Sus strozzii Biostratigraphical Importance

Sus strozzii was defined in the Upper Valdarno Formation in Italy (Forsyth Major 1881; Azzaroli 1954; Gallai 2007; Rook et al. 2013), whose age is late Villafranchian (MNQ 18-19). It first appears in the early Villafranchian (MNQ 16) site of Valdeganga II (loci 4, 8 and 16) in Spain (Mein et al. 1978; Opdyke et al. 1997). It is also present at Damartia on the island of Rhodes (Kostopoulos et al. 2002), which is about the same age. During the middle Villafranchian (MNQ 17), *S. strozzii* is known in France at Saint-Vallier (Drôme; Faure 2004); in Italy at Mugello (Azzaroli 1954; Faure & Guérin 1984), Olivola (Val di Magra, Azzaroli 1950; on the age of Olivola, see Guérin et al. 2004), Quercia (Tuscany, Azzaroli 1954, Iannucci 2023), Coste San Giacomo (Anagni, Latium; Florindo et al. 2021), Podere San Lorenzo (Perugia, Umbria; Cherin et al. 2019) and Torre di Picchio (Terni; Petronio et al. 2000–2002); in Great Britain in the Nodule Bed of the Red Crag (Stuart 1982); in the Netherlands in Oosterschelde estuary (Reumer et al. 2005); in Romania at Valea Grăunceanului (Dacic Basin; Radulesco 2005, Terhune et al. 2020); in Crimea in the Khapry Faunal Complex (Titov 2000; Bajgusheva & Titov 2004); and in Azerbaijan at Palan-Tyukan (Titov 2000). The species is more common during the late Villafranchian (MNQ 18 and 19). In France *S. strozzii* is known at Senèze and Le Coupet (Haute-Loire; Heintz et al. 1974), both of the MNQ 18 biozone (we do not agree with Martínez-Navarro et al. 2015, who consider Senèze as part of the MN17 biozone). Its most important sites are those of the Upper Valdarno, where Azzaroli (1954) listed no less than 14 localities (mainly in the Tasso "faunal unit"): Figline; Cetino presso Incisa; La Faella; Vacchereccia; Il Tasso; Poggiolino presso Montemarciano (le Ville); la Ginestra; le Valli presso Trigesimo; Tarchio; le Mignaie; S. Maria presso il Tasso; Colombaiolo presso S. Giovani Valdarno; Pian Rinaldi presso Viesca; and Viciali (Val d'Ambra). Petronio et al. (2000-2002) listed other sites from Central Italy: Villa S. Faustino (Terni, Umbria), Monte Riccio (Viterbo, Latium) and Fontana Acetosa (Anagni, Latium); a mandible was described from Pantalla (Latium; Cherin et al. 2018). In Greece it occurs in two sites, Gerakarou and Vassiloudi, of the Mygdonia Basin (Koufos 1986, Koufos et al. 1995; Kostopoulos & Athanassiou 2005); and in Croatia in the Sandalja I cave near Pula, Southern Istria, according to M. Malez. Radulesco (2005) also listed it at Fântâna lui Mitilan (Dacic Basin, Romania), but Terhune et al. (2020) did not include it in their updated faunal list. Recently it was reported for the first time in a Hungarian deposit (Iannucci et al. 2019).

Tegelen (Netherlands) is one of the rare North European sites where *S. strozzii* was cited (Bernsen 1931; van Regteren Altena 1951). The age of Tegelen was considered as earliest Middle Pleistocene by Guérin (1980: 978-979) owing to the unquestioned presence of the rhinoceros *Dicerorhinus mercki*; contrary to the statement of Van den Hoek Ostende and de Vos (2006: 298), Guérin was aware of the work of Bernsen (1931) and never affirmed that *Hippopotamus* was present in the site. Van den Hoek Ostende and de Vos (2006) suggested that the site may be constituted of several fossiliferous levels of different ages.

Sus strozzii is probably present at Kvabebi in Georgia, a site attributed by Tobien (1981) to the early Villafranchian (biozone MNQ 16a), but more recently considered MNQ 16b and dated ca. 2.8 Ma by Lazarev et al. (2021). *Sus* cf. *strozzii* is also reported from the Early Pleistocene of Dursunlu in Anatolia (Turkey; Güleç et al. 1999). In the Middle East, *S. strozzii* was found at Bethlehem (Palestine, late Villafranchian; Hooijer 1958) and in the latest Villafranchian site of 'Ubeidiya (Israel; Geraads et al. 1986) whose age is estimated around 1.5 Ma (Tchernov et al. 1986).

Azzaroli (1954) considered that, compared with the Valdarno specimens, the dentition of the Senèze *Sus* shows some primitive characters like the preservation of the I3 (which is observed on the Olivola cranium as well), the rudimentary state of the anterior part of the cingulum on the P3, and the *Sus palaeochoerus* shape of the p4, with a clearly bifid cuspid. Azzaroli thought that these differences may be of "racial" origin, or due to individual variability, or the expression of evolutionary trends. The last hypothesis seems to us the most probable but the scarcity of comparative material does not allow us to confirm it. Geraads et al. (1986) pointed out that 'Ubeidiya *S. strozzii* cheek teeth are narrower than the European ones; such a difference seems more linked to a geographical distinction than to an age difference.

Sus strozzii Paleoecological Interest

Faure and Guérin (1984) and Faure (2004) considered that *Sus strozzii* is a good environmental marker because it is anatomically similar to the present East Asian species *Sus verrucosus* and *Sus barbatus*, two long-snouted boars whose favorite biotope is dense forest under a wet and rather warm climate. Groves (1981) pointed out that presently the two *Sus* species of Java (*Sus verrucosus* and *S. scrofa*) inhabit fields, scrub, secondary, tidal and teak forests and montane regions and feed on roots, tubers, fallen fruits, vegetal shoots, small rodents, snakes, insect larvae and deer fawns. Groves presumed that long-snouted pigs, whose snout is more delicate but much more mobile, root more deeply and in softer, looser soil than the short-snouted species of the *Sus scrofa* type.

Conclusion

The Senèze suid is clearly *Sus strozzii*. Although not fully adult, the nearly complete male skeleton from Senèze improves our knowledge of the osteology and osteometry of the species. The dimensions and proportions of the cranium, the mandible, the teeth and the usable limb bones (those with fused epiphyses) of *S. strozzii* were compared with those of European and Middle Eastern recent and fossil *S. scrofa*, and with those of recent Southeast Asian *S. verrucosus* and *S. barbatus*. Some fossil *S. scrofa* are larger, but *S. strozzii* appears larger and stockier than the recent species. Its size is about the same as the slender *S. barbatus*, *S. verrucosus* being much smaller.

Our study confirms that *S. strozzii* presents, beyond its well known dental similarities with the two Southeast Asian long-snouted *Sus*, strong similarities to them in general proportions of the cranium and the mandible. Such features include the reduced length of the temporal fossa, the reduced width and the great length of the palate, the great relative antero-posterior diameter of the mandibular ramus, the height of the corpus, and the height of the coronoid apophysys and of the articular condyle. Our study also shows that *Sus strozzii* possesses some distinctive cranial and mandibular proportions: the height and the narrowness of the anterior part of the cranium, the height of its occipital face, the strong postorbital width, and a relatively advanced position of the orbit. On the mandible its main peculiarities are the relative shortness of the symphysis, the width of the corpus at the p4-m1 level, the weak transverse extension of the condyle, and the important width of the symphyseal constriction.

The Senèze *S. strozzii*, like that of Olivola, is a possibly more primitive form than the Upper Valdarno one. Such a supposition is impossible to test owing to the paucity of comparative material, but it fits well with the older age of Senèze and Olivola. Anatomical analogies of *S. strozzii* to the recent *S. verrucosus* and *S. barbatus* lead us to assign to it similar ecological requirements, i.e., a forested area under a wet and perhaps rather warm climate.

Acknowledgments For access to *S. strozzii* and other fossil suid specimens preserved in their collections, we acknowledge the late Professor Augusto Azzaroli of Firenze University; Loïc Costeur, Curator of the vertebrate collections of the Naturhistorisches Museum Basel; Andy Currant, Curator of the fossil mammal collection at the Natural History Museum, London; the late Professor Léonard Ginsburg of the Muséum national d'Histoire naturelle in Paris; and Abel Prieur, Curator of the paleontological collections of Université Claude Bernard-Lyon I. We thank Dimitris S. Kostopoulos, Martin Pickford and Chris Robinson, whose comments on an earlier version helped to improve our manuscript; and Eric Delson for assistance with language editing.

References

Arribas, A., & Garrido, G. (2008). Un nuevo jabalí del género *Potamochoerus* (Suidae, Artiodactyla, Mammalia) en el Plioceno superior terminal euroásiatico (Fonelas P-1, Cuenca de Guadix, Granada). In A. Arribas Herrera (Ed.), *Vertebrados del Plioceno superior terminal en el suroeste de Europa: Fonelas P-1 y el Proyecto Fonelas. Cuadernos del Museo Geominero*, Madrid, *10*, 337–364.

Azzaroli A. (1950). Osservazioni sulla formazione Villafranchiana di Olivola in Val di Magra. *Atti della Società Toscana di Scienze Naturali, Memorie*, serie A, 57, 104–111.

Azzaroli, A. (1954). Filogenesi e biologia di *Sus strozzii* e di *Sus minor*. *Palaeontographia Italica*, *48* (1952), 41–76.

Azzaroli, A. (1975). Remarks on the pliocene suidae of Europe. *Zeitschrift für Säugetierkunde*, *40*, 355–367.

Bajgusheva, V. S., & Titov, V. V. (2004). Results of the Khapry Faunal Unit Revision. In L.C. Maul & R.D. Kahlke (eds.), *Terra Nostra*, Berlin, Heft 2004/2, 18th International Senckenberg Conference & VI International Palaeontological Colloqium in Weimar "*Late Neogene and Quaternary biodiversity and evolution: Regional developments and interregional correlations*" (Weimar, April 2004), Conference Volume.

Berdondini, E. (1992). Suids of the Early Villafranchian of Villafranca d'Asti and China. *Rendiconti Lincei-Scienze Fisiche e Naturali, 3*, 109–124.

Bernsen, J. J. A. (1931). Eine Revision der fossilen Säugetierfauna aus den Tonen von Tegelen IV. *Natuurhistorisch Maandblad*, *20*, 104–108.

Cherin, M., Sorbelli, L., Crotti, M., Iurino, D. A., Sardella, R., & Souron, A. (2018). New material of *Sus strozzii* (Suidae, Mammalia) from the Early Pleistocene of Italy and a phylogenetic analysis of suines. *Quaternary Science Reviews*, *194*, 94–115.

Cherin, M., Azzarà, B., Breda, M., Ansoleaga, A.B., Buzi, C., Pandolfi, L., et al. (2019). Large mammal remains from the early Pleistocene site of Podere San Lorenzo (Perugia, Central Italy). *Rivista Italiana di Paleontologia e Stratigrafia, 125*, 489–515.

Faure, M. (2004). Le *Sus strozzii* du Pliocène final de Saint-Vallier (Drôme). In Faure, M., & Guérin, C. (Eds.), Le gisement pliocène terminal de Saint-Vallier (Drôme, France). *Geobios*, *37*, Mémoire spécial 26, 189–190.

Faure, M., & Guérin, C. (1984). *Sus strozzii* et *Sus scrofa*, deux mammifères artiodactyles marqueurs des paléoenvironnements. *Palaeogeography, Palaeoclimatology, Palaeoecology*, *48*, 215–228.

Faure, M., Sen, S., & Costeur, L. (2022). Pierre Philis *Fossil Hunter*: un demi-siècle de recherches de vertébrés fossiles (1892-1942). *Geodiversitas*, *44*, 321–383.

Florindo, F., Marra, F., Angelucci, D. E., Biddittu, I., Bruni, L., Florindo, F., et al. (2021). Environmental evolution, faunal and human occupation since 2 Ma in the Anagni basin, central Italy. *Scientific Reports*, *11*, 7056 (p. 17).

Forsyth Major, C. J. (1881). Studii sugli avanzi pliocenici del genere *Sus* (*Sus strozzii* Menegh.). *Atti della Società Toscana di Scienze Naturali*, Proc. Verb., *2*, 227.

Gallai, G. (2007). Sistematica, paleoecologia, paleogeografia dei Suidae fossili italiani. *PaleoItalia*, *17*, 17–22.

Geraads, D., Guérin, C., & Faure, M. (1986). Les Suidés du Pléistocène ancien d'Oubéidiyeh (Israël). In E. Tchernov & C. Guérin, (Eds), Les mammifères du Pléistocène inférieur de la vallée du Jourdain à Oubéidiyeh. *Mémoires et travaux du Centre de Recherche français de Jérusalem*, *5*, 93–105.

Gliozzi, E., Abbazzi, L., Argenti, P., Azzaroli, A., Caloi, L., Carpasso Barbato, L. et al. (1997). Biochronology of selected Mammals, Molluscs and Ostracods from the Middle Pliocene to the Late Pleistocene in Italy. The state of the art. *Rivista Italiana di Paleontologia e Stratigrafia*, *103*, 369–388.

Goldfuss, A. (1823). Osteologische beiträge zur kenntniss verschiedener säugthiere der vorwelt. *Verhandlungen der Kaiserlichen Leopoldinisch-Carolinischen Akademie der Naturforscher, 11*, 449–490.

Gray, J. E. (1873). Observations on pigs (Sus, Linnaeus; Setifera, Illiger) and their skulls, with the description of a new species. *The Annals and Magazine of Natural History, 4th Series, 11*, 431–441.

Groves, C. P. (1981). *Ancestors for the Pigs: Taxonomy and Phylogeny of the Genus Sus*. Canberra: Australian National University Press.

Guérin, C. (1980). Les rhinocéros (Mammalia, Perissodactyla) du Miocène terminal au Pléistocène supérieur en Europe occidentale. Comparaison avec les espèces actuelles. *Documents des Laboratoires de Géologie de Lyon*, *79*, 1-3, 1–1185.

Guérin, C. (1990). Biozones or mammal Units? Methods and Limits in Biochronology. In Lindsay, E. H., Fahlbusch, V., & Mein, P. (Eds.), *European Neogene Mammal Chronology*, pp. 119–130. New York: Plenum Press.

Guérin, C. (2007). Biozonation continentale du Plio-Pléistocène d'Europe et d'Asie occidentale par les mammifères: état de la question et incidence sur les limites Tertiaire/Quaternaire et Plio/Pléistocène. *Quaternaire*, 18, 23–33.

Guérin, C., & Faure, M. (1997). The Wild Boar (*Sus scrofa priscus*) from the Post-Villafranchian Lower Pleistocene of Untermassfeld. In R.D. Kahlke (Ed.), Das Pleistozän von Untermassfeld bei Meiningen (Thüringen), Teil 1, *Monographien Römisch-Germanisches Zentralmuseum für Vor- und Frühgeschichte*, Bonn, *40*, 375–383.

Guérin, C., & Faure, M. (2002). The Mid-Villafranchian mammals of Saint-Vallier (Drôme, France). *Proc. 1st International workshop on Late Plio/Pleistocene extinction and evolution in the Palearctic. The Vatera fauna.* Lesbos, Grèce, September - October 1999, *Annales géologiques des Pays helléniques*, *39*, A, 197–220.

Guérin C., Faure, M., & Sen, S. (1998). Le gisement de Vertébrés pliocènes de Çalta, Ankara, Turquie. 8. Suidae. *Geodiversitas*, *20*, 441–453.

Guérin, C., Faure, M., Argant, A., Argant, J., Crégut-Bonnoure, E., Debard, et al. (2004). Le gisement pliocène supérieur de Saint-Vallier (Drôme, France): synthèse biostratigraphique et paléoécologique. In Faure, M. & Guérin, C. (Eds.), Le gisement pliocène terminal de Saint-Vallier (Drôme, France), *Geobios*, *37*, Mémoire spécial 26, 349–360.

Güleç, E., Howell, F. C., & White, T. D. (1999). Dursunlu - A new lower Pleistocene faunal and artifact-bearing locality in Southern Anatolia. In H. Ullrich (Ed.) *Hominid Evolution – Lifestyles and Survival Strategies*, (pp. 349–364). Edition Archaea.

Hainard, R. (1988). *Mammifères sauvages d'Europe*, 4ème édition, t. II. Delachaux & Niestlé, Neuchâtel.

Heintz, E., Guérin, C., Martin, R., & Prat, F. (1974). Principaux gisements villafranchiens de France: listes fauniques et biostratigraphie. *Mémoires du Bureau de Recherches géologiques et minières*, *78*, 169–182.

Hooijer, D.A. (1958). An Early Pleistocene mammalian fauna from Bethlehem. *Bulletin of the British Museum (Natural History), Geology*, *3*, 265–292.

Iannucci, A. (2023). *Sus strozzii* (Suidae, Mammalia) from the historical locality of Quercia (Early Pleistocene, Italy). *Geobios*, *77*, 27–36.

Iannucci, A. (2024). *Sus arvernensis* (Suidae, Mammalia) from the early Villafranchian of Villafranca d'Asti (northern Italy): a revision of the largest sample of a key suid species. *Bollettino della Società Paleontologica Italiana, 63*, 45–81.

Iannucci, A., Gasparik, M., & Sardella, R. (2019). First report of *Sus strozzii* (Suidae, Mammalia) from the Early Pleistocene of Hungary (Dunaalmás) and species distinction based on deciduous teeth. *The Science of Nature*, *107*, 5. https://doi.org/10.1007/s00114-019-1661-6

Kostopoulos, D. S., Vassiliadou, K., & Koufos, J. D. (2002). The beginning of Pleistocene in the Balkan area according to the Mammal record; palaeozoogeographical approach. *Annales géologiques des Pays Helléniques*, *39*, A, 253–278.

Kostopoulos, D. S., & Athanassiou, A. (2005). In the shadow of Bovids: Suids, Cervids and Giraffids from the Plio-Pleistocene of Greece. In E. Crégut-Bonnoure (Ed.), Les Ongulés holarctiques du Pliocène et du Pléistocène, Actes Colloque international Avignon, 19-22 septembre 2000. *Quaternaire*, HS, *2*, 179–190.

Koufos, G. D. (1986). The presence of *Sus strozzii* in the Villafranchian (Villanyian) of Macedonia (Greece). *Paläontologische Zeitschrift*, *60*, 341–351.

Koufos, G. D., Syrides, G. E., Kostopoulos, D. S., & Koliadimou, K. K. (1995). Preliminary results about the stratigraphy and the palaeoenvironment of Mygdonia basin, Macedonia, Greece. *Geobios* Mémoire spécial *18*, 243–249.

Lazarev, S., Kuiper, K. F., Oms, O., Bukhsianidze, M., Vasilyan, D., Jorissen, E. L., et al. (2021). Five-fold expansion of the Caspian Sea in the late Pliocene: New and revised magnetostratigraphic and $^{40}Ar/^{39}Ar$ age constraints on the Akchagylian Stage. *Global and Planetary Change 206*, 103624.

Martínez-Navarro, B., Madurell-Malapeira, J., Ros-Montoya, S., Espigares, M. -P., Medin, T., Hortolà, P., et al. (2015). The Epivillafranchian and the arrival of pigs into Europe. *Quaternary International*, *389*, 131–138.

Mein, P., Moissenet, E., & Truc, G. (1978). Les formations continentales du Néogène supérieur des vallées du Júcar et du Cabriel au N. E. d'Albacete (Espagne). Biostratigraphie et environnement.

Documents des Laboratoires de Géologie de la Faculté des Sciences de Lyon, 72, 99–147.

Opdyke, N., Mein, P., Lindsay, E., Perez-Gonzales, A., Moissenet, A., & Norton, V. L. (1997). Continental deposits, magnetostratigraphy and vertebrate palaeontology, late Neogene of Eastern Spain. *Palaeogeography, Palaeoclimatology, Palaeoecology, 133*, 129–148.

Petronio, C., Argenti, P., Caloi, C., Esu, D., Girotti, O., & Sardella, R. (2000-2002). Updating Villafranchian mollusc and mammal faunas of Umbria and Latium. *Geologica Romana, 36*, 369–387.

Pickford, M. (2012). Ancestors of Broom's pigs. *Transactions of the Royal Society of South Africa, 67*, 17–35.

Radulesco, C. (2005). Artiodactyles du Pliocène et du Pléistocène inférieur de Roumanie. In E. Crégut-Bonnoure (Ed.), Les Ongulés holarctiques du Pliocène et du Pléistocène, Actes Colloque international Avignon, 19-22 septembre 2000. *Quaternaire*, HS *2*, 191–200.

Reumer, J. W. F., Mayhew, D. F., & van Veen, J. C. (2005). Small mammals of the late Pliocene oosterschelde dredgings. *Deinsea, 11*, 103–118.

Rook, L., Croitor, R., Delfino, M., Ferretti, M. P., Gallai, G., & Pavia, M. (2013). The Upper Valdarno Plio-Pleistocene vertebrate record : an historical overview, with notes on palaeobiology and stratigraphic significance of some important taxa. *Italian Journal of Geosciences, 132*, 104–125.

Schaub, S. (1943–1944). Die oberpliocaene Säugetierfauna von Senèze (Haute-Loire) und ihre verbreitungsgeschichtliche Stellung. *Eclogae Geologicae Helvetiae, 36*, 270–289.

Stehlin, H. G. (1899–1900). Über die Geschichte des Suiden Gebisses. *Abhandlungen Schweizerische Paläontologische Gesellschaft, 26*, 1–336; *27*, 337–527.

Stehlin, H. G. (1923). Die oberpliocaene fauna von Senèze (Haute-Loire). *Eclogae Geologicae Helvetiae, 18*, 268–281.

Stuart, A. J. (1982). *Pleistocene vertebrates in the British Isles*. Longmans: London.

Tchernov, E., Guérin, C., Ballesio, R., Bar-Yosef, O., Beden, M., Eisenmann, V., et al. (1986). Conclusion sur la faune du gisement pléistocène ancien d'Oubeidiyeh (Israël): implications paléoécologiques, biogéographiques et stratigraphiques. In Tchernov, E., & Guérin, C. (Eds.), "Les mammifères du Pléistocène inférieur de la vallée du Jourdain à Oubeidiyeh". *Mémoires et Travaux du Centre de Recherche français de Jérusalem, 5*, 351–398.

Terhune, C. E., Curran, S., Croitor, R., Drăgușin, V., Gaudin, T., Petculescu, A., et al. (2020). Early Pleistocene fauna of the Oltet River Valley of Romania: Biochronological and biogeographic implications. *Quaternary International, 553*, 14–33.

Titov, V. V. (2000). *Sus* (Suidae, Mammalia) from the Upper Pliocene of the northeastern part of the Azov region. *Paleontological Journal, 34*, 203–210.

Tobien, H. (1981). Mammals of the Neogene/Quaternary boundary in the Eastern Mediterranean area and from the Upper Siwaliks. In Sastry, M. V. A., Kurien, T. K., Dutta, A. K., Biswas, S. (Eds.), *Proceedings of the Field conference on Neogene/Quaternary boundary (India, 1979)* (pp.185–197). Calcutta: Sree Saraswary Press.

Van den Hoek Ostende, L. W., & de Vos, J. (2006). A century of research on the classical locality of Tegelen (province of Limburg, The Netherlands). *Courier Forschungs-Institut Senckenberg, 256*, 291–304.

Van der Made, J., & Moyà-Solà, S. (1989). European Suinae (Artiodactyla) from the Late Miocene onwards. *Bolletino della Società Paleontologica Italiana, 28*, 329–339.

Van Regteren Altena, C. O. (1951). Systematic catalogue of the palaeontological collection. *Archives du Musée Teyler*, III, *X*, 182–208.

Viret, J. (1954). Le loess à bancs durcis de Saint-Vallier (Drôme) et sa faune de mammifères villafranchiens. *Nouvelles Archives du Muséum d'Histoire naturelle de Lyon, 4*, 1–200.

Chapter 13
New Data on Cervidae (Mammalia, Artiodactyla) from the Villafranchian Locality of Senèze (Haute-Loire, France)

Andrea M. F. Valli

Abstract The cervid remains from Senèze recovered during the expeditions between 2001 and 2006 are studied and described here. Only two of the four previously known taxa are recognized in the new material: *Eucladoceros ctenoides senezensis* and *Metacervoceros rhenanus philisi*, the first one being more abundant than the second. Their morphology and dimensions are compared with those of the specimens already known from previous studies, and we found no significant differences. The impact of our findings for the problem of the existence, at Senèze, of more than one fauna is also discussed. The present results are compatible with the existence of just one fauna.

Résumé Les nouvelles fouilles du gisement paléontologique de Senèze (Haute-Loire, France), réalisées entre 2001 et 2006, n'ont livré que deux espèces de Cervidae parmi les quatre déjà signalées dans cette localité. Aucun élément attribuable à *Croizetoceros ramosus minor* ou à *Cervalces gallicus* n'a en effet été retrouvé. En revanche, de nombreux ossements appartenant à *Metacervoceros rhenanus philisi*, et surtout à *Eucladoceros ctenoides senezensis*, ont été mis au jour. Leur morphologie et leurs dimensions sont comparées à celles des spécimens recueillis lors des anciennes fouilles, et leur correspondent parfaitement. L'importance du nouveau matériel permet aussi de rediscuter l'hypothèse de l'existence de plusieurs faunes fossiles à Senèze. Les résultats de l'étude morphométrique des Cervidae sont d'une part en bon accord avec les données qu'il est possible d'obtenir des restes déjà présents dans les anciennes collections, et d'autre part sont compatibles avec l'existence dans ce site d'une seule faune fossile.

A.M.F. Valli (✉)
Société Scientifique du Bourbonnais, 03000 Moulins (Allier), France
e-mail: andrea_vallifr@yahoo.fr

Istituto Italiano di Paleontologia Umana, Anagni, Italy

Keywords *Eucladoceros ctenoides senezensis* • *Metacervoceros rhenanus philisi* • Morphometric and anatomical comparison • Pleistocene

Mots-clés Comparaison morphométrique et anatomique • *Eucladoceros ctenoides senezensis* • *Metacervoceros rhenanus philisi* • Pléistocène

Introduction

Senèze (Haute-Loire, France) is a well known fossiliferous locality belonging to the Late Villafranchian (*sensu* Heintz 1970, MNQ18 of Guérin 2007). Because of the richness and the quality of its preserved specimens, it has been designated as the reference locality for the Late Villafranchian in European mammalian biostratigraphy (see Delson et al. 2006; Crégut-Bonnoure et al. 2024). At present, four taxa belonging to the family Cervidae are recognized from this site: *Eucladoceros ctenoides senezensis* (Depéret, 1912 in Depéret & Mayet, 1912; = *Eucladoceros senezensis senezensis* for Heintz 1970), *Metacervoceros rhenanus philisi* (= *Cervus philisi* Schaub, 1941), *Cervalces gallicus* (Azzaroli, 1952; = *Libralces gallicus* = *Alces gallicus*), and *Croizetoceros ramosus minor* Heintz, 1970. The first three are represented by complete (or almost complete) skeletons (Azzaroli 1952; Breda 2001; Heintz 1970; Schaub 1941; Viret 1954). Cervid paleontological material is mainly conserved in the Naturhistorisches Museum of Basel and the collections of the Laboratoire de Géologie de l'Université Claude Bernard-Lyon 1, with some fossils housed in the Muséum national d'Histoire naturelle in Paris and in the Muséum d'Histoire naturelle of Lyon (now Musée des Confluences).

E. Delson et al. (eds.), *Senèze: Life in Central France Around Two Million Years Ago*, Vertebrate Paleobiology and Paleoanthropology, https://doi.org/10.1007/978-3-031-64415-3_13

Many scientists analyzing Quaternary fossil cervids have studied and described the Senèze species (e.g., Heintz 1970; Pfeiffer 1999, 2005), so many features of their anatomy and morphology are well known (but also see Mennecart et al. 2017). The aim of this paper is to report and describe the cervid fossil remains recovered by the French-American team led by M. Faure, C. Guérin and E. Delson, who excavated the locality between 2001 and 2006.

Cervid species were not equally common at Senèze: in the recent fieldwork, we were able to find fossils representing only two taxa, *Eucladoceros* and *Metacervoceros*, which are the most abundant deer at this locality. The newly recovered cervid fossils total more than 220 remains: 142 belong to *Eucladoceros*, 64 to *Metacervoceros* and some 20, mostly antler fragments, cannot be allocated to any specific taxon. Among them is an incomplete antler (FSL SEN 03–0044) belonging to a very young individual (Fig. 13.1). It lacks any burr and looks like a spike just tilted on the last third. Scars or grooves are very attenuated on the first 40 mm, but they are well developed on the distal part. It measures 88 mm in total length, 12 and 14 mm respectively in transverse and antero-posterior diameters at its base.

Material and Methods

All the analyzed specimens are housed in the collection of the Laboratoire de Géologie de l'Université Claude Bernard-Lyon 1 (FSL). They will be identified by the field numbers assigned during the work (expressed here by: FSL, the acronym of the locality, "SEN", the year, e.g.,"04–" and a progressive four-digit code). For fossils employed in comparative analyses, catalog numbers and institutions of storage will be given in the text.

The nomenclature for bones, teeth and anatomical features follows Desse et al. (1986), Heintz (1970; modified as in Valli 2001) and Smith and Dodson (2003). Details on measurements are given in Desse et al. (1986) and Valli (2001). The few abbreviations used in the text, figures, and tables are: DAP = antero-posterior diameter, DDV = dorso-ventral diameter, DT = transversal diameter. The measurement error is put at 0.5 mm, as greater precision is unnecessary when measuring the fossil bones of large mammals (Guérin & Faure 2002).

Cervalces gallicus (Azzaroli, 1952) and *Croizetoceros ramosus minor* Heintz, 1970

Cervalces gallicus is the rarest of the fossil cervids from Senèze. All the material is conserved in the collections of the Université Claude Bernard-Lyon 1, where an assembled skeleton is exhibited. In addition, an almost complete set of leg bones is stored separately. It is a relatively small moose characterized by its palmated antlers, which depart laterally from the skull: they are horizontal and their beam is very long. Other typical features are the contact between the nasals and the premaxillae (the lack of articulation between these bones is a diagnostic character of the genus *Alces*, following Breda & Marchetti 2005) and the proportionally very long leg bones, especially metapodials. Heintz (1970) included this species in the genus *Libralces* Azzaroli, 1952, but, except for the length of the beam, it presents the same main features as *Cervalces* Scott (1885). Moreover, according to Lister (1993), there are fewer phenetic differences among all the *Cervalces* species (including *C. gallicus*) than among the populations belonging to the modern moose, *Alces alces* (Linnaeus 1758).

Croizetoceros ramosus minor is a quite rare deer at Senèze. Its fossils are mainly conserved in Basel with a few in the Lyon and Paris collections. In Paris, there is a fragmented skull with the proximal part of the antlers, on which it is possible to observe the typical *Croizetoceros* first tine, hooked in shape, emerging on the beam after a certain distance from the burr and the piriform section of the beam. This specimen has been figured by Heintz (1970; plate VI.4). *Croizetoceros* is a medium-sized cervid with the body-mass of a small fallow deer. Its antlers are very characteristic: multi-pointed (up to six tines) and lyre-shaped. Its molars present a protocone fold, and P_4 is molarized (see the section concerning *M. rhenanus*). *C. ramosus minor* is a subspecies with an important role in biochronology, because it contributes, together with other taxa, to defining the Late Villafranchian interval (Guérin 2007; Heintz 1970). *C. ramosus* is considered a plesiometacarpal deer, because a fossil attributed to this species (from La Puebla de Valverde, Spain) displays the typical features of such animals (for the differentiation between plesiometacarpal and telemetacarpal cervids, see Heintz 1970; Groves & Grubb 1987; Geist 1999). However, a recent analysis of the bony labyrinth of one specimen from Les Étouaires (Puy-de-Dôme, France) and *C. pyrenaicus* (Depéret, 1884) from Perpignan (Pyrénées-Orientales, France) shows that, in this respect, the anatomy of *Croizetoceros* is more similar to that of telemetacarpal species than plesiometacarpal ones (Mennecart et al. 2017). It should be recalled that the ear region is considered crucial for the discrimination between these two cervid taxa (Bouvrain et al. 1989).

Metacervoceros rhenanus philisi (Schaub, 1941)

Taxonomy

For its synonymy, until 1970, see Heintz (1970). Afterward, see Spaan (1992), De Vos et al. (1995), Valli (2004), Pfeiffer (2005), and Valli and Palombo (2008). In 1941, Schaub described *Cervus philisi* from a complete skeleton exhibited in the Naturhistorisches Museum of Basel. Heintz, the author of the most important monograph on European Villafranchian cervids, recognized several subspecies in 1970, including the nominate "*Cervus*" *philisi philisi* for the Senèze cervid. He accepted Schaub's species, but he preferred to maintain the generic level in an uncertain status until more complete analyses were published. Spaan (1992) proposed that "*C.*" *philisi* was a junior synonym of a species described by Dubois (1904), *Cervus rhenanus*, from Tegelen (Netherlands). Both specific nomenclatures are in current alternative use (e.g., Azzaroli 1992; De Vos et al. 1995; Delpech & Guérin 1996; Kostopoulos 1997; Valli 2004). At present, Spaan's proposition is followed more frequently, even if the alternative can sometimes appear (e.g., Mennecart et al. 2017; Sirakov et al. 2010).

At the generic level, Azzaroli (1992) proposed a new name, *Pseudodama*, for the Villafranchian European *Dama*-like deer, including the Senèze species. Other suggestions were also advanced, ascribing the taxon either to a new genus or to modern ones (e.g., *Euraxis*, Di Stefano & Petronio 1998; *Dama*, Van der Made 1999; and *Rusa*, Di Stefano & Petronio 2002). Pfeiffer (1999, 2005) proposed that the Villafranchian European *Dama*-like taxa may constitute a phyletic lineage to *Dama* Frisch, 1775, and she suggested including all of these in the modern genus, despite the fact that the original generic definition mentions the peculiar shape of the fallow-deer's antlers. In fact, Dietrich (1938) had previously erected a new genus, *Metacervoceros*, for several Villafranchian deer, including taxa related to the Senèze species. Grubb (2000), in his assessment of the nomenclature of living and fossil deer, considered it valid, and thus the (sub)species from Senèze can be attributed to the genus *Metacervoceros* (e.g., Croitor 2001, 2006a, 2014; Kostopoulos & Athanassiou 2005; Valli & Palombo 2008).

Morphology, Antlers

M. rhenanus philisi is a fallow deer-sized plesiometacarpal cervid, with simple three-pronged antlers. In term of fossil remains, it is one of the most abundant taxa present at Senèze. Hundreds of specimens (teeth, bones and antlers) are stored in the Basel, Lyon and Paris collections, including the complete skeleton in Basel and at least 4 complete or sub-complete skulls housed in the three collections.

Of the 64 *Metacervoceros* remains recovered during the recent fieldwork, one is an antler fragment, another is the fragment of a left scapula, eight belong to the dentition, and the rest (54) are limb bones, half of these being phalanges. Because of their poor preservation, a number of fossils can be neither described nor measured.

The only portion of antler which definitively belonged to an adult *Metacervoceros* is a basal (right?) fragment (FSL SEN 03–0154; Fig. 13.1). The burr is quite developed, the beam is oval just over it, but is becomes very flat a few cm above, at the level of the dichotomy, about 5 cm above the burr. Several shallow groves are present from the base along the beam. Measurements include: DT at the burr = 36.5 mm, DAP at the burr = 40.0 mm, DT above the burr = 27.5 mm, DAP above the burr = 31.0 mm.

Dentition

Upper teeth (right and left series FSL SEN 01–0140; Fig. 13.1A). P2 is a wide tooth: on the labial side, the posterior style is turned at the back and the pillar of the cone forward; on the lingual side, a well-developed groove is present; in occlusal view, the hypoconal "fold" is present. P3 is wider but shorter than P2: on the labial side, the pillar of the cone is turned forward; on the lingual side, the groove is absent and there is a very slight cingulum in the anterior part of the tooth; in occlusal view, the hypoconal "fold" is also present. P4 is short but wide: on the lingual side, the groove is absent but the anterior and posterior cingula are present even if weak; in occlusal view, the pillar of the cone is in the middle of the wall and it is symmetric on the left tooth and almost symmetric on the right. A clearly forward turned pillar of the cone, in occlusal view, would be a *Croizetoceros* feature (De Giuli & Heintz 1974). The hypoconal "fold" is present. On the molars, the pillar of the paracone and the two first styles are well marked in occlusal view, the metastyle is well recognizable only on the M3 and it is turned at the back. The hypoconal "fold" is always present but on the right M1 it is only slightly expressed. The interlobal column is present as well as the anterior cingulum (but it is weak on the M1).

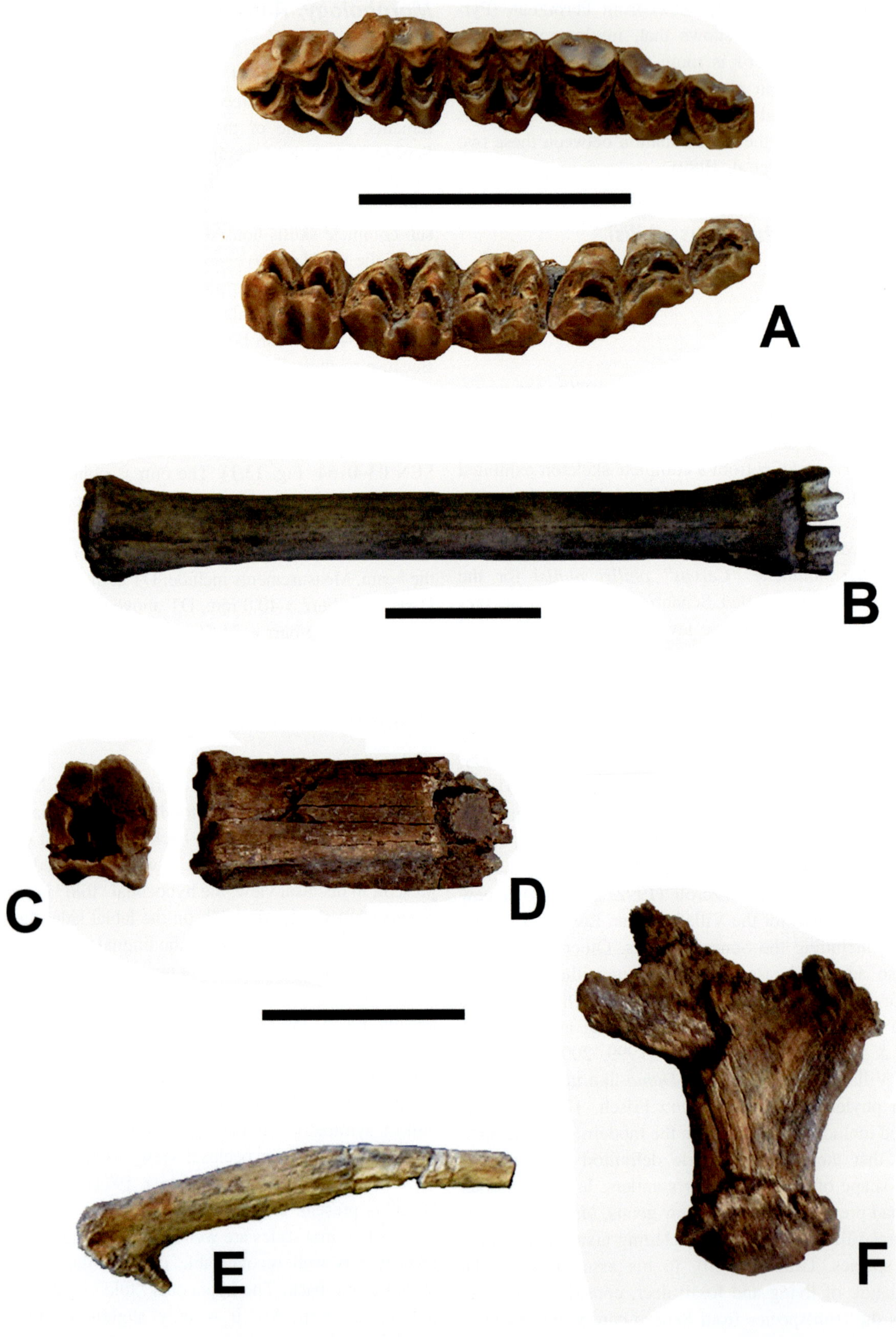

Fig. 13.1 *Metacervoceros rhenanus philisi*: **A**, right and left upper dental series (FSL SEN 01–0140), occlusal view; **B**, right metacarpal III and IV (FSL SEN 06–0079), dorsal view; **C-D**, left metatarsal III and IV (FSL SEN 03–0121), **C** proximal view, **D** medial view; **F**, fragment of (right?) antler (FSL SEN 03–0154), external view. Indeterminate cervidae: **E**, fragment of antler (FSL SEN 03–0049). Scale bars 50 mm

Lower teeth. No complete series of the lower dentition have been found in the recent fieldwork, only isolated teeth or fragments of lower jaw, so not all the lower teeth can be described. The only deciduous cervid tooth found during the 2001–2006 fieldwork is a right D_4 (FSL SEN 04–0303) belonging to *Metacervoceros*. It does not present any differences from normal cervid lacteal teeth. For a more complete description of this kind of tooth, the reader can see Valli (2004). P_4 (two specimens, FSL SEN 04–0413 and FSL SEN 06–0002): the tooth is not molarized (*sensu* Heintz 1970), on the labial side the valley between protoconid and hypoconid is deep but narrow; no cingulids are present on either side. M_3 (right tooth, FSL SEN 04 –0020): in occlusal view, the anterior entoconid wing and posterior metaconid wing overlap, on the upper half of the crown, no clear step is present between the 2nd and the 3rd lobe of the tooth; on the lingual side, the ridge between the 2nd and the 3rd lobe is clear (for more details on these features, see Lister 1996). Other teeth (e.g., P_3) are too worn or fragmented to be described. Measurements of the dentition are presented in Table 13.1.

Table 13.1 Measurements (in mm) of the teeth of *Metacervoceros rhenanus philisi* from Senèze, 2001–2006 collection

	N	Min	Mean	Max		
P^2						
Total length	2	13.0	13.75	14.5	–	–
Length at the base	2	12.0	12.00	12.0	–	–
Width	2	11.5	12.00	12.5	–	–
P^3						
Total length	2	13.0	13.25	13.5	–	–
Length at the base	2	12.0	12.00	12.0	–	–
Width	2	14.0	14.25	14.5	–	–
P^4						
Total length	2	13.0	13.25	13.5	–	–
Length at the base	2	12.0	12.00	12.0	–	–
Width	2	14.0	14.25	14.5	–	–
M^1						
Total length	2	16.5	16.75	17.0	–	–
Length at the base	2	15.0	15.00	15.0	–	–
Width of the first lobe	2	17.0	17.25	17.5	–	–
Width of the second lobe	2	16.5	16.50	16.5	–	–
M^2						
Total length	2	19.5	19.50	19.5	–	–
Length at the base	2	16.0	16.50	17.0	–	–
Width of the first lobe	2	18.5	18.75	19.0	–	–
Width of the second lobe	2	18.0	18.00	18.0	–	–
M^3						
Total length	2	18.5	18.50	18.5	–	–
Length at the base	2	17.5	17.50	17.5	–	–
Width of the first lobe	2	18.0	18.50	19.0	–	–
Width of the second lobe	2	17.0	17.25	17.5	–	–
P^2–M^3 length	2	85.5	85.75	86.0	–	–
P^2–P^4 length	2	38.0	38.50	39.0	–	–
P^3–P^4 length	2	25.0	25.00	25.0	–	–
M^1–M^3 length	2	53.00	53.00	53.0	–	–
D_4						
Total length	1	–	20.0	–	–	–
Length at the base	1	–	19.5	–	–	–
Width	1	–	8.0	–	–	–
P_3						
Total length	1	–	14.5	–	–	–
Length at the base	1	–	13.0	–	–	–
Width	1	–	8.50	–	–	–

(continued)

Table 13.1 (continued)

	N	Min	Mean	Max		
P_4						
Total length	2	15.0	15.00	15.0	–	–
Length at the base	2	13.0	13.50	14.0	–	–
Width	2	8.5	8.50	8.5	–	–
M_3						
Total length	1	–	25.0	–	–	–
Length at the base	1	–	24.0	–	–	–
Width of the first lobe	1	–	11.5	–	–	–
Width of the second lobe	1	–	11.5	–	–	–
Width of the third lobe	1	–	7.0	–	–	–
P_3–P_4 length	1	–	27.5	–	–	–

(Standard deviation and coefficient of variation are only calculated if N $\geq$ 3)

Postcranium

Among the *Metacervoceros* scapular remains are a distal end of a left element (FSL SEN 06–0024) and a damaged right one (FSL SEN 06–0191). The cavitas glenoidalis is oval, more elongated than wide and there is no V-shaped cavity in its articular surface. The tuberculum superglenoidalis is slender. A few measurements have been taken on the specimen FSL SEN 06–0024: minimal DAP = 28.0 mm; DAP at the articular processus = 48.0 mm; DAP of the cavitas glenoidalis = 37.5 mm and DT of the cavitas glenoidalis = 32.5 mm.

Humerus: the only available specimen for measurements and description is a left distal end (FSL SEN 01−0141). On the dorsal side, the external pit is deeper than the internal one, but both are wide.

Radius (three specimens: FSL SEN 04−0028, and two labelled FSL SEN 06−0002). On the proximal end, the attachment of the ligamentum collaterale laterale sticks out laterally, sloping a little downwards; the margo lateralis is relatively pronounced; the cranial facet shows one isolated circular break in its articular surface (two specimens) or one extending to the plantar side (just one specimen). On the distal end, the dorsal edge of the magnum's articular surface is more convex, the edge between the lunar and the scaphoid is more angled, and the plantar edge of the magnum shows a right angle. The distal part of the ulna seems well connected to that of the radius.

Metacarpal III and IV (right FSL SEN 06-0079, Fig. 13.1; also FSL SEN 04-0386 and FSL SEN 04-0403 fit together into another right element; and FSL SEN 06-0002[b] is a third right metacarpal). Proximal end (four specimens: FSL SEN 01-0022, FSL SEN 04-0386, FSL SEN 06-0002, FSL SEN 06-0079): the proximal facets project approximately equally in plantar view (three specimens); on the dorsal side, the tuberosity (Lister 1996) is well developed and bordered by a well-defined ridge; in three cases it is smoother, in one it is rougher. Distal end (six specimens: FSL SEN 01-0022, FSL SEN 03-0140, FSL SEN 04-0008, FSL SEN 04-0403, FSL SEN 06-0002b, FSL SEN 06-0079): on two specimens, a split between the articular ends is visible on the plantar side but not on the dorsal one; on four specimens the split is not visible either on the dorsal or the plantar side. On four specimens, the plantar face slopes anteriorly towards the two distal articular edges. In least two specimens the dorsal sulcus, crossing the dorsal side from one end of the bone to the other, is slight.

Tibia (only two distal ends, one right and the other left, both labelled FSL SEN 06-0004). The dorsal facet for the lateral malleolus bone is not downturned plantarly, it is smaller than the other in distal view and the contact between the two is very reduced. The lateral part of the cochlea tibiae is quite enlarged.

Talus. Only one damaged specimen (a left talus, FSL SEN 06-0103) is available. It can be associated with the left calcaneus with the same number. On the medial side, the proximal-medial ridge is greater in its proximal part than in its distal one and the bulge above the distal pulley is more pronounced. On the dorsal side, a V-shaped pit is present on the distal pulley.

Calcaneus. Three specimens have been discovered: two of them, apparently belonging to the same individual and labelled FSL SEN 06-0103, are damaged. The third, a right calcaneus (FSL SEN 03-0035) is in better condition. On two of these, the sustentaculum tali shows a step, more or less pronounced, on its dorsal ridge.

Cubo-navicular (two right specimens, FSL SEN 03-0110 and FSL SEN 03-0048, and a left one FSL SEN 06-0090). No gap is present in the articular surface for the talus. On the dorsal side, the bulge is quite sharp, at least in two specimens.

Metatarsal III and IV. Five specimens are available: a distal end (FSL SEN 04-0035), two proximal ones (right FSL SEN 03-0110 and left FSL SEN 01-0148), and two complete bones (right FSL SEN 06-0241 and left FSL SEN 06-0244, probably belonging to the same individual). Proximal end: the dorsal articular surfaces meet on a long space, a spur can be present (2 of 4) on the medial side. The posterior articular facet for the cubo-articular is well-developed, reaching the middle of the facet of the metatarsal III. Distal end: no split is visible either on the dorsal or on the plantar side. On the plantar side, the foramen is quite large and oval, and the rim of the distal articular surface of the metatarsal pulley shows a long extension.

Another fragment of left metatarsal III and IV (FSL SEN 03–0121; Fig. 13.1C–D) is considered here. It seems small for *Metacervoceros* (proximal DP = 26.5 mm, proximal DAP = 31.0 mm), but it is also quite large for *Croizetoceros ramosus minor*. It is damaged but the morphology of the posterior articulation for the cubo-navicular looks very similar to that of *Metacervoceros*, and the bone seems to have been not slender enough for *Croizetoceros*. Thus, it is tentatively attributed to *Metacervoceros*. For a more definitive identification, it would be necessary to have the complete bone in order to estimate the ratio of proximal DT/total length (Valli 2004).

The measures of the main limb bones (phalanges excluded) are given in Table 13.2.

Table 13.2 Measurements (in mm) of the limb bones (without the phalanges) of *Metacervoceros rhenanus* philisi from Senèze, Valli (2004–2006) collection

	N	Min	Mean	Max	S.D.[a]	C.V.[b]
Radius						
Proximal DT	3	37.5	42.67	45.5	4.48	10.50
Proximal DAP	3	20.5	21.83	22.5	1.15	5.29
Distal DT	2	39.5	39.50	39.5	–	–
Distal DAP	2	30.0	30.00	30.0	–	–
Metacarpal III–IV						
Total length	2	218.0	227.75	237.5	–	–
Proximal DT	4	31.5	33.63	35.5	1.75	5.20
Proximal DAP	4	23.0	25.00	26.5	1.47	5.89
Diaphysis DT	1	–	20. 5	–	–	–
Diaphysis DAP	1	–	20.0	–	–	–
Supra–articular DT	6	31.5	32.83	35.5	1.54	4.69
Distal DT	6	31.0	33.50	35.5	1.64	4.90
Distal DAP	6	21.5	23.00	24.5	1.26	5.50
Tibia						
Distal DT	2	38.0	38.00	38.0	–	–
Distal DAP	2	30.0	30.00	30.0	–	–
Talus						
Median height	1	–	41.5	–	–	–
Median DAP	1	–	24.5	–	–	–
Distal DT	1	–	27.5	–	–	–
Cubo–navicular						
Total height	1	–	35.5	–	–	–
DT	2	36.0	36.75	37.5	–	–
DAP	2	33.5	33.75	34.0	–	–
Metatarsal III–IV						
Total length	2	240.0	241.00	243.0	–	–
Proximal DT	4	31.5	32.00	32.5	0.41	1.28
Proximal DAP	4	33.5	33.50	37.0	1.47	4.15
Diaphysis DT	2	22.0	20. 25	22.5	–	–
Diaphysis DAP	2	24.5	25.50	26.5	–	–
Supra–articular DT	3	31.5	34.50	36.0	2.60	7.53
Distal DT	3	30.5	33.83	35.5	2.89	8.53
Distal DAP	3	19.5	23.17	25.0	3.18	13.71

[a] Standard deviation only calculated if N $\geq$ 3
[b] Coefficient of variation only calculated if N $\geq$ 3

Phalanx I. Pfeiffer (1999) explained how to distinguish fallow deer anterior phalanges I from posterior ones. Such features can, in certain cases, also be employed for *Metacervoceros*. The anterior phalanges have, in their proximal articular surface, an upper profile which is square while, on the posterior ones, it is quite trapezoidal. The anterior phalanges have the sides of the diaphysis straight and parallel. The sides of the posterior phalanges are not straight but slightly curved inward (see Fig. 80 in Pfeiffer 1999, page 99). Our collections of *Metacervoceros* first phalanges include one anterior, three posterior and seven indeterminate (it is not possible, despite the features listed above, to establish their exact anatomical position). The anterior phalanx is FSL SEN 6–0024b; the posterior ones are FSL SEN 03–0051, FSL SEN 06–0217 and FSL SEN 06–0103c. Eight of them are complete enough for description of their anatomical features. The shape of the articulation for the sesamoid bones is clearly squared, like that of the cervids (Heintz 1970). In plantar view, the lateral bulge is short and, in lateral view, the proximal part of the bone generally presents a U–shaped scar (all but one specimen). In plantar view, five phalanges, independent of their anatomical position, show the boundary of the distal articular surface with the diaphysis is quite straight: a slight V–shaped scar can be detected on some of them. On the other bones this feature is most variable.

Phalanx II. Nine phalanges II are present (FSL SEN 01–0072, FSL SEN 03–0053a, FSL SEN 04–0056, FSL SEN 04–0281, FSL SEN 04–0402, FSL SEN 04–0406, FSL SEN 06–0012, FSL SEN 06–0015 and FSL SEN 06–0018) but only seven are in good enough condition to be described. In proximal view, a post–articular plateau is present. In plantar view, the proximal attachments of tendons are expressed by bulges exceeding the articular surface (four specimens), and the distal trochlea forms no elongate peaks (four specimens).

Phalanx III (eight specimens, but not all complete, FSL SEN 03–0053b, FSL SEN 04–0404, FSL SEN 04–0405, FSL SEN 05–0062, FSL SEN 06–0002e, FSL SEN 06–0013, FSL SEN 06–0016 and FSL SEN 06–0019). The two articular surfaces are slightly different in size. The articulation between the II and the III phalanges allows a more extensive range of movement than that between the II and the I. There is no articulation for the sesamoid bones. In plantar view, the lateral edge is slightly but securely convex. No pyramidal eminence is present; the dorsal edge is straight.

The measures of *Metacervoceros* phalanges are shown in Table 13.3. For phalanx I, the measurement values are given separately for anterior, posterior or indeterminate specimens.

Table 13.3 Measurements (in mm) of phalanges of *Metacervoceros rhenanus philisi* from Senèze, 2001–2006 collection

	N	Min	Mean	Max	S.D.[a]	C.V.[b]
Phalanx I anterior						
Total length	1	–	51.0	–	–	–
Proximal DT	1	–	17.0	–	–	–
Proximal DAP	1	–	22.0	–	–	–
Diaphysis DT	1	–	13.0	–	–	–
Diaphysis DAP	1	–	13.0	–	–	–
Distal DT	1	–	15.5	–	–	–
Distal DAP	1	–	14.5	–	–	–
Phalanx I posterior						
Total length	3	52.5	53.00	53.5	0.50	0.94
Proximal DT	3	17.0	17.33	17.5	1.29	1.67
Proximal DAP	3	21.5	22.00	22.5	0.50	2.27
Diaphysis DT	3	12.5	13.00	13.5	0.50	3.85
Diaphysis DAP	3	16.5	16.67	17.0	0.29	1.73
Distal DT	3	14.5	14.83	15.0	0.29	1.95
Distal DAP	3	13.5	13.83	14.5	0.58	4.17
Phalanx I indeterminate						
Total length	5	46.0	50.60	55.0	4.26	8.43
Proximal DT	4	16.5	17.38	19.0	1.18	6.80
Proximal DAP	4	20.0	21.13	23.0	1.31	6.22
Diaphysis DT	5	12.0	12.60	13.0	0.40	3.32
Diaphysis DAP	5	14.5	15.90	18.0	1.39	8.73
Distal DT	5	14.0	14.90	16.0	0.82	5.51
Distal DAP	5	13.5	14.30	16.0	1.04	7.25
Phalanx II						

(continued)

Table 13.3 (continued)

	N	Min	Mean	Max	S.D.[a]	C.V.[b]
Total length	8	32.5	35.50	39.0	2.27	6.39
Proximal DT	7	15.0	15.93	17.5	0.84	5.26
Proximal DAP	8	20.5	21.50	23.5	0.93	4.31
Diaphysis DT	9	11.0	12.44	13.5	0.73	5.84
Diaphysis DAP	9	14.5	15.39	16.5	0.74	4.81
Distal DT	9	11.5	12.61	13.0	0.49	3.85
Distal DAP	9	16.5	17.89	19.5	1.11	6.22
Phalanx III						
Total length	7	36.0	37.29	39.5	1.11	2.98
Proximal DT	7	32.5	33.93	37.0	1.46	4.29
Proximal DAP	7	13.5	14.43	16.0	1.02	7.05
Distal DAP	7	21.5	22.86	25.5	1.31	5.75

[a] Standard deviation only calculated if N $\geq$ 3
[b] Coefficient of variation only calculated if N $\geq$ 3

Upper Canines

In order to complete the anatomical description of the taxon, it is interesting to remark that upper canines are not present (Croitor 2006a, 2006b; Pfeiffer 2005). Unfortunately, those authors do not report the specimens on which they determined this feature. Although no cranium was recovered in the recent fieldwork, specimens from Senèze were surveyed in Paris (Muséum national d'Histoire naturelle), Lyon (Université Claude Bernard) and Basel (Naturhistorisches Museum). Two of them (the skeleton described by Schaub in 1941 and exhibited in Basel; and the skull labeled "89 1922–5" and seen in a glass case in Paris) confirmed the absence of the upper canines and their alveoli.

Upper canines are missing on all modern telemetacarpal deer except *Hydropotes inermis* Swinhoe, 1870, which retain large ones (Geis 1999; Nowak 1999), and *Rangifer tarandus* (Linnaeus, 1758) where they can be present (Groves & Grubb 1987), even if, normally, they do not pierce the gums (Banfield 1961). Among the best known telemetacarpal taxa, *Odocoileus* Rafinesque, 1832 species do not have upper canines (except rare individuals: Brokx 1972; Heffelfinger 2006), and the same appears true in *Capreolus capreolus* (Linnaeus, 1758) (Chaplin & Atkinson 1968; Kierdorf et al. 2007).

On the contrary, plesiometacarpal deer generally present such teeth. In *Cervus elaphus* Linnaeus, 1758 and *C. canadensis* Erxleben, 1777, it is even possible to sex and age an individual thanks to these teeth (D'Errico & Vanhaeren 2002; Greer & Yeager 1967). *Dama dama* (Linnaeus, 1758) and *Axis axis* (Erxleben, 1777) are some of the few exceptions. In the fallow deer only milk upper canines may be present; in adulthood, it completely lacks canines and their alveoli (Geist 1999). The absence of such teeth in *M. rhenanus philisi* seems to agree with Pfeiffer's idea of a phylogenetic link to *Dama*, but the analysis of Mennecart et al. (2017, Figs. 13.3 and 13.4), shows a closer relationship between the species from Senèze and *A. axis* than with the fallow deer, the last being more closely related to *Megaloceros giganteus* (Blumenbach, 1799). In any case, it is worth noting that in the phylogenetic tree of selected Cervidae based on 26 characters of the bony labyrinth (Mennecart et al. 2017, Fig. 13.3), these four cervid taxa cluster together in the same clade.

Eucladoceros ctenoides senezensis (Depéret, 1912 in Depéret & Mayet 1912)

Taxonomy, Occurrence at Senèze

Concerning the taxonomy of the *Eucladoceros* from Senèze, we follow the proposition of de Vos et al. (1995), who synonymized *E. senezensis* (the specific name was presented for the first time at the Dijon congress in August 1911 but the paper was only published in 1912) with *E. ctenoides* (Nesti, 1841). Similarities between the two species were also noted by Stehlin (1923). Heintz (1970) recognized subspecific status for the Senèze taxon, which he considered significant for European biochronology. It is a red deer–sized plesiometacarpal cervid, characterized by the shape of its complex antlers, with several large tines placed in the anterior part of the beam.

It is one of the most common taxa at Senèze. Two superb skeletons are assembled in the (non–public) exhibition rooms of the Lyon (FSL) collection, and two other more or less complete skulls are conserved in Basel and Paris. Several antler fragments are known, belonging to adult and juvenile individuals. Many hundreds of *Eucladoceros* specimens, in good condition, are housed in Basel, Lyon and Paris.

With 142 fossil specimens, *E. ctenoides senezensis* is the most abundant cervid found during the recent fieldwork. Of these, 15 are fragments of antlers. There are also two incomplete crania and eight other dental remains. But the majority of the bones belongs to the postcranial skeleton. An almost complete individual has been unearthed (Parenti et al. 2024, Fig. 2.5B). The skull and most of the distal part of the legs were missing, but we recovered several ribs and vertebrae, for a total of 73 elements. Those mentioned in this article will be indicated by an asterisk (*) following the field number.

Of course, because of the quality of the preservation, not all the bones have been measured and described. We have concentrated our analysis on the cranial remains and the main leg bones recovered in 2001–2006.

Antlers and Crania

Antlers (15 fragments, FSL SEN 01–0002, FSL SEN 01–0105, FSL SEN 04–0006, FSL SEN 04–0082, FSL SEN 04–0085, FSL SEN 04–0310, FSL SEN 04–0343, FSL SEN 05–0006, FSL SEN 05–0028, FSL SEN 06–0033, FSL SEN 06–0106, FSL SEN 06–0123, FSL SEN 06–0126, FSL SEN 06–0154 and FSL SEN 06–0170). The attribution of these remains depends on their shape and size. The branching pattern indicates that they are not moose, as does the tendency of the beam to be flat at the divergence points. The beam is furrowed by grooves.

Two partial *Eucladoceros* crania are among our finds. One (FSL SEN 04–0349) is the posterior portion of a male cranium, with the bases of the antlers, but without the muzzle or any portion of the orbits (Fig. 13.2). The basioccipital bone is pentagonal in shape. The pharyngeal tuberosities are well marked and the foramen ovale is large. The profile of the upper margin of the foramen magum, in basal view, is convex (for an example of this shape on a *Eucladoceros* skull see Fig. 6a in Croitor 2006b, page 98). On the temporal bone, the temporal ridges are clearly visible, but not too protruding. The frontal bone is large and flattened, with a shallow depression in front of the pedicles. The letter are inclined posteriorly and sidewards, forming an almost right angle between their axes. The axes of the pedicles are almost cylindrical with an antero–posterior compression especially marked on the lateral sides. In frontal view, the space between the pedicles is horizontal.

The second cranium (FSL SEN 04–0342) is a partial muzzle, including only parts of the portion anterior to the orbits with nasals and palatines, the left maxilla with all teeth but only the right P^2-P^4 (Fig. 13.2), because of damage to the right maxilla, and the complete right premaxilla (the tip of the left one is missing; Fig. 13.2). The alveoli for the upper canines are visible. The area for the maxillary muscle is trapezoidal in shape and as long as the one of *C. elaphus* (for more details about this feature, see Valli & Palombo 2005). The surface for the origin of the masseter muscle does not go down to the upper teeth; its lower boundary is about 2 cm above their roots. Anteriorly, it reaches the posterior portion of the P^4. The anterior part of the orbit lies above the M^3. The complete right premaxilla shows that it is wide but lacks the notch typical of fallow deer, chital, sambar (*Cervus unicolor* [Kerr, 1792]), wapiti and Père David deer (*Elaphurus davidianus* Milne Edwards, 1866), all considered mixed-feeder cervids (Fortelius & Solounias 2000; Geist 1999; Hofmann 1989). Overall, the muzzle is similar to that of a red deer. This shape confirms Valli and Palombo's (2005) anatomical observations used to advance hypotheses on the paleodiet of *E. ctenoides senezensis*. They suggested that its alimentary regime was similar to the red deer's, possibly eating more leaves and buds than the modern species. On the other hand, a recent microwear analysis of a larger dental sample of *Eucladoceros* from Senèze claims that this deer grazed more than previously supposed (Berlioz et al. 2018).

The measures taken on *Eucladoceros* crania are given in Table 13.4.

Standard deviation and coefficient of variation are only calculated if $N \geq 3$.

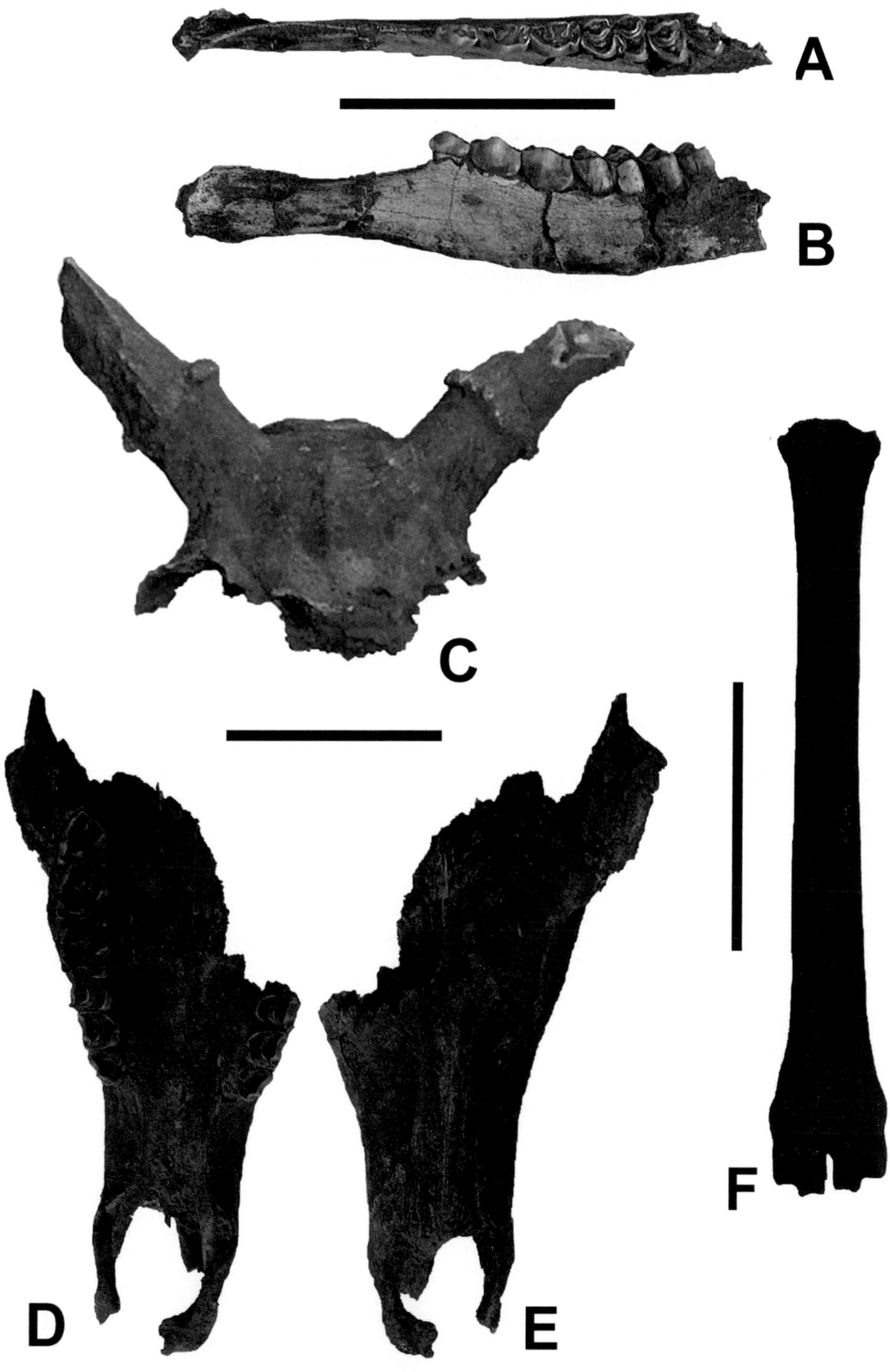

Fig. 13.2 *Eucladoceros ctenoides senezensis*: **A–B**, fragment of left half mandible (FSL SEN 04–0196), **A** occlusal view, **B** lateral view; **C**, partial skull (FSL SEN 04–0349), frontal view; **D–E**, partial skull (FSL SEN 04–0342), **D** occlusal view, **E** dorsal view; **F**, right metatarsal III and IV (FSL SEN 04–0305), dorsal view. Scale bars 100 mm

Table 13.4 Measurements[a] (in mm) of the skull of *Eucladoceros ctenoides senezensis* from Senèze, 2001–2006 collection

	N	Min	Mean	Max		
Nasion-prosthion length	1	–	203.0	–	–	–
Palationoral–prosthion length	1	–	171.5	–	–	–
Otion–otion DT	1	–	111.0	–	–	–
Occipital condyle DT	2	70.0	72.00	74.0	–	–
Paraccipital processes DT	2	101.0	102.50	104.0	–	–
Occipital hole DT	2	26.0	26.00	26.0	–	–
Occipital hole height	2	26.5	26.00	28.5	–	–
Minimum frontal DT	2	131.5	133.00	134.5	–	–
Maximum DT at nasal bones	1	–	40.0	–	–	–
Maximum DT at prenasal bones	1	–	68.0	–	–	–
Basion nuchal crest height	2	81.0	83.25	85.5	–	–
Muzzle minimum DT	1	–	64.5	–	–	–
Right pedicle DT	2	48.0	49.00	53.0	–	–
Right pedicle DAP	2	44.5	26.00	53.5	–	–
Left pedicle DT	1	–	48.0	–	–	–
Left pedicle DAP	1	–	44.5	–	–	–

[a] For more details about these measures, see Desse et al. (1986) and Valli (2001)

Dentition

Concerning the upper teeth, apart from those found on FSL SEN 04–0342, a few other specimens can be added (FSL SEN 04–0067, a right series M^1–M^3; FSL SEN 04–0221, a damaged left upper molar; FSL SEN 04–0346, a right P^3; and a left M^1–M^2, at present without label). P^2 is an asymmetrical tooth with a well marked lingual valley. P^3 is similar to P^2, but more symmetrical, with the lingual valley less pronounced than on P^2 and the pillar of the cone asymmetric, in vertical view. P^4 is almost symmetrical, without lingual valley and cingulum; the labial wall of the tooth is vertical. Molar teeth: the pillars of the cones and the styles are relatively well marked, the last do not bend at the base of the crown (see feature 5, Fig. 3.3 in Lister 1996, page 138). A cingulum can be present on the anterior part of M^3 or on the other molars, between the two lobes; the endostyles are present and well developed on all the molars.

Lower teeth (FSL SEN 04–0217 a damaged right lower molar; FSL SEN 04–0196, a fragment of left mandible; FSL SEN 06–0212, left M_2–M_3; FSL SEN 06–0215, right M_2–M_3, probably the same individual as the preceding specimen). The partial mandible holds P_2–M_2 (the last tooth is damaged; Fig. 13.2); the corpus is shallow under the teeth (Fig. 13.2). P_2 is a compact tooth with the conids little separated from the central body of the crown, the paraconid is absent and the anterior base not inflated (see, for an example of this kind of morphology, Fig. 10 in Valli 2004, page 200). On P_3, in occlusal view, the paraconid and the parastylid are not well separated but more than on P_4, the metaconid is deflected toward the rear part of the tooth, the lingual division between the entoconid and the endostylid, in lingual view, is not deep (see, for example, feature 7, Fig. 3. 1 in Lister 1996, page 136). On P_4, in occlusal view, the paraconid and the parastylid are not well separated, and, in lingual view, the division between entoconid and endostylid is not deep. This tooth is not molarised (*sensu* Heintz 1970). Molars: in occlusal view, on M_2 and M_3, the anterior entoconid wing overlaps with the posterior metaconid, and, on M_3, there is not a clear step between the 2nd and 3rd lobes.

Table 13.5 provides measures of the upper and lower teeth of *Eucladoceros*.

Table 13.5 Measurements (in mm) of the teeth of *Eucladoceros ctenoides senezensis* from Senèze, 2001–2006 collection

	N	Min	Mean	Max	S.D.[a]	C.V.[b]
P^2						
Total length	2	17.0	17.25	17.5	–	–
Length at the base	2	14.5	14.75	15.0	–	–
Width	2	16.5	16.50	16.5	–	–
P^3						
Total length	3	18.5	18.67	19.0	0.29	1.55
Length at the base	3	14.0	15.00	16.0	1.00	6.67
Width	3	18.0	18.67	19.5	0.76	4.09
P^4						
Total length	2	15.5	16.00	16.5	–	–
Length at the base	2	13.5	14.00	14.5	–	–
Width	2	19.5	19.75	20.0	–	–
M^{1c}						
Length at the base	3	21.5	21.50	21.5	0.00	0.00
Width of the first lobe	3	22.5	22.83	23.0	0.29	1.26
Width of the second lobe	3	22.5	22.33	24.0	0.76	3.27
M^{2c}						
Length at the base	3	20.5	21.83	22.5	1.15	5.29
Width of the first lobe	3	23.5	24.17	25.0	0.76	3.16
Width of the second lobe	3	23.0	23.67	24.5	0.76	3.23
M^3						
Total length	2	24.5	25.00	25.5	–	–
Length at the base	2	21.0	21.25	21.5	–	–
Width of the first lobe	2	24.0	24.00	24.0	–	–
Width of the second lobe	2	22.5	22.75	23.0	–	–
P^2–M^3 length	1	–	120.0	–	–	–
P^2–P^4 length	2	51.0	51.75	52.5	–	–
P^3–P^4 length	1	–	35. 0	–	–	–
M^1–M^3 length	2	71.0	71.75	72.5	–	–
P_2						
Total length	1	–	14.5	–	–	–
Length at the base	1	–	12. 5	–	–	–
Width	1	–	7.5	–	–	–
P_3						
Total length	1	–	18.5	–	–	–
Length at the base	1	–	14.0	–	–	–
Width	1	–	10.0	–	–	–
P_4						
Total length	1	–	18.5	–	–	–
Length at the base	1	–	17. 0	–	–	–
Width	1	–	11.0	–	–	–
$M_1{}^d$						
Total length	1	–	24.0	–	–	–
Length at the base	1	–	21.0	–	–	–
Width of the first lobe	1	–	13.0	–	–	–
M_2						
Total length	2	26.0	26.75	27.0	–	–
Length at the base	2	23.0	23.25	23.5	–	–
Width of the first lobe	3	14.5	14.83	15.0	0.29	1.95
Width of the second lobe	2	14.5	14.75	15.0	–	–
M_3						
Total length	3	33.5	34.83	36.5	1.53	4.39
Length at the base	3	32.0	33.33	35.5	1.89	5.68

(continued)

Table 13.5 (continued)

	N	Min	Mean	Max	S.D.[a]	C.V.[b]
Width of the first lobe	3	14.5	14.67	15.0	0.29	1.97
Width of the second lobe	3	14.0	14.17	14.5	0.29	2.04
Width of the third lobe	3	7.5	8.67	10.0	1.26	14.52
P_2–M_3 length	1	–	130.0	–	–	–
P_2–P_4 length	1	–	50.0	–	–	–
P_3–P_4 length	1	–	32.0	–	–	–
M_1–M_3 length	1	–	80. 5	–	–	–

[a] Standard deviation only calculated if N $\geq$ 3
[b] Coefficient of variation only calculated if N $\geq$ 3
[c] Teeth too worn to measure the total length
[d] The second lobe is damaged

Postcrania

Concerning the study of the postcranial skeleton, we concentrated on the scapula and the limb bones. We neglected the vertebrae, with the exception of the atlas, whose shape is similar to that of a red deer. The measures of the specimen FSL SEN 04–0036* are given in Table 13.6.

Scapula. Only one specimen has been found (FSL SEN 04–0277*), the left element of the almost complete skeleton (the right has not been found). The cavitas glenoidalis is oval, more elongated than wide and without a V–shaped cavity on its articular surface. The tuberculum superglenoidalis is slender and straight but shows an inclination with respect to the major articular surface axis (see, for example, Fig. 52 in Pfeiffer 1999, page 63). In lateral view, the cavitas glenoidalis is regularly concave and horizontal with respect to the scapular spine, as on the red deer. As the bone is incomplete, few measurements are possible. Minimal DAP = 43.5 mm; DAP at the articular processus = 70.0 mm; DAP of the cavitas glenoidalis = 51.5 mm and DT of the cavitas glenoidalis = 47.5 mm.

Humerus (FSL SEN 03–0010, FSL SEN 04–0326* and FSL SEN 06–0026, right distal fragments; FSL SEN 04–0322*, complete left bone). The proximal extremity can just be analyzed on FSL SEN 04–0322, the only complete bone. The tuberculum majus show a small indentation that is not deep or rounded in shape; the caput humeri forms a thin border on the diaphysis (see, for example, Fig. 53 in Pfeiffer, 1999, page 66). On the dorsal side of the distal epiphysis, the external pit is deeper than the internal one, but the internal is wider. On the plantar side, the border of the trochlea humeri is deeply tilted as in *C. elaphus*.

Radio–ulna (two complete specimens, a left, FSL SEN 04–0323*, and a right, FSL SEN 04–0363*). On the proximal end of the radius, the facet on the surface for the ulna is largely posterior (see, for example, Fig. 2.4 in Lister 1996, page 127); the attachment of the ligamentum collaterale laterale sticks out laterally, sloping a little downwards; the margo lateralis forms one lip. On the distal end, the dorsal edge of the magnum is flat, the edge between the lunar and the scaphoid is smooth, and the plantar edge of the magnum

Table 13.6 Measurements[a] (in mm) of the atlas (FSL SEN 04–0036*) of *Eucladoceros ctenoides senezensis* from Senèze, 2001–2006 collection

Total length	71.5
Total width	124.0
Total height	67.5
External proximal DT	98.5
Internal proximal DT	91.5
Right proximal DT	28.5
Left proximal DT	26.5
DT of the vertebral hole	37.0
DDV of the vertebral hole	36.5
External distal DT	90.0

[a] For more details about these measures, see Desse et al. (1986) and Valli (2001)

shows an angle wider than 90°. The edge of the tuber olecrani of the ulna has an angular, flat–topped shape; there are no distinct muscle attachments on the side.

The elements of the left carpus (pyramidal, FSL SEN 04–0337*; scaphoid, FSL SEN 04–0319*; lunar, FSL SEN 04–0339*; magnum, FSL SEN 04–0338*; capitato–trapezoidal, FSL SEN 04–0314*; rudimentary phalanges II and III, FSL SEN 04–0362* and FSL SEN 04–0318*) have been recovered. They do not present any feature different from the same bones in a red deer.

Metacarpal III and IV (only one specimen, a left, FSL SEN 04–0368*). On the proximal end, the two articular surfaces meet posteriorly but without constituting an articular facet, and, in dorsal view, a rough tuberosity is present. On the distal end, a split is visible on the plantar surface (but it is not possible to detect whether it is also present on the dorsal one) and the plantar face slopes anteriorly toward the distal edge. The bone is marked by a dorsal groove, the dorsal sulcus, continuing to the distal end.

The measures of the main forelimb bones are given in Table 13.7.

The innominates were recovered (FSL SEN 04–0373* and FSL SEN 04–0372*), but they are incomplete and damaged, so the observations are limited to the acetabular cavity. It is similar to that of a red deer, with a wrinkled ischium.

Femur (two complete bones, right FSL SEN 04–0365* and left FSL SEN 04–0367*, and a fragment of right distal end, FSL SEN 04–0298, too damaged to be described). Proximal end: in proximal view, the fossa trochanterica is "U"–shaped, as in red deer. In lateral view, the muscular impression on the trochanter is similar to that figured by Pfeiffer (1999; Fig. 66B, page 81) for *C. elaphus*. In dorsal view, between the femoral head and the trochanter major, the insertion for the musculus vastus intermedius is linear and directed cranio–laterally. Distal end: in plantar view, the insertion for the musculus gastrocnemius extends onto the fossa supracondylaris (see, for example, Fig. 68A in Pfeiffer 1999, page 82). This last feature could be observed only on one specimen because the second is damaged. In distal view, the fossa intercondylaris is very deep, and the two lips of the trochlea are divergent. Both femora from the skeleton hold

Table 13.7 Measurements (in mm) of the main forelimb bones of *Eucladoceros ctenoides senezensis* from Senèze, 2001−2006 collection

	N	Min	Mean	Max		
Humerus						
Total length	1	–	298.0	–	–	–
Proximal DAP	1	–	94.0	–	–	–
Diaphysis DT	1	–	34.0	–	–	–
Diaphysis DAP	1	–	43.5	–	–	–
Distal DT	2	66.5	67.75	69.0	–	–
Distal DAP	2	60.5	61.25	62.0	–	–
Radius						
Total length	2	305.0	307.25	309.5	–	–
Proximal DT	2	66.5	66.75	67.0	–	–
Proximal DAP	2	34.0	35.25	36.5	–	–
Diaphysis DT	2	35.5	35.75	36.0	–	–
Diaphysis DAP	2	24.5	24.75	25.0	–	–
Distal DT	2	56.5	56.50	56.5	–	–
Distal DAP	2	38.5	38.75	39.0	–	–
Metacarpal III–IV						
vcTotal length	1	–	279.0	–	–	–
Proximal DT	1	–	45.5	–	–	–
Proximal DAP	1	–	33.5	–	–	–
Diaphysis DT	1	–	26.0	–	–	–
Diaphysis DAP	1	–	29.0	–	–	–
Supra–articular DT	1	–	44.5	–	–	–
Distal DT	1	–	47.0	–	–	–
Distal DAP	1	–	31.5	–	–	–

Standard deviation and coefficient of variation are only calculated if N $\geq$ 3 the associated patellas. These bones show a pronounced oblique crest, going from the distal end to external corner.

Tibia (FSL SEN 04–0366*, left; FSL SEN 04–0295, right; FSL SEN 04–0360*, right, but only the distal end can be measured and described). Proximal end (feature observation possible only on one specimen), in plantar view: the plantar development of the intercondylar area is longer than wide, with a rounded tip (as shown in Fig. 71A of Pfeiffer 1999, page 86). Distal end (on all three specimens): in medial view, the insertion for the ligamentum collaterale tarsi mediale develops horizontally, drawing a slightly curved crest and ending with a slight bulge, and the edge of the medial trochlea is first horizontal then bends towards the malleolus medialis with an angle of about 115°. In distal view, the articular facets for the malleolus are well separated by a small "U"-shaped indentation and the medial trochlea is deep and narrow; the anterior articular facet for the malleolus is not downturned posteriorly. In dorsal view, the shape between the cochlea tibiae and the malleolus medialis is variable on the three available specimens.

Talus (two right and one left, FSL SEN 01–0074, FSL SEN 04–0300 and FSL SEN 04–0052; only the last two are well enough preserved to allow description). On the medial side, the proximal-medial ridge is larger in its proximal part than in its distal one and the bulge above the distal pulley is more pronounced. On the dorsal side, a V-shaped pit may be present on the distal pulley (only one specimen).

Calcaneus (right, FSL SEN 04–0299, and left, FSL SEN 04–0288*). They are recognized as cervid bones because the articular facet for the talus is vertical, not oblique as on bovids (Heintz 1970). In medial view, the shape of the sustentaculum tali shows a round step, as in Fig. 75A, page 92, of Pfeiffer (1999).

Cubo-navicular (four bones, a left, FSL SEN 04–0271*, and three right, FSL SEN 04–0296, FSL SEN 04–0374* and FSL SEN 04–0013, the last too damaged to be described or measured). In proximal view, there are no gaps on the articular surface for the talus. On the dorsal side, the bulge can be sharp or rounded (this feature could be analyzed on only two specimens).

Metatarsal III and IV (four specimens, two right, FSL SEN 04–0216 and FSL SEN 04–0305 [Fig. 13.2] and two left, FSL SEN 04–0286* and FSL SEN 04–0287). Proximal end, in medial view, one small tuberosity may be present between the articular facets for the tarsus (it has been detected on two specimens, the others presenting some damage at this site). In plantar view, the medial articular facet of the metatarsal IV does not cross the longitudinal groove between the metatarsal III and the IV. In proximal view, the anterior edges of the medial and lateral articular surfaces protrude approximately equally. The anteriormost portions of these two surfaces may contact each other for about 1 cm (see, for detail about this feature, Fig. 13.2.10 in Lister 1996, page 133), or do not meet (half of the specimens). On the distal extremity, in plantar view, the foramen is large and oval and the articular surface for the phalanges is long towards the back. The splits between the distal articular surfaces are generally present (three specimens of four).

The measures of the main hind limb bones are given in Table 13.8.

Table 13.8 Measurements (in mm) of the main hind limb bones of *Eucladoceros ctenoides senezensis* from Senèze, 2001–2006 collection

	N	Min	Mean	Max	S.D.[a]	C.V.[b]
Femur						
Total length	2	371.0	375.75	376.5	–	–
Proximal DT	2	96.5	97.00	97.5	–	–
Articular head DT	2	36.5	37.50	38.5	–	–
Articular head DAP	2	41.0	41.25	41.5	–	–
Diaphysis DT	2	31.0	33.00	35.0	–	–
Diaphysis DAP	2	36.5	38.00	39.5	–	–
Distal DT	2	83.5	84.25	85.0	–	–
Distal DAP	2	104.5	106.50	108.5	–	–
Patella[c]						
Total length	2	63.0	63.50	64.0	–	–
Total DT	2	48.0	48.00	48.0	–	–
Tibia						
Total length	2	333.5	347.75	362.0	–	–
Proximal DT	1	–	89.0	–	–	–
Diaphysis DT	2	33.5	33.50	33.5	–	–
Diaphysis DAP	2	25.5	27.00	28.5	–	–
Distal DT	3	52.0	52.50	53.0	0.50	0.95

(continued)

Table 13.8 (continued)

	N	Min	Mean	Max	S.D.[a]	C.V.[b]
Distal DAP	3	40.0	41.17	42.0	1.04	2.53
Talus						
Total height	3	55.0	58.50	61.0	2.78	4.76
Maximum DT	3	36.0	37.33	39.0	1.53	4.09
Lateral height	3	31.5	48.83	60.0	15.22	31.16
Lateral DAP	3	30.0	32.17	33.5	1.89	5.88
Median height	3	51.5	54.17	55.5	2.31	4.26
Median DAP	2	30.0	31.75	33.5	–	–
Height at the middle	3	43.5	45.83	47.5	2.08	4.54
Proximal DT	3	36.0	36.67	37.5	0.76	2.08
Distal DT	3	34.5	35.67	37.5	1.61	4.51
Calcaneum						
Total length	2	116.5	123.50	130.5	–	–
Proximal DT	2	26.0	27.50	29.0	–	–
Proximal DAP	2	34.0	35.25	36.5	–	–
Middle DT	2	15.5	16.00	16.5	–	–
Middle DAP	2	30.5	33.50	36.5	–	–
Maximum DT	2	40.0	42.25	44.5	–	–
Maximum DAP	2	45.5	49.75	54.0	–	–
Cubo–navicular						
Total height	2	38.0	39.75	41.5	–	–
DT	3	43.5	46.17	48.0	2.36	5.12
DAP	3	38.5	44.33	49.5	5.53	12.47
Metatarsal III–IV						
Total length	4	279.5	301.88	328.5	20.60	6.83
Proximal DT	4	38.0	40.75	42.5	2.02	4.96
Proximal DAP	4	40.0	42.25	45.0	2.63	6.22
Diaphysis DT	4	24.5	25.00	25.5	0.41	1.63
Diaphysis DAP	4	27.5	29.13	31.0	1.49	5.13
Supra–articular DT	4	42.5	44.63	46.0	1.70	3.81
Distal DT	4	43.0	45.38	46.5	1.65	3.64
Distal DAP	4	29.5	31.25	32.5	1.50	4.80

[a] Standard deviation only calculated if $N \geq 3$
[b] Coefficient of variation only calculated if $N \geq 3$
[c] For more details about these measures, see Desse et al. (1986)

Even though some features which differentiate the anterior phalanges I from the posterior ones in cervid species have been defined (Barbet 2006; Herrera 1990; Pfeiffer 1999), it was not possible to extend the method to *Eucladoceros*, so these elements are all considered as indeterminate. Five phalanges I are present in our collections (FSL SEN 04–0291*, FSL SEN 04–0309, FSL SEN 04–0330*, FSL SEN 04–0334, and FSL SEN 04–0380*). The articulation for the sesamoid bones is square in shape. In dorsal view, the axial protuberance is generally short and rounded and the lateral protuberance is small and sharp; one specimen presents a longer axial protuberance. In plantar view, the limit of the distal articular surface is almost horizontal except on one specimen, where it presents a small "V"–shaped prominence in the middle (for more details, see Figs. 82D and 82C in Pfeiffer 1999, page 101). In lateral view, the insertion of the ligamentum collaterale does not present any bulge, except on one specimen.

Phalanx II (six specimens, FSL SEN 04–0292, FSL SEN 04–0304, FSL SEN 04–0328*, FSL SEN 04–0335, FSL SEN 04–0378* and FSL SEN 04–0381*). In plantar view, the lateral protuberance is wide and developed, the axial one is flat without any indentation between the two, except on two specimens, where the axial protuberance is developed less than the lateral, and a narrow concavity is present between the two. The limit of the distal articular surface is variable: on the lateral side, it can reach to the middle part of the bone (without constituting a real articular surface until its end) but in one specimen it is very short.

Phalanx III (five specimens, FSL SEN 04–0294*, FSL SEN 04–0327*, FSL SEN 04–0336, FSL SEN 04–0361* and FSL SEN 04–0364*). The articular surfaces (the axial and the lateral ones) are different in size (but less different than those of the Senèze moose). The plantar surface is large and slightly concave in this posterior area. The facets for the articulation with the sesamoid bones are not marked. The pyramidal eminence is not present and the dorsal profile is straight rather than convex.

All the measures taken on the *Eucladoceros* phalanges are presented in Table 13.9.

Recently, several studies have investigated the paleoecology of this deer by analyzing some leg bones (femur, tibia, calcaneus and third phalanx) from the earlier collections. These studies consider that it was adapted to move in quite open habitats: it was an "intermediate open areas dweller" (*sensu* Curran 2015, 2018). This result is not necessarily at odds with those related to its ecology obtained with our analysis of the skull and dental wear (but also see Berlioz et al. 2018), because the latter suggest dietary preferences for this species but do not assess its preferred habitat.

Discussion

The fallow deer–sized remains found during the 2001–2006 fieldwork in Senèze agree, in size and anatomical features, with *M. rhenanus philisi* fossils known from the same locality. Our measurements are always within the range established by Heintz (1970) for the taxon or obtained by measuring specimens of the previous collections from Senèze, except for the distal metatarsal III–IV FSL SEN 04–0035 (see Table 13.2). This bone is slightly smaller than the other known specimens. However, it is too wide for a *Croizeroceros* (the DT of the metapodial extremities is an important feature for discerning between *Metacervoceros* and *Croizetoceros*) and morphologically very similar to the other *M. rhenanus philisi* specimens. Thus, like FSL SEN 03–0121 discussed above, it is considered to belong to *M. rhenanus philisi*. This result slightly enlarges the measurement range for metatarsals of this taxon.

Table 13.9 Measurements (in mm) of phalanges of *Eucladoceros ctenoides senezensis* from Senèze, 2001–2006 collection

	N	Min	Mean	Max	S.D	C.V
Phalanx I						
Total length	5	57.0	60.60	62.5	2.27	3.75
Proximal DT	5	22.0	23.40	24.5	1.14	4.87
Proximal DAP	5	27.5	28.80	30.5	1.35	4.69
Diaphysis DT	5	17.0	17.90	19.0	0.82	4.59
Diaphysis DAP	5	21.0	15.90	21.5	0.27	1.29
Distal DT	5	19.0	20.00	20.5	0.71	3.54
Distal DAP	5	18.0	18.70	19.5	0.67	3.59
Phalanx II						
Total length	6	43.5	45.67	48.0	1.57	3.44
Proximal DT	6	21.5	21.75	22.5	0.99	4.54
Proximal DAP	6	29.0	29.58	30.0	0.49	1.66
Diaphysis DT	6	16.5	17.00	17.5	0.32	1.86
Diaphysis DAP	6	19.5	20.50	21.5	0.71	3.45
Distal DT	6	17.0	18.75	19.5	1.17	6.25
Distal DAP	6	23.0	25.17	27.0	1.69	6.73
Phalanx III						
Total length	4	45.6	51.75	57.0	5.07	9.81
Proximal DT	4	42.5	46.00	48.0	2.42	5.25
Proximal DAP	5	17.5	19.00	20.0	0.94	4.92
Distal DAP	5	30.5	31.50	32.5	0.71	2.24

Our sample is too limited for a statistical analysis, but the diagrams of the measurements of the distal end of the metacarpal III + IV (Fig. 13.3) can help to clarify the dimensional equivalence between *M. rhenanus philisi* from the previous collections and the remains found during our excavations. For a more complete comparison, the measurements of *M. rhenanus valliensis* (Heintz 1970) from the locality of Saint–Vallier (Drôme, France) have been added. This taxon is slightly smaller than the form from Senèze, but some superpositions are possible. In Fig. 13.3, the samples from our collection match more clearly with the previously–known specimens from Senèze than with the fossils from Saint–Vallier, especially concerning the distal DAP.

Regarding the anatomical characters, the features recognized on our sample document the resemblance between the bones of *M. rhenanus philisi* and those of *Dama*. The distal shape of the cannon bones is more similar to that of red deer, but the features concerned are those established by Lister (1996) and that author explains that they can present some variability. The talus also could seem anomalous for the taxon. The study of the *M. rhenanus valliensis* population from Saint–Vallier, a different subspecies from ours and often considered its ancestor (Delpech & Guérin 1996; Heintz 1970), shows that, on the medial side of the talus, the ridge is generally greater (or equal) in the distal part than in the proximal one (Valli 2004). At least one specimen showing the more common red deer feature (a medial ridge larger in its proximal part than in its distal one) has been found at Saint-Vallier (Valli 2001). The small size of the sample from Senèze (only one specimen) can explain this anomaly.

The results of the morphological survey for *M. rhenanus philisi* from Senèze seem to agree with Pfeiffer's conclusions (1999, 2005). Using anatomical and morphometrical data, she argued for a relationship between this taxon (and the Italian Villafranchian forms) and the modern genus *Dama*. Our result was achieved mainly by comparing the post-cranial remains. On the other hand, Croitor (2014) considered the *Metacervoceros* skull too primitive (and different) to be included in *Dama* (or in *Cervus*). In turn, the analysis of the bony labyrinth (Mennecart et al. 2017) agrees with Croitor's claim, because it set apart *M. rhenanus philisi* from Senèze from *D. dama*. Thus, the morphology of the postcranium (and teeth) alone cannot resolve the phylogenetic relationship between these species.

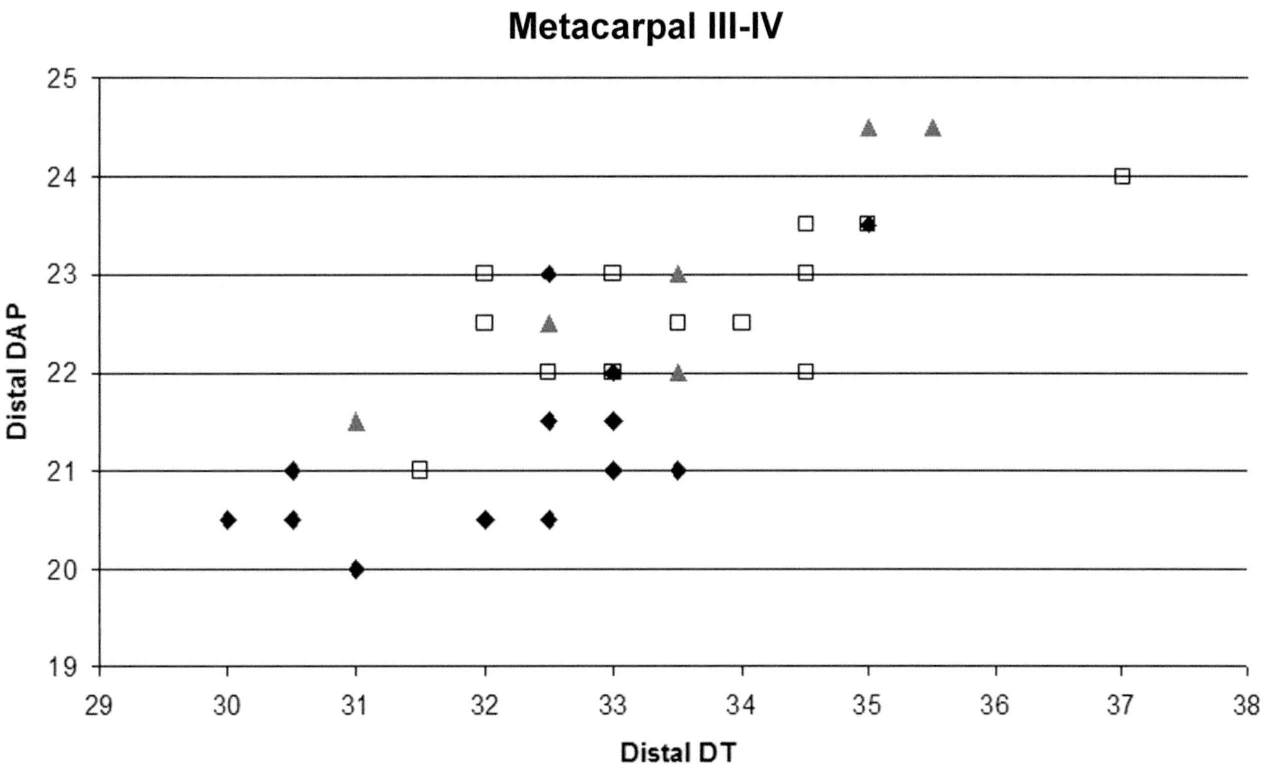

Fig. 13.3 Bivariate plot of metacarpal III + IV distal DT versus distal DAP for French *M. rhenanus*. Black diamonds: *Metacervoceros rhenanus valliensis* from Saint–Vallier; open squares: *M. rhenanus philisi* from older Senèze collections; gray triangles: *M. rhenanus philisi* from Senèze 2001–2006 fieldwork. Dimensions in mm

Perhaps the similarity between postcranial and dental remains from Senèze and those of *Dama* might be due to the actual occurrence of *Dama* in this locality, as recently proposed by Brugal et al. (2020). But, if this hypothesis is carefully examined, it does not seem really supported. In fact, Brugal et al. cited Mennecart et al. (2017) as the source of this idea, but the latter analysed two samples of bony labyrinth from Senèze and reported (pages 6–7) "The relationships of *Metacervocerus* with *Dama* or *Axis* are not solved in our phylogenetic analysis and more than one species attributed to *Metacervocerus* could be present in Senèze. Indeed, the two *Metacervocerus* bony labyrinths under study display significant differences in orientation of the vestibular aqueduct recalling the *Dama* lineage". Therefore, either the morphological variation of *Metacervoceros* bony labyrinths is more extensive than supposed or there are actually two species. This last hypothesis is really interesting, but so far, none of the previous studies (e.g., Heintz 1970; Pfeiffer 1999, 2005; Croitor 2006a) have corroborated this result and, unfortunately, during the last excavations, no ear bones were found with which to test this idea.

Most recently, Croitor and Robinson (2020; Fig. 10) reported the results of a cluster analysis of cranial, dental and antler characters among selected Eurasian cervids, including *Cervus, Rusa, Axis, Dama* and *Metacervoceros*. Their analysis recognized two main clusters, with *Cervus* and *Rusa* in one, and *Dama* plus *Metacervoceros* linked to *Axis* in the other.

The dimensions of the *Eucladoceros* bones found during our fieldwork are within the range established by previous authors (Barbet 2006; Heintz 1970) and my measurements for the taxon. The proximal DT of the femur exceeds that found on the classical remains, but, as the bone belongs to the almost complete skeleton, its determination is not in question. As for *M. rhenanus*, the sample is too limited for statistical comparison, but it is possible to compare the sample from the recent work to the older collection with bivariate plots. In Fig. 13.4, the two samples of *Eucladoceros* distal metatarsal III + IV are shown. The bones from the older collections are conserved at the Université Claude Bernard-Lyon 1 and the Muséum national d'Histoire naturelle in Paris. To these data, we have added the bones of the population from Saint-Vallier (fossils in FSL, NMB and Paläontologisches Institut Zürich). As shown in the diagram, the taxon from Senèze is slightly smaller than the population from Saint-Vallier, but the range of superposition is important. The bones from our 2001–2006 sample fall in the lower part of the Senèze *Eucladoceros*, with almost no overlap with Saint-Vallier.

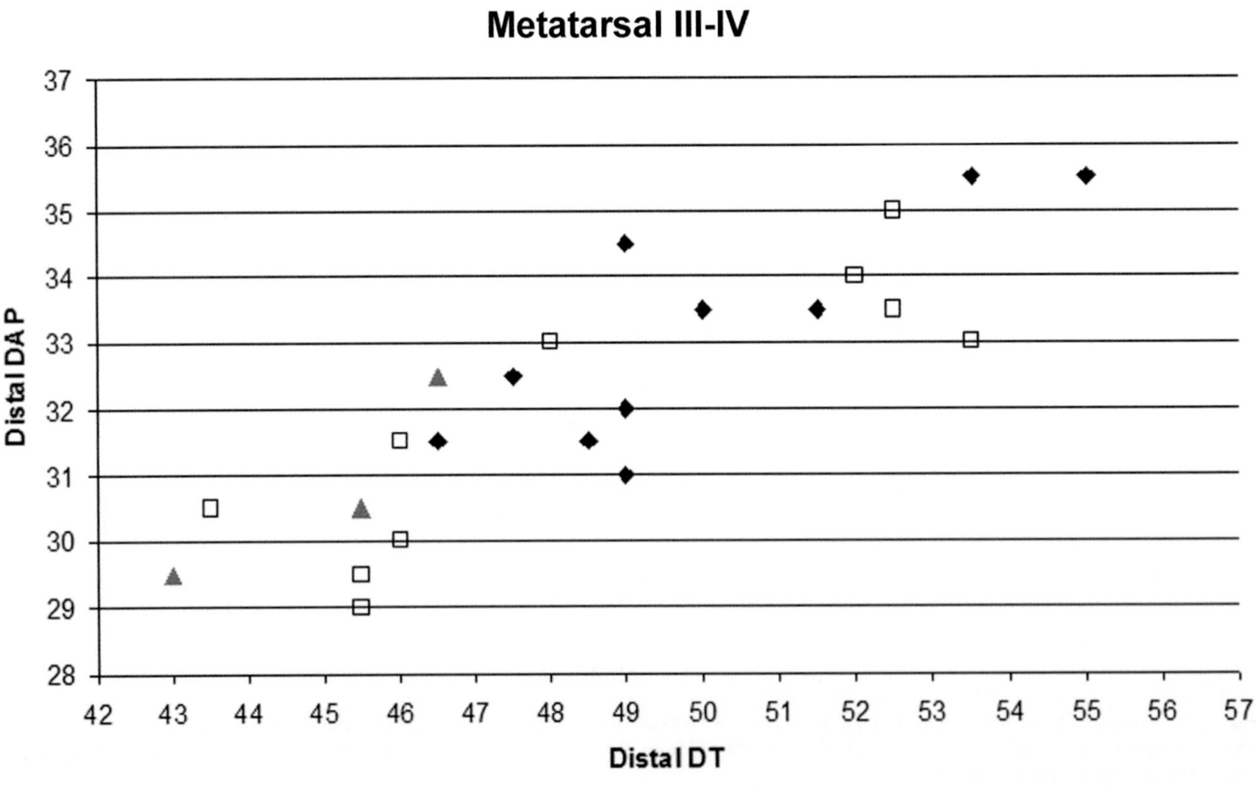

Fig. 13.4 Bivariate plot of metatarsal III + IV distal DT versus distal DAP for French *Eucladoceros ctenoides*. Black diamonds: *E. ctenoides vireti* from Saint-Vallier; open squares: *E. ctenoides senezensis* from older Senèze collections; gray triangles: *E. ctenoides senezensis* from Senèze, 2001–2006 fieldwork. Dimensions in mm

The features of the *Eucladoceros* skull are similar to the ones described by Valli and Palombo (2005) from the older collections of Senèze. Concerning the postcranial skeleton, the majority of the remains look more similar to those of red deer than those of fallow deer, although certain features set them apart (e.g., the shape of the distal tibia or the plantar proximal end of the metatarsal III + IV). Globally, the characters agree with the ones described by Pfeiffer (1999, 2005) and Barbet (2006) on the *Eucladoceros* bones from Senèze. Some differences can be found (e.g., the margo lateralis of the radius) but they are attributed to the limited sample from our fieldwork.

According to certain authors (e.g., Azzaroli et al. 1988) more than one fauna, of different ages, might be present at Senèze. One of our goals was to evaluate this suggestion. Several layers, all potentially fossiliferous, have been found during our surveys (Delson et al. 2006; Debard 2024). Are these nearly contemporaneous, or were they deposited over a large span of time? What do the new cervid fossils from Senèze tell us? Remains of *Eucladoceros*, the most abundant cervid, come from at least two different loci. According to the morphometric analysis they seem homogeneous, and they might belong to the same population. However, these bones might derive from only three animals, which is not enough to draw a definitive conclusion. The analysis of the *Metacervoceros* fossils leads to the same result. The fossils from the older collections are compatible with the presence of only one fauna at Senèze, but they do not necessarily refute the idea that the fossiliferous layers have been deposited during a significant span of time.

Conclusion

The study of the cervid fossils from the recent fieldwork at Senèze, between 2001 and 2006, led to identification of the remains of at least two species of deer, the most common taxa, *E. ctenoides senezensis* and *M. rhenanus philisi*. Both were already known by complete skeletons, and we recovered an almost complete *Eucladoceros* postcranial skeleton. The new cervid remains match very well, in morphology and dimension, with the fossils recovered previously.

Both these species have a wide diffusion in Europe during the Villafranchian. *M. rhenanus* fossils are signalled in France and Spain (Heintz 1970; Heintz & Dubar 1981; Azanza and Menéndez 1989–1990; Garrido 2008; Brugal et al. 2020), Great Britain (Lister 1993), Netherlands (De Vos et al. 1995; Spaan 1992), Yugoslavia (Malez 1986), Greece (Kostopoulos 1997; Koufos & Kostopoulos 1997) and Eastern Europe (Sirakov et al. 2010; Vislobokova 1992). *E. ctenoides* is also known from several European countries, from Spain through the Netherlands to Greece and Bulgaria (Azzaroli & Mazza 1992; De Vos et al. 1995; Heintz 1970; Koufos & Kostopoulos 1997; Spassov & Crégut-Bonnoure 1999).

The importance of the two subspecies found at Senèze is that they are key elements for European mammalian biochronology: they contribute to defining the Late Villafranchian, in Western Europe (France and Spain; Heintz 1970, 1972; Delpech & Guérin 1996), as well as in Eastern (especially Greece: Kostopoulos & Athanassiou 2005; Koufos 2001; Koufos & Kostopoulos 1997). Their accurate recognition is crucial for correctly placing the Quaternary fossil assemblages of this continent in a biochronological framework.

Acknowledgments I thank E. Delson, M. Faure and C. Guérin who have entrusted to me the study of the cervid fossil remains from Senèze. I am also very grateful to A. Prieur, E. Robert, P. Tassy and C. Sagne for allowing me to survey the fossil material conserved in the collections of the Laboratoire de Géologie de l'Université Claude Bernard-Lyon 1 and the Muséum national d'Histoire naturelle in Paris. L. Costeur, curator at Naturhistorisches Museum of Basel, kindly checked, on my request, the absence of the upper canines and their alveoli on the mounted skeleton exposed in the Basel Museum and sent me some important information about fossil cervids from Senèze conserved in the collections under his responsibility; he also contributed to the quality of my manuscript with his advice and comments. E. Delson provided help with language and other editorial suggestions. M. Breda and R. Croitor reviewed my manuscript and helped to increase its quality. To all these colleagues, I offer my deepest gratitude.

References

Azanza, B., & Menéndez, E. (1989–1990). Los Ciervos fósiles del neógeno español. *Paleontologia i Evolució, 23*, 75–82.

Azzaroli, A. (1952). L'Alce di Senèze. *Palaeontographia Italica, 47*, 133–141.

Azzaroli, A. (1992). The cervid genus *Pseudodama* n. g. in the Villafranchian of Tuscany. *Palaeontographia Italica, 79*, 1–41.

Azzaroli, A., & Mazza, P. (1992). The cervid genus *Eucladoceros* in the early Pleistocene of Tuscany. *Palaeontographia Italica, 79*, 43–100.

Azzaroli, A., De Giuli, C., Ficcarelli, G., & Torre, D. (1988). Late Pliocene to early mid-pleistocene mammals in Eurasia: Faunal succession and dispersal events. *Palaeogeography, Palaeoclimatology, Palaeoecology, 66*, 77–100.

Banfield, A. W. F. (1961). A revision of the Reindeer and Caribou, genus *Rangifer*. *Bulletin of the National Museum of Canada, 177*, 1–137.

Barbet, P. (2006). Approche taphonomique du site, pliocène terminal, de Chilhac (Haute-Loire, France) et étude paléontologique des Cervidae. Thèse de Doctorat, Muséum national d'Histoire naturelle, Paris.

Berlioz, É., Kostopoulos, D. S., Blondel, C., & Merceron, G. (2018). Feeding ecology of *Eucladoceros ctenoides* as a proxy to track regional environmental variations in Europe during the early pleistocene. *Comptes Rendus Palevol, 17*, 320–332.

Blumenbach, J. (1799). *Handbuch der Naturgeschichte*. J. C. Dieterich.

Bouvrain, G., Geraads, D., & Jehenne, Y. (1989). Nouvelles données relatives à la classification des Cervidae (Artiodactyla, Mammalia). *Zoologischer Anzeiger, 223*, 82–90.

Breda, M. (2001). The holotype of *Cervalces gallicus* (Azzaroli, 1952) from Senèze (Haute-Loire, France) with nomenclature implications and taxonomical-phylogenetic accounts. *Rivista Italiana di Paleontologia e Stratigrafia, 107*, 439–449.

Breda, M., & Marchetti, M. (2005). Systematical and biochronological review of Plio-Pleistocene Alceini (Cervidae; Mammalia) from Eurasia. *Quaternary Science Reviews, 24*, 775–805.

Brokx, P. A. (1972). The superior canines of *Odocoileus* and other deer. *Journal of Mammalogy, 53*, 359–366.

Brugal, J.-P., Argant, A., Boudadi-Maligne, M., Crégut-Bonnoure, É., Croitor, R., Fernandez, P., et al. (2020). Pleistocene herbivores and carnivores from France: An updated overview of the literature, sites and taxonomy. *Annales de Paléontologie, 106*, 102384.

Chaplin, R. E., & Atkinson, J. (1968). The occurrence of upper canine teeth in Roe deer (*Capreolus capreolus*) from England and Scotland. *Journal of Zoology, 155*, 141–144.

Crégut-Bonnoure, É., Argant, A., Argant, J., Debard, É., Delson, E., Eisenmann, et al. (2024). Biochronology of the Senèze faunal assemblage. In E. Delson, M. Faure & C. Guérin (Eds.), *Senèze: Life in Central France two million years ago. Paleontology, geochronology, stratigraphy and taphonomy* (pp. 633–652). Cham (Switzerland): Springer.

Croitor, R. (2001). Functional morphology of small-sized deer from the early and middle Pleistocene of Italy: Implication for paleolandscape reconstruction. In G. Cavarretta, P. Gioia, M. Mussi & M. R. Palombo (Eds.), *The World of Elephants* (pp. 97–102). Rome: CNR editions.

Croitor, R. (2006a). Early Pleistocene small-sized deer of Europe. *Hellenic Journal of Geosciences, 41*, 89–117.

Croitor, R. (2006b). Taxonomy and systematics of large-sized deer of the genus *Praemegaceros* Portis, 1920 (Cervidae, Mammalia). *Courier Forschungsinstitut Senckenberg, 256*, 91–116.

Croitor, R. (2014). Deer from late Miocene to Pleistocene of Western Palearctic: Matching fossil record and molecular phylogeny data. *Zitteliana Reihe B, 32*, 115–153.

Croitor, R., & Robinson, C. (2020). A revision of "*Cervus*" *punjabiensis* Brown, 1926 (Cervidae, Mammalia) from the Upper Siwaliks of Chandigarh, India. *Quaternary International, 550*, 147–158.

Curran, S. C. (2015). Exploring *Eucladoceros* ecomorphology using geometric morphometrics. *The Anatomical Record, 298*, 291–313.

Curran, S. C. (2018). Three-dimensional geometric morphometrics in paleoecology. In D. Croft, D. Su, & S. Simpson (Eds.), *Methods in Paleoecology: Reconstructing Cenozoic Terrestrial Environments and Ecological Communities* (pp. 319–337). Cham, Switzerland: Springer.

D'Errico, F., & Vanhaeren, M. (2002). Criteria for identifying red deer (*Cervus elaphus*) age and sex from their canines. Application to the study of upper Palaeolithic and Mesolithic ornaments. *Journal of Archaeological Science, 29*, 211–232.

De Giuli, C., & Heintz, É. (1974). *Croizetoceros ramosus* (Cervidae, Artiodactyla, Mammalia) de Montopoli, nouvel élément de la faune villafranchienne d'Italie. *Atti della Società Toscana di Scienze Naturali, Serie A, 81*, 241–251.

De Vos, J., Mol, D., & Reumer, J. W. F. (1995). Early Pleistocene Cervidae (Mammalia, Artiodactyla) from the Oosterschelde (the Netherlands), with a revision of the cervid genus *Eucladoceros* Falconer, 1868. *Deinsea, 2*, 95–121.

Debard, É. (2024) Geological study of the paleontological site of Senèze (Domeyrat, Haute-Loire, France). In E. Delson, M. Faure, & C. Guérin (Eds.), *Senèze: Life in Central France two million years ago. Paleontology, geochronology, stratigraphy and taphonomy.* Cham, Switzerland: Springer, 37–82.

Delpech, F., & Guérin, C. (1996). Ordre des Artiodactyles; Famille des Cervidae. In C. Guérin & M. Patou-Mathis (Eds.), *Les grands mammifères plio-pléistocènes d'Europe* (pp. 47–62). Paris: Masson.

Delson, E., Faure, M., Guérin, C., Aprile, L., Argant, J., Blackwell, B., et al. (2006). Franco-American renewed research at the late Villafranchian locality of Senèze (Haute-Loire, France). *Courier Forschungsinstitut Senckenberg, 256*, 275–290.

Depéret, C. (1884). Nouvelles études sur les Ruminants pliocènes et quaternaires d'Auvergne. *Bulletin de la Société géologique de France, 12*, 247–284.

Depéret, C., & Mayet, L. (1912). Le gisement de mammifères pliocènes de Senèze (Haute-Loire). *Association française pour l'Avancement des Sciences, Compte Rendu de la 40e session, Dijon, 1911, Mémoire hors série*, 138–151.

Desse, J., Chaix, L., & Desse-Berset, N. (1986). "Ostéo" Base-réseau de données ostéométriques pour l'archéozoologie. *Editions du CNRS-Notes et Monographies Techniques, 20*, 1–161.

Di Stefano, G., & Petronio, C. (1998). Origin and relationships among the *Dama*-like cervids in Europe. *Neues Jahrbuch für Geologie und Paläontologie, Abhandlungen, 207*, 37–55.

Di Stefano, G., & Petronio, C. (2002). Systematics and evolution of the Eurasian Plio-Pleistocene tribe Cervini (Artiodactyla, Mammalia). *Geologica Romana, 36*, 311–334.

Dietrich, W. O. (1938). Zur Kenntnis der Oberpliocänen echten Hirsche. *Zeitschrift der Deutschen Geologischen Gesellschaft, 90*, 261–267.

Dubois, E. (1904). On an equivalent of the Cromer forest bed in the Netherlands. *Koninklijke Nederlandse Akademie van Wetenschappen, Series B, Physical Sciences*, 214–222.

Fortelius, M., & Solounias, N. (2000). Functional characterisation of ungulate molars using the abrasion-attrition wear gradient: A new method for reconstructing paleodiets. *American Museum Novitates, 3301*, 1–36.

Garrido, G. (2008). La asociación de los géneros *Croizetoceros*, *Metacervoceros* y *Eucladoceros* (Cervidae, Artiodactyla, Mammalia) en el yacimento de Fonelas P-1 (Cuenca de Guadix, Granada). In A. Arribas (Ed.), *Vertebrados del Plioceno superior terminal en el suroeste de Europa: Fonelas P-1 y el projeto Fonelas* (pp. 365–396). Madrid: Cuaderno del Museo Geominero, *10*.

Geist, V. (1999). *Deer of the world-their evolution, behaviour and ecology*. Swan Hill Press.

Greer, K. R., & Yeager, H. W. (1967). Sex and age indications from upper canine teeth of elk (wapiti). *The Journal of Wildlife Management, 31*, 408–417.

Groves, C. P., & Grubb, P. (1987). Relationships of living deer. In C. S. Wemmer (Ed.), *Biology and Management of the Cervidae* (pp. 21–59). Washington, D.C.: Smithsonian Institution Press.

Grubb, P. (2000). Valid and invalid nomenclature of living and fossil deer, Cervidae. *Acta Theriologica, 45*, 289–307.

Guérin, C. (2007). Biozonation continentale du Plio-Pléistocène d'Europe et d'Asie occidentale par les mammifères: état de la question et incidence sur les limites Tertiaire/Quaternaire et Plio/Pléistocène. Colloque international INQUA/AFEQ-Q5-Le Quaternaire, limites et spécificités, Paris Février 2006. *Quaternaire, 18*, 23–33.

Guérin, C., & Faure, M. (2002). Grands Mammifères. In J. C. Miskovsky (Ed.), *Géologie de la Préhistoire: Méthodes, techniques, applications* (pp. 879–887). Paris: Géopré édit., Maison de la géologie.

Heffelfinger, J. (2006). *Deer of the Southwest*. Texas A&M University Press.

Heintz, É. (1970). Les Cervidés villafranchiens de France et d'Espagne. *Mémoires du Muséum national d'Histoire naturelle, 22*(I and II), 1–317 and 1–206.

Heintz, É. (1972). Apports de la biométrie à la stratigraphie des Mammifères fossiles. *Mémoires du Bureau de recherches géologiques et minières, 77*, 395–399.

Heintz, É., & Dubar, M. (1981). Place et signification des dépôts villafranchiens de Moustiers-Ségriès et faune de Mammifères de Cornillet (Alpes-de-Haute-Provence). *Bulletin du Muséum national d'Histoire naturelle*, 4e sér., *3*, section C (2), 363–397.

Herrera, P. L. (1990). Critères pour la détermination des phalanges du cerf élaphe (*Cervus elaphus* L.). *Revue de Paléobiologie, 9*, 1–7.

Hofmann, R. R. (1989). Evolutionary steps of ecophysiological adaptation and diversification of ruminants: A comparative view of their digestive system. *Oecologia, 78*, 443–457.

Kierdorf, U., Kahlke, R.-D., & Egerer, U. (2007). Obere Eckzähne des Europäischen Rehs *Capreolus capreolus* L. (Cervidae, Mammalia) aus ontogenetischer und phylogenetischer Sicht. *Terra Praehistorica, 70*, 111–115.

Kostopoulos, D. S. (1997). The Plio-Pleistocene Artiodactyls from Macedonia, Greece: 2. the fossiliferous locality of Volakas, VOL (Volakas Basin, Drama, NE Greece). *Paleontologia i Evolució, 30–31*, 83–92.

Kostopoulos, D. S., & Athanassiou, A. (2005). In the shadow of bovids: suids, cervids and giraffids from the plio-pleistocene of Greece. In É. Crégut-Bonnoure (Ed.), *Les Ongulés holarctiques du Pliocène et du Pléistocène. Actes Colloque International, Avignon, 19–22 septembre 2000* (pp. 179–190). *Quaternaire*, Hors-série, 2.

Koufos, G. D. (2001). The Villafranchian mammalian faunas and biochronology of Greece. *Bollettino della Società Paleontologica Italiana, 40*, 217–223.

Koufos, G. D., & Kostopoulos, D. S. (1997). Biochronology and succession of the plio-pleistocene macromammalian localities of Greece. In J.-P. Aguilar, S. Legendre & J. Michaux (Eds.), *European Neogene Mammal Chronology* (pp. 619–634). Montpellier: Mémoires et Travaux E.P.H.E.

Linnaeus, C. (1758). *Systema naturae*. Holmiae (Laurentii Salvii).

Lister, A. M. (1993). The stratigraphical significance of deer species in the cromer forest-bed formation. *Journal of Quaternary Science, 8*, 95–108.

Lister, A. M. (1996). The morphological distinction between bones and teeth of fallow deer (*Dama dama*) and red deer (*Cervus elaphus*). *International Journal of Osteoarchaeology, 6*, 119–143.

Malez, M. (1986). Die quartären Vertebraten-Faunen in der SFR Jugoslawien. *Quartärpaläontologie, 6*, 101–117.

Mennecart, B., DeMiguel, D., Bibi, F., Rössner, G. E., Métais, G., Neenan, J. M., et al. (2017). Bony labyrinth morphology clarifies the origin and evolution of deer. *Scientific Reports, 7*, 1–11.

Nowak, R. M. (1999). *Walker's Mammals of the World* (6th Ed.). Baltimore and London: Johns Hopkins University Press.

Parenti, F., Aprile, L., Candelato, F., Monguillon, A., Natali, L. et al. (2024). Surveying and excavations at Senèze: 2001–2006. In E. Delson, M. Faure and C. Guérin (Eds.), *Senèze: Life in Central France two million years ago. Paleontology, geochronology, stratigraphy and taphonomy*. Cham, Switzerland: Springer, 27–35.

Pfeiffer, T. (1999). Die Stellung von *Dama* (Cervidae, Mammalia) im System plesiometacarpaler Hirsche des Pleistozäns. *Courier Forschungsinstitut Senckenberg, 211*, 1–218.

Pfeiffer, T. (2005). The position of *Dama* (Cervidae, Mammalia) in the system of fossil and living deer from Europe-phylogenetical analysis based on the postcranial skeleton. In É. Crégut-Bonnoure (Ed.), *Les Ongulés holarctiques du Pliocène et du Pléistocène. Actes Colloque International, Avignon, 19–22 septembre 2000* (pp. 39–57). *Quaternaire*, Hors-série, 2.

Schaub, S. (1941). Die kleine Hirschart aus dem Oberpliocaen von Senèze (Haute-Loire). *Eclogae Geologicae Helvetiae, 34*, 264–271.

Scott, W. B. (1885). *Cervalces americanus*, a fossil moose, or elk, from the Quaternary of New Jersey. *Proceedings of the Academy of Natural Sciences of Philadelphia, 37*, 181–202.

Sirakov, N., Guadelli, J.-L., Ivanova, S., Sirakova, S., Boudadi-Maligne, M., Dimitrova, I., et al. (2010). An ancient continuous human presence in the Balkans and the beginnings of human settlement in western Eurasia: A lower pleistocene example of the lower pleistocene levels in Kozarnika cave (North-western Bulgaria). *Quaternary International, 223–224*, 94–106.

Smith, J. B., & Dodson, P. (2003). A proposal for a standard terminology of anatomical notation and orientation in fossil vertebrate dentitions. *Journal of Vertebrate Paleontology, 23*, 1–12.

Spaan, A. (1992). A revision of the deer from Tegelen (province of Limburg, The Netherlands). *Scripta Geologica, 98*, 1–85.

Spassov, N., & Crégut-Bonnoure, É. (1999). Premières données sur les Bovidae Villafranchiens de Bulgarie. *Comptes Rendus de l'Académie des Sciences, 328*, 493–498.

Stehlin, G. (1923). Die oberpliocaene Fauna von Senèze (Haute Loire). *Eclogae Geologicae Helvetiae, 18*, 268–281.

Valli, A. M. F. (2001). Le gisement villafranchien moyen de Saint-Vallier (Drôme): Nouvelles données paléontologiques (Cervidae, Bovinae) et taphonomiques. *Documents des Laboratoires de Géologie, Lyon, 153*, 1–275.

Valli, A. M. F. (2004). Les Cervidae du gisement pliocène supérieur (Villafranchien moyen) de Saint-Vallier (Drôme, France). In M. Faure & C. Guérin (Eds.), *Le gisement pliocène final de Saint-Vallier (Drôme, France)* . *Geobios*, Mémoire spécial 26, *37*, 191–232.

Valli, A. M. F., & Palombo, M. R. (2005). Le régime alimentaire du Cervidae (Mammalia) *Eucladoceros ctenoides* (Nesti, 1841) reconstitué par la morphologie du crâne et par l'usure dentaire. *Eclogae Geologicae Helvetiae, 98*, 133–143.

Valli, A. M. F., & Palombo, M. R. (2008). Feeding behaviour of middle-size deer from the upper pliocene site of Saint-Vallier (France) inferred by morphological and micro/mesowear analysis. *Palaeogeography, Palaeoclimatology, Palaeoecology, 257*, 106–122.

Van Der Made, J. (1999). Ungulates from Atapuerca TD6. *Journal of Human Evolution, 37*, 389–413.

Viret, J. (1954). Le loess à bancs durcis de Saint-Vallier (Drôme) et sa faune de mammifères villafranchiens. *Nouvelles Archives du Muséum d'Histoire naturelle de Lyon, 4*, 1–266.

Vislobokova, I. (1992). Neogene deer in Europa. *Paleontologia i Evolució, 24–25*, 149–154.

Chapter 14
The Bovidae (Mammalia, Cetartiodactyla) from Senèze

Evelyne Crégut-Bonnoure

Abstract The new field research at Senèze has yielded seven incomplete remains belonging to bovids, but there are hundreds from previous work. Since the last century, eight species have been described from the locality. A revision was undertaken of *Gazellospira torticornis*, *Procamptoceras brivatense*, *Gallogoral meneghinii*, *Pliotragus ardeus*, *Megalovis latifrons* and *Ovis* sp., which are represented by relatively abundant remains. A comparison with taxa from other localities in the Massif Central and across Europe has allowed the detection of a number of mistakes in the specific attribution of some remains and to discover new material. Today eleven species are identified at Senèze: *G. torticornis*, *P. brivatense*, cf. *Hemitragus* sp., *G. meneghinii*, *P. ardeus*, *M. latifrons*, *Ovis claudiusguerini* nov. sp., *Leptobos etruscus*, *L. furtivus*, *Bison* (*Eobison*) sp. and Bovidae indet. Of these, one is newly named here and two others (cf. *Hemitragus* sp. and Bovidae indet.) are new for the site. A fourth taxon, *Bison* (*Eobison*) sp., is known from a single specimen donated to the team and collected from an unknown horizon; it is younger than the other species and must date from the end of the Early Pleistocene. Morphological variability is observed in samples of *Pliotragus*, *Gallogoral* and *Megalovis*, which suggests evolutionary change in these species through time. This could imply a mixing of the species and possibly indicate more than one fossiliferous level. Unfortunately, there are no details concerning the stratigraphical position of the material from previous field research in the maar.

Résumé Les nouvelles fouilles du site de Senèze ont fourni sept restes incomplets de Bovidae. Au siècle dernier, huit espèces ont été décrites. La révision de six de ces taxons a été réalisée: *Gazellospira torticornis*, *Procamptoceras brivatense*, *Gallogoral meneghinii*, *Pliotragus ardeus*, *Megalovis latifrons* et *Ovis* sp. Ils sont représentés par un matériel relativement abondant. Leur comparaison avec le matériel d'autres localités du Massif Central et d'Europe a permis de relever des erreurs d'attribution spécifique pour quelques restes et de découvrir de nouvelles pièces. Aujourd'hui onze espèces sont identifiées à Senèze: *G. torticornis*, *P. brivatense*, cf. *Hemitragus* sp., *G. meneghinii*, *P. ardeus*, *M. latifrons*, *Ovis claudiusguerini* nov. sp., *Leptobos etruscus*, *L. furtivus*, *Bison* (*Eobison*) sp. et Bovidae indet. Parmi ces taxons, un est défini pour la première fois et deux autres (cf. *Hemitragus* sp. et Bovidae indet.) sont nouveaux pour le site. *Bison* (*Eobison*) sp., est connu par un seul fossile donné à l'équipe de fouille et trouvé dans un niveau incertain; il est chronologiquement plus jeune que les autres taxons et doit dater de la fin du Pléistocène inférieur. Une variabilité morphologique est observable chez *Pliotragus*, *Gallogoral* et *Megalovis*, ce qui suggère une évolution morphologique de ces taxons dans le temps. Ce constat suggère un mélange des espèces et pourrait indiquer plusieurs niveaux fossilifères. Malheureusement, nous n'avons aucun détail sur la position stratigraphique du matériel provenant des anciennes fouilles dans le maar.

Keywords Antilopinae • Caprinae • Morphometry • MNQ 18

Mots-clés Antilopinae • Caprinae • Morphométrie • MNQ 18

This chapter includes one or more new nomenclatural-taxonomic actions, registered in Zoobank, the ICZN online registration system, and for such purposes the official publication date is March 2025. ZooBank Life Science Identifiers (LSIDs) can be accessed through any standard web browser by appending the LSID to the prefix 'http://zoobank.org/'. The LSID for this publication is urn:lsid:zoobank.org:pub: 004CD4B9-C1AF-46FD-B3B0-0F9E3A863504.

E. Crégut-Bonnoure (✉)
Muséum Requien, Avignon, France
e-mail: evelyne.cregut1313@orange.fr

TRACES, UMR 5608 (CNRS—Université Toulouse 2 Jean-Jaurès), Toulouse, France

E. Delson et al. (eds.), *Senèze: Life in Central France Around Two Million Years Ago*,
Vertebrate Paleobiology and Paleoanthropology, https://doi.org/10.1007/978-3-031-64415-3_14

Introduction

The bovid material from Senèze is abundant, with crania, mandibles, isolated teeth and postcranial remains. Over the last century, eight bovid taxa have been studied and described from this locality (Schaub 1922, 1923, 1943; Stehlin 1923, Pilgrim & Schaub 1939; Guérin 1965; Duvernois & Guérin 1989; Duvernois 1989, 1990; Crégut-Bonnoure 2002, 2020): *Gazellospira torticornis*, *Procamptoceras brivatense*, *Gallogoral meneghinii*, *Pliotragus ardeus*, *Megalovis latifrons*, *Ovis* sp., *Leptobos etruscus* and *Leptobos furtivus*. This association has been dated to the Late Villafranchian, and the locality is considered as the reference locality for the biozone MNQ 18 (Delson et al. 2006; Guérin 1980, 1982, 1996). According to the $^{40}Ar/^{39}Ar$ dates obtained on tephra from the Mont-Dore strato-volcano and the U/Pb geochronology on zircon for fossiliferous slope deposits on the edge of the crater, the age proposed for the site is now ca. 2.2–2.1 Ma (Roger et al. 2000; Nomade et al. 2014; Pastre et al. 2015; Paquette et al. 2021; Pastre 2024; Delson et al. 2024). Given that the base of the Quaternary is now defined as ca. 2.6 Ma (=Gelasian; Gibbard & Head 2009), this date corresponds to the Early Pleistocene and to the transition between Middle and Late Villafranchian.

The new field research directed by Faure, Guérin and Delson has yielded only seven bovid remains (plus two donated by a local collector). They were compared with the previously recovered material from this locality and from other Villafranchian French and European sites preserved at the Laboratoire de Géologie, Université Claude Bernard – Lyon 1 (FSL; previously Faculty of Science, Lyon), the Crozatier Museum of Le Puy-en-Velay (MC), the national Natural History Museum of Paris (MNHN-F), the Emil Racoviţă Institute of Speleology in Bucharest (ISER), the Laboratory of Geology and Paleontology of the University of Thessaloniki (LGPUT), the national Natural History Museum of Sofia (NNHM-S) and the Natural History Museum of Basel (NMB). Other than *Gallogoral*, which is known by three skeletons, all the Senèze species are represented by isolated elements. In the present study, all available material is described and taxonomically assessed, with some new interpretations proposed. Heintz's (1970) terminology is used for dental description, and the veterinary one for the cranium and postcranial remains (Barone 1976). Measurements are defined in the Appendix, following Crégut-Bonnoure (1987, 1995, 2020) and Driesch (1976). Upper teeth are indicated by uppercase letters, lowers by lowercase. The stratigraphic scheme for the Late Pliocene to Early Pleistocene large mammal record and chronostratigraphic framework for the French MNQ biozones based on new $^{40}Ar/^{39}Ar$ ages provided by Nomade et al. (2014) is used.

Bovidae Gray, 1821

Antilopinae Gray, 1821

Tribe Antilopini Gray, 1821

Genus *Gazellospira* Pilgrim & Schaub, 1939

Species *Gazellospira torticornis* (Aymard, 1854)

G. torticornis is a medium sized antelope with heteronymously twisted horn cores. In Western Europe its chronological range and geographical distribution are wide, the species being indicated from MN 15 to MNQ 19 from France to Russia (Crégut-Bonnoure 2007; Hermier et al. 2020).

The first appearance of *Gazellospira* occurs at the end of the Pliocene, during MN 15 (Bouvrain 1998; de Bruijn et al. 1992; Hermier et al. 2020): Çalta, Alçoköy, and Gülyazi in Turkey. The genus is identified during MN 16 with two species: *G. torticornis* (Aymard, 1854) and *G. gromovae* Dmitrieva, 1975 (see Dmitrieva 1977) (defined at Kuruksaj in Tajikistan). This species has been recognized at Liventsovka in Russia (MNQ 17 or MNQ 18; Titov 2008) and at Bolgrad in Ukraine (MNQ 18; Alexeeva & Baygucheva 1976). Recently Hermier et al. (2020) proposed to synonymize *G. gromovae* with *Antilospira? zdanskyi* Teilhard de Chardin and Trassaert, 1938 from Shanxi (China) and to include all the specimens under *Gazellospira zdanskyi*.

G. torticornis has been recovered in many European localities:

- MN 15b: in Hungary at Csarnóta 1 and 2 (Kretzoï 1956; Schaub 1932).
- MN 16: in France at Vialette (Guérin *in* Duvernois & Guérin 1989; MN16a), Perrier-Rocca-Neyra (Nomade et al. 2014; Stehlin 1904) and Vals-Le-Crozas (Bout 1960); in Italy at Montopoli (Pilgrim & Schaub 1939); in Spain at Villarroya (Schaub 1943); and in Greece at Seskló, lower level (Athanassiou 2018).
- MN(Q)16–18: in Hungary at Beremend 4 (Kretzoï 1956; Schaub 1932).
- MNQ 17: in France at Le Coupet (Aymard 1854), Perrier-Pardines (Pilgrim & Schaub 1939), Saint-Vallier (Heintz 1966), Chilhac II (Boeuf 1983; Paquette et al. 2021) and perhaps at Cornillet (Heintz & Dubar 1981); in Spain at Casablanca 1 (Soto & Morales 1985); in Hungary at Villány 3 (=Villány-Kalkberg) (Kretzoï 1956; Schaub 1932); in Greece at Seskló, upper level (Athanassiou 2005, 2018; Symeonidis 1992), Dafneró (Kostopoulos & Koufos 1994; Hermier et al. 2020; Bennami et al. 2020), Volax (or Volakas) (Sickenberg 1968) and Vaterá F (De Vos et al. 2002); in Bulgaria at

Varshets (Spassov 2005; Spassov & Crégut-Bonnoure 1999); in Romania at Valea Grăunceanului (Rădulescu et al. 2003; Rădulesco 2005; Terhune et al. 2020; Croitor et al. 2023); and in Turkey at Sarikol Tepe (Kostopoulos & Sen 1999).

- MNQ 18: in France at Senèze (Schaub 1923); in Italy at Olivola (Schaub 1928), Costa San Giacomo (Cassoli & Segre Naldini 1994; Belluci et al. 2014) and Montecarlo (Belluci & Sardella 2015); in Spain at Huélago (Azanza & Morales 1989), Fonelas-P1 and Fonelas SCC-1 (1.8–1.7 Ma; Garrido 2006, 2008a) and at La Puebla de Valverde (Andrés Rodrigo 2011; Gautier & Heintz 1974; Nomade et al. 2014); in Germany at Erpfinger Höhle (or Erpfingen; Abel et al. 2006; Lehmann 1957; Ufrecht 2008); in Bulgaria at Slivnitsa (Spassov & Crégut-Bonnoure 1999); and in Greece at Karnezeika (Sianis et al. 2022) and perhaps at Gerakaroú (Kostopoulos & Koufos 1994).
- MNQ 18–19: in Hungary at Villány-Nagyharsányhegy 4 (=Villány-Nagyharsányberg 4) (Kretzoï 1956; Schaub 1932).

In southern Europe, *Gazellospira* is reported in Spain, Italy and the Balkans (Greece, Montenegro) during MNQ 19 and the beginning of MNQ 20, however a doubt remains as this material could belong to *Pontoceros*: in Spain at Orce-2 (Agusti et al. 1987; Maldonado-Garrido et al. 2017); in Italy in the Tasso FU (Caloi et al. 1986); in Greece at Krímni 1 (Kostopoulos & Koufos 1994), Alykés (Athanassiou 1996, 1998), and Pyrgos (Kostopoulos et al. 2002); and in Montenegro at Trlica (Vislobokova & Agadjanian 2015). At the same time, the species inhabited the Middle East: in Turkey at Yassigüme (Burdur Basin; Demirel & Mayda 2014; 1.5 Ma). Its possible presence at ʻUbeidiya (Israel) was suggested by Geraads (1986), but the material has been reallocated to cf. *Pontoceros ambiguus/Spirocerus* sp. (Martínez-Navarro et al. 2012).

During MNQ 17 and MNQ 18, some remains could belong to another species (at Varshets; Spassov 2005; Spassov & Crégut-Bonnoure 1999) or subspecies (La Puebla de Valverde; Andrés Rodrigo 2011). At Fonelas, a new subspecies has been clearly identified: *G. torticornis hispanica*. Recently, another subspecies has been proposed (by Hermier et al. 2020) for the Late Pliocene specimen from Kvabebi (Georgia) previously attributed to *Parastrepsiceros sokolovi* Vekua, 1968: *G. torticornis sokolovi*.

The holotype of *G. torticornis*, a right horn core recovered from Le Coupet (MC 2003-14-26 COU), is preserved in the Crozatier Museum (Aymard *in* Dorlhac 1854: Fig. 8; Lacombat 2004–2005: Fig. 62 p. 78; Lacombat et al. 2010: pp. 82–83). It has been described by Pilgrim and Schaub (1939) and a neodiagnosis is given by Guérin (*in* Duvernois & Guérin 1989). The main morphological features of the species have been presented by: Viret 1961; Heintz 1966; Crégut-Bonnoure & Guérin 1996; Andrés Rodrigo 2011. *G. torticornis* has been identified at Senèze by Schaub (1923). All the material is from previous field research.

Description

Crania and Horn Cores

There are 20 incomplete specimens which represent a minimum of 12 individuals. The most complete element is a partial cranium belonging to a young animal (NMB Se 1704; Pilgrim & Schaub 1939: pl. 1, Figs. 5 and 6) (Fig. 14.1: D). The first twist of the horn cores is preserved with the frontals between them and 5 cm in front of them, along with the parietal, the left temporal and a rostral fragment of the supra-occipital. The external acoustic meatus and the base of the auditory bulla are preserved on the left side. The transverse section of the frontals is convex above the orbits and concave in the sagittal plane.

There are three frontlets in Basel belonging to adults: one of them has been described by Pilgrim and Schaub (NMB Se 1739; Fig. 2 pl. 1); another one (NMB Se 200; Fig. 14.1: C) shows the frontal sinuses which are wide; the third (NMB Se 1) has quite complete horn cores (Fig. 14.1: A). In addition, there are five more or less complete horn cores: an isolated complete right horn core (NMB Se 1452; Pilgrim & Schaub 1939: pl. 1, Fig. 3) which is slender, the first twist of a right horn core (NMB Se 199), two apical parts (NMB Se 1593 and another specimen without number) and several fragments belonging to a single horn core (NMB Se 1865).

In the Lyon collection there are six isolated horn cores. The specimens FSL 16902 and FSL 16898 belong to young animals. The first one is a frontlet with the two horn cores (Fig. 14.1: B). The frontal suture is open and the two horn cores, without the tips, are slender. The fronto-parietal suture is 2 cm from the caudal side of the horn core. The second one is a right specimen with dorsoventrally compressed base; the rostral side is less convex in comparison to the other specimen and the anterior keel is weak. Three other specimens belong to adults: FSL 16899 is a fragmentary frontal with complete elongated horn core (Fig. 14.1: G); FSL 16900 is a right horn core with the caudal part of the frontal and a rostral fragment of the parietal, with the position of the fronto-parietal suture at the same distance from the horn cores as on FSL 16902 (Fig. 14.1: B, E); FSL 16901, a well preserved left specimen, could be the same individual as the previous (Fig. 14.1: F). The last one, FSL 211088, is a base of a right horn core.

In Paris, there are two quite complete horn cores belonging to adults: a left one (MNHN-F-1929-9) with the inside base of the frontal and the inner surface of the orbit; and a right (MNHN-F-1923-4) with the base of the frontal showing the sinuses. There are also three fragments: an

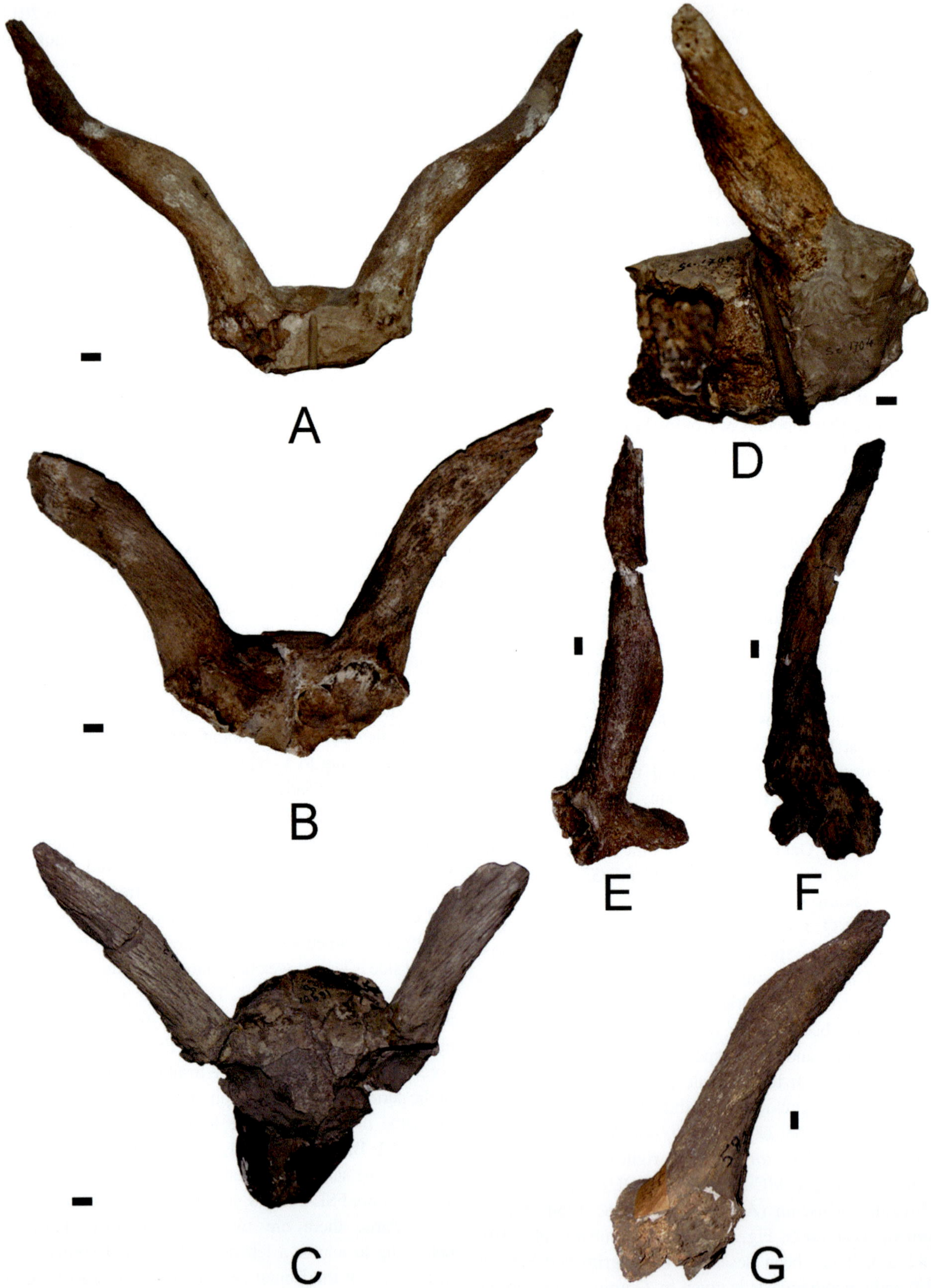

Fig. 14.1 Cranium and horn cores of *Gazellospira torticornis* from Senèze. A, frontlet NMB Se 1; B, frontlet FSL 16902; C, frontlet NMB Se 200; D, cranium NMB Se 1704; E, right horn core FSL 16900; F, left horn core FSL 16901; G, FSL 16899. A, B, G, *norma frontalis*; C, *norma dorsalis*; D, E, F, *norma lateralis*. Scale 1 cm

incomplete right base (MNHN-F-unnumbered) and two apical fragments (a right MNHN-F-1929-9 and MNHN-F-unnumbered).

The horn cores are well spaced apart, heteronymously twisted and divergent. There are no morphological differences from the type, and the general proportions of the horn cores are the same (Table 14.1). However, the Senèze specimens are very slightly smaller than those of Le Coupet, which may be related to sexual dimorphism (Fig. 14.2). The morphology of the fragmentary horn core from Villány-Nagyharsányhegy 4 is similar (NMB cast UP 795). The scatter-plot (Fig. 14.3) separates the young animals from the adults and males (n = 3) from females (n = 6).

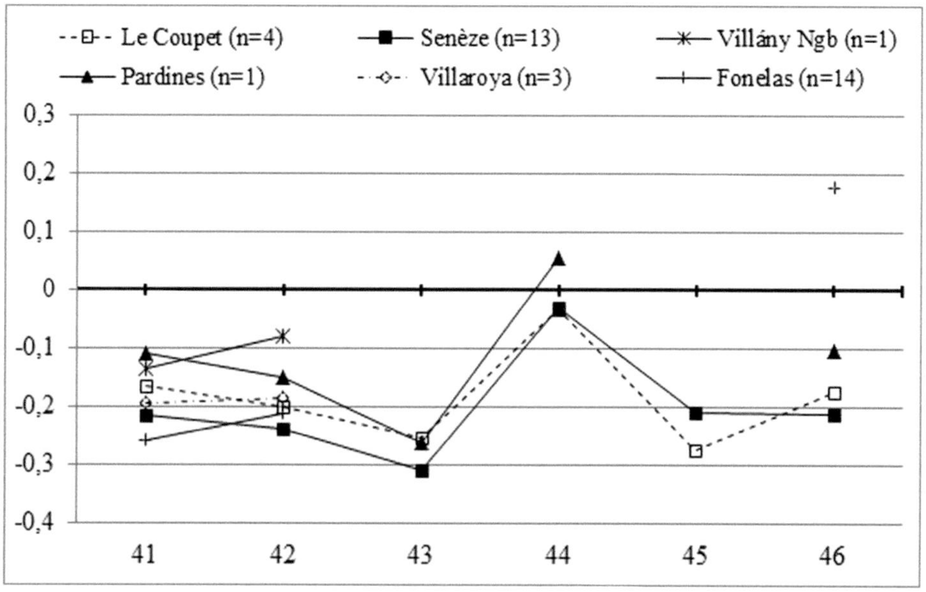

Fig. 14.2 *Gazellospira torticornis*. Ratio-diagram of the horn cores. Reference axis: recent *Capra ibex* ♂ (in Crégut-Bonnoure 2020). 41, rostrocaudal diameter at the base; 42, dorsolateral diameter at the base; 43, rostrocaudal diameter at the middle; 44, dorsolateral diameter at the middle; 45, length; 46, circumference at the base. References of the measurements in Fig. 14.A.2 and Measurements dimensions in Table 14.1

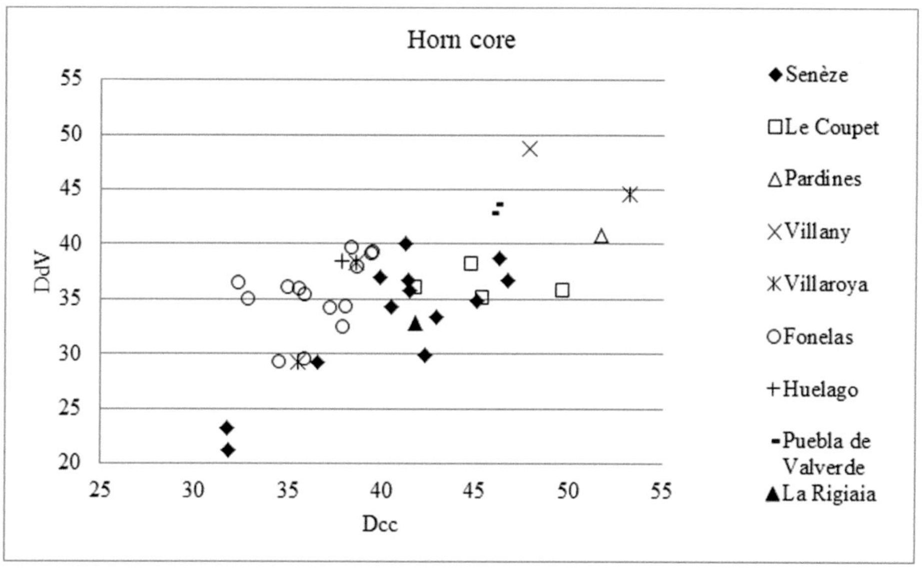

Fig. 14.3 *Gazellospira torticornis*. Scatter diagram of the horn cores. Dcc (=measurement 41), rostrocaudal diameter at the base. Ddv (= measurement 42), dorsoventral diameter at the base. References of the measurements in Fig. 14.A.2 and selected dimensions in Table 14.1

Table 14.1 Comparative dimensions (in mm) of the parietals and horn cores in *Gazellospira torticornis* from Senèze and other localities

Variable	25	26	37	38	39	41	42	43	44	45	46
Senèze N	1	1	4	4	4	13	13	11	11	4	12
Mean	46.1	81.4	126.5	65.7	180.8	40.6	33.1	24.9	31.3	228.5	120.5
Minimum			117.4	60.1	134.1	31.7	21.3	15.5	21.0	216.5	96.5
Maximum			135.8	68.3	201.9	46.7	40.0	29.4	35.6	255.0	140.0
S.D			8.62	3.87	32.07	4.76	5.74	4.69	4.79	17.84	13.29
Pardines (NMB LP 166)			136.9	58.8	282.2	51.7	40.8	27.7	38.2		
Le Coupet											
MC 2003-14-27-Cou (Holotype)						41.8	36.1	28.1	35.0	173.8	127.0
NMB Cp 62						45.4	35.1				135.0
NMB Cp 59						49.6	38.8	28.3	26.9	220.0	130.0
NMB Cp 60						44.8	38.2				135.0
Villány-Nagyharsànyberg (NMB UP 795–cast)						48.7	47.9				

Variable definitions (from Crégut-Bonnoure 2002, 2020) (Figs. 14.A.1 and 14.A.2). Parietal: 25, dorsal length; 26, maximal width; Horn cores: 37, width between the lateral sides at the base; 38, width between the internal sides at the base; 39, width between the internal sides in the middle; 40, width between the internal sides at the tip; 41, rostrocaudal diameter at the base; 42, dorsolateral diameter at the base; 43, rostrocaudal diameter in the middle of complete horn core; 44, dorsolateral diameter in the middle; 45, rostral length; 46, circumference at the base

Teeth

A description of the teeth is presented by Heintz (1966) and Guérin (*in* Duvernois & Guérin 1989: pp. 341–342). Some missing details are presented below. The teeth are moderately hypsodont.

Upper Teeth

There are seven upper teeth belonging to three individuals (one adult, two juvenile) (Tables 14.2, 14.3 and 14.4): a fragmentary maxilla with left P2-4 (MNB Se 1727; Pilgrim & Schaub 1939: pl. 2, Figs. 13 and 14); the right DP2-4 from a single maxilla (NMB Se 307; Pilgrim & Schaub 1939: pl. 2, Figs. 3 and 4); and an isolated left DP4 (FSL 16893, intially attributed to *Procamptoceras*; Fig. 14.4: A to D).

The upper teeth are characterized by the lack of the occlusal islet (Heintz 1966). The crown is broad and the inflation of the paracone is particularly strong, including at the base; between it and the mesostyle, the surface of the crown is broad and concave. The M3 lacks a metastylar wing. The ribs between the styles are well inflated, especially the paracone, and the width of styles is moderate.

Three specimens previously identified as *Gazellospira* are reallocated to *Gallogoral*:

- NMB Se 539 (on label *Gazellospira* right M2) which has an occlusal islet and an important entostyle (height: 7.7 mm).
- The two M1 NMB Se 387 (left; Pilgrim & Schaub 1939: pl. 2, Figs. 10–12) and NMB Se 1437 (right) are relatively more hypsodont and less broad at their base. There is an occlusal islet, the inflation of the paracone is weak and between it and the mesostyle the surface is straight and slightly concave.

Table 14.2 Comparative dimensions (in mm) of the upper premolars in *Gazellospira torticornis* from Senèze and other localities

P2 variable	Ht	1	2	3	4	
Senèze (NMB Se 1727)	10.9				8.5	
Le Coupet (NMB Cp 71)	11.2	11.6	11.6	10.8	10.8	
Pardines						
NMB Lp 43 R	15.1	11.8	11.3	8.4	8.7	
NMB Lp 43 L	15.5	11.8	11.3	8.3	8.7	
Rocca-Neyra						
NMB Rn 179 (cast) R	9.8	11.4	10.9	9.6	9.5	
NMB Rn 179 (cast) L	12.4	11.3		9.6		
P3 variable	Ht	1	2	3	4	
Senèze (NMB Se 1727)		11.4		10.0		
Le Coupet (NMB Cp 66)	10.4	11.6		9.5	9.7	
Pardines						
NMB Lp 42	10.5	11.0	10.7	10.4	10.3	
NMB Lp 43 R	16.5	11.9	11.5	9.6	10.6	
Rocca-Neyra						
FSL 211708 R	13.9	11.3		11.6	12.6	
FSL 211708 L	13.4	12.2		10.8	12.4	
P4 variable	Ht	1	2	3	4	5
Senèze (NMB Se 1727)		12.1		11.0	10.6	
Le Coupet						
NMB Cp 66	12.4	11.6		11.0	10.8	12.6
NMB Cp 137	14.2	11.0	10.2	10.0	10.0	12.0
Pardines						
NMB Lp 42	9.7	10.1		12.4	12.4	
NMB Lp 43 R	19.2	13.0	11.5			
Rocca-Neyra						
FSL 211708 R	14.4	11.9		11.4	11.4	13.5
FSL 211708 L	13.9	11.3		11.6	11.6	14.0

Variable definitions (from Crégut-Bonnoure, 1995, 2002, 2020) (Fig. 14.A.4: A): Ht, height; 1, occlusal length 2, basal length. 3, occlusal width total. 4, occlusal width in the middle. 5, basal width. L, left; R, right

Table 14.3 Comparative dimensions (in mm) of the upper molars in *Gazellospira torticornis* from Senèze and other localities

M1 variable	Ht	1	2	3	4	5	6	7
Senèze (FSL 211708)		17.9		16.0	16.3	15.2	14.5	17.2
Le Coupet								
NMB Cp 66	7.8	16.7		14.5	14.5	13.5	13.5	15.5
NMB Cp 46	15.7	20.9	18.5			12.9	11.8	18.2
NMB Cp 73	12.5	18.8	15.3			13.0	11.3	15.2
Pardines (NMB LP 42)	6.2	15.4		16.0		16.9		
Rocca-Neyra								
FSL 211708 (L)	11.3	18.3				14.0	14.0	16.8
FSL 211708 (R)	9.0	17.7		16.1	16.2	14.9	14.8	17.3
M2 variable	Ht	1	2	3	4	5	6	7
Senèze (FSL 211708)		24.1		16.9	16.3	14.4	13.4	19.0
Chilhac (NMB PP 27)	22	23.6	21.4	15.4	14.7	12.7	11.6	18.2
Le Coupet (NMB Cp 47)	28.8	22.6	21.2	13.8	12.4	12.6	11	17.2
Vialette (FSL 211178)	37.2	22.8	20.7	15.7	14.6	12.5	11.8	20.6
Pardines (NMB LP 42)	11.2	21.3		17.9	17.6	17.2	16.7	
Rocca-Neyra								
FSL 211708 (L)	17.2	23.0		16.6	16.6	140.0	13.1	18.6
FSL 211708 (R)	15.1	23.7		16.2	16.2	15.1	13.7	18.2

(continued)

Table 14.3 (continued)

M3 variable	Ht	1	2	3	4	5	6	7
Senèze (FSL 211708)		24.2		14.7	14.7		12.6	12.0
Le Coupet (NMB Cp 72)	17.5	21.7	21.9	13.2	13.0	17.8	11.3	11.1
Pardines (NMB LP 42)	12.4	24.3		16.6	16.5		13.4	13.2
Rocca-Neyra								
FSL 211708 (L)		24.8		14.2	14.2		12.5	11.7
FSL 211708 (R)		24.2		14.3	14.3		12.2	11.3
M3 variable	8	9	10	11	12			
Senèze (FSL 211708)		9.8	9.8					
Le Coupet (NMB Cp 72)	17.3	8.7	9.2	19.7	21.1			
Pardines (NMB LP 42)		9.38	9.38					
Rocca-Neyra								
FSL 211708 (L)		9.5	8.7					
FSL 211708 (R)		10.8	10.8					
Toothrow variable	LP+M	LP	LM					
Pardines (NMB LP 42)			59.1					
Rocca-Neyra (NMB RN 179–cast)	100.5	36.5	65.6					

Variable definitions (from Crégut-Bonnoure 1995, 2002, 2020) (Fig. 14.A.4: B, C): Ht, height, LP+M, length of the premolar and molars, LP: length of the premolars; LM: length of the molars. M1 and M2: 1, occlusal length; 2, basal length. 3, occlusal width total, first lobe; 4, occlusal width in the middle, first lobe; 5, occlusal width total, second lobe; 6, occlusal width in the middle, second lobe; 7, basal width. M3: 1, occlusal length; 2, basal length. 3, occlusal width total, first lobe; 4, occlusal width in the middle, first lobe; 5, basal width, first lobe; 6, occlusal width total, second lobe; 7, occlusal width in the middle, second lobe; 8, basal width, second lobe; 9, length of the interstylar surface, mesial; 10, length of the interstylar surface, distal; 11, occlusal length of the two lobes at the lingual side; 12, basal length of the two lobes at the lingual side; L, left; R, right

Table 14.4 Comparative dimensions (in mm) of the upper deciduous teeth in *Gazellospira torticornis* from Senèze

DP2 variable	Ht	1	2	3	5			
Senèze (NMB Se 307)	10.7	12.8	12.5	8.3	7.9			
DP3 variable	Ht	1	2	3	4	5	6	7
Senèze (NMB Se 307)	10.1	17.6	15.4	8.0	8.3	8.9	8.8	11.4
DP4 variable	Ht	1	2	3	4	5	6	7
Senèze								
FSL 16893	18.4	20.4	17.4	12.0	11.7	11.1	9.7	15.2
NMB Se 307	14.1	19.6	16.1	11.8	11.3	10.8	10.3	13.6

Variable definitions (from Crégut-Bonnoure, 1995, 2002, 2020): Ht, height; DP2 (Fig. 14.A.4: A): 1, occlusal length; 2, basal length; 3, occlusal width; 5, basal width. DP3 and DP4 (Fig. 14.A.4: B): 1, occlusal length; 2, basal length. 3, occlusal width total, first lobe; 4, occlusal width in the middle, first lobe; 5, occlusal width total, second lobe; 6, occlusal width in the middle, second lobe; 7, basal width.

Lower Teeth

There are 18 lower teeth belonging to five individuals (four adult, one juvenile) (Tables 14.5 and 14.6).

There are three fragmentary mandibles; two of them belong to adults (NMB Se 1434, left with m1-3, Fig. 14.4: I to K); NMB Se 1734, right with p4-m3) and one to a juvenile (NMB Se 403 with right dp2-4 and m1; Pilgrim & Schaub 1939: pl. 2, Fig. 5). In all, there are seven isolated teeth: left m1-2 (MNHN-F-unnumbered); four other m2s (two left, FSL 16896 (Fig. 14.4: E to H); and two right, NMB Se 1439 and MNHN-F-1922-5); and one right m3 (NMB Se 1435).

The crown of p4 is bilobed with a developed second lobe. The buccal sides are rounded. The lingual valley is shallow and "V" shaped. The molars are characterized by a prominent and well developed goat fold which runs from the top to the base of the crown. On the isolated specimens, the mesial side is flat and the shape of the cemento-enamel junction is quadrangular and higher on the mesial side than on the distal one. The lobes are narrow and pinched. As noticed by Guérin (*in* Duvernois & Guérin 1989), the inflation of the lingual cuspids is moderate; generally, this inflation disappears towards the base of the crown. The lingual valley is shallow and broad, and the mesial valley behind the parastylid is short and does not reach the base of the crown as shown in the illustration of Pilgrim and Schaub (1939: Fig. 9 plate II). On NMB Se 1434, the distal lobe of the m3 is displaced labially (Fig. 14.4: K). It is slightly convex whereas it is rectilinear on an isolated molar from Saint-Vallier (Heintz 1966: Fig. 3c pl. I). The inflation of the cuspids is stronger on the isolated m2 from Chilhac (NMB PP 156; identified as "*Antilope* cf. *Nemorhoedus*" on label) whose goat fold is not well developed. A small interlobar cingulum is occasionally present (height: 5.3 mm), for example on the m3 NMB Se 1435 and on the m2 of the mandible NMB Se 1734. The proportions of the m2s seem to have similarities with those of Perrier-Pardines (Fig. 14.5) whereas those of the m3 are similar to the Le Coupet population (Fig. 14.6).

The isolated right m2 and m3 from Villány-Nagyharsányhegy 4 (NMB UP 486 and 485) do belong to *Gazellospira*.

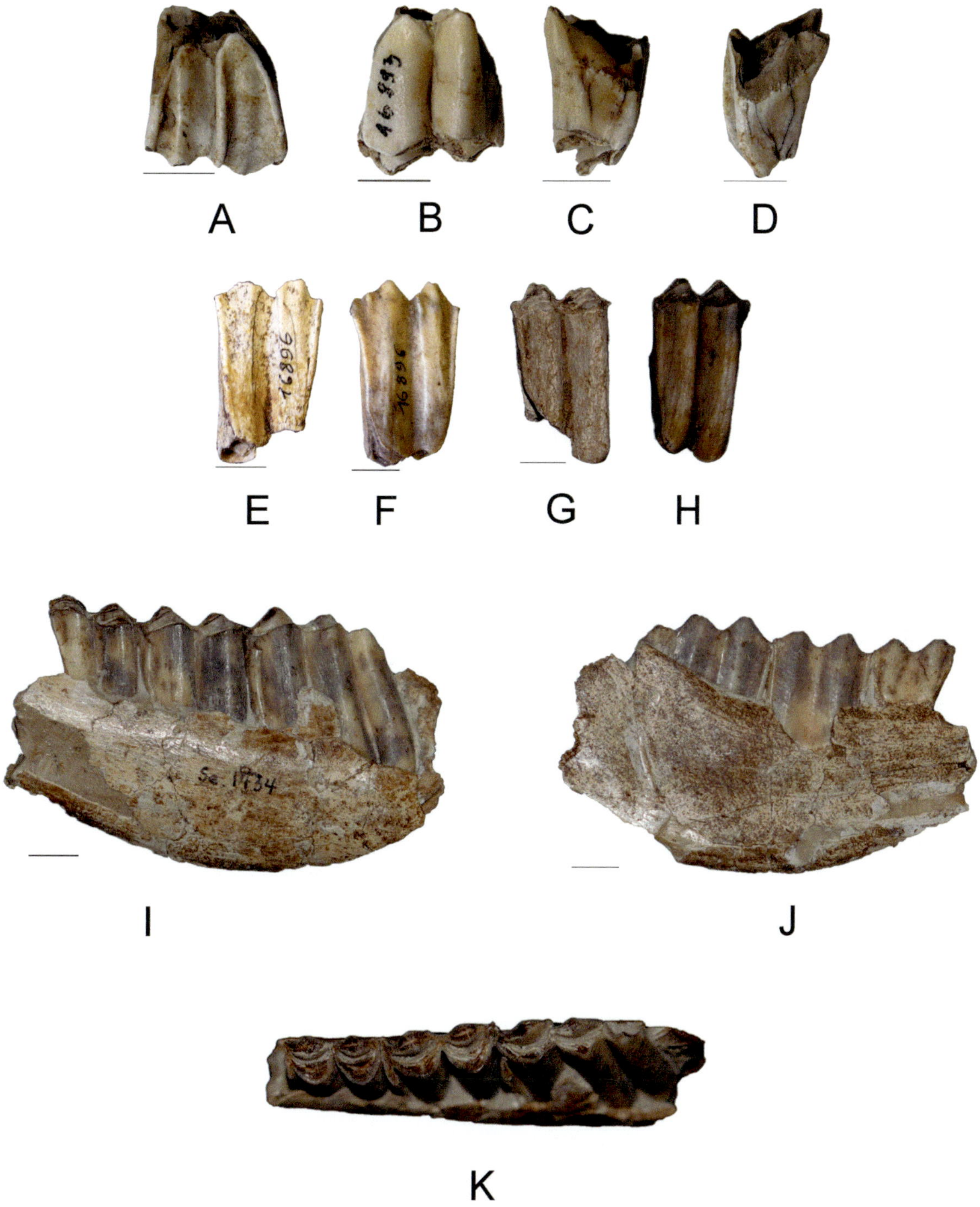

Fig. 14.4 Mandible and teeth of *Gazellospira torticornis* from Senèze. FSL 16893: A to D, left DP4; E, G, left m2; F, H, left m2. I, J, K, left mandible NMB Se 1434. A, G, H, I, buccal view; B, E, F, J, lingual view; C, mesial view; D, distal view; K, occlusal view. Scale 1 cm

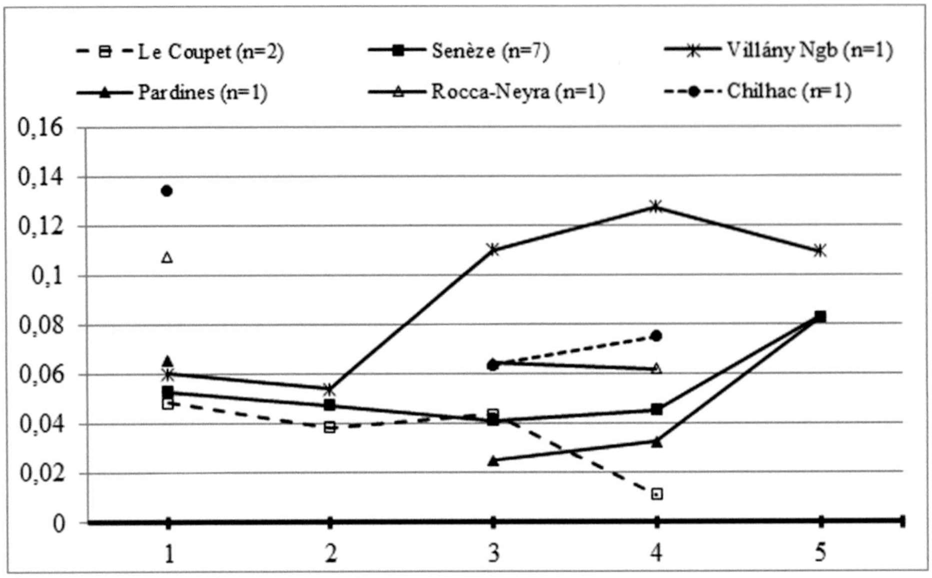

Fig. 14.5 *Gazellospira torticornis*. Ratio-diagram of m2. Reference axis: recent *Capra ibex* from Le Taï cave (in Crégut-Bonnoure 2020). Ngb, Villány-Nagyharsányhegy 4. Length: 1, occlusal; 2, basal; Width: 3, occlusal, first lobe; 4, occlusal, second lobe; 5, basal. References of the measurements in Fig. 14.A.4 and dimensions in Table 14.6

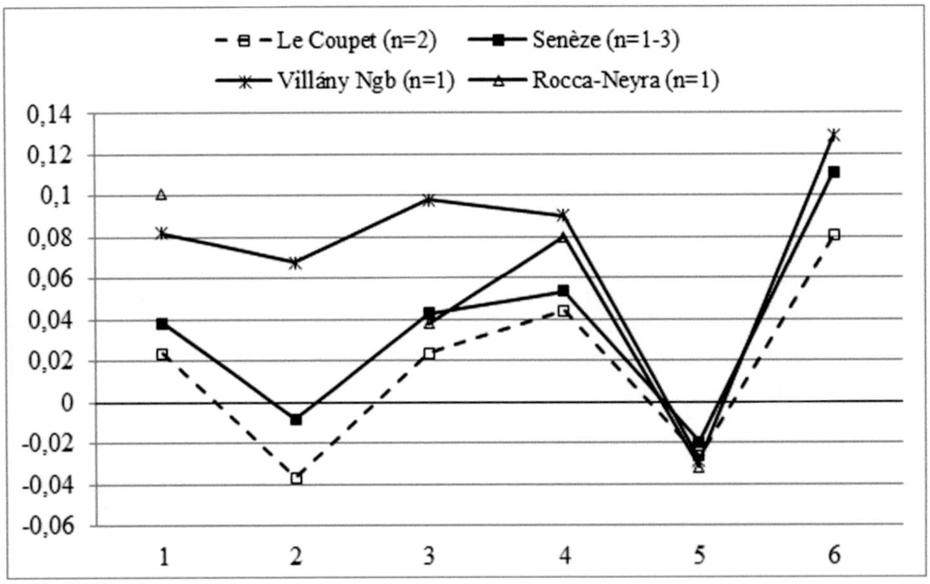

Fig. 14.6 *Gazellospira torticornis*. Ratio-diagram of m3. Reference axis: recent *Capra ibex* from Le Taï cave (in Crégut-Bonnoure 2020). Ngb, Villány-Nagyharsányhegy 4. Length: 1, occlusal; 2, basal. Width: 3, occlusal, first lobe; 4, occlusal, second lobe; 5, occlusal, third lobe; 6, basal. References of the measurements in Fig. 14.A.4 and dimensions in Table 14.6

Table 14.5 Comparative dimensions (in mm) of the lower premolars in *Gazellospira torticornis* from Senèze and Pardines

p2 variable	Ht	1	2	5		
Pardines (NMB LP 41)		9.4	9.8	6.28		
p3 variable	Ht	1	2	3	5	
Pardines (NMB LP 41)	11.8	13.3	11.9	7	8	
p4 variable	Ht	1	2	3	4	5
Senèze (NMB Se 1734)	7.0	13.4	13.4	8.2	8.5	8.3
Pardines (NMB LP 41)	16.1	15.7	15.0	8.0	8.3	9.3

Variable definitions (from Crégut-Bonnoure, 1995, 2002, 2020) (Fig. 14.A.4: D): Ht, height; 1, occlusal length; 2, basal length; 3, occlusal width, first lobe; 4, occlusal width, second lobe; 5, basal width

Table 14.6 Comparative dimensions (in mm) of the lower molars in *Gazellospira torticornis* from Senèze and other localities

m1 variable	Ht	1	2	3	4	5	
Senèze							
NMB Se 1434	11.5	17.6	14.2	9.2	9.8	10.5	
NMB Se 1734	6.7	14.1	14.1	10.2	11.2	11.0	
NMB Se 403		18.7		8.4	8.2		
MNHN-F-?	22.4	19.0		10.1			
Pardines (NMB LP 41)	10.5	17.1	14.6	9.7	10.3	11.5	
m2 variable	Ht	1	2	3	4	5	
Senèze N	7	7	5	7	7	6	
Mean	22.4	20.7	18.4	10.4	10.3	12.3	
Minimum	10.4	19.8	16.2	9.5	9.3	11	
Maximum	32.1	21.5	20.6	11.9	12.0	13.3	
S.D	7.6	0.6	2.0	0.1	0.9	0.8	
Le Coupet							
NMB Cp 51	13.1	19.6	17.4	10.7			
NMB Cp 56	19.6	21.5	18.7	10.1	9.5		
Pardines (NMB LP 41)	23.7	21.3		10.0	10.0	12.3	
Chilhac (NMB PP 156)		25.0		10.9	11.0		
Villány-Kalkberg (NMB UP 486)	14.7	21.1	18.7	12.2	12.4	13.1	
m3 variable	Ht	1	2	3	4	5	6
Senèze							
NMB Se 1435	16.5		28.8	12.1	11.5	6.4	13.0
NMB Se 1434		27.9		8.8	8.7	4.6	
NMB Se 1734		29.7		11.1	10.9	5.6	
Le Coupet							
NMB Cp 52	17.0	28.8	27.1	11.0	10.8	5.9	12.4
NMB Cp 49	22.2	26.8	26.9	9.5	9.5	4.9	11.7
Rocca-Neyra (NMB Rn 27)		33.2		10.6	11.0	5.4	
Villány-Nagyharsànyberg (NMB UP 485)	25.0	31.8	34.3	12.1	11.3	5.4	13.5

Variable definitions (from Crégut-Bonnoure, 1995, 2002, 2020): Ht, height; m1 and m2 (Fig. 14.A.4: E): 1, occlusal length; 2, basal length; 3, occlusal width, first lobe; 4, occlusal width, second lobe; 5, basal width. m3 (Fig. 14.A.4: F): 1, occlusal length; 2, basal length; 3, occlusal width, first lobe; 4, occlusal width, second lobe; 5, occlusal width, third lobe; 6, basal width

Postcranium

There are 30 postcranial elements: one atlas (NMB Se 1517), four humeri (complete left NMB Se 1095 [Pilgrim & Schaub 1939: pl. 3, Figs. 1–3]; left NMB Se 1441, right NMB Se 1443, left NMB Se 1675), four more or less complete radii (left NMB Se 1095 [Pilgrim & Schaub 1939: pl. 3, Fig. 4], NMB Se 1232 [left and right], left NMB Se 1943 belonging to a young animal), two left ulnae (NMB Se 1095 [Pilgrim & Schaub 1939: pl. 3, Fig. 4], NMB Se 1232), two left scaphoids (NMB Se 1095 [Pilgrim & Schaub 1939: pl. 3, Fig. 5], FSL 16899), one left semi-lunar (NMB Se 1095, Pilgrim & Schaub 1939: pl. 3, Fig. 5); one left pisiform (FSL 16899), one left unciform (FSL 16899), two left capitato-trapezoids (NMB Se 1095 [Pilgrim & Schaub 1939: pl. 3, Fig. 5], FSL 16899), two left metacarpal III+IV (NMB Se 1095 [Pilgrim & Schaub 1939, pl. 3, Fig. 5 and 7], FSL 210827), two first anterior phalanges (NMB Se 1095, Pilgrim and Schaub: 1939, pl. 3, Fig. 6), two second anterior phalanges (NMB Se 1095: Pilgrim and Schaub: 1939, pl. 3, Fig. 6), two tibiae (right NMB Se 1084 [Pilgrim and Schaub: 1939, pl. 3, Fig. 11], left NMB Se 1472), three calcanei (two left: NMB Se 1152, NMB Se 1153 and one right: NMB Se 1136), one right metatarsal III +IV (FSL 210827).

The carpals FSL 16899 and the metacarpal FSL 210827 belong to the same individual, while the carpals and the metacarpal NMB Se 1095 belong to another. It is not possible to tell if the metatarsal of FSL 210827 derives from the same individual as the metacarpal, nor if the frontal and postcrania of FSL 16899 represent a single individual.

This material has been compared with the rich samples from Le Coupet, Perrier-Pardines, Perrier-Rocca-Neyra and Chilhac.

Atlas

Two incomplete atlases have been considered to be *Gazellospira*. NMB Se 235 is similar to the *Gallogoral* atlases NMB Se 313 and NMB Se 550 from skeletons: consequently, I reallocate it to this genus. NMB Se 1517 appears to be corrrectly identified. On the label, Schaub wrote that the proportions are a little different from the Le Coupet *Gazellospira* NMB Cp 211; however, their morphology is quite similar. On these two vertebrae, the atloidian fossa is deep and its connection with the lateral sides of the ventral arch is abrupt, marked by a rectilinear keel. On NMB Cp 211 this connection is less abrupt. On the dorsal side, the tubercle is weakly developed and in front of it there are two concavities delimited laterally by two crests. The cranial notch is eroded in NMB Cp 311; in NMB Se 1517 from Senèze it is widely “V shaped”.

Humerus

The verticillius of the trochlea is up to the capitulum-trochlea base when the bone is put on its distal surface (Fig. 14.7: A). In medial view, the trochlea is well developed and the caudal epicondyle is projected downward. The trochlear fossa, deep and rounded, is bordered caudaly by a pinched crest which runs up and transversely to the diaphysis; this crest stops 6 mm above the proximal side of the trochlea. The surface between the crest and the proximal side of the trochlea is quadrangular shaped and deep. The coronoid fossa is elongated; nevertheless, it stops at the level of the crest. A small and vertical septum, located more or less in the middle of the bone. It is well developed on NMB Se 1095 as on the humeri from Chilhac (NMB PP 137; Fig. 14.7: C) and Le Coupet (NMB Cp 80), less on the other specimens from Senèze. The depth of the olecranon fossa is medially important.

The small specimen NMB Se 1441 has been assigned on the label to *Procamptoceras*. Its morphology is similar to *Gazellospira*. It belongs to a female, like the Le Coupet specimen NMB Cp 80. In addition, the specimen FSL 16899 (a left humerus) labeled as *Gazellospira* in fact belongs to *Metacervoceros rhenanus (=Cervus philisi)*.

If I compare the general proportions of the distal epiphysis between populations, they are fairly close but not identical (Fig. 14.8; Table 14.7).

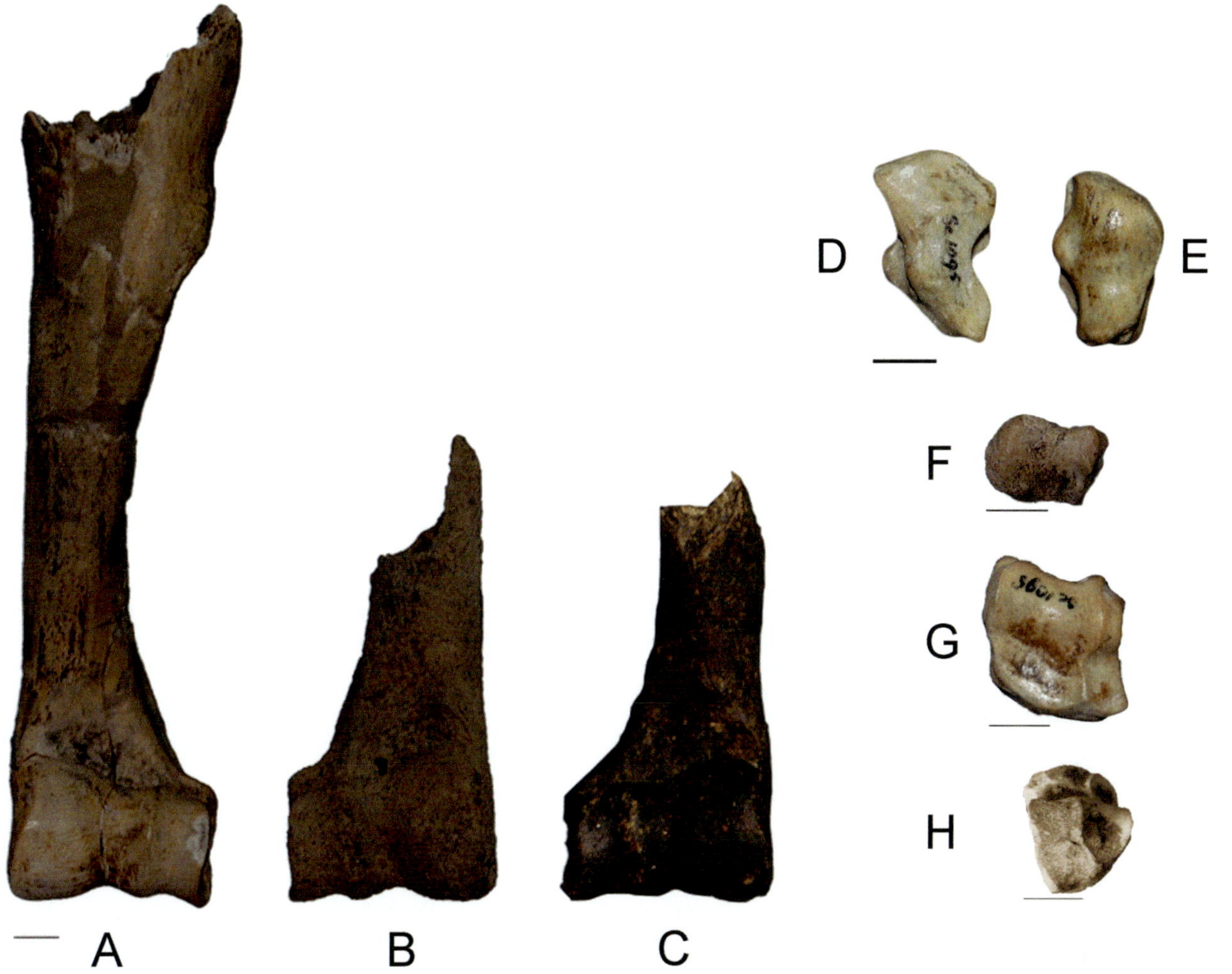

Fig. 14.7 Postcranium of *Gazellospira torticornis* from Senèze and other localities. A, left humerus NMB Se 1095 from Senèze; B, right humerus NMB Rn 158 from Rocca-Neyra; C, right humerus NMB Ch 137 from Chilhac; D to G, left carpals NMB Se 1095 from Senèze; D, semi-lunar; E, scaphoid; F, pisiform; G, capitato-trapezoid; H, left unciform FSL 16899 from Senèze. A, B, C, cranial view; D, E, G, H, proximal view; F, lateral view. Scale 1 cm

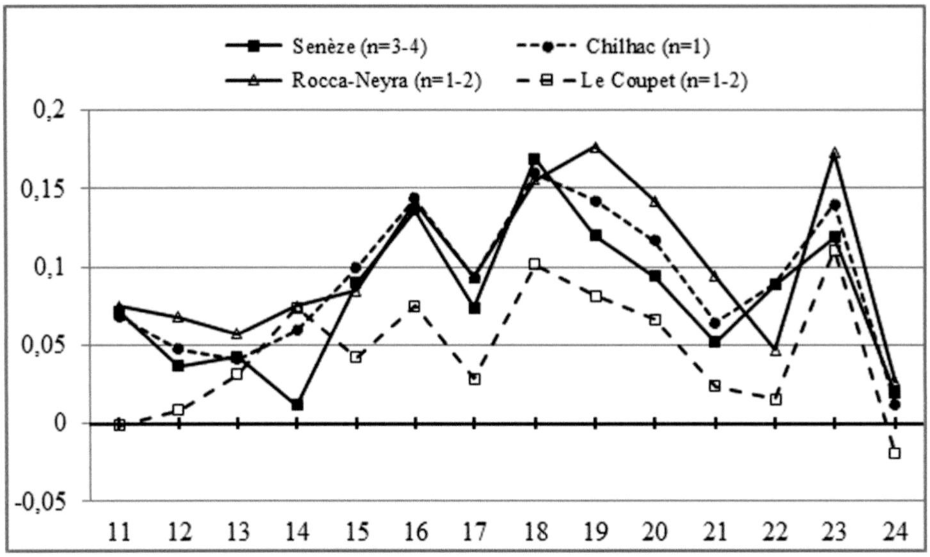

Fig. 14.8 *Gazellospira torticornis*. Ratio-diagram of the humerus. Reference axis: recent *Capra ibex* ♂+♀ (in Crégut-Bonnoure 2020). ccd, craniocaudal diameter; td, transversal diameter. Humerus: Td distal: 11, maximal; 12, articular; 13, trochlea; 14 capitulum. Ccd distal: 15, medial side, total; 16, medial side, articulation; 17, lateral side, total; 18, lateral side, articulation; 19, trochlea, minimum; 20, trochlea, median guiding crest. Diaphysis, sus articular: 21, transversal diameter; 22, ccd. Olecranon fossa: 23, transversal diameter; 24: height. References of the measurements in Fig. 14.A.6 and dimensions in Table 14.7

Table 14.7 Comparative dimensions (in mm) of the humerus and ulna in *Gazellospira torticornis*

Humerus variable	2	9	10	11	12	13	14	15	16
Senèze									
NMB Se 1095	195.5	24.5	30.3	45.7	43.1	29.7	12.9	42.4	31.4
NMB Se 1441					38.2	27.5	11.69	37.67	28.1
NMB Se 1443		26.2	35.2	46.4	42.5	29.1	12.12	42.0	33.4
NMB Se 1675		25.8	32.5	51.0	45.6	31.0	13.58	44.1	34.4
Rocca-Neyra									
NMB Rn 158				48.1	45.5	30.3	14.5	43.7	34.2
NMB Rn 163								38.5	30.1
Chilhac (NMB PP 137)				47.4	43.4	29.2	14.0	42.5	32.4
Humerus variable	17	18	19	20	21	22	23	24	
Senèze									
NMB Se 1095	34.5	26.3	22.5	26. 6	25.7	28.9	15.2	20.9	
NMB Se 1441			20.6	24.0	23.1	23.1		15.4	
NMB Se 1443	34.0		23.2	25.4	26.1	29.4	15.54		
NMB Se 1675	38.5	25.8	24.5	27.7	26.1	28.1	15. 9	18.1	
Rocca-Neyra									
NMB Rn 158	37.3	25.2	25.8	29.0	27.8	24.8	18.7	20.3	
NMB Rn 163							16.6	16.6	
Chilhac (NMB PP 137)	37.3	25.5	23.8	27.3	26.0	27.4	16.4	17.8	
Ulna variable	2	3	4	5	7	8	10		
Senèze									
NMB Se 1095					11.6		24.4		
NMB Se 1232	36.4	18.0	74.1	33.2	10.7	16.5	23.9		

Variable definitions (from Crégut-Bonnoure 1995, 2002, 2020) (Figs. 14.A.6 and 14.A.8): ccd, craniocaudal diameter; dpd, dorso-palmar diameter; td, transversal diameter. Humerus (Fig. 14.A.6): Length: 2, from proximal articulation. Middle of the diaphysis: 9, td; 10, ccd. Td distal: 11, maximal; 12, articular; 13, trochlea; 14 capitulum. Ccd distal: 15, medial side, total; 16, medial side, articulation; 17, lateral side, total; 18, lateral side, articulation; 19, trochlea, minimum; 20, trochlea, median guilding crest. Diaphysis, supra-articular: 21, td; 22, ccd. Olecranon fossa: 23, transversal diameter; 24: height. Ulna (Fig. 14.A.8): 2, dpd of the *tuber olecrani*; 3, td of the *tuber olecrani*; 4, length from the top of the *tuber olecrani* to the proximal angle of the anconeal process; 5, dpd minimal between the top of the *tuber olecrani* and the proximal angle of the anconeal process; 7, minimal td perpendicular to measurement 5; 8, td of trochlea incisura; 10: td of the articulation for radius

Radius

One of the five specimens belongs to an immature individual (NMB Se 1943). It was assigned initially to *M. rhenanus*; its re-attribution to a bovid was suggested by T. Pfeiffer (in a note on the label from 1996).

The diaphysis is rectilinear and enlarged distal to the proximal third. At this level the medial side is pinched and acute. The proximo-lateral tuberosity is pinched below the articulation, and its surface constitutes a planum of about 1 cm. The dorsal side for the capitulum is especially projected upward and acute at its top. On the palmar side, at 5 mm from the base of the lateral facets for the ulna, there is a more or less transverse and oblique elongated tuberosity. The facet for the scaphoid is slighly concave but less than in cervids. Its lateral side, in dorsal and palmar views, is acute and clearly projected down. At the level of the epiphyseal line, a distinct tuberosity is visible on the palmar side between the proximal reliefs for the scaphoid and semi-lunar facets. In distal view, it is rounded. The area for the ulnar styloid process constitutes an important planum (2 cm length, 1 cm width), directed obliquely upwards.

The proportions of the proximal articulation of the Senèze specimens are similar to those of Pardines and Perrier-Rocca-Neyra (measurements 3–9) (Fig. 14.9; Table 14.8), except the length of the facet for the capitulum (measurement 6). The proportions of the distal articulation are still very close to those of Perrier-Rocca-Neyra (measurements 16 to 23).

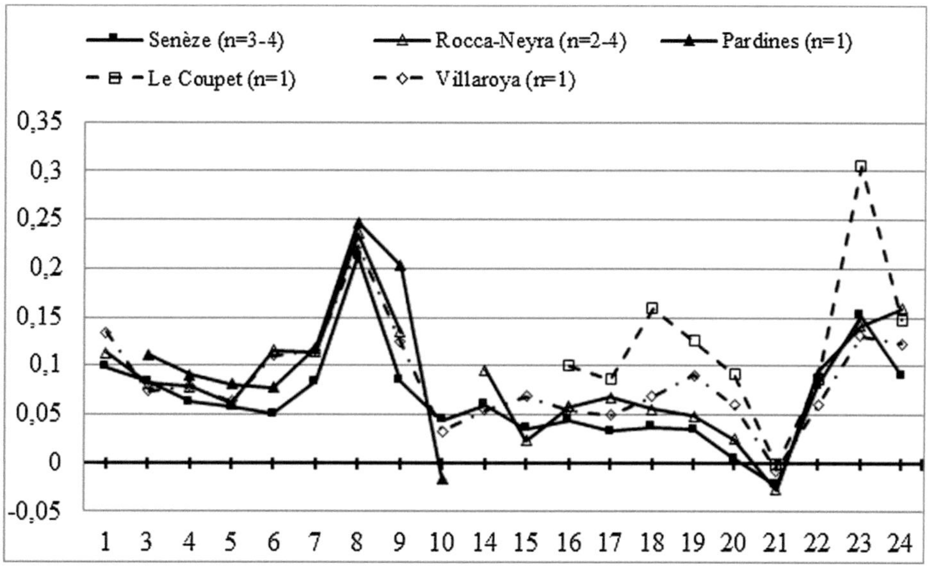

Fig. 14.9 *Gazellospira torticornis*. Ratio-diagram of the radius. Reference axis: recent *Capra ibex* ♂+♀ (in Crégut-Bonnoure 2020). dpd, dorsopalmar diameter; td; transversal diameter. Length: 1, total; 2, median. Proximal epiphysis: 3, td total; 4, td articular; 5, td of the facet for the trochlea articulation; 6, td of the facet for the capitulum articulation; 7, dpd of the facet for the trochlea articulation; 8, dpd of the facet for the capitulum articulation; 9, dpd minimum; 10, td of the facets for the ulna; 11, td of the lateral facet for the ulna, 12; td of the median facet for the ulna; 13, td of the medial facet for the ulna. Middle of the diaphysis: 14, td; 15, dpd. Distal epiphysis: 16, td total; 17, dpd; 18, td articular; 19, td of the facets for the scaphoid and the semi-lunar; 20, td of the facet for the scaphoid; 21, td caudal of the facet for the semi-lunar; 22, longest length of the facet for the scaphoid; 23, length of the ridge separating the facets for the scaphoid and semi-lunar; 24, td of the facet for the pyramidal. References of the measurements in Fig. 14.A.7 and dimensions in Table 14.8

Table 14.8 Comparative dimensions (in mm) of the radius in *Gazellospira torticornis* from Senèze and other localities

Variable	1	2	3	4	5	6	7	8
Senèze								
NMB Se 1095	236.8	236.3	50.5	44.1	30.6	12.6	24.2	17.2
NMB Se 1232	240.5	244.5	51.3	45.0	31.0	13.4	22.8	19.8
NMB Se 1943	230.6	231.8	46.5	42.8	30.1	11.7	23.7	18.2
Pardines (NMB LP 56)			52.6	46.9	32.2	13.4	25.5	19.9
Rocca-Neyra								
NMB Rn 158	231.7		51.2	48.1	31.7	16.5	25.4	21.1
NMB Rn 148	256.9	235.0	46.7	42.9	29.2	12.5	22.6	18.6
NMB Rn ?			49.8	45.7	31.6	14.8	27.7	
NMB Rn 30a							25.5	18.5
Villaroya (NMB Spa 190)	256.1	257.4	48.4	45.7	31.0	14.4	25.4	18.9
Variable	9	10	11	12	13	14	15	16
Senèze								
NMB Se 1095	19.5	29.0	7.4	5.0	7.4	29.0	17.1	
NMB Se1232	17.6	27.5	8.9	6.4	7.7	29.3	17.8	44.0
NMB Se 1943	16.9	29.4	10.1	6.7	8.9	25.7	15.4	40.5
Pardines (NMB LP 56)	23.6	24.9	7.0	5.5				48.1
Rocca-Neyra								
NMB Rn 158	21.6					27.3	13.9	41.1
NMB Rn 148	19.0					33.4	18.6	46.8
NMB Rn 30a	20.1							43.1
Le Coupet (NMB Cp 268)								48.1
Villaroya (NMB Spa 190)	19.8	27.9	6.4	5.2	10.4	27.7	18.0	43.3
Variable	17	18	19	20	21	22	23	24
Senèze								
NMB Se 1095				16.1		27.9		
NMB Se1232	28.7	36.2	34.2	17.0	11.4	28.0	15.5	4.4
NMB Se 1943	28.0	33.5	31.6	14.3	11.8	27.2	14.5	4.4
Pardines (NMB LP 56)	32.1	46.1	40.6	19.2	12.2	28.1	21.3	5.0
Rocca-Neyra					3	3	3	3
NMB Rn 158	30.0	34.0	304	15.8	10.2	27.0	13.5	4.4
NMB Rn 148	33.3	39.4	38.3	18.5	13.0	30.8	14.3	5.4
NMB Rn 30a	28.9	35.6	33.3	15.2	11.2	28.8	16.1	5.6
Le Coupet (NMB Cp 268)	32.1	46.1	40.6	19.2	12.2	28.0	21.3	5.0
Villaroya (NMB Spa 190)	29.4	37.4	37.4	17.9	12.0	26.5	14.3	4.7

Variable definitions (from Crégut-Bonnoure 1995, 2002, 2020) (Fig. 14.A.7): ccd, craniocaudal diameter; td; transversal diameter. Length: 1, total; 2, median. Proximal epiphysis: 3, td total; 4, td articular; 5, td of the facet for the trochlea articulation; 6, td of the facet for the capitulum articulation; 7, ccd of the facet for the trochlea articulation; 8, ccd of the facet for the capitulum articulation; 9, ccd minimum; 10, td of the facets for the ulna; 11, td of the lateral facet for the ulna, 12; td of the median facet for the ulna; 13, td of the medial facet for the ulna. Middle of the diaphysis: 14, td; 15, dpd. Distal epiphysis: 16, td total; 17, dpd; 18, td articular; 19, td of the facets for the scaphoid and the semi-lunar; 20, td of the facet for the scaphoid; 21, td caudal of the facet for the semi-lunar; 22, longest length of the facet for the scaphoid; 23, length of the ridge separating the facets for the scaphoid and semi-lunar; 24, td of the facet for the pyramidal

Carpals

The morphology of the carpals is succinctly described by Pilgrim and Schaub (1939). Their dimensions are given in Table 14.9.

Scaphoid. In proximal view, the medial side has a broad median expansion, prominent especially in palmar view (Fig. 14.7: E). In lateral view, the distal side is slightly concave. A small tubercle is present on FSL 16899, below the proximo-dorsal facet. In medial view, an important tubercle is present palmo-distally and also visible in proximal view.

Semi lunar. In proximal view, the dorsomedial and palmolateral crests are perpendicular to the dorsopalmar plane (Fig. 14.7: D). The palmar angle is very sharp and in palmar view a longitudinal tubercle runs from it to the base. Behind the palmo-lateral tubercle, whose surface is acute, the articulation is elongated and moderately broad. In lateral and medial views, the distal side is distinctly concave at the middle of the bone. The proximo-lateral facet is triangularly shaped. There is laterally a distopalmar tubercle, elongated proximo-distally. On the medial side, the proximodistal facet is roundish, the distal one is quadrangular.

Pisiform. The pisiform has a roundish shape (Fig. 14.7: F). The two facets are small (6.7 × 3.9 mm and 7.5 × 2.4 mm).

Capitato-trapezoid. The general shape is quadrangular (Fig. 14.7: G). The lateral side is straight. In proximal view, the lateral crest, more or less rectilinear, is thin and prominent. The palmar osseous bulge is weakly developed and, in palmar view, the surface before the lateral crest is obliquely directed. It is possible to compare the proportions of this carpal with the specimen from Le Coupet (Fig. 14.10). Although the morphology is the same, the proportions are distinct.

Unciform. The medial side is straight (Fig. 14.7: H). In dorsal view, the median crest is very prominent. On each side of it, the facets are very deep. In palmar view, the lateral facet is lower than the medial one with a deep and transverse pit below the latter. The shape of the lateral facet is triangular and the lateral side is riding up. There are two facets on the medial side, separated by a thin transverse crest.

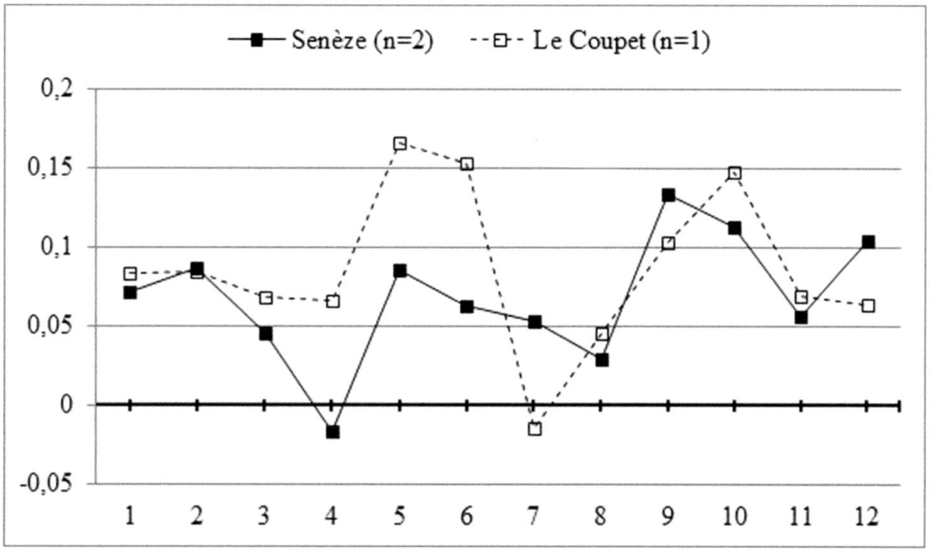

Fig. 14.10 *Gazellospira torticornis*. Ratio-diagram of the capitato-trapezoid. Reference axis: recent *Capra ibex* ♂+♀ (in Crégut-Bonnoure 2020). dpd, dorsopalmar diameter; td, transversal diameter. 1, td in anatomical position; 2, dpd in anatomical position; 3, total height; 4, height of the dorsal rim; 5, td of the lateroproximal facet, dorsal side; 6, td of the medioproximal facet, dorsal side; 7, td of the lateroproximal facet, palmar side; 8, td of the medioproximal facet, palmar side; 9, td of the lateroproximal facet (at the base of the dorsal rim); 10, dpd of the proximal ridge; 11, td maximum of the distal articulation; 12, dpd maximum of the distal articulation. References of the measurements in Fig. 14.A.9 and dimensions in Table 14.9

Table 14.9 Comparative dimensions (in mm) of the carpals in *Gazellospira torticornis* from Senèze and Le Coupet

Scaphoid variable	1	2	3	4	5	6	7	8	9	10
NMB Se 1095	20.1	20.3	25.6	18.2	20.8	14.0	13.9	6.8	6.2	3.8
FSL 16899	18.9	26.5	23.8	15.4	23.0					
Semi-lunar variable	1	2	3	4	5	6	7	8	9	
NMB Se 1095	19.0	23.6	30.8	29.0	19.6	14.2	10.3	15.1	15.7	
Semi-lunar variable	10	11	12	13	14	15	16	17	18	
NMB Se 1095	8.4	8.4	6.9	8.3	9.7	14.5	6.2	8.0	12.9	
Semi-lunar variable	19	20	21	22	23	24				
NMB Se 1095	3.4	8.1	4.0	10.8	9.2	18.0				
Pisiform variable	1	2	3	4	5					
FSL 16899	20.2	14.8	14.6	9.0	11.3					
Unciform variable	1	2	3	4	5	6	7			
FSL 16899	14.8	20.3	15.9	17.7	10.4	9.3	9.6			
Unciform variable	8	9	12	13	14	15				
FSL 16899	8.4	16.6	18.2	14.6	16.0	14.7				

(continued)

Table 14.9 (continued)

Capitato-trapezoid variable	1	2	3	4	5	6
Senèze (NMB Se 1095)	23.5	24.6	15.7	5.9	7.8	19.0
Senèze (FSL 16899)	23.2	22.0	13.7	6.3	6.6	17.6
Le Coupet (NMB Cp 161)	23.9	23.2	15.5	7.3	8.6	22.6
Capitato-trapezoid variable	7	8	9	10	11	12
Senèze						
NMB Se 1095	5.0	16.9	6.7	20.8	29.7	18.2
FSL 16899	5.2	16.5	6.3	18.9	28.9	21.4

Variable definitions (from Crégut-Bonnoure, 1987, 2002, 2020) (Fig. 14.A.9): dpd, dorsopalmar diameter; td, transversal diameter. Scaphoid (Fig. 14.A.9: A to C): Height: 1, total; 8, proximal articular facet, palmar side; 10, distal articular facet, dorsal side. Dpd: 2, total; 3, proximal articulation; 5, distal articulation; 7, proximal articular facet, palmar side; 9, distal articular facet, dorsal side. Td: 4: proximal articulation; 6, distal articulation

Semi-lunar (Fig. 14.A.9: D to H): 1, height. Dpd proximal: 2, in the middle; 3, from dorsolateral angle to palmolateral angle; 4, in anatomical position. Td proximal: 5: dorsal side; 6, palmar side; 7, minimum, in the middle. Td distal: 8, dorsal side; 9, palmar side; 10, dorsolateral articulation; 11, dorsomedial articulation; 12, palmomedial articulation; 13, palmolateral articulation. Dpd: 14, distal crest. Td: 15, minimum dorsal side. Medial side: 16, height of the proximo-dorsal articulation; 17, td of the proximodorsal articulation; 18, td of the proximopalmar articulation; 19, minimum height of the proximopalmar articulation; 20, td of the dorsodistal articulation; 21, height of the dorsodistal articulation. Lateral side: 22, td of the proximodorsal articulation; 23, minimum height of the proximodorsal articulation. Dpd: 24, lateral side

Pisiform (Fig. 14.A.9: L): 1, total height; 2, dpd. Pyramidal articulation (Fig. 14.A.9: I to K): 3, proximodistal diameter; 4, td; 5, total lateromedial diameter

Unciform (Fig. 14.A.9: M to P): 1, total height. 2, dpd. Td: 3, total; 4, proximal articulation, dorsal side; 5, lateroproximal articulation, dorsal side; 6, medio proximal articulation, dorsal side; 7, lateroproximal articulation, palmar side; 8, medioproximal articulation, palmar side; 9, dorsal side, minimum. Articulation of the medial side: 10, proximodistal diameter; 11, dpd. Distal articulation: 12, dpd; 13, td. Proximal articulation: 14, td, palmar side; 15, dpd, medial side; 16, dpd of the median crest

Capitato-trapezoid (Fig. 14.A.9: Q to S): dpd, dorso-palmar diameter; td; transversal diameter. 1, td in anatomical position; 2, dpd in anatomical position; 3, total height; 4, height of the dorsal rim; 5, td of the lateroproximal facet, dorsal side; 6, td of the medioproximal facet, dorsal side; 7, td of the latero-proximal facet, palmar side; 8, td of the medioproximal facet, palmar side; 9, td of the lateroproximal facet (at the base of the dorsal rim); 10, dpd of the proximal ridge; 11, td maximum of the distal articulation; 12, dpd maximum of the distal articulation

Metacarpal III+IV

The proximal articular surface of NMB Se 1095 (Pilgrim & Schaub, 1939: Fig. 7, pl. 3), in comparison with the specimen preserved in Lyon (FSL 210827), has an unciform facet less extended laterally. However, in *Gazellospira* from Perrier-Pardines and Perrier-Rocca-Neyra (n = 7), the shape of this facet is variable. The proximopalmar side is less flat than on the Perrier-Rocca-Neyra and Perrier-Pardines metapodials. On FSL 210827 a small tubercle exists at the middle of this side; it is absent on the other specimens. On the Senèze specimens, a small longitudinal groove is present at the middle of the lateral side, immediately below the articulation as on the Perrier-Rocca-Neyra and Perrier-Pardines bones; nevertheless, in these two last populations, the groove is bordered on its palmar side by a triangular tubercle which is absent on the Senèze specimens. The capitato-trapezoid articular facet is medially undulating; in proximal view, the medial side shows a recess; the dorsal side is rounded and the palmar one is more or less straight. On the dorsal side, the proximal interosseous canal is present; however, it is weak and disappears rapidly on the shaft, only remaining distally (Fig. 14.11: A). The foramen of the interosseous canal is relatively large compared to that of the Perrier-Rocca-Neyra and Perrier-Pardines metacarpals (Fig. 14.11: B, C); on each side smooth crests, weakly developed, are present and the dorsal groove is not very deep. The keels are parallel on each side. They are thinner on FSL 210827 than on NMB Se 1095. The lateral keel is well developed and its connection with the shaft in lateral view is abrupt because the dorsal fossa is deep.

The length of the Senèze specimens is similar to the Perrier-Rocca-Neyra and Perrier-Pardines bones (measurements 1–3; Fig. 14.12; Table 14.10); however, the epiphysis and articular surfaces (measurements 4–11 and 14–24) and the shaft at the middle (measurement 12) are smaller. The differences among the three populations are very slight.

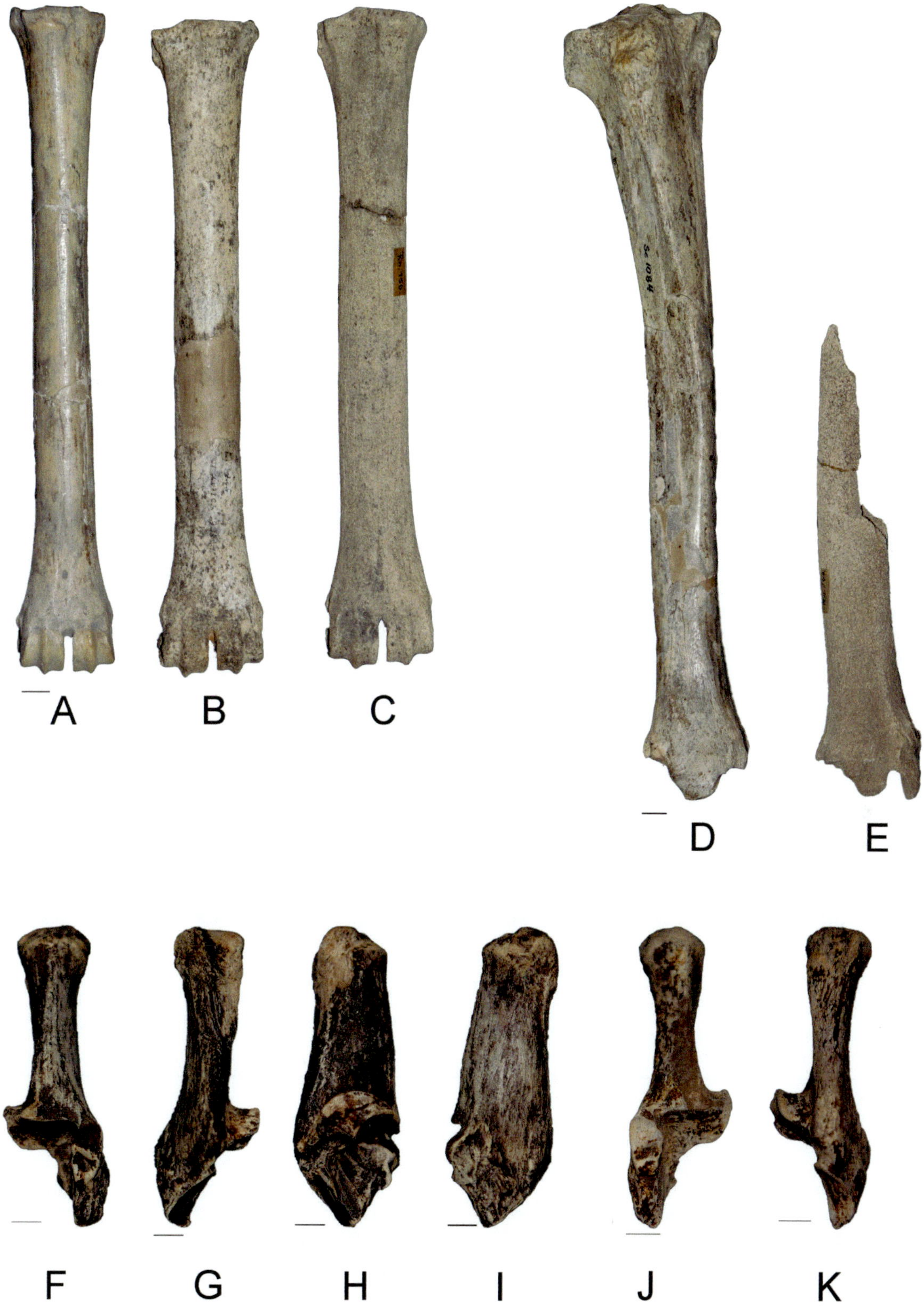

Fig. 14.11 Postcranium of *Gazellospira torticornis* from Senèze and other localities. A, left metacarpal III+IV NMB Se 1095 from Senèze; B, right metacarpal III+IV NMB LP 125 from Perrier-Pardines; C, right metacarpal III+IV NMB Rn 156 from Rocca-Neyra; D, right tibia NMB Se 1084 from Senèze; E, right tibia NMB Rn 144 from Rocca-Neyra; F to I, left calcaneus NMB Se 1152 from Senèze; J, K, right calcaneus NMB Se 1136 from Senèze. A, B, C, F, J, dorsal view; D, E, cranial view; G, K, plantar view; H, medial view; I, lateral view. Scale 1 cm

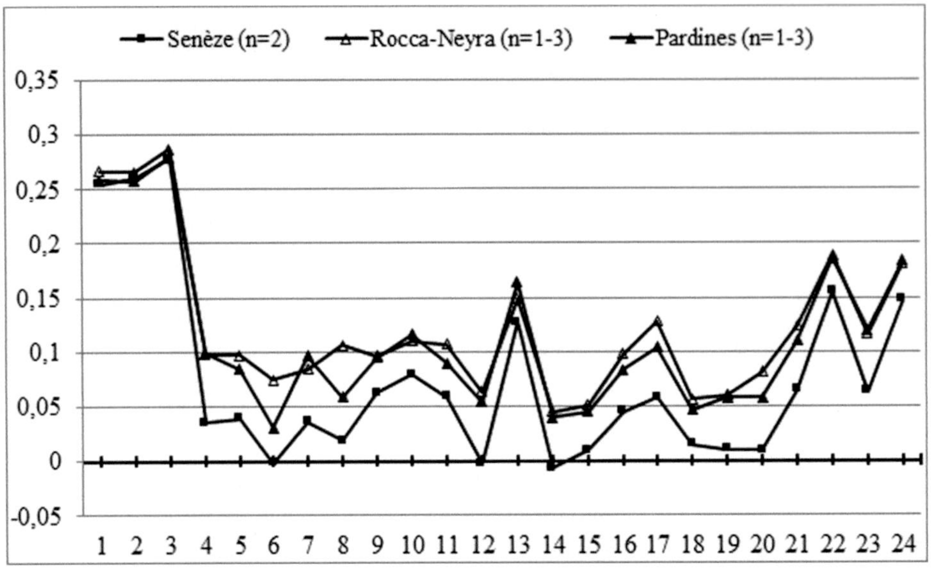

Fig. 14.12 *Gazellospira torticornis*. Ratio-diagram of the metacarpal III+IV. Reference axis: recent *Capra ibex* ♂+♀ (in Crégut-Bonnoure 2020). dpd, dorsopalmar diameter; td, transversal diameter. Length: 1, total; 2, minimum at the middle; 3: minimum of the medial side. Proximal epiphysis: 4, td total; 5, td of the articulation; 6, td of the unciform facet; 7, td of the capitato-trapezoid facet; 8, oblique length of the unciform facet; 9, oblique length of the capitato-trapezoid facet; 10, dpd; 11: dpd of the articulation. Middle of the diaphysis: 12, td; 13, dpd. Distal epiphysis: 14, td of the epiphysis; 15, td supra-articular; 16, dpd of the epiphysis; 17, dpd of the epiphysis at the foramen level; 18, distance between the keels on the palmar side; 19, distance between the keels at their base; 20, distance between the keels at their middle; 21, dpd of the medial condyle; 22, dpd of the medial keel; 23, dpd of the lateral condyle; 24: dpd of the lateral keel. References of the measurements in Fig. 14.A.10 and dimensions in Table 14.10

Table 14.10 Comparative dimensions (in mm) of the metacarpal III+IV in *Gazellospira torticornis* from Senèze and other localities

Variable	1	2	3	4	5	6	7	8	9	10	11	12
Senèze												
FSL 210827	230.2	215.2	225.1	33.4	32.7	12.7	20.6	19.7	27.7	25.9	21.6	20.3
NMB Se 1095	239.8	224.1	234.6	36.2	35.9	13.8	23.0	22.2	30.5	27.0	24.4	21.7
Rocca-Neyra												
NMB Rn 156	239.1	219.4	232.8	41.1	39.9	16.3	24.1	25.4	31. 7	28.8	25.9	25.9
NMB Rn 24 a				40.4	39.1	14.5	25.8	25.3	32.0	29.4	26.1	
NMB Rn 149	244.1	225.6	236.7	39.0	38.7	16.5	23.2	26.1	30.8	27.0	25.3	23.0
Pardines												
NMB LP 733	237.3	218.4	230.8	40.4	38.1	14.3	25.0	23.1	32.5	28.5	25.0	24.6
NMB LP 59				40.1	38.0	14.2		22.9	30.2	29.1	24.4	23.3
NMB LP 168												
NMB LP 60				27.7	27.2	11.8	18.5	16.8	22.4	20.9	20.1	16.8
Variable	13	14	15	16	17	18	19	20	21	22	23	24
Senèze												
FSL 210827	19.0	34.3	35.3	18.4	16.4	21.0	20.0	17.0	23.2	17.8	22.8	16.8
NMB Se 1095	19.4	36.2	36.6	19.4	17.2	21.9	21.0	17.3	24.6	19.8	24.0	18.8
Rocca-Neyra												
NMB Rn 156	20.2	39.6	39.6	21.4	19.7	23.6	23.0	20.13	27.4	20.3	26.4	19.3
NMB Rn 24 a	19.4	39.3	38.9	21.2	19.1	23.1	22.6	19.7	27.2	20.0	26.2	18.9
NMB Rn 149	20.9	40.2	40.2	21.8	20.7	24.0	23.6	27.6	27.65	20.5	26.6	19.4
Pardines												
NMB LP 733	20.7	38.9	38.8	20.7	18.7	23.1	22.9	19.1	26.5	19.6	25.7	18.7
NMB LP 59	21.1											
NMB LP 168		39.5	39.3							20.8	27.7	20.1
NMB LP 60	16.3											

Variable definitions (from Crégut-Bonnoure, 1995, 2002, 2020) (Fig. 14.A.10): dpd, dorsopalmar diameter; td; transversal diameter. Length: 1, total; 2, minimum at the middle; 3, minimum of the medial side. Proximal epiphysis: 4, td total; 5, td of the articulation; 6, td of the unciform facet; 7, td of the capitato-trapezoid facet; 8, length of the unciform facet, diagonally from lateropalmar angle to dorsal point of the proximal crest; 9, length of the capitato-trapezoid facet, diagonally from medio-palmar angle to dorsal point of the proximal crest; 10, dpd of the epiphysis;11, dpd of the articulation. Middle of the diaphysis: 12, td;13, dpd. Distal epiphysis: 14, td of the articulation; 15, supra-articular td;16, supra-articular dpd;17, dpd at the level of the dorsal foramen;18, distance between the lateral edges of the keels at the palmar top; 19, distance between the lateral edges of the keels at the palmar base; 20, distance of the keels at their middle; 21, td of the medial keel; 22, td of the medial condyle; 23, td of the lateral keel; 24, td of the lateral condyle.

Anterior Phalanges

There are two first and two second phalanges associated with the metacarpal III–IV (Table 14.11). In palmar view, the first phalanx has a short and wide abaxial muscular insertion (potato-shaped) whereas the axial one is longer, with a sigmoid internal border and a distal point forming a sharp edge. The sesamoid articular facets are directed downward. On each side, the supra-condylar fossae are surrounded by an edge.

The second phalanx has proximal sides more or less straight and parallel. The axial side is longer than the abaxial. On the palmar side, there is a pit in the median plane, below the proximal articulation, and a continuous and pinched bulge.

Tibia

The tibial crest runs from the proximal articulation to the medial malleolus (Fig. 14.11: D). On the caudal side, there is a rounded tubercle below the proximomedial articulation, more or less near the middle of the bone. The fibular groove is deep. The dorsal facet for the malleolar bone is ovoid with a flat surface. In plantar view, the lateral side of the distal articulation is truncated. Distally, the border makes a deep depression followed by an acute projection.

The two specimens can only be compared with the tibia from Perrier-Rocca-Neyra (NMB Rn 144), which is incomplete (Fig. 14.11: E). The morphology of the distal articulation is the same and the proportions are more or less similar (Fig. 14.13; Table 14.12).

Table 14.11 Dimensions (in mm) of the first and second anterior phalanges in *Gazellospira torticornis* from Senèze

Phal. I ant. variable	1	2	3	4	5	6	7	8	9	10	11	12	13
Senèze													
NMB Se 1095 medial	49.6	49.8	50.3	18.5	23.4	17.5	19.9	5.7	7.4	5.9		17.8	17.2
NMB Se 1095 lateral	50.5	50.4	50.8	18.1	21.1	17.1	18.4	5.6	8.2	4.7	5.6	20.5	16.7
Phal. I ant. variable	14	15	16	17	18	19	20	21	22	23	24	25	26
Senèze													
NMB Se 1095 medial	5.5		17.0		14.1	15.8	16.3	17.5	13.9	15.6	16.5	14.0	27.8
NMB Se 1095 lateral	4.8	3.6	21.2	19.3	14.0	15.8	16.3	17.6	14.1	15.6	16.7	14.0	27.9
Phal. II ant. variable	1	2	3	4	5	6	7	8	9				
Senèze													
NMB Se 1095 medial	29.8	28.3	26.9	16.3	21.3	15.6	13.1	7.8	9.6				
NMB Se 1095 lateral	29. 8	28.0	27.0	16.6	21.1	16.0	13.1	7.8	9.1				
Phal. II ant. variable	10	11	12	13	14	15	16	17	18				
Senèze													
NMB Se 1095 medial	14.0	16.2	12.8	15.2	15.1								
NMB Se 1095 lateral	13.0	16.5	12.8	14.3	15.6	14.3	17.0	14.0	16.2				

Variable definitions (from Crégut-Bonnoure, 1995, 2002, 2020) (Fig. 14.A.11): td: transversal diameter. dpd: dorsopalmar diameter. first and second phalanges: Length: 1, middle; 2, axial; 3, abaxial. Proximal epiphysis: 4, td; 5, dpd. Proximal articulation: 6, td of the proximal articulation; 7, dpd in the middle; 8, td of the proximoaxial facet; 9, td of the proximoabaxial facet. First phalanx: 10, td of proximoaxial sesamoid facet; 11, td of the proximo-abaxial sesamoid facet, 12, dpd of the proximoaxial facet; 13, dpd of the proximoabaxial facet; 14, dpd of the proximoaxial sesamoid facet; 15, dpd of the proximo-abaxial sesamoid facet; 16, dpd of the proximoaxial articulation; 17, dpd of the proximoabaxial articulation. Middle of the diaphysis: 18, td; 19, dpd. Distal articulation: 20, td; 21, td of the distal epiphysis; 22, dpd minimum; 23, dpd of the disto-axial condyle; 24, dpd of the disto-abaxial condyle. 25, dpd sub-articular distal; 26, height on the palmar side, from the base of the proximoaxial tubercle to the top of the distoaxial condyle. Second phalanx: Proximal articulation: 10, dpd of the proximoaxial facet; 11, dpd of the proximo-abaxial facet. Middle of the diaphysis: 12, td; 13, dpd. Distal articulation: 14, td; 15, dpd of the axial condyle; 16, dpd of the abaxial condyle; 17, dpd minimum; 18, dorsal height from the median top to the proximal side of the distal articulation

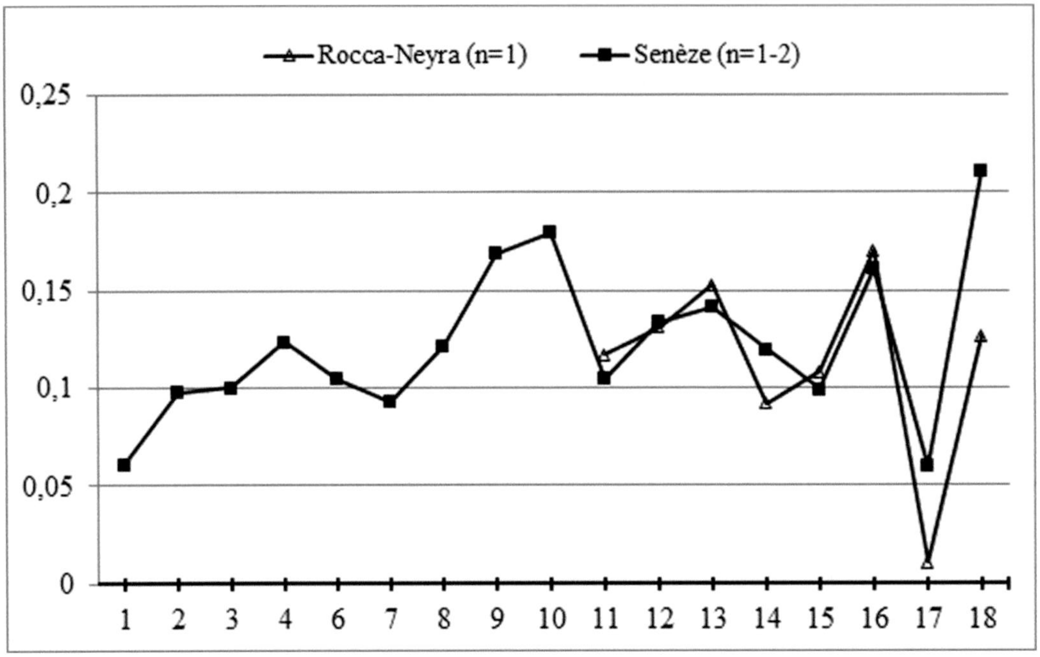

Fig. 14.13 *Gazellospira torticornis.* Ratio-diagram of the tibia. Reference axis: recent *Capra ibex* ♂+♀ (in Crégut-Bonnoure 2020). ccd, craniocaudal diameter; td, transversal diameter. 1, total length. Proximal epiphysis: 2, td; 3, ccd diameter; 4, ccd minimal. Proximal articulation: 6, td total; 7, td of the td medial facet; 8, td of the lateral facet. Middle of the diaphysis: 9, td; 10, ccd. Distal epiphysis: 11, ccd 12, td. Distal articulation: 13, td of the articulation for the talus; 14, ccd of the middle of the epiphysis; 15, ccd of the lateral side of the epiphysis; 16, ccd of the articulation for malleolar bone; 17, ccd of the cranial facet for malleolar td bone; 18, ccd of the caudal facet for malleolar bone. References of the measurements in Fig.14.A.14 and dimensions in Table 14.12

Table 14.12 Comparative dimensions (in mm) of the tibia and malleolus in *Gazellospira torticornis* from Senèze and other localities

Tibia variable	1	2	3	4	6	7	8	9	10
Senèze (NMB Se 1084)	314.0	67.0	63.0	30.3	65.5	27.1	33.1	28.2	26.8
Tibia variable	11	12	13	14	15	16	17	18	
Senèze									
NMB Se 1084	32.7	43.3	29.0	27.4	26.2		7.9		
NMB Se 1472	33.4	43.0	34.2	27.6	27.3	20.9	6.3	12.4	
Rocca-Neyra (NMB Rn 144)	34.0	42.9	32.5	25.8	27.3	21.3	6.3	10.2	
Malleolar bone variable	1	2	3	4	5				
Rocca-Neyra (NMB Rn 144)	19.6	18.0	16.6	17.9	14.5				

Variable definitions (from Crégut-Bonnoure 1995, 2002, 2020) (Figs. 14.A.13 and 14.A.14): ccd: craniocaudal diameter; td: transversal diameter. Tibia (Fig. 14. A.14): 1, total length. Proximal epiphysis: 2, td; 3, ccd diameter; 4, ccd minimal. Proximal articulation: 6, td total; 7, td of the td medial facet; 8, td of the lateral facet. Middle of the diaphysis: 9, td; 10, ccd. Distal epiphysis: 11, ccd 12, td. Distal articulation: 13, td of the articulation for the talus; 14, ccd of the middle of the epiphysis; 15, ccd of the lateral side of the epiphysis; 16, ccd of the articulation for malleolar bone; 17, ccd of the cranial facet for malleolar td bone; 18, ccd of the caudal facet for malleolar bone. Malleolar bone (Fig. 14.A.13): 1, total height; 2, height of the cranial side; 3, height of the caudal side; 4, ccd diameter; 5, td

Tarsals

Three calcanei were initially identified as cervid and then excluded from this family by Pfeiffer (1996–on label). One of them (NMB Se 1136; Fig. 14.11: J, K) is bigger and probably corresponds to a male, the other two to females (NMB Se 1152; Fig. 14.11: F to I), NMB Se 1153). They differ in some ways from cervids, although they are similar because the body between the *tuber calcanei* and the coracoid process is pinched in dorsal view and, in plantar view, the groove of the *sustentaculum tali* is clearly concave. They present a *tuber calcanei* with a transverse groove more or less in a median position as on the *Gazellospira* from Le Coupet (NMB Cp 153); on the Perrier-Pardines calcaneus NMB Lp 48, the groove is longer and reaches the medial side. An important groove is located laterally more or less at the middle of the distal border. In front of it, there is a strong tubercle. The articular facet for the malleolar bone is characterized by a distal part inclined downward. The navicular-cuboid articular facet is more or less straight. The distal base is truncated and broad in dorsal view and acute in lateral view. These calcanei show some similarities with the Perrier-Pardines specimen (measurements 3–5; Fig. 14.14; Table 14.13) and with those from Perrier-Rocca-Neyra (measurements 8 and 9) and Le Coupet (measurements 1–5 and 8–14). However, each population has its own characteristics.

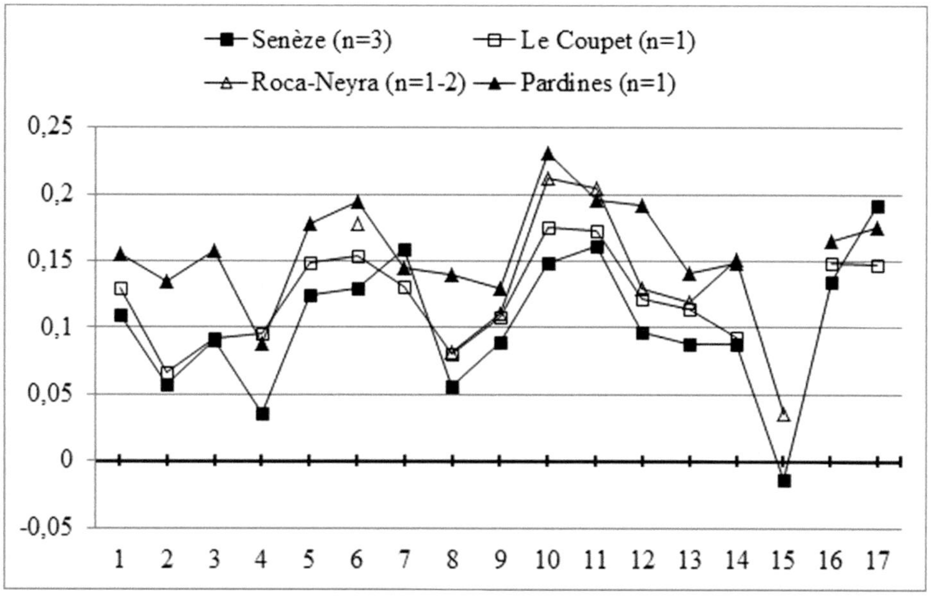

Fig. 14.14 *Gazellospira torticornis*. Ratio-diagram of the calcaneus. Reference axis: recent *Capra ibex* ♂+♀ (in Crégut-Bonnoure 2020). dpd, dorsoplantar diameter; td, transversal diameter. 1, total length. *Tuber calcanei*: 2, td; 3, dpd. Proximal body: 4, td minimum; 5, dpd. Total: 6, dpd; 7, td. *Sustentaculum tali*: 8, td. Navicular cuboid facet: 9, length; 10, td. Fibular trochlea facet: 11, length; 12, td. Length: 13, from the fibular trochlea facet at its middle to the distal point of the bone; 14, from the fibular trochlea facet at its proximal side to the distal point of the bone. Width: 15, *sustentaculum tali* furrow. Length: 16, from the dorsal *tuber calcanei* apex to the coracoid process top; 17, from the palmar *tuber calcanei* apex to the coracoid process tip. References of the measurements in Fig. 14.A.15 and dimensions in Table 14.13

Table 14.13 Comparative dimensions (in mm) of the calcaneus in *Gazellospira torticornis* from Senèze and other localities

Variable	1	2	3	4	5	6	7	8	9
Senèze									
NMB Se 1152	97.1	23.8	26.4	13.8	26.4	38.0	34.1	29.5	23.5
NMB Se 1153	88.8	19.9	24.8	11.3	25.8	35.4	30.0	27.5	23.7
NMB Se 1136	89.9	20.0	24.2	11.0	24.9	34.0	30.0	27.5	21.9
Pardines (NMB LP 48)	102.1	25.4	29.3	13.6	29.1	41.5	30.4	34.1	25.3
Rocca-Neyra									
NMB Rn 144						40.0		29.8	24.6
NMB Rn 173									23.8
Le Coupet (NMB CP 153)	96.2	21.7	25.2	13.8	27.2	37.8	29.5	29.7	24.1
Vialette (MC 2003-5-893-VIA)	87.7	24.2	26.4	14.2	25.5	32.5	31.7	25.8	19.9
Variable	10	11	12	13	14	15	16	17	
Senèze									
NMB Se 1152	8.8	22.3	10.3	32.4	35.1	11.7	64.7	97.1	
NMB Se 1153	7.7	18.6	9.0	29.4	30.9	9.3	57.6	61.9	
NMB Se 1136	9.0	18.9	9.7	31.6	34.3	10.1	57.8	61.9	
Pardines (NMB LP 48)	10.3	21.6	12.0	35.2	38.5		64.4	70.8	
Rocca-Neyra									
NMB Rn 144	9.6	25.6	11.0	33.3	38.7	11.6			
NMB Rn 173	10.2	18.5	9.9	33.7					
Le Coupet (NMB CP 153)	9.1	20.4	10.2	33.0	33.9		62.1	66.4	
Vialette (MC 2003-5-893-VIA)	7.2	18.8	9.5	29.3	31.5	11.5	58.2	62.5	

Variable definitions (from Crégut-Bonnoure 1987, 2002, 2020) (Fig. 14.A.15: G to F): td: transversal diameter. dpd: dorsoplantar diameter.1, total length; *Tuber calcanei*: 2, td; 3, dpd. Proximal body: 4, td minimum; 5, dpd. Total: 6, dpd; 7, td. *Sustentaculum tali*: 8, td. Navicular cuboid facet: 9, length; 10, td. Fibular trochlea facet: 11, length; 12, td. Length: 13, from the fibular trochlea facet in its middle to the distal point of the bone; 14, from the fibular trochlea facet at its proximal side to the distal point of the bone. Width: 15, *sustentaculum tali* furrow. Length: 16, from the dorsal *tuber calcanei* apex to the coracoid process top; 17, from the palmar *tuber calcanei* apex to the coracoid process tip

Metatarsal III+IV

The metatarsal is slender with its dorsal longitudinal groove constituting, in the distal quarter of the length, a gutter surrounded by two sharp crests. On the plantar side, the groove is present on the second quarter of the shaft. Its lateral border is more developed than the medial one and constitutes a thin bladelike crest whereas the medial border is rounder. The proximal articulation is asymmetric with the medial side dorsoplantarly longer than the lateral one. The plantar border of the small medioplantar cuneiform facet is acute and directed upward. The small plantar facet for the navicular-cuboid is double: the dorsal facet is very small and in a vertical position, whereas the larger plantar facet is directed downward. The dorsomedial cuneiform and the navicular-cuboid facets are elongated. The keels are parallel on each side. On the plantar side, they are thin and sharp, and they run to the base of the shaft.

The Senèze metapodial shows analogies with Le Coupet concerning the enlargement of its distal articulation (measurements 19 to 21) and on the other hand with Perrier-Rocca-Neyra concerning the thickness (measurements 22 and 23) (Fig. 14.15; Table 14.14). The proportions of the distal epiphysis distinguish Varshets (MNQ 17) and Villány-Nagyharsányhegy 4 (MNQ 18–19) from the group of French localities, which may be related to the more continental position of these two localities.

Discussion

At least 12 individuals are present, and sexual dimorphism is perceptible for the horn cores. The general proportions of the horn cores and the m3s from Le Coupet and Senèze are similar and some similarities are also visible for the postcrania. However, the number of specimens is insufficient

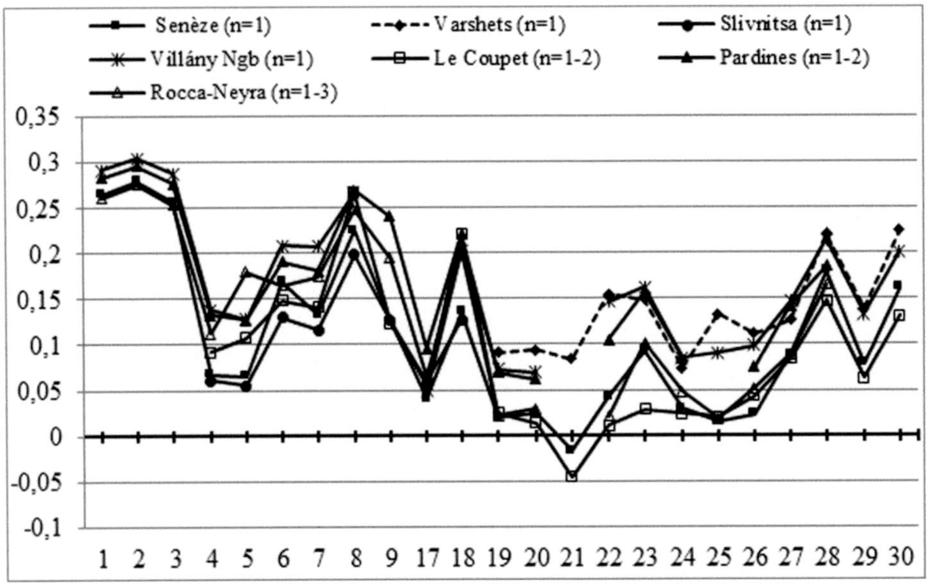

Fig. 14.15 *Gazellospira torticornis*. Ratio-diagram of the metatarsal III+IV. Reference axis: recent *Capra ibex* ♂+♀ (in Crégut-Bonnoure 2020). Ngb, Villány-Nagyharsányhegy 4; dpd, dorsoplantar diameter; td, transversal diameter. Length: 1, dorsal side; 2, minimum in the middle; 3, plantar side. Proximal epiphysis: 4, td total; 5, td of the articulation; 6, dpd total; 7, dpd of the articulation; 8, dpd of the small medioplantar facet; 9, td of the small medioplantar facet; 10, td of the mediodorsal facet; 11, dpd of the mediodorsal facet; 12, td of the dorsal facet for the navicular-cuboid; 13, dpd of the dorsal facet for the navicular-cuboid; 14, length of the connection of the dorsal facets. 15: td of the plantar facet for the navicular-cuboid; 16: dpd of the plantar facet for the c navicular-cuboid. Middle of the diaphysis: 17, td; 18: dpd. Distal epiphysis: 19, td maximal of the articulation; 20, td minimal of the articulation; 21, td supra articular; 22, dpd of the epiphysis; 23, dpd of the epiphysis at the foramen level; 24, distance between the keels on the plantar side; 25, distance between the keels at their base; 26, distance between the keels in their middle; 27, dpd of the medial condyle; 28, dpd of the medial keel; 29, dpd of the lateral condyle; 30, dpd of the lateral keel. References of the measurements in Fig. 14.A.16 and selected dimensions in Table 14.14

Table 14.14 Comparative dimensions (in mm) of the metatarsal III+IV in *Gazellospira torticornis* from Senèze and other localities

Variable	1	2	3	4	5	6	7	8	9	10
Senèze (FSL 210827)	245.7	229.3	249.1	30.7	29.4	34.3	30.5	10.7	10.2	9.1
Rocca-Neyra										
NMB Rn 175	244.0	227.7	247.7	30.3		34.2				
NMB Rn 150				32.1		31.5				
NMB Rn 146				40.0	38.3	36.5	33.6	11.3	12.0	16.1
Le Coupet (NMB Cp 152)				32.5	32.4	32.8	31.1	11.8	10.2	12.0
Pardines (NMB LP 57+58)	256.6	238.4	261.3	35.8	33.9	36.1	34.1	11.8	13.3	13.2
Villány-Nagyharsànyberg (NMB UP 487 cast)	261.5	243.1	267.9	36.2	34.0	37.6	36.2	11.8		14.3
Slivnitsa (NNHM-S SL 50)				30.3	28.7	31.4	29.3	10.0	10.2	11.1
Variable	11	12	13	14	15	16	17	18	19	20
Senèze (FSL 210827)	23.4	12.1	19.9	6.4		6.1	19.2	21.1	33.5	32.5
Perrier (FSL 211713)							21.6	26.3	41.2	39.6
Rocca-Neyra										
NMB Rn 175									22.5	33.7
NMB Rn 150								19.1	21.5	
NMB Rn 146	25.9	16.0	22.7	12.5	25.9		6.1	21.4	29.8	
Le Coupet										
NMB Cp 152	23.1	11.8	24.4	8.2		6.1	19.9	25.7		
NMB Cp 167									33.9	32.5
NMB Cp 192										31.1
Pardines										
NMB LP 57+58	24.7	14.7	23.7	9.3	12.2	6.3	21.1	24.7	37.6	35.5
NMB LP 164							22.4	26.5		
Villàny-Nagyharsànyberg (NMB UP 487 cast)	27.6	12.3	23.4	11.2	15.9	6.6	19.6	24.5	37.8	36.1
Varshets (NNHM-S VR2)									39.5	38.2
Slivnitsa (NNHM-S SL 50)	21.9	11.7	20.2	8.6	8.7	5.4	20.0	20.7		
Variable	21	22	23	24	25	26	27	28	29	30
Senèze (FSL 210827)	30.1	19.1	19.1	20.4	19.4	16.5	24.4	18.6	23.1	16.9
Perrier (FSL 211713)	39.5	23.9	23.2	24.9	24.7	20.8	28.4	20.7	27.1	18.6
Rocca-Neyra (NMB Rn 175)		18.2	19.5	21.2	19.5	17.7	24.4	18.0		
Le Coupet										
NMB Cp 167	28.1	17.4	16.5	20.1	19.6	17.3	24.1	17.3	22.3	15.7
NMB Cp 192		18.15								
Pardines (NMB LP 57+58)		22.1	22.1	23.1		18.6	27.6	18.9		
Villány-Nagyharsányberg (NMB UP 487 cast)		24.3	22.4	23.1	23,0	19.7	27.9	20.1	26.2	18.5
Varshets (NNHM-S VR2)	38.0	24.8	21.7	22.5	25.4	20.3	26.6	20.5	26.5	19.5

Variable definitions (from Crégut-Bonnoure 2002, 2020) (Fig. 14.A.16): dpd: dorsoplantar diameter; td: transversal diameter. Length: 1, dorsal side; 2, minimum in the middle; 3, plantar side. Proximal epiphysis: 4, td total; 5, td of the articulation; 6, dpd total; 7, dpd of the articulation; 8, dpd of the facet for the small cuneiform; 9, td of the facet for the small cuneiform; 10, td of the great cuneiform facet; 11, dpd of the great cuneiform facet; 12, td of the dorsal facet for the navicular-cuboid; 13, dpd of the dorsal facet for the navicular-cuboid; 14, length of the connection of the dorsal facets. 15: td of the plantar facet for the navicular-cuboid; 16: dpd of the plantar facet for the navicular-cuboid. Middle of the diaphysis: 17, td; 18: dpd. Distal epiphysis: 19, td maximal of the articulation; 20, td minimal of the articulation; 21, td supra-articular; 22, dpd of the epiphysis; 23, dpd of the epiphysis at the foramen level; 24, distance between the keels on the plantar side; 25, distance between the keels at their base; 26, distance between the keels in their middle; 27, dpd of the medial condyle; 28, dpd of the medial keel; 29, dpd of the lateral condyle; 30, dpd of the lateral keel

to check for real statistical differences. The teeth and bones are smaller in the Senèze specimens compared to those from Le Coupet, Perrier-Pardines and Perrier-Rocca-Neyra: this could be related to different sex-ratios in the Senèze and Le Coupet samples (more females at Senèze), and/or to chronological differences concerning Senèze and the two Perrier localities, which are older. In addition, the morphology of the m3, the metacarpal III+IV and the calcaneus is not exactly the same at Senèze and the other sites.

Caprinae Gray, 1821

Tribe Caprini Gray, 1821

Genus *Procamptoceras* Schaub, 1923

Species *Procamptoceras brivatense* Schaub, 1923

This medium sized antelope has long been considered as a member of Rupicaprinae, which was reclassified as tribe Rupicaprini of the Caprinae (Guérin *in* Duvernois & Guérin 1989; Crégut-Bonnoure & Guérin 1996) and later included in Caprini. Remains of *Procamptoceras* are scarce and only one species is known: *P. brivatense*. Its temporal range is mainly MNQ 17 to MNQ 19; in the literature it is indicated in 11 localities:

- MN 15b: in Hungary at Csarnóta 1 (Kretzoï 1956; Schaub 1932).
- MN(Q)16–18: in Hungary at Beremend 4 (Kretzoï 1956; Schaub 1932).
- MNQ 17: in Greece at Volax (Kostopoulos 1996); in Hungary at Villány 3 (=Villány-Kalkberg; Kretzoï 1956; Schaub 1932).
- MNQ 18: in France at Senèze (Schaub 1923), perhaps at Montoussé 5 (Clot et al. 1976); in Italy at Olivola (Schaub 1928); perhaps in Greece at Vassiloudi (Kostopoulos 1996, 2022; Kostopoulos & Koufos 1994); in Bulgaria at Slivnitsa (Spassov & Crégut-Bonnoure 1999) and at Kozarnika (Biostratigraphic unit B2-2; Fernandez & Crégut-Bonnoure 2007).
- MNQ 18–19: in Hungary at Villány-Nagyharsányhegy 4 (=Villány-Nagyharsányberg 4) (Kretzoï 1956).

The holotype is a quite complete cranium from Senèze preserved in Basel (NMB Se 555). A synthetic diagnosis is given by Viret (1961: p. 1067) and a neodiagnosis by Guérin (*in* Duvernois & Guérin 1989: p. 344).

Description

The material is mainly from the previous excavations (Fig. 14.15).

Crania and Horn Cores

A total of three more or less complete crania and two frontlets are known (Table 14.15; Figs. 14.16, 14.17 and 14.18): two incomplete crania belonging to males (FSL 210768 [Duvernois & Guérin 1989: pl. 1 Figs. 4–7 p. 373] and FSL 16861) and the female holotype (NMB Se 555 [Schaub 1923: Fig. 2 p. 284; Viret 1961: Fig. 158 A et B p. 284]); one frontlet belongs to an adult male (NMB Se 789; Schaub 1923: Fig. 1 p. 282) and a second to a juvenile male (MC 2007-2-1-Sen; Couthures et al. 1991). Sexual dimorphism may be discerned through the robusticity of the horn cores as indicated by Schaub (1923).

The holotype is well preserved, except for the right side which is broken at the orbit level (Fig. 14.16). The main morphological feature of the species is its relatively short face; nevertheless, the nasal area is reconstructed and there is no contact between the left premaxilla (the right is missing) and the maxilla: the total length as the length of the muzzle is an estimation. The frontal on this specimen is slightly concave in the sagittal plane whereas on the frontlet NMB Se 789 (Fig. 14.17: D, E), which belongs to a male, the concavity is more pronounced on each side of the prominent sagittal suture: this difference reflects sexual dimorphism. On the holotype, the caudal side of the infraorbital foramen opens above the middle of the P3. The facial tubercle is located above the middle of M1-M2, as on FSL 210768 (Fig. 14.17: A to C). The palatine notch is "U" shaped and its rostral side is located at the level of the second lobe of M3. The braincase has convex parietals. Guérin indicates that the angle between parietal and frontal is about 103° and between the parietal and the occipital about 127° (Table 14.18 Duvernois & Guérin 1989). The temporal ridges are prominent, also on the adult female cranium. The convexity of the frontals is pronounced and begins immediately below the base of the horn cores. The supra-orbital foramen is small. The orbits are very large. The occipital condyles are large and projected caudally. The basioccipital shape is rectangular as in Caprinae, there is no sagittal groove (as indicated by Duvernois & Guérin 1989) and the muscular tubercles are relatively small and rounded.

The horn cores, inserted just above the orbits, are parallel, weakly diverging laterally at the tips, elongated, with a concave rostral side, and the thickness decreases regurlarly (Fig. 14.18: D, E). The surface is covered with thin grooves. On the female cranium the horn cores are slender and spaced apart, whereas they are thick and set close together and close to the sagittal suture in males. The fronto-parietal suture is located ca. 2 cm behind the horn core; it is rectilinear. The parietal crests form an hourglass shape. On the adult male horn cores, (such as NMB Se 789; Fig. 14.17: D, E), a strong groove occupies the rostral side. It runs from the base to the first third of the length and shallows toward the tips. Two other grooves are present on the lateral side; however, there is only one on the female cranium. For all the samples, the thickness of the posterior side decreases regularly, being very narrow above the half of the length. A crest is visible on NMB Se 789. The section at the base is oval on the female cranium; it is more rounded for the males and the rostral side is flat.

The specimen preserved in the Crozatier Museum belongs to an immature male according to the open frontal

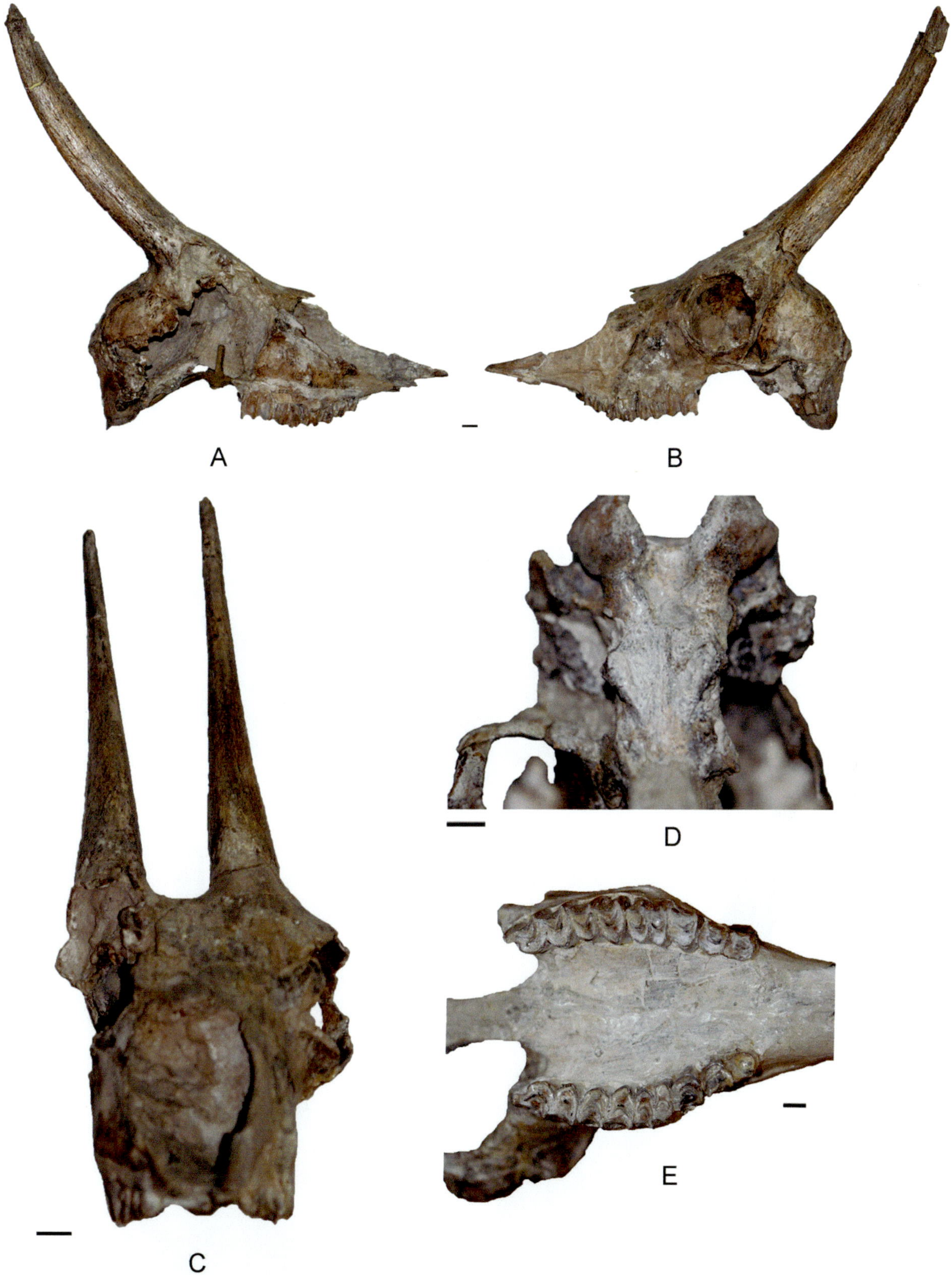

Fig. 14.16 Cranium of *Procamptoceras brivatense* from Senèze. A to E, holotype NMB Se 555. A, B, *norma lateralis*; C, *norma frontalis*; D, E, *norma basalis*. Scale 1 cm

A B C D E

Fig. 14.17 Cranium and frontlet of *Procamptoceras brivatense* from Senèze. A to C, cranium FSL 210768; D, E, frontlet NMB Se 789. A, D, *norma lateralis*; B, *norma dorsalis*; C, *norma basalis*; E, *norma frontalis*. Scale 1 cm

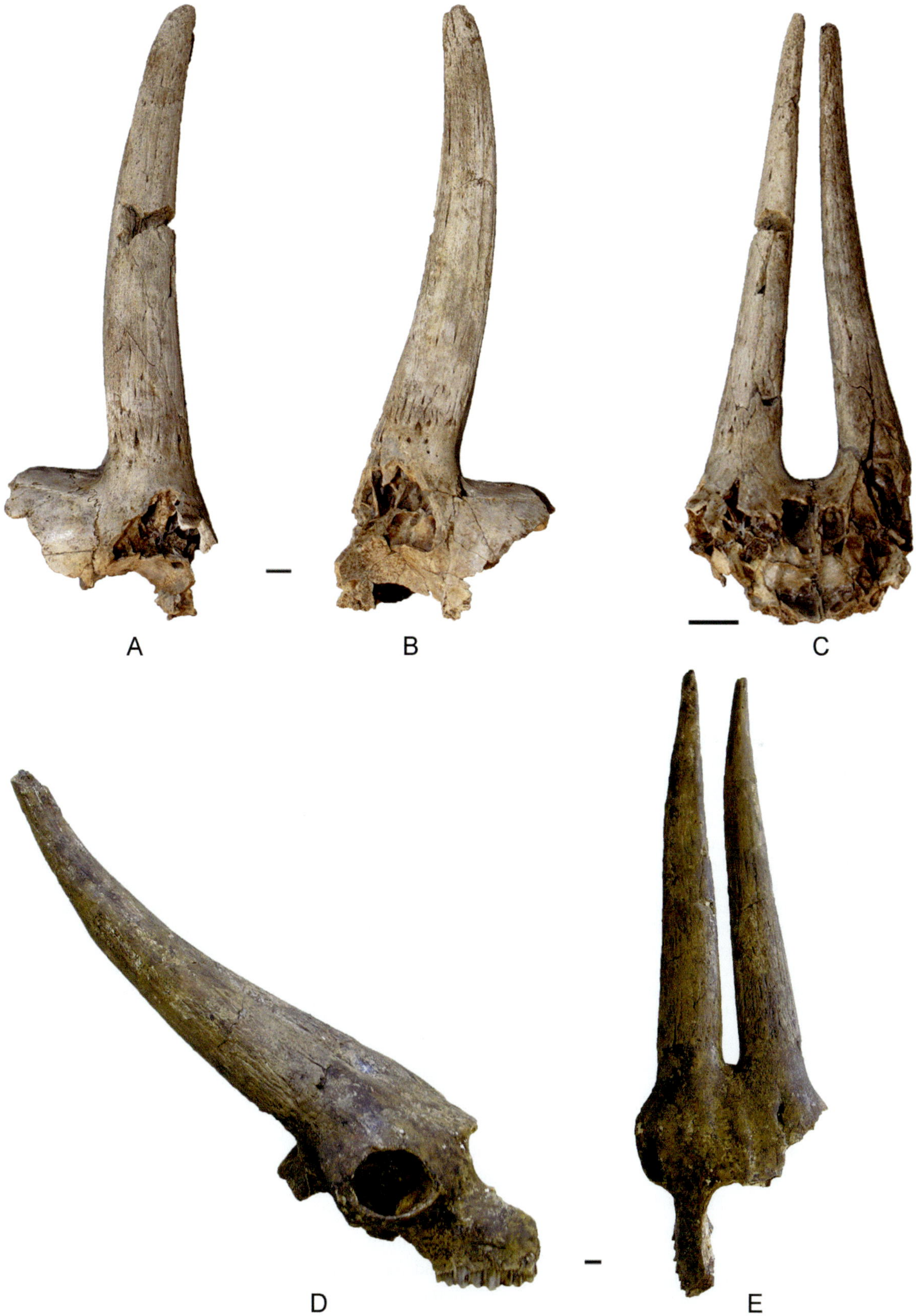

Fig. 14.18 Cranium and frontlet of *Procamptoceras brivatense* from Senèze. A to C, frontlet MC 2007-2-1-Sen; D, E, cranium FSL 16861. A, B, D, *norma lateralis*; C, *norma frontalis*; E, *norma dorsalis*. Scale 1 cm

Table 14.15 Dimensions (in mm) of the cranium in *Procamptoceras brivatense* from Senèze

Variable	1	2	3	4	5	6	9	10	11	12
FSL 210768				167.7			81.8	28.9	55.3	
FSL 16888							87.1	33.0	53.9	86.2
NMB Se 555 ♀	250.6	243.3	222.8	155.0	70.2	146.1	78.2	28.0	53.0	
Variable	13	14	15	17	18	19	20	21	23	24
FSL 210768	66.4	34.5	23.0	165.0	110.4	66.7	129.0			
NMB Se 555 ♀	71.8	40.2	27.6	144.8	91.6	65.5	114.3	173.7	158.8	175.8
Variable	25	26	27	28	29	30	31	32	35	37
FSL 16861 ♂				106.6	137.6					104.3
FSL 210768	40.2	87.1	31.3	112.4	136.0	106.24				109.7
NMB Se 555 ♀	43.8	77.2	29.9	94.1			77.4	76.8	16.9	83.2
NMB Se 789 ♂				111.8						98.5
MC 2007-2-1 sen			74.1							98.5
Variable	38	39	40	41	42	43	44	45	46	47
FSL 16861 ♂	12.9	14.0	45.1	51.4	46.5	36.2	41.4	268.5	152.5	
FSL 210768	22.8	14.4		51.7	42.7	44.6	35.4		155.0	187.2
FSL 16888	27.8	31.6	37.1	32.4	28.1	24.5	13.8	185.0	100.0	164.5
NMB Se 555 ♀	14.1	12.6	25.1	47.9	42.4	39.4	31.7	240.0	150.0	
NMB Se 789 ♂	21.7	14.6	6.1	29.5	24.9	24.5	16.5	167.2	84.91	
MC 2007-2-1 sen	12.9	14.0	45.1	51.4	46.5	36.2	41.4	268.5	152.5	
Variable	48	49	53	54	55	56	57	58	59	60
FSL 16888	160.0		35.0	38.4	79.0		85.0	66.7	23.2	29.3
NMB Se 555 ♀	136.4	132.2	43.9	46.3	73.8	68.1		54.9	23.2	18.3
Variable	61	62	63	64	65	66	67	68		
FSL 16888	23.8	17.4	42.0	31.8	42.3	66.5	130.0	157.4		
NMB Se 555 ♀	30.1	19.4	50.4	32.5	31.4	103.0	134.0			

Variable definitions (from Crégut-Bonnoure 2002, 2020) (Figs. 14.A.1 and 14.A.2): Length: 1, acrocranion-prosthion; 2, from occipital condyles to prosthion; 3, basion-prosthion; 4, from basion to the middle of the line of the P2; 5, from prosthion to the middle of the line of the P2; 6, from prosthion to the middle of the line of the M3; 9, P2 mesial-M3 distal; 10, P2 mesial-P4 distal; 11, M1 mesial-M3 distal; 17, acrocranion-nasion; 18, frontal from supra-orbital foramens; 19, acrocranion-bregma; 20, from supra-orbital foramens to acrocranion; 21, from supra-orbital foramens to prosthion; 23, entorbital-prosthion; 24, from infra-orbital foramen to acrocranion; 47, from occipital condyle to infra-orbital foramen; 48, basion-nasion. 49, nasion-prosthion; 68, from the top of the frontal to the tip of the occipital condyles. Transversal diameter:12, otion-otion; 13, tooth row at the labial side;14, between the lingual side of the M3 second lobe;15, between the lingual side of the P2; 27, between the parietal ridges; 28, post-orbital; 29, ectorbital-ectorbital; 30, entorbital-entorbital; 31, between orbit and facial tuberosities. 32, at the facial tuberosities level; 35, caudal of the nasal; 57, between jugular processes; 58, between the external side of the occipital condyles; 59, of the foramen magnum; 60, of the occipital condyle. Parietal: 25, dorsal length; 26, width. Horn cores: 37, width between the lateral sides; 38, width between the internal sides at the base; 39, width between the internal sides in the middle; 40: width between the internal sides at the tip; 41, rostrocaudal diameter at the base; 42, dorsolateral diameter at the base; 43, rostrocaudal diameter in the middle of complete horn core; 44, dorsolateral diameter in the middle of complete horn core; 45, rosral length; 46, circumference at the base. Orbit: 53, rostrocaudal diameter; 54, dorsoventral diameter. Height: 55, of the braincase from basioccipital to the top of the parietal; 56, from the middle of the palatine suture to the nasal surface; 61, of the occipital condyle; 62, of the foramen magnum; 63, from opisthion to fronto-parietal suture. Basi-occipital: 64, transversal diameter of the caudal tuberosities; 65, transversal diameter of the rostral tuberosities; 66, length from basion to basisphenoid; 67, frontal curvature

suture and the relatively wide spacing at the base of the complete horn cores (Fig. 14.18: A to C). The frontal is broken below the left. When the specimen was first recovered, the horn cores were convergent from the base to the tips in frontal view and consequently distinct from the other members of this species (Fig. 14.18: C). Recently, this specimen has been restored: the sagittal suture was opened and the frontals bent laterally. As a result, today its appearance is similar to that of other specimens (Lacombat et al. 2010: Fig. p. 77).

A fragmentary braincase from Perrier-Rocca-Neyra (NMB Rn 26) is identified as *Procamptoceras* on the label, but it actually belongs to a small cervid.

Upper Teeth

The teeth are scarce (NR = 30; Tables 14.16 and 14.17; Figs. 14.16: E, 14.17: C and 14.19: A to G): there are the upper toothrows of the crania NMB Se 555, FSL 210748 and FSL 16861; a fragmentary right maxilla (without M2; FSL 16888); a fragmentary right M1 (NMB Se 1536); and three specimens labelled FSL 16893: the worn left M1 and M2 from one individual, a right M1 and a right m3 (another upper tooth labelled FSL 16893 belongs to *Gazellospira* (left DP4; (see section "Upper Teeth" under *Gazellospira*). The lingual side of P2 and P3 is rounded and on P2 it is bilobed. The rib is broad and more projected outwards than the styles. The lingual side of P4 is rough mesially and flat distally. On the molars, the labio-lingual plane is perpendicular to the mesio-distal one. The styles are strong but the mesostyle is less prominent by comparison to the other styles and directed forward. The valley distal to the parastyle is deep and reaches the crown base. The paracone is strongly inflated, the metacone less so. The mesial lobe is lingually pinched, the distal one rounded. There is a small and rounded occlusal islet. If I exclude the tooth from the holotype cranium (NMB Se 555), the metastylar wing on M3 is not well developed; however, the metastyle is wide and prominent in occlusal view. The distal side is concave due to a longitudinal groove. This groove is intensified on the male cranium FSL 210768 by a linguodistal stylid (Fig. 14.17: C).

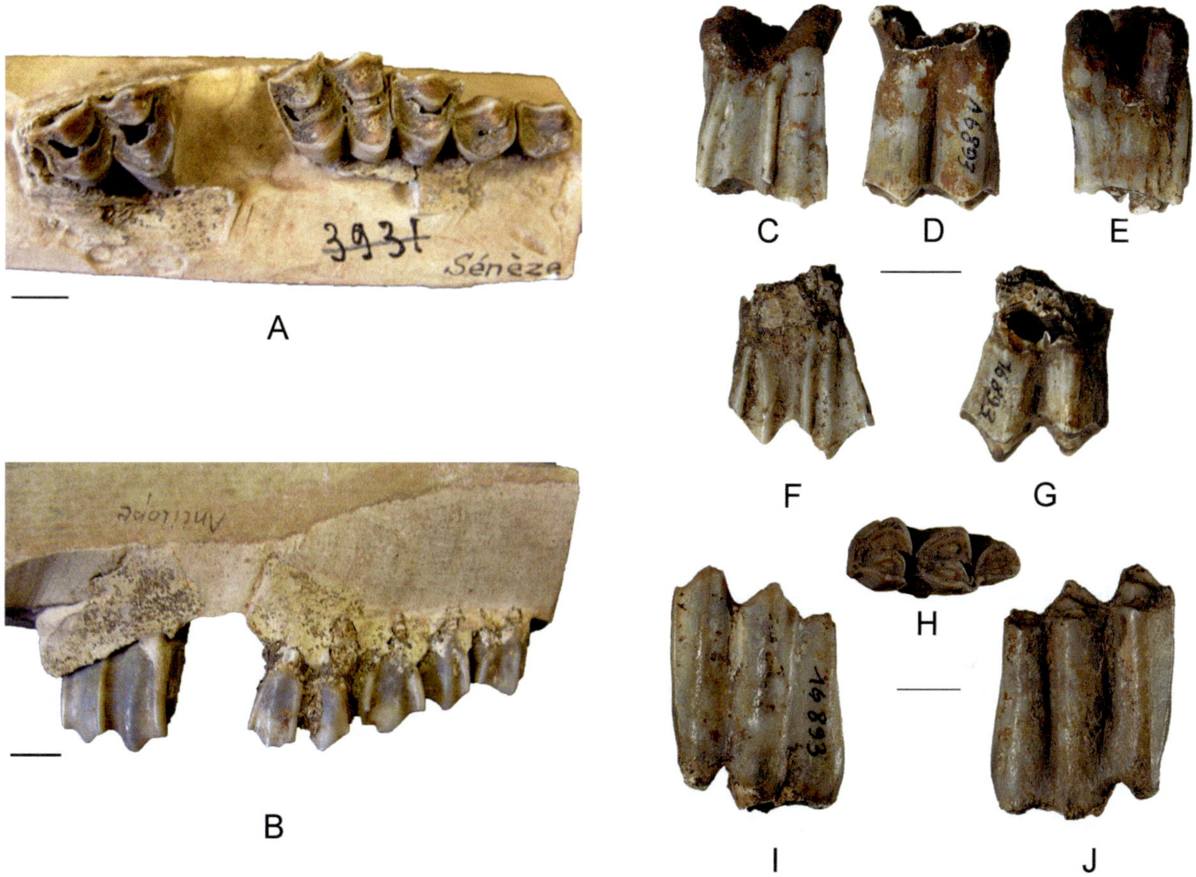

Fig. 14.19 Teeth of *Procamptoceras brivatense* from Senèze. A, B, right maxilla FSL 16888; C to E, left M2 FSL 16893; F, G, left M1 FSL 16893; H to J, right m3 FSL 16893. A, H, occlusal view; B, C, F, J, buccal view; D, G, I, lingual view; E, mesial view. Scale 1 cm

Table 14.16 Dimensions (in mm) of the upper premolars in *Procamptoceras brivatense* from Senèze

P2 variable	Ht	1	2	3	5	
FSL 16888	11.9	10.0	9.8	8.7	9.0	
FSL 21748	11.4	9.7	8.8	8.5	8.9	
NMB Se 555 (L)	8.6	8.9		8.4	9.2	
NMB Se 555 (R)	9.0	10.7		8.9	8.9	
P3 variable	Ht	1	2	3	5	
FSL 16888	12.8	11.5		9.4	10.5	
FSL 21748		9.7		9.6		
NMB Se 555 (L)	10.1	9.1		9.3	9.9	
NMB Se 555 (R)	10.6	9.2		9.5	9.9	
P4 variable	Ht	1	2	3	4	5
FSL 16888	14.5	12.0	10.7	11.1	11.0	12.9
FSL 21748		10.0		10.3	10.0	
NMB Se 555 (L)	11.4	10.5		10.4	9.8	10.5
NMB Se 555 (R)	11.0	9.4		11.3	11.1	12.0

Variable definitions (from Crégut-Bonnoure 1995, 2002, 2020) (Fig. 14.A.4: A): Ht, height: 1, occlusal length; 2, basal length. 3, occlusal width total. 4, occlusal width in the middle. 5, basal width. L, left; R, right

Table 14.17 Dimensions (in mm) of the upper molars in *Procamptoceras brivatense* from Senèze

M1 variable	Ht	1	2	3	4	5	6	7
N	5	5	2	5	5	6	6	4
Mean	10.5	14.0	14.0	12.9	12.9	12.3	12.1	14.2
Minimum	8.9	12.4	12.4	12.4	12.8	11.4	10.8	13.1
Maximum	14.3	17.3	15.6	13.4	13.0	14.4	15.1	15.2
SD	2.28	1.92	2.29	0.44	0.09	1.20	1.53	0.87
M2 variable	Ht	1	2	3	4	5	6	7
FSL 16893	17.0	17.2	15.4	13.4	12.9	11.8	11.0	15.1
FSL 21748		18.6		13.4	13.1	11.9	11.0	
NMB Se 555 (L)		18.8		13.8	12.5	12.4	10.8	
NMB Se 555 (R)	14.5	18.6	14.7	13.0	12.4	11.4		
M3 variable	1	3	4	6	7	9	10	11
FSL 16893	22.7	14.3	13.4	11.0	10.7	9.1	8.0	19.4
FSL 21748	22.9	13.1	12.6	11.2	10.7	6.9	10.5	19.1
NMB Se 555 (L)	21.8	13.0	11.8	11.4	10.0	6.8	7.8	18.0
NMB Se 555 (R)	22.2	13.3	11.7	11.8	9.9	7.1	7.6	17.3

Variable definitions (from Crégut-Bonnoure 1995, 2002, 2020): Ht, height; M1 and M2 (Fig. 14.A.4: B): 1, occlusal length; 2, basal length. 3, occlusal width total, first lobe; 4, occlusal width in the middle, first lobe; 5, occlusal width total, second lobe; 6, occlusal width in the middle, second lobe; 7, basal width. M3 (Fig. 14.A.4: C): 1, occlusal length; 3, occlusal width total, first lobe; 4, occlusal width in the middle, first lobe; 6, occlusal width total, second lobe; 7, occlusal width in the middle, second lobe; 9, length of the interstylar surface, mesial; 10, length of the interstylar surface, distal; 11, occlusal length of the two lobes at the lingual side; L, left; R, right

Mandible and Lower Teeth

There is no mandible associated with crania. Three specimens have been considered to be *Procamptoceras*.

The mandible FSL 16892 is a left specimen belonging to an adult (Duvernois & Guérin 1989: pl. 1 Fig. 8 p. 373). The toothrow length is 82 mm. This value is less than in the upper toothrow FSL 16861, which belongs to a male (87.1 mm). It is the same order of magnitude as the NMB Se 555 and FSL 210768 upper toothrows which belong to females (respectively 78.2 and 81.8 mm). However, the premolar segment of FSL 16892 is shorter than on the latter specimens and the molar segment longer: the index of premolar length versus the total lower toothrow is 30.72, the molar index is 69.55; in *Procamptoceras* upper toothrows the premolar index variation is 35.32–37.87 (n = 3) and the molar one is 61.85–67.66 (n = 3). Furthermore, the morphology of the m3 doesn't correspond to that of the isolated m3 of FSL 16893 whose length corresponds exactly to the last lobe of M2 and the two lobes of M3 (Table 14.18) (Fig. 14.19: H to J). Furthermore, on this isolated tooth, the valley behind the goat fold is shallow and extends towards the base, being deep and short on the m3 from the mandible. Finally, the ramus is short and cannot be connected with the mandibular fossa of the crania, which is in a relatively high position. Consequently, the mandible FSL 16892 cannot be a *Procamptoceras* specimen: the morphological characteristics suggest its attribution to cf. *Hemitragus* (see section "Mandible and Lower Teeth" under cf. *Hemitragus*). Two other mandibles are small sized, one of them with deciduous teeth (NMB Se 1718) and the other one p3 and p4 erupting (NMB Se 402). They were never described but are assigned to *Procamptoceras* on their labels. I reallocate them as cf. *Hemitragus* mandibles also (see section "Mandible and Lower Teeth" under cf. *Hemitragus*).

Table 14.18 Comparative dimensions (in mm) of the lower molars in *Procamptoceras brivatense* from Senèze and Olivola, and in Caprini indeterminate from Csarnóta

p3 variable	Ht	1	2	3	4	5	
Csarnóta (NMB UP 796–cast)	9.7	12.5	11.5			9.4	
p4 variable	Ht	1	2	3	4		
Csarnóta (NMB UP 796–cast)		13.9		7.7	7.9		
m1 variable	Ht	1	2	3	4	5	
Olivola (NMB Ol 6–cast)		10.7					
Csarnóta (NMB UP 796–cast)		15.4			9.8		
m2 variable	Ht	1	2	3	4	5	
Olivola (NMB Ol 6–cast)	5.8	14.8	9.9	9.7	10.3	10.2	
Csarnóta (NMB UP 796–cast)		18.5		9.7	8.6		
m3 variable	Ht	1	2	3	4	5	6
Senèze (FSL 16893)	21.8	24.3	24.6	9.3	8.5	5.1	9.3
Olivola (NMB Ol 6–cast)	16.7	23.3		10.1	9.0	5.4	
Csarnóta (NMB UP 796–cast)		22.9		8.0	7.9	4.9	

Variable definitions (from Crégut-Bonnoure 1995, 2002, 2020) (Fig. 14.A.4): Ht, height; p3 and p4 (Fig. 14.A.4: D): 1, occlusal length; 2, basal length; 3, occlusal width, first lobe; 4, occlusal width, second lobe; 5, basal width. m1 and m2 (Fig. 14.A.4: E): 1, occlusal length; 2, basal length; 3, occlusal width, first lobe; 4, occlusal width, second lobe; 5, basal width. m3 (Fig. 14.A.4: F): 1, occlusal length; 2, basal length; 3, occlusal width, first lobe; 4, occlusal width, second lobe; 5, occlusal width, third lobe; 6, basal width

In Europe two other adult mandibles are attributed to *Procamptoceras*: one of them is from Csarnóta (Schaub 1932: 323), the other from Olivola (*Antilope* cf. *Procamptoceras brivatense*; Schaub 1928: 265) (Fig. 14.20). Their casts are preserved in the Basel Museum. The Csarnóta specimen (right p3-m3 with p2 alveolus; NMB UP 796; Fig. 14.20: A to C) belongs to a young individual. The lobes of the molars are very pinched, especially the second lobe on p4, they are directed obliquely backward on the molars, a goat fold exists and an interlobar column is present on m1 and m2. The stylids are narrow and the inflation of the conids is important. The buccal valley is very deep and narrow. The index of premolar length versus the total lower toothrow is 35.4 (31.4 mm/88.7 mm), the molar index is 63.92 (56.7 mm/88.7 mm). The Olivola mandible (NMB Ol 9; Fig. 14.20: D to F) is incomplete (left p4-m3) and belongs to an old individidual. The pinching of the molars, which lack any goat fold, is less significant. It is clear that the Olivola and Csarnóta mandibles do not belong to the same species. Although the ratio of premolar-molar segments of the Csarnóta teeth is consistent with *Procamptoceras*, because of the strong pinching of the molars, I can assume that it is not a *Procamptoceras* mandible: the pinching of the upper teeth is moderate in this genus. The Olivola mandible is a better candidate, and the m3 is more in accordance with the FSL 16893 specimen from Senèze. The third lobe of this isolated specimen is convex and its shape is rounded buccally and directed outwards (Fig. 14.19: H). An elongated stylid, not clearly visible, occupies the distal side.

Fig. 14.20 Mandibles of Bovidae indeterminate from Csarnóta and *Procamptoceras brivatense* from Olivola. A to C, Bovidae indeterminate, right mandible NMB UP 79; D to F *Procamptoceras brivatense*, left mandibule NMB Ol 9. A, D: lingual view; B, E: buccal view; C, F: occlusal view; phot. C. Triat. Scale 1 cm

Postcrania

There is no skeleton associated with the skulls. Consequently, the attribution of isolated bones was previously done taking into account only their middle sized proportions and not their morphological features. Some mistakes resulted from this approach.

Vertebrae

There are an incomplete atlas (NMB Se 576), a seventh cervical (NMB Se 555) and a thoracic vertebra (also NMB Se 555). All of them are smaller than in *Gallogoral*. The width of the cranial side of the atlas corresponds to the width of the occipital condyle of the *Procamptoceras* cranium. This vertebra is a good candidate for *Procamptoceras*. It is compared to *Pliotragus* and other taxa in section "Atlas" under *Pliotragus*. The two other vertebrae could also be *Procamptoceras*, but it has not been possible to compare them with *Hemitragus* vertebrae. Consequently, their attribution is still in question.

Scapula

Two ventral articulations belonging to two individuals are attributed to *Procamptoceras*: NMB Se 1450 (left and right) (Table 14.19; Fig. 14.21: A to D). Their morphology is similar. On the right specimen the lateral side of the glenoid fossa is relatively flat, on the left one it is more rounded; however, it is less rounded than in cervids. These two articulations have a hooked supraglenoidal tubercle. On the lateral side, between the acromion and the glenoid fossa, there is a curvilinear crest, more acute on the left specimen. In ventral view, the lateral border of the glenoid fossa is weakly bilobed.

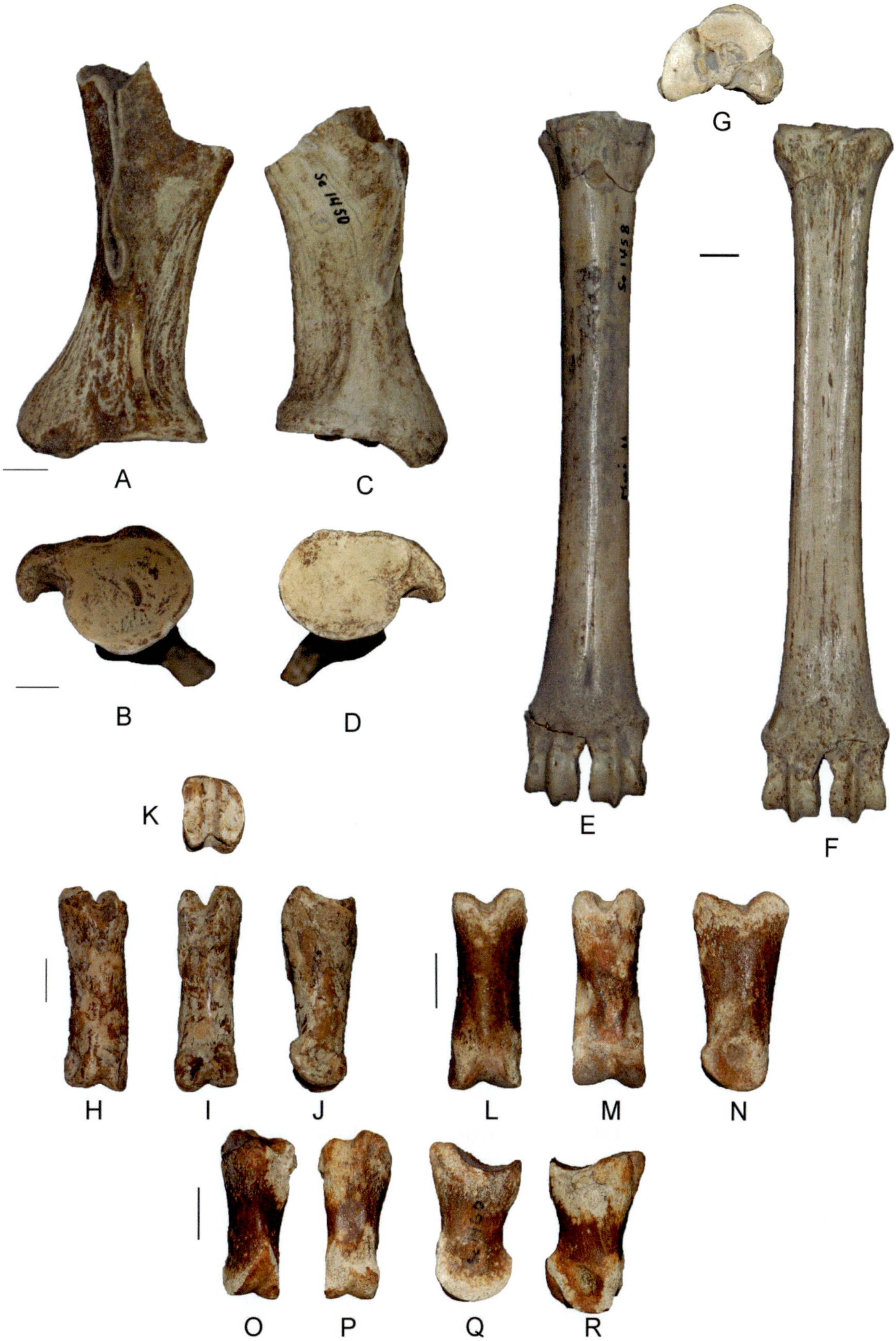

Fig. 14.21 Postcranium of *Procamptoceras brivatense* and cervid from Senèze. *Procamptoceras*: A to D, left (A, B) and right (C, D) scapula NMB Se 1450; E to G, left metacarpal III+IV NMB Se 1458; H to K, first anterior phalanx NMB Se 755. Cervid: L to M, first anterior phalanx NMB Se 1460; O to R, second phalanx NMB Se 1460. A, C, lateral view; B, D, ventral view; E, H, L, O, dorsal view; F, I, M, P, palmar view; G, K, proximal view; J, N, Q, abaxial view; R, axial view. Scale 1 cm

Table 14.19 Dimensions (in mm) of the scapula in *Procamptoceras brivatense* from Senèze

Variable	5	6	7	8	9	10	11
NMB Se 1450 (L)	23.9	12.5	30.6	31.0	29.0	41.0	28.3
NMB Se 1450 (R)	24.1	14.7	30.9	31.2	28.8	39.7	26.2

Variable definitions (from Crégut-Bonnoure 2002, 2020) (Fig. 14.A.5): 5, minimal length of the neck; 6, minimal breath of the neck; 7, length from the base of the acromion to the lateral side of the glenoidal cavity; 8, articular length of the lateral side of the glenoidal cavity; 9, articular length of the medial side of the glenoidal cavity; 10, total length of the ventral articulation; 11, thickness of the ventral articulation; L, left; R, right

Humerus

Two left distal epiphyses have been considered to be *Procamptoceras*: NMB Se 1441 and NMB Se 1792. NMB Se 1441 is instead a *Gazellospira* (see section "Humerus" under *Gazellospira*) close to the Le Coupet right humeri NMB Cp 80 and NMB Cp 114 and NMB Pn 163 from Perrier-Pardines (left). The sole distal humerus that could be from *Procamptoceras* is NMB Se 1792 (Table 14.20): compared to *Gazellospira*, the radial fossa is higher, and the crest located caudally is vertical rather than transverse in *Gazellospira*. The coronoid fossa is without any vertical septum and reaches the lateral side. In distal view, the olecranon fossa is wide and the medial epicondyle is perpendicular to the transverse plane whereas in *Gazellospira* the olecranon fossa is less wide because the medial epicondyle is obliquely directed to the middle of the bone. The lateral lip of the trochlea is prominent, being flatter and broader in *Gazellospira*. The epicondyle is caudally less projected down than in *Gazellospira*.

Table 14.20 Dimensions (in mm) of the humerus in *Procamptoceras brivatense* from Senèze

Variable	11	12	13	14	15	16	17
NMB Se 1792	40.8	39.2	29.7	10.7	37.3	25.8	29.9
Variable	18	19	20	21	22	23	24
NMB Se 1792	18.8	20.8	23.9	25.5	22.8	14.8	16.0

Variable definitions (from Crégut-Bonnoure 1995, 2002, 2020) (Fig. 14.A.6): Transversal diameter distal: 11, maximal; 12, articular; 13, trochlea; 14, capitulum. Craniocaudal diameter distal: 15, medial side, total; 16, medial side, articulation; 17, lateral side, total; 18, lateral side, articulation; 19, trochlea, minimum; 20, trochlea, median guilding crest. Diaphysis, supra-articular: 21, transversal diameter; 22, craniocaudal diameter. Olecranon fossa: 23, transversal diameter; 24, height

Metacarpal III–IV

There is a left metacarpal (NMB Se 1458) which is considered to belong to *Procamptoceras*. It is a juvenile specimen, the distal epiphysis being unfused. It is slender (Table 14.21; Fig. 14.21: E to G). The dorsomedial tuberosity is developed and prominent. Two other tuberosities occupy the palmar sides below the facets; the lateral is prominent and developed longitudinally. The palmar part of the capitato-trapezoid articular facet is broken. The lateral side of the unciform articular facet is curved. The palmar side of the proximal articulation is angled (125°). The palmar groove is shallow and disappears distally. The keels are parallel on each side.

A metacarpal from Villány 3 (cast NMB UP 483; right) was previously assigned to this species. It is a half shaft with the proximal articulation. It is smaller than the Senèze bone, and its morphology is different: the palmar side is rectilinear, the unciform articular facet has its palmar area directed downward, and the capitato-trapezoid articular facet presents an undulating medial surface and a concave palmar side. As indicated by Spassov and Crégut-Bonnoure (1999) it belongs to *M. rhenanus*.

Phalanges

Three specimens have been assigned to this species: the first and second anterior phalanges NMB Se 1460 and a first phalanx NMB Se 755 ("Ruminantier" on label).

Table 14.21 Dimensions (in mm) of the metacarpal III+IV in *Procamptoceras brivatense* from Senèze

Variable	1	2	3	4	5	6	7	8	9	10	11	12
NMB Se 1458	166.7	152.5	162.3	29.5	27.8	11.2	18.7	17.9	22.7	22.4	20.1	17.3
Variable	13	14	15	16	17	18	19	20	21	22	23	24
NMB Se 1458	14.2	30.5	31.5	18.3	16.5	17.8	15.5	15.3	20.2	13.9	19.7	14.0

Variable definitions (from Crégut-Bonnoure 1995, 2002, 2020) (Fig. 14.A.10): dpd, dorsopalmar diameter; td; transversal diameter. Length: 1, total; 2, median; 3, medial. Proximal epiphysis: 4, td of the epiphysis; 5, td of the articulation; 6, td of the unciform facet; 7, td of the capitato-trapezoid facet; 8, length of the unciform facet, diagonally from lateropalmar angle to dorsal point of the proximal crest; 9, length of the capitato-trapezoid facet, diagonally from mediopalmar angle to dorsal point of the proximal crest; 10, dpd of the epiphysis; 11, dpd of the articulation. Middle of the diaphysis: 12, td; 13, dpd. Distal epiphysis: 14, td of the articulation;15, supra-articular transversal diameter;16, dpd supra-articular;17, dpd at the level of the dorsal foramen;18, distance between the lateral edges of the keels at the palmar top; 19, distance between the lateral edges of the keels at the palmar base; 20, distance of the keels at their middle; 21, td of the medial keel; 22, td of the medial condyle; 23, td of the lateral keel; 24, td of the lateral condyle

The first anterior phalanx NMB Se 755 has a wide proximal epiphysis with clear sesamoid articular facets (Table 14.22) (Fig. 14.21: H to K). The axial muscular insertion is longer than the abaxial and slender. The proximal articulation corresponds to the metacarpal condyle. It is a good candidate for *Procamptoceras*.

The specimens numbered NMB Se 1460 are not bovid phalanges (Fig. 14.21: L to R): the first phalanx, without its proximal articulation (it belongs to a juvenile), has a dorsal side less convex than on NMB Se 755; the upper side of the distal articulation on the plantopalmar face is straight; on the diaphysis, the two ligament tubercles are not at the same level, the axial being in a more distal position. The proximal articulation of the second phalanx is elongated dorsopalmarly, with an abaxial side more developed than the axial one. Distally, the axial condyle is slightly extended downward and the fossa is very deep. All these features characterize a cervid.

species. The attribution of two vertebrae is under question. A revision of specimens referred to *Procamptoceras* from other sites shows that remains are generally scarce and in some cases (such as the mandible from Villány 3, and a metacarpal from Csarnóta 1) belong to other species. It is possible to confirm the presence of the species only at Olivola. Its occurrence in other European localities needs confirmation.

Genus Cf. *Hemitragus* Hodgson, 1841 sp. indet

Hemitragus is present in Europe from MNQ 15b to the beginning of the Late Pleistocene (MIS 5e) (Crégut-Bonnoure 1989, 2002, 2006, 2020; Crégut-Bonnoure & Spassov 2002). During the Early Pleistocene, *H. orientalis* Crégut-Bonnoure & Spassov, 2002 is found in three localities: in Hungary at Csarnóta 1 (MN 15b) (Crégut-Bonnoure & Spassov 2002; Kretzoï 1956; Schaub 1932), at Villány 3 (MNQ 17) (Crégut-Bonnoure & Spassov 2002; Kretzoï 1956; Schaub 1932), and at Villány-Nagyharsányhegy 4 (MNQ 18–19) (Schaub 1932); and two localities from MNQ 18 in Bulgaria: at Slivnitsa (Crégut-Bonnoure & Spassov 2002) and at Kozarnika (Biostratigraphic unit B2-2;

Table 14.22 Dimensions (in mm) of the proximal anterior phalanx in *Procamptoceras brivatense* from Senèze

Variable	1	2	3	4	5	6	7	8	9	10	11	12	13
NMB Se 755	40.3	41.6	42.2	15.0	18.4	14.7	16.6	4.6	6.1	4.1	4.1	13.8	11.4
Variable	14	15	16	17	18	19	20	21	22	23	24	25	26
NMB Se 755	3.8	3.3	16.5	13.9	12.6	13.2	14.8	13.8	11.3	12.2	14.3	11.4	15.4

Variable definitions (from Crégut-Bonnoure 1995, 2002, 2020) (Fig. 14.A.11: A to F): td, transversal diameter; dpd, dorsopalmar diameter. Length: 1, middle; 2, axial; 3, abaxial. Proximal epiphysis: 4, td; 5, dpd. Proximal articulation: 6, td of the proximal articulation; 7, dpd in the middle; 8, td of the proximoaxial facet; 9, td of the proximoabaxial facet. 10, td of proximoaxial sesamoid facet; 11, td of the proximoabaxial sesamoid facet, 12, dpd of the proximoaxial facet; 13, dpd of the proximoabaxial facet; 14, dpd of the proximoaxial sesamoid facet; 15, dpd of the proximoabaxial sesamoid facet; 16, dpd of the proximoaxial articulation; 17, dpd of the proximoabaxial articulation. Middle of the diaphysis: 18, td; 19, dpd. Distal articulation: 20, td articular; 21, td of the distal epiphysis; 22, dpd minimum; 23, dpd of the distoaxial condyle; 24, dpd of the distoabaxial condyle. 25, dpd supra-articular distal; 26, height on the palmar side, from the base of the proximoaxial tubercle to the top of the disto-axial condyle

Discussion

At least six individuals are present. The *Procamptoceras* cranium is well known, but not the mandible or the skeleton: the three mandibles and six of the eleven postcranial bones from Senèze previously referred to this taxon belong to other

Fernandez & Crégut-Bonnoure 2007). It might persist in Greece during MNQ 20 at Apollonia 1 (Kostopoulos 1996: Caprinae indet.; Crégut-Bonnoure 2020).

Some elements have been assigned to *Capra* in Spain, at Fonelas P-1 (MNQ 18; Garrido 2006; Arribas & Garrido 2008), and in Georgia, at Dmanisi (MNQ 18; Bukhsianidze 2005, 2006). I interpret the morphological features of this material differently: it does not have the characteristics of *Capra* from the Middle and Late Pleistocene or the extant species, and its dental morphology is the same as in *Hemitragus* (Crégut-Bonnoure 2009).

Description

Mandible and Lower Teeth

Three mandibles from Senèze have been referred previously to *Procamptoceras*. As indicated above (see section "Mandible and Lower Teeth" under *Procamptoceras.*), these include the left mandible FSL 16892, the left mandible NMB Se 402 and the right mandible NMB Se 1718 (Table 14.23), all of which are characterized by more hypsodont teeth than in *Procamptoceras* uppers. A note from Sickenberg on the 1972 label indicates that NMB Se 402 could be *Gazellospira* sp., however the goat fold is less developed than in this genus, in which it is very strong.

The NMB Se 402 mandible, with p3-m2, belongs to a juvenile: p3 and p4 are not fully erupted, and m1 and m2 are in an early state of wear (Tables 14.24 and 14.25; Fig. 14.22: D to E). The p3 is well molarized, with a relatively wide second lobe. Buccally, the lobes are rounded at the base, slightly pinched at the top. The parastylid is vertical, perpendicular to the sagittal plane, and wide. The metaconid is inflated in a small column in median position. The entostylid is vertical, perpendicular to the sagittal plane, and its thickness is the same as that of the parastylid. Only the top of the p4 can be observed. The parastylid is vertical and wide, the metaconid is widely inflated and the valley behind it is flat. The entoconid is restricted to the top of the crown. The parastylid is similar to that of the p3. It is impossible to see the base of the molars which are still embedded within the mandible. The goat fold is moderate on m1, stronger on m2, but much less than in *G. torticornis*. The lobes are buccally pinched. The stylids are well differentiated and the protoconid and the hypoconid are significantly inflated; consequently, the valley between them is not very wide. On m2 there is an oblique and thin metastylid which extends into the mandibular bone. At Slivnitsa, there is an incomplete mandible with molarized and unworn p3 and p4; it was assigned initially to an indeterminate bovid (Spassov & Crégut-Bonnoure 1999) and later, it was considered as an *H. orientalis* specimen (Crégut-Bonnoure 2009). The p3 parastylid is open and the paraconid is inflated. The inflation of the metaconid is restricted to the top of the crown. The entostylid is thin. The buccal valley is deep and broad. On the p4, a short valley, restricted to the top, separates the paraconid and the metaconid. The morphology of the Senèze premolars in NMB Se 402 is distinct from the Slivnitsa teeth, and both are different from the Middle Pleistocene *H. bonali*.

The base of the corpus is convex in FSL 16892, as noticed by Guérin (*in* Duvernois & Guérin 1989) (Fig. 14.22: A to C). The attachment of the masseter is small. Two crests originate from it and occupy the rostral part of the mandibular corpus fossa which is deep. The *foramen mandibulare*, located below the p4, is small: 2.8 mm length, 1.4 mm height. The height of the ramus is short. The teeth are worn but it is clear that they are hypsodont. The p2 and p3 are missing. The p4 is molarized, with its buccal side rounded, and the first lobe is broader than the second one. It is not possible to assess the morphology of the m1 which is too worn. There is a moderate goat fold on m2 which is stronger on m3. The distal lobe on m2 appears narrower than the mesial one and more obliquely directed backward. The stylids of m2 are wide (0.7 mm) and the lingual valley is not deep. The valley behind the parastylid doesn't reach the base of the crown as on m3. The third lobe of the m3 has its distal side flat and wide (4.5 mm at the top, 5 mm at the base), and the surface is obliquely directed anteriorly. The flatness results from the connection of a small stylid on the lingual side, whose top shows an early state of wear, and an elongated stylid on the distal side. This m3 has a quite similar morphology to that of the NNHM-S 1782/SL 63 specimen from Slivnitsa (Crégut-Bonnoure & Spassov 2002: Fig. 3a; Crégut-Bonnoure 2020: Fig. 3.88, E, F). More or less at the same level, the Slivnitsa molar shows a small lingual stylid and the distal stylid; the space between these two structures is flat and wide (7.1 mm at the top, 4.9 mm at the base).

The NMB Se 1718 mandible preserves the deciduous teeth and the isolated m1 (Tables 14.25 and 14.26; Fig. 14.22: F to I). The dp2 is relatively elongated with a midline cuspid. The dp3 has a thin parastylid which is open. The inflation of the paraconid is moderate, but more than that of the metaconid. An interlobar column is present between the second and the third dp4 lobes (height: 8.9 mm). The stylids are thin and the cuspids inflated. The valleys are relatively broad. The morphology of the m1 is similar to that of NMB Se 402, except for the presence of a metastylid at the top of the crown. There is no basal depression in the middle of the lingual side which is characteristic for Pleistocene *Hemitragus* (Crégut-Bonnoure 1995). However, its presence is variable on the m1s from Slivnitsa

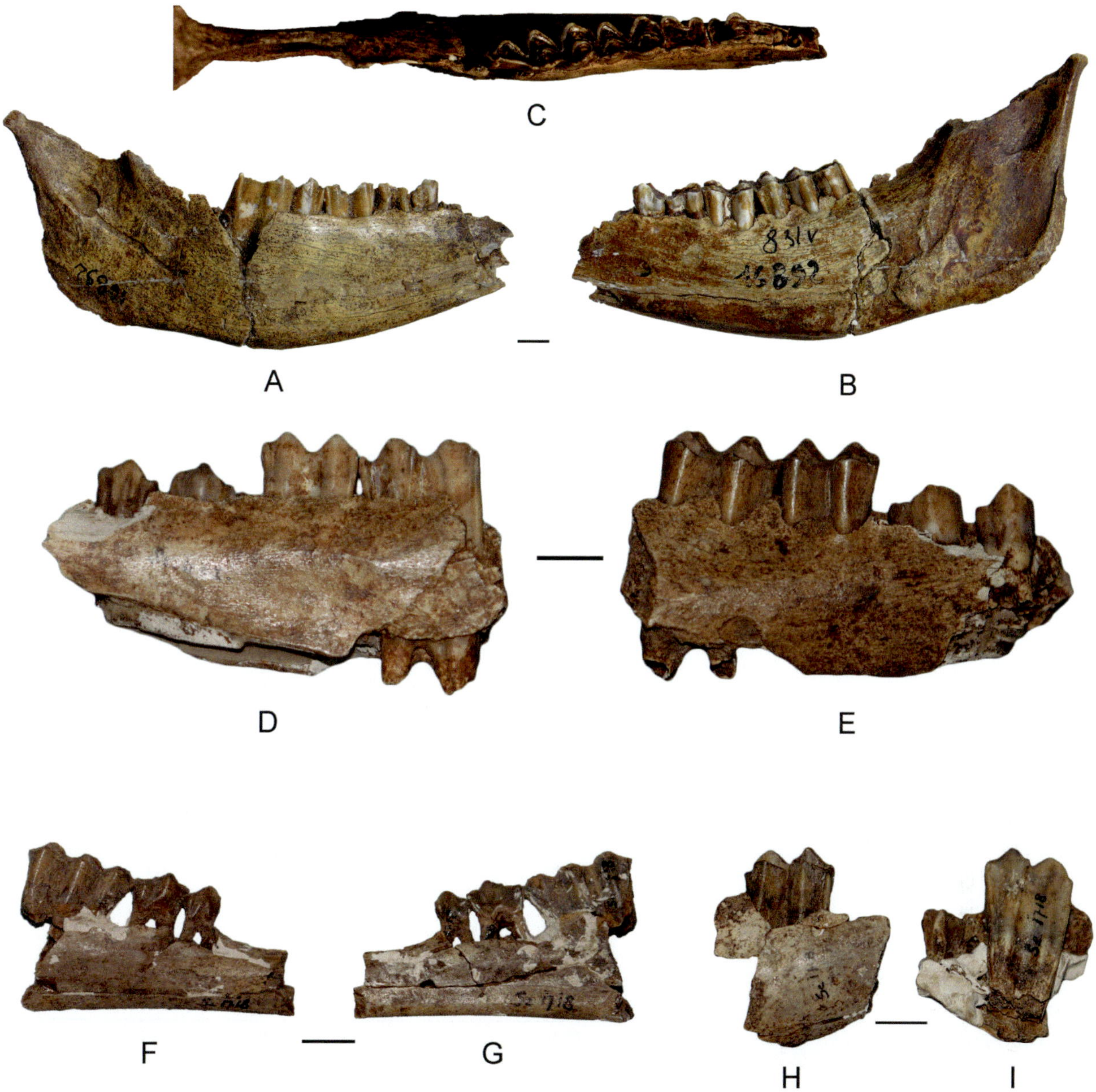

Fig. 14.22 Mandibles of cf. *Hemitragus* from Senèze. A to C, left FSL 16892; D, E, left NMB Se 402; F to I, right NMB Se 1718. A, D, G, I, lingual view; B, E, F, H, buccal view; C occlusal view. Scale 1 cm

Table 14.23 Dimensions (in mm) of the mandible in cf. *Hemitragus* from Senèze

Variable	3	4	6	8	9	10	12	15	16	17	18
FSL 16892	63.1	60.9	142.1	82.0	25.2	57.0	62.2	19.3	24.7	31.9	39.6

Variable definitions (from Crégut-Bonnoure 2002, 2020) (Fig. 14.A.3): Length: 3, from m3 distal side to the ramus; 4, from m3 distal side to the angle of the mandible; 6, from p2 mesial side to the posterior margin of the *processus angularis*; 8, from p2 mesial side to m3 distal side; 9, from p2 mesial side to p4 distal side; 10; from m1 mesial side to m3 distal side; ramus: 12, height of the ramus from condylar process to the base; 15, transversal diameter of the mandibular condyle; Height of the corpus: 16, at the p2 level. 17, at the p4-m1 level; 18, at the m3 level; 19, at the diastema level

Table 14.24 Dimensions (in mm) of the lower premolars in cf. *Hemitragus* from Senèze

p3 variable	Ht	2	5			
NMB Se 402	11.8	13.3	11.9	7.0	7.9	
p4 variable	Ht	1	2	3	4	5
FSL 16892	7.7	9.4	8.4	7.5	6.9	6.8
NMB Se 402		11.4		6.6		

Variable definitions (from Crégut-Bonnoure 1995, 2020) (Fig. 14.A.4: D): Ht, height; 1, occlusal length; 2, basal length; 3, occlusal width, first lobe; 4, occlusal width, second lobe; 5, basal width

Table 14.25 Comparative dimensions (in mm) of the lower molars in cf. *Hemitragus* from Senèze and *Hemitragus orientalis* from other localities

m1 variable	Ht	1	2	3	4	5	
Senèze							
FSL 16892	5.8	11.6	9.7	8.7	9.2	8.8	
NMB Se 402		14.9		8.1	7.8		
NMB Se 1718	26.5	16.8	13.7	7.2	7.0	8.2	
Slivnitsa (NNHM-S SL 95)	19.9	16.3	14.6	8.3	8.6	9.0	
Villány-Nagyharsànyberg 4 (NMB UP 500)	36.4	19.9	17.0	8.6	8.8	10.0	
Villány 3 (NMB UP 477)	34.9	18.3	15.6	8.2	7.8	9.8	
Dmanisi (D 1717–cast)	20.1	17.3	13.8	9.2	9.8	9.1	
m2 variable	Ht	1	2	3	4	5	
Senèze							
FSL 16892		15.4		9.7	9.2		
NMB Se 402	27.4	17.5	14.8	8.2	7.6	10.4	
Slivnitsa							
NNHM-S SL 82	25.1	18.2	15.5	10.0	9.8	10.7	
NNHM-S SL 78	24.7	19.3	17.6	10.2	9.6	10.8	
NNHM-S SL 75	34.9	19.7	18.5	9.7		11.7	
NNHM-S SL 73	39.0	22.2	18.0	9.4	9.0	113	
Villány-Nagyharsànyberg 4 (NMB UP 503)	37.8	21.6	18.5	11.7	11.4	11.6	
Villány 3							
NMB UP 479	19.9	17.7	15.7	9.3			
NMB UP 480	20.3	18.2	17.0	9.4	8.6	10.6	
Dmanisi (D 1716–cast)	35.5	20.2	16.5	9.6	9.8	10.7	
m3 variable	Ht	1	2	3	4	5	6
Senèze FSL 16892		27.1		10.3	9.8	6.1	
Slivnitsa	7	5	5	6	7	6	6
Mean	32.4	28.6	32.1	9.8	9.3	5.6	11.2
Minimum	25.0	25.3	31.1	8.6	8.0	4.9	10.3
Maximum	39.5	30.3	33.4	10.5	10.4	6.3	11.6
SD	6.22	2.09	0.93	0.71	0.85	0.50	0.50
Villány-Nagyharsànyberg 4 (NMB UP 506)	31.7	36.0	10.2		6.3		
Villány 3 (NMB UP 499)	29.0	27.8	27.8	10.5	9.7	5.5	11.1
Dmanisi (casts)							
D 2397	43.4	25.6		9.3	8.0	4.5	10.7
D 1715	44.5	28.9	29.9	9.6	8.8	4.8	10.4

Variable definitions (from Crégut-Bonnoure 1995, 2002, 2020): Ht: height. m1 and m2 (Fig. 14.A.4: E): 1, occlusal length; 2, basal length; 3, occlusal width, first lobe; 4, occlusal width, second lobe; 5, basal width. m3 (Fig. 14.A.4: F): 1, occlusal length; 2, basal length; 3, occlusal width, first lobe; 4, occlusal width, second lobe; 5, occlusal width, third lobe; 6, basal width

Table 14.26 Dimensions (in mm) of the lower deciduous teeth in cf. *Hemitragus* from Senèze

dp2-dp4 variable	L						
NMB Se 1718	37.3						
dp2 variable	Ht	1	3				
NMB Se 1718	7.0	6.7	3.8				
dp3 variable	Ht	1	2	3	5		
NMB Se 1718	8.5	10.3	8.9	5.7	5.6		
dp4 variable	Ht	1	2	3	4	5	6
NMB Se 1718	9.6	20.8	16.2	5.6	6.8	6.8	7.4

Variable definitions (from Crégut-Bonnoure, 1995, 2002, 2020): Ht, height; L, length. dp2 and dp3 (Fig. 14.A.4: D); 1, occlusal length; 2, basal length; 3, occlusal width, first lobe; 4, occlusal width, second lobe; 5, basal width. dp4 (Fig. 14.A.4: F); 1, occlusal length; 2, basal length; 3, occlusal width, first lobe; 4, occlusal width, second lobe; 5, occlusal width, third lobe; 6, basal width

Tooth Proportions

The Senèze molars are smaller than those of *H. orientalis*, except the m3. The general proportions of the molars show some similarities with Slivnitsa and Villány 3 *Hemitragus* (Figs. 14.23, 14.24 and 14.25). The material from Dmanisi, which has the morphological features of *Hemitragus* (Crégut-Bonnoure 2009), is proportionally distinct.

H. orientalis. However, the morphology of the lower premolars is not the same as the fossils from Slivnitsa, a locality considered as contemporaneous to Senèze (Spassov & Crégut-Bonnoure 1999), whereas m3 shows affinities. The morphological differences could result from geographical or chronological differences. In sum, this material remains insufficient to decide whether the differences are due to these factors or if there are two distinct species. Consequently, I attribute these mandibles cautiously to cf. *Hemitragus* sp., without proposing a new taxon at the species or subspecies rank.

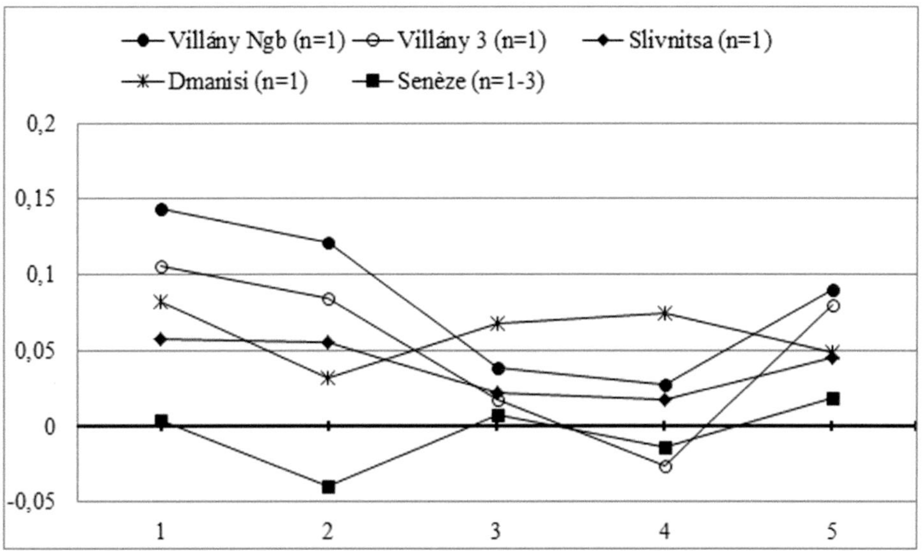

Fig. 14.23 A cf. *Hemitragus* and *Hemitragus orientalis*. Ratio diagram of the m1. Reference axis: *Hemitragus cedrensis* (in Crégut-Bonnoure 2020). Ngb, Villány-Nagyharsányhegy 4. Length: 1, occlusal; 2, basal. Width: 3, occlusal, first lobe; 4, occlusal, second lobe; 5, basal. References of the measurements in Fig. 14.A.4 and dimensions in Table 14.25

Discussion

There are three individuals. Taking into account the size, the hypsodonty and the morphology of the teeth, these elements cannot belong to *Procamptoceras* or *Gazellospira* as previously suggested on their labels. Although the number of specimens is insufficient to check for real statistical differences, the proportions of the molars are similar to

A number of specimens from Villány 3 have been re-identified as *Hemitragus*. Four teeth previously identified as *Procamptoceras* (Schaub 1932) have been reallocated to *H. orientalis* (Crégut-Bonnoure & Spassov 2002): right M2 NMB UP 478, left m1 NMB UP 477, right m2 NMB UP

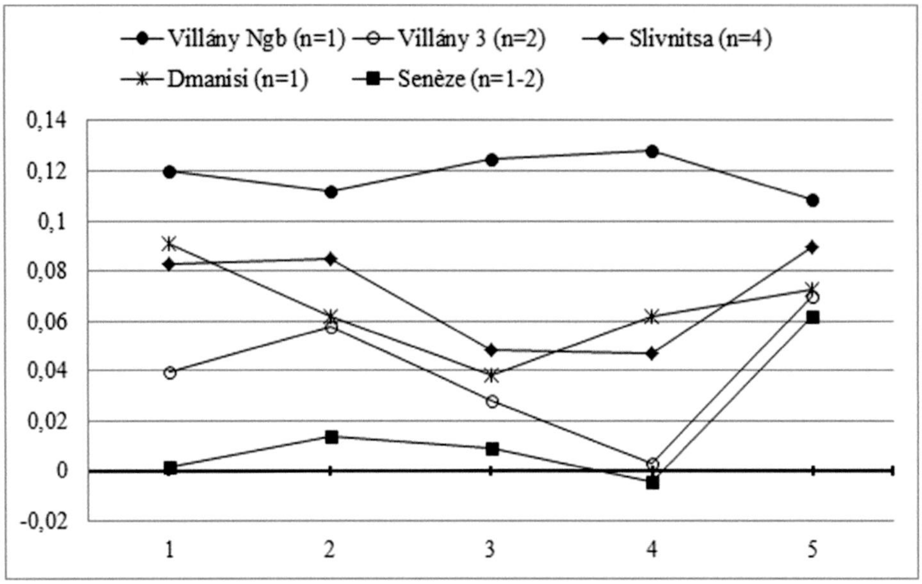

Fig. 14.24 A cf. *Hemitragus* and *Hemitragus orientalis*. Ratio diagram of the m2. Reference axis: *Hemitragus cedrensis* (in Crégut-Bonnoure 2020). Ngb, Villány-Nagyharsányhegy 4. Length: 1, occlusal; 2, basal. Width: 3, occlusal, first lobe; 4, occlusal, second lobe; 5, basal. References of the measurements in Fig. 14.A.4 and dimensions in Table 14.25

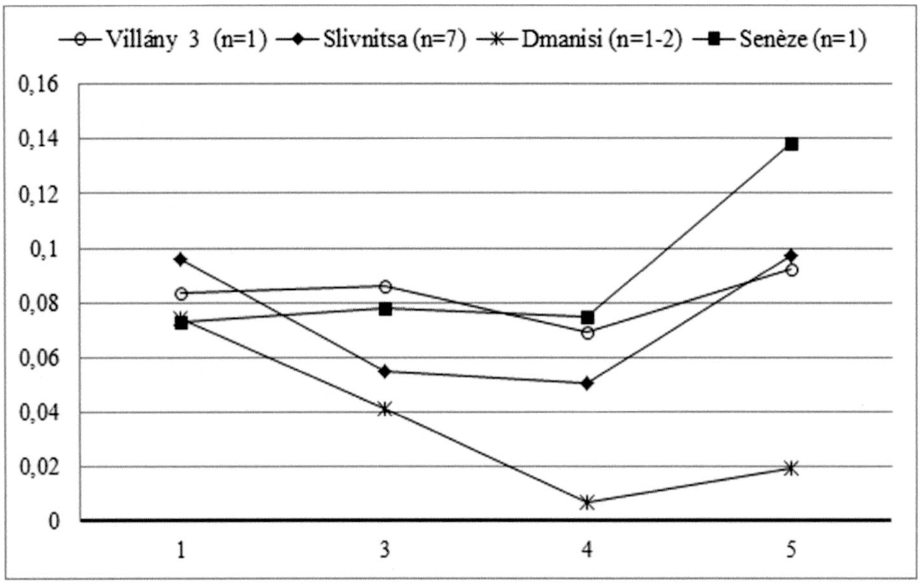

Fig. 14.25 A cf. *Hemitragus* and *Hemitragus orientalis*. Ratio diagram of the m3. Reference axis: *Hemitragus cedrensis* (in Crégut-Bonnoure 2020). Length: 1, occlusal. Width: 3, occlusal, first lobe; 4, occlusal, second lobe; 5, occlusal, third lobe. References of the measurements in Fig. 14.A.4 and dimensions in Table 14.25

479, right m2 NMB UP 480. The left scaphoid NMB UP 488 and the left semi lunar NMB UP 489 labeled as *Tragelaphus* (=*Gazellospira*) belong to cf. *Hemitragus*. Some phalanges initially assigned to *Gazellospira* were reallocated to *Hemitragus* (Crégut-Bonnoure 2020; Crégut-Bonnoure & Spassov 2002): anterior second phalanx NMB UP 492, posterior first phalanx NMB UP 490.

In addition, a right metacarpal III-IV from Csarnóta 1 (cast NMB UP 797) was originally identified as *Procamptoceras*. The distal articulation is missing and the proximal articulation lacks the palmar side. It is larger than the Senèze *Procamptoceras* metapodial and the morphology is not the same. The lateral side of the unciform articular facet is straight. The palmar groove is restricted to the proximal third of the diaphysis. This morphology is the same as in *Hemitragus* (Crégut-Bonnoure 1995; Crégut-Bonnoure & Spassov 2002).

Genus *Pliotragus* Kretzoi, 1941

Species *Pliotragus ardeus* (Depéret, 1883)

The genus was first described as *Deperetia* by Schaub (1923). This name having already been applied to a bivalve mollusc in 1922, it was changed by Kretzoï in 1941. The genus is indicated at the end of the Miocene (Stratzing in Austria; Gentry 2001), however, this attribution is debatable; there is no record in the early Pliocene. Two species are recognized: *P. ardeus* (Depéret, 1883) and *P. macedonicus* Crégut-Bonnoure & Tsoukala, 2005.

P. ardeus is a large-sized antelope as described by Depéret in 1883 (*Antilope ardea*). A neodiagnosis was proposed by Viret (1961) and a more developed one by Guérin (*in* Duvernois & Guérin 1989). It is known from MN 16 to MNQ 18 in 9–10 localities:

- MN 16: in France at Perrier-Etouaires (Depéret 1884; Nomade et al. 2014), at Perrier-Rocca-Neyra (Heintz et al. 1974; Nomade et al. 2014) and perhaps at Vialette (Heintz et al. 1974).
- MNQ 17: in Hungary at Villány 3 (this paper); in Romania at Valea Grăunceanului (Bolomey 1965; Terhune et al. 2020) but probably not at La Pietriş (Terhune et al. 2020) despite an earlier suggestion (Rădulesco & Samson 1990); perhaps in the Middle East at Sarikol Tepe (Kostopoulos & Sen 1999).
- MNQ 18: in France at Senèze (Schaub 1923); in Romania at Leu (Popescu 2004; Rădulesco 2005); in Bulgaria at Slivnitsa (Spassov & Crégut-Bonnoure 1999; this could be a new subspecies or even a new species).

In Greece, *Pliotragus* sp. is reported from the fauna of Karnezeika in the Peloponnesus (MNQ 17?; Kokotini et al. 2019).

The right M2 from Erpfingen in Germany assigned to *Deperetia* (NMB D461; Lehmann 1957) is a *Megalovis* DP4 (Crégut-Bonnoure 2002, 2020); the incomplete right M3 was not studied but, according to Lehmann's Fig. 20, its morphology matches better with *Gallogoral*.

P. macedonicus is identified in Greece at Petralona, probably at the beginning of the Middle Pleistocene (Crégut-Bonnoure & Tsoukala 2005). Guérin (*in* Duvernois & Guérin 1989) thought that *Hesperoceras merlae* Villalta and Crusafont Pairó 1956, from Villarroya in Spain (MN 16 or perhaps MNQ 17) is a synonym of *P. ardeus*. Because this genus name was used for a cephalopod mollusc in 1947, Crusafont Pairó (1965) created the new genus *Hesperidoceras*. *Hesperidoceras merlae* is now considered a valid genus and species (Andrés Rodrigo, 2011).

Description

No complete associated skeleton has been found at Senèze or anywhere else in Europe. Because the cranium is bigger than that of *Gallogoral*, all previous authors have considered that the teeth and postcrania were also. The revision below proposes some corrections.

Crania and Horn Cores

There are three incomplete crania. NMB Se 1636 is a braincase with the right horn core; NMB Se 1755 is a left lateral fragment with the base of the horn core, the parietal with the temporal and the beginning of the occipital protuberance; MNHN-F-1923-8 is a complete braincase with the horn cores; the right is nearly complete, while the left is broken about one-third of the way from the base to the tip (see Duvernois & Guérin 1989: pl. 2. *Nota bene*: their figure is of the cast MNHN-F-1958-12; two other casts are present in the Basel Museum).

There are five isolated horn cores: two were recovered during the recent fieldwork–FSL SEN 06-0001, P172 (right) and FSL SEN 04-0084, P233 (fragmentary); and also, NMB Se 1483 (left), NMB 1726 (left), and NMB Se 1725 (right).

The MNHN-F-1923-8 braincase is wide, larger than NMB Se 1636, which belongs to a young individual (Table 14.27) (Figs. 14.26 and 14.27: A, B). The parietal and the frontal are slightly convex transversely and rostrocaudally. In lateral view, the supra-occipital is obliquely directed backward and the condyles are projected backward. The horn cores, inserted just behind the orbits, are directed upward and laterally: in caudal view they form part of a circle. Their divergence at the base is about 120°. In lateral view, they are directed forward. Three grooves are visible on the rostral side, and a large one is present rostrally on the dorsal side. The section at the base is more or less triangular, the lateral face being convex; the section is flatter above. The basioccipital is broad and grooved, with very prominent tubercles: consequently, the caudal tubercles are visible in caudal view. The area between the rostral and caudal tubercles is concave in lateral view.

The left NMB Se 1483 specimen, without the tip (Fig. 14.27: F, G), was initially considered as the horn core of a *Megalovis* cranium (Schaub 1943: Fig. 5 p. 280, Fig. 6 p. 281; Viret 1961: Fig. 165 p. 1073), but it was reallocated to *Pliotragus* by Spassov and Crégut-Bonnoure (1999). Its surface is longitudinally grooved. The section at the base is rounded, and then laterally compressed. The dorsal side is concave. The rostral side, convex, shows a deep groove on its dorsal part. Another groove, less deep, runs on the ventral side. The angle with the frontal is about 148°. It is directed upwards and laterally. This morphology is the same as on cranium MNHN-F-1923-8 and on the isolated horn cores. On NMB Se 1483 and NMB Se 1725 (which belongs to a male) the sagittal suture is preserved; it is thick (40 mm) on Se 1483. The gap between the horn core base and the suture is respectively 23.5 and 26 mm; therefore, the distance between the two horn cores is respectively 47 and 52 mm. These values are comparable to MNHN-F-1923-8 (46.4 mm).

The better-preserved horn core from the new field research (FSL SEN 06-0001) is broken with a preserved length of about 6 cm (Fig. 14.27: C to E). Laterally, a part of the fronto-parietal suture is present. There is a significant sinus (length 24 mm) which occupies the horn core base. The section is elliptical. The dimensions decrease rapidly from the base to the top: at the base, rostrocaudal diameter is 81 mm, dorsoventral diameter is 34.5 mm; at 50 mm from the base, respectively 29.6 and 24.6 mm. There is no pedicle. The rostral side is convex, the ventral side is slightly convex, and the caudal and dorsal sides are slightly concave. The surface is covered by small grooves at the base, which deepen toward the tips. The deepest one is located on the rostral side. This specimen is smaller than the previous horn cores and could belong to a female.

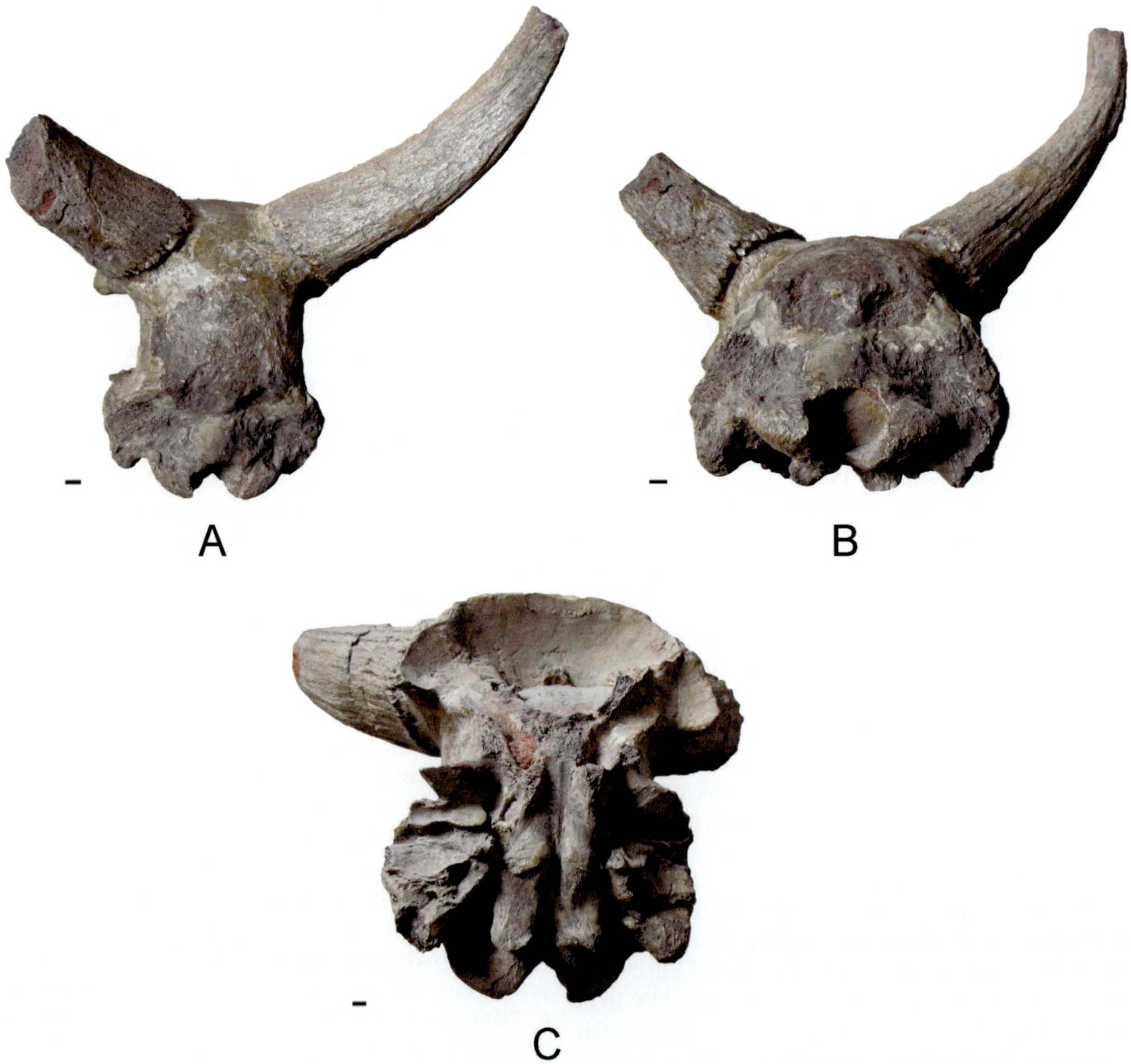

Fig. 14.26 Cranium of *Pliotragus ardeus* from Senèze. A to C, MNHN-F-1923-8. A, *norma dorsalis*; B, *norma caudalis*; C, *norma basalis*. Scale 1 cm

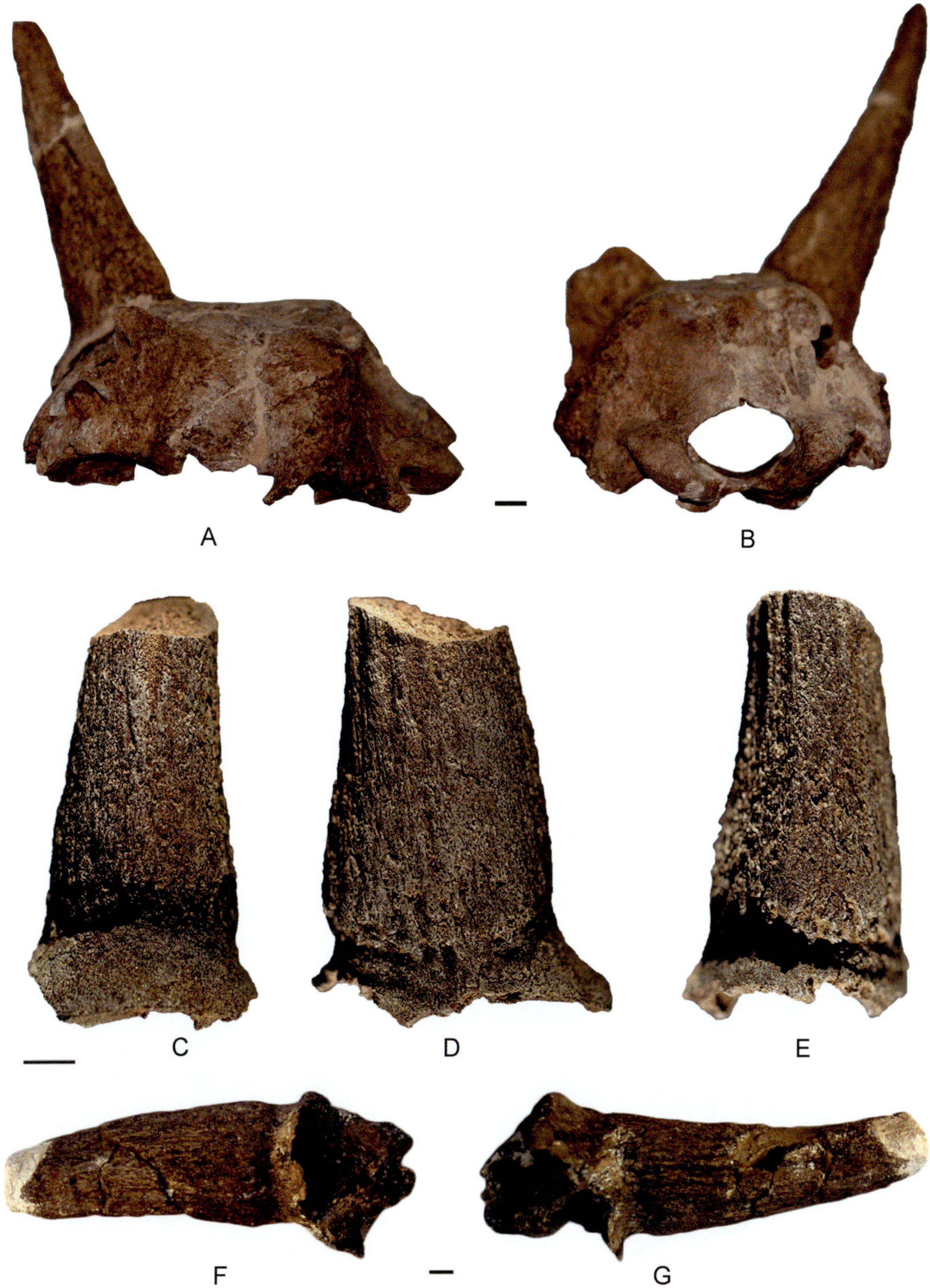

Fig. 14.27 Cranium and horn cores of *Pliotragus ardeus* from Senèze. A, B, cranium NMB Se 1636; C to E, horn core FSL SEN 06 001; F, G, horn core NMB Se 1483. A, D and F, *norma lateralis*; B and E, *norma caudalis*; C, G, *norma dorsalis*. Scale 1 cm

Comparison with a Dacian Basin Cranium

The Senèze specimens can be compared with a braincase from the Dacian Basin preserved in the Emil Racoviţă Institute of Speleology in Bucharest (Romania). A quite complete specimen (Gr. 454/207) was described by Bolomey (1965: pl. I Figs. 1 and 2; pl. II Fig. 1) and a frontlet was also reported (Gr. 393). The specimen studied here is unnumbered and was damaged (Fig. 14.28): the whole face is absent, having been broken through the orbits (maxilla and teeth are missing); it is the same for the occipital area which is now without the *foramen magnum* and the condyles; laterally, the left parietal is broken; the tip of the right horn core is missing and a recent fracturing is visible at its dorsal base, which has been reattached; the left horn core is incomplete. This cranium could be perhaps the specimen Gr. 454/207: the left horn core was detached at the level of the crack that is visible on Bolomey's Fig. 1 (1965: p. 317), and there is a triangular hole at the dorsocaudal base of the right horn core that is visible on Bolomey's Fig. 3 (1965: p. 318). However, the dimensions provided by this author (p. 321) do not correspond to what I measured, which are larger. According to Croitor et al. (2023) the cranium described by Bolomey could not be identified in the

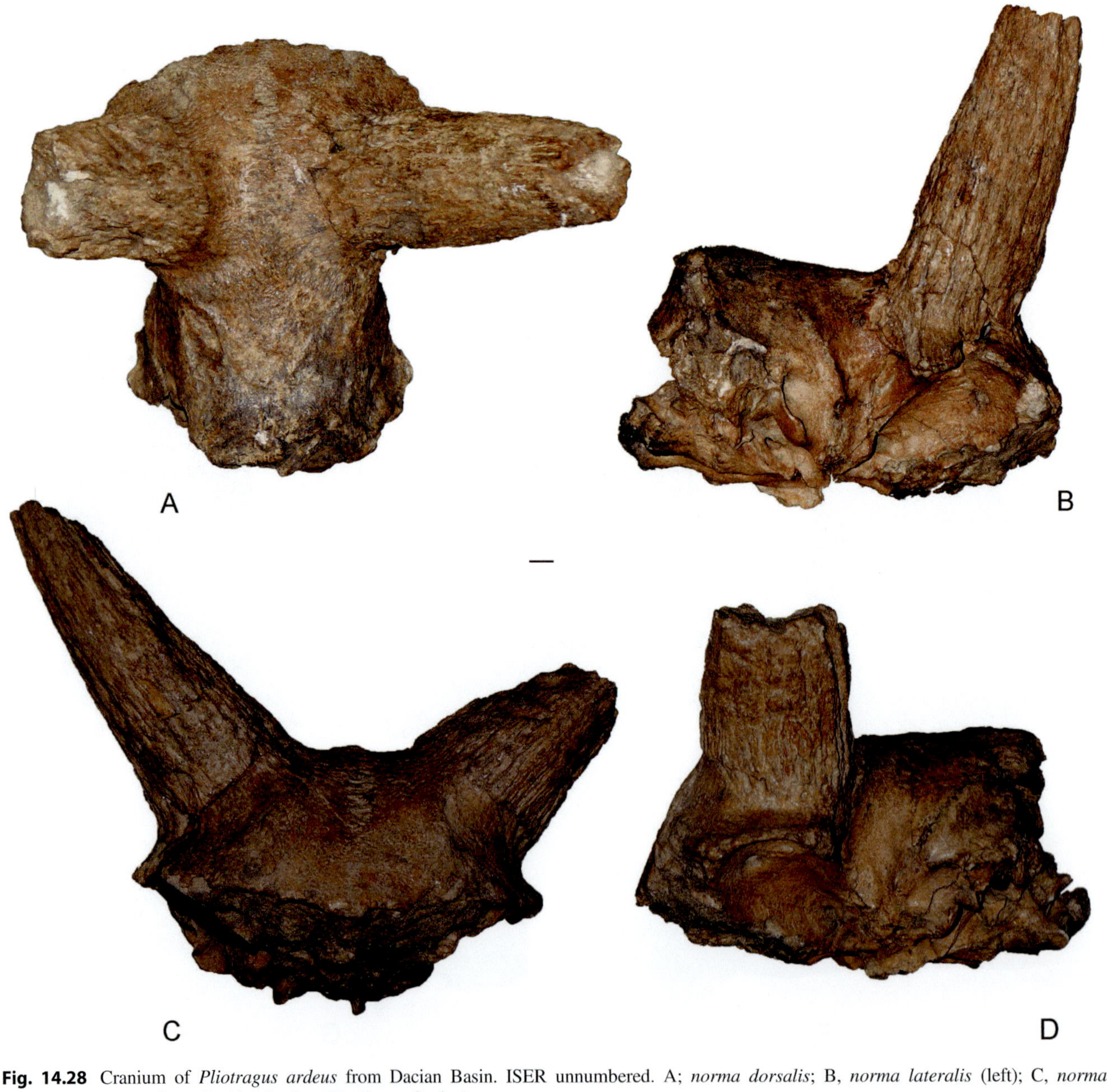

Fig. 14.28 Cranium of *Pliotragus ardeus* from Dacian Basin. ISER unnumbered. A; *norma dorsalis*; B, *norma lateralis* (left); C, *norma frontalis;* D, *norma lateralis* (right); phot. Ph. Fosse. Scale 1 cm

Table 14.27 Comparative dimensions (in mm) of the cranium in *Pliotragus ardeus* from Senèze and Dacian Basin

Variable	9	10	11	12	17	19	25	26	27	28
Senèze										
NMB Se 1636				111.9		69.6	33.8	84.9	44.3	
FSL 16895	119.3	51.3	68.2							
MNHN-F-1923-8				128.8		57.0	36.0	91.1		126.9
Dacian Basin				121.5	165.8	81.4	50.4	100.2	42.0	132.0
Variable	29	37	38	39	40	41	42	43	44	45
Senèze										
NMB Se 1636		107.8	49.3		170.5	44.5	36.4	33.7	24.6	125.0
MNHN 1923-8		148.6	49.3	179.0	200.0	67.9	58.9	50.6	34.3	195.0
NMB Se 1725						52.5	50.0			
NMB Se 1483						55.6	52.9	51.7	42.0	180.0
NMB Se 1726						52.1	49.8			165.0
NMB Se 1755						56.7	57.6			170.0
Dacian Basin	152.5	152.0	54.9	196.0	255.0	59.0	56.2	36.8	47.6	~188.0
Variable	46	55	57	58	59	60	61	62	63	64
Senèze										
NMB Se 1636	130.0	73.8	101.0	77.19	31.8	24.7	22.3	24.2	60.1	46.2
MNHN-F-1923-8	197.0		115.1	74.6	25.8	29.9	26.3	24.9		48.5
NMB Se 1725	158.0									
Dacian Basin	140.0	91.0								41.8
Variable	65	66	68							
Senèze										
NMB Se 1636			125.5							
MNHN-F-1923-8	45.8	81.8								
Dacian Basin	48.2	82.0								

Variable definitions (from Crégut-Bonnoure 2002, 2020) (Figs. 14.A.1 and 14.A.2): Length: 9, P2 mesial-M3 distal; 10, P2 mesial-P4 distal; 11, M1 mesial-M3 distal; 17, acrocranion-nasion; 18; frontal from supra-orbital foramina; 19, acrocranion-bregma; 68, from the top of the frontal to the tip of the occipital condyles. Transversal diameter: 12, otion-otion; 27, between the parietal ridges; 28, post-orbital; 29: ectorbital-ectorbital; 57, between jugular processes; 58, between the external side of the occipital condyles; 59, of the foramen magnum; 60, of the occipital condyle. Parietal: 25, dorsal length; 26, width. Horn cores: 37, width between the lateral sides; 38, width between the internal sides at the base; 39, width between the internal sides in the middle; 40: width between the internal sides at the tip; 41, rostrocaudal diameter at the base; 42, dorsolateral diameter at the base; 43, rostrocaudal diameter in the middle of complete horn core; 44, dorsolateral diameter in the middle of complete horn core; 45, rostral length; 46, circumference at the base. Height: 55, of the braincase from basioccipital to the top of the parietal; 61, of the occipital condyle; 62, of the foramen magnum; 63, from opisthion to fronto-parietal suture. Basioccipital: 64, transversal diameter of the caudal tuberosities; 65, transversal diameter of the rostral tuberosities; 66, length from basion to basisphenoid

collection of the Institute of Speleology. It seems that the material in question is curated in another collection (Croitor, pers. comm. 2023), and the unnumbered specimen I examined does not correspond to Gr. 454/207. Its morphology is distinct from those of Senèze. The parietal is flat, the temporal ridge is strong, and on each side of the parietal there is a diamond shaped ridge originating from the rear of the horn cores and running obliquely to the sagittal plane. The horn cores are slightly curved. The rostral surface is well grooved, and there is a large groove laterally on the caudal side. These differences could be related to sexual dimorphism.

The proportions of the *Pliotragus ardeus* cranium from Senèze are quite similar to those of *Gallogoral*, except for the horn cores (Fig. 14.29: measurements 37–46), whereas the Dacian Basin cranium's proportions are more distinct: the parietal and occipital are larger (measurements 25–28), the orbit is caudally more projecting laterally (measurement 29), the horn core is shorter and thicker in the middle (measurements 43 and 44), the width between the rostral tubercles of the basioccipital is slightly smaller (measurement 65), and the basioccipital is slightly longer (measurement 66).

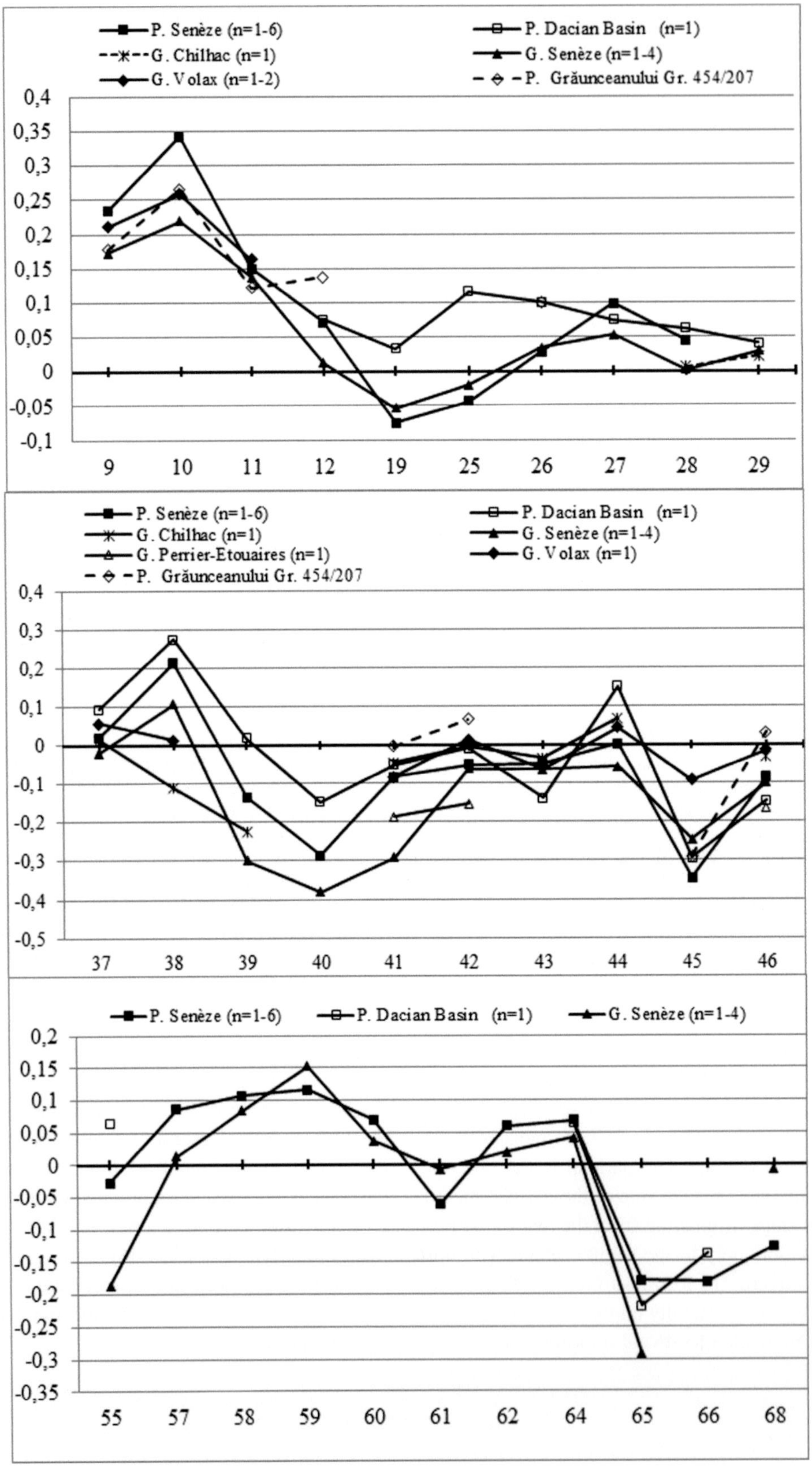
P. Senèze (n=1-6)
P. Dacian Basin (n=1)
G. Chilhac (n=1)
G. Senèze (n=1-4)
G. Volax (n=1-2)
P. Grăunceanului Gr. 454/207
0,4
0,35
0,3
0,25
0,2
0,15
0,1
0,05
0
-0,05
-0,1
9 10 11 12 19 25 26 27 28 29
P. Senèze (n=1-6)
P. Dacian Basin (n=1)
G. Chilhac (n=1)
G. Senèze (n=1-4)
G. Perrier-Etouaires (n=1)
G. Volax (n=1)
P. Grăunceanului Gr. 454/207
0,4
0,3
0,2
0,1
0
-0,1
-0,2
-0,3
-0,4
-0,5
37 38 39 40 41 42 43 44 45 46
P. Senèze (n=1-6)
P. Dacian Basin (n=1)
G. Senèze (n=1-4)
0,2
0,15
0,1
0,05
0
-0,05
-0,1
-0,15
-0,2
-0,25
-0,3
-0,35
55 57 58 59 60 61 62 64 65 66 68

◀ **Fig. 14.29** *Pliotragus ardeus* and *Gallogoral meneghinii*. Ratio diagram of the cranium. G: *Gallogoral*. P: *Pliotragus*. Reference axis: recent *Capra ibex* ♂ (in Crégut-Bonnoure 2020). Length: 9, P2 mesial-M3 distal; 10, P2 mesial-P4 distal; 11, M1 mesial-M3 distal; 17, acrocranion-nasion; 19, length acrocranion-bregma; 68, from the frontal convexity to the caudal side of the occipital condyle. Transversal diameter: 12, otion-otion; 27, between the parietal ridges; 28, post-orbital; 29: ectorbital-ectorbital; 57, between jugular processes; 58, between the external side of the occipital condyles; 59, of the foramen magnum; 60, of the occipital condyle. Parietal: 25, length; 26, width. Horn cores: 37, width between the lateral sides; 38, width between the internal sides at the base; 39, width between the internal sides in the middle; 40: width between the internal sides at the tip; 41, rostrocaudal diameter at the base; 42, dorsolateral diameter at the base; 43, rostrocaudal diameter in the middle; 44, dorsolateral diameter in the middle; 45, length; 46, circumference at the base. Height: 55, of the braincase from basioccipital to the top of the parietal; 61, of the occipital condyle; 62, of the foramen magnum. Basioccipital: 64, transversal diameter of the caudal tuberosities; 65, transversal diameter of the rostral tuberosities; 66, length from basion to basisphenoid. References of the measurements in Figs. 14.A.1 and 14.A.2 and dimensions in Tables 14.27 and 14.42 and in Croitor et al. (2023)

Maxilla

There is only one fragmentary maxilla which preserves its lateral surface (FSL 16867; Fig. 14.30: A, B), the others being reduced to the toothrow (FSL 16887, FSL 16895, NMB Se 38, NMB Se 1936, MNHN-F-1923-3; Fig. 14.31: A). The infraorbital foramen is in the axis of the middle of P3-4. Two small foramina are present in front of it, and the bone is slightly depressed. The maxilla is convex from M1 to M3, but less than in *Megalovis*.

Mandible

There are 13 more or less complete mandibles belonging to at least eight adults and two juveniles (Table 14.28; Figs. 14.31: D to H, 14.32: A to D): FSL 3935 (left), FSL 3935bis (right), FSL 16886 (right), FSL 16887 (right), FSL 16891 (left), NMB Se 383 (left, incomplete), NMB Se 562 (left and right), NMB Se 1733 (right, incomplete), NMB Se 1459 (left, juvenile), NMB Se 1719 (left, juvenile), MNHN-F-1922-3 (right, incomplete), and MNHN-F-cast 2659 (=NHMUK-P 12009; right, incomplete). FSL 16886 was initially attributed to *Gallogoral*, but it was reallocated to *Pliotragus* by Maria Andrès Rodrigo (note on label).

On the complete specimens, the buccal side of the corpus is convex as is the lingual side but only at the alveolar margin, below which it is slightly concave. The bone is thick. The ramus is only present on NMB Se 562 (Fig. 14.31: D, E), NMB Se 1719, and MNHN-F-2659. It is relatively low, obliquely directed backward and long at the level of m3 (NMB Se 562, length from m3 to caudal side: 78.7 mm). Two partial mandibles belonging to a single individual (FSL 3935 and NMB Se 562; Duvernois & Guérin 1989: pl. 3 Fig. 3 p. 350) are the most complete of the series; unfortunately, the alveoli of the p2 are missing on FSL 3935 (oligodonty) and are broken on NMB Se 562. On FSL 3935, the masseter attachment is preserved and elongated rostrocaudally, however its base is broken. On FSL 16891, the cranial part of the masseter attachment is preserved: it is convex and broad (17.3 mm).

The bone at the diastema level is slender on the specimens belonging to young and immature animals, more robust on the adult ones, relatively short, pinched with a crest. The *foramen mandibulare* is in a rostral position and it is well developed (13 mm long, 7 mm high). Compared to *Gallogoral*, the corpus of the mandible is deeper, especially at the diastema level (measurements 16, 19; Fig. 14.33).

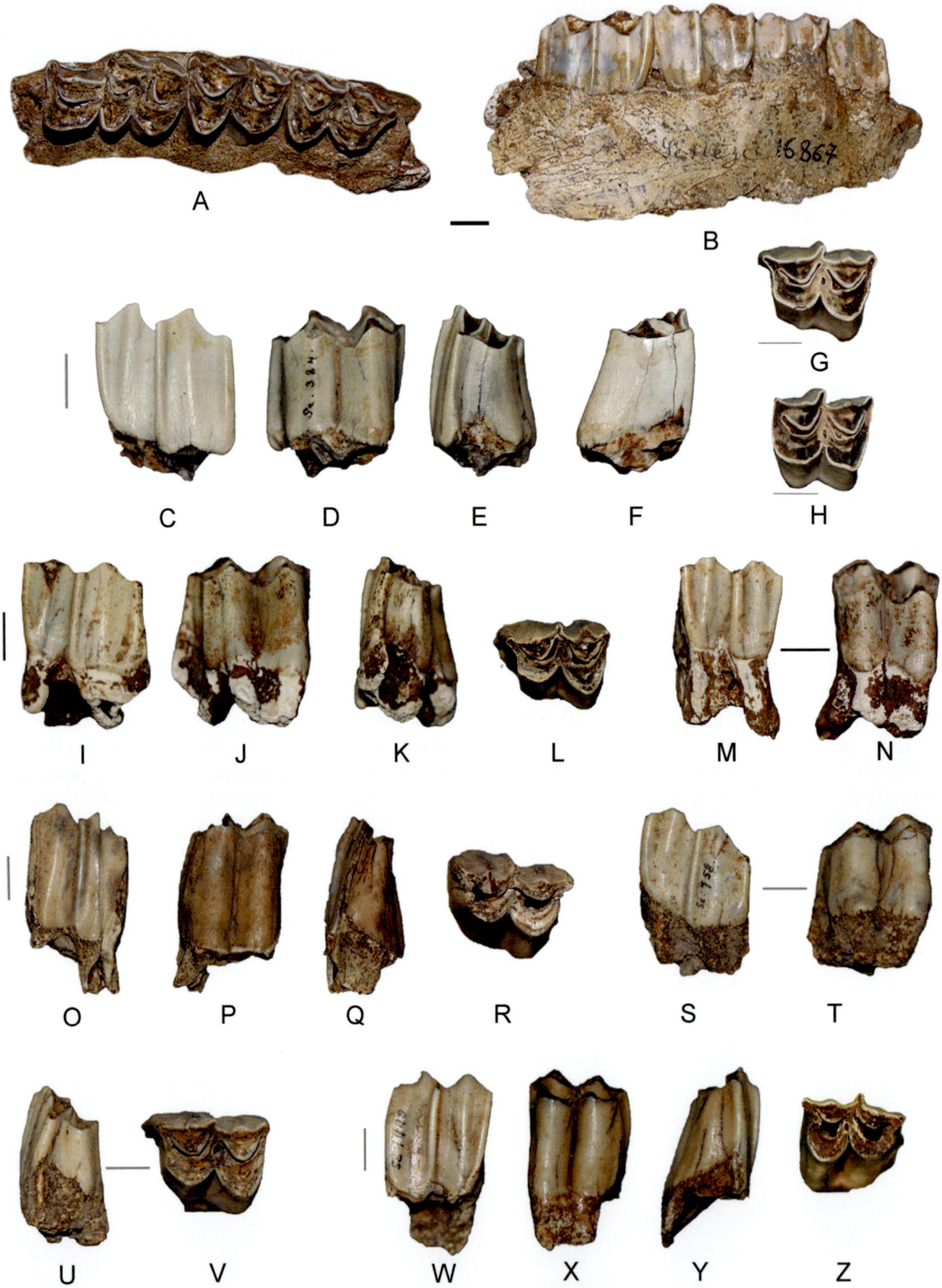

Fig. 14.30 Maxilla and upper teeth of *Pliotragus ardeus* and *Gallogoral meneghinii* from Senèze. *Pliotragus ardeus*, A to R and S to Z: A, B, left maxilla FSL 16867; C to G right M3 NMB Se 384; H, right M2 NMB Se 384; I to L, right M3 NMB Se 385; right M2 NMB Se 385; S to V, right M3 NMB Se 758; W to Z, left M3 NMB Se 1440. *Gallogoral meneghinii*, O to R, right M3 NMB Se 388. A, G, H, L, R, V, Z, occlusal view; B, C, I, M, O, S, W, buccal view; D, J, N, P, T, X, lingual view; E, K, Q, V, Y, distal view; F, mesial view. Scale 1 cm

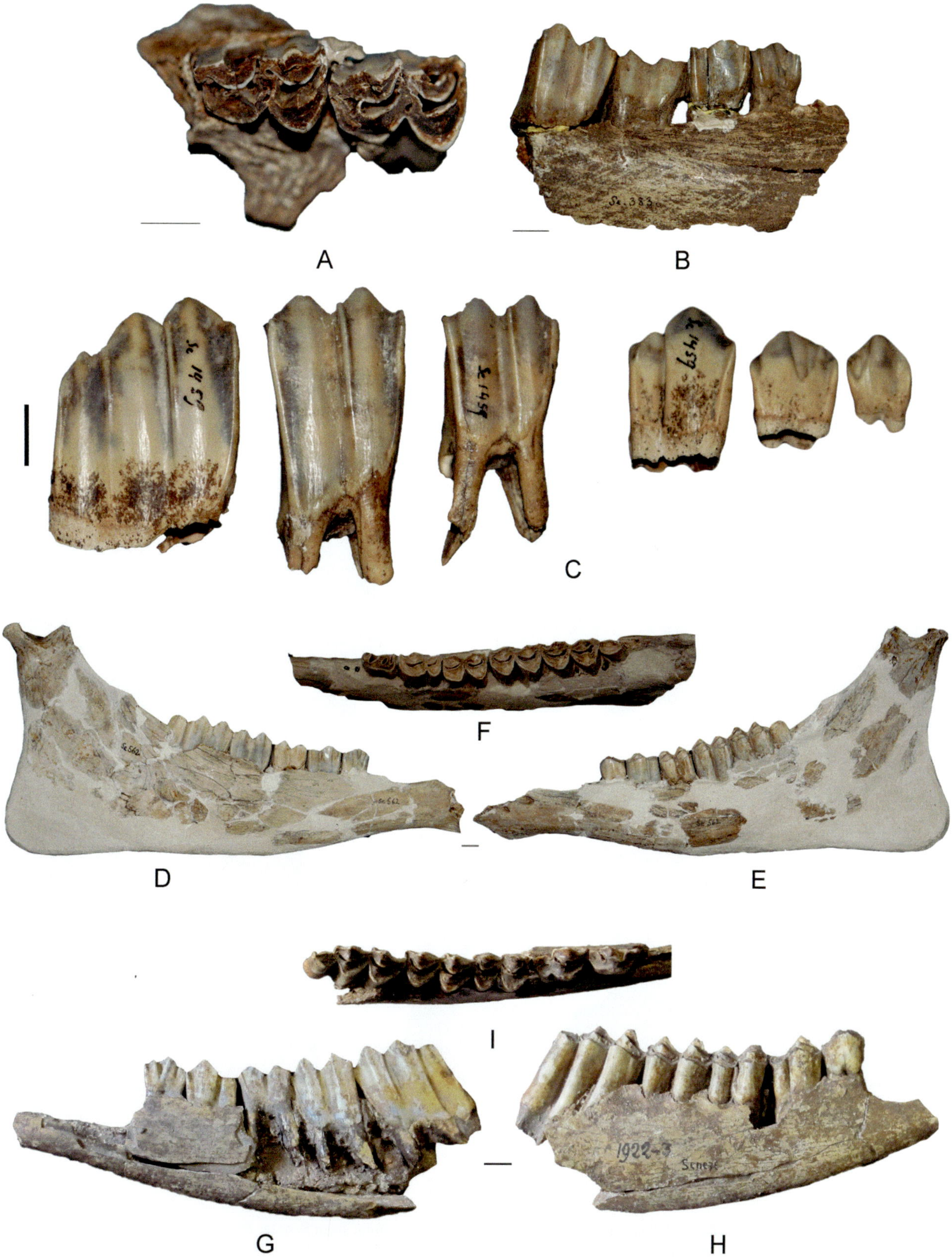

Fig. 14.31 Maxilla, mandibles and lower teeth of *Pliotragus ardeus* from Senèze. A, right maxilla NMB Se 1936; B, left mandible NMB Se 383; C, left p2-m3 NMB Se 1459; D to F, left mandible NMB Se 562; G to I, right mandible MNHN-F-1922-3. A, F, I, occlusal view; B, C, D, G, lingual view; E, H, buccal view. Scale 1 cm

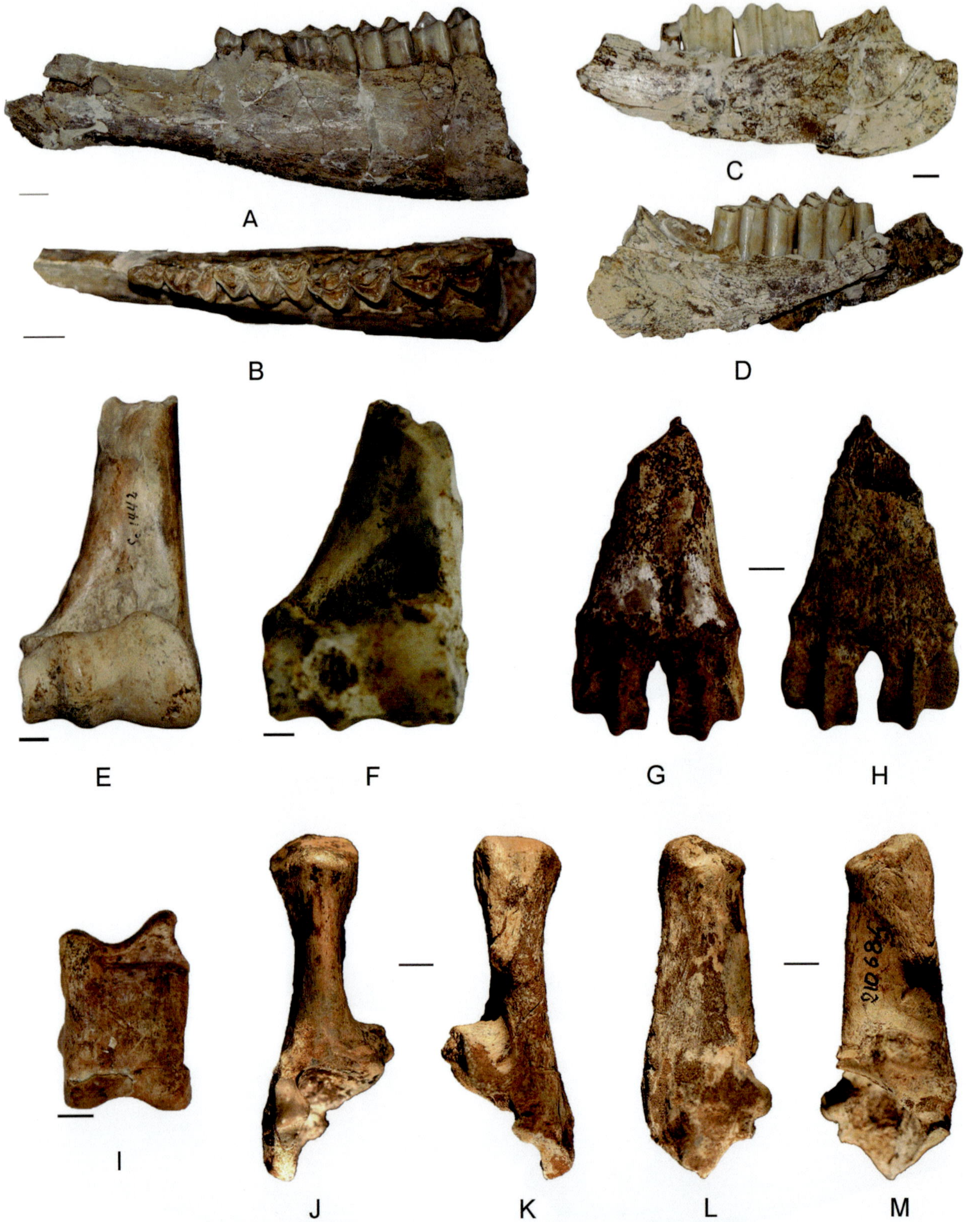

Fig. 14.32 Mandibles and post-cranium of *Pliotragus ardeus* and *Gallogoral meneghinii* from Senèze. *Pliotragus ardeus*, A to E, J to H: A, B, left mandible NMB Se 1719; C, D, right mandible NMB Se 1733; E, right humerus NMB Se 1442; J to M, right calcaneus FSL 210684; G, H, left metatarsal III+IV NMB Se 201. *Gallogoral meneghinii*, I, right talus NMB Se 1672. Cf. *Pliotragus*, F, right humerus NMB Se 1361. A, D, buccal view; B, C, lingual view; E, F, cranial view; H, I, K, plantar view; G, J, dorsal view; L, lateral view; M, medial view. Scale 1 cm

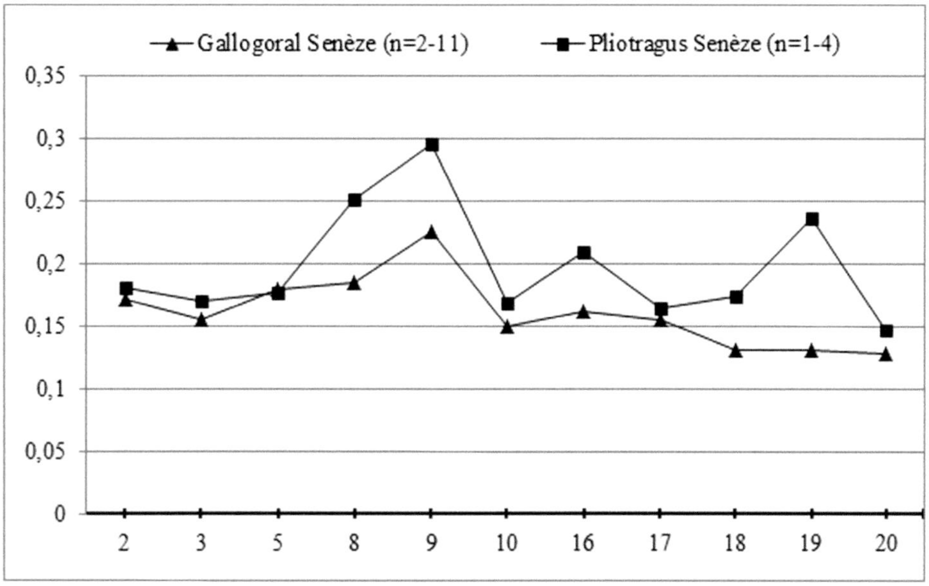

Fig. 14.33 *Pliotragus ardeus* and *Gallogoral meneghinii*. Ratio diagram of the mandible. Reference axis: recent *Capra ibex* ♂ (in Crégut-Bonnoure 2020). Length: 2, from infradental to condyle; 3, horizontally, from m3 posterior side to the distal side of ramus; 5, from infradental to m3 posterior side 8, total row teeth; 9, premolar row; 10, molar row; 20: from the foramen posterior side to the p2 anterior side. Height: 16, horizontal ramus, p2 level; 17, horizontal ramus, p4-m1 level; 18, horizontal ramus, m3 level; 19, horizontal ramus, diastema level. References of the measurements in Fig. 14.A.3 and dimensions in Table 14.28

Table 14.28 Comparative dimensions (in mm) of the mandible in *Pliotragus ardeus* and *Gallogoral meneghinii* from Senèze

Variable	1	2	3	4	5	6	7	8	9	10
P. ardeus										
FSL 3935 bis										74.0
FSL 16887									42.8	
FSL 16886										73.5
FSL 16891										66.8
NMB Se 562		300.0	78.6		197.5			128.4	44.6	83.4
G. meneghinii										
N	4	4	8	6	4	6	2	9	11	9
Mean	273.6	293.4	76.1	76.52	198.9	185.6	19.5	110.5	37.2	71.2
Minimum	243.2	262.3	67.8	75.3	173.2	178.5	15.1	104.6	24.6	67.3
Maximum	280	305	84.8	87.2	207.8	193.4	23.9	119.1	46.4	73.8
S.D	9.11	16.35	5.82	1.15	12.59	6.66	6.26	5.37	6.50	2.37
Variable	11	12	13	14	15	16	17	18	19	20
P. ardeus										
FSL 3935 bis						31.9	38.5		26.0	32.3
FSL 16891							30.2	46.3		
NMB Se 562						29.1		59.2	25.3	41.3
NMB Se 1719						31.9	38.5		26.0	32.3
G. meneghinii										
N	3	6	5	2	5	10	10	8	10	7
Mean	76.4	97.3	96.8	145.3	26.9	28.2	35.0	47.7	20.1	35.5
Minimum	70.6	93.1	90.3	141.5	22.7	23.3	33.1	43.0	18.4	29.4
Maximum	82.2	103.1	102.2	149.0	30.6	33.3	38.7	60.1	22.7	41.3
S.D	8.20	4.68	5.97	5.25	3.97	3.43	1.26	3.12	1.61	5.73

Variable definitions (from Crégut-Bonnoure 2002, 2020) (Fig. 14.A.3): Length: 1, from infradental to the posterior margin of the *processus angularis*; 2, from infradental to the caudal side of the condyle; 3, horizontally from m3 distal side to distal side of ramus; 4, from m3 distal side to the angle of the mandible; 5, from infradental to the distal side of m3; 6, from p2 mesial side to the posterior margin of the *processus angularis*; 7; from mesial side of I1 to distal side of canine; 8, from p2 mesial side to m3 distal side; 9, from p2 mesial side to p4 distal side; 10; from m1 mesial side to m3 distal side; 11, of diastema, from distal side of canine to mesial side of p2; Height of the ramus: 12, from the base to the condylar process; 13, from the base to the mandibular notch; 14: from the base to the coronoid process. Transversal diameter: 15, of the mandibular condyle. Height of the corpus: 16, at the p2 level. 17, at the p4-m1 level; 18, at the m3 level; 19, at the diastema level; 20, from the distal side of the mandibular foramen to the mesial side of p2

Teeth

Upper Teeth

Because of the great resemblance with *Gallogoral*, some specimens have previously been attributed to that genus. Correctly identified *Pliotragus* upper teeth are: those from the maxilla FSL 16867 (left P4-M3; initially assigned to *Gallogoral*; Fig. 14.28: A, B), FSL 16887 (right M2-3), the six isolated teeth of FSL 16868 (left M1; left and right M2 from a single individual; three M3 [one left, two right], initially assigned to *Gallogoral*), FSL 16895 (right P3-M3), NMB Se 1936 (right M2-3; initially assigned to *Gallogoral*), NMB Se 197 (left P4; initially assigned to *Antilope* sp.), NMB Se 384 (right M1-3; initially assigned to *Gallogoral*), NMB Se 385 (right M2-3; initially assigned to *Gallogoral*), NMB Se 1431 (left M2-M3; initially assigned to *Gallogoral*), NMB Se 71 (left M3; initially assigned to *Gallogoral*), NMB Se 390 (left M3; initially assigned to *Gallogoral*), NMB Se 758 (right M3; initially assigned to *Gallogoral*), the teeth from maxilla MNHN-F-1923-3 (right P3-M3), MNHN-F-1921-9 (right M2), a right M2 (MNHN-F-unnumbered) and a right M3 (MNHN-F-unnumbered). There are a total of 38 teeth attributable to at least 11 individuals (Tables 14.29, 14.30 and 14.31).

On P3 (n = 2), the buccal valley is deep and doesn't reach the base of the crown, stopping at more or less 6 mm. The parastyle is stronger than the metastyle, respectively 2.3 and 1.2 mm for MNHN-F-1923-3. Lingually, there is a longitudinal valley which distinguishes the mesial part from the distal one, this last forming as a small column. On P4 (n = 4) (Fig. 14.30: A, B), the styles, divergent at the top, are well projected outside and asymmetric, the parastyle being stronger than the metastyle (respectively, on NMB Se 197: 2.9 and 1.9 mm; on FSL 16867: 3.6 and 1.3 mm; on MNHN-F-1923-3: 3.1 and 1.1 mm). The buccal surface is in a more internal position and the rib is inflated but with unclear edges. On each side of the rib, there is a valley, the mesial shorter than the distal one; however, they are only weakly developed on MNHN-F-1923-3. The lingual side is flat at the top, with the mesial angle rounded, especially at the base, whereas the distal angle is less developed; consequently, the lingual side is asymmetrical, the mesial side being wider than the distal. On MNHN-F-1923-3, there is a longitudinal valley separating the mesial part from the distal side.

There are five M1, 12 M2 and 15 M3 (Fig. 14.30). The first lobe of M1 and M2 is pinched, but less than in *Gallogoral*. The second lobe is more or less rounded, sometimes slightly pinched. The crown is straight with a mesostyle which is particularly prominent from the top to the base. The parastyle is strong, the metastyle is weak. The valley behind the parastyle doesn't reach the base of the crown. The metacone is narrow and well inflated whereas the paracone is diffuse. There is an occlusal islet, small and rounded. It could be elongated as on NMB Se 385. The M3 has a straight metastylar wing (Fig. 14.30: D, J, P, T, X); however, on MNHN-F-1923-3 it is not developed. The first lobe is lingually pinched; the second lobe is slightly pinched near the lingual valley. The distal side of the second lobe, anterior to the metastyle, is rectilinear, perpendicular to the sagittal plane. The linguo-distal angle of the second lobe is pinched

Table 14.29 Comparative dimensions (in mm) of the upper premolars in *Pliotragus ardeus* from Senèze and other localities

P2 variable	Ht	1	2	3	5	
Perrier (NMB P 522–cast)		12.9		12.3	11.7	
P3 variable	Ht	1	2	3	5	
Senèze						
FSL 16895		15.5		12.5		
MNHN-F-1923-3	18.2	16.5		12.8		
Perrier (NMB P 522–cast)		15.3			14.3	
P4 variable	Ht	1	2	3	4	5
Senèze						
MNHN-F-1923-3		16.1		13.7	12.2	
FSL 16867	17.7	14.8		16.7	14.2	19.3
FSL 16895		15.9		13.2	11.4	
NMB Se 197	24.6	15.7	13.4	12.5	14.9	
Slivnitsa						
NNHM-S SL 31		18.0	16.3	16.6	15.0	20.5
NNHM-S SL 33	28.8	17.9	16.8	16.2	13.9	19.7
Perrier						
NMB Perrier 522 (cast–L)		14.9		16.1	14.2	
NMB Perrier 522 (cast–R)	15.4	16.8		15.4	14.0	16.6
Olivola (NMB M8–cast)		14.1				

Variable definitions (from Crégut-Bonnoure 1995, 2002, 2020) (Fig. 14.A.4: A): 1, occlusal length; 2, basal length. 3, occlusal width total. 4, occlusal width in the middle. 5, basal width; L, left; R, right

Table 14.30 Comparative dimensions (in mm) of the M1 and M2 in *Pliotragus ardeus* from Senèze and other localities

M1 variable	Ht	1	2	3	4	5	6	7
Senèze								
MNHN 1923-3	16.3	20.8	15.6	16.5	16.1	14.6	14.5	18.8
FSL 16895		20.3		15.2	14.9	13.1	12.8	19.3
FSL 16867	10.2	19.3	18.1	18.6	18.3	18.6	18.2	20.6
FSL 16868	7.5	17.6	17.6	18.1	17.3	16.1	16.1	18.9
NMB Se 384	13.7	17.9	17.9	17.5	16.6	16.0	15.1	20.4
Perrier								
NMB 522 (L–cast)		19.6		17.9	17.3	15.8	15.5	
NMB 522 (R–cast)	11.8	19.1	15.5	17.2	17.1	16.6	16.5	18.3
M2 variable	Ht	1	2	3	4	5	6	7
Senèze N	11	11	11	11	11	11	11	9
Mean	22.41	23.42	21.45	17.31	16.41	14.94	14.04	20.87
Minimum	15.8	20.9	19.9	14.9	13.2	12.5	12.0	19.5
Maximum	30.4	26.7	24.1	18.7	17.8	16.0	15.9	22.3
SD	4.66	1.68	1.42	1.14	1.35	1.02	1.22	0.98
Slivnitsa								
NNHM-S SL 30	28.5	25.9	24.1	17.1	17.2	14.1	13.5	
NNHM-S SL 26	22.9	26.7		18.19	17.7	15.1	14.8	
NNHM-S SL 39	23.1	26.5						
Perrier								
NMB 522 (L–cast)		21.8		18.1	17.3			
NMB 522 (R–cast)		22.5		18.7	18.1	16.6	15.5	22.5
Chilhac (NMB PP 27)	21.2	23.9	20.9	15.0	14.6	13.0	12.6	17.6

Variable definitions (from Crégut-Bonnoure 1995, 2002, 2020) (Fig. 14.A.4: B): Ht, height; L, length; O, occlusal; B, basal; m, in the middle; W, width; 1, first lobe; 2, second lobe. M1 and M2: 1, occlusal length; 2, basal length. 3, occlusal width total, first lobe; 4, occlusal width in the middle, first lobe; 5, occlusal width total, second lobe; 6, occlusal width in the middle, second lobe; 7, basal width; L, left; R, right

Table 14.31 Comparative dimensions (in mm) of the M3 in *Pliotragus ardeus* from Senèze and other localities

M3 variable	Ht	1	2	3	4	5	6
Senèze N	11	14	10	14	14	10	13
Mean	24.1	24.7	25.3	15.9	15.2	21.1	13.3
Minimum	18.0	23.5	23.0	12.6	12.1	18.9	10.6
Maximum	32.1	27.8	27.3	17.5	17.1	23.1	15.2
SD	4.00	1.06	1.28	1.32	1.33	1.28	1.09
Slivnitsa N	6	7	6	7	7	6	7
Mean	27.0	26.2	26.32	17.3	16.2	22.1	14.3
Minimum	19.2	22.3	19.7	15.3	14.9	15.5	12.7
Maximum	33.5	28.9	31.5	18.9	18.0	25.8	15.5
SD	5.26	2.29	4.24	1.32	1.17	3.62	1.16
M3 variable	7	8	9	10	11	12	
Senèze N	13	10	13	13	13	13	
Mean	12.8	18.0	8.9	9.8	22.2	21.5	
Minimum	10.2	15.3	8.0	8.0	21.2	20.1	
Maximum	14.9	20.0	10.4	11.3	23.4	23.4	
SD	1.28	1.56	0.75	0.93	0.75	1.14	
Slivnitsa N	7	5	6	6	5	5	
Mean	14.7	20.1	9.5	11.1	24.2	24.1	
Minimum	11.6	17.7	8.5	9.6	21.1	21.2	
Maximum	22.3	22.9	9.8	12.2	28.1	26.8	
SD	3.60	2.29	0.49	1.20	2.52	2.44	

Variable definitions (from Crégut-Bonnoure 1995, 2002, 2020) (Fig. 14.A.4: C): Ht, height.1, occlusal length; 2, basal length. 3, occlusal width total, first lobe; 4, occlusal width in the middle, first lobe; 5, basal width, first lobe; 6, occlusal width total, second lobe; 7, occlusal width in the middle, second lobe; 8, basal width, second lobe; 9, length of the interstylar surface, mesial; 10, length of the interstylar surface, distal; 11, occlusal length of the two lobes at the lingual side; 12, basal length of the two lobes at the lingual side with a vertical crest. The surface between the pinching and the crest is more or less flat and obliquely directed backward; in distal view the surface is oblique. Lingually the distal root is separated into two distinct units, the mesial being shorter and joined with the first lobe root. The separation is wide and prolonged on the base of the distal lobe by a kind of notch. On FSL 16887 the distal interstylar surface is concave; on the other specimens it is more or less flat.

Lower Teeth

The lower teeth are represented both by isolated specimens and those from mandibles. After reassigning several specimens previously allocated to *Gallogoral*, the sample of *Pliotragus* lower teeth includes: those from the mandibles FSL 3935 (left p4-m3 and right [3935bis] p3-m3), FSL 16886 (left m1-3; initially assigned to *Gallogoral*), FSL 16887 (right p2-m2), FSL 16891 (left m1-3), NMB Se 383 (left p3-m3), NMB Se 562 (right p3-m3; left p3-p4, m2-3), NMB Se 1459 (left dp2-4, p2-m3), NMB Se 1733 (right m2-3), NMB Se 1719 (left dp2-4, m1-2), MNHN-F-1922-3 (right p3-m3; right m3) and MNHN-F-cast 2659 (=NHMUK-P 12009) (right p2-3, m1-3); and isolated teeth: NMB Se 1357 (left p3), NMB Se 1929 (left p2-4), NMB Se 399 (right m2; initially assigned to *Gazellospira*), NMB Se 1668 (right m3, fragmentary), NMB Se 1725 (right m3, unworn) and MNHN-F-1922-? (right m1). A total of 63 teeth are available which represent at least eight individuals (Tables 14.32 and 14.33).

Two p2 (n = 4) are unworn (NMB Se 1459, NMB Se 1719) (Fig. 14.31: C). The buccal side is convex, without bilobation on NMB Se 1459 but bilobed on the other specimens. The parastylid is convex. The inflation of the metaconid is moderate and the surface from it to the distal valley is flat. This valley is short, its base stopped at 6/7 mm above the collar. The entoconid fold is found at the bottom of the valley. The p3 (n = 10), is weakly bilobed buccally except on NMB Se 1719 and on the cast MNHN-F-2659. The distal side is rectilinear without any enlargement of its base, except on NMB Se 562 and NMB Se 1357 on which the buccal side is distally rounded. There is a deep lingual valley, which stops at 7/8 mm above the collar (Fig. 14.31: C). Below this valley, the surface of the lingual side constitutes a flat wall, without any inflation. The mesial side of the valley is sinuous, except on MNHN-F-1922-3, the distal side is rectilinear. The parastylid is moderate (length 2 mm) and separated from the paraconid by a longitudinal valley, except on NMB Se 1459 and NMB Se 1357 on which the valley is restricted to the top. In occlusal view, the paraconid and the parastylid are distinct, the latter forming a small loop. The metaconid is little inflated, but more on FSL 16887. At the bottom of the valley, the protoconid is inflated. In occlusal view, on worn specimens, the metaconid forms a column fused with the entoconid. On the unworn NMB Se 1459, the hypoconid is projected backward; on the other specimens it is perpendicular to the sagittal plane. On p4 (n = 9), the lobes are clearly distinct (Fig. 14.31: C). The first lobe is longer and wider than the second. The first lobe is buccally rounded, especially at the base (and consequently when the tooth is worn). The second lobe is buccally pinched. The distal side is straight without any basal enlargement. The lingual valley is deeper than the buccal one. The metaconid is widely inflated. The entoconid is obliquely directed: consequently, the lingual wall of the second lobe is oriented obliquely inside the crown and this lobe is in a more internal position than the first which is projected outside. On the unworn NMB Se 1459 a small islet, originating from the protoconid, is visible on the buccal side of the first lobe.

Twelve m1 and 14 m2 have been identified (Fig. 14.31: C). The crown is rectilinear. There is no goat fold on the Lyon specimens or NMB Se 1733, but it is well developed on NMB Se 1459, NMB Se 1719 and the isolated m2 MNHN-F-1922-?. Its presence is correlated with tooth wear: on worn molars, it seems to disappear at the base of the crown. The interlobar column is absent except on six molars (4 m1, 1 m2, 1 m3): on the worn m1 from the mandible MNHN-F-1922-3 (9.3 mm high; Fig. 14.31: H), on the worn m1 from the mandible FSL 16887 (more than 4.7 mm high), on the m1 FSL 16886 (6.9 mm high), on the m1 from the mandible NMB Se 383 (11.3 mm high), on the m2 FSL 16887 (4.1 mm high) and on the incomplete m3 NMB Se

1668 (11.8 mm high).The lobes are lingually pinched and they are obliquely directed backward. However, the pinching is less important on NMB Se 1733. The cementoenamel junction is low on the isolated m2 FSL 16887. The base of the second lobe is prominent and the thickness is greater than that of the first lobe: on NMB Se 1459 the width is 14.4 mm for the first lobe and 15.1 mm for the second; on NMB Se 383 the values are respectively 14.8 and 17.2 mm. The width and depth of the lingual valley is variable. The parastylid and the metastylid are weak. The mesostylid is narrow (generally 1.3 mm), quite prominent and present on the top two third of the crown, disappearing at the base; on the unworn molars of NMB Se 1459 it runs to the base. Distal to the parastylid there is a relatively wide longitudinal valley which reaches the base of the crown. The inflation of the conids is moderate.

Fourteen m3 are known (Fig. 14.31: C). The third lobe is located in the sagittal plane of the crown, and its shape is polymorphic. In some specimens, it is triangular with a pinched lingual side and a wide distal side (7.2 mm wide on FSL 16886, 5.7 mm wide on FSL 16891 and 5.1 mm wide on FLS 3935 bis) with a longitudinal groove and a stylid lingually. Or it could be more rounded and more elongated as on NMB Se 562, NMB Se 1733, and MNHN-F-1922-3; however, these teeth are characteristic of the species (Fig. 14.31: G to I). Moreover, the distal side is rectilinear

Table 14.32 Comparative dimensions (in mm) of the lower premolars in *Pliotragus ardeus* from Senèze

p2 variable	Ht	1	3	5		
Senèze						
MNHN 2659 (cast)	11.2	11.0		7.1		
NMB Se 1459	13.8	11.0	9.9	7.1		
NMB Se 1929	11.3	11.6	11.2	5.6		
FSL 16887	11.1	12.9	12.5	8.3		
p3 variable	Ht	1	2	3	5	
Senèze N	10	10	10	10	10	
Mean	15.5	14.8	13.9	8.8	9.6	
Minimum	12.4	13.7	13	7.8	7.6-	
Maximum	19.0	16.2	15.1	10.4	10.7	
SD	2.28	0.81	0.64	0.72	0.86	
p4 variable	Ht	1	2	3	4	5
Senèze N	4	9	5	9	9	7
Mean	17.7	18.0	17.0	10.7	9.1	11.87
Minimum	9.7	15.3	16.3	9.8	6.9	11.0
Maximum	25.7	19.9	18.5	11.7	10.8	12.9
SD	6.52	1.35	0.96	0.67	1.29	0.56
Chilhac (NMB PP 32)	13.9	17.8	16.1	10.7	10.7	10.9

Variable definitions (from Crégut-Bonnoure 1995, 2002, 2020): (Fig. 14.A.4: D): Ht, height; 1, occlusal length; 2, basal length; 3, occlusal width, first lobe; 4, occlusal width, second lobe; 5, basal width

Table 14.33 Comparative dimensions (in mm) of the lower molars in *Pliotragus ardeus* from Senèze and other localities

m1 variable	Ht	1	2	3	4	5
Senèze N	10	11	4	11	10	6
Mean	14.2	18.8	17.7	12.0	12.2	13.8
Minimum	6.8	17.2	16.1	10.8	10.1	13.0
Maximum	24.9	21.5	19.2	13.2	14.6	14.9
SD	5.83	1.47	1.30	0.88	1.67	0.75
Slivnitsa (NNHM-S SL 40)	29.2	25.4	21.9	11.7	12	15.1
m2 variable	Ht	1	2	3	4	5
Senèze N	7	16	6	15	15	8
Mean	25.0	22.4	19.8	12.3	12.0	14.9
Minimum	14.6	20.4	18.7	10.5	10.1	13.1
Maximum	33.6	25.4	20.8	14.2	13.6	16.9
SD	7.86	1.27	0.91	1.02	1.17	0.93
Slivnitsa						
NNHM-S SL 28	32.4	25.3	23.3	12.4	12.3	16.7
NNHM-S SL 41	33.5	25.9	23.6	12.6	11.9	16.9

(continued)

Table 14.33 (continued)

m3 variable	Ht	1	2	3	4	5	6
Senèze N	5	12	3	13	11	11	3
Mean	31.1	31.7	31.7	11.8	11.0	6.6	13.8
Minimum	27.4	27.3	30.3	10.2	9.3	5.0	12.1
Maximum	34.3	36.0	34.6	13.6	12.7	7.5	15.0
SD	2.67	2.70	2.51	1.10	1.05	0.78	1.49

Variable definitions (from Crégut-Bonnoure, 1995, 2002, 2020): Ht, height; m1 and m2 (Fig. 14.A.4: E): 1, occlusal length; 2, basal length; 3, occlusal width, first lobe; 4, occlusal width, second lobe; 5, basal width. m3 (Fig. 14.A.4: F): 1, occlusal length; 2, basal length; 3, occlusal width, first lobe; 4, occlusal width, second lobe; 5, occlusal width, third lobe; 6, basal width

on NMB Se 1459 and convex on the isolated specimen MNHN-F-1922-3 with a stylid.

The proportions of the upper teeth have been compared across several populations of *Pliotragus* and *Gallogoral*. For all the molars, there are few differences between Senèze *Gallogoral* and *G. meneghinii sickenbergii* Kostopoulos, 1997a from Volax.[1]

Pliotragus M1s from Senèze and Perrier-Pardines do not have exactly the same proportions, but the variation is not important (Fig. 14.34). *Gallogoral* from Volax is more distinct. The proportions in *Pliotragus* and *Gallogoral* are quite comparable while the morphology is not the same (Fig. 14.34).

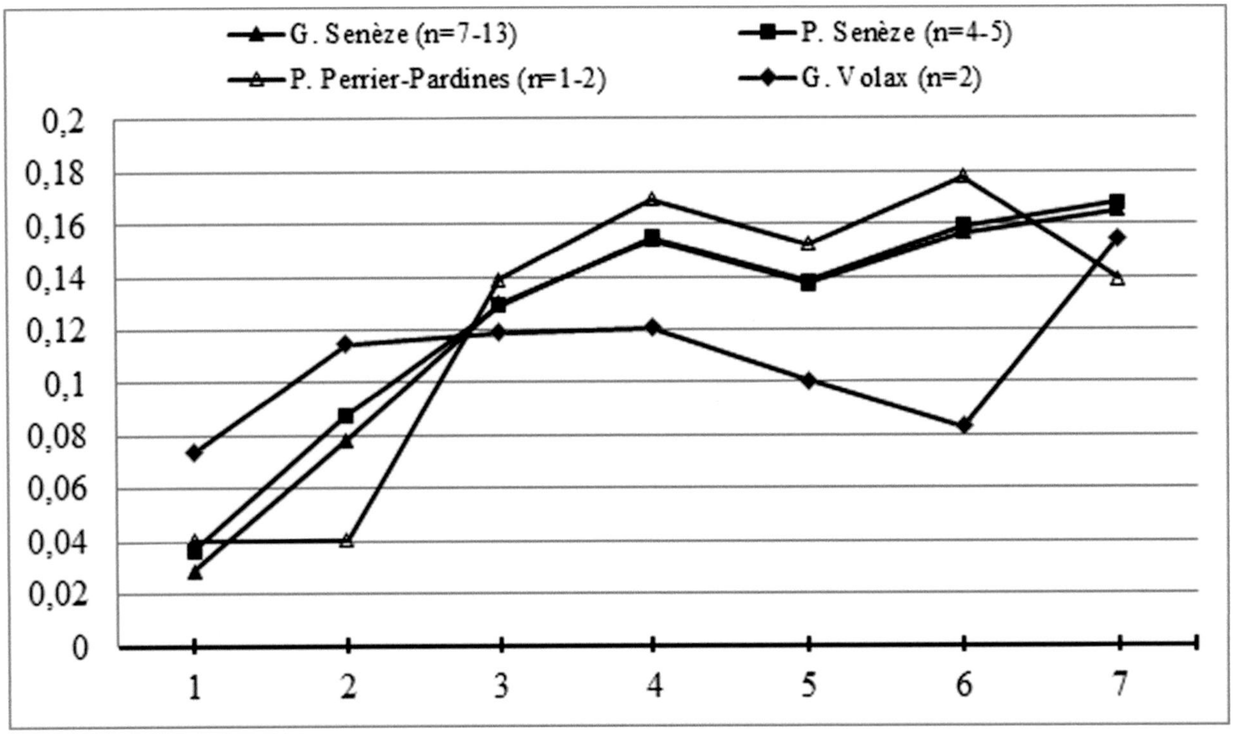

Fig. 14.34 *Pliotragus ardeus* and *Gallogoral meneghinii*. Ratio diagram of the M1. G: *Gallogoral*. P: *Pliotragus*. Reference axis: *Capra ibex* from the Observatoire cave (Monaco principality) (in Crégut-Bonnoure 2020). Length: 1, occlusal; 2, basal. Width: 3, occlusal, first lobe; 4, occlusal in the middle, first lobe; 5, occlusal, second lobe; 6, occlusal in the middle, second lobe; 7, basal. References of the measurements in Fig. 14.A.4 and dimensions in Tables 14.30 and 14.44

For M2 (Fig. 14.35), the Slivnitsa molars are longer than the Senèze ones (measurements 1 and 2) with a narrower second lobe (measurements 5 and 6) and a wider base (measurement 7). Compared to *Gallogoral*, the Senèze *Pliotragus* teeth are longer with lobes relatively narrower at

[1]Kostopoulos (1996, 1997a) named the new subspecies *G. m. sickenbergii* after O. Sickenberg, but he added two i's at the end, perhaps thinking of the species name *G. meneghinii*. That name, however, was in honor of Prof. Meneghini, whose name ended in an i. According to the rules of ICZN, the original subspecies name should be maintained even if it was created wrongly or misspelled.

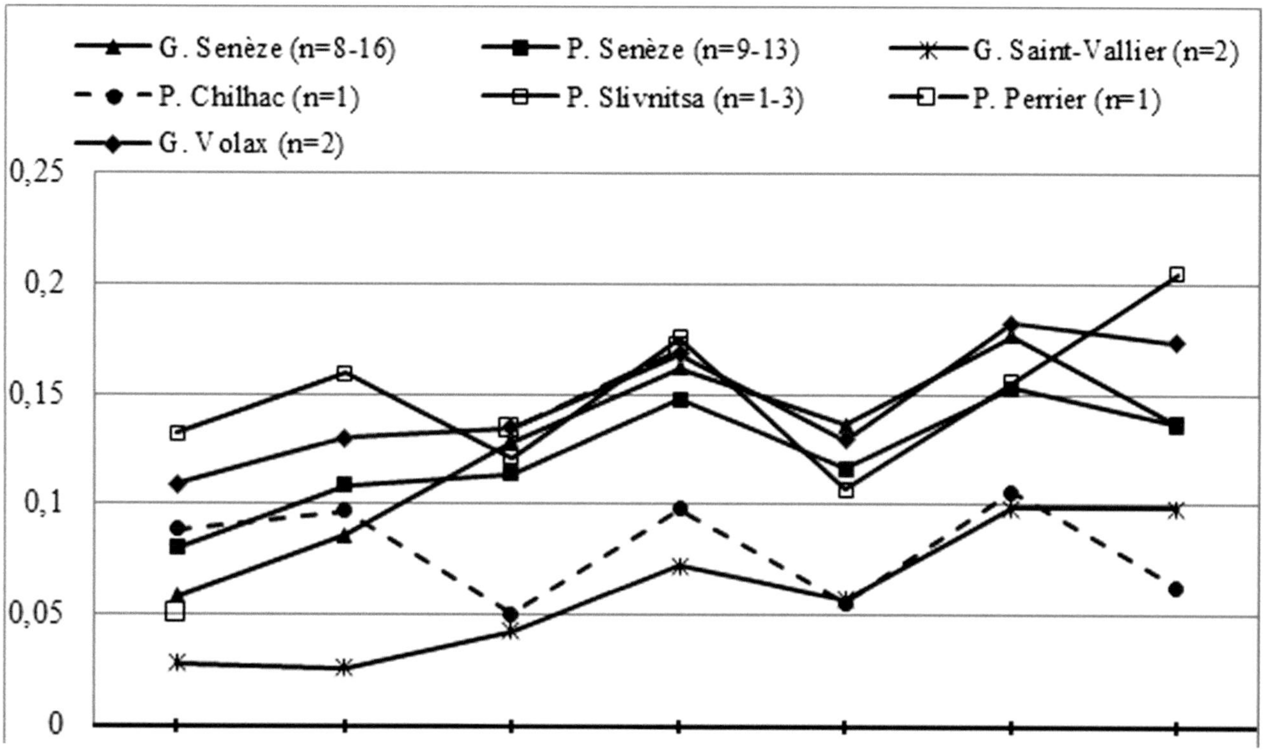

Fig. 14.35 *Pliotragus ardeus* and *Gallogoral meneghinii*. Ratio diagram of the M2. G: *Gallogoral*. P: *Pliotragus*. Reference axis: *Capra ibex* from the Observatoire cave (Monaco principality) (in Crégut-Bonnoure 2020). Length: 1, occlusal; 2, basal. Width: 3, occlusal, first lobe; 4, occlusal in the middle, first lobe; 5, occlusal, second lobe; 6, occlusal in the middle, second lobe; 7, basal. References of the measurements in Fig. 14.A.4 and dimensions in Tables 14.30 and 14.44

the occlusal level (measurements 3–6) but similar at the base (measurement 7).

The *Pliotragus* M3s from Senèze are slightly smaller than those of Slivnitsa (Fig. 14.36). The lobes are narrower and the distal interstylar surface is shorter (measurement 10). In comparison with *Gallogoral* from Senèze, the teeth are shorter at the occlusal level (measurement 1) with a wider base (measurements 5 and 8) and a first lobe wider in the middle (measurement 4), whereas the second lobe is narrower (measurements 6 and 7). The interstylar surfaces are slightly longer (measurements 9 and 10).

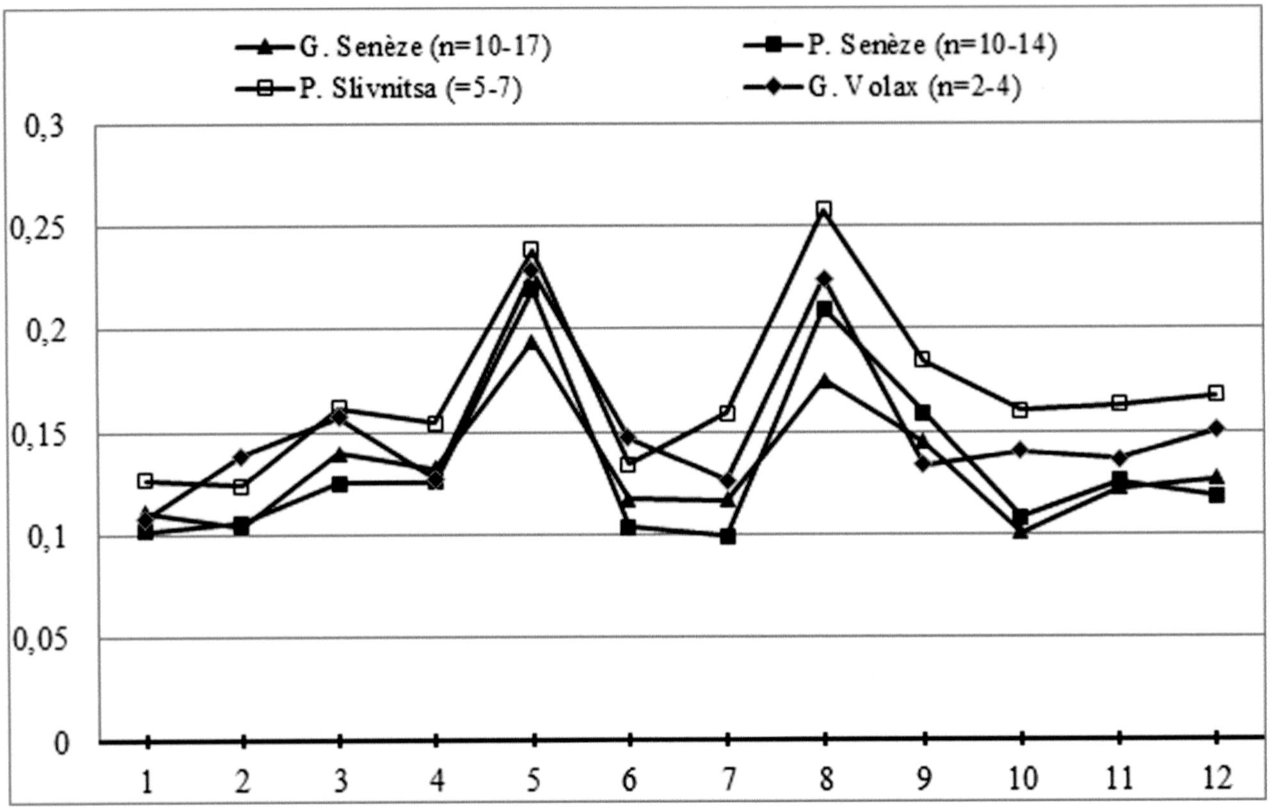

Fig. 14.36 *Pliotragus ardeus* and *Gallogoral meneghinii*. Ratio diagram of the M3. G: *Gallogoral*. P: *Pliotragus*. Reference axis: recent *C. pyrenaica* from Tortosa (Spain) (in Crégut-Bonnoure 2020). Length: 1, occlusal. 2, basal. 9, anterior interstylar surface. 10, posterior interstylar surface. 11, occlusal, first+second lobes (lingual side). 12, basal, first+second lobes (lingual side). Width: 3, occlusal, first lobe. 4, occlusal in the middle, first lobe. 5, basal, first lobe. 6, occlusal, second lobe. 7, occlusal in the middle, second lobe. 8, basal, second lobe. References of the measurements in Fig. 14.A.4 and dimensions in Tables 14.31 and 14.45

It is more difficult to compare the lower molars due to their scarcity in the localities. The proportions of m1s are quite similar in *Pliotragus* and *Gallogoral* from Senèze, but the formers are larger. The Slivnitsa tooth is longer than those of the Senèze population (measurements 1 and 2), with lobes slightly narrower at the occlusal level (measurements 3 and 4), but wider at the base (measurement 5) (Fig. 14.37).

The Slivnitsa m2s are longer than those of Senèze (measurements 1 and 2), whereas the lobes are similar at the occlusal level (measurements 3 and 4), while the base is wider (measurement 5) (Fig. 14.38). The teeth of these two populations differ from the Senèze *Gallogoral* in having a wider first lobe and base (measurements 3 and 5).

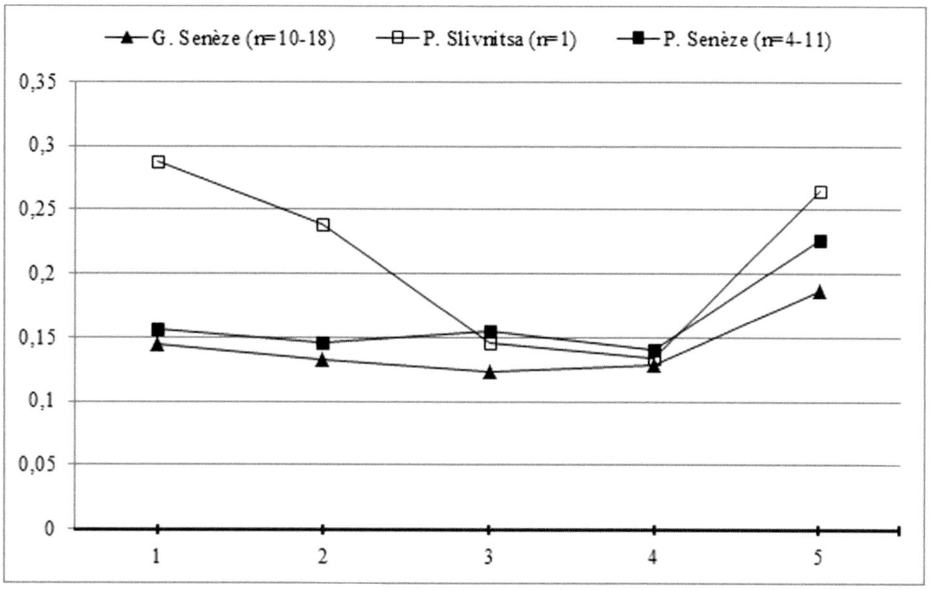

Fig. 14.37 *Pliotragus ardeus* and *Gallogoral meneghinii*. Ratio diagram of the m1. G: *Gallogoral*. P: *Pliotragus*. Reference axis: *Capra ibex* from the Observatoire cave (Monaco principality) (in Crégut-Bonnoure 2020). Length: 1, occlusal; 2, basal. Width: 3: occlusal, first lobe; 4, occlusal, second lobe; 5, basal. References of the measurements in Fig. 14.A.4 and dimensions in Tables 14.33 and 14.47

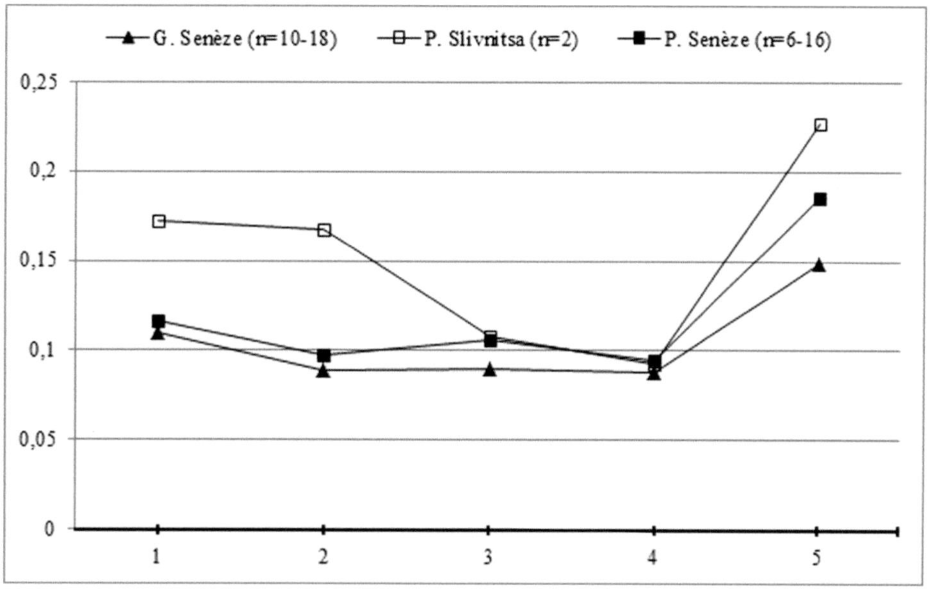

Fig. 14.38 *Pliotragus ardeus* and *Gallogoral meneghinii*. Ratio diagram of the m2. G: *Gallogoral*. P: *Pliotragus*. Reference axis: *Capra ibex* from the Observatoire cave (Monaco principality) (in Crégut-Bonnoure 2020). Length: 1, occlusal; 2, basal. Width: 3, occlusal, first lobe; 4, occlusal, second lobe; 5, basal. References of the measurements in Fig. 14.A.4 and dimensions in Tables 14.33 and 14.47

Pliotragus and *Gallogoral* m3s from Senèze are very similar (Fig. 14.39). The *Gallogoral* molars from Saint-Vallier and Volax have a third lobe which is thicker at the occlusal level (measurement 5) This difference is not significant because the number of specimens in each of these two localities is reduced to one.

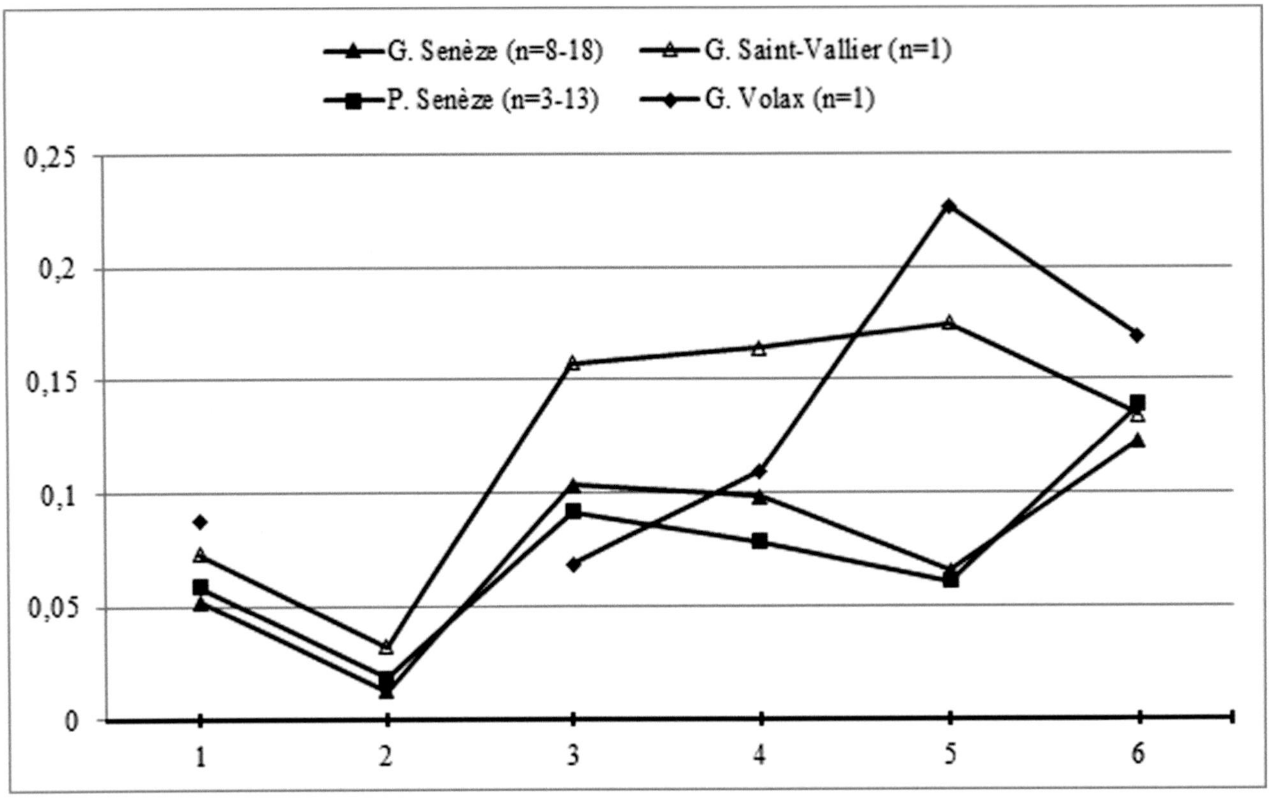

Fig. 14.39 *Pliotragus ardeus* and *Gallogoral meneghinii*. Ratio diagram of the m3. G: *Gallogoral*. P: *Pliotragus*. Reference axis: *Capra ibex* from the Observatoire cave (Monaco principality) (in Crégut-Bonnoure 2020). Length: 1, occlusal; 2, basal. Width: 3, occlusal, first lobe; 4, occlusal, second lobe; 5, occlusal, third lobe; 6, basal. References of the measurements in Fig. 14.A.4 and dimensions in Tables 14.33 and 14.45

Postcrania

The material is scarce, and there is no associated skeleton. Consequently, the postcrania were identified by comparison with *Gallogoral*.

Atlas

One vertebra, the atlas NMB Se 1515, has been attributed to *Megalovis* (Schaub, 1923: 294). This specimen is a little larger than in *Gallogoral*, smaller than the atlas of *Eucladoceros ctenoides senezensis* (=*Cervus senezensis*), *Leptobos etruscus* and the extant *Ovibos moschatus*. This specimen has been compared with these species and with the smaller *Procamptoceras* atlas.

Dorsal side:

- In front of the dorsal tubercle there are two concavities. It is the same in *Gallogoral* where they are less deep. In *Ovibos* they are missing and the tubercle is less developed. In *Leptobos* they are little developed and the tubercle is more prominent. In *Eucladoceros ctenoides senezensis* they are missing and the tubercle is reduced. This area is broken on *Procamptoceras*.
- The *foramina alare* are rounded, as in *Gallogoral* and *Leptobos*. In *Ovibos* they are transversally elongated as in *Procamptoceras*. In *Cervus* they are rostrocaudally elongated.
- The caudal angle of the wing is partially broken, but it seems that it was transversally directed. In *Gallogoral* as in *Leptobos* it is directed backward. In *Ovibos* there is no transversal development. In *Eucladoceros*, the backward development is important whereas it is weak in *Procamptoceras*.

Ventral side:

- The ventral tubercle is elongated, not prominent and located in the plane of the inter- articular notch. This morphology is similar to that of *Gallogoral*. In *Ovibos* the tubercle is rounded and very prominent; in *Leptobos* it is also prominent, but it is offset from the notch as in *Eucladoceros ctenoides senezensis* and *Procamptoceras*.
- On each side of the tubercle, there is a planum delimited rostrally by a crest. It is missing in the other species.
- The atlantal fossa is deep and its connection with the ventral side is abrupt, bounded by a crest which is rectilinear as in *Procamptoceras*. The fossa is less deep in

Gallogoral, there is no abrupt connection and there is no crest. In *Ovibos* there is no atlantal fossa and the ventral side is regularly convex. In *Leptobos*, the fossa is deep but the connection is less abrupt and the shape of the borders is curvilinear. In *Eucladoceros ctenoides senezensis* the fossa is deep and its connection with the ventral side is more abrupt than on NMB Se 1515, and the borders are curvilinear.

- The rostroventral notch is doubly curved as in *Procamptoceras* and *Gallogoral*. It is the same in *Ovibos* but the two concavities are deeper and a supplementary facet occupies the bottom. In *Leptobos*, the notch is "U" shaped, with a very large base. In *Eucladoceros ctenoides senezensis* it forms an arc of a circle.

To conclude, this atlas doesn't belong to a cervid or to *Leptobos*. It shows similarities with *Gallogoral* but the morphology of the ventral side differs. It is distinct from the extant musk ox. In the locality of Trlica where *M. balcanicus* has been identified, the axis is similar to *Ovibos* as are the ribs, the carpals and the tarsals (Crégut-Bonnoure & Dimitrijević 2006). I would think that the *Megalovis* atlas could be similar to that of the extant musk ox, which is not the case for the Senèze specimen. The width of the rostral articulation (79.80 mm) is not so different from the width of the *Pliotragus* condyles (77.1 and 74.6 mm). Consequently, I identify this vertebra as *Pliotragus*.

A second atlas, NMB Se 1516, is labeled ?*Deperetia ardea*. It is smaller than NMB Se 1515 but the morphology is similar. It could represent a female *Pliotragus* specimen.

Table 14.34 Comparative dimensions (in mm) of the scapula in *Pliotragus ardeus* and *Gallogoral meneghinii* from Senèze

Variable	1	2	3	4	5	6	7	8	9	10	11
P. ardeus											
Senèze (NMB Se 1637)					33.5	20.5	38.6	43.3	40	51.7	40.1
G. meneghinii											
Senèze N	4	2	2	4	5	6	6	2	3	4	5
Mean	256.0	150.4	47.0	118.9	35.9	22.6	39.4	41.8	38.9	53.6	38.4
Minimum	245.3	146.6	46.0	111.1	33.3	20.7	38.2	41.2	37.4	49.4	36.4
Maximum	273.8	154.2	48.0	129.2	39.1	25.6	40.5	42.4	39.9	57.1	40.4
SD	12.33	5.36	1.44	8.87	2.41	1.91	0.86	0.83	1.29	3.90	1.62

Variable definitions (from Crégut-Bonnoure 2002, 2020) (Fig. 14.A.5): 1, length; 2, total width; 3, supraspinatus width; 4, infraspinatus width; 5, minimal length of the neck; 6, minimal breath of the neck; 7, length from the base of the acromion to the lateral side of the glenoidal cavity; 8, articular length of the lateral side of the glenoidal cavity; 9, articular length of the medial side of the glenoidal cavity; 10, total length of the ventral articulation; 11, thickness of the ventral articulation

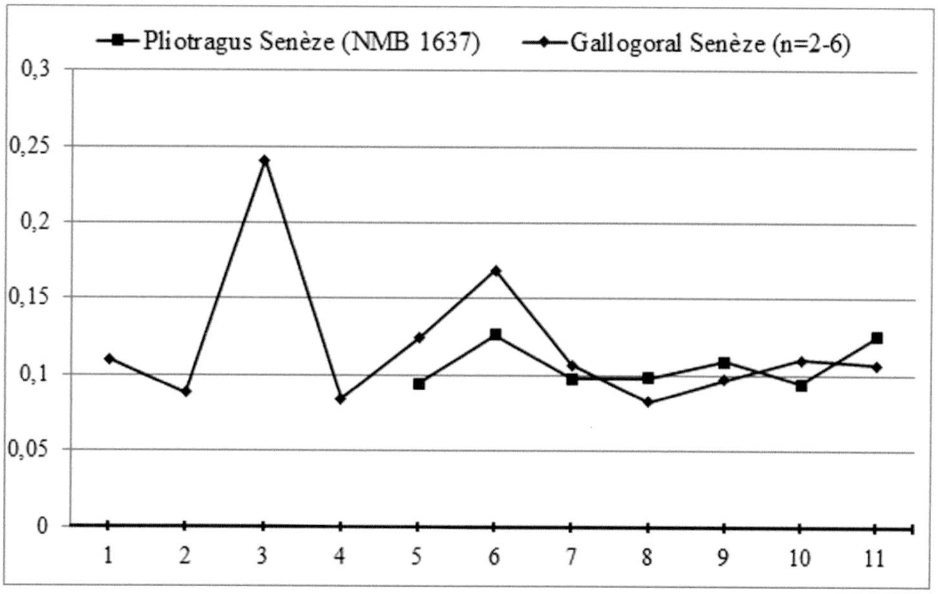

Fig. 14.40 *Pliotragus ardeus* and *Gallogoral meneghinii*. Ratio diagram of the scapula. Reference axis: recent *Capra ibex* ♂+♀ (in Crégut-Bonnoure 2020). 1, total length. Width: 2, total; 3, supraspinatus; 4, infraspinatus. Neck: 5, minimal length; 6, minimal width. Glenoid cavity: 7, length from the base of the acromion to the lateral side; 8, articular length of the lateral side; 9, articular length of the medial side. Ventral articulation: 10, total length; 11, thickness. References of the measurements in Fig. 14.A.4 and dimensions in Table 14.34

Scapula

NMB Se 1637 (right) is incomplete (cf. *Deperetia ardea* on label). It is bigger than in *Gallogoral* and doesn't have the same morphology (Table 14.34). Compared to *Gallogoral*, the lateral side, in the cranial area, is more concave at the level of the neck; the ventral tubercle is triangularly shaped with a medial development, whereas it is elongated in *Gallogoral* and located on the cranial side. The lateral side of the glenoid is pinched as in *Gallogoral*. This pinching is at 21.2 mm from the caudal angle of the cavity whereas in *Gallogoral* this distance is greater: 30.8 mm on NMB Se 1449. In cranial view, up the glenoid cavity, there is a kind of edge which is absent in *Gallogoral*. The supra-glenoid tubercle is well-developed, unlike in *Gallogoral*. The proportions between this specimen and *Gallogoral* are distinct (Fig. 14.40).

Humerus

Two incomplete left humeri have been attributed initially to the species (cf *Deperetia ardea* on label): NMB Se 1442 and NMB Se 1361 (Table 14.35). They preserve approximately the distal quarter (Fig. 14.32: E, F). The medial side of the distal articulation and the lateral lip of the trochlea (cranial side) are damaged in NMB Se 1361. They are similar in size, larger than in *Gallogoral* and *Gazellospira* but smaller than in *Megalovis*. Some morphological differences are visible between them. When the bone is placed on its distal articulation, the diaphysis of NMB Se 1442 is straight and stays in a vertical position because the base of the epicondyle is not angular; this part is broken on NMB Se 1361. However, the diaphysis, when the bone is on the ventral side, is more vertical on NMB Se 1442 than on NMB Se 1361, which is bent laterally. In lateral view, the caudal crest, back to the fossa, is pinched on NMB Se 1442, more rounded on NMB Se 1361; on the cranial side, the crest is thin and not prominent on Se 1442, being broader and more prominent on NMB Se 1361; the articular surface of the capitulum is rectilinear on NMB Se 1442, oblique on NMB Se 1361. The base of the lateral side is preserved only on NMB Se 1442: the lateral epicondyle is angled just before its junction with the capitulum. The coronoid fossa does not reach the medial side, and it is elongated in the longitudinal plane of the diaphysis on NMB Se 1442; it is elongated to the medial side on NMB Se 1361. The olecranon fossa is triangular shaped on NMB Se 1442, asymmetrical on NMB Se 1361 with an oblique lateral edge and a more elongated fossa. The base of this side is delimited by a transverse groove on NMB Se 1442; it is oblique on NMB Se 1361. In cranial view, the lateral side of the articulation below the olecranon fossa, at the junction point with the base of the lateral side, is pinched and acute on NMB Se 1361, not on NMB Se 1442. The epicondylar crest is less developed on NMB Se 1361 and

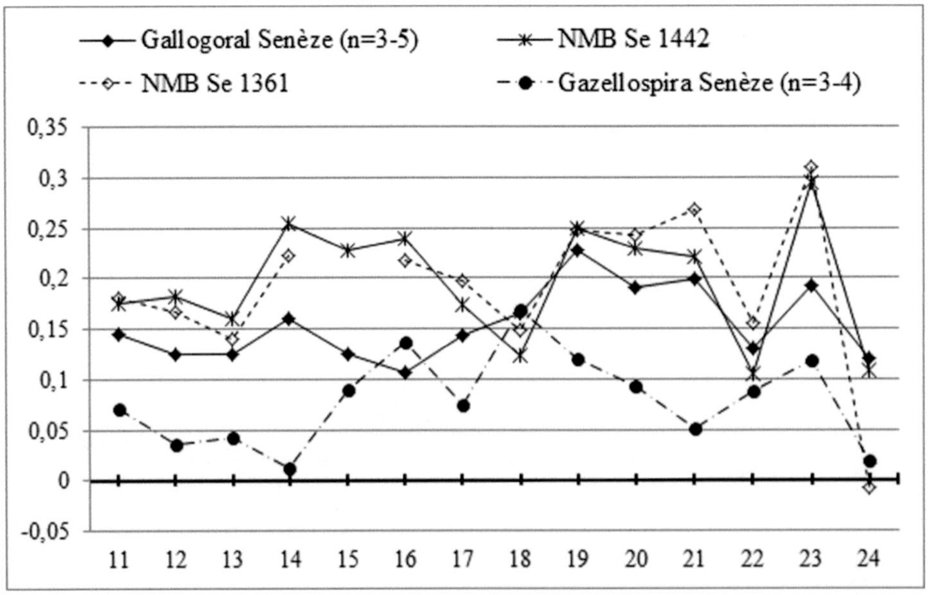

Fig. 14.41 Cf. *Pliotragus ardeus* (NMB Se 1442 and 1361), *Gallogoral meneghinii* and *Gazellospira torticornis*. Ratio diagram of the humerus. Reference axis: recent *Capra ibex* ♂+♀ (in Crégut-Bonnoure 2020). ccd, craniocaudal diameter; td, transversal diameter. Length: 1, total; 2, from proximal articulation. Proximal epiphysis: 3: td total; 4: ccd. Proximal articulation: 5, td; 6, ccd; 7, dt minimum, at the intertubercular furrow level; 8, td of the major tuber. Middle of the diaphysis; 9, td; 10, ccd. Transversal diameter distal: 11, maximal; 12, articular; 13, trochlea; 14, capitulum. Ccd distal: 15, medial side, total; 16, medial side, articulation; 17, lateral side, total; 18, lateral side, articulation; 19, trochlea, minimum; 20, trochlea, median guiding crest. Diaphysis distal, sus articular: 21, td; 22, ccd. Olecranon fossa: 23, td; 24: height. References of the measurements in Fig. 14.A.6 and dimensions in Tables 14.7 and 14.35

Table 14.35 Comparative dimensions (in mm) of the humerus in cf. *Pliotragus ardeus* and *Gallogoral meneghinii* from Senèze

Variable	1	2	3	4	5	6	7	8	9	10	11	12
cf. *P. ardeus*												
Senèze (NMB Se 1442)											60.6	59.2
Senèze (NMB Se 1361)											61.3	57.3
G. meneghinii												
Senèze												
NMB Se 313 (R)	280.2	249.0	73.1	83.1	63.5	53.0	54.1	58.7	30.7	35.6	56.4	52.0
NMB Se 313 (L)	279.2	241.9	69.4	78.4	62.8	49.7	53.3	65.8	28.1	35.2	53.4	50.1
NMB Se 361											61.3	55.4
NMB Se 1626												49.8
FSL 5008											53.4	52.2

Variable	13	14	15	16	17	18	19	20	21	22	23	24
cf. *P. ardeus*												
Senèze (NMB Se 1442)	38.5	22.0	57.1	40.3	44.9	23.5	30.6	35.4	37.2	28.3	23.3	22.2
Senèze (NMB Se 1361)	36.7	20.4		38.5	47.5	24.8	30.5	36.5	41.5	31.8	24.2	17.0
G. meneghinii												
Senèze												
NMB Se 313 (R)	35.2	17.3	46.0	26.51	39.9	25.7		30.8	35.8	29.5	16.0	22.3
NMB Se 313 (L)	36.9	17.6	46.1	28.87	40.5	27.4		31.6	37.6	31.5	16.8	
NMB Se 361	36.9	19.6			48.0		30.6	35.4	38.6	31.5	23.8	24.5
NMB Se 1626	33.5	16.2	42.0				29.8	33.3				
FSL 5008	35.0	17.6	46.2	33.72	39.3	24.2	26.6	30.7	29.7	27.7	17.2	21.9

Variable definitions (from Crégut-Bonnoure 1995, 2002, 2020) (Fig. 14.A.6): ccd, craniocaudal diameter; td, transversal diameter. Length: 1, total; 2, from proximal articulation. Proximal epiphysis: 3: td total; 4: ccd. Proximal articulation: 5, td; 6, ccd; 7, dt minimum, at the intertubercular furrow level; 8, td of the major tuber. Middle of the diaphysis; 9, td; 10, ccd. Transversal diameter distal: 11, maximal; 12, articular; 13, trochlea; 14, capitulum. Craniocaudal diameter distal: 15, medial side, total; 16, medial side, articulation; 17, lateral side, total; 18, lateral side, articulation; 19, trochlea, minimum; 20, trochlea, median guilding crest. Diaphysis distal, supra-articular: 21, td; 22, ccd. Olecranon fossa: 23, td; 24: height; L, left; R, right

disappears distally. On NMB Se 1442 it is prominent and curved distally. The diaphysis is slender on NMB Se 1442, and the articulation is more elongated. It is massive on NMB Se 1361, and the trochlea is shorter than on NMB Se 1442.

These two specimens are characterized by a capitulum longer than in *Gallogoral* and *Gazellospira* (Fig. 14.41: measurement 14). NMB Se 1361 has a distal epiphysis a little longer than NMB Se 1442 whereas its distal articulation is a little smaller (measurements 11 and 12). Its trochlear lateral lip is bigger (measurement 20) and its diaphysis is broader up the distal articulation (measurement 21). The height of its olecranon fossa is smaller (measurement 24).

To conclude, these two humeri are of similar size but with some distinct morphology and proportions. NMB Se 1361 shows some similarities with the extant *Ovis* with regard to the structure of the epicondylar crest and the obliquity of the articular surface of the capitulum. However, it is larger than the humerus of the present *Ovis ammon* and Middle Pleistocene *O. a. antiqua*, and its proportions are different. Without any skeleton associated with a *Pliotragus* skull, it is impossible to know the morphology of the humerus of this species; these two specimens are insufficient to check for real statistical differences and to decide whether the differences are due to sexual dimorphism or to an evolutionary factor or if there are two distinct species. As the morphological and the size differences between these two specimens are not very important, I cautiously refer them to *Pliotragus* as on the label.

Radius

A complete radius (FSL 16889, left; Guérin 1965: pl. XLIII p. 199) and a fragment belonging to an immature individual (a distal articulation with pyramidal and pisiform facets) (NMB Se 1761, right), were assigned to the genus by Schaub, (1943, pp. 280–281). A comparison with *Gallogoral* is provided by Guérin (1965). Their dimensions are in Table 14.36.

Pliotragus and *Gallogoral* radii have morphological similarities but also differences. In *Pliotragus,* the tuberosity for the lateral ligament of the humero-radial articulation is prominent laterally and the surface between it and the lateral side of the proximal articulation is more or less concave. On the *Gallogoral* radius the tuberosity is less prominent and its surface is flatter. On the dorsal side, near this tuberosity, the *Pliotragus* bone has an ovoid depression which is more marked than in *Gallogoral*. The articulation for the capitulum is in a higher position and palmarly pinched in *Pliotragus*; it is less high and there is no pinching in *Gallogoral*. Distally, the lateral side in contact with the ulna shows a crest in *Pliotragus*, but not in *Gallogoral*.

Valea Grăunceanului is the only site with radii attributed to *Pliotragus*. A complete specimen was assigned to this genus by Rădulesco (2005) but no figures or measurements are presented. Bolomey (1965: p. 319) described two other specimens that were later assigned to the giraffid *Mitilanotherium inexspectatum* Samson and Rădulesco, 1966 (see Rădulesco 2005). Recently, Croitor et al. (2023) reported the presence of three radii, of which one is complete and may correspond to the specimen reported by Rădulesco. All the other radii of the collection belong to the new cervid *Rucervus radulescui* Croitor, 2018. The morphology of the complete radius from Valea Grăunceanului, which can be identified as a male individual by its dimensions, is similar to that of the Senèze specimen (Croitor et al. 2023: Fig. 7, p. 7).

Compared to *Gallogoral*, the *Pliotragus* specimens are characterized by a broader transverse diameter of the articulation for the ulna (measurement 10) (Fig. 14.42). The pyramidal articular facet is also broader (measurement 24) than the caudal part of the facet for the semi-lunar (measurement 21).

Table 14.36 Comparative dimensions (in mm) of the radius and ulna in *Pliotragus ardeus* and *Gallogoral meneghinii* from Senèze and Slivnitsa

Radius variable	1	2	3	4	5	6	7	8	9	10	11	12
P. ardeus												
Senèze (FSL 16889)	263.8	269.5	60.1	58.0	41.7	14.3	30.1	19.1	23.5	40.0	13.5	10.7
G. meneghinii												
Senèze N	5	5	6	6	6	6	6	4	6	4	4	4
Mean	237.9	236.9	53.4	50.6	34.5	15.7	27.2	18.5	21.5	30.8	9.3	7.9
Minimum	135.0	231.0	50.7	49.1	32.3	13.5	25.1	16.4	20.9	28.6	8.3	6.8
Maximum	240.7	243.3	55.9	52.7	37.5	17.7	29.7	20.8	22.3	33.2	10.4	10.1
SD	2.22	5.66	1.96	1.35	1.69	1.41	1.58	1.82	0.49	1.88	0.89	1.53
Radius variable	13	14	15	16	17	18	19	20	21	22	23	24
P. ardeus												
Senèze												
FSL 16889	11.4	36.5	22.1	55.8	37.5	49.4	43.5	21.2	15.2	34.6	15.7	7.3
NMB Se 1761				51.6	39.8							6.9
Slivnitsa (NNHM-S SL8)				53.8	38.5	44.1	40.1	15.9	11.3	32.4	16.9	4.6
G. meneghinii												
Senèze N	4	5	5	5	5	3	3	3	3	3	3	3
Mean	9.5	32.9	20.3	49.6	35.3	40.8	38.4	18.18	11.2	30.1	14.4	4.6
Minimum	8.6	31.9	19.3	48.0	34.8	40.3	37.8	17.3	10.4	28.9	12.7	3.4
Maximum	11.0	34.2	21.0	51.0	35.8	41.2	39.5	18.6	12.6	30.8	16.1	5.8
SD	1.09	0.89	0.67	1.25	0.40	0.47	0.99	0.72	1.26	1.03	1.71	1.18
Ulna variable	1	2	3	4	5	6	7	8	9	11	12	
G. meneghinii												
Senèze												
FSL 16879		38.2	16.8		36.4	43.8	16.5	10.0	25.1		27.2	
NMB Se 313 (L)	320.0	41.9	21.6		38.7		17.5			16.7	15.9	
NMB Se 313 (R)	310.0	42.1	21.9	69.4	38.3	44.1	15.8	19.3				

Variable definitions (from Crégut-Bonnoure 1995, 2002, 2020): ccd: craniocaudal; dpd, dorsopalmar diameter; td; transversal diameter. Radius (Fig. 14.A.7): Length: 1, total; 2, median. Proximal epiphysis: 3, td total; 4, td articular; 5, td of the facet for the trochlea articulation; 6, td of the facet for the capitulum articulation; 7, dpd of the facet for the trochlea articulation; 8, dpd of the facet for the capitulum articulation; 9, dpd minimum; 10, td of the facets for the ulna; 11, td of the lateral facet for the ulna, 12; td of the median facet for the ulna; 13, td of the medial facet for the ulna. Middle of the diaphysis: 14, td; 15, dpd. Distal epiphysis: 16, td total; 17, dpd; 18, td articular; 19, td of the facets for the scaphoid and the semi-lunar; 20, td of the facet for the scaphoid; 21, td caudal of the facet for the semi-lunar; 22, maximal dpd of the facet for the scaphoid; 23, length of the crest between the facets for the scaphoid and the semi-lunar; 24, td of the facet for the pyramidal. Ulna (Fig. 14.A.8): 1, length; 2, dpd of the tuber olecrani; 3, td of the tuber olecrani; 4, length from the top of the tuber olecrani to the proximal angle of the anconeal process; 5, dpd minimal between the top of the tuber olecrani and the proximal angle of the anconeal process; 7, td minimal perpendicular to measurement 5; 8, td of trochlea incisura; 10: td of the articulation for radius; L, left; R, right

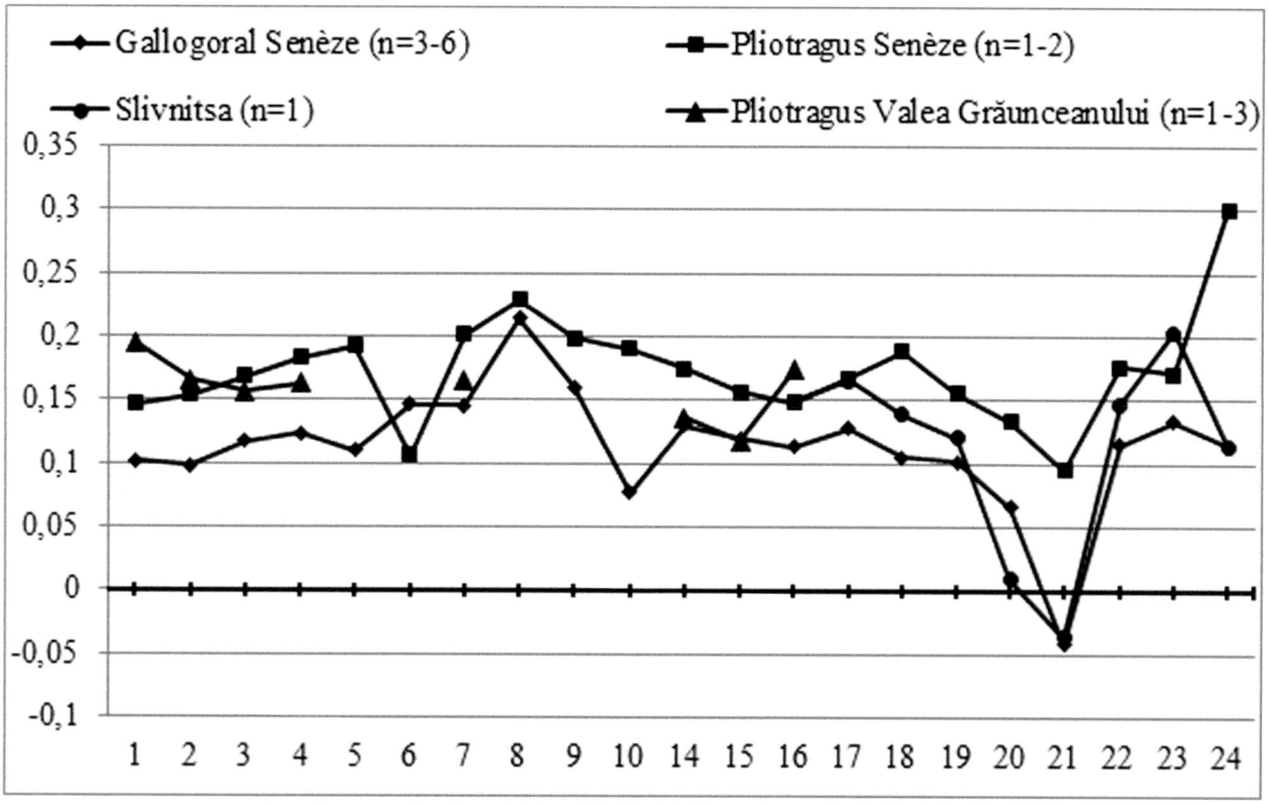

Fig. 14.42 *Pliotragus ardeus* and *Gallogoral meneghinii*. Ratio diagram of the radius. Reference axis: recent *Capra ibex* ♂+♀ (in Crégut-Bonnoure 2020). dpd, dorsopalmar diameter; td; transversal diameter. Length: 1: total. 2: median. Proximal epiphysis: 3, td total; 4, td articular; 5, td of the facet for the trochlea articulation; 6, td of the facet for the capitulum articulation; 7, dpd of the facet for the trochlea articulation; 8, dpd of the facet for the capitulum articulation; 9, dpd minimum; 10, td of the facets for the ulna; 11, td of the lateral facet for the ulna, 12; td of the median facet for the ulna; 13, td of the medial facet for the ulna. Middle of the diaphysis: 14, td; 15, ccd. Distal epiphysis: 16, td total; 17, dpd; 18, td articular; 19, td of the facets for the scaphoid and the semi-lunar; 20, td of the facet for the scaphoid; 21, td caudal of the facet for the semi-lunar; 22, maximal ccd of the facet for the scaphoid; 23, length of the crest between the facets for the scaphoid and the semi-lunar; 24, td of the facet for the pyramidal. References of the measurements in Fig. 14.A.7 and dimensions in Table 14.36

Tibia

There is one distal articulation (NMB Se 1677; left) (Table 14.37). The medial malleolus is broken, so the thickness is approximate, but it seems less than in *Gallogoral* whereas the articulation is more elongated. The fibular groove is deep with a lateral crest as in *Gallogoral*. It is distinct from the latter in having a medial and acute edge or process caudally, whereas in *Gallogoral* it is more or less rounded. The epiphysis is thinner (measurement 11; Fig. 14.43) and broader (measurement 12) than in *Gallogoral*, with a longer cranial facet for the malleolar bone (measurement 17).

Table 14.37 Comparative dimensions (in mm) of the tibia in *Pliotragus ardeus* and *Gallogoral meneghinii* from Senèze

Variable	1	2	3	4	5	6	7	8	9
G. meneghinii									
FSL 5008 (L)	305.6	71.4	62.6	30.2	41.5	65.9	28.0	32.5	29.1
FSL 5008 (R)	306.0	71.5	64.9	31.2	41.3	67.2	28.6	31.1	28.5
NMB Se 313 (L)	310.0	69.9	66.1	32.9	41.7	71.2	29.4	39.7	26.5
NMB Se 313 (R)	315.0	71.2	66.2	36.1	43.9	67.2	29.9	34.5	27.0

(continued)

Table 14.37 (continued)

Variable	10	11	12	13	14	15	16	17	18
P. ardeus									
NMB Se 1677		33.0	47.8	37.6	28.8	28.9	21.7	9.0	14.2
G. meneghinii									
FSL 5008 (L)	28.8	33.2	42.8	31.1	26.0	27.1	20.3	7.2	12.3
FSL 5008 (R)		43.6	33.1	30.5	26.7	27.3	19.8	7.2	12.8
NMB Se 313 (L)	29.5	35.4	45.4						
NMB Se 313 (R)	28.0	34.5	43.2						

Variable definitions (from Crégut-Bonnoure 1995, 2002, 2020) (Fig. 14.A.14): ccd, craniocaudal diameter; td, transversal diameter. 1, total length. Proximal epiphysis: 2, dt; 3, ccd; 4, ccd minimal; 5, ccd of the lateral side. Proximal articulation: 6, td total; 7, td of the medial facet; 8, td of the lateral facet. Middle of the diaphysis: 9, td; 10, ccd. Distal epiphysis: 11, ccd; 12, td. Distal articulation: 13, td of the articulation for the talus; 14, ccd of the middle of the epiphysis; 15, ccdof the lateral side of the epiphysis; 16, ccd of the articulation for the malleolar bone; 17, ccd of the dorsal facet for the malleolar bone; 18, ccd of the plantar facet for the malleolar bone; L, left; R, right

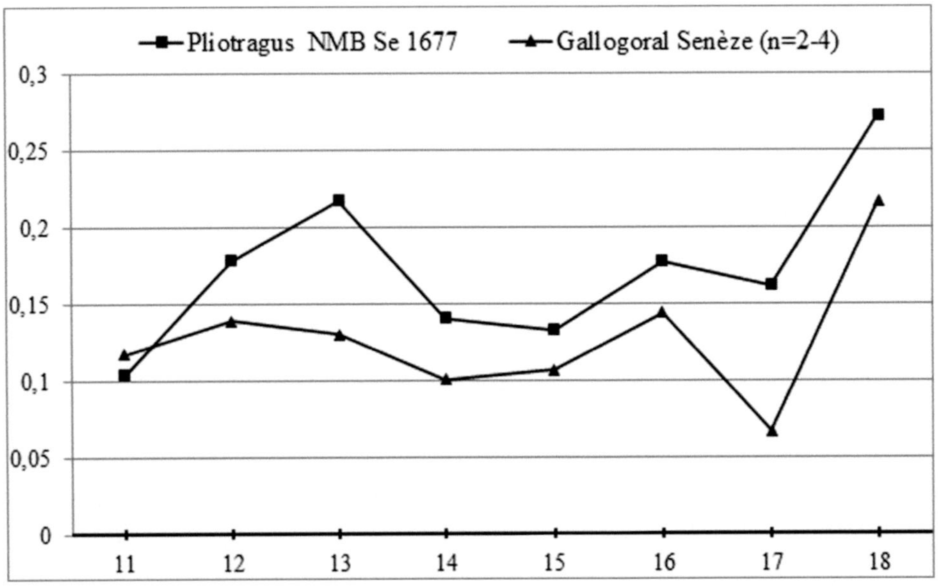

Fig. 14.43 *Pliotragus ardeus* and *Gallogoral meneghinii*. Ratio diagram of the tibia. Reference axis: recent *Capra ibex* ♂+♀ (in Crégut-Bonnoure 2020). ccd: craniocaudal diameter; td; transversal diameter. Distal epiphysis: 11, ccd; 12, td. Distal articulation: 13, td of the articulation for the talus; 14, ccd of the middle of the epiphysis; 15, ccd of the lateral side of the epiphysis; 16, ccd of the articulation for the malleolar bone; 17, ccd of the cranial facet for the malleolar bone; 18, ccd of the caudal facet for the malleolar bone. References of the measurements in Fig. 14.A.14 and dimensions in Table 14.37

Talus

A right talus (NMB Se 1672) is assigned to *Pliotragus* on the label (Fig. 14.32: I). The medial condyle has been reconstructed, as has a part of the disto-lateral condyle. The proximal trochlea has no median tubercle and the lips are not at the same level. At the base of the plantar trochlea, the lateral recess for the small navicular-cuboid articular facet is shallow and straight. The plantar surface is wide and relatively flat, with a slightly divergent lateral side and the medial one sub-vertical and in the plane of the proximal trochlear lip. The base of the latter is deep and pinched; in dorsal view, this base is abruptly inclined downward and oblique. The distal articulation has the same proportions as the proximal articulation of the navicular-cuboid NMB Se 1673. Only the proximal part of the articulation for the medial process of the navicular-cuboid is preserved. It is at the level of the lateral recess of the plantar articulation for the calcaneus: the structure corresponds to the low position of the medial process of the navicular-cuboid. The base of the medial articulation for the tibia is dorsally pinched. The general proportions are comparable to those of *Gallogoral* (Table 14.38; Fig. 14.44).

Table 14.38 Comparative dimensions (in mm) of the talus in *Pliotragus ardeus* and *Gallogoral meneghinii* from Senèze and Slivnitsa

Variable	1	2	3	4	5	6	7	8	9	10	11	12
P. ardeus												
Senèze (NMB Se 1672)	34.6	29.2	12.8	34.5	15.5	19.1	28.7	45.9	27.8	50.9	10.2	
Slivnitsa (NNHM-S SL 15)	34.5	29.9	12.4	32.5	14.0	17.9		48.2	27.1	53.6	9.4	17.7
G. meneghinii												
Senèze												
FSL 5008	31.3	26.9	10.0	31.3	14.3	17.3	26.0	43.5	26.7	47.9	11.3	14.9
FSL 16900 (L)	32.8	25.5	10.5	30.5	14.1	17.5	27.2	42.0	24.6	45.4	10.7	14.8
FSL 16900 (R)	32.8	25.5	10.7	30.6	13.8	17.6	27.2	41.9	24.6	45.7	10.7	15.0
Variable	13	14	15	16	17	18	19	20	21	22	23	
P. ardeus												
Senèze (NMB Se 1672)	39.5	28.5		35.7	24.0	26.3	18.1	26.0	28.4			
Slivnitsa (NNHM-S SL 15)	41.7	29.2	30.0	34.0	22.3	25.4	19.3	24.0	21.8			
G. meneghinii												
Senèze												
FSL 5008	37.3	27.1	31.3	31.3	24.5	21.8	17.4	26.0	21.7	30.8	30.8	
FSL 16900 (L)	36.1	28.5	29.9	31.1	24.4	20.9	16.6	24.5	21.7	30.6	31.0	
FSL 16900 (R)	36.1	28.4	30.1	31.3	24.4	19.6	16.8	21.8	24.3	30.7	31.1	

Variable definitions (from Crégut-Bonnoure 1987, 2002, 2020) (Fig. 14.A.15: A to F): dpd, dorsoplantar diameter; td, transversal diameter.1, td total. Proximal trochlea: 2, td; 3, dpd minimum. Distal trochlea: 4, td total; 5, td medial condyle; 6, td lateral condyle. Medial side: 7, dpd; 8, height. Lateral side: 9, dpd; 10, height. Laterodistal facet for the calcaneus: 11; height, 12, dpd. Height: 13, in the middle; 14, of the plantar trochlea for calcaneus; 15, trochlear lip of the proximolateral side; 16, trochlear lip of the proximomedial side. Distal trochlea: 17, dpd medial condyle; 18, dpd lateral condyle;19, dpd in the middle. Td: 20, of the plantar trochlea for calcaneus in the middle; 21, of the plantar trochlea for calcaneus at the proximal edge; 22, of the dorsal side; 23, of the proximal trochlea at its dorsal base; L, left; R, right

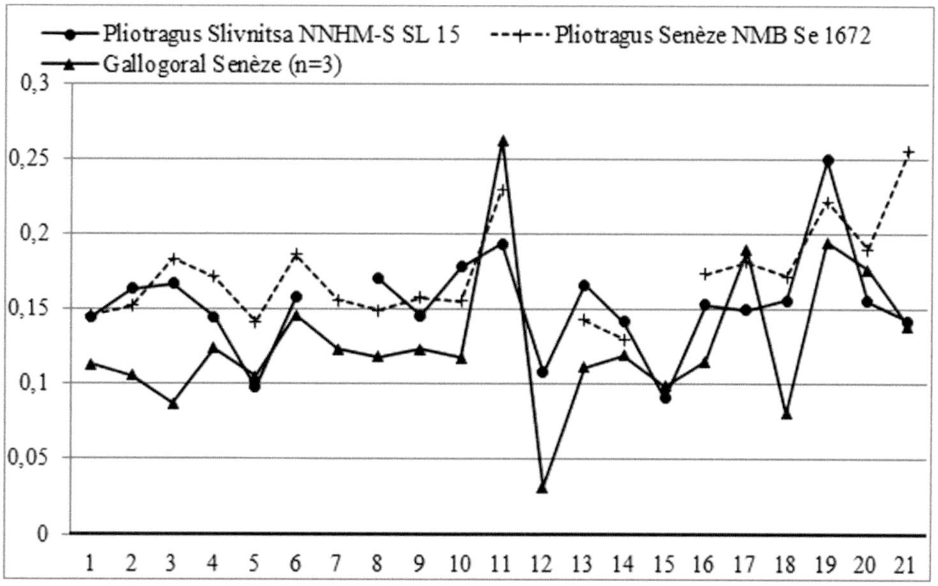

Fig. 14.44 *Pliotragus ardeus* and *Gallogoral meneghinii*. Ratio diagram of the talus. Reference axis: recent *Capra ibex* ♂+♀ (Crégut-Bonnoure 2020). dpd, dorso-plantar diameter; td, transversal diameter.1, td total. Proximal trochlea: 2, td; 3, dpd minimum. Distal trochlea: 4, td total; 5, td medial condyle; 6, td lateral condyle. Medial side: 7, dpd; 8, height. Lateral side: 9, ddp; 10, height. Laterodistal facet for the calcaneus: 11; height, 12, ddp. Height: 13, in the middle; 14, of the plantar trochlea for calcaneus; 15, trochlear lip of the proximo-lateral side; 16, trochlear lip of the proximo-medial side. Distal trochlea: 17, dpd medial condyle; 18, dpd lateral condyle; 19, dpd in the middle. Td: 20, of the plantar trochlea for calcaneus in the middle; 21, of the plantar trochlea for calcaneus at the proximal edge. References of the measurements in Fig. 14.A.15 and dimensions in Table 14.38

Calcaneus

One right calcaneus, FSL 210684, could be of *Pliotragus* (Fig. 14.32: J to M). It is similar in size to that of *Gallogoral*, but the morphology is not the same. In proximal view, the dorsal side of the *tuber calcanei* is rounded, rather than acute in *Gallogoral*. The transverse groove doesn't reach the lateral side as in *Gallogoral*; however, it is longer and deeper than in *Gallogoral*. On the lateral side there is a small notch, which is rounded in *Gallogoral* with a tubercle is in front of it that is absent on FSL 210684.The plantar groove is enlarged and shallow, while in *Gallogoral* it is narrow and deep. In plantar view, the base of the tuber is not clearly visible; it is well defined and shorter in *Gallogoral*. The lateral side is less sinuous than in *Gallogoral*. The medial side between the *tuber calcanei* and the *sustentaculum tali* is straight and well developed. In the last third of its length, it narrows rapidly. In *Gallogoral* the medial side is weakly concave and in the last third of its length weakly convex. The medial side of the *sustentaculum tali* has a tubercle, which is absent in *Gallogoral*. The tendinous furrow of the *sustentaculum tali* is broader and more concave than in *Gallogoral*. At the junction with the body, in *Gallogoral* there is a blade-like crest which is absent on FSL 210684. In dorsal view, the lateral side of the malleolar facet is straight, but sinuous in *Gallogoral*. The proximal area of this facet is flat whereas in *Gallogoral* it is concave with its proximal side pinched. The distal area is narrow and sharp, whereas it is enlarged and rounded in *Gallogoral*. In medial view, the surface of the *sustentaculum* is flat. In *Gallogoral* the surface is convex.

The proportions also are different from those of *Gallogoral* (Table 14.39; Fig. 14.45): the *tuber calcanei* is broader (measurement 2) but thinner (measurement 3) with a broader *sustentaculum tali* (measurement 8) and distal facet (measurement 10).

Table 14.39 Comparative dimensions (in mm) of the calcaneus in *Pliotragus ardeus* and *Gallogoral meneghinii* from Senèze and other localities

Variable	1	2	3	4	5	6	7	8	9
cf. P. ardeus									
Senèze (FSL 210684)	92.0	24.8	23.4	14.0	27.2	38.6	37.1	32.0	25.2
Slivnitsa (NNHM-S SL 20)		28.8	32.4	16.9					
G. meneghinii									
Senèze									
FSL 5008	93.0	23.8	25.8	13.1	26.4	37.0	32.6	27.6	24.5
FSL 16900 (L)	92.6	23.7	23.1	13.0	24.6	35.8	32.3	30.0	25.5
FSL 16900 (R)	92.0	24.8	24.2	12.7	25.0	35.7	33.3	29.0	24.6
Chilhac (NMB PP 134)	88.2	20.0	24.6	11.4	26.6	36.3	26.3		
Variable	10	11	12	13	14	15	16	17	
cf. P. ardeus									
Senèze (FSL 210684)	9.0	18.7	10.1	30.0	33.0	12.9	58.1	63.8	
Slivnitsa (NNHM-S SL20)						16.0	72.9	77.0	
G. meneghinii									
Senèze									
FSL 5008	7.3	19.1	10.0	34.0	36.1	11.0	57.7	61.6	
FSL 16900 (L)	8.6	19.7	11.2	33.3	36.9	11.8	56.1	61.8	
FSL 16900 (R)	8.3	19.6	11.4	33.2	36.1	11.7	56.8	61.7	
Chilhac (NMB PP 134)	7.5	18.8	10.2				58.8	63.8	

Variable definitions (from Crégut-Bonnoure 1987, 2002, 2020) (Fig. 14.A.15: G to F): dpd, dorso-plantar diameter; td, transversal diameter. 1, total length. *Tuber calcanei*: 2, td; 3, dpd. Proximal body: 4, td minimum; 5, dpd. Total: 6, dpd; 7, td. *Sustentaculum tali*: 8, td; Navicular-cuboid facet: 9, length; 10, td. Fibular trochlea facet: 11, length; 12, td. Length: 13, from the fibular trochlea facet in its middle to the distal point of the bone; 14, from the fibular trochlea facet at its proximal side to the distal point of the bone. Width: 15, *sustentaculum tali* furrow. Length: 16, from the dorsal *tuber calcanei* apex to the coracoid process tip; 17, from the palmar *tuber calcanei* apex to the coracoid process tip; L, left; R, right

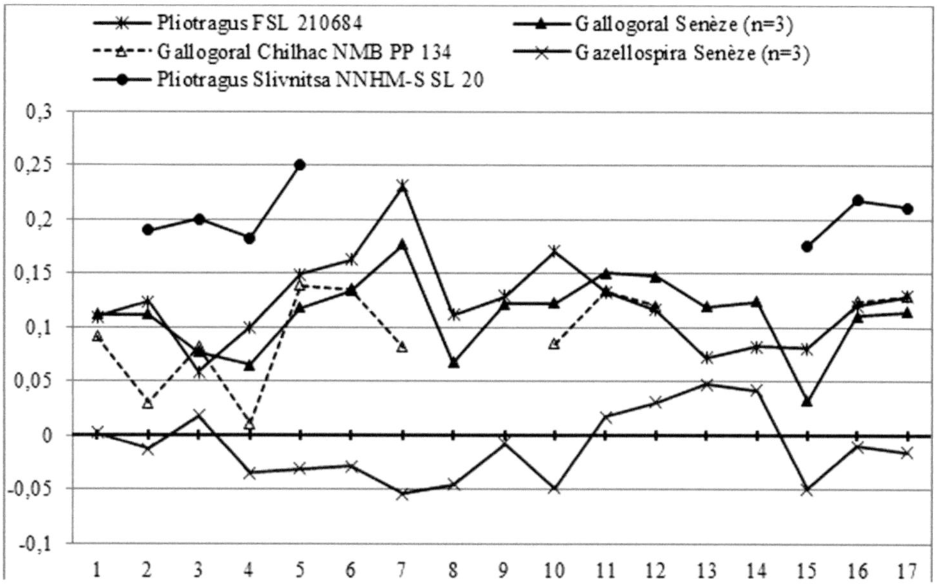

Fig. 14.45 *Pliotragus ardeus*, *Gazellospira torticornis* and *Gallogoral meneghinii*. Ratio diagram of the calcaneus. Reference axis: recent *Capra ibex* ♂+♀ (in Crégut-Bonnoure 2020). dpd, dorsoplantar diameter; td, transversal diameter. 1, total length. *Tuber calcanei*: 2, td; 3, dpd. Proximal body: 4, td minimum; 5, dpd. Total: 6, dpd; 7, td. *Sustentaculum tali*: 8, td; Navicular-cuboid facet: 9, length; 10, td. Fibular trochlea facet: 11, length; 12, td. Length: 13, from the fibular trochlea facet at its middle to the distal point of the bone; 14, from the fibular trochlea facet at its proximal side to the distal point of the bone. Width: 15, *sustentaculum tali* furrow. Length: 16, from the dorsal *tuber calcanei* apex to the coracoid process tip; 17, from the palmar *tuber calcanei* apex to the coracoid process tip. References of the measurements in Fig. 14.A.15 and dimensions in Tables 14.13 and 14.39

Navicular-Cuboid

The specimen NMB Se 1073 (right) is assigned to *Pliotragus* on the label. The morphology is distinct from Perrier-Rocca-Neyra *Gazellospira* (NMB Rn 144, a quite complete tarsal articulation) and from Senèze *Gallogoral* (FSL 5008).

The proximal articulation is asymmetrical, with the lateral side wider than the medial. The calcaneal articular facet is elongated and rises in contact with the facet for the lateral recess of the talus. The small lateral facet for the base of the plantar trochlea of the talus is eroded. On the plantar side the medial process is very low, its top being below the level of the median process; it is above this level in *Gallogoral*. Distally, the metatarsal III+IV dorsal articular facet is elongated, with the medial side bilobed; its surface is undulating. The metatarsal III–IV plantar articular facet is triangular and obliquely directed downward. The dorsomedial cuneiform facet is elongated, equally enlarged. On the plantar side, the neck between the medial process and the calcaneal articular facet is broad and "U" shaped. It is continued on the plantar surface by a deep "V" shaped fossa which reaches almost to the distal side. In *Gallogoral* the fossa is less deep and its shape is rounded. In plantar view, the metatarsal III+IV process is acute distally and projects inferiorly; in *Gallogoral* it is less developed, rounded and stops before the base. In lateral view, the groove for the tendon to *musculus peroneus longus* is deeper in *Gallogoral*.

Compared to *Gallogoral*, the bone is broader (measurement 1), the facet for the calcaneal articular facet is longer (measurement 6), and the small medio-plantar cuneiform articular facet is thinner (measurement 17) (Table 14.40; Fig. 14.46). The proportions of the plantar side are in accordance with the morphology (measurements 22 and 23).

Table 14.40 Comparative dimensions (in mm) of the navicular-cuboid in *Pliotragus ardeus*, *Gazellospira torticornis* and *Gallogoral meneghinii*

Variable	1	2	3	4	5	6	7	8	9	10	11	12
P. ardeus												
Senèze (NMB Se 1073)	35.7	41.5	33.7	19.2	15.1	35.5	9.5			26.5	15	8.1
G. meneghinii												
Senèze												
FSL 5008	31.7	38.5	31.1	17.3	13.5	26.9	7.2	9.1	3.7	24.0	13.7	8.0
FSL 16900 (L)	31.7	39.0	32.1	17.4	14.4	28.9	8.3	10.4	3.5	24.2	13.1	6.8
FSL 16900 (R)	31.8	38.9	31.8	17.7	14.4	27.6	7.8	12.0	3.4	25.0	14.3	8.4
G. torticornis												
Rocca-Neyra												
NMB Rn 144	37,0	40.1	30.5	17.4	14.1	28.6	10.8	4.1	9.5	24.5	14.2	9.4
NMB Rn 29	38.0	39.4	30.5	16.8	14.1	25.8	11.3	3.1	8.0	25.5	15.2	10.4
Variable	13	14	15	16	17	18	19	20	21	22	23	
P. ardeus												
Senèze (NMB Se 1073)	21.1	14.9	8.6	13.7	8.97	11.1	17.4	11.9	10.8	23.3	25.9	
G. meneghinii												
Senèze												
FSL 5008	21.1	15.8	6.4	12.1	10.4	10.8	15.7	10.2	9.2	24.0	21.9	
FSL 16900 (L)	21.1	14.6	6.2	10.9	8.2	10.2	17.3	8.5	10.4	23.1	20.6	
FSL 16900 (R)	19.9	14.9	6.2	11.7	9.7	8.8	15.8	11.0	8.9	21.6	20.1	
G. torticornis												
Rocca-Neyra												
NMB Rn 144	23.1	16.3	6.7	11.9	11.1	12.2	16.9	12.2	10.8	25.2	23.3	
NMB Rn 29	23.8	16.1	7.2	12.4	10.9	14.0	18.5	12.4	11.7	26.4	23.7	

Variable definitions (from Crégut-Bonnoure 1987, 2002, 2020) (Fig. 14.A.15: J to M): dpd, dorsoplantar diameter; td, transversal diameter. Total: 1, dpd; 2, td. Td, proximal articulation for the talus: 3, total; 4, lateral; 5, medial. Articulation for the calcaneus: 6, dpd; 7, td. Small dorsomedial articulation for the talus: 8, dpd; 9, td. Articulation for the great medial cuneiform facet: 10, dpd; 11, td. Dorsal side: 12, height between great cuneiform surface and metatarsal III+IV dorsoateral surface. Articulation for the metatarsal III +IV: 13, dpd of the dorsolateral facet; 14, td of the dorsolateral facet; 15, dpd of the palmolateral facet; 16, td of the plantar articulation of the palmolateral facet. Articulation for the small medial cuneiform facet: 17, dpd; 18, td. Dorsal side: 19, maximal lateral height; 20, maximal medial height; 21, maximal median height. Plantar side: 22, maximal medial height; 23, maximal lateral height; L, left; R, right

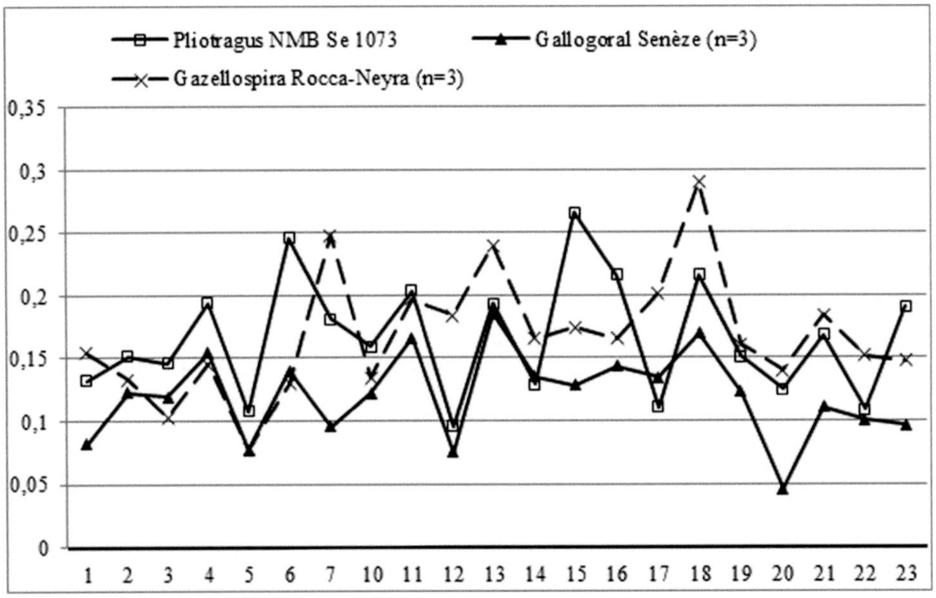

Fig. 14.46 *Pliotragus ardeus* (NMB Se 1073), *Gazellospira torticornis* and *Gallogoral meneghinii*. Ratio diagram of the navicular-cuboid. Reference axis: recent *Capra ibex* ♂+♀ (in Crégut-Bonnoure 2020). dpd, dorsoplantar diameter; td, transversal diameter. Total: 1, dpd; 2, td. Td, proximal articulation for the talus: 3, total; 4, lateral; 5, medial. Articulation for the calcaneus: 6, dpd; 7, td. Articulation for the mediodorsal facet: 10, dpd; 11, td. Dorsal side: 12, height between mediodorsal surface and metatarsal III+IV dorsolateral surface. Articulation for the metatarsal III +IV: 13, dpd of the dorsolateral facet; 14, td of the dorsolateral facet; 15, dpd of the palmolateral facet; 16, td of the plantar articulation of the palmo-lateral facet. Articulation for the small medioplantar facet: 17, dpd; 18, td. Dorsal side: 19, maximal lateral height; 20, maximal medial height; 21, maximal median height. Plantar side: 22, maximal medial height; 23, maximal lateral height. References of the measurements in Fig. 14.A.15 and dimensions in Table 14.40

Metatarsal III+IV

A distal articulation, belonging to a juvenile individual (NMB Se 201; left), is assigned to *Gallogoral* on the label (Fig. 14.32: G, H). Compared to other elements of that genus, this specimen is larger. In medial and lateral views, the keel is more developed and more prominent. In distal view, the sides of the articular condyles are obliquely directed, whereas they are parallel in *Gallogoral*. In dorsal view, the sulcus is broad, shallow and at the foramen level there is no crest. In *Gallogoral* it is narrow, deep, bordered by two crests slightly lower than the foramen level. Left metatarsal FSL 16877 has also been assigned to *Gallogoral* on the label, but as with the previous metatarsal, its morphology does not agree with this genus. The shaft is slender with a distal enlargement; it is shorter without the distal enlargment in *Gallogoral*. The proximal tubercle is triangular, rather than rounded in *Gallogoral*. The dorsal sulcus is deep and the distal foramen is bordered by the two acute sides.

The proportions of these two metapodials are clearly distinct from *Gazellospira*, *Gallogoral* and *Ovis* (Table 14.41; Fig. 14.47). The main difference between these two specimens concerns the thickness of the distal epiphysis (measurement 23) which could be due to sexual dimorphism.

Table 14.41 Comparative dimensions (in mm) of the metatarsal III+IV in *Pliotragus ardeus* from Senèze

Variable	1	2	17	18	19	20	21	22
NMB Se 201					46.8	44.9	45.2	25.4
FSL 16877	176.9	160.0	23	21.6	45.2	43.5	43.4	23.7
Variable	23	24	25	26	27	28	29	30
NMB Se 201	21.8	28.1	28.0	23.1	27.7	19.3	26.9	17.6
FSL 16877	23.0	26.2	27.8	22.6	26.2	18.0	25.0	16.6

Variable definitions (from Crégut-Bonnoure 1995, 2002, 2020) (Fig. 14.A.16): td: transversal diameter. dpd: dorsoplantar diameter. Length: 1, dorsal side; 2, minimum in the middle. Middle of the diaphysis: 17, td; 18, dpd. Distal epiphysis: 19, td maximal of the articulation; 20, td minimal of the articulation; 21, td supra-articular; 22, dpd of the epiphysis; 23, dpd of the epiphysis at the foramen level; 24, distance between the keels on the plantar side; 25, distance between the keels at their base; 26, distance between the keels in their middle; 27, dpd of the medial condyle; 28, dpd of the medial keel; 29, dpd of the lateral condyle; 30, dpd of the lateral keel

Discussion

According to the teeth, there are 11 individuals in the Senèze series studied. The cranium (N = 3), the mandibles (N = 14) and the teeth (N = 101) have morphological features that distinguish *Pliotragus* from *Gallogoral*. However, the proportions of the teeth are quite similar in these two genera. The postcranial material is scarce, represented only by isolated bones (N = 10). Without a complete associated skeleton, it is impossible to be sure that all of them belong to this species. The comparison between the Senèze and Dacian Basin crania shows that they differ. This could be due to sexual dimorphism or a (sub)species-specific characteristic.

Genus *Gallogoral* Guérin, 1965

Species *Gallogoral meneghinii* (Rütimeyer, 1877-78)

Subspecies *Gallogoral meneghinii meneghinii* (Rütimeyer, 1877-78)

Gallogoral was considered as a rupicaprin (now a caprin) and is monospecific. The species was first described at Senèze by Schaub (1922) as *Nemorhoedus philisi*, which was synonymized with *Palaeoryx meneghinii* Rütimeyer, 1877-78 (from Olivola) by Schaub (1928). Guérin (1965) created the new genus *Gallogoral* for this medium-sized antelope.

The species is known in Western Europe from MNQ 17 to MNQ 18:

- MNQ 17: in France at Le Coupet, Perrier-Pardines, Saint Vallier (Heintz 1964, 1966; Crégut-Bonnoure *in* Crégut-Bonnoure & Valli 2004), Chilhac I (Schaub 1943) and Chilhac II (Boeuf 1983; Paquette et al. 2021); in Italy at Coste San Giacomo (Bellucci et al. 2014; Florindo et al. 2021); in Greece at Sésklo, upper level (Athanassiou 2002, 2018), Dafneró (Kostopoulos 1996) and Volax (Kostopoulos 1996, 1997a).
- MNQ 18: in France at Senèze (Schaub 1922, 1923); in Spain at La Puebla de Valverde (Crusafont et al. 1964; Sinusía et al. 2004); in Italy at Olivola (Rütimeyer 1877-78; Schaub 1928); in Bulgaria at Slivnitsa (Spassov & Crégut-Bonnoure 1999); in Greece at Karnezeika (Sianis et al. 2022); and in Georgia at Dmanisi (Bukhsianidze 2005).

The subspecies *G. meneghinii sickenbergii* Kostopoulos, 1997a, 1997b was identified at Volax and Sésklo. This subspecies may also occur at Dmanisi and Slivnitsa (Spassov 2005). Recently, another subspecies was

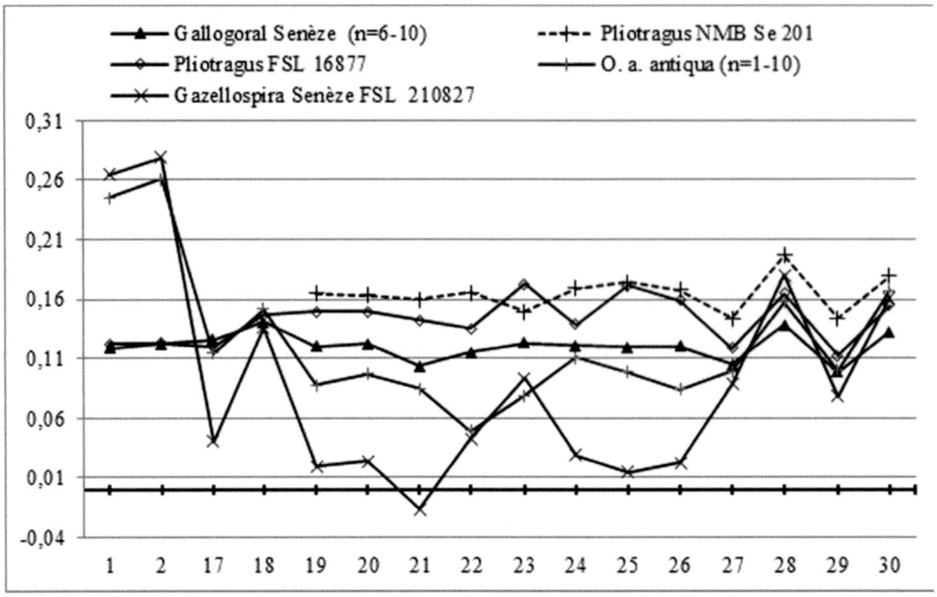

Fig. 14.47 *Pliotragus ardeus* (NMB Se 201 and FSL 16877), *Gallogoral meneghinii*, *Gazellospira torticornis*, *Ovis ammon antiqua*. Ratio diagram of the metatarsal III+IV. Reference axis: recent *Capra ibex* ♂+♀ (in Crégut-Bonnoure 2020). td: transversal diameter. dpd: dorsoplantar diameter. Length: 1, dorsal; 2, minimum at the middle. Middle of the diaphysis: 17, td; 18, dpd. Distal epiphysis: 19, td maximal of the articulation; 20, td minimal of the articulation; 21, td supra-articular; 22, dpd of the epiphysis; 23, dpd of the epiphysis at the foramen level; 24, distance between the keels on the plantar side; 25, distance between the keels at their base; 26, distance between the keels in their middle; 27, dpd of the medial condyle; 28, dpd of the medial keel; 29, dpd of the lateral condyle; 30, dpd of the lateral keel. References of the measurements in Fig. 14. A.16 and dimensions in Tables 14.14, 14.41 and 14.55 and in Crégut-Bonnoure (2020)

recognized at La Puebla de Valverde: *G. meneghinii heintzi* Andrés Rodrigo, 2011.

Description

Guérin (1965) gave a complete description of the Senèze material, followed by a synthesis with a neodiagnosis in 1989 (*in* Duvernois & Guérin 1989). Here I add some morphological details and present a new exhaustive list.

Crania and Horn Cores

Four crania are preserved in the collection (all of them are incomplete): FSL 5008, FSL 5009 (Guérin 1965: pl. 1 p. 31; pl. II p. 33; pl. III p. 35; pl. V p. 47; pl. VI p. 49; pl. VII p. 51; pl. IX p. 61; pl. X p. 63; pl. XI p. 65) (Fig. 14.48), NMB Se 313 and NMB Se 550. There are also a frontlet NMB Se 1755 (Fig. 14.49: A) and three isolated horn cores: NMB Se 1482 (Fig. 14.49: B, C), NMB Se 1666 (Fig. 14.49: D, E) and NMB Se 1695 (identified as "Antelope?" on the label; Fig. 14.49: F, G).

The cranium NMB Se 313 belongs to a female according the slenderness of the horn cores (Fig. 14.48). It has been partially reconstructed, and thus it is impossible to fully assess its morphology. The other crania are more or less distorted as indicated by Guérin (*in* Duvernois & Guérin 1989). The relevant morphological criteria are the significant length and divergence of the horn cores, a high temporal fossa with an only slightly convex parietal, and the strong convexity of the basioccipital tuberosities which are separated by a wide groove. The dimensions are given in Table 14.42. The comparative proportions between *Pliotragus* and *Gallogoral* crania are presented in Fig. 14.29.

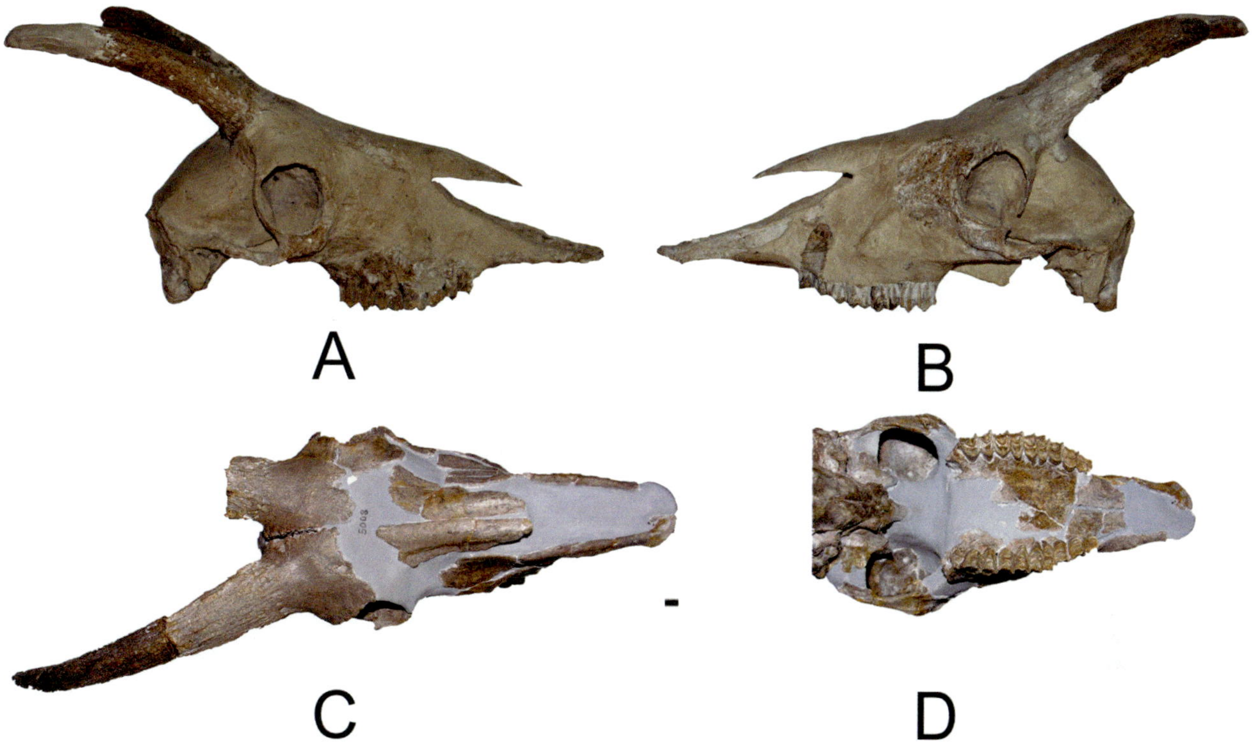

Fig. 14.48 Cranium of *Gallogoral meneghinii* from Senèze. A to D, NMB Se 313. A, B: *norma lateralis*; C, *norma dorsalis*; D, *norma basalis*. Scale 1 cm

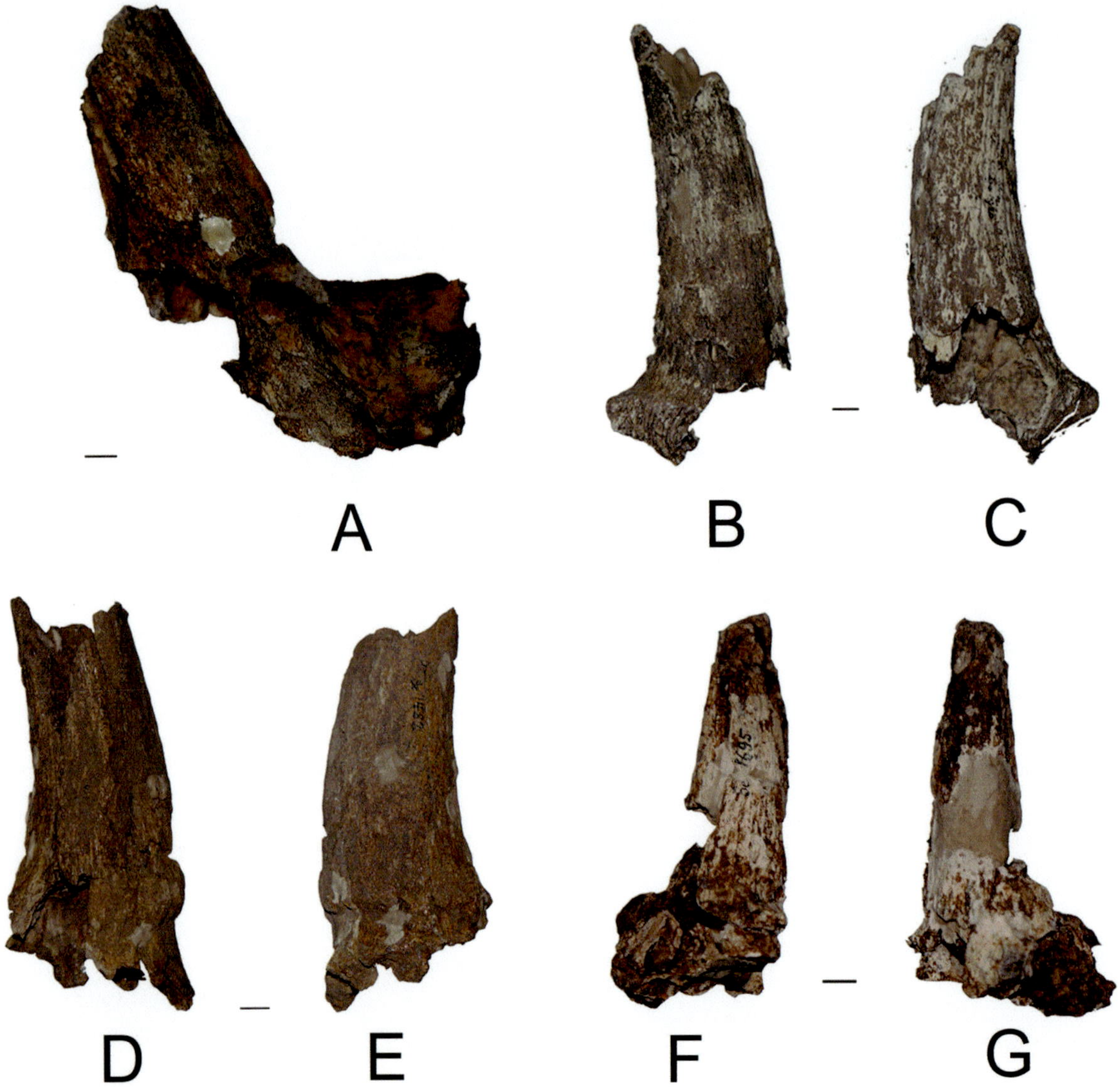

Fig. 14.49 Frontlet and horn cores of *Gallogoral meneghinii* from Senèze. A, frontlet NMB Se 1755; B, C, left horn core NMB Se 1482; D, E, right horn core NMB Se 1666; F, G, horn core NMB Se 1695. A, *norma frontalis*; B, E, *norma dorsalis*; C, D, *norma lateralis*. Scale 1 cm

Mandible

Eighteen more or less complete mandibles are available (Table 14.28): FSL SEN 06-0238 (P 172) from the new fieldwork (left), FSL 5008 (left and right), FSL 16866 (left), FSL 16862 (left and right; Guérin 1965, Pl. XV to XVIII), NMB unnumbered (indicated "neotype"[2]; right), NMB Se 103 (right), NMB Se 195 (right), NMB Se 313 (left and right; Figs. 14.50 and 14.51), NMB Se 550 (left and right), NMB Se 491 (right, with deciduous teeth), NMB Se 563 (left), NMB Se 1359 (right), NMB Se 538 (left, with deciduous teeth) and NMB Se 1438 (right; Fig. 14.52: A, B).

[2]It is unclear why this specimen is marked "neotype" on the label. A neotype can only be designated by an author when the holotype or any syntypes cannot be located. In fact, Rütimeyer (1877-78) figured a partial cranium from Olivola in the Pisa collection, and Schaub (1928) illustrated the same specimen as the "type skull" (*Typusschädel*) along with a metatarsal and upper toothrow in Florence which he termed topotypes (material from the type locality). Schaub (1922) named *Nemorhoedus philisi* from Senèze, indicating the mounted skeleton NMB Se 550 as type, but in 1928 he recognized the synonymy of these two taxa. Guérin (1965, p. 316) indicated that he thought the three specimens mentioned by Schaub were "very insufficient" and thus designated three "paratypes": FSL 5008 and NMB Se 550 from Senèze and NMB PP 133 from Chilhac. These are not formally paratypes, as they were not part of the type series, but merely important members of Guérin's updated hypodigm. Guérin (1965) does not appear to include any mention of this "neotype" mandible, and it is unclear when or by whom this annotation was added to the specimen label.

Table 14.42 Comparative dimensions (in mm) of the cranium in *Gallogoral meneghinii* from Senèze and other localities

Variable	1	2	3	4	5	6	9	10	11	12
Senèze										
NMB Se 313	360.0	345.0	325.3	219.8	101.6	210.2	103.3	37.7	65.9	
FSL 5008	338.2	330.5	308.7	213.9	96.3	201.1	103.9	39.6	66.8	105.1
Volax										
LGPUT Vol 1	300.0						115.3	43.5	71	
LGPUT Vol 2							111.2	41.2	69.5	
Variable	13	14	15	17	18	19	20	21	23	24
Senèze										
NMB Se 313	88.3	44.0	34.0							
FSL 5008	99.8	61.5	44.9	21.8		66.6	138.3		251.1	221.8
Chilhac (NMB PP 133)					119.3					
Olivola (NMB Ol 10)					99.6					
Volax										
LGPUT Vol 1		51.0	36.0		123.0					
LGPUT Vol 2	88.5	56.5	39.0		113.0					
Variable	25	26	27	28	29	30	31	32	33	34
Senèze										
NMB Se 313					152.0	107.7	102.7	96.5		46.8
FSL 5008	36.9	86.1	39.9	115.1	144.7	122.3	114.2	104.2	104.2	48.2
Chilhac (NMB PP 133)				116.1						
Variable	37	38	39	40	41	42	43	44	45	46
Senèze										
NMB Se 1482						62. 6	48.0			183.0
NMB Se 1666						50.7	50.2			166.0
NMB Se 313	113.6	30.7	76.4		38.0	42.5	29.0	29.6	210.0	130.0
FSL 5008	120.2	43.8	112.2	149.4	30.0	43.6	47.1	29.22		145.0
Chilhac (NMB PP 133)	126.3	22.6	112.3		59.8	56.9	46.7	39.3		182.0
Olivola (NMB Ol 10)						42.6				
Volax (LGPUT Vol 1)	140.0	30.0			55.0	59.2	43.7	37.1	300.0	190.0
Perrier-Etouaires (NMB Prr 523)					43.3	40.3			134.0	
Variable	47	48	51	52	53	54	55	56	57	58
Senèze										
NMB Se 313					42.1	51.8				73.9
FSL 5008	290.0	214.9	122.7	57.3	77.3	47.8	51.2	85.4	91.4	69.9
Volax (LGPUT Vol 1)				70.5	45.0					
Senèze	59	60	61	62	63	64	65	68		
NMB Se 313	31.6	26.9	25.9	21.6		46.4				
FSL 5008	30.9	23.7	29.2	23.3	66.0	44.7	35.3	164.7		

Variable definitions (from Crégut-Bonnoure 2002, 2020) (Fig. 14.A.1): Length: 1, acrocranion-prosthion; 2, from occipital condyles to prosthion; 3, basion-prosthion; 4, from basion to the middle of the line of the P2; 5, from prosthion to the middle of the line of the P2; 6, from prosthion to the middle of the line of the M3; 9, P2 mesial-M3 distal; 10, P2 mesial-P4 distal; 11, M1 mesial-M3 distal; 17, acrocranion-nasion; 18, frontal from supra-orbital foramens; 19, acrocranion-bregma; 20, from supra-orbital foramens to acrocranion; 21, from supra-orbital foramens to prosthion; 23, entorbital-prosthion; 24, from infra-orbital foramen to acrocranion; 25, parietal dorsal length. Transversal diameter: 12, otion-otion; 13, tooth row at the labial side;14, between the lingual side of the M3 second lobe;15, between the lingual side of the P2; 26, parietal width 27, between the parietal ridges; 28, post-orbital; 29, ectorbital-ectorbital; 30, entorbital-entorbital; 31, between orbit and facial tuberosities. 32, at the facial tuberosities level; Parietal: 25, length; 26, width. Incisal bone: 33, maximal length; 34, rostral width. Variable definitions (from Crégut-Bonnoure 2002, 2020) (Fig. 14.A.2): Horn cores: 37, width between the lateral sides; 38, width between the internal sides at the base; 39, width between the internal sides in the middle; 40: width between the internal sides at the tip; 41, rostrocaudal diameter at the base; 42, dorsolateral diameter at the base; 43, rostrocaudal diameter in the middle; 44, dorsolateral diameter in the middle; 45, length; 46, circumference at the base. Length: 47, from occipital condyle to infra-orbital foramen; 48, basion-nasion; 51, base of the incisal bone; 52, lacrimal bone from the entorbital; 68, from the top of the frontal to the tip of the occipital condyles. Orbit: 53, rostrocaudal diameter; 54, dorsoventral diameter. Height: 55, of the braincase from basioccipital to the top of the parietal; 56, from the middle of the palatine suture to the nasal surface. Transversal diameter: 57, between jugular processes; 58, between the external sides of the occipital condyles; 59, of the foramen magnum; 60, of the occipital condyle. Height: 61, of the occipital condyle; 62, of the foramen magnum; 63, from opisthion to fronto-parietal suture. Basioccipital: 64, transversal diameter of the caudal tuberosities; 65, transversal diameter of the rostral tuberosities.

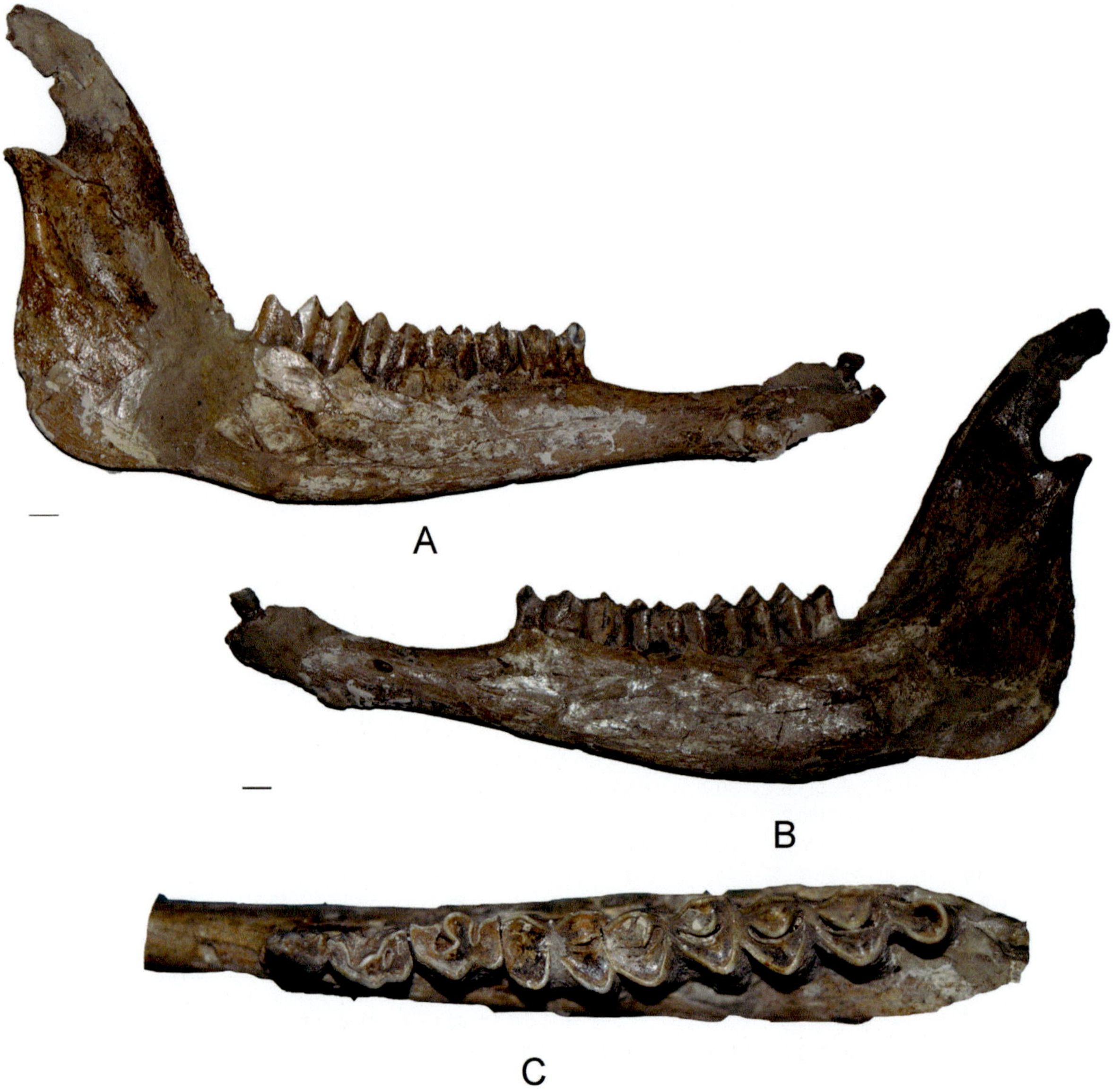

Fig. 14.50 Mandible of *Gallogoral meneghinii* from Senèze. A, B, C, left NMB Se 313 (from skeleton). A, lingual view; B, buccal view; C, occlusal view. Scale 1 cm

Compared to *Pliotragus*, the buccal side is more flattened than the lingual one from p2 to the distal part of m1; consequently, in occlusal view, the corpus is less visible than in *Pliotragus*. From m2 to the masseter insertion, the corpus is buccally inflated, except on FSL 16866. The masseter insertion is elongated rostrocaudally. The corpus at the diastema level is thin, as is its crest, without any pinching. The mandibular foramen is in a more distal position than in *Pliotragus*, and it leads into a longitudinal groove. On NMB Se 195 and NMB Se 563, which are from immature individuals, the diastema is shorter than on the adult mandibles. The comparative proportions between *Pliotragus* and *Gallogoral* mandibles are presented in Fig. 14.33.

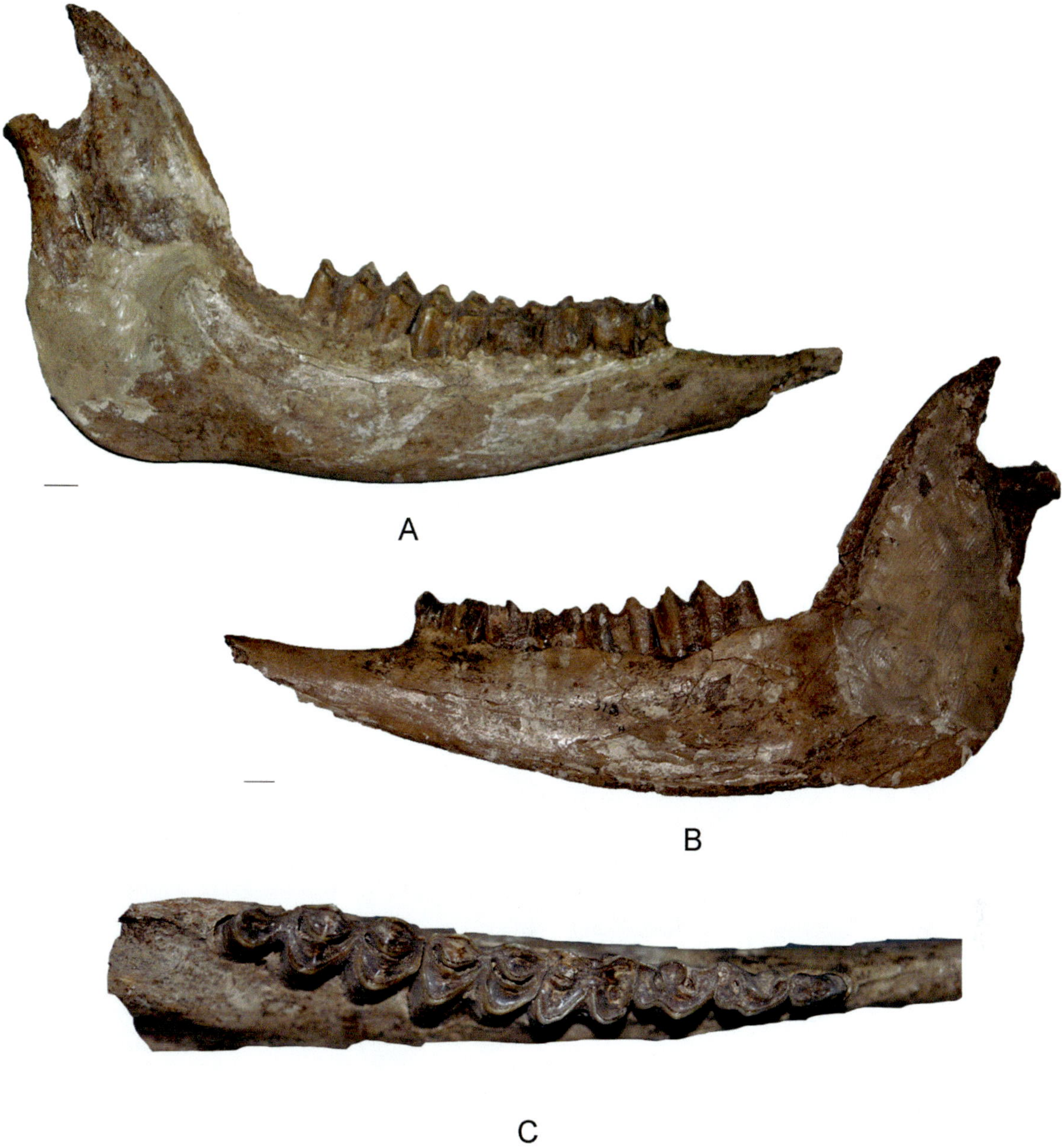

Fig. 14.51 Mandible of *Gallogoral meneghinii* from Senèze. A, B, C, right NMB Se 313 (from skeleton). A, buccal view; B, lingual view; C, occlusal view. Scale 1 cm

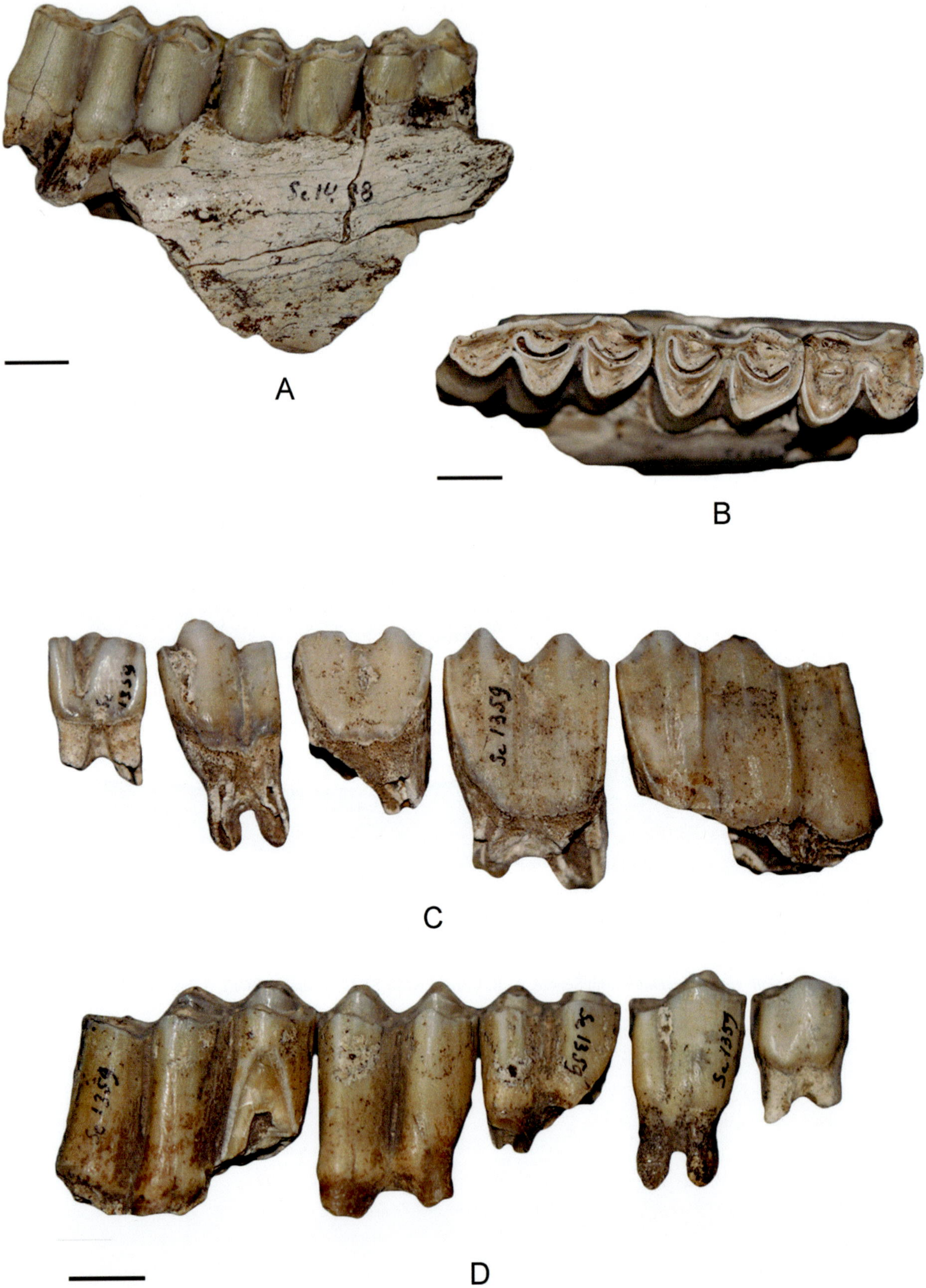

Fig. 14.52 Mandible and lower teeth of *Gallogoral meneghinii* from Senèze. A, B, right mandible NMB Se 1438; C, D, right p3-m3 NMB Se 1359. A, D, buccal view; B, occlusal view; C, lingual view. Scale 1 cm

Teeth

Upper Teeth

As indicated under *Pliotragus*, the morphological differences with *Gallogoral* are minimal and some mistakes of attribution exist. Currently, upper teeth of *Gallogoral* include (Tables 14.43, 14.44 and 14.45): FSL 3610 (left P3-4 and right M1), FSL 3933 (right M1), FSL 5008 and FSL 5009 (right and left P2-M3 from the crania), FSL 16890 (right M2-3; initially assigned to *Pliotragus*), NMB Se 313 (right and left P2-M3 from the cranium), NMB Se 779 (left P3-M3), NMB Se 198 (right M2), NMB Se 387 (left M1; initially assigned to *Gazellospira;* Pilgrim & Schaub 1939: pl. 2, Figs. 10–12), NMB Se 388 (right M2), NMB Se 539 (right M2; initially assigned to *Gazellospira*), NMB Se 1355 (right M2), NMB Se 1359 (left M2), NMB Se 1549 (left M2), NMB Se 1667 (left D3-M1; juvenile), NMB Se 1670 (left M2), NMB Se 1437 (right M1; initially assigned to *Gazellospira*), MNHN-F-1921-9 (left M2-3), MNHN-F-unnumbered (right P3-4, juvenile; left M2), and a group of isolated M3: NMB Se 91 (right), NMB Se 196 (right), NMB Se 386 (right), NMB Se 388 (left; initially assigned to *Pliotragus*), NMB Se 391 (right), NMB Se 1432 (left), NMB Se 1440 (left; initially assigned to *Pliotragus*), NMB Se 1669 (right). There are also the teeth from the skeleton NMB Se 550, not studied here.

On P2 (n = 6), the lingual side is bilobed. The parastyle and the metastyle are weak; the inflation of the paracone is moderate and behind it the valley is broad and shallow. The P3 (n = 9) and the P4 (n = 9) present a rounded lingual side. The parastyle is stronger than the metastyle but the latter is relatively broad: P3 FSL 5009: respectively 2.3 and 1.7 mm; P3 FSL 5008: 2.6 and 2.5 mm; P3 FSL 3610: 3.0 and 2.2 mm; unworn P3 MNHN-F-unnumbered: 2.4 and 2.3 mm; P4 FSL 5009: 3.0 and 2.3 mm; P4 FSL 5008: 2.4 and 2.2 mm; P4 FSL 3610: 3.1 and 2.2 mm; unworn P3 MNHN-F-unnumbered: 3.2 and 3.1 mm. The inflation of the ribs is less extended than in *Pliotragus*. The valleys on each side of the cusp have the same development. In relation with the development of the styles, the interstylar surface seems to be in a more internal position; the styles are parallel.

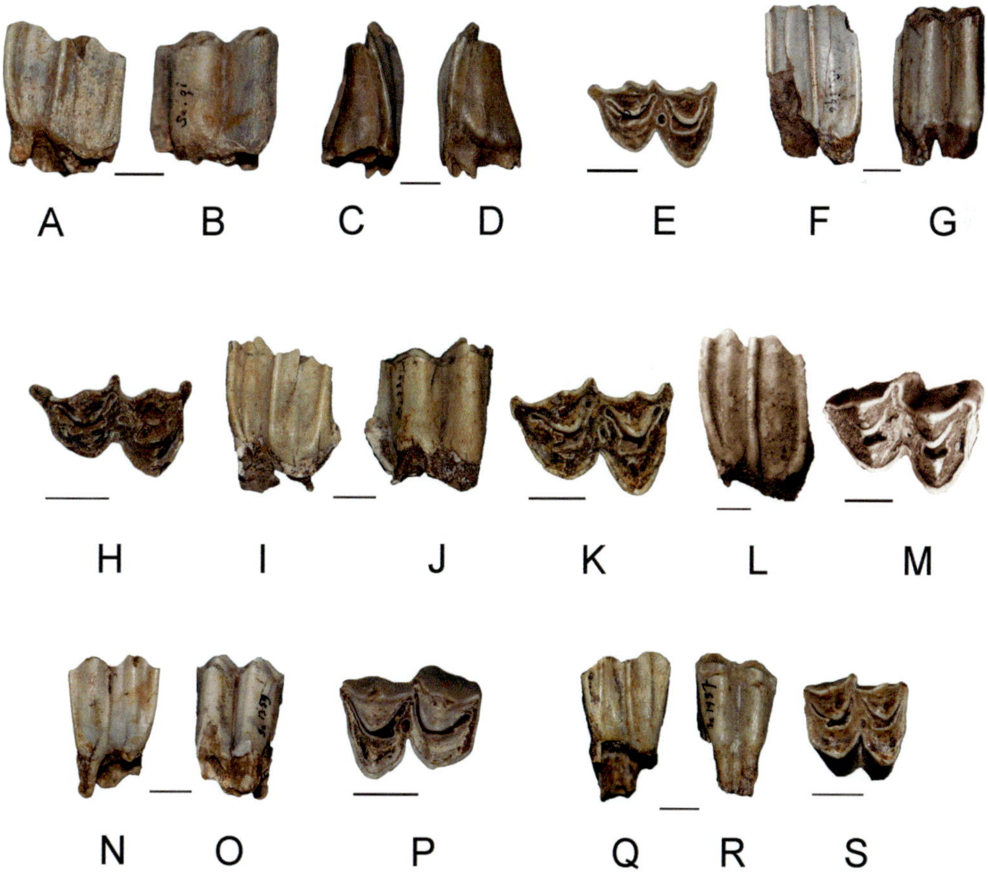

Fig. 14.53 Upper molars of *Gallogoral meneghinii* from Senèze. A, B, right M3 NMB Se 91; C, D, right M3 NMB Se 71; F to H, right M3 NMB Se 196; I to K, right M3 NMB Se 386; L, M, right M3 FSL 16890; N to P, left M2 Se 1359; Q to S, NMB Se 1937. A, F, I, L, N, Q, buccal view; B, G, J, O, R, lingual view; C, distal view; D, mesial view; E, H, K, M, P, S, occlusal view. Scale 1 cm

Table 14.43 Comparative dimensions of the upper premolars (in mm) in *Gallogoral meneghinii* from Senèze and Volax

P2 variable	Ht	1	2	3	5	
Senèze N	4	6	2	6	6	
Mean	11.6	11.9	12.1	11.8	12.1	
Minimum	9.6	10.3	11.5	11.3	11.0	
Maximum	13.1	12.8	12.7	12.5	13.1	
SD	1.45	0.93	0.87	0.59	0.77	
Volax						
LGPUT Vol 142	18.3	13.8	14.0			
LGPUT Vol 1	16.4	14.4	14.1	11.9	12.5	
P3 variable	Ht	1	2	3	5	
Senèze N	5	9	3	9	9	
Mean	16.7	14.1	12.9	14.0	15.0	
Minimum	12.5	12.5	11.2	12.3	14.5	
Maximum	22.6	15.5	14.0	14.7	16.1	
SD	4.01	0.89	1.48	0.73	0.49	
Volax						
LGPUT Vol 142	20.0	14.2	13.2			
LGPUT Vol 1	17.0	13.7	13.7	14.6	16.0	
P4 variable	Ht	1	2	3	4	5
Senèze N	6	9	3	9	9	9
Mean	16.7	13.8	12.8	14.6	13.7	15.5
Minimum	11.1	13.1	11.8	13.7	11.3	14.5
Maximum	23.2	15.3	14.7	14.9	14.8	17.6
SD	4.46	0.65	1.65	0.48	1.0	0.83
Volax						
LGPUT Vol 142	22.6	14.0	12.8			
LGPUT Vol. 155	20.3	14.8	13.7	15.2	14.0	16.5
LGPUT Vol 142	22.6	14.0				
LGPUT Vol 1		14.6		15.4	14.6	

Variable definitions (from Crégut-Bonnoure 1995, 2002, 2020) (Fig. 14.A.4: A): Ht, height; 1, occlusal length; 2, basal length. 3, occlusal width total. 4, occlusal width in the middle. 5, basal width

Table 14.44 Comparative dimensions of the upper M1 and M2 (in mm) in *Gallogoral meneghinii* from Senèze and other localities

M1 variable	Ht	1	2	3	4	5	6	7
Senèze N	10	13	7	12	12	13	13	11
Mean	17.9	18.9	17.0	17.3	16.6	15.7	15.3	19.4
Minimum	8.9	16.5	15.8	13.2	11.8	11.3	10.9	16.9
Maximum	32.6	22.2	18.6	19.7	19.5	18.6	18.1	21.1
SD	8.31	2.08	0.94	2.46	2.84	2.02	2.23	1.37
Volax N	6	6	6	3	3	3	2	2
Mean	21.3	20.9	18.4	16.8	15.4	14.4	12.9	18.9
Minimum	14.4	17.8	17.1	13.9	13.6	11.9	11.4	18.8
Maximum	28.7	23.0	20.2	18.8	16.4	16.6	14.4	19.0
SD	5.02	1.77	1.35	2.54	1.54	2.36	2.11	0.13
Olivola (NMB Ol 8–cast)		17.6		18.5	18.5	17.8	18.3	

(continued)

Table 14.44 (continued)

M2 variable	Ht	1	2	3	4	5	6	7
Senèze N	13	16	10	14	14	15	15	8
Mean	23.2	22.3	20.4	17.94	17.0	15.6	14.8	20.8
Minimum	16.8	19.4	18.5	14.0	14.0	12.9	12.5	19.1
Maximum	41.5	24.4	22.4	19.6	18.7	17.3	16.7	24.0
SD	7.34	1.33	1.30	1.83	1.58	1.22	1.45	1.51
Saint-Vallier								
MHNL 163-352	21.9	21.2	18.4	15.5	14.9	13.6	12.7	20.4
FSL 94027	25.0	20.4	17.1	13.9	12.7	12.4	12.1	17.8
Volax N	5	5	4	4	4	4	4	2
Mean	27.2	25.0	22.5	18.2	17.2	15.4	15.0	22.7
Minimum	23.8	24.2	20.1	17.0	14.9	13.6	12.9	21.4
Maximum	30.6	25.8	24.4	19.0	18.4	17.3	16.5	23.9
SD	3.11	0.66	2.00	0.85	1.56	1.57	1.52	1.77
Olivola (NMB Ol 8–cast)		23.4		20.7	19.0	18.4	16.0	

Variable definitions (from Crégut-Bonnoure 1995, 2002, 2020) (Fig. 14.A.B): Ht, height; 1, occlusal length; 2, basal length. 3, occlusal width total, first lobe; 4, occlusal width in the middle, first lobe; 5, occlusal width total, second lobe; 6, occlusal width in the middle, second lobe; 7, basal width

Table 14.45 Comparative dimensions of the upper M3 and upper deciduous teeth (in mm) in *Gallogoral meneghinii* from Senèze and other localities

M3 variable	Ht	1	2	3	4	5	6	7	8	9	10	11	12
Senèze N	12	17	10	16	16	10	15	16	10	13	13	11	11
Mean	25.1	25.3	25.2	16.4	15.4	20.0	13.7	13.3	16.6	8.6	9.7	22.1	21.9
Minimum	17.3	21.4	23.2	14.1	13.4	17.6	11.8	11.7	15.3	8.0	8.4	20.5	20.0
Maximum	35.4	28.1	28.1	18.3	17.5	21.9	15.9	15.0	17.6	9.4	11.2	23.9	24.0
SD	5.58	1.78	1.96	1.36	1.25	1.25	1.18	1.07	0.88	0.49	0.99	1.26	1.04
Volax													
LGPUT Vol 142 (L)	30.6	22.6											
LGPUT Vol 142 (R)		26.0	27.5	16.0	14.3	21.6	13.9	13.7	19.4	8.0	10.9	22.0	22.1
LGPUT Vol 1		26.8		18.3	16.3		15.9	13.9					
LGPUT Vol 155	35.3	24.9	27.0				14.4	13.3	17.8	8.8	10.3	23.6	24.3
Olivola (NMB Ol 8–cast)		26.0		17.7	16.9					10.2	11.0	21.9	
DP3 variable	Ht	1	2	3	4								
Senèze (NMB Se 1667)	12.0	17.6	15.8	9.9	9.4								

Variable definitions (from Crégut-Bonnoure, 1995, 2002, 2020): Ht, height; M3 (Fig. 14.A.4: C): 1, occlusal length; 2, basal length. 3, occlusal width total, first lobe; 4, occlusal width in the middle, first lobe; 5, basal width, first lobe; 6, occlusal width total, second lobe; 7, occlusal width in the middle, second lobe; 8, basal width, second lobe; 9, length of the interstylar surface, mesial; 10, length of the interstylar surface, distal; 11, occlusal length of the two lobes at the lingual side; 12, basal length of the two lobes at the lingual side. DP3 (Fig. 14.A.4: A): 1, occlusal length; 2, basal length. 3, occlusal width; 4: basal width

There are 13 M1, 16 M2 and 17 M3. The lobes are pinched on M1 and M2 (Fig. 14.53). There is a more or less elongated occlusal islet; a smaller one, rounded, is present on the M1 FSL 3610. On this specimen and another M1 an entostyle is present (7.8 and 7.5 mm height). The crown is curved with a mesostyle prominent especially at the top, less at the base because of the curvature of the crown. The parastyle is strong, and the valley behind it reaches the base of the crown as in *Pliotragus*. Unlike in the latter genus, the metastyle is strong. The metacone and the paracone are narrow and well inflated. There is no metastylar wing on the M3s from the Lyon collection (except on FSL 5008, but it is restricted to the base) nor on the isolated specimen from Paris (MNHN-F-1921-9). On the other specimens a more or less important metastylar wing exists (Fig. 14.53). It is straight and less developed than in *Pliotragus*. The first lobe is lingually pinched, as is the second lobe whose distal side is rounded. At the lingual base, unlike *Pliotragus* the separation of the root is not prolonged on the base of the distal lobe.

Lower Teeth

The lower teeth are less numerous than the upper ones (Tables 14.46 and 14.47). After revision, *Gallogoral* lower teeth include: the teeth from the four mandibles: FSL 3611 (right m3), FSL 5008 (left p4-m3; right p2-m3), FLS 16862 (left and right p2-m3), FSL 16866 (left p2-m3), also FSL 16870 (left m3), FSL 16891 (left m1-3), NMB unnumbered "neotype" (right p3-m3), NMB Se 103 (right p2-m3), NMB Se 195 (right p3-m1), NMB Se 313 (left and right p2-m3), NMB Se 563 (left p3-m2), NMB Se 1359 (right p3-m3), NMB Se 1438 (right m1-3), MNHN-F-1954 11-8+1954 11-10 (right m2-3) and MNHN-F-1921-9 (left m2). In addition, there are the teeth from the skeleton NMB Se 550, not studied here, and two mandibles belonging to young animals: NMB Se 538 (left dp2-4) and NMB Se 491 (left dp2-m1). The isolated teeth are: NMB Se 382 (right m2), NMB Se 389 (right m2), NMB Se 392 (left m1), NMB Se 393 (left m3), NMB Se 394 (left m3, fragmentary), NMB Se 396 (right m2). NMB Se 1360 (right m1), NMB Se 1433 (right p4-m1), NMB Se 1439 (right m1), NMB Se 1671 (right m2), NMB Se 1676 (left p2, p4-m3), MNHN-F-1954 11-8+1954 11-10 (right m2-3), MNHN-F-1921-9 (left m2).

On p2 (n = 8), the lingual conid is broad, well inflated and in median position. Its base is connected with the two stylids. The crown of p3 (n = 13) is buccally bilobed. On the unworn NNB Se 195 the buccal valley is "V" shaped. The parastylid and the paraconid are fused in a vertical wall, less strong than in *Pliotragus*, where the two structures are well separated; in occlusal view, it is directed forward and the mesio-buccal angle is rounded. The lingual valley is deep and extended practically to the base. Its distal branch is obliquely directed. In occlusal view, the metaconid is directed obliquely backwards; in lingual view, it is inflated in the upper third but the inflation disappears at the base. At this level, the conid is fused with the entoconid and the entostylid. In occlusal view, the entoconid is directed obliquely backwards. An occlusal islet exists at the level of the lingual fusion of the metaconid and the entoconid spur. The entostylid is well developed and prominent. On p4 (n = 15), the buccal side is bilobed, the distal lobe being shorter than the mesial but equally wide, whereas the lingual side is molarized; consequently, there is no valley on the lingual side and on the worn specimens the side is flat (Figs. 14.52: C, D and 14.53 A to C). However, on the worn p4 NMB Se 1459 and on the unworn teeth NMB Se 195, NMB Se 563 and NMB Se 1676, the metaconid is dilated but not clearly and a longitudinal notch separates the lobes; the entoconid is inflated and in occlusal view it constitutes a spur directed backward. In occlusal view, on FSL 5008, FSL 16862 and NMB Se 313, which are worn, the metaconid forms a loop. In front of it, the parastylid and the paraconid are fused, constituting a vertical wall. On NMB Se 313, more worn than the previous premolars, the lingual side is vertical, without any inflation; however, in occlusal view, a small rounded islet corresponds to the base of the area behind the loop.

Twenty m1, 18 m2 and 18 m3 are present (Fig. 14.54: D to O). As in *Pliotragus*, the goat fold disappears at the base of the crown and consequently it is absent on the worn specimens. The pinching of the lobes is less acute than in *Pliotragus*, but sometimes it is similar. As in *Pliotragus*, the lingual valley is broad but less deep; the parastylid is weak. The valley behind it doesn't reach the base of the crown. The metastylid is less developed than in *Pliotragus,* not prominent. The enamel has a rugose aspect. The third lobe of m3 is buccally rounded or oval. It is more or less elongated, directed outwards. On the isolated molars, the crown is curved. The cemento-enamel junction on the isolated m2s is tall. The second lobe is swollen at the base, wider than the first lobe, as in *Pliotragus*: on NMB Se 389 the width is 12.9 mm for the first lobe and 14.1 mm for the second, on NMB Se 1671 the values are respectively 13.0 and 14.4 mm and on NMB Se 396 they are 13.9 and 14.6 mm.

I have already discussed the different proportions between *Pliotragus* and *Gallogoral* molars (Figs. 14.34, 14.35, 14.36, 14.37, 14.38 and 14.39). The m3s from Senèze are smaller than the *Gallogoral* teeth from Saint-Vallier and Volax, and their proportions are distinct.

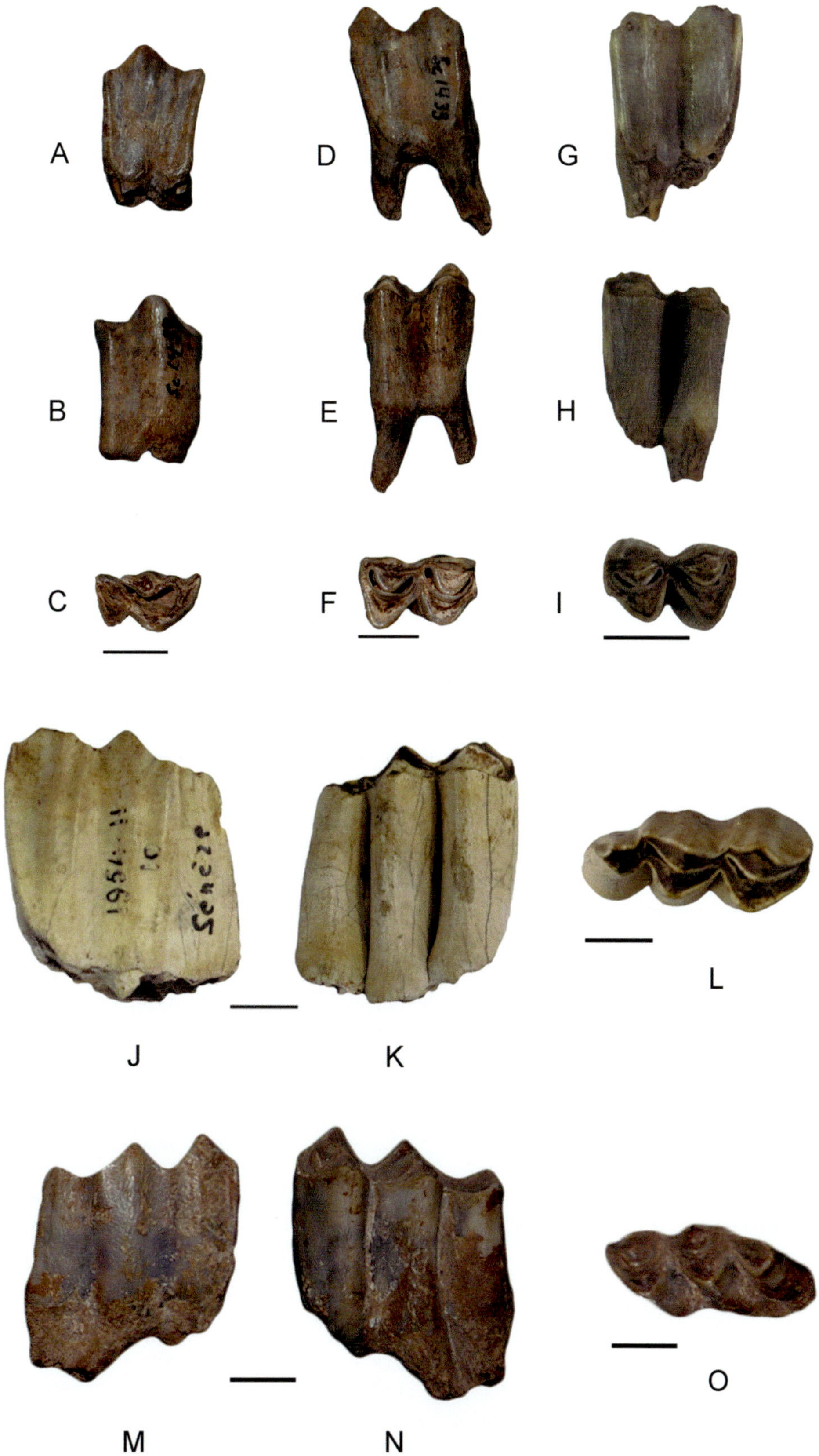

Fig. 14.54 Lower premolar and molars of *Gallogoral meneghinii* from Senèze. A to C, right p4 Se 1433; D to F, right m1 NMB Se 1439; G to I, left m2 MNHN-F-1921-9; J to L, right m3 MNHN-F-11; M to O, left m3 FSL 16870. A, D, G, J, N, lingual view; B, E, H, K, M, buccal view; C, F, I, L, O, occlusal view. Scale 1 cm

Table 14.46 Comparative dimensions of the lower premolars (in mm) in *Gallogoral meneghinii* from Senèze and other localities

p2 variable	Ht	1	5			
Senèze N	7	7	7			
Mean	9.6	9.6	6.7			
Minimum	8.3	8.8	6.0			
Maximum	11.4	10.3	7.5			
SD	1.16	0.56	0.53			
Saint-Vallier (MHNL 163–351)	9.5	10.7	7.6			
p3 variable	Ht	1	2	3	5	
Senèze N	9	11	5	11	10	
Mean	11.9	13.4	12.7	8.6	9.5	
Minimum	10.2	12.7	12.5	7.8	8.9	
Maximum	13.7	14.2	13.1	9.1	9.5	
SD	1.16	0.48	0.24	0.46	0.2	
Saint-Vallier (MHNL 163–351)	5.1	13.5	13.5	9.5	9.6	
Volax						
LGPUT Vol 22	20.8	16.2	15.3	9.7	9.8	
LGPUT Vol 10		18.9		10.3		
p4 variable	Ht	1	2	3	4	5
Senèze N	10	15	7	15	15	11
Mean	14.5	15.8	14.8	9.8	8.5	10.0
Minimum	10.0	14.6	14.1	9.0	7.2	8.8
Maximum	21.0	16.9	15.9	11.3	9.5	10.8
SD	3.28	0.66	0.63	0.67	0.72	0.62
Volax						
LGPUT Vol 21	29.7	19.2	18.4	10.0	7.5	10.8
LGPUT Vol 10		16.7		9.8		

Variable definitions (from Crégut-Bonnoure 1995, 2002, 2020) (Fig. 14.A.4: D): Ht, height; 1, occlusal length; 2, basal length; 3, occlusal width, first lobe; 4, occlusal width, second lobe; 5, basal width

Table 14.47 Comparative dimensions of the lower molars (in mm) in *Gallogoral meneghinii* from Senèze and other localities

m1 variable	Ht	1	2	3	4	5	
Senèze N	14	18	10	18	15	12	
Mean	17.4	18.3	17.1	11.2	11.9	12.6	
Minimum	8.3	13.8	15.3	10.3	9.6	11.6	
Maximum	29.1	22.8	19.1	12.6	13.5	13.6	
SD	8.52	2.32	1.28	0.72	1.03	0.69	
m2 variable	Ht	1	2	3	4	5	
Senèze N	12	18	10	17	17	12	
Mean	27.4	22.1	19.6	12.0	12.0	14.0	
Minimum	12.7	20.0	17.3	10.6	10.8	12.9	
Maximum	38.6	25.0	20.5	13.6	13.6	16.2	
SD	8.95	1.41	0.97	0.88	0.95	0.99	
Saint-Vallier (MHNL 163-351)		17.8		13.0	14.6	14.0	
m3 variable	Ht	1	2	3	4	5	6
Senèze N	9	16	8	18	17	15	8
Mean	25.5	31.2	31.3	12.2	11.5	6.6	13.3
Minimum	14.6	27.2	27.4	9.9	10.1	5.5	10.6
Maximum	43.4	33.9	33.8	14.3	13.5	7.6	15.1
SD	8.55	1.86	1.88	1.17	1.01	0.69	1.39
Saint-Vallier (MHNL 163-351)		32.8	32.8	13.8	13.4	8.5	13.6
Volax (LGPUT Vol 223)	45.8	33.9		11.2	11.8	9.6	14.8

Variable definitions (from Crégut-Bonnoure, 1995, 2002, 2020): Ht, height; m1 and m2 (Fig. 14.A.4: E): 1, occlusal length; 2, basal length; 3, occlusal width, first lobe; 4, occlusal width, second lobe; 5, basal width. m3 (Fig. 14.A.4: F): 1, occlusal length; 2, basal length; 3, occlusal width, first lobe; 4, occlusal width, second lobe; 5, occlusal width, third lobe; 6, basal width

Postcrania

Most bones have been studied by Guérin (1965) and described with numerous drawings and photographs; the vertebrae, ribs and sesamoids were not studied here.

Scapula

There are nine scapulae (Table 14.34) (Fig. 14.40): one left isolated scapula from the FSL 5008 skeleton (Guérin 1965: pl. XXXV p. 179, pl. XXXVI p. 181, pl XXXVII p. 183), the right and left from the skeletons NMB Se 313 and NMB Se 550, three right isolated (NMB Se 1449, NMB Se 1453, NMB Se 1473) and one left belonging to a neonatal individual (NMB Se 343a).

The glenoid fossa is oval. The coracoid apophysis is not strong and weakly curved backward. The acromion is poorly developed, but the infraglenoidal tubercle is prominent.

Humerus

There are seven humeri (Table 14.35) (Fig. 14.41): the left from the FSL 5008 skeleton (Guérin 1965: pl XXXVIII p. 185, pl XXXVIX p. 187), the right and left from the skeletons NMB Se 313 and NMB Se 550 and two right isolated and incomplete specimens (NMB Se 361, NMB Se 1626).

The diaphysis is nearly straight in cranial view, sinuous in lateral view; if the bone is set on its distal articulation, it stays in a vertical position. The greater tubercle is higher than the head. In lateral view, the capitulum is rounded. The fossa is deeper cranially. Its caudal side is bordered by a weakly developed bulge. In medial view, there is a shallow longitudinal fossa, above the limit between the trochlea and the medial epicondyle. In cranial view, the coronoid fossa is triangular and elongated. It is deeper from the middle of the trochlea to the lateral lip of the trochlea. The trochlear lip is pinched in its proximal border. In caudal view, the lateral epicondyle is shorter than the medial one, with a truncated base. The olecranon fossa is very deep and broad.

Radius

There are eight radii (Table 14.36) (Figs. 14.42 and 14.55: A): the right and left from the FSL 5008 skeleton (Guérin 1965: pl. XL p. 193, pl. XLI p. 195, pl. XLIII p. 199), the right and left from the NMB Se 313 and NMB Se 550 skeletons, a left FSL 16869 and a proximal right NMB Se 1674.

The diaphysis is straight in dorsal view whereas it is convex in lateral and medial views. The bone is flat except at the base where it is thick, and the diaphysis is pinched medially. In dorsal view, the lateral tuberosity is not very prominent and lower than the upper side of the proximal articulation (more or less 13 mm). Near this tuberosity, the bone has an ovoid depression. The dorsal border of the articulation for the capitulum is much higher than the coronoid process. The surface of the articulation is extremely concave. In palmar view, the facets for the ulna are rounded and their bases are at the same level. At the distal epiphysis, the attachment for the base of the ulna is prominent and the lateral border is pinched. The dorsal groove for the *extensor carpi radialis* is well defined. On the palmar side, the synostosed area for the fusion of the diaphysis and the epiphysis is identifed by a thick bulge. The facet for the scaphoid is dorsally deep to receive its proximal tubercle. The crest separating this facet from that for the semi-lunar is pinched, especially in its palmar part.

Ulna

There are seven ulnae (Table 14.36): the right and left from the FSL 5008 skeleton (Guérin 1965: pl. XLIV p. 203), the right and left from the NMB Se 313 and NMB Se 550 skeletons and one left (FSL 16879).

The olecranon process is strong with a straight medial side. Its tuberosity is weakly developed. In lateral view, the general shape between the olecranon and the anconeal processes is rectangular, with a concave cranial side. However, the cranio-palmar diameter at the level of the anconeal process is greater than at the olecranon. The troclear incisure is broad and the facet for the radius is rounded. The surface of the radio-ulnar synostosis is short and triangular (Figs. 14.52, 14.53 and 14.54)

Carpals

It was not possible to extract the carpals of the skeletons in Basel (NMB Se 313 and NMB Se 550). Consequently, the observations concern the bones of the FSL 5008 skeleton and the isolated specimens (Fig. 14.55). Their dimensions are given in Table 14.48.

There are seven scaphoids (Fig. 14.55: F): the right and left from the FSL 5008 skeleton (Guérin 1965: pl. XLVII Figs. 6–9 p. 211), the right and left from the NMB Se 313 and NMB Se 550 skeletons, and a left specimen (NMB Se 431). In proximal view, the lateral angle is very important, acute, with sides more or less perpendicular each other. In distal view, this lateral angle is clearly visible. The proximal tubercle is large and prominent. Behind it, the surface of the articulation is clearly concave. The distal articulation is more or less rectangular with a concave medial side and a broad dorsolateral angle. The medial tubercle is well defined.

There are six semi-lunars (Fig. 14.55: G): the right and left from the FSL 5008 (Guérin 1965: pl. XLVI Figs. 1–5

p. 209), NMB Se 313 and NMB Se 550 skeletons. The proximal articulation has a thin dorsomedial tubercle, elongated and perpendicular to the dorsopalmar axis. The dorsolateral tubercle is obliquely directed to the center of the bone, more developed than the dorsal one. The medial notch for the lateral protrusion of the scaphoid is widely open and deep. The palmomedial angle is broad and more or less rectangular. The palmar side has a strong tubercle laterally. On the medial side, the dorsodistal facet for the scaphoid is well developed and quadrangular. In distal view, the dorsal side is shorter than the palmar one, with a pinched palmar border. At its middle, the bone is narrower than on the dorsal and palmar sides and its surface is concave. The two dorsal facets are elongated and well differentiated.

There are five pyramidals (Fig. 14.55: H): the right from the FSL 5008 skeleton (Guérin 1965: pl. XLV Figs. 1–5 p. 205), and the right and left from the NMB Se 313 and NMB Se 550 skeletons. The proximal surface is weakly concave. The surface of the articular facet for the pisiform is rectangular, more or less flat; however, in lateral view the palmar side is convex. The disto-palmar angle is rounded and there is a tubercle on the lateral side. The medial facet for the semi-lunar is well developed and triangular. An elongated pit is located medially on the dorsal side.

There are six pisiforms (Fig. 14.55: I): the right and left from the FSL 5008 (Guérin 1965: pl. XLV Figs. 8–9 p. 205), NMB Se 313 skeleton and NMB Se 550 skeletons. The palmar side is well rounded. The facet for the pyramidal is flat.

There are seven capitato-trapezoids (Fig. 14.55: J): the right and left from the FSL 5008 (Guérin 1965: pl. XLVII Figs. 7–10 p. 211), NMB Se 313 and NMB Se 550 skeletons and a left (FSL 16865). The lateral side is weakly angled with an elongated palmar angle: at this level, the lateral facet is enlarged and projected laterally. This angle in palmar view is projected laterally, more than the lateral angle of the distal articulation. The proximal surface is dorsally shallow on the medial side of the dorsopalmar crest, which is prominent. The dorsolateral facet is oblique and directed downward. In palmar view, the surface between the medial side and the dorsopalmar crest is clearly concave. In medial view, the palmar side constitutes distally a kind of process directed downward.

There are six unciforms (Fig. 14.55: K): the right and left from the FSL 5008 (Guérin 1965: pl. XLVII Figs. 1–4 p. 211), NMB Se 313 and NMB Se 550 skeletons. The proximal facets are strongly concave and the median crest is prominent. The dorsomedial facet has its medial side obliquely directed. The palmar tubercle is compressed on each side and projected palmarly.

Metacarpal III+IV

There are 11 metacarpals III+IV (Table 14.49) (Fig. 14.55: B to E): the right and left from the FSL 5008 (Guérin 1965: pl. XLVIII p. 215, pl. XLIX p. 217, pl. L p. 219), NMB Se 313 and left from NMB Se 550 skeletons, two other right (NMB Se 1791, FSL 16878) and three left (NMB Se 400 [juvenile], FSL 16873, FSL 16874). On the proximal epiphysis, the dorsal tubercle is weak. At its base, it forms a transverse bulge; above it, the surface constitutes a slight cavity. At the proximal articulation, the contact between the unciform and the capitato-trapezoid surfaces is pinched palmarly on NMB Se 400 and NMB Se 1791. On FSL 5008, a supplementary facet exists at the palmar side of these two proximal facets. There is a single synovial fossa on the facet for the capitato-trapezoid; however, it is missing on FSL 16878. The dorsal sulcus is obliterated, but distally its foramen (interosseous canal) is relatively broad and deep. The palmar groove is proximally deep and narrow. The distal tubercles are faint. The keels on each side are parallel.

A left metacarpal from "Montagne de Perrier" ("niveau supérieur") was first attributed to *Gazellospira torticornis* (*Paleoras torticornis*; Depéret 1884: pl. VIII, Fig. 5; MNHN-F-53, Bravard collection and cast NMB Prr 525), then to *Gallogoral* (*Nemorhoedus*: Schaub 1922; Guérin 1965). Its fossilization is distinct from those of the Villafranchian Perrier-Pardines and Perrier-Rocca-Neyra samples. The shaft is shorter than in *Gallogoral*, the supra-articular tubercles are well developed and project on the palmar side. In fact, it is a *Capra* specimen dating from the Late Pleistocene (Crégut-Bonnoure 2002, 2020). The proportions of the Senèze, Volax and Olivola metapodials are not exactly the same but the Perrier specimen MNHN-F-53 is clearly distinct (Fig. 14.56).

Table 14.48 Dimensions of the carpals (in mm) in *Gallogoral meneghinii* from Senèze

Scaphoid variable	1	2	3	4	5	6	7	8	9	10		
FSL 5008 (R)	21.0	30.6	28.4	21.2	24.8	17.5	18.3	5.7	18.7	8.9		
FSL 5008 (L)	22.0	31.2	28.4	21.2	26.8	16.7	18.8	5.8	17.4	9.2		
NMB Se 431	22.0	32.8	28.3	20.4	26.3	16.3			19.6	8.1		
Semi-Lunar variable	1	2	3	4	5	6	7	8	9	10		
FSL 5008 (R)	21.8	25.5		29.7		16.5		13.3	17.1	7.5		
FSL 5008 (L)	21.7	26.7	33.4	20.8	21.1	16.7	11.0	13.6	16.8	8.5		
Semi-Lunar variable	11	12	13	14	15	16	17	18	19	20		
FSL 5008 (R)	7.7	10.2	7.7	9.5		6.0	11.3	16.7	5.1	10.2		
FSL 5008 (L)	7.0	8.2	9.3	10.4	21.0	5.6	10.0	16.1	4.8	10.2		
Semi-Lunar variable	21	22	23	24								
FSL 5008 (R)	6.0											
FSL 5008 (L)	5.3	12.3	6.8	21.6								
Pyramidal variable	1	2	3	4	5	6	7	8	9	10		
FSL 5008 (R)	19.4	17.5	18.4	14.9	8.0	22.6	12.1	13.0	12.8	9.0		
Pisiform variable	1	2	3	4	5							
FSL 5008 (R)	18.5	25.0	17.3	7.6	13.2							
FSL 5008 (L)	18.3	24.5	17.4	9.1	12.8							
NMB Se 313	19.0	27.4			13.7							
Unciform variable	1	2	3	4	5	6	7	8				
FSL 5008 (R)	18.0	26.7	18.7	23.7	17.8	9.3	10.7	8.8				
FSL 5008 (L)	17.8	28.0	18.6	23.5	17.2	10.7	12.0	9.5				
Unciform variable	9	10	11	12	13	14	15	16				
FSL 5008 (R)	21.3	6.2	6.0	23.2	17.1	19.2	17.1	15.5				
FSL 5008 (L)	21.3			22.1	16.5	18.7	18.1	16.0				
Capitato-trapezoid variable	1	2	3	4	5	6	7	8	9	10	11	12
FSL 5008 (L)	27.7	25.7	19.1	9.3	7.3	19.9	7.16	19.9	6.8	22.7	32.9	21.0
FSL 5008 (R)	27.3	26.8	19.6	8.8	7.8	22.3	7.6	21.4	6.1	24.3	33.9	22.2
FSL 16865	26.1	26.2	19.6	8.2	6.5	21.1	6.73	20.3	6.5	23.2	33.7	21.8

Variable definitions (from Crégut-Bonnoure 1987, 2002, 2020): dpd, dorsopalmar diameter; td, transversal diameter. Scaphoid (Fig. 14.A.9: A to C): Height: 1, total; 8, proximal articular facet, palmar side; 10, distal articular facet, dorsal side. Dpd: 2, total; 3, proximal articulation; 5, distal articulation; 7, proximal articular facet, palmar side; 9, distal articular facet, dorsal side. Td: 4: proximal articulation; 6, distal articulation

Semi-lunar (Fig. 14.A.9: D to H): 1, height. Dpd proximal: 2, in the middle; 3, from dorsolateral angle to palmolateral angle; 4, in anatomical position. Td proximal: 5: dorsal side; 6, palmar side; 7, minimum, in the middle. Td distal: 8, dorsal side; 9, palmar side; 10, dorsolateral articulation; 11, dorsomedial articulation; 12, palmomedial articulation; 13, palmolateral articulation. Dpd: 14, distal crest. Td: 15, minimum dorsal side. Medial side: 16, height of the proximodorsal articulation; 17, td of the proximodorsal articulation; 18, td of the proximopalmar articulation; 19, minimum height of the proximopalmar articulation; 20, td of the dorsodistal articulation; 21, height of the dorsodistal articulation. Lateral side: 22, td of the proximodorsal articulation; 23, minimum height of the proximodorsal articulation. Dpd: 24, lateral side

Pyramidal (Fig. 14.A.9: I to K): Height: 1, total; 2, minimum of the dorsal side; 3, medial of the dorsal side; 4, of the articulation for the pisiform; 10, of the proximodorsal articulation of the medial side. Td: 5, of the articulation for the pisiform; 7, of the distal articulation; dpd: 6: of the distal articulation; 8, proximodorsal of the lateral side; 9, of the proximodorsal articulation of the medial side

Pisiform (Fig. 14.A.9: L): 1, total height; 2, dpd. Pyramidal articulation: 3, proximodistal diameter; 4, td; 5, total lateromedial diameter

Variable definitions (from Crégut-Bonnoure 1987, 2002, 2020): dpd, dorsopalmar diameter; td, transversal diameter. Unciform (Fig. 14.A.9: M to P): 1, total height. 2, dpd. Td: 3, total; 4, proximal articulation, dorsal side; 5, lateroproximal articulation, dorsal side; 6, medio proximal articulation, dorsal side; 7, lateroproximal articulation, palmar side; 8, medio proximal articulation, palmar side; 9, dorsal side, minimum. Articulation of the medial side: 10, proximodistal diameter; 11, dpd. Distal articulation: 12, dpd; 13, td. Proximal articulation: 14, td, palmar side; 15, dpd, medial side; 16, dpd of the median crest

Capitato-trapezoid Fig. 14.A.9: Q to S): dpd, dorsopalmar diameter; td; transversal diameter. 1, td in anatomical position; 2, dpd in anatomical position; 3, total height; 4, height of the dorsal rim; 5, td of the latero-proximal facet, dorsal side; 6, td of the medioproximal facet, dorsal side; 7, td of the lateroproximal facet, palmar side; 8, td of the medioproximal facet, palmar side; 9, td of the lateroproximal facet (at the base of the dorsal rim); 10, dpd of the proximal ridge; 11, td maximum of the distal articulation; 12, dpd maximum of the distal articulation

L, left; R, right

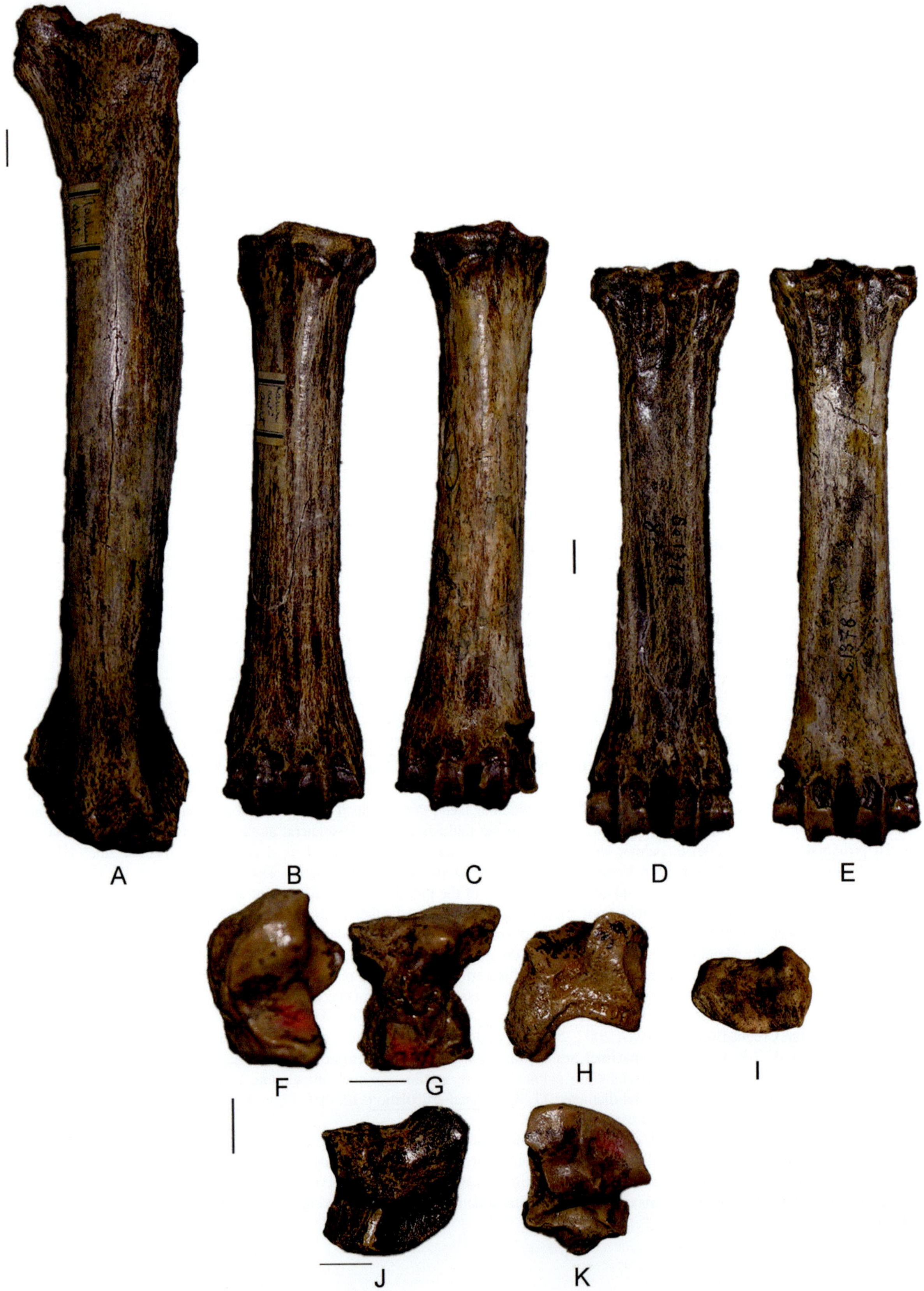

Fig. 14.55 Post-cranium of *Gallogoral meneghinii* from Senèze. A, right radius FSL 5008; B, D, right metacarpal III+IV FSL 5008; C, E, left metacarpal III+IV FSL 5008; F, left scaphoid FSL 5008; G, left semi-lunar FSL 5008; H, left pyramidal FSL 5008; I, left pisiform J, left capitato-trapezoid FSL 5008; K, left unciform FSL 5008. A, B, C, dorsal view; D, E, palmar view; F, G, J, K, proximal view; H, lateral view; I, medial view. Scale 1 cm

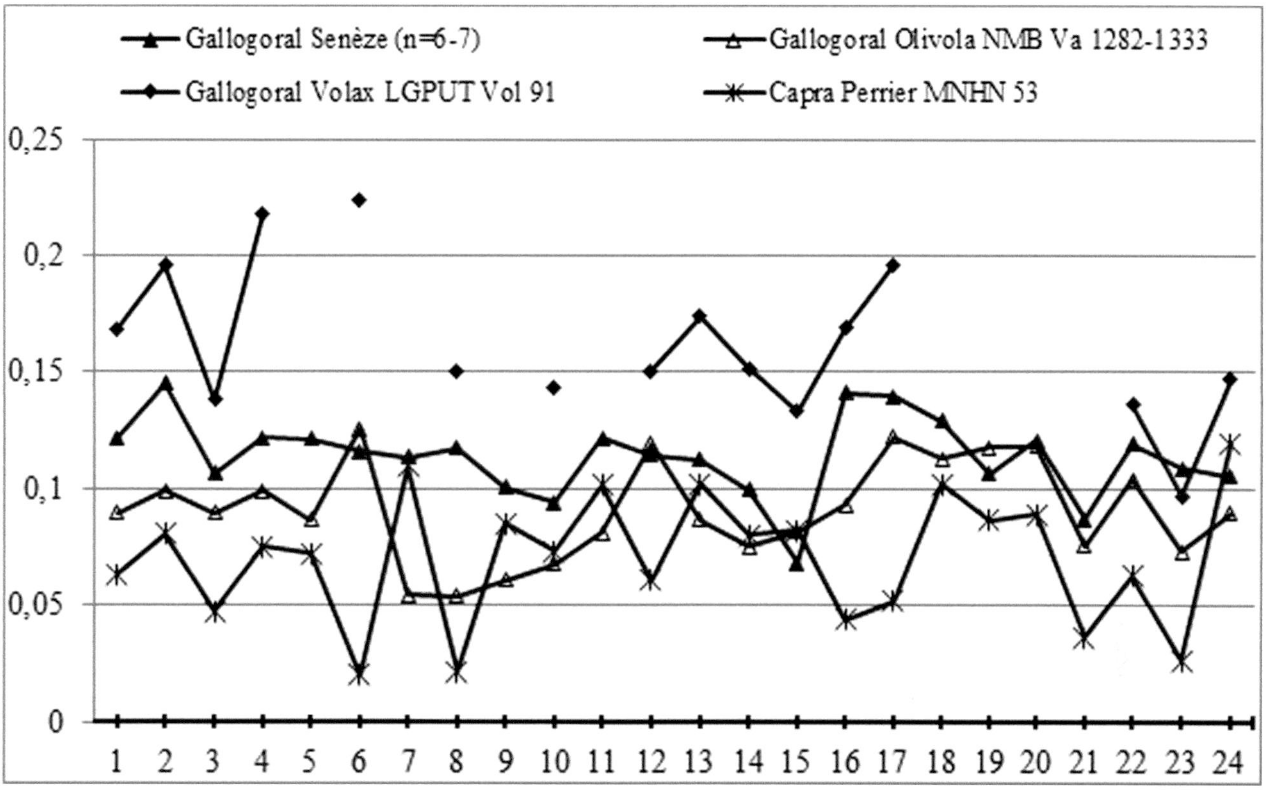

Fig. 14.56 *Gallogoral meneghinii* from Senèze and other localities. Ratio diagram of the metacarpal III+IV. Reference axis: recent *Capra ibex* ♂+♀ (in Crégut-Bonnoure 2020). dpd, dorsopalmar diameter; td, transversal diameter. Length: 1, total; 2, minimum at the middle; 3: minimum of the medial side. Proximal epiphysis: 4, td total; 5, td of the articulation; 6, td of the unciform facet; 7, td of the capitato-trapezoid facet; 8, oblique length of the unciform facet; 9, oblique length of the capitato-trapezoid facet; 10, dpd; 11: dpd of the articulation. Middle of the diaphysis: 12, td; 13, dpd. Distal epiphysis: 14, td of the epiphysis; 15, td supra-articular; 16, dpd of the epiphysis; 17, dpd of the epiphysis at the foramen level; 18, distance between the keels on the palmar side; 19, distance between the keels at their base; 20, distance between the keels at their middle; 21, dpd of the medial condyle; 22, dpd of the medial keel; 23, dpd of the lateral condyle; 24: dpd of the lateral keel. References of the measurements in Fig. 14.A.10 and dimensions in Table 14.49

Table 14.49 Comparative dimensions of the metacarpal III+IV (in mm) in *Gallogoral meneghinii* from Senèze and other localities, and in *Capra* sp. from Perrier

Variable	1	2	3	4	5	6	7	8	9	10	11	12
G. meneghinii												
Senèze N	7	6	7	7	7	7	7	7	7	7	7	7
Mean	173.2	156.8	165.2	42.4	41.4	17.3	26.0	26.3	31.8	27.4	26.6	27.5
Minimum	167.9	152.6	160.5	41.7	40.4	16.1	25.2	25.3	26.7	26.5	24.9	26.9
Maximum	177.9	160.8	170.3	43.3	42.5	18.7	27.7	28.2	33.5	28.3	32.0	28.6
SD	3.33	2.92	3.02	0.55	0.81	0.92	0.90	0.99	2.3	0.67	2.48	0.61
Val d'Arno (NMB Va 1282-1333–cast)	160.8	147.2	153.3	40.3	38.3	17.7	22.7	22.7	29.0	25.8	24.3	27.8
Volax (LGPUT Vol 91)	192.8	171.3	184.0	53.0		22.2		28.4		30.7		29.8
Capra sp.												
Perrier (MNHN-F-53)	151.4	145.3	135.3	38.1	37.0	13.9	25.8	21.1	30.7	26.1	25.4	24.3

(continued)

Table 14.49 (continued)

Variable	13	14	15	16	17	18	19	20	21	22	23	24
Senèze N	7	8	8	8	7	7	7	6	8	7	7	7
Mean	18.5	45.0	41.2	23.6	20.3	27.9	25.6	22.0	25.2	17.3	26.0	16.12
Minimum	16.8	41.6	25.6	21.9	16.1	27.0	25.0	21.6	20.3	16.7	25.1	15.2
Maximum	20.0	46.3	45.3	26.1	22.4	30.0	26.5	22.6	20.7	18.4	26.7	16.8
SD	0.98	1.49	6.58	1.52	2.08	0.98	0.51	0.45	2.18	0.66	0.57	0.62
Val d'Arno (NMB Va1282-1333–cast)	17.5	42.5	42.5	21.2	19.5	26.9	26.3	21.9	24.5	16.7	23.9	15.6
Volax (LGPUT Vol 91)	21.4	50.7	47.8	25.2	23.1					18.0	25.3	17.8
Capra sp.												
Perrier (MNHN-F-53)	18.1	43.0	42.5	18.9	16.6	26.2	24.5	20.5	22.4	15.2	21.5	16.7

Variable definitions (from Crégut-Bonnoure 1995, 2002, 2020) (Fig. 14.A.10): dpd, dorsopalmar diameter; td, transversal diameter. Length: 1, total; 2, minimum at the middle; 3, minimum of the medial side. Proximal epiphysis: 4, td of the epiphysis; 5, td of the articulation; 6, td of the unciform facet; 7, td of the capitato-trapezoid facet; 8, length of the unciform facet, diagonally from lateropalmar angle to dorsal point of the proximal crest; 9, length of the capitato-trapezoid facet, diagonally from mediopalmar angle to dorsal point of the proximal crest; 10, dpd of the epiphysis; 11, dpd of the articulation. Middle of the diaphysis: 12, td; 13, dpd. Distal epiphysis: 14, td of the articulation;15, td supra-articular;16, dpd supra-articular;17, dpd at the level of the dorsal foramen;18, distance between the lateral edges of the keels at the palmar top; 19, distance between the lateral edges of the keels at the palmar base; 20, distance of the keels at their middle; 21, td of the medial keel; 22, td of the medial condyle; 23, td of the lateral keel; 24, td of the lateral condyle

Anterior Phalanges

There are 15 first phalanges (Table 14.50) (Fig. 14.57: E to H): four from the FSL 5008 skeleton (Guérin 1965: pl. LI p. 225), one from the NMB Se 313 skeleton, one each from NMB Se 752, NMB Se 1382, and NMB 1752, four numbered NMB 1832a, an isolated specimen also labeled FSL 5008 and FSL 16863 (two specimens). The sesamoid articular facets are well differentiated, the axial one being rounded and the abaxial rectangular. Palmarly, the proximal tubercles are short. The axial one is more or less triangular with a distal point. In palmar view, the abaxial side is higher than the axial one, and the distal condyles are very asymmetrical.

There are 11 second phalanges (Table 14.51) (Fig. 14.57: K, M): four from the FSL 5008 skeleton (Guérin 1965: pl. LII p. 229), one from the NMB Se 313 skeleton, NMB Se 1392, NMB 1832a (two specimens), FSL 210686 and FSL 16863 (two specimens). The dorsal side is pinched and there is a dorsal tubercle in the axis of the dorsoproximal extensor protuberance. In proximal view, the abaxial side is more developed than the axial one, and the palmar angle is strongly projected palmarly. In palmar view, this angularity constitutes a pinched crest which runs to the distal articulation. Another one, in symmetrical position, exists on the axial side. In axial view, the distal fossa is deep. On the abaxial side, there is a kind of vertical groove in the middle of the diaphysis.

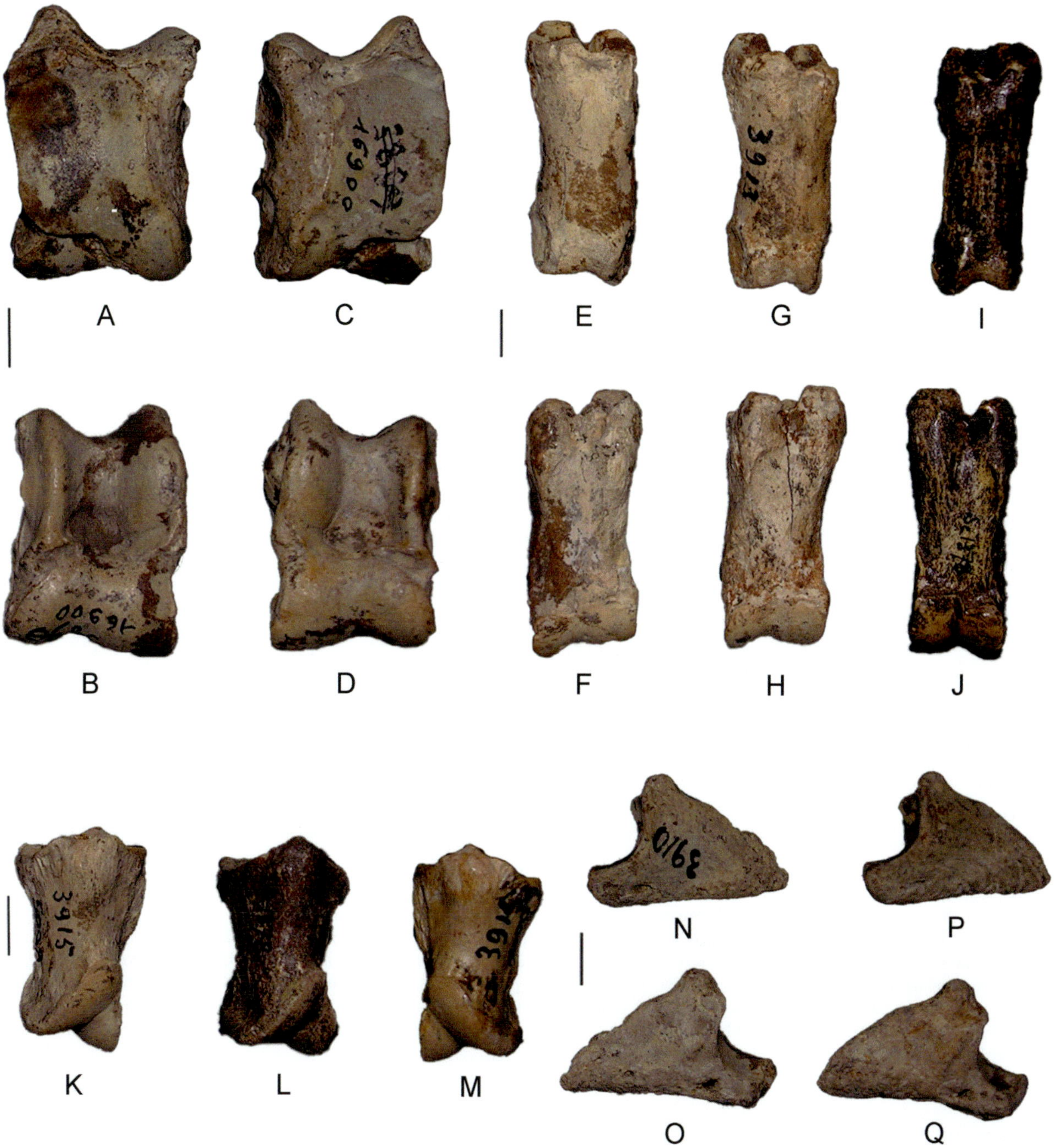

Fig. 14.57 Post-cranium of *Gallogoral meneghinii* from Senèze. A, B, left talus FSL 16900; C, D, right talus FSL 16900; E to H, anterior first phalanges FSL 5008 (=3913); I, J, posterior first phalanx FSL 5008; K, M, anterior second phalanges FSL 5008 (=3915); L, posterior second phalanx FSL 5008 (=3978); N, O, anterior third phalanx FSL 5008 (=39150); P, Q, posterior third phalanx FSL 5008. A, C, plantar view, B, D,; E, G, I, K, L, M, dorsal view; F, H, J, palmar/plantar view; N, P, abaxial view; O, Q, axial view. Scale 1 cm

There are 10 third phalanges (Table 14.52) (Fig. 14.57: N, O): four from the FSL 5008 skeleton (Guérin 1965: pl. LIII p. 251), an isolated specimen also numbered FSL 5008, NMB Se 1392, NMB 1832a and FSL 16863 (three specimens). The dorsal extensor protuberance is well developed and prominent, rounded in lateral view.

Innominate

There are four innominates: the right and left from the NMB Se 313 and NMB Se 550 skeletons (Guérin 1965: pl. LV p. 243). It was not possible to study them because the skeletons are mounted. However, there is a cast of the left

Table 14.50 Dimensions of the first phalanges (in mm) in *Gallogoral meneghinii* from Senèze

Anterior variable	1	2	3	4	5	6	7	8	9	10	11	12	13
Senèze N	15	15	14	14	15	15	15	15	15	14	14	14	15
Mean	46.2	47.3	48.7	22.5	23.7	21.1	19.9	6. 5	9.5	6.3	7.62	18.6	15.1
Minimum	41.4	45.9	46.9	21.4	22.6	19.4	18.5	5.9	7.0	5.2	6.6	16.1	13.5
Maximum	49.4	49.4	50.7	24.9	25.2	22.1	23.1	7.3	11.0	7.4	8.9	21.0	16.1
SD	2.24	1.43	1.26	0.92	0.82	0.66	1.31	0.42	0.93	0.61	0.71	1.35	0.86
Anterior variable	14	15	16	17	18	19	20	21	22	23	24	25	26
Senèze N	14	14	13	14	15	15	15	14	15	15	15	15	15
Mean	4.9	5.6	22.7	20.0	18.6	17.8	20.7	20.0	15.5	16.6	18.6	14.7	18.2
Minimum	4.1	4.4	19.5	18.1	17.2	16.9	18.4	19.0	14.3	13.9	17.3	13.9	14.6
Maximum	5.8	6.8	23.7	21.5	20.0	20.3	22.7	21.6	17.4	17.6	19.1	16.0	21.6
SD	0.47	0.67	1.09	1.16	0.73	0.67	0.98	0.73	0.88	1.11	0.59	0.58	2.11
Posterior variable	1	2	3	4	5	6	7	8	9	10	11	12	13
Senèze N	8	8	8	10	10	10	10	11	11	10	10	11	9
Mean	44.8	44.8	47.5	20.6	23.5	19.3	19.0	6.2	8.4	5.9	6.6	18.1	14.7
Minimum	43.5	43.1	46.6	19.9	22.8	18.6	18.0	5.0	7.5	4.7	6.0	13.5	13.2
Maximum	48.4	48.6	51.8	21.9	24.4	19.9	20.3	7.0	9.3	6.7	7.1	19.7	18.2
SD	1.55	1.61	1.75	0.7	0.55	0.41	0.69	0.55	0.53	0.58	0.34	1.76	1.73
Posterior variable	14	15	16	17	18	19	20	21	22	23	24	25	26
Senèze N	10	10	10	9	8	8	8	8	8	8	8	8	8
Mean	4.6	4.3	21.9	18.7	16.4	16.5	18.1	17.8	13.6	15.5	15.5	13.5	21.3
Minimum	3.8	3.1	20.6	17.0	15.3	16.0	17.4	16.9	13.1	14.9	14.9	12.9	16.9
Maximum	5.6	5.0	23.4	22.0	17.3	17.0	18.5	18.5	14.1	15.8	15.8	14.0	23.7
SD	0.61	0.52	0.83	1.42	0.8	0.45	0.35	0.59	0.33	0.29	0.26	0.29	2.02

Variable definitions (from Crégut-Bonnoure 1995, 2002, 2020) (Fig. 14.A.11: A to F): dppd: dorsopalmar/plantar diameter; td: transversal diameter. Length: 1, middle; 2, axial; 3, abaxial. Proximal epiphysis: 4, td; 5, dppd. Proximal articulation: 6, td of the proximal articulation; 7, dppd in the middle; 8, td of the proximo-axial facet; 9, td of the proximoabaxial facet. 10, td of proximoaxial sesamoid facet; 11, td of the proximoabaxial sesamoid facet, 12, dppd of the proximoaxial facet; 13, dppd of the proximoabaxial facet; 14, dpd of the proximoaxial sesamoid facet; 15, dppd of the proximoabaxial sesamoid facet; 16, dppd of the proximoaxial articulation; 17, dppd of the proximoabaxial articulation. Middle of the diaphysis: 18, td; 19, dppd. Distal articulation: 20, td; 21, td of the distal epiphysis; 22, dppd minimum; 23, dppd of the distoaxial condyle; 24, dppd of the distoabaxial condyle. 25, dppd sub-articular distal; 26, height on the palmar side, from the base of the proximo-axial tubercle to the top of the disto-axial condyle

Table 14.51 Dimensions of the second phalanges (in mm) in *Gallogoral meneghinii* from Senèze

Anterior variable	1	2	3	4	5	6	7	8	9
Senèze N	11	11	11	11	10	11	11	11	11
Mean	34.5	33.3	29.6	22.6	24.3	21.5	15.2	10.6	11.5
Minimum	31.9	32.4	28.6	21.9	22.0	19.9	13.6	9.7	10.5
Maximum	35.5	35.1	31.4	23.3	25.3	22.6	16.7	11.2	12.3
SD	0.97	0.89	0.82	0.41	0.93	0.65	0.87	0.51	0.63
Anterior variable	10	11	12	13	14	15	16	17	18
Senèze N	11	10	11	11	11	11	11	11	11
Mean	16.0	19.5	15.8	17.9	17.4	15.1	22.2	15.2	21.0
Minimum	15.0	16.1	15.2	16.7	16.4	14.2	20.2	14.3	19.5
Maximum	16.9	22.1	16.4	19.7	18.6	16.0	23.6	16.5	22.2
SD	0.58	1.49	0.38	0.86	0.70	0.56	0.84	0.63	0.74

(continued)

Table 14.51 (continued)

Posterior variable	1	2	3	4	5	6	7	8	9
Senèze N	7	6	9	8	7	7	7	8	8
Mean	34.0	31.2	29.6	19.5	21.6	18.6	13.0	9.2	10.2
Minimum	32.9	29.7	28.3	19.1	21.1	18.0	12.5	8.6	9.3
Maximum	35.6	32.3	31.8	20.0	22.3	18.9	13.5	9.7	10.8
SD	0.93	0.88	1.08	0.33	0.5	0.37	0.45	0.43	0.6
Posterior variable	10	11	12	13	14	15	16	17	18
Senèze N	7	7	9	9	7	7	9	9	7
Mean	14.0	16.3	13.5	14.2	15.4	13.5	17.5	13.8	21.6
Minimum	13.1	15.7	12.4	13.3	14.4	12.8	16.7	12.6	20.5
Maximum	14.8	17.3	14.3	14.9	16.2	13.9	18.5	15.4	23.1
SD	0.68	0.64	0.6	0.48	0.65	0.39	0.58	0.97	1.12

Variable definitions (from Crégut-Bonnoure 1995, 2002, 2020) (Fig. 14.A.11: G to K): dppd: dorsopalmar/plantar diameter; td: transversal diameter. Length: 1, middle; 2, axial; 3, abaxial. Proximal epiphysis: 4, td; 5, dppd. Proximal articulation: 6, td of the proximal articulation; 7, dppd in the middle; 8, td of the proximoaxial facet; 9, td of the proximoabaxial facet. 10, dppd of the proximoaxial facet; 11, dppd of the proximoabaxial facet. Middle of the diaphysis: 12, td; 13, dppd. Distal articulation: 14, td; 15, dppd of the axial condyle; 16, dppd of the abaxial condyle; 17, dppd minimum; 18, dorsal height from the median top to the proximal side of the distal articulation

Table 14.52 Comparative dimensions of the third phalanges (in mm) in *Gallogoral meneghinii* from Senèze

Anterior variable	1	2	3	4	5	6	7	8	9	10	11	12
Senèze N	10	10	8	10	10	10	10	10	10	10	9	10
Mean	26.3	23.2	19.2	14.7	8.4	12.0	8.9	7.0	16.4	20.8	44.4	17.1
Minimum	24.3	20.2	16.8	13.1	6.3	9.7	7.3	6.0	15.0	18.0	40.7	15.6
Maximum	27.7	25.5	21.5	15.9	10.4	14.4	10.7	7.7	19.0	22.4	50.6	18.8
SD	1.29	1.84	1.46	0.88	1.21	1.5	1.11	0.57	1.22	1.87	3.31	1.06
Posterior variable	1	2	3	4	5	6	7	8	9	10	11	12
Senèze N	4	5	5	5	4	5	5	5	5	5	5	5
Mean	24.9	21.8	17.4	14.1	7.8	10.8	8.4	6.7	16.9	20.2	44.6	15.6
Minimum	23.4	20.3	15.9	12.6	7.2	9.8	7.2	6.1	14.9	16.8	38.1	12.8
Maximum	27.2	24.2	19.7	15.7	9.3	11.5	9.9	8.7	18.9	23.2	50.5	18.0
SD	1.68	1.65	1.41	1.27	1.01	0.59	1.13	1.14	1.92	2.71	5.7	2.17

Variable definitions (from Crégut-Bonnoure 1995, 2002, 2020) (Fig. 14.A.11: L to M): dppd: dorsopalmar/plantar r diameter; td: transversal diameter. Height: 1, total; 2, at the level of the articulation; 3, of the distal part. Td: 4, of the articulation; 5, dorsal of the axial facet; 6, dorsal of the abaxial facet; 7, palmoplantar of the axial facet; 8, palmoplantar of the abaxial facet. Dppd: 9, of the articular crest; 10, of the abaxial facet; 11, total length; 12, total width

specimen from NMB Se 550 in the Lyon collection. The insertion point of the straight thigh muscle is convex.

Femur

There are 10 femora (Table 14.53): the right and left from the FSL 5008 (Guérin 1965: pl. LVII p. 249, pl. LVIX p. 251, pl. LX p. 253), NMB Se 313 and NMB Se 550 skeletons, three left (NMB Se 1451, FSL 16871, MNHN-F-1922-5) and one right (FSL 16872). The greater trochanter is a little higher than the head (more or less 2 cm). The lesser trochanter is prominent. The diaphysis is enlarged in its distal third, just above the supracondylar pit. Distally, between the lateral condyle and the trochlea, the *fossa extensoria* is deep and elongated craniocaudally. There is an elongated fossa for ligament insertion on the lateral epicondyle, whose axis is more or less perpendicular to the *fossa extensoria*. It is as deep as the latter but wider and elongated.

Tibia

There are six tibiae (Table 14.37) (Fig. 14.43): the right and left from the FSL 5008 (Guérin 1965: pl. LXI p. 261, pl.

LXII p. 263, pl. LXIII p. 265), NMB Se 313 and NMB Se 550 skeletons. In lateral view, the proximal articulation is convex and its plantar angle is projected downward. In proximal view, the tibial crest is enlarged and the dorsal side is truncated. The lateral articulation has an acute plantar side; a rounded tubercle occupies the middle of this side. Distally the malleolar facets have prominent and pinched borders. The surface of the cranial facet is flat whereas the plantar one is concave.

Patella

There are six patellae (Table 14.54): the right and left from the FSL 5008 (Guérin 1965: pl. LX p. 257), NMB Se 313 and NMB Se 550 skeletons. The dorsal tubercle is rounded. In proximal view the medial side is curvilinear, less developed than the lateral side (NMB Se 313 (left): respectively 22.3 and 23.7 mm). Its connection with the lateral side is a right angle.

Malleolar Bone

There are two specimens (Table 14.54): the right from the FSL 5008 skeleton (Guérin 1965: pl. LXII, Figs. 5–7, p. 257) and the left from the NMB Se 313 skeleton. The cranial and the caudal sides have the same development on NMB Se 313; they are more asymmetrical on FSL 5008. The tubercle of the medial side is located at the center of the bone. The cranial angle of the articulation for the talus is prominent and rounded.

Table 14.53 Dimensions of the femur (in mm) in *Gallogoral meneghinii* from Senèze

Variable	1	2	3	4	5	6	7	8	9	10	11
Senèze N	4	4	5	5	5	4	4	4	5	5	7
Mean	295.2	290.1	78.2	41.3	36.0	36.5	16.7	61.3	27.6	30.6	65.8
Minimum	276.7	268.0	70.0	36.9	32.5	34.2	15.6	51.5	26.0	28.4	59.7
Maximum	317.5	310.5	93.3	48.7	43.5	39.0	18.5	77.4	32.1	33.1	78.8
SD	16.84	17.42	8.92	4.39	4.4	2.31	1.27	11.28	2.57	2.02	6.21
Variable	12	13	14	15	16	17	18	19	20	21	
Senèze N	7	7	7	7	7	7	7	7	7	7	
Mean	65.3	24.5	24.3	69.0	55.8	45.1	76.1	56.5	42.6	32.3	
Minimum	61.5	22.0	22.6	63.9	51.7	42.6	71.2	52.7	40.4	31.0	
Maximum	77.6	29.3	27.2	83.4	63.0	52.0	92.0	63.3	49.2	37.4	
SD	2.17	2.57	1.8	6.71	3.77	3.52	7.14	4.01	3.01	2.3	

Variable definitions (from Crégut-Bonnoure 2002, 2020) (Fig. 14.A.12): td, transversal diameter; ccd, craniocaudal diameter. 1, total length; 2, length from the femoral head to the base; 3, td of the proximal epiphysis; Femoral head: 4, td; 5, ccd. Ccd: 6, greater trochanter; 7, femoral neck; 8, lesser trochanter. Middle of the diaphysis: 9, td; 10, ccd. Td distal: 11, epiphysis; 12, articulation; 13, lateral condyle; 14, medial condyle. Ccd distal: 15, lateral side. 16, lateral lip of the trochlea; 17: lateral condyle; 18, medial side; 19, medial lip of the trochlea; 20, medial condyle. Td: 21, trochlea on the dorsal side

Table 14.54 Dimensions of the patella and the malleolar bone (in mm) in *Gallogoral meneghinii* from Senèze

Patella variable	1	2	3	4	5	
NMB Se 313 (L)	51.6	38.1	39.4	33.3	29.0	
NMB Se 313 (R)	51.2	38.6	36.3	35.6	28.5	
Malleolar bone variable	1	2	3	4	5	6
NMB Se 313 (L)	22.9	17.4	16.9	24.1	14.4	20.4

Variable definitions (from Crégut-Bonnoure 2002, 2020): td, transversal diameter; ccd, craniocaudal diameter. Patella (Fig. 14.A.13: A and B): 1, height; 2, td; 3, height of the caudal articulation; 4, td of the caudal articulation; 5, ccd. Malleolar bone (Fig. 14.A.13: C): 1, total height; 2, height of the cranial side; 3, height of the caudal side; 4, ccd; 5, td; 6, length of the medial articulation; L, left; R, right

Tarsals

There are seven tali (Table 14.38) (Figs. 14.44 and 14.57: A to D): the right from the FSL 5008 skeleton (Guérin 1965: pl. LXIV p. 269), the right and left from the NMB Se 313 and NMB Se 550 skeletons and a right and a left of FSL 16900 (a third specimen with this number belongs to a cervid). The proximal trochlea lacks a median tubercle and the lips are not at the same level. At the base of the plantar trochlea, the protrusion for the navicular-cuboid medio-plantar facet is larger than on the specimen attributed to *Pliotragus* (see section "Talus" under *Pliotragus*). It is straight and below it there is a large oval synovial fossa. The lateral sides of the plantar articulation for the calcaneus are obliquely directed upward; the medial one is angled farther, laterally.

There are seven calcanei (Table 14.39) (Fig. 14.45): the right from the FSL 5008 skeleton (Guérin 1965: pl. LXV p. 271), the right and left from the NMB Se 313 and NMB Se 550 skeletons and a right and a left of FSL 16900. Their description is given in section "Calcaneus" concerning the distinction with *Pliotragus*.

There are seven navicular-cuboids (Table 14.40) (Fig. 14.46): the right from the FSL 5008 skeleton (Guérin 1965: pl. LXVI p. 273), the right and left from the NMB Se 313 and NMB Se 550 skeletons and a right and a left of FSL 16900. Even if the morphology shows some similarities with *Pliotragus*, this bone is distinct as discussed in section "Navicular-Cuboid".

Metatarsal III+IV

There are 12 metatarsal III+IV (Table 14.55) (Fig. 14.58): the right and left from the FSL 5008 (Guérin 1965: pl. LXVIII p. 281, pl. LXVIX p. 283, pl. LXX p. 285), NMB Se 313 and NMB Se 550 skeletons, the right and left from two other individuals (left FSL 16875 and right FSL 16876 from a single individual and both sides of NMB Se 1382), one left (NMB Se 1381) and one right (MNHN-F-1921-9). The proximal articulation has a broad dorsal side. The dorso-medial articular facet is triangular with a median concavity; consequently, the surface is undulating. The dorsal articular facet for the navicular-cuboid is quadrangular with a prominent planto-median angle. The plantar facet for the navicular-cuboid is obliquely directed upward and clearly visible in plantar view. The proximo-lateral tendinous grooves are shallow. Behind them, there is a projecting transverse tubercle, and below it there is a kind of basin. The dorsal groove is shallow and its sides are rounded. Up the distal articulation, slightly lower than the foramen which is small, the sides of the groove are acute. The space between the distal condyles is broad plantarly. As for the metacarpal III+IV, the proportions of the Senèze, Volax and Olivola metatarsals are not exactly the same (Fig. 14.59).

Posterior Phalanges

There are:

- 19 first phalanges (Table 14.50) (Fig. 14.57: I, J): four each from the FSL 5008, NMB Se 313 and NMB Se 550 skeletons and several sets from different individuals (three from FSL 16864 and four more from NMB Se 1010). The phalanx NMB Se 1592, initially considered as *Gallogoral*, belongs to a cervid.
- 17 second phalanges (Table 14.51) (Fig. 14.57: L): three from the FSL 5008 skeleton, four each from the NMB Se 313 and NMB Se 550 skeletons, plus four from FSL 16864 and two from another individual (NMB Se 1010).
- five third phalanges (Table 14.52) (Fig. 14.57: P, Q): two from the FSL 5008 skeleton and three from another individual (NMB Se 1010). The morphology is similar to the anterior ones.

Discussion

The remains are abundant, with three complete skeletons, which is a considerable advantage for our knowledge of the anatomy. There are four crania, three isolated horn-cores, one frontlet and 18 mandibles. The number of bones from the scapular and pelvic girdles, limbs, carpals, tarsals, metapodials and phalanges is 218. There is a minimum of 15 individuals: ten adults, three juveniles, two very young

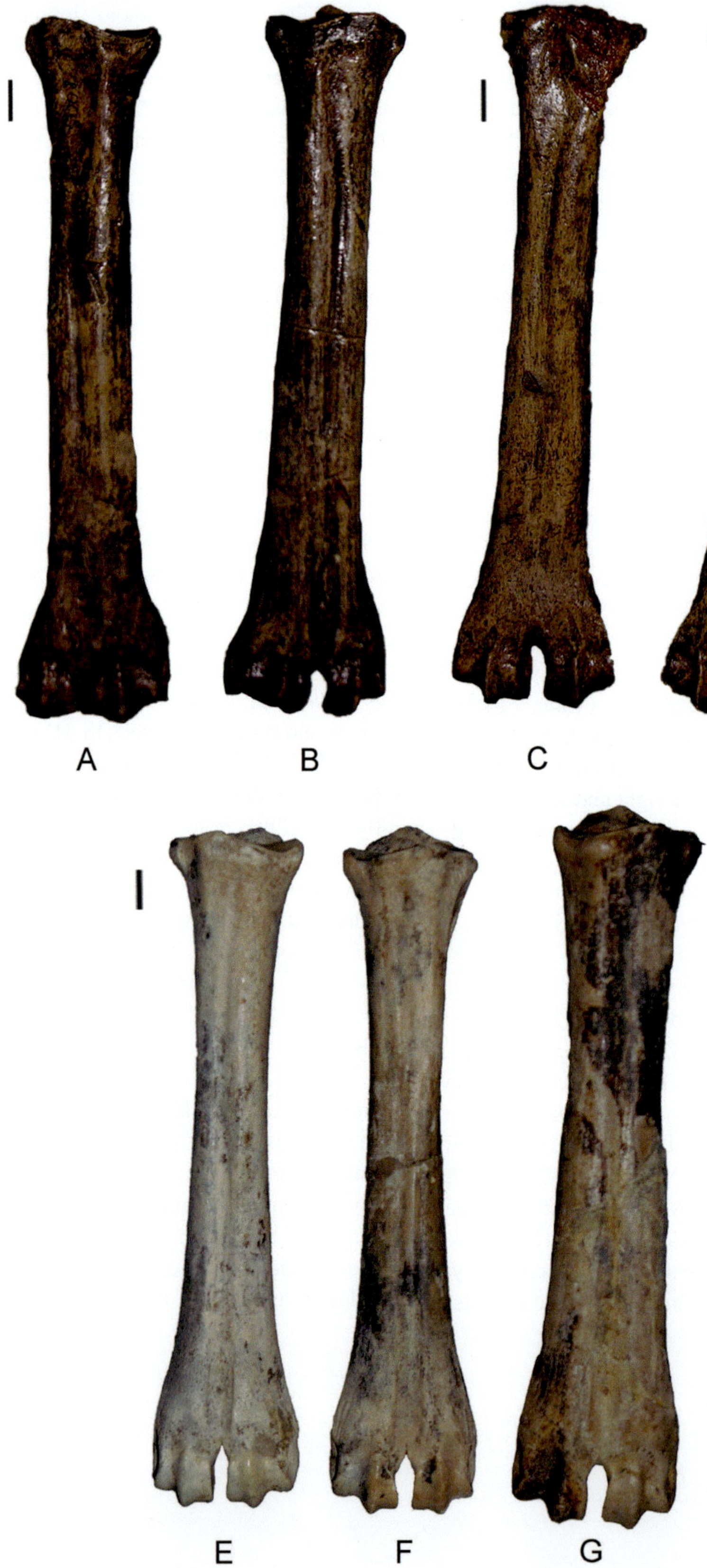

Fig. 14.58 Post-cranium of *Gallogoral meneghinii* from Senèze. A, C, left metatarsal III+IV FSL 5008; B, D, right metatarsal III+IV FSL 5008; E, right metatarsal III+IV NMB Se 1382; F, left metatarsal III+IV; G, left metatarsal III+IV NMB Se 1381. A to G, dorsal view. Scale 1 cm

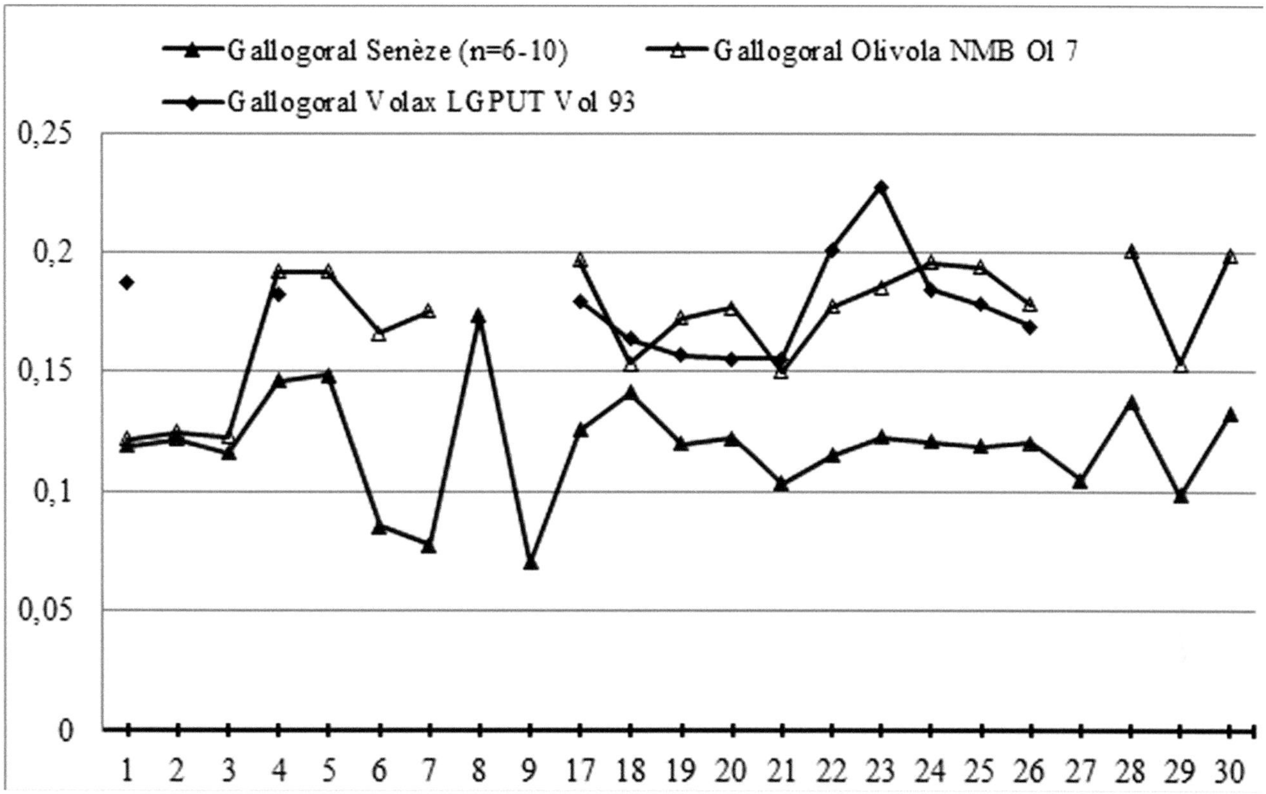

Fig. 14.59 *Gallogoral meneghinii* from Senèze and other localities. Ratio diagram of the metatarsal III+IV. Reference axis: recent *Capra ibex* ♂+♀ (in Crégut-Bonnoure 2020). td: transversal diameter. dpd: dorsoplantar diameter. Length: 1, dorsal; 2, minimum at the middle; 3, plantar side. Proximal epiphysis: 4, td total; 5, td of the articulation; 6, dpd total; 7, dpd of the articulation; 8, dpd of the small medio-plantar articular facet; 9, td of the small medio-plantar articular facet. Middle of the diaphysis: 17, td; 18, dpd. Distal epiphysis: 19, td maximal of the articulation; 20, td minimal of the articulation; 21, td supra-articular; 22, dpd of the epiphysis; 23, dpd of the epiphysis at the foramen level; 24, distance between the keels on the plantar side; 25, distance between the keels at their base; 26, distance between the keels in their middle; 27, dpd of the medial condyle; 28, dpd of the medial keel; 29, dpd of the lateral condyle; 30, dpd of the lateral keel. References of the measurements in Fig. 14.A.16 and dimensions in Table 14.55

Table 14.55 Comparative dimensions of the metatarsal III+IV (in mm) in *Gallogoral meneghinii* from Senèze and other localities

Variable	1	2	3	4	5	6	7	8	9	10	11	12
Senèze N	8	10	6	10	8	6	6	8	8	7	8	8
Mean	175.66	159.67	180.46	36.87	35.6	33.08	31.33	9.5	8.98	14.92	25.15	14.47
Minimum	171.3	156.5	174.8	35.6	34.2	29.8	29.5	6.8	7.1	13.1	23.7	12.1
Maximum	184.7	165.9	189.4	38.9	37.9	37.0	34.5	12.5	9.8	17.2	29.6	17.8
SD	4.8	3.06	6.05	1.06	1.09	2.66	1.83	1.79	0.97	1.39	1.86	1.60
Olivola (NMB Ol 7–cast)	176.8	160.8	183.4	40.9	39.3	34.1	33.6			14.8	27.4	15.2
Volax (LGPUT Vol 93)	205.8			40.1								

(continued)

Table 14.55 (continued)

Variable	13	14	15	16	17	18	19	20	21	22	23	24
Senèze N	8	7	6	6	10	10	10	8	8	9	9	8
Mean	22.57	8.38	13.32	5.9	23.33	21.33	42.21	40.78	39.67	22.62	20.5	25.13
Minimum	20.7	6.5	11.5	5.2	21.4	20.0	40.4	38.5	36.9	20.8	18.4	23.3
Maximum	24.5	9.7	15.2	6.7	27.4	23.6	45.2	43.4	44.0	25.6	22.6	26.9
SD	1.24	1.18	1.43	0.6	1.67	1.21	1.58	1.71	2.36	1.6	1.41	1.08
Olivola (NMB Ol 7–cast)	24.4	10.4	11.6	6.8	27.5	21.9	47.6	46.3	44.2	26.1	23.7	29.9
Volax (LGPUT Vol 93)					26.4	22.5	46.0	44.0	44.7	27.6	26.0	29.1
Variable	25	26	27	28	29	30						
Senèze N	8	8	8	8	9	9						
Mean	24.61	20.7	25.32	16.91	24.24	15.79						
Minimum	23.5	19.6	24.3	15.9	23.1	14.9						
Maximum	26.7	22.1	27.2	18.7	26.2	26.2						
SD	0.98	0.78	1.02	0.87	1.03	0.83						
Olivola (NMB Ol 7–cast)	29.2	23.7		19.6	27.5	18.4						
Volax (LGPUT Vol 93)	28.2	23.2										

Variable definitions (from Crégut-Bonnoure 2002, 2020) (Fig. 14.A.16): dpd, dorsoplantar diameter; td, transversal diameter. Length: 11, dorsal side; 2, minimum in the middle; 3, plantar side. Proximal epiphysis: 4, td total; 5, td of the articulation; 6, dpd total; 7, dpd of the articulation; 8, dpd of the facet for the small cuneiform; 9, td of the facet for the small cuneiform; 10, td of the great cuneiform facet; 11, dpd of the great cuneiform facet; 12, td of the dorsal facet for the navicular-cuboid; 13, dpd of the dorsal facet for the navicular-cuboid; 14, length of the connection of the dorsal facets. 15: td of the plantar facet for the navicular-cuboid; 16: dpd of the plantar facet for the navicular-cuboid. Middle of the diaphysis: 17, td; 18: dpd. Distal epiphysis: 19, td maximal of the articulation; 20, td minimal of the articulation; 21, td supra-articular; 22, dpd of the epiphysis; 23, dpd of the epiphysis at the foramen level; 24, distance between the keels on the plantar side; 25, distance between the keels at their base; 26, distance between the keels in their middle; 27, dpd of the medial condyle; 28, dpd of the medial keel; 29, dpd of the lateral condyle; 30, dpd of the lateral keel

animals. I have mentioned already that a similarity of proportions exists between the teeth of *Gallogoral* and *Pliotragus*, whereas their morphology is not the same. I have discerned some morphological variability for the M3 and the p4. The comparison of the Senèze metapodials with those from Volax and Olivola highlights differences in their proportions. In the current state of our knowledge on the species, they could reflect further subspecific distinction.

A number of specimens previously identified as other taxa can be reallocated to *Gallogoral*. A fragmentary left horn core (a base with a part of the frontal) from Perrier-Etouaires was assigned to *Antilope ardea* by Depéret (1884: p. 253). From an examination of a cast (NMB Pr 523), I reallocate it to *Gallogoral*: the base of the horn core is well rounded, rather than being compressed dorsoventrally in *Pliotragus*. According to its morphology, the mandible NMB Cp 65 from Le Coupet doesn't belong to *Gazellospira*: it is instead a *Gallogoral*. The radius NNHM-S SL 8 from Slivnitsa was originally referred to *Pliotragus* (Spassov & Crégut-Bonnoure 1999), but its proportions are similar to those of *Gallogoral*, and SL 8 could belong to this genus.

Genus *Ovis* Linnaeus, 1758

Species *Ovis claudiusguerini* nov. sp.

The earliest *Ovis* known is represented by a fragmentary horn core from Russian Siberia (Trans-Baikal region), dating from the Early Pliocene (Kalmykov 2013). *Ovis* from the Villafranchian of China was first described as *Ovis ammon shantungensis* Matsumoto, 1926 and then termed *O. shantungensis* (Teilhard de Chardin and Piveteau 1930). Its dispersal into Europe could probably originate from a species closely related to *O. zdanskyi* Bohlin, 1938 from locality 18 of Honan (Crégut-Bonnoure 2002, 2020). In Europe the genus is identified from biozone MNQ 18: in France at Senèze (Schaub 1943); in Bulgaria at Slivnitsa (Spassov & Crégut-Bonnoure 1999; Crégut-Bonnoure 2002, 2020; *Ovis ammon* aff. *antiqua*) and at Kozarnika (Biostratigraphic Unit B2-2; Fernandez & Crégut-Bonnoure 2007; *Ovis* sp.). It is still present during the Middle Pleistocene (Crégut-Bonnoure 2002, 2020) with *O. ammon antiqua* Pommerol, 1880 and *O. argaloides* Nehring, 1891.

Diagnosis

<u>Type specimen</u>: Male skull NMB Se 302; Fig. 14.60.

<u>*Derivatio nominis*</u>: Named in honor of the late Dr. Claude Guérin, who was a leader of the new excavations in Senèze.

<u>*Locus typicus*</u>: Senèze (Domeyrat, Haute-Loire, France).

<u>*Age:*</u> Early Pleistocene, late Villafranchian, MNQ 18.

<u>*Diagnosis*</u>: Male braincase smaller than in the extant and fossil *O. ammon*. Horn cores with a short pedicle and widely separated at the base. Surface of the temporal fossa directed obliquely downward and temporal lines concave. Parietal without fossa in the axis of the frontal suture. Frontal relatively short in comparison to *O. shantungensis* and *O. ammon antiqua*.

Simplified Synonymy

Ovis spec., Schaub, 1943, p. 282–283, Fig. 8
"*Ovis*" sp., Duvernoy and Guérin, 1989, p. 355.
Ovis sp., Crégut-Bonnoure, 1992, p. 180.
Ovis sp., Crégut-Bonnoure, 2002, p. 270–271, Fig. 3.330 B, pl. XXVII Fig. 2
Ovis from Senèze, Crégut-Bonnoure, 2002, p. 159.
Ovis sp., Crégut-Bonnoure, 2020, p. 456, Fig. 3.331 B p. 433, pl. 21 Fig. 3.

Description

The record of *Ovis* at Senèze is limited to the fragmentary male cranium NMB Se 302, which was described by Schaub (1943: pp. 282–283, Fig. 8) (Table 14.56; Fig. 14.60). The caudal side of the left horn core, the caudal base of the right one, the frontal between the horn cores, the parietal and the upper part of the supra-occipital are preserved. The horn core surface is eroded. The dorsoventral diameter preserved is 72 mm. The parieto-occipital suture is not synostosed, and the ligamentary insertions on the external occipital protuberance are moderately developed: these features indicate an adult of moderate age. This specimen is larger than extant *O. orientalis* Gmelin, 1774 but smaller than extant and fossil *O. ammon* ssp. The horn cores are more widely spaced at the base than in extant and fossil *Ovis ammon* as in *O. zdanskyi*. The base is clear, weakly prominent above the frontal, and less elongated than in the two previous species. The parietal, inside the *fossa temporalis*, is directed obliquely downward, rather than vertical as in extant and fossil *O. ammon* subspecies. Compared to *O. shantungensis* (Teilhard de Chardin and Piveteau 1930: pl. XIV, Fig. 1) and *O. a. antiqua* (Pommerol 1880: pl. III), the parietal surface is flat, without median fossa or depression. The temporal lines are concave. The proportions of the parietal and the horizontal area of the occipital are distinct from *O. a. antiqua* and other *O. ammon* sspp. (Fig. 14.61).

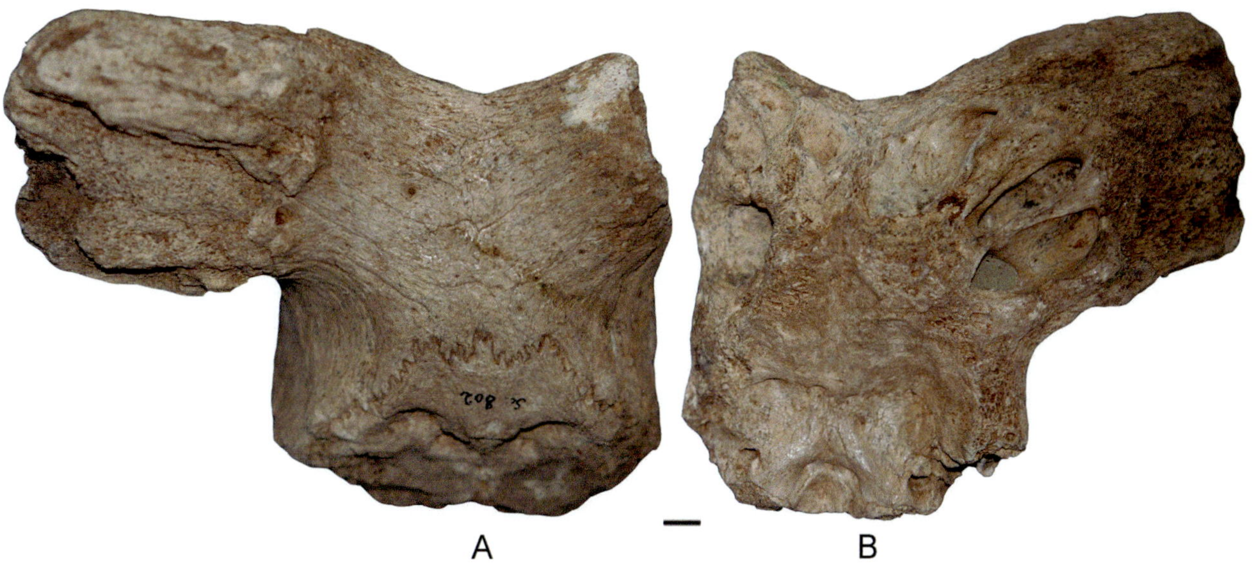

Fig. 14.60 Cranium of *Ovis claudiusguerini* n. sp. from Senèze. Holotype NMB Se 802. A, *norma dorsalis*; B, *norma basalis*. Scale 1 cm

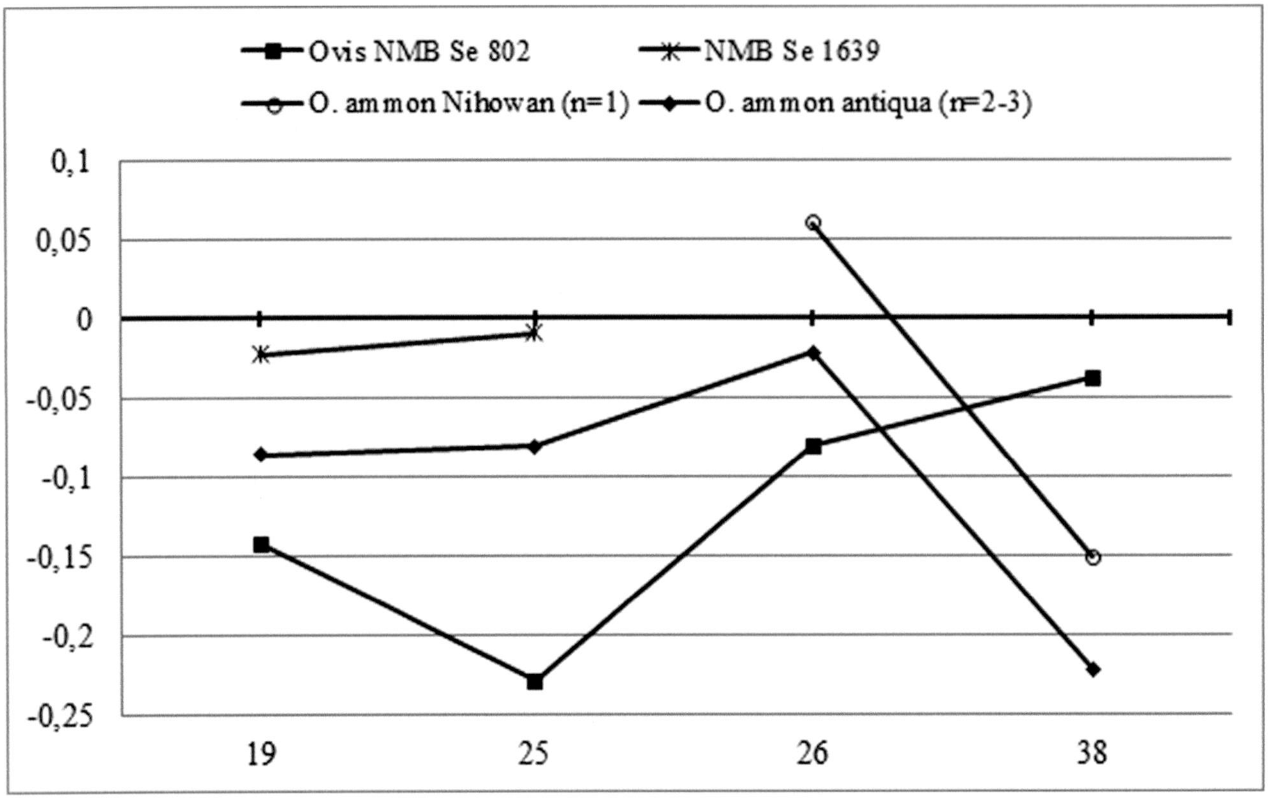

Fig. 14.61 *Ovis claudiusguerini* n. sp. from Senèze, *O. ammon antiqua* from France, *O. ammon* from Nihowan and indeterminate bovid (NMB Se 1639) from Senèze. Ratio diagram of the cranium. Reference axis: recent *Ovis ammon ammon* ♂ (in Crégut-Bonnoure 2020). 19, length acrocranion-bregma; 25, length of the parietal; 26, width of the parietal; 38, width of the horn cores between the internal sides at the base. References of the measurements in Fig. 14.A.1 and dimensions in Table 14.56

Table 14.56 Comparative dimensions of the cranium (in mm) in *Ovis claudiusguerini* n. sp. and Bovid indeterminate from Senèze and in *Ovis* from other localities

Variable	19	25	26	27	38
Ovis claudiusguerini n. sp.					
Senèze (NMB Se 802 ♂)	49.2	25.9	91.8	54.4	66.6
Ovis ammon ssp.					
Nihowan (MNHN without reference)			126.7		51.3
O. ammon antiqua					
Pont-du-Château	59.4	35.0	112.6		51.6
Arago					
E 14 EFN1 961	52.0	35.0	102.0		40.1
I18 IFY0 207	56.7	39.2	100.5		39.3
O. ammon ammon extant N	4	4	4	5	2
Mean	68.25	43.87	110.49	90.90	72.69
Minimum	63.1	39.8	96.4	78.6	72.2
Maximum	74.0	47.3	129.7	108.0	73.2
SD	5.49	3.35	14.01	11.12	1.00
Bovid indet					
Senèze (NMB Se 1639)	64.8	43.9			

Variable definitions (from Crégut-Bonnoure 2002, 2020) (Fig. 14.A.1): 19, length acrocranion-bregma; 25, dorsal length of the parietal; 26, width of the parietal; 38, width of the horn cores between the internal sides at the base

Discussion

The morphology and the proportions of this cranium are clearly distinct from the extant and fossil *O. ammon* sspp. It clearly represents a new species. However, the age of this specimen is uncertain. It has been considered as more recent than the bovids discussed above (Crégut-Bonnoure 1992; Crégut-Bonnoure & Guérin 1996; Schaub 1943). Its color is yellowish rather than being blackish-brown as is the great majority of the other bovid remains from Seneze. On the other hand, examination of the material from the Basel, Lyon, Paris and Le Puy-en-Velay collections reveals the existence of additional yellowish specimens belonging to the other taxa described in this analysis. Consequently, the color of the *Ovis* cranium is not a valid criterion to suggest that the species is younger than the other Senèze taxa. In addition, the genus is known in Europe during MNQ 18 from two other localities (see above).

Tribe Ovibovini Gill, 1872

Genus *Megalovis* Schaub, 1923

Species *Megalovis latifrons* Schaub, 1923

Megalovis is found in several localities dating to the Early Pleistocene, from Europe to Asia. At least six species exist: *M. latifrons* Schaub, 1923; *M. balcanicus* Crégut-Bonnoure & Dimitrijević, 2006; *M. piveteaui* Schaub, 1937; *M. wimani* Schaub, 1937; *M. guangxiensis* Han, 1987 and *Megalovis* sp..

In Europe, *M. latifrons* is now identified in four localities:

- MNQ 17: in Bulgaria at Varshets (Spassov & Crégut-Bonnoure 1999).
- MNQ 18: in France at Senèze (Schaub 1923) and in Germany at Erpfinger Höhle (or Erpfingen; Abel et al. 2006; Crégut-Bonnoure 2002, 2020; Lehmann 1957).
- MNQ 18–19: in Hungary at Villány-Nagyharsányhegy 4 (Crégut-Bonnoure this paper).

In Germany, some teeth from the Swabian Jura (probably Erpfinger Höhle) and from Stetten were identified as *Antilope jaegeri* by Rütimeyer (1867). Their casts are preserved in the Basel Museum. Their morphology is similar to those of *M. latifrons*. According to the teeth and forelimb metapodial, whose morphology is distinct from *M. latifrons*, another species could be present in Bulgaria at Slivnitsa (Spassov & Crégut-Bonnoure 1999; MNQ 18) and perhaps a subspecies in Romania at Fântâna lui Mitilan (Crégut-Bonnoure 2002, 2020; Rădulesco & Samson 1962; Rădulescu & Samson 1995).

M. balcanicus is recognized in Montenegro at Trlica (Crégut-Bonnoure & Dimitrijević 2006; MNQ 19–20), in Italy at Pirro Nord (de Giuli et al. 1987, 1990; Crégut-Bonnoure & Dimitrijević 2006; MNQ 19) and in Bulgaria at Kozarnika (biostratigraphic unit B2-2; Fernandez & Crégut-Bonnoure 2007; MNQ 18). In China, *M. piveteaui* was described from the localities of Sangkan-ho (Nihowan; Teilhard de Chardin and Piveteau 1930) and Shanshenmiaozui (Nihowan; Tong et al. 2022), *M. wimani* from the locality of Yang-Shao-Tsun (site A, Honan) and *M. guangxiensis* from several localities of Guangxi (Jin et al. 2009a). This last is still present at the boundary between Middle and Early Late Pleistocene from *Homo sapiens* Cave (Chongzuo; Jin et al. 2009b).

The inclusion of *M. latifrons* among the Ovibovini has been discussed by many authors. Schaub (1937) rejected it from the Ovibovini. The species was considered as a member of Caprini by Viret (1961). Dechaseaux (1961) thought its brain is distinct from the Ovibovinés (*sic*), but more similar to the Ovinés (*sic*). Because of the incorrect attribution of the *Pliotragus* horn core NMB Se 1483 to *Megalovis*, as discussed above (see section "Crania and Horn Cores" under *Megalovis*), the two genera were wrongly synonymized (Schaub 1937; Gentry 1970, 1978, 1980, 1996, 2001; and McKenna & Bell 1997). The holotype cranium of *Megalovis* (NMB Se 556), which could belong to a female, comprises a partially preserved braincase, but a complete one of *Pliotragus* exists (MNHN-F-1923-8). Gentry (1992) initially referred *Megalovis* to the Ovibovini, due to the synonymy of *Megalovis* and *Pliotragus* resulting from the reconstruction of the holotype cranium with the *Pliotragus* horn core (NMB Se 1483). Some years later, Gentry (1996) excluded the species from this tribe after studying the basioccipital of the *Pliotragus* skull, which doesn't have a supplementary facet near the condyle. Following the latest revision, I can safely include *Pliotragus* in Caprini and *Megalovis* in Ovibovini (Crégut-Bonnoure 2002, 2020; Crégut-Bonnoure & Dimitrijević 2006; Crégut-Bonnoure & Guérin 1996).

M. latifrons was described by Schaub (1923) based on the incomplete cranium from Senèze. Later, other remains were discovered in this locality (Schaub 1943). Guérin (1965) has given some anatomical details, and new remains from other localities were described (Lehmann 1957; Rădulesco & Samson 1962; Spassov & Crégut-Bonnoure 1999). Some corrections have been made (Crégut-Bonnoure 2002, 2006, 2020; Duvernois & Guérin 1989; Spassov & Crégut-Bonnoure 1999), and a neodiagnosis has been proposed (Crégut-Bonnoure 2002, 2020). The Senèze remains come mainly from the previous field research; only a fragmentary tooth and a fragmentary metatarsal III+IV come from the recent fieldwork.

Description

Crania

Two fragmentary crania are preserved.

The type cranium NMB Se 556 was reconstructed from several fragments (Schaub 1923: Fig. 5; Schaub 1943: Figs. 5 and 6) (Figs. 14.62 and 14.63: A): the frontal with the base of the horn cores and the parietal, the caudal part of the palatal, the left and right maxillae. The dimensions are given in Table 14.57. It belongs to a young adult; the suture being incompletely closed and the wear of the teeth being slight. Only the bases of the horn cores are preserved. The distance between them is 80 mm. They insert well posterior to the orbit. The base shows a slight circle of pearls and the rostral side of the right horn core has a deep and longitudinal groove. The cross section is roundish. The frontal bone is flattened and the post-orbital constriction is moderate (161 mm). It is transversely convex behind the supra-orbital foramina. The frontal suture forms a longitudinal bulge. As Guérin (1965) indicated, this bone is low. The fronto-parietal suture is "V" shaped, and it is located immediately behind the caudal side of the horn cores. The parieto-occipital suture is straight. The parietal is obliquely directed and its angle of connection with the frontal is about 120°. The supra-orbital foramina are well developed (5 mm in the longitudinal plane). On the right frontal, a supplementary foramen seems to exist. The lachrymal constitutes a very deep and wide (40 mm) fossa extending to P4 and caudally delimited by a frontal crest. Its rostral extension is unknown, the bone being missing. The lachrymal caudal rim, which is a part of the orbit, is similar to that of *Ovibos*. The maxilla is strongly convex from the facial tubercle (which is at the level of M1) to the M3. The convexity forms a more or less triangular protuberance toward the facial tubercle, which is more strongly developed than in *Praeovibos* and *Ovibos*. It is not possible to obtain a value for the palatal distance between the teeth because a part of the palate was reconstructed. The horizontal blade of the palatine initially was isolated. Consequently, its position with respect to M3 cannot be specified. The choanae are wide and similar to those of *Ovibos*. In front of the infraorbital foramen a grooved area runs toward P2.

The cranium fragment NMB Se 1638 is composed of the right premaxilla and maxilla, with P2, P4-M3 (Fig. 14.63: B, C). The ventral rim of the orbit and the zygomatic arch are supposedly associated with it, but there is no connection with these elements. The rostral side of the premaxilla is not preserved. The bone is thick and convex as in *Ovibos*. The morphology of the maxilla is similar to the type specimen. The ventral rim of the orbit is moderately prominent.

The horn core NMB Se 1483, used in a reconstruction of the type cranium (Schaub 1943: Fig. 5 p. 280, Fig. 6 p. 281; Viret 1961: Fig. 165 p. 1073) is now reallocated to *Pliotragus* (see section "Crania and Horn Cores" under *Pliotragus*). On the *M. balcanicus* cranium, the horn cores show similarities with the North American ovibovin *Euceratherium* Sinclair and Furlong, 1904 (=*Preptoceras* Furlong, 1905) (Crégut-Bonnoure & Dimitrijević 2006). On NMB Se 1483, the preserved base of the horn core is rounded and not rostrocaudally elongated as in *M. balcanicus*: it could be due to sexual dimorphism or a species-specific characteristic (Crégut-Bonnoure & Dimitrijević 2006).

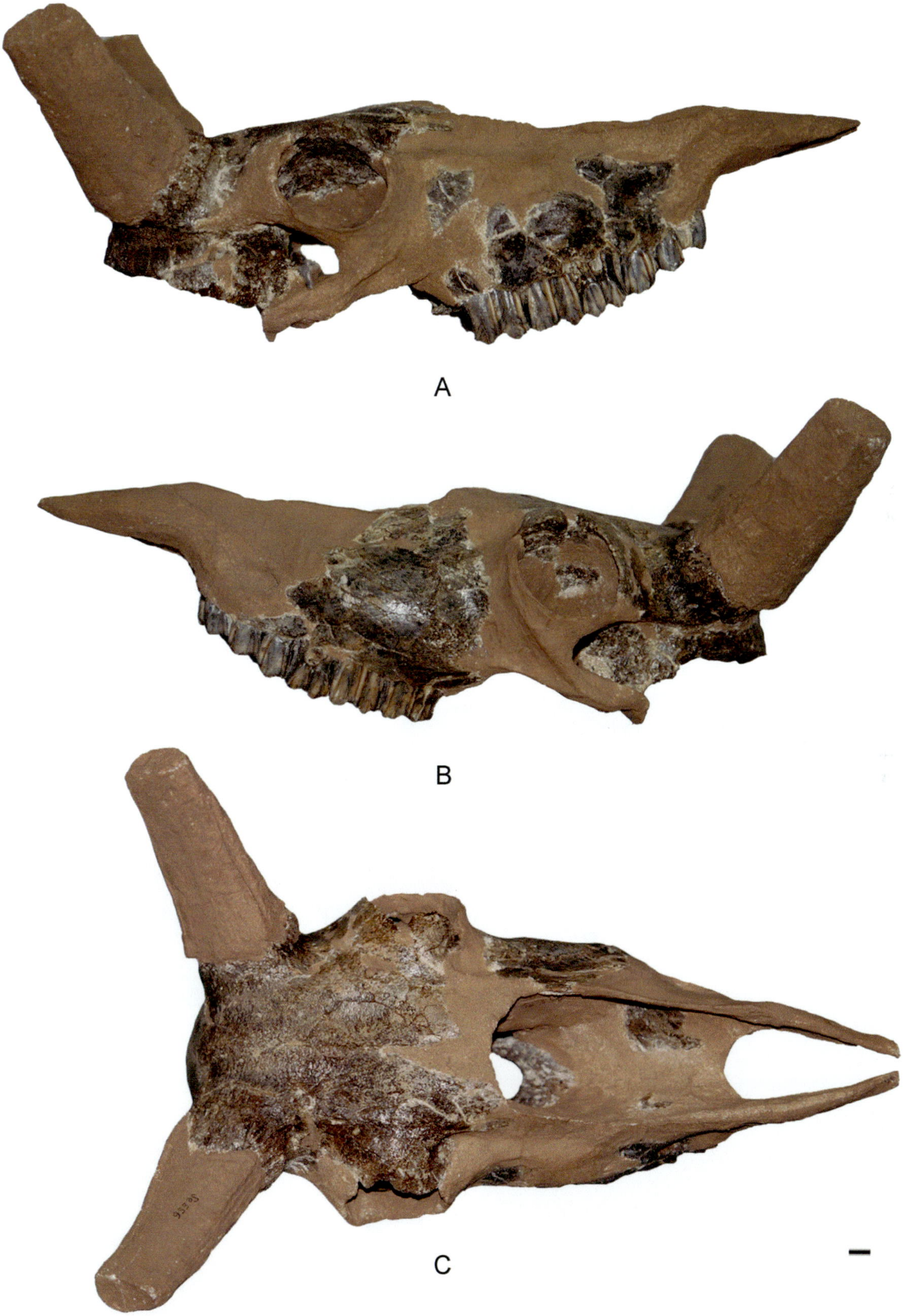

Fig. 14.62 Cranium of *Megalovis latifrons* from Senèze. Holotype NMB Se 556. A, B, *norma lateralis*; C, *norma dorsalis*. Scale 1 cm

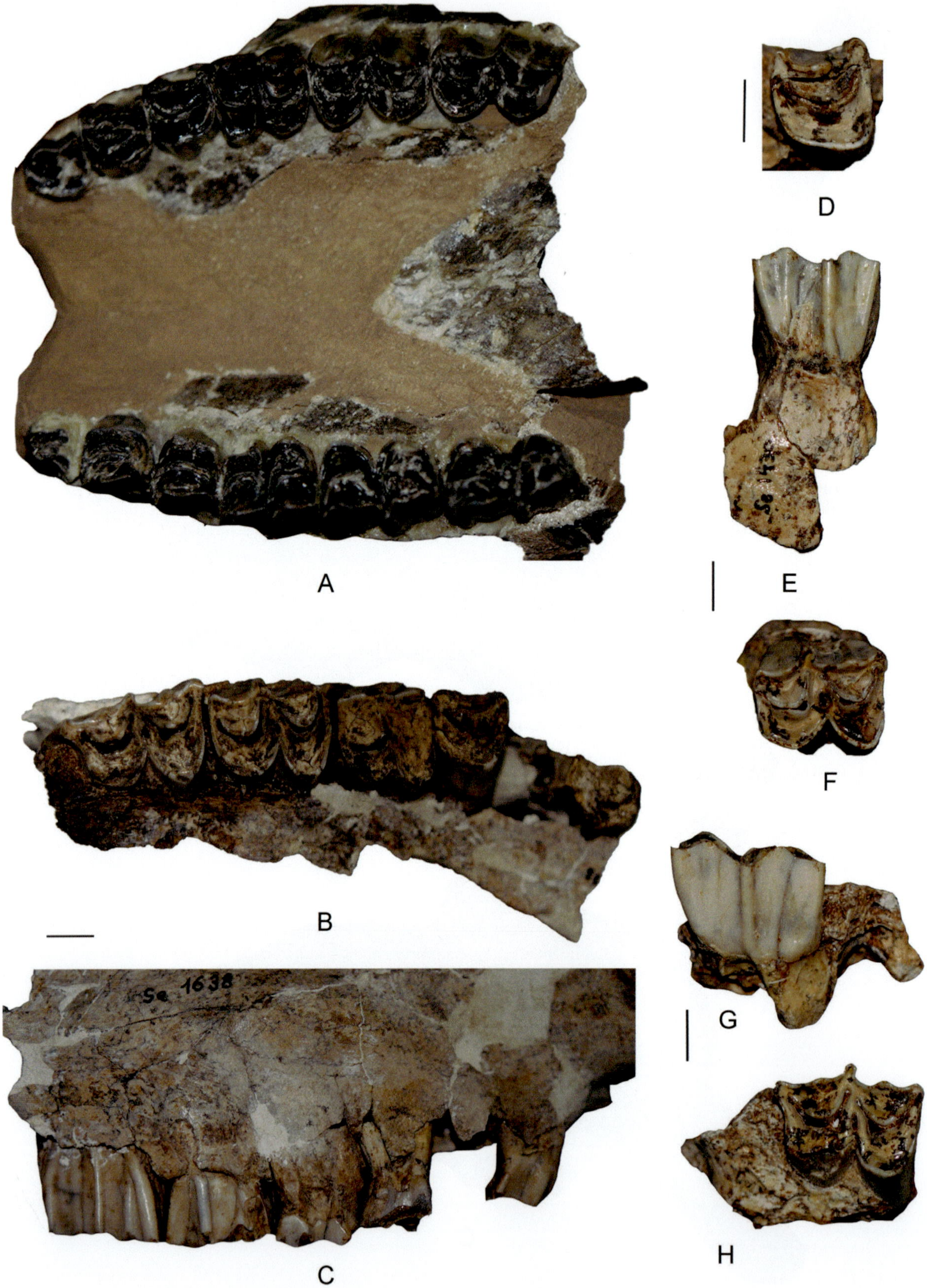

Fig. 14.63 Cranium and upper teeth of *Megalovis latifrons* from Senèze. A, maxilla of holotype NMB Se 556; B, C, right maxilla NMB 1638; D, right P4 NMB 1430; E, F, right M2 NMB Se 1430; G, H, right M3 NMB Se 1430. A, B, D, F, H, occlusal view; C, E, G, buccal view. Scale 1 cm

Table 14.57 Dimensions of the cranium (in mm) in *Megalovis latifrons* from Senèze

Variable	9	10	11	13	14	15	25	28
NMB Se 556	131.9	53.9	77.8	110.1	69.7	48.4	43.3	133.9
Variable	29	30	32	37	38	41	42	
NMB Se 556	171.0	151.1	131.6	152.0	151.0	61.0	50.9	

Variable definitions (from Crégut-Bonnoure 2002, 2020) (Figs. 14.A.1 and 14.A.2): td: transversal diameter. 25, dorsal length of the parietal; 9, length P2 mesial-M3 distal; 10, length P2 mesial-P4 distal; 11, length M1 mesial-M3 distal; 13, td of the tooth row at the labial side;14, td between the lingual side of the M3 second lobe;15, td between the lingual side of the P2; 28, td post-orbital; 29, td ectorbital-ectorbital; 30, td entorbital-entorbital; 32, td at the facial tuberosities;37, width between the lateral sides of the horn cores at the base; 38, width between the internal sides of the horn cores at the base; 41, rostrocaudal diameter of a horn core at its base; 42, dorsolateral diameter of a horn core at its base

Mandible

Two left mandibles are known (Table 14.58). The corpus of specimen NMB Se 401 (Schaub 1923: Fig. 5) is thick (Fig. 14.64: A to C). Guérin (1965) indicated that the corpus and ramus meet in an obtuse caudal angulation and that the caudal angle of the ramus is marked. However, the caudal side has been reconstructed (Schaub 1923: Fig. 5; Schaub 1943: Fig. 6 p. 281). Nonetheless, according to the preserved morphology, this extrapolation must be close to correct. At the level of the third lobe of m3, the masseter attachment constitutes a well developed tubercle even though the teeth indicate that the specimen belongs to a young adult. The toothrow is about the same size as the extant muskox *Ovibos*, but the diastema is shorter. The *foramen mandibulare* is similar to that of *Ovibos*: "V" shaped pointing caudally, very wide and extended forwards by a longitudinal groove. In comparison to *Ovibos*, the incisor region is not pinched below the alveoli and consequently it is not prominent.

The mandible NMB Se 1732 belongs to an old adult (Fig. 14.64: D, E). It is incomplete, lacking the diastema, the ramus and the base of the corpus. The bone is thick as in NMB Se 401.

Teeth

There are 22 upper teeth (Tables 14.59, 14.60 and 14.61): the right and left P2-M3 from the type cranium NMB Se 556, the right P4, M2-3 belonging to an old individual (NMB Se 1430), the right P2, P4-M3 from another maxilla (NMB Se 1638) and one little worn left M1 (FSL 16897). From the new field research there is a buccal face of a P4 (FSL SEN 01-0053, from Trench 3).

These teeth are less robust than those of *Ovibos*, and their morphology is distinct (Figs. 14.63 and 14.64: A to C). The cross section of the premolars is quadrangular, the lingual side being straight. P2 and P3 are bilobed as is P4. However, the bilobation decreases from P2 to P4. On P4, the bilobation is in a more distal position. On P2 (n = 2), the parastyle is weak. The paracone rib is very broad and prominent, flanked distally by a deep valley. On P3 (n = 2), the median rib is flanked by two valleys, the mesial one being deeper than the distal. There is a small occlusal islet in a disto-lingual position. On P4 (n = 5), the crown is broader at the front than the back. Consequently, in occlusal view the lingual side is directed obliquely and outward (Fig. 14.63: D).

There are four M1, four M2 and four M3 (Figs. 14.63 and 14.65: A to C). The buccal side is higher than the lingual one, due to wear. The styles of the molars are narrower than in *Ovibos* and there is neither an interlobar column nor cingulum. The mesial lobes are lingually pinched but less than in *Pliotragus*, in which this lobe is relatively shorter and the crown less hypsodont. The distal lobe is rounded. An occlusal islet is present, more or less rounded, which is less developed than in *Ovibos* and *Praeovibos*. The paracone rib is well developed, but not the metacone rib. The interstylar surfaces are less deep than in *Praeovibos* and *Ovibos*. On M3, the metastyle is straight without any metastylar wing.

Lower Teeth

There are 23 lower teeth (Tables 14.62 and 14.63) (Figs. 14.65: D to N, 14.66: A to C): the toothrows from the left mandibles NMB Se 401 and NMB Se 1732, the left m1-2 of one individual (FSL 16897), the left p3-4, m2-3 from another animal (MNHN-F-1922-5), the right m1-2 of another one (MNHN-F-1907-28), a fragmentary left m1 (NMB Se 1641) and a left m2 (NMB Se 1726). This material belongs to at least six individuals.

The buccal side of the premolars is bilobed, the bilobation decreasing from p2 to p4, and p3 and p4 have a variable

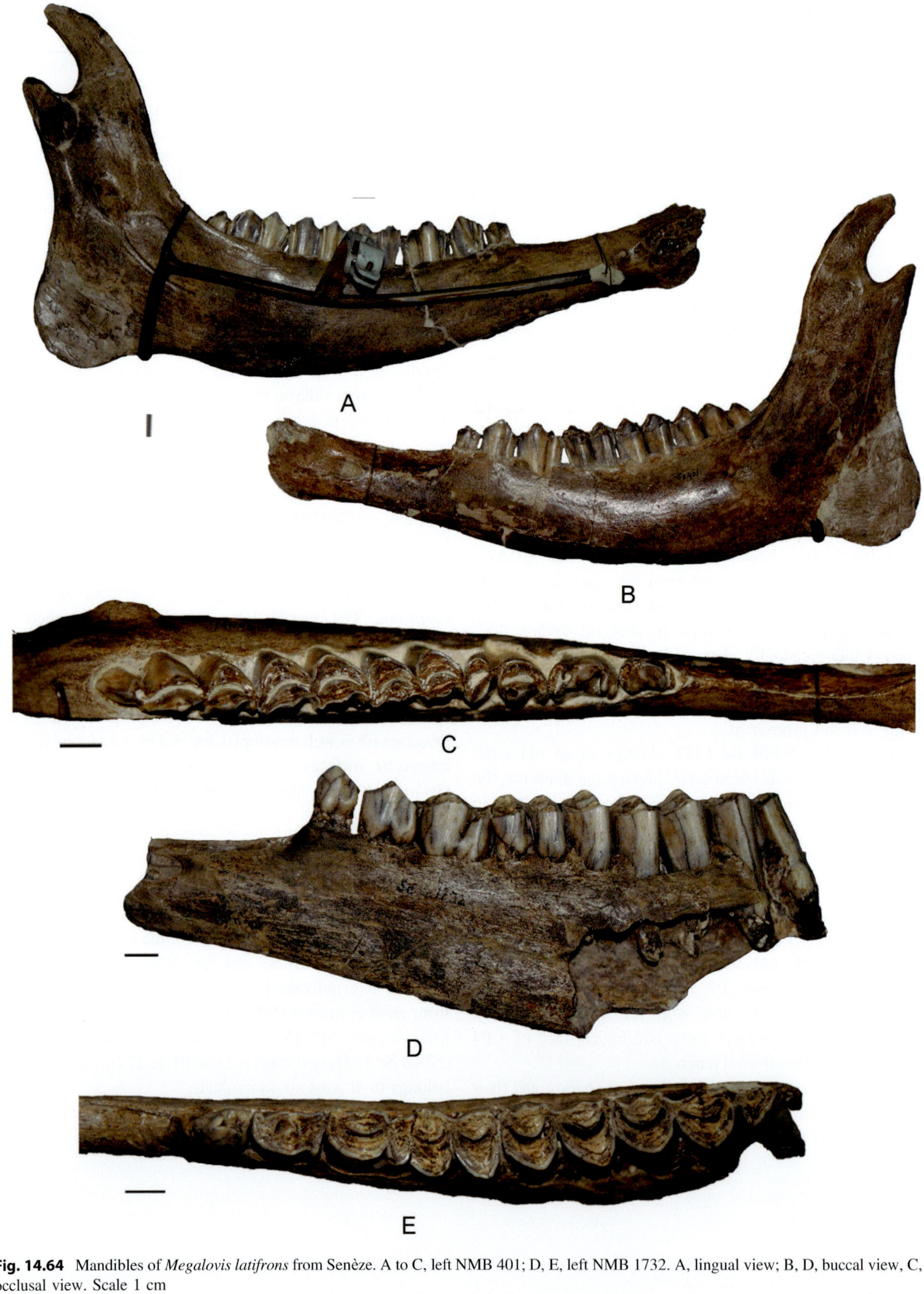

Fig. 14.64 Mandibles of *Megalovis latifrons* from Senèze. A to C, left NMB 401; D, E, left NMB 1732. A, lingual view; B, D, buccal view, C, E, occlusal view. Scale 1 cm

Table 14.58 Dimensions of the mandible (in mm) in *Megalovis latifrons* from Senèze

Variable	2	3	4	6	7	8	9	10
NMB Se 401	320.0	68.9	231.5	23.4	140.4	52.4	88.6	77.7
NMB Se 1732					146.4	54.7	93.0	
Variable	12	13	14	16	17	18	19	20
NMB Se 401	120.5	162.7	32.6	35.4	43.2	58.0	30.4	45.1
NMB Se 1732				40.2	47.6		31.6	39.3

Variable definitions (from Crégut-Bonnoure 2002, 2020) (Fig. 14.A.3): Length: 2, from infradental to the caudal side of the condyle; 3, L. horizontally from m3 distal side to the ramus; 4, from m3 distal side to the angle of the mandible; 6, from p2 mesial side to the posterior margin of the *processus angularis*; 7; from mesial side of I1 to distal side of canine; 8, from p2 mesial side to m3 distal side; 9, from p2 mesial side to p4 distal side; 10; from m1 mesial side to m3 distal side

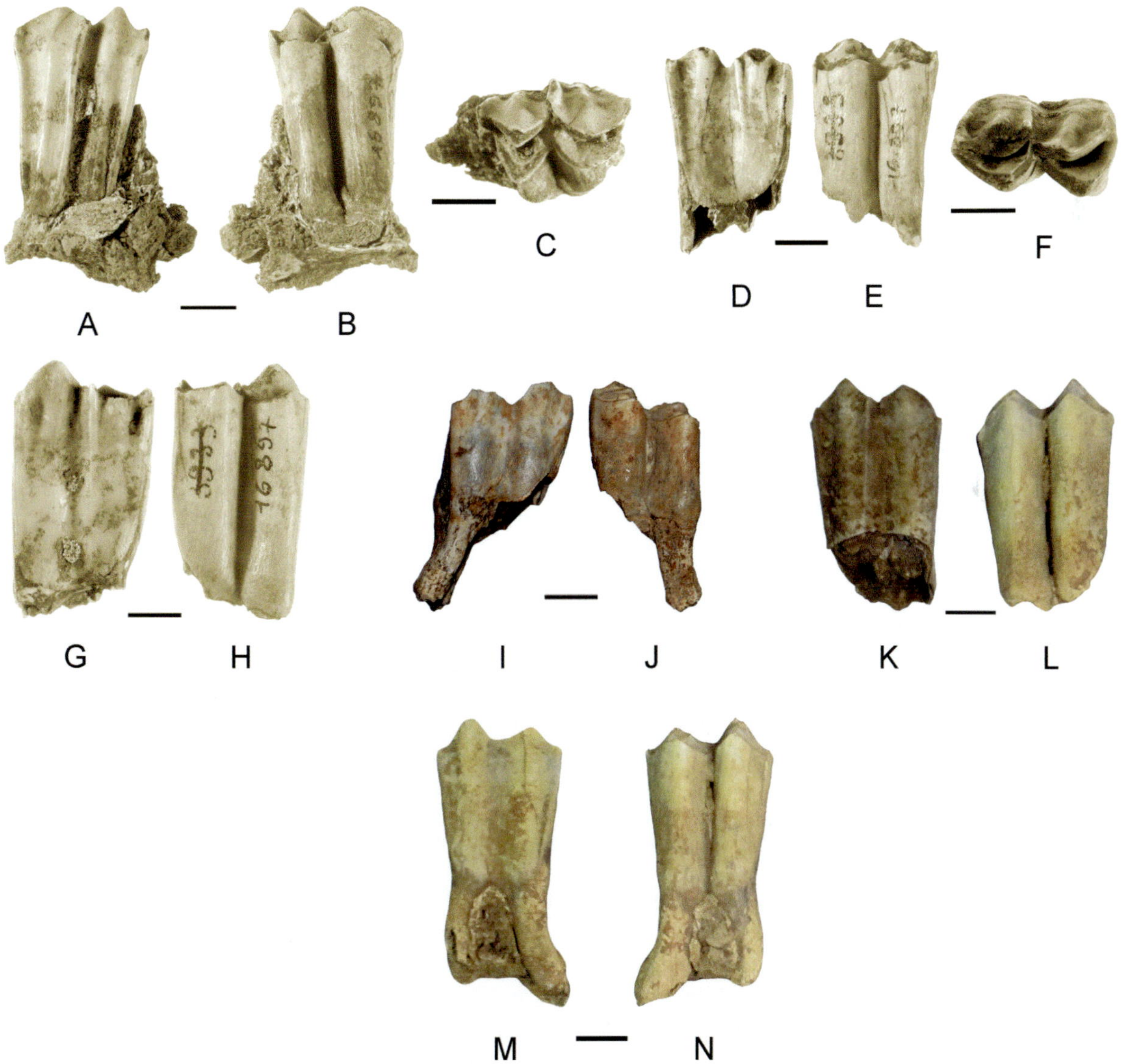

Fig. 14.65 Upper and lower teeth of *Megalovis latifrons* from Senèze. A to C, left M1 FSL 16897; D to F, right m1 FSL 16897; G, H, left m2 FSL 16897; I, J, left m1 NMB Se 1726; K, L, right m2 MNHN-F-1907-28; M, N, right m1 MNHN-F-1907-28. A, E, H, J, L, N, buccal view; B, D, G, I, K, M, lingual view; C, F, occlusal view; Scale 1 cm. A to H, phot. N. Podevigne

Table 14.59 Comparative dimensions of the upper premolars (in mm) in *Megalovis* from Senèze and other localities

P2 variable	Ht	1	2	3	5	
Megalovis latifrons						
Senèze						
NMB Se 556 (L)		16.4		13.4		
NMB Se 556 (R)		15.4		15.0		
Erpfinger Höhle? (NMB DP 33 cast)	28.7	15.9	15.5	13.7	16.9	
P3 variable	Ht	1	2	3	5	
Megalovis latifrons						
Senèze						
NMB Se 556 (L)		17.2		14.7		
NMB Se 556 (R)		16.6		14.5		
Megalovis balcanicus						
Trlica						
TRL 01/3/1		17.2		14.9		
TRL 01/21/7	21.3	16.8	15.5	15.9		
P4 variable	Ht	1	2	3	4	5
Megalovis latifrons						
NMB Se 556 (L)		15.8		15.6	13.8	
NMB Se 556 (R)		16.7		15.8	13.8	

Variable definitions (from Crégut-Bonnoure 1995, 2002, 2020) (Fig. 14.A.4: A): Ht, height; 1, occlusal length; 2, basal length. 3, occlusal width total. 4, occlusal width in the middle. 5, basal width. L, left; R, right

Table 14.60 Comparative dimensions of the M1 and M2 (in mm) in *Megalovis* from Senèze and other localities

M1 variable	Ht	1	2	3	4	5	6	7
Senèze								
NMB Se 556 (L)		23.1		18.8	18.2	17.3	16.8	
NMB Se 556 (R)		22.6				17.4	16.9	
FSL 16897 (juvenile)	33.8	22.4	17.7	13.9	13.3	12.0	11.6	19.4
Erpfinger Höhle?								
NMB DP 31 (cast)	20.6	26.8	25.5			16.7	15.4	
NMB DP 32 (cast)	20.1	24.4	22.8	20.5	17.1	19.8	18.0	25.0
Stetten (NMB DP 25–cast)	30.7	28.7	26.1	18.6	18.6	17.2	15.6	23.5
M2 variable	Ht	1	2	3	4	5	6	7
Senèze								
NMB Se 556 (L)		26.9		17.8	16.1	15.5	14.6	
NMB Se 556 (R)		26.5		17.8	16.7	15.8	14.6	
NMB Se 1430	18.4	25.7	24.6	20.1	19.1	18.3	17.9	23.7
Erpfinger Höhle (NMB DP 559–cast)	33.6	29.6	27.9	19.2	17.7	17.0	15.2	25.6
Erpfinger Höhle? (NMB DP 30–cast)	38.9	31.6	27.2	17.9	16.3	15.2	13.5	24.3

Variable definitions (from Crégut-Bonnoure 1995, 2002, 2020) (Fig. 14.A.4: B): Ht, height; 1, occlusal length; 2, basal length. 3, occlusal width total, first lobe; 4, occlusal width in the middle, first lobe; 5, occlusal width total, second lobe; 6, occlusal width in the middle, second lobe; 7, basal width; L, left; R, right

molarization. The p2s (n = 2) have a well developed parastylid. The protoconid is inflated, located in the middle of the crown, and flanked by two valleys, the distal being the deeper. The metastylid is prominent. The entoconid constitutes an oblique wall. The p3 morphology is variable (n = 3): on NMB Se 401 and MNHN-F-1922-5, the buccal bilobation is more important than on NMB Se 1732. On NMB Se 401, quite unworn, the paraconid and the parastylid are distinct. The inflation of the metaconid is weak, and behind it the wall is obliquely directed to the center of the crown. The entoconid constitutes a spur perpendicular to the longitudinal plane. Its fusion with the metaconid occurs at the base. The metastylid is thin. On NMB Se 1732, which is worn, the lingual side is molarized. A poorly defined valley exists behind the parastylid. On MNHN-F-1922-5, even more worn than NMB Se 1732, the parastylid and the paraconid are fused in a vertical wall. However, at the top, there is a small valley whose base is at 8.6 mm above the cemento-enamel junction: this morphology is similar to that of *Pliotragus*. A deep valley separates the paraconid and the metaconid. This last is not inflated but is fused with the entoconid (vertical wall). The

Table 14.61 Comparative dimensions of the M3 (in mm) in *Megalovis* from Senèze and other localities

M3 variable	Ht	1	2	3	4	5	6
Megalovis latifrons							
Senèze							
NMB Se 556 (L)		28.3		15.7	15.1		15.9
NMB Se 556 (R)		28.9		15.8	15.3		14.0
NMB Se 1430	21.5	31.0		20.5	19.6		
Erpfinger Höhle (NMB D 559–cast)	30.2	30.6	32.4	17.9	16.6	24.2	15.7
Erpfinger Höhle? (NMB DP 29–cast)	32.8	31.7	33.0	19.8	18.8	24.9	16.3
Megalovis sp							
Villány 3 (NMB UP 504)	34.9	27.1	30.6	16.5	15.2	20.4	13.3
Varshets (NNHM-S VR 3)	37.7	24.1	24.7	16.0	14.4	20.5	13.3
M3 variable	7	8	9	10	11	12	
Megalovis latifrons							
Senèze							
NMB Se 556 (L)	15.1		10.2	10.8			
NMB Se 556 (R)	13.3		10.7	10.9			
NMB Se 1430			10.4	11.1			
Erpfinger Höhle (NMB DP 558–cast)	15.6	22.8	11.8	12.7	27.5	28.7	
Erpfinger Höhle? (NMB DP 29–cast)	15.8	21.6	12.0	13.5	26.8	27.6	
Megalovis sp							
Villány 3 (NMB UP 504)	12.2	17.6	11.9	11.9	25.6	27.8	
Varshets (NNHM-S VR 3)	12.0	18.0	10.0	8.0	22.1	22.8	

Variable definitions (from Crégut-Bonnoure 1995, 2002, 2020) (Fig. 14.A.4: C): Ht, height; 1, occlusal length; 2, basal length. 3, occlusal width total, first lobe; 4, occlusal width in the middle, first lobe; 5, basal width, first lobe; 6, occlusal width total, second lobe; 7, occlusal width in the middle, second lobe; 8, basal width, second lobe; 9, length of the interstylar surface, mesial; 10, length of the interstylar surface, distal; 11, occlusal length of the two lobes at the lingual side; 12, basal length of the two lobes at the lingual side; L, left; R, right

entostylid is prominent. The p4s (n = 3) are characterized by a first lobe longer and wider than the second one, being isolated from each other by a deep buccal valley. The entoconid is obliquely directed and its lingual wall is oriented backward. Other characteristics are variable. The unworn p4 NMB Se 401 is bilobed. The parastylid is flanked by a narrow and deep valley. Behind this last, the paraconid is inflated into a wide column (8 mm). A lingual valley, very deep and narrow, reaches the buccal valley, separating the two lobes. The metaconid and metastylid are fused, constituting an oblique wall. The buccal side of the second lobe is more pinched than the first one and it is more prominent. On NMB Se 1732, the lingual side is molarized. The lingual valley is less deep than on NMB Se 401. The buccal side is similar to the latter specimen. On MNHN-F-1922-5, the paraconid is not inflated. A small and shallow valley is visible behind the parastylid. The lingual valley is similar to that of NMB Se 401. The metaconid and the metastylid are fused thus forming a flat surface less obliquely directed than on NMB Se 401, and the second lobe is less pinched.

There are six m1, six m2 and three m3. The lingual side is higher than the buccal one. The lobes are pinched buccally, more than in *Praeovibos* and *Ovibos*, and directed backward. The stylids are also more prominent and the conids more inflated. The valley between the parastylid and the paraconid is short and does not reach the base of the crown. The goat fold is present only at the top of the crown, except on the isolated m1 MNHN-F-1907-28 on which it reaches the base. The metaconid rib rejoins the parastylid, which is narrow, well above the base of the crown. The metastylid is well developed in the upper part of the crown, not at the base. The paraconid and metaconid are inflated, and the lingual valley is wide, "U" shaped. The base of the crown is wide and the cemento-enamel junction is low on the lateral sides, being higher on m1 than on m2.

The section of the third lobe of the m3 is triangular with some variation. On NMB Se 401 (unworn) it is lingually pinched and the distal side forms a longitudinal crest (Fig. 14.64: C). On NMB Se 1732, the shape is also triangular with a pinched lingual side; the distal side is without any crest (Fig. 14.64: E). On MNHN-F-1907-8, the section is rectangular: the distal side is flat, the buccal lobe being more or less rounded with a distal angle forming a longitudinal crest (Fig. 14.66: C). This shape is similar to that of the *Praeovibos priscus* Staudinger, 1908 specimen from Arago cave (Crégut 1979: Fig. 35 p. 216; Crégut-Bonnoure 2020: Fig. 3.48 p. 227). Nevertheless, despite this analogy, the morphology of p4, m1 and m2 lobes is similar to *Megalovis* and not to *Praeovibos*. The *Praeovibos* p4 has a distal lobe as long as in *Megalovis*, but narrower bucco-lingually, and the molar lobes are rounded buccally as in *Ovibos* (Crégut 1979; Crégut-Bonnoure 2002, 2006, 2020).

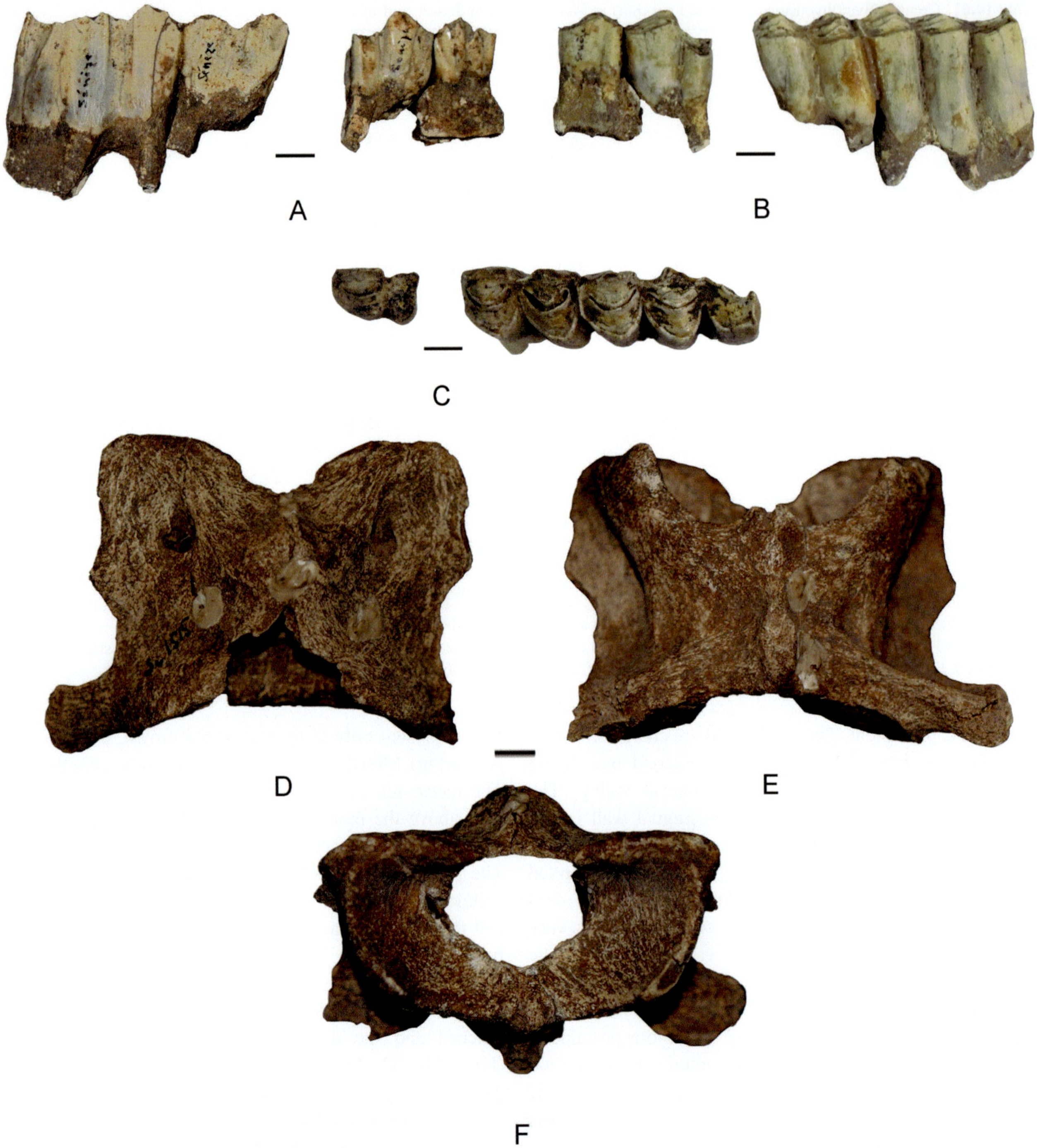

Fig. 14.66 Lower teeth of *Megalovis latifrons* and atlas of *Pliotragus ardeus* from Senèze. A to C, left p3, p4, m1-3 MNHN-F-1922-5; D to F, atlas MNB Se 151. A, lingual view; B, buccal view; C, occlusal view; D, dorsal view; E, ventral view; F, cranial view. Scale 1 cm

Table 14.62 Comparative dimensions of the lower premolars (in mm) in *Megalovis* from Senèze and other localities

p2 variable	Ht	1	5			
M. latifrons						
Senèze						
NMB Se 401		12.1	8.4			
NMB Se 1732		13.4	9.1			
p3 variable	Ht	1	2	3	5	
M. latifrons						
NMHN 1922-5	16.0	17.0	16.7	12.1	12.3	
NMB Se 401		16.7	10.9	9.4		
NMB Se 1732	12.5	18.7	11.9	11.2		
p4 variable	Ht	1	2	3	4	5
M. latifrons						
Senèze						
NMHN 1922-5	24.6	22.7	22.1	12.4	13.5	13.3
NMB Se 401		20.9		10.3	10.8	
NMB Se 1732	17.0	20.8		12.2	11.9	13.3
Erpfinger Höhle? (NMB DP 36–cast)	13.2	19.5	17.6	11.0	13.6	
M. wimani						
NMB Ca 147 (cast)		18.9	19.9	11.9	10.8	13.6

Variable definitions (from Crégut-Bonnoure 1995, 2002, 2020) (Fig. 14.A.4: D): Ht, height; 1, occlusal length; 2, basal length; 3, occlusal width, first lobe; 4, occlusal width, second lobe; 5, basal width

Table 14.63 Comparative dimensions of the lower molars (in mm) in *Megalovis* from Senèze and other localities

m1 variable	Ht	1	2	3	4	5
Senèze N	5	5	3	6	5	3
Mean	23.85	23.40	20.58	13.42	13.41	16.07
Minimum	12.5	21.6	19.9	11.5	11.6	15.2
Maximum	31.3	25.3	21.2	16.5	16.9	
SD	7.47	1.41	0.66	1.78	2.18	0.81
m2 variable	Ht	1	2	3	4	5
Megalovis latifrons						
Senèze N	3	5	3	5	5	3
Mean	34.79	27.57	23.74	13.79	14.1	17.82
Minimum	21.9	26.4	23.1	12.0	11.9	16.9
Maximum	46.2	29.7	24.6	16.4	16.8	18.7
SD	12.23	1.24	0.79	2.24	2.35	0.9
Stetten						
NMB DP 24	36.1	26.9	23.1	11.8	12.2	15.3
Varshets						
NNHM-S VR1	28.4			12.3	12.5	
NNHM-S VR4	26.0	24.9	23.2	12.41	12.4	14.7
Megalovis sp						
Slivnitsa						
NNHM-S SL 1		25.7		15.5	15.2	
NNHM-S SL2		25.6		15.1	15.2	

(continued)

Table 14.63 (continued)

m3 variable	Ht	1	2	3	4	5	6
Megalovis latifrons							
Senèze							
NMHN 1922-5	29.9	41.9	41.2	16.2	17.0	9.5	17.4
NMB Se 401		37.3		12.2	12.0	7.6	
NMB Se 1732		40.0		15.4	15.4	8.1	
Erpfinger Höhle?							
NMB DP 35 (cast)	32.8				14.7	10.5	17.2
NMB DP 34 (cast)	38.8	34.4	36.0	14.0	12.5	6.7	15.3
Stetten (NMB DP 23–cast)	41.7	37.0	37.4		13.6	8.0	16.1
Megalovis sp							
Slivnitsa							
NNHM-S SL 1		39.9		15.5	14.5	8.2	
NNHM-S SL2				15.1			
M. wimani							
NMB Ca 147 (cast)		35.4		13.2	12.6	6.6	

Variable definitions (from Crégut-Bonnoure 1995, 2002, 2020): Ht, height; m1 and m2 (Fig. 14.A.4: E): 1, occlusal length; 2, basal length; 3, occlusal width, first lobe; 4, occlusal width, second lobe; 5, basal width. m3 (Fig. 14.A.4: F): m3, occlusal view (above), lingual view (below left), mesial view (below right): 1, occlusal length; 2, basal length; 3, occlusal width, first lobe; 4, occlusal width, second lobe; 5, occlusal width, third lobe; 6, basal width

The proportions of the upper teeth have already been compared with the European and Asian *Megalovis* populations (Crégut-Bonnoure 2002, 2020; Crégut-Bonnoure & Dimitrijević 2006). All of them show similarities whereas the morphology is distinct. All the *M. latifrons* populations have similar proportions, and the material from Erpfinger Höhle and Stetten is relatively more robust (Figs. 14.67, 14.68 and 14.69).

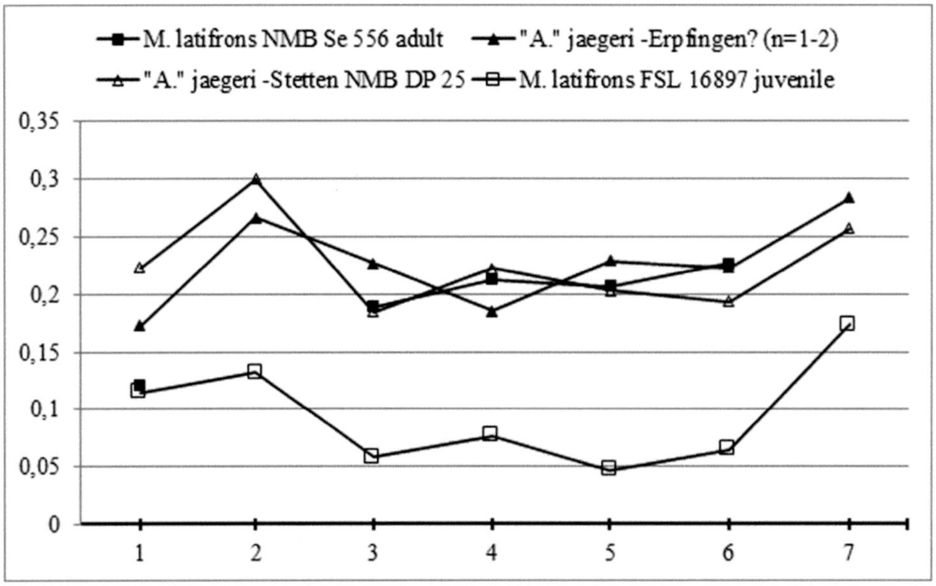

Fig. 14.67 *Megalovis latifrons* from Senèze and Germany. Ratio diagram of the M1. Reference axis: *Capra ibex* from Le Taï cave (in Crégut-Bonnoure 2020). Length: 1, occlusal; 2, basal. Width: 3, occlusal, first lobe; 4, occlusal in the middle, first lobe; 5, occlusal, second lobe; 6, occlusal in the middle, second lobe; 7, basal. References of the measurements in Fig. 14.A.4 and dimensions in Table 14.60

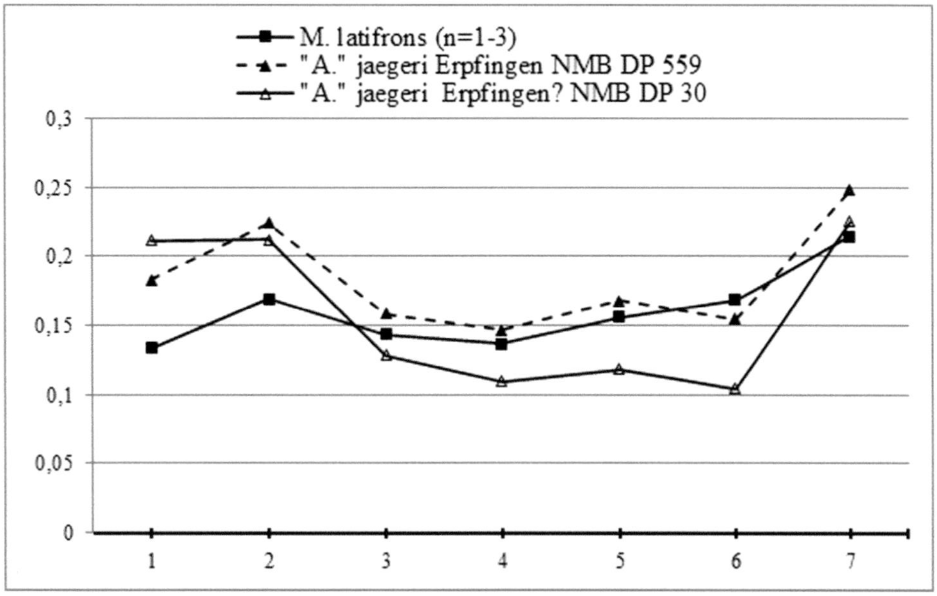

Fig. 14.68 *Megalovis latifrons* from Senèze and Germany. Ratio diagram of the M2. Reference axis: *Capra ibex* from Le Taï cave (in Crégut-Bonnoure 2020). Length: 1, occlusal; 2, basal. Width: 3, occlusal, first lobe; 4, occlusal in the middle, first lobe; 5, occlusal, second lobe; 6, occlusal in the middle, second lobe; 7, basal. References of the measurements in Fig. 14.A.4 and dimensions in Table 14.60

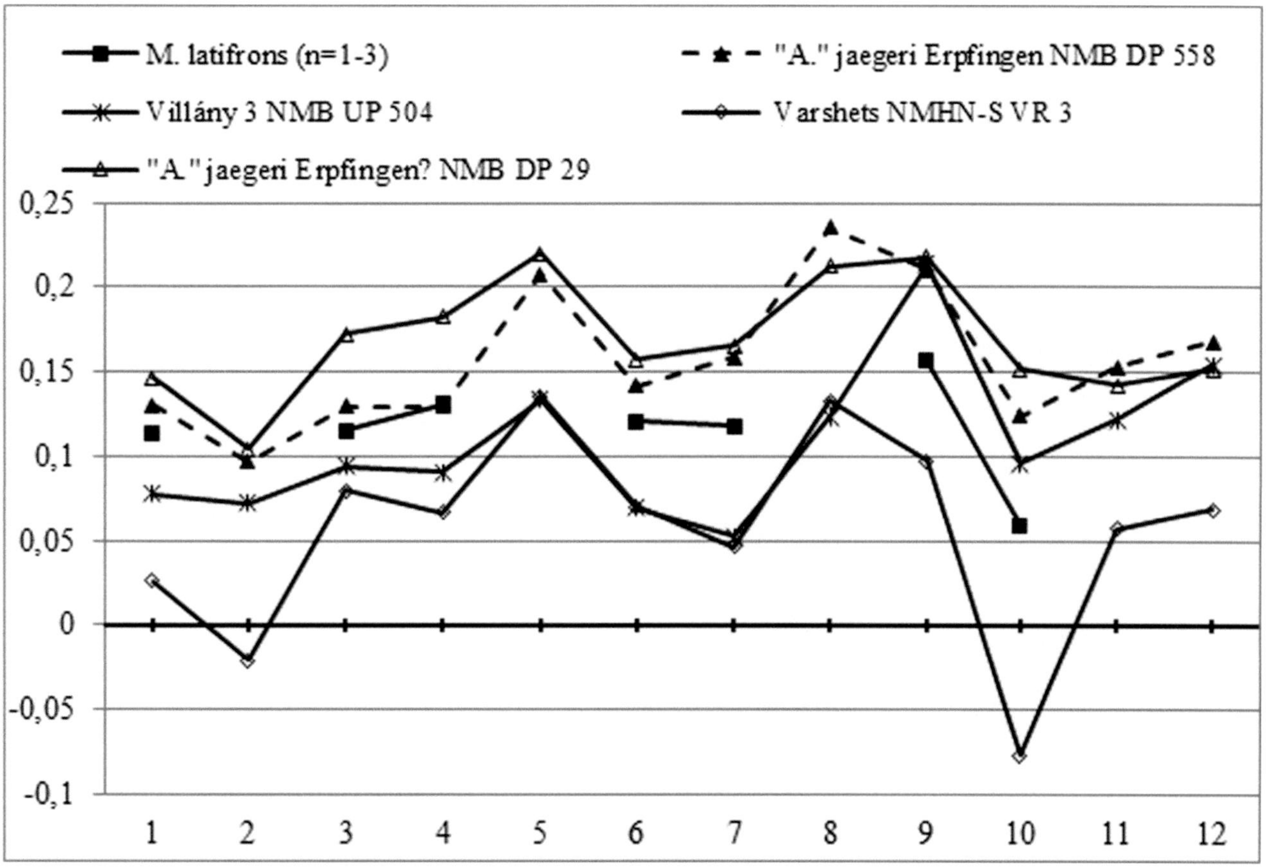

Fig. 14.69 *Megalovis latifrons* from Senèze and other localities. Ratio diagram of the M3. Reference axis: *Capra ibex* from Le Taï cave (in Crégut-Bonnoure 2020). Length: 1, occlusal. 2, basal. 9, anterior interstylar surface. 10, posterior interstylar surface. 11, occlusal, first+second lobes (lingual side). 12, basal, first+second lobes (lingual side). Width: 3, occlusal, first lobe. 4, occlusal in the middle, first lobe. 5, basal, first lobe. 6, occlusal, second lobe. 7, occlusal in the middle, second lobe. 8, basal, second lobe. References of the measurements in Fig. 14.A.3 and dimensions in Table 14.61

Postcrania

The postcranial material is scarce and some of the previous attributions must be changed.

Atlas

The atlas NMB Se 1515 is now considered as *Pliotragus*, as discussed in section "Atlas" (Fig. 14.66: D to F).

Humerus and Radius

Guérin (*in* Duvernois & Guérin 1989) indicated that the right humerus (NMB Se 263) and radius (NMB Se 272) belonging to a single leg (assigned to this species by Schaub, 1923, 1943) actually correspond to *Leptobos furtivus* (Fig. 14.70: A, B). The humerus proportions are clearly distinct from those of *M. balcanicus* and from the FSL 211101 specimen which is described below (Fig. 14.71).

FSL 211101 is a right humerus assigned initially to a cervid (on the label) (Fig. 14.72). It is not complete: more or less half its length and the distal articulation are preserved (Table 14.64). However, the trochlea is broken caudally. Its morphology is quite similar to the *M. balcanicus* specimen (Crégut-Bonnoure & Dimitrijević 2006: pl. IV Fig. 1 a to c). Because the trochlea is broken, it is not possible to see if the bone stays vertical when it is put on its distal articulation.

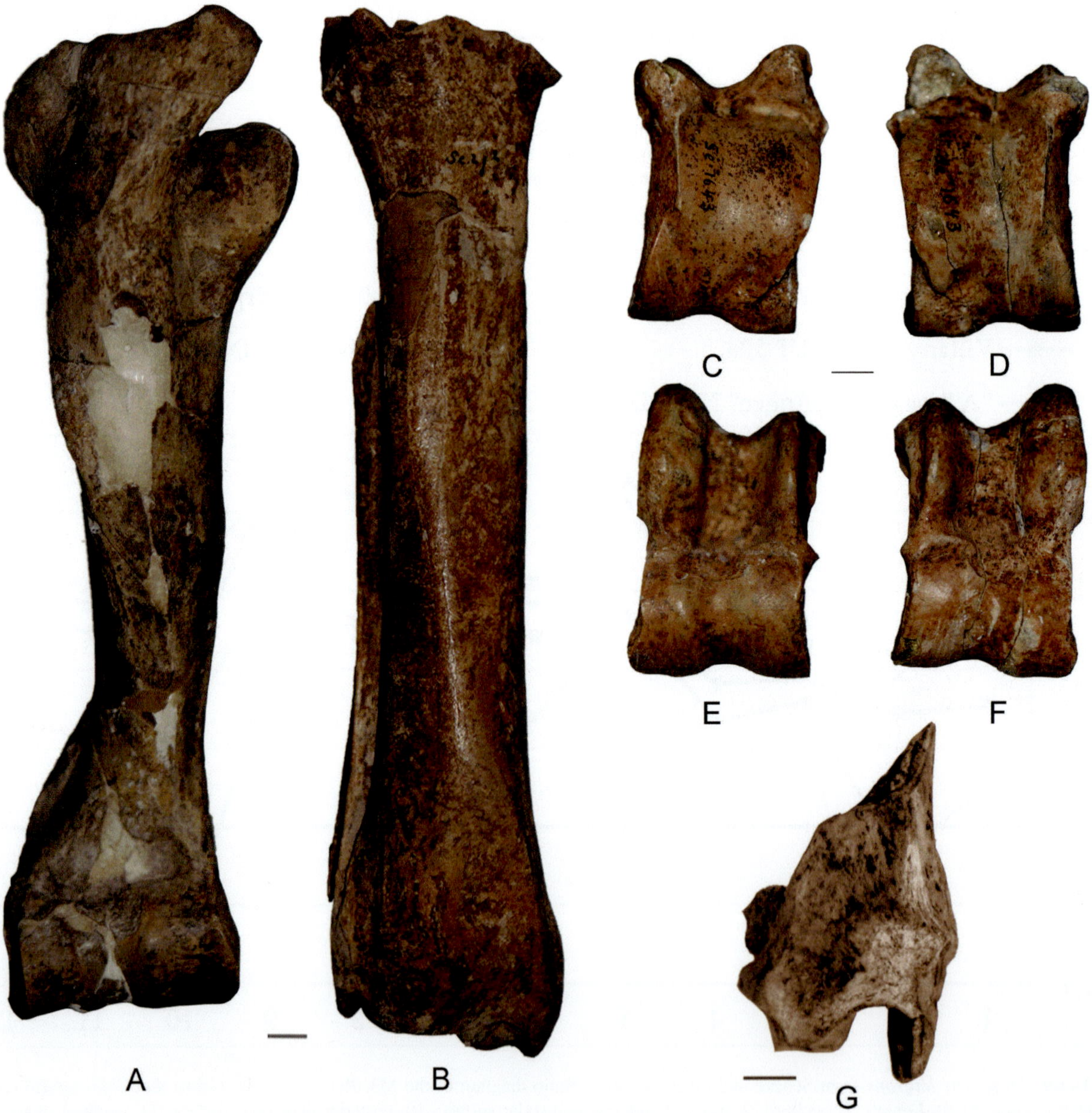

Fig. 14.70 Post-cranium of *Megalovis latifrons*, *Leptobos furtivus* and *Eucladoceros ctenoides senezensis* from Senèze. A, *Leptobos* right humerus NMB Se 263; B, *Leptobos* right radius NMB Se 264; C, E, *Eucladoceros* left talus NMB Se 1643; D, F, *Eucladoceros* right talus NMB Se 1643; G, *Megalovis* left tibia NMB Se 1535. A, G, cranial view; B, E, F, dorsal view; C, D, plantar view. Scale 1 cm

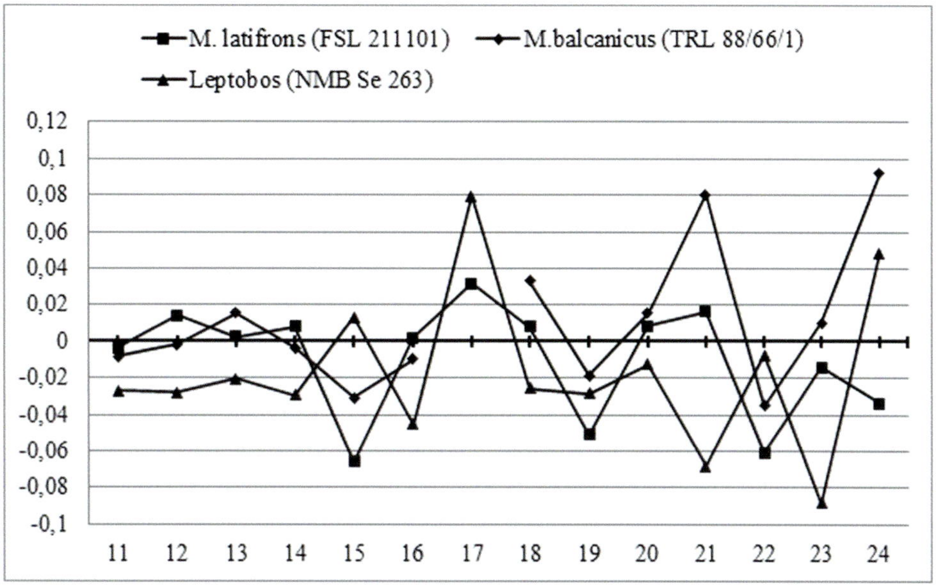

Fig. 14.71 *Megalovis latifrons* from Senèze, *M. balcanicus* from Trlica and *Leptobos* from Senèze. Ratio diagram of the humerus. Reference axis: recent *Ovibos moschatus* ♂ (in Crégut-Bonnoure 2020). Transversal diameter distal: 11, maximal; 12, articular; 13, trochlea; 14 capitulum. Craniocaudal diameter distal: 15, medial side, total; 16, medial side, articulation; 17, lateral side, total; 18, lateral side, articulation; 19, trochlea, minimum; 20, trochlea, median guiding crest. Diaphysis, sus articular: 21, transversal diameter; 22, craniocaudal diameter. Olecranon fossa: 23, transversal diameter; 24: height. References of the measurements in Fig. 14.A.6 and dimensions in Table 14.64 and in Crégut-Bonnoure and Dimitrijević (2006) for *M. balcanicus*

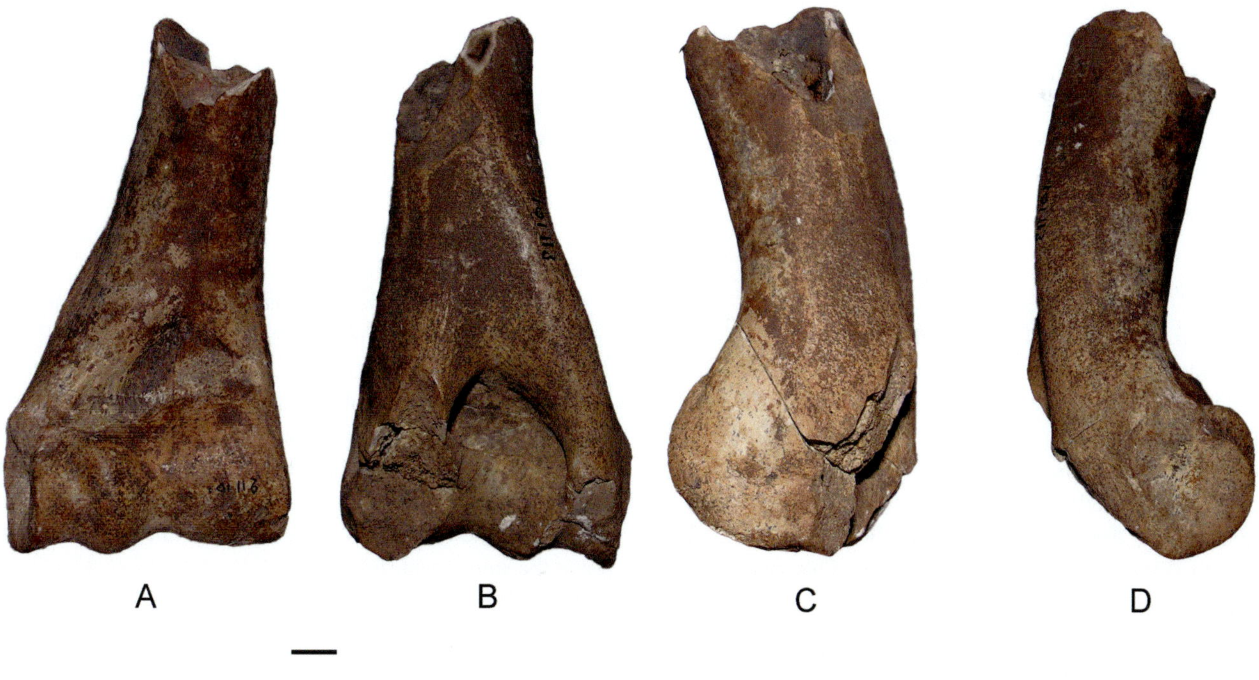

Fig. 14.72 Humerus of *Megalovis latifrons* from Senèze. Right FSL 211101. A, cranial view; B, caudal view; C, medial view; D, lateral view. Scale 1 cm

Table 14.64 Comparative dimensions of the humerus (in mm) in *Megalovis* and *Leptobos* from Senèze

Variable	11	12	13	14	15	16	17
M. latifrons							
Senèze (FSL 211101)	69.8	69.0	45.6	21.8	56.5	45.2	45.9
Leptobos furtivus							
Senèze (NMB Se 263)	66.1	62.6	43.3	20.0	68.0	40.5	51.2
Variable	18	19	20	21	22	23	24
M. latifrons							
Senèze (FSL 211101)	29.9	29.4	38.6	41.5	34.6	25.8	23.4
Leptobos furtivus							
Senèze (NMB Se 263)	27.6	30.9	36.8	34.2	39.1	21.7	28.2

Variable definitions (from Crégut-Bonnoure 1995, 2002, 2020) (Fig. 14.A.6): transversal diameter distal: 11, maximal; 12, articular; 13, trochlea; 14 capitulum. Craniocaudal diameter distal: 15, medial side, total; 16, medial side, articulation; 17, lateral side, total; 18, lateral side, articulation; 19, trochlea, minimum; 20, trochlea, median guilding crest. Diaphysis, supra-articular: 21, transversal diameter; 22, craniocaudal diameter. Olecranon fossa: 23, transversal diameter; 24: height

However, when the bone is placed on its cranial side, as in *M. balcanicus*, the olecranon fossa shows a medial side more vertical and thicker than in *Ovibos* and *Praeovibos*. The coronoid fossa is elongated toward the trochlear side, being shorter in the musk ox. Compared to *M. balcanicus*, the bone is slender with a shorter trochlea (measurement 13; Fig. 14.71). It may be identified as *Megalovis latifrons*.

Tibia

A fragmentary right tibia, NMB Se 1535, has been assigned to the species by Schaub, (1923) (Fig. 14.70: G). It is a distal fragment without the malleolar facets. The morphology is similar to that of *Ovibos* except that the articulation for the talus is visible on each side of the medial malleolus in cranial view, unlike in *Ovibos*.

Talus

Two tali (NMB Se 1643 right and left) belonging to a single individual have been assigned to the species by Schaub (1923). Their morphology is not ovibovin (Fig. 14.70: C to F). In *Ovibos*, *Praeovibos* and *Soergelia*, as in the *Megalovis* from Trlica, Slivnitsa and Pirro Nord, the main characteristic consists of an ellipsoidal synovial fossa at the base of the trochlea for the calcaneus (Crégut-Bonnoure 2002, 2020; Crégut-Bonnoure & Dimitrijević 2006). This synovial facet is absent on these two specimens. They are characterized by proximal trochlear lips only slightly offset to one another and a narrow intertrochlear space. Distally, the lateral condyle is flat at its base and its connection to the median groove is sharp. The medial condyle is shorter than the lateral one. There is also an inverted-comma shape of the protrusion for the navicular-cuboid medio-plantar facet at the base of the the plantar trochlea. All these features characterize a cervid (Heintz 1970: Fig. 31). According to their size, they belong to *Eucladoceros ctenoides senezensis* (Heintz 1970: pl. XL Fig. 6) (Crégut-Bonnoure 2002, 2020).

Metapodials

There are three metapodials (Tables 14.65 and 14.66): a right metacarpal III+IV NMB Se 1731 (Fig. 14.73: A to C), a right metatarsal III+IV FSL 16894 (Guérin 1965: pl. LXX phot. A; Fig. 14.73: E) and an incomplete right metatarsal III +IV FSL SEN 02-0007 given to the scientific team by local amateur A. Consigny, who collected it in parcel 263, in an area not studied by the team but altitudinally below excavated finds (Fig. 14.73: D).

The shaft of the metapodials is longer and slender in comparison to *Ovibos* and *Praeovibos* (Schaub 1943: Fig. 7). The shape of the distal part is hourglass-like, less than in *Ovibos*, and asymmetrical: the lateral side is more divergent than the medial one. The metacarpal III+IV proximal articulation has articular facets angled to one another, the palmar side of the capitato-trapezoid articular facet being obliquely directed to the dorsal side: consequently, the two facets are not at the same level (Rădulesco & Samson 1962: Fig. 2; De Giuli & Masini 1983: pl.2, Fig. 2d) (Fig. 14.73: C). The shaft is relatively longer than in the bone from Fântâna lui Mitilan and in the Chinese *Megalovis*, especially the median length (measurement 2, Fig. 14.74).

On the two metatarsal III+IV, the proximal articulation is partially broken. However, the main features of the genus are present (Crégut-Bonnoure, 2002, 2006, 2020). Two proximal tubercles are visible on the sides. On the medial side, it is located at the level of the small plantar cuneiform articular facet as in *Ovibos*. Its length is about 40 mm (37 mm on FSL 1743; the proximal side is broken on FSL SEN 02-0007 so this measurement is impossible). On the lateral side, which is flat, the tubercle is located at the level of the navicular-cuboid articular facet. It is absent in *Ovibos*. It is triangular and shorter than the medial one on FSL 1743 (25 mm); on FSL SEN 02-0007 the base is at the same level as the medial tubercle. The proportions are similar to those of *M. piveteaui* (Fig. 14.75).

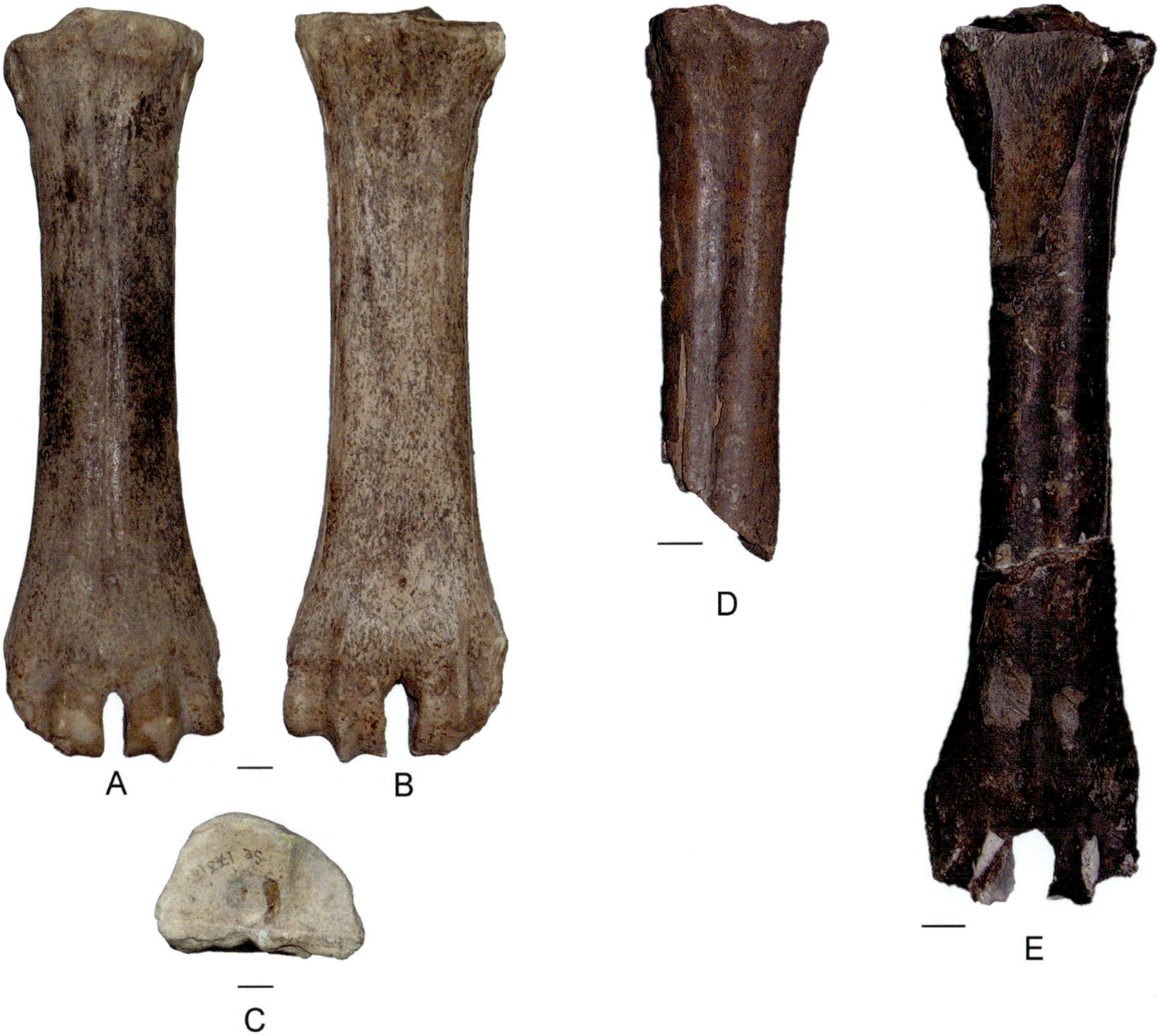

Fig. 14.73 Metapodials of *Megalovis latifrons* from Senèze. A to C, right metacarpal III+IV NMB Se 1731; D, right metatarsal III+IV FSL SEN-02-0007; E, right metatarsal III+IV FSL 16894. A, D, E, dorsal view; B, palmar view; C, proximal view. Scale 1 cm

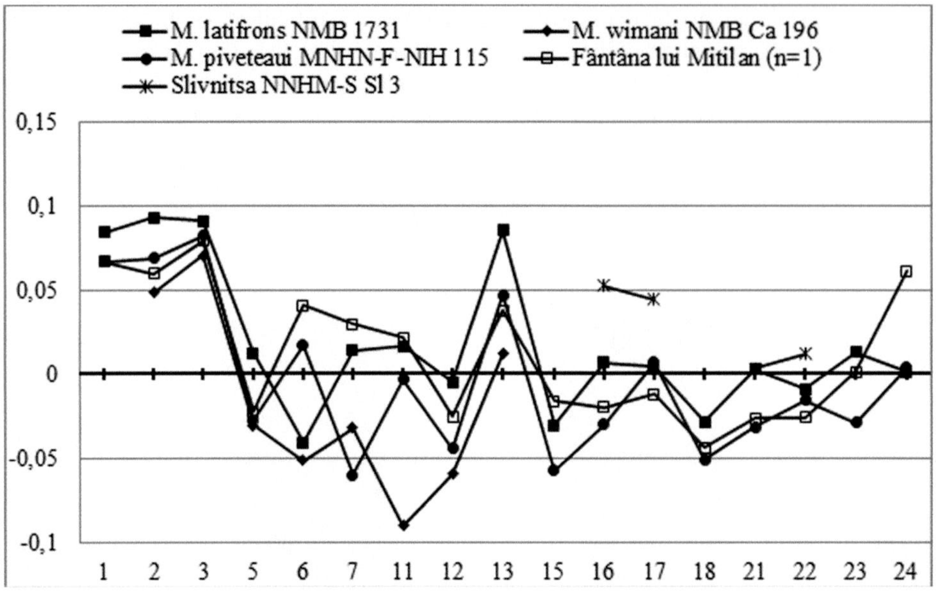

Fig. 14.74 *Megalovis latifrons* from Senèze and *Megalovis* spp. from other localities. Ratio diagram of the metacarpal III+IV. Reference axis: recent *Ovibos moschatus* ♂ (in Crégut-Bonnoure 2020). dpd, dorsopalmar diameter; td, transversal diameter. Length: 1, total; 2, minimum at the middle; 3: minimum of the medial side. Proximal epiphysis: 5, td of the articulation; 6, td of the unciform facet; 7, td of the capitato-trapezoid facet; 11, dpd of the articulation. Middle of the diaphysis: 12, td; 13, dpd. Distal epiphysis: 15, td supra-articular; 16, dpd supra-articular; 17, dpd at the level of the dorsal foramen; 18, distance between the lateral edges of the keels at the palmar top; 21, td of the medial keel; 22, td of the medial condyle; 23, td of the lateral keel; 24, td of the lateral condyle. References of the measurements in Fig. 14.A.10 and dimensions in Table 14.65

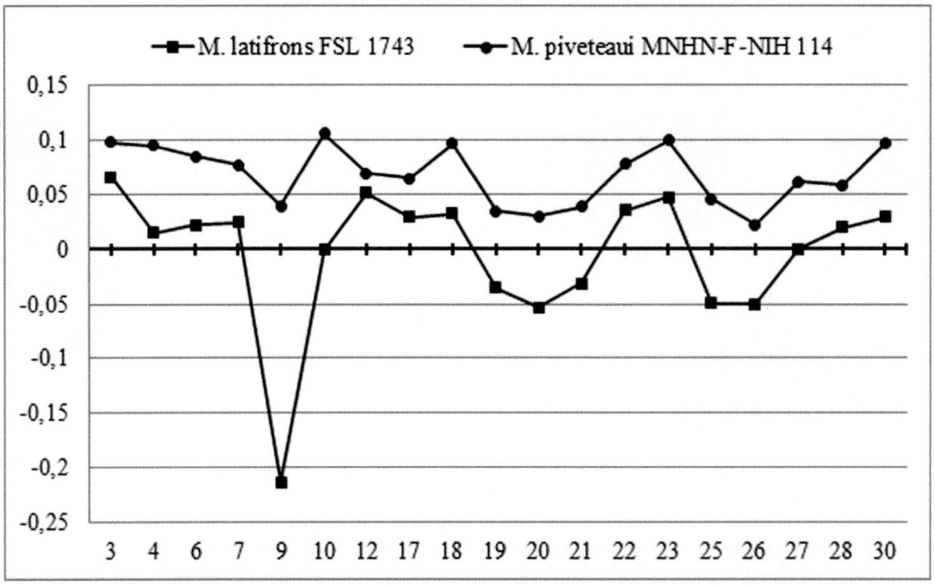

Fig. 14.75 *Megalovis latifrons* from Senèze and *Megalovis piveteaui* from Nihowan. Ratio diagram of the metatarsal III+IV. Reference axis: recent *Ovibos moschatus* ♂ (in Crégut-Bonnoure 2020). dpd: dorsoplantar diameter; td: transversal diameter. 3, length of the plantar side. Proximal epiphysis: 4, td total; 6, dpd total; 7, dpd of the articulation; 9, td of the small medioplantar articular facet; 10: td of the mediodorsal articular facet; 12, td of the dorsal facet for the navicular-cuboid. Middle of the diaphysis: 17, td; 18: dpd. Distal epiphysis: 19, td maximal of the articulation; 20, td minimal of the articulation; 21, td supra-articular; 22, dpd of the epiphysis; 23, dpd of the epiphysis at the foramen level; 25, distance between the keels at their base; 26, distance between the keels in their middle; 27, dpd of the medial condyle; 28, dpd of the medial keel; 30, dpd of the lateral keel. References of the measurements in Fig. 14.A.16 and dimensions in Table 14.63

Table 14.65 Comparative dimensions of the metacarpal III+IV (in mm) in *Megalovis* from Senèze and other localities. Measurements from Fântâna lui Mitilan provided by the late Dr. Costin Rădulesco

Variable	1	2	3	4	5	6	7	8	9	10	11
M. latifrons											
Senèze (NMB Se 1731)	200.0	180.0	187.0	56.4	55.2	20.7	35.7	35.4	43.4	37	32.6
Megalovis sp											
Fântâna lui Mitilan	192.2	166.7	182.0	56.0	51.1	25.0	37.0	32.0	44.0	35.0	33.0
M. wimani											
NMB Ca 196		162.5	178.5	52.8	50.0	20.2	32.1	27.7	37	31	25.5
M. piveteaui											
MNHN NIH 115	192.2	170.3	183.4	51.3	50.4	23.7	30.1	35.1	38.3	35.0	31.2

Variable	12	13	15	16	17	18	21	22	23	24
M. latifrons										
Senèze (NMB Se 1731)	37.4	24.9	59.0	29	23.9	36.2	34	21.5	33.4	20.0
Megalovis sp.										
Fântâna lui Mitilan	35.7	22.3	64.0	27.3	23.0	33.8	31.8	20.7	32.5	23.0
Slivnitsa (NNHM-S SL 3)				32.3	26.2		34.1	22.6		
M. wimani										
NMB Ca 196 (cast)	33.0	21.0								
M. piveteaui										
MNHN NIH 115	34.2	22.8	55.4	26.7	24.0	34.4	31.4	21.2	30.3	20.1

Variable definitions (from Crégut-Bonnoure 2002, 2020) (Fig. 14.A.10): dpd, dorsopalmar diameter; td, transversal diameter. Length: 1, total; 2, minimum at the middle; 3, minimum of the medial side. Proximal epiphysis: 4, td of the epiphysis; 5, td of the articulation; 6, td of the unciform facet; 7, td of the capitato-trapezoid facet; 8, length of unciform facet, diagonally from latero-palmar angle to dorsal point of the proximal crest; 9, length of capitato-trapezoid facet, diagonally from medio-palmar angle to dorsal point of the proximal crest; 10, dpd of the epiphysis; 11, dpd of the articulation. Middle of the diaphysis: 12, td; 13, dpd. Distal epiphysis: 14, td of the articulation;15, td supra-articular;16, dpd supra-articular;17, dpd at the level of the dorsal foramen;18, distance between the lateral edges of the keels at the palmar top; 19, distance between the lateral edges of the keels at the palmar base; 20, distance of the keels at their middle; 21, td of the medial keel; 22, td of the medial condyle; 23, transversal diameter of the lateral keel; 24, td of the lateral condyle

Table 14.66 Comparative dimensions of the metatarsal III+IV (in mm) in *Megalovis latifrons* from Senèze and in *M. piveteaui*

Variable	3	4	6	7	9	10	12	17	18
M. latifrons									
Senèze (FSL 1743)	198.5	45.3	41.7	40.8	7.2	15.7	18.4	30.1	25.5
M. piveteaui									
MNHN NIH 114	214.0	54.5	48.2	46.1	12.9	20.1	19.2	32.6	29.6

Variable	19	20	21	22	23	25	26	27	28	30
M. latifrons										
Senèze (FSL 1743)	51.7	48.9	49.4	27.9	25.4	29.6	25.4	32.4	22.7	19.6

Variable definitions (from Crégut-Bonnoure 1995, 2002, 2020) (Fig. 14.A.16): dpd: dorsoplantar diameter; td: transversal diameter. 3, length of the plantar side. Proximal epiphysis: 4, td total; 6, dpd total; 7, dpd of the articulation; 9, td of the facet for the small cuneiform; 10: td the facet for the great cuneiform; 12, td of the dorsal facet for the navicular-cuboid. Middle of the diaphysis: 17, td; 18: dpd. Distal epiphysis: 19, td maximal of the articulation; 20, td minimal of the articulation; 21, td supra-articular; 22, dpd of the epiphysis; 23, dpd of the epiphysis at the foramen level; 25, distance between the keels at their base; 26, distance between the keels in their middle; 27, dpd of the medial condyle; 28, dpd of the medial keel; 30, dpd of the lateral keel

Discussion

There are six individuals. Two of them are old, two others are young adults, and two are full adults. Several mistakes of attribution occurred, and consequently there are few specimens: the cranium, an incomplete maxilla, two mandibles, three incomplete toothrows, nine isolated teeth and five bones. The molarization of p3 and p4 is not homogeneous as is true also for the shape of the m3 third lobe and the tubercle of the proximo-lateral side of the metatarsal III+IV. A fragmentary metatarsal III+IV was presented by a local person from an area whose stratigraphic position is not known with certainty; it appears from its altitude to lie ca. 30 m lower than the team's western trenches (1–3), although if the bone were found on the surface that may not be meaningful.

Bovinae Gray, 1821

Tribe Bovini Gray, 1821

Genus Leptobos Rütimeyer, 1877-78

Species *Leptobos furtivus* Duvernois, 1990 and *Leptobos etruscus* (Falconer, 1868).

Leptobos is a Villafranchian bovid (Duvernois 1989, 1990, 1992; Dimitrijevič and Knezevic 1996). It appears at the beginning of this Land Mammal Age (Azzaroli et al. 1988; Guérin 1980, 1982; Heintz et al. 1974; Kurtén 1963). Six large species are recognized in Europe, all of which are more slender than *Bos* or *Bison*; among them are *L. etruscus* and *L. furtivus*, both present at Senèze. *L. furtivus* is considered by some authors as a *L. etruscus* female (Andrés Rodrigo 2011; Garrido 2006, 2008b) or a species related to *L. merlai* (Cherin et al., 2017).

Description

The first analysis was done by Duvernois (1990; and *in* Duvernois & Guérin 1989). The material is relatively abundant. Here I will discuss only the remains from the new fieldwork: a metapodial condyle (FSL SEN 01-0051, Trench 3), a left M2 (FSL SEN 04-0217) and a distal epiphysis from a left metatarsal III+IV (FSL SEN 04-0182) (Fig. 14.76: B, E; both the last two fromTrench 5). The molar is incomplete; the lingual face of the second lobe is missing (Table 14.67). It is not possible to be sure of its specific atribution. The same is true for the isolated condyle. Regarding the metatarsal, its proportions are more similar to those of *L. furtivus* than to *L. etruscus* (Table 14.68; Fig. 14.77).

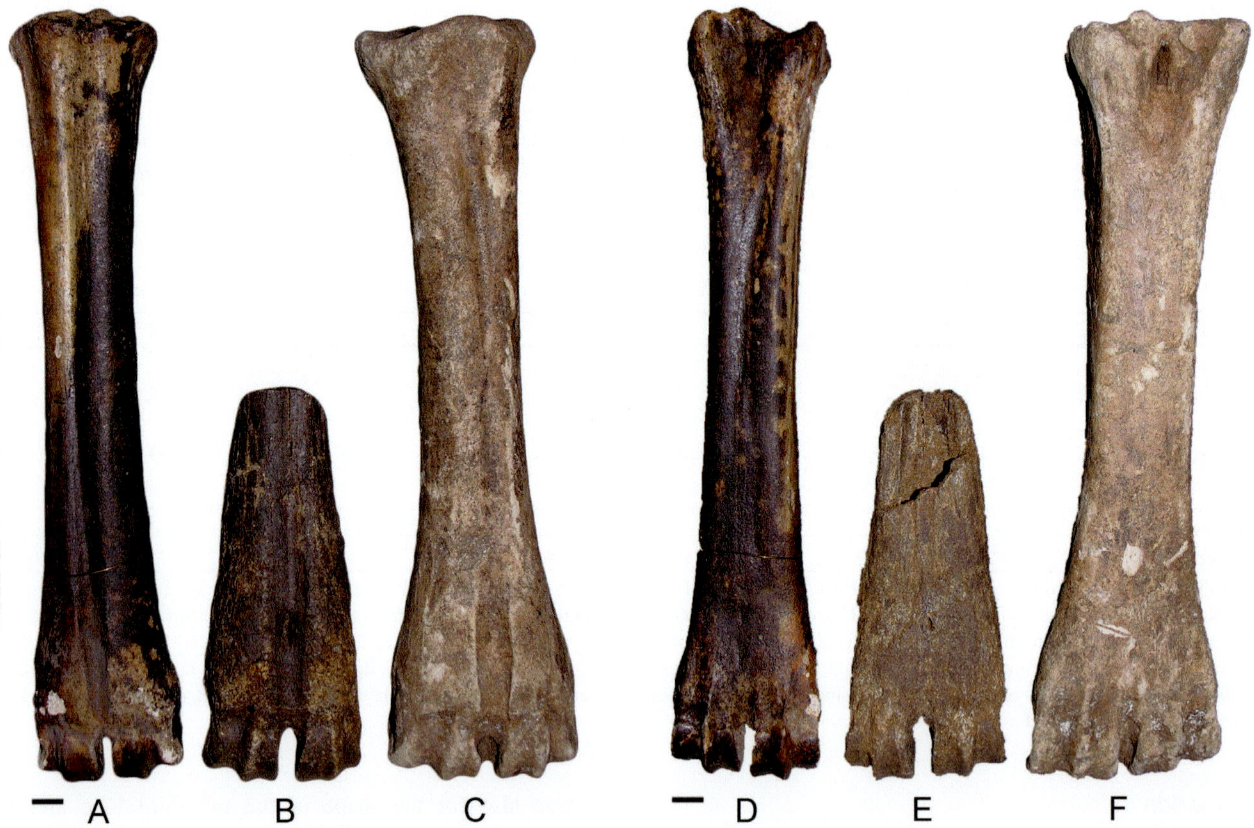

Fig. 14.76 Metatarsal III+IV of *Leptobos furtivus* and *Bison* (*Eobison*) sp. from Senèze. A, D, *Leptobos* left FSL 210779; B, E, *Leptobos* left FSL SEN 04-0182; C, F, *Bison* left FSL SEN 02-0020. A to C, dorsal view; D to E, plantar view. Scale 1 cm

Discussion

The question of the validity of *L. furtivus* is not addressed here. Some specimens were measured for comparison. I can only say that the proportions of the *L. furtivus* metatarsals are not exactly the same as for those of *L. etruscus*.

Table 14.67 Dimensions of the M2 (in mm) in *Leptobos* sp. from Senèze

Variable	Ht	1	2	3	4	7
Senèze (FSL SEN 04-0217)	20.3	32.7	31	32.4	20.6	24.4

Variable definitions (from Crégut-Bonnoure 1995, 2020) (Fig. 14.A.4: B): Ht, height; 1, occlusal length; 2, basal length. 3, occlusal width total, first lobe; 4, occlusal width in the middle, first lobe; 7, basal width

Table 14.68 Comparative dimensions of the metatarsal III+IV (in mm) in *Bison* (*Eobison*) sp. and *Leptobos* from Senèze

Variable	1	2	3	4	6	10	11	12	17	18	19
Bison (Eobison)											
FSL SEN 02-0020	235.8	218.4	238.9	55.9	56.3	18.2	40.0	18.9	33.2	34.2	64.8
L. etruscus											
FSL SEN 04-0182											53.2
FSL 213125	277.3	254.3	283.7	55.7	53.6	18.2	40.6	20.4	36.7	36.6	57.8
L. furtivus N	4	4	4	4	4	4	4	4	4	4	4
Mean	239.33	221.64	245.45	46.15	44.4	17.16	29.04	17.55	29.72	28.7	49.38
Minimum	238.7	221.2	244.7	45.1	43.5	16.4	27.4	17.1	28.2	28.4	48.3
Maximum	239.8	222.0	246.8	47.2	43.8	18.6	30.5	17.8	30.1	29.1	50.3
SD	0.5	0.36	0.93	0.86	1.11	1.02	1.6	0.3	1.39	0.27	0.88
Variable	20	21	22	23	24	25	26	27	28	29	30
Bison (Eobison)											
FSL SEN 02-0020	64.8	61.0	35.2	33.7	37	37.2	30.9		28.4		25.2
L. etruscus											
FSL SEN 04-0182		52.4	31.6	30.5	32.5	34.1	30.6	34.9	28.9	34.5	26.5
FSL 213125	58.6	57.5	35.7	32.5	35.0	34.8	31.0	38.0	29.1	36.7	27.3
L. furtivus N	4	4	4	4	4	4	4	4	4	4	4
Mean	49.14	49.55	29.08	27.57	29.52	29.54	25.27	30.58	24.04	30.24	21.89
Minimum	48.2	48.7	28.3	26.9	28.4	28.4	23.8	30.0	23.8	29.4	21.1
Maximum	49.8	50.3	29.7	29.0	30.2	30.2	26.5	31.2	24.4	31.0	22.7
SD	0.68	0.65	0.59	1.00	0.8	0.81	1.12	0.69	0.26	0.82	0.81

Variable definitions (from Crégut-Bonnoure 1995, 2002, 2020): dpd: dorsoplantar diameter; td: transversal diameter. Length: 1, dorsal side; 2, minimum in the middle; 3, plantar side. Proximal epiphysis: 4, td total; 6, dpd total; 9, td of the facet for the small cuneiform; 10: dpd of the facet for the great cuneiform; 11, dt of the facet for the great cuneiform; 12, td of the dorsal facet for the navicular-cuboid. Middle of the diaphysis: 17, td; 18: dpd. Distal epiphysis: 19, td maximal of the articulation; 20, td minimal of the articulation; 21, td supra-articular; 22, dpd of the epiphysis; 23, dpd of the epiphysis at the foramen level; 24, distance between the keels on the plantar side; 25, distance between the keels at their base; 26, distance between the keels in their middle; 27, dpd of the medial condyle; 28, dpd of the medial keel; 29, dpd of the lateral condyle; 30, dpd of the lateral keel

Genus *Bison* C.H. Smith, 1827

Subgenus *Eobison* Flerov, 1972

Species indeterminate

The small-sized and primitive *Bison* (*Eobison*) is known in southern Eurasia during the Early Pleistocene. The reality of the subgenus has been subject to debate (Bukhsianidze 2005; Kostopoulos et al. 2018), but its validity is now established (Sorbelli et al. 2022). Two archaic lineages are recognized: a northern and a southern (Croitor 2016). In Eastern Europe, the species closest to those of Western Europe is *Bison* (*Eobison*) *georgicus* Burchak-Abramovich & Vekua, 1994 (see also Bukhsianidze 2005) from Dmanisi, Georgia. In Western and Mediterranean Europe, the oldest populations are identified in Montenegro at Trlica [*Bison* (*Eobison*) sp.; Crégut-Bonnoure & Dimitrijević 2006], in Italy at Pirro Nord [*Bison* (*Eobison*) *degiulii* Masini 1989] and at Pietrafitta [*Bison* (*Eobison*) *degiulii*; Sorbelli et al. 2022]; in Greece at Apollonia I (*Bison* (*Eobison*) sp.; Kostopoulos 1996, 1997b) and in Germany at Untermaßfeld

(*Bison* (*Poephagus*) *menneri* Sher, 1997; Bukhsianidze 2020; and *Bison* sp.; Bukhsianidze 2020). All these localities are younger than MNQ 18.

Description

There is a complete left metatarsal III+IV assigned to *L. etruscus* on the label (FSL SEN 02-0020, parcel 164) (Table 14.68) (Fig. 14.76: C, F). It was given to the scientific team by local amateur A. Consigny, who collected it on the surface of a large parcel whose altitude overlaps or partly underlies the area of our southeastern trenches, but the stratigraphic level is essentially unknown.

This metapodial is more robust than those of *Leptobos* (Fig. 14.76: A, B, D, E). The shaft's sides are regularly concave, rather than parallel as in *Leptobos*. The lateral reliefs for the extensors, distal to the proximal articulation, are weakly developed, while in *Leptobos* they constitute a deep furrow flanked by two crests. The dorsal groove is wide, especially near the distal condyles, and shallow; in *Leptobos* it is even shallower. The dorsal foramen is smaller in comparison to *Leptobos*. The keels are broad and not acute; in *Leptobos* they are more narrow and acute. The plantar side is without any longitudinal groove; in *Leptobos* the middle of the diaphysis is grooved. Its proportions are distinct from those of *Leptobos* (Fig. 14.78). It is best identified as *Bison* (*Eobison*) sp. indet.

Discussion

I cannot identify this single specimen to species. At Dmanisi, *B.* (*Eobison*) is associated with *Hemitragus* (Crégut-Bonnoure 2009), at Trlica and Pirro Nord it is associated with *Megalovis*; however, these occurrences do not involve the same species as at Senèze. Moreover, these localities are younger than MNQ 18. Accordingly, the Senèze *B.* (*Eobison*) may suggest the presence of a younger level than those yielding the other bovids.

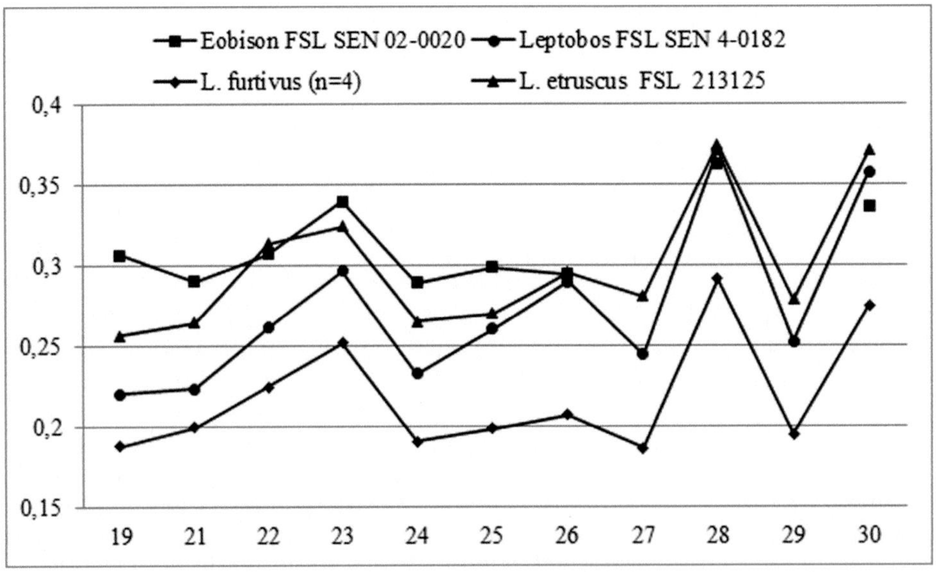

Fig. 14.77 *Leptobos* and *Bison*. Ratio diagram of the metatarsal III+IV. Reference axis: recent *Capra ibex* ♂+♀ (in Crégut-Bonnoure 2020). dpd: dorsoplantar diameter; td: transversal diameter. Distal epiphysis: 19, td maximal of the articulation; 21, td supra-articular; 22, dpd of the epiphysis; 23, dpd of the epiphysis at the foramen level; 24, distance between the keels on the plantar side; 25, distance between the keels at their base; 26, distance between the keels in their middle; 27, dpd of the medial condyle; 28, dpd of the medial keel; 30, dpd of the lateral keel. References of the measurements in Fig. 14.A.16 and dimensions in Table 14.68

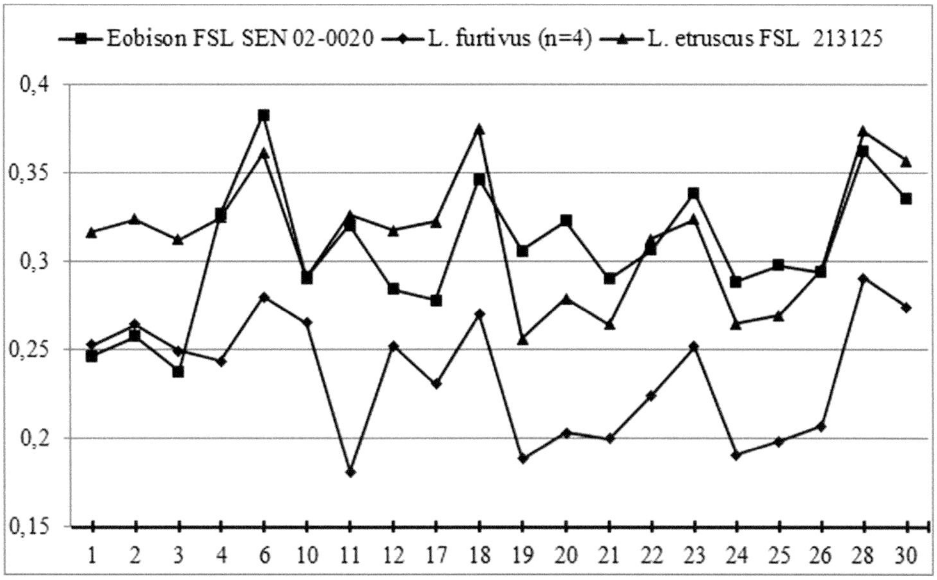

Fig. 14.78 *Leptobos* and *Bison*. Ratio diagram of the metatarsal III+IV. Reference axis: recent *Capra ibex* ♂+♀ (in Crégut-Bonnoure 2020). dpd: dorsoplantar diameter; td: transversal diameter. Length: 1, dorsal side; 2, minimum in the middle; 3, plantar side. Proximal epiphysis: 4, td total; 6, dpd total; 10: dpd of the medio-dorsal articular facet; 11, dt of the medio-dorsal articular facet; 12, td of the dorsal facet for the navicular-cuboid. Middle of the diaphysis: 17, td; 18: dpd. Distal epiphysis: 19, td maximal of the articulation; 20, td minimal of the articulation; 21, td supra-articular; 22, dpd of the epiphysis; 23, dpd of the epiphysis at the foramen level; 24, distance between the keels on the plantar side; 25, distance between the keels at their base; 26, distance between the keels in their middle; 28, dpd of the medial keel; 30, dpd of the lateral keel. References of the measurements in Fig. 14.A.16 and dimensions in Table 14.68

Bovidae Indeterminate

Two specimens correspond to indeterminate bovids.

Cranium NMB Se 1639

This fragment was assigned to an indeterminate antelope by Schaub (on the label) (Table 14.56) (Fig. 14.79: A). The parietal and the occipital are preserved from the fronto-parietal suture to the external occipital protuberance. The fronto-parietal suture is weakly concave laterally and more or less convex in the middle. Consequently, it is not an *Ovis* or *Megalovis* cranium, where the fronto-parietal suture is "V" shaped and whose proportions are distinct (Fig. 14.60). The parietal is flat with a median bulge and two small foramina on each side. The left temporal line is weakly developed (the right one is eroded). The parieto-occipital suture is straight. The crests of the external occipital protuberance are well developed. It could be cf. *Hemitragus*, but it is impossible at present to confirm this hypothesis.

Calcaneus MNHN-F-1926-?

Another bone reveals the existence of an indeterminate bovid. It is a left calcaneus whose catalog number is unreadable (Fig. 14.79: B, C). The *tuber calcanei* is triangular but broader than in cervids. At its top, there is a transverse groove stopping before the lateral side. At this level, the surface of the bone is incurved distally. The plantar groove is broad. The dorsal side is relatively pinched. The tendinous furrow of the *sustentaculum tali* is concave with a medial side pinched and raised up. The fibular trochlea is broad at the middle and the distal surface is convex. This specimen belongs to a medium sized caprine. Its morphology and proportions are distinct from those of *Gazellospira*, *Gallogoral* and cf. *Pliotragus* (Table 14.69; Fig. 14.80). The navicular-cuboid articular facet is thin (measurement 10), and the malleolar bone articular facet is longer (measurement 11). The bone is short from the malleolar surface to the distal apophysis (measurements 13 and 14), whereas it is longer from the *tuber calcanei* to the coracoid process (measurements 16 and 17). The dimensions are of the same order of magnitude as in *Capra* and *Hemitragus*, but the morphology is clearly distinct (Crégut-Bonnoure 1987). It could correspond to *Procamptoceras* but it cannot be proven, no associated skeleton of that genus being known.

Metatarsal III+IV NMB Se 751

This distal epiphysis is assigned to an antelope on the label. It is typical of cervids with a deep dorsal groove stopping at 15 mm before the articulation. It is smaller than in *Metacervoceros rhenanus*, and very similar to the extant *Capreolus* and to *Croizetoceros ramosus*. I attribute it to *Croizetoceros*.

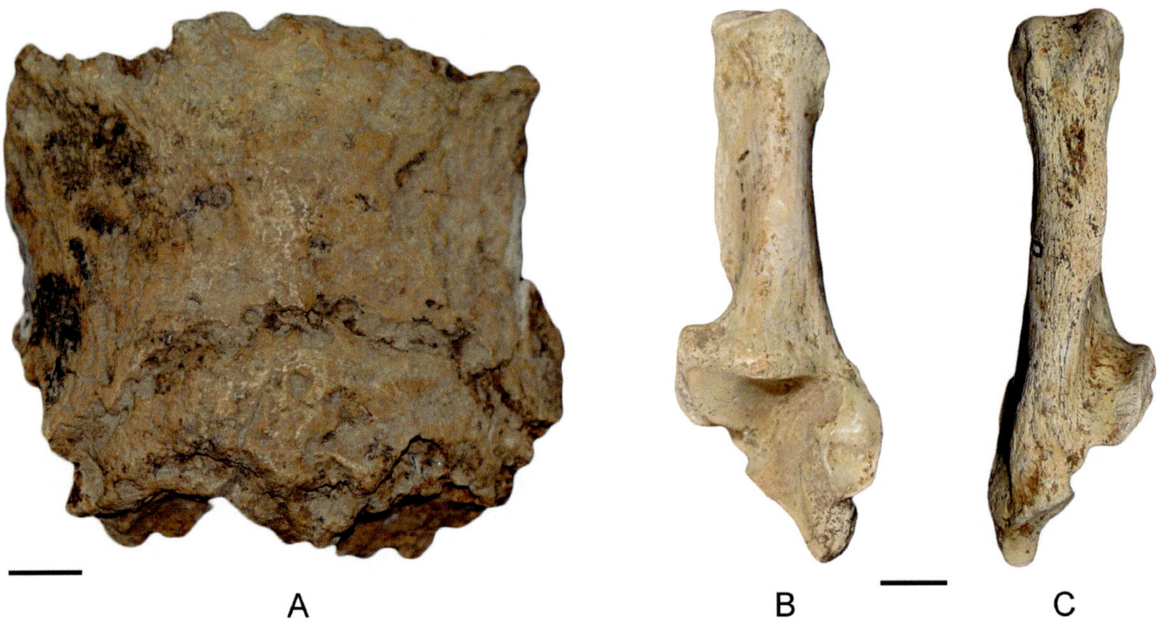

Fig. 14.79 Indeterminate bovid from Senèze. A, braincase NMB Se 1639; B, C, left calcaneus MNHN-F-1926-?. A, *norma dorsalis*; B, dorsal view; C, plantar view. Scale 1 cm

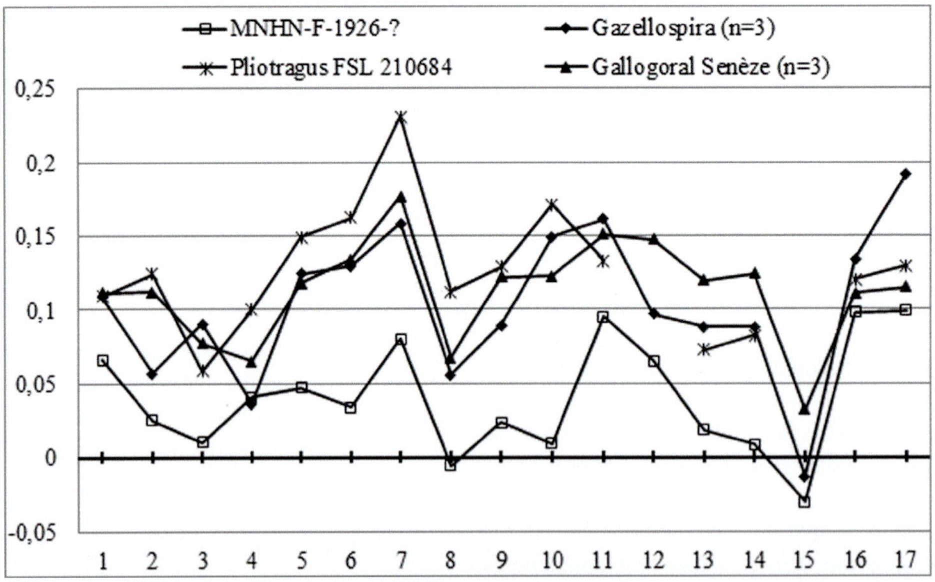

Fig. 14.80 Indeterminate bovid (MNHN-F-1926–?), *Gazellospira*, *Gallogoral* and cf. *Pliotragus* (FSL 210684). Ratio diagram of the calcaneus. Reference axis: recent *Capra ibex* ♂+♀ (in Crégut-Bonnoure 2020). dpd, dorsoplantar diameter; td, transversal diameter. 1: total length. *Tuber calcanei*: 2: td. 3: dpd. Proximal body: 4: td minimum. 5: dpd. Total: 6: dpd. 7: td. *Sustentaculum tali*: 8: td. Navicular-cuboid facet: 9: length. 10: td. Fibular trochlea facet: 11: length.12: td. Length: 13: from the fibular trochlea facet at its middle to the distal point of the bone. 14: from the fibular trochlea facet at its proximal side to the distal point of the bone. Width: 15: *sustentaculum tali* furrow. Length: 16: from the dorsal *tuber calcanei* apex to the coracoid process tip. 17: from the palmar *tuber calcanei* apex to the coracoid process tip. References of the measurements in Fig. 14.A.15 and dimensions in Tables 14.13, 14.39 and 14.69

Table 14.69 Dimensions of the calcaneus (in mm) in indeterminate Bovidae from Senèze

Variable	1	2	3	4	5	6	7	8	9
Senèze (MNHN-F-1926–?)	83.2	19.7	20.9	12.2	21.5	28.7	26.2	24.4	19.8
Variable	10	11	12	13	14	15	16	17	
Senèze (MNHN-F-1926–?)	6.2	17.1	9.0	26.5	27.8	9.9	55.2	59.5	

Variable definitions (from Crégut-Bonnoure 1987, 2002, 2020) (Fig. 14.A.15: G to F): dpd, dorsoplantar diameter; td, transversal diameter. 1: total length. *Tuber calcanei*: 2: td. 3: dpd. Proximal body: 4: td minimum. 5: dpd. Total: 6: dpd. 7: td. *Sustentaculum tali*: 8: td. Navicular-cuboid facet: 9: length. 10: td. Fibular trochlea facet: 11: length.12: td. Length: 13: from the fibular trochlea facet in its middle to the distal point of the bone. 14: from the fibular trochlea facet at its proximal side to the distal point of the bone. Width: 15: *sustentaculum tali* furrow. Length: 16: from the dorsal *tuber calcanei* apex to the coracoid process tip. 17: from the palmar *tuber calcanei* apex to the coracoid process tip

Discussion

Hundreds of bovid specimens have been recovered from Senèze and examined for this study. Of these, only seven were recovered during the recent fieldwork and two more were collected by local people from stratigraphically uncertain locations, at least one of them on the surface (*Bison*). As a result, 51 specimens from the initial investigations carried out in the early twentieth century by Pierre Philis (Table 14.70) and 37 from other French and European localities (Table 14.71) have been reidentified.

Three species are relatively abundant: *Gazellospira torticornis* (minimum number of individuals, MNI = 13), *Pliotragus ardeus* (MNI = 8), and *Gallogoral meneghinii meneghinii* (MNI = 15). The latter is especially well known thanks to three complete skeletons. *Pliotragus* and *Gallogoral* teeth have similar proportions but their morphology is distinct, as is true of their crania and limb bones. The remains previously referred to *Procamptoceras brivatense* belong mainly to other species; only 12 specimens can be attributed to this taxon with confidence (MNI = 6). It is the same problem with *Megalovis latifrons* (MNI = 6). For the first time, a caprin close to *Hemitragus* is identified from this locality (MNI = 3). The morphometric characteristics of *Ovis* (MNI = 1) lead me to attribute it to a new species dedicated to the late Claude Guérin. The *Leptobos* material was not revised but a metatarsal from the new field research has the same proportions as *Leptobos furtivus*. A primitive bison is identified (MNI = 1). Finally, two specimens can only be identified as indeterminate bovid. The analysis also revealed morphometric differences between *P. ardeus* crania from Senèze and from the Dacian Basin, probably related to sexual dimorphism.

The faunal assemblage from Senèze has been suggested to come from more than one fossiliferous level (e.g., Azzaroli et al. 1997). There is no indication of the stratigraphic origin of the remains from the previous excavations. The color of the *Ovis* bone is not the same as that of the majority of the other remains. However, even if there are few, some other teeth and bones do have the same coloring. This genus is present in Slivnitsa along with *Hemitragus*, which could also be the case in Senèze (cf. *Hemitragus*), and the two localities are considered to be contemporaneous (Spassov & Crégut-Bonnoure 1999). Consequently, the color of the *Ovis* cranium is not an adequate criterion to indicate another fossiliferous level, especially as the process of fossilization depends on sedimentary conditions. Only the presence of *Bison* (*Eobison*) sp. suggests a younger level, which could date to the end of the Early Pleistocene, and in fact, this specimen was collected by a local person on the surface.

Pliotragus shows a variability of the p3 bilobation, the development of the molar and of the third lobe shape of m3. Some morphological variation can be observed in the *G. meneghinii* M3, p4 and proximal articulation of metacarpal III+IV: the inflation of the p4 metaconid is variable; the metastylar wing of M3, the proximal synovial fossa of the metacarpal III+IV, the pinching of the contact between the unciform and the capitato-trapezoid surfaces and the palmar supplementary facet of these two facets all could be present or absent. *Megalovis* shows an important variation in the morphology of some of its lower teeth (p3, p4, m3). The molarization of p3 could be variable in some bovids as in *O. ammon antiqua* and *Soergelia minor* (Crégut-Bonnoure 2002, 2006, 2020). Other than the Trlica *M. balcanicus* specimen, no p3 is available in the European localities where *Megalovis* is present. Consequently, it is difficult to assess the morphological variation of this premolar in a single population. At Senèze, the morphology of the three p4s is also variable. NMB Se 401 is similar to the Erpfinger Höhle tooth (Lehmann 1957: Fig. 15), whereas NMB Se 1732 is similar to *M. wimani*. There is no variation of the morphology of this tooth in *M. balcanicus* p4s (n = 4). In summary, the morphology of the remains is not homogeneous in *Pliotragus*, *Gallogoral* and *Megalovis*. The variability observed in these bovids could be related to evolution of these species over time, but it is imposible to appreciate the span of this time. Furthermore, the material is not extensive enough to capture the real range of the variation.

Table 14.70 Change of taxonomic attribution of Senèze cetartiodactyls

Catalog reference	Specimen	Prior determination	New determination
NMB Se 71	Left m3	*Gallogoral*	*Pliotragus*
NMB Se 197	Left p4	*Antilope* sp.	*Pliotragus*
NMB Se 201	Left metatarsal III+IV	*Gallogoral*	*Pliotragus*
NMB Se 235	Atlas	*Gazellospira*	*Gallogoral*
NMB Se 263	Right humerus	*Megalovis*	*Leptobos*
NMB Se 264	Right radius	*Megalovis*	*Leptobos*
NMB Se 384	Right M1 to M3	*Gallogoral*	*Pliotragus*
NMB Se 385	Right M2 and M3	*Gallogoral*	*Pliotragus*
NMB Se 387	Left M1	*Gazellospira*	*Gallogoral*
NMB Se 388	Right M3	*Pliotragus*	*Gallogoral*
NMB Se 390	Left M3	*Gallogoral*	*Pliotragus*
NMB Se 399	Right m2	*Gazellospira*	*Pliotragus*
NMB Se 402	Right mandible	*Procamptoceras* and *Gazellospira* sp.	*cf. Hemitragus*
NMB Se 539	Right M2	*Gazellospira*	*Gallogoral*
NMB Se 555	Seventh cervical and thoracic vertebrae	*Procamptoceras*	cf. *Procamptoceras*
NMB Se 751	Metatarsal II+IV, distal	Antelope	cf. *Croizetoceros*
NMB Se 755	First anterior phalanx	Ruminantier	cf. *Procamptoceras*
NMB Se 758	Right M3	*Gallogoral*	*Pliotragus*
NMB Se 936	Right M2-3	*Gallogoral*	*Pliotragus*
NMB Se 1136	Right calcaneus	*Cervus philisi*	*Gazellospira*
NMB Se 1152	Left calcaneus	*Cervus philisi*	*Gazellospira*
NMB Se 1153	Left calcaneus	*Cervus philisi*	*Gazellospira*
NMB Se 1431	Left M2-3	*Galllogoral*	*Pliotragus*
NMB Se 1436	Right maxilla	*Galllogoral*	*Pliotragus*
NMB Se 1437	Right M1	*Gazellospira*	*Gallogoral*
NMB Se 1441	Left humerus	*Procamptoceras*	*Gazellospira*
NMB Se 1460	First and second phalanges	*Procamptoceras*	cervid
NMB Se 1483	Left horn core	*Megalovis*	*Pliotragus*
NMB Se 1515	Atlas	*Megalovis*	*Pliotragus*
NMB Se 1516	Atlas	*? Deperetia*	*Pliotragus*
NMB Se 1592	First posterior phalanx	*Gallogoral*	cervid
NMB Se 1637	Right scapula	*Cf. Deperetia*	*Pliotragus*
NMB Se 1639	Cranium	*Antilope*	Bovidae indeterminate
NMB Se 1643	Right and left tali	*Megalovis*	*Eucladoceros senezensis*
NMB Se 1718	Right mandible	*Procamptoceras*	*cf. Hemitragus*
NMB Se 1792	Left humerus	*Procamptoceras*	*cf. Procamptoceras*
NMB Se 1936	Right M2-3	*Galllogoral*	*Pliotragus*
NMB Se 1943	Left radius	*Cervus philisi*	*Gazellospira*
FSL 16866	Right mandible	*Galllogoral*	*Pliotragus*
FSL 16867	Left maxilla	*Galllogoral*	*Pliotragus*
FSL 16868	Three M3: one left, two right	*Galllogoral*	*Pliotragus*
FSL 16877	Left calcaneus	*Gallogoral*	*Pliotragus* or Bovidae indet
FSL 16886	Left m1-3	*Galllogoral*	*Pliotragus*
FSL 16890	Right M2-3	*Pliotragus*	*Gallogoral*
FSL 16892	Left mandible	*Procamptoceras*	*cf. Hemitragus*
FSL 16893	Left DP4	*Procamptoceras*	*Gazellospira*
FSL 16899	Left humerus	*Gazellospira*	*Metacervoceros*
FSL 16900	One of the three tali	*Gallogoral*	cervid
FSL 211101	Right humerus	cervid	*Megalovis*
FSL SEN 02-0020	Left metatarsal III+IV	*Leptobos*	*Bison (Eobison)*
MNHN-F-unreadable	Left calcaneus	*?*	Bovidae indeterminate

Table 14.71 Change of specific attribution of some French and Hungarian cetartiodactyls

Locality/catalog reference	Specimen	Prior determination	New determination
Chilhac			
NMB PP 156	Left m2	*Antilope* cf. *nemorhaedus*	*Gazellospira*
Le Coupet			
NMB Cp 65	Mandible	*Gazellospira*	*Gallogoral*
NMB Cp 80	Right humerus	*Procamptoceras*	*Gazellospira*
NMB Cp 114	Right humerus	*Procamptoceras*	*Gazellospira*
Pardines			
NMB Pn 163	Left humerus	*Procamptoceras*	*Gazellospira*
Perrier-Rocca-Neyra			
NMB Rn 26	Fragmentary braincase	*Procamptoceras*	small cervid
Perrier-Etouaires			
NMB Pr 523	Horn core	*Pliotragus*	*Gallogoral*
"Montagne de Perrier"			
MNHN-F unnumbered	Metacarpal III+IV	*Gazellospira/Gallogoral*	*Capra sp.*
Olivola			
NMB Ol 9 cast	Left mandible	*Antilope* cf. *Procamptoceras*	*Procamptoceras*
Erpfingen			
NMB D 461–cast	Left DP4 (ex M2)	*Deperetia*	*Megalovis*
Lehmann's Fig. 20	Right M3	*Deperetia*	*Gallogoral*
Erpfingen ?			
NMB DP 29 cast	Right M3	*Antilope jaegeri*	*Megalovis*
NMB DP 30 cast	Left M2	*Antilope jaegeri*	*Megalovis*
NMB DP 31 cast	Left M1	*Antilope jaegeri*	*Megalovis*
NMB DP 32 cast	Left M1	*Antilope jaegeri*	*Megalovis*
NMB DP 33 cast	Right P2	*Antilope jaegeri*	*Megalovis*
NMB DP 36 cast	Right p4	*Antilope jaegeri*	*Megalovis*
Stetten			
NMB DP 23 cast	Left m3	*Antilope jaegeri*	*Megalovis*
NMB DP 24 cast	Left m2	*Antilope jaegeri*	*Megalovis*
NMB DP 25 cast	Right M1	*Antilope jaegeri*	*Megalovis*
Villány Nagyharsányberg			
NMB UP 485	Right m3	no indication	*Gazellospira*
NMB UP 486	Right m2	no indication	*Gazellospira*
NMB UP 487	Right metacarpal III+IV	*Gazellospira*	*Metacervoceros rhenanus*
NMB UP 504	Left M3	*Tragelaphus*	*Megalovis*
Villány 3 = Villány kalkberg			
NMB UP 477	Left m1	*Procamptoceras*	*Hemitragus*
NMB UP 478	Right M2	*Procamptoceras*	*Hemitragus*
NMB UP 479	Right m2	*Procamptoceras*	*Hemitragus*
NMB UP 480	right m2	*Procamptoceras*	*Hemitragus*
NMB UP 483	Right metacarpal III+IV	*Procamptoceras*	*Metacervoceros rhenanus*
NMB UP 488	Left scaphoid	*Tragelaphus*	*Hemitragus*
NMB UP 489	Left semi-lunar	*Tragelaphus*	*Hemitragus*
NMB UP 490	Posterior first phalanx	*Gazellospira*	*Hemitragus*
NMB UP 492	Anterior second phalanx	*Gazellospira*	*Hemitragus*
NMB UP 493	Anterior second phalanx	*Gazellospira*	*Hemitragus*
NMB UP 510	Anterior second phalanx	*Tragelaphus*	*Hemitragus*
Csarnóta			
NMB UP 796	Right mandible	*Procamptoceras*	Bovidae indeterminate
NMB UP 797–cast	Right metacarpal III+IV		*Hemitragus*

Conclusions

The faunal assemblage of Senèze has been known for a long time. Bovids are diverse and relatively abundant. Unfortunately, few specimens were found during the new field research: they belong to *Pliotragus ardeus* (N = 2), *Gallogoral meneghinii* (N = 1), *Megalovis latifrons* (N = 2, of which one was donated to the team), *Leptobos furtivus* (N = 3). *Bison* (*Eobison*) sp. (N = 1, donated to the team) was not excavated but collected on the surface by a local person. With the exception of the bison, this association does not differ fundamentally from that resulting from the initial investigations carried out in the early twentieth century by Pierre Philis. The revision undertaken here led to new interpretations for some of the remains, and it has permitted me to discover new taxa. Today nine genera, eleven species and one or more indeterminate bovid(s) are identified at Senèze, three of them being new for the locality: *Gazellospira torticornis*, *Procamptoceras brivatense*, cf. *Hemitragus*, *Gallogoral meneghinii*, *Pliotragus ardeus*, *Megalovis latifrons*, *Ovis claudiusguerini* nov. sp., *Leptobos etruscus*, *L. furtivus*, *Bison* (*Eobison*) sp. and Bovidae sp. indet. Of these taxa only *Bison* (*Eobison*) (known only from a single metatarsal III+IV collected on the surface) appears to originate from a younger faunal assemblage. The morphological variation especially in teeth of *Pliotragus*, *Gallogoral* and *Megalovis* from Senèze could be populational but it could also be the result of an evolution of these species and of more than one fossiliferous level for the bovid assemblage found during the previous excavations. However, in this hypothesis, considering the results of the stratigraphic and geochronological analyses (Debard 2024; Delson et al. 2024) these possible distinct levels would not be very distant in time (the entire sequence appears to span about 100 kyr or less), and it is not possible to distinguish successive bovid assemblages. The Senèze bovid community is dated to the beginning of the Late Villafranchian (Crégut-Bonnoure et al. 2024). Its comparison with bovids from other European localities led to new attributions of some of them.

Acknowledgments I am very grateful to the late Dr. Claude Guérin who had proposed that I collaborate on the Senèze project. The discussions that we had on the Senèze bovids and their complexity led to some useful suggestions. He read the first manuscript, and his comments greatly improved this text; I dedicate this article to him. Dr. Eric Delson edited the language of the final version, and I sincerely thank him. I am also very grateful to Dr. Martine Faure who financially supported my visits to Basel. I thank Dr. Sevket Sen for his assistance in my bibliographical research on the Turkish localities. I am also grateful to Dr. Jan Wagner for information on the chronology of the various localities from the Hungarian Karst. I thank Dr. Burkhart Engesser, Dr. Clemens Model and Dr. Loïc Costeur from the Naturhistorische Museum of Basel; Dr. Abel Prieur and Dr. Emmanuel Robert from the University Claude-Bernard-Lyon I; Dr. Frédéric Lacombat from the Musée Crozatier of Le Puy-en-Velay; Prof. Pascal Tassy, Dr. Daniel Goujet, Dr. Hervé Lelièvre and Dr. Christine Argot from the Muséum national d'Histoire naturelle in Paris; Dr. Dimitris Kostopoulos from the University of Thessaloniki; and Dr. Silviu Costantin from the "Emil Racoviță" Institute of Speleology of Bucharest for granting me access to the collections at their disposal and for their hospitality at each of my visits. I sincerely thank Dr. Emmanuel Desclaux and Jessica Cohen from the Departmental Laboratory of Lazaret, Noël Podevigne from the Université Claude Bernard-Lyon I, Dr. Pierre Moulet and Cédric Triat from Museum Requien for their logistical support. I offer my sincere thanks to Dr. Maïa Bukhsianidze, Dr. Roman Croitor and Dr. Dimitris Kostopoulos who reviewed this manuscript. Their relevant comments greatly improved the text. I am very grateful to them.

Appendix

The measurements used in the text and tables are defined here, following Crégut-Bonnoure (1987, 1995, 2020) and Driesch (1976), with illustrations.

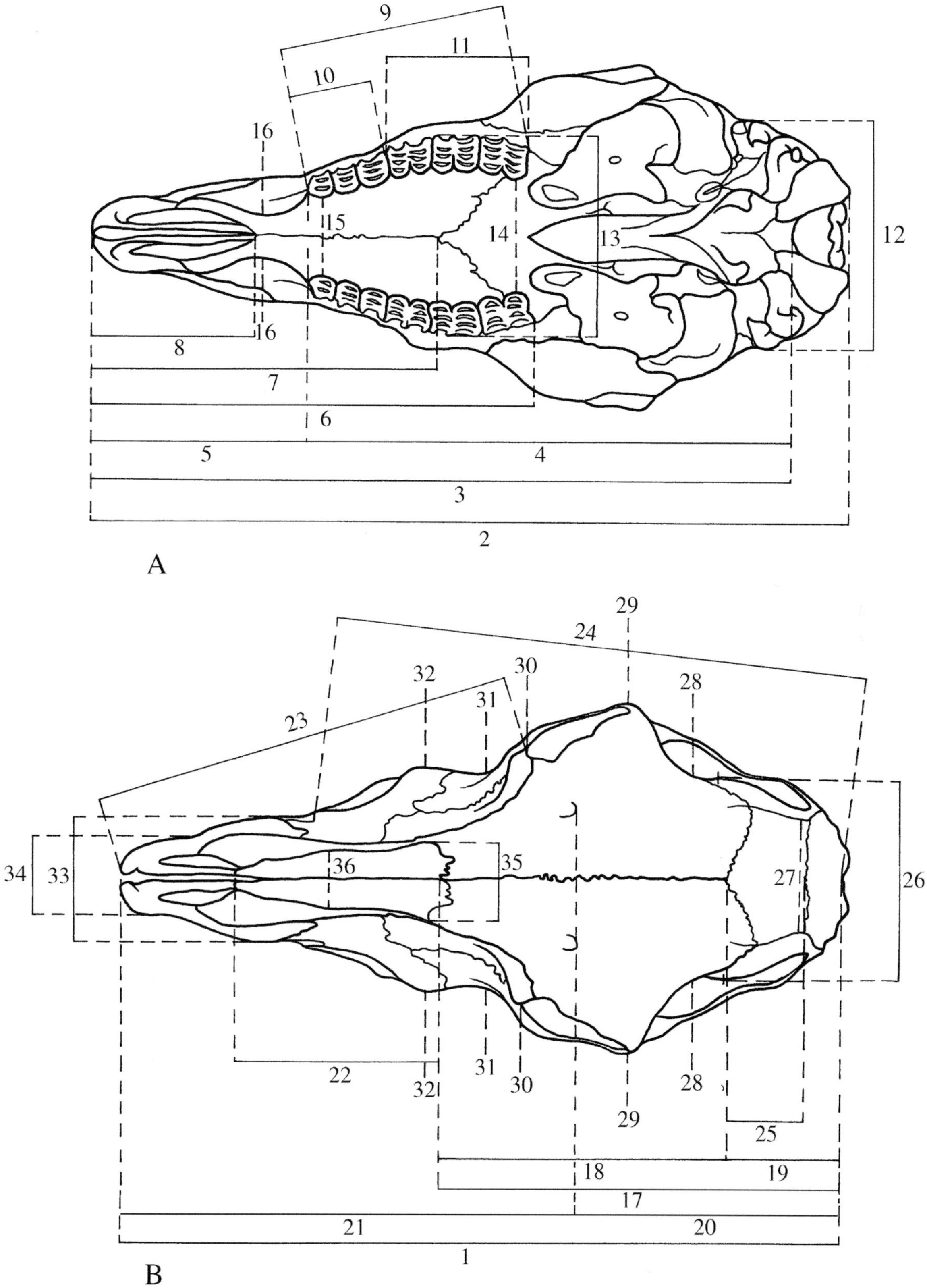

◀ **Fig. 14.A.1** Variable definitions on the cranium (from Crégut-Bonnoure 2020, modified from Driesch 1976). Abbreviations: L.: length; td: transverse diameter; Ht: Height. **A**) basal view; **B**) dorsal view. 1, L. acrocranion-prosthion; 2, L. from occipital condyles to prosthion; 3, L. basion-prosthion; 4, L. from basion to a line tangent to the mesial edge of both P2s; 5, L. from prosthion to a line tangent to the mesial edge of both P2s; 6, L. from prosthion to a line tangent to the distal edge of both M3s; 7, L. from palatine suture in midline to prosthion; 8, L. from posterior edge of palatal process of the premaxilla (anterior end of intermaxillary suture);, L. P2 mesial-M3 distal; 10, L. P2 mesial-P4 distal; 11, L. M1 mesial-M3 distal; 12, td otion-otion; 13, td (maximal) across tooth rows on the labial side; 14, td minimal between the lingual side of the M3 second lobe; 15, td minimal between the lingual side of the P2; 16, td minimal between the inter-alveolar margins. 17, L. acrocranion-nasion; 18, L. of the frontal from bregma to nasion; 19, L. acrocranion-bregma; 20, L. from line tangent to supraorbital foramina to acrocranion; 21, L. from line tangent to supraorbital foramina to prosthion; 22, L. from nasion to rhinion; 23, L. entorbital-prosthion; 24, L. from infraorbital foramen to acrocranion; 25, dorsal length of the parietal; 26, maximal width of the parietal; 27, td minimal between the parietal ridges; 28, td post-orbital; 29, td ectorbital-ectorbital; 30, td entorbital-entorbital; 31, td between orbit and facial tuberosities. 32, td at the facial tuberosities; 33, td maximal of the premaxilla; 34, td rostral of the premaxilla; 35, td caudal of the two nasals; 36, td rostral of the two nasals

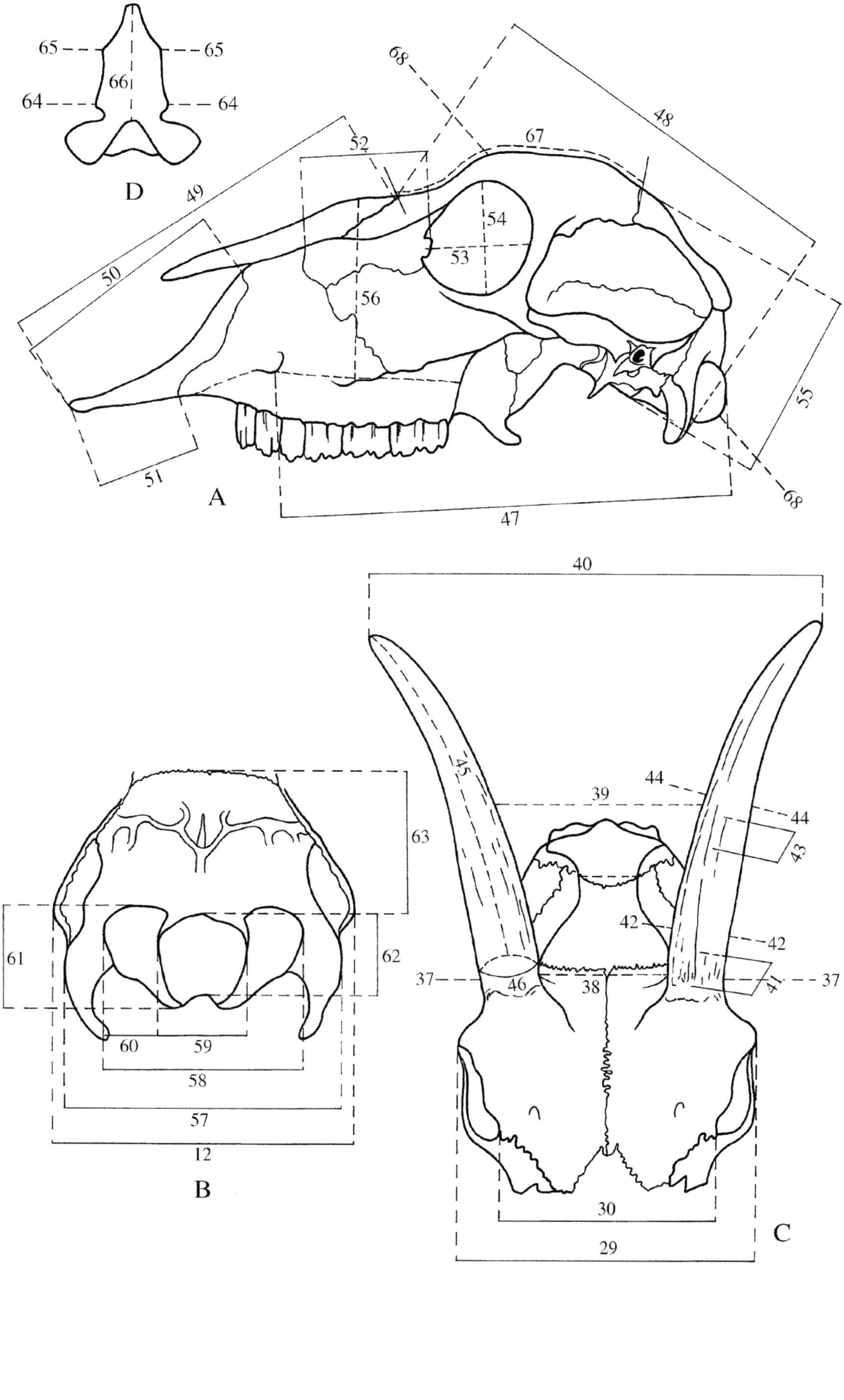

D
65
64
66
A
47
48
49
50
51
52
53
54
55
56
67
68
B
12
57
58
59
60
61
62
63
C
29
30
37
38
39
40
41
42
43
44
45
46

◀ **Fig. 14.A.2** Variable definitions of the cranium (from Crégut-Bonnoure 2020, modified from Driesch 1976), continued. Abbreviations: L.: length; td: transverse diameter; Ht: Height. **A**) left lateral view; **B**) occipital view; **C**) anterior view of neurocranium; **D**) inferior view of basioccipital. 12, td otion-otion; 37, width between the lateral sides of the horn cores at the base; 38, width between the internal sides of the horn cores at the base; 39, width between the internal sides of the horn cores in the middle; 40: width between the internal sides of the horn cores at the tip; 41, rostrocaudal diameter of a horn core at its base; 42, dorsolateral diameter of a horn core at its base; 43, rostrocaudal diameter in the middle of a complete horn core; 44, dorsolateral diameter in the middle of a complete horn core; 45, rostral length of a complete horn core; 46, circumference at the base of a horn core; 47, L. from posterior edge of occipital condyle to posterior edge of infraorbital foramen; 48, L. basion-nasion. 49, L. nasion-prosthion; 50, maximal L. of the premaxilla; 51, basal L. of the premaxilla; 52: L. of the lacrimal; 53, rostrocaudal diameter of the orbit; 54, dorsoventral diameter of the orbit; 55, Ht of the braincase from basioccipital to the top of the parietal; 56, Ht from the middle of the palatine suture to the nasal surface; 57, td between jugular processes; 58, td maximal between the external sides of the occipital condyles; 59, td maximal of the foramen magnum; 60, td maximal of an occipital condyle; 61, Ht maximal of an occipital condyle; 62, midline Ht of the foramen magnum; 63, midline Ht from opisthion to frontoparietal suture; 64, td of the caudal tuberosities of the basioccipital; 65, td of the rostral tuberosities of the basioccipital; 66, L from basion to basisphenoid; 67, frontal curvature arc from bregma to nasion; 68, L. from the top of the frontal (vertex) to the posterior (caudal) tip of the occipital condyles

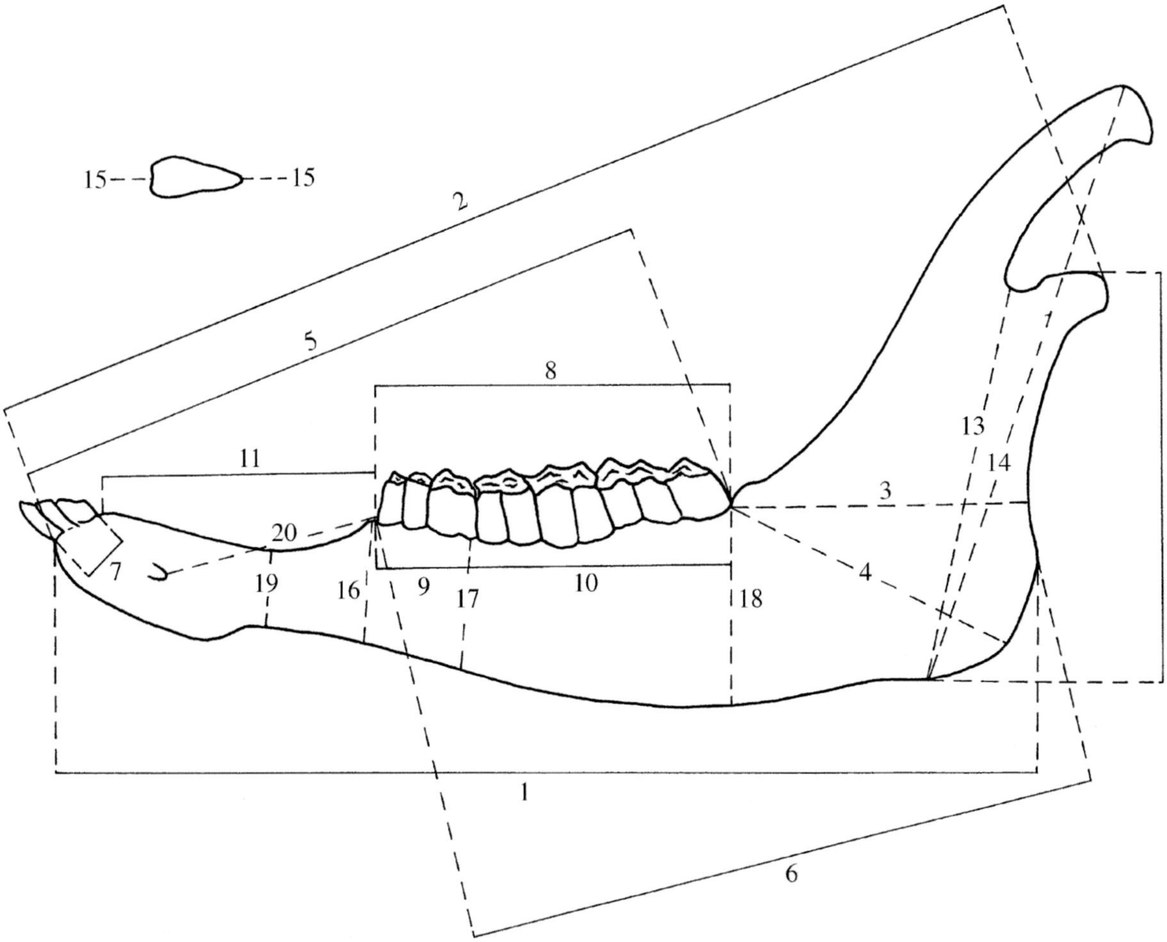

Fig. 14.A.3 Variable definitions on the mandible, with occlusal plane horizontal (from Crégut-Bonnoure 2020, modified from Driesch 1976): Abbreviations: L.: length; td: transverse diameter; Ht: height. 1, L. from infradentale to gonion; 2, L. from infradentale to the caudal side of the condyle; 3, L. from m3 distal side to ramus; 4, L. from m3 distal side to angle of the mandible; 5, L. from infradentale to the distal side of m3; 6, L. from p2 mesial side to gonion; 7; L. from mesial side of I1 to distal side of canine; 8, L. from p2 mesial side to m3 distal side; 9, L. from p2 mesial side to p4 distal side; 10; L. from m1 mesial side to m3 distal side; 11, L. of diastema, from distal side of canine to mesial side of p2; 12, Ht of the ramus from the base to the condylar process; 13, L. from the posterior base of the ramus to the mandibular notch; 14: L. from the posterior base of the ramus to the coronoid process; 15, td of the mandibular condyle; 16, Ht of the corpus at the mesial p2 level. 17, Ht of the corpus at the p4-m1 level; 18, Ht of the corpus at the distal m3 level; 19, minimal Ht of the corpus at the diastema level; 20, L. from the distal side of the mandibular foramen to the mesial side of p2

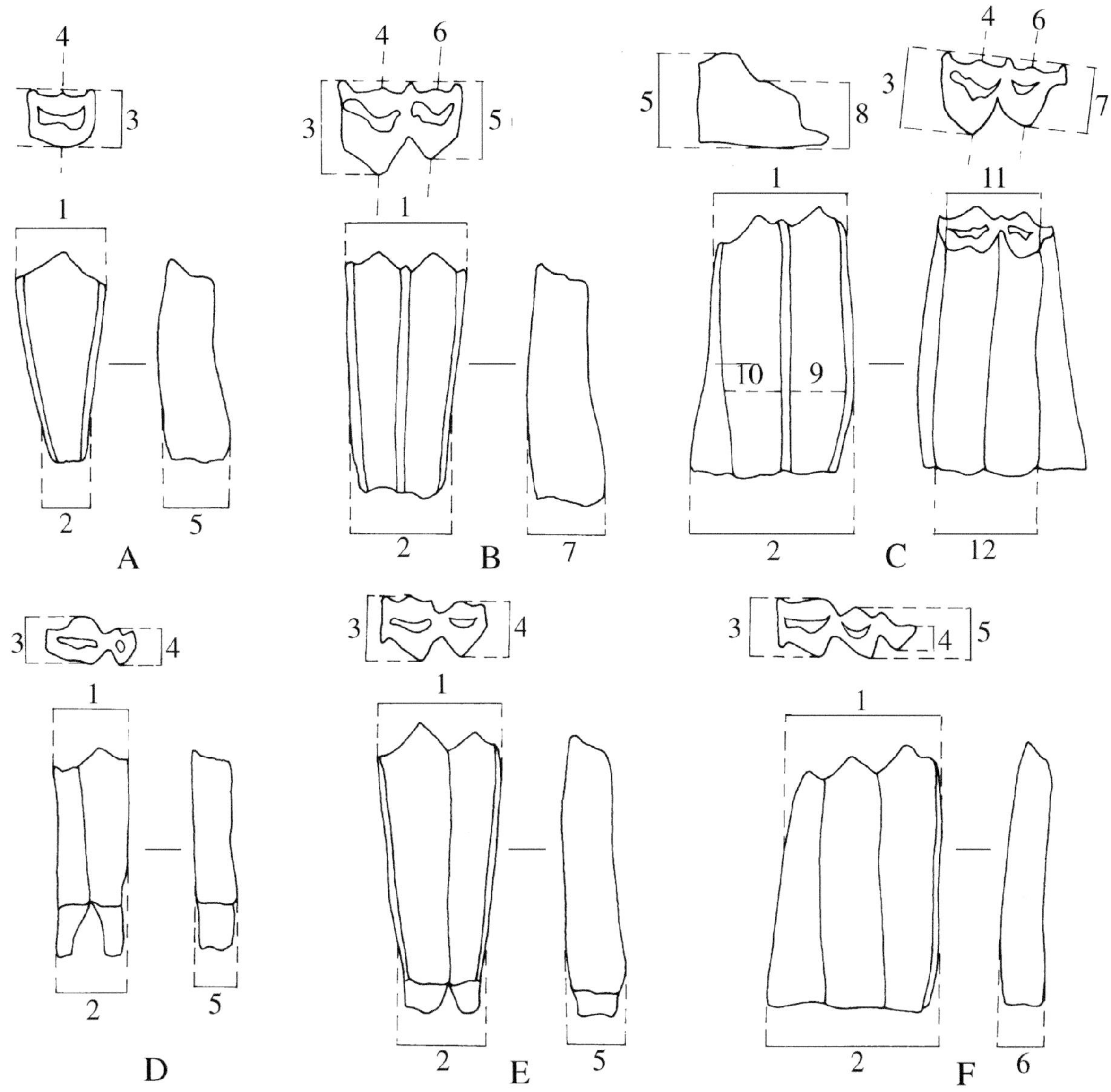

Fig. 14.A.4 Variable definitions of the teeth (from Crégut-Bonnoure 1995, 2020). upper (**A** to **C**) and lower teeth (**D** to **F**). **A**) P4, occlusal view (above), labial view (below left), mesial view (below right): 1, occlusal length; 2, basal length. 3, occlusal width total. 4, occlusal width in the middle. 5, basal width. **B**) M1 and M2, occlusal view (above), labial view (below left), mesial view (below right): 1, occlusal length; 2, basal length. 3, occlusal width total, first lobe; 4, occlusal width in the middle, first lobe; 5, occlusal width total, second lobe; 6, occlusal width in the middle, second lobe; 7, basal width. **C**) M3, basal view (above left), occlusal view (above right), labial view (below left), lingual view (below right): 1, occlusal length; 2, basal length. 3, occlusal width total, first lobe; 4, occlusal width in the middle, first lobe; 5, basal width, first lobe; 6, occlusal width total, second lobe; 7, occlusal width in the middle, second lobe; 8, basal width, second lobe; 9, length of the interstylar surface, mesial; 10, length of the interstylar surface, distal; 11, occlusal length of the two lobes at the lingual side; 12, basal length of the two lobes at the lingual side. **D**) p4, **E**) m1 and m2, occlusal view (above), lingual view (below left), mesial view (below right): 1, occlusal length; 2, basal length; 3, occlusal width, first lobe; 4, occlusal width, second lobe; 5, basal width. **F**) m3, occlusal view (above), lingual view (below left), mesial view (below right): 1, occlusal length; 2, basal length; 3, occlusal width, first lobe; 4, occlusal width, second lobe; 5, occlusal width, third lobe; 6, basal width

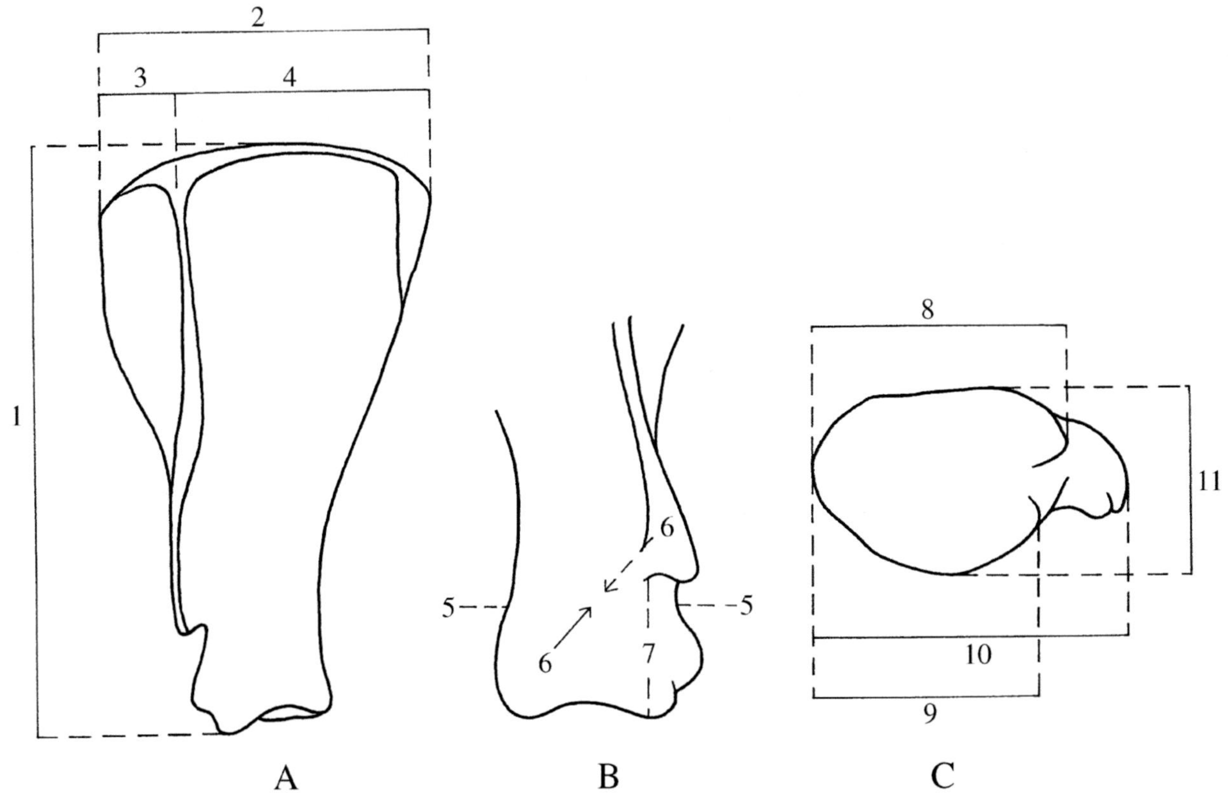

Fig. 14.A.5 Variable definitions on the scapula (from Crégut-Bonnoure 2020, modified from Driesch 1976). **A, B**) lateral view; **C**) ventral view. 1, length; 2, total width; 3, supraspinatus width; 4, infraspinatus width; 5, minimal length of the neck; 6, minimal breath of the neck; 7, length from the base of the acromion to the lateral side of the glenoid cavity; 8, articular length of the lateral side of the glenoid cavity; 9, articular length of the medial side of the glenoid cavity; 10, total length of the ventral articulation; 11, thickness of the ventral articulation

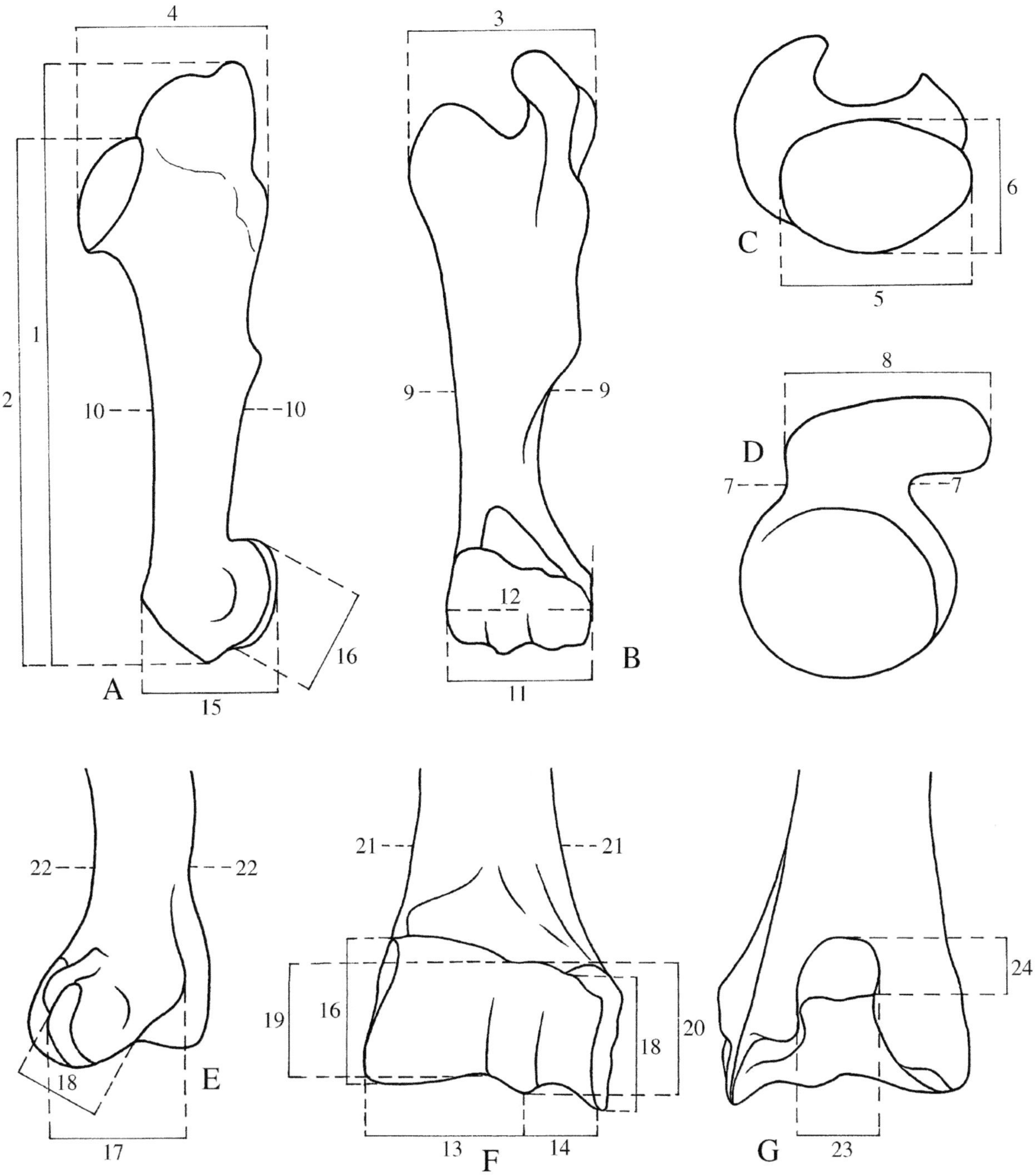

Fig. 14.A.6 Variable definitions on the humerus (from Crégut-Bonnoure 1995, 2020, modified from Driesch 1976). **A**), medial view; **B** and **F**), cranial view; **C** and **D**), proximal view; **E**), lateral view; **G**), caudal view. Abbreviations: L., length; ccd, craniocaudal diameter; td, transverse diameter; ht: height. 1, total L.; 2, L. from proximal articulation. 3: td total of the proximal epiphysis; 4: ccd, of the proximal epiphysis. 5, td of the proximal articulation; 6, ccd of the proximal articulation; 7, dt minimum, at the intertubercular furrow level, of the proximal articulation; 8, td of the major tuber. 9, td in the middle of the diaphysis; 10, ccd in the middle of the diaphysis; 11, td distal maximal; 12, td distal articular; 13, td of the trochlea; 14, td of the capitulum; 15: ccd médial of the distal epiphysis; 16, ht of the trochlea; 17, ccd lateral of the distal epiphysis; 18, ht of the capitulum; 19, ht minimum of the trochlea; 20, ht of the lateral lip of the trochlea; 21, td supraarticular of the distal epiphysis; 22, ccd of the distal epiphysis. 23, td olecranon fossa; 24: height olecranon fossa

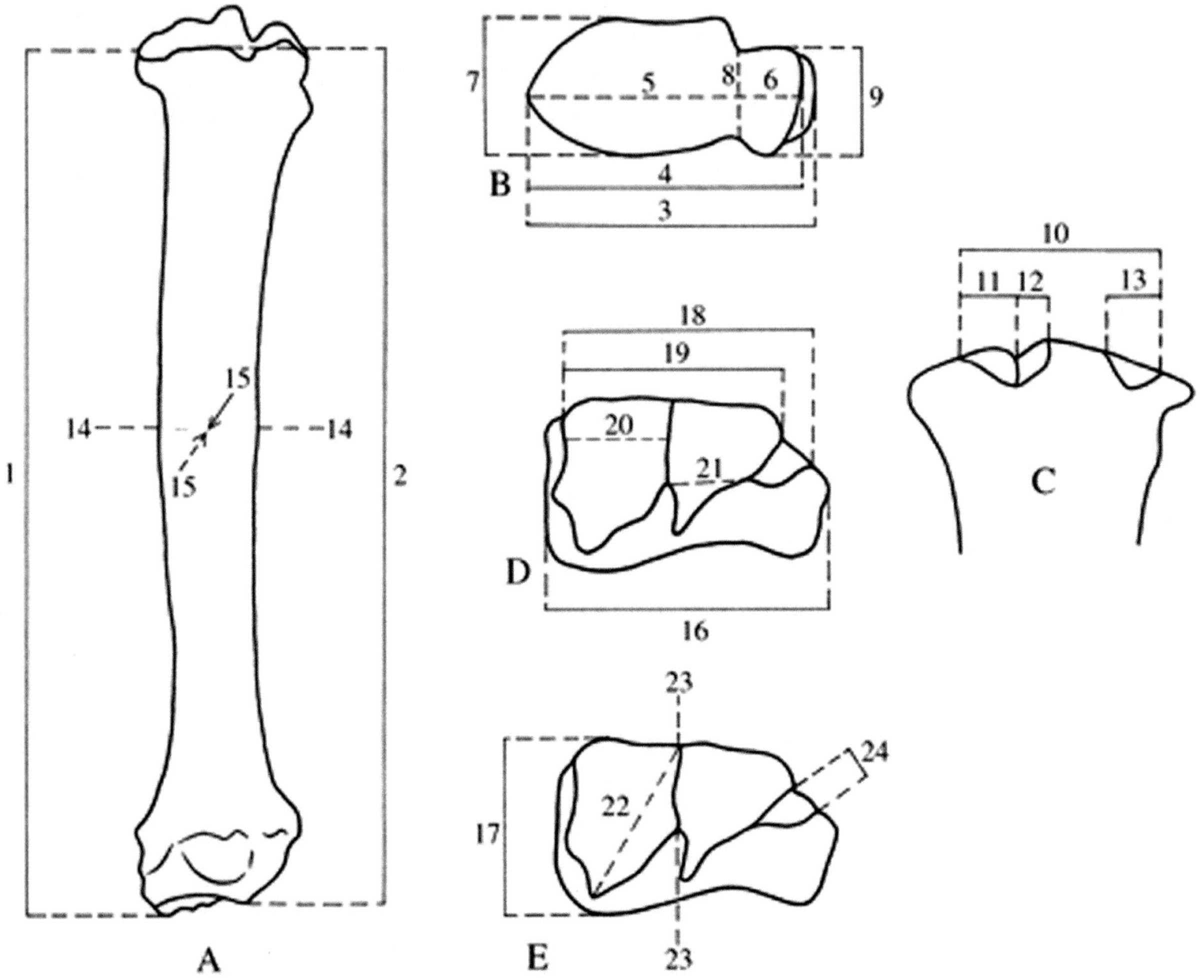

Fig. 14.A.7 Variable definitions on the radius (from Crégut-Bonnoure 1995, 2020, modified from Driesch 1976). **A**) dorsal view; **B**) proximal view; **C**) palmar view; **D** and **E**) distal view. Abbreviations: L., length; dpd, dorsopalmar diameter; td; transverse diameter. 1, L. total; 2, L. medial; 3, td total of the proximal epiphysis; 4, td articular of the proximal epiphysis; 5, td of the facet for the trochlea articulation; 6, td of the facet for the capitulum articulation; 7, dpd of the facet for the trochlea articulation; 8, dpd of the facet for the capitulum articulation; 9, dpd minimum; 10, td of the facets for the ulna; 11, td of the lateral facet for the ulna, 12; td of the median facet for the ulna; 13, td of the medial facet for the ulna; 14, td in the middle of the diaphysis; 15, dpd in the middle of the diaphysis; 16, td total of the distal epiphysis; 17, dpd of the distal epiphysis; 18, td articular of the distal epiphysis; 19, td of the facets for the scaphoid and the semi-lunar; 20, td of the facet for the scaphoid; 21, td caudal of the facet for the semi-lunar; 22, maximal dpd of the facet for the scaphoid; 23, length of the crest between the facets for the scaphoid and the semi-lunar; 24, td of the facet for the pyramidal

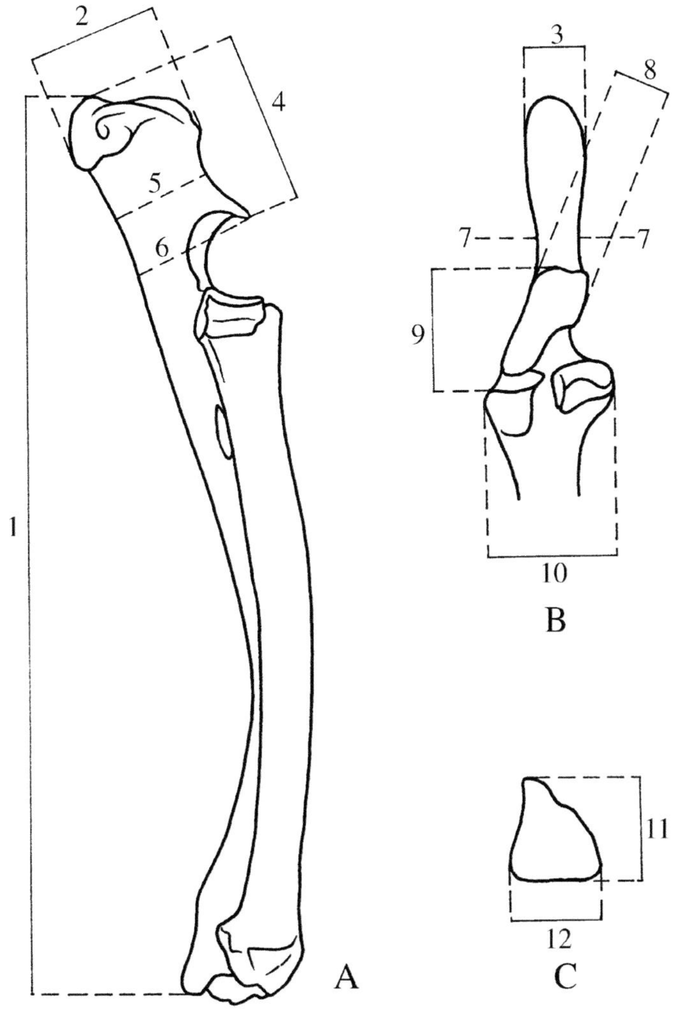

Fig. 14.A.8 Variable definitions on the ulna (from Crégut-Bonnoure 2020, modified from Driesch 1976). **A**) medial view; **B**) dorsal view; **C**) distal view. Abbreviations: dpd, dorsopalmar diameter; td; transverse diameter. 1, maximum length; 2, dpd of the tuber olecrani; 3, td of the tuber olecrani; 4, length from the top of the tuber olecrani to the proximal angle of the anconeal process; 5, dpd minimal between the top of the tuber olecrani and the proximal angle of the anconeal process; 6, dpd at the level of the anconeal process; 7, td minimal perpendicular to measurement 5; 8, td of trochlea incisura; 9, height of the trochlea incisura; 10: td of the articulation for the radius; 11: dpd of the distal articulation; 12: dt of the distal articulation

Fig. 14.A.9 Variable definitions on the carpals (from Crégut-Bonnoure 1987, 2020). **A** to **C**, scaphoid; **D** to **H**, semi-lunar; **I** to **K**, pyramidal; **L**, pisiform; **M** to **P**, unciform; **Q** to **S**, capitato-trapezoid. **A**, **D**, **M**, **Q**: proximal views; **B**, **E**, **I**, **N**, **R**: distal views; **C**, **F**, **J**, **S**: lateral views; **G**, **K**, **L**, **O**: medial views; **H**, **P**: dorsal views. Abbreviations: ht: height; dpd, dorsopalmar diameter; td, transverse diameter. Scaphoid: 1, total ht; 2, dpd total; 3, dpd of the proximal articulation; 4: td of the proximal articulation; 5, dpd of the distal articulation; 6, td of the distal articulation; 7, dpd of the proximal articular facet, palmar side; 8, ht of the proximal articular facet, palmar side; 9, dpd of the distal articular facet, dorsal side; 10, ht of the distal articular facet, dorsal side. Semi-lunar: 1, height; 2, dpd proximal in the middle; 3, dpd proximal from dorsolateral angle to palmolateral angle; 4, dpd proximal in anatomical position; 5: td proximal of the dorsal side; 6, td proximal of the palmar side; 7, td proximal of the minimum, in the middle; 8, td distal of the dorsal side; 9, td distal of the palmar side; 10, td distal of the dorsolateral articulation; 11, td distal of the dorsomedial articulation; 12, td distal of the palmomedial articulation; 13, td distal of the palmolateral articulation; 14, dpd of the distal crest; 15, td minimum of the dorsal side; 16, height of the proximodorsal articulation, medial side; 17, td of the proximodorsal articulation, medial side; 18, td of the proximopalmar articulation, medial side; 19, minimum height of the proximopalmar articulation, medial side; 20, td of the dorsodistal articulation, medial side; 21, height of the dorsodistal articulation, medial side. 22, td of the proximodorsal articulation, lateral side; 23, minimum height of the proximodorsal articulation, lateral side; 24, dpd of the lateral side. Pyramidal: 1, total ht; 2, minimum ht of the dorsal side; 3, medial ht of the dorsal side; 4, ht of the articulation for the pisiform; 5, td of the articulation for the pisiform; 6: ddp of the distal articulation; 7, td of the distal articulation; 8, ddp proximodorsal of the lateral side; 9, ddp of the proximodorsal articulation of the medial side; 10, ht of the proximodorsal articulation of the medial side. Pisiform: 1, total height; 2, dpd; 3, proximodistal diameter of the articulation for the pyramidal; 4, td; 5, total latéro-medial diameter of the articulation for the pyramidal. Unciform: 1, total height. 2, dpd; 3, td total; 4, td of the proximal articulation, dorsal side; 5, td of the lateroproximal articulation, dorsal side; 6, td of the medio proximal articulation, dorsal side; 7, td of the lateroproximal articulation, palmar side; 8, td of the medio proximal articulation, palmar side; 9, td of the dorsal side, mínimum; 10, proximodistal diameter of the articulation of the medial side; 11, dpd of the articulation of the medial side; 12, dpd of the distal articulation; 13, td of the distal articulation; 14, td of the proximal articulation, palmar side; 15, dpd of the proximal articulation, medial side; 16, dpd of the median crest of the proximal articulation. Capitato-trapezoid: 1, td in anatomical position; 2, dpd in anatomical position; 3, total height; 4, height of the dorsal rim; 5, td of the lateroproximal facet, dorsal side; 6, td of the medioproximal facet, dorsal side; 7, td of the lateroproximal facet, palmar side; 8, td of the medioproximal facet, palmar side; 9, td of the lateroproximal facet (at the base of the dorsal rim); 10, dpd of the proximal ridge; 11, td maximum of the distal articulation; 12, dpd maximum of the distal articulation

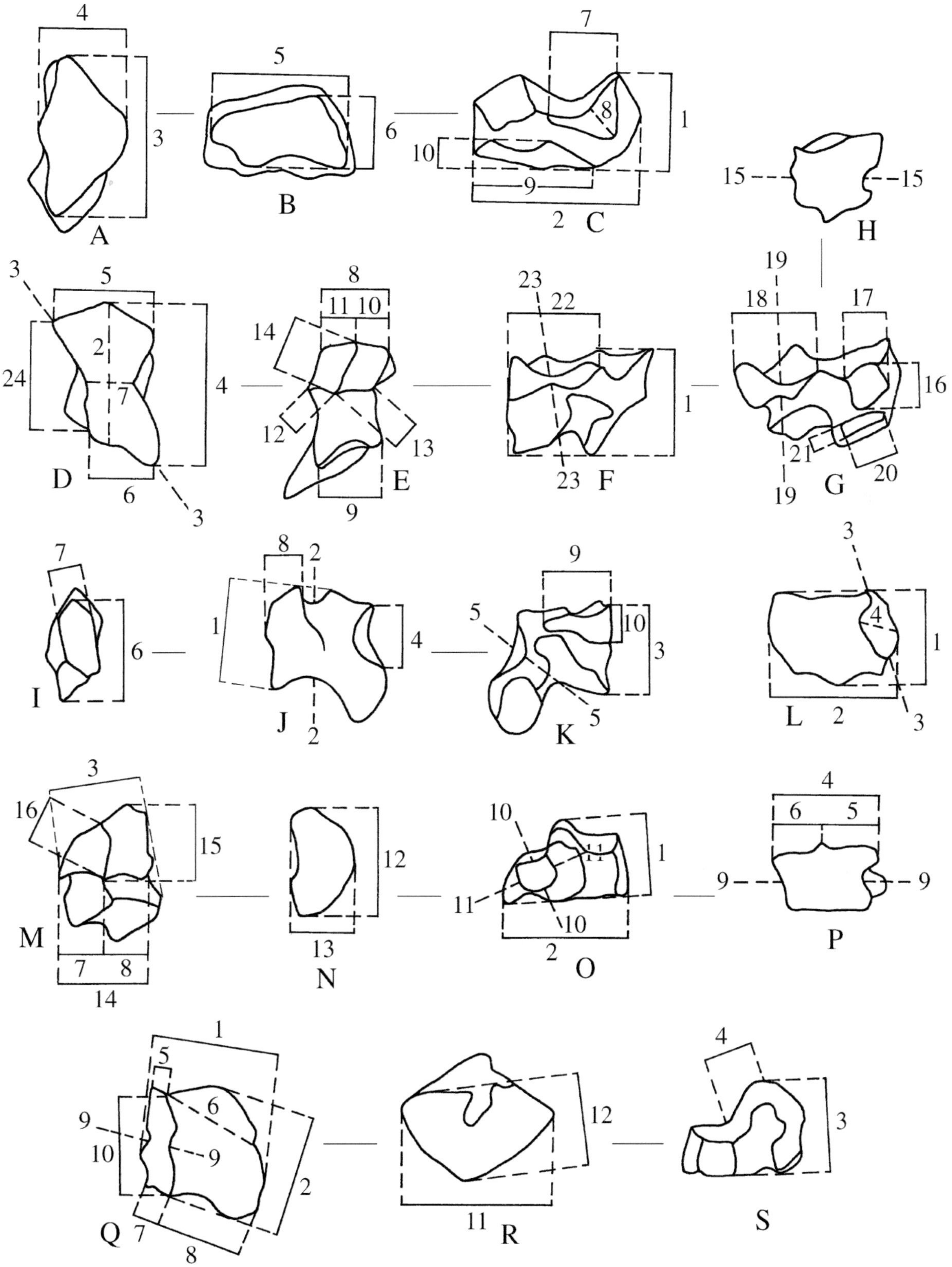
A 4 3
B 5 6
C 7 8 1 10 9 2
H 15 15
D 3 5 2 24 7 4 6 3
E 8 11 10 14 12 13 9
F 23 22 1 23
G 19 18 17 16 21 20 19
I 7 6
J 8 2 1 4 2
K 9 5 10 3 5
L 3 4 1 2 3
M 3 16 15 7 8 14
N 12 13
O 10 11 1 11 10 2
P 4 6 5 9 9
Q 1 5 6 9 10 9 2 7 8
R 12 11
S 4 3

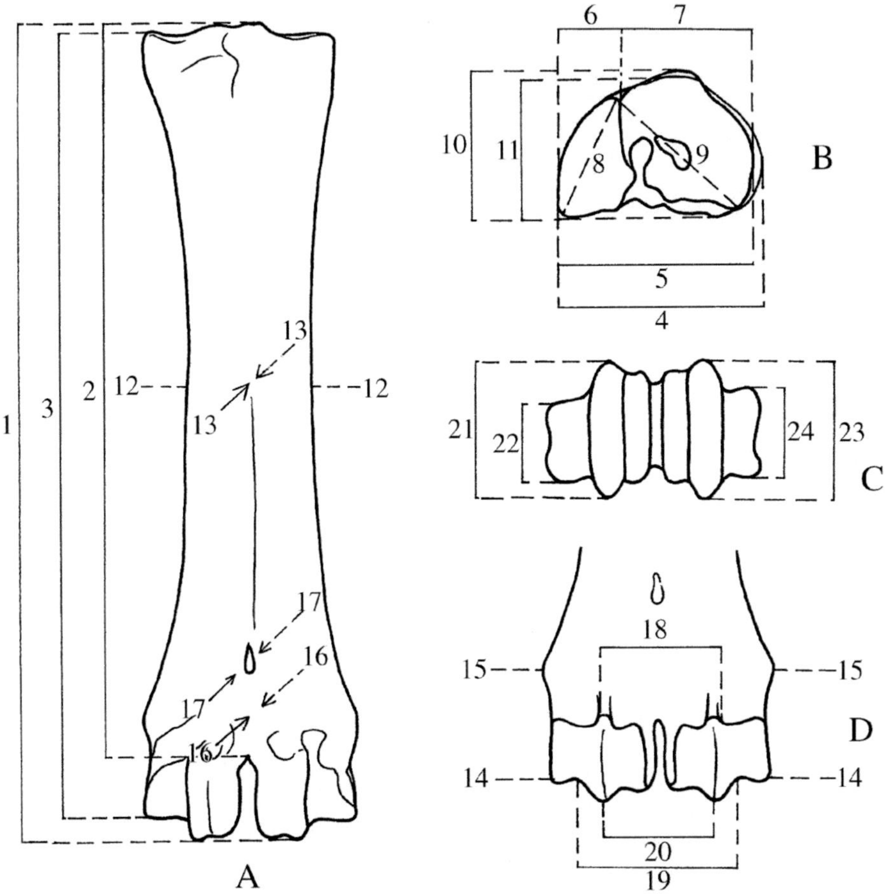

Fig. 14.A.10 Variable definitions on the metacarpal III+IV (from Crégut-Bonnoure 1995, 2020, modified from Driesch 1976). **A**) dorsal view; **B**) proximal view; **C**) distal view; **D**) palmar view. Abbreviations: L.: length; dpd, dorsopalmar diameter; td, transverse diameter. 1, total L.; 2, minimum L. at the middle; 3, minimum L. of the medial side; 4, td of the proximal epiphysis; 5, td of the proximal articulation; 6, td of the unciform facet; 7, td of the capitato-trapezoid facet; 8, length of the unciform facet, diagonally from lateropalmar angle to dorsal point of the proximal crest; 9, length of the capitato-trapezoid facet, diagonally from mediopalmar angle to dorsal point of the proximal crest; 10, dpd of the epiphysis; 11, dpd of the articulation; 12, td in the middle of the diaphysis; 13, dpd in the middle of the diaphysis; 14, td of the distal articulation; 15, td supraarticular; 16, dpd supraarticular; 17, dpd at the level of the dorsal foramen; 18, distance between the lateral edges of the keels at the palmar top; 19, distance between the lateral edges of the keels at the palmar base; 20, distance between the keels at their middle; 21, td of the medial keel; 22, td of the medial condyle; 23, td of the lateral keel; 24, td of the lateral condyle

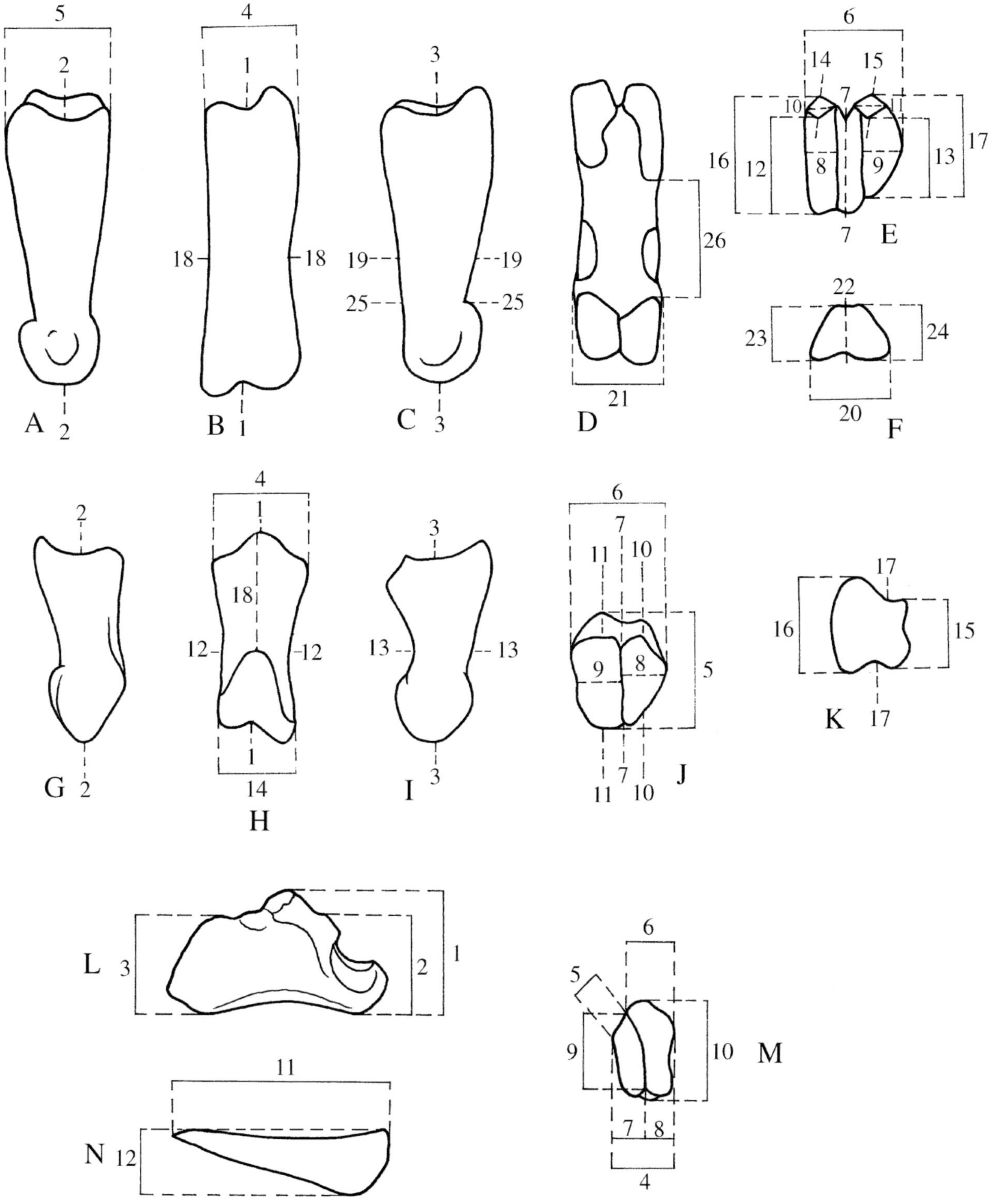
A
B
C
D
E
F
G
H
I
J
K
L
M
N

◄ **Fig. 14.A.11** Variable definitions on the phalanges (from Crégut-Bonnoure 1995, 2020). **A** to **F**) first phalanx; **G** to **K**) second phalanx; **L** to **N**) third phalanx. **A, G, L**) axial view; **B, H**) dorsal view; **C, I**) abaxial view; **D**) palmar/plantar view; **E, J, M**) proximal view; **F, K, N**) distal view. Abbreviations: L.: length; dppd: dorsopalmar-plantar diameter; td: transverse diameter. First phalanx: 1, L. in the middle; 2, axial L.; 3, abaxial L.; 4, td of the proximal epiphysis; 5, dppd of the proximal epiphysis: 6, td of the proximal articulation; 7, dppd of the proximal articulation in the middle; 8, td of the proximoaxial facet; 9, td of the proximoabaxial facet. 10, td of proximoaxial sesamoid facet; 11, td of the proximoabaxial sesamoid facet, 12, dppd of the proximoaxial facet; 13, dppd of the proximoabaxial facet; 14, dppd of the proximoaxial sesamoid facet; 15, dppd of the proximoabaxial sesamoid facet; 16, dppd of the proximoaxial articulation; 17, dppd of the proximoabaxial articulation; 18, td in the middle of the diaphysis; 19, dppd in the middle of the diaphysis; 20, td of the distal articulation; 21, td of the distal epiphysis; 22, dppd minimum of the distal articulation; 23, dppd of the distoaxial condyle; 24, dppd of the distoabaxial condyle. 25, dppd sub-articular distal; 26, height on the palmar side, from the base of the proximoaxial tubercle to the top of the distoaxial condyle. Second phalanx: 1, L. in the middle; 2, axial L.; 3, abaxial L.; 4, td of the proximal epiphysis; 5, dppd of the proximal epiphysis; 6, td of the proximal articulation; 7, dppd of the proximal articulation in the middle; 8, td of the proximoaxial facet; 9, td of the proximoabaxial facet; 10, dppd of the proximoaxial facet; 11, dppd of the proximoabaxial facet; 12, td in the middle of the diaphysis; 13, dppd in the middle of the diaphysis; 14, td of the distal articulation; 15, dppd of the axial condyle; 16, dppd of the abaxial condyle; 17, dppd minimum of the distal articulation; 18, dorsal height from the median top to the proximal side of the distal articulation. Third phalanx: 1, total height; 2, height at the level of the articulation; 3, height of the distal part; 4, td of the articulation; 5, td dorsal of the axial facet; 6, td dorsal of the abaxial facet; 7, td palmoplantar of the axial facet; 8, td palmoplantar of the abaxial facet; 9, dppd of the articular crest; 10, dppd of the abaxial facet; 11, total length; 12, total width

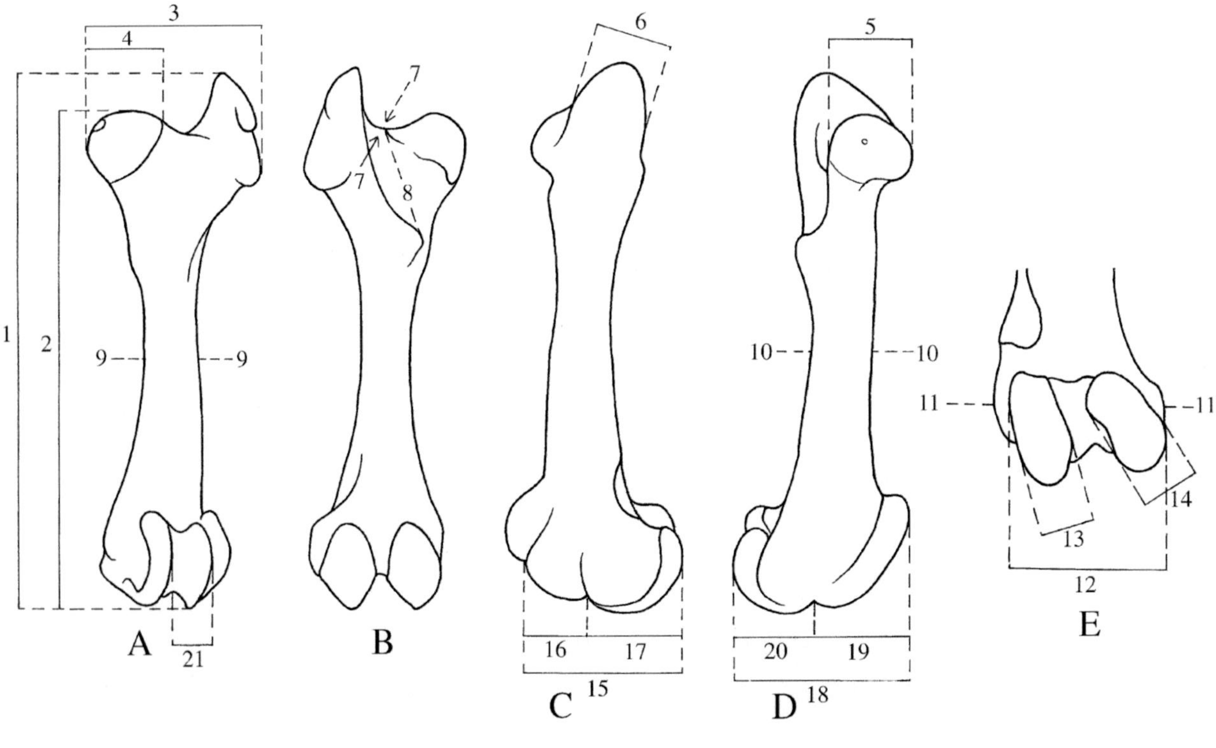

Fig. 14.A.12 Variable definitions on the femur (from Crégut-Bonnoure 2020, modified from Driesch 1976). **A**) cranial view; **B, E**) caudal view; **C**) lateral view; **D**) medial view. Abbreviations: td, transverse diameter; ccd, craniocaudal diameter. 1, total length; 2, length from the femoral head to the base; 3, td of the proximal epiphysis; 4, td of the femoral head; 5, ccd of the femoral head; 6, ccd of the greater trochanter; 7, ccd of the femoral neck; 8, ccd of the lesser trochanter; 9, td in the middle of the diaphysis; 10, ccd in the middle of the diaphysis; 11, td of the distal epiphysis; 12, td of the distal articulation; 13, td of the distal lateral condyle; 14, td of the distal medial condyle; 15, ccd of the distal lateral side. 16, ccd of the distal lateral lip of the trochlea; 17: ccd of the distal lateral condyle; 18, ccd of the distal medial side; 19, ccd of the distal medial lip of the trochlea; 20, ccd of the distal medial condyle; 21, td of the trochlea on the dorsal side

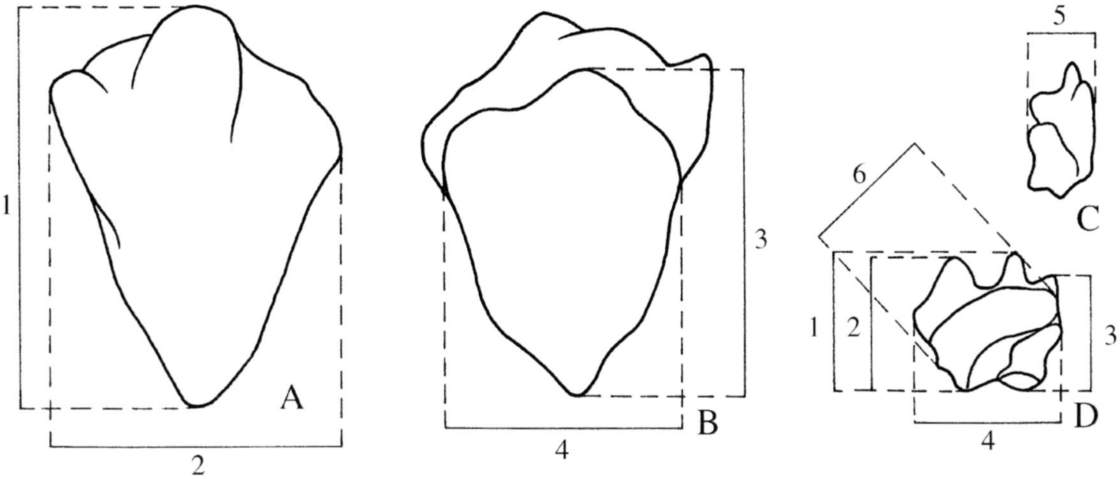

Fig. 14.A.13 Variable definitions on the patella and the malleolar bone (from Crégut-Bonnoure 2020, modified from Driesch 1976). **A**, **C**) cranial view; **B**) caudal view; **D**) medial view. Abbreviations: td, transverse diameter; ccd, craniocaudal diameter. Patella (**A**, **B**): 1, height; 2, td; 3, height of the caudal articulation; 4, td of the caudal articulation; 5, ccd (not in the figure). Malleolar bone (**C**, **D**): 1, total height; 2, height of the cranial side; 3, height of the caudal side; 4, ccd; 5, td; 6, length of the medial articulation

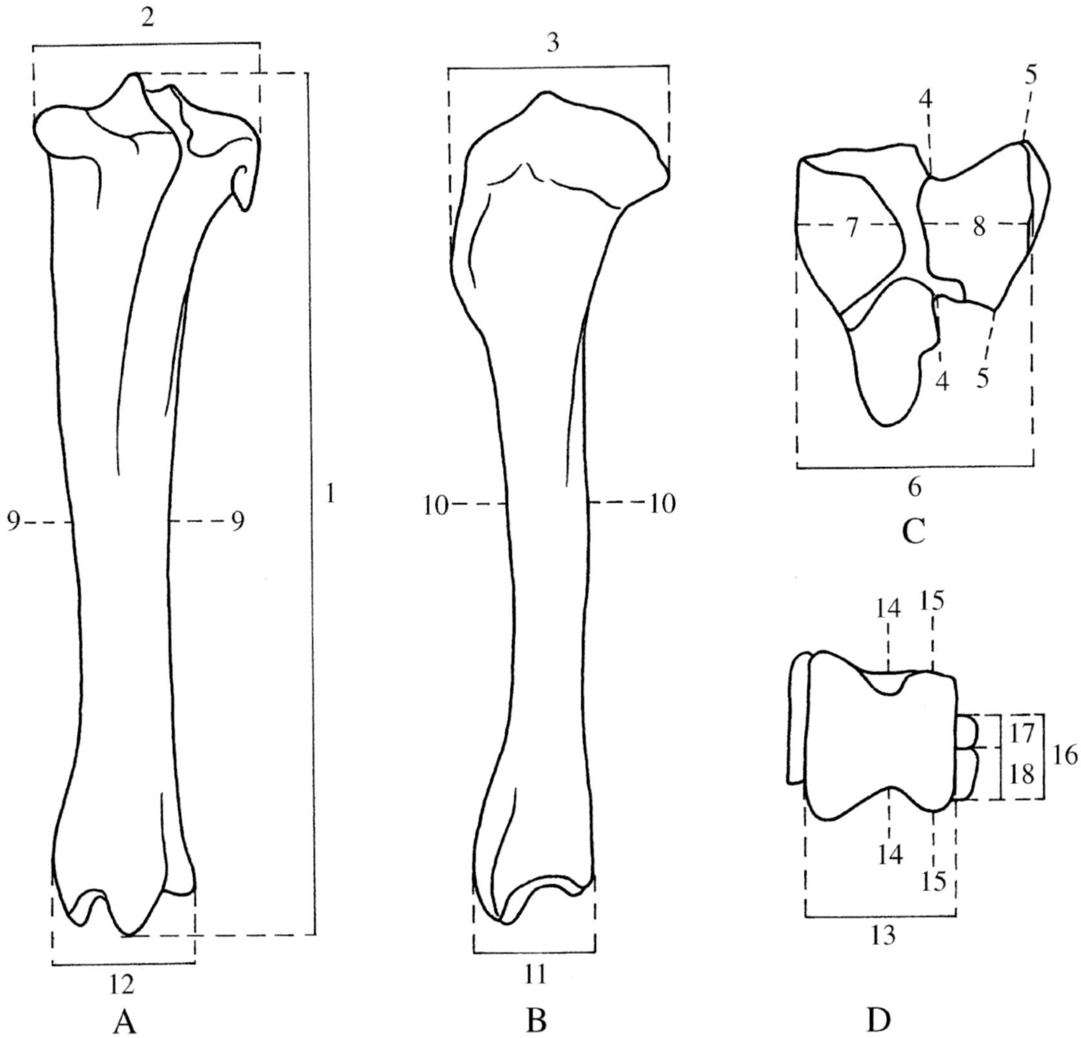

Fig. 14.A.14 Variable definitions on the tibia (from Crégut-Bonnoure 1995, 2020, modified from Driesch 1976). **A**) dorsal view; **B**) lateral view; **C**) proximal view; **D**) distal view. Abbreviations: dpd: dorsoplantar diameter; td: transverse diameter. 1, total length; 2, td of the proximal epiphysis; 3, dpd of the proximal epiphysis; 4, dpd minimal of the proximal epiphysis; 5, dpd of the lateral side of the proximal epiphysis; 6, td of the proximal articulation; 7, td of the medial facet of the proximal articulation; 8, td of the lateral facet of the proximal articulation; 9, td in the middle of the diaphysis; 10, dpd in the middle of the diaphysis; 11, dpd of the distal epiphysis; 12, td of the distal epiphysis; 13, td of the articulation for the talus; 14, dpd of the middle of the distal epiphysis; 15, dpd of the lateral side of the distal epiphysis; 16, dpd of the articulation for malleolar bone; 17, dpd of the dorsal facet for malleolar bone; 18, dpd of the plantar facet for malleolar bone

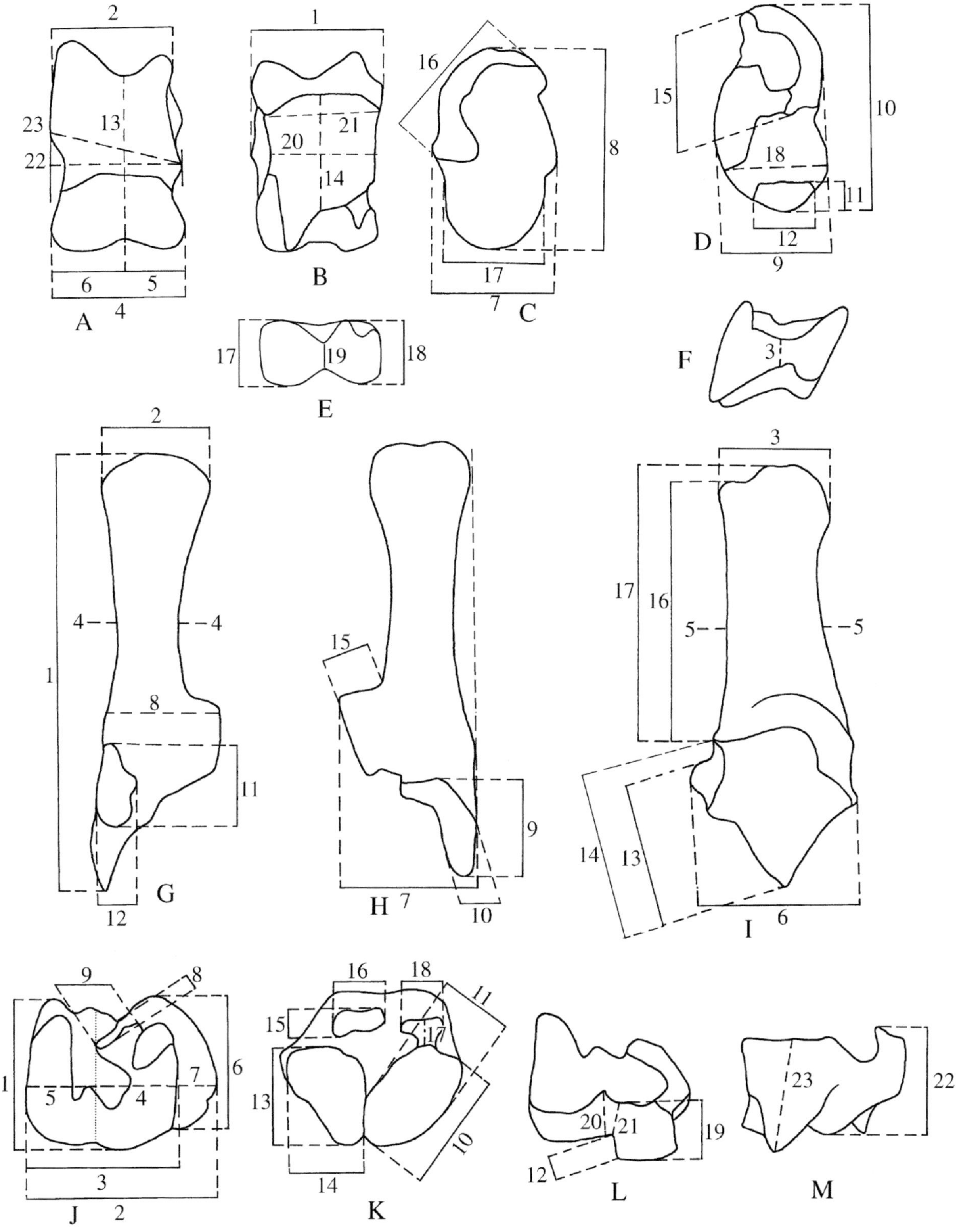
A
B
C
D
E
F
G
H
I
J
K
L
M

◀ **Fig. 14.A.15** Variable definitions on the tarsals (from Crégut-Bonnoure 1987, 2020). **A** to **F**) right talus; **G** to **I**) right calcaneus; **J** to **M**) left navicularcuboid. **A**, **G**, **L**) dorsal view; **B**, **H**, **M**) plantar view; **C**, **I**) medial view; **D**) lateral view, **E**, **J**) proximal view; **F**, **K**) distal view. Abbreviations: dpd, dorsoplantar diameter; td, transverse diameter. **Talus**: 1, td total; 2, td of the proximal trochlea; 3, dpd minimum of the proximal trochlea. 4, td total of the distal trochlea; 5, td medio distal condyle; 6, td latero distal condyle; 7, dpd of the medial side; 8, height of the medial side; 9, dpd of the lateral side; 10, height of the lateral side; 11; height of the laterodistal facet for the calcaneus, 12, dpd of the laterodistal facet for the calcaneus. Height: 13, in the middle; 14, of the plantar trochlea for calcaneus; 15, trochlear lip of the proximolateral side; 16, trochlear lip of the proximomedial side. Distal trochlea: 17, dpd medial condyle; 18, dpd lateral condyle;19, dpd in the middle. Td: 20, of the plantar trochlea for calcaneus in the middle; 21, of the plantar trochlea for calcaneus at the proximal edge; 22, of the dorsal side; 23, of the proximal trochlea at its dorsal base. **Calcaneus**: 1, total length. *Tuber calcanei*: 2, td; 3, dpd. Proximal body: 4, td minimum; 5, dpd. Total: 6, dpd; 7, td. *Sustentaculum tali*: 8, td; Navicular-cuboid facet: 9, length; 10, td. Fibular trochlea facet: 11, length; 12, td. Length: 13, from the fibular trochlea facet at its middle to the distal point of the bone; 14, from the fibular trochlea facet at its proximal side to the distal point of the bone. Width: 15, *sustentaculum tali* furrow. Length: 16, from the dorsal *tuber calcanei* apex to the coracoid process tip; 17, from the palmar *tuber calcanei* apex to the coracoid process tip. **Navicular-cuboid**: Total: 1, dpd; 2, td. Td, proximal articulation for the talus: 3, total; 4, lateral; 5, medial. Articulation for the calcaneus: 6, dpd; 7, td. Small dorsomedial articulation for the talus: 8, dpd; 9, td. Articulation for the great medial cuneiform facet: 10, dpd; 11, td. Dorsal side: 12, height between great cuneiform surface and metatarsal III+IV dorsolateral surface. Articulation for the metatarsal III+IV: 13, dpd of the dorsolateral facet; 14, td of the dorsolateral facet; 15, dpd of the palmolateral facet; 16, td of the plantar articulation of the palmolateral facet. Articulation for the small medial cuneiform facet: 17, dpd; 18, td. Dorsal side: 19, maximal lateral height; 20, maximal medial height; 21, maximal median height. Plantar side: 22, maximal medial height; 23, maximal lateral height

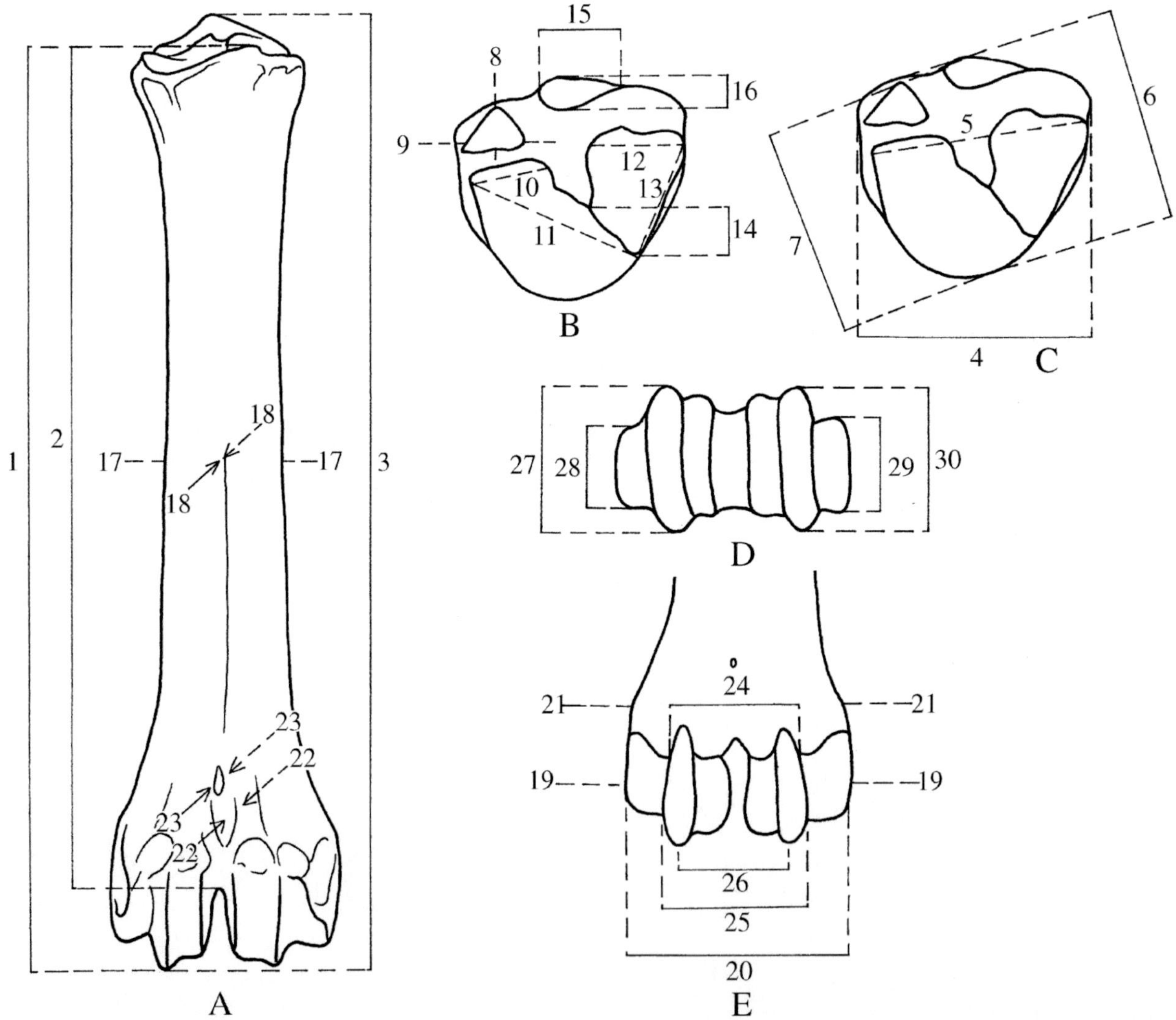

Fig. 14.A.16 Variable definitions on the metatarsal III+IV (from Crégut-Bonnoure 2020): Abbreviations: dpd, dorsoplantar diameter; td, transverse diameter. Length: 1, dorsal side; 2, minimum in the middle; 3, plantar side. Proximal epiphysis: 4, td total; 5, td of the articulation; 6, dpd total; 7, dpd of the articulation; 8, dpd of the facet for the small cuneiform; 9, td of the facet for the small cuneiform; 10, td of the great cuneiform facet; 11, dpd of the great cuneiform facet; 12, td of the dorsal facet for the navicular-cuboid; 13, dpd of the dorsal facet for the navicular-cuboid; 14, length of the connection of the dorsal facets. 15: td of the plantar facet for the navicular-cuboid; 16: dpd of the plantar facet for the navicular-cuboid. Middle of the diaphysis: 17, td; 18: dpd. Distal epiphysis: 19, td maximal of the articulation; 20, td minimal of the articulation; 21, td supraarticular; 22, dpd of the epiphysis; 23, dpd of the epiphysis at the foramen level; 24, distance between the keels on the plantar side; 25, distance between the keels at their base; 26, distance between the keels in their middle; 27, dpd of the medial condyle; 28, dpd of the medial keel; 29, dpd of the lateral condyle; 30, dpd of the lateral keel

References

Abel, T., Harlacher, C., & Ufrecht, W. (2006). Zur erkarstungsgeschichte der Bären- und Karlshöhle bei Erpfingen (Schwäbische Alb) im Plio-Pleistozän unter Berücksichtigung von Sinterchronologie und Paläontologie. *Jahresberichte und Mitteilungen des Oberrheinischen Geologischen Vereins, 88*, 9–51.

Agustí, J., Moyá-Solá, S, Martín-Suárez, E., & Marin, M. (1987). Faunas de mamíferos en el Pleistoceno inferior de la región de Orce (Granada, España). *Paleontologia i Evolució*, mem. esp., *1*, 73–86.

Alexeeva, L. I., & Baygucheva, V. S. (1976). New data on the horn cored antelope in the territory of southern Eastern Europe. *Bulletin of the Commission for the Study of the Quaternary Period, 45*, 109–112.

Andrés Rodrigo, M. (2011). Los bóvidos Villafranquienses de La Puebla de Valverde y Villarroya: sistemática, filogenia y paleobiología. Ph.D. Dissertation, Universidad de Zaragoza.

Arribas, A., & Garrido, G. (2008). Los representantes más antiguos del género *Capra* (Bovidae, Artiodactyla, Mammalia) en el registro euroasíatico –Fonelas P-1, Cuenca de Guadix, España). In A. Arribas (Ed.), Vertebrados del Plioceno superior terminal en el suroeste de Europa: Fonelas P-1 y el Proyecto Fonelas. *Cuadernos del Museo Geominero* (Vol. 10, pp. 461–473).

Athanassiou, A. (1996). Contribution to the study of the fossil Mammals of Thessaly. Thesis, University of Athens (in Greek).

Athanassiou, A. (1998). Contribution to the study of the fossil mammals of Thessaly. *Gaia, 5*, 1–354. (in Greek).

Athanassiou, A. (2002). Neogene and Quaternary mammal faunas from Thessaly. *Annales géologiques des Pays helléniques*, 1° série, XXXIX (A), 279–293.

Athanassiou, A. (2005). *Gazellospira torticornis* (Aymard, 1854) from the late Pliocene locality of Sésklo (Thessaly, Greece). In E. Crégut-Bonnoure (Ed.), *Les ongulés holarctiques du Pliocène et du Pléistocène*. Actes colloque international Avignon, 10–22 septembre 2000. *Quaternaire, hors-série* (Vol. 2, pp. 137–144).

Athanassiou, A. (2018). A Villafranchian *Hipparion*-bearing mammal fauna from Sésklo (E. Thessaly, Greece): Implications for the question of *Hipparion–Equus* sympatry in Europe. *Quaternary, 1*, 12. https://doi.org/10.3390/quat1020012

Aymard, A. (1854). Ossements fossiles de Sainzelle, commune de Polignac. *Annales de la Société d'Agriculture, Sciences, Arts et Commerce du Puy, 18*, 51–54.

Azanza, B., & Morales, J. (1989). Los artiodáctylos de Huélaguo, Huescar-1 y Cúllar de Baza-1 (Cuenca de Guadix-Baza, Granada). *Trabajos sobre Neógeno-Cuaternario, 11*, 289–314.

Azzaroli, A., Colalongo, M. L., Nakagawa, H., Pasini, G., Rio, D., Ruggeri, G., et al. (1997). The Plio-Pleistocene boundary in Italy. In J.A. Van Couvering (Ed.), *The Pleistocene boundary and the beginning of the Quaternary* (pp. 141–155). Cambridge: Cambridge University Press.

Azzaroli, A., de Giuli, C., Ficcarelli, G., & Torre, D. (1988). Late Pliocene to early Mid-Pleistocene mammals in Eurasia: Faunal succession and dispersal events. *Paleogeography, Paleoclimatology, Paleoecology, 66*, 77–100.

Barone, R. (1976). *Anatomie comparée des mammifères domestiques. Ostéologie*. Vigot frères.

Bellucci, L., Bona, F., Corrado, P., Magri, D., Mazzini, I., Parenti, F., et al. (2014). Evidence of late Gelasian dispersal of African fauna at Coste San Giacomo (Anagni Basin, central Italy): Early Pleistocene environments and the background of early human occupation in Europe. *Quaternary Science Reviews, 96*, 72–85.

Belluci, L., & Sardella, R. (2015). The last antilopini bovids from the Early Pleistocene of Italy. *Quaternary International, 357*, 245–252.

Boeuf, O. (1983). Le site villafranchien de Chilhac (Haute-Loire, France). Thèse de Doctorat, Université de Paris VII.

Bohlin, B. (1938). Einige Jungtertiäre und Pleistozene Cavicornier aus Nord-China. *Nova Acta Regiae Societatis Scientarum Upsaliensis, 11*, 1–49.

Bolomey, A. (1965). Contribution à la connaissance de la morphologie de *Pliotragus ardeus*. *Revue roumaine de Biologie, Zoologie, 10*, 315–323.

Bout, P. (1960). *Le Villafranchien du Velay et du bassin hydrographique moyen et supérieur de l'Allier. Corrélations françaises et européennes*. Imprimerie Jeanne d'Arc.

Bouvrain, G. (1998). Bovidae. In S. Sen (Ed.), *Le gisement de vertébrés pliocènes de Çalta, Ankara, Turquie*. *Geodiversitas* (Vol. 20, pp. 467–485).

Bruijn, H. de, Daams, R., Daxner-Hock, G., Fahlbusch, L., Ginsburg, L., Mein, P., & Morales, J. (1992). Report on the RCMNS working group on fossil mammals, Reisenburg 1990. *Newsletters on Stratigraphy, 16*, 65–118.

Bukhsianidze, M. (2005). Fossil Bovidae of Dmanisi. Ph.D. dissertation, University of Ferrara.

Bukhsianidze, M. (2006). *Capra dalii* nov. sp. (Caprinae, Bovidae, Mammalia) at the limit of Plio-Pleistocene from Dmanisi (Georgia). In R. D. Kahlke, L. C. Maul & P. Mazza (Eds.), *Late Neogene and quaternary biodiversity and evolution: Regional developments and interregional correlations*. 18th International Senckenberg Conference, VI International Paleontological Colloquium in Weimar, *Courier Forschungsinstitut Senckenberg* (Vol. 256, pp. 159–171).

Bukhsianidze, M. (2020). New results on bovids from the Early Pleistocene site of Untermaasfeld. In R.D. Kahlke (Ed.), *The Pleistocene of Untermassfeld near Meiningen (Thüringen, Germany) – Part 4. Monographien des Römisch-Germanischen Zentralmuseums, 40*, 1169–1196.

Burchak-Abramovich, N. I., & Vekua, A. K. (1994). Novy predstavitel pleiotsenovykh bykov iz vostovhnoi Gruzii. In L. P. Tatarinow (Ed.), *Paleoteriologia* (pp. 253–261).

Caloi, L., Gliozzi, E., Kotzakis, T., Malatesta, A., & Palombo, M. R. (1986). Osservazioni sulla paleobiogeografia dei mammiferi del Pleistoceno italiano. *Hystrix, 1*, 7–23.

Cassoli, P. F., & Segre Naldini, E. (1994). Some observations on Latium "Villafranchian" layers. In *Il Significato del Villafranchiano nella stratigrafia del Plio-Pleistocene*. INQUA congress, Peveragno, Abstract.

Cherin, M., D'Allestro V., & Masini, F. (2017). New bovid remains from the Early Pleistocene of Umbria (Italy) and a reappraisal of *Leptobos merlai*. *Journal of Mammalian Evolution, 26*, 201–224 (2019). https://doi.org/10.1007/s10914-017-9421-x

Clot, A., Chaline, J., Heintz, E., Jammot, D., Mourer-Chauviré, C., & Rage, J.-C. (1976). Montoussé 5 (Hautes-Pyrénées), un nouveau remplissage de fissure à faune de vertébrés du Pléistocène inferieur. *Geobios, 9*, 511–514.

Couthures, J., Ablin, D., & Hadjouis, D. (1991). Le site villafranchien de Senèze (Haute-Loire), Rapport des fouilles 1991. Service Régional de l'Archéologie, Direction Régionale des Affaires Culturelles d'Auvergne, 55 p.

Crégut, E. (1979). La faune de Mammifères du Pléistocène moyen de la Caune de l'Arago à Tautavel (Pyrénées-Orientales). *Travaux du laboratoire de Paléontologie humaine et de Préhistoire, 3*, 1–381.

Crégut-Bonnoure, E. (1987). Caractères distinctifs du carpe et du tarse d'*Hemitragus* et de *Capra* (*Mammalia, Bovidae, Caprinae*). *Bulletin du Musée d'anthropologie Préhistorique de Monaco, 30*, 43–94.

Crégut-Bonnoure, E. (1989). Un nouveau Caprinae, *Hemitragus cedrensis* nov. sp. (*Mammalia, Bovidae*), des niveaux pléistocènes moyens de la grotte des Cèdres (Le Plan d'Aups, Var). Intérêt biogéographique. *Geobios, 22*, 653–663.

Crégut-Bonnoure, E. (1992). Les Caprinae (Mammalia, Bovidae) du Pléistocène d'Europe. Intérêt biostratigraphique, paléoécologique et archéozoologique. *Mémoire de la Société Géologique de France, 160*, 85–93.

Crégut-Bonnoure, E. (1995). Les grands mammifères. In A. Defleur & E. Crégut-Bonnoure (Eds.), Le gisement Paléolithique moyen de la grotte des Cèdres (Var), *Documents d'Archéologie française, 49*, 54–142.

Crégut-Bonnoure, E. (2002). Les Ovibovini, Caprini et Ovini (Mammalia, Artiodactyla, Bovidae, Caprinae) du Plio-Pléistocène d'Europe: Systématique, évolution et biochronologie. Thèse de Doctorat d'État, Université Claude Bernard-Lyon 1.

Crégut-Bonnoure, E. (2006). European Ovibovini, Ovini, Caprini (Caprinae, Mammalia) from the Plio-Pleistocene: New interpretations. *Courier Forschungsinstitut Senckenberg, 256*, 139–158.

Crégut-Bonnoure, E. (2007). Apport des Caprinae et Antilopinae (Mammalia, Bovidae) à la biostratigraphie du Pliocène terminal et du Pléistocène d'Europe. *Quaternaire, 18*, 73–97.

Crégut-Bonnoure, E. (2009). Biochronologie et grands Mammifères au Pléistocène moyen et supérieur en Europe occidentale: l'apport des genres *Hemitragus* et *Capra*. *Quaternaire, 20*, 481–508.

Crégut-Bonnoure, E. (2020). Les Ovibovini, Caprini et Ovini (Mammalia, Artiodactyla, Bovidae, Caprinae) du plio-pléistocène d'Europe. Systématique, évolution et biochronologie. *British Archaeological Reports, S2975*.

Crégut-Bonnoure, E., & Dimitrijević, V. (2006). *Megalovis balcanicus* sp. nov. and *Soergelia intermedia* sp. nov. (Mammalia, Bovidae, Caprinae), new Ovibovini from the Early Pleistocene of Europe. *Revue de Paléobiologie, 25*, 423–473.

Crégut-Bonnoure, E., & Guérin, C. (1996). Ordre des Artiodactyles; Famille des Bovidae. In C. Guérin & M. Patou-Mathis (Eds.), *Les grands Mammifères plio-pléistocènes d'Europe* (pp. 62–95). Paris: Masson.

Crégut-Bonnoure, E., & Spassov, N. (2002). *Hemitragus orientalis* nov. sp. (*Mammalia, Artiodactyla, Bovidae*): Un nouveau taxon d'Europe orientale. *Revue de Paléobiologie, 21*, 553–573.

Crégut-Bonnoure, E., & Tsoukala, E. (2005). The Pleistocene *Caprinae (Mammalia, Bovidae)* from the Petralona Cave (Macedonia, Greece): new interpretation and biogeographical implications. In E. Crégut-Bonnoure (Ed.), *Les ongulés holarctiques du Pliocène et du Pléistocène*, Actes colloque international Avignon, 10–22 septembre 2000. *Quaternaire*, hors-série (Vol 2, pp. 161–178).

Crégut-Bonnoure, E., & Valli, A. (2004). Les Bovidae du site Villafranchien supérieur de Saint-Vallier (Drôme, France). Mémoire spécial 26. *Geobios, 37*, 233–258.

Crégut-Bonnoure, E., Guérin, C., Argant, A., Argant, J., Debard, E., Delson, E., et al. (2024). Biochronology of the Senèze faunal assemblage. In E. Delson, M. Faure & C. Guérin (Eds), *Senèze: Life in Central France two million years ago. Paleontology, geochronology, stratigraphy and taphonomy* (pp. 633–652). Cham (Switzerland): Springer.

Croitor, R. (2016). Genus *Bison* (Bovidae, Mammalia) in Early Pleistocene of Moldova. Materialele conferinței naționale cu participarea internațională "Mediul și dezvoltarea durabilă". In *Ediția a III-a/3rd National Scientific Conference with international Participation "Environment and Sustainable Development"*, October 2016, Chisinau, Moldova (pp. 15–20). https://hal.archives-ouvertes.fr/hal-01766112

Croitor, R. (2018). A Description of two new species of the Genus Rucervus (Cervidae, Mammalia) from the Early Pleistocene of Southeast Europe, with comments on Hominin and South Asian Ruminants dispersals. *Quaternary, 1*, 17; https://doi.org/10.3390/quat1020017

Croitor, R., Robinson, C., Curran, S., Terhune, C., Drăgușin, V., Pobiner, B., et al. (2023). Early Pleistocene ruminants (Artiodactyla, Mammalia) from the Dacian Basin (South Romania) before and after the *Pachycrocuta* event: implications for hominin dispersals in Western Eurasia. *Historical Biology, 36*, 485–533.

Crusafont Pairó, M. (1965). *Hesperidoceras*, nuevo nombre para *Hesperoceras* Vill. et Crus, 1953. *Boletin de la Real Sociedad Española de Historia Natural (Geologia), 63*, 223–224.

Crusafont Pairó, M., Hartenberger, J.-L., & Heintz, E. (1964). Un nouveau gisement de mammifères fossiles d'âge villafranchien à la Puebla de Valverde (Province de Teruel, Espagne). *Comptes Rendus hebdomadaires des séances de l'Académie des Sciences, 258*, 2869–2871.

Debard, E. (2024) Geological study of the Early Pleistocene site of Senèze (Domeyrat, Haute-Loire, France). In E. Delson, M. Faure, & C. Guérin (Eds.), *Senèze: Life in Central France two million years ago. Paleontology, geochronology, stratigraphy and taphonomy* (pp. 37–82). Cham (Switzerland): Springer.

Dechaseaux, C. (1961). Encéphales d'Artiodactyles fossiles. In J. Piveteau (Ed.), *L'origine des Mammifères et les aspects fondamentaux de leur évolution. Mammifères, origine reptilienne, évolution* (pp. 1085–1104). Masson.

De Giuli, C., & Masini F. (1983). A new element of the Late Villafranchian (Tasso Unit) faunas of Italy : occurrence of Ovibovini (Bovidae, Artiodactyla, Mammalia) in the fauna of Casa Frata (Upper Valdarno, Tuscany). *Bolletino della Società Paleontologia Italiana, 22*, 271–280.

De Giuli, C., Masini, F., & Torre, D. (1987). The latest villafranchian faunas in Italy: The Pirro Nord fauna (Apricena, Gargano). *Palaeontographia Italica, LXXIV*, 51–62.

De Giuli, C., Masini, F., & Torre, D. (1990). The latest villafranchian faunas in Italy: The Pirro Nord fauna (Apricena, Gargano). *Quartärpaläontologie, 8*, 29–34.

Delson, E., Faure, M., Guérin, C., Aprile, L., Blackwell, B. A. B., Debard, E., et al. (2006). Franco-American renewed research at the Late Villafranchian locality of Senèze (Haute-Loire, France). *Courier Forschungs-institut Senckenberg, 256*, 275–290.

Delson, E., Nomade, S., Sen, S., Debard, E., Pastre, J.-F., Bahain, J.-J., et al. (2024). Geochronology of Senèze: ^{40}Ar/^{39}Ar Dating and Magnetostratigraphy, with Notes on an ESR/U-series dating attempt. In E. Delson, M. Faure, & C. Guérin (Eds.), *Senèze: Life in Central France two million years ago. Paleontology, geochronology, stratigraphy and taphonomy* (pp. 99–122). Cham (Switzerland): Springer.

Demirel, A., & Mayda, S. (2014). A new Early Pleistocene mammalian fauna from Burdur Basin, SW Turkey. *Russian Journal of Theriology, 13*, 55–63.

Depéret, C. (1883). Nouvelles études sur les Ruminants fossiles d'Auvergne. *Comptes Rendus de l'Académie des Sciences, 97*, 866–869.

Depéret, C. (1884). Nouvelles études sur les Ruminants pliocènes et quaternaires d'Auvergne. *Bulletin de la Société géologique de France, 12*, 247–284.

De Vos, J., van der Made, J., Athanassiou, A., Lyras, G., Sondaar, P. Y., & Dermitzakis, M. D. (2002). Preliminary note on the Late Pliocene fauna from Vaterá (Lesvos, Greece). *Annales géologiques des Pays helléniques, 39*, 37–70.

Dimitrijevic, V., & Knezevic, S. (1996). *Leptobos* (Bovidae, Mammalia) from "Paludina Beds" at Cortanovci, Srem. *Annales Géologiques de la Péninsule Balkanique, 60*, 219–228.

Dmitrieva, E. L. (1975). Fossil Tragelaphinae from western Mongolia and certain central-Asian. "Fauna, flora and biostratigraphy of the Mesozoic and Cenozoic of Mongolia". The joint Soviet Mongolian

Paleontological expedition. *Transactions of the Paleontological Institute of the Academy of Sciences of the USSR, 2*, 76–83.

Dmitrieva, E. L. (1977). *Neogene Antelopes from Mongolia and adjacent regions*. Nauka. (in Russian).

Dorlhac, J. (1854). Notice géologique sur le cratère du Coupet et sur son gisement de gemmes et d'ossements fossiles. *Annales de la Société d'Agriculture, Sciences, Arts et Commerce du Puy, 19*, 497–517.

Driesch, A. von den (1976). A guide to measurement of animal bones from archaeological sites. *Peabody Museum Bulletin, 1*, 1–137 (Peabody Museum of Archaeology and Ethnology Harvard University).

Duvernois, M.-P. (1989). Les *Leptobos* (Mammalia, Artiodactyla) du Villafranchien inférieur d'Europe occidentale. *Comptes Rendus de l'Académie des Sciences, 309*, 769–775.

Duvernois, M.-P. (1990). Les *Leptobos* (Mammalia, Artiodactyla) du Villafranchien d'Europe Occidentale. *Documents des Laboratoires de Géologie de Lyon, 113*, 213 p.

Duvernois, M.-P. (1992). Mise au point sur le genre *Leptobos* (Mammalia, Artiodactyla, Bovidae); implications biostratigraphiques et phylogénétiques. *Geobios, 25*, 155–166.

Duvernois, M.-P., & Guérin, C. (1989). Les Bovidae (Mammalia, Artiodactyla) du Villafranchien supérieur d'Europe Occidentale. *Geobios, 22*, 339–379.

Falconer, H. (1868). Note of an undescribed species of bos in the Florence Museum (Bos etruscus). In C. Murchison (Ed.), *Paleontological memoirs and notes of the late Hugh Falconer with a biographical sketch of the author, vol. 2, Mastodon, elephant, Rhinoceros, ossiferous caves, primeval man and his contemporaries* (p. 481). Hardwicke.

Fernandez, P., & Crégut-Bonnoure, E. (2007). Les *Caprinae* (Rupicaprini, Ovibovini, Ovini et Caprini) de la séquence pléistocène de Kozarnika (Bulgarie du Nord): Morphométrie, biochronologie et implications phylogéniques. *Revue de Paleobiologie, 26*, 425–503.

Flerov, C. C. (1972). *Dryevnyeyshiye pryedstavityelya i istoriya roda Bison [The most ancient bisons and the history of genus Bison (in Russian)]*. Teriologia 1. Novosibirsk: Nauka.

Florindo, F., Marra, F., Angelucci, D. E., Biddittu, I., Bruni, L., Florindo, F., et al. (2021). Environmental evolution, faunal and human occupation since 2 Ma in the Anagni basin, central Italy. *Scientific Reports, 11*, 7056.

Garrido, G. (2006). Paleontología sistemática de grandes mamíferos del yacimiento del Villafranquiense superior de Fonelas P-1 (Cuenca de Guadix, Granada). Ph.D. thesis, Facultad de Ciencias Geológicas de la Universidad Complutense de Madrid.

Garrido, G. (2008a). La muestra más moderna y completa de *Gazellospira torticornis* (Bovidae, Artiodactyla, Mammalia) en el Plioceno superior terminal de Europa occidental (Fonelas P-1 (Cuenca de Guadix, Granada). In A. Arribas (Ed.), *Vertebrados del Plioceno superior terminal en el suroeste de Europa: Fonelas P-1 y el Proyecto Fonelas. Cuadernos del Museo Geominero* (Vol. 10, pp. 413–460).

Garrido, G. (2008b). Primera cita de *Leptobos etruscus* (Bovidae, Artiodactyla, Mammalia) en la Península Ibérica (Fonelas P-1 (Cuenca de Guadix, Granada). In A. Arribas (Ed.), *Vertebrados del Plioceno superior terminal en el suroeste de Europa: Fonelas P-1 y el Proyecto Fonelas. Cuadernos del Museo Geominero* (Vol. 10, pp. 489–516).

Gautier, F., & Heintz, E. (1974). Le gisement villafranchien de La Puebla de Valverde (Province de Teruel, Espagne). *Bulletin du Muséum national d'Histoire naturelle, 228*, 113–136.

Gentry, A. W. (1970). Revised classification from *Makapania broomi* WELLS AND COOKE (Bovidae, Mammalia). *Palaeontologia Africana, 3*, 63–67.

Gentry, A. W. (1978). Bovidae. In V. J. Maglio & H. B. S. Cooke (Eds.), *Evolution of African Mammals* (pp. 540–572). Cambridge: Harvard University Press.

Gentry, A. W. (1980). Fossil Bovidae (Mammalia) from Langebaanweg, South Africa. *Annals of the South African Museum, 79*, 213–337.

Gentry, A. W. (1992). The subfamilies and tribes of the family Bovidae. *Mammal Review, 22*, 1–32.

Gentry, A. W. (1996). A fossil *Budorcas* (Mammalia, Bovidae) from Africa. In K. M. Stewart & K. L. Seymour (Eds.), *Paleoecology and palaeoenvironments of Late Cenozoic Mammals* (pp. 571–587). Toronto: University of Toronto Press.

Gentry, A. W. (2001). An Ovibovine (Mammalia, Bovidae) from the Neogene of Stratzing, Austria. *Annales des Naturhistorische Museum in Wien, 102*, 189–199.

Geraads, D. (1986). Les Ruminants du Pléistocène d'Oubéidiyeh (Israël). In E. Tchernov and coll. (Eds), *Les Mammifères du Pléistocène inférieur de la vallée du Jourdain à Oubéidiyeh. Mémoires et travaux du Centre de Recherche français de Jérusalem* 5, 143–181.

Gibbard, P. L, & Head, M. J. (2009). IUGS ratification of the Quaternary System/Period and the Pleistocene Series/Epoch with a base at 2.58 Ma. *Quaternaire, 20*, 411–412. https://doi.org/10.4000/quaternaire.5289

Gill, T. (1872). Arrangement of the families of mammals with analytical tables. *Smithsonian Miscellaneous Collections, 11*, 1–98.

Gmelin, S. G. (1774). *Reise durch Russland zur Untersuchung der drey Natur-Reiche. Vol. 3*. St Petersbourg.

Gray, J. E. (1821). On the natural arrangement of vertebrose animals. *London Medical Repository and Record, 15*, 296–310.

Guérin, C. (1965). *Gallogoral* (nov. gen.) *meneghinii* (Rütimeyer, 1878), un Rupicapriné du Villafranchien d'Europe Occidentale. *Documents du Laboratoire de Géologie de la Faculté des Sciences de Lyon, 11*, 1–353.

Guérin, C. (1980). Les rhinocéros (Mammalia, Perissodactyla) du Miocène terminal au Pléistocène supérieur en Europe occidentale. Comparaisons avec les espèces actuelles. *Documents des Laboratoires de Géologie de Lyon, 79*, 1–1185.

Guérin, C. (1982). Première biozonation du Pléistocène européen, principal résultat biostratigraphique de l'étude des Rhinocerotidae (Mammalia, Perissodactyla) du Miocène terminal au Pléistocène supérieur d'Europe occidentale. *Geobios, 15*, 593–598.

Guérin, C. (1996). Limites et problèmes de chronologie. In C. Guérin & M. Patou-Mathis (Eds.), *Les grands mammifères plio-pléistocènes d'Europe* (pp. 1–11). Masson.

Han, D. F. (1987). Artiodactyla fossil from Liucheng *Gigantopithecus* cave in Guangxi. *Memoirs of the Institute of Vertebrate Paleontology and Paleoanthropology Academia Sinica, 18*, 135–208.

Heintz, E. (1964). Un nouveau Rupicaprinae fossile du gisement villafranchiern de Saint-Vallier (Drôme). *Comptes-Rendus sommaires de la Société Géologique de France, 9*, 373–374.

Heintz, E. (1966). La présence de *Gazellospira torticornis* AYMARD (Ruminant) dans le gisement villafranchien de Saint-Vallier (Drôme). Essai de répartition géographique et stratigraphique. *Bulletin de la Société Géologique de France, 8*, 25–30.

Heintz, E. (1970). Les Cervidés villafranchiens de France et d'Espagne. *Mémoires du Muséum national d'Histoire naturelle*, n.s., *22*, volumes 1 and 2, 1–317, 1–206.

Heintz, E., & Dubar, M. (1981). Place et signification des dépôts villafranchiens de Moustiers-Ségriès et faune de Mammifères de Cornillet (Alpes de Haute-Provence). *Bulletin du Muséum national d'Histoire naturelle, 4*, 363–397.

Heintz, E., Guérin, C., Martin, P., & Prat, F. (1974). Principaux gisements villafranchiens de France: Listes fauniques et biostratigraphie. *Mémoires du Bureau de Recherches Géologiques et Minières, 78*, 169–182.

Hermier, R., Merceron, G., & Kostopoulos, D. S. (2020). The emblematic Eurasian Villafranchian antelope *Gazellospira* (Mammalia: Bovidae): New insights from the Lower Pleistocene Dafnero fossil sites (Northern Greece). *Geobios, 61*, 11–29. https://doi.org/10.1016/j.geobios.2020.06.006

Jin, C. Z., Qin, D. G., Pan, W. S., et al. (2009a). A newly discovered *Gigantopithecus* fauna from Sanhe Cave, Chongzuo, Guangxi, South China. *Chinese Science Bulletin, 54*, 788–797.

Jin, C. Z., Pan, W. S., Zhang, Y. Q., et al. (2009b). The *Homo sapiens* Cave hominin site of Mulan Mountain, Jiangzhou District, Chongzuo, Guangxi with emphasis on its age. *Chinese Science Bulletin, 54*, 8483–8856. https://doi.org/10.1007/s11434-009-0641-1

Kalmykov, N. P. (2013). The most ancient finding of mountain sheep (Mammalia, Artiodactyla: *Ovis*) in the Holarctic (western Transbaikalia). *Doklady Biological Sciences, 448*, 25–27.

Kokotini, M., Kargopoulos, N., Iliopoulos, G., Roussiakis, S., Skandalos, P., Michailidis, D., et al. (2019). Karnezeika (Argolis, Peloponnese): preliminary data concerning a new Villafranchian locality of southern Greece. 15th International Congress of the Geological Society of Greece, Athens, 2019. *Bulletin of the Geological Society of Greece, Sp. Pub. 7*, 104–105.

Kostopoulos, D. (1996). The Plio-Pleistocene Artiodactyls of Macedonia (Northern Greece). Systematic, biochronology, biostratigraphy. Doctoral Thesis, University of Thessaloniki (in Greek).

Kostopoulos, D. (1997a). The Plio-Pleistocene artiodactyls from Macedonia, Greece: 2. The fossiliferous locality of Volakas, VOL (Volakas Basin, Drama, NE Greece). *Paleontologia i Evolució 30–31*, 83–92.

Kostopoulos, D. (1997b). The Plio-Pleistocene Artiodactyls (Vertebrata, Mammalia) of Macedonia. 1. The fossiliferous site "Apollonia - 1" Mygdonia Basin of Greece. *Geodiversitas, 19*, 845–875.

Kostopoulos, D. S., Maniakas, I., & Tsoukala, E. (2018). Early bison remains from Mygdonia Basin (Northern Greece). *Geodiversitas, 40*, 283–319.

Kostopoulos, D. (2022). The Fossil Record of Bovids (Mammalia: Artiodactyla: Ruminantia: Pecora: Bovidae) in Greece. In E. Vlachos (Ed.), *Fossil Vertebrates of Greece* (Vol. 2, Springer Nature Switzerland, pp. 113–203). https://doi.org/10.1007/978-3-030-68442-6_5

Kostopoulos, D., & Koufos, G. (1994). The Plio-Pleistocene Artiodactyls of Macedonia (Northern Greece) and their biostratigraphic significance. Preliminary report. *Comptes Rendus de l'Académie des Sciences, 318*, 1267–1272.

Kostopoulos, D., & Sen, S. (1999). Late Pliocene (Villafranchian) mammals from Sarikol Tepe, Ankara, Turkey. *Mitteilungen der Bayerischen Staatssammlung für Paläontologie und historische Geologie, 39*, 165–202.

Kostopoulos, D., Vassiliadou, K., & Koufos, G. (2002). The beginning of the Pleistocene in the Balkan area according to the mammals record: palaeozoogeographical approach. *Annales géologiques des Pays helléniques, 39*(A), 253–278.

Kretzoï, M. (1941). Präokkupierte Namen im Saügetiersystem. *Földtani Közlöny, 71*, 349–350.

Kretzoï, M. (1956). Die altpleistozänen Wirbeltierfaunen des Villányer Gebirges. *Geologica Hungarica, Series Palaeontologica, 27*, 1–264.

Kurtén, B. (1963). Villafranchian faunal evolution. *Societas Scientiarum Fennici, Commentationes Biologicae, 26*(3), 1–18.

Lacombat, F. (2004–2005). (Ed.). *Les grands mammifères fossiles du Velay. Les collections paléontologiques du Plio-Pléistocène du Musée Crozatier–Le Puy-en-Velay. Annales des amis du Musée Crozatier*, n° *13–14*, 1–206.

Lacombat, F., Mol, D., Borget, J.-N., & Magne, E. (2010). *Haute-Loire: des volcans, des mammouths et des hommes*. Jeanne-d'Arc.

Lehmann, U. (1957). Weitere Fossilfunde aus dem ältesten Pleistozän der Erpfinger Höhle (Schwäbische Alb.). *Mitteilungen Geologisches Staatsinstitut Hamburg, 26*, 60–99.

Linnaeus, C. (1758). *Systema naturæ per regna tria naturæ, secundum classes, ordines, genera, species, cum characteribus, differentiis, synonymis, locis. Editio decima reformata*. Holmiæ: Laurentii Salvii.

Maldonado-Garrido, E., Piñero, P., & Agustí, J. (2017). A catalogue of the vertebrate fossil record from the Guadix-Baza Basin (SE Spain). [Catálogo del registro fósil de vertebrados de la Cuenca de Guadix-Baza (SE de España)]. *Spanish Journal of Palaeontology, 32*(1), 207–236.

Martínez-Navarro, B., Belmaker, M., & Bar-Yosef, O. (2012). The Bovid assemblage (Bovidae, Mammalia) from the Early Pleistocene site of 'Ubeidiya, Israel: Biochronological and paleological implications for the fossil and lithic bearing strata. *Quaternary International, 267*, 78–97.

Masini, F. (1989). I bovini Villafranchiani dell'Italia. Doctoral Thesis, University of Florence.

Matsumoto, H. (1926). On a new fossil race of bighorn sheep from Shantung, China. *Science Reports of the Tohoku Imperial University, 10*, 38–41.

McKenna, M. C., & Bell, S. K. (1997). *Classification of mammals above the species level*. New York: Columbia University Press.

Nehring, A. (1891). Diluviale Reste von *Cuon, Ovis, Saiga, Ibex* und *Rupicapra* aus Mähren. *Neues Jahrbuch für Mineralogie, Geologie und Paläontologie, II*, 107–155.

Nomade, S., Pastre, J. F., Guillou, H., Faure, M., Guérin, C., Delson, E., et al. (2014). $^{40}Ar/^{39}Ar$ constraints on some French landmark Late Pliocene to Early Pleistocene large mammalian paleofaunas: Paleoenvironmental and paleoecological implications. *Quaternary Geochronology, 21*, 2–15.

Paquette, J.-L., Médard, E., Poidevin, J.-L., & Barbet, P. (2021). Precise dating of middle to late Villafranchian mammalian paleofaunae from the Upper Allier River valley (French Massif Central) using U-Pb geochronology on volcanic zircons. *Quaternary Geochronology, 65*(101198), 1–12.

Pastre, J.-F. (2024). The Senèze maar and paleontological site (Massif Central, France): their volcanic context and studies of related tephras. In E. Delson, M. Faure & C. Guérin (Eds), *Senèze: Life in Central France two million years ago. Paleontology, geochronology, stratigraphy and taphonomy* (pp. 83–98). Cham (Switzerland): Springer.

Pastre, J.-F., Debard, E., Nomade, S., Guillou, H., Faure, M., Guérin, C., & Delson, E. (2015). Nouvelles données géologiques et téphrochronologiques sur le gisement paléontologique du maar de Senèze (Pléistocène inférieur, Massif Central, France). *Quaternaire, 26*, 225–244.

Pilgrim, G. E., & Schaub, S. (1939). Die schraubenhörnige Antilope des europäischen Ober, pliocaens und ihre systematische Stellung. *Abhandlungen der Schweizerischen Palaeontologischen Gesellschaft, 62*, 1–30.

Pommerol, F. (1880). Le mouflon quaternaire. *Congrès de l'Association française pour l'avancement des Sciences*, 8° session. *Montpellier, 1879*, 600–609.

Popescu, A. (2004). Sur le gisement paléontologique Leu (Mammalia, Pléistocène). *Acta Paleontologica Romanae, 4*, 369–372.

Rădulesco, C. (2005). Artiodactyles du Pliocène et du Pléistocène inférieur de Roumanie. In E. Crégut-Bonnoure (Ed.), *Les ongulés holarctiques du Pliocène et du Pléistocène. Quaternaire*, hors-série *2*, 191–200.

Rădulesco, C., & Samson, P. (1962). Sur la présence de *Megalovis latifrons* dans le Villafranchien d'Olténie (Roumanie). *Vertebrata Palasiatica, 9*, 262–269.

Rădulesco, C., & Samson, P. (1990). The Plio-Pleistocene Mammalian Succession of the Olteţ Valley, Dacic Basin, Romania. *Quartärpaläontologie, 8*, 225–232.

Rădulescu, C., & Samson, P. (1995). Review of the villafranchian *s.s.* Faunas of Romania. *Il Quaternario, 8*, 377–382.

Rădulescu, C., Samson, P., Petculescu, A., & Ştiucă, E. (2003). Pliocene large mammals of Romania. *Coloquios de Paleontología, 53*, 549–558.

Roger, S., Coulon, C., Thouveny, N., Féraud, G., Van Velzen, A., Fauquette, S., et al. (2000). $^{40}Ar/^{39}Ar$ dating of a tephra layer in the Pliocene Senèze maar lacustrine sequence (French Massif Central): Constraint on the age of the Reunion-Matuyama transition and implication on paleoenvironmental archives. *Earth and Planetary Science Letters, 138*, 431–440.

Rütimeyer, L. (1867). Versuch einer natürlichen Geschichte des Rindes in seinen Beziehungen zu den Wiederkäunern im Allgemeinen. *Neue Denkschriften der Allgemeinen Schweizerischen Gesellschaft für die Gesammten Naturwissenschaften Gessellschaft, 22*, 1–102.

Rütimeyer, L. (1877-78). Die Rinder der Tertiär Epoche. In: *Abhandlungen der Schweizerischen Paläontologischen Gesellschaft, 4-5*, 1-208.

Samson, P., & Rădulesco, C. (1966). Sur la présence des Girafidés dans le Villafranchien supérieur de Roumanie. *Neues Jahrbuch für Geologie und Paläontologie, Monatshefte, 1966*, 588–594.

Schaub, S. (1922). *Nemorhoedus philisi* nov. spec. ein fossiler Goral aus dem Oberpliocaen der Auvergne. *Eclogae Geologicae Helvetiae, 16*, 558–563.

Schaub, S. (1923). Neue und wenig bekannte Cavicornier von Senèze. *Eclogae Geologicae Helvetiae, 18*, 281–295.

Schaub, S. (1928). Die Antilopen des toskanischen Oberpliocaens. *Eclogae Geologicae Helvetiae, 21*, 260–266.

Schaub, S. (1932). Die Ruminantier des ungarischen Praeglacials. *Eclogae Geologicae Helvetiae, 25*, 319–330.

Schaub, S. (1937). Ein neuer Cavicornier aus dem Oberpliocaen von Honan. *Bulletin of the Geological Institutions of the University of Uppsala, 27*, 25–31.

Schaub, S. (1943). Die Oberpliocaene Säugetierfauna von Senèze (Haute-Loire) und ihre veibreitungsgeschichtliche Stellung. *Eclogae Geologicae Helvetiae, 36*, 270–289.

Sher, A. V. (1997). An early Quaternary bison population from Untermaßfeld: *Bison menneri* sp. nov. In R. D. Kahlke (Ed.). *Das Pleistozän von Untermaßfeld bei Meiningen (Thuringen). Monographien des Römisch-Germanisches Zentralmuseum, 40*, 101–180.

Sianis, P. D., Kostopoulos, D. S., Roussiakis, S., Athanassiou, A., & Iliopoulos, G. (2022). The bovids (Artiodactyla) from the Lower Pleistocene locality of Karnezeika (Southern Greece). *Historical biology, 35*, 661–677.

Sickenberg, O. (1968). Die pleistozän knochenbrekzien von Vólax (Griech. Mazedonien). *Geologische Jahrbuch, 85*, 33–54.

Sinusía, C., Pueya, E. L., Azanza, B., & Pocoví, A. (2004). Datación magnetoestratigráfica del yacimiento paleontológico de la Puebla de Valverde (Teruel). *Geotemas, 6*(4), 339–342.

Smith, C. H. (1827). Genus Bison. In E. Griffith (Ed.), *The animal kingdom arranged in conformity with its organization. Class Mammalia, Volume 5* (p. 373). London: Geo. B. Whittaker.

Sorbelli, L., Cherin, M., Kostopoulos, D., Sardella, R., Mecozzi, B., Plotnikov, V., Prat-Vericat, M., Azzarà, B., Bartolini-Lucenti, S., & Madurell Malapeira, J. (2022). Earliest bison dispersal in Western Palearctic: Insights from the *Eobison* record from Pietrafitta (Early Pleistocene, central Italy). *Quaternary Science Reviews, 301*, 1–2. https://doi.org/10.1016/j.quascirev.2022.107923

Soto, E., & Morales, J. (1985). Grandes mammíferos del yacimiento villafranquiense de Casablanca 1, Almenara (Castellón). *Estudios Geológicos, 41*, 243–249.

Spassov, N. (2005). Brief review of the Pliocene ungulate fauna of Bulgaria. In E. Crégut-Bonnoure (Ed.), *Les ongulés holarctiques du Pliocène et du Pléistocène. Quaternaire*, h.s., *2*, 201–212.

Spassov, N., & Crégut-Bonnoure, E. (1999). Premières données sur les Antilopinae et les Caprinae villafranchiens de Bulgarie. Richesse faunique et signification biozoogéographique. *Comptes Rendus de l'Académie des Sciences, 328*, 493–498.

Stehlin, H. G. (1904). Une faune à Hipparion à Perrier. *Bulletin de la Société Géologique de France, 4*, 432–444.

Stehlin, H. G. (1923). Die oberpliocaene Fauna von Senèze (Haute-Loire). *Eclogae geologicae Helvetiae, 18*, 268–281.

Symeonidis, N. (1992). Fossil mammals of Lower Pleistocene (Villafranchian age) of the Sesklo basin (Volos). *Annales Géologiques des Pays hélléniques, 35*, 1–21.

Teilhard de Chardin, P., & Piveteau, J. (1930). Les mammifères fossiles de Nihowan (Chine). *Annales de Paléontologie, 9*, 1–134.

Teilhard de Chardin, P., & Trassaert, M. (1938). Cavicornia of southeastern Shansi. *Paleontologia sinica, 6*, 1–101.

Terhune, C. E., Curran, S., Croitor, R., Drăguşin, V., Gaudin, T., Petculescu, A., et al. (2020). Early Pleistocene fauna of the Olteţ River Valley of Romania: Biochronological and biogeographic implications. *Quaternary International, 550*, 14–33.

Titov, V.V. (2008). Late Pliocene large mammals from Northeastern Sea of Azov Region (In Russian). *Southern Scientific Centre RAS*, Rostov-on-Don, 1–262.

Tong, H.-W., Zhang, B., Cheng, X., & Wang, X.-M. (2022). New fossils of small and medium-sized bovids from the Early Pleistocene Site of Shanshenmiaozui in Nihewan Basin, North China. *Vertebrata PalAsiatica, 60*, 134–168.

Ufrecht, W. (2008). Evaluating landscape development and karstification of the Central Schwäbische Alb (Southwest Germany) by fossil record of karst fillings. *Zeitschrift Für Geomorphologie (neue Folge), 52*(4), 417–436. https://doi.org/10.1127/0372-8854/2008/0052-0417

Villalta, J. F. de, & Crusafont Pairo, M. (1956). Un nuevo Ovicaprino en la fauna villafranquiense de Villaroya (Logroño). In: *Actes du IV Congrès international du Quaternaire, Rome-Pise, Août-Septembre 1953* (pp. 426-432)

Viret, J. (1961). Artiodactyla. In J. Piveteau (Ed.), *Traité de Paléontologie, 6* (pp. 887–1084). Paris: Masson.

Vislobokova, I. A., & Agadjanian, A. K. (2015). New data on large mammals from the Trlica Pleistocene fauna (Montenegro, the Central Balkans). *Paleontological Journal, 49*(6), 651–667.

Chapter 15
Mammuthus meridionalis of the Late Villafranchian Site of Senèze (Domeyrat, Haute-Loire, Central France)

Bernard Ménouret and Claude Guérin

Abstract The fossiliferous site of Senèze has yielded 36 remains of the elephantid *Mammuthus meridionalis meridionalis*, all but one found during the old excavations but never studied in detail nor published. The best preserved specimens are a cranium, two mandibles with their teeth, two isolated teeth and some limb bones. The subspecies is known in Western Europe from the middle (biozone MNQ 17) to the late (biozone MNQ 18) Villafranchian, and its occurrence helps to date the site to the late Villafranchian. The presence of the Southern mammoth is a clear indication of a wooded savanna under a temperate or warm climate.

Résumé Le site de Senèze a fourni 36 restes de l'éléphantidé *Mammuthus meridionalis meridionalis*, tous, sauf un, proviennent d'anciennes fouilles mais n'avaient jamais été étudiés en détails ni publiés. Les pièces les mieux conservées sont un crâne, deux mandibules avec leurs dents, deux dents isolées et des os des membres. La sous-espèce est connue en Europe de l'Ouest du Villafranchien moyen (biozone MNQ 17) au Villafranchien supérieur (biozone MNQ 18), sa présence contribue à dater le site du Villafranchien supérieur et montre une savane arborée sous un climat tempéré à chaud.

Keywords Anatomy • Biochronology • Biometry • Elephantidae • Paleoenvironment.

Mots-clés Anatomie • Biochronologie • Biométrie • Elephantidae • Paléoenvironnement

Claude Guérin—Deceased 2016

B. Ménouret (✉)
3 rue de l'Amouroy, 13680 Lançon-de-Provence, France
e-mail: menouret@gmail.com

C. Guérin
Université Claude Bernard-Lyon 1, Villeurbanne, France

Introduction

The late Villafranchian site of Senèze (Haute-Loire) is known since the end of the nineteenth century. The first mammalian fossil found in the locality was a complete proboscidean skeleton discovered while plowing a cultivated field in 1892 and identified by Boule (1892) as *Elephas meridionalis*; only three teeth were collected and sent to the Muséum national d'Histoire naturelle in Paris: two M3s and a lower molar figured in Gaudry (1896, Fig. 190, p. 189, not seen by us). Some of the limb bones were salvaged in 1909 by Depéret (Depéret & Mayet 1912).

The most important part of the material previously found in Senèze was excavated between 1911 and 1939 by Pierre Philis, a farmer who owned most of the site. He gathered a huge quantity of fossils which were sold to the Naturhistorisches Museum of Basel, the Faculté des Sciences de Lyon and the Muséum national d'Histoire naturelle in Paris, and sometimes to private collections (Roux-Vernière and De Brun). For more information about fieldwork at the site before the Second World War, see Faure et al. (2022). Later, some small scientific excavations were conducted but yielded no proboscidean remains. The 2000–2006 excavations of the Franco-American team co-directed by Delson, Faure and Guérin was the first real paleontological research ever made on a large scale at Senèze. During this excavation work only one proboscidean fossil, an incomplete tusk, was found. After the first indication by Boule (1892), several authors pointed out the presence of an elephant species in Senèze. Depéret and Mayet (1912), Stehlin (1923), and Schaub (1943) published the first faunal lists of the site, including short enumerations of the remains of *Elephas meridionalis* housed in the Faculté des Sciences de Lyon and the Naturhistorisches Museum Basel, but those fossils were never described in detail.

Depéret and Mayet (1912) enumerated four different fossiliferous points, including the one excavated by Boule,

E. Delson et al. (eds.), *Senèze: Life in Central France Around Two Million Years Ago*,
Vertebrate Paleobiology and Paleoanthropology, https://doi.org/10.1007/978-3-031-64415-3_15

where *Elephas meridionalis* remains were found at Senèze. In a general work on Pliocene elephants, Depéret et al. (1923) very briefly studied and figured some remains of Senèze under the name *Elephas planifrons*. Viret (1954) noted that the *Elephas meridionalis* from Senèze is very similar to that of Saint-Vallier (Drôme, France, middle Villafranchian). Lastly Heintz et al. (1974) provided a critical revision of the faunal lists of the main Villafranchian deposits of France and confirmed the presence of what they termed *Archidiskodon meridionalis* from Senèze and five other classical French sites. The purpose of the present work is to describe all the proboscidean material found at Senèze.

The following abbreviations are used: ant. = anterior; APD = antero-posterior diameter; dia. = diaphyse; dist. = distal; h = height; hy = hypsodonty index (h/w); L = length; M = upper molar; m = lower molar; Max = maximum; Mc = metacarpal; Mt = metatarsal; prox. = proximal; RI = robustness index for metapodial (TD dia. /L) × 100; TD = transverse diameter; w = width. The hypsodonty index of a tooth is the ratio height of the tooth by its width. We used the biochronological scheme based on biozones (MNQ, Mammal Neogene Quaternary), first developed by Mein (1975) and later expanded for the Pleistocene units by Guérin (1980, 1982, 1990, 2007).

Systematic Paleontology

Order PROBOSCIDEA Illiger, 1881
Family ELEPHANTIDAE Gray, 1821
Subfamily ELEPHANTINAE Gray, 1821
Genus *Mammuthus* Brookes, 1828

Type species: *Mammuthus primigenius* (Blumenbach, 1799).

The genus is known in Africa, Eurasia and North America. The oldest record of *Mammuthus* is about 6.2–5.6 Ma (Sanders 2007) in the Lukeino Fm. (Tugen Hills, Kenya) with Late Miocene to Early Pliocene distribution of the species *M. subplanifrons* known in South and East Africa. The genus appeared in Eurasia at about 3.5–3.0 Ma in Cernatesti and Tulucesti (Romania) with the species *M. rumanus* (Konidaris et al. 2020; Kostopoulos & Koulidou 2015; Lister et al. 2005; Markov 2012). The genus became extinct during the Late Pleistocene around 10 ka in Siberia (Stuart and al. 2002; Mol et al. 2003; Kuzmin 2010; Nikolskiy et al. 2011), but survived on some islands; Santa Rosa Island (California), with the pygmy mammoth *M. exilis* (Semprebon et al. 2016); Pribilof Islands (Alaska), where *M. primigenius* survived until ca. 5,600 BP (Veltre et al. 2008); and Wrangel Island (Russia), in the Arctic Sea, where it became extinct about 3,900 BP (Tikhonov et al. 2003; Vartanyan et al. 1993).

Several hypotheses have been proposed for the systematics of the Eurasian mammoths. For some authors (Dubrovo & Jakubowski 1988; Reggiani & Sala 1994; Garutt 1998a,b; Kosintsev & Vorobiev 2001; Titov 2001; Maschenko et al. 2011; Baygusheva & Titov 2012; Golovachev & Titov 2018 and others), the more primitive species *M. rumanus* and *M. meridionalis* belong to the genus *Archidiskodon* Pohlig, 1885. An alternative classification, more widely accepted, sees a succession of chronospecies: *M. rumanus*, *M. meridionalis*, *M. trogontherii*, and *M. primigenius*. Labe and Guérin (2005) moved the taxon *Elephas intermedius* into the genus *Mammuthus* as chronologically intermediate between *M. trogontherii* and *M. primigenius* and considered *M. chosaricus* as a junior synonym of *M. intermedius*. We follow this classification and consider *Archidiskodon* also a junior synonym.

The principal anatomical trends within this genus are, change in body size, increasing in height and shortening of the cranium, shortening of the mandible, increasing crown height of molars, thinning of their enamel and increasing in the number of lamellae. From the beginning to the end of the *Mammuthus* lineage, for the M3, the mean enamel thickness changes from 5 mm to 1.5 mm, the number of the laminae from 8–10 to 20–28, the lamellar frequency from 3 up to 9–10 and the hypsodonty index from 0.9 up to 3 (Ferretti 1999; Konidaris et al. 2020; Labe & Guérin 2005; Lister et al 2005; Palombo & Ferretti 2005; Wei et al. 2010).

Species: *Mammuthus meridionalis* (Nesti, 1825).

Elephas meridionalis was defined from the Upper Valdarno Formation (Italy, ca. 2.0–1.75 Ma) probably in the Montevarchi Group.

Mammuthus meridionalis is known in France from the middle Villafranchian (MNQ 17 zone, with Saint-Vallier as its oldest site, according to Guérin 2004) up to the early Middle Pleistocene (MNQ 21 zone). It is present during the Villafranchian (Boeuf 1983; Heintz et al. 1974) at Chagny (Sâone-et-Loire, MNQ 17), Saint-Vidal (Haute-Loire, MNQ 17), Saint-Vallier (Drôme, MNQ 17), Le Coupet (Haute-Loire, MNQ 17), Chilhac (Haute-Loire, MNQ 17), Senèze (Haute-Loire, MNQ 18) and Blassac-La-Girondie (Haute-Loire, MNQ 19); and in the early Middle Pleistocene (Aouadi 2001; Guérin et al. 2003) at Saint-Prest (Eure-et-Loir, MNQ 20), Sainzelles (Haute-Loire, MNQ 20), Ceyssaguet (Haute-Loire, MNQ 20), Soleilhac (Haute-Loire, MNQ 20), Malbattu (Puy-de-Dôme, MNQ 20) and Durfort (Gard, MNQ 20). Some other sites from Haute-Loire also yielded *M. meridionalis*; La Rochelambert, Pranlary, La Malouteyre and Vals Le Crozas (middle

Villafranchian), Communac and Saint-Privat-d'Allier (MNQ 20). (Laurent 1907; Ferretti & Palombo 2005; Ros-Montoya 2010).

Its spatial distribution extended from Europe to Siberia, and it perhaps migrated into North America about 1.5 Ma, where it could have been the earliest species of *Mammuthus* (McDaniel and Jefferson 2003; Enk et al. 2016; Lucas et al. 2017). Lister and Sher (2015) and Lister (2017) disagree with this proposal: for these authors *M. meridionalis* never entered North America, it was *M. trogontherii* which reached North America. New findings are necessary to support one of these two hypothesis.

Mammuthus meridionalis is a tall species, the shoulder height of adult males varying from 3.5 to 4.2 m. The estimated male weight ranged between 7 and 10 tons, or more for the largest individuals (Christiansen 2004; Larramendi 2016). The skull has a rather sharp occipital crest, inflated parietals and a concave forehead, the mandible is long and rather low, with the symphysis elongated as a "beak" (Lister 1996). The tusks are very stocky with a small upward and inward curvature. The M3 is thick and moderately hypsodont (hy = 0.96–1.76), with a small number of laminae (10–15), a low lamellar frequency (3.75–6.5) and weakly folded thick enamel (2.7–5 mm) (Baygusheva & Titov 2012; Konidaris et al. 2020; Palombo & Ferretti 2005).

Material

The Senèze fossils of *Mammuthus* are stored at the Naturhistorisches Museum Basel (NMB, numbers beginning with Se), in the collections of the geology laboratory of Université Claude Bernard - Lyon 1 (UCBL, numbers beginning with FSL) and at the Muséum national d'Histoire naturelle in Paris (MNHN).

- 1 cranium and mandible NMB Se 547.
- 1 mandible FSL 211130.
- 2 M3 of the same individual MNHN.F.SEN 98 & SEN 99 (1892–15).
- 1 tusk NMB Se 193.
- 1 partial tusk (found during the new excavations) FSL 456286.
- 1 right humerus (distal part) FSL 210990.
- 2 right scaphoids FSL 211125, NMB Se 1770.
- 2 right lunars FSL 210962 a, FSL 211123.
- 1 left trapezium Se 194.
- 3 left trapezoids FSL 211121, NMB Se 194, NMB Se 1494.
- 2 left magnums FSL 211126, NMB Se 1770.
- 1 right unciform NMB Se 194.
- 1 right Mc III (distal fragment) NMB Se 194.
- 1 left Mc V NMB Se 1770.
- 1 right Mc V (fragment) NMB Se 194.
- 1 patella NMB Se 1424.
- 1 left cuboid Se 1699.
- 1 left lateral (first) cuneiform FSL 210961 a.
- 1 left Mt III FSL 210962 b.
- 1 left Mt V FSL 210962 c.
- 1 phalanx I FSL 211111.
- 1 phalanx I digit IV, right foot NMB Se 767.
- 1 phalanx fragment digit III right foot NMB Se 1425.
- 1 phalanx I (fragment) of the digit V of a left foot NMB Se 1770.
- 1 phalanx I digit IV right NMB Se 1770.
- 1 phalanx II digit IV, left foot NMB Se 1427.
- 3 distal phalanges FSL 210965, FSL 211120, NMB Se 1425.
- 1 sesamoid NMB Se 1777.

Description

Cranium (Figs. 15.1, 15.2, 15.3).

The skull found in 1913 was prepared and partially reconstructed by K. Rothpletz and A. Schenker, preparators of the Basel Museum, from thousands of fragments; the restoration took more than one year (1946–1947). The cranium and mandible are mounted together and cannot be separated so we were unable to observe the ventral part of the cranium and the tooth rows. The specimen was figured by Schaub (1948, p. 93 fig. 1-2) in lateral and ventral views, in lateral view by Maccagno (1962, p. 27, fig. 12c and in lateral and frontal views by Azzaroli (1966, p. 543 fig. 1, p. 545 fig. 1a), but the drawing of the lateral face is not very faithful.

The cranium is compressed antero-posteriorly and greatly elevated with long recurved tusks. The distance between the top of the cranium (nuchal crest) and the anterior margin of the premaxillae is 1.40 m. The transverse distance between the posterior parts of the last upper molars is 118 mm.

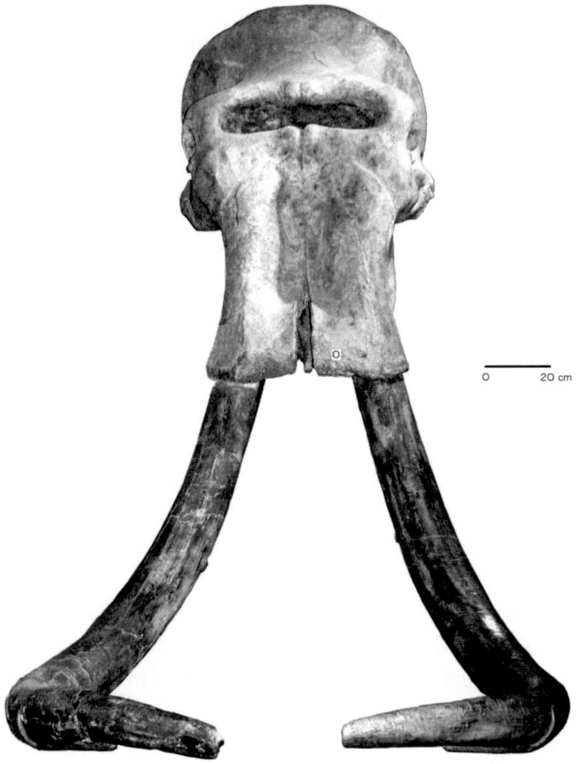

Fig. 15.1 Cranium NMB Se 547 of *M. meridionalis* from Senèze, anterior view

In anterior view (Fig. 15.1) the nasal cavity is wide (670 mm) and low (103 mm in the middle) with slightly rounded edges. On the lower edge the suture between the premaxillae is obvious; on both sides of the suture we observed strong tuberosities, on the upper edge it is the same with stronger tuberosities. The level of the upper border is higher than the upper edge of the orbit; the lower border is at the level of the processus lacrymalis. The fossa incisiva is deep. The lateral border of the rostrum is weakly concave. The tusks alveoli are almost parallel, faintly divergent in their distal part. The nasal deviated in its ventral part during the fossilization.

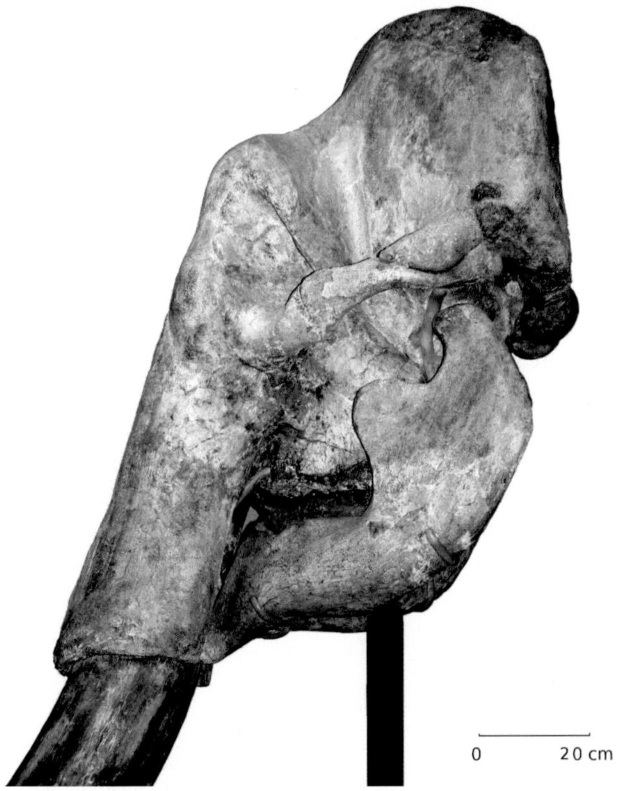

Fig. 15.2 Cranium NMB Se 547 of *M. meridionalis* from Senèze, left lateral view

In lateral view (Fig. 15.2), the top of the cranium is above the front of the external auditory meatus. The outline of the parietal is round in its upper part and concave in its frontal part. The premaxillary rostrum is strongly inclined downward: its angle with the vertical is about 15°. Its dorsal border is straight.

The eye socket, more or less triangular, is elongated vertically (237 mm), its anterior border is narrow in the middle and enlarged both dorsally and ventrally. The posterior border has a moderate thickness and becomes quickly thinner. The temporal fossa is narrow and deep.

The jugal is higher anteriorly than posteriorly. The ventral edge is angled downward in front, then slightly convex and parallel to the occlusal plane. The dorsal edge is concave at the insertion with the processus zygomaticus of the squamosal, which is strong, inclined downward and frontally offset. The processus zygomaticus of the maxillae is big and twisted up, under it the maxillae is quite deep. The zygomatic arch is 457 mm long.

The lacrimal apophysis is nearly in the middle of the anterior border of the orbit. The external auditory meatus is closed ventrally by the processus posttympanicus of the squamosal, it is situated at the same level as the processus postorbitalis of the frontal bone and is half-moon shaped. The occipital face is flat and vertical.

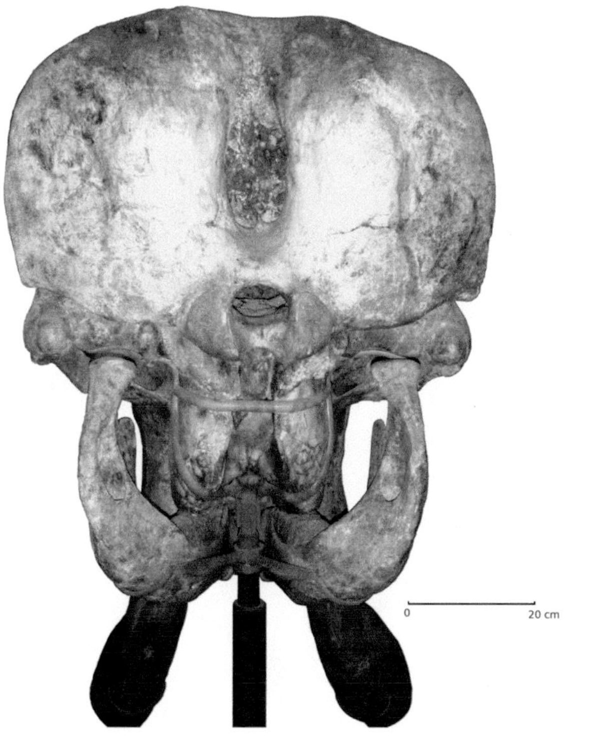

Fig. 15.3 Cranium NMB Se 547 of *M. meridionalis* from Senèze, posterior view

In posterior view (Fig. 15.3), the occipital area is rounded and wider than high, its height, excluding the foramen magnum, is 520 mm. The largest width (868 mm) is at the level of the external auditory meatus. On the supraoccipital, the median fossa is very deep, drop shaped and elongated dorso-ventrally; in the middle of its ventral border we can see the beginning of a small vertical crest. In its upper extension we observe a slight depression at the cranial vertex. The foramen magnum has an ellipsoid shape and is wider (101 mm) than high (82 mm). The occipital condyles are semi-elliptic in shape, higher than broad, the TD of the occipital processes measured between their external borders is 228 mm.

The holotype skull of *M. meridionalis* from Upper Valdarno illustrated by Palombo and Ferretti (2005, p. 110, fig. 2A) has a concave forehead, wide parietals, flat and vertical occipital, and its apex is above the external auditory meatus as on the cranium from Senèze.

Another partial skull from Upper Valdarno, also illustrated by Palombo and Ferretti (2005, p. 110, fig. 2B-C), has a wide nasal cavity with upper and lower edges consisting of two arcuate parts separated by strong tuberosities, while on the Senèze skull the outline is more rectangular. A third skull from Upper Valdarno (Weithofer 1890, T. I, fig. 1) shows a nasal opening with an upper border in two arcuate parts separated by strong tuberosities, the lateral borders making an acute angle with the lower one. The skeleton of *M. meridionalis* from Madonna della Strada (Scoppito, Italy, 1.5–1.4 Ma) studied by Maccagno (1962) has been restored (Rossi et al. 2017); the restoration changed the shape of the nuchal fossa, and thus it is now wider and shallower than in the previous restoration and it looks more like that of the Senèze skull. The skull differs from that of Senèze by a wider temporal fossa, an external auditory meatus not half-moon shaped but rounded, a nasal cavity which looks like those from Upper Valdarno, a foramen magnum higher than wide and occipital processes wider than high: the TD of the occipital processes measured between their external borders is 280 mm (Maccagno 1962). The incomplete skull of *M. meridionalis* from Oriolo (Italy, ca. 1.0 Ma) shows long and parallel tusk alveoli faintly divergent in their distal part with a deep fossa incisiva, the premaxillae are not completely fused distally and the preserved part of the frontal is concave in lateral view (Ferretti 1999). The height of the maxillae under the processus zygomaticus seems to be less than those of the Senèze cranium (Ferretti 1999, p. 505, Fig. 15.3a).

One skull of male *M. trogontherii* from Kagal'nik sand pit (Sea of Azov Region, Russia, early Middle Pleistocene) is almost complete. The forehead is concave, the premaxillae are long and nearly vertical with a weakly convex dorsal border, lateral sides concave and there is a wide fossa incisiva. The nasal foramen is low and wide (ca. 700 mm), its upper border is higher than the upper edge of the orbit and its lower border at the level of the mid orbit. It differs from the Senèze cranium by pointed lateral edges making an acute angle with the lower edge. The height of the maxillae under the processus zygomaticus is less than on the cranium of Senèze, the occipital is wider (1056 mm) and the foramen magnum (wxh = 100 × 72 mm) has a more elliptical shape (Baygusheva et al. 2012). Another partial skull of male *M. trogontherii* from Kostolac (Danube valley, Serbia, ca. 0.8 Ma) has the same morphology for the nasal foramen as that from Kagal'nik but its estimated width is less (600 mm), the premaxillae are long, and the estimated width of the cranium around the occiput is 1080 mm (Lister et al. 2012). A skull of a male *M. trogontherii* from Ust'-Tarka (Novosibirsk Region, Russia, ca. 0.4 Ma) has nearly the same distance between the top of the cranium and the anterior part of the premaxillae (I450 mm) as that of Senèze. In lateral view the outline of the forehead is flat, the external auditory meatus is round and the apex of the skull is located forward the front of the auditory meatus. The occipital is weakly convex. The morphology of the nasal cavity is the same as those of Kostolac and Kagal'nik with a width of 630 mm and a lower edge at the level of the orbital midline (Shpansky et al. 2015, p. 309, pl. 10, fig. 1-2). A partial skull of *M.* cf. *trogontherii* from Monte Spaccato (Italy, ca. 0.7 Ma) shows long, downward pointed premaxillae with a wide fossa incisiva and deep maxilla (Palombo & Ferretti 2005).

Mandible (Figs. 15.4, 15.5).

A photograph of the Basel mandible (taken before its restoration and thus without its ramus) is given in Depéret et al. (1923, pl. II, fig. 6), together with a drawing of the same piece (text-fig. 6 bis, p. 109); this drawing is reproduced in Lister (1996, p. 207, text-Fig. 19.6) and Konidaris et al. (2020, p. 84, text-fig. 9). The dorsal and ventral borders of the corpus are parallel under the tooth row, then in their anterior part they bend down at an obtuse angle of about 120° for the dorsal and 150° for the ventral one, so that the height of the rostrum decreases rapidly. The rostrum is elongated. On its upper border, on each side, there is a crest extending from the fore part of the tooth up to the fore part of the rostrum. The mandible height taken under the anterior part of the m3 is 175 mm and the TD is 115 mm. The length from the extremity of the rostrum to the mandibular angle is more than 800 mm. A mental foramen is located in the upper part of the corpus, at the beginning of the anterior part. The ramus is broad, its height from the mandibular angle to the coronoid process is 330 mm. The anterior border of the ramus is sinuous, first slightly concave and then convex. The posterior border is slightly convex after the mandibular angle which measures about 110°. The coronoid process is wide. On the medial side of the ramus the foramina mandibulae open at the level of the mandibular angle. In posterior view the two rami do not diverge, the span at the exterior part of the articular condyles is 610 mm. The length of the symphysis is 142 mm, and the external TD at the back of the symphysis is 270 mm.

Another mandible from Senèze, with broken rami and two teeth, belonged to a private collection when illustrated in Depéret et al. (1923 pl. II fig. 1-3, and text fig. 13 III p. 81) but is now held by Université Claude Bernard-Lyon 1 (Figs. 15.4, 15.5). The minimal distance between the right and left teeth is 77 mm at the level of lamina 3, the maximal distance between the posterior ends of the teeth is 206 mm. The TD of the corpus is 96 mm beneath the anterior part of the molar and 145 mm beneath the middle of the molar, the heights at the same levels are respectively 215 and 177 mm. There are two mental foramina, one at the level of the front of the molar, another lower and in a more rostral position. The level of the posterior part of the symphysis is that of the lamina 3. On each side of the mandible, a descending crest joins the front part of the tooth with the base of the symphyseal beak; its length is about 240 mm. The beak, oriented downward, is 253 mm long and its TD at its anterior extremity is 55 mm.

The mandible of the *M. meridionalis* of Chagny (Sâone-et-Loire, France, FSL 211776), dated to biozone MNQ 17 (Guérin et al. 2004), is very similar, but it possesses a shorter and slimmer beak (length 197.5 mm, TD 40 mm), stronger descending anterior crests and a single mental foramen. The minimal distance between the right and left teeth has the same value as that of Senèze FSL 211130, i.e., 77 mm, but at the level of lamina 4 instead of lamina 3; the maximal distance between the posterior ends of the teeth

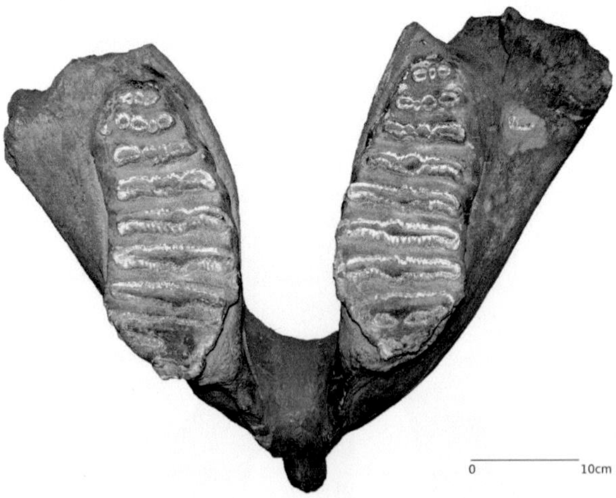

Fig. 15.4 Mandible FSL 211130 of *M. meridionalis* from Senèze, occlusal view

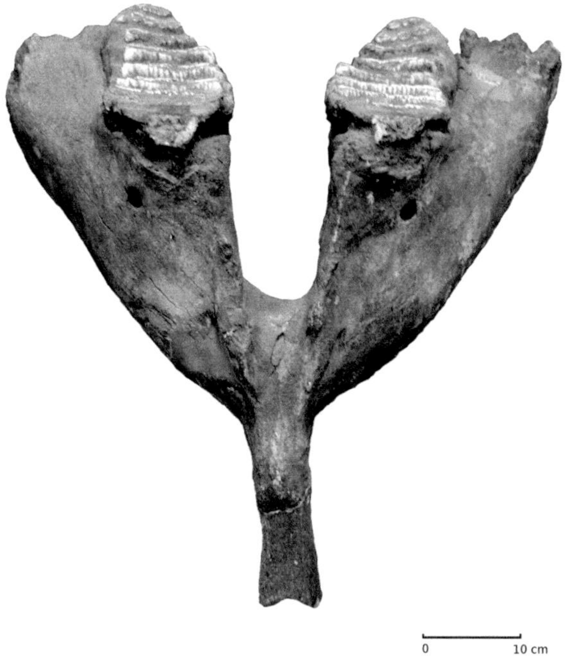

Fig. 15.5 Mandible FSL 211130 of *M. meridionalis* from Senèze, anterior view

is 211 mm, nearly the same. The TD of the corpus under the anterior part of the molar is 109 mm and 173 mm at the middle of the same tooth, the heights at the same level are 191 and 165 mm respectively. The ramus is inclined backward.

Three mandibles of *M. meridionalis* from Upper Valdarno [one termed *Elephas lyrodon,* a young female with m1-m2 in use in Weithofer (1890, pl. VI, fig. 1-2), one of an old male in Azzaroli (1966, p. 545, fig. 7) and one illustrated in Ferretti (1999, p. 507, fig. 4d) have an elongated rostrum with a dorsal border making an obtuse angle with the occlusal plane and with the ramus inclined backward like those from Senèze. One from Apollonia-1 (Northern Greece, 1.6–1.3 Ma) has the same morphology (Konidaris et al. 2020). Two mandibles of *M. meridionalis* from Incarcal-1 and Venta Micena (Spain, 1.6–1.4 Ma) also have a long and narrow symphyseal beak, that of Incarcal is better preserved, the symphyseal rostrum makes an obtuse angle with the occlusal plane, the dorsal and ventral borders of the corpus are parallel under the tooth as on those from Senèze but the ramus is not backward shifted (Mazo et al. 2003; Ros-Montoya et al. 2012). The mandible of Scoppito differs from those of Senèze in having a nearly right angle between the tooth row and the dorsal border of the symphysis and the height of the corpus at the front of the molar (240 mm) is much greater than that at the back (151 mm). The broad ramus is inclined forward, its anterior border, straight on its lower part, makes an acute angle with the occlusal plane (Maccagno 1962, Pl. III, fig. 3). The minimal distance between the right and left teeth is 95 mm and at the posterior ends of the teeth is 203 mm (Maccagno 1962). That of *M. meridionalis* from St-Prest differs from those of Senèze by dorsal and ventral borders of the corpus not parallel but as in the Senèze mandibles the angle between the tooth row and the dorsal border of the symphysis is obtuse (Coppens & Beden 1982, p. 292, fig. 1). A mandible of *M. meridionalis* from Oriolo, probably a male, has an obtuse angle between the tooth row plane and the dorsal border of the rostrum, but the anterior margin of the ramus is not concave in its lower part (Ferretti 1999). That of *M. meridionalis* from Huéscar-1 (Spain, ca. 0.9 Ma) has, like in Senèze, a sinuous anterior border of the ramus, the caudal border of the ramus gently inclined posteriorly, an obtuse angle between the occlusal plane and the dorsal border of the symphysis, but the dorsal and ventral borders of the corpus are divergent, the heights anteriorly and posteriorly to the tooth alveoli are respectively 201.5 mm and 146 mm (Ros-Montoya et al. 2018).

Eight mandibles of *M. trogontherii* from West Runton (United Kingdom), Loussiká (Southern Greece), the Sea of Azov Region (Russia), Drmno (Serbia), Mavrodin (Romania), Zhalainuoer III (Inner Mongolia, China) and Ust'Tarka (Novosibirsk Region, Russia) were recently studied; all except one of the two from the Sea of Azov Region, are of male individuals (Athanassiou 2012; Baygusheva et al. 2012; Larramendi 2014; Lister & Stuart 2010; Lister et al. 2012; Shpansky et al. 2015; Vasile et al. 2012). They differ from those of Senèze by a higher corpus, the height at the anterior border of the tooth is 235–271 mm, a broader corpus (TD max. 192–205 mm), an almost right angle between the dorsal border of the symphysis and the tooth row, the ramus inclined forward, its anterior border making an acute angle with the occlusal teeth plane and the symphysal rostrum shorter when it's present.

Golovachev and Titov (2018) studied two mandibles of *Mammuthus intermedius* from the lower Volga Region (Russia), both differ from those of Senèze by a lower height of the corpus under the tooth (138–137 mm), a lower TD (TD max 134.5 mm), a more rounded mandibular angle, and a weak beak at the end of the symphysis (Golovachev & Titov 2018, p. 272, fig. 1). Another one from La Fage (Corrèze, France; MNQ 24), with a small beak, has a corpus height of 166 mm in front of the tooth and 136 mm at its back (Beden & Guérin 1975). The last three mandibles differ from those of *M. trogontherii* by a more gracile corpus.

Petrova et al. (2017) give measurements of eleven mandibles of *M. primigenius* from Russia and one from Condover (England); all those mandibles have a lower corpus, 112–155 mm at posterior edge of alveolus (mean 135 mm) and are narrower than those of Senèze. The range of the maximal width between the outermost points of the articular processes is 411–494 mm. Narrow symphyseal beaks, short or long, are present on male or female mandibles (Álvarez-Lao & Méndez 2011).

Dentition

Upper Molars. The Senèze M2 is known from the Basel cranium Se 547, nine plates are still present. The lamellar frequency (number of laminae in 10 cm, counted in the middle of the occlusal surface) is 5. Maximum crown width is 83 mm on lamina 6. The length is about 166 mm. Van Essen (2003) gives observed ranges for M2 of *M. meridionalis* length 174–248 mm, width 79–104.5 mm, number of laminae 8–10 and lamellar frequency 4.7–6.1.

We studied a cast of the left M3 from the Basel cranium Se 547. This tooth was not in occlusion along its full length, only 7 laminae showing traces of wear. The lamellar frequency is 4 (4.5 in Titov 2001, but this author erroneously noted 15 laminae), the maximal enamel thickness is 3.5 mm and maximal crown width is 103 mm at the lamina 4 level; the maximal width of a lamina is 91 mm. Using the age groups proposed by Laws (1966), the estimated age at death was about 35–40 years.

For the two M3 collected by Boule (1892), the maximal crown width is 122 mm at the lamina 5 level, the maximal width of a lamina is 104 mm at the same level, the enamel maximal thickness is 4.5 mm and the lamellar frequency 4.5. The length is 247 mm for the left one and 261 mm for the right one. The two have twelve laminae, eight of each showing traces of wear. Only the right one has roots preserved, at lamina 8 the height is 202 mm and its hypsodonty index is 1.66. This is just slightly greater than those of the M3s of *M. meridionalis* from Upper Valdarno, the maximal value being 1.60 (Palombo & Ferretti 2005). The features of the M3 from Chagny are nearly the same as those of Senèze. It possesses 10 laminae, the laminar frequency is 4.5, the enamel maximal thickness is 3.5 mm for a length of 265 mm and a maximal width of 123 mm. Two M3s from Colle Mancino (L'Aquila, Italy, estimated to date ca. 1.2–1.1 Ma by Agostini et al. 2012, Fig. 15.12) have nearly the same dimensions as those from Senèze (L = 259–258 mm, w = 108–109 mm, h = 200–213 mm), with greater hypsodonty indexes (hy = 1.85–1.95) both with 11 laminae, a lamellar frequency of 5, an average enamel thickness of 3.7 (Maccagno 1965). Another one from Imola (Italy, ca 0.8 Ma) also has an elevated hypsodonty index of 1.76 but with more advanced features, 14 laminae, enamel thickness 2.4, lamellar frequency 6 and a height of 158 mm (Palombo & Ferretti 2005).

Even if they are unusual, high crowned molars with high hypsodonty indexes and archaic features are present in two sites, Senèze and Colle Mancino. The high crowned M3 of Senèze is in agreement with the height of the maxilla of the skull Se 547.

Tusks. The tusk is straight in its proximal part, then recurved, slightly twisted outward and finally twisted inwards and upwards. The section is almost round; maximal diameter x minimal diameter are 166 × 165 mm for the right one of the Se 547 cranium, 170 × 170 mm for the left one and, 210 × 180 mm for Se 193. The tusks are relatively long for an individual of about 35–40 years (as suggested above): the length, following the curve, is about 2.30 m for the left one of Se 547, and the length of Se 193 is 2.80 m, almost the same as that of the big old individual from Scoppito (see below). A large tusk fragment (FSL 456,286) was discovered in 2003 during our excavations: its length is 1.5 m but proximal and distal extremities are missing, its maximal diameter is ca. 200 mm, and the frontal side is slightly flattened (Fig. 15.6).

Lister and Stuart (2010) gave diameters for tusks of *M. meridionalis* from the Upper Valdarno; for females the range is 95–110 mm and for males 170–237 mm. The left tusk of a male *M. meridionalis* from Georghievsk is 3.20 m long, basal diameters are 300 × 238 mm (Garutt 1998b); this tusk is even larger than that of Scoppito. Another male tusk from Georghievsk is smaller, length 2.80 m and diameter 220 mm

Fig. 15.6 Tusk FSL 456286 of *M. meridionalis* from Senèze

(Shvyreva & Titov 2018). That of a male from Rodionovo (North Caucasus, Russia, Early Pleistocene) is short, length 2.05 m, with quite large basal diameters 216 × 184 mm and, that of a male from Nogaisk (Ukraine, 1.3–1.0 Ma), has smaller basal diameters 207 × 138 mm (Maschenko et al. 2011). The right tusk of Scoppito belongs to a very old male, its estimated length is 3.10 m and basal diameters are 238 × 204 mm (Maccagno 1962). A partial tusk of *M. meridionalis* from Colle Mancino is almost round (250 × 230 mm) near the alveolus (Maccagno 1965).

The tusk of Campo di Pile (Italy, ca. 1.0 Ma), from a young adult female, is more compressed near the alveolus (167 × 88 mm) and rounded after (110 × 110 mm), its double curvature is less accentuated than in adult male (Agostini et al. 2012). The basal diameters of a female tusk from Savignano sul Panaro (Northeastern Italy, ca. 1.8 Ma), identified as *Archidiskodon* cf. *gromovi* by Reggiani and Sala (1994) and attributed to *M. meridionalis* by Palombo & Ferretti 2005), are 173 × 100 mm.

Tusks of male *M. trogontherii* from West Runton (Lister & Stuart 2010), Kostolac (Lister et al. 2012), Nosak (Dimitrijević et al. 2015) and Loussiká (Athanassiou 2012) have maximal basal diameters ranging from 200 to 250 mm. Basal diameters for female tusks of *M. trogontherii* from Edersleben (Germany) and from Novogeorgievsk (Ukraine) are respectively 138 and 149 mm (Lister & Stuart 2010). A partial tusk, distal end missing, of male *M.* cf. *intermedius* from Fourcaud (Aude, MNQ 24) is 1.95 m with a basal diameter of approximately 160 mm and a sub-circular cross section (Filoux et al. 2014). The tusks of *M. primigenius* are slender, in females basal diameter has a range of 47–90 mm for a length of 1.60–2.47 m, and for the male ranges are respectively 89–180 mm and 2.40–3.80 m (Vereshchagin & Tikhonov 1986, 1999).

The length, the basal diameter and the curvature of the tusks make it possible to distinguish male and female mammoths (Agostini et al. 2012; Averianov 1996; Azzaroli 1966; Lister & Stuart 2010; Palombo & Ferretti 2005). The shape and dimensions of the two isolated tusks and those of the skull unearthed at Senèze allow an attribution to male individuals for these specimens.

Lower teeth. The mandible of Se 547 is mounted against the cranium, so we were not able to observe the lower tooth. Depéret et al. (1923, pl. II Fig. 15.6) published a photograph of that mandible in occlusal view, before restoration, and provided a short description of the tooth. There are 11 laminae excluding the posterior talon; the lamellar frequency of the m3 is about 4. The molars of the mandible FSL 211,130 from Senèze (Figs. 15. 4–5) have 8 laminae (without the talon), the last two without wear, and probably 1 or 2 lost, the lamellar frequency is 4. The enamel, strongly folded for the laminae 3–6, has a thickness of 4–5 mm. The lengths are 253 mm for the right and 257 mm for the left, the maximal width is 110 mm for the right and 108 mm for the left, at a level between the laminae 3 and 4. Using the data given by some authors (Baygusheva & Titov 2012; Ferretti 1999; Palombo & Ferretti 2005) for the molars of *M. meridionalis*, the observed ranges of the length and the width are for the m3 250–330 mm and 75–113 mm and for the m2 181–220 mm and 76.5–92.5 mm. According to these data the two teeth of the mandible FSL 211,130 are probably m3s (see also the test elaborated by Rousselières 2006). After Laws' age groups (Laws 1966), we can estimate an age of about 45 years. The m3's of the Chagny mandible possess 10 laminae, their lengths are 284 and 288 mm, the maximal width 112.5 mm; the lamellar frequency is 4 and the enamel thickness 3.5 mm. The mandibles and teeth from Senèze appear to be very similar to those of Chagny and Upper Valdarno.

Postcranium

Humerus. Only the distal part of the right humerus FSL 210,990 is preserved. The dimensions of the articulation are TD x APD = 300 × 234 mm.

Scaphoid (Fig. 15.7). The bone looks like an elevated pyramid with an almost round articular facet in the proximal part of the lateral face; this articular facet is divided in two, the upper part corresponding to the radius and the inferior corresponding to the medial side of the lunar. On the distal part of the lateral side there is a second articular facet for the lunar: arcuate, higher in the posterior part, it joins the distal articular facet which is strongly convex transversally. The shape of the distal articular facets seems to be a good criterion to distinguish *Mammuthus* from *Elephas* (Beden & Guérin 1975).

Dimensions

FSL 211,125: L = 135 mm; w = 75 mm; h = 144 mm;

Se 1770: L = 126 mm; w = 70.5 mm; h > 143 mm.

The two scaphoids of *M. meridionalis* from Leu (Romania, Early Pleistocene) have nearly the same dimensions and morphology (Popescu 2008). That of Leffe (Italy, ca. 1.5 Ma) is higher and a little shorter (Vialli 1956) and that of Scoppito is larger with the same morphology (Maccagno 1962). The scaphoid of *M. trogontherii* from Loussiká has nearly the same dimensions (L = 119 mm, h = 142 mm) as those of *M. meridionalis* but the radial facet is higher and ends proximally in a pointed top (Athanassiou 2012). The scaphoid of *M. trogontherii* from Zhalainuoer III has a greater height (176 mm) but there is no description of the bone (Larramendi 2014). Following Dubrovo and Jakubowski (1988), the scaphoid of *Palaeoloxodon antiquus* is relatively higher than in *Mammuthus*, with a nearly

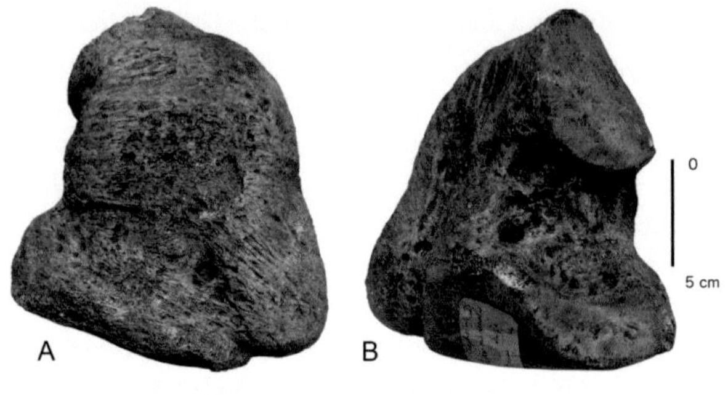

Fig. 15.7 Right scaphoid FSL 211125 of *M. meridionalis* from Senèze; **A** medial, **B** lateral, views

triangular upper facet for the lunar and a lower facet for the lunar reduced in height anteriorly.

Lunar (Fig. 15.8). On the anterior side the medial border is strongly convex, higher than the lateral one which is concavo-convex; proximal and distal edges are concave. The anterior face is pentagonal, with depressed upper and lower borders, and higher medially. The fully articular proximal face is trapezoidal in shape and narrower on its posterior part, with a large notch on its lateral border. The distal face is also fully articular and trapezoidally shaped, with only a small notch on its lateral border. On the medial side there are two articular facets for the scaphoid, a proximal one in the posterior part, oval shaped and elongated but remote from the anterior border of the bone, and a distal one in the anterior part, parallelogram in form, very elongated (3/4 of the face length) on FSL 211,123. On the lateral side there are two facets: a proximal one elongated and very low, continuing in the upper facet for the ulna on the proximal face; and a distal one, elliptical, joining with the pyramidal; a deep and wide groove separates the two facets.

Dimensions

FSL 210962: L = 122 mm; w = 122 mm; h = 71.5 mm;

FSL 211,123: L = 152 mm; w = 140.5 mm; h = 99.5 mm.

Mol et al. (1999) gave ranges for three specimens of *M. meridionalis* from the Oosterschelde Estuary and North Sea (L = 162–196 mm, w = 157–173 mm, h = 93–114 mm); the dimensions of those from Scoppito (Maccagno, 1962) and Leffe (Vialli, 1956) fall into these ranges. The dimensions of the lunar of *M. meridionalis* from Leu are nearly the same (L = 152.9 mm; w = 149.2 mm; h = 93.7 mm) than those of FSL 211,123 from Senèze with very similar articular facets (Popescu 2008). The two lunars of *M. trogontherii* from Loussiká have intermediate measurements between the two of Senèze (Athanassiou 2012). We agree with Rousselières (2006) that the posterior tuberosity is much more developed in *Palaeoloxodon* than in *Mammuthus*.

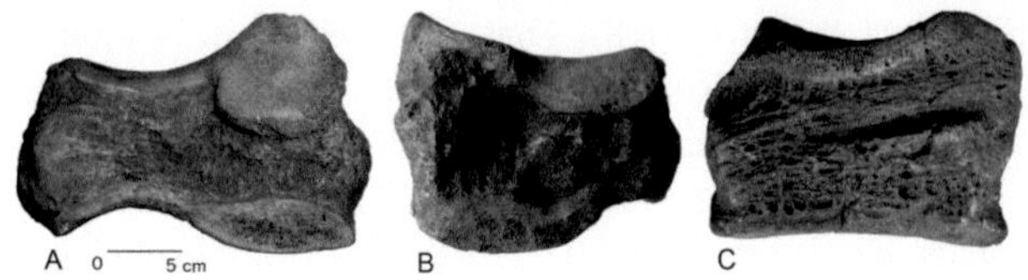

Fig. 15.8 Right lunar FSL 211123 of *M. meridionalis* from Senèze; **a** lateral, **b** medial, **c** anterior views

Trapezium. On the proximal part of the anterior side there is a trapezoidal articular facet for the scaphoid, it joins the proximal articular facet for the trapezoid at a right angle. The proximal articular facet occupies the whole proximal side and continues on the posterior side at an obtuse angle. The distal face is totally utilized for the articulation with the Mc I. The medial face is pentagon shaped, more extended distally than proximally.

Dimensions

Se 194: APD = 108 mm; TD = 62 mm; h = 96 mm.

The trapezium of *M. meridionalis* from Scoppito is wider than higher. The trapezium of *M. trogontherii* from Loussiká has a greater TD (95 mm) and a similar height (Athanassiou 2012).

Trapezoid (Fig. 15.9). The front face is trapezoidal, much higher laterally than medially. The lower edge is an obtuse dihedron whose vertex is laterally offset; the upper edge is weakly convex, the lateral one is convex and the medial one weakly concave. On the medial side there is one articular facet for the trapezium. In front it occupies the whole height and in the back only the upper part of the bone. This articular facet is almost flat on Se 1494 and concave on Se 194 and FSL 211,121. On the lateral side, there are three articular facets for the magnum; two are on the anterior side, a very large upper trapezoidal facet and a small lower facet fused with it. A third facet is in a postero-superior position. The two upper facets are separated by a notch, the anterior one occupying almost the whole height of the bone, the posterior one (only on the upper half) is of semicircular shape and medially offset. The distal facet, low and relatively short, joins in front the proximal anterior facet. A strong depression separates it from the posterior upper facet. The convexity of the anterior and posterior edges is very variable. The proximal face is fully articular, longitudinally concave, transversely flat or weakly convex for FSL 211,121. On Se 194 the articulation is strongly concave, and on Se 1494 nearly flat. The proximal articulation joins the proximo-lateral facets. The distal face is fully articular with the Mc II. It is triangle shaped, and convex.

Dimensions
Se 194: L = 135 mm; ant. w = 98 mm; ant. h = 82.5 mm;

Se 1494: L = 130 mm; ant. w = 92.5 mm; ant. h = 73.5 mm;

FSL 211,121: L = 132 mm; ant. w = 89 mm; ant. h = 76 mm.

Mol et al. (1999) give measurements for one specimen of *M. meridionalis* from the North Sea: L = 130 mm; ant. w = 70; ant. h = 70 mm. That of Scoppito (Maccagno 1962) has a greater anterior width (110 mm) and anterior height (93 mm). Two trapezoids of *M. trogontherii* from Loussiká have nearly the same dimensions as those of Senèze (Athanassiou 2012). In *Palaeoloxodon antiquus* the two articular facets for the magnum may be fused or separated in front of the lateral face of the trapezoid (Dubrovo & Jakubowski 1988).

Magnum. The front side is trapezoidal, with an upper edge much longer than the distal one, both being convex; the medial is concave, the lateral more or less right. The proximal face seems fully articular but is damaged laterally and anteriorly on Se 1770. The proximal articular facet protrudes on the back side of the bone. On the medial side there are two proximal articular facets for the trapezoid separated by a notch, and a distal one occupies the entire edge of the lateral side; that distal facet makes an obtuse dihedron with the proximal articular facet. On the lateral side the articular facet looks like an hourglass, higher on its posterior part. The upper edge is strongly concave. The proximal margin of the posterior side is regularly convex.

Fig. 15.9 Left trapezoid FSL 211121 of *M. meridionalis* from Senèze, lateral view

Dimensions:
Se 1770: L = 169 mm; ant. w = 113.5 mm; ant. h = 96 mm; post. w = 119.5 mm; post. h = 129.5 mm;

FSL 211,126: ant. w = 123 mm; ant. h = 102.5 mm.

Mol et al. (1999) give measurements for two magnums of *M. meridionalis* from the Oosterschelde Estuary and the North Sea: L = 149–159 mm; post. w = 124–135 mm; h = 129–135 mm.

Two magnums of *M. meridionalis* from Leu are described by Popescu (2008), only one is complete: L = 152.5 mm; ant. w = 94–113.7; post. w = 105.6–106.4 mm; h max. = 97–104.7 mm. That of Scoppito (Maccagno 1962) is longer (185 mm) and has an anterior height (110 mm) a little bit greater than those of Senèze. The magnum of *M. trogontherii* from Loussiká is high (123 mm) with the same length as that from Scoppito (Athanassiou 2012). The 27 specimens of *M. primigenius* from the North Sea and Padul (Southern Spain) are smaller, the means are 70.6 mm for ant. h, 90.7 mm for ant. w and 127 mm for the length (Álvarez-Lao et al. 2009).

Unciform. The anterior side is trapezoidal, its anterior medial height is 115 mm and its lateral height 73 mm. On the medial side, the articular facet for the magnum is almost pear-shaped, much higher on the back, stretched from the anterior to the posterior part of the side. The lower part of the face is damaged. On the lateral side, the posterior part of the articular facet for the Mc V occupies the full height of the bone, while the anterior part occupies only the inferior half; this facet joins together with the distal articular facet. The upper face articulates mostly with the pyramidal and to a lesser extent with the lunar. A blunt crest separates the two facets. The distal face is fully articular with the Mc V.

Dimensions:
Se 194: L = 161 mm; l = 139 mm; ant. w = 115 mm; post. h = 139 mm.

Mol et al. (1999) give values for one specimen of *M. meridionalis* from the Oosterschelde Estuary: L = 143 mm; l = 136 mm; ant. w = 130 mm. That from Scoppito is somewhat larger, its anterior width is 164 mm and its anterior medial height (123 mm) is greater than its lateral one (Maccagno 1962). That of Leffe has the same morphology of the anterior face, the medial height (103 mm) is greater than the lateral one (Vialli 1956). The unciform of *M. trogontherii* from Loussiká is a little bit lower than that of Senèze, h max. = 123 mm (Athanassiou 2012). Those of *M. primigenius* from North Sea and Padul are smaller (Álvarez-Lao et al. 2009).

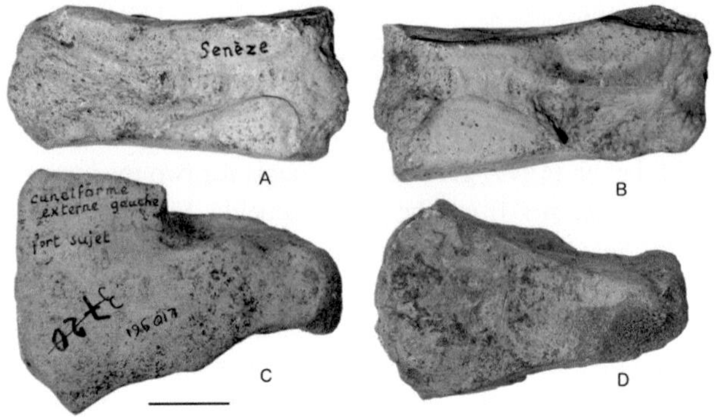

Fig. 15.10 Left lateral (first) cuneiform FSL 210961a of *M. meridionalis* from Senèze; **A** medial, **B** lateral, **C** distal, **D** proximal views

Mc III. The only specimen is a proximal epiphysis whose posterior part is damaged. On the medial and lateral sides the articular facets for the Mc II and the Mc IV occupy the full depth of the bone.
Dimensions:
Se 194: prox TD > 95 mm.

Mol et al. (1999) give a prox TD of 87 mm for a *M. meridionalis* Mc III dredged up from the Oosterschelde Estuary and Ros-Montoya (2010) gives 92.84 mm for one from Venta Micena (Spain, 1.6–1.4 Ma). Athanassiou (2012) gives a measurement of 82 mm for the proximal TD of an Mc III of *M. trogontherii* from Loussiká.

Mc V (Table 15.1). The Mc V is stocky and well developed antero-posteriorly. The proximal articulation corresponds medially with the unciform, laterally with the pyramidal. In frontal view the proximal edge is oblique toward the lateral side and above; there are two tuberosities along the lateral edge, a proximal and a distal one; the distal breadth is much greater than the proximal one. In lateral view there is a proximal facet for a sesamoid. Its robustness index is 46.6. Two Mc V of *M. meridionalis*, one from Leffe (L = 175 mm) and one from Scoppito (L = 209 mm) have slightly lower robustness indexes, respectively 42.3 and 45.6 (Maccagno 1962; Vialli 1956).

Two Mc V of an adult male of *M. trogontherii* from Loussiká, both have the same length, 181 mm. Using a picture in Athanassiou (2012, p. 20, Fig. 17) we estimated the TD of the diaphysis at 68 mm, hence the RI at 37.6. Dietrich (1912) gives measurements of a Mc V from Steinheim an der Murr (Germany) (L = 220 mm), its RI = 35.5 is low. A Mc V of *M. intermedius* from Durfort has a length of 185 mm and RI = 47.6 (Beden & Guérin 1975). That of *M. primigenius* is smaller. For 17 specimens of *M. primigenius* from Padul and North Sea the average length is 162.1 mm and, for two specimens from Padul the RI are 46.8 and 32.5 (Álvarez-Lao et al. 2009). Kulczycki (1955) gave measurements of four Mc V of *M. primigenius*, the lengths are from 110 to 172 mm (mean 141.5 mm) and, the RIs from 36.0 to 41.8 (mean 38.8 mm).

Patella. The bone is elongated transversely, the TD being just a little lower than the height. The maximal width is almost situated in the superior quarter of the height. The posterior side is divided into two parts by a vertical flexure, the medial portion is clearly smaller than the lateral one.
Dimensions:
Se 1424: L = 164 mm; TD = 139 mm; APD = 108 mm.

We have also seen, in the Basel Museum, a patella of *M. meridionalis* from Upper Valdarno (NMB V.A. 1363) of identical morphology and with nearly the same dimensions: L = 174 mm; TD = 131 mm; APD = 102 mm. Two patellae of *M. meridionalis*, one of a female from Incarcal-1 and one from Fonelas P-1 (Spain, MNQ 18) are smaller (Garrido 2008; Mazo et al. 2003). That from Scoppito (Maccagno 1962) is larger (L = 174 mm; TD = 165 mm; APD = 120 mm) than that of Senèze. Shpansky et al. (2015) give the dimensions of five patellae of *M. trogontherii* from different European sites and Mol and Lacombat (2009) of one from Nolhac (Haute-Loire); the ranges of the length, width and APD are respectively 143–189 mm, 112–154 mm, and 84–117 mm; the measurements of the patella from Senèze fall within these ranges. The patella of *M. primigenius* is on average smaller; for 11 specimens, the means of L, TD and, APD are respectively 136.3 mm, 109.2 mm and 73.2 mm (Álvarez-Lao et al. 2009).

Cuboid. On the anterior side, the proximal margin is regularly convex, the distal one is concave in its middle part and convex in the extremities. On the medial side, a semicircular shaped postero-proximal articular facet joins the proximal facet in a right angle. A second articular facet in antero-distal position joins the distal facet also in a right angle. The lateral face is damaged. The proximal side is triangular in shape. The articular facet for the calcaneum is flat and larger than the one for the navicular, which is concave antero-posteriorly and transversely. The distal side is not articular in its posterior part.
Dimensions:
Se 1699: L = 133 mm; w = 129.5 mm; ant. h = 67 mm.

The cuboids from Savignano sul Panaro, Incarcal and Leu are smaller (Mazo et al. 2003; Popescu 2011; Reggiani & Sala 1994). The cuboid of *M. trogontherii* from Loussiká (Athanassiou 2012) has nearly the same height (58 mm) as that of Leu. The two cuboids of *M. intermedius* from La Fage are smaller, the heights are only 41–45.5 mm (Beden & Guérin 1975). For 6 specimens of *M. primigenius* from North Sea, the means of L, w and maximal height are respectively 110 mm, 116 mm and 49 mm (Álvarez-Lao et al. 2009).

Lateral (first) Cuneiform (Fig. 15.10). The proximal articulation corresponds to the navicular, it is very wide in its front part and presents a notch on its medial edge. It is concave longitudinally and almost flat transversally. The anterior face has a trapezoidal outline; the proximal edge is shorter than the distal one; the proximal and distal borders are convex, and the lateral and medial ones are concave. The distal articulation is triangular, it corresponds to the Mt III and the Mt IV. Its lateral edge presents a right angle. The face is transversely convex and longitudinally flat. In lateral view there are two articular facets for the cuboid, one antero-distal half-elliptical and one postero-proximal, more elongated. In medial view there are two articular facets for the intermediate cuneiform, one antero-distal drop shaped and one postero-proximal, half-elliptical and elongated.
Dimensions:
FSL 210,961: L = 105 mm; w = 75 mm; h = 45 mm.

That from Scoppito has greater length and width but the same height (Maccagno 1962). The two of *M. intermedius* from La Fage are smaller (Beden & Guérin 1975).

Mt III (Fig. 15.11). This bone is rather stout (Table 15.1). The proximal articular facet is triangular in shape. Its anterior border is uniformly convex, the lateral one is slightly concave after the articular facet for the Mt IV. This articular facet is rectangular and elongated antero-posteriorly, it joins the proximal facet at an acute, nearly right angle. On the medial side the articular facet for the Mt II is also rectangular but higher than that for the Mt IV; like the latter, it joins the proximal facet but the angle is slightly obtuse. Its robustness index is 42.2.

The Mt IIIs of *M. meridionalis* from Scoppito (L = 174 mm) and Leffe (L = 146 mm) are longer than that from

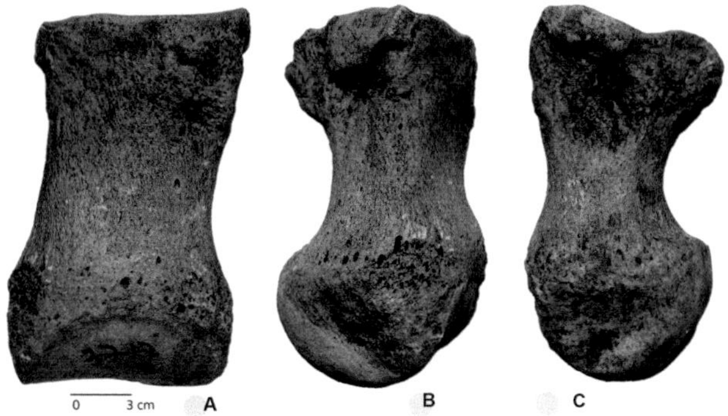

Fig. 15.11 Left Mt III FSL 210962b of *M. meridionalis* from Senèze, **A** anterior, **B** medial, **C** lateral views

Senèze, but with very close robustness indexes, respectively 44.3 and 41.8 (Maccagno 1962; Vialli 1956). Two Mt IIIs of females, one from Savignano sul Panaro and one from Incarcal-1 (Mazo et al. 2003) are shorter, L = 135 mm and 130 mm, but with very close robustness indexes, respectively 39.3 and 41.5. Four Mt III of *M. trogontherii*, one from Zhalainuoer, two from the Sea of Azov Region and one from Steinheim are longer (L range 167–180 mm) than that of *M. meridionalis* but the robustness index of that from Steinheim (RI = 40.4) is the same (Baygusheva et al. 2012; Dietrich 1912; Larramendi 2014).

A Mt III of *M. primigenius* from Ilskaya 2, Caucasus (Russia) has a length of 125 mm and its robustness index is 43.2 (Baryshnikov 2003) and for three from Poland and Ukraine the lengths are 130 mm, 127 mm and 115 mm, with respective RI of 39.2, 44.1 and 47.8 (Kulczycki 1955). The Mt III of *M. primigenius* is smaller than that of *M. meridionalis* and *M. trogontherii* with similar RI.

Table 15.1 Measurements (in mm) of the metapodials of *Mammuthus meridionalis* from Senèze

	Mc V	Mt III	Mt V
	NMB Se 1770	FSL 210,962	FSL 210,962
L	192	141	109
Prox. TD	99	74	65
Prox. APD	104	85.5	85.5
TD dia	89.5	59.5	82
APD dia	81	47	87
Dist. TD	113	77	83
Dist. APD	117	76.5	91

Fig. 15.12 Left Mt V FSL 210962c of *M. meridionalis* from Senèze, anterior view

Mt V (Fig. 15.12). The Mt V is very stout and all its APD are superior to the corresponding TD (Table 15.1). The proximal articular facet is oval in shape. On the anterior side the proximal border is slightly concave, very oblique medially and to the top. On the medial side, on the anterior half of the upper border, the articular facet for the Mt IV is half-moon shaped and joins the proximal facet at an acute angle. On the lateral side there is a strong infero-posterior tuberosity. The Mt V of a male *M. meridionalis* from Rodionovo, as nearly the same measurements (L = 106 mm), that from Scoppito is longer (L = 120 mm), that of a female from Incarcal-1 is smaller (L = 90 mm) but the robustness indices are close to each other; RI = 75.2 for Senèze, 79.2 for Scoppito, 80.2 for Rodionovo, 81.1 for Incarcal-1 (Maccagno 1962; Maschenko et al. 2011; Mazo et al. 2003).

Two Mt Vs of a probably female *M. trogontherii* from the Azov Sea Region are small (L = 105.2–106.2 mm), that from Steinheim is large (L = 131 mm) with a RI of 67.9 (Baygusheva et al. 2012; Dietrich 1912). A Mt V of a young *M. intermedius* from La Fage is very short (L = 67 mm) with a RI of 74.6 (Beden & Guérin, 1975). Two Mt Vs of a young adult female of *M. primigenius* from Kastykhtakh, Taimyr Peninsula (Russia, ca. 0.31 Ma) has a length of only 83 mm and a RI of 44.0 (Kirillova et al. 2012), another one from Starunia (Ukraine) has a length of 93 mm and a RI of 69.9 (Kulczycki 1955). The Mt V of *M. meridionalis* are larger than that of *M. primigenius* with a proportionally more robust diaphysis.

Table 15.2 Measurements (in mm) of the phalanges of *Mammuthus meridionalis* from Senèze

	Phalanx I			Phalanx II
	NMB Se 1770 Digit IV	NMB Se 767 Digit IV	FSL 211,111	NMB Se 1427
L	106	100	98	72
Prox. TD	100	81	93.5	55
Prox. APD	75.5	70	82.5	49
TD dia	89.5	67	79	47.5
APD dia	57.5	49.5	55	42
Dist. TD	87	66	79	40
Dist. APD	55.5		52	33.5

Phalanges (Table 15.2). The first phalanges are the thickest, the length is just a little bit greater than the proximal TD. In anterior view, there are four tuberosities, two proximal and two distal, the lateral seeming to be larger than the medial ones. The proximal articular facet is oval-shaped, concave anterio-posteriorly and almost flat medio-laterally. The second phalanx is proportionally more elongated than the first. The third phalanges are short and elongated transversely. The lengths of the first phalanges of the hand from Leffe are close to those of Senèze. The length for that of the digit IV is 98 mm and its proximal TD is 64 mm, smaller than the distal one, unlike those of Senèze (Viali 1956). Those of the hand from Scoppito are longer, from 107 to 118 mm, and slender, proximal and distal TDs are smaller than those of Senèze (Maccagno 1962). The first phalanges of the manus (digit IV) of *M. trogontherii* from Loussiká have nearly same lengths (93–100 mm) and proximal TDs (94-ca. 97 mm) than those of Senèze (Athanassiou 2012) and those from Zhalainuoer are a little bit longer, L = 109 mm for the digit IV (Larramendi 2014). The second phalanges of the hand of *M. meridionalis* from Scoppito are short, except for the digit I (Maccagno 1962), and those of the foot from Savignano sul Panaro are very short with much larger TD (Reggiani & Sala 1994). Those of the manus of *M. trogontherii* from Zhalainuoer are very short 55–57 mm (Larramendi 2014).

Conclusions of the descriptive study

The main features for the skull of Senèze are a concave forehead in lateral view, apex of the skull located over the external auditory meatus, a narrow temporal fossa, a long and strong zygomatic arch, a flat vertical occipital, a rectangular outline of the nasal cavity and a great height of the maxillae which make possible high crowned molars. The skull Se 547 and the two isolated tusks belong to male individuals. For Garutt (1998b) the gender of an individual can be determined by the shape of the nasal opening, a wide nasal opening with sharpened lateral edges directed downward indicates a male, while a rounded one occurs in females. We do not agree with this hypothesis, since the skulls from Senèze and Scoppito, both of male individuals with different shaped lateral edges, contradict it.

The mandibles from Senèze with primitive features, a low and elongated corpus, an obtuse angle between the occlusal tooth plane and the dorsal border of the rostrum, a weakly rounded mandibular angle, and a sinuous anterior border look like those of Upper Valdarno, contrary to that of Scoppito which has more derived characters, in having a right angle between the occlusal plane and dorsal border of the rostrum, a less sinuous anterior border of the ramus and a height of the corpus under the anterior border of the tooth greater than that under the posterior border; these characters remind us of a "trogontheroid' jaw.

For some authors the length of the symphyseal process is useful in assessing the gender of mammoths (Deperet et al. 1923; Erdbrink & Augustinus 1973; Mazo et al. 2003; Ros-Montoya et al. 2012; Vasile et al. 2019). We have seen above that mandibles of *M. meridionalis, M. trogontherii* and *M. primigenius* with elongated processes can be of male or female individuals. We think, like Álvarez-Lao and Méndez (2011) and Athanassiou (2012), that the length of the symphyseal process is not diagnostic for gender determination.

The M3 shows archaic features such as low lamellar frequency, reduced number of laminae and thick enamel mixed with more advanced ones, great height and high hypsodonty index. A tooth with large roots requires a fairly high maxilla. We did not find significant differences in the post-cranial bones studied here between *M. meridionalis* and *M. trogontherii,* as has already been reported (Mol et al. 1999; van Essen 2011).

Biochronology and Paleoenvironment

Proboscideans of the *Mammuthus* lineage are especially important for the biozonation of the Eurasian Plio-Pleistocene (Guérin 1980, 1982, 1990, 2007). The most ancient species in Europe is *M. rumanus*, defined at Tulucesti (Late Pliocene, 3.5 to 3 Ma), also known from Cernatesti (both in Romania), Bossilkovtsi in Bulgaria, Novotroitsk (Stavropol Region, Russia) and Liyucun, Yushe Basin, Shanxi (China, 3.4–2.8 Ma), and perhaps from the Red Crag of Great Britain (2.6–2.5 Ma) and Montopoli (2.6–2.5 Ma) and Laiatico (Lower Valdarno) in Italy (Baygusheva & Titov 2012; Kahlke 2014; Markov 2012; Palombo & Ferretti 2005; Wei et al. 2006;). In Romania and maybe Bulgaria it could be sympatric with *Anancus arvernensis* and *Zygolophodon borsoni* (Lister & van Essen, 2003; Markov & Spassov 2003). The next species *M. meridionalis* was defined from the Upper Valdarno Formation, in the Montevarchi Group (about 1.8 Ma, Napoleone et al. 2003). This species has a long "lifespan" from about 2.4 Ma to 0.7 Ma and a wide geographical extension. Based on the degree of dental evolution numerous subspecies have been erected. The most widely used subspecies are *M. m. gromovi* defined in the Khapry Faunal Complex (Crimea), *M. m. meridionalis* the nominotypical subspecies from Upper Valdarno (Olivola and Tasso Faunal Units), *M. m. vestinus* defined at Scoppito (Farneta Faunal Unit), *M. meridionalis tamanensis* of Crimea (Taman Faunal Complex) and *M. m. depereti* defined at Saint-Prest (MNQ 20). The validity of these subspecies is subject to many assumptions and is beyond the scope of this study. Heintz (1969) established that, in Western Europe, *Mammuthus meridionalis* can be sympatric with *Anancus arvernensis* during the middle and the early late Villafranchian, but it is never sympatric with *Mammut borsoni*, which became extinct during the early Villafranchian, i.e., before the appearance of *M. meridionalis*.

Mammuthus trogontherii, first appearing at Majuangou-3, Nihewan Basin, China, around 1.66 Ma, spread to North Siberia (ca. 1.2 Ma) and slightly later to Europe around 0.8 Ma (Wei et al. 2006; Virag 2009; Wei et al. 2010; van Essen 2011; Konidaris et al. 2020). Shoulder height and body mass estimates have the same range as *M. meridionalis* (Athanassiou 2012; Larramendi 2016; Lister & Stuart 2010). At Dorn-Durkheim-3 (Germany, ca. 0.8 Ma) the two species *M. meridionalis* and *M. trogontherii* are present (Lister et al. 2012; van Essen 2011). The species is still present in Europe ca. 0.19 Ma at Nosak, Kostolac Basin (Serbia) and Marsworth (England). In this latter site, *M. trogontherii* is present together with *M. primigenius* (Dimitrijević et al. 2015; Lister & Sher 2001; Muttoni et al. 2015). The latest record is in China (Shangmadu, Shaanxi Province) dating to 33.9–24.9 ka (Wei et al. 2010).

M. intermedius whose locus typicus is the karst filling of Aven I des Abîmes de La Fage, at Noailles, Corrèze (France, MNQ 24, 0.3–0.24 Ma), considered as an intermediate link in the transition from *M. trogontherii* to *M. primigenius*, is known in numerous sites from the Lyon region (France), notably Choulans (Camaret et al. 2011; Labe & Guérin 2005), in Russia (Baygusheva & Titov 2021; Foronova 2014a; Golovachev & Titov 2018), in Ukraine (Foronova 2014b) and in Moldova (Obada 2014).

M. primigenius is first known from North Eastern Siberia, then it migrated west to Europe ca. 190 ka and later reached North America (McDaniel and Jefferson 2003; Lister & Sher 2015). There are two chronological overlaps, the first between *M. meridionalis* and *M. trogontherii* and the second between *M. trogontherii* and *M. intermedius*/*M. primigenius*. During these two time intervals the transition is more complex than previously assumed, intermediate or mosaic forms suggest a possibility of hybridization (Lister & Sher 2001; Lister et al. 2005; Virag, 2009; Wei et al. 2010; Lister & Stuart 2010; van Essen 2011; Palombo 2017; Konidaris et al. 2020).

M. meridionalis paleoenvironmental preference

Pollen analysis from Senèze (Delson et al. 2006; J. Argant 2024) shows temperate conditions with possible beginning of a cooler phase, the coexistence of deciduous trees and conifers and the presence of open spaces near the lake. For Garutt (1998b) a palynological analysis of the deposits from Georghievsk indicates that *Mammuthus meridionalis* lived in open steppe and forest-steppe environments. A dental microwear study of *M. meridionalis* from North Sea indicates that it was a leaf browser (Rivals et al. 2012). These results indicate that the favorite biotope of *M. meridionalis* was wooded savanna under temperate or cooler conditions in a relatively dry environment (Ferretti & Croitor 2001; Lister 1996). From *M. trogontherii* onwards, species of *Mammuthus* appear to have preferred a cold and dry steppe biotope.

Conclusion

The evolutionary level of the molars and mandible from Senèze is comparable to the *Mammuthus meridionalis* from Upper Valdarno and should accordingly be attributed to the nominotypical subspecies *M. meridionalis meridionalis*. Its presence is in agreement with the age of Senèze, one of the earliest sites allocated to MNQ 18. An M3 with primitive features, mixed with more advanced ones, great height and elevated hypsodonty index, like those of Colle Mancino, is present at Senèze. Such M3s with primitive and more progressive features appeared earlier than previously thought: that of Senèze is the oldest. The high crowned M3 of Senèze is in agreement with the height of the maxilla. The measurements and descriptions of the remains from Senèze, especially for the limb bones, complement those previously published for *M. meridionalis*.

Acknowledgments We acknowledge Abel Prieur, Curator of the Paleontological collections of Université Claude Bernard-Lyon 1 and Loïc Costeur, Curator of the Vertebrate collections of the Naturhistorisches Museum Basel, who provided access to specimens in their care, and Dick Mol (Museum of Rotterdam) who gave us a very valuable bibliographical contribution. We thank the municipality of Domeyrat which provided us a place to store the bulky tusk before its transport to Lyon, Abel Prieur and Emmanuel Robert who helped us to transport it and Abel who meticulously prepared it in the laboratory. We thank our colleagues Martine Faure and Sevket Sen for help during the measurement and photography of the fossils and Eric Delson for revising the English text. We thank Dick Mol, Maria Rita Palombo, William Sanders and Haowen Tong for their reviews of a previous version, which helped to improve this contribution.

References

Agostini, S., Palombo, M. R., Rossi, M. A., Di Canzio, E., & Tallini, M. (2012). *Mammuthus meridionalis* (Nesti, 1825) from Campo di Pile (L'Aquila, Abruzzo, Central Italy). *Quaternary International, 276–277*, 42–52.

Álvarez-Lao, D. J., & Méndez, M. (2011). Ontogenetic changes and sexual dimorphism in the mandible of adult woolly mammoths (*Mammuthus primigenius*). *Geobios, 44*, 335–343.

Álvarez-Lao, D. J., Kahlke, R.-D., García, N., & Mol, D. (2009). The Padul mammoth finds—On the southernmost record of *Mammuthus primigenius* in Europe and its southern spread during the Late

Pleistocene. *Palaeogeography, Palaeoclimatology, Palaeoecology, 278*, 57–70.

Aouadi, N. (2001). New data on the diversity of Elephants (Mammalia, Proboscidea) in the Early and early Middle Pleistocene of France. In G. Cavarretta, P. Gioia, M. Mussi, & M. R. Palombo (Eds.), *The world of Elephants* (pp. 81–84). Consiglio Nazionale delle Ricerche.

Argant, J. (2024). Palynological analysis of the fossiliferous layers of the late Villafranchian deposit of Senèze (Domeyrat, Haute-Loire, France). Taphonomy and palaeoenvironmental data. In E. Delson, M. Faure & C. Guérin (Eds), *Senèze: Life in Central France two million years ago. Paleontology, geochronology, stratigraphy and taphonomy* (pp. 123–140). Cham (Switzerland): Springer.

Athanassiou, A. (2012). A skeleton of *Mammuthus trogontherii* (Proboscidea, Elephantidae) from NW Peloponnese, Greece. *Quaternary International, 255*, 9–28.

Averianov, A. O. (1996). Sexual dimorphism in the mammoth skull, teeth, and long bones. In In J. Shoshani & P. Tassy (Eds.), *The Proboscidea. Evolution and Palaeoecology of Elephants and Their Relatives* (pp. 260–267). New York: Oxford University Press.

Azzaroli, A. (1966). La valeur des caractères crâniens dans la classification des éléphants. *Eclogae Geologicae Helvetiae, 59*, 541–564.

Baryshnikov, G. (2003). *Mammuthus primigenius* from the Crimea and the Caucasus. In: J.W.F. Reumer, J. De Vos & D. Mol (Eds.), *Advances in Mammoth Research (Proceedings of the Second International Mammoth Conference, Rotterdam, May 16–20 1999). Deinsea, 9*, 41–56.

Baygusheva, V., & Titov, V. V. (2012). The evolution of Eastern European meridionaloid elephants' dental characteristics. *Quaternary International, 255*, 206–216.

Baygusheva, V., & Titov, V. V. (2021). Late Middle-early late Pleistocene mammoths from the lower Don River region (Russia). *Quaternary, 4*, 5. https://doi.org/10.3390/quat4010005.

Baygusheva, V. S., Titov, V. V., & Timonina, G. I. (2012). Two skeletons of *Mammuthus trogontherii* from the Sea of Azov Region. *Quaternary International, 276–277*, 242–252.

Beden, M., & Guérin, C. (1975). Les proboscidiens (Mammalia) du gisement pléistocène moyen des Abîmes de La Fage à Noailles (Corrèze). *Nouvelles Archives du Muséum d'histoire naturelle de Lyon, 13*, 69–87.

Blumenbach, J. F. (1799). Blumenbach's Handbuch der Naturgeschichte, Sechste Auflage. Göttingen: Dietrich.

Boeuf, O. (1983). Le site villafranchien de Chilhac (Haute-Loire) France. Thèse de Doctorat, Université Paris VII.

Boule, M. (1892). Découverte d'un squelette d'*Elephas meridionalis* dans les cendres basaltiques du volcan de Senèze (Haute-Loire). *Comptes Rendus de l'Académie des Sciences, 115*, 624–626.

Brookes, J. (1828). *A catalogue of the anatomical and zoological museum of Joshua Brookes, museum of Joshua Brookes...* London: R. Taylor.

Camaret, J., George, I., Audibert, C., & Berthet, D. (2011). Expertise sur l'état de fossilisation et de conservation du mammouth de Choulans (*Mammuthus intermedius*). *Les Cahiers du Musée des Confluences. Etudes Scientifiques, 2*, 5–21.

Christiansen, P. (2004). Body size in proboscideans, with notes on elephant metabolism. *Zoological Journal of the Linnean Society, 140*, 523–549.

Coppens, Y., & Beden, M. (1982). *Mammuthus meridionalis depereti* n. subsp. (Mammalia, Elephantidae), nouveau Mammouth du Pléistocène inférieur d'Europe occidentale. *Comptes rendus de l'Académie des Sciences*, II, *294*, 291–294

Delson, E., Faure, M., Guérin, C., Aprile, L., Argant, J., Blackwell, B. A. B., et al. (2006). Franco-American renewed research at the Late Villafranchian locality of Senèze (Haute-Loire, France). *Courier Forschungsinstitut Senckenberg, 256*, 275–290.

Depéret, C., & Mayet, L. (1912). Le gisement de mammifères pliocènes de Senèze (Haute-Loire). *Association Française pour l'Avancement des Sciences, Comptes rendus de la 40e session Congrès de Dijon* 1911, *Notes et Mémoires, Mémoire hors volume*, 138-151

Depéret, C., Mayet, L., & Roman, F. (1923). Les éléphants pliocènes. *Annales de l'Université de Lyon*, N.S., *42*, p. 221.

Dietrich, W. O. (1912). Elephas primigenius Fraasi, eine schwäbische Mammutrasse. *Jahreshefte des Vereins für Vaterländische Naturkunde in Württemberg, 68*, 42–106.

Dimitrijević, V., Mrdjić, N., Korać, M., Chu, S., Kostić, D., Jovičić, M., & Blackwell, B. A. B. (2015). The latest steppe mammoths (*Mammuthus trogontherii* (Pohlig)) and associated fauna on the Late Middle Pleistocene steppe at Nosak, Kostolac Basin, Northeastern Serbia. *Quaternary International, 379*, 14–27.

Dubrovo, I.A., & Jakubowski, G. (1988). The carpus morphology of the forest elephant (*Palaeoloxodon*) and its significance for taxonomy. *Prace Muzeum Ziemi*, Warszawa, *40*, 65–83.

Enk, J., Devault, A., Widga, C., Saunders, J., Szpak, P., Southon, J., et al. (2016). *Mammuthus* Population Dynamics in Late Pleistocene North America: Divergence, Phylogeography, and Introgression. *Frontiers in Ecology and Evolution, 4*, 1–13.

Erdbrink, D. P., & Augustinus, P. G. E. F. (1973). Two fragmentary mandibles of Archidiskodon elephants from the Netherlands. *Geologie en Mijbouw, 52*, 351–358.

Faure, M., Sen, S., & Costeur, L. (2022). Pierre Philis Fossil Hunter: Un demi-siècle de recherches de vertébrés fossiles (1892–1942). *Geodiversitas, 44*, 321–383.

Faure M., Delson E. & Guérin C. (2024). Introduction and History of Research in the Senèze Maar. In E. Delson, M. Faure & C. Guérin (Eds), *Senèze: Life in Central France two million years ago. Paleontology, geochronology, stratigraphy and taphonomy* (pp. 1–25). Cham (Switzerland): Springer

Ferretti, M. P. (1999). *Mammuthus meridionalis* (Mammalia, Proboscidea, Elephantidae) from the "Sabbie Gialle" of Oriolo (Cava La Salita, Faenza, Northern Italy) and other European late populations of southern mammoth. *Eclogae Geologicae Helvetiae, 92*, 503–515.

Ferretti, M. P., & Croitor, R. V. (2001). Functional morphology and ecology of Villafranchian Proboscideans from Central Italy. In G. Cavarretta, P. Gioia, M. Mussi, & M. R. Palombo (Eds.), *The world of Elephants* (pp. 103–108). Rome: Consiglio Nazionale delle Ricerche.

Ferretti, M. P., & Palombo, M. R. (2005). *Mammuthus meridionalis* (Nesti 1825). In F. Lacombat (Ed.), *Les grands Mammifères fossiles du Velay. Les collections paléontologiques du Plio-Pléistocène du musée Crozatier, le Puy-en-Velay. Annales des amis du Musée Crozatier, 13/14*, 66, 70-71.

Filoux, A., Magniez, P., Moigne, A. -M., Perrenoud, C., & Le Loeuff, J. (2014). The Fourcaud mammoth: A discovery in an ancient Aude terrace (Quillan, Aude, France). *Quaternaire, 25*, 3–8.

Foronova, I. V. (2014a). *Mammuthus intermedius* (Proboscidea, Elephantidae) from the late Middle Pleistocene of the southern Western and Central Siberia, Russia: The problem of intermediate elements in the mammoth lineage. *Russian Journal of Theriology, 13*, 71–82.

Foronova, I. V. (2014b). Mammoths of the Molodova V Paleolithic site (Dniester Basin): The case of dental thin-enamel specialization and paleoecological adaptation. *Quaternary International, 326–327*, 235–242.

Garrido, G. (2008). Presencia de *Mammuthus meridionalis* (Nesti, 1825) en el Villafranquiense superior de Fonelas P-1 (Cuenca de Guadix, Granada). In Arribas, A. (Ed.), Vertebrados del Plioceno superior terminal en el suroeste de Europa: Fonelas P-1 y el Proyecto Fonelas. (pp. 597–607). *Instituto Geológico y Minero de España, serie Cuadernos del Museo Geominero, 10*, 597-607.

Garutt, W. E. (1998a). Is there a genus *Archidiskodon* Pohlig, 1885, of the family Elephantidae Gray, 1821? *Cranium, 15*, 15–20.

Garutt, W. E. (1998b). A skeleton of the Southern Elephant, *Archidiskodon meridionalis* (Nesti, 1825), from a sand-pit near Georghievsk, Northern Caucasus, Russia. *Cranium, 15*, 33–38.

Gaudry, A. (1896). *Essai de paléontologie philosophique*. Paris: Masson et Cie.

Golovachev, M. V., & Titov, V. V. (2018). Remains of *Mammuthus intermedius* Jourdan, 1861 from the type localities of the Khazar Fauna of the lower Volga Region. (In Russian with English summary). *Proceedings of the Zoological Institute of the Russian Academy of Sciences, 322*, 268–284.

Gray, J. E. (1821). On the natural arrangement of vertebrose animals. *London Medical Repository, 15*, 296–310.

Guérin, C. (1980). Les rhinocéros (Mammalia, Perissodactyla) du Miocène terminal au Pléistocène supérieur en Europe occidentale. Comparaison avec les espèces actuelles. *Documents des Laboratoires de Géologie de Lyon, 79*, 1–1185.

Guérin, C. (1982). Première biozonation du Pléistocène européen, principal résultat biostratigraphique de l'étude des Rhinocerotidae (Mammalia, Perissodactyla) du Miocène terminal au Pléistocène supérieur d'Europe occidentale. *Geobios, 15*, 593–598.

Guérin, C. (1990). Biozones or mammal Units? Methods and Limits in Biochronology. In E. H. Lindsay, V. Fahlbusch, & P. Mein (Eds.), *European Neogene Mammal Chronology* (pp. 119–130). New York: Plenum Press.

Guérin, C. (2004). Les Proboscidiens (Mammalia) du gisement villafranchien moyen de Saint-Vallier (Drôme). In Faure, M., & Guérin, C. (Eds.), *Le gisement pliocène terminal de Saint-Vallier (Drôme, France)*. Mémoire spécial 26, *Geobios, 37*, 306–317.

Guérin, C. (2007). Biozonation continentale du Plio-Pléistocène d'Europe et d'Asie occidentale par les mammifères: état de la question et incidence sur les limites Tertiaire/Quaternaire et Plio/Pléistocène. *Quaternaire, 18*, 23–33.

Guérin, C., Dewolf, Y., & Lautridou, J.-P. (2003). Révision d'un site paléontologique célèbre: Saint-Prest (Chartres, France). *Geobios, 36*, 55–82.

Guérin, C., Faure, M., Argant, A., Argant, J., Crégut-Bonnoure, E., Debard, E. et al. (2004). Le gisement pliocène supérieur de Saint-Vallier (Drôme, France): synthèse biostratigraphique et paléoécologique. In M. Faure & C. Guérin (Eds.), *Le gisement pliocène terminal de Saint-Vallier (Drôme, France)*, Mémoire spécial 26, *Geobios, 37*, 349–360.

Haynes, G. (1991). *Mammoths, Mastodonts & Elephants. Biology, Behavior, and the Fossil Record*. Cambridge: Cambridge University Press.

Heintz, E. (1969). Mise au point sur les Proboscidiens des gisements villafranchiens de Perrier-Etouaires (Puy-de-Dôme) et de Vialette (Haute-Loire). *Comptes Comptes rendus sommaires de la Société géologique de France, 2*, 56–58.

Heintz, E., Guérin, C., Martin, R., & Prat, F. (1974). Principaux gisements villafranchiens de France: listes fauniques et biostratigraphie. *Mémoires du Bureau de Recherches géologiques et minières, 78*, 169-182

Illiger, C. (1811). *Prodromus systematis mammalium et avium additis terminis zoographicis utriusque classis*. Berlin, C. Salfeld.

Kahlke, R.-D. (2014). The origin of Eurasian Mammoth Faunas (*Mammuthus-Coelodonta* Faunal Complex). *Quaternary Science Reviews, 96*, 32–49.

Kirillova, I. V., Shidlovskiy, F. K., & Titov, V. V. (2012). Kastykhtakh mammoth from Taimyr (Russia). *Quaternary International, 276–277*, 269–277.

Konidaris, G., Kostopoulos, D. S., & Koufos, G. D. (2020). *Mammuthus meridionalis* (Nesti, 1825) from Apollonia-1 (Mygdonia Basin, Northern Greece) and its importance within the Early Pleistocene mammoth evolution in Europe. *Geodiversitas, 42*, 69–91.

Kosintsev, P. A., & Vorobiev, A. A. (2001). Ancient elephants of West Siberia. In G. Cavarretta, P. Gioia, M. Mussi, & M. R. Palombo (Eds.), *The world of elephants* (pp. 710–713). Rome: Consiglio Nazionale delle Ricerche.

Kostopoulos, D. S., & Koulidou, I. (2015). An early mammoth maxilla from north-western Greece. *Quaternary International, 379*, 155–163.

Kulczycki, J. (1955). Les ossements des mammouths. *Palaeontologia Polonica, 7*, 1–65.

Kuzmin, Y. V. (2010). Extinction of the woolly mammoth (*Mammuthus primigenius*) and woolly rhinoceros (*Coelodonta antiquitatis*) in Eurasia: Review of chronological and environmental issues. *Boreas, 39*, 247–261.

Labe, B., & Guérin, C. (2005). Réhabilitation de *Mammuthus intermedius* (Jourdan, 1861), un mammouth (Mammalia, Elephantidae) du Pléistocène moyen récent d'Europe. *Comptes Rendus Palevol, 4*, 235–242.

Larramendi, A. (2014). Skeleton of a Late Pleistocene steppe mammoth (*Mammuthus trogontherii*) from Zhalainuoer, Inner Mongolian Autonomous Region, China. *Paläontologische Zeitschrift, 89*, 229–250.

Larramendi, A. (2016). Shoulder height, body mass, and shape of proboscideans. *Acta Palaeontologica Polonica, 61*, 537–574.

Laurent, A. (1907). Quelques observations nouvelles sur les terrains sédimentaires du Velay; avec un complément par M. Boule. *Bulletin de la Société géologique de France, 4e série, 7*, 386–391.

Laws, R. M. (1966). Age criteria for the African elephant *Loxodonta a. africana*. *East African Wildlife Journal, 4*, 1–37.

Lister, A. M. (1996). Evolution and taxonomy of Eurasian mammoths. In J. Shoshani & P. Tassy (Eds.), *The Proboscidea, evolution and palaeoecology of Elephants and their relatives* (pp. 203–213). Oxford: Oxford University Press.

Lister, A. M. (2017). On the type material and evolution of North American mammoths. *Quaternary International, 443*, 14–31.

Lister, A. M., & Sher, A. V. (2001). The origin and evolution of the woolly mammoth. *Science, 294*, 1094–1097.

Lister, A. M., & Sher, A. V. (2015). Evolution and dispersal of mammoths across the Northern Hemisphere. *Science, 350*, 805–809.

Lister, A. M., & Stuart, A. J. (2010). The West Runton mammoth (*Mammuthus trogontherii*) and its evolutionary significance. *Quaternary International, 228*, 180–209.

Lister, A. M., & Van Essen, H. (2003). *Mammuthus rumanus* (Stefanescu), the earliest mammoth in Europe. In A. Petculescu & E. Ştiucă (Eds.), *Advances in Vertebrate Paleontology "Hen to Panta"*, (pp. 47–52). Bucharest: Romanian Academy.

Lister, A. M., Sher, A. V., Van Essen, H., & Wie, G. (2005). The pattern and process of mammoth evolution in Eurasia. *Quaternary International, 126–128*, 49–64.

Lister, A. M., Dimitrijević, V., Marković, Z., Knežević, S., & Mol, D. (2012). A skeleton of "steppe" mammoth (*Mammuthus trogontherii* (Pohlig)) from Drmno, near Kostolac, Serbia. *Quaternary International, 276–277*, 129–144.

Lucas, S. G., Morgan, G. S., Love, D. W., & Connell, S. D. (2017). The first North American mammoths: Taxonomy and chronology of early Irvingtonian (early Pleistocene) *Mammuthus* from New Mexico. *Quaternary International, 443*, 2–13.

Maccagno, A.M. (1962). L'*Elephas meridionalis* Nesti di Contrada "Madonna della Strada" Scoppito (L'Aquila). *Atti dell'Accademia di Scienze fisiche e matematiche di Napoli, Memoria*, 4, 1–132.

Maccagno, A. M. (1965). Nuovi ritrovamenti di resti elefantini nel Villafranchiano della Conqua aquilana. *Bollettino della Società dei Naturalisti, 74*, 231–242.

Markov, G. N. (2012). *Mammuthus rumanus*, early mammoths, and migration out of Africa: Some interrelated problems. *Quaternary International, 276–277*, 23–26.

Markov, G. N., & Spassov, N. (2003). Primitive mammoths from Northeast Bulgaria in the context of the earliest mammoth migrations in Europe. In A. Petculescu & E. Ştiucă (Eds.), *Advances in Vertebrate Paleontology "Hen to Panta"* (pp. 53–58), Bucharest: Romanian Academy.

Maschenko, E. N., Schvyreva, A. K., & Kalmykov, N. P. (2011). The second complete skeleton of *Archidiskodon meridionalis* (Elephantidae, Proboscidea) from the Stavropol Region, Russia. *Quaternary Science Reviews, 30*, 2273–2288.

Mazo, A., Galobart, À., & Colomer, F. (2003). *Mammuthus meridionalis* (Nesti, 1825) de Incarcal (Girona, NE de la Península Ibérica): Descripción e identificación. *Paleontologia i Evolució, 34*, 185–209.

McDaniel, G. E. Jr., & Jefferson, G. T. (2003). *Mammuthus meridionalis* (Proboscidea: Elephantidae) from the Borrego Badlands of Anza-Borrego Desert State Park, California: phylogenetic and biochronologic implications. In: J.W.F. Reumer, J. De Vos & D. Mol (Eds.), *Advances in Mammoth Research (Proceedings of the Second International Mammoth Conference, Rotterdam, May 16–20 1999). Deinsea, 9*, 239-252.

Mein, P. (1975). Résultats du groupe de travail des vertébrés. In J. Senes (Ed.), *Report on Activity of the R.C.M.N.S. Working Groups*, pp. 78-81. Bratislava: Regional Committee on Mediterranean Neogene Stratigraphy.

Mol, D., & Lacombat F. (2009). *Mammuthus trogontherii* (Pohlig, 1885), the steppe mammoth of Nolhac. Preliminary report of a left and right upper M3, excavated at the ancient maar of Nolhac, Haute-Loire, Auvergne, France. *Quaternaire, 20*, 569-574.

Mol, D., van den Bergh, G. D., & de Vos, J. (1999). Fossil proboscideans from the Netherlands, the North Sea and the Oosterschelde Estuary. In G. Haynes, J. Klimowicz & J. W. F. Reumer (Eds), *Mammoths and the Mammoth fauna: Studies of an Extinct Ecosystem. Deinsea, 6*, 119–145.

Mol, D., Tikhonov, A. N., van der Plicht, J., & Bolshiyanov, D. Y. (2003). Discoveries of woolly mammoth, *Mammuthus primigenius* (Proboscidea: Elephantidae) and some other Pleistocene mammals on the Taimyr Peninsula. *Russian Journal of Theriology, 2*, 77–95.

Muttoni, G., Scardia, G., Dimitrijević, V., Kent, D. V., Monesi, E., Mrdjić, N., & Korać, M. (2015). Age of *Mammuthus trogontherii* from Kostolac, Serbia, and the entry of megaherbivores into Europe during the Late Matuyama climate revolution. *Quaternary Research, 84*, 439–447.

Napoleone, G., Albianelli, A., Azzaroli, A., Bertini, A., Magi, M., & Mazzini, M. (2003). Calibration of the Upper Valdarno basin to the Plio-Pleistocene for correlating the Apennine continental sequences. *Italian Journal of Quaternary Sciences, 16*, 131–166.

Nesti, P. (1825). Lettere sopra alcune ossa fossili del Valdarno non per anco descritte. Sulla nuova specie di elefante fossile del Valdarno all'illustrissimo Sig. Dott. Prof. Ottaviano Targioni Tozzetti. Lettera prima. *Nuovo Giornale dei Letterati, 11*, 195–216.

Nikolskiy, P. A., Sulerzhitsky, L. D., & Pitulko, V. V. (2011). Last straw versus Blitzkrieg overkill: Climate-driven changes in the Arctic Siberian mammoth population and the Late Pleistocene extinction problem. *Quaternary Science Reviews, 30*, 2309–2328.

Obada, T. F. (2014). Evidence of contact between *Mammuthus intermedius* (Jourdan, 1861) and ancient humans from Duruitoarea Veche, Republic of Moldova: preliminary data. In D.S. Kostopoulos, E. Vlachos & E. Tsoukala (Eds.), *VIth International Conference on Mammoths and their Relatives. Scientific Annals, School of Geology, Aristotle University of Thessaloniki, Greece. Grevena-Stiatista, 102*, 145–146.

Palombo, M. R. (2017). Discrete dispersal bioevents of large mammals in Southern Europe in the post-Olduvai Early Pleistocene: A critical overview. *Quaternary International, 431*, 3–19.

Palombo, M. R., & Ferretti, M. P. (2005). Elephant fossil record from Italy: Knowledge, problems, and perspectives. *Quaternary International, 126–128*, 107–136.

Pastre, J.-F., Debard, E., Nomade, S., Guillou, H., Faure, M., Guérin, C., et al. (2015). Nouvelles données géologiques et téphrochronologiques sur le gisement du maar de Senèze (Pléistocène inférieur, massif central, France). *Quaternaire, 26*, 225–244.

Petrova, E. A., Masutin, V. V., & Zhuykova, I. A. (2017). Two incomplete skeletons of woolly mammoth (*Mammuthus primigenius*) from the late Pleistocene in the Kirov Region, European Russia. *Russian Journal of Theriology, 16*, 157–175.

Pohlig, H. (1885). Über eine Hipparionen-Fauna von Maragha in Nordpersien, über fossile Elephantenreste Kaukasiens und Persiens und über die Resultate einer Monographie der fossilen Elephanten Deutschlands und Italiens. *Zeitschrift der Deutschen Geologischen Gesellschaft, 37*, 1022–1027.

Popescu, A. (2008). Carpal bones of *Mammuthus meridionalis* (Nesti, 1825) in Leu (Dolj district). *Oltenia Studii şi Comunicări Ştiinţele Naturii, 24*, 255–261.

Popescu, A. (2011). The tarsals of *Mammuthus meridionalis* (Nesti, 1825) from Leu (Dolj County, Romania). *Geo-Eco-Marina, 17*, 211–217.

Reggiani, P., & Sala, B. (1994). *Archidiskodon* cf. *gromovi* (Proboscidea, Elephantidae) from Savignano sul Panaro, Modena (Northeastern Italy). *Memorie di Scienze Geologiche, 46*, 313–327.

Rivals, F., Semprebon, G., & Lister, A. (2012). An examination of dietary diversity patterns in Pleistocene proboscideans (*Mammuthus*, *Palaeoloxodon*, and *Mammut*) from Europe and North America as revealed by dental microwear. *Quaternary International, 255*, 188–195.

Ros-Montoya, S. (2010). Los Proboscídeos del Plio-Pleistoceno de las cuencas de Guadix-Baza y Granada. Tesis, Universidad de Granada.

Ros-Montoya, S., Madurell-Malapeira, J., Martínez-Navarro, B., Espigares, M.-P., & Palmqvist, P. (2012). Late Villafranchian *Mammuthus meridionalis* (Nesti, 1825) from the Iberian Peninsula: Dentognathic remains from Incarcal-I (Crespià, Girona) and Venta Micena (Orce, Granada). *Quaternary International, 276–277*, 17–22.

Ros-Montoya, S., Palombo, M. R., Espigares, M. P., Palmqvist, P., & Martínez-Navarro, B. (2018). The mammoth from the archeo-paleontological site of Huéscar-1: A tile in the puzzling question of the replacement of *Mammuthus meridionalis* by *Mammuthus trogontherii* in the late Early Pleistocene of Europe. *Quaternary Science Reviews, 197*, 336–351.

Rossi, M. A., Agostini, S., Palombo, M. R., Angelini, I., Caramiello, S., Casarin, F., et al. (2017). *Mammuthus meridionalis* from Madonna della Strada (Scoppito, L'Aquila): Diagnostics and restoration. *Bollettino della Società Paleontologica Italiana, 56*, 359–373.

Rousselières, F. (2006). Proboscidiens du Pléistocène moyen et supérieur d'Europe sud-occidentale: Anatomie et biométrie statistique. Thèse, Université Claude Bernard-Lyon I.

Sanders, W. J. (2007). Taxonomic review of fossil Proboscidea (Mammalia) from Langebaanweg, South Africa. *Transactions of the Royal Society of South Africa, 62*, 1–16.

Schaub, S. (1943). Die oberpliocäne Säugetierfauna von Senèze (Haute-Loire) und ihre verbreitungsgeschichtliche Stellung. *Eclogae Geologicae Helvetiae, 36*, 270-289.

Schaub, S. (1948). Das Gebiss der Elephanten. *Verhandlungen der Naturforschenden Gesellschaft Basel, 59*, 89–112.

Semprebon, G. M., Rivals, F., Fahlke, J. M., Sanders, W. J., Lister, A. M., & Göhlich, U. B. (2016). Dietary reconstruction of pygmy mammoths from Santa Rosa Island of California. *Quaternary International, 406*, 123–136.

Shpansky, A. V., Vasiliev, S. K., & Pecherskaya, K. O. (2015). The Steppe Elephant *Mammuthus trogontherii* (Pohlig) from the Irtysh Region Near Omsk. *Paleontological Journal, 49*, 304–325.

Shvyreva, A. K., & Titov, V. V. (2018). About cases of pathology of the dental system in *Archidiskodon meridionalis meridionalis* (Nesti, 1825) from Georgievsk sand pit (Stavropol Territory, Russia). *Proceedings of the Zoological Institute of the Russian Academy of Sciences, 322*, 259–267.

Stehlin, H. G. (1923). Die oberpliocäne Fauna von Senèze (Haute-Loire). *Eclogae Geologicae Helvetiae, 38*, 268–281.

Stuart, A. J., Sulerzhitsky, L. D., Orlova, L. A., Kuzmin, Y. V., & Lister, A. M. (2002). The latest woolly mammoths (*Mammuthus primigenius* Blumenbach) in Europe and Asia: A review of the current evidence. *Quaternary Science Reviews, 21*, 1559–1569.

Tikhonov, A., Agenbroad, L., & Vartanyan, S. (2003). Comparative analysis of the mammoth populations on Wrangel Island and the Channel Islands. In: J.W.F. Reumer, J. De Vos & D. Mol (Eds.), *Advances in Mammoth Research (Proceedings of the Second International Mammoth Conference, Rotterdam, May 16–20 1999). Deinsea, 9*, 415–420.

Titov, V. V. (2001). Most ancient elephants from the south of Russia. In G. Cavarretta, P. Gioia, M. Mussi, & M. R. Palombo (Eds.), *The World of Elephants* (pp. 152–156). Rome: Consiglio Nazionale delle Ricerche.

Van Essen, H. (2003). Tooth morphology of *Mammuthus meridionalis* from the southern bight of the North Sea and from several localities in the Netherlands. *Deinsea, 9*, 453–511.

Van Essen, H. (2011). Tracing transitions. an overview of the evolution and migrations of the genus *Mammuthus* Brookes, 1828 (Mammalia, Proboscidea). Ph.D. Dissertation, Leiden University.

Vartanyan, S. L., Garutt, V. E., & Sher, A. V. (1993). Holocene dwarf mammoths from Wrangel Island in the Siberian Arctic. *Nature, 362*, 337–340.

Vasile, S., Panaitescu, D., Ştiucă, E., & Virág, A. (2012). Additional Proboscidean Fossils from Mavrodin (Teleorman County, Romania). *Oltenia. Studii şi comunicări. Ştiinţele Naturii, 28*, 211–218.

Vasile, S., Cojocaru, V., & Torcărescu, B. A. (2019). A partial *Mammuthus meridionalis* skeleton from Brebina (Teleorman County, Dacian Basin). *Buletinul Muzeului Judeţean Teleorman, 11*, 5–14.

Veltre, D. W., Yesner, D. R., Crossen, K. J., Graham, R. W., & Coltrain, J. B. (2008). Patterns of faunal extinction and paleoclimatic change from mid-Holocene mammoth and polar bear remains, Pribilof Islands, Alaska. *Quaternary Research, 70*, 40–50.

Vereshchagin, N. K., & Tikhonov, A. N. (1986). A Study on Mammoth Tusks. *Proceedings of the Zoological Institute. USSR Academy of Sciences, 149*, 3–14.

Vereshchagin, N. K., & Tikhonov, A.N. (1999). Exterior of the mammoth. *Cranium, 16*, 4–44.

Vialli, V. (1956). Sul Rinoceronte e l'Elefante dei livelli superiori della serie lacustre di Leffe (Bergamo). *Memorie della Società Italiana di Scienze Naturale del Museo Civico di Storia Naturale di Milano, 12*, 1–70.

Virág, A. (2009). Revision of the "*Archidiskodon meridionalis ürömensis*" Vörös, 1979 in the context of the mammoth evolution in Eurasia. *Central European Geology, 52*, 405–419.

Viret, J. (1954). Le loess à bancs durcis de Saint-Vallier (Drôme) et sa faune de mammifères villafranchiens. *Nouvelles Archives du Muséum d'histoire Naturelle de Lyon, 4*, 1–200.

Wei, G. B., Taruno, H., Kawamura, Y., & Jin, C. (2006). Pliocene and early pleistocene primitive mammoths of Northern China: Their revised taxonomy, biostratigraphy and evolution. *Journal of Geosciences, 49*, 59–101.

Wei, G. B., Hu, S. M., Yu, K. F., Hou, Y. M., Li, X., Jin, C. Z., et al. (2010). New materials of the steppe mammoth, *Mammuthus trogontherii*, with discussion on the origin and evolutionary patterns of mammoths. *Science China Earth Sciences, 53*, 956–963.

Weithofer, K. A. (1890). Die fossilen Proboscidier des Arnothales in Toskana. *Beiträge zur Paläontologie von Österreich-Ungarns, 8*, 107–240.

Chapter 16
Primates from Senèze

Eric Delson

Abstract Senèze has yielded two primate fossils, neither from the recent field work. A partial ulna can be identified as cf. *Macaca sylvanus*. A nearly complete female skull is the holotype of *Dolichopithecus arvernensis* Depéret, 1928, now included in *Paradolichopithecus*. Emended and differential diagnoses are provided for this genus and species for the first time, and the Senèze skull is described and briefly compared to those of other species of the genus. Dental and cranial measurements are provided, along with comparative data. The macaque and *Paradolichopithecus* are moderately to highly terrestrial, respectively, and they suggest a woodland to more open environment. Senèze may be the youngest known occurrence of *Paradolichopithecus arvernensis*. A comparison of previous analyses results in retention of a *Paradolichopithecus*-*Macaca* link and of *Paradolichopithecus* and *Procynocephalus* as distinct genera.

Résumé Deux restes de Cercopithécidés fossiles sont connus à Senèze, ni l'un ni l'autre ne proviennent des fouilles récentes. Un fragment de cubitus est attribué à cf. *Macaca sylvanus*. Un crâne femelle sub-complet est l'holotype de *Paradolichopithecus arvernensis* (Depéret 1928). Nous retraçons un bref historique de la recherche sur le genre *Paradolichopithecus*. Nous présentons des diagnoses amendées de *Paradolichopithecus* et de *P. arvernensis*, accompagnées de comparaisons avec d'autres genres papionins. Le crâne de Senèze est décrit en détail. En se basant sur du matériel provenant d'autres sites nous pouvons dire que les squelettes postcrâniens de *Macaca* et de *Paradolichopithecus* sont adaptés à un habitus en partie ou totalement terrestre; cela suggère un paysage boisé relativement ouvert autour du maar de Senèze. Dans l'état actuel des connaissances, Senèze est peut-être le gisement le plus tardif ayant livré le genre *Paradolichopithecus,* vers 2,2–2,1 Ma. Compte tenu de sa taille légèrement plus petite, les spécimens de Grăunceanu sont probablement un peu plus ancien. Le spécimen d'individu femelle subadulte écrasé de Dafnero paraît de taille similaire, et une réévaluation des données paléomagnétiques publiées, suggèrent un âge compris entre 2,58 et 2,2 Ma.

Deux questions sur la systématique de *Paradolichopithecus* sont particulièrement discutées: est-il un synonyme de *Procynocephalus*, et est-il plus proche de *Macaca* ou de *Papio* ? La plupart des caractéristiques communes aux deux genres fossiles eurasiens sont basées sur une grande taille et une adaptation locomotrice terrestre, bien qu'elles diffèrent par la forme du dos du museau; de ce fait les deux genres sont maintenus comme distincts. De même, les caractéristiques suggérant un lien entre *Paradolichopithecus* et *Papio* (ou d'autres Papionina) semblent principalement liées à leur grande taille. Les arguments de distinctions ou de relations fondés sur le rapport de la surface de P4/M1 et à la forme du labyrinthe osseux ne résistent pas à un examen minutieux. Des analyses inédites de la morphométrie géométrique 3D et de cladistique des caractères du crâne tendent à montrer une relation entre *Paradolichopithecus* et *Macaca*.

Keywords *Paradolichopithecus arvernensis* • *Macaca sylvanus* • Cercopithecinae • Papionini • Macacina

E. Delson (✉)
Department of Vertebrate Paleontology, American Museum of Natural History, New York, NY, USA
e-mail: eric.delson@lehman.cuny.edu

Department of Anthropology, Lehman College of the City University of New York, Bronx, NY, USA

PhD Program in Anthropology, The Graduate Center of the City University of New York, New York, NY, USA

New York Consortium in Evolutionary Primatology, New York, NY, USA

Institut Català de Paleontologia Miquel Crusafont, Universitat Autònoma de Barcelona, Cerdanyola del Vallès, Barcelona, Spain

E. Delson et al. (eds.), *Senèze: Life in Central France Around Two Million Years Ago*,
Vertebrate Paleobiology and Paleoanthropology, https://doi.org/10.1007/978-3-031-64415-3_16

Introduction

Only two primate specimens have been recovered at Senèze: one is a fragment of ulna, the other is a nearly complete skull, the holotype of a widespread species. This chapter will review the two specimens and discuss their taxonomic position and implications for paleoenvironment and relative age.

Note: upper teeth are indicated by uppercase letters, lower teeth by lowercase letters.

Institutional abbreviations: AMNH-M, American Museum of Natural History, Department of Mammalogy: AP, Sciences de la Terre, Faculté des Sciences, Université d'Alger, Algiers, Algeria: CMAI, Catedra de Morfologie si Antropologie, Universitatea 'Al. I. Cuza', Iasi, Romania; ERIS, Emil Racoviţă Speleological Institute, Romanian Academy, Bucharest, Romania; FSL, Laboratoire de Paléontologie, Université Claude Bernard (UCB)-Lyon 1 (previously Faculty of Sciences, Lyon); ICP, Institut Català de Paleontologia Miquel Crusafont, Barcelona, Spain; LGPUT, Laboratory of Geology and Paleontology, University of Thessaloniki, Greece; MAP, Museo dell'Accademia del Poggio, Montevarchi, Italy; MNHN-Z-AC, Muséum national d'Histoire naturelle, Zoologie, Labo. Anatomie Comparée, Paris, France; NHCV, Natural History Collection, Museum of Vrissa, Lesvos, Greece; (now housed in Athens Museum of Paleontology and Geology; NMB, Naturhistorisches Museum Basel, Switzerland; NMP, Narodni Muzeum, Prague, Czech Republic; PIN-M, Institute of Paleontology, Russian Academy of Sciences, Moscow, Russia).

Previous Work on *Paradolichopithecus*

Delson and Plopsor (1975) and Szalay and Delson (1979) summarized the taxonomic history of this genus, including its frequent confusion with the colobine *Dolichopithecus*, which Depéret (1889) had described from Perpignan. In brief (updated here), Depéret (1928) reported the discovery of a "giant" relative of *Dolichopithecus* from Senèze, which he named *D. arvernensis*. This paper was delivered at the 1926 meeting of the International Geological Congress; it was rarely cited previously and never indicated as the original description of this species. The following year, Depéret (1929) described in detail the nearly complete skull and mandible of an adult female from Senèze, still considering that it was a later stage in a lineage including his Ruscinian species *D. ruscinensis*; in fact, Depéret linked the two more by size-related characters than by distinctive morphology. Verheyen (1962) first recognized that the narrow interorbital space, especially by comparison to the long muzzle, suggested that "*D.*" *arvernensis* was likely not a colobine but a cercopithecine. Vogel (1966, 1968) noted other features which placed this species in the Cercopithecinae, close to *Macaca*. Jolly (1967) reviewed the evolution of "baboons" (which he defined as large-bodied terrestrial cercopithecids) and followed Vogel and Verheyen, suggesting that *Dolichopithecus arvernensis* was a cercopithecine similar in cranial morphology to some macaques or possibly to the Asian *Procynocephalus*. Simons (1970) went a step further, suggesting that "*D.*" *arvernensis* should be included in *Procynocephalus*, perhaps in the type species *Pro. wimani* Schlosser (1924).

Meanwhile a partial face of a large monkey was one of the first fossils to be recovered from the Romanian site of Valea Grăunceanului (hereafter Grăunceanu), and Necrasov et al. (1961) described this as a new form similar to Depéret's *D. arvernensis*. They argued that it was probably colobine, but with some cercopithecine dental features, as well as the unique character of possessing a double supraorbital notch. This supposed double notch is, in fact, a single supraorbital notch, typical of larger cercopithecines, next to the broken region where the medial orbit margin turns down along the nasal bones; the damage probably confused Necrasov and colleagues, who were not primate specialists. They named the new genus and species *Paradolichopithecus geticus*, with this specimen as holotype. Simons (1970) recognized the similarity of this new species to "*D.*" *arvernensis* and suggested placing it also in *Procynocephalus*.

In 1971, Delson described a fragmentary juvenile mandible from the Spanish site of Cova Bonica and compared its large papionin dentition to those of extant *Papio* as well as "*D.*" *arvernensis* and *Paradolichopithecus geticus*, but only identified it as a large cercopithecine (cf. *Papio*). Delson and colleagues (Delson 1975; Delson & Plopsor 1975; Szalay and Delson 1979; Delson et al. 2000, 2014; Delson and Frost 2004; Ting et al. 2004; Frost et al. 2005) reviewed these and other finds, terming them *Paradolichopithecus arvernensis*, and employing *Paradolichopithecus* sp. for slightly smaller-bodied and geologically older specimens (see list below). Additional fossils of *P. arvernensis* were found at the Vatera-F site on Lesvos island (Greece; van der Geer & Sondaar 2002) and most recently at Dafnero-3 (Greece; Kostopoulos et al. 2018) and probably Karnezeika, Greece (two teeth and a fragment of radius; Sianis et al. 2023).

Meanwhile, Trofimov (1977) described several crania and mandibles from Kuruksay, Tajikistan, as *Paradolichopithecus sushkini*. These were discussed in more detail by

Maschenko (1994, 2005), who argued that *Paradolichopithecus* was phylogenetically close to or congeneric with *Papio*, while previous authors (e.g., Szalay and Delson 1979, Delson et al. 2000, Frost et al. 2005) had linked it with *Macaca* on the basis of facial shape. Qiu et al. (2004) named *Paradolichopithecus gansuensis* on the basis of a partial maxilla and mandible from Longdan (Gansu, China), suggesting that the teeth of *Paradolichopithecus* differed from those of *Procynocephalus*. These six localities (Grăunceanu, Vatera-F, Dafnero-3, Karnezeika, Kuruksay and Longdan) all appear to be similar in age (middle Villafranchian, MNQ 17, ca. 2.6–2.2 Ma), somewhat older than Senèze (dated 2.2–2.1 Ma, see Delson et al. 2024).

Another series of mostly fragmentary finds come mainly from still older sites, of late early Ruscinian to early Villafranchian age (ca. 4.3–2.6 Ma). Delson (1974) noted in a list (see also Szalay and Delson 1979 and Delson et al. 2000) that "*Dolichopithecus*" *arvernensis* was known from Malusteni (Romania) and Vialette (France). The former specimen was discussed as *D. ruscinensis* by Simionescu (1930), the latter was mentioned as an intermediate member of the "*Dolichopithecus*" lineage by Schaub (1943), but neither has yet been described in detail. Aguirre and Soto (1974, 1978) reported a juvenile partial mandible from La Puebla de Valverde (Spain). Moyà Solà et al. (1990) mentioned similar material from Moreda-1 (Spain), and Vlačiky (2009; see also Šujan et al. 2023) noted its presence at Nová Vieska (Slovakia). The most recent report of new material concerns three teeth from Ridjake, Serbia (Radović et al. 2019; 2024). All of these and the mandible and new specimens from Cova Bonica are attributed here to *Paradolichopithecus* sp. The origin of this lineage is unclear, whether from a European macaque or as the sister taxon to *Macaca*.

Meanwhile, in a series of papers, Nishimura and colleagues (2007, 2009, 2010, 2014) examined the facial morphology of *Paradolichopithecus* specimens from Kuruksay, Senèze and Longdan and the maxilla of *Procynocephalus wimani* from Xinan (Henan, China). They focused on the presence and extent of the maxillary sinus, which had been recognized in all extant species of *Macaca*, alone among modern cercopithecids. Of these, only "*P.*" *sushkini* presented a clear maxillary sinus. Nishimura et al. (2014) suggested that the maxillary sinus might be lost in some large macacinans due to development of the maxillary fossa.

In a brief abstract, Plastiras et al. (2017) termed the Dafnero specimen *Procynocephalus* without comment. Kostopoulos et al. (2018) suggested that the Dafnero cranium could be identified as either *Paradolichopithecus* or *Procynocephalus* and proposed that *Paradolichopithecus* is likely a junior synonym of *Procynocephalus*, following Jolly (1967) and Simons (1970). They also implied that *Paradolichopithecus* was more likely related to *Papio* than to *Macaca* (following Maschenko 1994, 2005).

Le Maître et al. (2023) analyzed the bony labyrinth morphology of the Dafnero specimen and other cercopithecines. Their results were not definitive, but they concluded that *Paradolichopithecus* showed similarities to Cercopithecini and Papionina, although not many to macaques. They suggested that *Paradolichopithecus* might be a stem papioninan or perhaps a basal papionin closer to Papionina than to *Macaca*.

Here I provide revised diagnoses of *Paradolichopithecus* and *P. arvernensis*, describe the Senèze specimen in detail and briefly analyze the fragmentary macaque ulna from the site. I also consider some of the above arguments in more detail (see Discussion below). Delson et al. (in preparation) will describe and analyze the fossils from Grăunceanu and Lesvos and place *Paradolichopithecus* in a broader context.

Systematic Paleontology

Order Primates Linnaeus, 1758
Semiorder Euprimates Hoffstetter, 1977
Hyporder Anthropoidea Mivart, 1864
Infraorder Catarrhini E. Geoffroy St. Hilaire, 1812
Parvorder Eucatarrhini Delson, 1977
Superfamily Cercopithecoidea Gray, 1821
Family Cercopithecidae Gray, 1821
Subfamily Cercopithecinae Gray, 1821
Tribe Papionini Burnett, 1828
Subtribe Macacina Owen, 1843
Genus *Macaca* Lacépède, 1799
cf. *M. sylvanus*, Linnaeus, 1758

Material: left partial ulna, NMB Se 1534, discovered in 1914, first reported by Stehlin (1923) as *Macaca* sp. Fragmentary specimen, with most of the proximal end and distal half broken away, but preserving the radial and distal part of the humeral articular surfaces. Figure 16.1 shows it in lateral view compared to a slightly larger ulna from Fornace RDB/Villafranca d'Asti (earliest Villafranchian) and a male extant *Macaca sylvanus*. The Senèze ulna is smaller than both others but may be a female (or possibly subadult if the distal epiphysis was unfused). The shaft appears straight and not curved, as also seen in the other two specimens, but the two fossils appear relatively more robust distal to the coronoid process compared to the extant specimen. Se 1534 presents less excavation or muscle markings laterally than the Fornace specimen or partial ulnae from the Upper Valdarno and Ain Mefta, or medially (posterior to the trochlear notch) than in the latter two ulnae or one from Zlaty Kun (see Table 16.1). This gracility again suggests a more lightly built, possibly female individual. The preserved length of the bone is 93 mm, about half the length of a complete male or 55% of a complete female element of *Macaca sylvanus* (Table 16.1). Only two standard measurements can be taken on the Senèze ulna. Table 16.1 presents those values in comparison to extant and fossil *Macaca sylvanus* and the much larger *Paradolichopithecus arvernensis*. Those are the only two cercopithecid species known in Europe around 2 Ma, implying that the Senèze ulna can be identified as cf. *Macaca sylvanus*.

Genus *Paradolichopithecus* Necrasov, Samson and Rădulesco (1961).

Synonymy: *Dolichopithecus* Depéret, 1889, in part: Depéret (1928), Simionescu (1930); *Papio* Erxleben, 1777, in part: Delson (1971) (cf.), Aguirre and Soto (1974) (cf.), Maschenko (1994, 2005); *Procynocephalus* Schlosser, 1924, in part: Jolly 1967 (tentative), Simons (1970) (tentative), Kostopoulos et al. (2018) (tentative).

Type species: *Paradolichopithecus geticus* Necrasov, Samson and Rădulesco, 1961 (now generally accepted as a junior synonym of *Dolichopithecus arvernensis* Depéret, 1928).

Other included species: *Paradolichopithecus* sp. Delson and Plopsor (1975) (Vialette, France; Moreda 1A, Cova Bonica, La Puebla de Valverde, Spain; Malusteni, Romania; Nová Vieska, Slovakia; Ridjake, Serbia); possibly also

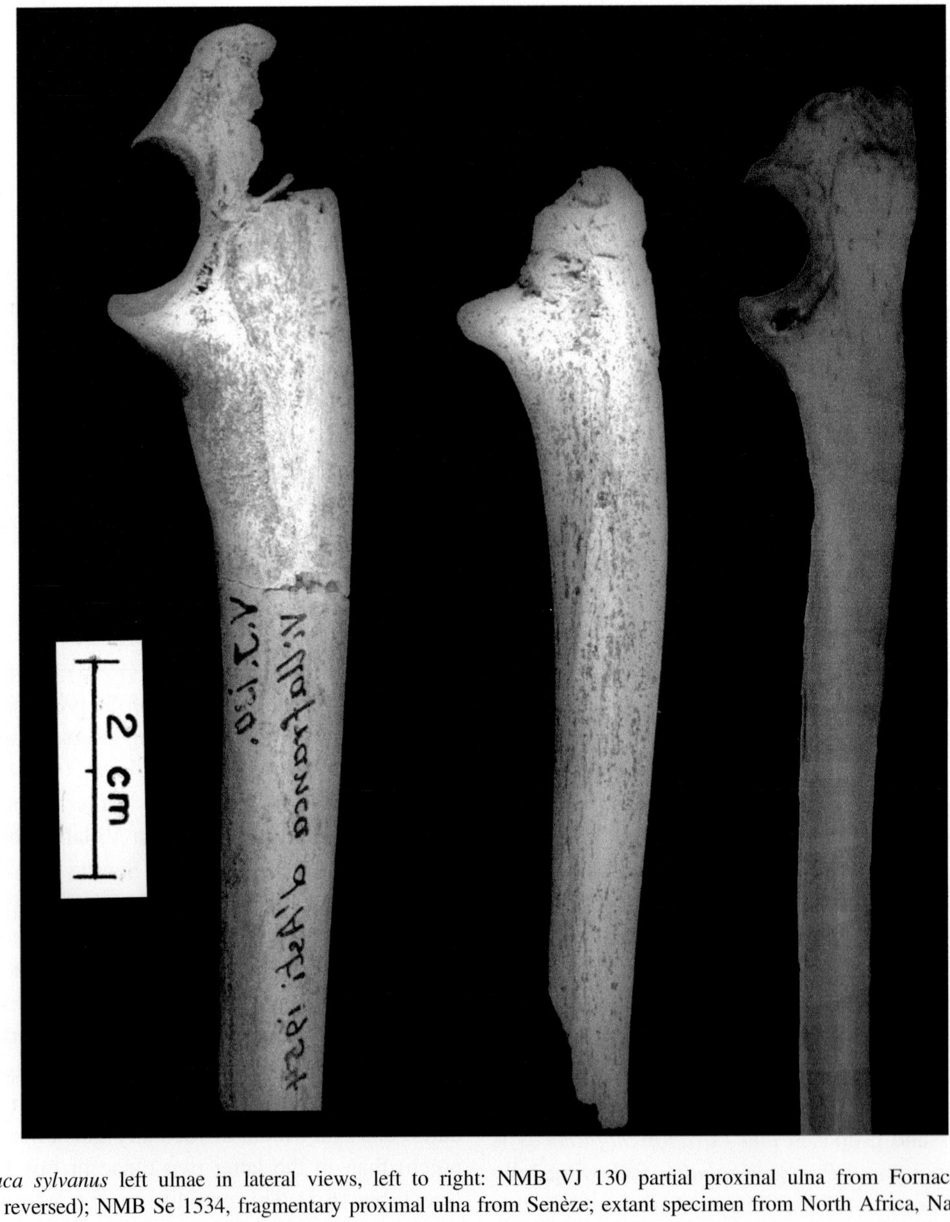

Fig. 16.1 *Macaca sylvanus* left ulnae in lateral views, left to right: NMB VJ 130 partial proxinal ulna from Fornace RDB (right side photographically reversed); NMB Se 1534, fragmentary proximal ulna from Senèze; extant specimen from North Africa, Naturalis (Leiden)

"*Paradolichopithecus*" *sushkini* Trofimov, 1977 (Kuruksay, Tajikistan) and "*Paradolichopithecus*" *gansuensis* Qiu, Deng and Wang, 2004 (Longdan, Gansu, China); see discussion below.

Emended diagnosis (formalized and modified in part after Szalay and Delson 1979): Large papionin comparable in size to larger subtaxa of extant *Papio*. Distinguished by combination of smooth median dorsal profile without anteorbital drop, rounded and slightly peaked (not flattened) transverse muzzle dorsum and nasals, little to no facial fossae or maxillary ridges and known postcranial elements (especially forelimb and phalanges) indicative of highly terrestrial locomotion.

Differential diagnosis (see also Gilbert 2013): *Paradolichopithecus* is distinguished from *Dinopithecus, Gorgopithecus*, *Mandrillus*, *Papio*, *Pliopapio*, *Soromandrillus* and *Theropithecus* by the lack of a defined anteorbital drop in the midline facial profile; *Cercocebus, Lophocebus, Macaca, Parapapio, Procercocebus* and *Rungwecebus* also lack an anteorbital drop. Distinguished from *Dinopithecus, Gorgopithecus*, *Mandrillus*, *Papio* and *Soromandrillus* by rounded (and slightly "peaked") rather than flattened transverse muzzle dorsum; similar in this region to *Pliopapio* and to some *Cercocebus, Lophocebus, Macaca, Parapapio, Procercocebus*, *Rungwecebus* and *Theropithecus* (all variable); differs from *Procynocephalus*, which has a more flattened middle part of muzzle dorsum (breakage prevents observation of anteorbital region). Similar to *Macaca* and *Procynocephalus* (and variably *Dinopithecus, Parapapio* and *Pliopapio*) but distinguished from *Cercocebus, Gorgopithecus, Lophocebus, Mandrillus*, *Papio*, *Procercocebus*, *Rungwecebus*, *Soromandrillus* (possibly variable in later Omo and Angolan samples) and *Theropithecus* (other than most *T. oswaldi*) by the absence of a maxillary fossa and mandibular corpus fossa. Distinguished from *Macaca* (and "*P.*" *sushkini*) by lack of a maxillary sinus. Distinguished from *Mandrillus*, *Papio*, *Procercocebus*, *Soromandrillus*, most fossil *Theropithecus* and (morphology variable) *Cercocebus, Lophocebus* and *Parapapio* in lacking male maxillary ridges. Distinguished from *Cercocebus, Parapapio, Pliopapio*, *Procercocebus*, *Rungwecebus* and (variably) *Lophocebus* and *Macaca* by prominent and horizontally oriented supraorbital torus in males; in *Gorgopithecus* also prominent but anteroinferiorly oriented, downturned in midline. Distinguished from *Cercocebus, Mandrillus*, *Procercocebus* and *Soromandrillus* by anteriorly converging temporal lines, rather than more divergent, more posteriorly converging lines. Distinguished from *Gorgopithecus* by vertical rather than anteroposteriorly curved upper incisor roots. Distinguished from *Theropithecus* by simpler molar structure lacking elevated and complex enamel folding. The p4/P4 is not typically enlarged relative to the m1/M1 as it is in *Cercocebus, Mandrillus*, and some (geochronologically younger) *Soromandrillus*, but more as in *Lophocebus, Papio* and at least some *Macaca* (although there is some overlap of ranges in extant taxa, see Table 16.2 and Fig. 16.2). Distinguished from *Cercocebus, Lophocebus, Macaca, Parapapio, Pliopapio* and *Rungwecebus* but broadly similar to *Mandrillus*, *Papio*, *Procercocebus* (probably), *Procynocephalus* and *Theropithecus* by strongly terrestrial/cursorial adaptations of the forelimb such as robust humerus with medial epicondyle well retroflexed and medial trochlear

Table 16.1 Comparative ulnar measurements (in mm) of extant and fossil *Macaca sylvanus* and *Paradolichopithecus arvernensis*

Specimen, locality, reference	Age	Sex	Length	APThTN	WRA
Macaca sylvanus					
NMB Se 1534, Senèze	eEPl	f?	93 (b)	10.5	14.0
AP 1A7 Ain Mefta, Algeria (Pomel, 1896)	lMPl	?		13.0	15.0
NMP unnumbered, Zlaty Kun Cave 718, Czech Republic (Vlcek, 1961)	mMPl	f?		10.5	13.2
ICP 14956, Terrassa, Spain (Alba et al. 2008; specimen not yet published)	lEPl	m?		11.3	14.0
MAP 12, Upper Valdarno (Terranova Bracciolini), Italy (Gentili et al., 1998; not yet published)	mEPl	m?		11.7	14.4
NMB VJ 130, Fornace RDB (Villafranca d'Asti), Italy (Rook et al., 2001)	LP	m?		13.4	15.8
MNHN-Z-AC, 1910.166, extant	modern	m	185.1	14.5	15.8
MNHN-Z-AC 1900.244, extant	modern	f	167.4	11.4	14.1
MNHN-Z-AC 1899.52, extant	modern	m	186.0	14.2	17.2
AMNH-M 185,277, extant	modern	f	168.9	11.6	12.5
Paradolichopithecus arvernensis					
ERIS VGr/350 b, Grăunceanu, Romania (Szalay and Delson 1979)	eEPl	m?		20.0	21.0
NHCV PO 059 F, Vatera-F, Lesvos, Greece (van der Geer and Sondaar 2002)	eEPl	m?		20.3	–
NHCV PO 229 F, Vatera-F, Lesvos, Greece (van der Geer and Sondaar 2002)	eEPl	m?		20.6	23.4

Age abbreviations: l,. L–late; m, M–middle; e, E–early; Pl–Pleistocene; P–Pliocene. Sex: f, female; m, male. All measurements by ED on originals.
Length: (proximal surface of olecranon to distal tip of styloid process ("head"); b: measurement of broken length as preserved.
APThTN: anteroposterior thickness of the trochlear notch in the (proximo-distal) middle of the notch;
WRA: width across the radial articulation from the coronoid tip to the lateral margin

"crest" expanded distally; ulnar olecranon large and somewhat tilted posteriorly; partial femora robust; and phalanges stout (relevant postcrania unknown in *Dinopithecus, Gorgopithecus* and *Soromandrillus*). Distinguished from *Cercocebus, Lophocebus, Macaca, Parapapio, Pliopapio, Procercocebus, Rungwecebus* and smaller subtaxa of *Papio* and *Theropithecus* in size and broadly comparable to *Dinopithecus, Gorgopithecus, Mandrillus*, larger subtaxa of *Papio, Soromandrillus,* and mid-sized subtaxa of *Theropithecus* (as indicated by tooth and skull size and estimated mass: females ca. 15–23 kg, males ca. 18–40 kg, from Delson et al. 2000).

Description: The cranial midline profile descends from glabella in a smooth, hollow curve to rhinion, which projects slightly, and then more steeply and almost linearly to prosthion. The flanks of the maxilla and mandibular corpus are

Table 16.2 Comparative dental measurements (in mm) of *Paradolichopithecus*, *Papio* and *Mandrillus*

Taxon/specimen	I1 W	I1 L	I1 H	I2 W	I2 L	I2 H	C W f	C L f	C H f	P3 W f	P3 L f	P4 W	P4 L
P. arvernensis Senèze FSL 41333	7.1	7.7	6.9	6.6	5.5	9.8	7.9	8.2	13.2	9.0	7.0	9.9	7.3
P. arvernensis Grăunceanu CMAI unnumbered	8.5	8.2	10.7	7.8	8.2	11.6	6.5	9.0	14.1	7.9	7.4	9.0	8.3
P. arvernensis Dafnero-3 LGPUT DFN-3 150 (right side)										7.1	6.9	7.9	7.8
"*P.*" *sushkini* Kuruksay PIN-M 3120–523				7.2	7.3	9.5	7.0	8.6	16.3	8.0	7.4	9.6	7.6
Papio h. ursinus Mean	9.28	9.63		7.91	7.2		6.65	8.54		7.23	6.99	8.72	7.85
Papio h. ursinus N	130	132		116	117		57	57		61	61	144	146
Papio h. ursinus Max	11.9	11.6		10.0	8.7		8.1	12.3		8.3	8.6	10.3	9.8
Papio h. ursinus Min	6.8	6.4		5.6	5.1		4.8	5.9		6.1	5.7	7.0	6.1
	M1AW	M1PW	M1 L	M2AW	M2PW	M2 L	M3AW	M3PW	M3 L	I1 A	P4 A	M1 A	P4/M1 A
P. arvernensis Senèze FSL 41333	10.2	9.6	10.8	12.8	11.0	13.3	12.8	10.7	13.7	54.7	72.27	110.16	0.66
P. arvernensis Grăunceanu CMAI	10.0	9.7	12.1	12.1	10.9	12.9	12.4		14.3	69.7	74.7	121.0	0.62
P. arvernensis Dafnero-3 LGPUT	9.4	8.9	10.8	10.6	11.4	13.4	11.7	10.2	13.5		61.6	101.5	0.61
"*P.*" *sushkini* Kuruk PIN-M 3120–523	11.7	10.7	13.3	14.0	12.4	14.7	14.8	12.2	14.2		72.96	155.61	0.47
Papio h. ursinus Mean	10.0	9.34	11.13	11.87	10.95	13.48	12.34	10.69	13.91	89.1	68.82	111.47	0.62
Papio h. ursinus N	171	170	173	160	159	160	141	141	141		143	171	145
Papio h. ursinus Max	12.0	11.0	13.1	13.7	12.9	15.5	14.2	13.1	16.4		94.9	140.2	0.81
Papio h. ursinus Min	8.7	8.0	8.35	9.5	9.0	10.4	10.2	8.2	10.1		43.9	72.8	0.48
	i1 W	i1 L	i1 H	i2 W	i2 L	i2 H	c W f	c L f	c H f	p3 W f	p3 Lf	p4 W	p4 L
P. arvernensis Senèze FSL 41333	7.0	4.8	6.8	3.6	7.5	6.4	9.8	6.4	12.4	6.4	10.9	7.2	8.4
P. arvernensis Grăunceanu CMAI							8.2	4.9	11.7	5.3	10.1	8.7	9.3
P. arvernensis Vatera NHCV PO 170 F	6.67	5.39	10.2	6.09	5.53	9.5	7.7	5.57	14.13	5.37	9.9	7.11	8.67
Papio h. ursinus Mean	8.64	7.28		7.99	6.23		8.23	5.01		5.12	12.17	7.00	8.88
Papio h. ursinus N	118	120		120	119		59	59		61	61	133	137
Papio h. ursinus Max	10.3	8.6		9.8	8.0		9.7	5.8		6.2	16.1	8.4	10.6
Papio h. ursinus Min	6.5	4.3		5.9	3.6		7.1	4.3		4.4	9.0	5.7	7.0

(continued)

Table 16.2 (continued)

Taxon/specimen	m1AW	m1PW	m1 L	m2AW	m2PW	m2 L	m3AW	m3PW	m3 L	p4 A	m1 A	p4/m1 A
P. arvernensis Senèze FSL 41333	8.6	8.5	10.1	10,5	10.0	12.4	10,8	10.4	17.8	60.48	86.86	0.70
P. arvernensis Grăunceanu CMAI	8.6	8.7	11.3	11.0	10.9	13.3	10.6	10.1	16.5	80.91	97.18	0.83
P. arvernensis Vatera PO 170 F	8.2	8.09	9.87	10.5	9.44	13.25	11.08	9.48	17.84	61.64	80.93	0.76
Papio h. ursinus Mean	8.2	8.56	10.68	10.38	9.97	13.09	11.29	10.22	16.74	62.71	87.95	0.72
Papio h. ursinus N	157	158	164	146	148	150	134	133	133	132	157	127
Papio h. ursinus Max	9,6	9.9	12.4	12.5	11.9	15.3	13.2	11.6	19.4	84.8	112.2	0.95
Papio h. ursinus Min	6.4	6.8	8.0	8.7	7.9	10.1	9.2	8.1	13.2	43.4	55.2	0.51
	P4 A	M1A	P4/M1 A							p4 A	m1A	p4/m1 A
Mandrillus sp. Mean	71.97	92.4	0.78							62.24	74.31	0.85
Mandrillus sp. N	48	52	48							66	69	46
Mandrillus sp. Max	109.2	125.9	0.95							90.4	102.0	1.31
Mandrillus sp. Min	48.9	72.4	0.53							34.2	53.5	0.43

Fossil specimens all judged female. *P. arvernensis* Senèze and Grăunceanu, *P. sushkini* Kuruksay original measurements; Dafnero from Kostopoulos et al. (2018)

Papio hamadryas ursinus and *Mandrillus* sp. from PRIMO (http://primo.nycep.org; original measurements by Delson, Eisenhart and colleagues); all mixed-sex other than canine and P3/p3 values limited to female specimens (f).

Upper teeth indicated by uppercase letters (I, C, P, M), lower teeth by lowercase letters (I, c, p, m), L = maximum mesiodistal length: measured on c perpendicular to W; measured on p3 as projected maximum parallel to alveolar plane from distal margin to mesialmost end of flange. W = buccolingual width: measured on incisors at cervix; measured on C perpendicular to L; on c, this is maximum dimension at cervix due to rotation of canine in jaw. AW = buccolingual width across mesial loph(id); PW = buccolingual width across distal loph(id). A = area; premolar area is product of L x W; molar area is product of L x AW (following Fleagle and McGraw 1999, 2002). N, number of specimens measured; Max, maximum; Min, minimum.

smooth and generally unhollowed by fossae, but this is variable, especially on the mandible. The muzzle is grooved below the orbits for vessels exiting the infraorbital foramina. There is no maxillary sinus (except in "*P.*" *sushkini*, which might suggest inclusion in a distinct genus); the maxillary body is thin at the level of P3-M1 or M2, but may be thicker in males at M3; the zygomatico-maxillary suture occurs at the level of M3. Cranial sexual dimorphism may be low compared to extant African papionins of similar size, but lack of both male and female specimens from the same locality makes this difficult to demonstrate. The supraorbital torus is strongly built in both sexes, running nearly horizontally above the orbits before turning rather sharply inferiorly at the lateral margins; glabella is slightly protruding. The neurocranium is somewhat more globular in females and more elongate in males. The temporal lines run medially and posteriorly from the lateral end of the supraorbital torus. In females, they are nearly parallel and diverge slightly as they merge into the moderate nuchal crest. In males, they meet near the inferred position of bregma to form a low sagittal crest which again diverges slightly as it merges with the nuchal crest, which is relatively flat, not upturned in the midline. The endocast was described as typically cercopithecine in sulcal patterns, if with a slightly lower occipital region (Radinsky 1974).

The mandibular corpus is of moderate and constant depth; the ramus is somewhat posteriorly inclined. The dentition is typically papionin, with large front teeth and moderate molar flare. Lower molars show a variable degree of "crestiness" of lophids, even at similar stages of wear. The ratio of fourth premolar to first molar area is comparable to values in *Papio, Macaca* and *Lophocebus*, somewhat lower than the means for *Mandrillus* and *Cercocebus*. The humerus is robust with a well retroflexed medial epicondyle and distally expanded medial trochlear "crest"; the ulnar olecranon is large and somewhat tilted posteriorly; partial femora are robust; and phalanges are stout.

Paradolichopithecus arvernensis (Depéret, 1928)

Synonymy: *Dolichopithecus arvernensis* Depéret, 1928: *Paradolichopithecus geticus* Necrasov, Samson and Rădulesco, 1961; "*Dolichopithecus*" *arvernensis* Depéret, 1928: Simons (1970), Delson (1974); *Procynocephalus* sp. cf. *P. wimani* Schlosser, 1924: Simons (1970); *Paradolichopithecus arvernensis* (Depéret, 1928): Delson (1975).

Holotype: FSL 41336, nearly complete female skull.

Type locality: Senèze (Haute-Loire, France), ca. 2.2–2.1 Ma, late Villafranchian, MNQ 18 (see Delson et al. 2024, Crégut-Bonnoure et al. 2024).

Hypodigm: numerous specimens from Grăunceanu (Romania) and Vatera-F (Lesvos, Greece) and probably one or two specimens each from Dafnero-3 and Karnezeika (Greece); not listed here as not from Senèze.

Emended diagnosis: A large species of *Paradolichopithecus* broadly comparable in size to larger subtaxa of extant *Papio* (e.g., samples of *P. h. anubis* and *P. h. ursinus*). Estimated mass (Delson et al. 2000) is ca. 25–40 kg for males (combining dental, cranial and postcranial estimates) and around 15–23 kg for females (dental only). These values are similar to those for "*P.*" *sushkini* from Kuruksay but larger than for male *Paradolichopithecus* sp. from Cova Bonica (and probably other sites noted above) at ca. 18–25 kg. Selected dental measurements for female specimens of *P. arvernensis* and "*P.*" *sushkini* are provided in Table 16.2, along with comparative values for a large sample of *Papio hamadryas ursinus* and area ratios of P4/M1 and p4/m1 for these taxa and also *Mandrillus sphinx* which has relatively large fourth premolars (Fleagle and McGraw 1999, 2002; Gilbert 2007, 2013).

Figure 16.2 is a box plot of P4/M1 area ratios in large samples of three mixed-sex groups of mainly extant papionins: *Cercocebus* and *Mandrillus*; *Papio*, *Lophocebus* and *Theropithecus* (including some fossils of the latter); and *Macaca*; plus selected individuals of *Paradolichopithecus* and *Procynocephalus* (details in figure caption). The boxes for the *Cercocebus* and *Mandrillus* (C-M) and *Papio*, *Lophocebus* and *Theropithecus* (P-L–T) groups do not overlap; the former have larger values indicating their relatively larger P4s. Macaques mainly fall between those two groups, the boxes overlapping slightly. *Paradolichopithecus arvernensis* overlaps macaques and the P-L–T group. "*Paradolichopithecus*" *sushkini* has quite small P4s, overlapping only the P-L–T lower whisker. *Procynocephalus subhimalayanus* lines up with the macaques but also the lower C-M and upper P-L–T whiskers, while "*Paradolichopithecus*" *gansuensis* lies lower on the plot. The fossils mostly align with macaques (and P-L–T), indicating moderately-sized P4 compared to M1.

Figure 16.3 is a box plot of M3 area in a selection of mixed-sex extant and fossil larger papionins. *Paradolichopithecus* and *Procynocephalus* species overlap extant large *Papio*, are generally larger than *Mandrillus* and overlap the smaller individuals of the large African fossil species. *Macaca sylvanus* is clearly the smallest of these taxa, much smaller than the larger extant *Papio* subspecies and the

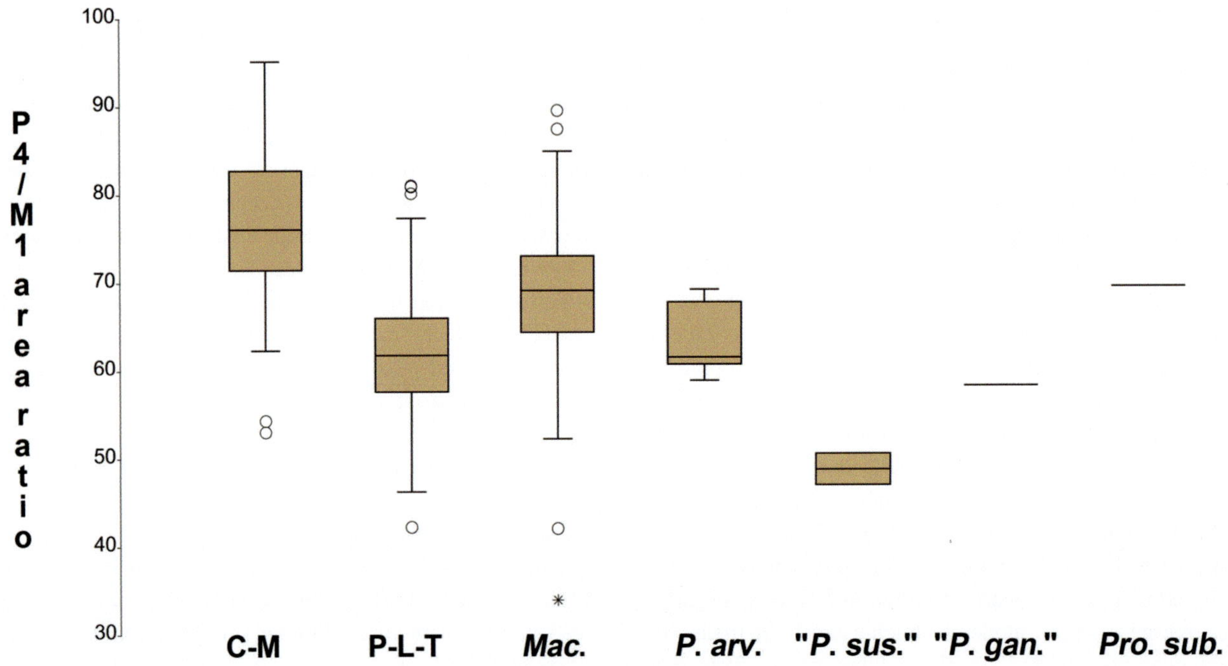

Fig. 16.2 Box plot of P4/M1 area ratio in percent (100 × P4 width x P4 length/M1 mesial width x M1 length) in selected mixed-sex mainly extant papionins: C-M, *Cercocebus* spp. (40 specimens) and *Mandrillus* spp. (48 specimens); P-L–T, *Papio* sspp. (211 specimens), *Lophocebus* spp. (51 specimens), *Theropithecus* spp. (32 T. *gelada*, 12 fossil *T. oswaldi* sspp.); *Mac.*, *Macaca* spp. (48 M. *sylvanus* specimens, 50 specimens of various other species). The focal fossils include: *P. arv.*, *Paradolichopithecus arvernensis* from Senèze (1 specimen), Grăunceanu (6 specimens) and Dafnero-3 (1 specimen); "*P.*" *sus.*, "*Paradolichopithecus*" *sushkini* (2 specimens); "*P.*" *gan.*, "*Paradolichopithecus*" *gansuensis* (1 specimen); *Pro. sub.*, *Procynocephalus subhimalayanus* (1 specimen). Boxes show 25% and 75% quartiles, horizontal line inside the box is the median; whiskers are drawn from the top (or bottom) of the box to the largest (smallest) data point less than 1.5 times the box height outside the box; outlier values beyond the whiskers are indicated as circles, while those more than 3 times the box height outside the box are indicated as asterisks. Data from PRIMO (http://primo.nycep.org) and Table 16.2

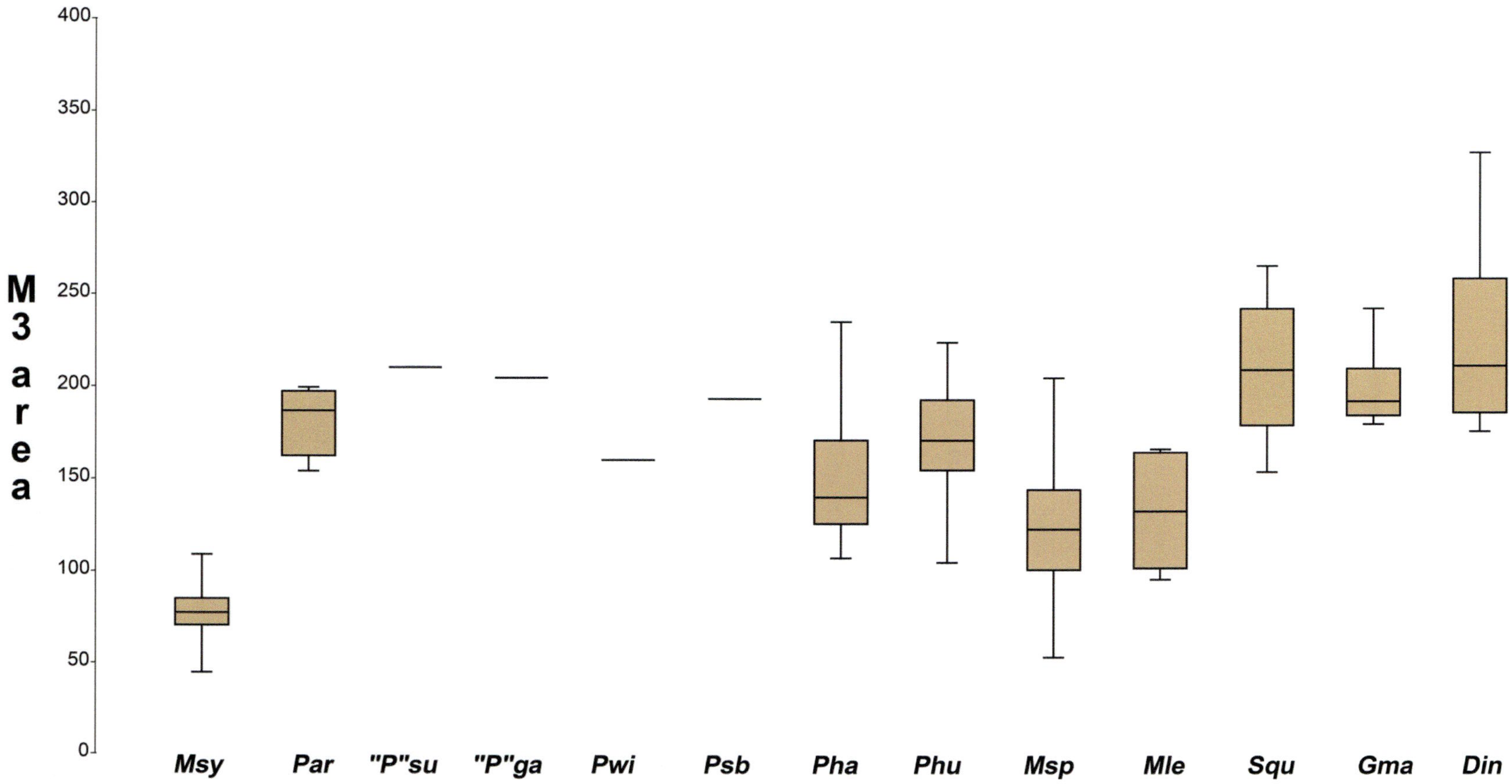

Fig. 16.3 Box plot of M3 area in cm^2 (M3 mesial width x M3 length) in selected mixed-sex mainly extant larger papionins (fossil taxa marked with †): *Msy*, *Macaca sylvanus* (50 specimens); *Par*, *Paradolichopithecus arvernensis* † from Senèze (1 specimen), Grăunceanu (6 specimens) and Dafnero-3 (1 specimen); *"P"su*, *"Paradolichopithecus" sushkini* † (1 specimen); *"P"ga*, *"Paradolichopithecus" gansuensis* † (1 specimen); *Pwi*, *Procynocephalus wimani* † (1 specimen); *Psb*, *Procynocephalus subhimalayanus* † (1 specimen); *Pha*, *Papio hamadryas anubis* (30 specimens); *Phu*, *Papio hamadryas ursinus* (141 specimens); *Msp*, *Mandrillus sphinx* (26 specimens); *Mle*, *Mandrillus leucophaeus* (11 specimens); *Squ*, *Soromandrillus quadratirostris* † (12 specimens); *Gma*, *Gorgopithecus major* † (8 specimens); *Din*, *Dinopithecus ingens* † (7 specimens). Boxes show 25% and 75% quartiles, horizontal line inside the box is the median; whiskers are drawn from the top (or bottom) of the box to the largest (smallest) data point. Data from PRIMO (http://primo.nycep.org) and Table 16.2

fossils plotted, overlapping only the smallest *Mandrillus sphinx*; it is among the largest macaques and shows that a significant increase in size must have accompanied the origin of the Eurasian fossil taxa.

Table 16.3 provides selected measurements (defined in Table 16.4) on female skulls of the same taxa. The original Grăunceanu female face (Necrasov et al. 1961) does not permit many skull measurements to be taken, but two are essentially identical to the Senèze specimen: upper molar row length 40.6 mm and buccal jaw depth below m3 at 29 mm. The cranium from Senèze does not have a maxillary sinus (those from Grăunceanu are under study), nor does that of "*P.*" *gansuensis* from Longdan, as opposed to those of "*P.*" *sushkini* which have sinuses. The maxillary body is thin at the level of P3-M2 but moderately thick at M3 (in the Senèze holotype but thin in males from Grăunceanu), again similar to "*P.*" *gansuensis* but different from "*P.*" *sushkini* (which is thick at P3-M2 but thin at M3). Neither females nor males present more than minimal maxillary or mandibular corpus fossae, as opposed to moderately developed ones in "*P.*" *sushkini* and "*P.*" *gansuensis*.

Detailed Description of Senèze Skull

General. FSL 41336 was partly reconstructed from numerous fragments by Depéret (1929). This reconstruction was taken apart and re-restored after cleaning and more careful alignment by Battetta (1969), who made a cast of the restoration (Fig. 16.4) and then disarticulated the specimen into fourteen fragments, most of which do not cleanly contact one another. The canine roots (absolutely short but longer than the crown) combined with the small p3 confirm identification as a female.

The largest cranial fragment comprises the nearly complete braincase and part of the upper face (Figs. 16.5 and 16.6). Lacking from this region are: part of the right temporal (including the petrous region, the external auditory meatus and zygomatic process); the right occipital condyle; and the antero-lateral portion of the right frontal, behind which the bone is damaged along the right side of the coronal suture. Most of the left malar is present on this piece, completing the lateral half of the orbit rim. Another fragment includes almost all of the right malar and the adjacent anterolateral part of the frontal (a space remains near the external corner of the orbit, although a carved piece of plaster fills it in casts). The fronto-malar segment makes a loose contact with a fragment comprising the posterior portion of the temporal process of the malar firmly sutured to the anterior two-thirds of the malar process of the temporal (together making up about half of the right zygomatic arch). Two yet smaller elements almost complete the connection to the neurocranium at the root of the malar process. Another small fragment nearly closes the right orbit inferiorly, between the malar and the maxilla.

The second largest cranial fragment is composed of the majority of the right maxilla and teeth, part of the left maxilla and portions of the premaxillae and nasals (Fig. 16.7). The right maxilla extends almost to the malar and ethmoid margins along the inferior border of the orbit; there are two areas of damage inferiorly, above the M1-2 contact and on the inferior part of the premaxillary suture alongside the nasal aperture. The palatal roof is almost entirely missing on both sides. The inferior two-thirds of the nasal bones (probably damaged at rhinion) contact a small slip of the right premaxilla next to the nasal aperture. Another segment of the premaxilla contains the right I2 and abuts the maxillary toothrow with C-M3. The superior portion of the left maxilla is present as an elongate strip along the nasal bones and aperture to above the canine alveolus, but a large region is missing from above the molars past the zygo-maxillary suture up to the inferior orbit margin and frontal. The three left molars are conjoined in one fragment of alveolar margin, while the anterior segment of premaxilla with I1-2 and a portion of internal maxillary wall with P3-4 form another. The right I1 and left C are isolated.

The mandible is almost completely preserved, except for the left i2 and an area of external surface bone posterior and inferior to the right canine root (Figs. 16.8 and 16.9). It is broken into two portions between the right canine and p3. A piece of external corpus below the p3 and abutting (and covering) the canine root was never recovered, while some of the corpus base inferior to this was previously attached to the larger segment but is currently missing. The two parts of the corpus make poor contact and have not been glued together.

Dentition. The teeth are heavily worn but clearly papionin in morphology (Figs. 16.7, 16.8 and 16.9). The upper canine is short from apex to cervix, and the root of the isolated left C is taller than the crown. This is typical of papionin females. The P3 flange is slightly longer than that of the P4. The molar buccal notches are shallow. The lower canine crown is slender and looks like those of males, but the root (clearly visible on the left where the corpus is broken off, see Fig. 16.9) is still taller and curved distally. This shape is different from the smaller crowns of the *P.* "*geticus*" holotype and Vatera PO 170 F. The p3 flange (in all three mandibles) extends just slightly below the alveolar plane. The p4-m3 lingual notches are worn but shallow.

Face. The reconstructed face has a moderately steep but curving anteorbital drop and a tall maxilla above the alveolar plane (Figs. 16.4 and 16.10). The orbits have a straight superior margin below fairly thick brows, then curving smoothly down both medially and laterally to zygomax superior, where the zygomaticomaxillary suture crosses the

Table 16.3 Comparative skull measurements (in mm) of female *Paradolichopithecus arvernensis* and *Papio hamadryas anubis* (variables defined in Table 16.4)

Taxon/Specimen	NAIN	NABA	NABR	NAPR	NARH	BAIN	BABR	BAPR	PRIN	PRBR	PRRH	PROB	INOB	ZMOB
P. arvernensis Senèze FSL 41333	115.0	78.5	68.0	96.5	46.0	52.0	64.0	128.0	180.0	155.0	54.0	85.5	116.0	27.5
P. arvernensis Dafnero-3 LGPUT DFN3-150	120.1 d			84 d	45.5				171.3 d					
Papio h. anubis females Mean	99.49	73.32	57.07	79.97	45.40	48.73	62.00	111.53	154.48	133.83	35.83	70.93	94.19	30.75
Papio h. anubis females N	13	11	2	2	2	2	2	2	2	2	2	2	2	2
Papio h. anubis females Maximum	107.6	79.5	57.9	83.9	52.8	50.9	63.4	111.8	158.1	138.2	39.2	72.7	98.3	31.8
Papio h. anubis females Minimum	91.7	66.3	56.3	76.0	38.0	46.5	60.6	111.2	150.9	129.4	32.5	69.2	90.1	29.7
	BIPG	MAXW	FACH	BIZY	TMIN	PORB	BIOR	INOR	ORBH	ORBW	SUPO	FORW	FORL	NASW
P. arvernensis Senèze FSL 41333	76.7	81.8	82.5	118.0	12.0	48.2	101.0	14.0	27.8	32.5	8.1	15.0	15.0	22.0
P. arvernensis Dafnero-3 DFN3-150						39.8 d	63.4 d	9.0 d	26.5 d	26.7 d		12.8 d	18.5 d	
Papio h. anubis females Mean	74.56	73.87	62.52	97.33	28.73	54.09	68.11	4.51	23.43	28.99	3.77	16.69	17.76	19.10
Papio h. anubis females N	2	2	2	2	2	4	4	2	4	4	2	2	2	2
Papio h. anubis females Maximum	77.1	78.5	64.3	100.3	38.7	55.6	75.6	5.2	24.8	30.2	3.9	17.4	18.4	20.2
Papio h. anubis females Minimum	72.1	69.2	60.7	94.4	18.8	52.5	59.9	3.8	21.3	27.3	3.7	16.0	17.1	18.0
	NASL	BMEU	I1MU	I1AU	M3IU	M3PU	M3MU	ALWU	PDEP	AFAC	IDCO	SYML	PLAL	GNAL
P. arvernensis Senèze FSL 41333	35.5	61.5	15.5	14.0	75.0	50.8	38.6	15.8	10.5	40.0	128.5	40.5	29.5	22.5
P. arvernensis Dafnero-3 DFN3-150						50.9	37.7							
P. arvernensis Vatera NHCV PO 170 F												37.2	21.8	19.1
Papio h. anubis females Mean	26.84	50.29	16.18	17.46	70.98	47.61		13.52	7.16	36.00	112.09	35.61	22.49	19.84
Papio h. anubis females N	2	4	2	2	2	2		4	2	2	2	2	2	2
Papio h. anubis females Maximum	29.0	56.8	18.0	19.7	74.3	49.8		15.7	7.6	36.0	115.7	39.7	23.6	20.6
Papio h. anubis females Minimum	24.7	45.1	14.4	15.2	67.6	45.4		12.3	6.7	36.0	108.5	31.5	21.4	19.1
	BICO	RAML	CORH	BJD3	BJD4	ALWL	M3IL	I1AL	I1ML	ARAM	ASYM	ORBW/NAIN	FACH/ BIOR	BIOR/ NAIN
P. arvernensis Senèze FSL 41333	80.0	35.5	70.0	34.4	34.2	14.6	76.5	9.5	10.3	115.0	130.0	28.3	81.7	87.8
P. arvernensis Dafnero-3 DFN3-150														
P. arvernensis Vatera PO 170 F				28.1	29.9	13.1	70.4							
Papio h. anubis females Mean	72.09	33.37	57.14	21.81	26.06	12.02	71.45	11.19	12.53	105.50	131.50	30.6	83.3	72.0
Papio h. anubis females N	2	2	2	2	2	4	2	2	2	2	2	4	2	4
Papio h. anubis females Maximum	72.8	34.2	59.9	23.0	28.1	14.0	75.9	12.7	14.2	108.0	135.0	32.0	85.0	81.2
Papio h. anubis females Minimum	71.4	32.5	54.4	20.7	24.0	10.3	67.0	9.6	10.9	103.0	128.0	28.9	81.5	63.4

Data sources: *P. arvernensis* Senèze and Vatera original measurements; Dafnero from Kostopoulos et al. (2018; specimen crushed bilaterally, values with "d" are distorted from original values); all female individuals.

Papio hamadryas anubis from PRIMO (http://primo.nycep.org; original measurements by Delson and colleagues); limited to female specimens

Table 16.4 Definition of skull measurements in Table 16.3

NAIN	Nasion to Inion
NABA	Nasion to Basion
NABR	Nasion to Bregma
NAPR	Nasion to Prosthion (= Alveolare)
NARH	Nasion to Rhinion
BAIN	Basion to Inion
BABR	Basion to Bregma
BAPR	Basion to Prosthion
PRIN	Prosthion to Inion
PRBR	Prosthion to Bregma
PRRH	Prosthion to Rhinion
PROB	Prosthion to Zygoorbitale
INOB	Inion to Zygoorbitale
ZMOB	Zygomaxillare to Zygoorbitale
ZMPG	Zygomaxillare to Postglenoid
BIPG	Bi-Postglenoid Breadth (centers of process tips)
MAXW	Maximum Vault Width (on temporal dorsal to porion or zygomatic arch)
FACH	Face Height (top of brow ridge to alveolar margin; cranium in alveolar plane)
BIZY	Maximum Bi-Zygomatic Breadth
TMIN	Minimum Temporal Line Width (between medialmost edges; if sagittal crest exists, defined as 1 mm)
PORB	Minimum Postorbital Width
BIOR	Bi-Orbital Breadth (taken on lateral margin of frontal/zygoma; near Fronto-malar temporale)
INOR	Minimum Interorbital Distance (close to Bi-Dacryon; but often not as deep into orbit)
ORBH	Orbit Height (taken near center of orbit)
ORBW	Orbit Width (taken near center of orbit)
SUPO	Supraorbital Torus Thickness (taken near center of orbit; maximum away from notch)
FORW	Maximum Foramen Magnum Width
FORL	Foramen Magnum Length (Basion to Opisthion)
NASW	Maximum Nasal Aperture Width
NASL	Maximum Nasal Aperture Length (Rhinion to Nasospinale)
BMEU	Breadth across upper molars (on alveolar margin between M1 & M2)
I1MU	Maximum breadth across upper I1 (Chord across 2 incisors near tips)
I1AU	Breadth across upper I1 at alveoli (chord across 2 incisors at alveolar margin)
M3IU	Upper I1 to M3 (I1 incisal edge to M3 distal face)
M3MU	Upper M1 to M3 (M1 mesial face to M3 distal face)
ALWU	Width of maxillary alveolar process (near middle of M2)
PDEP	Depth of palate (near middle of M2; from alveolar margin projected to midline)
AFAC	Facial angle (Approximate angle of alveolar margin and Glabella to Prosthion chord)
IDCO	Infradentale to Condylion (midpoint of condyle)
SYML	Symphyseal Length (Infradentale to Gnathion)
PLAL	Alveolar Planum Length (Infradentale to Inferior Endpoint of planum alveolare)
GNAL	Gnathion to Inferior Endpoint of PLAL
GNGO	Gnathion to Gonion
RAML	Length (M-D) of mandibular ramus (minimum anterior–posterior [=mesial-distal] length of ramus above alveolar plane and below incisura; often called "width")
CORH	Height of coronoid process (with mandible resting on flat support; maximum height above support of coronoid process)
BJD3	Buccal corpus depth below m3 (depth from alveolar margin in middle of m3 to inferior border of corpus)
BJD4	Buccal corpus depth below p4 (depth from alveolar margin in middle of p4 to inferior border of corpus)
ALWL	Width of mandibular alveolar process (near middle of m2)
M3IL	Lower i1 to m3 (i1 incisal edge to m3 distal end)
I1AL	Breadth across lower i1 at alveoli (chord across 2 incisors at alveolar margin)
I1ML	Maximum breadth across lower i1 (chord across 2 incisors near tips)
ARAM	Ramus-corpus angle (approximate angle of corpus inferior margin and [upper] posterior margin of ramus at gonion)
ASYM	Symphyseal angle (approximate angle of corpus inferior margin and anterior face of symphysis at gnathion)

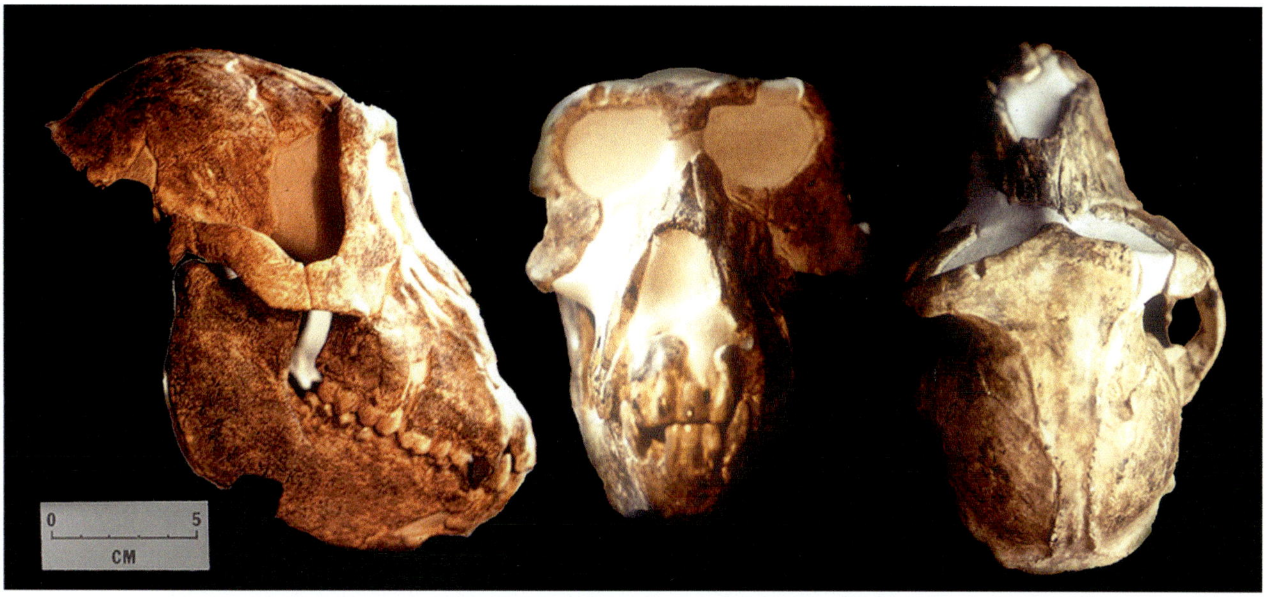

Fig. 16.4 *Paradolichopithecus arvernensis*, cast of Battetta (1969) reconstruction of FSL 41336 Senèze female holotype skull. Left to right: left lateral (Frankfurt plane), frontal and dorsal (occlusal plane) views, photos by F.S. Szalay

inferior margin, producing a "D" shape (Figs. 16.4 and 16.11). The nasal bones are smoothly curved and not raised. Several infraorbital foramina lead into shallow grooves on the maxilla, which shows no sign of a fossa or other excavation. The nasal aperture is oval in anterior view, with maximum width about halfway down, but the inferior portion is damaged. The maxillary root of the zygomatic arch arises above the distal part of M3. The arch itself is very deep, giving the face great lateral robusticity. As Simons (1970) illustrated and discussed, the overall shape in lateral view is similar to that in males of several macaque species (e.g., *M. nemestrina* or *M. nigra*), whose longer faces compare better to *Paradolichopithecus* of both sexes than those of female macaques.

The palate is poorly preserved, but there is a slight shallowing mesially and not much elongation due to the small canines. The toothrow widens slightly from M3 to M2, then narrows mesially to C, as in female macaques but unlike females of *Papio*, *Mandrillus*, *Soromandrillus* or *Dinopithecus*, which are of comparable palate length but narrower and more straight-sided.

FSL 41336 has a shallow supraorbital notch on the right with a slight projection medially; the left side is damaged and unclear (Fig. 16.6). Based on the analysis of a CT scan by Nishimura et al. (2014; Fig. 16.12 here), an inferior concha likely descended from the superior portion of the nasal cavity, which would have separated a large inferior meatus from the main portion of the nasal cavity at the level of M1 to mesial M2. There is no indication of a maxillary sinus, a conservative trait shared with most extant cercopithecoids excluding macaques.

Calvaria. Behind the tall but not protruding supraorbital tori, there is a small ophryonic depression. The temporal lines arise near the lateral ends of the tori and turn fairly sharply medially before curving posteriorly well medial to the postorbital constriction. The lines converge to a closest approach (12 mm) on the posterior parietals before diverging slightly and eventually meeting the strong nuchal crest, which turns neither up nor down at inion but remains level. This pattern is most similar to that seen in male macaques. The occipital planum is fairly smooth and angled from the nuchal crest to the foramen magnum. The postglenoid process is robust and leans anteriorly only slightly; the mandibular fossa is wide.

Mandible. The Senèze mandible is well preserved. The anterior symphyseal surface is robust, with relatively strong muscle scars bounding the smooth midline region. The scars are less posterolateral than in female *Papio* or *Mandrillus* mandibles of nearly similar size, but more pronounced than in those of large female macaques (which are far smaller overall). Large male macaques (e.g., *M. nemestrina leonina*, AMNH-M 11090) are perhaps most similar in shape. The planum alveolare is long, the superior torus slight and the inferior torus barely present. There is a slight corpus fossa (really just a shallowing) below the p4-m1 contact, also as in male macaques but not female papioninans, which usually have significant fossa development (see e.g., Gilbert 2013). Corpus depth remains constant from m1 to m3 and beyond, the margin curves smoothly through gonion and up the distal edge of the ramus to the condyle. A strong buccinator groove precedes the anterior edge of the ramus, which rises subvertically and thickens slightly below the coronoid. The

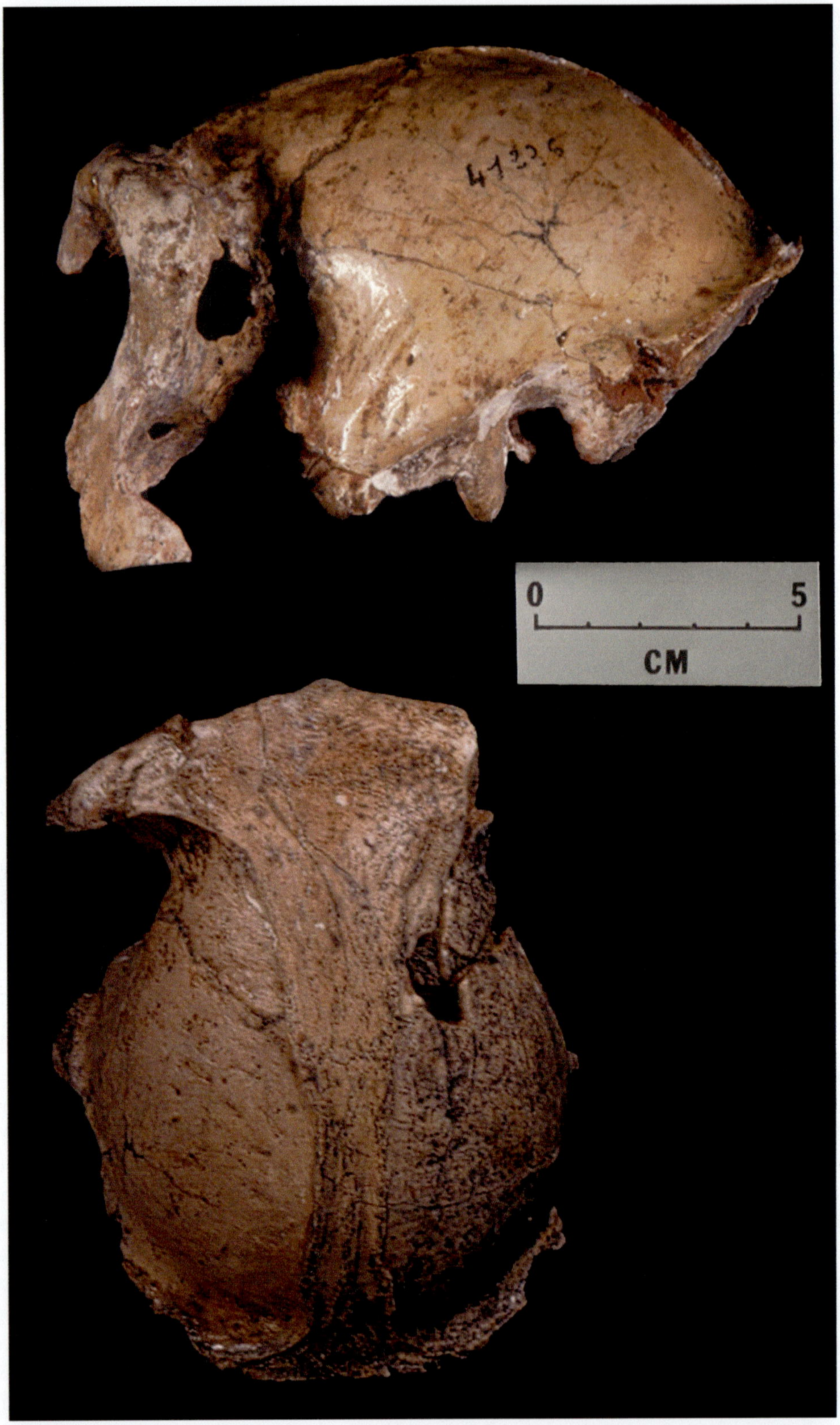

Fig. 16.5 *Paradolichopithecus arvernensis*, FSL 41336, Senèze holotype skull, neurocranial element. Top to bottom: left lateral and dorsal views (near Frankfurt plane)

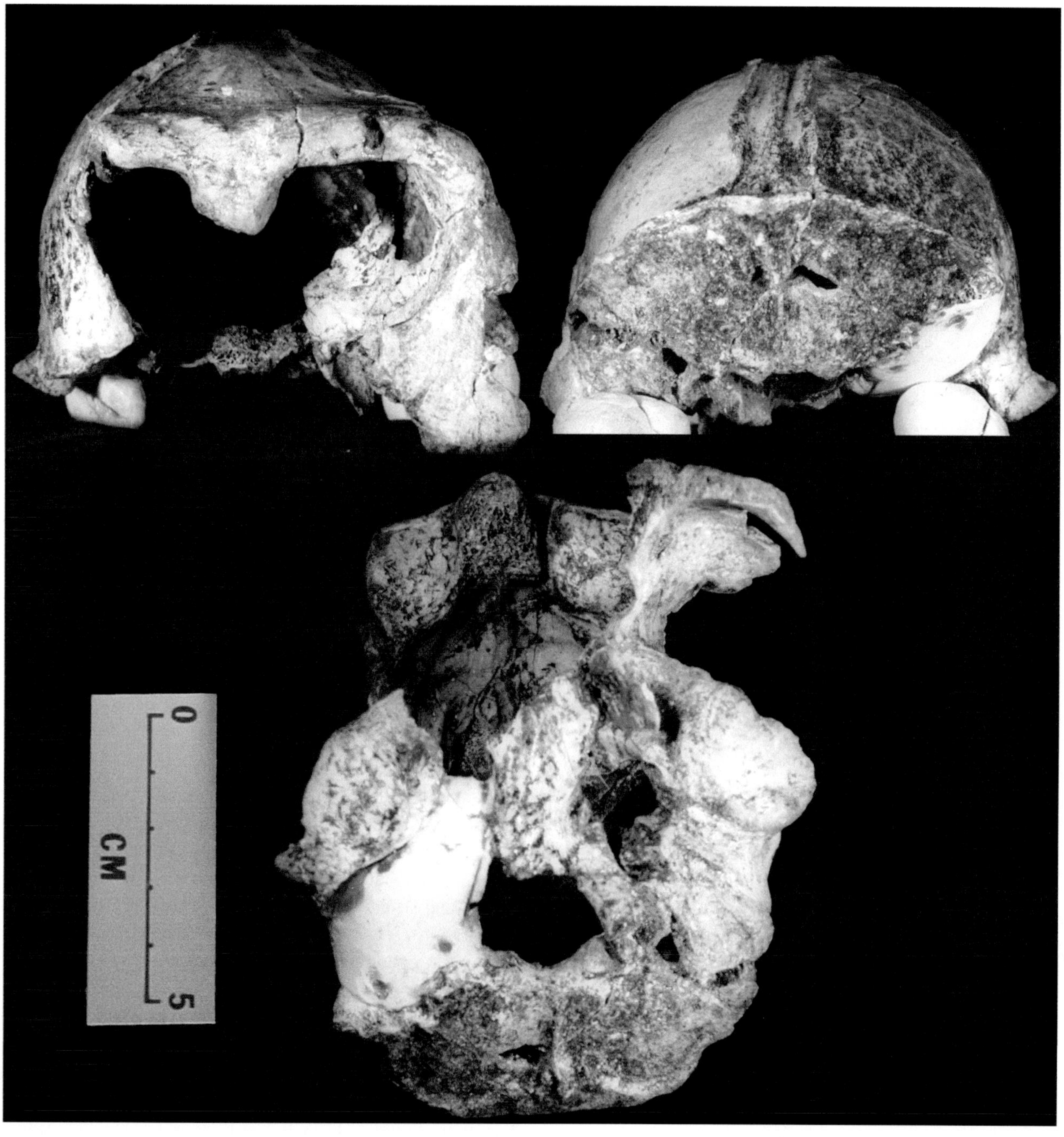

Fig. 16.6 *Paradolichopithecus arvernensis*, FSL 41336, Senèze holotype skull, neurocranial element in frontal (top left), occipital (top right) and basal (below) views (near Frankfurt plane)

Fig. 16.7 *Paradolichopithecus arvernensis*, FSL 41336, Senèze holotype skull, right maxilla with separated partial right premaxillary fragment and isolated I1 attached, in lateral (left) and occlusal (right) views (occlusal plane)

posterior slope of the coronoid is nearly linear, and the notch is acute but restricted. The ramus is angled slightly posteriorly relative to the corpus, most comparable to that of female *M. tonkeana hecki* (e.g., AMNH-M 152897) or *Mandrillus sphinx* (e.g., AMNH-M 89357), rather than "leaning" farther posteriorly as in female *Papio* spp. or male macaques, or being nearly vertical as in smaller papioninans or some female macaques (e.g., *M. thibetana*, AMNH-M 83994).

Implications for Paleoenvironment and Relative Age

The two primate taxa known from Senèze provide only slight information about the local environment or age of the site. *Macaca sylvanus* is known across much of Europe in small numbers throughout the Pliocene and Pleistocene. Based on the extant populations of the Maghreb, the species often feeds and travels on the ground in open to mixed woodland and into grassland but sleeps in trees (Fooden 2007; Rowe and Myers 2016). Environments are variable, often wet and cold (even snowy) in winter, with warm and dry summers. Its postcranium is moderately adapted to terrestriality, among the most cursorial of macaques. *Paradolichopithecus arvernensis* is known by postcrania from Grăunceanu and Vatera which are similar to those of *Papio* and *Mandrillus* (Ting et al. 2004, Delson et al. in preparation). They suggest a high degree of terrestriality, perhaps in open terrain to woodland.

European macaques do not yet provide any direct indication of geological age. Delson (1980; Szalay and Delson 1979) suggested that several chronosubspecies could eventually be distinguished, but later studies (e.g., Alba et al. 2008, 2011) have not confirmed his proposal; further analysis is in progress. Delson and Plopsor (1975) and Aguirre and Soto (1978) suggested that *Paradolichopithecus arvernensis* might serve as a marker for the beginning of the Calabrian Stage, then recognized as the base of the Pleistocene. Current assessments of age for localities including the genus (listed above) in fact reveal a range from at least 4 Ma to 2 Ma, with no known specimens in Calabrian-age deposits. Detailed analysis is still ongoing, but the Senèze population appears to be slightly larger than that from Grăunceanu in terms of cranial size, which led to previous suggestions that female crania (then known only from Senèze) were little smaller than males (known from

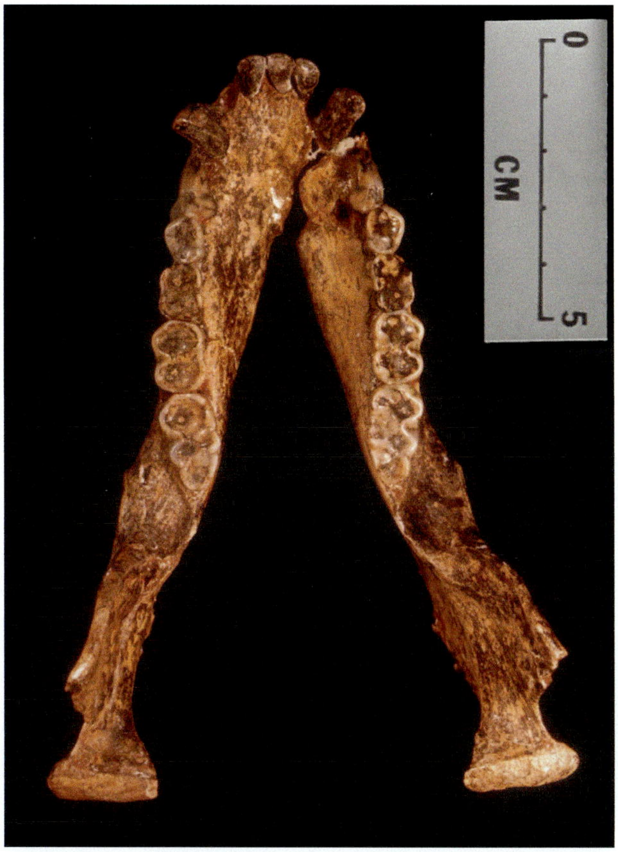

Fig. 16.8 *Paradolichopithecus arvernensis*, FSL 41336, Senèze holotype skull, mandible in two pieces loosely attached in occlusal view (occlusal plane)

Grăunceanu) and thus that sexual dimorphism was less than expected in a cercopithecine of this size. The subadult female Dafnero cranium is crushed bilaterally and distorted, but it appears similar in size to the Senèze specimen. It may be that Senèze is younger than Grăunceanu (whose precise age is still unknown, see Terhune et al. 2020), given that the still older populations termed *Paradolichopithecus* sp. appear even smaller.

Recently Benammi et al. (2020) have proposed on the basis of magnetostratigraphy that Dafnero, faunally similar to Vatera (and Grăunceanu), might date to ca 2.3 Ma. They studied two neighboring sections, both of which contained a normal magnetozone between reversed intervals, with the faunal level near the bottom of each section. They estimated the age of the fauna between 2.5 and 2.0 Ma and thus correlated the normal magnetozone with either the Olduvai or the "Réunion" (now Feni) subchron. They compared the thickness of the normal magnetozone to the length of those two normal intervals to approximate the sedimentation rate for each correlation hypothesis and in turn to estimate the duration of the interval between the base of the normal and the fossiliferous level. This yielded an estimated age of 2.4 Ma if the normal correlated to the Olduvai and 2.2 Ma if it was the Feni (based on a 20 kyr duration of the latter).

However, there are a number of problems in their approach. First, in the two magnetostratigraphic sections studied, there are only 2 normal samples over a span of 5–10 m, and the stratigraphic length of the normal was calculated by placing the estimated end points almost randomly between the normal samples and the nearest reversed sample (rather than midway between the nearest sample levels as is usually done). Moreover, the sampling interval was not tight enough to delineate the polarity zones that closely; on average there are several meters between samples. It is not clear if the normal or reversed zones were significantly shorter or longer than estimated. Given the dates for Senèze in Nomade et al. (2014), which they cited, the normal is most likely to be the Feni, not the Olduvai. But Benammi et al. (2020) used an older age range for the Feni of 2.148–2.128 Ma, rather than the current 2.14–2.116 Ma (26 not 20 kyr; see Channell et al. 2020). Therefore, the age can only be estimated as older than Feni and younger than the Gauss-Matuyama boundary, i.e., 2.58–2.14 Ma. A date between 2.6 and 2.2 Ma is the most reasonable approximation, perhaps older in that range as the fossils come from low in the section. The Senèze specimen of *Paradolichopithecus arvernensis*, dated between 2.2 and 2.1 Ma (Delson et al. 2024) may be one of the youngest known representatives of the genus.

Discussion

There are two major questions about *Paradolichopithecus* which have been raised in the literature: distinction from *Procynocephalus* and affinity to Papionina or Macacina. Delson and colleagues (e.g., Delson and Plopsor 1975; Szalay and Delson 1979; Delson et al. 2000, 2014; Delson and Frost 2004) agreed with previous work by Vogel (1966, 1968), Jolly (1967) and Simons (1970) in arguing for morphological similarity to *Macaca*. They also opted to retain the two Plio-Pleistocene forms as distinct genera, pending the possibility of more direct comparison. Maschenko (1994, 2005) suggested that *Paradolichopithecus* (especially "*P.*" *sushkini*) was instead close to or even congeneric with *Papio*. Kostopoulos et al. (2018) also suggested a close relationship with *Papio* and further argued that *Paradolichopithecus* could be included in *Procynocephalus* as there were few if any features that separated the two genera. Le Maître et al. (2023) supported the possibility that *Paradolichopithecus* (or *Procynocephalus*) aligned more closely with Papionina than Macacina based on morphology of the bony labyrinth.

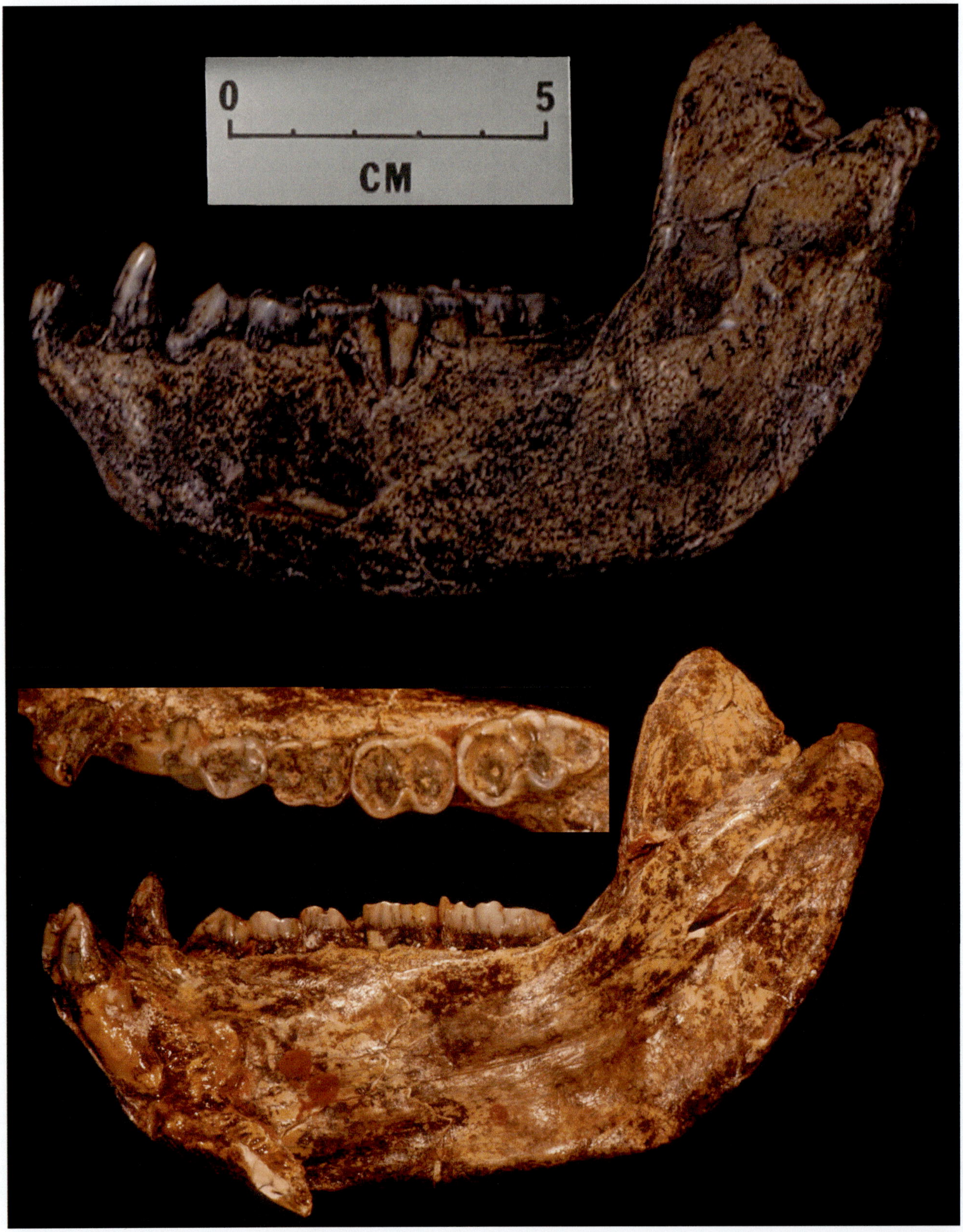

Fig. 16.9 *Paradolichopithecus arvernensis*, FSL 41336, Senèze holotype skull, left mandibular segment in lateral above internal views (occlusal plane, interval view photographically reversed); left c-m_3 inset at 1.3 × main scale

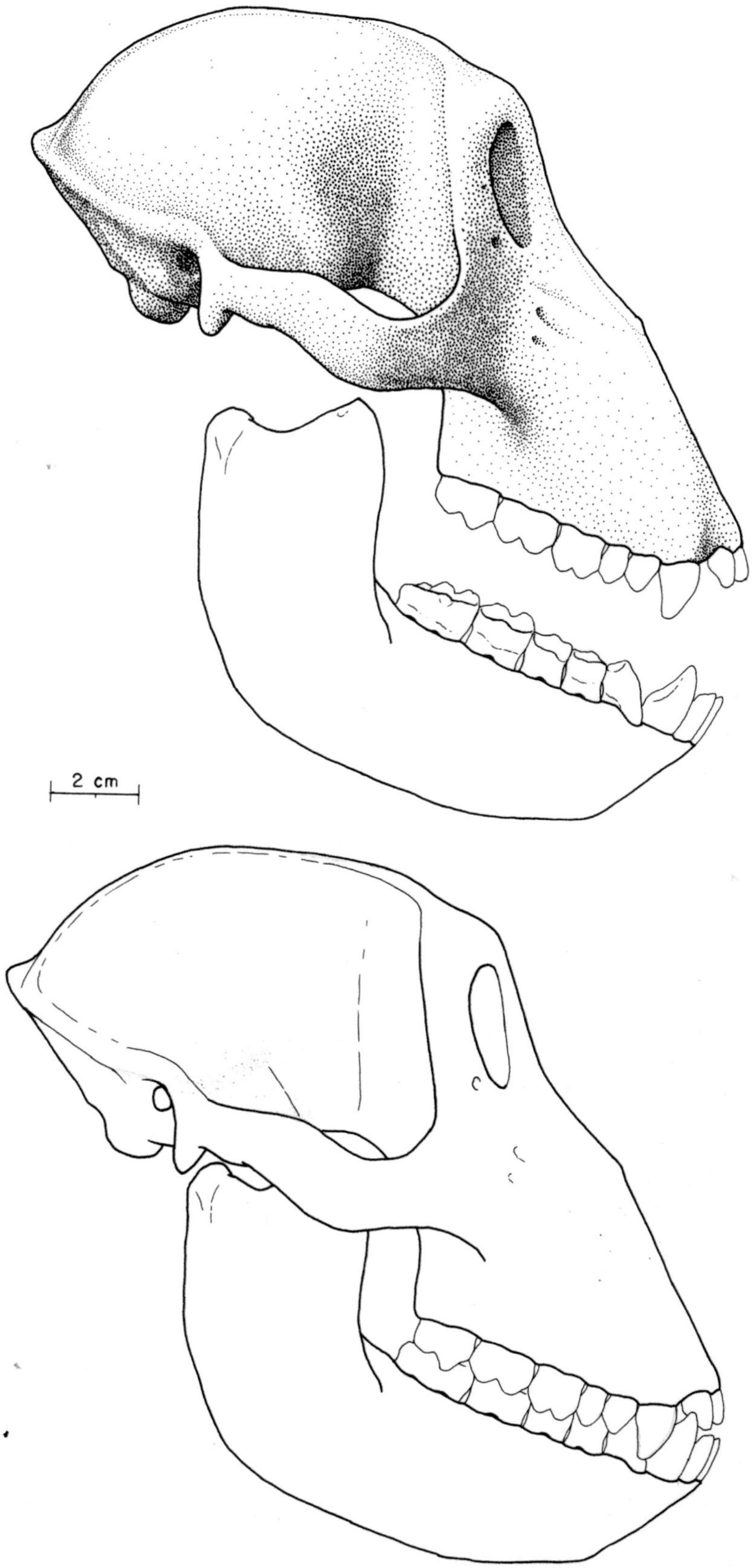

Fig. 16.10 *Paradolichopithecus arvernensis*, FSL 41336, Senèze holotype skull, right lateral drawings in Frankfurt plane: cranium and mandible separated above the two elements in occlusion, artwork by B. Akerbergs

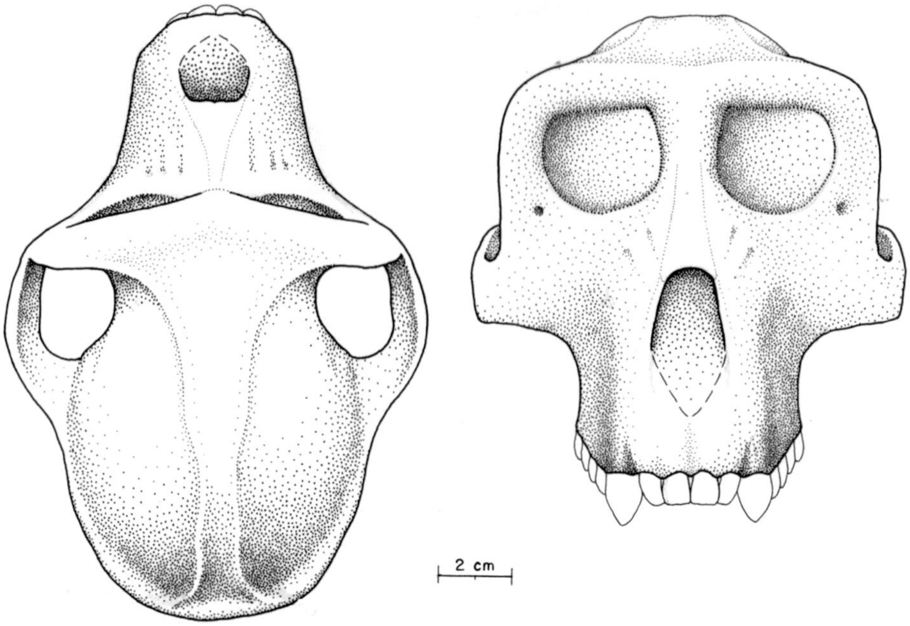

Fig. 16.11 *Paradolichopithecus arvernensis*, FSL 41336, Senèze holotype skull, drawings in Frankfurt plane: cranium in dorsal (left) and frontal (right) views, artwork by B. Akerbergs

Are Paradolichopithecus *and* Procynocephalus *synonymous or distinct genera?*

Kostopoulos et al. (2018: 188–189) listed a number of features they claimed were shared by *Paradolichopithecus* and *Procynocephalus*:

(a) similar size and body mass;
(b) similar dental morphology, low fourth premolar to first molar area ratio and a strong molar flare;
(c) females with weak to moderate anteorbital drop (comparison between DFN3-150, UCBL-FSL 41336, and PIN 3120–523 of *Paradolichopithecus* and the partial face of female *Pro. subhimalayanus* illustrated by Szalay and Delson (1979: Fig. 181A);
(d) premaxillary bones marginally reaching nasal bones;
(e) females with a smoothly rounded anterior muzzle traveling upward towards the nasion in a gentle curve (comparison based on the same specimens as (c); more vertical in male *Paradolichopithecus*);
(f) females with weak to absent maxillary fossae and weak to moderate but always present (n = 7) fossae of the mandibular corpus;
(g) narrow parabolic upper and lower toothrows with parallel molar series;
(h) small (compared to I2) and non-shovel shaped first incisors;
(i) a gradually descending capitulum of the distal humerus (shared between *Pro. wimani* and *Paradolichopithecus* from Vatera, Greece; e.g., van der Geer and Sondaar 2002);
(j) a protruding (in dorsal view) lateral suspensory facet for the fibular malleolus of the talus and parallel trochlea tali (shared between *Pro. wimani* and *Paradolichopithecus* from Valea Grăunceanului, Romania and Vatera, Greece; e.g., van der Geer and Sondaar (2002) and Sondaar et al. (2006);
(k) a narrow coronoid process of the proximal ulna (shared between *Pro. wimani* and *Par. geticus*; compare Fig. 182f and 186f in Szalay and Delson (1979) in relation to Fleagle and McGraw (2002: Fig. 10).

Of these 11 points, restricting the comparison to *Paradolichopithecus arvernensis*, *Procynocephalus wimani* and *Pro. subhimalayanus* (see Fig. 16.13), a, b, d and probably g and h are just a matter of similar large size (equivalent to *Papio* and other large genera unrelated to *Mandrillus*); *c* may be correct but linked to relationship with *Macaca*, although breakage mostly prevents observation of the anteorbital region in NHMUK-P M37157; *e* is untrue, *Pro. subhimalayanus* has a more flattened muzzle dorsum, confirmed by a CT scan which shows that this shape is not due to crushing; *f* again is expected for relatives of *Macaca*, and the mandibular corpus fossa is not as extensive as suggested (see above); *i*, *j* and *k* are correct, due to similar locomotor adaptation (a Grăunceanu humerus, which is less

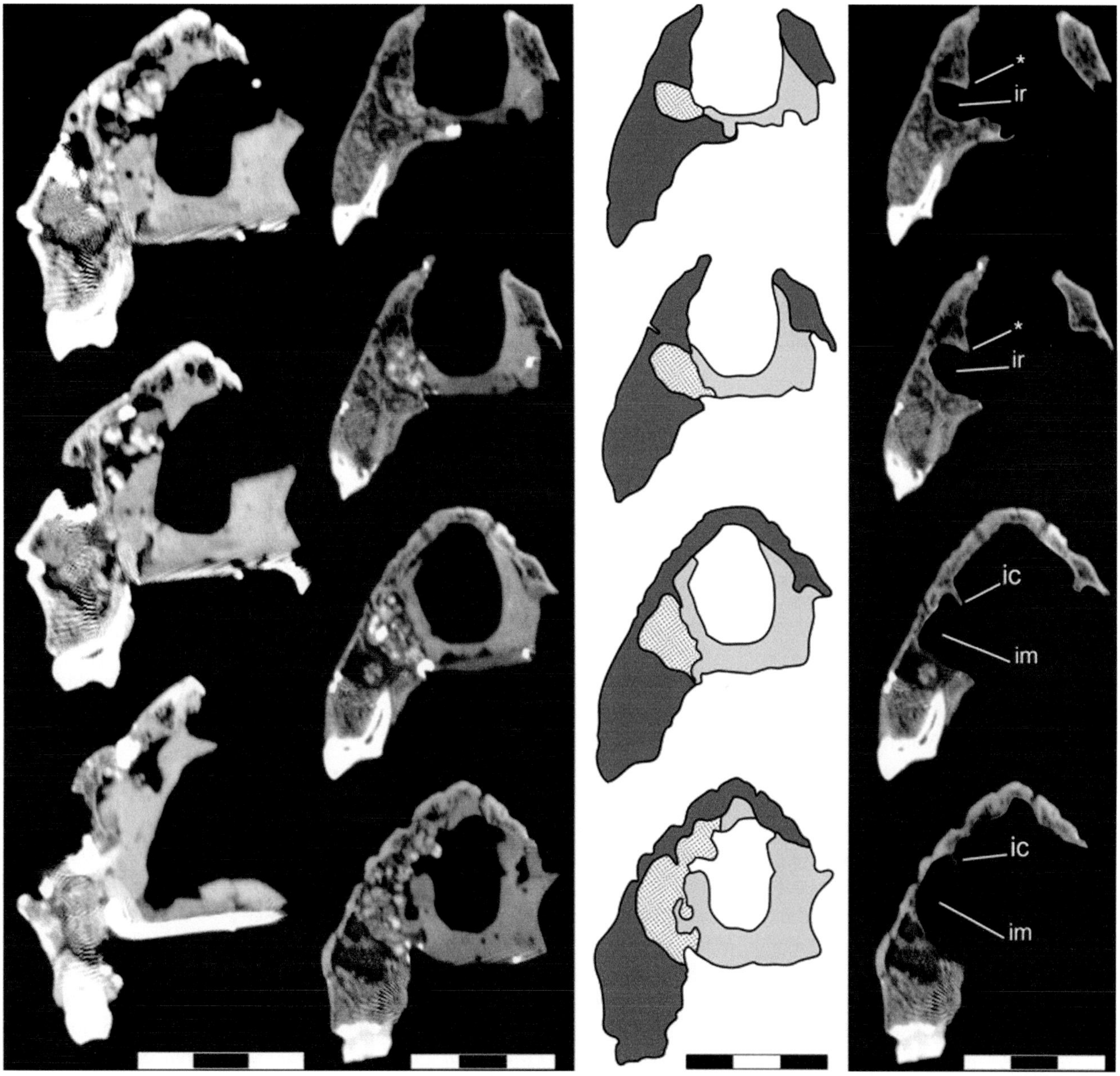

Fig. 16.12 *Paradolichopithecus arvernensis*, FSL 41336, Senèze holotype skull, coronal CT scan slices from Nishimura et al. (2009), with permission: left column, top to bottom, slices through mesial, middle and distal regions of M3; second column, top to bottom, slices through M1, M1-2 contact area, M2, M2-3 contact area; third column, top to bottom, interpretive schematic drawings of previous slices; right column, top to bottom, images virtually excluding the bony structure. The scales are in centimeters. The areas filled in dark gray, light gray, and dots on the schematic drawings indicate bone, cement or plaster, and matrix, respectively. At the level of M1 and M1–2 contact on the right (left on the image), a small inferior recess (ir) accompanies a thick bony ridge projecting inferiorly and medially (*). At the level of M2 and M2–3 contact, a thin basal portion of the inferior concha (ic) descends from the superior portion of the muzzle, and the inferior meatus (im) expands laterally and superiorly to occupy a large area of the nasal cavity. At the level of mesial M3, the inferior concha (ic) can be observed in the superior portion of the muzzle and the maxillary body faces just the inferior meatus. The nasolacrimal canal probably extended anteriorly through the lateral-superior region (**, area in white) of the muzzle at the level of middle M3, opening towards the inferior meatus (*) at the level of mesial M3. The inferior concha can be observed from the middle portion of the maxillary body at the level of distal M3 and the superior portion of the body faces the middle meatus. The medial bony wall (bw) separates the maxillary body continuously from the middle meatus of the nasal cavity

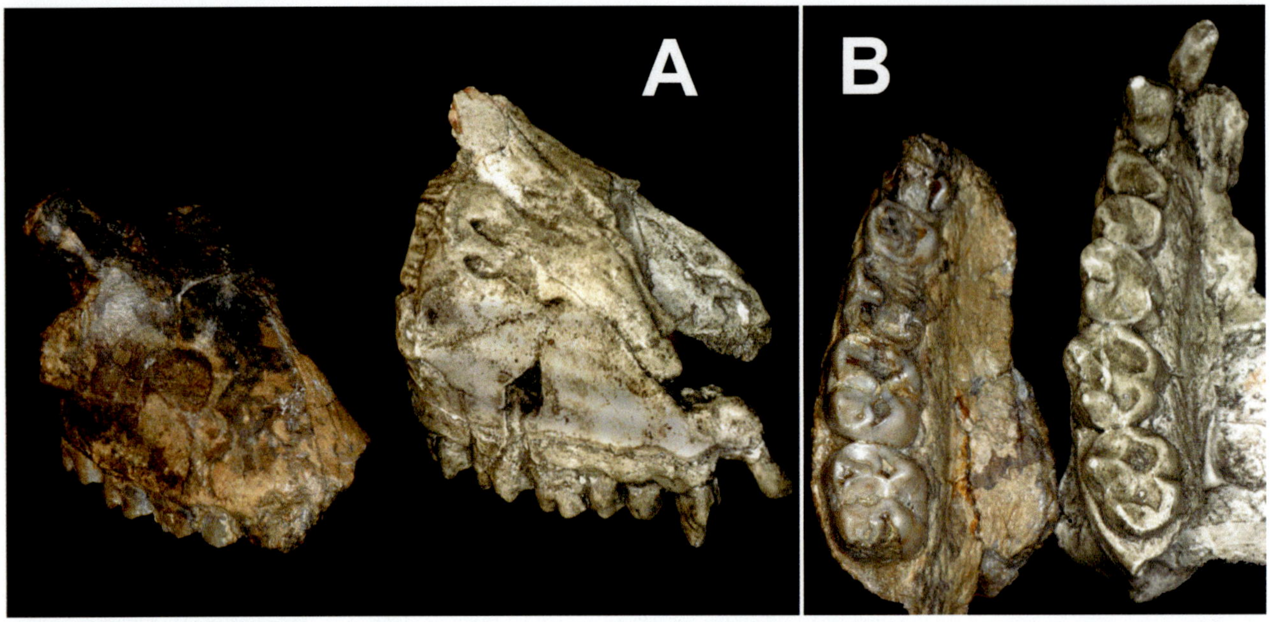

Fig. 16.13 Maxillae of female large cercopithecines: right, FSL 41336, the Senèze holotype of *Paradolichopithecus arvernensis* (cast) and left, NHMUK-P M37157, the Siwalik holotype of *Procynocephalus subhimalayanus*; (**A**) right lateral view, slightly oblique, showing difference in elevation of muzzle dorsum; (**B**) occlusal view, showing similarity in size

distorted, is illustrated in Szalay and Delson 1979 and in Sondaar et al. 2006; *Pro. wimani* [illustrated by Teilhard de Chardin 1938] and Vatera are similar, but there is no talus known from Grăunceanu).

Kostopoulos et al. (2018) also noted the presence of *Paradolichopithecus* in China, but in fact the generic identity of "*P.*" *gansuensis* is moot. Qiu et al. (2004, p. 170), in part following Trofimov (1977), suggested that the cheek teeth of *Paradolichopithecus* are more lophodont than those of *Procynocephalus* and the distal margin of the M3 is more rounded, with a larger fovea. These distinctions are not supported by an examination of all available *Paradolichopithecus arvernensis*, which show some variability in the "crestiness" of the lower molars especially. Dental features do not seem to distinguish between the two genera, and thus the Longdan species is only questionably indicated here as possibly *Paradolichopithecus*. Kostopoulos et al. (2018, p. 188) further suggested that the Dafnero specimen could be included just as easily in *Procynocephalus wimani* due to shared morphology, such as "weak-moderate anteorbital drop", but that feature is not visible in the holotype maxilla of *P. wimani* (see Nishimura et al. 2014).

Thus, all the points where the two genera are similar can be explained by comparable size, linkage to *Macaca* and/or locomotor adaptation. One character does appear to separate the genera, namely the shape of the muzzle dorsum, which is more elevated and "peaked" in *Paradolichopithecus arvernensis* but relatively flattened in *Procynocephalus subhimalayanus* (Fig. 16.13 A), although not at all as much so as in large Papionina. Pending further analysis and additional material of *Procynocephalus*, the two genera should be kept distinct.

Macacina or Papionina

Determining the subtribal allocation of *Paradolichopithecus*, whether Macacina or Papionina, is also fraught. Maschenko (1994, 2005) discussed the large primate from Kuruksay as *Papio* (*Paradolichopithecus*) *sushkini* in some places and *Papio sushkini* in others. It is not clear if he meant to use *Paradolichopithecus* as a subgenus or was merely indicating the "original" name parenthetically. He also mentioned *Paradolichopithecus arvernensis* without employing *Papio* as part of the name.

In a summary of his views, Maschenko (2005, pp. 111–112) wrote: "*Papio sushkini* … is similar to recent *Papio* in morphology of the infraorbital foramina, ratio of lengths of the facial and cranial parts of the skull, the short and robust alveolar process of the upper jaw, the large anteroposterior diameter of foramen magnum, the shape of the nasal bones, position of the fossa glandula lacrimalis (only on os lacrimale), shape of the cecal foramen, morphology of the CP3 complex molar morphology, and relation of lengths of the premolar and molar tooth rows. Specialization of *Papio sushkini* in comparison with modern representatives of the genus is found in a greater robustness of molars relative to premolars, greater enamel thickness, greater length of the zygomatic arch relative to total length of the skull, and a more robust M^3 in comparison with M^2."

Once again, most of these similarities are linked to the large size of the Eurasian fossils. Maschenko (1994, 2005) did not mention the anteorbital drop characterizing *Papio* or its lack in the fossils, which may be the most diagnostic distinction of *Macaca* and *Paradolichopithecus* from *Papio*.

Kostopoulos et al. (2018) also plotted the P4/M1 area ratio, using a much smaller sample and both sides of the jaw for fossil specimens. They found that the *Paradolichopithecus* box overlapped the P-L–T range but only the whisker of *Macaca*, while the *Paradolichopithecus* upper whisker overlapped the *Macaca* box and the lower end of the C-M range. From this pattern, they suggested another similarity to Papionina rather than Macacina, but their results are vitiated with the larger samples used here (Fig. 16.2, see above).

As noted previously, Le Maître et al. (2023) analyzed the inner ear bony labyrinth morphology of the Dafnero specimen and other cercopithecines. They concluded with some hesitancy that *Paradolichopithecus* showed similarities to Cercopithecini and Papionina, but not to macaques. They suggested that *Paradolichopithecus* might be a stem papioninan or perhaps a basal papionin closer to Papionina than to *Macaca*. The data they presented do not strongly support these conclusions.

In their introductory remarks, Le Maître et al. (2023, p. 210), reported the disagreement between those who considered *Paradolichopithecus* (and *Procynocephalus*) macaque-related and those who "propose closer phylogenetic affinities with the African baboons". The first author they cited in the latter group is Jolly (1967), who in fact clearly linked both taxa with Asian macaques (p. 45). They may have been confused by his broad use of the term "baboon", by which he meant terrestrial cercopithecines (and possibly colobines), usually large and long-faced.

Le Maître et al. (2023, p. 218) studied both labyrinths of the Dafnero cranium, noting that the left one was undistorted, while the right was damaged and somewhat distorted. Overall shape was most similar to those of *Erythrocebus patas* and *Mandrillus sphinx*. Some features were similar to those seen in Papionina, but some also occurred with Macacina. Typicality probabilities determined from their bgPC analysis indicated that the left labyrinth was most likely to be a cercopithecin, while the (distorted) right grouped with papioninans; linkage to macaques was low (p. 224).

In a plot of regression score vs. log centroid size (Fig. 16.3, p. 220), macaques fell mostly below the regression line for the whole sample, along with some *Chlorocebus* and two of the three *E. patas*, with the Dafnero fossil falling near one of the latter, on the Cercopithecini line. Few papioninans fell below the overall line. Plots of the first four PCs from a standard PCA (Fig. 16.4, p. 221) revealed that the fossil has no clear affinities with any clade but was said to weakly group with Papionina (p. 223).

Three neighbor-joining cluster analyses based on Euclidean distances (p. 225) include the Dafnero specimen but do not capture known phylogenetic relationships (as they mentioned on p. 226). The fossil often links with *Mandrillus sphinx*, but also with *Macaca nigra* and a *Papio*; and centroid size (allometry) is important when included. None of the three tribal-group taxa (Macacina, Papionina and Cercopithecini) is consistently monophyletic; this does not lead to much faith in the systematic placement of the Dafnero fossil. A discussion of phylogenetic signal across these features also finds only a weak signal and suggestions of hesitancy (p. 227): "interpreting these results is tricky, because the null hypothesis of no pattern of similarity among relatives is very unlikely to be true in any biological organism."

Overall, the discussion is rather convoluted, noting some links but ignoring others. Similarities to macaques are mostly suggested to be ancestral retentions from a basal papionin or cercopithecine ancestry (p. 228). Le Maître et al. (2023, p. 229) wrote: "we can not exclude that the shape affinities of *Paradolichopithecus* with the baboon-related clade are the result of a parallel evolution for large, terrestrial species that would arise from a common allometric pattern across all cercopithecines." They noted (p. 230) that similarities of cranial shape between young *Paradolichopithecus* (i.e., Dafnero) and *Macaca* could be related to similar growth trajectories in all papionins, but they make no mention of adult similarity. In sum, they conclude that *Paradolichopithecus* is phylogenetically unrelated to macaques but either a papioninan or a basal papionin closer to papioninans. They wrote (p. 231): "Our findings do not provide a clear answer to the taxonomic classification of *Paradolichopithecus*, though the traditional hypothesis of a *Paradolichopithecus*–Macacina relationship appears to be the least supported."

Two incompletely published analyses presented arguments for the opposite view, namely that *Paradolichopithecus* is a phylogenetic sister to *Macaca*. Delson and Frost (2004) mentioned the results of a 3D geometric morphometric analysis of the face which linked *Paradolichopithecus* from Senèze (and Grăunceanu) to macaques rather than *Papio* when allometry (centroid size) was included; the details of this study were presented in the talk and will be included in Delson et al. (in preparation). An analysis of cranial landmarks placed the *Paradolichopithecus* specimens within the *Macaca* scatter but also showed some overlap with *Papio*. However, analyses of the lateral muzzle profile and the transverse curvature of the dorsal rostrum more clearly placed *Paradolichopithecus* with *Macaca* to the exclusion of *Papio*. Delson and Frost (2004) also noted that *Paradolichopithecus* had relatively small I1 area compared to M1 area (this is indicated in Table 16.2), but this pattern might in part be related to greater wear of the *Paradolichopithecus* incisors in older individuals.

O'Shea et al. (2016) presented preliminary results of a cladistic analysis also placing *Paradolichopithecus* with macaques. Details will be included in Delson et al. (in preparation), but their poster listed a number of synapomorphies with *Macaca*: intermediate pinching of the anterior temporal lines (in females); tympanic unfused and moderately separated from the postglenoid process (both sexes); maxillary fossa and maxillary ridge absent (in males); and glabella moderately prominent (both sexes). This resulted in a macacinan clade sister to Papionina (crown *Papio-Lophocebus-Theropithecus* and *Cercocebus-Mandrillus* groups plus several stem taxa), with some *Parapapio* and the outgroups "*Parapapio*" *lothagamensis* and *Victoriapithecus macinnesi* outside the crown.

For the moment, at least, it appears that a *Paradolichopithecus-Macaca* relationship is better supported than one linking the large Eurasian fossils with *Papio* or Papionina as a whole.

Summary

Two cercopithecid fossils are known from Senèze. A partial ulna represents cf. *Macaca sylvanus*. The holotype of *Paradolichopithecus arvernensis* (Depéret, 1928) is a nearly complete female skull. A brief history of research on *Paradolichopithecus* is provided. Emended diagnoses of both *Paradolichopithecus* and *P. arvernensis* are presented, with comparisons to other papionin genera. The Senèze skull is described in detail. *Macaca* and *Paradolichopithecus* are postcranially adapted to a partly or fully terrestrial habitus (based on material from other sites); this supports a relatively open wooded landscape around the Senèze maar. *Paradolichopithecus arvernensis* from Senèze may be one of the youngest known occurrence of the genus, at 2.2–2.1 Ma. The Grăunceanu sample is possibly somewhat older based on a slightly smaller size. The crushed subadult female from Dafnero appears similar in size, and a reassessment of the reported paleomagnetic data suggests an age between 2.58 and 2.2 Ma.

Two questions regarding the systematics of *Paradolichopithecus* are actively discussed: is it a synonym of *Procynocephalus* and is it more closely related to *Macaca* or *Papio*? Most features shared by the two Eurasian fossil genera are based on large size and terrestrial locomotor adaptation, although they differ in the shape of the muzzle dorsum; thus both genera are retained as distinct. Similarly, features suggesting a link between *Paradolichopithecus* and *Papio* (or other papioninans) mainly appear related to shared large size. Claimed distinctions or links related to the P4/M1 area ratio and the shape of the bony labyrinth do not stand up to scrutiny. Unpublished 3D geometric morphometric and cladistic analyses of the cranium do support a *Paradolichopithecus-Macaca* relationship.

Acknowledgments I thank Prof. L. David and Drs. M. Faure, C. Guérin, P. Mein, A. Prieur, E. Robert (UCB-Lyon 1) and J. Hürzeler (NMB) for facilitating my access to the Senèze fossil primates over many decades. I thank many curators and collection managers/associates of the Department of Mammalogy, American Museum of Natural History, for allowing me to make use of their incomparable collection of cercopithecid and other primates over an even longer period. Many other curators of fossil and extant primate collections across Europe permitted me to study specimens in their care. My travel to Senèze and to Lyon and other collections was supported, in part, by research grants from many sources, including Columbia University, the Leakey Foundation, the National Science Foundation (grants BCS-0083166 and -0104781), the National Geographic Society (grants 944, 7050-01 and 7434-03), the PSC-CUNY faculty research program (awards 11152, 6-69381, 6-64333, 6-65407, 63528-32. 64542-33, 65509-34, 66578-35. 67356-36 and 68016-37) and the Wenner-Gren Foundation for Anthropological Research (grant no. 2810).

I acknowledge Ms. Biruta Ackerbergs for her drawings of the Senèze cranium and mandible (from casts), originally made for inclusion in Szalay and Delson (1979). I thank Drs. Fred Szalay and Takeshi Nishimura for permission to use their images in Figs. 16.2 and 16.10. I thank Dr. Martine Faure for locating and showing me Depéret's (1928) article and for writing the résumé in French.

I thank Julia L. Arenson, Dr. Martine Faure and Dagmawit Abebe Getahun for comments on an early version of this manuscript. I thank Dr. Eric J. Sargis and especially reviewers Drs. David Alba, Stephen Frost and Christopher Gilbert for their helpful suggestions. All of these have improved the manuscript significantly.

References

Aguirre, E., & Soto, E. (1974). Nuevo fósil de Cercopitécido en el Pleistoceno inferior de Puebla de Valverde (Teruel). *Estudios Geológicos, 30*, 213–217.

Aguirre, E., & Soto, E. (1978). *Paradolichopithecus* in La Puebla de Valverde, Spain: Cercopithecoidea in European Neogene stratigraphy. *Journal of Human Evolution, 7*, 559–565.

Alba, D. M., Moyà-Solà, S., Madurell, J., & Aurell, J. (2008). Dentognathic remains of *Macaca* (Primates, Cercopithecidae) from the late early Pleistocene of Terrassa (Catalonia, Spain). *Journal of Human Evolution, 55*, 1160–1163.

Alba, D. M., Carlos Calero, J. A., Mancheño, M. Á., Montoya, P., Morales, J., & Rook, L. (2011). Fossil remains of *Macaca sylvanus florentina* (Cocchi, 1872) (Primates, Cercopithecidae) from the Early Pleistocene of Quibas (Murcia, Spain). *Journal of Human Evolution, 61*, 703–718.

Battetta, J. (1969). Compte rendu du remaniement de la reconstitution du crâne de *Dolichopithecus arvernensis* Depéret, type. *Bulletin mensuel de la Société Linnéenne de Lyon, 38ème année*, 279–284.

Benammi, M., Aidona, E., Merceron, G., Koufos, G.D. & Kostopoulos, D.S. (2020). Magnetostratigraphy and chronology of the Lower Pleistocene primate bearing Dafnero fossil site, N. Greece. *Quaternary, 3*, 22 [16 pp.]; https://doi.org/10.3390/quat3030022.

Burnett, G. T. (1828). Illustrations of the Manupeda or apes and their allies: Being the arrangement of the Quadrumana or

Anthropomorphous beasts indicated in the outline. *Quarterly Journal of Science, Literature and Art, 26*, 300–330.

Channell, J. E. T., Singer, B. S., & Jicha, B. R. (2020). Timing of Quaternary geomagnetic reversals and excursions in volcanic and sedimentary archives. *Quaternary Science Reviews, 228*, 106114.

Crégut-Bonnoure, E., Guérin, C., Argant, A., Argant, J., Debard, E., & Delson, E., et al. (2024). Biochronology of the Senèze faunal assemblage. In E. Delson, M. Faure & C. Guérin (Eds.), *Senèze: Life in Central France two million years ago. Paleontology, geochronology, stratigraphy and taphonomy* (pp. 633–652). Cham (Switzerland): Springer.

Delson, E. (1971). Estudio preliminar de unos restos de simios pliocenicos procedentes de "Cova Bonica" (Gava) (Prov. Barcelona). *Acta Geológica Hispánica, 6*, 54–57.

Delson, E. (1974). Preliminary review of cercopithecid distribution in the circum-Mediterranean region. *Mémoires du Bureau de Recherches Géologiques et Minières, 78*, 131–135.

Delson, E. (1975). Evolutionary history of the Cercopithecidae. *Contributions to Primatology, 5*, 167–217.

Delson, E. (1977). Catarrhine phylogeny and classification: Principles, methods and comments. *Journal of Human Evolution, 6*, 433–459.

Delson, E. (1980). Fossil macaques, phyletic relationships and a scenario of deployment. In D. G. Lindburg (Ed.), *The Macaques: Studies in Ecology, Behavior and Evolution* (pp. 10–30). New York: Van Nostrand.

Delson, E., & Frost, S. R. (2004). *Paradolichopithecus*: A large-bodied terrestrial papionin (Cercopithecidae) from the Pliocene of western Eurasia (abstract). *American Journal of Physical Anthropology, Supplement 38*, 85.

Delson, E. & Plopsor, D. (1975). *Paradolichopithecus*, a large terrestrial monkey (Cercopithecidae, Primates) from the Plio-Pleistocene of southern Europe and its importance for mammalian biochronology. In *Proceedings VIth Congress of the Regional Committee on Mediterranean Neogene Stratigraphy*, Bratislava, 1975 (pp. 91–96). Slovak Academy of Sciences.

Delson, E., Terranova, C. J., Jungers, W. L., Sargis, E. J., Jablonski, N. G., & Dechow, P. C. (2000). Body mass in Cercopithecidae (Primates, Mammalia): Estimation and scaling in extinct and extant taxa. *Anthropological Papers of the American Museum of Natural History, 83*, 1–159.

Delson, E., Alba, D. M., Frost, S. R., Harcourt-Smith, W. E., Martín Suárez, E., Mazelis, E. J., et al. (2014). *Paradolichopithecus*, a large terrestrial Pliocene cercopithecine from Europe: new remains and an update (abstract). In M. Delfino, G. Carnevale & M. Pavia (Eds.), *Abstract book and field trip guide*, 12th Annual Meeting of the European Association of Vertebrate Palaeontologists, Turin, Italy, June 2014 (p. 50).

Delson, E., Nomade, S., Sen, S., Debard, E., Pastre, J-F., Bahain, J-J. et al. (2024). Geochronology of Senèze: ^{40}Ar/^{39}Ar Dating and Magnetostratigraphy, with Notes on an ESR/U-Series Dating Attempt. In E. Delson, M. Faure & C. Guérin (Eds), *Senèze: Life in Central France two million years ago. Paleontology, geochronology, stratigraphy and taphonomy* (pp. 99–122). Cham (Switzerland): Springer.

Depéret, C. (1889). Sur le *Dolichopithecus ruscinensis* nouveau singe fossile du Pliocène du Roussillon. *Comptes Rendus de l'Académie des Sciences* (Paris), *109*, 982–983.

Depéret, C. (1928). Les singes fossiles du Pliocène de France. *Comptes-Rendus du XIVe Congrès Géologique International, 1926* (pp. 3–4). Madrid: Gráficas Reunidas.

Depéret, C. (1929). *Dolichopithecus arvernensis* Depéret: Nouveau singe du Pliocène supérieur de Senèze (Haute-Loire). *Travaux du Laboratoire de Géologie de la Faculté des Sciences de Lyon*, fascicule *15* (mémoire 12), 5–12.

Erxleben, J. C. P. (1777). *Systema regni animalis. Classis 1, Mammalia*. Lipsiae: Weygand.

Fleagle, J. G., & McGraw, W. S. (1999). Skeletal and dental morphology supports diphyletic origin of baboons and mandrills. *Proceedings of the National Academy of Sciences of the USA, 96*, 1157–1161.

Fleagle, J. G., & McGraw, W. S. (2002). Skeletal and dental morphology of African papionins: Unmasking a cryptic clade. *Journal of Human Evolution, 42*, 267–292.

Fooden, J. (2007). Systematic review of the Barbary macaque, *Macaca sylvanus* (Linnaeus, 1758). *Fieldiana Zoology, 113*, 1–60.

Frost, S. R., Ting, N., Harcourt-Smith, W., & Delson, E. (2005). Positional and locomotor behavior of *Paradolichopithecus arvernensis* as inferred from the functional morphology of the postcrania (abstract). *American Journal of Primatology, 66*, 134.

Geoffroy Saint-Hilaire, E. (1812). Tableau des quadrumanes, 1: Ord. Quadrumanes. *Annales du Muséum d'Histoire naturelle, Paris 19*, 85–122.

Gilbert, C. C. (2007). Craniomandibular morphology supporting the diphyletic origin of mangabeys and a new genus of the *Cercocebus/Mandrillus* clade, *Procercocebus*. *Journal of Human Evolution, 53*, 69–102.

Gilbert, C. C. (2013). Cladistic analysis of extant and fossil African papionins using craniodental data. *Journal of Human Evolution, 64*, 399–433.

Gray, J. E. (1821). On the natural arrangement of vertebrose animals. *London Medical Repository, 15*, 296–310.

Hoffstetter, R. (1977). Phylogénie des primates. *Bulletin et Mémoires de la Société d'Anthropologie de Paris, 13*, 327–346.

Jolly, C. J. (1967). The evolution of the baboons. In H. Vagtborg (Ed.), *The Baboon in medical research* (Vol. 2) (pp. 23–50). Austin: University of Texas Press.

Kostopoulos, D. S., Guy, F., Kynigopoulou, Z., Koufos, G. D., Valentin, X., & Merceron, G. (2018). A 2Ma old baboon-like monkey from northern Greece and new evidence to support the *Paradolichopithecus - Procynocephalus* synonymy (Primates: Cercopithecidae). *Journal of Human Evolution, 121*, 178–192.

Lacépède, B. G. E. (1799). Tableau des divisions, sous-divisions, ordres et genres des mammifères. In G. L. L. de Buffon (Ed.), *Histoire naturelle* (vol. 38, Quadrupèdes t. 14, pp. 144–195). Paris: P. Didot L'Ainé et Firmin Didot.

Le Maître, A., Guy, F., Merceron, G., & Kostopoulos, D. S. (2023). Morphology of the bony labyrinth supports the affinities of *Paradolichopithecus* with the Papionina. *International Journal of Primatology, 44*, 209–236.

Linnaeus, C. (1758). *Systema naturae per regna tria naturae, secundum classes, ordines, genera, species cum characteribus, differentiis, synonymis, locis* (10th Edn.) Holmiae: Laurentii Salvii.

Maschenko, E. N. (1994). *Papio* (*Paradolichopithecus*) *suschkini* (Trofimov) - revision of systematics, morphofunctional peculiarities of the skull and lower jaw (in Russian). In L. P. Tatarinov (Ed.), *Paleotheriology* (pp. 15–57). Nauka.

Maschenko, E. N. (2005). Cenozoic primates of Eastern Eurasia (Russia and adjacent territories). *Anthropological Science, 113*, 103–115.

Mivart, St. G. (1864). Notes on the crania and dentition of the Lemuridae. *Proceedings of the Zoological Society of London, 1864*, 611–648.

Moyà Solà, S., Pons Moyà, J., & Köhler, M. (1990). Primates catarrinos (Mammalia) del Neógeno de la península Ibérica. *Paleontologia i Evolució, 23*, 41–45.

Necrasov, O., Samson, P., & Rădulesco, C. (1961). Sur un nouveau singe catarrhinien fossile, découvert dans un nid fossilifère d'Olténie (R. P. R.). *Analele Stiintifice Universitatii "Al. I. Cuza" din Iaşi*, Sec. *II, Stiinte Naturale, 7*, 401–416.

Nishimura, T. D., Takai, M., & Maschenko, E. (2007). The maxillary sinus of *Paradolichopithecus sushkini* (late Pliocene, southern Tajikistan) and its phyletic implications. *Journal of Human Evolution, 52*, 637–646.
Nishimura, T. D., Senut, B., Prieur, A., Treil, J., & Takai, M. (2009). Nasal architecture of *Paradolichopithecus arvernensis* (late Pliocene, Senèze, France) and its phyletic implications. *Journal of Human Evolution, 56*, 213–217.
Nishimura, T. D., Zhang, Y., & Takai, M. (2010). Nasal anatomy of *Paradolichopithecus gansuensis* (Early Pleistocene, Longdan, China) with comments on phyletic relationships among the species of this genus. *Folia Primatologica, 81*, 53–62.
Nishimura, T. D., Ito, T., Yano, W., Ebbestad, J. O. R., & Takai, M. (2014). Nasal architecture in *Procynocephalus wimani* (Early Pleistocene, China) and implications for its phyletic relationship with *Paradolichopithecus*. *Anthropological Science*, *122*, 101–113.
O'Shea, N., Delson, E., Pugh, K. D., & Gilbert, C. C. (2016). Phylogenetic analysis of *Paradolichopithecus*: Fossil baboon or macaque? (abstract). *American Journal of Physical Anthropology, S62*, 244.
Owen, R. (1843). Report on the British fossil Mammalia (Part I: Unguiculata and Cetacea). In *Report on the l2th meeting of the British association for the advancement of science for 1842* (pp. 54–74). London: John Murray.
Plastiras, C., Kostopoulos, D. S., & Merceron, G. (2017). Tooth morphology, molar enamel thickness and dental microwear textural analysis with application on European *Paradolichopithecus/Procynocephalus* and comparisons with Pleistocene papionins (abstract). In *2nd international meeting of early-stage researchers in paleontology, 2017, Book of Abstracts* (pp. 161–162). http://www.imerp2.upatras.gr/files/mariatest/2IMERP_BOOK_OF_ABSTRACTS.pdf.
Qiu, Z., Deng, T., & Wang, B. (2004). Early Pleistocene mammalian fauna from Longdan, Dongxiang, Gansu, China. *Palaeontologica Sinica, 191. New Series C, 27*, 1–198.
Radinsky, L. (1974). The fossil evidence of anthropoid brain evolution. *American Journal of Physical Anthropology, 41*, 15–27.
Radović, P., Lindal, J., Marković, Z., Alaburić, S., & Roksandic, M. (2019). First record of a fossil monkey (Primates, Cercopithecidae) from the Late Pliocene of Serbia. *Journal of Human Evolution, 137*, 102681.
Radović, P., Marković, Z., Alaburić, S., & Roksandic, M. (2024) A new papionin molar (Primates, Cercopithecidae) from the Pliocene of Serbia. *Paläontologische Zeitschrift*, *98*, 637–646.
Rowe, N., & Meyers, M. (Eds.). (2016). *All the World's primates.* Pogonias Press.
Schaub, S. (1943). Die oberpliocaene Säugetierfauna von Senèze (Haute-Loire) und ihre verbreitungsgeschichtliche Stellung. *Eclogae Geologicae Helvetiae*, *36*, 270–289.
Schlosser, M. (1924). Fossil primates from China. *Palaeontologica Sinica, 1* (2), 1–16.
Sianis, P. D., Athanassiou, A., Kostopoulos, D. S., Roussiakis, S., Kargopoulos, N., & Iliopoulos, G. (2023). The remains of a large cercopithecid from the Lower Pleistocene locality of Karnezeika (southern Greece). *Earth and Environmental Science Transactions of the Royal Society of Edinburgh, 114*, 177–182.
Simionescu, I. (1930). Vertebratele Pliocene dela Măluşteni (Covurlui). *Academia Romána, Publicaţiunile Fondului Vasile Adamachi,, 9*, 83–136.
Simons, E. L. (1970). The deployment and history of Old World monkeys (Cercopithecidae, Primates). In J. R. Napier & P. H. Napier (Eds.), *Old World Monkeys* (pp. 97–137). London: Academic Press.
Sondaar, P. Y., van der Geer, A. A. E., & Dermitzakis, M. D. (2006). The unique postcranial of the Old World monkey *Paradolichopithecus*: More similar to *Australopithecus* than to baboons. *Hellenic Journal of Geosciences, 41*, 19–28.
Stehlin, H. G. (1923). Die oberpliocaene Fauna von Senèze (Haute-Loire). *Eclogae Geologicae Helvetiae, 18*, 268–281.
Šujan, M., Braucher, R., Chyba, A., Vlačiky, M., Aherwar, K., Rózsová, B., et al. (2023). Mud redeposition during river incision as a factor affecting authigenic ^{10}Be/^{9}Be dating: Early Pleistocene large mammal fossil-bearing site Nová Vieska, eastern Danube Basin. *Journal of Quaternary Science, 38*, 347–364.
Szalay, F. S., & Delson, E. (1979). *Evolutionary History of the Primates*. New York: Academic Press.
Teilhard de Chardin, P. (1938). The fossils from locality 12 of Choukoutien. *Palaeontologica Sinica*, *new series C, 5*, 1–50.
Terhune, C. E., Curran, C., Croitor, R., Drăguşin, V., Gaudin, T., Petculescu, A., et al. (2020). Early Pleistocene fauna of the Olteţ River Valley of Romania: Stratigraphic and biogeographic implications. *Quaternary International, 553*, 14–33.
Ting, N., Harcourt-Smith, W. E. H., Frost, S. R., & Delson, E. (2004). Description and analysis of postcranial elements of *Paradolichopithecus arvernensis*: A large bodied papionin from the Pliocene of Eurasia (abstract). *American Journal of Physical Anthropology, S38*, 195.
Trofimov, B. A. (1977). Primate *Paradolichopithecus sushkini* sp. nov. from the Upper Pliocene of the Pamirs Piedmont (preliminary communication). *Journal of the Palaeontological Society of India* (for 1975), *20*, 26–32.
Van der Geer, A. A. E., & Sondaar, P. Y. (2002). The postcranial elements of *Paradolichopithecus arvernensis* (Primates, Cercopithecidae, Papionini) from Lesvos, Greece. *Annales Géologiques des Pays Helléniques, 39*, 71–86.
Verheyen, W. N. (1962). Contribution à la craniologie comparée des primates. *Annales du Musée Royale de l'Afrique Centrale, série in 8vo, Sciences Zoologiques, 105*, 1–247.
Vlačiky, M. (2009). Výskumy kvartérnych paleontologických lokalít na Slovensku v roku 2009. In Z. Németh, D. Plašienka, L. Šimon, M. Kohút, and Ľ. Iglárová (Eds.) *Nové poznatky o stavbe a vývoji Západných Karpát* (Vol. 41, pp. 549–550). Bratislava: Mineralia Slovaca, Geovestník, ŠGÚDŠ. (in Slovak).
Vogel, C. (1966). Morphologische Studien an der Gesichtsschädel Catarrhiner Primaten. *Bibliotheca Primatologica, 4*, 1–226.
Vogel, C. (1968). The phylogenetical evaluation of some characters and some morphological trends in the evolution of the skull in catarrhine primates. In B. Chiarelli (Ed.), *Taxonomy and Phylogeny of Old World Primates with References to the Origin of Man* (pp. 21–55). Turin: Rosenberg and Sellier.

Chapter 17
Biochronology of the Senèze Faunal Assemblage

Evelyne Crégut-Bonnoure, Claude Guérin, Alain Argant, Jacqueline Argant, Evelyne Debard, Eric Delson, Véra Eisenmann, Martine Faure, Bernard Ménouret, Cécile Mourer-Chauviré, and Andrea M. F. Valli

Abstract Our fieldwork at Senèze from 2000–2006 was designed to place the fauna from this important site in a precise stratigraphic and geochronologic framework, as well as to seek new elements of known taxa and possibly additional taxa. The analyses reported in this volume led to the recognition of six new taxa: *Dinofelis* sp., *Canis* sp., cf. *Hemitragus*, two indeterminate bovids and an indeterminate hyaenid; *Ovis claudiusguerini* n. sp. was newly named for a specimen previously indicated as *Ovis* sp. *Hystrix refossa* and Hyaenidae indet. (both recovered in our excavations) are new taxa for the site, as is *Bison* (*Eobison*) sp., a specimen of which was collected by A. Consigny on the surface and presented to the team. Senèze is the type-locality for 13 mammalian taxa: three genera, seven species and three subspecies. Fifty-eight vertebrates have now been recorded for the site: 39 mammals (one lagomorph, four rodents, 12 carnivores, two primates, 16 cetartiodactyls, three perissodactyls and one proboscidean), 17 birds and two fishes; three additional rare equids, if accepted, would raise the number of vertebrates to 61 and of mammals to 42. The major mammalian and avian taxa in the assemblage are reviewed, especially in terms of their chronological implications. The faunal list is compared to those from other sites in the Massif Central and the nearby Saint-Vallier, resulting in Senèze being distinguished from all others as representing the MNQ 18 biochronological unit, for which it is considered to be a reference locality. The definition of that unit is discussed and clarified. Further comparison was made with selected MNQ 17–18 localities from Spain, Italy, the Netherlands and Germany. The Senèze fauna (other than *Bison* and perhaps some rare equids) is relatively homogeneous, with no

Claude Guérin—Deceased 2016.

E. Crégut-Bonnoure (✉)
Muséum Requien, Avignon, France
e-mail: evelyne.cregut1313@orange.fr

E. Crégut-Bonnoure
TRACES, UMR 5608 (CNRS-Université Toulouse 2 Jean-Jaurès), Toulouse, France

C. Guérin
Université Claude Bernard – Lyon 1, Villeurbanne, France

A. Argant · J. Argant
Aix Marseille Université, CNRS, Minist. Culture, Lampea, UMR 7269, Aix-en-Provence, France

E. Debard
Université de Lyon, UCBL, ENSL, CNRS, LGL-TPE, Villeurbanne, France

Saint-Didier-au-Mont-d'Or, France

E. Delson
Department of Vertebrate Paleontology, American Museum of Natural History, New York, NY, USA

Department of Anthropology, Lehman College of the City University of New York, Bronx, NY, USA

Ph.D. Program in Anthropology, The Graduate Center of the City University of New York, New York, NY, USA

New York Consortium in Evolutionary Primatology, New York, NY, USA

Institut Català de Paleontologia Miquel Crusafont, Barcelona, Spain

V. Eisenmann
CR2P (Paléontologie), Muséum National d'Histoire Naturelle, Paris, France

M. Faure
Université Lumière Lyon 2, Lyon, France

Université de Lyon, UCBL, ENSL, CNRS, LGL-TPE, Villeurbanne, France

B. Ménouret
Lançon-de-Provence, France

C. Mourer-Chauviré
Université de Lyon, UCBL, ENSL, CNRS, LGL-TPE, Villeurbanne, France

A.M.F. Valli
Société Scientifique du Bourbonnais, Moulins (Allier), France

Istituto Italiano di Paleontologia Umana, Anagni, Italy

E. Delson et al. (eds.), *Senèze: Life in Central France Around Two Million Years Ago*,
Vertebrate Paleobiology and Paleoanthropology, https://doi.org/10.1007/978-3-031-64415-3_17

evidence for a second assemblage as suggested by some previous authors. Its base age of ca. 2.20 Ma suggests this date for the beginning of MNQ 18, but the age of the end of this unit is still under discussion.

Résumé Nos travaux de fouille sur le site de Senèze, de 2000 à 2006, avaient pour objectif de situer la faune de ce gisement important dans un cadre stratigraphique et géochronologique précis, ainsi que de retrouver de nouveaux restes d'espèces connues, et éventuellement de nouvelles espèces. Nos recherches ont conduit à la reconnaissance de six nouveaux taxons: *Dinofelis* sp., *Canis* sp., cf. *Hemitragus*, deux bovidés indéterminés et un hyaenidé indéterminé; *Ovis claudiusguerini* n. sp. a été nouvellement nommé à partir d'un spécimen précédemment indiqué comme *Ovis* sp. *Hystrix refossa* et Hyaenidae indet. (trouvés dans nos fouilles) sont de nouveaux taxons pour le site, comme le *Bison* (*Eobison*) sp., dont un fossile a été découvert par A. Consigny en surface et donné à l'équipe. Senèze est la localité type de 13 taxons de mammifères: trois genres, sept espèces et trois sous-espèces. Cinquante-huit vertébrés sont désormais recensés sur le site: 39 mammifères (un lagomorphe, quatre rongeurs, 12 carnivores, deux primates, 16 cetartiodactyles, trois périssodactyles et un proboscidien), 17 oiseaux et deux poissons; l'existence de trois autres espèces rares d'Equidae reste incertaine, mais augmenterait le nombre total des vertébrés à 61 et celui des mammifères à 42. Les principales espèces de mammifères et d'oiseaux de l'assemblage sont passées en revue, notamment pour leurs indications biochronologiques. La liste faunique est comparée à celles d'autres sites paléontologiques du Massif Central, et du gisement proche de Saint-Vallier (Drôme). Depuis plus de 40 ans Senèze est le site de référence pour l'unité biochronologique MNQ 18. La définition de cette unité est discutée et clarifiée. Une comparaison plus approfondie a été effectuée avec des sites paléontologiques attribués aux unités biochronologiques MNQ 17–18 en Espagne, en Italie, aux Pays-Bas et en Allemagne. La faune de Senèze (en dehors du bison et peut-être de quelques rares équidés) est relativement homogène, sans aucune preuve d'un deuxième assemblage comme ont pu le suggérer par le passé certains auteurs. Son âge de base est d'env. 2,20 Ma, correspondant au début de MNQ 18, mais l'âge de la fin de cette unité est encore en discussion.

Keywords MNQ 17 • MNQ 18 • Late Villafranchian • Middle Villafranchian • Mammals • Birds • Faunal lists

Mots-clés MNQ 17 • MNQ 18 • Villafranchien supérieur • Villafranchien moyen • Mammifères • Oiseaux • Listes fauniques

Introduction

The Massif Central and its margins are home to a remarkable concentration of late Pliocene and early Quaternary paleontological deposits, of which Senèze is an important reference in Plio-Pleistocene paleontology. Before the Second World War, thanks to Pierre Philis, the site yielded a series of fossil mammals of remarkable quality and diversity (see Faure et al. 2024). However, most of this material suffers from a lack of spatial and, above all, stratigraphic localization, due to the field methods of his time. As a result, the age and the homogeneity of the Senèze fauna have been debated at length. The chronostratigraphic setting of Senèze is therefore of great importance. With this in mind, the excavations of 2001–2006[1] were aimed, among other things, at finding additional fossil remains and locating them as precisely as possible in the stratigraphy, in order to propose a reliable dating.

Geological Context

The Senèze site is located in a maar, on the edge of the lake that once occupied its bottom. Here the surrounding fauna, particularly herbivores and their predators, regularly came to drink, probably over a very long period of time. The exceptional wealth of vertebrate fossils is therefore entirely logical and explainable. This maar corresponds to a highly volcanic zone, which is not without consequences for the stratigraphy of the deposit. The analysis of sedimentology and stratigraphy (Debard 2024) shows how difficult it is to establish stratigraphic correlations between the various excavation sites and suggests the sliding of slope deposits into the lake, or even of lake sediment flowing into the water. As a result, it is difficult, if not impossible, to correlate stratigraphically the various faunal elements from the Philis collections and those from the new excavations, and also to establish relative dates. In the southeast sector (Trenches 5–7), fossils are associated with coarse deposits. These are the result of the reworking of slope formations that slid into the lake during cold periods when water levels dropped. In this sector, numerous faults affect both slope deposits and lake sediments, accentuating their plunge towards the center of the maar. In contrast, in the western sector (Trenches 1 and 2, zone H8), fossils are found in clayey or loamy deposits corresponding to lacustrine sediments, marking a high lake level. These different sedimentary dynamics from one side of the maar to the other should be taken into consideration to

[1]The 2000 season was dedicated to survey and prospecting, without any excavation or collecting of fossils.

explain the presence of the oldest tephras in the southeastern sector, whereas they are more recent in the western sector, despite an altitudinal shift of a few meters.

Delson et al. (2024) interpreted $^{40}Ar/^{39}Ar$ ages (recalculated following Schaen et al. 2021) and linked paleomagnetic "columns" to date the finds from these two sectors. Fossils from the western sector ranged in age between 2.10 and 2.08 Ma, while those from the southeastern sector dated between ca. 2.20–2.18 Ma. It is likely that the fossils recovered by Philis dated within the broader 2.20–2.08 Ma interval as well, but no closer estimate can be provided.

Paleoflora

During the latest excavations, all sediment types directly related to the fauna discovered were sampled for pollen analysis, to which we must add 11 hyaenid coprolites. Out of 208 samples, only 24 preserved pollens, often in insufficient quantity for statistical analysis, so these results are essentially qualitative. Nevertheless, they provide important palynological information, as they are directly related to the fauna, without the classic risks of disturbance of lake sediments at the bottom of a maar. Palynological analysis of the paleontological sediments and hyaenid coprolites demonstrates the non-contemporaneity of these sediments with those of the Elhaï core (see J. Argant 2024).

Mammalian and Avian Fauna

Fauna from our excavations provide the essential data for precise spatial location of remains in a reliable chronostratigraphic context. However, the fauna from Pierre Philis's earlier discoveries is much more extensive and must be included in any analysis of the site's relative age. The analyses in this volume have revised several mammal groups, leading to the discovery of six new (but not fully identified) taxa (*Dinofelis* sp., *Canis* sp., cf. *Hemitragus*, two indeterminate bovids and an indeterminate hyaenid), and the better characterization of several others, including a new species dedicated to the late Claude Guérin (*Ovis claudiusguerini*). A total of 58 vertebrates has been recorded for the combined collection: 39 mammals (one lagomorph, four rodents, 12 carnivores, two primates, 16 cetartiodactyls, three perissodactyls and one proboscidean), 17 birds and two fishes (Table 17.1). Three additional equids represented only by one or two speciens each have also been tentatively recognized (Eisenmann & Delson 2024); if they are included, the totals would rise to 61 vertebrates and 42 mammals. The fauna discovered during our excavations includes 16 species: two rodents, two carnivores, 7 cetartiodactyls, three perissodactyls, one proboscidean and one fish. Three of these are new to Senèze (*Hystrix refossa*, Hyaenidae indet. and *Bison* (*Eobison*) sp.).

Table 17.1 Faunal list with indications of sources of the fauna from Senèze

Taxon	FGD[1] 2001–2006	Previous finds
Mammalia		
Lagomorpha		
Leporidae		
Oryctolagus cf. *lacosti*		x
Rodentia		
Sciuridae		
Sciuridae indet. aff. *Eutamias*		x
Arvicolidae		
Mimomys pliocaenicus–M. ostramosensis group		x
Mimomys pitymyoides (group)	x	x
Hystricidae		
Hystrix refossa	x	
Carnivora		
Felidae		
Dinofelis sp.		x
Acinonyx pardinensis	x	x
Homotherium crenatidens		x
Megantereon cultridens		x
Hyaenidae		
Chasmaporthetes lunensis		x
Pachycrocuta perrieri		x
Hyaenidae indet.	x	
Canidae		
Vulpes alopecoides		x
Nyctereutes megamastoides		x
Canis arnensis		x
Canis sp. cf. *C. etruscus*		x
Ursidae		
Ursus etruscus		x
Primates		
Cercopithecidae		
Paradolichopithecus arvernensis		x
cf. *Macaca sylvanus* (cf. *M. s. florentina*)		x
Cetartiodactyla		
Suidae		
Sus strozzii		x
Cervidae		
Metacervoceros rhenanus philisi	x	x
Croizetoceros ramosus minor		x
Eucladoceros ctenoides senezensis	x	x
Cervalces gallicus		x
Bovidae		
Antilopinae		
Gazellospira torticornis		x
Caprinae		

(continued)

Table 17.1 (continued)

Taxon	FGD[1] 2001–2006	Previous finds
Procamptoceras brivatense		x
Ovis claudiusguerini		x
cf. *Hemitragus* sp		x
Megalovis latifrons	x[2]	x
Pliotragus ardeus	x	x
Gallogoral meneghinii	x	x
Bovinae		
Leptobos furtivus	cf	x
Leptobos etruscus		x
Bison (*Eobison*) sp	x[3]	
Bovidae indet.		x
Perissodactyla		
Equidae		
Allohippus senezensis senezensis	x	x
Allohippus major	x[3]	x
Possible additional equids known from phalanges and/or metapodials:		
(1) larger than the average for *Allohippus stenonis vireti*		x
(2) cf. Valdarno *Allohippus stehlini*		x
(3) cf. small ?*Allohippus* of Pyrgos		x
Rhinocerotidae		
Dicerorhinus etruscus etruscus[4]	x	x
Proboscidea		
Elephantidae		
Mammuthus meridionalis meridionalis	x	x
Aves		
Ciconiiformes		
Ciconiidae		
Ciconia nigra		x
Anseriformes		
Anatidae		
Tadorna cf. *ferruginea*		x
Anas clypeata		x
Aythya sp.		x
Mergus cf. *merganser*		x
Mergellus sp. size of *M. albellus*		x
Anatidae indet.		x
Accipitriformes		
Accipitridae		
Haliaeetus albicilla		x
Buteo sp., size of *Buteo buteo* or *Buteo lagopus*		x
Galliformes		
Tetraonidae		
Tetrao cf. *partium*		x
Phasianidae		
Pavo bravardi		x
Alectoris cf. *graeca*		x
Gruiformes		
Otididae		
Otis sp. size of *Otis tarda*		x
Strigiformes		
Strigidae		
Bubo sp.		x
Surnia robusta		x
Passeriformes		
Turdidae		
Cf. Turdidae, size of *Turdus merula*		x
Corvidae		
Corvus corax antecorax		x
Osteichthyes		
Cypriniformes		
Cyprinidae		
Tinca sp.		x
Perciformes		
Percidae		
Perca fluviatilis	x	

[1] FGD: the Faure/Guérin/Delson fieldwork of 2001–2006

[2] One specimen of *Megalovis latifrons* was presented to our team in 2002 by A. Consigny, who recovered the fossil in parcel 263, whose stratigraphic position is unknown

[3] One specimen each of *Allohippus major* and *Bison* (*Eobison*) sp. were presented to our team in 2002 by A. Consigny, who recovered the fossils on the surface in parcel 164, from a horizon whose stratigraphic position is unknown. The latter taxon usually occurs in much younger sites

[4] Since 1993, most authors include this species in the genus *Stephanorhinus*, but we retain Guérin's allocation to *Dicerorhinus* as discussed in Chap. 10

This fauna is considered to belong to the MNQ 18 unit. Extending the biozonation of the Mediterranean Neogene based on mammals by Mein (1975, 1989), the foundations of Villafranchian biochronology, defined by Guérin (1980, 2007), are based on the analysis and comparison of mammalian faunal assemblages whose composition and evolution in different European deposits allow us to characterize several such units (MNQ = Mammal Neogene/Quaternary).

There has been significant disagreement over the distinctiveness and content of the MNQ 18 unit. We argue that there is sufficient faunal turnover between Saint-Vallier and Senèze to recognize a full MNQ unit here. Senèze is universally accepted as a reference locality for this unit.

Senèze is the type-locality for 13 mammalian taxa, three genera, seven species and three subspecies (underlined in the list that follows): *Gallogoral* Guérin, 1965; *Procamptoceras brivatense* Schaub, 1923; *Megalovis latifrons* Schaub, 1923; *Paradolichopithecus arvernensis* (Depéret, 1928); *Cervalces gallicus* (Azzaroli, 1952); *Allohippus senezensis* Prat, 1964; *Leptobos furtivus* Duvernois, 1990; *Ovis claudiusguerini* Crégut-Bonnoure, 2024; *Eucladoceros ctenoides senezensis* (Depéret, 1912 in Depéret & Mayet, 1912); *Metacervoceros rhenanus philisi* (Schaub, 1941); and *Croizetoceros ramosus minor* Heintz, 1970. *Pliotragus* Kretzoï, 1941 was named to replace *Deperetia* Schaub, 1923 (preoccupied), which was in turn named for a species (*Antilope ardea* Depéret, 1883) typified at Les Etouaires. Schaub (1923) discussed a specimen from Senèze (NMB Se 1636) which he termed the "Typus der neuen Genus" (p. 288) or "Genustypus" (p. 289), but of course no such term exists in formal nomenclature. This genus is not among those for which Senèze is the type locality.

Azzaroli et al. (1988) proposed that the Senèze maar had yielded two associations of different ages, on the basis of an analysis of Schaub's (1943) faunal list plus original studies. These authors, however, did not cite and may not have been aware of the revision by Heintz et al. (1974). Of the taxa listed in Table 17.1, Azzaroli et al. (1988, p. 82) considered that *Cervalces* (= *Libralces*) *gallicus*, *Megalovis* and *Allohippus major* (= *Equus bressanus* for them) were indicative of a younger age than the "main" fauna. They also included in that category *Canis senezensis* (included by them in *C. arnensis*) and "a small equid which may possibly be *Equus stehlini*." Azzaroli et al. (1988) suggested that the larger and more speciose assemblage might date to their middle Villafranchian, while the less extensive faunule might be placed near the end of the Villafranchian as they conceived of it.

According to sites in France and across Europe (modified after Guérin 2007), MNQ 18 is characterized by:

(a) Four associations: *Gallogoral* + *Metacervoceros rhenanus philisi*; *Acinonyx* + *Megalovis latifrons*; *Pliotragus* + *Canis arnensis*; *Cervalces gallicus* + *Sus strozzii*.
(b) The presence of six well-defined evolutionary stages in diverse mammalian lineages: *Mimomys pliocaenicus*, *Allohippus senezensis senezensis*, *Metacervoceros rhenanus philisi*, *Eucladoceros ctenoides senezensis*, *Croizetoceros ramosus minor* and *Dicerorhinus etruscus etruscus* (stage II).
(c) The first appearance of seven new taxa: *Canis arnensis; Cervalces gallicus; Leptobos etruscus; Procamptoceras brivatense; Megalovis latifrons; Pliotragus ardeus;* and *Ovis claudiusguerini*. *Dinofelis* sp. and cf. *Hemitragus* sp. are also new but not well defined. Contrary to Guérin (2007), *Mimomys pliocaenicus* is common in MNQ 17, while *Felis lunensis* (= *Felis sylvestris lunensis* for some authors) only appears later in MNQ 18. The European jaguar, *Panthera gombaszogensis*, is first known in Eurasia at Slivnitsa (Bulgaria), a site which Spassov (e.g., 1998, 2000, 2023) considered equivalent in age to Senèze; this taxon's next occurrences are at Dmanisi (Georgia) dated at 1.80–1.76 Ma and Olivola (Italy, estimated to date ca. 1.9 to 2.0 Ma). It is also present at Pirro Nord (Italy, probably dating between 1.6 and 1.3 Ma, following Napoleone et al. 2003, but see below on this date). It is not known at Senèze.

Following Mein (1975), the associations are pairings of taxa characteristic of the MN(Q) unit, combining a first appearance with a last appearance. The importance of this approach has changed over time, and some researchers (e.g., Delson) think it may actually be better just to consider a group of taxa typical of the unit (combining the "associations" and the lineage stages). The new paleontological analyses confirm Senèze's attribution to MNQ 18 and provide new information for the definition of this biochron.

Review of Senèze Taxa

The carnivore fauna, very limited in the fossils from the 2001–2006 excavations, is remarkable for the presence of the complete skeletons of *Homotherium crenatidens* and *Megantereon cultridens* from earlier collections. A new taxon, *Dinofelis* sp., is added to the list (A. Argant 2024); it is present in several European Villafranchian deposits but cannot provide biochronological information due to its extreme rarity. *Acinonyx pardinensis*, already reported from the older material, is confirmed by our recent excavation. Among Hyaenidae, the giant hyena *Pachycrocuta brevirostris* is not represented at Senèze despite the presence of three different species (*Chasmaporthetes lunensis*, *Pachycrocuta perrieri*, Hyaenidae indet.). It only appeared later, after 2 Ma. *Megantereon cultridens* from Senèze was later replaced in Europe by the African *Megantereon whitei* around 1.8 Ma. *Homotherium crenatidens* from Senèze is a middle to late Villafranchian form (Sotnikova et al. 2002). It differs from the Ceyssaguet *Homotherium* (ca. 1.2 Ma) but it is little different from that at Olivola (Olivola faunal unit ca. 2.0–1.8 Ma). As for *Canis arnensis*, its arrival in Italy was around 1.9 Ma (Bartolini-Lucenti & Rook 2016). A second species of *Canis* is less well-represented at Senèze but might indicate the presence of *C. etruscus*. Compared with the Saint-Vallier assemblage, the carnivore fauna at Senèze shows some similarities. It is, however, more recent, as attested by the presence of *Nyctereutes megamastoides* instead of *Nyctereutes vulpinus* in Saint-Vallier and by levels of evolution of *Ursus etruscus* and *Homotherium crenatidens*, more recent at Senèze than at Saint-Vallier.

Senèze has yielded two primate fossils, cf. *Macaca sylvanus* and *Paradolichopithecus arvernensis*, neither from the current field work (Delson 2024). These two primates provide only slight information about the local environment or age of the site. *Macaca sylvanus* is known across much of Europe in small numbers throughout the Pliocene and Pleistocene. It may be that Senèze *Paradolichopithecus* is younger than the form at Grăunceanu (Romania, whose precise age is still unknown: MNQ 17 or MNQ 18). The macaque and *Paradolichopithecus* are moderately to highly terrestrial, respectively, and they suggest a woodland to more open environment. Senèze may be one of the youngest known occurrences of *Paradolichopithecus arvernensis*.

Only two species of Cervidae were found during our excavations: *Metacervoceros rhenanus philisi* and *Eucladoceros ctenoides senezensis* (Valli 2024). With *Croizetoceros ramosus minor*, they are characteristic of the MNQ 18 unit, and this is even truer if we limit ourselves only to cervids from deposits in the Massif Central as well as those from Saint-Vallier, which have been assigned to MNQ 17 and are at different evolutionary stages. *Cervalces gallicus* and the elks in general make their appearance in MNQ 18; it is perhaps one of the few taxa that is only present from this time onwards. In France, it is known only from MNQ 18, but it seems to persist up to MNQ 19 in the Balkans and Caucasus.

Seven incomplete remains of Bovidae have been found during the new fieldwork, representing four species: *Pliotragus ardeus*, *Gallogoral meneghinii*, *Megalovis latifrons* and *Leptobos furtivus*. All of them are already known from Senèze, but their fragmentary state makes it impossible to compare them with material from previous collections (Crégut-Bonnoure 2024). The review of Senèze's bovid assemblage highlights the morphological variability of the samples of *Pliotragus, Gallogoral* and *Megalovis*, which suggests an evolutionary change in these species over time and more than one fossil level. However, these possible distinct levels would not be very distant in time (the whole sequence seems to extend over about 100 kyr or less), and it is not possible to distinguish successive bovid assemblages. *Bison* (*Eobison*), known from a single specimen donated to the team by A. Consigny, is younger than the other species. It was discovered on the surface and must date from the end of the Early Pleistocene.

Only one species of suid, *Sus strozzii*, is known from Senèze, and it has little biochronological impact given its long temporal range (Faure & Guérin 2024).

Two partial skeletons of equids and a few other possibly associated hindlimb elements were recovered in the new excavations close to the bottom of the local sequence, thus ca 2.2 Ma, and are referred to *Allohippus senezensis senezensis* (Eisenmann & Delson 2024). Another specimen donated to the team by A. Consigny belongs to a large-bodied *Allohippus* termed *A. major*. The rich material of equids collected by Philis is not homogeneous. *Allohippus senezensis senezensis* is by far the most frequent, and in size and proportions it is intermediate between *Allohippus senezensis guthi* of La Puebla de Valverde (Spain; MNQ 17) and *Allohippus senezensis mygdoniensis* of Gerakarou (Greece; MNQ 18). *Allohippus major* is also represented in the older collection by a moderate number of postcranial elements and one tooth. There are moreover a few fossils which might represent up to three additional taxa: 2 specimens larger than the average of *A. stenonis vireti* of Saint-Vallier (MNQ 17), one similar to *Allohippus stenonis* from the Upper Valdarno and two others as small as ?*Allohippus* of Pyrgos (Greece; MNQ 19). It appears that some caballine specimens may also be present in the Senèze collections, either because they came from as yet unrecognized younger levels or because they were erroneously thought to come from Senèze; they are not included in our analysis.

The new excavations yielded many remains of rhinoceros. They belong to *Dicerorhinus etruscus etruscus* and are similar to the specimens coming from Philis' collections (Guérin 2024; following his view, this genus name is preferred to *Stephanorhinus*). Senèze is by far the richest locality for *Dicerorhinus etruscus*. Owing to its size and proportions, all the Senèze material is identified as belonging to evolutionary stage II of *Dicerorhinus etruscus etruscus*. Evolutionary stage I of this subspecies is known from the MN 16 locality of Les Étouaires and from Saint-Vallier (MNQ 17). Stage I thus corresponds to the early and middle Villafranchian (MN 16 and MNQ 17), while stage II corresponds to the late and final (or Epivillafranchian) Villafranchian (MNQ 18–20).

Only one proboscidean fossil, an incomplete tusk, was found during the new excavation. The revision of the material previously found at Senèze allowed the attribution of all the remains to *Mammuthus meridionalis* (Ménouret & Guérin 2024). The anatomical trends within this species distinguish several stages, recognized as subspecies. The evolutionary level of the molars and mandible from Senèze is comparable to the *Mammuthus meridionalis* from Upper Valdarno (Italy) and should accordingly be attributed to the nominotypical subspecies *Mammuthus meridionalis meridionalis*. However, the M3's collected by Boule (1892) are somewhat different. With a mixture of archaic and evolved characteristics, they are similar to the later Italian mammoth of Colle Mancino. It is not clear if this similarity implies a younger age for the Boule specimen. There is only one other M3 from Senèze, in a skull whose cranium is joined to the mandible, so the teeth are not easily seen.

Several taxa were not considered in detail in this volume. There are few rodents or other micromammals in the Senèze fauna, probably because Philis mainly searched by eye for larger specimens. As discussed briefly by Faure et al. (2024), Maul (2004) re-examined previous finds and identified the few arvicolid teeth as belonging to the *Mimomys pliocaenicus*—*M. ostramosensis* and *M. pitymyoides* groups. In the 2003 and 2004 field campaigns, Elvira Martín Suárez (Granada, Spain) processed quantities of sediment through a specialized screen-washing apparatus and recovered two teeth: a molar of *Mimomys pitymyoides* and an incisor of *Hystrix refossa* (which was reported by Mörs & Hugueney, 2017). Schaub (1943) also noted the presence of a sciurid (aff. *Eutamias*) and the leporid *Oryctolagus* cf. *lacosti*; there have been no recent studies of these specimens. The arvicolids indicate an Early Pleistocene age (effectively MNQ 18), following Maul and Markova (2007), although Minwer-Barakat et al. (2008) and Fejfar et al. (2011) discuss *M. pliocaenicus* in MNQ 17 (see also Tables 17.1 and 17.2).

One taxon which does not figure in our faunal list is *Hippopotamus*. Jung (1946) first suggested its presence at Senèze, but Faure and Guérin (1979) rejected that idea. They reported that five specimens could represent this taxon (four teeth or fragments and a right IVth metacarpal), but that the preservation was unlike that of any Senèze specimen. Faure (1985) suggested that they in fact came from Chambezon, another site in Haute-Loire exploited by P. Philis in 1939 (Faure, 1982; Faure et al., 2022, p. 355). Unfortunately, the idea of a hippopotamid at Senèze continues to reappear. Mazza and Rustioni (1994) discussed a first phalanx (FSL 211082) which had previously been identified as *Equus*, but which they suggested to represent *Hippopotamus* cf. *antiquus*. Most recently, Iannucci et al. (2023) repeated that identification. However, two posterior first phalanges catalogued as FSL 211082 are discussed as *Allohippus senezensis* by Eisenmann and Delson (2024), which suggests that Mazza and Rustioni (1994) were incorrect in their allocation. In sum, there is no evidence of *Hippopotamus* in the Senèze assemblage.

As far as birds are concerned, it is not possible to consider only material from recent excavations, as these yielded only two remains, which were poorly preserved and could not be identified. The biochronological aspect can therefore only be based on material from earlier excavations (Mourer-Chauviré, 2024). The avifauna of Senèze is particularly close to that of Saint-Vallier, with several taxa in common, particularly *Pavo bravardi*. This species is also known from other, older deposits, and Senèze is the last time this large peacock was present in Europe. The avifauna at Senèze appears to be more recent than that at Saint-Vallier, with the presence of a tetraonid, *Tetrao* cf. *partium*. Tetraonidae are known in Central Europe as early as the Early Pliocene (MN 15), but in Western Europe they first appear at Senèze. They subsequently became very abundant in all Middle and Late Pleistocene deposits. Tyrberg (2022) provided additional records for birds from western and central European sites.

Comparison of Senèze with the Villafranchian Localities of the Massif Central and Saint-Vallier

Western Europe is rich in Villafranchian sites, most of which were excavated in the late nineteenth and early twentieth centuries, and their chronostratigraphic data is often imprecise. In addition, the faunas sometimes come from distinct levels that are not always identified. Some localities, such as Senèze, remain paleontological references. Analyses and syntheses based on geochronology (including paleomagnetic correlation and chronometric dating when such data are available) and biochronology provide a framework for positioning Villafranchian reference localities. We begin this analysis of relative Villafranchian biochronology at the local scale, comparing the Senèze faunal assemblage to those of the Massif Central and the nearby Saint-Vallier (Drôme) fauna. This avoids potential problems of diachronicity in the appearance and disappearance of species on a Eurasian scale.

Comparison of Senèze with these nearby Villafranchian localities (see Table 17.2) highlights the differences among the three oldest sites (from MN 16) due to the presence of taxa which are not represented in Senèze: *Mammut borsoni, Dicerorhinus jeanvireti, Procapreolus cusanus, Cervus perrieri* and *Leptobos bravardi* from Vialette (3.2–3.0 Ma; see notes to Table 17.2); *Procapreolus cusanus, Cervus perrieri, Tapirus arvernensis* from Les Étouaires (ca. 2.75 Ma, or broadly 3.1–2.6 Ma); and *Anancus arvernensis* from Rocca-Neyra (2.56 Ma). The association *Anancus arvernensis-Mammuthus meridionalis* can be seen in MNQ 17 at Saint-Vallier (approx. 2.4 Ma), Chilhac (2.28 Ma) and Le Coupet (2.27 Ma) but is not present at Senèze or in the younger sites of MNQ 19–20: Le Creux de Peyrolles (1.45 Ma), Sainzelles (1.3 Ma) and Ceyssaguet (1.2 Ma). At these, *M. meridionalis* is the only proboscidean present. Given the abundance of fossils at Senèze and the wide exploitation of the maar by Philis and our fieldwork, the absence of *Anancus* seems significant to us. As discussed above, *Pachycrocuta brevirostris* and *Panthera gombaszogensis* are absent from Senèze; the former only occurs in much later sites, while the latter occurs at Slivnitsa, close in age to Senèze but farther east. Despite some similarities, evolutionary stages of *Ursus etruscus, Homotherium crenatidens, Allohippus senezensis*, cervids and rhinoceros are different at Senèze in comparison to those of MNQ 17 localities. Furthermore five species disappear whereas nine new mammalian taxa appear: *Trogontherium cuvieri, Gazella borbonica, Anancus arvernensis, Tapirus arvernensis* and *Allohippus stenonis* (which appears again in MNQ 19-20) do not occur in MNQ 18, while *Dinofelis* sp., *Canis arnensis, Cervalces gallicus, Procamptoceras brivatense, Pliotragus ardeus*, cf. *Hemitragus* sp., *Megalovis latifrons, Ovis claudiusguerini* and *Leptobos etruscus* (plus *Tetrao* cf. *partium* among the birds) are new arrivals in MNQ 18 (at Senèze). These differences are significant enough to place Senèze in a MNQ unit different from MNQ 17, namely MNQ 18. The age range determined here for Senèze (ca. 2.20–2.08 Ma; Delson, 2024) suggests that the "boundary" between MNQ 17 and MNQ 18 is close to 2.2 Ma (see Fig. 17.1).

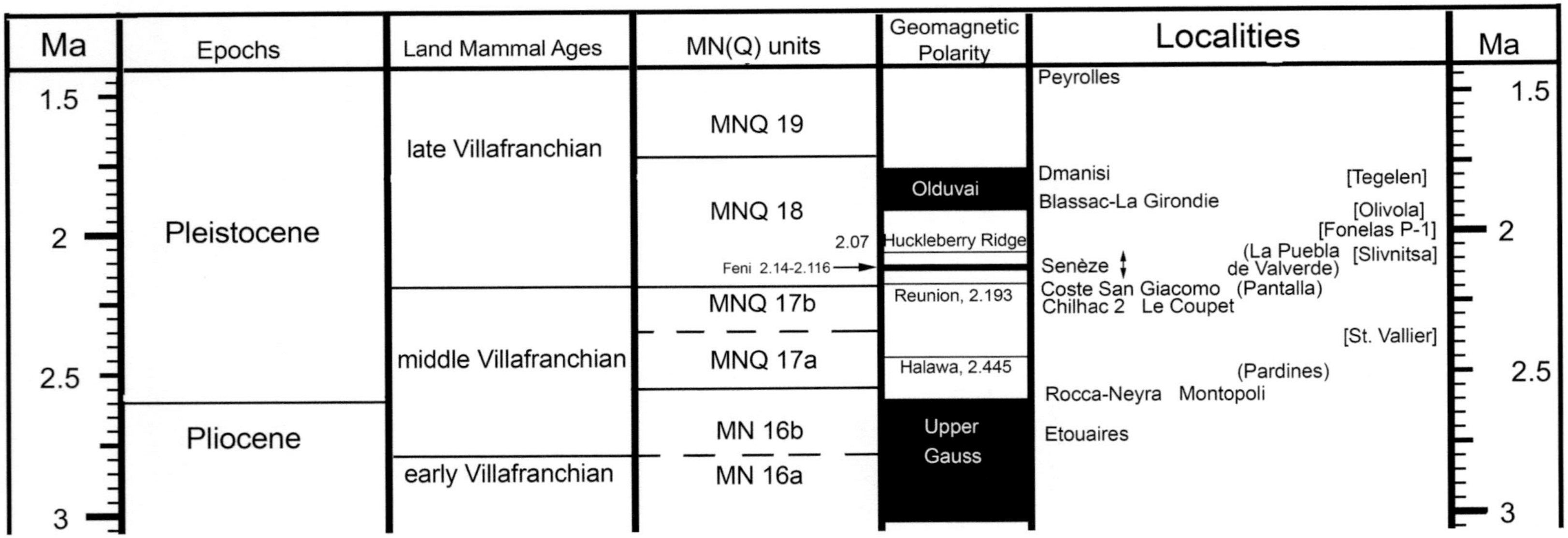

Fig. 17.1 Selected Villafranchian localities discussed in the text, placed on a magneto-biochronologic time scale. Sites indicated only by name and on the left side have chronometric dates; double-headed arrow indicates range of time represented at Senèze. Sites in parentheses and in the center are approximately or questionably dated; sites in square brackets and to the right side are placed in estimated position, as discussed in the text

Rook and Martínez-Navarro (2010, Figs. 2–3) have used the appearances and disappearances of the main lineages of large mammals to better characterize the early, middle and late Villafranchian intervals, particularly in Italy and France. In this context, they placed Senèze in the late Villafranchian, which agrees with Guérin (2007) as well. More recently, the Montopoli site has been the subject of a detailed study of large mammal biochronology in order to situate it among Pliocene and Early Pleistocene sites, using several variants of the GFRI (Genus-rank Faunal Resemblance Index) (Bartolini-Lucenti et al. 2022). The BCA (Bootstrapping Cluster Analysis) dendrogram of 58 Eurasian and African deposits shows, among other things, the close association of Saint-Vallier with Senèze and Le Coupet (Bartolini-Lucenti et al. 2022: Fig. 5, node 20). In this analysis Senèze was considered to belong in the middle Villafranchian whereas Le Coupet was considered as late Villafranchian. These decisions do not agree with our assessments above, but their work only considers genera, not species or evolutionary stages within them. Moreover, those a priori allocations do not affect the analysis. The BCA dendrogram also reveals an important geographical component among the European sites which in some cases outweighs the temporal sequencing.

In addition to providing zircon U–Pb dates for Chilhac and Le Coupet (see Table 17.2 notes), Paquette et al. (2021) published dates for the fossiliferous horizons of Blassac-La Girondie and Vazeilles. Blassac is usually considered to be an MNQ 19 site (although Kahlke et al. 2011 considered it to lie close to the MN 16/MNQ 17 boundary), with three cervids which separate it from Senèze: *Metacervoceros rhenanus* perolensis, *Eucladoceros tetraceros* (also at Vazeilles) and *Alces* cf. *carnutorum*. Paquette et al.'s (2021) date for Blassac was 1.946 ± 0.029 Ma, and on that basis they suggested the MNQ 18/MNQ 19 boundary was close to 2 Ma, much older than usually thought. Vazeilles was dated slightly younger, but its fauna is very limited and adds little to this analysis. Paquette et al. (2021) also suggested that the MNQ 17/MNQ 18 boundary could be located at 2.22 Ma, but they did not have the latest values for Senèze and only considered central French localities. It is interesting to take our analysis one step further and consider some of the most comparable localities of MNQ 17–19 age from Spain and Italy, as well as other western and central European sites.

Table 17.2 Comparison of Senèze taxa with those of Massif Central sites and Saint-Vallier

Site	Vialette	Les Etouaires	Rocca-Neyra	Pardines	Saint-Vallier	Chilhac	Le Coupet	La Rochelambert	Saint-Vidal	Senèze	Blassac-La Girondie	Peyrolles	Sainzelles	Ceyssaguet
MNQ unit	MN 16a	MN 16b	MN 16b	MNQ 17a	MNQ 17	MNQ 17b	MNQ 17b	MNQ 17	MNQ 17	MNQ 18	MNQ 18 or 19	MNQ 19	MNQ 19	MNQ 19 or 20
Age (chronometric or estimated)	ca. 3.2–3.0 Ma[1]	ca. 2.75 Ma[2]	ca. 2.56 Ma[2]	ca 2.5 Ma[3]	ca. 2.4 Ma[3]	2.285 Ma[4]	2,274 Ma[4]			2.20–2.08 Ma[5]	1.946 ± 0.029 Ma[4]	1.5–1.4 Ma[2]	1.3 Ma[6]	1.2 Ma[7]
Leporidae														
Lepus lacosti					+					cf				
Arvicolidae														
Mimomys pitymyoides										+				
Mimomys pliocaenicus					+					+				
Hystricidae														
Hystrix refossa					+	+				+				
Hystrix sp.		+												
Felidae														
Felis lunensis											sp.			
Lynx issiodorensis	+	+	+	+	+									
Lynx lynx														+
Acinonyx pardinensis					+					+				+
Homotherium crenatidens		+		+	+	+	+	+		+				+
Homotherium latidens													+	
Megantereon cultridens		+		+	+	+				+	+			+
Puma pardoides		+			+									
Dinofelis sp										+				
Panthera gombaszogensis														+
Hyaenidae														
Chasmaporthetes lunensis			+	+	+					+				
Pachycrocuta perrieri	+	+		?	+	+	+			+	+			
Pachycrocuta brevirostris													+	+
Hyaenidae indet										+				
Canidae														
Nyctereutes megamastoides		+	+	+		+				+				
Nyctereutes vulpinus					+									
Vulpes alopecoides					+	cf				+				+
Canis arnensis (=C. senezensis)										+				+
Canis sp.	+				+	+				cf. *etruscus*	*etruscus*		+	
Xenocyon lycanoides													+	

(continued)

Table 17.2 (continued)

Site	Vialette	Les Etouaires	Rocca-Neyra	Pardines	Saint-Vallier	Chilhac	Le Coupet	La Rochelambert	Saint-Vidal	Senèze	Blassac-La Girondie	Peyrolles	Sainzelles	Ceyssaguet
Ursidae														
Agriotherium sp.	+													
Ursus etruscus	+	+		+	+	+				+	+			+
Cercopithecidae														
Paradolichopithecus arvernensis	+									+				
Macaca sylvanus					+					**cf**				
Suidae														
Sus arvernensis	+	+					+							
Sus strozzii					+					+				
Hippopotamidae														
Hippopotamus major													+	
Cervidae														
Procapreolus cusanus	+	+												
Croizetoceros ramosus	*ramosus*	*ramosus*		*medius*	*medius*	+	+	*medius*	*medius*	*minor*				+
Metacervoceros pardinensis	+	+												
Metacervoceros rhenanus				*valliensis*	*valliensis*	+	*valliensis*			*philisi*	*perolensis*	*perolensis*		+
"Cervus" perrieri	+	+												
Arvenoceros ardei	+	+												
Eucladoceros sp.	?										*tetraceros*	+		
Eucladoceros ctenoides			*vireti*	*vireti*	*vireti*	+	*senezensis*	+		*senezensis*				+
Praemegaceros obscurus													cf	+
Cervalces gallicus										+			+	
Alces cf. *carnutorum*											+			
Bovidae					+									
Gazellospira torticornis	+		+	+	+	+	+			+				
Gazella borbonica	+	+	+	+	+			+	+					
Procamptoceras brivatense										+				
cf. *Hemitragus*										+				
Pliotragus ardeus	?	+	+							+				
Gallogoral meneghinii				+	+	+	+			+				
Ovis claudiusguerini										+				
Megalovis latifrons										+				
Leptobos bravardi		+												
Leptobos elatus		*elatus*	*elatus*		*merlae*									
Leptobos furtivus							+			+				
Leptobos etruscus										+	+	+		
Equidae														
Hipparion crusafonti			+											
Allohippus sp.			+											
?*Allohippus* sp. 1				+										

(continued)

Table 17.2 (continued)

Site	Vialette	Les Etouaires	Rocca-Neyra	Pardines	Saint-Vallier	Chilhac	Le Coupet	La Rochelambert	Saint-Vidal	Senèze	Blassac-La Girondie	Peyrolles	Sainzelles	Ceyssaguet
Allohippus stenonis	cf. *vireti*				*vireti* (LD3)		*vireti*					cf. *vireti*	cf.	several subsp.
Allohippus senezensis						*guthi*				*senezensis*	+			
Allohippus other							+			1–3 Taxa				
Allohippus major and/or very large *Equus*										+	+			
Equus cf. *granatensis*							+							
Rhinocerotidae														
Dicerorhinus jeanvireti	+	+												
Dicerorhinus etruscus etruscus		stage I		+	stage I	+		+	+	stage II	+		+	
Tapiridae														
Tapirus arvernensis	*arvernensis*	*arvernensis*					+							
Mammutidae														
Mammut borsoni	+	+												
Gomphotheriidae														
Anancus arvernensis	+	+	+	+	*falconeri*	*chilhiacensis*	*falconeri*	+	+					
Elephantidae														
Mammuthus meridionalis					+	+	+		+	*meridionalis*	+		+	+
Aves														
Phalacrocoracidae														
Phalacrocorax carbo														+
Ciconiidae														
Ciconia nigra					+					+				
Anatidae														
Cygnus cf. *cygnus*					+									
Anser sp.						+								
Tadorna cf. *tadorna*						+								
Tadorna cf. *ferruginea*										+				
Anas clypeata										+				
Aythya sp.										+				
Bucephala cereti						+								
Mergus cf. *merganser*										+				
Mergellus sp. size of *M. albellus*										+				
Accipitridae														
Aquila claudeguerini														+
Accipiter gentilis														+
Haliaeetus albicilla					+					+				

(continued)

Table 17.2 (continued)

Site	Vialette	Les Etouaires	Rocca-Neyra	Pardines	Saint-Vallier	Chilhac	Le Coupet	La Rochelambert	Saint-Vidal	Senèze	Blassac-La Girondie	Peyrolles	Sainzelles	Ceyssaguet
Buteo sp.										+				
Tetraonidae														
Tetrao cf. *partium*										+				
Phasianidae														
Pavo bravardi					+					+				
Alectoris cf. *graeca*										+				
Otididae														
Otis sp					+					+				
Strigidae														
Bubo sp										+				
Surnia robusta										+				
cf. Turdidae, size of *Turdus merula*										+				
Corvidae														
Corvus pliocaenus														+
Corvus corax antecorax					+					+				

[1] Vialette is a complex maar with a possibly long interval represented (as at the nearby Senèze). Bandet et al. (1978) and Couthures (1979) provide a simplified stratigraphic overview. Savage and Curtis (1970) offered an early K–Ar date of 3.8 Ma for a lava said to overlie the fauna; this date was only given in a table with no further details. Bandet et al. (1978) dated a number of carefully selected samples from across the Vialette region. Five of these yielded a mean age of 3.3 ± 0.11 Ma for a basalt flow underlying the fauna. Two others from a flow overlying the mammals provided a mean date of 2.6 ± 0.2 Ma. Biquand et al. (1981) reported a normal geomagnetic polarity near the level with fossils, as well as a fission-track date of 3.14 ± 0.6 Ma from the fossiliferous horizon (note the large error range); they also dated an overlying basalt at 2.9 ± 0.05 Ma but noted a probable loss of potassium and suggested the date was not reliable. Fouris et al. (1991) dated a basanite flow underlying the fossils at 3.07 ± 0.08 Ma. These early K–Ar dates cannot readily be converted to modern standards (see Delson et al. 2024), and the ages must be considered approximations. Lacombat et al. (2008) updated the Vialette fauna and accepted the 3.3 Ma date as preceding the fauna while 3.14 Ma was thought to date it; no other dates were mentioned. Taken together, these data suggest an age for the Vialette fauna roughly between 3.3 and 3.0 Ma, with the normal interval perhaps the middle Gauss normal (C2An.2n), 3.21–3.12 Ma. Other evidence suggests that MN 16 begins after 3.2 Ma, which shortens the age range for Vialette

[2] The Les Étouaires fauna on the Perrier plateau is usually considered a reference fauna for the 16b unit. The stratigraphy and history of study of the Perrier region is discussed in detail by Pastre (2004). Savage and Curtis (1970) reported a K–Ar date of 3.4 Ma on an ash supposedly below the Les Etouaires fauna and a whole-rock date of 3.5 Ma said to be correlative with the fauna. Numerous other dates were published by various authors with conflicting links to the faunas. Biquand et al. (1990) summarized these to show the Les Etouaires fauna lay between dates of 2.4 and 2.5 Ma, just above a normal to reversed paleomagnetic transition (documented by only a few samples) correlated to the Gauss/Matuyama boundary. Pastre (2004) reviewed additional dates and suggested that the Les Etouaires fauna was in fact younger than 2 Ma. The Perrier sequence was re-dated by Nomade et al. (2014), who (in collaboration with Pastre) obtained a concordant series of dates (recalculated here following Schaen et al. 2021) including 2.74 ± 0.02 Ma underlying or equivalent to the Les Etouaires faunal level. Iannucci and Sardella (2023) have questioned this date, noting that the Les Etouaires fauna probably derived from several different horizons, not just the Côte d'Ardé dated by Nomade et al., and thus could only be bounded by an underlying date of 3.07 ± 0.05 Ma and the overlying date of 2.56 ± 0.02 Ma (both recalculated) for the Rocca-Neyra faunal assemblage. Pending further analysis of the provenance of the Les Etouaires fossils, the fauna can be dated between 3.1 and 2.6 Ma, perhaps close to 2.75 Ma. Few if any European sites with large mammals are solidly dated between 3.1 and 2.8 Ma, so no further comparisons are possible. The highest faunal level of the Perrier complex, le Creux de Peyrolles, was dated by Nomade et al., (2014, recalculated) at 1.45 ± 0.03 Ma

[3] The Perrier-Pardines fauna lies slightly above that of Rocca-Neyra, suggesting an age close to 2.5 Ma. In turn, Saint-Vallier is usually thought to be faunally younger than or close to Pardines but older than Chilhac (Guérin et al. 2004). Its age may be estimated close to 2.4 Ma

[4] Chilhac (localities 2 and 3) has been dated several times with improving results. Boivin et al. (2010) K–Ar dated an underlying lava at 2.47 ± 0.1 Ma and an overlying lava at 1.8 ± 0.2 Ma. Nomade et al. (2014) dated an overlying tephra at 2.344 ± 0.03 Ma (recalculated). Paquette et al. (2021) U–Pb dated zircons from the fossiliferous layer of Chilhac 2 at 2.285 ± 0.046 Ma. Le Coupet was first dated by Savage and Curtis (1970), who reported a K–Ar date of 1.9 Ma, presumably on an overlying horizon. Fouris et al. (1991) dated an underlying lava not directly linked to the fauna at 2.12 ± 0.09 Ma (this date cannot be recalculated due to lack of details provided. Paquette et al. (2021) dated zircons from the Le Coupet fossiliferous slope deposits at 2.274 ± 0.032

[5] As discussed by Delson et al. (2024), Senèze faunal elements recovered by the FGD team date between 2.2–2.08 Ma, with fossils known between 2.2–2.18 Ma and 2.1–2.08 Ma and probably inbetween

[6] Thouveny and Bonifay (1984) placed the Sainzelles fauna in a reversed interval between the top of the Olduvai subchron and a K/Ar date of 1.4–1.3 Ma. Bonifay and Brugal (1996) erroneously reported that as 1.4 Ma. Kahlke et al., (2011, Fig. 17.1) charted the site close to 1.3 Ma (just older than Ceyssaguet), but without any discussion in the text. Palombo (2016, Fig. 17.1), on the other hand, charted it possibly slightly younger than Ceyssaguet, ca. 1.1 Ma, but again without discussion

[7] Croitor and Bonifay (2001) stated that Ceyssaguet is located on the slopes of a volcano whose last eruption was K–Ar dated at 1.3 Ma and estimated that the overlying faunal level might date to 1.2 Ma. Later, Mourer-Chauviré and Bonifay (2018) reported an unpublished K–Ar age of 1.4 Ma for the underlying basalt and suggested that the maar infilling might have taken 200 kyr, resulting in an approximate date for the fauna of 1.2 Ma

Table 17.3 Comparison of Senèze taxa with those of similar age (MNQ 17–18) from Spain, Italy, Netherlands and Germany

Site	Coste San Giacomo	Pantalla	La Puebla de Valverde	Senèze	Huélago	Olivola	Fonelas P1	Tegelen	Erpfingen
MNQ unit	MNQ 17b	MNQ 17b	MNQ 17	MNQ 18	MNQ 18	MNQ 18	MNQ 18	MNQ 18	MNQ 18
Age (chronometric or estimated)	2.23 ± 0.03 Ma	2.23 ± 0.18 Ma	ca. 2.0 or 2.3	2.20–2.08 Ma		??2.0–1.8 Ma	ca. 2 Ma	ca. 1.8 Ma	
Arvicolidae									
Mimomys pitymyoides				+				+	
Mimomys pliocaenicus	+			+				+	+
Hystricidae									
Hystrix refossa	+			+					+
Felidae									
Felis lunensis						+			
Lynx issiodorensis		+	+			+			
Lynx pardinus spelaeus							+		
Acinonyx pardinensis		+	+	+		+	+		
Homotherium crenatidens				+			+		
Homotherium latidens	sp.		+			+			
Megantereon cultridens			+	+		+	+		
Puma pardoides			+						
Dinofelis sp				+					
Panthera gombaszogensis						+		+	+
Hyaenidae									
Chasmaporthetes lunensis			+	+		+			*lunensis*
Pachycrocuta perrieri	?		+	+				+	+
Pachycrocuta brevirostris						+	+		
Hyaenidae indet				+					
Hyaena (Parahyaena) brunnea							+		
Canidae									
Nyctereutes megamastoides			+	+					
Nyctereutes vulpinus			+						
Vulpes alopecoides	cf	sp.		+		+	+		
Canis arnensis (=C. senezensis)				+					
Canis sp.	sp.								
Canis etruscus	cf	+		cf		+	+		+
Canis accitanus							+		
Xenocyon falconeri			cf				+		
Ursidae									
Ursus etruscus	cf		+	+		+			
Cercopithecidae									
Paradolichopithecus arvernensis			cf	+					
Macaca sylvanus	+		+	cf				+	
Suidae									
Sus strozzii	+	+		+		+		+	
Potamochoerus magnus							+		
Hippopotamidae									
Hippopotamus sp.	sp								
Cervidae									
Croizetoceros ramosus	cf		*pueblensis*	*minor*	+				+
"Cervus" lyra	+								
Metacervoceros rhenanus			+	*philisi*			cf. *philisi*	+	cf. *philisi*
Eucladoceros sp.	+					+	+		
Eucladoceros ctenoides			*vireti*	*senezensis*	*senezensis*			*tegulenis*	*tegulenis*
Cervalces gallicus				+		+			+
Cervus nestii		+				+			+
Giraffidae									
Mitilanotherium sp.					+		+		
Bovidae									
Gazellospira torticornis	+		+	+	+		*hispanica*		+
Gazella borbonica	+		+		+				
Gazella							+		
Procamptoceras brivatense				+		+			

(continued)

Table 17.3 (continued)

Site	Coste San Giacomo	Pantalla	La Puebla de Valverde	Senèze	Huélago	Olivola	Fonelas P1	Tegelen	Erpfingen
cf. *Hemitragus*				+					
?Capra/Hemitragus							+		
Hesperidoceras merlae			+						
Pliotragus ardeus				+					
Gallogoral meneghinii	+		*pueblensis*	+		+			cf
Ovis claudiusguerini				+					
Megalovis latifrons				+					+
Praeovibos sp.							+		
Parabos soriae					+				
Leptobos elatus		*merlai*			+	*merlai*	*merlai* (or *furtivus*)	+	
Leptobos furtivus				+					
Leptobos etruscus	sp.			+			+		
Equidae									
Allohippus stenonis	+	+	*guthi*						
Allohippus senezensis				*senezensis*					
Allohippus other				1–3 Taxa					
Allohippus major and/or very large *Equus*				+			cf		+
Allohippus large and slender								+	
Onager-like equid					+				
Rhinocerotidae									
Dicerorhinus etruscus etruscus	sp.		+	Stage II			+	+	+
Gomphotheriidae									
Anancus arvernensis	+								
Elephantidae									
Mammuthus meridionalis	+	cf	+	*meridionalis*	+		+	+	+
Aves									
Ciconiidae									
Ciconia nigra				+					
Anatidae									
Tadorna cf. *ferruginea*				+					
Anas clypeata				+					
Aythya sp.				+					
Mergus cf. *merganser*				+					
Mergellus sp., size of *M. albellus*				+					
Accipitridae									
Aquila cf. *chrysaetos*			+						
Haliaeetus albicilla				+				+	
Buteo sp				+					
Tetraonidae									
Tetrao cf. *partium*				+					
Tetrao praeurogallus									+
Phasianidae									
Pavo bravardi				+					
Alectoris cf. *graeca*				+					
Gruiformes sp.							+		
Gruidae									
Grus cf. *grus*			+						
Otididae									
Otis sp				+					
Columbidae sp.							+		
Strigidae									
Bubo sp				+					
Surnia robusta				+					
cf. **Turdidae**, size of *Turdus merula*				+					
Corvidae									
Corvus corax antecorax				+					
Corvus cf. *corax*			+						

Comparison of Senèze with the Later Villafranchian Localities of Italy, Spain, the Netherlands and Germany

As indicated in Table 17.3, several of these sites have yielded chronometric age determinations. Coste San Giacomo (Italy) is clearly in MNQ 17, with *Anancus* and *Mammuthus* alongside *Axis (=?Cervus) lyra* and a range of mostly fragmentary remains of MNQ 17–18 taxa (Bellucci et al. 2014). Spassov (e.g., 2016, 2023) has proposed to include Coste San Giacomo in an early MNQ 18a subunit, along with Senèze and Slivnitsa, but the date and presence of *Anancus* keep Coste San Giacomo in MNQ 17. *Allohippus stenonis* might also suggest MNQ 17 but could just be a marker of an Italian site. Florindo et al. (2021) reported an argon-argon age of 2.233 ± 0.032 Ma, which is recalculated here to 2.228 ± 0.032 Ma, rounded to 2.23 ± 0.03 Ma. This seems to be the youngest date for any definitive MNQ 17 assemblage and fits well with the idea of beginning MNQ 18 at 2.2 Ma.

Pantalla (Italy) has an even more limited fauna, including a number of "Italian" taxa in common with Olivola and other cosmopolitan species. Cherin et al. (2023) placed the assemblage close to the middle-late Villafranchian "boundary" but included it as well as Senèze in MNQ 17b. It seems most likely an MNQ 17 locality as interpreted in this chapter. It was ESR/U-series dated to 2.227 ± 0.177 Ma by Cherin et al. (2023), with the possibility of a slightly older age from another sample. This date corresponds closely to Coste San Giacomo.

Olivola (Italy) has yielded a more extensive fauna but not a clear date. The assemblage is broadly similar to that from Senèze and lacks definitive MNQ 17 taxa such as *Anancus*. Following detailed paleomagnetic calibration of the Upper Valdarno faunal sequence (e.g., Napoleone et al. 2003), Olivola was considered to predate the oldest Valdarno assemblage, from Matassino (within the Olduvai subchron) and estimated to date roughly between 2 and 1.8 Ma. It would thus fall later in MNQ 18. It is worth noting that Bartolini-Lucenti et al. (2022) erroneously included *Anancus* at Olivola in their supplementary table of genera at sites in their analysis. Recently, however, Mattei et al. (2023) have suggested that the Matassino and other Upper Valdarno mammal paleofaunas might be less definitively dated by paleomagnetic correlation, due to problems with the mineral carrying the magnetic signal. Their arguments appear solid, but further analysis is required: for the moment, precise ages of the later Upper Valdarno faunal units must be reconsidered, although they probably span 2.0–1.3 Ma.

La Puebla de Valverde (Spain) has long been considered an MNQ 17 locality with a fauna similar to that of Saint-Vallier and Chilhac (Sinusía et al, 2004). In an analysis of their paleomagnetic results, Sinusía et al. (2004) suggested that the fossil mammals occurred above a short normal interval correlated to what is now the Feni subchron, dated ca. 2.140–2.116 Ma and below the base of the (not seen) Olduvai at 1.935 Ma. This would be younger than Senèze and far too young if the fauna were equivalent to Chilhac. One possibility is that some of the Spanish taxa persisted longer than in southern France and Italy , such as *Eucladoceros ctenoides vireti*, *Allohippus stenonis guthi*, *Metacervoceros rhenanus ?valliensis* and *Croizetoceros ramosus pueblensis* (a form close to that from Saint-Vallier). Alternatively, the short normal could be correlated to the Halawa (= X) excursion at ca. 2.445 Ma (see Ogg 2020), in which case the fauna would fall between that and the Feni, around 2.3 Ma, well within MNQ 17.

Still younger sites usually included in MNQ 18 (Huélago and Fonelas P-1, Spain) yield species not found at Senèze or other localities, such as *Pachycrocuta brevirostris*, *Hyaena brunnea*, *Potamochoerus magnus*, *Gazella* spp. and *Praeovibos* sp. Fonelas P-1 was suggested to date ca. 2 Ma (Arribas et al. 2009), thus close to Olivola. The age of Tegelen (Netherlands) has been contentious, perhaps implying several different assemblages among the collections under this name, but some of the taxa are surely MNQ 18, as is also true of the comparably-aged Erpfingen (Germany). They are often suggested to date close to 1.8 Ma, but that would contradict Paquette et al.'s (2021) contention that MNQ 19 began close to 2.0 Ma. An alternative interpretation could place Blassac-La-Girondie later in MNQ 18 than Senèze, but not in MNQ 19; its three "late" cervids are unique to this site and are not found in younger MNQ 19 assemblages. The youngest locality often included in MNQ 18 may be Dmanisi (Georgia). Despite some analogies of the fauna (Kahlke et al. 2011), it is now dated firmly at 1.80–1.76 Ma (Ferring et al. 2022), which would suggest MNQ 19 might begin between 1.75–1.70 Ma. In Bulgaria, Slivnitsa is an important site of the Balkans which is close to Senèze in terms of fauna, perhaps the most similar of any known site (Spassov 1997, 2003, 2016; Spassov & Crégut-Bonnoure 1999). The two localities belong to the same biozone, but the evolutionary differences/stages of the bovids indicate a small biochronological difference or a geographical difference (Crégut-Bonnoure 2024). Spassov (2003, 2016) proposed the creation of subzone MNQ 18a including these two localities, but we do not recognize this distimction here. Further analyses of a wider range of sites than examined here (and additional dates) are required to resolve these questions.

Conclusions

The systematic study of the Senèze fauna collected by Pierre Philis led to the discovery of seven new species: *Canis* sp., *Dinofelis* sp., an indeterminate hyaenid, cf. *Hemitragus* sp., *Ovis claudiusguerini* (previously *Ovis* sp.) and two indeterminate bovids. The mammalian fossils recovered during our excavations, although few in number, mostly represent taxa known from the earlier collections: *Mimomys pitymyoides*, *Acinonyx pardinensis*, *Allohippus senezensis senezensis*, *Metacervoceros rhenanus philisi*, *Eucladoceros ctenoides senezensis*, *Pliotragus ardeus*, *Gallogoral meneghinii*, *Megalovis latifrons*, *Leptobos furtivus*, *Dicerorhinus etruscus* and *Mammuthus meridionalis*. Three new taxa for Senèze come from our fieldwork: *Hystrix refossa,* Hyaenidae indet. (distinct from the *Chasmaporthetes* and *Pachycrocuta* lineages) and *Bison* (*Eobison*) sp. (the last is a surface find presented to the team by A. Consigny and probably representing a younger horizon than the rest of the fauna).

Despite similarities with MNQ 17 faunas from the Massif Central and Saint-Vallier, there are some major differences at Senèze (as detailed above): disappearance of five mammalian species and appearance of nine new ones (plus one bird). In addition, the "evolutionary stages" (mainly subspecies) of five cervid and perissodactyl species confirm the attribution of the Senèze fauna to MNQ 18, at the beginning of this unit.

The bulk of the fauna is relatively homogeneous, with no evidence for more than one assemblage (contrary to the proposal of Azzaroli et al. 1988). Tentative evidence suggests some intraspecific evolutionary change in *Pliotragus*, *Gallogoral* and *Megalovis*. The site saw several episodes of sedimentation between ca. 2.20 and 2.08 Ma, resulting in a pattern of isolated bones on the slopes of the maar or clusters of elements on the lakeshore. The western sector preserves a lakeshore with terrestrial elements, hyena coprolites and fish remains in the excavated Trenches 1 and 2. The southeastern sector (Trenches 5–7) represents slope deposits from a colder interval of low lake level. In both areas, animals likely slipped into the lake and drowned when unable to climb out (Debard 2024; Fernández-Jalvo et al. 2024). The result was a 100+ kyr interval as a natural trap for the mammals and birds living around the lake, characterized by rapid sedimentation allowing quick fossilization and good preservation of this remarkable fauna, which corresponds to the beginning of MNQ 18 (and the late Villafranchian) about 2.2 Ma in the Early Pleistocene around the Feni subchron.

Acknowledgments We again thank the landowners who gave permission to work in their fields and the many funding agencies who supported our research (see details in Faure et al. 2024). We thank Eric Sargis for supervising the review process and Ralf D. Kahlke, Bienvenido Martínez-Navarro and Nikolai Spassov for reviews which improved an earlier version of the manuscript. Figure 17.1 was made with the TimeScale Creator program (https://timescalecreator.org/index/index.php).

References

Argant A. (2024). The carnivores of Senèze. In E. Delson, M. Faure & C. Guérin (Eds.), *Senèze: Life in Central France two million years ago. Paleontology, geochronology, stratigraphy and taphonomy* (pp. 165–243). Cham (Switzerland): Springer.

Argant, J. (2024). Palynological analysis of the fossiliferous layers of the late Villafranchian deposit of Senèze (Domeyrat, Haute-Loire, France). Taphonomy and paleoenvironmental data. In E. Delson, M. Faure & C. Guérin (Eds.), *Senèze: Life in Central France two million years ago. Paleontology, geochronology, stratigraphy and taphonomy* (pp. 123–140). Cham (Switzerland): Springer.

Arribas, A., Garrido, G., Viseras, C., Soria, J. M., Pla, S., Solano, J. G., et al. (2009). A mammalian lost world in southwest Europe during the Late Pliocene. *PLoS ONE, 4*(9), e7127.

Azzaroli, A. (1952). L'Alce di Senèze. *Palaeontographia Italica, 47*, 133–141.

Azzaroli, A., De Giuli, C., Ficcarelli, G., & Torre, D. (1988). Late Pliocene to early Mid-Pleistocene mammals in Eurasia: Faunal succession and dispersal events. *Palaeogeography, Palaeoclimatology, Palaeoecology, 68*, 77–100.

Bandet, Y., Donville, B., & Michaux, J. (1978). Étude géologique et géochronologique du site villafranchien de Vialette (Puy-de-Dôme). *Bulletin de la Société géologique de France,* Série, *7*(20), 245–251.

Bartolini-Lucenti, S., & Rook, L. (2016). A review on the Late Villafranchian medium-sized canid *Canis arnensis* based on the evidence from Poggio Rosso (Tuscany, Italy). *Quaternary Science Reviews, 151*, 58–71.

Bartolini-Lucenti, S., Cirilli, O., Pandolfi, L., & Rook, L. (2022). The Montopoli site, a reference Local Fauna in the Pliocene and Pleistocene European large mammals biochronology, first discovered by Giovanni Capellini (1833–1922). *Bollettino della Società Paleontologica Italiana, 61*, 207–222.

Bellucci, L., Bona, F., Corrado, P., Magri, D., Mazzini, I., Parenti, F., et al. (2014). Evidence of late Gelasian dispersal of African fauna at Coste san Giacomo (Anagni Basin, central Italy): Early Pleistocene environments and the background of early human occupation in Europe. *Quaternary Science Reviews, 96*, 72–85.

Biquand, D., Dubar, M., & Sémah, F. (1981). Paleomagnetic correlation of the Mediterranean Upper Neogene biochronology and Villafranchian vertebrate sites of the Massif Central, France. *Quaternary Research, 33*, 241–252.

Bonifay, M.-F., & Brugal, J.-P. (1996). Biogéographie et biostratigraphie des grandes faunes du Pléistocène inférieur et moyen en Europe du sud: apport des gisements français. *Paléo, 8*, 19–30.

Boule, M. (1892). Découverte d'un squelette d'*Elephas meridionalis* dans les cendres basaltiques du volcan de Senèze (Haute-Loire). *Comptes Rendus de l'Académie des Sciences, 115*, 624–626.

Cherin, M., Basilici, G., Duval, M., Shao, Q., Sier, M. J., Parés, J. M., et al. (2023). The dawn of the Late Villafranchian: Paleoenvironment and age of the Pantalla paleontological site (Italy; Early Pleistocene). *Quaternary Science Reviews, 316*, 108279.

Couthures, J. (1979). La stratigraphie du gisement plio-pléistocène de Vialette (Haute-Loire). *Bulletin de l'Association française pour l'Etude du Quaternaire, 4*, 171–173.

Crégut-Bonnoure, E. (2024). The Bovidae (Mammalia, Cetartiodactyla) from Senèze. In E. Delson, M. Faure & C. Guérin (Eds.), *Senèze: Life in Central France two million years ago. Paleontology, geochronology, stratigraphy and taphonomy* (pp. 433–585). Cham (Switzerland): Springer.

Croitor, R., & Bonifay, M. F. (2001). Étude préliminaire des cerfs du gisement pléistocène inférieur de Ceyssaguet (Haute-Loire). *Paléo, 13*, 129–144.

Debard, E. (2024). Geological study of the paleontological site of Senèze (Domeyrat, Haute-Loire, France). In E. Delson, M. Faure & C. Guérin (Eds.) *Senèze: Life in Central France two million years ago. Paleontology, geochronology, stratigraphy and taphonomy* (pp. 37–82). Cham, Switzerland: Springer.

Delson, E. (2024). Primates from Senèze. In E. Delson, M. Faure & C. Guérin (Eds.), *Senèze: Life in Central France two million years ago. Paleontology, geochronology, stratigraphy and taphonomy* (pp. 607–632). Cham (Switzerland): Springer.

Delson, E., Nomade, S., Sen, S., Debard, E., Pastre, J.-F., Bahain, J.-J., et al. (2024). Geochronology of Senèze: $^{40}Ar/^{39}Ar$ dating and magnetostratigraphy, with notes on an ESR/U-series dating attempt. In E. Delson, M. Faure & C. Guérin (Eds.), *Senèze: Life in Central France two million years ago. Paleontology, geochronology, stratigraphy and taphonomy* (pp. 99–122). Cham (Switzerland): Springer.

Depéret, C. (1928). Les singes fossiles du Pliocène de France. In *Comptes-Rendus du XIVe Congrès Géologique International, 1926* (pp. 3–4). Madrid: Gráficas Reunidas.

Depéret, C., & Mayet, L. (1912). Le gisement de mammifères pliocènes de Senèze (Haute-Loire). In *Association française pour l'Avancement des Sciences, Compte Rendu de la 40e session, Dijon, 1911, Mémoire hors série*, pp. 138–151.

Duvernois, M.-P. (1990). Les *Leptobos* (Mammalia, Artiodactyla) du Villafranchien d'Europe Occidentale. *Documents des Laboratoires de Géologie de Lyon, 113*, 1–213.

Eisenmann, V., & Delson, E. (2024). The Senèze equids. In E. Delson, M. Faure & C. Guérin (Eds.), *Senèze: Life in Central France two million years ago. Paleontology, geochronology, stratigraphy and taphonomy* (pp. 307–385). Cham (Switzerland): Springer.

Faure, M. (1982). Les hippopotames fossiles de la région d'Issoire (Puy-de-Dôme, France). *Revue des Sciences naturelles d'Auvergne, NS, 48*, 11–21.

Faure, M. (1985). Les hippopotames Quaternaires non insulaires d'Europe occidentale. *Nouvelles Archives du Museum d'Histoire naturelle de Lyon, 23*, 13–79.

Faure, M., & Guérin, C. (1979). Le problème de l'hippopotame villafranchien de Senèze (Haute-Loire, France) et ses implications biostratiqraphiques. In *Septième Réunion annuelle des Sciences de la Terre, Lyon, Avril 1979* (p. 184). Paris: Société Géologique de France.

Faure, M., & Guérin, C. (2024). *Sus strozzii* (Mammalia, Suidae) of the Late Villafranchian site of Senèze (Domeyrat, Haute-Loire, France). In E. Delson, M. Faure & C. Guérin (Eds.), *Senèze: Life in Central France two million years ago. Paleontology, geochronology, stratigraphy and taphonomy* (pp. 387–407). Cham (Switzerland): Springer.

Faure, M., Sen, S., & Costeur, L. (2022). Pierre Philis *Fossil Hunter*: un demi-siècle de recherches de vertébrés fossiles (1892–1942). *Geodiversitas, 44*, 321–383.

Faure, M., Delson, E., & Guérin, C. (2024). Introduction and history of research in the Senèze maar. In E. Delson, M. Faure & C. Guérin (Eds.), *Senèze: Life in Central France two million years ago. Paleontology, geochronology, stratigraphy and taphonomy* (pp. 1–25). Cham (Switzerland): Springer.

Fejfar, O., Heinrich, W.-D., Kordos, L., & Maul, L. C. (2011). Microtoid cricetids and the early history of arvicolids (Mammalia, Rodentia). *Palaeontologia Electronica, 14*(3), 27A:38p. https://palaeo-electronica.org/2011_3/6_fejfar/index.html

Fernández-Jalvo, Y., Valli A. M. F., Marín-Monfort, M. D., & Pesquero-Fernández, M. D. (2024). The taphonomy of Senèze. In E. Delson, M. Faure & C. Guérin (Eds.), *Senèze: Life in Central France two million years ago. Paleontology, geochronology, stratigraphy and taphonomy* (pp. 653–581). Cham (Switzerland): Springer.

Ferring, R., Oms, O., Nomade, S., Humphrey, J. D., Tappen, M., Coil, R., et al. (2022). Early Pleistocene stratigraphy, sedimentary environments, and formation contexts at Dmanisi in the Georgian Caucasus. *Journal of Human Evolution, 172*, 103254.

Florindo, F., Marra, F., Angelucci, D. E., Biddittu, I., Bruni, L., Florindo, F., et al. (2021). Environmental evolution, faunal and human occupation since 2 Ma in the Anagni basin, central Italy. *Scientific Reports, 11*, 7056.

Fouris, M., Cantagrel, J.-M., Poidevin, J.-L., & Mergoil, J. (1991). Le Plio-Pléistocène du Velay: Volcanologie et chronologie K/Ar des gisements fossilifères. Données actuelles, problèmes et hypothèses. In J. P. Raynal & D. Miallier (Eds.), *Datation et Caractérisation des milieux Pléistocènes.* Actes des Symposiums 11 et 17 de la 11ème R. S.T., Clermont-Ferrand, 1986. *Cahiers du Quaternaire, 16*, 401–416.

Guérin, C. (1965). *Gallogoral* (nov. gen.) *meneghinii* (Rütimeyer, 1878), un Rupicapriné du Villafranchien d'Europe Occidentale. *Documents des Laboratoires de Géologie de la Faculté des Sciences de Lyon, 11*, 1–353.

Guérin, C. (1980). Les rhinocéros (Mammalia, Perissodactyla) du Miocène terminal au Pléistocène supérieur en Europe occidentale. Comparaison avec les espèces actuelles. *Documents des Laboratoires de Géologie de Lyon, 79*, 1–1185.

Guérin, C. (2007). Biozonation continentale du Plio-Pléistocène d'Europe et d'Asie occidentale par les mammifères: état de la question et incidence sur les limites Tertiaire/Quaternaire et Plio/Pléistocène. *Quaternaire, 18*, 23–33.

Guérin, C. (2024). *Dicerorhinus etruscus etruscus* from the Late Villafranchian site of Senèze (Domeyrat, Haute-Loire, Central France). In E. Delson, M. Faure & C. Guérin (Eds.), *Senèze: Life in Central France two million years ago. Paleontology, geochronology, stratigraphy and taphonomy* (pp. 245–306). Cham (Switzerland): Springer.

Guérin, C., Faure, M., Argant, A., Argant, J., Crégut-Bonnoure, E., Debard, E., et al. (2004). Le gisement pliocène supérieur de Saint-Vallier (Drôme, France): Synthèse biostratigraphique et paléoécologique. *Geobios, 37*, S349–S360.

Heintz, É. (1970). Les Cervidés villafranchiens de France et d'Espagne. *Mémoires du Muséum national d'Histoire naturelle, 22*, Volume I and II, 1–317 and 1–206.

Heintz, É., Guérin, C., Martin, R., & Prat, F. (1974). Principaux gisements villafranchiens de France: Listes fauniques et biostratigraphie. *Mémoires du Bureau de Recherches Géologiques et Minières, 78*, 169–182.

Iannucci, A., & Sardella, R. (2023). What does the "Elephant-*Equus*" event mean today? Reflections on mammal dispersal events around the Pliocene-Pleistocene boundary and the flexible ambiguity of biochronology. *Quaternary, 6*, 16.

Iannucci, A., Mecozzi, B., & Sardella, R. (2023). Beware of the 'Wolf event'—Remarks on large mammal dispersals in Europe and the

Late Villafranchian faunal turnover. *Alpine and Mediterranean Quaternary, 36*, 75–90.

Jung, J. (1946). Géologie de l'Auvergne et de ses confins bourbonnais et limousins. *Mémoires pour servir à l'explication de la carte géologique détaillée de la France,* Paris, 372 p.

Kahlke, R. D., García, N., Kostopoulos, D. S., Lacombat, F., Lister, A. M., Mazza, P. P. A., et al. (2011). Western Palaearctic palaeoenvironmental conditions during the Early and early Middle Pleistocene inferred from large mammal communities, and implications for hominin dispersal in Europe. *Quaternary Science Reviews, 30*, 1368–1395.

Kretzoï, M. (1941). Präokkupierte Namen im Saügetiersystem. *Földtani Közlöny, 71*, 349–350.

Lacombat, F., Abbazzi, L., Ferretti, M. P., Martínez-Navarro, B., Moullé, P.-E., Palombo, M.-R., et al. (2008). New data on the Early Villafranchian fauna from Vialette (Haute-Loire, France) based on the collection of the Crozatier Museum (Le Puy-en-Velay, Haute-Loire, France). *Quaternary International, 160*, 64–71.

Mattei, M., Fioramonti, V., Cavinato, G. P., Cifelli, F., Mazzini, I., Parotto, M., et al. (2023). Magnetostratigraphy of the Collepardo Late Pliocene faunal assemblage (Early Villafranchian Triversa Faunal Unit), Ernici Mts., Central Italy. *Italian Journal of Geosciences, 142*, 398–410.

Maul, L. C., & Markova, A. K. (2007). Similarity and regional differences in quaternary arvicolid evolution in Central and Eastern Europe. *Quaternary International, 160*, 81–99.

Maul, L.C. (2004). The arvicolids of the old collection of Senèze reconsidered. In L. C. Maul & R.-D. Kahlke (Eds.), Conference volume for 18th international Senckenberg conference—VI International palaeontological colloquium in Weimar. Late Neogene and Quaternary biodiversity and evolution: Regional developments and interregional correlations. Weimar (Germany), 25–30 April 2004. *Terra Nostra, 2004*, 168–169.

Mazza, P., & Rustioni, M. (1994). The fossil bear from Senèze (Southern France). *Rendiconti Lincei, 5*, 17–26.

Mein, P. (1975). Résultats du groupe de travail des vertébrés. In J. Senes (Ed.), *Report on activity of the R.C.M.N.S. working groups* (pp. 78–81). Regional Committee on Mediterranean Neogene Stratigraphy.

Mein, P. (1989). Updating of MN zones. In E. H. Lindsay, V. Fahlbusch, & P. Mein (Eds.), *European Neogene Mammal Chronology* (pp. 73–90). New York: Plenum Press.

Ménouret, B., & Guérin, C. (2024). The *Mammuthus meridionalis* of the Late Villafranchian site of Senèze (Domeyrat, Haute-Loire, Central France). In E. Delson, M. Faure & C. Guérin (Eds.), *Senèze: Life in Central France two million years ago. Paleontology, geochronology, stratigraphy and taphonomy* (pp. 587–606). Cham (Switzerland): Springer.

Minwer-Barakat, R., García-Alix, A., Martín-Suárez, E., & Freudenthal, M. (2008). The latest Ruscinian and early Villanyian Arvicolinae from southern Spain re-examined: Biostratigraphical implications. *Journal of Vertebrate Paleontology, 28*, 841–850.

Mörs, T., & Hugueney, M. (2017). A *Hystrix* Linnaeus, 1758 incisor (Mammalia, Rodentia) from the Early Pleistocene of Senèze, France. *Geodiversitas, 39*, 797–802.

Mourer-Chauviré, C., & Bonifay, M.-F. (2018). Les oiseaux du Pléistocène inférieur de Ceyssaguet (Lavoûte-sur-Loire, Haute-Loire, France): description d'une nouvelle espèce du genre Aquila. *Quaternaire, 29*, 183–194.

Mourer-Chauviré, C. (2024). The birds (aves) of the Early Pleistocene Locality of Senèze (Haute-Loire, France). In E. Delson, M. Faure & C. Guérin (Eds.), *Senèze: Life in Central France two million years ago. Paleontology, geochronology, stratigraphy and taphonomy* (pp. 145–163). Cham (Switzerland): Springer.

Napoleone, G., Albianelli, A., Azzaroli, A., & Mazzini, M. (2003). Dating the late Villafranchian (Pliocene) vertebrate collections of the Upper Valdarno (Tuscany, Italy) by the magnetochronologic framework of the basin fill. *Bollettino della Società Paleontologica Italiana, 42*, 401–413.

Nomade, S., Pastre, J. F., Guillou, H., Faure, M., Guérin, C., Delson, E., et al. (2014). $^{40}Ar/^{39}Ar$ constraints on some French landmark Late Pliocene to Early Pleistocene large mammalian paleofaunas: Paleoenvironmental and paleoecological implications. *Quaternary Geochronology, 21*, 2–15.

Ogg, J. (2020). Geomagnetic polarity time scale. In F. M. Gradstein, J. G. Ogg, M. D. Schmitz, & G. M. Ogg (Eds.), *Geologic time scale 2020* (pp. 159–192). Elsevier.

Palombo, M. R. (2016). Large mammals faunal dynamics in southwestern Europe during the late Early Pleistocene: Implications for the biochronological assessment and correlation of mammalian faunas. *Alpine and Mediterranean Quaternary, 29*, 143–168.

Paquette, J.-L., Médard, E., Poidevin, J.-L., & Barbet, P. (2021). Precise dating of middle to late Villafranchian mammalian paleofaunae from the Upper Allier River valley (French Massif Central) using U-Pb geochronology on volcanic zircons. *Quaternary Geochronology, 65*, 101198.

Pastre, J.-F. (2004). The Perrier Plateau: A plio-pleistocene long fluvial record in the River Allier Basin, Massif Central, France. *Quaternaire, 15*, 87–101.

Prat, F. (1964). Contribution à la classification des équidés Villafranchiens. *Procès-verbaux de la Société linnéenne de Bordeaux, 101*, 14–32.

Rook, L., & Martínez-Navarro, B. (2010). Villafranchian: The long story of a Plio-Pleistocene European large mammal biochronologic unit. *Quaternary International, 219*, 134–144.

Savage, D. E., & Curtis, G. H. (1970). The Villafranchian stage-age and its radiometric dating. *Geological Society of America Special Paper, 124*, 207–231.

Schaen, A. J., Jicha, B. R., Hodges, K. V., Vermeesch, P., Stelten, M. E., Mercer, C. M., et al. (2021). Interpreting and reporting $^{40}Ar/^{39}Ar$ geochronologic data. *GSA Bulletin, 133*, 461–487.

Schaub, S. (1923). Neue und wenig bekannte Cavicornier von Senèze. *Eclogae Geologicae Helvetiae, 18*, 281–295.

Schaub, S. (1941). Die kleine Hirschart aus dem Oberpliocaen von Senèze (Haute-Loire). *Eclogae Geologicae Helvetiae, 34*, 264–271.

Schaub, S. (1943). Die oberpliocaene Säugetierfauna von Senèze (Haute-Loire) und ihre verbreitungsgeschichtliche Stellung. *Eclogae Geologicae Helvetiae, 36*, 270–289.

Sinusìa, C., Pueya, E. L., Azanza, B., & Pocoví, A. (2004). Datación magnetoestratigráfica del yacimiento paleontológico de la Puebla de Valverde (Teruel). *Geotemas, 6*, 339–342.

Sotnikova, M. V., Baigusheva, V. S., & Titov, V. V. (2002). Carnivores of the Khapry faunal assemblage and their stratigraphic implications. *Stratigraphy and Geological Correlation, 10*, 375–390.

Spassov, N. (1997). Villafranchian succession of mammalian megafaunas from Bulgaria and the biozonation of South-East Europe. In J. P. Aguilar, S. Legendre & J. Michaux (Eds.), *Actes du Congrès Biochrom'97. Mémoires et travaux de l'Institut de Montpellier de l'École Pratique des Hautes Études, 21*, 669–676.

Spassov, N. (1998). A New Late Villafranchian locality of vertebrate Fauna—Slivnitsa (Bulgaria) and the carnivore dispersal events inEurope on the Pliocene/Pleistocene boundary. *Historia Naturalis Bulgarica, 9*, 101–113.

Spassov, N. (2000). Biochronology and Zoogeographic affinities of theVillafranchian faunas of South Europe. *Historia Naturalis Bulgarica, 12*, 89–128.

Spassov, N. (2003). The Plio-Pleistocene vertebrate fauna in South-Eastern Europe and the megafaunal migratory waves from the east to Europe. *Revue de Paleobiologie, 22*(1), 197–229.

Spassov, N. (2016). Southeastern Europe as a route for the earliest dispersal of *Homo* toward Europe: Ecological conditions and the timing of the first human occupation of Europe. In K. Harvati & M. Roksandic (Eds.), *Paleoanthropology of the Balkans and Anatolia* (pp. 281–290). Dordrecht: Springer.

Spassov, N. (2023). The Final Pliocene and Early Pleistocene faunal dispersals from east to Europe and correlation of the Villafranchian biochronology between eastern and western Europe. XXI INQUA Congress 2023, Rome, 13–20 July 2023, session 135; abstract (Programme | XXI INQUA Congress 2023 | Ex Ordo).

Spassov, N., & Crégut-Bonnoure, E. (1999). Premières données sur les Bovidae Villafranchiens de Bulgarie. *Comptes Rendus de l'Académie des Sciences, Sciences de la Terre et des Planètes, 328*, 493–498.

Thouveny, N., & Bonifay, E. (1984). New chronological data on European Plio-Pleistocene faunas and hominid occupation sites. *Nature, 308*, 355–358.

Tyrberg, T. (2022). *Pleistocene birds of the palearctic* (2nd ed.). Privately pubished by the author.

Valli, A. M. F. (2024). New data on Cervidae (Mammalia, Artiodactyla) from the Villafranchian Locality of Senèze (Haute-Loire, France). In E. Delson, M. Faure & C. Guérin (Eds.), *Senèze: Life in Central France two million years ago. Paleontology, geochronology, stratigraphy and taphonomy* (pp. 409–431). Cham (Switzerland): Springer.

Chapter 18
The Taphonomy of Senèze

Yolanda Fernández-Jalvo, Andrea M. F. Valli, María Dolores Marín-Monfort, and María Dolores Pesquero-Fernández

Abstract The later Villafranchian reference site of Senèze (Haute-Loire, France) has yielded not only valuable specimens of this period for taxonomic and paleobiological analyses, but also an interesting taphonomic problem. Several nearly complete or partial skeletons of fossil mammals have been found in anatomical position. In the first half of the twentieth century, thousands of fossils (including numerous partial skeletons) were exhumed through excavation over an extensive area of this paleo-lake. The high number of skeletons recovered from this maar suggested a catastrophic hypothesis about the site's origin. The survey and systematic excavations carried out from 2001 to 2006 by the Franco-American research team were focused on a more restricted area, providing more detailed information that makes possible a better interpretation of site formation at Senèze. A detailed taphonomic study has been undertaken to investigate this exceptional site. The analysis reported here is restricted to bone surface modifications of large-mammal fossils recovered from the systematic excavations. Some specimens from the older collections and mounted skeletons housed at the University Claude Bernard-Lyon 1 were also analyzed to confirm taphonomic modifications and pathologies. As a result of this taphonomic study of Senèze fossils, we demonstrate that carnivore activity has been mimicked by subaquatic compression marks and soil corrosion, and we establish criteria to distinguish each taphonomic agent. This investigation concludes that catastrophic events were not part of the principal taphonomic process at Senèze, nor was there any reworking (i.e., transportation or mixing of fossils from different time periods or environments). The fossil assemblage of Senèze is an accumulated association of animals (biocoenosis) that regularly visited the site in life to drink, where they eventually died and their carcasses were slowly buried, resting on the floor of the lake under water.

Résumé Le site de Senèze (Haute-Loire, France), référence internationale pour le Villafranchien supérieur, n'est pas seulement riche en vertébrés fossiles, mais se révèle aussi intéressant du point de vue taphonomique. Les restes de plusieurs grands mammifères ont été trouvés en connexion anatomique. Dans la première moitié du XXe siècle, des milliers de fossiles ont été exhumés, dont plusieurs squelettes plus ou moins complets. Cela aurait pu suggérer l'hypothèse d'évènements catastrophiques ayant eu lieu dans le maar. Les fouilles menées par une équipe franco-américaine, entre 2001 et 2006, bien que concernant une surface plus réduite que celle des anciennes fouilles, ont permis d'obtenir des informations précieuses pour l'interprétation de la formation du site fossilifère. Une étude taphonomique soigneuse a été réalisée sur les vestiges de ce gisement exceptionnel. Nous avons analysé les modifications sur la surface osseuse des restes de grands mammifères qui ont été retrouvés à partir des fouilles rigoureuses avec repérage des positions des pièces. Les marques de corrosion du sol et celles dues à la compression des sédiments sont parfois similaires aux traces caractéristiques de l'action des carnivores, bien que l'analyse effectuée a permis d'établir des critères pour reconnaître chaque agent taphonomique. Un contrôle réalisé sur les spécimens des collections anciennes, notamment les squelettes montés conservés à l'Université Claude Bernard-Lyon 1, a permis de confirmer les modifications taphonomiques et les stigmates de pathologies articulaires déjà mis en évidence. Le résultat de ces observations confirme qu'à Senèze des événements catastrophiques n'ont pas été à l'origine du

Y. Fernández-Jalvo (✉) · M.D. Marin-Monfort ·
M.D. Pesquero-Fernández
Museo Nacional de Ciencias Naturales (CSIC), Madrid, Spain
e-mail: yfj@mncn.csic.es

A.M.F. Valli
Société Scientifique du Bourbonnais, Moulins (Allier), France

M.D. Marin-Monfort
INGEOSUR Dept. Geology, Universidad Nacional del Sur (UNS)-CONICET, Bahía Blanca, Argentina

E. Delson et al. (eds.), *Senèze: Life in Central France Around Two Million Years Ago*,
Vertebrate Paleobiology and Paleoanthropology, https://doi.org/10.1007/978-3-031-64415-3_18

principal processus taphonomique et que la plupart des restes fossiles n'ont pas subi de transport ni de remaniement: il s'agit d'une accumulation osseuse (biocénose) produite à partir de cadavres d'animaux, morts en venant se désaltérer dans le lac.

Keywords Villafranchian • Large mammals • Maar • Volcanic crater lake • Mimic tooth marks

Mots-clés Villafranchien • grands mammifères • maar • cratère volcanique • paléo-lac • traces imitant des morsures de carnivores

Introduction

Taphonomy investigates "site formation" as the first step in interpreting the transition from the biosphere to the lithosphere (Efremov 1940, 1953). This discipline was, in fact, originally created due to the difficulties of studying faunal remains which could be found at places where they were not living, i.e., thanatocoenosis (Efremov 1940). In this respect, taphonomic research is not only focused on fossil specimens, but also on trace fossils and signs of activity of contemporaneous life forms of the past (e.g., footprints, coprolites, pellets, tooth marks, digestion traces, insect damage and root marks) and abiotic agents (e.g., environmental agents and weathering process fluctuations, transport, exhumation and mixture of fossils) that might influence site formation. These sources of information tell us about the life and behaviors of past animals which could never be recovered without taphonomy (Fernández-López 1981, 2000). Fossil assemblages are susceptible to being affected by the way they were accumulated and the preservation processes they underwent (Behrensmeyer 1975, 1978; Behrensmeyer & Kidwell 1985; Behrensmeyer et al. 2000; Fernández-López 1981, 1991, 1995, 2006; Shipman 1981). Taphonomy is, therefore, an integrative and multidisciplinary investigation that aims to reconstitute the past in all details, providing a more dynamic view of the past rather than a simplistic view of a discipline just involved with burial processes, evidence of bias or destructive processes (Fernández-Jalvo et al. 2011; Fernández-López 1981, 1991).

A detailed taphonomic study has been undertaken here to investigate the exceptional nature of the fossil locality of Senèze. This later Villafranchian (sensu Heintz 1970) site yielded one of the most important vertebrate fossil collections of this period, including holotypes and complete skeletons of large mammals. Surveys and excavations have been performed since the end of the nineteenth century, and the most important collections are conserved in the Paleontology collection of the Laboratoire de Géologie, Université Claude-Bernard Lyon 1 (code FSL) and the Naturhistorisches Museum of Basel (NMB). Recently, an international Franco-American research team, led by Martine Faure, Claude Guérin and Eric Delson, undertook new surveys and programmed field research between 2000 and 2006.

Several hypotheses have been proposed to explain the richness and completeness of fossils from Senèze. Couthures (1989) reported that the mammalian fauna was derived from the upper levels of the volcanic crater lake (maar) infilling as well as from the slope deposits. This author also proposed that poisonous fumaroles could produce a catastrophic mass death of fauna in the area, but evidence is lacking. An alternative idea (Delson et al. 2006) envisaged mud or debris that slid down the steep slopes of this maar and engulfed large animals either while still alive or soon after death and washed them into the lake margin before they could be preyed upon by any of the local carnivores. In this situation, skeletons would have remained in anatomical connection at least during the early phases before burial.

Debard (2024, p. 81) expanded this idea as follows: "The geomorphological and sedimentological context provides a better understanding of the formation conditions of the fossiliferous assemblages. During the cold phases, dead or still living animals could be dragged by solifluction to be trapped in the icy waters of the lake. The association of remains with pebbles and blocks (frequent in the south-eastern sector) may reflect very cold conditions where the surface of the lake, partly frozen and covered with stony pavement, could yield under the weight of heavier animals, causing their death by drowning. During the warmer phases, as seen in the western lake beds, they could simply slip on the slope while coming to drink; because the stiffness of the sedimentary layers resulted in bluffs 0.5 to 1.5 m high at the edge of the lake and due to the instability of the waterlogged sediments, the animals could not get back up onto the bank and drowned. Subsequently, the collapse of sediment panels, delimited by faults, could further break up the resulting skeletons."

These hypotheses can be tested through taphonomic analyses investigating bone surface modifications (including pathologies produced in life), alterations recorded on the fossil bone surfaces due to biotic and abiotic taphonomic agents, sediment-fossil relationships, context and sequence of deposition. The aim of this taphonomic study is to obtain a conclusive understanding of how this reference site for the Late Villafranchian (MNQ 18), with multiple fossiliferous horizons between 2.20 and 2.08 Ma (Nomade et al. 2014, Delson et al. 2024) was formed and the origin and nature of this fossil assemblage.

Material and Methods

Fossils analyzed here come from systematic excavations from 2001 to 2006 conducted in two areas or parcels (P172 and P233). Both of these areas are located near Senèze village at the paleo-lakeshore (Fig. 18.1; see also Figs. 2.2 and 2.3). Fossils from P172 and P233 have been excavated from trenches and were spatially located using laser EDM (electronic distance measuring or Total Station) coordinate equipment (Parenti et al. 2024). Fossils were then photographed in situ, and location, numbering and specific remarks on the fossil were noted on the label accompanying each. Specimens are numbered sequentially by year of recovery and currently stored in the FSL collection; for example, the 9th specimen collected in 2006 is thus labeled FSL SEN 06–0009. The sediment is not cemented, so fossils once recovered from the sediment needed only to be washed in water avoiding any mechanical cleaning procedure or chemical product application; some fossils were not even washed.

The available material to be taphonomically analyzed comprises a total of 527 fossils from the 2003–2006 seasons. Approximately 140 fossils of *Allohippus senezensis* come from the mostly articulated skeletons recovered from parcel 172 (Fig. 18.2), and about 129 fossils of *Eucladoceros ctenoides senezensis* from parcels 172 and 233, with most postcranial skeletons in anatomical connection (mainly recovered from parcel 233). *Metacervoceros rhenanus philisi* fossils have also been studied.

Fossil dimensions (length, width and thickness) have been measured with a caliper and analyzed with the naked eye and under a portable binocular light microscope from 7 × to 40 × magnification. Most fossils were photographed for specific taphonomic traits. All taphonomic observations were systematically recorded in a database.

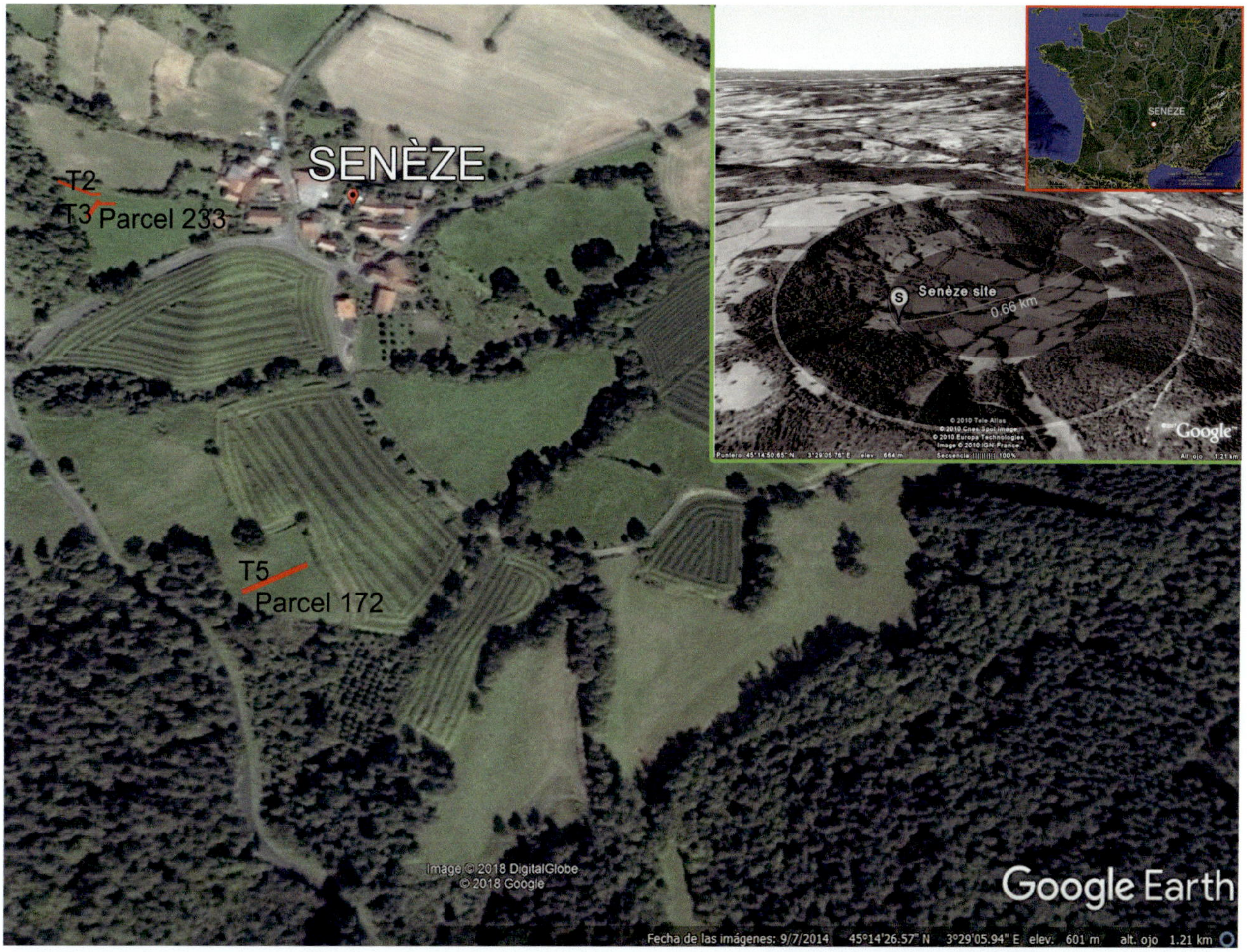

Fig. 18.1 Satellite view of Senèze town and location of trenches T5 and T2 from where fossil assemblages were recovered and studied in this chapter. Top right B/W inset of an aerial photograph of the maar (outer circle) indicating the edge of the crater. **S** marks the area of the Senèze site, while the inner greyish circle shows the possible extension of the paleolake, with an approximate diameter of 660 m. Map of France and location of Senèze (small red spot, top right inset box) sourced from Google Earth

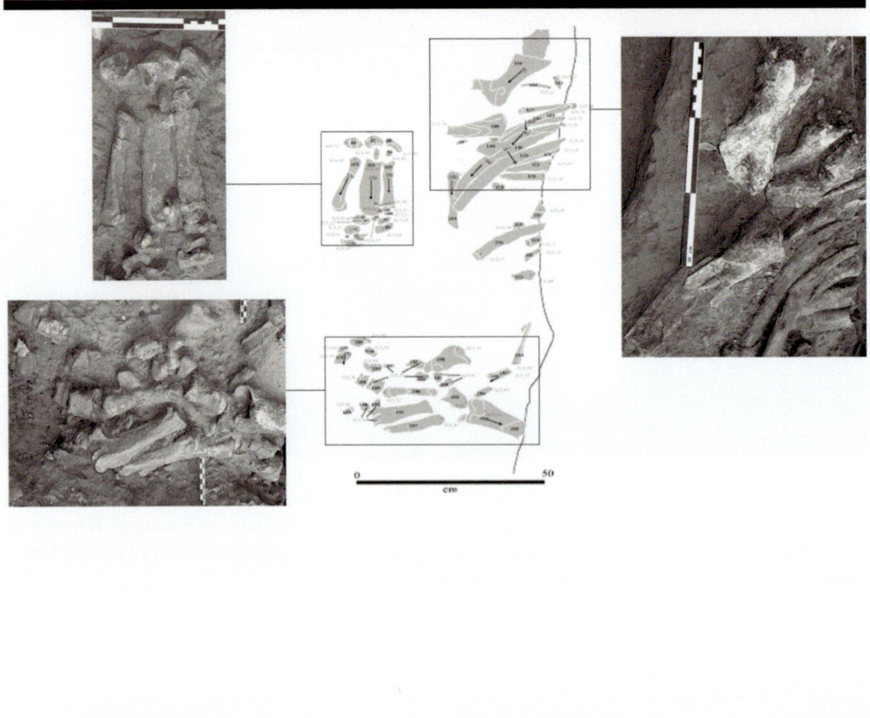
0
50
cm

◀ **Fig. 18.2** Top: partial *Allohippus senezensis* skeleton excavated in parcel 172 near the previous emplacement of trench 5 during the 2005 field season, almost in anatomical connection. The limb bones are close to each other and to the shoulder and pelvic girdles, but the vertebrae and ribs are dispersed. The orientation of these ribs is in agreement with a gravitational organization of isolated long but light bones (photo A. Monguillon, see also Fig. 2.5). Bottom: partial skeleton of the rhinoceros *Dicerorhinus etruscus etruscus* excavated in 2001 from trench 2 of parcel 233; additional remains of the same adult individual recovered in 2003 and 2004 are not shown. Most of the postcranial skeleton (especially the limb elements) is present, in addition to a fragmentary mandibular body with tooth-row and an anterior fragment of nasal bone (photos by E. Delson, drawings by F. Parenti, see also Fig. 2.7)

Identification of Anatomical Elements and Dimensions

- Body part: segment and portion (e. g., diaphysis, proximal end, and distal end; complete; lateral; body; process; arch);
- Age (adult/immature) determined from epiphyseal fusion and bone texture.
- Taxonomic identification to level of group or species (Valli 2024) could mostly be possible if fossils were complete. Fragments were also allocated to size classes: small (<115 kg), medium (115–350 kg), large (>350 kg) following Blumenschine (1986). Sometimes this was the only identification if fragments could not be identified taxonomically.
- Dimensions were classified according to Cáceres (2002), and surface (mm^2) and volumes (mm^3), according to Fernández-Jalvo and Andrews (2011). Complete bones vs. fragments were also distinguished.

Fracture. All fractures were described according to Villa and Mahieu's (1991) nomenclature and methodology for long bone shaft fragments bigger than 4 cm:

- Fracture outline; transverse, curved/V-shaped, longitudinal.
- Fracture angle; oblique, right, mixed.
- Fracture edge; smooth, jagged.
- Shaft circumference; C1: < ½; C2: > ½; C3: whole circumference.
- Shaft fragmentation. L1: < ¼; L2: between ¼ and ½; L3: between ½ and ¾; L4: > ¾ of the total length

Surface modifications related to fracture. Peeling was defined as a roughened surface with parallel grooves or fibrous texture produced "when fresh bone is fractured and peeled apart similar to bending a small fresh twig from a tree branch between two hands" (White 1992: page 140). Deformation is the response of bone under the long-term action of an applied force (the weight of sediment, as in Fig. 18.3A or a stone) producing an irreversible plastically deformed bone shape (Briggs & Williams 1981; Müller et al. 2018; Webster & Hughes 1999). A shorter action (e.g., a falling block striking the bone) may produce a fragile response, i.e. breakage (Fig. 18.3B). Deformation is, therefore, related to the length of time of an applied force and/or to the elasticity of the bone due to collagen content or external water, such as damp environments (Kirchner 2006; Marin-Monfort et al. 2014; Samuel et al. 2016). We have distinguished between impact marks (pits at the broken edge, related to breakage) and percussion marks (pits on the shafts that are not necessarily related to breakage).

Fig. 18.3 **A**: ribs of *Sivatherium hendeyi* from Langebaanweg site in South Africa (5.2–5 Ma). The site is a paleo-flood plain. and the sub-aquatic conditions deformed these skeletonized bones that adapted to the uneven surface and the shape of stones deposited on the riverbed. **B**: Fragment of human long bone from Tianyuan Cave Site (Zhoukoudian. China) broken by gravitational impact of falling roof blocks when already skeletonized, affecting both sides (medial and lateral). **C**: Tooth mark on a *Hipparion* vertebra matching with a hyena *Lycyaena chaeretis* premolar, both found in the site of Cerro de la Garita (Teruel. Spain). **D**: Transverse scratches from a monitored cow skeleton in Draycot, UK (Andrews & Cook 1985) produced by trampling. **E**: Extreme to heavy surface bone corrosion on salient angles of long bones from a horse skeleton that was resting on highly humid moorland acid soil (pH 4) covered by rich grass, heather and bracken; from left to right: femur after 13 years. femur after 11 years and humerus after only 3 years. Pictures from Fernández-Jalvo and Andrews (2016)

Other Surface Modifications Identified on Senèze Fossils

Rodent gnawing marks: grooves or wide striations with U-shaped cross section along anatomical or broken edges produced on bone surfaces scored with the incisors. The width of the incisor marks depends on the size of the animal (Tong et al. 2008). These marks are unbranched although they may be superimposed leaving a characteristic shape.

Tooth marks were measured under the microscope using a digital caliper (minimum length and width for punctures, and minimum width for grooves) and followingAndrews and Fernández-Jalvo's (1997) classification, which provides indications of the type of score and location.

a. punctures on bone surface: pits anywhere on the surface of shafts but not on articular or broken ends
b. grooves on bone surface: scores anywhere on the surface of shafts but not on articular or broken ends
c. punctures on articular ends of bones up to and including epiphyseal lines
d. punctures on the edges of spiral breaks exposed in profile and probably associated with the break
e. puncture marks on the edges of transverse breaks exposed in profile and probably associated with the break
f. punctures on split shafts: puncture marks along the edges of split bones exposed in profile and probably associated with the break
g. puncture print made by multi-cuspid mark (Fig. 18.3C): molar punctures made by multi-cusped teeth usually on flat areas of bone
h. punctures on intact bone edges: puncture marks on the angular edges of some elements (e.g., on ribs and on scapular spines)
i. punctures on crenulated/broken edges (Fernández-Jalvo & Andrews 2011).

Surface weathering has a gradual increment of splitting, flaking and exfoliation of the bone surface due to changes in humidity, temperature and sun exposure. Bones exposed on the ground to weather or atmospheric agents are classified in 5 stages (from 1 = light cracking to 5 = extreme exfoliation and destruction) of increasing damage on the bone surfaces (Behrensmeyer 1978, Andrews & Whybrow 2005, Andrews & Fernández-Jalvo 2019).

We also distinguished cracks that show raised or warped up ridges like mud-cracks in contrast to weathering cracks whose edges are just apart. These cracks are due to humidity (Díez et al. 1999) and were recorded as present or absent.

Sediment friction marks refer to processes that entail movement of bone against a rocky/sandy substrate or friction of rocks falling on bones causing multiple randomly dispersed scratch marks, usually transverse to the length of the bone (Fig. 18.3D), as the bones are rubbed against the stones. These marks are also described in archaeological sites as *trampling* marks by animals (Andrews & Cook 1985), including humans (Domínguez-Rodrigo et al. 2009) due to bones pressed into the rocky substrate. In archaeological contexts, trampling marks strongly mimic cut marks made by stone tools, and distinction between them is especially relevant. A new type of marks has been described in similar environments (lakeshores) as result of animals trampling on bones when deposited in damp environments (Pesquero 2006; Pesquero et al. 2013). This modification is described as "punctures surrounded by plastically deformed bone", and they have also been experimentally reproduced (Marin-Monfort et al. 2014). More details about these marks and their relevance at Senèze are discussed below. Abrasion (by water or wind) may also produce scratches, although the striations are microscopic in size (Fernández-Jalvo & Andrews 2003).

Rounding may be the result of trampling and abrasion (by water or wind), as well as by digestion (Andrews 1990; Brain 1967; Fernández-Jalvo & Andrews 2016; Korth 1979; Shipman & Rose 1983). The difference between these processes is referred to macro- and microscopic scratches and chemical corrosion on the bone surface that distinguishes each process. Rounding affects broken edges as well as anatomical protuberances.

Root-marks (linear marks on bones) are the result of roots of vascular plants in symbiosis with fungi or bacteria, and marks are commonly divided into branches and show signs of chemical corrosion at their interior (Behrensmeyer 1978; Morlan 1980; Johnson 1985; Grayson 1988; Macho-Callejo et al. 2023).

Soil corrosion on bone surfaces (Fernández-Jalvo et al. 2002, 2010; Gordon & Buikstra 1981) indicates the side of the bone that has been in direct contact with acidic ground under constant humidity (Fig. 18.3E).

Manganese and iron oxide stains (Bao et al. 1998; López-González et al. 2006; Marín-Arroyo et al. 2008; Shahack-Gross et al. 1997). Distribution of manganese stains on fossils is described as isolated (I: single stains), clustered (C: in patches or on a particular area on the fossil), dispersed (D: small stains all over the fossil) and widespread (W: almost covering the whole fossil surface). Manganese deposits and iron oxides have been distinguished in the Senèze fossil assemblage based on color (blackish vs. reddish, respectively) as we could not make EDS chemical element analyses. Sediment attached to fossils (included into cancellous bone tissue and/or covering compact bone), as well as a sequence of sediments (clay-sand-silt) recorded on the fossils has been described in order to find criteria congruent with the site's sedimentation.

The taphonomic study of the fossil bone surfaces have shown frequent pathological modifications (Fernández-Jalvo & Andrews 2016) sometimes severe, which have been distinguished and described separately.

Experimental work. Breakage, deformation and gravel/stone compressive puncture marks were reproduced at the Laboratory of Experimental Taphonomy at the Museo Nacional de Ciencias Naturales (Madrid, Spain). A Zwick/Roell Z2.5 "materials testing device" with a single column (uniaxial) for tensile, compression and cyclic testing from 0.5 to 2.5 kN (562 lbf) was used to reproduce deformations on dry and wet bones, and breakage by instantaneous efforts using different kiloNewtons of force, tools of compression and time of action (Fig. 18.4). The equipment operates with a standard PC fully compatible with *testXpert®* software. Experiments were made with bones of cow, sheep, fallow-deer and red-deer, as well as small mammals. Bones used for the experiments were dry and wet (after several weeks immersed in water, acidic pH 4.5, neutral pH 6.5 and alkaline pH 10). Compression was applied on bones lying on fine to coarse sediment or single stones (Fig. 18.4A and Fig. 18.4B), as well as immersed in water (Fig. 18.4C), applying low to high forces (500 to 2500 N), a sudden force (Fig. 18.4D) or more constant and slow compression. Results from all these experiments under different compressive–environmental conditions have been published in Marin-Monfort et al. (2014) and include mechanical deformations related to modifications found at Senèze, mainly associated with trampling and falling blocks.

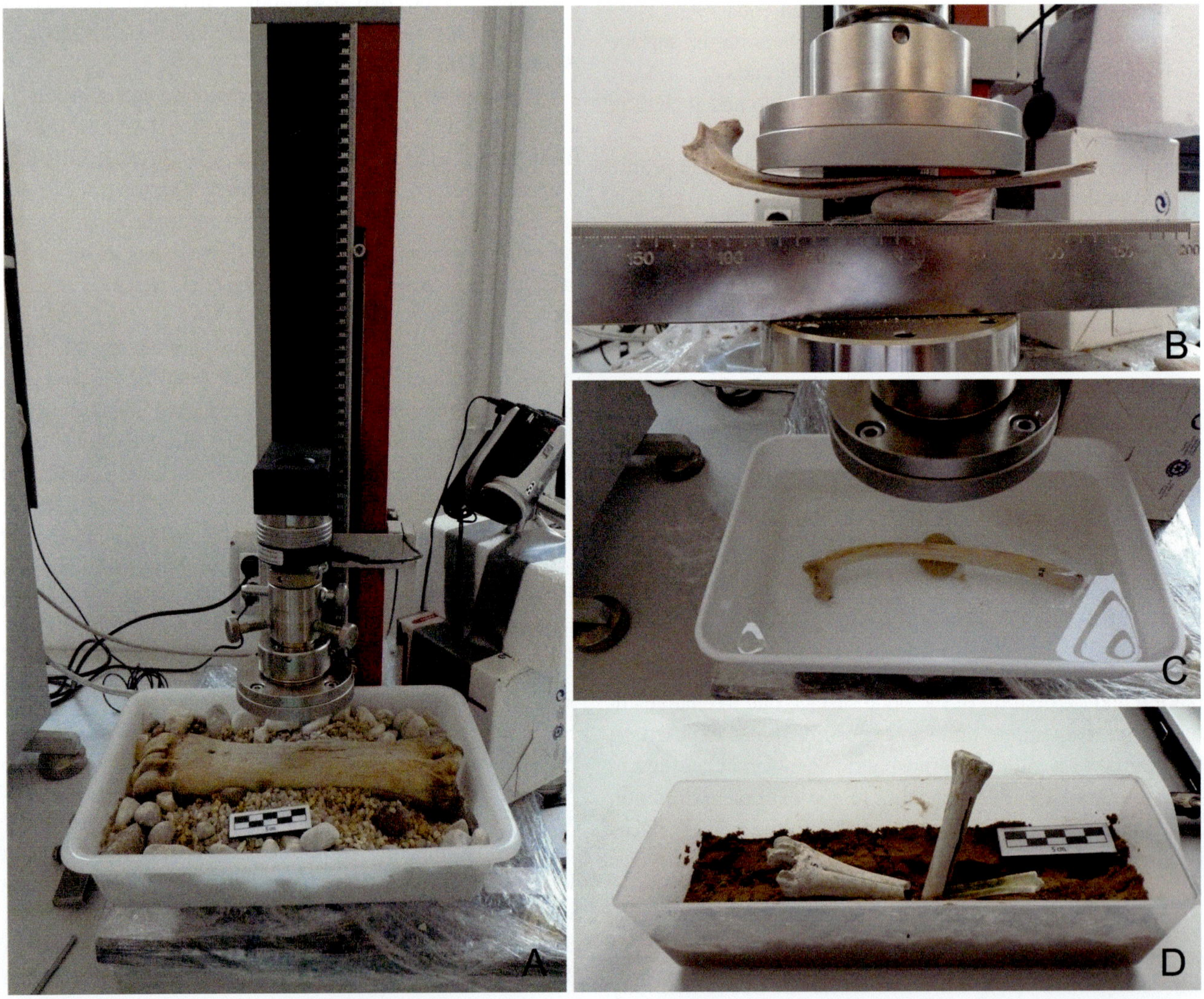

Fig. 18.4 Experiments using static materials testing equipment (Zwick/Roell Z2.5KN). A: Cow metapodial, previously immersed in water for several weeks, going to be compressed against coarse gravel and stones. B: Rib of a sheep previously immersed in water for several weeks compressed against a single stone and deformed during the process. C: Photograph before starting an experiment using a sheep rib immersed in water. D: Damage to a dry fallow deer metapodial resting on silty sediment, broken as result of an instant strike using a narrowed tool of compression

Results

Fossils from two nearly complete skeletons of *Allohippus senezensis* (from parcel 172), part of one almost complete skeleton of *Eucladoceros ctenoides senezensis* (P233) and some bones of *Metacervoceros rhenanus philisi* have been taphonomically analysed here, together with remains of limbs or cranial portions found in anatomical connection and/or dispersed of these taxa. In trench 5 of parcel 172, the team also found remains of *E. ctenoides,* i.e., the posterior part of a cranium with antlers, and another partial skeleton with face, mandible, pelvis and some long bones, which belong to two different individuals. In trench 2 of parcel 233, a partial skeleton of the rhinoceros *Dicerorhinus etruscus etruscus* was recovered in 2001 (Fig. 18.2), with additional remains of the same adult individual coming to light in 2003 and 2004. Most of the postcranial skeleton (especially the limb elements) is present, in addition to a fragmentary mandibular body with toothrow and an anterior fragment of nasal bone. Unfortunately, most fossils from the rhino skeleton were not available when the taphonomic study was undertaken. A total of 527 fossils have been taphonomically analyzed here. The percentage of complete or almost complete anatomical elements is close to 40% of the total number of fossils studied (56% of the taxonomically identified fossils). Seventy percent of the bones in the two partial skeletons of *Allohippus* are complete anatomical elements. Most fossils of *Eucladoceros*, of which 52% are complete elements, were found in parcel 233 (103 fossils), and fewer from parcel 172 (26 fossils). *Metacervoceros* remains are mainly hindlimbs from parcel 172 (13 fossils), and scattered parts of another skeleton were recovered from parcel 233 (25 fossils). In general, 45% of *Metacervoceros* fossils are complete elements.

Table 18.1 Senèze bone fractures classified according to the Villa and Mahieu (1991) nomenclature and methodology for shafts of long bone fragments bigger than 4 cm

Type of fracture	Number	Percentage (%)
Fracture outline	**245**	
Long	61	24.9
Curved	117	47.8
Transverse	67	27.3
Fracture angle	**245**	
Mixed	116	47.3
Right	60	24.5
Oblique	69	28.2
Fracture edge	**245**	
Jagged	58	23.7
Smooth	187	76.3
Total broken edges	**245**	

With regard to breakage, the total number of broken edges analyzed was restricted to 245 specimens given that Villa and Mahieu's (1991) methodology can only be applied on shaft fragments larger than 4 cm. The results of this analysis (Fig. 18.5A) provide indications of a high rate of complete bones (variables C3-L4); most fossils have a complete circumference, although the length is variable. Fracture outline, angle and edge values are displayed in Table 18.1 and Fig. 18.5B.

Deformations and percussion marks are not always linked to breakage (Fig. 18.6A, B, C). Most bones affected by deformation are thin ones like ribs or vertebral apophyses, while long bones have been observed to be deformed only by large stones (e.g., Fig. 18.6C) found underneath the fossil (A. Monguillon pers. comm.). The very fine layer of plastic brown clay sediment, typical of lake deposits, that covers several Senèze fossils favored the preservation of all broken fragments together. Indications of wet environments are also shown by a particular cracking (Fig. 18.6D) recorded at Senèze affecting almost 14% of these fossils (Table 18.2). These cracks have curled up edges and are frequently present in caves and damp environments (lakeshores). These warped cracks are related to prolonged immersion in water or high relative humidity. Weathering may also produce cracks and fissures, especially at low stages of alteration. Surface weathering on bones at Senèze has only reached incipient stages (1–2).

Fissures on Senèze fossils are related to breakage and impact marks (Fig. 18.6E) or present at the other side of percussion punctures or compression marks. Peeling has been recorded only rarely at Senèze (Fig. 18.6F, Table 18.2). During the analyses of these fossils, percussion pits seen at Senèze have been observed to strongly mimic carnivore tooth marks. Tooth marks have a conical puncture shape (Fig. 18.3C) produced when tooth cusps impact on the bone surface, in contrast to the irregular shapes and size of punctures made by percussion. Some of these irregular impact marks at the edge of breakage and linked to percussion marks on surfaces (Fig. 18.6G) mimicking tooth marks will be discussed below.

Taphonomic modifications on the bone surfaces of Senèze fossils are presented in Tables 18.1 and 18.2 according to excavation area (parcel 172 and parcel 233). Weathering has been seen affecting very few remains from Senèze fossils and only at low degrees. The degree of rounding observed at Senèze is also low, and the number of fossils affected by root-marks is very low. As a whole, surface modifications are rare (less than 10% of the fossils at Senèze) and have similar percentages between excavation areas, except for a slightly more abundant number of fossils in P172 that are affected by cracking (due to both weathering and humidity). Evidence of biotic agents of modification (both animal and plant) is practically absent. Evidence of acid soil corrosion has been observed and described as chemical corrosion in Table 18.2; the incidence is low but present at the site, indicating rather acidic soils.

A

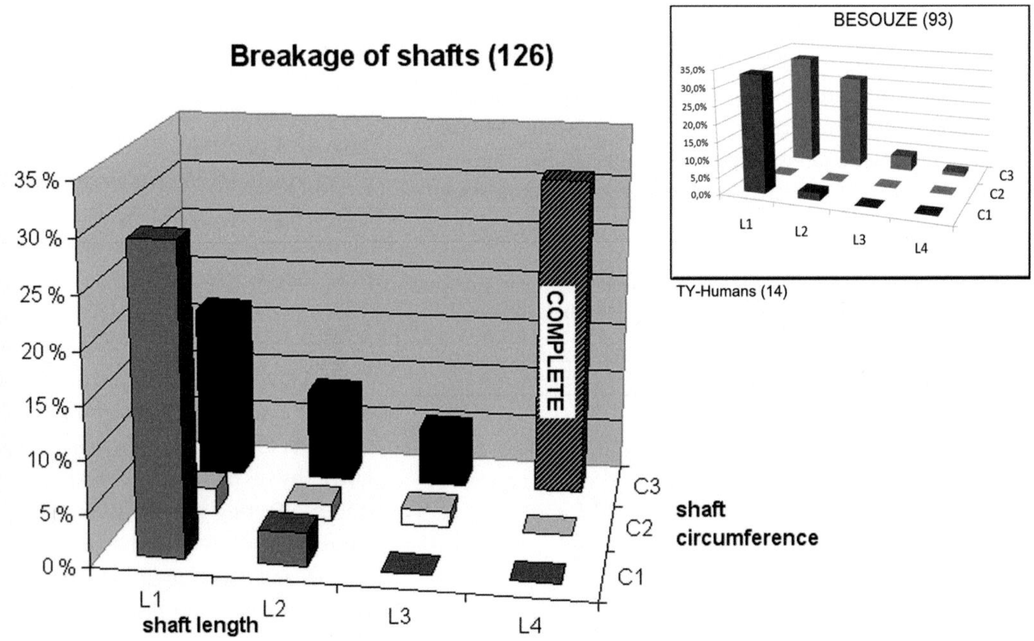

	L1	L2	L3	L4
■ C1	29.37%	3.17%	0.00%	0.00%
□ C2	2.38%	1.59%	1.59%	0.00%
■ C3	16.67%	8.73%	5.56%	30.95%

Fig. 18.5 **A** Long bone breakage of Senèze shafts analyzed according to the methodology of Villa and Mahieu (1991), considering shaft length (L1-L4) and shaft circumference (C1-C3). The inset at top right shows one of the sites (Besouze) analyzed by these authors with close similarities to the Senèze results, except for a higher number of complete specimens (L4-C3) at Senèze. **B** Diagrams of the different types of breakage established by Villa and Mahieu (1991) compared to fossils from Senèze (P233 and P172) analyzed here

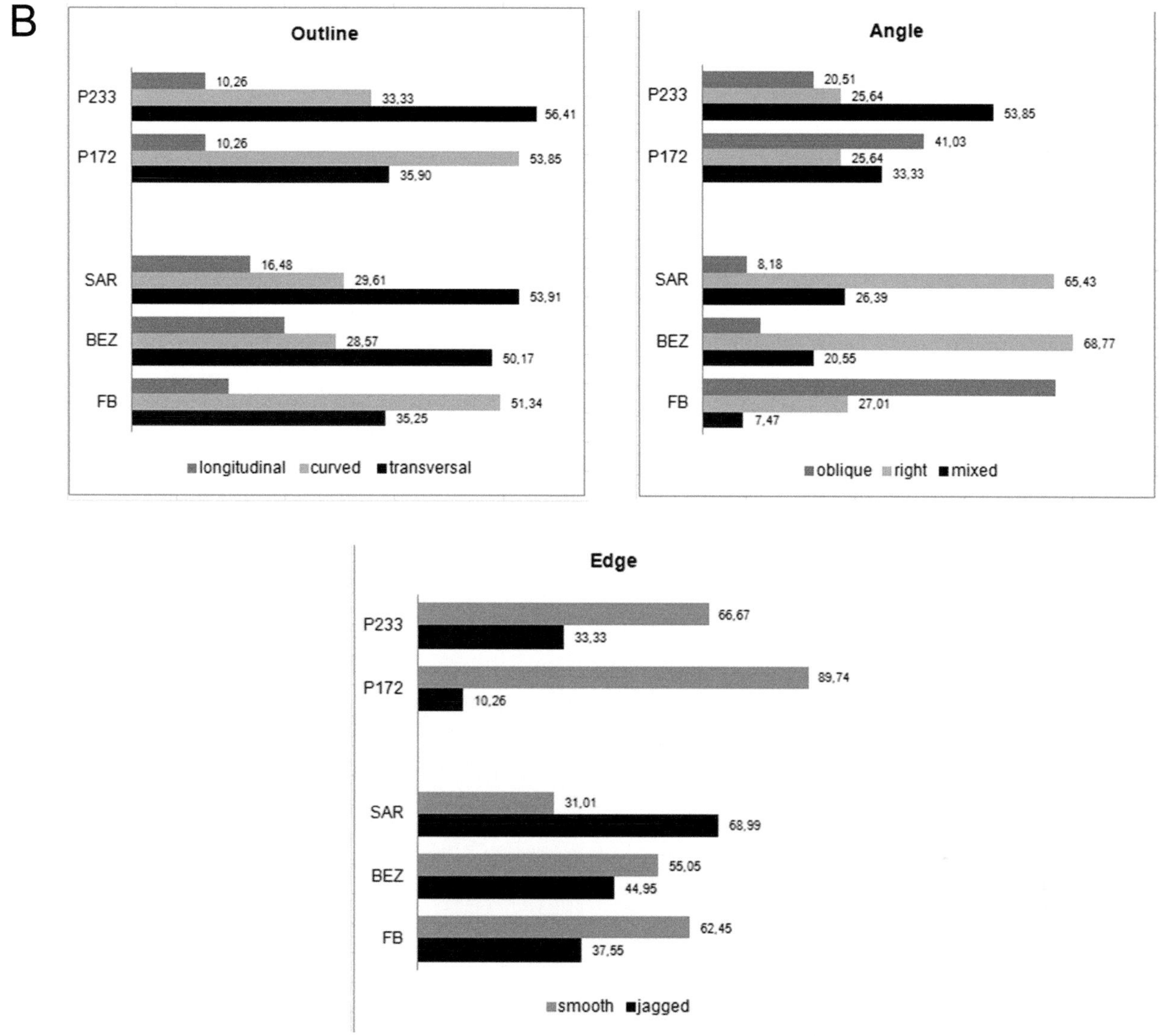

Fig. 18.5 (continued)

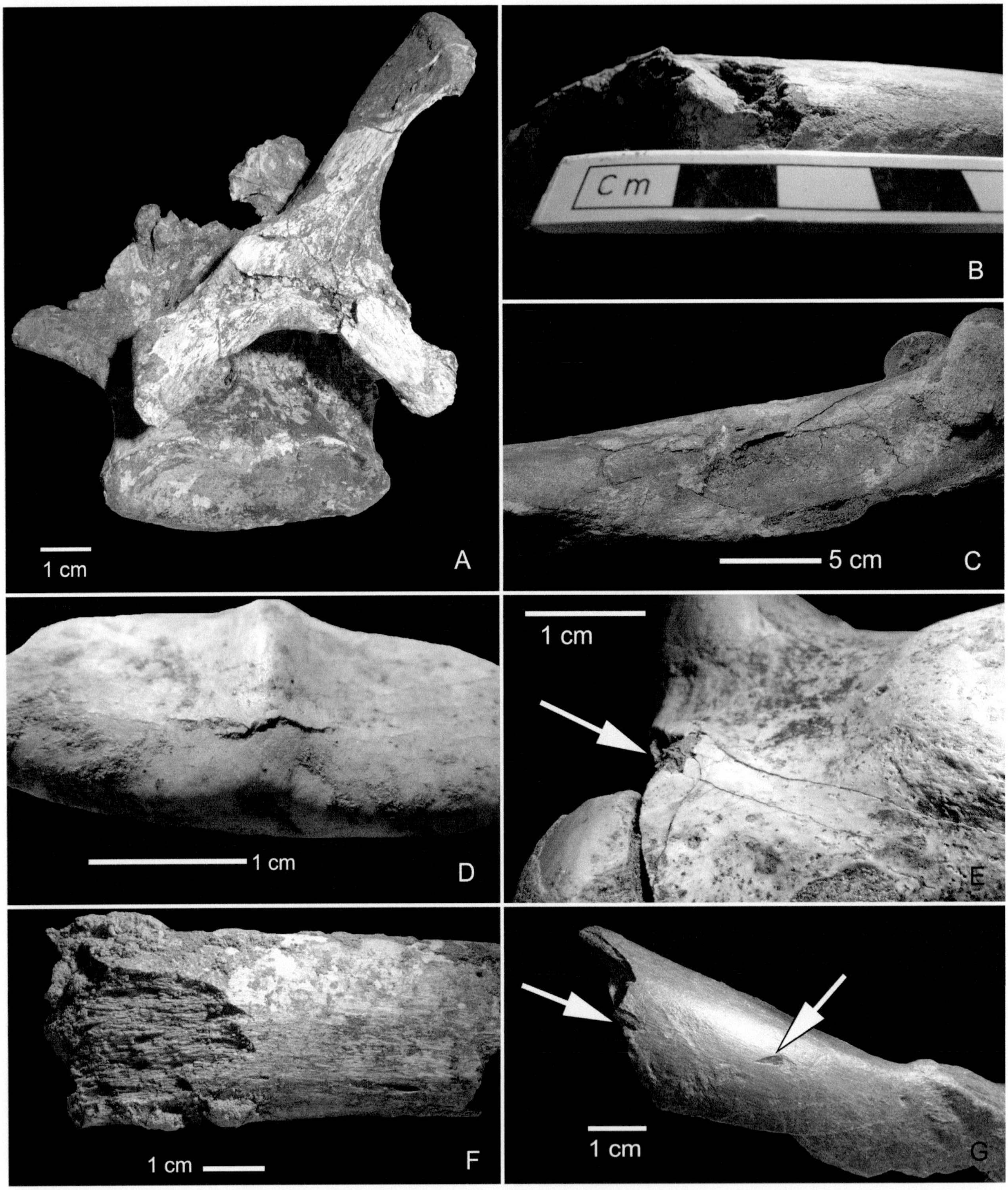

Fig. 18.6 Deformation and breakage of Senèze fossils. A: Lumbar vertebra of *Eucladoceros* (FSL SEN 04–0372); the dorsal apohysis appears laterally bent and the vertebral body deformed, probably as result of long term sediment compression under aquatic conditions. B: Impact deformation on a radio-ulna of *Allohippus senezensis* (FSL SEN 06–0288), probably due to a falling block that struck the surface when the bone was wet or immersed in water; the rapid action produces a fragile response of the bone breaking at the area where the block struck the bone. C: Deformation by compression of a left femur of *Allohippus senezensis* recovered in 2005 (FSL SEN 05–0152) resting on a stone. D: Equid sesamoid (FSL SEN 06–0312) showing cracking by humidity characterized by curled up edges. E: Fissures related to impact marks (white arrow) on an *Allohippus senezensis* first phalanx (FSL SEN 06–0248). F: Bone fragment (FSL SEN 03–0232) showing extensive peeling (roughened surface with exposure of bone fibrous texture) produced by bending; the fossil is also affected by impact and percussion marks and, therefore, peeling may be the result of falling blocks or trampling. G: Right femur of *Eucladoceros* (FSL SEN 04–0298) bearing impact marks associated with breakage (white arrow) and percussion marks on surface (black-outlined white arrow) that may mimic carnivore tooth marks

Table 18.2 Modifications observed on the bone surfaces of Senèze fossils

Type of modification	Number	%	Parcel 172 (%)	Parcel 233 (%)
Impact marks	120	22.8	13.3	9.3
Percussion	86	16.3	10.8	5.3
Deformation	129	24.5	12.3	12.0
Peeling	11	2.1	1.7	0.4
Tooth-marks	5	0.9	0.4	0.6
Gnawing	1	0.2	0.2	0.0
Root-marks	6	1.1	0.8	0.2
Rounding	13	2.5	1.7	0.4
Weathering	44	8.3	7.2	1.3
Cracking-humidity	72	13.7	10.2	3.4
Fissure deformation	89	16.9	8.2	8.5
Surface exfoliation	15	2.8	4.0	0.6
Chem. corrosion	32	6.1	3.6	2.5
Sediment friction/trampling	20	3.8	2.1	1.7
Manganese	479	90.9	51.6	38.5
Mn homogeneous	362	68.7	42.1	26.8
Mn heterogeneous	112	21.3	9.5	11.8
Mn arborescent	465	88.2	49.5	38.0
Mn-isolated	15	2.8	1.5	1.3
Mn-clustered	78	14.8	8.0	6.8
Mn-dispersed	201	38.1	21.1	16.5
Mn-generalized	180	34.2	20.5	13.5
whitish fossil	99	18.8	4.6	14.0
iron oxides	144	27.3	21.3	5.9
Total No. Fossils	527		297	230

Fossils are often covered by a dark coating of manganese oxides. Staining by manganese dioxide is common at Senèze and suggests total immersion in damp sediment. Differences in manganese distribution on fossils have been observed and compared between fossils recovered from parcel 172 and parcel 233. Parcel 172 yields a high number of fossils darkened by manganese stains (52%) and homogeneous manganese coverage (42%) mainly dispersed or all over the surface (Fig. 18.7). Some fossils have been observed to have a bluish color. Discoloration has not been chemically analyzed, and the origin remains unknown, but it also has been seen to be useful to distinguish between excavation areas at Senèze. Fossils from P233 have a more whitish general color than fossils from parcel 172 (Table 18.2).

The location and type of attached sediment on bone surfaces or inside these fossils was described. We have distinguished dark-brown plastic clay, sands and mixed sandy-silty sediments. This sequence suggests different phases of sedimentary deposition and energy. Table 18.3a shows the result of these observations on 278 taxonomically identified fossils (equids and cervids). Table 18.3b compares complete vs. broken fossils (independently of their taxonomic identification) that still had sediment attached on 423 specimens. Some fossils show a consistent sequence of sediment attached to fossils at the cancellous tissues, and clay or sand superimposed to sand and clay respectively. The different excavation areas (parcels 233 and 172) have also been distinguished in order to find criteria that may allow us to better interpret the sedimentation sequence and fossil site formation.

Table 18.3 Location and type of attached sediment on bone surfaces or at the interior of Senèze fossils. (18.3a: distribution observed on taxonomically identified fossils. 18.3b: distribution observed on complete vs. broken fossils.)

Sediment distribution–18.3a		*Allohippus senezensis*		*Eucladoceros ctenoides*	
Location		Parcel 172	Parcel 233	Parcel 172	
Total no. ID fossils:271		142 (%)	103 (%)	26 (%)	
Brown plastic clay	On bone surface	8.4	3.2	0.6	
	In cancellous	3.9	9.1	0.3	
	In and on bone	1.3	0.0	0.0	
	Clay/sand-clay sequence	0.0	5.2	0.0	
Sand	On bone surface	16.5	20.7	2.3	
	In cancellous	0.3	0.0	0.0	
	In and on bone	1.9	0.3	0.0	
	Sand on plastic clay	6.5	5.2	1.0	
Sediment distribution–18.3b		Parcel 172		Parcel 233	
Location		Complete	Broken	Complete	Broken
Total no. fossils: 527		133	164	73	157
Brown plastic clay	On bone surface	3.6	4.0	2.1	1.7
	In cancellous	2.1	4.6	1.7	8.4
	In and on bone	0.2	0.8	0.0	1.5
	Clay/sand-clay sequence	0.0	0.0	0.0	3.6
Sand	On bone surface	9.5	11.6	6.9	11.3
	In cancellous	0.0	1.3	0.0	0.6
	In and on bone	0.0	5.5	0.0	1.7
	Sand on plastic clay	3.2	4.0	1.9%	3.6
Mixed sand-clay	On bone surface	0 .2	0.6	0.4	0.2
	In cancellous	0.0	0.6	0.0	0.6
	In and on bone	0.0	0.0	0.0	0.2

Another modification observed on these fossils, which is not taphonomic but pathological in origin (osteoarthritis), also has relevance in the interpretation of the site (Fig. 18.8). The incidence may vary from mild to heavy arthritis (Table 18.4) and affects limbs and extremities (radio-ulna, tibia, metapodials, phalanges, sesamoids and tarsal-carpal bones). Each of the complete skeletons recovered in recent seasons has at least one element which is pathologically deformed. The taxa affected are *Allohippus senezensis senezensis*, *Eucladoceros ctenoides senezensis*, *Metacervoceros rhenanus philisi* and *Dicerorhinus etruscus etruscus*. Several phalanges, carpal-tarsals and metapodials of these species from the older collections have also been deformed by arthritis of different degrees. For the mounted skeletons of the paleontology collection of the Université Claude Bernard-Lyon 1 only the exposed sides of articulations could be analyzed. An *Allohippus senezensis* skeleton shows heavy arthritis on the distal radius, carpals and proximal metacarpal of the right forelimb (Fig. 18.8F). A *Cervalces gallicus* skeleton has arthritis on phalanges I and II (medial side) of the left forelimb. A *Eucladoceros ctenoides senezensis* mounted skeleton bears signs of pathology on the lateral side of the right distal metapodial, and on the left distal tibia (lateral side as well). One specimen apparently bears an injury due to a blow of some kind, but the bone tissue appears not fully healed, suggesting a perimortem lesion (Fig. 18.8G).

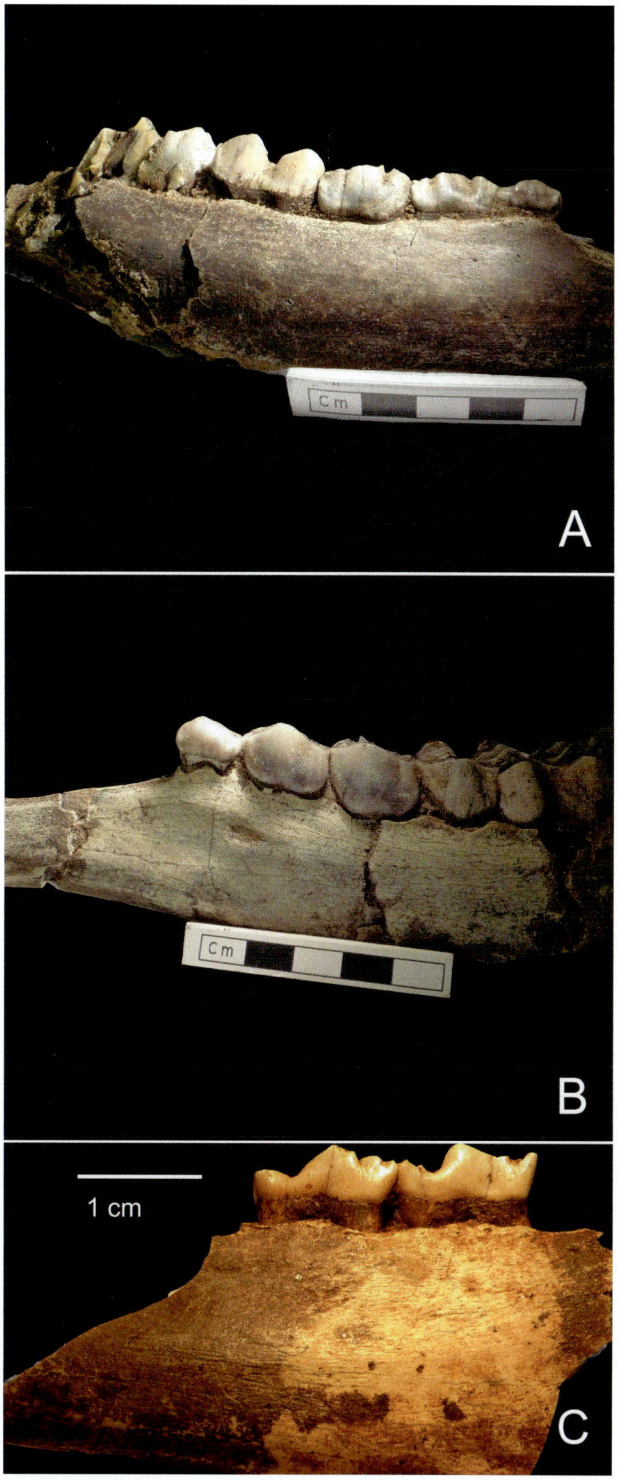

Fig. 18.7 Manganese deposits heterogeneously dispersed. **A** and **B** right mandible of *Eucladoceros* (FSL SEN 04–0196): A: buccal side completely covered by manganese and extensive scratches produced by friction against the sediment (see Fig. 18.10B). In contrast, the lingual side **B** is not stained by manganese deposits. **C**: *Metacervoceros* mandible (FSL SEN 04–0413) heterogeneously stained by manganese

Table 18.4 List of Senèze fossils from 2003–2006 excavations showing bone pathologies

FSL SEN	Species	Anatomical element	Location of pathologies (x = arthritis)
03–0011	*Metacervoceros rhenanus*	Phalanx	x heavy on articulation distal + palmar
03–0053 a	*Metacervoceros rhenanus*	Phalanx II	x heavy on palmar proximal. contact with phalanx I and lateral
03–0053 b	*Metacervoceros rhenanus*	Phalanx III	x heavy on dorsal side
04–0015	*Dicerorhinus etruscus*	Metacarpal IV	x light on distal articulation/metaphysis
04–0132	*Allohippus senezensis*	Metatarsal left	perimortem lesion
04–0133	*Allohippus senezensis*	Phalanx I left	x light on dorsal
04–0134	*Allohippus senezensis*	Phalanx II left	x light on dorsal
04–0216	*Eucladoceros ctenoides*	Metatarsal right	x light
04–0292	*Eucladoceros ctenoides*	Phalanx II	x articulation
04–0294	*Eucladoceros ctenoides*	Phalanx III	x possible metaphysis prox
04–0295	*Eucladoceros ctenoides*	Tibia right	x light on proximal metaphysis
04–0296	*Eucladoceros ctenoides*	Cubonavicular right	x on posterior medial
04–0298	*Eucladoceros ctenoides*	Femur right	possible tumor on medial side
04–0299	*Eucladoceros ctenoides*	Calcaneus right	x moderate on posterior medial
04–0309	*Eucladoceros ctenoides*	Phalanx I	x heavy on distal end
04–0327	*Eucladoceros ctenoides*	Phalanx III	x heavy on proximal and lateral side
04–0328	*Eucladoceros ctenoides*	Phalanx II	x light
04–0361	*Eucladoceros ctenoides*	Phalanx III	x light
04–0363	*Eucladoceros ctenoides*	Radio-ulna right	x light distal radial metaphysis and very light distal ulna
04-0366c	*Eucladoceros ctenoides*	Medial cuneiform	x light on lateral edge
04–0381	*Eucladoceros ctenoides*	Phalanx II	x light on proximal palmar
04–0392	indet	Phalanx III	x heavy
04–0402	*Eucladoceros ctenoides*	Phalanx II	x moderate. proximal dorsal
04–0405	*Metacervoceros rhenanus*	Phalanx III right	x heavy on proximal dorsal
05–0098	*Allohippus senezensis*	Radio-ulna	x distal end
06–0217	*Eucladoceros ctenoides*	Phalanx I	x light on ventral side
06–0261	*Allohippus senezensis*	Calcaneus right	x moderate on distal articular cavity
06–0264	*Allohippus senezensis*	Phalanx II left	light-moderate on posterior side
06–0276	*Allohippus senezensis*	Phalanx II left	x light-moderate
06–0281	*Allohippus senezensis*	Metatarsal III right	x light

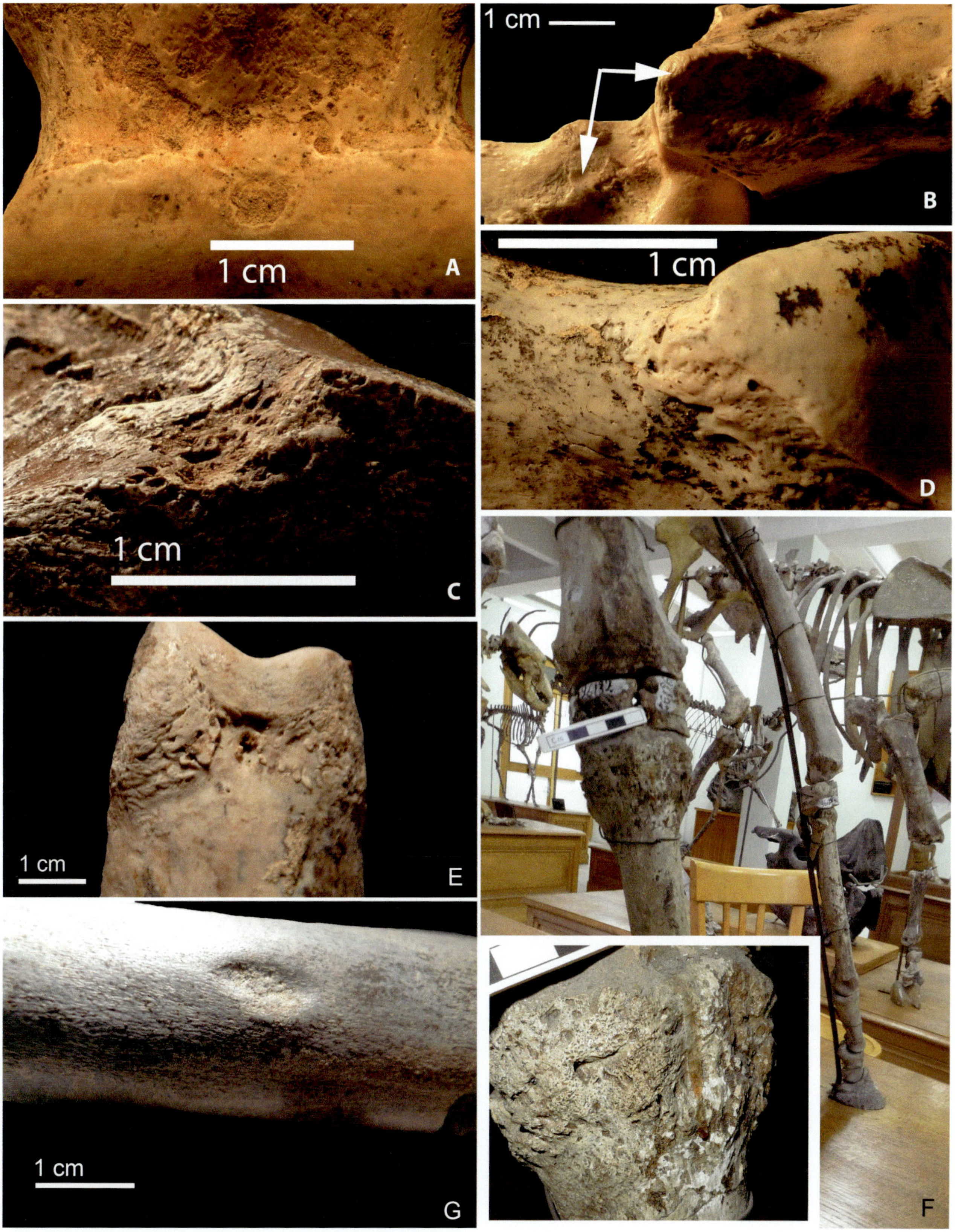
1 cm
A
1 cm
B
1 cm
C
1 cm
D
1 cm
E
F
1 cm
G

◀ **Fig. 18.8** Pathologies at Senèze affect all ungulate taxa studied from different excavation seasons (previous and recent) and mounted skeletons, as well. Bone overgrowth is an attempt to improve the articular effectiveness. These pathologies would have produced difficulties and pain when walking. **A**: Second phalanx of *Allohippus senezensis* (FSL SEN 06–0264) has anomalous bone growth on the palmar side. **B**: Strong osteoarthritis on the joint between phalanx I and phalanx II of *Metacervoceros rhenanus*. Both phalanges have strong outgrowths on the margins of the joints (white arrows) and the distal end of phalanx I (on the left). **C**: Mild osteoarthritis on the distal end of a fourth metacarpal of *Dicerorhinus etruscus* (FSL SEN 04–0015). **D**: Second phalanx of *Eucladoceros ctenoides* (FSL SEN 04–0402) having moderate bone overgrowth on the dorsal side of the distal end in connection with the third phalanx (FSL SEN 04–0405), heavily affected by osteoarthritis on the proximal end, but not shown in this figure. **E**: Distal end of phalanx I of left hindlimb identified as "Eland" from old collections (FSL 210517) having anomalous bone overgrowth on the dorsal side. **F**: Mounted skeleton of horse from Senèze on exhibit in the paleontology collection of the Université Claude Bernard-Lyon 1; the proximal end of the right metacarpal has heavy osteoarthritis (see inset). **G**: *Allohippus senezensis* left metatarsal (FSL SEN 04–0132) shows perimortem injury that started remodeling bone at the interior of the scar, but the animal died before the scar could be healed

Discussion

Distribution and Completeness of Skeletons

The first and most relevant characteristic of the Senèze fossil assemblage is the presence of several individuals whose skeletons are in anatomical connection and almost complete. In 2005, a nearly complete and mostly articulated skeleton of *Allohippus senezensis* (Fig. 18.2) was excavated in Parcel 172. Limb bones of the carcass appeared very close to anatomical position, but the vertebral column was disturbed, and the ribs were all dispersed. This fossil distribution suggests that carcasses were skeletonized and resting on the lake floor permanently covered by water for some time, and bones could rearrange according to the slope and gravitational forces. Apparently, bodies immersed in water (Payne 1965) suffer a quicker decay of ligaments from the vertebral column than from the limbs, making ribs and vertebrae detach first. Similarly, limbs of the large deer *Eucladoceros ctenoides* recovered in 2004 from Parcel 233 were found in anatomical connection with minimal dispersion of bones (except for the head and part of the limbs that were missing). All this is in agreement with dead bodies that decay under permanent immersion in water and sediment without any sign of subaerial exposure.

During the 2006 season, work was concentrated in Parcel 172, extending the excavation area laterally to the previous trenches 5 and 6. The excavation yielded fossil remains all along the new trench 7 both on top and, especially, at the bottom of the slope, where a new partly complete horse skeleton was recovered (see Fig. 11.25). Fossils from the top part of the excavation area at P172 appeared accompanied by stones (Fig. 18.9A), and a carnivore coprolite was found next to fossils (Fig. 18.9B, white arrow). Fossils from this top part of the excavation appear fragile, some also show superficial exfoliation, cracking and splitting (weathering), some others have lost or suffered flaking of the cortical compact bone, leaving the cancellous tissue exposed (Fig. 18.9C). Longitudinal breakage and cracking has been found frequently among these "upper" fossils, in contrast to those recovered from the middle and lower parts of the slope that are almost complete and had almost no surface modifications. Nonetheless, some of these bones have also been broken diagenetically (Fig. 18.9D, black arrows), and fragments remained together indicating an absence of reworking processes. Fossils from the uppermost topographic zone (Fig. 18.9E white arrow) are probably from the edge of the lake shore, i.e., the minimum water level of the lake.

Fig. 18.9 Signs of subaerial exposure at the topographically upper part of deposits in Parcel 172 (trench 7, 2006 excavation season). A: Stones grouped together with fossils which are damaged by subaerial exposure (exfoliation, cracking and splitting). B: Hyena coprolites (white arrow) found next to fossils which have impact marks (black arrows) that caused longitudinal breakage. C: Fossil bones in this area appear fragile, with pitted and flaked cortex probably caused by cyclic subaerial exposure and water immersion. D: Rib with transverse breakage (black arrows) broken diagenetically (post-burial) with all fragments in situ indicating the absence of reworking processes (post-burial transport). E: location of fossils from the top of the excavation that show signs of sub-aerial exposure (as in A–C)

Breakage, Compression and Deformations

Breakage has been studied following the methodology of Villa and Mahieu (1991) (Table 18.1) to typify the shape of fractures. Villa and Mahieu (1991) used this method to compare three different types of sites: Fontbrégoua, a Neolithic site (ca. 4000 BC), and the collective burials of Besouze and Sarrians (Late Neolithic, ca. 2500 BC). Fontbrégoua has evidence of cannibalism (Villa et al. 1986) and hence frequent traces of long bone breakage by humans while bones were still green or fresh. At Besouze, the sub-fossil bones were broken by impact (equivalent to falling blocks), and at Sarrians fossil bones were broken by sediment pressure. The shaft length of fragments compared to circumference completeness analyzed at Senèze (Fig. 18.5A) shows similarities with the site of Besouze except for a higher percentage of complete elements (C3-L4) found at Senèze. However, the number of Senèze fragments that show curved outlines and mixed angles exceeds the frequency in Besouze; this may be caused by wetter conditions (dry in Besouze) and the higher incidence of trampling observed at Senèze (absent at Besouze) that also produces breakage (Andrews 1990; Olsen & Shipman 1988).

Impact marks on the edges of breakage (Table 18.2) are related more directly to breakage by falling blocks, and they are more abundant than percussion marks (on the bone surface). Deformation of fossils by sediment and stone compression producing plastically deformed fossils is also high, indicating a wet environment of burial in plastic sediment such as clay (still attached to fossils). Similar deformation has also been observed at Langebaanweg (Pliocene, South Africa, Fig. 18.3A), and Batallones (Miocene, Madrid, Spain; Domingo 2009); both of these sites are under water. We have also experimentally observed (Fig. 18.4B) that bones immersed in water are much more plastic and deform under compression forces more readily than when they are dry; dry bones react in a more fragile or brittle way and easily break.

Surface Modifications

The presence of rodent gnawing is usually related to dry bones or bones that lost their greasy periosteum in relatively dry environments. Rodents may chew to wear out their ever-growing teeth (Brain 1981), gain minerals from the bone (Laudet & Fosse 2001) or, among rats, to feed on bone nutrients (Rabinovich & Horwitz 1994; Klippel & Synstelin 2007). A low incidence of rodent gnawing is common in highly humid climates (Andrews & Armour-Chelu 1998) or wet caves (Marin-Monfort et al. 2016). High humidity is also suggested by cracks on bone surfaces (as shown in Fig. 18.6D), that affects around 14% of Senèze fossils analyzed. These cracks were formerly identified in caves with a high relative humidity (Díez et al. 1999; Fernández-Jalvo & Andrews 2016) and recognized on fossils from environments under permanent water (lakeshores, Fernández-Jalvo & Andrews 2016).

Evidence of carnivore chewing or of any other contemporaneous biological agent that could cause death and accumulate these individuals is rare (e.g., tooth marks, corrosion by digestion or fracture by chewing), in spite of the high diversity of carnivore taxa (Delson et al. 2006; Argant 2024) and coprolites found at the site, at the edge of the water level. However, physical modifications by abiotic agents observed on Senèze fossils may strongly mimic carnivore activity. We will discuss here features which can distinguish the action of abiotic agents that mimic the effects of biological taphonomic agents. We correlate the type of score and location of tooth marks according to categories established by Andrews and Fernández-Jalvo (1997, described here in methods). We describe them below as the "pseudo-category" of the corresponding modification as they are not real carnivore tooth marks. Differences are described through specific examples of Senèze fossils.

This is the case of punctures on bone surfaces (pseudo-category a, according to the classification of Andrews & Fernández-Jalvo 1997, FSL SEN 06–0009, Fig. 18.10A), grooves on diaphyses (pseudo-category b, FSL SEN 04–0196 Fig. 18.10B), punctures on articular surfaces (pseudo-category c, FSL SEN 06–0087, Fig. 18.10C), punctures on a spiral break (pseudo-category d, FSL SEN 05–0139, Fig. 18.10D, see also Fig. 18.6G), and multi-cuspid tooth-mark prints (pseudo category g, FSL SEN 06–0243, Fig. 18.10E-G). Punctures can be produced by falling blocks, trampling and soil corrosion (Fig. 18.11A). Soil corrosion may also mimic carnivore "scoop out" at the ends of long bones (FSL SEN 04–0322, FSL SEN 04–0298, Fig. 18.11B-C). In fact, only 5 fossils from the new Senèze collection could be confirmed to bear tooth marks on their surface (Fig. 18.11D). We have also found mimics of stone tool-cutmarks (Fig. 18.11E) and peeling (Fig. 18.6F), although the origin is not restricted to human activity as originally defined by White (1992). Both modifications that mimic human action are the result of trampling on stony substrates on bones free of soft tissues, but still rich in collagen content.

Carnivore Activity Mimicked by Compressive Marks and Soil Corrosion

Punctures on bone surfaces, except for articular ends (Fig. 18.10A). Compressive marks can be mistaken for carnivore action similar to category a of chewing marks

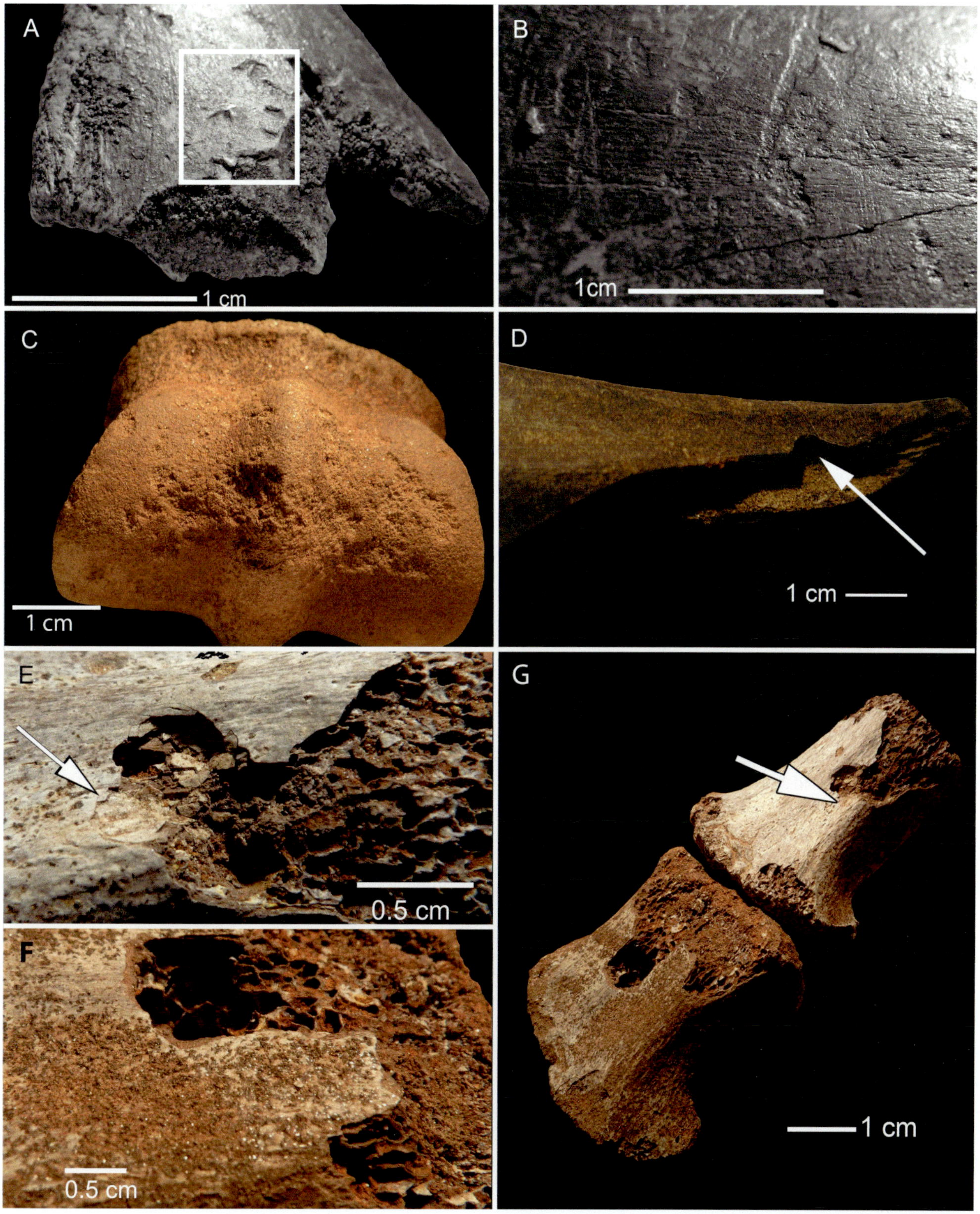
A
1 cm
B
1cm
C
1 cm
D
1 cm
E
0.5 cm
F
0.5 cm
G
1 cm

◀ **Fig. 18.10** Trampling, percussion and impact marks mimicking tooth marks. **A**: Long bone fragment (FSL SEN 06–0009) showing puncture marks on bone surface. These marks have been interpreted as result of trampling against sand or gravel grains of sediment when bones were immersed in water. **B**: Right mandible of *Eucladoceros* (FSL SEN 04–0196. detail of Fig. 18.7A) with distinct grooves produced by friction against the sediment. **C**: Distal metapodial of an immature individual of *Allohippus senezensis* (FSL SEN 06–0087) with puncture marks on articular surface produced by pressure of gravel. **D**: Fossil bone fragment (FSL SEN 05–0092) showing a conchoidal scar on spiral breakage produced by an impact mark (white arrow). **E**: Vertebra (FSL SEN 06–0243) bearing a mimicked multi-cuspid tooth mark that was produced by percussion; the white arrow shows a rounded and flat-bottomed print accompanying a conic print that could be mistaken for a tooth cusp print (see text). **F**: Percussion marks on a vertebra (FSL SEN 06–0242), the depressions have not compressed cancellous tissue in the interior. **G**: Vertebrae FSL SEN 06–0243 and FSL SEN 06–0242 found in anatomical connection (details of these two vertebrae are shown in images **E** and **F**). As in image **E**, the white arrow points to a rounded and flat-bottomed print next to the apparent multi-cuspid mark; the recovery of this vertebra in anatomical connection with the following vertebra. bearing percussion marks. dismisses the identification of a possible tooth mark on FSL SEN 06–0243

described by Andrews and Fernández-Jalvo (1997). A similar case has been observed at the Miocene paleo-lakeshore site of Cerro de la Garita (Teruel, Spain), where the activity of carnivores was clear (Fig. 18.3C), but other puncture marks were surrounded by plastically deformed bone (Fig. 18.11F). These punctures were proposed to be caused by trampling against sand or gravel grains of sediment when bones were wet or immersed in water (Marin-Monfort et al. 2014; Pesquero 2006; Pesquero et al. 2013). The action was reproduced in the laboratory using a "materials testing device". Experiments were conducted with bones immersed in water for several weeks and compressed against coarse sediment (Fig. 18.4A). Results of the experiment (Fig. 18.11G) showed grains of sediment penetrating the bone shaft and articular surfaces, producing similar punctures to those observed at Cerro de la Garita (Fig. 18.11F) and at Senèze (Fig. 18.11H). The main trait which distinguishes these compressive marks against coarse sediment from carnivore tooth marks on bone surfaces is the presence of plastically deformed bone around the depression instead of a more conical shaped perforation compressing the cortical bone inwards (Fig. 18.3C). However, care must be taken with crocodile toothmarks (Baquedano et al. 2012; Njau & Blumenschine 2006), which have strong similarities to marks described in this maar site as the result of trampling. Some fossils appear tilted at strong angles and others remain horizontal.

Grooves on bone surfaces. Striations made by trampling may occur isolated as in Figs. 18.10B and 18.11E, but frequently there are shallow scratches randomly dispersed on the bone surface as in Fig. 18.3D. Difficulties in differentiating trampling marks from cut marks have been extensively documented in the archaeological literature (Andrews & Cook 1985; Behrensmeyer et al. 1986; Olsen & Shipman 1988), most recently by McPherron et al. (2010) and the reply by Domínguez-Rodrigo et al. (2010). Trampling marks are randomly dispersed, mainly affecting shafts and salient angles of bones and usually transversely oriented to the length of the bone (Andrews & Cook 1985; Fernández-Jalvo & Andrews 2016). Tooth marks, as well as cut marks, located near insertion areas aim to detach ligaments or muscle insertions to dismember and/or obtain meat. The striation observed in Fig. 18.11E and peeling shown in Fig. 18.6F mimic hominin activity, but both of these traces are isolated and are not accompanied by any other indication of butchery activities (e.g., scraping marks or breakage to extract the marrow). On the contrary, most fossils are complete and show superficial marks dispersed on the bone surface (Fig. 18.10B), impacts (Fig. 18.6B, Fig. 18.6E and Fig. 18.6G), deformations (Fig. 18.6C) or breakage caused by gravitational rock fall (Figs. 18.6B and 18.10D). We conclude that these two exceptional modifications (mimicking anthropic activity) are responses to friction against the sediment (landslides) and falling blocks, which characterize the fossil assemblage of Senèze.

Puncture marks on articular surfaces. The mean sizes of punctures on articular ends are generally larger and deeper than those on more compact bone (Andrews & Fernández-Jalvo 1997; Haynes 1983; Selvaggio & Wilder 2001), as would be expected in the softer bone of the articular surfaces. Figure 18.10C shows shallow punctures, apparently incongruent with carnivore activity. This is a distal epiphysis of a juvenile metapodial which was still unfused and, therefore, very soft. A carnivore would have damaged this bone more heavily to access the marrow cavity or to obtain the marrow and blood contained in the cancellous tissue. The lack of heavy damage suggests that this is the effect of another taphonomic agent, apparently not related to the nutritional value of the bone. The shape and traits of punctures recorded on the articular end are not those of tooth cusps, but instead the punctures have irregular shapes that may be more congruent with compression of the bone against gravel or coarse sand, probably due to sediment compactness.

Puncture marks on broken edges. Figure 18.10D shows a small conchoidal scar (white arrow) on a spiral broken edge. Spiral breaks indicate that breakage occurred when the bone was fresh. In this particular case (a horse rib), the lack of more chewing marks by carnivores on this fossil and the high frequency of impact marks in this fossil assemblage as a whole make the identification of this damage as the result of chewing by carnivores unconvincing. Such breakage is more likely to be the result of an impact by a stone or block rather than a carnivore tooth puncture. The criteria to distinguish the origin of this particular case are given by the taphonomic context in which the fossil has been found.

Similarly, a mimic multi-cuspid mark was found on specimen FSL SEN 06–0243 (Fig. 18.10E). This fossil was found next to FSL SEN 06–0242 (Fig. 18.10F) in anatomical connection (Fig. 18.10G). The latter specimen also has pits on its surface that expose the trabecular bone, and the edges of the cancellous tissue are fresh and not compressed. This is also observed in the former fossil, next to the puncture, where the thin compact bone was flaked, probably by the percussion mark. Punctures caused by tooth cusps are conical in shape, compressing the cortical bone covering the bottom and walls of the puncture. Looking more carefully at the bone surface of FSL SEN 06–0243, shallower pits can be distinguished at the top of the picture, and a rounded bottom flat print (Fig. 18.10E, white arrow) next to the "pseudo multi-cuspid" mark does not match a tooth print. In this case, the accompanying fossil provides the criterion to confirm/reject a tooth mark.

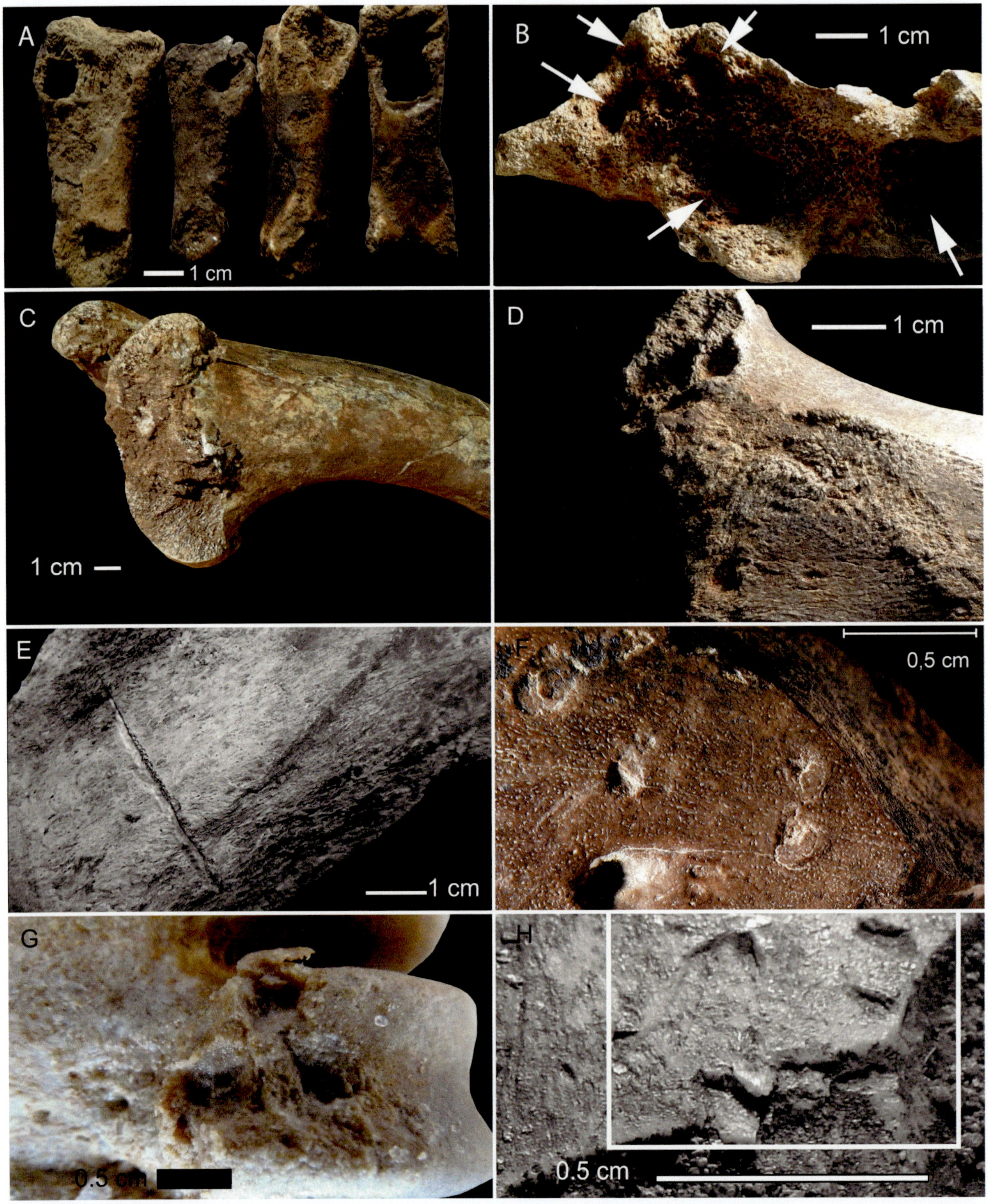

Fig. 18.11 **A**: Several phalanges from the older collection showing perforations produced by soil corrosion indicated by the irregular texture and chemical damage of bone around these holes. **B**: Soil acidity corrosion affecting a fragment of *Eucladoceros* right femur (FSL SEN 04–0298) mimicking scooping out damage by carnivores (see text) **C**: Proximal end of *Eucladoceros* left humerus (FSL SEN 04–0322) corroded by soil acidity. **D**: Right tibia of *Eucladoceros* (FSL SEN 04–0295) on which punctures have been identified as tooth marks, located at a heavily muscled attachment area. **E**: Bone fragment of rhinoceros from Senèze bearing a deep striation caused by friction against the sediment (trampling) that mimics a human-made cut mark. **F**: Detail of a bone fragment from the site of Cerro de la Garita (Spain) with punctures surrounded by plastically deformed bone. **G**: Experimental bone compressed against coarse sediment after several weeks immersed in water. **H**: Detail of puncture marks shown in Fig. 18.10A (FSL SEN 06–0009) showing plastically deformed bone around those punctures (see text)

Surface corrosion on fossil bones has been observed at Senèze producing perforations (Fig. 18.11A) and corroded surfaces (Figs. 18.11B-C). At first glance, the surface modifications in Fig. 18.11B could be mistaken for carnivore activity classified as hollowing out or scooping out of articular ends of limb bones (Binford 1981; Haynes 1980, 1983; Sutcliffe 1970). However, punctures of different sizes in the cancellous tissue (white arrows) and the chemical damage observed on the compact and cancellous tissues indicate that soil corrosion caused this damage and suggest that punctures in Fig. 18.10G were likely made by stones in contact with the cancellous tissue. A modern case of soil corrosion has been monitored over a period of 13 years in the fields of Neuadd, Wales, UK (Andrews & Armour-Chelu 1998; Andrews and Fernández-Jalvo 2019); see Fig. 18.3E). Horse bones resting on the soil surface showed progressive corrosion on salient angles that have been in direct contact with the ground (Fernández-Jalvo et al. 2010). Areas of Neuadd where bone surface corrosion appears more severe have pH values near 4 and have constant high humidity. Histological studies of the Neuadd bones showed that when corrosion was active, bacterial attack was absent (Fernández-Jalvo et al. 2010). The absence of tooth marks on these bones and observations of progressive increase in "corrosion-scooping out" of salient angles and areas in direct contact with the ground, observed over a period of years, excluded the possibility of carnivore action on these Neuadd specimens. The main differences are that salient angles are the most affected by the corrosion, while carnivores may alter any surface and leave tooth marks on any part of the bone. Similarly, and in an aquatic environment as observed in Senèze, some bones (around 5%) were affected by soil corrosion indicating acidity, which is in agreement with traits of volcanic sediments. Histological analyses have not been done on Senèze fossils, but the external aspect of these fossils does not show evidences of bacterial attack. If conditions at Senèze were similar to those measured and analyzed at Neuadd, the fossil site would have had a low rate of bacterial activity. The reduced or absent bacterial activity in turn indicates a much slower decay process of soft tissues, such as ligaments and cartilaginous joints, which would tend to preserve bones in anatomical position. This situation would explain the high number of complete skeletons, almost undisturbed, at Senèze.

The Site Formation of Senèze

Sedimentological and stratigraphic studies by Debard (2024; summarized in Delson et al. 2006) suggest that the lakeshore of this maar had abrupt slopes and steep meter-high banks. Sediments are characterized by lacustrine deposits of upgrading sequence (from plastic and slippery clay to loose sand to fine and coarse gravels), alternating in the sequence with volcanic tuffs, now known from studies by Pastre (2024) to be wind transported ash from different volcanic sources. The lake water level could fluctuate seasonally and annually. This situation enabled sediment slumping and mud flows due to gravitational instability of plastic clay units.

Analyzing sediments attached to bones and the general aspect of fossils from Senèze, we may see slight differences in the location of sediment inside and on the surface of fossils, oxide deposits and the general aspect and coloration of fossils. Sediment distribution is shown in Table 18.3. The *Allohippus* skeletons from parcel 172 have both clay and sands mainly attached on their surface, with sands covering clay layers (not the inverse), and some fossils having gravels and coarse sands on bone surfaces (upgrading sequence). The almost complete skeleton (lacking only the head and part of the limbs) of *Eucladoceros* found in parcel 233 in layer "H inférieur 5" shows a distinctive sedimentary sequence of brown plastic clay (in cancellous bone) and sand covered by brown/reddish plastic clay on bone surfaces only present in broken bones, suggesting a more cyclic sedimentation perhaps related to seasonal changes. Fossils from P172 have more oxides (ferric or manganese) that may have bacterial origin and/or suggest a more oxidizing environment and changes in oxygen content from anoxic to oxidation conditions compared to P233. These indications, together with signs of subaerial surface modifications at the top of the trench of P172 suggest that this area could be closer to the edge of the lakeshore than P233. Debard (2024), however, observes coarser sediments in parcel 172 than in parcel 233, indicating landslide deposits. These coarser sediments would correspond to lower levels in the lake than sediments in parcel 233 (fine lake sediments). Debard (2024) also suggests that these two deposits may not be strictly contemporaneous, and our observations on the sediment infilling cavities support subsequent likely seasonal events.

Manganese occurs in some fossils with a differential distribution, with one side completely covered and the opposite side almost free of manganese deposit (Fig. 18.7). Sometimes shafts are completely covered, but articular ends have no remains of manganese. This pattern of articular joints being free of manganese deposit has been considered to be indicative that bones were in anatomical connection when manganese deposited (López-González et al. 2006). This distinctive pattern is congruent with the characteristically high number of complete skeletons in Senèze, with several skeletal elements in anatomical connection when buried.

The general indication of the sedimentary sequences recorded on fossils suggests calm depositional conditions interrupted by slightly higher energy streams that deposited sand, sometimes followed by calmer clay episodes. This is in agreement with lake deposits, which often have minor variations of energy due probably to seasonal changes. No hydrodynamic abrasion has been observed on these fossils, except for a few affected by rounding that seems more likely to have been produced by trampling/sediment friction. No size, volumetric or shape selection of fossils has been observed, and most remains appear in anatomical connection. No piles of fossils have been observed as should be expected if mud-flows had concentrated previously deposited bones or carcasses. Diagenetic fragmentation in situ (Fig. 18.9D) is frequent, indicating buried fossils were never re-exposed or mixed. In summary, we have not found evidence of re-sedimentation by transport of bones from environments other than lake taphonomic environments, or reworking of fossils producing time–space mixtures.

The main force that acted at Senèze has been gravitational, with blocks falling down the inner slope of the maar and hitting already skeletonized carcasses causing breakage and impacts on bones immersed in water (Fig. 18.6B-C). Skeletons may have slowly slid along the slope while disarticulating as shown in Fig. 18.2 for the horse skeleton recovered in 2005, with ribs dispersed along the slope), sometimes affected by sediment friction/trampling while immersed in water, and sediment compaction and/or sudden mud flows and rock slumps that compressed and deformed buried bones.

Other sites related to maars, such as Eckfeld and Messel (Germany) and Camp dels Ninots (Spain) have yielded a large number of skeletons in anatomical connection. Based on the catastrophic event which occurred in the volcanic Lake Nyos (Cameroon) in 1986, investigators of these fossil sites have recently proposed the possibility of gas emissions from the volcano lake that could kill the animals recorded in these sites (Lutz & Kaulfuß 2006; Oms & Vehí 2010). At Lake Nyos, sudden emission of a large cloud of CO_2 suffocated 1,700 people and 3,000 livestock in nearby towns and villages (Rouwet et al. 2016; Tchindjang 2020). Corpses of these humans and cattle were dispersed on the ground some hundreds of meters away from the lake. The most similar site to Senèze is Camp dels Ninots (Girona, Spain) that yielded fully articulated skeletons of large sized mammals (rhinos, tapirs and large bovids). Systematic excavations of the lake interior started at Camp dels Ninots in 2003, revealing that the strata have a relatively low tilt. So far, the number of skeletons recovered from Camp dels Ninots (26 up to now, 21 recovered in the last 10 years) is lower than at Senèze, but the Spanish site has a high potential of fossil content. At Camp dels Ninots (Oms & Vehí 2010), several events have been distinguished with skeletons recorded in different strata. The process of bringing the large mammals into the lake has been proposed to be the result of a mass death caused by gas emissions. Only those animals that died close to the water's edge could float (after bloating by microbial putrefaction) on the water and later sink to the interior of the lake. If carnivorous mammals or birds scavenged a body (feeding on the viscera and meat), endogenous gut bacteria are prevented from spreading through the body (Bell et al. 1996) and, consequently, the corpses would not be bloated. None of the skeletons recovered from Camp dels Ninots have signs of chewing by carnivores. Oms and Vehí (2010) suggest that carnivores could have been present, but the gas emissions killed so many individuals that scavengers could not access all of them. It is also feasible that those recorded in the site are the bodies that were not scavenged, those that could float and sink to the interior of the maar.

A similar scenario could be considered to explain some aspects of the site formation of Senèze, where the high inclination of the strata could help dead animals killed by gas emission suffocation to roll into the lake. Perhaps such gas emissions could have been released periodically over a long time span, and even the lake waters could have been poisoned by volcanic products such as sulfur or methane. However, it is difficult, at least in the context of Senèze, to explain the low incidence of carnivore activity, with the apparent record of carnivore coprolites high in the trenches of Parcel 172, which could be the lakeshore edge. Various potential scavengers are recorded at Senèze and, therefore, were present in the environment. Catastrophic causes, such as poisoning or suffocation, could have occurred on occasion, but the slope instability and landslides would have played a more important role and caused more frequent casualties than the relatively sporadic volcanic gas emission. We propose that the observed number of skeletons come from animals that regularly visited the site in life to drink, where they eventually died onshore or drowned in the lake. This was especially true for animals that had some pathological problems in the extremities (e.g., arthritis) which made it difficult to easily escape the waters.

Conclusions

Surface modifications recorded on fossils from Senèze have provided taphonomic indications of permanent water coverage in a calm and low energy deposit, congruent with the lake paleo-environment in which fossils have been recovered. Only some fossils from the top of the slope at P172 bear surface modifications indicating local sub-aerial exposure.

The lack of numerous chewing marks on many bones or multiple marks on the same fossils indicates a low level of carnivore activity at the site, including access to carcasses (e.g., scavenging). So far, carnivore tooth marks or other damage are very rare (only 5 fossils bear convincing tooth marks), in spite of coprolites found dispersed at the site, on top of the slope at P172. Physical damage by block and stone impact, compression or sediment friction, as well as chemical soil corrosion, strongly mimics carnivore action.

No signs of re-sedimentation or reworking have been observed on this fossil assemblage. Some rearrangement of certain anatomical elements of complete skeletons (e.g., the horse skeleton from P172 recovered in 2005) indicates that the carcass was resting on the floor of the lake under water for a relatively long time period, long enough to become completely free of soft tissues. Complete skeletons and fossils in anatomical connection were not exposed to the open air, but were permanently under the water lying on the lake floor. This situation ensured that the herbivore remains could decay without emission of any smell that could attract a potential scavenger. Relatively frequently, falling blocks rolled down the slope and fell on skeletonized bones immersed in water, producing bone breakage patterns different from those caused by falling blocks in sub-aerial environments. Prolonged immersion is also indicated by surface bone modifications due to impact strikes, peeling, sediment friction and soil corrosion. Slumps of sediment also could suddenly bury bones and increase deformation of these fossils.

Indications from sedimentological and geomorphological traits of the site, bone surface modifications and fossil distribution suggest that falling blocks and landslides were frequent. The strong inclination of strata alongside the Senèze crater lake became a natural trap for animals that came to the lake to drink water. The nature of these species is definitively autochthonous. This fossil assemblage should not be characterized as a thanatocoenosis (death assemblage) despite the maar having acted as a natural trap. We consider, however, that the faunal assemblage of Senèze is a close image of the past biocoenosis, as these animals (herbivores and carnivores) were not selected or collected by any biotic or abiotic external agent, but by their metabolic necessity of drinking and their own capacity of escaping. This trap acted for several decades or centuries at a time during multiple intervals over perhaps 100,000 yr at the broad lakeshore area, which explains the high number of complete skeletons, without the need of invoking catastrophic events.

Previously proposed hypotheses requiring only catastrophic events can be discarded due to the geological and biological context of Senèze. Instead, the scenario determined after this taphonomic study reinforces the hypothesis proposed by Delson et al. (2006): *mud or debris that slid down and engulfed large animals either while still alive or soon after death and washed into the lake margin before they could be preyed upon by any of the local carnivores. In this situation skeletons would have remained in anatomical connection at least during the early phases of fossilization.* Incidental gas emissions could have occurred, but their effects would have been limited in comparison to the results of lakeside deaths at Senèze.

Acknowledgments We are grateful to Martine Faure, Claude Guérin and Eric Delson who invited us to do this research. Thanks also to Eric Delson and Evelyne Debard for discussions and comments that have greatly improved this paper and to Eric Delson for revising the language. Thanks to the excavation team that worked at Senèze and greatly contributed by providing information and the fossil collections that allowed us to do this study. We are especially pleased we could share time and experiences with Claude Guérin. YFJ offers special thanks to Martine Faure, Claude Guérin, Eric Delson, Angelique Monguillon and Erwan Ermel for information on the excavation fossil recovery and especially for looking after her so nicely, both in the city of Lyon and at Senèze. This research by YFJ has been financially supported in part by Lehman College (indirect cost recovery funds made available to E. Delson). The Spanish authors could undertake this research thanks to the Spanish Ministry of Science and Innovation (PGC2007-66231), DPF has been under a Juan de la Cierva contract (JCI-2007-132-565) and DMM a CSIC Intramural Project (RF2004OE519). We thank Peter Andrews, Isabel Cáceres and an anonymous reviewer for their helpful comments which allowed us to improve an earlier version of the manuscript.

References

Andrews, P. (1990). *Owls, caves and fossils*. London: Natural History Museum, and Chicago: University of Chicago Press.

Andrews, P., & Armour-Chelu, M. (1998). Taphonomic observations on a surface bone assemblage in a temperate environment. *Bulletin de la Société Géologique de France, 169*, 433–442.

Andrews, P., & Cook, J. (1985). Natural modifications to bones in a temperate setting. *Man (N.S.), 20*, 675–691.

Andrews, P., & Fernández-Jalvo, Y. (1997). Surface modifications of the Sima de los Huesos fossil humans. *Journal of Human Evolution, 33*, 191–217.

Andrews, P., & Fernández-Jalvo, Y. (2019). *Understanding time in Taphonomy: A 30-year field study in Wales*. Hauppage (USA): Nova Science Publishers.

Andrews, P., & Whybrow, P. (2005). Taphonomic observations on a camel skeleton in a desert environment in Abu Dhabi. *Paleontologia Electronica, 8*(1), 23A, 17.

Argant, A. (2024). The Carnivores of Senèze (Haute-Loire). In E. Delson, M. Faure & C. Guérin (Eds), *Senèze: Life in Central France two million years ago. Paleontology, geochronology, stratigraphy and taphonomy* (pp. 165–243). Cham (Switzerland): Springer.

Bao, H., Koch, P. L., & Hepple, R. P. (1998). Hematite and Calcite coatings on fossil vertebrates. *Journal of Sedimentary Research, 68*, 727–738.

Baquedano, E., Domínguez-Rodrigo, M., & Musiba, C. (2012). An experimental study of large mammal bone modification by crocodiles and its bearing on the interpretation of crocodile predation at FLK Zinj and FLK NN3. *Journal of Archaeological Science, 39*, 1728–1737.

Behrensmeyer, A. K. (1975). The taphonomy and paleoecology of Plio-Pleistocene vertebrate assemblages east of Lake Rudolf, Kenya. *Bulletin of the Museum of Comparative Zoology, 146*, 473–578.

Behrensmeyer, A. K. (1978). Taphonomic and ecologic information on bone weathering. *Paleobiology, 4*, 150–162.

Behrensmeyer, A. K., & Kidwell, S. M. (1985). Taphonomy's contribution to paleobiology. *Paleobiology, 11*, 105–119.

Behrensmeyer, A. K., Gordon, K. D., & Yanagi, G. T. (1986). Trampling as a cause of bone surface damage and pseudo cutmarks. *Nature, 319*, 768–771.

Behrensmeyer, A. K., Kidwell, S. M., & Gastaldo, R. A. (2000). Taphonomy and paleobiology. *Paleobiology, 26*, 103–147.

Bell, L. S., Skinner, M. F., & Jones, S. J. (1996). The speed of post mortem changes to the human skeleton and its taphonomic significance. *Forensic Science International, 82*, 129–140.

Binford, L. R. (1981). *Bones. Ancient men and modern myths*. New York: Academic.

Blumenschine, R. J. (1986). *Early hominid scavenging opportunities: Implications of carcass availability in the Serengeti and Ngorongoro Ecosystems*. British Archaeological Reports, S283: 1-176.

Brain, C. K. (1967). Bone weathering and the problem of bone pseudo-tools. *South African Journal of Science, 63*, 97.

Brain, C. K. (1981). *The hunters or the hunted? An introduction to African Cave Taphonomy*. Chicago: University of Chicago Press.

Briggs, D. E. G., & Williams, S. H. (1981). The restoration of flattened fossils. *Lethaia, 14*, 157–164.

Cáceres, I. (2002). Tafonomía de yacimientos antrópicos en Karst. Complejo Galería (Sierra de Atapuerca, Burgos), Vanguard Cave (Gibraltar) y Abric Romaní (Capellades, Barcelona). Ph.D. Dissertation, Universitat Rovira i Virgili.

Couthures, J. (1989). The Senèze maar (French Massif Central): hypothesis regarding a catastrophe occurring about 1.5 million years ago. *Journal of Volcanology and Geothermal Research, 39*, 207–210.

Debard, E. (2024). Geological study of the Early Pleistocene site of Senèze (Domeyrat, Haute-Loire, France). In E. Delson, M. Faure & C. Guérin (Eds), *Senèze: Life in Central France two million years ago. Paleontology, geochronology, stratigraphy and taphonomy* (pp. 37–82). Cham (Switzerland): Springer.

Delson, E., Faure, M., Guérin, C., Aprile, L., Argant, J., Blackwell, B., et al. (2006). Franco-American renewed research at the Late Villafranchian locality of Senèze (Haute-Loire, France). *Courier Forschungsinstitut Senckenberg, 28*, 297–317.

Delson, E., Nomade, S., Sen, S., Debard, E., Pastre, J-F., Bahain, J-J., et al. (2024). Geochronology of Senèze: ^{40}Ar/^{39}Ar Dating and Magnetostratigraphy, with Notes on an ESR/U-series dating attempt. In E. Delson, M. Faure & C. Guérin (Eds), *Senèze: Life in Central France Two million years ago. Paleontology, geochronology, stratigraphy and taphonomy* (pp. 99–122). Cham (Switzerland): Springer.

Díez, J. C., Fernández-Jalvo, Y., Rosell, J., & Cáceres, I. (1999). Zooarchaeology and taphonomy of Aurora Stratum (Gran Dolina, Sierra de Atapuerca, Spain). *Journal of Human Evolution, 37*, 623–652.

Domingo, S. (2009). Estudio Tafonómico del Yacimiento de Vertebrados Vallesienses (Mioceno Superior) de Batallones 1, Torrejón de Velasco, Madrid. Ph.D. Dissertation, Universidad Complutense de Madrid.

Domínguez-Rodrigo, M., de Juana, S., Galán, A., & Rodríguez, M. (2009). A new protocol to differentiate trampling marks from butchery cut marks. *Journal of Archaeological Science, 36*, 2643–2654.

Domínguez-Rodrigo, M., Pickering, T. R., & Bunn, H. T. (2010). Configurational approach to identifying the earliest hominin butchers. *Proceedings of the National Academy of Sciences USA, 107*, 20929–20934.

Efremov, I. A. (1940). Taphonomy: New branch of paleontology. *Pan-American Geologist, 74*, 81–93.

Efremov, I. A. (1953). Taphonomie et annales géologiques. *Annales du Centre d'Études et de Documentation paléontologiques, 4*, 1–196.

Fernández-Jalvo, Y., & Andrews, P. (2003). Experimental effects of water abrasion on bone fragments. *Journal of Taphonomy, 1*, 147–163.

Fernández-Jalvo, Y., & Andrews, P. (2011). When humans chew bones. *Journal of Human Evolution, 60*, 117–123.

Fernández-Jalvo, Y., & Andrews, P. (2016). *Atlas of Taphonomic Identifications: 1001+ images of Fossil and recent mammal bone modification*. Dordrecht: Springer.

Fernández-Jalvo, Y., Andrews, P., Pesquero, D., Smith, C., Marin-Monfort, S., B., et al. (2010). Early bone diagenesis in temperate environments, Part I: Surface features and histology. *Palaeogeography, Palaeoclimatology, Palaeoecology, 288*, 62–81.

Fernández-Jalvo, Y., Sánchez-Chillón, B., Andrews, P., Fernández-López, S., & Alcalá Martínez, L. (2002). Morphological taphonomic transformations of fossil bones in continental environments, and repercussions on their chemical composition. *Archaeometry, 44*, 353–361.

Fernández-Jalvo, Y., Scott, L., & Andrews, A. (2011). Taphonomy in palaeoecological interpretations. *Quaternary Science Reviews, 30*, 1296–1302.

Fernández-López, S. R. (1981). La evolución tafonómica (un planteamiento neodarwinista). *Boletín de la Real Sociedad Española de Historia Natural, 79*, 243–254.

Fernández-López, S. R. (1991). Taphonomic concepts for a theoretical biochronology. *Revista Española De Paleontología, 6*, 37–49.

Fernández-López, S. R. (1995). Taphonomie et interprétation des paléoenvironnements. *Geobios, 18*, 137–154.

Fernández-López, S. R. (2000). *Temas de Tafonomía*. Departamento de Paleontología, Universidad Complutense de Madrid.

Fernández-López, S. R. (2006). Taphonomic alteration and evolutionary taphonomy. *Journal of Taphonomy, 4*, 111–142.

Gordon, C. C., & Buikstra, J. E. (1981). Soil pH, bone preservation, and sampling bias at mortuary sites. *American Antiquity, 46*, 566–571.

Grayson, D. K. (1988). Danger Cave, Last Supper Cave, and Hanging Rock Shelter: The faunas. *Anthropological Papers of the American Museum of Natural History, 66*, 1–130.

Haynes, G. (1980). Evidence of carnivore gnawing on Pleistocene and recent mammalian bones. *Paleobiology, 6*, 341–351.

Haynes, G. (1983). A guide for differentiating mammalian carnivore taxa responsible for gnaw damage to herbivore limb bones. *Paleobiology, 9*, 164–172.

Heintz, E. (1970). Les Cervidés villafranchiens de France et d'Espagne. *Mémoires du Muséum national d'Histoire naturelle, 22*, volumes 1 and 2, 1–317, 1-206.

Johnson, E. (1985). Current developments in bone technology. In M.B. Schiffer. (Ed.), *Advances in archaeological method and theory* (pp. 157–235). Nueva York: Academic Press.

Kirchner, H. (2006). Ductility and brittleness of bone. *International Journal of Fracture, 139*, 509–516.

Klippel, W. E., & Synstelien, J. A. (2007). Rodents as taphonomic agents: Bone gnawing by brown rats and gray squirrels. *Journal of Forensic Sciences, 52*, 765–773.

Korth, W. W. (1979). Taphonomy of microvertebrate fossil assemblages. *Annals of the Carnegie Museum, 48*, 235–285.

Laudet, F., & Fosse, P. (2001). Un assemblage d'os grignoté par les rongeurs au Paléogène (Oligocène supérieur, phosphorites du Quercy). *Comptes Rendus de l'Académie des Sciences : Sciences de la Terre et des Planètes, 333*, 195–200.

López-González, F., Grandal-d'Anglade, A., & Vidal-Roman, J. R. (2006). Deciphering bone depositional sequences in caves through the study of manganese coatings. *Journal of Archaeological Science, 33*, 707–717.

Lutz, H., & Kaulfuß, U. (2006). A dynamic model for the meromictic lake Eckfeld Maar (Middle Eocene, Germany). *Zeitschrift der Deutschen Gesellschaft für Geowissenschaften, 157*, 433–450.

Macho-Callejo, A., García-Morato, S., Gutiérrez, A., Marin-Monfort, D., & Fernández-Jalvo, Y. (2023). Put down roots and find the plant!: Preliminary results of root etching and its implications.

Historical Biology. https://doi.org/10.1080/08912963.2023.2263865

Marín-Arroyo, A. B., Landete, D., Vidal, G., Seva, R., González Morales, M., & Straus, L. G. (2008). Archaeological implications of human-derived manganese coatings: A study of blackened bones in El Mirón Cave, Cantabrian Spain. *Journal of Archaeological Science, 35*, 801–813.

Marín-Monfort, M. D., Pesquero, M. D., & Fernández-Jalvo, Y. (2014). Compressive marks from gravel substrate on vertebrate remains: A preliminary experimental study. *Quaternary International, 330*, 118–125.

Marín-Monfort, M. D., Cáceres, I., Andrews, P., Pinto-Llona, A. C., & Fernández-Jalvo, Y. (2016). Taphonomy and site formation of Azokh 1. In Y. Fernández-Jalvo, T. King, L. Yepiskoposyan, & P. Andrews (Eds.), *Azokh Cave and the Transcaucasian Corridor* (pp. 211–249). Dordrecht: Springer.

McPherron, S.P., Alemseged, Z., Marean, C.W., Wynn, J.G., Reed, D., Geraads, D., et al. (2010). Evidence for stone-tool-assisted consumption of animal tissues before 3.39 million years ago at Dikika, Ethiopia. *Nature, 466*, 857-860.

Morlan, R. E. (1980). Taphonomy and archaeology in the upper Pleistocene of the northern Yukon Territory: A glimpse of the peopling of the New World. *Mercury Series, 94*, 1–380.

Müller, R. T., García, M. S., Da-Rosa, Á. A. S., & Dias-da-Silva, S. (2018). Under pressure: Effect of sedimentary compression on the iliac morphology of early sauropodomorphs. *Journal of South American Earth Sciences, 88*, 345–351.

Njau, J. K., & Blumenschine, R. J. (2006). A diagnosis of crocodile feeding traces on larger mammal bone, with fossil examples from the Plio-Pleistocene Olduvai Basin, Tanzania. *Journal of Human Evolution, 50*, 142–162.

Nomade, S., Pastre, J.F., Guillou, H., Faure, M., Guérin, C., & Delson, E., et al. (2014). $^{40}Ar/^{39}Ar$ constraints on some French landmark Late Pliocene to Early Pleistocene large mammalian paleofaunas: Paleoenvironmental and paleoecological implications. *Quaternary Geochronology, 21*, 2–15.

Olsen, S. L., & Shipman, P. (1988). Surface modification on bone: Trampling versus butchery. *Journal of Archaeological Science, 15*, 535–553.

Oms, O., & Vehí, M. (2010). La geologia del Camp del Ninots. In G. Campeny Vall-Llosera, and B. Gómez de Soler (Eds.), *Camp dels Ninots, Rastres de l'Evolució* (pp. 91–103). Girona: Ajuntament de Calde de Malavella.

Parenti, F., Valli, A., Aprile, L., Candelato, F., Monguillon, A., Natali, L., et al. (2024). Surveying and excavations at Senèze: 2001–2006. In E. Delson, M. Faure & C. Guérin (Eds), *Senèze: Life in Central France two million years ago. Paleontology, geochronology, stratigraphy and taphonomy* (pp. 27–35). Cham (Switzerland): Springer.

Pastre, J.-F. (2024). The Seneze maar and paleontological site (Massif Central, France): their volcanic context and studies of related tephras. In E. Delson, M. Faure & C. Guérin (Eds), *Senèze: Life in Central France two million years ago. Paleontology, geochronology, stratigraphy and taphonomy* (pp. 83–98). Cham (Switzerland): Springer.

Payne, J. A. (1965). A summer carrion study of the baby pig *Sus scrofa* Linnaeus. *Ecology, 46*, 592–602.

Pesquero, M.D. (2006). Tafonomía del yacimiento de vertebrados miocenos de Cerro de la Garita (Concud, Teruel). Ph.D Dissertation, Universidad Complutense de Madrid.

Pesquero, M.D., Alcalá, L., & Fernández-Jalvo, Y. (2013). Taphonomy of the reference Miocene vertebrate mammal site of Cerro de la Garita, Spain. *Lethaia, 46*, 378–398.

Rabinovich, R., & Horwitz, L. K. (1994). An experimental approach to the study of porcupine damage to bones. In M. Patou-Mathis (Ed.), *Taphonomy, Bone Modification* (pp. 99–118). Treignes: Editions du CEDARC.

Rouwet, D., Tanyileke, G., & Costa, A. (2016). Cameroon's Lake Nyos gas burst: 30 years later. *Eos, 97*. https://doi.org/10.1029/2016EO055627.

Samuel, J., Park, J. S., Almer, J., & Wang, X. (2016). Effect of water on nanomechanics of bone is different between tension and compression. *Journal of the Mechanical Behavior of Biomedical Materials, 57*, 128–138.

Selvaggio, M. M., & Wilder, J. (2001). Identifying the involvement of multiple carnivore taxa with archaeological bone assemblages. *Journal of Archaeological Science, 28*, 465–470.

Shahack-Gross, R., Bar-Yosef, O., & Weiner, S. (1997). Black-coloured bones in hayonim cave, Israel: Differentiating between burning and oxide staining. *Journal of Archaeological Science, 24*, 439–446.

Shipman, P. (1981). *Life history of a Fossil: An introduction to Taphonomy and paleoecology*. Cambridge: Harvard University Press.

Shipman, P., & Rose, J. (1983). Early hominid hunting, butchering, and carcass-processing behaviors: Approaches to the fossil record. *Journal of Anthropological Archaeology, 2*, 57–98.

Sutcliffe, A. J. (1970). Spotted hyaena: crusher, gnawer, digester and collector of bones. *Nature, 227*, 1110–1113.

Tchindjang, M. (2020). Lake Nyos, a multirisk and vulnerability appraisal. In L. Chenming, (Ed.), *Earth and its atmosphere* (pp.1–55). Hyderabad: Vide Leaf.

Tong, H. W., Zhang, S. Q., Chen, F. Y., & Li, Q. (2008). Rongements sélectifs des os par les porcs-épics et autres rongeurs: Cas de la grotte Tianyuan, un site avec des restes humains fossiles récemment découvert près de Zhoukoudian (Choukoutien). *L'Anthropologie, 112*, 353–369.

Valli, A. M. F. (2024). New Data on Cervidae (Mammalia, Artiodactyla) from the Villafranchian Locality of Senèze (Haute-Loire, France). In E. Delson, M. Faure & C. Guérin (Eds), *Senèze: Life in Central France two million years ago. Paleontology, geochronology, stratigraphy and taphonomy* (pp. 409–431). Cham (Switzerland): Springer.

Villa, P., & Mahieu, E. (1991). Breakage patterns of human long bones. *Journal of Human Evolution, 21*, 27–48.

Villa, P., Bouville, C., Courtin, J., Helmer, D., Mahieu, E., Shipman, P., et al. (1986). Cannibalism in the Neolithic. *Science, 233*, 431–437.

Webster, M., & Hughes, N. C. (1999). Compaction-related deformation in Cambrian Olenelloid trilobites and its implications for fossil morphometry. *Journal of Paleontology, 73*, 355–371.

White, T. D. (1992). *Prehistoric Cannibalism at Mancos 5MTUMR-2346*. Princeton: Princeton University Press.

Chapter 19
Summary and Conclusions of the Senèze Research Project

Eric Delson, Claude Guérin, Martine Faure, Luca Aprile, Alain Argant, Jacqueline Argant, Jean-Jacques Bahain, Federica Candelato, Evelyne Crégut-Bonnoure, Evelyne Debard, Véra Eisenmann, Christophe Falguères, Yolanda Fernández-Jalvo, Jean Gaudant, María Dolores Marín-Monfort, Bernard Ménouret, Angélique Monguillon, Cécile Mourer-Chauviré, Luca Natali, Sébastien Nomade, Fabio Parenti, Jean-François Pastre, María Dolores Pesquero-Fernández, Vittorio Rioda, Sevket Sen, Qingfeng Shao, and Andrea M. F. Valli

Abstract The new fieldwork at Senèze ran from 2000 to 2006, with the goals of clarifying the age, stratigraphy and taphonomy of Senèze, as well as finding additional remains, especially of the less well-known taxa. Here we summarize the findings of each chapter and discuss their broader implications. Four geological chapters consider field methods, stratigraphy, volcanology and tephra mineralogy and dating. Following a chapter on palynology, two chapters discuss non-mammalian paleontology: ichthyology and ornithology. Eight chapters cover work on the fossil mammals. The chapter on biochronology places Senèze among other sites at the start of MNQ 18. Based on that work and the mammal chapters, it is possible to review the relative frequency of mammalian families in the total Senèze assemblage. Of some 2200 specimens, over half are cervids, with bovids, rhinocerotids and equids far behind. According to data from palynology and the habitat preferences of the more common mammals, the paleoenvironment around the Senèze maar would have included forest, woodland and grassland, perhaps in a warmer and moister climate than

Claude Guérin—Deceased 2016.
Jean Gaudant—Deceased 2015.

E. Delson (✉)
Department of Vertebrate Paleontology, American Museum of Natural History, New York, NY, USA
e-mail: eric.delson@lehman.cuny.edu

Department of Anthropology, Lehman College of the City University of New York, Bronx, NY, USA

Ph.D. Program in Anthropology, The Graduate Center of the City University of New York, New York, NY, USA

New York Consortium in Evolutionary Primatology, New York, NY, USA

Institut Català de Paleontologia Miquel Crusafont, Universitat Autònoma de Barcelona, Cerdanyola del Vallès, Barcelona, Spain

C. Guérin
Université Claude Bernard-Lyon 1, Villeurbanne, France

M. Faure
Université Lumière-Lyon 2, Lyon, France

M. Faure · E. Debard · C. Mourer-Chauviré
Université de Lyon, UCBL, ENSL, CNRS, LGL-TPE, Villeurbanne, France

L. Aprile
Associazione Aditus in Rupe, Floridia, SR, Italy

A. Argant · J. Argant
LAMPEA, UMR 7269 CNRS, MMSH, Université Aix-Marseille, Aix-en-Provence, France

J.-J. Bahain · C. Falguères
Département Homme et Environnement, UMR 7194 HNHP (MNHN-CNRS-UPVD), Muséum National d'Histoire Naturelle, Paris, France

F. Candelato
3A soc. coop, via Campofiore 33, 37129 Verona, Italy

E. Crégut-Bonnoure
Muséum Requien, Avignon, France

TRACES, UMR 5608 (CNRS—Université Toulouse 2 Jean-Jaurès), Toulouse, France

E. Debard
25 rue Paul Chevrel, 69370 Saint-Didier-au-Mont-d'Or, France

V. Eisenmann · S. Sen
CR2P (Paléontologie), Muséum National d'Histoire Naturelle, Paris, France

Y. Fernández-Jalvo · M.D. Marín-Monfort · M.D. Pesquero-Fernández
Museo Nacional de Ciencias Naturales (CSIC), Madrid, Spain

J. Gaudant
Université Paris 7, Paris, France

B. Ménouret
3 rue de l'Amouroy, 13680 Lançon-de-Provence, France

A. Monguillon
Réserve Naturelle Régionale du site géologique de Limay, Limay, France

L. Natali · F. Parenti · V. Rioda · Andrea M. F.Valli
Istituto Italiano di Paleontologia Umana, Anagni, FR, Italy

E. Delson et al. (eds.), *Senèze: Life in Central France Around Two Million Years Ago*,
Vertebrate Paleobiology and Paleoanthropology, https://doi.org/10.1007/978-3-031-64415-3_19

today. Taphonomic studies revealed that bones often rested a long time under water, lacked any indication of carnivore attack and often displayed pathologies in their joints. Combining these results and those from stratigraphy, it is proposed that most associated skeletons were preserved after large mammals fell into the lake and drowned without being disturbed.

Résumé Les nouveaux travaux de recherche à Senèze se sont déroulés de 2000 à 2006, avec pour objectifs de clarifier la stratigraphie, l'âge et la taphonomie du gisement, ainsi que de mettre au jour de nouveaux fossiles, plus particulièrement ceux des espèces les moins bien représentées. Nous résumons ici les conclusions de chaque chapitre et discutons de leurs implications plus larges. Quatre chapitres présentent tour à tour nos méthodes de fouille et de relevés topographiques, la stratigraphie, la volcanologie et la minéralogie des téphras, et les datations. Un chapitre traite ensuite de palynologie, deux autres de paléo-ichtyologie et paléo-ornithologie. Huit chapitres concernent les mammifères fossiles. L'étude biochronologique nous conduit à placer le gisement de Senèze, parmi d'autres sites, au début de MNQ 18. À partir de ce travail et des études des mammifères, il est possible de revoir la fréquence relative des différentes familles de mammifères dans l'assemblage faunique total du gisement. Sur quelques 2200 spécimens, plus de la moitié sont des Cervidés suivis, loin derrière, par les Bovidés, les Rhinocérotidés et les Equidés. Selon les données palynologiques et les préférences d'habitat des mammifères les plus communs, le paléoenvironnement autour du maar aurait été un paysage en mosaïque avec des zones boisées et de prairies, peut-être dans un climat plus chaud et plus humide que celui d'aujourd'hui. L'étude taphonomique a révélé que les os sont restés souvent assez longtemps dans l'eau, que certains présentaient des stigmates de pathologies articulaires, et qu'ils ne portaient aucune trace de carnivore. En combinant ces données et ceux de la stratigraphie, nous en déduisons que la plupart des squelettes retrouvés en connexion anatomique sont ceux d'animaux noyés dans le lac, qui ont été conservés grâce à l'absence de perturbations post-mortem.

Keywords Dating • Faunal composition • Mammal paleontology • Stratigraphy • Site formation • Taphonomy

Mots-clés Datation • Composition faunique • Paléontologie des mammifères • Stratigraphie • Formation du gisement • Taphonomie

S. Nomade
Laboratoire des Sciences du Climat et de l'Environnement, CEA Saclay, LSCE/IPSL, UMR CEA-CNRS-UVSQ, 8212 Gif sur Yvette, France

F. Parenti
Universidade Federal do Paraná, Curitiba, PN, Brazil

J.-F. Pastre
23, rue de l'Eglise, 93370 Montfermeil, France

Laboratoire de Géographie Physique, Environnements quaternaires et actuels, UMR 8591, CNRS, Thiais, France

Q. Shao
College of Geography Science, Nanjing Normal University, Nanjing, China

Andrea M. F.Valli
Société scientifique du Bourbonnais, Moulins (Allier), France

Background

The Early Pleistocene fossil mammal site of Senèze (Haute-Loire, France) was first discovered in 1892, when scholars from the region heard about a chance find and contacted Marcellin Boule of the Muséum national d'Histoire naturelle, in Paris. Boule (1892) reported the excavation of a proboscidean skeleton and other mammalian fossils. A young peasant from Senèze, Pierre Philis, was drawn by this find to a search for mammal fossils on his land and over the next half-century recovered many fossils including numerous partial skeletons which he sold to curators in Paris, Lyon and Basel. The details of Philis' life were presented by Faure et al. (2022) and are summarized in Chap. 1 of this book (Faure et al. 2024). After Philis' death in 1942, only a few short field campaigns were undertaken at Senèze in the twentieth century, although many fossils from the site were described and analyzed.

In 1999, Martine Faure, Claude Guérin and Eric Delson met at a field conference on Lesvos (Greece; see the preface) and agreed to collaborate on new field and laboratory research in hopes of clarifying the age, stratigraphy and taphonomy of Senèze, as well as finding additional remains, especially of the less well-known taxa. A short exploratory season of survey and introduction in 2000 was followed by six seasons of intensive excavation, stratigraphic analysis and sampling for dating, paleomagnetic records, palynology and taphonomy (details in Faure et al. 2024). Chapters 2–18 report the results of this work, and we summarize the major findings and implications here.

Physical (Abiotic) Geology

In Chap. 2, Parenti et al. (2024) discuss the approach to survey and excavation techniques employed during our fieldwork. Recent widespread adoption of digital surveying and data recording techniques in paleontological and archaeological sites has transformed not only the daily routine of fieldwork but also the relationship among excavation, observation and interpretation. In the Senèze campaigns the field team included some specialists who already had worked on the middle Villafranchian site of Saint-Vallier between 1995 and 1999 (Faure & Guérin 2004), with some technical improvements, such as automatic recording of polar coordinates by total station and GIS elaboration in the field by digital camera. The Senèze maar was documented with good topographic and cadastral maps, but they were not adequately detailed for microrelief, vegetation cover, rural paths and local power lines. Stratigraphy and taphonomy require a very careful description of each sedimentary body, with a centimetric tolerance. Therefore, each targeted parcel (hereditary land unit) was surveyed at a 1:500 scale, drawing it at a 1:1,000 scale. For altimetric reference, they chose the 606 m a.s.l. point close to the village of Senèze; for planimetry a polygonal net was established with a Topcon GTS 229 total station and for topographic description a Breithaupt transit alidade with plane table was utilized, allowing manual drawing in the field. Digitization of contour lines was performed with *Topografia & Strade* software from Software Tecnico Scientifico. The positions of fossils and samples were plotted by total station and transferred to *ArcGIS* mapping software (ESRI). Detailed topographic recording of each fossil accumulation or remarkable isolated pieces was performed by direct drawing at a 1:10 scale and vertical digital photographs rectified by a GIS georeferencing module. Hand excavation and macrovertebrate collecting by "natural layers" followed removal of the topsoil with a backhoe, creating trenches. Students and other volunteer excavators were trained to draw the stratigraphy of these trenches to scale by using a grid placed against the exposed trench surface and drawing the visible contents of each grid square. Because of the good preservation of fossils, short or massive bones were recovered without any additional treatment. Fragile specimens were removed in plaster jackets. Based on the Senèze experience, Parenti et al. evaluated pros and cons of multiple techniques for recording the precise position of any piece of evidence, no matter its size, for the future understanding of its natural history. Two main methodological conclusions are: (1) optical and digital devices (levels, theodolites, tacheometers and total stations) are very precise and sophisticated but insufficient for the description of the landscape. (2) The growing availability of computer-assisted drawing (designed for engineering) is rapidly blurring any residual capacity of observation and understanding via the manual reproduction of forms and dimensions. As a synthesis, they present an evaluation of data recording techniques and sampling strategies in different kinds of unlithified fossil deposits.

Geological study (Chap. 3, Debard 2024) focused on the excavated trenches, enabling a detailed stratigraphic study of the layers exposed at Senèze, which clarified the nature and geometry of the deposits. Work was concentrated on two sectors of the maar, a western area excavated in the first years of study and a southeastern zone emphasized in the last three years. The trenches were numbered T1 to T9, and the layers distinguished were named alphabetically by trench (with numeric subdivisions in some cases), e.g., T2h, [T2] hi2; groups of such layers were combined into numbered "sets" with common origin. Some trenches were excavated in two sections, an upper ("sup") and lower ("inf"), but those indications did not necessarily agree with relative stratigraphic position, given local dip and faulting. Additional work was carried out on naturally exposed sections, some named (such as the "Philis ravine" where P. Philis recovered numerous fossils) and others identified by letter (e.g., section A). Detailed scaled drawings are presented for each trench or section studied, including positions of samples taken for palynology, magnetostratigraphy and tephra mineralogy (including dating).

After the emplacement of phreatomagmatic products linked to the formation of the maar, slope deposits, incorporating elements of the basement and products ejected by the maar, are organized in prograding lenses towards the center of the crater, following a slope of 15–20°. They form sedimentary bevels characteristic of maar lake margins. In the western sector, the first lacustrine deposits appear at elevation 625 m, in T2, where they thicken rapidly (set 2 = T2h) and have been recognized down to elevation 620 m, the lower limit of this trench. Trench T1 shows mainly slope deposits; lacustrine levels appear at around 621 m, with a thickness less than that of T2 (less than one meter); they only become more widespread towards the 616 m mark. In the south-east sector, slope formations predominate in T4, T8, T9 and T5 sup, at the same height as the lacustrine levels in trench T2. The highest lacustrine level was observed in T7 sup at 619.50 m, i.e., practically at the same altitude as the base of T2. Lake sedimentation becomes continuous at around 617.50 m, i.e., at about the same height as in trench T1. Trenches T6 and T7 inf show only lacustrine sediments, as they were dug below this elevation, marking the generalization of lacustrine levels.

From one sector to the next, and even from one trench to the next, there is great variability in the organization and nature of the deposits. Several hypotheses can be proposed to explain these rapid facies changes.

The morphological context has clearly led to an evolution in sedimentation around the crater rim. Coastal bevels were built in the less steep areas. Slightly marked thalwegs have channelled colluvium, as in the case of Trench 4.

The formations outcropping around the crater also influenced the nature of the deposits below. Those in the western sector were largely fed by a slag cone in the form of stones, gravel and red grains, present in all layers. The phreatomagmatic tuffs produced much of the fine material, but also boulders, such as the one visible in the western part of section A. On the southeastern side, on the other hand, the sediments were largely fed by erosion of the gneissic basement.

Syn- and post-sedimentary reworking resulted in several types of modifications. Load, stretch and boudinage patterns are visible on many sections, revealing subaqueous landslides. In some cases, such as in Trenches 2 and 5, movements, perhaps linked to sediment desiccation, have induced settlement faults, resulting in a panel-like lowering of the deposits towards the center of the maar. Other, larger-scale subsidence is also apparent. Thus, the deposits in the area of section A appear to have been affected by mass detachment, revealed by the lack of deformation of the sediments beneath the tuff block. Slippage also affected the upper sections of trench T5, causing them to fold. Observations of present-day maars still holding water, such as Lac d'Issarlès and Lac Pavin, show that such gravity-driven reworking is a frequent occurrence. Several hypotheses can be put forward to explain these sublacustrine collapses: disruption of the equilibrium of sediments deposited in great thickness on the flats, seismic tremors, fluctuations in lake level, rapid drops in lake level induced by the incision of the outlet. At Senèze, this may have started at the very beginning of the formation of the fossiliferous deposits, perhaps in connection with the evolution from 2.5 to 2.0 Ma of the Sénouire river, into which the Senèze lake drained.

The interweaving of lacustrine deposits and slope formations indicates the emplacement of fossiliferous levels in the zone of water level variations in the paleolake, which may reflect climatic changes. Thus, the arrival of slope deposits may reflect reduced vegetation cover on the crater slopes, linked to periglacial conditions favoring solifluction flows and runoff during ground thaw. The slope deposits in Trenches 5–7 (the oldest layers) may represent a colder interval of lower lake level. Given the absolute dates, which place the bulk of the deposits between 2.19 and 2.07 Ma, this climatic deterioration could correspond to the marked cooling evident in the Tiglian or Pretiglian. Lake levels would have risen as a result of heavy precipitation, leading to a positive water balance. This change, linked to wetter climatic fluctuations, would initiate the transition to warmer conditions in the later Tiglian. These varied sedimentary dynamics show that the fossiliferous deposits developed in different physical and temporal contexts. It is therefore difficult to establish precise correlations between the two main excavated areas, although they are possible within each sector.

Building on the foregoing stratigraphic analysis and detailed mapping, Pastre (2024) was able to clarify the extent and succession of the products of the Senèze volcano and provide a more precise map of the entire area in Chap. 4. A first phreatomagmatic episode precedes the basanite flow which seems closely accompanied by the deposition of scoriae. In a later phase, the maar crater formed and the main phreatomagmatic tuffs were emitted. They are followed by a long lacustrine deposition phase in the maar crater.

The mineralogical study of several thin layers interbedded in the lacustrine and slope deposits containing the fauna on the shore of the maar lake led to the discovery of 10 tephra layers and 6 reworked tephric deposits coming from the Mont-Dore strato-volcano 60 km to the NW. These tephras were discovered in our paleontological excavations (mostly in trenches T2 and T5) and in two previously known sections (section A and the Philis Ravine). The sandy or silty-clayey texture of these beds results from weathering which has systematically destroyed the glass initially present. Because of the absence of ash, the tephra layers can only be characterized by their feldspars and heavy mineral content (ferromagnesians, oxides and accessory minerals) which are typical of each tephra.

Heavy mineral spectra are dominated either by brown amphiboles (kaersutite & pargasite), brown clinopyroxenes (Al_2O_3 and TiO_2 rich brown diopside), green clinopyroxenes (green diopside poor in Al_2O_3 and TiO_2, typical of the Mont-Dore) or by titanite. Among the accessory minerals, zircons and apatite are present in low percentages in most of the samples. Biotite and Fe–Ti oxides (magnetite, ilmenite) complete the assemblages. All these spectra are typical of the Mont-Dore pyroclastites, mostly from trachytic ones. Feldspars are more frequent minerals. Their compositions are typical of each sample. Several samples are dominated by alkali feldspars (anorthoclase, sodic sanidine) and less by plagioclases (oligoclase, andesine, labrador).

The different mineralogical compositions of the tephras indicate distinct ages for each deposit sequence (section) in which they occur. These sequences characterize several evolutionary stages of the lacustrine shore and of the crater slopes featured by the various stratigraphic sequences observed in the paleontological excavations. The frequency of the discovered tephra indicate the high pyroclastic activity of the Mont-Dore around 2.1–2.0 Ma. If no known pyroclastic unit of the strato-volcano can be directly correlated with a Senèze tephra bed, several in the Guéry area in its NE part may be contemporaneous: for example, the Guéry pumice flow dated at 2.09 ± 0.03 Ma.

The heavy mineral assemblages of sedimentary beds precise their origin. Brown clinopyroxenes and olivine which dominate several samples are basaltic minerals and can come from scoriae. The olivine, enstatite and chromian diopside which are common in many samples are mantellic minerals issued from lherzolite (peridotite) nodules. The basement minerals are dominated by garnets and sillimanite with some staurolite and andalousite. Like a part of the brown amphiboles, most of these minerals derive from the phreatomagmatic tuffs dominating the crater where they have been pulverized by phreatomagmatic explosions. The green automorphic diopside and/or titanite which are present in several samples, like other minerals of the Mont-Dore, show the reworking of tephras from the strato-volcano which fell on the crater rim.

In Chap. 5, Delson et al. (2024) analyze data from two complementary sources bearing on the age of the Senèze deposits and fauna. Nomade et al. (2014) published five $^{40}Ar/^{39}Ar$ ages for Senèze tephra, but their stratigraphic positions were not completely known then. These ages were recalculated (based on the guidelines in Schaen et al. 2021) and placed in precise stratigraphic position following Debard (2024). Two ages were obtained on tephra from Trench 2, two from Trench 5 and one from section A (far from any fossils but on the stratigraphically highest horizon at Senèze).

In addition, S. Sen collected samples for paleomagnetic analysis from five sequences across the site. These came from Trenches 1, 2 and 5 (both upper and lower portions), zone H8 (near Trench 2) and a pit dug for sampling at the north end of parcel 233. Virtual geomagnetic polarity was calculated for each sample and the sequences examined for magnetostratigraphic zonation. Most of the polarity was reversed (Figs. 5.5 and 5.6), but there were normal zones at the top of Trench 2, in the lower half of the pit and in the upper part of Trench 5 inf (lower; in fact stratigraphically above T5 sup). The two T2 normal zones were correlated to the Huckleberry Ridge excursion (ca. 2.07 Ma) because they lie above the SEN 1 dated tephra (2.10 ± 0.03 Ma) located in a reversed zone (Fig. 5.7). The Trench 5 normal is most likely correlated to the Feni subchron (2.140–2.116 Ma), although there is a slight possibility it equates to the Huckleberry Ridge (Fig. 5.8). The section A dated tephra is the youngest of the five, and its age fits well with its inferred position near the top of the Senèze sequence (Fig. 5.9). Two other dates have also been published, one from an unknown level in section A (Paquette et al. 2021) and one on a normally magnetized tephra in a core drilled into the maar in 1989 (Roger et al. 2000). Both ages agree with their correlated position in the stratigraphy (Fig. 5.9). The fauna recovered during our fieldwork occurs in two relatively restricted intervals: 2.10–2.08 Ma in the western sector and roughly 2.20–2.18 Ma in the southeastern sector. It is not known if the fossils collected by Philis before 1942 also occurred in these intervals or throughout the 2.20–2.07 span of the Senèze sediments we studied, or even through a longer interval. The rare co-occurrence of argon-argon ages and geomagnetic polarity zones allows the Senèze sequence to be dated with great precision.

In a brief appendix, Bahain, Shao and Falguères report their attempt to obtain comparable dates using the ESR/U-series approach. The combination at Senèze of relatively early U-uptake history reconstructed from the analytical data and high dose rates linked to the volcano-sedimentary context yielded ages which were far too young by comparison to the dates discussed above.

Palynology

J. Argant (2024) reviewed earlier palynology at Senèze and analyzed new samples taken during our fieldwork in Chap. 6. The first palynological analysis was carried out by Elhaï (1969), on a 175 m borehole drilled in 1965 into the lacustrine deposits at the center of the maar (altitude 590 m). The first five meters were sterile. The following 120 m yielded pollen and spores. Due to the size of the column, the analysis was carried out at a sampling interval of 40–60 cm. The top of this core is 30 m below the fauna from the nearby western sector. It was therefore necessary to carry out systematic sampling in conjunction with the new paleontological excavations in order to avoid the disadvantages associated with deep lake deposits, which may have suffered from sediment mixing over time.

Two hundred eight samples were taken between 2001 and 2006, to test all sediment types: coarse sands and gravels from slope deposits, fine sands and clays from lacustrine deposits collected in columns on the sections of trenches T1 to T6 or near the bones as they were cleared by the excavators. Eleven hyaenid coprolites, four of which were found in the vicinity of the *Dicerorhinus etruscus* skeleton, were also analyzed, along with the sediment on which they were found. This would avoid some of the drawbacks associated with deep lake deposits. Only 24 samples retained pollen grains, generally in excellent condition, but unfortunately often not very abundant. The other samples were sterile.

For the most part, therefore, the data remain qualitative. Forty-two taxa were identified: 18 trees, 22 herbaceous plants, one pteridophyte (monolete spore) and one moss (*Sphagnum*). Deciduous trees are constantly present: *Quercus*, *Corylus*, *Alnus* and, more sporadically, *Carpinus*, *Castanea*, *Fagus*, *Tilia*, *Betula*, *Fraxinus*, *Juglans*, *Ligustrum* and *Platanus*. *Hedera* is noted once. Conifers include *Pinus*, *Abies*, *Cedrus* and *Juniperus*. The presence of *Alnus*, *Fraxinus* and *Ligustrum* evokes the wettest substrates along the lake shore. Although the number of tree pollen grains is often higher than that of herbaceous plants, the role of grass should not be underestimated, as evidenced by the ever-present Poaceae, *Plantago*, *Rumex* and heliophilous trees such as *Betula*, *Juniperus*, *Alnus*, *Platanus* which take advantage of the openness of the landscape. Hygrophilous plants are rare. Only *Filipendula*, Cyperaceae and Typhaceae are regularly noted, but the lake shore seems to have been unfavorable to the establishment of an extensive megaphorbia. Climatic variations probably existed during the deposition of the layers containing the paleontological remains, but it is impossible to clearly identify and quantify them in a diagram.

Comparison with the Elhaï results remains problematic due to the sterility of the sediments in the upper five meters of the deep borehole. However, it appears that the pollen spectra of the fossiliferous sediments can only be more recent than those of the upper part of the Elhaï diagram. The climatic conditions encountered by the animals living in the vicinity of the Senèze maar were probably cool and temperate.

Non-mammalian Paleontology

The few fishes found at Senèze were reported by Gaudant (2024) in Chap. 7. A partial skeleton of a tench (*Tinca*) was found in the 1965 core, 12 m below the top. Our fieldwork recovered two partial skulls and several other fragments of the modern perch, *Perca fluviatilis*.

Birds are more frequent among the fossils from Senèze, as discussed in Chap. 8 by Mourer-Chauviré (2024). Seventeen species are represented, all from the older collections. Most of these are extant taxa, known from one to three specimens. Four extinct taxa are recognized: *Pavo bravardi*, *Tetrao* cf. *partium*, *Surnia robusta* and *C. corax antecorax*. Of these, the first is known by two specimens, the others only by one each. *P. bravardi* is a peafowl, closely related to the modern southeast Asian species *Pavo muticus*, the Green Peafowl. It is known from seven Plio-Pleistocene localities in France, Bulgaria, Greece and Moldavia; Senèze appears to be the youngest of these. *Tetrao partium* may be the direct ancestor of *Tetrao tetrix*, the Black Grouse, although known elements differ in proportions. The modern species inhabits the northern Palearctic, but not in the tundra. *T. partium* has been recovered from many European localities of Late Pliocene to Middle Pleistocene age. *Surnia robusta* is a very large hawk-owl, probably not the ancestor of the modern *Surnia ulula*. The fossil species is known only from a few localities in Hungary (MN 16-MNQ 20) and Ahl al Oughlam in Morocco (age estimated at 2.5 Ma). *Corvus corax*, the Common Raven, is widely distributed across the Holarctic province (and in the northern Ethiopian province), in a great range of climatic zones. *C. c. antecorax* was originally described as a distinct species but is now usually recognized as a subspecies of the modern taxon, smaller than most extant subspecies. It is known from MNQ 17 at Saint-Vallier to the late Middle Pleistocene, mainly in warmer intervals. The overall avifauna from Senèze is most similar to that from Saint-Vallier. They share five taxa: *Ciconia nigra*, *Haliaeetus albicilla*, *Pavo bravardi*, *Otis* sp. (size of *O. tarda*) and *C. corax antecorax*. The paleoenvironmental implications of the Senèze birds will be discussed with that of the mammals, below.

Mammalian Paleontology and Systematics

Drawing on the results from Chaps. 9–16 (see below), the Senèze mammalian fauna includes 39–42 species (one lagomorph, four rodents, 12 carnivores, two primates, 16 cetartiodactyls, 3–6 perissodactyls and one proboscidean). One of the cetartiodactyls, the bovid *Bison* (*Eobison*) sp., is known from a single donated specimen and must derive from a significantly younger horizon. Three of the equid perissodactyls are represented by only one or two specimens each and may not document distinct species or might derive from other sites. Less than half this number of mammalian species were recovered during our fieldwork: two rodents, two carnivores, seven cetartiodactyls (including *Bison*), three perissodactyls and one proboscidean; many taxa are known from only one to three specimens. We can examine the material of each order in more detail.

In Chap. 9, A. Argant (2024) revised the Senèze carnivores. The fossil carnivore remains unearthed during our excavations are extremely limited, in contrast to the material from the last century. The study of this recent material provided an opportunity to review all the carnivore remains at Senèze and to attempt to date the deposit using biochronology. Twelve species were identified: four Canidae (*Nyctereutes megamastoides*, *Vulpes alopecoides*, *Canis arnensis* and *Canis* sp. cf. *C. etruscus*); one Ursidae (*Ursus etruscus*); three Hyaenidae (*Pachycrocuta perrieri*, *Chasmaporthetes lunensis*, Hyaenidae indet.); and four Felidae (*Acinonyx pardinensis*, *Megantereon cultridens*, *Homotherium crenatidens* and *Dinofelis* sp.). Most of these were known from previous finds, including skeletons of *Homotherium crenatidens* and *Megantereon cultridens*. One tooth and a radius suggest a second canid, possibly referable to *C. etruscus*, while a single calcaneus indicates the presence of a *Dinofelis*. These are old finds but new taxa compared to the previous faunal list. Among the few specimens recovered recently is a small hyaenid fragmentary half mandible with p3. This specimen differs from the other two hyaenids known at Senèze and also from *Hyaena* and

Parahyaena; it must be identified for now simply as Hyaenidae indet., another new taxon for the site.

Our large-scale fieldwork has provided invaluable data on the fossiliferous levels in place, enabling their stratigraphy and paleoenvironments to be clarified. The Senèze carnivores probably span a long interval, but there is no evidence to suggest two distinct chronological periods. This assemblage of carnivores corresponds perfectly to the usual guild of Villafranchian predators. They regularly visited the banks of the Senèze maar to hunt the herbivores that came to the lake to drink. Carnivores put the Senèze deposit chronologically after that of Saint-Vallier and before those of Val d'Arno and Dmanisi, at around 2.2–2.1 Ma.

Guérin (2024) analyzed the remains of *Dicerorhinus etruscus etruscus* in Chap. 10. As documented here, over 530 fossils are known, representing at least 19 individuals (based on the number of mandibles recognized). Senèze is thus the richest locality for this species. Following a complete list of every Senèze specimen by element, each element is described, illustrated and measured, with comparisons to other species of the genus.

D. etruscus ranged from the second half of MN 16 (MN 16b: Les Etouaires, Montopoli) to about 0.5 Ma or later. Two subspecies are recognized: *D. etruscus etruscus* (MN 16b to MNQ 19) and the larger *D. etruscus brachycephalus* from the Epivillafranchian (e.g., Le Vallonnet) to the later Middle Pleistocene (e.g., Lunel-Viel and Mosbach). In turn *D. etruscus etruscus* is divided into two informal stages: stage I (MN 16b-MNQ 17 [Saint-Vallier]) and stage II (MNQ 18–19: Senèze, Upper Valdarno, Cueva Victoria). These two stages are distinguished by a series of morphological trends in the cranium (receding of the orbit and widening of the posterior part); mandible (thinning and shallowing of the corpus); dentition (reduction in width of the upper cheek-teeth, slight increase of hypsodonty); and post-cranium (e.g., slight reduction of the length of the long bones and the metapodials, slight changes in the proportions of the carpal and tarsal bones). *D. etruscus* was more cursorial than extant species of rhinoceros and probably ate soft foods like leaves, buds, young bark and twigs.

The Senèze equids are described, illustrated, documented with measurements and interpreted in Chap. 11 by Eisenmann and Delson (2024). First, they distinguish three genera of monodactyl equines by cranial proportions: *Equus*, *Plesippus* and *Allohippus*. The Senèze sample comprises over 170 fossils (from complete skeletons to isolated teeth) found by Philis, as well as two partial skeletons and some possibly associated hindlimb elements recovered during our excavations from stratigraphically low horizons in the southeastern sector. The great majority of them belong to *Allohippus senezensis senezensis*, which is intermediate in size and proportions between *A. senezensis guthi* from La Puebla de Valverde and *A. senezensis mygdoniensis* of Gerakarou.

Despite suggestions of similarity, it does not seem that *A. senezensis* and *A. stehlini* are closely related, even if one MC III does resemble *A. stehlini* from the Upper Valdarno. The geographic and possible temporal occurrences of these taxa and other forms of *Allohippus* are shown in a chart; *A. senezensis* ranges from MNQ 17b through MNQ 18.

One tooth and 16 postcranial elements (including one donated to the team in 2002 by A. Consigny) are identified as the much larger *A. major*, although the lack of information on its cranial proportions renders the generic attribution uncertain. This taxon is rare but has been recognized from early MNQ 17 through MNQ 19 and perhaps in the Epivillafranchian. It is broadly similar in size to but morphologically distinct from *Equus (Sussemionus)* sp.

In addition, the Senèze equid collection includes two first phalanges larger than the average for *A. stenonis vireti* of Saint-Vallier; one MT III and one first phalanx resembling the small ?*Allohippus* sp. of Pyrgos; the *A. stehlini*-like MC III from the Upper Valdarno (both the latter sites are significantly younger than Senèze) and some caballine teeth and metapodials. The caballine material is presumably from a potential younger horizon at Senèze such as that which yielded the specimen of *Bison* (*Eobison*) or are wrongly labeled as to the locality of origin. The five distinctive *Allohippus* specimens might also derive from other localities or represent extremely rare Senèze taxa.

In Chap. 12, Faure and Guérin (2024) discuss the rare fossils of *Sus strozzii* from Senèze. No material was recovered during our fieldwork, but the nearly complete skeleton of a young male was found in 1937 and is exhibited in the Naturhistorisches Museum Basel (which also has a partial scapula of an adult). The skeleton was studied and illustrated by Azzaroli (1954), and this chapter provides measurements and scatter-plots with comparison to congeneric species. *Sus strozzii* is large and morphologically more similar to the modern southeast Asian long-snouted *S. verrucosus* and *S. barbatus* than to the extant Eurasian *S. scrofa*. The cheek teeth of *S. strozzii* are wider and lower, its molars less complicated and covered with thicker enamel than those of *S. scrofa*, and its limb bones are stockier and relatively short [which Azzaroli (1954) suggested could be an adaptation to soft and marshy soils]. The Senèze specimen retains I3, which has been lost in the younger sample from the Upper Valdarno, suggesting a possible temporal trend. *S. strozzii* is known across Europe and in the Middle East from MNQ 16 up to the early Middle Pleistocene.

As reviewed by Valli (2024) in Chap. 13, four species of deer are known from Senèze: *Eucladoceros ctenoides senezensis*, *Metacervoceros rhenanus philisi*, *Croizetoceros ramosus minor* and *Libralces gallicus*. All four taxa have Senèze as their type locality. The remains of the last two are quite rare, while those of the other two are by far the most common mammals of the locality; they were also the only cervids recovered during our fieldwork. Valli discusses the morphology and morphometrics of these remains, as well as their taxonomy.

M. rhenanus philisi is a cervid slightly larger than a European fallow deer, but its antlers have only three points on their beams. Like the fallow deer, the Philis deer lacks upper canines. Specimens found by our team belong to at least three different individuals. Their size and morphology corresponds to that described in Heintz (1970), a monograph devoted to the Villafranchian cervids of France and Spain. *E. ctenoides senezensis* is larger; its size is comparable to that of a modern red deer, from which it differs mainly in its antlers. The numerous fossils excavated during our research include one almost complete skeleton and remains of at least two other individuals. Both these subspecies help to define the late Villafranchian in several European countries, including France, Spain and Greece.

As to the two other species, *L. gallicus* represents the oldest elk currently known from Europe. It is characterized by slender limbs and, above all, by its beams, very elongated and horizontally arranged on both sides of the animal. In France, it is only known from the late Villafranchian. *C. ramosus minor* is the smallest deer from Senèze; its dimensions are a little larger than those of a roe deer. It is characterized by its antlers; all the tines grow in the frontal part of the beam and face forward.

Crégut-Bonnoure (2024) analyzed the bovids from Senèze in Chap. 14. This diverse group is the second most common family among the Senèze mammals. The Bovidae have undergone multiple radiations over geological time, making them particularly interesting for biochronological analysis. In the Plio-Pleistocene, the paleoenvironmental changes linked to the onset of climatic instability led to new radiations and major faunal movements, resulting in the establishment of new guilds. The family is represented at Senèze by four tribes (Antilopini, Caprini, Ovibovini and Bovini), and the aim of this analysis is to present and discuss their richness and diversity. Our excavations yielded only seven fragmentary remains, and unfortunately their incomplete nature makes it impossible to compare them significantly with the material from the previous collections.

Eight bovid species were described in the last century. In this chapter, they were all revised, except *Leptobos*, which allowed the correction of some identifications, including many from other localities. As a result, eleven taxa are now identified: *Gazellospira torticornis*, *Procamptoceras brivatense*, cf. *Hemitragus* sp., *Gallogoral meneghinii*, *Pliotragus ardeus*, *Megalovis latifrons*, *Ovis claudiusguerini* nov. sp., *Leptobos etruscus*, *L. furtivus*, *Bison* (*Eobison*) sp. and Bovidae indeterminate. Cf. *Hemitragus* sp. and Bovidae indeterminate are new for the locality, while the previous *Ovis* sp. is now identified as a new species dedicated to the late Claude Guérin. *Bison* (*Eobison*), known from a single specimen donated to the team and whose stratigraphic position is unknown, is younger than the other species and must derive from an otherwise unknown horizon dating to the end of the Early Pleistocene.

Morphological variability can be observed in *Pliotragus*, *Gallogoral* and *Megalovis*, suggesting an evolution of these taxa over time and the possibility of more than one fossiliferous level. However, these possible distinct levels would not be very distant in time (as opposed to that yielding the *Bison*), and it is not possible to distinguish successive bovid assemblages. The bovid fauna from Senèze is clearly younger than those from MN 16 and MNQ 17 localities from the Massif Central.

In Chap. 15, Ménouret and Guérin (2024) discuss the proboscidean from Senèze, *Mammuthus meridionalis meridionalis*. They begin with a taxonomic review of the genus and the species, which ranges in France from MNQ 17 into the early Middle Pleistocene. It is a tall species, with male height at the shoulder between 3.5 and 4.2 m and weight mostly 7–10 tons. The skull presents a concave forehead, inflated parietals, a sharp occipital crest, long and low mandibular corpus and elongated symphysis. The tusks are stocky, curving slightly upward and inward; the M3 is thick and moderately hypsodont.

Following a full list of specimens (including one partial tusk recovered in our fieldwork), each element is described in some detail and compared morphologically and morphometrically to those of other populations of the species (and in some cases to other species of the genus). There is good agreement in general between the Senèze fossils and those of this subspecies from other middle-late Villafranchian sites. Interestingly, two M3s from Colle Mancino (estimated age ca. 1.2 Ma) are the most similar to those from the original skull excavated by Boule (1892) and on exhibit in Paris (see Faure et al. 2024), the only remaining elements of that specimen. The implications of this similarity are unclear, but it seems that this morphology first appeared at Senèze.

Delson (2024) reviews the few Senèze primates in Chap. 16. There are only two specimens known, both cercopithecids found by Philis: a fragmentary ulna and a nearly complete skull. The ulna is referred to *Macaca sylvanus*, of which several subspecies may have sequentially inhabited Europe from the Late Miocene into the Late Pleistocene. The female skull was first described by Depéret (1928), who made it the holotype of *Dolichopithecus arvernensis*, which he thought was a colobine related to *D. ruscinensis* Depéret, 1889 from the mid-Pliocene of Perpignan. Necrasov, Samson and Rădulesco (1961) described a fragmentary face of a cercopithecid from the newly-studied Romanian site of Valea Grăunceanului (usually abbreviated as Grăunceanu) as *Paradolichopithecus geticus*. They thought this taxon also represented a colobine, similar to *D. arvernensis*. Later in the 1960s, several authors recognized that these taxa were synonymous and actually cercopithecines. Thus the Senèze skull is now termed *Paradolichopithecus arvernensis*; additional samples of this species have been described from Grăunceanu, and from Vatera-F, Karnezeika and Dafnero-3 (all in Greece), dating to MNQ 17 (and perhaps 18). Other specimens have been referred to further species of this genus. Emended and differential diagnoses and a general description are presented for *Paradolichopithecus*, and an emended diagnosis for *P. arvernensis*. The Senèze skull is described and illustrated in detail.

There are two major systematic questions regarding *Paradolichopithecus*: is it more closely related to *Macaca* or *Papio*?; and is it a synonym of the east Asian *Procynocephalus* Schlosser, 1924? The arguments on both sides of these questions are presented, and (at least in a preliminary decision), *Paradolichopithecus* is considered a distinct genus of the Macacina. A longer analysis of these problems will be presented along with descriptions of the Grăunceanu and Vatera-F samples in a forthcoming study.

Analytical Interpretations of the Senèze Fauna

Two chapters analyze broader aspects of the Senèze assemblage, taphonomy and biochronology. In addition, three other related topics can be discussed on the basis of the entire volume's content: composition and proportions of the fauna; assessment of paleoenvironment; and site formation.

Biochronology is discussed by a group of authors in Chap. 17 (Crégut-Bonnoure et al. 2024). Following a brief review of stratigraphy, geochronology and palynology (as discussed above), the implications for relative dating of each mammalian group and the birds are presented, along with a full list of taxa present, separated by origin (Philis and other twentieth century finds vs. our collection). The faunal list is then compared to those from twelve other Massif Central localities dating from MN 16 through MNQ 20, plus Saint-Vallier. It is clear that Senèze falls between sites placed in MNQ 17 and MNQ 19. The lack of *Mammut borsoni* (only MN 16), *Tapirus arvernensis* (mainly MN 16) and *Anancus arvernensis* (MN 16 and MNQ 17) places Senèze younger than MNQ 17. The carnivores *Panthera gombaszogensis* and *Pachycrocuta brevirostris* also do not occur at Senèze: the former appears at Slivnitsa (Bulgaria), which may be slightly younger than Senèze, while the latter only occurs in the Epivillafranchian. Subspecies or "stages" in the lineages of ursids, felids, cervids, bovids, equids and rhinocerotids also set Senèze apart from earlier and later sites. A further faunal comparison was made with a selection of Spanish and Italian sites of MNQ 17 and MNQ 18, as well as with Tegelen (the Netherlands) and Erpfingen (Germany), both MNQ 18. Senèze again shares taxa with early MNQ 18 sites, especially.

A number of the studied localities have been dated recently, so their ages could be compared with those for Senèze. Sites yielding *Anancus arvernensis* are considered to be placed in MNQ 17. These include Chilhac and Le Coupet (France), dating to 2.285 ± 0.046 Ma and 2.274 ± 0.032 Ma and Coste San Giacomo and Pantalla (Italy), with ages of 2.23 ± 0.03 Ma and 2.23 ± 0.18 Ma, respectively. These dates, all preceding 2.20 Ma, combined with the estimate of 2.20 Ma for the base of the Senèze sequence, suggest that the boundary between MNQ 17 and MNQ 18 is close to 2.2 Ma. In turn, Blassac-La Girondie (France) was dated to 1.946 ± 0.029 Ma by Paquette et al. (2021), who considered it to fall in MNQ 19 on the basis of three distinctive cervids not found at Senèze. But those species are not diagnostic of MNQ 19, leading Crégut-Bonnoure et al. (2024) to suggest the site might be of mid MNQ 18 age. They accepted Dmanisi (Georgia) as very late in MNQ 18, with a date of 1.80–1.76 Ma, suggesting that the end of MNQ 18 was ca. 1.75–1.70 Ma.

The composition of the Senèze mammalian assemblage has not been discussed earlier in this volume. Many authors provide lists of all specimens in their taxa of interest, allowing the number of separate fossils to be estimated. Specimens with the same catalogue number or those known to be part of a single individual were counted only once. Valli (2024) did not review the large number of cervids recovered prior to our research, so Heintz (1970) was consulted to estimate similar counts for *Eucladoceros ctenoides senezensis* and *Metacervoceros rhenanus philisi*. The total summed the number of specimens with antlers (and separate skulls), number of teeth (sum of largest numbers for any position in upper and lower permanent and deciduous rows), and half the sum of estimated numbers of each major postcranial element tabulated by Heintz. The resulting estimates by species, family and order are presented in Table 19.1, and values by family are illustrated in Fig. 19.1. The total of 2200+ mammal specimens is clearly dominated by cervids (over 63%). *Metacervoceros rhenanus philisi* and *Eucladoceros ctenoides senezensis* each are represented by more specimens than any other order. Bovidae, Rhinocerotidae and Equidae are well-represented (with roughly 13.8, 9.3 and 8.9 %, respectively), but far behind the two cervids. Other families and orders are rare at Senèze, never reaching 2% of the mammal specimens. Birds account for slightly <1% of the tetrapod total. These frequencies will be discussed again when considering paleoenvironmental indications below.

An alternative way of looking at the size of the total Senèze collection is to consider the number of catalogue numbers at each major institution. In some cases multiple

Table 19.1 List of numbers per taxon for the fauna from Senèze

Taxon	FGD 2001–06	Previous finds	Total by species	FGD 2001–06	Previous finds	Total by family	Percent by family	FGD 2001–06	Previous finds	Total by order	Percent by order
Mammalia								140	2098	2238	
Lagomorpha									10	10	0.45
Leporidae					10	10	0.45				
Oryctolagus cf. *lacosti* (number estimated—see Masson 1943)		10	10								
Rodentia								2	22	24	1.07
Sciuridae					1	1	0.04				
Sciuridae indet. aff. *Eutamias*		1	1								
Arvicolidae				1	21	22	0.98				
Mimomys pliocaenicus-M. ostramosensis group		10?	10								
Mimomys pitymyoides (group)	1	11?	12								
Hystricidae				1		1	0.04				
Hystrix refossa	1		1								
Carnivora								2	42	44	1.97
Felidae				1	16	17	0.76				
Dinofelis sp.		1	1								
Acinonyx pardinensis	1	3	4								
Homotherium crenatidens		7	7								
Megantereon cultridens		5	5								
Hyaenidae				1	8	9	0.40				
Chasmaporthetes lunensis		3	3								
Pachycrocuta perrieri		5	5								
Hyaenidae indet. (not counting coprolites)	1		1								
Canidae					12	12	0.54				
Vulpes alopecoides		3	3								
Nyctereutes megamastoides		4	4								
Canis arnensis		3	3								
Canis sp.		2	2								
Ursidae					6	6	0.27				
Ursus etruscus		6	6								
Primates									2	2	0.09
Cercopithecidae					2	2	0.09				
Paradolichopithecus arvernensis		1	1								
Cf. *Macaca sylvanus*		1	1								
Cetartiodactyla								126	1598	1724	77.03
Suidae					2	2	0.09				
Sus strozzii		2									
Cervidae				119	1295	1414	63.18				
Metacervoceros rhenanus philisi	40	640	680								
Croizetoceros ramosus minor		12	12								
Eucladoceros ctenoides senezensis	59	441	500								
Cervalces gallicus		2	2								
Cervidae indet	20	200	220								
Bovidae				7	301	308	13.76				
Gazellospira torticornis		37	37								
Procamptoceras brivatense		16	16								
Ovis claudiusguerini		1	1								
Cf. *Hemitragus* sp.		3	3								
Megalovis latifrons	2	13	15								

(continued)

Table 19.1 (continued)

Taxon	FGD 2001–06	Previous finds	Total by species	FGD 2001–06	Previous finds	Total by family	Percent by family	FGD 2001–06	Previous finds	Total by order	Percent by order
Pliotragus ardeus	1	55	56								
Gallogoral meneghinii	1	90	91								
Leptobos	3	84	87								
Bovidae indet		2	2								
Perissodactyla								9	398	407	18.19
Equidae				4	195	199	8.89				
Allohippus senezensis senezensis	3	174	177								
Allohippus major	1	16	17								
Possible additional equids known from phalanges and/or metapodials:											
(1) Larger than the average for *Allohippus stenonis vireti*		2	2								
(2) Cf. Valdarno *Allohippus stehlini*		1	1								
(3) Cf. small ?*Allohippus* of Pyrgos		2	2								
Rhinocerotidae				5	203	208	9.29				
Dicerorhinus etruscus etruscus	5	203	208								
Proboscidea								1	26	27	1.21
Elephantidae				1	26	27	1.21				
Mammuthus meridionalis meridionalis	1	26									
Aves									21	21	[0.93][1]
Ciconiiformes											
Ciconiidae											
Ciconia nigra		1									
Anseriformes											
Anatidae											
Tadorna cf. *ferruginea*		1									
Anas clypeata		2									
Aythya sp.		1									
Mergus cf. *merganser*		1									
Mergellus sp. size of *M. albellus*		1									
Anatidae indet		1									
Accipitriformes											
Accipitridae											
Haliaeetus albicilla		1									
Buteo sp., size of *Buteo buteo* or *Buteo lagopus*		1									
Galliformes											
Tetraonidae											
Tetrao cf. *partium*		1									
Phasianidae											
Pavo bravardi		1									
Alectoris cf. *graeca*		3									
Gruiformes											
Otididae											
Otis sp. size of *Otis tarda*		1									
Strigiformes											
Strigidae											
Bubo sp.		2									
Surnia robusta		1									

(continued)

Table 19.1 (continued)

Taxon	FGD 2001–06	Previous finds	Total by species	FGD 2001–06	Previous finds	Total by family	Percent by family	FGD 2001–06	Previous finds	Total by order	Percent by order
Passeriformes											
Turdidae											
Cf. Turdidae, size of *Turdus merula*		1									
Corvidae											
Corvus corax antecorax		1									
Osteichthyes								5	1	6	
Cypriniformes											
Cyprinidae											
Tinca sp.		1									
Perciformes											
Percidae											
Perca fluviatilis	5										

[1]Percentage of birds among all tetrapods (birds + mammals). FGD: the Faure/Guérin/Delson fieldwork of 2001–2006

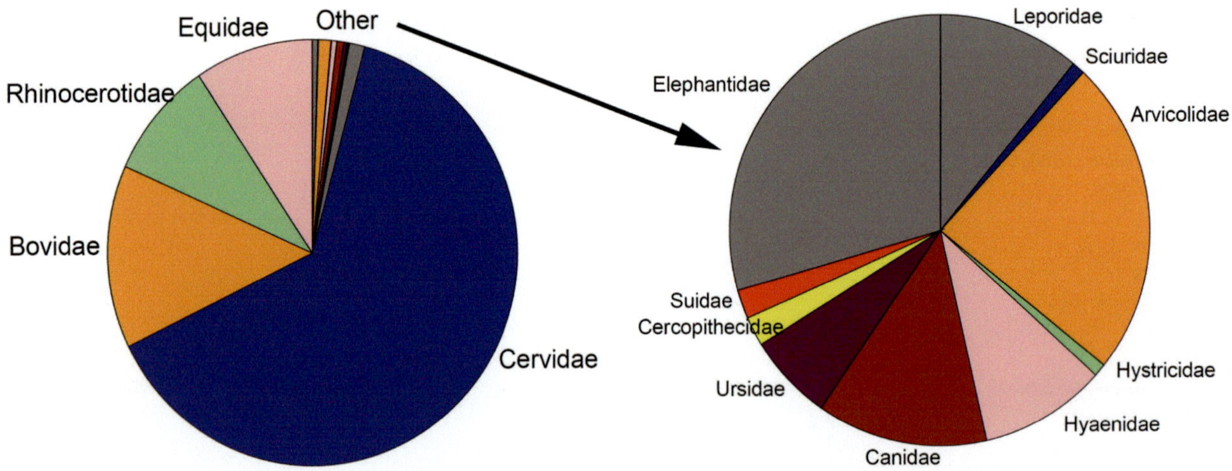

Fig. 19.1 Frequency of occurrence of mammalian families and two higher taxa in the Senèze assemblage. Data from Table. 19.1. This figure was produced with the PAST statistical program (v 4.03), see Hammer et al. (2001)

numbers were given to single fossil individuals, so the total is greater than that given in Table 19.1. There are 1860 numbers for Senèze mammals (and 20 for birds) in the Naturhistorisches Museum Basel; 826 (of which 63 are indeterminate) in the Université Claude Bernard-Lyon 1 and about 100 in the Institut de Paléontologie, Muséum national d'Histoire naturelle, Paris. Our sample from 2001–2006 includes 1314 numbers for mammals (identified or indeterminate), plus 20 coprolites and 5 fish; every element of a partial skeleton has a separate number.

There are also mounted skeletons (see Faure et al. 2024) of a number of taxa in the Basel and Lyon collections: *Homotherium crenatidens*, *Megantereon cultridens* (now dismounted), *Sus strozzii*, *Metacervoceros rhenanus philisi* (2), *Eucladoceros ctenoides senezensis* (2), *Cervalces gallicus*, *Gallogoral meneghinii* (2), *Leptobos etruscus, Allohippus senezensis* (3, plus one unmounted) and *Dicerorhinus etruscus* (2). Our fieldwork recovered four additional partial skeletons (with more than a dozen elements each): *Eucladoceros ctenoides senezensis, Allohippus senezensis* (2) and *Dicerorhinus etruscus*.

Paleoenvironment

Most of the paleontological contributions have provided some information on the paleoenvironmental implications of the taxa studied. We will bring these together and then attempt to summarize an overall conclusion.

J. Argant (2024) analyzed the pollen collected at various points across the site, especially in contact with skeletons, confirmed by the content of hyaenid coprolites. Leafy trees dominated the pollen assemblages, especially *Quercus*, *Corylus* and *Alnus*; conifers were less common, including *Pinus*, *Abies*, *Cedrus* and *Juniperus*. Non-arboreal pollen was often less frequent than arboreal, but Poaceae (grasses) is always present and *Plantago* and *Rumex* are also found on occasion. This led her to interpret an environment with both open spaces and wooded areas, with climate relatively temperate or cool temperate.

Of the fish, assuming the fossils had similar adaptations to their extant congeners, *Tinca* (found in the core) lives on muddy and grassy bottoms in quiet rivers and ponds, while *Perca* (excavated in lakeshore deposits) is found in water plant communities of rivers and lakes.

A number of birds provide information about their habitat preferences; each species is known by only 1–3 specimens. Many taxa live on the water or its banks, in forests with lakes or gallery forest along watercourses. *Tetrao* and *Surnia* inhabit forest edges or lower altitude forests. *Alectoris* and *Otis* are found in more open environments, grassy plains or bushland. Some forms are migratory and winter in the tropics, south and southeast Asia and eastern Africa. Senèze would have been more humid and wooded than Saint-Vallier, with no indication of a cooler climate than today's.

Among the mammals, deer, especially elk, generally prefer wooded environments, although *Eucladoceros*, like the red deer, had a flexible diet which allowed it to also inhabit more open environments. *Sus strozzii* indicates a forested area under a wet and probably warm climate. Bovids are almost entirely open-country species, with a wide range of climatic preferences. Short muzzles and deep, not very robust, metapodials like those of *Allohippus senezensis* are usually found in equids living in rather dry conditions like extant *Equus hemionus* and *E. przewalskii*. *Dicerorhinus etruscus* probably preferred wooded savanna and moist climate. *Mammuthus meridionalis* also preferred wooded savanna, but under temperate or cooler conditions in a relatively dry environment. The modern *Macaca sylvanus* feeds and travels on the ground in open to mixed woodland and into grassland but sleeps in trees. Its environments are wet and cold in winter, with warm and dry summers. *Paradolichopithecus arvernensis* postcrania (from other sites) are similar to those of *Papio* and *Mandrillus*, suggesting a high degree of terrestriality, perhaps in open terrain to woodland. Carnivores often have a wide range of environmental tolerance, although the felids found at Senèze also occur in other sites which were close to large streams, lakes or fluvio-paludal environments.

As discussed, deer are the most common mammals at Senèze, with antelope, rhino and horse far behind. This suggests that the environments preferred by these animals would likely be the most common around the maar lake. Deer prefer wooded habitats and rhino wooded savanna, but bovids are open-country animals, and *Allohippus* is suggested to inhabit dry (and open) areas. Of the rarer species, *Sus* prefers forest, *Mammuthus* wooded savanna, the primates open country but perhaps with trees in which to sleep. This suggests a mixture of habitats with forest to woodland dominant. Perhaps more open steppe/savanna existed outside the maar itself. This agrees well with the palynological results, with leafy trees dominant, conifers and grasses present but less frequent. The climate would presumably have been warmer than today, and moist especially close to the lake.

Several MIS cycles spanned the 2.20–2.08 interval, and they would also have impacted the climate and vegetation around the maar. According to Channell et al. (2020) following dates in Lisiecki and Raymo (2005), the Feni correlated to MIS 81 (quite warm), MIS 80 (cooler) and into MIS 79 (warm). The broadly warm interval continued to the Huckleberry Ridge, when a sharp cooling occurred. The Réunion was also cool (MIS 84), followed by a warmer interval (MIS 83) and a cold phase (MIS 82) before the Feni. This implies that the two intervals when we suggest fossils were found in our fieldwork, ca. 2.20–2.18 Ma and 2.10–2.08 Ma, were both warmer times linked to MIS 83 and MIS 81, with the strongly cold MIS 82 between them and the cool MIS 84 and MIS 80 on either side. Debard (2024) has proposed that the slope deposits in Trenches 5–7 represented a colder interval, which might instead be correlated to MIS 84 and the Réunion excursion, which was not captured in the paleomagnetic sampling.

Taphonomy

Fernández-Jalvo et al. (2024, Chap. 18) report research on the taphonomic history of the Senèze large-mammal assemblage recovered in our excavations and some older finds housed in the paleontology collection of the Laboratoire de Géologie, Université Claude Bernard-Lyon 1. Five hundred twenty-seven fossils from the 2003–2006 seasons of our work were studied and classified according to taxon, age, body part and size class. Fractures, peeling and other surface modifications such as punctures, gnaw marks, cracks, friction marks (such as trampling), soil corrosion and pathology were noted for each element. Experiments were undertaken to replicate deformation and compression damage.

Sediment analysis on fossils revealed slight differences in sediment distribution, oxide deposits, and coloration, suggesting varying depositional conditions, possibly related to seasonal changes. Manganese staining on many bones indicates immersion in damp sediment. The distribution of this manganese on fossils, with most bones completely black, except for the stain-free articular surfaces, indicated that skeletal elements were in anatomical connection during deposition and possibly burial.

Several individuals exhibit almost complete and anatomically connected skeletons, suggesting a fast burial and/or prolonged period of permanent water immersion. This environment likely facilitated quicker decay of ligaments, resulting in rib and vertebrae detachment before limbs. A number of individuals showed significant pathology indicating arthritis in the limbs and especially distal extremities (carpals, tarsals, metapodials and phalanges).

The formation of the Senèze site was attributed mainly to gravitational forces, with blocks falling down the inner slope, and colliding with skeletonized carcasses, causing fracture and compression. The absence of hydrodynamic abrasion, size selection, or piles of fossils, together with diagenetic breakage and fragments still in anatomical connection reinforces the idea that re-sedimentation or reworking did not occur. There is often minimal bone dispersion, indicative of underwater decay. Plastically deformed long bone shapes, cracks with curled edges (associated with humid environments), manganese and iron oxides, fossils arranged on site both horizontally or almost vertically, together with the absence of rodent chewing or weathering support an underwater burial process. Only in very rare cases some fossils from the upper part of the excavation (upper part of the slopes) show signs of weathering, rodent chewing and the presence of hyena coprolites, indicating that these bones were exposed subaerially on the shore of the lake.

Patterns of bone breakage, compression and deformation, most likely caused by falling blocks and boulders due to sediment compaction and mudflows, were investigated. There was no clear evidence for carnivore activity, with potential mimics caused by abiotic means. Pseudo-carnivore tooth marks caused by falling block impacts on the bone surfaces, soil corrosion and pressure marks on coarse sediments were discussed in detail, emphasizing the need for careful differentiation of such marks according to their origin. There is also no evidence for human activity at the site, which is reasonable as it is older than Dmanisi and all sites in western Europe with signs of human occupation.

Site Formation

One of the primary goals of the taphonomic research was to determine the most likely processes which formed the Senèze fossil accumulations. Paleontologists often assume that assemblages of large mammal bones may be the result of carnivore kills or geological events such as flooding. There do not seem to be any specific suggestions about the origin of the Senèze assemblage until 1989, when Couthures (1989: 210) proposed that "Only sudden and repeated asphyxia could have been responsible for the particular ecological disaster of Senèze and for the abnormal mixing of animal fossils." It is not clear why he thought this or what he meant by "abnormal mixing".

The next alternative was proposed by Debard in Delson et al. (2006, p. 287): "mud or debris slides down the inner slopes from the local summits (possibly related to local faulting evidenced in the trench sequences) which engulfed large animals either while still alive or soon after death. The resulting cadavers could have been washed into the lake margin before they could be preyed upon by any of the local carnivores and would have remained in anatomical connection at least during the early phases of preservation."

The big question is why so many almost complete skeletons have been recovered from the crater lake at Senèze, especially without any indication of carnivore ravaging. A. Argant (2024) has suggested that predators attacked herbivores coming to drink at the lake, either on the shore or as they moved down the hillside. But this idea is unlikely given the lack of carnivore damage (rather only pseudo-carnivore tooth marks as noted above) on any bones studied by Fernández-Jalvo et al. (2024).

Instead, both Debard (2024) and Fernández-Jalvo et al. (2024) have proposed that the larger mammals fell into the lake and drowned when they were unable to climb out. Debard offered slightly different scenarios in colder or warmer intervals. During the cold phases, dead or still living animals could be dragged by solifluction to be trapped in the icy waters of the lake. The association of remains with pebbles and blocks (frequent in the southeastern sector, Trenches 5–7) may reflect very cold conditions where the surface of the lake, partly frozen and covered with stony pavement, could yield under the weight of heavier animals, causing their death by drowning. During the warmer phases, as seen in the western lake beds (Trenches 1–3), they could simply slip on the slope while coming to drink; because the stiffness of the sedimentary layers resulted in bluffs 0.5–1.5 m high at the edge of the lake and due to the instability of the waterlogged sediments, the animals could not get back up onto the bank and drowned. Fernández-Jalvo et al. (2024) emphasized the common presence of joint pathologies on many of the associated skeletons, which would have made it harder for such animals to extract themselves from the lake.

Among his several suggestions, A. Argant (2024) proposed a slightly different version of the same general idea, opining that some herbivores fled instinctively away from carnivores down the slopes and into the water. Unlike in Africa, there was no fear of crocodiles to deter them. Some would have swum to safety, others would have drowned. After a while their corpses would have risen back to the surface and drifted according to the direction of wind and current. Some of these would end up on the shore and be accessible to scavengers. Others would have sunk once again to the bottom of the lake and been preserved.

Fernández-Jalvo et al. (2024) compared Senèze with other maar-related sites such as Eckfeld, Messel and Camp dels Ninots, where gas emissions from volcanic lakes and fumaroles have been suggested as a possible cause of mass mortality and subsidence. In the case of Senèze, rare gas emissions could have had an influence and could have caused some of the deaths, but its maar was characterized by abrupt slopes, highly plastic sediments that allow slumping and mudflows and most of the skeletons with varying degrees of pathology. This pattern suggests that at Senèze, the instability of the sediment could have caused animals to fall into the water and slide down the slope, making it difficult for them to escape from sinking underwater. The herbivore remains decomposed under water without emitting odors that could attract scavengers. Falling boulders and landslides were frequent, as evidenced by the steep inclination of the strata. The site acted as a natural trap, active for decades or centuries at a time, during multiple intervals over perhaps 100,000 yr at the broad lakeshore area, which explains the high number of complete skeletons, without the need of invoking catastrophic events. The faunal assemblage represents a past biocoenosis shaped by the metabolic necessity of drinking and the (in)ability to escape, rather than by selective collection. This scenario is also consistent with the complementary hypothesis that mud or debris engulfed large animals at Senèze, preserving skeletons in anatomical association during early phases of fossilization.

Taken together, these suggestions envisage various large mammals, especially individuals with limb joint pathologies, drowning under a range of circumstances and their cadavers being preserved in anatomical connection to be discovered by P. Philis and later paleontologists searching at Senèze.

Epilogue

The Franco-American Senèze Research Project began as the result of a recognition among the three team leaders (and now editors of this book) that Senèze was a famous but poorly documented site typical of the western European late Villafranchian. We planned the research to throw greater light on the local stratigraphy, provide chronometric dates, recover additional specimens of rare taxa and revise the systematics of known species, understand the taphonomy and site formation processes and reconstruct the paleoenvironment. The chapters in this book, summarized above, show that almost all these goals were realized. The stratigraphy of two main sectors of the maar was clarified, and a combination of argon-argon ages and magnetostratigraphic sequences allowed the fossiliferous layers sampled by the team to be precisely dated. No new primates were found, and despite screen washing of several tons of sediment, only two micromammal teeth were recovered. But a number of well-preserved partial skeletons were excavated and two taxa new to the site were identified: *Hystrix refossa* and Hyaenidae indet. The fauna was demonstrated to be homogeneous, refuting the previous suggestion that two distinct assemblages could be identified. On the other hand, one specimen of *Bison* (*Eobison*) sp. collected on the surface from an unknown horizon by local amateur A. Consigny did demonstrate the local presence of Epivillafranchian or later sediments. Five other taxa were identified through systematic revision, and a previously known but indeterminate species was named for one of the team leaders: *Ovis claudiusguerini*. The total assemblage fits neatly between those of MNQ 17 (middle Villafranchian) and later MNQ 18-MNQ 19. Senèze is well justified to be considered the primary reference locality for the earlier part of MNQ 18 and the start of the late Villafranchian.

Findings from taphonomy and stratigraphy clarified that the many associated skeletons were probably preserved as a result of animals falling into the lake under different circumstances and drowning before their bones could be disturbed. Cervids were by far the most common animals recovered at Senèze, followed more distantly by bovids, rhinocerotids and equids. Combining information about the paleoenvironment from the likely habitat preferences of these groups and the results of palynological analysis indicates that a wide range of habitats were present in the Senèze neighborhood: forest, woodland, gallery forest and open plains, but where each of these were located is not yet known.

In sum, the project has been highly successful, justifying the financial and logistical support provided by many agencies and people. We hope that we have made Senèze more accessible and comprehensible and that our data will be of use to many colleagues in the years to come.

Acknowledgments Andrea Valli supervised the development of our catalogue and database of all specimens collected during our fieldwork and worked to integrate them into the paleontology collection of the Laboratoire de Géologie, Université Claude Bernard-Lyon 1. Abel Prieur and Emmanuel Robert provided a catalogue of the Senèze fossils already in that collection. Loïc Costeur (Naturhistorisches Museum Basel) and Christine Argot (Institut de Paléontologie, Muséum national d'Histoire naturelle, Paris) provided helpful assistance with material in the collections in their care. We are grateful to all of them. A full list of acknowledgments for the Senèze fieldwork is presented in Chap. 1.

References

Argant, A. (2024). The Carnivores of Senèze (Haute-Loire) . In E. Delson, M. Faure, & C. Guérin (Eds.), *Senèze: Life in Central France two million years ago. Paleontology, geochronology, stratigraphy and taphonomy* (pp. 165–243). Cham (Switzerland): Springer.

Argant, J. (2024). Palynological analysis of the fossiliferous layers of the late Villafranchian deposit of Senèze (Domeyrat, Haute-Loire, France). Taphonomy and paleoenvironmental data. In E. Delson, M. Faure, & C. Guérin (Eds.), *Senèze: Life in Central France two million years ago. Paleontology, geochronology, stratigraphy and taphonomy* (pp. 123–140). Cham (Switzerland): Springer.

Azzaroli, A. (1954). Filogenesi e biologia di *Sus strozzii* e di *Sus minor*. *Palaeontographia Italica, 48*(1952), 41–76.

Boule, M. (1892). Découverte d'un squelette d'Elephas meridionalis dans les cendres basaltiques du volcan de Senèze (Haute-Loire). *Comptes Rendus hebdomadaires des séances de l'Académie des sciences, 115*, 624–626.

Channell, H. E. T., Singer, B. S., & Jicha, B. R. (2020). Timing of Quaternary geomagnetic reversals and excursions in volcanic and sedimentary archives. *Quatermary Science Reviews, 228*, 106114.

Couthures, J. (1989). The Senèze maar (French Massif Central): Hypothesis regarding a catastrophe occurring about 1.5 million years ago. *Journal of Volcanology and Geothermal Research, 39*, 207–210.

Crégut-Bonnoure, E. (2024). The Bovidae (Mammalia, Cetartiodactyla) from Senèze. In E. Delson, M. Faure, & C. Guérin (Eds.), *Senèze: Life in Central France two million years ago. Paleontology, geochronology, stratigraphy and taphonomy* (pp. 433–585). Cham (Switzerland): Springer.

Crégut-Bonnoure, E., Guérin, C., Argant, A., Argant, J., Debard, E., Delson, E., et al. (2024). Biochronology of the Senèze faunal assemblage. In E. Delson, M. Faure, & C. Guérin (Eds.), *Senèze: Life in Central France two million years ago. Paleontology, geochronology, stratigraphy and taphonomy* (pp. 633–652). Cham (Switzerland): Springer.

Debard, E. (2024). Geological study of the paleontological site of Senèze (Domeyrat, Haute-Loire, France). In E. Delson, M. Faure, & C. Guérin (Eds.), *Senèze: Life in Central France two million years ago. Paleontology, geochronology, stratigraphy and taphonomy* (pp. 37–82). Cham, Switzerland: Springer.

Delson, E., Faure, M., Guérin, C., Aprile, A., Argant, J., Blackwell, B., et al. (2006). Franco-American renewed research at the Late Villafranchian locality of Senèze (Haute-Loire, France). *Courier Forschunginstitut Senckenberg, 256*, 275–290.

Delson, E. (2024). Primates from Senèze. In E. Delson, M. Faure, & C. Guérin (Eds.), *Senèze: Life in Central France two million years ago. Paleontology, geochronology, stratigraphy and taphonomy* (pp. 607–632). Cham (Switzerland): Springer.

Delson, E., Nomade, S., Sen, S., Debard, E., Pastre, J.-F., Bahain, J.-J., et al. (2024). Geochronology of Senèze: $^{40}Ar/^{39}Ar$ dating and magnetostratigraphy, with notes on an ESR/U-series dating attempt. In E. Delson, M. Faure, & C. Guérin (Eds.), *Senèze: Life in Central France two million years ago. Paleontology, geochronology, stratigraphy and taphonomy* (pp. 99–122). Cham (Switzerland): Springer.

Depéret, C. (1889). Sur le *Dolichopithecus ruscinensis* nouveau singe fossile du Pliocène du Roussillon. *Comptes Rendus de l'Académie des Sciences* (Paris), *109*, 982–983.

Depéret, C. (1928). Les singes fossiles du Pliocène de France. In *Comptes-Rendus du XIVe Congrès Géologique International*, 1926 (pp. 3–4). Madrid: Gráficas Reunidas.

Elhaï, H. (1969). La flore sporo-pollinique du gisement villafranchien de Senèze (Massif Central, France). *Pollen et Spores, 9*, 127–139.

Eisenmann, V., & Delson, E. (2024). The Senèze equids. In E. Delson, M. Faure, & C. Guérin (Eds.), *Senèze: Life in Central France two million years ago. Paleontology, geochronology, stratigraphy and taphonomy* (pp. 307–385). Cham (Switzerland): Springer.

Faure, M., & Guérin, C. (2024). *Sus strozzii* (Mammalia, Suidae) of the Late Villafranchian site of Senèze (Domeyrat, Haute-Loire, France). In E. Delson, M. Faure, & C. Guérin (Eds.), *Senèze: Life in Central France two million years ago. Paleontology, geochronology, stratigraphy and taphonomy* (pp. 387–407). Cham (Switzerland): Springer.

Faure, M., Delson, E., & Guérin, C. (2024). Introduction and history of research in the Senèze maar. In E. Delson, M. Faure, & C. Guérin (Eds.), *Senèze: Life in Central France two million years ago. Paleontology, geochronology, stratigraphy and taphonomy* (pp. 1–25). Cham (Switzerland): Springer.

Fernández-Jalvo, Y., Valli, A. M. F., Marín-Monfort, M. D., & Pesquero-Fernández, M. D. (2024). The taphonomy of Senèze. In E. Delson, M. Faure, & C. Guérin (Eds.), *Senèze: Life in Central France two million years ago. Paleontology, geochronology, stratigraphy and taphonomy* (pp. 653–681). Cham (Switzerland): Springer.

Gaudant, J. (2024). The fishes from the Maar of Senèze. In E. Delson, M. Faure & C. Guérin (Eds), *Senèze: Life in Central France two million years ago. Paleontology, geochronology, stratigraphy and taphonomy* (pp. 141–144). Cham (Switzerland): Springer.

Faure, M., & Guérin, C. (eds.). (2004). *Le Gisement pliocène final de Saint-Vallier (Drôme, France)*. Mémoire spécial 26, *Geobios, 37*, S1–S360.

Guérin, C. (2024). *Dicerorhinus etruscus etruscus* from the Late Villafranchian site of Senèze (Domeyrat, Haute-Loire, Central France). In E. Delson, M. Faure, & C. Guérin (Eds.), *Senèze: Life in Central France two million years ago. Paleontology, geochronology, stratigraphy and taphonomy* (pp. 245–306). Cham (Switzerland): Springer.

Hammer, Ø., Harper, D.A.T., & Ryan, P. D. (2001). PAST: Paleontological statistics software package for education and data analysis. *Palaeontologia Electronica*, *4*, 9 pp.

Heintz, E. (1970). Les Cervidés villafranchiens de France et d'Espagne. *Mémoires du Muséum national d'Histoire naturelle, 22*, Volume I and II, 1–317 and 1–206.

Lisiecki, L. E., & Raymo, M. E. (2005). A Pliocene-Pleistocene stack of 57 globally distributed benthic $\delta^{18}O$ records. *Paleoceanography, 20*, PA 1003, pp. 1–17.

Ménouret, B., & Guérin, C. (2024). The *Mammuthus meridionalis* of the Late Villafranchian site of Senèze (Domeyrat, Haute-Loire, Central France). In E. Delson, M. Faure, & C. Guérin (Eds.), *Senèze: Life in Central France two million years ago. Paleontology, geochronology, stratigraphy and taphonomy* (pp. 587–606). Cham (Switzerland): Springer.

Mourer-Chauviré, C. (2024). The birds (Aves) of the Early Pleistocene locality of Senèze (Haute-Loire, France). In E. Delson, M. Faure, & C. Guérin (Eds.), *Senèze: Life in Central France two million years ago. Paleontology, geochronology, stratigraphy and taphonomy* (pp. 145–163). Cham (Switzerland): Springer.

Necrasov, O., Samson, P., & Rădulesco, C. (1961). Sur un nouveau singe catarrhinien fossilifère, découvert dans un nid fossilifère d'Olténie (R. P. R.). *Analele Stiintifice Universitatii "Al. I. Cuza" din Iaşi, Sec. II, Stiinte Naturale, 7*, 401–416.

Nomade, S., Pastre, J. F., Guillou, H., Faure, M., Guérin, C., Delson, E., et al. (2014). $^{40}Ar/^{39}Ar$ constraints on some French landmark Late Pliocene to Early Pleistocene large mammalian paleofaunas: Paleoenvironmental and paleoecological implications. *Quaternary Geochronology, 21*, 2–15.

Paquette, J.-L., Médard, E., Poidevin, J.-L., & Barbet, P. (2021). Precise dating of middle to late Villafranchian mammalian paleofaunae from the Upper Allier River valley (French Massif Central) using U-Pb geochronology on volcanic zircons. *Quaternary Geochronology, 65*(101198), 1–12.

Parenti, F., Valli, A., Aprile, L., Candelato, F., Monguillon, A., Natali, L., et al. (2024). Surveying and excavations at Senèze: 2001–2006. In E. Delson, M. Faure, & C. Guérin (Eds.), *Senèze: Life in Central France two million years ago. Paleontology, geochronology, stratigraphy and taphonomy* (pp. 27–35). Cham (Switzerland): Springer.

Pastre, J.-F. (2024). The Senèze maar and paleontological site (Massif Central, France): their volcanic context and studies of related tephras. In E. Delson, M. Faure, & C. Guérin (Eds.), *Senèze: Life in Central France two million years ago. Paleontology, geochronology, stratigraphy and taphonomy* (pp. 83–98). Cham (Switzerland): Springer.

Roger, S., Coulon, C., Thouveny, N., Féraud, G., Van Velzen, A., Fauquette, S., et al. (2000). $^{40}Ar/^{39}/Ar$ dating of a tephra layer in the Pliocene Senèze maar lacustrine sequence (French Massif Central): Constraint on the age of the Réunion-Matuyama transition and implications on paleoenvironmental archives. *Earth and Planetary Science Letters, 183*, 431–440.

Schaen, A. J., Jicha, B. R., Hodges, K. V., Vermeesch, P., Stelten, M. E., Mercer, C. M., et al. (2021). Interpreting and reporting $^{40}Ar/^{39}Ar$ geochronologic data. *GSA Bulletin, 133*, 461–487.

Schlosser, M. (1924). Fossil primates from China. *Palaeontologica Sinica, 1* (2), 1–16.

Valli, A. M. F. (2024). New Data on Cervidae (Mammalia, Artiodactyla) from the Villafranchian locality of Senèze (Haute-Loire, France). In E. Delson, M. Faure, & C. Guérin (Eds.), *Senèze: Life in Central France two million years ago. Paleontology, geochronology, stratigraphy and taphonomy* (pp. 409–431). Cham (Switzerland): Springer.

Taxonomic Index

E. Delson et al. (eds.), *Senèze: Life in Central France Around Two Million Years Ago*,
Vertebrate Paleobiology and Paleoanthropology, https://doi.org/10.1007/978-3-031-64415-3

General Index

E. Delson et al. (eds.), *Senèze: Life in Central France Around Two Million Years Ago*,
Vertebrate Paleobiology and Paleoanthropology, https://doi.org/10.1007/978-3-031-64415-3

MIX
Papier aus verantwortungsvollen Quellen
Paper from responsible sources
FSC® C105338

If you have any concerns about our products,
you can contact us on
ProductSafety@springernature.com

In case Publisher is established outside the EU,
the EU authorized representative is:
Springer Nature Customer Service Center GmbH
Europaplatz 3, 69115 Heidelberg, Germany

Printed by Libri Plureos GmbH
in Hamburg, Germany